Wildland Plants: Physiological Ecology and Developmental Morphology

Donald J. Bedunah & Ronald E. Sosebee
Editors

Society for Range Management
1839 York Street
Denver, Colorado 80206
1995

FRONT & BACK COVER SKETCHES: The Society for Range Management extends its appreciation to Bridgett Crane and Mary Gilbert for their original sketches specifically designed for *Wildland Plants: Physiological Ecology and Developmental Morphology.*

First Edition

ISBN 1-884930-02-6

Library of Congress Catalog Card No. 95-069505

Printed in the United States of America

Table of Contents

Preface

Wildlands are a complex and diverse resource that provide for human needs and satisfy social and spiritual values as well. The values attributed to wildlands varies with the user. For land managers responsible for the sustainability of wildlands, a sound understanding of how abiotic and biotic interactions and genetic potential control plant community composition and plant quality are of primary importance. How the physiology of plants are affected by their environment has been termed ecophysiology, environmental physiology, stress physiology, or physiological ecology. The books thus far published in this field have not had great appeal for land managers because of their theoretical approach.

The publication of *Wildland Plants: Physiological Ecology and Developmental Morphology* is an attempt to provide land managers with an ecophysiology text with more applied information. It is not an all-inclusive text on plant ecophysiology and morphology, but a review of selected topics on how genetic, morphological, and environmental stresses affect rangeland communities. As such, this book is unique in its approach to plant ecophysiology and morphology information—it is designed to provide an understanding of rangelands for land managers, and is a source-book on genetic variation in plants, developmental morphology, carbon and nitrogen dynamics, processes controlling seed germination and seedling establishment, the influences and adaptations to stresses such as drought, nutrient deficits, salinity, and fire, the physiological and ecological aspects of allelopathy, and the influence of defoliation on the range ecosystem. Each author has provided insight into the knowledge gaps in each subject area. The immense literature on these subjects has also been reviewed and compiled, making it of value to upper-division undergraduate and graduate range management students. It should also prove to be an invaluable resource for all wildland management workers.

The first chapter, "Genetic Variation in Plants," provides an overview of ecotypic variation. It describes how within a biological species there may be both morphological and physiological variations. A knowledge and understanding of ecotypic variation is critical in plant selection for revegetation or restoration and to our understanding of the variations found in managing wildland ecosystems. Chapter 2, "Developmental Morphology," provides an overview of how plants grow and develop as an aid to understanding individual plant and community response to management. Growth is defined as a quantitative change in a plant; development is a series of qualitative changes in form, structure, function, and complexity. It is essential that wildland managers understand the derivation and arrangement of plant structure; in this chapter Bill Dahl expands on the principles of developmental morphology with examples in grazing management, undesirable plant control, and forage quality and how quality varies phenologically.

Chapter 3, "Carbon and Nitrogen Dynamics in Range Plants," reviews carbon and nitrogen acquisition and utilization. The manner in which plants gain and allocate carbon and nitrogen—and how these processes are affected by environmental factors—is fundamental to the development of management strategies and to the dynamics of the range ecosystem. This chapter provides an in-depth examination of the dynamics of carbon and nitrogen in range plants and the ecological implication these have for the management of rangelands.

The chapters "Seed Physiology and Germination" and "Seedling Morphology and Establishment" provide thorough reviews of the physiological and morphological principles controlling seed germination and seedling establishment. Seed germination, seedling emergence, and early seedling growth are all critical to plant and stand establishment and the revegetation and restoration of plant communities and natural community establishment.

Water is the most important environmental factor affecting the distribution, growth, and survival of rangeland plants. Chapter 6, "The Water Relations of Range Plants: Adaptations to Water Stress," summarizes the principles of plant water relations that have significance to range managers and describes plant adaptations to water deficits that may be useful in making land management decisions. Ray Brown specifically discusses land management practices and plant water relations including grazing and rangeland revegetation and plant water relations. In his discussion he uses literature pertaining to plant water relations irrespective of plant type because there is little data to suggest that the water relations of rangeland plant species are any more unique than those of other groups of plants, aside from perhaps specific adaptation to and interactive effects of grazing. Drought is the norm on range ecosystems and there is evidence that grazing can accentuate the effects of drought. However, Brown points to the fact that the water relations of many important range species have never been studied will continue to reduce our ability to best manage our rangelands.

Fire is an important factor affecting many wildland ecosystems. In chapter 7, "Fire As a Physical Stress," Allen Steuter and Guy McPherson examine fire as a physical stress and discuss mechanisms to facilitate survival and resilience of individual plants to fire and how fire interacts with other disturbances to effect individual plants and plant communities.

Plants adapt to nutrient stress and may have to cope with toxic levels of ions. Chapters 8 and 9, "Effects of Nutrient Stress" by James Klemmedson and Arthur Tiedemann and "Plant Adaptations to Saline Environments" by Richard Miller and Paul Doescher, discuss how plants respond and adapt to soil nutrient deficiencies and how plants cope with excess soluble ions of saline environments.

In general, ecophysiology texts stress environmental influences including chemical stresses. Rick Kelsey and Richard Everett state that biological life is chemistry—dynamic reactions and equilibria, bound and mediated by the physical laws of nature. It has become increasingly clear that chemicals produced by plants (phytochemicals) can have ecological functions that influence plant interactions and thus community function and it is relatively safe to say that the influences of allelopathy are not being addressed directly in the management of rangeland today. Furthermore, it is not likely to become a critical management criterion until it is more thoroughly understood. The chapter on allelopathy provides an excellent and thorough review of its potential relevance to the range ecosystem.

Herbivory by invertebrates and vertebrates are primary stresses on plant and plant communities that are rarely addressed in ecophysiology texts. Herbivory rarely directly kills individual plants, but does influence competitive interactions between plants thereby lowering the ability of a plant to compete for water and nutrients. Rangeland managers most often are more directly involved with the control of domestic vertebrate herbivores however, invertebrates, especially insects subject to outbreaks such as grasshoppers and locusts, may have major impacts on rangeland plants. Invertebrate herbivory is a chronic and universal phenomena on rangelands that is poorly understood by most range managers. Chapters 11 and 12, "The Relations of Phytophagous Invertebrates and Rangeland Plants" by Edward Evans and Tim Seasteadt and "Plant Responses to Defoliation: A Physiologic, Morphologic and Demographic Evaluation" by David Briske and Jim Richards, provide individual plant and community response information on the role of invertebrates and vertebrates in rangeland systems.

The complex interactions associated the interrelations of a plant to its environment, abiotic and biotic, are quite complex and so therefore is the management of wildlands complex. Future advances in management of wildlands will likely come from an integration of research from multiple disciplines. It is my hope that this text provides literature and ideas to land managers for more effective wildland management and provides research scientists with information on current research needs.

Don Bedunah

Genetic Variation in Plants: Environmental and Biotic Effects

Ronald E. Sosebee and David B. Wester

Introduction

Mayr (1940) defined a biological species as "a group of populations which replace each other geographically or ecologically and of which the neighboring ones intergrade ... or are potentially capable of doing so." Of the many concepts of species that have been proposed, this biological species concept (BSC) is most widely accepted by systematists today (Stussey 1990). In his original formulation, Mayr (1940) recognized that "most species occupy distinct ecological niches and that this ecological distinctiveness is the keystone of evolution" (O'Brien and Mayr 1991). In fact, Mayr (1982) suggested an amendment to the BSC, wherein a species is defined as "a reproductive community of populations (reproductively isolated from others) that occupies a specific niche in nature", that explicitly incorporates an ecological component into the definition of a biological species.

Two implications of the BSC are noteworthy. First, the BSC specifically states that reproductive isolation is the only discriminator among species. Second, the BSC acknowledges that there can be considerable genetic diversity within a species (O'Brien and Mayr 1991). It is this genetic diversity within a species that is explored in this chapter. It is this same genetic diversity that allows plants to exploit certain habitats and to fill selected niches. Plants are uniquely adapted to the environment in which they live and even though a species might be ubiquitous and extend across many different environments (e.g. blue grama [*Bouteloua gracilis (H.B.K.) Lag. ex Steud.*]), all of the populations are not exactly the same physiologically nor do they respond to the various environments in the same way. This variability results in ecotypic variation and ultimately in ecotypes. We shall attempt to define and describe the concepts of ecotype and ecotypic variation, and cite examples that occur in nature. In our discussion, we attempt to present variation that results from environmental differences and also from biotic effects within a population. Finally, we hope to show some implications of ecotypic variation and the value of ecotypes to range management.

Authors are professor and associate professor, Range and Wildlife Department, Texas Tech University, Lubbock, Texas 79409.

We wish to acknowledge preliminary review by B. E. Dahl and E. B. Fish and for their helpful comments on an earlier version of this paper.

This is contribution number T-9-626, College of Agricultural Sciences and Natural Resources, Texas Tech University.

Overview of Ecotypes and Ecotypic Variation

Genetic variation in plants occurs at several levels. *Ecotypic variation* refers to genetically-based variation within a species that is related to habitat (or "external" environment in the sense discussed by Antonovics et al. 1988). Application of the term "ecotype" usually refers to relatively distinct differences within a species. Along environmental gradients, however, plant variation may be relatively gradual or continuous; under these conditions, the term *ecocline* is appropriate. It is important that both ecotypic and ecoclinal (or clinal) variation refer to genetically-based variation that is related to environment. It may be reasonable to suggest that because ecotypic variation is a response to different environments, ecotypic variation is largely a "between population" phenomenon (but see below).

Variation within a population, wherein environmental differences are less striking, occurs on a smaller scale than variation between populations. Variation at this level may be attributable to very fine scale environmental variation, or to biotic interactions among neighboring plants. Terminology at this scale of focus varies among authors. Harper (1977) used the terms "within-population variation", or "polymorphism". Turkington and Aarssen (1984) used "polymorphism" to described a variety of "patterns of local population differentiation." Jain (1990) discussed "mosaic or patchwork in localized variation [that] is based on nonrandom occurrence of two or more genotypes in a population", or simply "mosaic variation". Turkington and Harper (1979) used the term "biotic ecotype". It might be noted that the term *biotype*, as defined by Heslop-Harrison (1967), refers to several individuals that are produced from one individual through vegetative reproduction; these individuals possess the same genotype, and are thus defined as belonging to the same biotype. According to this definition, the term "biotype" appears most appropriate for situations involving vegetative reproduction or apomixis (e.g., Solbrig and Simpson 1974, 1977; Hancock and Wilson 1976). This usage of the term biotype is more restrictive than the meaning implied by Turesson (1925), who used it in the context of "the mass of genetically different biotypes which make up the Linnaean species [which] do not distribute themselves indiscriminately over an area comprising different types of localities", but which tend to be grouped according to habitat type. Clausen et al. (1939) used the term biotype in the context of local variation within populations "in one kind of environment"; these variations are not correlated with environmental effects.

Plant Variation Related to Environment (Ecotypes and Ecotypic Variation)

Concepts and Definitions

Harper (1982) distinguished between two interpretations for the statement that a particular species has a "wide distribution." On the one hand, this could mean that "individuals of the species, though genetically narrowly based, have wide tolerances or plasticity so that individuals sampled over the range of the species will behave equally well over that range, and are mutually exchangeable." Distributional information about these so-called "all-purpose genotypes" also connotes physiological implications (Harper 1982).

On the other hand, what may be meant by a species having a wide distribution is "that the individuals have very narrow tolerances but the nature of the species is such that the taxonomist includes a wide range of locally specialized genotypes within one taxon." Distributional information in this context conveys information about genetic variation within and between populations (Harper 1982).

It was this kind of variation that Turesson (1922) recognized in his pioneering research on ecotypes. For example, *Hieracium umbellatum* is a widely distributed plant in Sweden. Despite considerable morphological variation displayed by this species throughout its range, taxonomists recognized it as a single taxon. Turesson (1922) grew plants collected from different habitats under the common environmental conditions of a nursery garden, and found that the morphological differences displayed by plants in their original habitats persisted under a common environment: these differences had a genetic basis (i.e., they were not phenotypic responses that disappeared under common environmental conditions) that reflected adaptation to different habitats. In other words, these plants were ecologically distinct in spite of the fact that they were interfertile: their ecological distinctiveness would suggest to an ecologist *different groups* of plants that a taxonomist would accept simply as the genetic variability within a *single group*.

Similar observations "work in reverse" as well. Turesson (1922) noted that what taxonomists recognized as three different species of *Atriplex* (*A. praecox, A. longipes and A. hastifolium* Salisb.) on the basis of morphological features were in fact all interfertile. In this instance, ecological adaptations resulted in genetically-based morphological differences that were sufficient to suggest to the taxonomist different species; despite these differences, however, these three different "species" were capable of interbreeding, and hence actually represented a single biological species.

These observations laid the foundation for genecology, the study of genetically-based differences within a (biological) species that are habitat- (or environment-) related. Turesson (1922) defined an *ecotype* as "the ecological unit to cover the product arising as a result of the genotypical response of an ecospecies to a particular habitat", where the term "ecospecies" refers to the Linnaean species as they are realized in nature (i.e., members of an ecospecies are not restricted to a particular habitat).

Ecotypes initially were delineated on the basis of growth response to climate, edaphic, or geographic distribution (Gregor 1944) including altitudinal differences (Clausen et al. 1940). [Gregor (1944) briefly discussed ecotypes related to biotic environment; this topic is pursued in a later section]. A definitional requirement of ecotypes includes genetically-based variation that represents adaptation to environmental factors. This adaptation need not, however, be morphologically based; i.e., ecotypes are not necessarily morphologically distinct, but they are physiologically distinct in the way that they relate to their environment. [For the sake of completeness, it may be interesting at this point to recognize the phenomenon of "sibling species." There are numerous examples of plants that are virtually indistinguishable morphologically. Although plants of these groups belong to the same morphological (or Linnaean) species, they do not belong to the same biological species by virtue of their intersterility.]

Examples of Ecotypic Variation

Historical Description of Ecotypic Variation

Clausen et al. (1940, 1948) studied numerous plant groups including genera with few to many species growing in various habitats throughout California. Populations within genera studied occurred along a transect from the Pacific Ocean at Point Montara (San Mateo Co.) across the Sierra Nevada to the Great Basin at Benton (Mono Co.). This transect included habitats as diverse as "Mediterranean" on the western extreme to "alpine" on the eastern extreme. The genera that initially displayed the most ecotypic variation were *Potentilla, Zauschneria, Achillea and Artemisia* (Clausen et al. 1940). Perhaps the classic example from their results is *Potentilla* (Table 1), particularly *P. glandulosa* Lindl. and its ecotypes (Clausen et al. 1940, Daubenmire 1974).

The classic examples of ecotypes discussed by Turesson (1922, 1923) and Clausen et al. (1940, 1948) suggested that ecotypes were discrete entities, and this was a key component in Turesson's development of the ecotype concept (Barbour et al. 1980). However, Gregor (1944) suggested that ecotypic differentiation occurs along an environmental gradient rather than as disjunct entities. He employed the term "ecocline" "to denote a series of habitat populations showing genotypical gradation related to a particular environmental gradient." He referred to ecotypes as populations that were different from each other and occurred at specific points along the ecocline. Heslop-Harrison (1965) concluded that Turesson's ecotypes might be an artifact of sampling procedures since only remote populations were sampled. More recently, Ehrlich and Raven (1969) indicated that local races within a species are commonplace; they might be widely separated or they might be in the proximity of other races within the same species. Therefore, Stebbins (1979) recommended that reference be made to ecotypic variation rather than to ecotypes.

Since the advancement of the ecotype concept in the 1920's, the questions have been "Are ecotypes equivalent to subspecies?" and "Are ecotypes equivalent to varieties?" Clausen et al. (1940) used ecotype and subspecies synonymously when the ecotypes were morphologically distinct. Taxonomically, varieties generally infer morphological differences among plants within a species. Agronomically, varieties often infer genetic differences among accessions within a species. Ecotypes usually are not synonymous with subspecies, because subspecies are based primarily on phenotypic or morphologic characteristics (Stebbins 1979). Stebbins suggested that there are some similarities between plant ecotypes and animal subspecies, but that there are also some distinctions.

Ecotypes have been used synonymously with ecological races (Hiesey and Milner 1965), physiological races or ecodemes (Daubenmire 1974). Ecotypes are not synonymous with ecophenes. Ecophenes are plants differing in morphological and phenotypical characteristics that are an expression of phenotypic plasticity. Neither is ecotype necessarily synonymous with genotype or biotype. Every plant within a population is potentially a different genotype as a result of fertilization and meiosis. Biotypes are local genetic variants within a species in the same environment; ecotypes connote genetic variability between different environments (Clausen et al. 1939).

Table 1.

Genus	*Species*	*Ecotype**	*Habitat*	*Subecotype***
Potentilla (Section Drymocallis)	glandulosa	glabrata (Rydb.) Keck	Northern Rocky Mountains	
		pseudorupestris (Rydb.) Keck		
		nevadensis (S. Wats.) Keck	Cascade and Blue Mts. (WA) south to San Bernadino and San Jacinto Mts. (CA); 1500-3500m	alpine >2600m; subalpine <2600m
		ashlandica (Greene) Keck	Cascade Mts. (OR) south to Siskiyou Mts.; 1500-2000m	
		arizonica (Rydb.) Keck	2400-3500m (Central and southern UT); <2000m (AZ)	
		Hanseni (Greene) Keck	1200-1800m; mid-altitude slope of Sierra Nevada	
		typica Keck	<2400m; southern B.C. to northern Baja, CA (mostly near coast)	
		globosa Keck	Chaparral (OR) 1950m; Siskiyou Mts. to S. Fork Mts. (CA); 1280-2150m	
		micropetala (Rydb.) Keck	Elevations 1500-2900m (Wasatch Mts.)	
		reflexa (Greene) Keck	250-2000m; well-drained slopes of Sierra NV Mts.	
		Ewanii Keck	San Gabriel Mts. (CA)	

* Clausen et al. (1940) referred to this group equivalent to subspecies, because they are morphologically distinct ecotypes.

** Clausen et al. (1940) referred to this group as ecotypes since their primary subdivision was called subspecies. This taxonomic subdivision was referred to ecotypes within subspecies because they were also distinguishable by growth habit.

Verification of Ecotypic Differentiation

In addition to the classic examples of ecotypic variation that have already been cited, the reader is referred to Tisdale (1962) and Hiesey and Milner (1965) for a review of early research pertaining to ecotypic variation. Sideoats grama [*Bouteloua curtipendula* (Michx.) Torr.] exhibits ecotypic variation along a latitudinal gradient from Texas to North Dakota with respect to several growth and development characteristics (internode elongation, height, tiller and rhizome production) (Olmstead 1944) and flowering (Olmstead 1945) in response to photoperiod. Likewise, Larsen

(1947) demonstrated ecotypic variation with respect to flowering response to photoperiod in little bluestem [*Schizachyrium scoparium* (Michx.) Nash] from the coastal bend of Texas to North Dakota and eastward to North Carolina and Connecticut. Subsequent transplant garden research revealed ecotypic variation in little bluestem as well as big bluestem (*Andropogon gerardii* Vitman), switchgrass (*Panicum virgatum* L.), and Indiangrass [*Sorghastrum nutans* (L.) Nash], particularly in relation to onset of spring growth, flowering, and dormancy in response to latitudinal, longitudinal, and altitudinal origins (McMillan 1964, 1965a). McMillan (1965b) obtained similar results from growth chamber experiments utilizing the same species listed above as well as blue grama and sideoats grama. Although ecotypic variation exists in little bluestem, forage quality is not necessarily distinguishable, especially in pinehill bluestem [*S. scoparium* var. *divergens* (Hack.) Gould] (Grelen 1974). However, Brown and Kalmbacher (1988) found significant differences in forage quality among ecotypes of creeping bluestem [*S. scoparium* var. *polycladus* (Schriber & Ball) Bruner (*S. stoloniferum* Nash)]. Likewise, ecotypic variation is exhibited in forage quality of red fescue (*Festuca rubra* L.), particularly as it relates to winter hardiness (Klebesadel 1985). Although distinct palatability preferences of strains (ecotypes) of Indian ricegrass [*Oryzopsis hymemoides* (R.& S.) Ricker] have not been shown for livestock, Bohmont and Lang (1957) demonstrated distinct differences in preference by rabbits. They also found distinct morphological characteristics in Indian ricegrass relative to the environment where it was selected.

Ecotypic variation has also been shown in annuals such as cheatgrass (*Bromus tectorum* L.) (Hulbert 1955). Whalley and Burfitt (1972) suggested that ecotypic differentiation occurs in wild oats (*Avena sterilis*, L. subsp. *ludoviciana* (Dur.) G.& M., *A. fatua*, L., and possibly *A. barbata* Pott.). Ecotypic variation exists in rosering Gaillardia (*Gaillardia pulchella* Foug.) adapted to calcareous and noncalcareous soils (Heywood 1986). These edaphic ecotypes have different frequencies of occurrence of isozymic forms of malate dehydrogenase (MDH) and a difference in the pigmentation pattern of the ray corollas. Most early-flowering strains (ecotypes) of perennial ryegrass (*Lolium perenne* L.) have no obligatory vernalization requirements, whereas all late-flowering strains (ecotypes) have a definite low-temperature requirement and a critical photoperiod of at least 11 hours (Cooper 1951). Cooper indicated that these requirements were related to the geographical and ecological origin of the plant populations.

In the early 1960s, a renewed emphasis was placed on physiological ecology and ecotypic differentiation studies. In 1961, Mooney and Billings compared physiological characteristics of arctic and alpine ecotypes of alpine sorrel [*Oxyria digyna* (L.) Hill]. They found a number of differences including increased flower production along a north to south gradient, increased rhizome production along a south to north gradient, and leaf chlorophyll increased with latitude. Also, the arctic plants had higher photosynthetic rates and attained maximum photosynthetic rates at lower temperatures than alpine plants. Arctic plants had higher respiration rates at all temperatures than alpine plants. High elevation, low latitude populations attained photosynthesis light compensation at higher light intensities than low elevation, high latitude plants. Similar ecotypic differentiation was found in arctic and alpine populations of alpine meadow-rue (*Thalictrum alpinum* L.), an herbaceous perennial (Mooney and Johnson 1965).

In recent years, there has been a number of studies involving gas exchange between species and among ecotypes within a species. Clough et al. (1980) found sun- and shade-ecotypes within bitter nightshade (*Solanum dulcamara* L.). Light saturated photosynthetic rate in the shade-ecotype was significantly reduced when grown in full sunlight. Therefore, without shade, its growth can be impaired and it then becomes a poor competitor with the sun-ecotype. However, when bitter nightshade is grown with adequate supplies of water and nitrogen, both ecotypes acclimated to various irradiance levels during growth (Osmund 1983). But Osmund (1983) showed that if both nitrogen and water become limiting, then the photosynthetic response curve was depressed in the shade-ecotype.

Ecotypes resistant and sensitive to sulfur dioxide have been identified in Carolina geranium (*Geranium carolinianum* L.) (Taylor et al. 1986). When exposed to chronic levels of SO_2, photosynthesis and biomass in the sensitive ecotype were substantially reduced, whereas in the resistant ecotype both photosynthesis and biomass were greater than in plants not exposed to chronic levels of SO_2. Increased atmospheric SO_2 resulting from air pollution is a recent environmental parameter to which plants have been exposed. Whether large numbers of plants can evolve rapidly enough to become adapted to air pollutants is yet to be determined.

In contrast, it is interesting to note that plants living in "less hospitable" environments are well adapted to the "inhospitable" environmental regime. High altitude ecotypes of sheep sorrel (*Rumex acetosella* L.) (Venezuela) and buckhorn plantain (*Plantago lanceolata* L.) (Peru) were unaffected by high levels of UV-B radiation whereas low altitude ecotypes from Seattle, Washington (buckhorn plantain) and northern Utah (sheep sorrel) absorbed much more UV-B radiation (Barnes et al. 1987).

Differences in several physiological processes have been shown in various ecotypes of the geographically widespread broad-leaved cattails. McNaughton (1966a) demonstrated that ecotypes of broad-leaved cattails are enzymatically distinct and the enzyme activities reflect the environment in which the plants grow. Ecotypes growing in a cool or semi-maritime environment had much higher glycolic acid oxidase (GAO) activity in the chloroplast than ecotypes growing in a warm or continental environment (McNaughton 1965, McNaughton 1969). Malate dehydrogenase (MDH) was more resistant to inactivation from high temperatures when obtained from plants endogenous to hot environments (McNaughton 1966b) and was more temperature sensitive when obtained from plants growing in a cool environment (McNaughton 1972). The Michaelis constant (Km, an inverse measure of enzyme substrate affinity) was three-fold greater in MDH obtained from plants endogenous to a cool environment (McNaughton 1972). Photochemical efficiency of Photosystem II was more efficient for ecotypes from high altitudes as compared to those from a maritime climate (McNaughton 1967). Likewise, ecotypes from the cool environs of Quebec demonstrated a greater dependence on temperature for net photosynthesis of young leaves and a decline in temperature dependence of older leaves when compared to ecotypes from the very warm environs of Red Bluff, California (McNaughton 1973).

Simon (1979, 1986) found similar responses in the NAD-MDH isozymes in ecotypic differentiation of maritime and continental populations of sweetpea (*Lathyrus japonicus* Willd.). Both apparent free energy (Ea) and free energy of activation (ΔG) of MDH were greater in the continental ecotype than in the maritime ecotype. Thermostability of MDH was also greater in the continental ecotype than in the maritime ecotype (Simon 1979). Kinetic behavior of mitochondrial MDH in four populations of sweetpea reflects the temperature regime of the site from which the various populations were collected (Simon 1986).

Mesquite (*Prosopis glandulosa* Torr.) often is divided into varieties *torreyana* (Benson) M. C. Johnst. and *glandulosa* (Benson) M. C. Johnst. However, Peacock and McMillan (1965) identified ecotypic variation within and between these varieties with respect to onset of dormancy, tolerance to cold temperatures, and response to photoperiod variation along a south to north gradient. Ecotypic variation occurs in *Juniperus communis* L. in Finland as indicated by a decrease in volatile terpenes along a gradient from a distant maritime site toward the coast (Kallio and Junger-Mannerma 1989).

The "Lost Pines" of Texas includes two "islands" of loblolly pine (*Pinus taeda* L.) that occur in a drier environment about 100 miles west of the rest of the loblolly pines of east Texas and Louisiana (Bastrop-Fayette County) (McClurkin et al. 1971). McClurkin et al. (1971) found an interesting physiological difference (presumed to be genetically based) in the "Lost Pines" ATPase activity and the relative degree of activation. ATPase activity in the root tips of the "Lost Pines" was enhanced by a potassium-calcium substrate (calcareous soil is characteristic of Bastrop-Fayette Co.), whereas a sodium-magnesium substrate was the most effective ATPase activator in the population of loblolly pines tested from Mississippi.

Edaphic factors are extremely influential in the kind of vegetation that inhabits an area (Jenny 1980). Edaphic ecotypes of bentgrass (*Agrostis tenuis* Sibth.) were identified within 50 feet (or less) of each other on lead-free soil and lead-toxic mine spoils in Central and North Wales (Jowett 1964). Likewise, ecotypic variation was evident in water sedge (*Carex aquatilis* Wahlenb.) growing in ice-wedge polygon microhabitats in the wet meadow tundra at Barrow, Alaska (Shaver et al. 1979). Field-grown plants collected from these microhabitats differed in above-ground weight/tiller; leaf-production rate; root:leaf weight ratio; rate of ^{32}P uptake; and concentration of sugar, nitrogen, and phosphorus in roots and rhizomes. Even though these microhabitats occurred within 3 feet of each other, they differed in soil water content, availability of nitrogen and phosphorus, and depth of thaw (Shaver et al. 1979). These authors found significant differences in phosphorus-uptake kinetics in nutrient concentrations in *Carex* populations occupying microhabitats with different soil characteristics. Similarly, ecotypic variation exists in mountain avens (*Dryas octapetala* L.) between snowbeds and fellfields in the arctic tundra. The two ecotypes are morphologically distinct, but they also differ in time of flowering (therefore, they seem to be reproductively isolated even though they are in the same proximity) (McGraw and Antonovics 1983) and competitive ability (McGraw 1985a; 1985b). A bermuda grass (*Cynodon dactylon* L.) ecotype adapted to alkaline soil did not respond to N-P-K fertilization, whereas the ecotype adapted to nonalkaline soil did (Ramakrishnan and Kumar 1976). Similarly, Main (1974) found that populations (ecotypes) of bluebunch wheatgrass (*Agropyron spicatum* Rydb.) growing on serpentine and nonserpentine soils responded differently to calcium and magnesium fertilization.

Ecotypic variation often occurs along a salinity gradient in Nuttall's saltbush (*Atriplex nuttallii* Wats.) and possibly in shadscale [*A. confertifolia* (Torr. & Frem.) Wats.] (Goodman and Caldwell 1971) as well as in red fescue (Rozema et al. 1978, Rhebergen and Nelissen 1985). Ecotypic variation has been shown in *Ephedra distachya* and *E. lometolepis* growing along a salinity gradient from the Aral Sea (Tkachenko 1985). The ephedra ecotypes were associated with the different salts along the gradient. Likewise, broad-leaved cattails endogenous to saline habitats were more tolerant of salinity than those growing in nonsaline habitats (McNaughton 1966c).

Ecotypic variation resulting from edaphic features has also been found in the arrowhead complex [*Sagittaria graminea* var. *graminea* Michx., *S. graminea* var. *chapmani* J. G. Sm. and *S. graminea* var. *weatherbiana* (Fern.) Bogin], an aquatic plant (Wooten 1973). Saxifrage (*Saxifraga hirculus*) growing within the mires of Sweden has evolved ecotypes from different habitats in very short distances on the same mire (Ohlson 1989). Ohlson (1989) also found saxifrage ecotypes between mires. Poorly drained soil often causes toxicosis in plants. However, plants persistent in these habitats evolve adaptations that allow them to survive there. Etherington and Thomas (1986) found clones (ecotypes) of orchard grass (*Dactylis glomerata* L.) growing in well-drained and poorly drained soils. The clones exhibited distinct environmental adaptation to the wet and dry soils.

Very often it is difficult to separate the effects of edaphic factors and salinity from water stress in plants.

Eickmeier et al. (1975) identified two physiological races (ecotypes) in eastern hemlock [*Tsuga canadensis* (L.) Carr.] in Wisconsin according to two temperature regimes and water stress. However, plants from southwestern Wisconsin had a more favorable carbon balance at warmer temperatures and higher water stress than plants from northeastern Wisconsin. The northeastern plants were better adapted to lower temperatures and lower water stress. Musselman et al. (1975) demonstrated ecotypic differentiation in response to soil water regimes between populations of northern white cedar (*Thuja occidentalis* L.) grown in lowland and on upland sites. Likewise, Sowell and Spomer (1986) found ecotypic variation within Engelmann spruce (*Picea engelmannii* Parry) and subalpine fir [*Abies lasiocarpa* (Hook.) Nutt.]. Plants within both species exhibited higher root respiration rates near timberline than at lower elevations. They concluded that higher respiration rates seemed to compensate for the low soil temperatures at the higher elevations.

Big sagebrush (*Artemisia tridentata*) has been subdivided into four subspecies [*A. tridentata* Nutt. ssp. *tridentata*, basin big sagebrush; *A. tridentata* Nutt. ssp. *vaseyana* (Rydb.) Beetle, mountain big sagebrush; *A. tridentata* Nutt. ssp. *wyomingensis*, Beetle & Young, Wyoming big sagebrush] (McArthur 1979, 1983; Winward 1983); and *A. tridentata* ssp. *xericensis* (West 1988). They are morphologically distinct; therefore, they have been classified as subspecies. However, their habitat type is very specific according to edaphic conditions (Hironaka 1979, West 1988); therefore, they exhibit ecotypic variation. In fact, many of the early settlers recognized that basin big sagebrush was a good indicator of arable land (Winward 1983), although the big sagebrush complex had not been taxonomically divided into subspecies.

Plant Variation Related to Biotic Effects (Within-Population or Polymorphic Variation)

Concepts and Definitions

Harper (1977) briefly but elegantly encapsulated several key elements in the theory of evolution through natural selection by observing that (1) some organisms leave more descendants than others, and that some characteristics of "successful" ancestors are heritable, and (2) the fact that most organisms leave more progeny than can survive in a given habitat insures not only that there is an excess of individuals on which natural selection can act, but that pressures of overpopulation are likely to be especially important in the process.

Those who envision natural selection as a " 'struggle' against the elements, the physical forces of the environment" follow what Harper (1977) called a "Wallacian" approach to plant ecology. According to Coleman (1986), however, Darwin himself shifted his (early) emphasis of natural selection and evolutionary change *away from* abiotic factors (an "early Darwinian" or "Wallacian" approach) and *toward* biotic factors after 1856:

> "In 1856 Darwin found a solution to perhaps the decisive evolutionary problem — divergence, the production of two or more new species from a preexisting stock. Adaptation, when effective, tuned an old species to new conditions; divergence assured the production of new multiple species. Divergence was Darwin's answer to the pressing question of evolutionary innovation. What is ecologically noteworthy in this advance is that Darwin's eye and mind now largely abandoned the abiotic environment as an interesting (that is to say, participatory) feature of the evolutionary process. The ultimate Darwinian explanation, population pressure (a strictly biotic matter), came to dominate the argument … The implications of Darwin's new understanding of divergence are remarkable, not least for the history of ecology. In the Darwinian scheme … the abiotic environment offered the background against which was played the central evolutionary event, speciation. It provided, however, only an abstract background; for speciation was the result of divergence, and divergence was the product of population pressure and the ensuing struggle for a place in nature, both understood to be predominantly biotic phenomena." (Coleman 1986).

Gould (1989) succinctly summarized this idea with the statement "In Darwinian theory, competition is the great regulator."

It is noteworthy that this same emphasis was a focal point for Haeckel (1866) when he first used the term "ecology" in reference to the "economy of nature":

> "… far more important [than the abiotic factors of the environment such as light, heat, moisture, etc], however, and exerting a much more powerful influence on the transformation and adaptation of species are organic conditions of existence, that is, the mutual relations of organisms to one another … the interactions between all neighboring organisms are extraordinarily important and … they exert far more influence on the change

and adaptation of species than do the inorganic conditions of existence ... " [Haeckel (1866) quoted from Coleman (1986)].

We wish to change our focus here as well, from one emphasizing genetic variation related to environment, to genetic variation related to biotic factors. We focus on neighbor-to-neighbor effects among plants.

By definition, ecotypic variation is genetically-based variation within a biological species that is related to habitat differences. The considerable body of ecological research that has focused on plant adaptations to physical (or abiotic) features of the environment reveals a "Wallacian approach" to plant ecology. In contrast, a "Darwinian approach" focuses on the evolutionary response of plants to biotic pressure (Harper 1977). This approach recognizes that organisms form a part of each other's environment; because genetically-based variation that is related to biotic neighbors is a "within-environment, within-population" phenomenon, we use the term *polymorphic* variation.

Examples of Polymorphic Variation

In the same classic paper in which he contributed so much toward clarifying the ecotype concept, Gregor (1944) also briefly discussed an example of polymorphic variation. He cited Sinskaja (1931) to illustrate what the latter called *synecotypes*, or phytosocial ecotypes, wherein there is plant differentiation in response to biotic environment. In this example, two types of black mustard (*Brassica nigra*) occupy the same climatic region. One form is widespread in pure crops, whereas the second form only occurs mixed with *Brassica campestris*. Gregor (1944) remarks that "Here we have a phytosocial ecotype selectively adapted to live in association with *B. campestris*, and a phytosocial environment sufficiently strong to overcome the effect of climatic factors."

One of the most intensively studied species with respect to polymorphic variation is clover (*Trifolium repens* L.). Turkington and Harper (1979) investigated the effects of biotic neighbors on clover yield when this plant was grown in association with four grasses (perennial ryegrass, *Holcus lanatus*, L., *Cynosurus cristatus* L., and bentgrass). Samples of clover were selected from within patches of each of these grasses, propagated in the greenhouse, and transplanted back into the same field from which they were collected. Each clover population was transplanted not only with the grass with which it was originally associated, but also with each other grass. Major results were (1) yield of the four clover populations (corresponding to populations collected from patches with the four grasses) differed among grass associates; (2) grasses affected clover growth differently; and (3) a significant interaction between the four clover populations and the four grasses was attributed to the tendency of each clover population to grow better with the grass with which it was originally associated. In a second experiment the four populations of clover were grown in all combinations with the four grasses under greenhouse conditions wherein a standard soil medium eliminated any edaphic variation that may have influenced field results. With one exception, each clover population grew better with the grass with which it was originally associated.

Turkington (1985) concluded that

> "This is a fine-scale level of differentiation within the clover population, selected by different forces imposed by different grass neighbors. This leads to the hypothesis that the mosaic of neighbors, mostly grasses, with which *T. repens* coexists, constitutes a prominent element of biotic heterogeneity that sorts genotypes of *T. repens* on the basis of neighbor-specific compatibilities."

In a similar study, Turkington (1979) removed alfalfa (*Medicago sativa* L.) and clover plants from 10-year old pastures where they were growing with smooth brome (*Bromus inermis* Leyss.), timothy (*Phleum pratense* L.) or orchard grass (*Dactylis glomerata* L.), propagated them in a greenhouse, and then transplanted them into pure stands of each of these grasses. Results for clover were generally similar to those found by Turkington and Harper (1979). Alfalfa plants that originally were associated with timothy had both higher yield and survival when planted with timothy than when planted with either of the other two grasses. Also, alfalfa plants that originally had been associated with smooth brome or orchard grass had higher survival when grown with these respective species than when grown with a grass that had not been an original associate. In these respects, results for alfalfa paralleled results for clover. However, alfalfa plants that originally had been associated with either orchard grass or smooth brome yielded more when planted with timothy than when planted with the originally associated grasses. Turkington (1979) concluded that, at least with respect to yield, if alfalfa developed in timothy mixtures it was "in some way 'better equipped'" to coexist with smooth brome or orchard grass than if it had originally developed with these latter species. It may be that timothy is a weaker competitor than either smooth brome or orchard grass, but Turkington (1979) could offer no explanation why selection followed a different course for plant survival than it did for yield.

Linhart (1974) studied variation in *Veronica peregrina* L., an annual that often dominates in areas where temporary pools of water dry. Plants were sampled in central portions of localized populations, where populations were relatively dense, as well as in peripheral areas where populations were less dense. Seeds from plants in both locales were collected and planted under common greenhouse conditions. Plants of central and peripheral origins were different with regard to a number of features (e.g., germination rate, variability in extent of germination, plant size, and number and size of seeds produced) that were interpreted in the context of localized adaptation. Linear distances separating central from peripheral areas were only 2 to 5 m. Linhart (1974) suggested that disruptive selection [wherein individuals of different genotypes may be at a selective advantage within a given area occupied by a breeding population, giving rise to genetic polymorphism in patchy environments (Briggs and Walters 1984)] may be the mechanism producing localized differentiation. In this instance, there were important environmental differences between central and peripheral zones (e.g., duration and predictability of soil moisture), and so it may be reasonable to apply the concept of ecotype. However, given the extremely short distances involved, and the fact that gene flow by both pollen and seed no doubt occurs between areas, it may also be reasonable to suggest that biotic specialization is partly involved in producing the observed patterns of localized population differentiation.

Kelley and Clay (1987) studied competitive interactions between *Danthonia spicata* L. and *Anthoxanthum odoratum* L. Eight individuals of each species were collected from a field where both species overlapped spatially and transferred to a greenhouse for propagation. The eight genotypes (individuals) of *Anthoxanthum* were planted with each of the eight genotypes of *Danthonia*. Results indicated that interspecific competition was related to genotypic composition of the populations. *Danthonia* was sensitive to the genotypic identity of *Anthoxanthum*, but *Anthoxanthum* performance did not depend on the genotypic identity of *Danthonia*. Kelley and Clay (1987) concluded that genotypic-specific interactions must be taken in account in an attempt to fully understand evolution in natural plant populations.

Taylor and Aarssen (1990) studied within-species variability of quackgrass (*Agropyron repens* L.), Kentucky bluegrass (*Poa pratensis* L.), and timothy. Genotypes of each species were collected in the field and transferred to a greenhouse for propagation; each genotype was then grown in monoculture and in pairwise combinations with all other collected genotypes. Genotypic variability within species was detected by a significant "target plant" site-of-origin x "associate plant" site-of-origin effect. Additionally, intransitive competitive relationships were detected at the genotype level whereas competitive relationships at the species level were transitive. (In a *transitive* competitive relationship, if Species A competitively excludes Species B, and Species B competitively excludes Species C, then Species A will also competitively exclude Species C. In an *intransitive* competitive relationship, it might be that Species A competitively excludes Species B, and Species B competitively excludes Species C, but Species C competitively excludes Species A). These results provide evidence for the hypothesis of competitive combining ability as a mechanism of species coexistence (see below), a key component of which is biotic specialization that results from neighbor-to-neighbor interactions (Aarssen 1983).

It is clear that, given the nature of the processes involved, the presence and degree of biotic specialization depend on the species involved in the interactions as well as on the environment in which the interactions take place. For example, whereas *Danthonia* performance depended on genetic identity of *Anthoxanthum*, the converse was not true (Kelly and Clay 1987). Goldberg (1988) failed to detect differences in competitive ability among clones of *Solidago canadensis* var. *scabra* (*Muhl.*) T.& G. Similarly, Platenkamp and Foin (1990) failed to detect evidence of specialization of genotypes relative to neighbors. In this study, clones of *Anthoxanthum odoratum* were reciprocally transplanted between a xeric site (dominated by *Anthoxanthum*) and a mesic site (dominated by *Holcus lanata*). *Anthoxanthum* transplants ("target" plants) were surrounded by (1) no neighbors, (2) *Anthoxanthum* neighbors transplanted from the mesic site, (3) *Anthoxanthum* transplanted from the xeric site, and (4) *Holcus* neighbors. Under the hypothesis of local biotic specialization, one would anticipate that target *Anthoxanthum* plants from the mesic site would perform relatively better with *Holcus* neighbors, and target plants from the xeric site would perform relatively better with *Anthoxanthum* neighbors. However, responses of target plants did not depend on target origin. Also, neighbor plants from the mesic site performed relatively better at the xeric site, and neighbor plants from the xeric site performed relatively better at the mesic site, a pattern just opposite of what would be anticipated if plants were locally adapted relative to their origin. Antonovics et al. (1988) collected 22 genotypes of *Anthoxanthum odoratum* from a mown field, propogated them in a greenhouse, and transplanted them into the field, where growth responses were monitored for three years. Although different vegetation types affected the reproductive output of *Anthoxanthum*, there was less evidence that this was so for different genotypes. That is, different genotypes responded differently to particular community components in some areas of the field but not in other areas.

General conclusions from research detecting local population differentiation suggest that (1) it may take place spatially over scales of a few meters or less; (2) it often can be attributed to neighbor effects in addition to (or in combination with) small scale environmental (i.e., abiotic) differences; and (3) it depends on the species involved and on the environmental conditions in which the interaction takes place.

Neighbor-to-neighbor interactions that involve biotic specialization have obvious implications with respect to speciation. This scale of local population differentiation may be the first step in divergence of species into subspecies or into new species (Turkington 1979; also see Ehrlich and Raven 1969). The time scale for this process is important. Turkington (1979) detected clear patterns among populations of alfalfa and associated grasses (timothy, smooth brome and orchard grass) within 10 years (also see Turkington et al. 1977). The polymorphic variability of clover detected by Turkington and Harper (1979) apparently developed in less than 50 years. Aarssen and Turkington (1985a) detected development of competitive combining ability (see below and Aarssen 1983) among plants that had been associated for 40 years. Additional studies reviewed by Turkington and Aarssen (1984) indicate that such rapid evolutionary change is not unusual.

In addition to possible consequences involving speciation, local patterns of population differentiation involving biotic specialization have clear implications with respect to patterns and processes of community organization and development. Aarssen and Turkington (1985b) studied patterns of species associations in fields that were seeded in 1939, 1958, and 1977, and their results led to a model of community development. Younger communities had more between-species associations that were temporary [i.e., between-species contacts (via analysis of contact samplng) temporarily deviated from random expectation with a seasonal pattern], whereas older communities had more stable species associations (i.e., non-random associations that persisted). These latter associations often involved stable negative associations between four different pairs of grasses and the development of more stable positive associations between clover and several grasses. Aarssen and Turkington (1985b) offered an interpretation of grassland development involving:

> "a transition from an essentially unorganized assemblage of neighborhoods in which significant deviations from random associations are in a constant state of flux, to a more stable community where neighbors of different species maintain a more permanent association."

The development of these associations takes place in the context of different species, and different genotypes of different species, continually encountering each other through growth and reproduction. These encounters allow for the possibility that natural selection may be reciprocal between competitors, "continually favoring different adaptive genetic variants as species track one another in a continuing process of coevolution which maintains a balance of relative competitive abilities"

(Aarssen and Turkington 1985b). Aarssen (1983) elaborated the distinction between ecological combining ability (essentially niche differentiation) and competitive combining ability, and it is the development among plants of the latter type of combining ability (a key component of which is biotic specialization) that is critical to the model of community development of Aarssen and Turkington (1985b).

Implications to Range Management

Tisdale (1962) noted that the ecotype concept had not been actively pursued as a tool for applied ecology. This specific remark regarding ecotypes can be extended to genetic variation in plants in general. In view of the accumulating information on ecotypic and polymorphic variation, ecologists can no longer regard all plants or populations of plants within a species as uniform. This view has implications in all areas of applied ecology.

For example, consider the area of selection of grasses for revegetating old fields or deteriorated rangelands. Coyne et al. (1982) evaluated four accessions of the Old World Bluestems (*Bothrichloa*). Three of the four accessions were varieties of *B. ischaemum* [WW-477 (*B. ischaemum* var. *songarica* (Rupr.) Celar & Harl.), WW-573 and WW-604 (*B. ischaemum* var. *ischaemum* Keng.)], WW-573 (WW-SPAR) produced the greatest amount of forage and remained green through both high temperature and water stress during the summer. It also exhibited the lowest transpiration rate, the lowest stomatal conductance and was among the highest in water use efficiency (g CO_2/g H_2O) (Coyne et al. 1982). In contrast, WW-535 (also *B. ishaemum* var. *ischaemum* Keng.) (Ironmaster) is best adapted to iron-deficient soils (Berg et al. 1986, Dewald et al. 1988). Ironmaster is a valuable forage species, but it is also useful for stabilizing erodible soils and for wildlife cover on iron deficient soils (Dewald et al. 1988). Although these accessions of Old World Bluestem are morphologically distinct (varieties), they are also physiologically distinct and respond to the environment differently according to their adaptations and these differences have a genetic basis; therefore, they exhibit ecotypic variation. Similarly, genotypes of weeping lovegrass [*Eragrostis curvula* (Schrad.) Nees] exhibit ecotypic variation to iron chlorosis on calcareous soils (Foy et al. 1981). Ecotypic differentiation also occurs in eastern gamagrass [*Tripsacum dactyloides* (L.) L.] with some populations having greater adaptability to calcareous soils and also some populations have higher forage quality than others (Schliesing and Dahl 1983).

Similarly, reforestation success could depend upon selection of tree seedlings or saplings that are not preferred by wildlife. Select genotypes of Douglas-fir [*Pseudotsuga menziesii* (Mirb.) Franco] resistant to browsing by black-tailed deer produced secondary chemicals (phenols, flavanols, and leucoanthocyanins) as well had lower dry matter and cellulose digestibilities (Radwan 1972). Revegetation of drastically disturbed lands, such as mine spoils, oil exploration, oil spills, etc, will dictate selection of ecotypes adapted to extreme environmental conditions.

An area in which little attention has been directed is the impact of ecotypic variation in weed control. Canada thistle [*Cirsuim arvense* (L.) Scop.] control with 2,4-D (2,4-Dichlorophenoxy acetic acid) is often related to ecotypic variation (Hodgson 1973). Hodgson found that three ecotypes with the greatest amount of lipid deposition on the leaves were resistant to 2,4-D, whereas four ecotypes with the least amount of lipid deposition were the most susceptible. Wilkenson (1980) found ecotypic variation in epicuticular wax alkane and fatty acid contents in salt cedar (*Tamarix pentandra* Pall.). He suggested this ecotypic variation might affect salt cedar response to postemergence-applied herbicides.

A number of examples of genetic variation related to environment (ecotypic variation) and to biotic effects (biotic specialization) have been discussed. It is clear that both types of variation in what a classical taxonomist might consider a single species are critically important to the ecologist who must understand how plants interact with their environment. When we consider that plant variation is related not only to abiotic but also to biotic components of the environment, it is clear that both "Wallacian" and Darwinian" approaches to plant ecology are necessary to the practicing range ecologist. These considerations apply not only when one regards plants individually in their autecological perspective, but also plants collectively in their synecological perspective. When we begin to incorporate state-of-the-art understanding of plant variation into issues of reseeding deteriorated rangelands, rehabilitation of drastically disturbed lands, reforestation, erosion control and land stabilization, forage quality for grazing management, wildlife habitat management, etc, then the success ratio of our management schemes should improve.

To the extent that range management involves understanding both aspects of plant ecology, there can be little doubt that failure to recognize and appreciate variability within and between plant populations can be disastrous:

> "One very important consequence of recognizing that ecology is evolution in action is that it becomes extremely dangerous to generalize about the ecological properties of a species. There is no meaningful way in which a study made of one sample from a species can be said to represent that species until the range of variation within and between its populations has been established" (Harper 1977).

Literature Cited

Aarssen, L. W. 1983. Ecological combining ability and competitive combining ability: toward a general evolutionary theory of coexistence in systems of competition. Am. Nat. 122:707-731.

Aarssen, L. W. and R. Turkington. 1985a. Competitive relations among species from pastures of different ages. Can. J. Bot. 63:2319-2325.

Aarssen, L. W. and R. Turkington. 1985b. Vegetation dynamics and neighbour associations in pasture-community evolution. J. Ecol. 73:585-603.

Antonovics, J., N. C. Ellstrand and R. N. Brandon. 1988. Genetic variation and environmental variation: expectations and experiments, pp. 275-303. *In*: L. D. Gottlieb and S. K. Jain (eds.), Plant Evolutionary Biology. Chapman and Hall, London, 414 pages.

Barbour, M. G., J. H. Burk and W. D. Pitts. 1980. Terrestrial Plant Ecology. The Benjamin/Cummings Publ. Co., Inc., Menlo Park, CA., 604 pages.

Barnes, P. W., S. D. Flint, and M. M. Caldwell. 1987. Photosynthesis damage and protective pigments in plants from a latitudinal arctic/alpine gradient exposed to supplemental UV-B radiation in the field. Arctic and Alpine Res. 19:21-27.

Berg, W. A., C. L. Dewald, and P. I. Coyne. 1986. Selection for iron-efficient old world bluestems. J. Plant Nut. 9:453-458.

Bohmont, B. L. and R. Lang. 1957. Some variations in morphological characteristics among geographic strains of Indian ricegrass. J. Range Manage. 10:127-134.

Briggs, D. and S. M. Walters. 1984. Plant Variation and Evolution, second ed., Cambridge Univ. Press, Cambridge, 412 pages.

Brown, W. F. and R. S. Kalmbacher. 1988. Variation in forage component composition and *in vitro* digestion of creeping bluestem. Crop Sci. 28:851-855.

Clausen, J., D. D. Keck, and W. M. Hiesey. 1939. The concept of species based on experiment. Am. J. Bot. 26:103-106.

Clausen, J., D. D. Keck, and W. M. Hiesey. 1940. Experimental studies on the nature of species. I. Effect of varied environments on western North American plants. Carnegie Inst. of Washington Pub. No. 520. Washington, D.C. 436p.

Clausen, J., D. D. Keck, and W. M. Hiesey. 1948. Experimental studies on the nature of species. III. Environmental responses of climatic races of achillea. Carnegie Inst. of Washington Pub. No. 581. Washington, D.C. 123p.

Clough, J. M., R. S. Aberte, and J. A. Teeri. 1980. Photosynthetic adaptation of *Solanum dulcamara* L. to sun and shade environments. III. Characterization of genotypes with differing photosynthetic performance. Oecologia 44:221-225.

Coleman, W. 1986. Evolution into ecology? The strategy of Warming's ecological plant geography. J. Hist. Biol. 19:181-196.

Cooper, J. P. 1951. Studies on growth and development in *Lolium*. II. Pattern of bud development of the shoot apex and its ecological significance. J. Ecol. 39:228-270.

Coyne, P. I., J. A. Bradford, and C. L. Dewald. 1982. Leaf water relations and gas exchange in relation to forage production in four Asiatic bluestems. Crop Sci. 22:1036-1040.

Daubenmire, R. F. 1974. Plants and environment, third ed. John Wiley and Sons. New York, New York.

Dewald, C. L., P. L. Sims, W. A. Berg, and L. M. White. 1988. Registration of 'WW-Ironmaster' Old World Bluestem. Crop Sci. 28:189-190.

Ehrlich, P. R. and P. H. Raven. 1969. Differentiation of populations. Science 165:1228-1232.

Eickmeier, W., M. Adams, and D. Lester. 1975. Two physiological races of *Tsuga canadensis* in Wisconsin. Can. J. Bot. 53:940-951.

Etherington, J. R. and O. M. Thomas. 1986. Response to waterlogging and differing sensibility to divalent iron and manganese in clones of *Dactylis glomerata* L. derived from well-drained and poorly-drained soils. Ann. Bot. 58:109-119.

Foy, C. D., A. L. Fleming, J. W. Schwartz. 1981. Differential resistance of weeping lovegrass genotypes to iron-related chlorosis. J. Plant Nutr. 3:537-550.

Goldberg, D. E. 1988. Response of *Solidago canadensis* clones to competition. Oecologia 77:356-364.

Goodman, P. J. and M. M. Caldwell. 1971. Shrub ecotypes in a salt desert. Nature 232:571-572.

Gould, S. J. 1989. Wonderful Life. W. W. Norton Co., New York.

Gregor, J. W. 1944. The ecotype. Biol. Rev. 19:20-30.

Grelen, H. E. 1974. Variation in pinehill bluestem, a Southern ecotype of the *Andropogon scoparius* complex. J. Range Manage. 27:283-284.

Haeckel, E. 1866. Generelle Morphologie der Organismen. Berlin: Georg Reimer.

Hancock, J.F., Jr. and R. E. Wilson. 1976. Biotype selection in *Erigeron annuus* during old field succession. Bull. Torr. Bot. Club 103:122-125.

Harper, J. L. 1977. Population Biology of Plants. Academic Press, New York, 892 pp.

Harper, J. L. 1982. After description, pp. 11-25. *In:* E. I. Newman (ed.), The Plant Community as a Working Mechanism. Spec. Publ. Ser., British Ecol. Soc. No. 1, Blackwell Sci. Publ., 128 pages.

Heslop-Harrison, J. 1965. Forty years of genecology. Adv. Ecol. Research. 2:159-247.

Heslop-Harrison, J. 1967. New Concepts in Flowering-Plant Taxonomy. Harvard Univ. Press, Cambridge, MA. 134 pages.

Heywood, J. S. 1986. Edaphic races of *Gaillardia pulchella* in central Texas. J. Heredity 77:146-150.

Hiesey, W. M. and H. W. Milner. 1965. Physiology of ecological races. Ann. Rev. Plant Physiol. 16:203-216.

Hironaka, M. 1979. Basic synecological relationships of the Columbia River sagebrush type, pp. 27-32. *In*: The Sagebrush Ecosystem: A Symp. Utah State University, College of Natural Resources. Logan, Utah.

Hodgson, J. M. 1973. Lipid deposition on leaves of Canada thistle ecotypes. Weed Sci. 21:169-172.

Hulbert, L. C. 1955. Ecological studies of *Bromus tectorum* and other annual bromegrasses. Ecol. Monogr. 25:181-213.

Jain, S. K. 1990. Variation and selection in plant populations, pp. 199-130. *In*: K. Wohrmann and S. K. Jain (eds.), Population Biology. Springer-Verlag, Berlin, 456 pages.

Jenny, H. 1980. The Soil Resource. Ecological Studies. Vol. 37. Springer-Verlag, New York.

Jowett, D. 1964. Population studies on lead-tolerant *Agrostis tenuis*. Evolution 18:70-80.

Kallio, H. and K. Junger-Mannerma. 1989. Maritime influence on the volatile terpenes in the berries of different ecotypes of juniper (*Juniperus communis* L.) in Finland. J. Agr. Food Chem. 37:1013-1016.

Kelley, S. E. and K. Clay. 1987. Interspecific competitive interactions and the maintenance of genotypic variation within two perennial grasses. Evolution 41:92-103.

Klebesadel, L. J. 1985. Hardening behavior, winter survival, and forage productivity of *Festuca* species and cultivars in subarctic Alaska. Crop Sci. 25:441-447.

Larsen, E. C. 1947. Photoperiodic responses of geographical strains of *Andropogon scoparius*. Bot. Gaz. 109:132-149.

Linhart, Y. B. 1974. Intra-population differentiation in annual plants. I. *Veronica peregrina* L. raised under non-competitive conditions. Evolution 28:232.

Main, J. L. 1974. Differential responses to magnesium and calcium by native populations of *Agropyron spicatum*. Am. J. Bot. 61:931-937.

Mayr, E. 1982. The Growth of Biological Thought: Diversity, Evolution, and Inheritance. Harvard Univ. Press, Cambridge, MA.

Mayr, R. 1940. Speciation phenomena in birds. Am. Nat. 74:249-278.

McArthur, E. D. 1983. Taxonomy, origin, and distribution of big sagebrush (*Artemisia tridentata* and allies (*Subgenus Tridentatae*), pp. 3-11. *In*: K. L. Johnson (ed.), First Utah Shrub Ecology Workshop, Proc. (Ephraim, Utah). Utah State University, College of Natural Resources. Logan, Utah.

McArthur, E. D. 1979. Sagebrush systematics and evolution, pp. 14-22. *In*: The Sagebrush Ecosystem: A Symp. Utah State University, College of Natural Resources. Logan, Utah.

McClurkin, D. C., I. T. McClurkin, and T. J. Culpepper. 1971. Cytochemical and tissue homogenate analysis of adenosine triphosphate in root tips of Texas "Lost Pines." Forest Sci. 17:446-451.

McGraw, J. B. 1985a. Experimental ecology of *Dryas octapetala* ecotypes: Relative response to competitors. New Phytol. 100:233-241.

McGraw, J. B. 1985b. Experimental ecology of *Dryas octapetala* ecotypes. III. Environmental factors and plant growth. Arctic and Alpine Res. 17:229-239.

McGraw, J. B. and J. Antonovics. 1983. Experimental ecology of *Dryas octapetala* ecotypes. I. Ecotypic differentiation and life-cycle stages of selection. J. Ecol. 71:879-897.

McMillan, C. 1964. Ecotypic differentiation within four North American prairie grasses. I. Morphological variation within transplanted community fractions. Am. J. Bot. 51:1119-1128.

McMillan, C. 1965a. Ecotypic differentiation within four North American prairie grasses. II. Behavioral variation within transplanted community fractions. Am. J. Bot. 52:55-65.

McMillan, C. 1965b. Grassland community fractions from central North America under simulated climates. Am. J. Bot. 52:109-116.

McNaughton, S. J. 1973. Comparative photosynthesis of Quebec and California ecotypes of *Typha latifolia*. Ecology 54:1260-1270.

McNaughton, S. J. 1972. Enzymic thermal adaptations: The evolution of homeostasis in plants. Am. Nat. 106:165-172.

McNaughton, S. J. 1969. Genetic and environmental control of glycolic acid oxidase activity in ecotypic populations of *Typha latifolia*. Am. J. Bot. 56:37-41.

McNaughton, S. J. 1967. Photosynthetic system II: Racial differentiation in *Typha latifolia*. Science 156:1363.

McNaughton, S. J. 1966a. Oxidase activity in ecotypic populations of *Typha latifolia* L. Nature 211:1377-1379.

McNaughton, S. J. 1966b. Thermal inactivation properties of enzymes from *Typha latifolia* L. ecotypes. Plant Physiol. 41:1736-1738.

McNaughton, S. J. 1966c. Ecotype function in the *Typha* community-type. Ecol. Monogr. 36:296-325.

McNaughton, S. J. 1965. Differential enzymatic activity in ecological races of *Typha latifolia* L. Science 150:1829-1830.

Mooney, H. A. and W. D. Billings. 1961. Comparative physiological ecology of arctic and alpine populations of *Oxyria digyna*. Ecol. Monogr. 31:1-29.

Mooney, H. A. and A. W. Johnson. 1965. Comparative physiological ecology of an arctic and an alpine population of *Thalictrum alpinum* L. Ecology 46:721-727.

Musselman, R. C., D. T. Lester, and M. S. Adams. 1975. Localized ecotypes of *Thuja occidentalis* L. in Wisconsin. Ecology 56:647-655.

O'Brien, S. J. and E. Mayr. 1991. Bureaucratic mischief: recognizing endangered species and subspecies. Science 251:1187-1188.

Ohlson, M. 1989. Ecotypic differentiation and phenotypic plasticity in *Saxifraga hirculus* populations in central and southern Sweden. Holarctic Ecol. 12:46-53.

Olmstead, C. E. 1945. Growth and development in range grasses. V. Photoperiodic responses of clonal divisions of three latitudinal strains of side-oats grama. Bot. Gaz. 106:382-401.

Olmstead, C. E. 1944. Growth and development in range grasses. IV. Photoperiodic responses in twelve geographic strains of sideoats grama. Bot. Gaz. 106:46-74.

Osmund, C. B. 1983. Interactions between irradiance, nitrogen nutrition, and water stress in the sun-shade responses of *Solanum dulcamara*. Oecologia 57:316-321.

Peacock, J. T. and C. McMillan. 1965. Ecotypic differentiation in *Prosopis* (mesquite). Ecology 46:35-51.

Plantenkamp, G. A. J. and T. C. Foin. 1990. Ecological and evolutionary importance of neighbors in the grass *Anthoxanthum odoratum*. Oecologia 83:201-208.

Radwan, M. A. 1972. Differences between Douglas-fir genotypes in relation to browsing preferences by black-tailed deer. Can J. For. Research 2:250-255.

Ramakrishnan, P. S. and R. Kumar. 1976. Adaptive responses of an alkaline soil population of *Cynodon dactylon* L. Pers. to NPK nutrition. J. Ecol. 64:187-193.

Rhebergen, L. J. and H. J. M. Nelissen. 1985. Ecotypic differentiation within *Festuca rubra* L. occurring in a heterogeneous coastal environment. Vegetatio 61:197-202.

Rozema, J., E. Rozema-Dijst, A. H. J. Freijsen, and J. J. L. Huber. 1978. Population differentiation within *Festuca rubra* L. with regard to soil salinity and soil water. Oecologia 34:329-341.

Schliesing, T. G. and B. E. Dahl. 1983. Ecotypic variation in *Tripsacum dactyloides* evaluated in Texas. J. Range Manage. 36:665-668.

Shaver, G. R., F. S. Chapin, III, and W. D. Billings. 1979. Ecotypic differentiation in *Carex aquatilis* on ice-wedge polygons in the Alaskan coastal tundra. J. Ecol. 67:1025-1046.

Simon, J. P. 1986. Molecular forms and kinetic properties of the mitochondrial fraction of NAD-malate dehydrogenase in four clones of *Lathyrus japonicus* (Leguminosae) from sites of contrasting climates. Can. J. Bot. 64:2675-2681.

Simon, J. P. 1979. Differences in thermal properties of NAD malate dehydrogenase in genotypes of *Lathyrus japonicus* Willd. (Leguminosae) from maritime and continental sites. Plant, Cell and Environment 2:23-33.

Sinskaja, E. N. 1931. Bull. Appl. Bot. Genet. Pl. Breed. 25:1-97.

Solbrig, O. T. and B. B. Simpson. 1974. Components of regulation of a population of dandelions in Michigan. J. Ecol. 63:473-486.

Solbrig, O. T. and B. B. Simpson. 1977. A garden experiment on competition between biotypes of the common dandelion (*Taraxacum officinale*). J. Ecol. 65:427-430.

Sowell, J. B. and G. G. Spomer. 1986. Ecotypic variation in root respiration rate among elevational populations of *Abies lasiocarpa* and *Picea engelmannii*. Oecologia 68:375-379.

Stebbins, G. L. 1979. Fifty years of plant evolution, pp. 18-41. *In*: O. Solbrig, S. Jain, G. B. Johnson, and P. H. Raven (eds.), Topics in Plant Biology. Columbia University Press. New York, New York.

Stussey, T. F. 1990. Plant Taxonomy, The Systematic Evaluation of Comparative Data. Columbia Univ. Press, New York. 514 pages.

Taylor, D. R. and L. W. Aarssen. 1990. Complex competitive relationships among genotypes of three perennial grasses: Implications for species coexistence. Am. Nat. 136:305-327.

Taylor, G. E., Jr., D. T. Tingey, and C. A. Gunderson. 1986. Photosynthesis, carbon allocation, and growth of sulfur dioxide ecotypes of *Geranium carolinianum* L. Oecologia 68:350-357.

Tisdale, E. W. 1962. The ecotype concept in range management, pp. 61-68. *In*: J. W. Gregory (Dir.), Scottish Plant Breeding Station Record. William Blackwood and Sons, Ltd. Edinburgh, Scotland.

Tkachenko, K. G. 1985. *Ephedra* cenoecotypes on the shore of the Aral Sea. Prob. of Desert Dev. 1985(2):77-80.

Turesson, G. 1922. The species and the variety as ecological units. Hereditas 3:100-113.

Turesson, G. 1923. The scope and import of genecology. Hereditas 4:171-176.

Turesson, G. 1925. The plant species in relation to habitat and climate. Hereditas 4:147-236.

Turkington, R. A., P. B. Cavers, and L. W. Aarssen. 1977. Neighbor relationships in grass-legume communities. I. Interspecific contacts in four grassland communities near London, Ontario. Can. J. Bot. 55:2701-2711.

Turkington, R. 1979. Neighbor relationships in grass-legume communities. IV. Fine scale biotic differentiation. Can. J. Bot. 57:2711-2716.

Turkington, R. 1985. Variation and differentiation in populations of *Trifolium repens* in permanent pastures, pp. 69-82. *In*: J. White (ed.), Studies on Plant Demography, A Festschrift for John L. Harper. Academic Press, New York, 393 pages.

Turkington, R. and J. L. Harper. 1979. The growth, distribution and neighbor relationships of *Trifolium repens* in a permanent pasture. J. Ecol. 67:245-254.

Turkington, R. and L. W. Aarssen. 1984. Local-scale differentiation as a result of competitive interactions, pp. 107-127. *In*: R. Dirzo and J. Sarukhan (eds.), Perspectives on Plant Population Ecology. Sinauer Assoc. Inc. Publ., Sunderland, MA 478 pages.

West, N. E. 1988. Intermountain deserts, shrub steppes, and woodlands, pp. 209-230. *In*: M. G. Barbour and W. D. Billings (eds.), North American Terrestrial Vegetation. Cambridge University Press. New York, New York.

Whalley, R. D. B. and J. M. Burfitt. 1972. Ecotypic variation in *Avena fatua* L., *A. sterilis*, (*A. ludoviciana*), and *A. barbata* Pott. in New South Wales and southern Queensland. Aust. J. Agric. Res. 23:799-810.

Wilkenson, R. E. 1980. Ecotypic variation of *Tamarix pentandra* epicuticular wax and possible relationship with herbicide sensitivity. Weed Sci. 28:110-113.

Winward, A. H. 1983. Using sagebrush ecology in wildland management, pp. 15-19. *In*: K. L. Johnson (ed.), First Utah Shrub Ecology Workshop, Proc. (Ephraim, Utah). Utah State University, College of Natural Resources. Logan, Utah.

Wooten, J. W. 1973. Edaphic factors in species and ecotype differentiation of *Sagittaria*. J. Ecol. 61:151-156.

Developmental Morphology of Plants

Bill E. Dahl

Introduction

Plant morphology is a branch of botany that deals with how plants grow. Developmental morphology, or morphogenesis, relates to structural changes during the development of an organism (Hyder 1974). *Growth* is a quantitative change in a plant; in contrast, *development* is a series of qualitative changes in form, structure, function, and complexity. Developmental processes involve cell division, cell enlargement, and differentiation of functionally separate parts.

In range management, interest essentially begins with shoot and root apices (growing points), and proceeds to a full array of developmental parts, especially photosynthetic tissues. It is essential that range managers understand the derivation and arrangement of plant structures. Fundamental interest begins with apical meristems of shoots and roots. Although this review omits most of the historical theories about the progression of cell division from a single apical cell or a group of apical cells, it is important to appreciate that apical meristems are the ultimate source of all tissues in a plant body, and that apical meristems produce small masses of meristematic cells that become specialized into the kinds of tissues that they in turn produce.

Mature tissues that arise from enlargement and differentiation of cells derived from the apical meristem are considered *primary* tissues. Primary growth produces roots, stems, leaves, fruits, and seeds as well as the primary xylem and primary phloem. Tissues resulting from vascular cambium activity (secondary xylem and secondary phloem) are *secondary* tissues. Secondary xylem is preserved and eventually forms the bulk of woody plants (Esau 1965). Thus, cambium activities are responsible for the wood used by man. However, primary tissues constitute the bulk of the plant body in herbaceous plants. In monocotyledons, a cambium is absent and all tissues are primary in origin (Wilson and Loomis 1962; Kozlowski 1971).

The importance of understanding plant growth and development in range management lies in the realization that primary objectives in grazing management include maximum production and harvest of digestible material and protein consistent with optimum animal response, sustained plant production and community composition, and soil conservation. In each of these considerations, managing for the expansion of photosynthetic tissues, and their replacement after removal, is a key to success.

In this chapter, appropriate terminology related to plant growth and development is developed. Then, basic plant growth patterns are described. A strong foundation

---

Author (deceased) was professor, Range and Wildlife Management, Texas Tech University, Lubbock, Texas 79409. This is contribution number T-9-651, College of Agricultural Sciences and Natural Resources, Texas Tech University.

in plant developmental morphology will help our understanding of individual plant and community response to management.

Developmental Morphology

Regions of the Shoot Apex

Whether plants develop primarily long shoots (shoots whose internodes elongate while the apex remains vegetation; see below) or primarily short shoots (shoots which elongate very little during vegetative growth; see below) impacts management decisions of grazing managers, weed control specialists, park managers, foresters (Kramer and Kozlowski 1979) and probably most others who deal in plant manipulation. Relatively little botanical literature elaborates on the importance of long *versus* short shoot development except to acknowledge that such differences do exist. Romberger (1963), following the lead of Wardlaw (1957), describes shoot development and specifically describes the shoot apex segment responsible for long *versus* short shoot development, particularly in woody plants. Wardlaw's scheme (Fig. 1) for naming morphologically distinguishable areas of shoot apices as reworded by Romberger follows.

"*Distal region.*– The distal region comprises the summit of the apical dome. It includes the single apical cells of lower plants and the apical initial cell group (or zone) of higher plants.
Subdistal region.– Inception of growth centers (or loci), the sites of subsequent primordial initiation, occurs in the subdistal region. Growth centers are

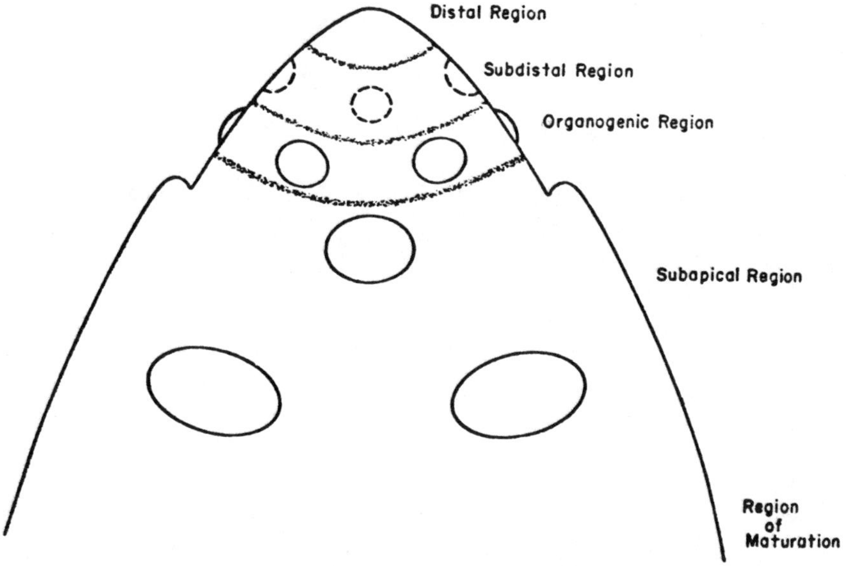

Figure 1. A shoot apex with whorled phyllotaxy, as a system of interrelated regions. Boundaries between regions are diffuse (Romberger 1963).

groups of cells in which concentrations of metabolites conducive to growth have presumably accumulated, but in which no obvious morphological changes can yet be detected.

Organogenic region.– In the organogenic region obvious outgrowth of foliar primordia occurs from the growth centers and internal tissue differentiation becomes detectable or even conspicuous. The boundary between this region and that below is not sharp.

Subapical region.– Characteristically the subapical region exhibits conspicuous primoridal enlargement, considerable widening of the axis, continued differentiation of the vascular tissue, and elongation of internodes resulting from cell division and extension in the rib meristem.

Region of maturation.– In the diffuse lower limits of the subapical region, meristematic activity declines. There is a gradual transition to the region of maturation. In this region the morphogenic patterns initiated and developed in the upper regions are finally fixed."

Similarly, other botanists refer to shoot segments simply as regions of cell division, regions of cell elongation, and regions of maturation (Kaufman et al. 1989).

Special Significance of the Subapical Region

Special significance has always been attached to apical meristems, i.e., the distal region, in plant management because of the phenomenon of apical dominance and because removal of the apical meristem stops generation of new organs on that shoot. However, without an understanding of the role of growth in the subapical region, one can little understand the sources of shoot extension, grazing resistance, susceptibility to herbicides, etc. Romberger (1963) stated that such problem areas as formation and breaking of buds, alterable long shoot *versus* short shoot growth habit in woody plants, rosette *versus* cauline habit in herbaceous plants, physiological dwarfing of plants grown from embryos with unsatisfied chilling requirements, and lengthy "grass stages" in some pine seedlings are only different aspects of the one problem of control of cell division and elongation in the subapical meristem.

The subapical meristem region is of primary importance in normally caulescent plants because once restraint upon its activity has been released, it generates most of the cells that make up the mature internodes. Because of cell elongation in the subapical region, movement of cells and regions is usually upward, i.e., youngest cells occur usually at the bottom or base of internodes. Internodes elongate because their original cells elongate. They also elongate because there is a great increase in cell numbers (Romberger 1963). For further discussion of this point, see the section "Growth of Forbs" below.

Intercalary Meristems

Intercalary meristems are merely portions of apical meristems separated from the apex during development by layers of more mature or permanent tissues (Esau 1960), e.g., at the base of internodes, at the base of monocotyledon leaves and at the petiole base of dicot leaves (Evans 1958; Jewiss 1966; Youngner 1969;

Kozlowski 1971). The term *rib meristem* is commonly used for this growth region when applied to trees (Wilson 1984). These more mature areas are the nodal regions, so the intercalary meristems are "internodal." When the youngest region is basal, the order of maturation in internodal tissues is acropetal, i.e., the most mature section is at the top of the internode. Grasses, other monocots, and *Equisetum* sp. L. are examples of acropetal maturation. Also, the youngest portions of the leaves of grasses, most other monocots, pines (*Pinus* sp. L.), etc., occur basally. Rarely, the youngest region is at the top, and then the order of maturation is basipetal.

Lateral Meristems

Lateral meristems are situated laterally in an organ. They are composed of initials that divide chiefly in one plane (periclinally) and increase the diameter of an organ, e.g., cambium and cork cambium of trees and shrubs (Wilson and Loomis 1962; Northington and Goodin 1984; Salisbury and Ross 1985). While we think of lateral meristems as primarily occurring in woody plants, the majority of dicotyledons and gymnosperms have secondary growth resulting from the vascular (lateral) cambium. Only small dicotyledonous annuals and most monocotyledons have only primary growth (Esau 1960).

Growth of Leaves

The leaf primordium is first evident externally as a rounded or wedge-like protuberance on the side of the shoot apex promeristem, i.e., the organogenic region of the apex. At the tip of the protuberance is a group of meristematic cells that constitutes the leaf meristem. If the leaf is stipulate, stipular meristems arise early as swellings at the base of this structure. Lateral ridges, which are the marginal meristems that build the leaf blade, appear next, except at the base (Esau 1965; Kramer and Kozlowski 1979). Thus, leaf growth is at first apical and marginal, but cell division is soon restricted to the base and perhaps the margin of a leaf. In most dicots the base of the finger-like structure remains unexpanded and forms an intercalary meristem that builds the petiole (Romberger 1963; Esau 1960). In grasses, which usually have no petiole, a collar develops before the blade attains much length and separates the intercalary meristems of blade and sheath. Also, early in the development of a phytomer (shoot segment), intercalary meristems separate from the leaf and internode. Thus, intercalary meristems in grasses occur primarily at the base of the blade, the base of the sheath, and the base of the internode (Esau 1960).

Cell division in leaves ceases while the leaves are minute, and growth consists largely of rapid increase in cell size and maturation of the mesophyll cells and vascular tissues (Romberger 1963; Esau 1960). In grasses, the intercalary meristem of leaf blades ceases activity by the time of leaf collar exertion (Sharman 1942). Appearance of the ligule heralds the final leaf extension of most grasses (Sharman 1942; Langer 1954).

Growth of Buds

A mature bud is a rudimentary shoot, or part of a shoot, bearing at its tip the apical meristem from which it originated (Kozlowski 1971). For convenience the initiation of buds will be considered the initial phase of shoot growth. Most buds are initiated in leaf axils, where some remain dormant and others develop into shoots (Romberger 1963) depending largely on growth control mechanisms referred to as apical dominance. According to Esau (1960) the apical meristem of the bud produces leaf primordia before tissue specialization in the central axis.

Woody plants generally have a wider variety of buds than herbaceous plants; hence, a relatively complex classification scheme has evolved to characterize them. Because shoot origin and form is important to plant management and control, a brief description of the more important and more common buds is given. According to Kramer and Kozlowski (1979) buds of woody plants are classified according to location, content, or activity. They may be terminal, lateral, axillary, vegetative, flower, mixed, adventitious, etc. Any of these may be further classified as active or dormant. A vegetative bud contains a small mass of meristematic tissue, nodes, internodes, small rudimentary leaves with miniature buds in their axils, and all are usually enclosed in bud scales. Cataphylls, the largest and oldest leaf primordia, are located at the bud base and the small rudimentary leaves occur toward the growing point. Flower buds contain embryonic flowers and most of them contain rudimentary leaves. Mixed buds contain both flowers and leaves. A mixed bud type that terminates in a flower cluster in leaf axils is found in *Diospyros* sp. L., *Quercus* sp. L., and *Fagus* sp. L. (Kramer and Kozlowski 1979). *Prosopis* sp. L., common on southwestern rangelands, have mixed buds (Meyer et al. 1971).

Adventitious buds of woody plants form irregularly in the cambium of older portions of a plant rather than in leaf axils near an apical meristem. These form on parts of the root or stem that have no connection to apical meristems (Kozlowski 1971). Whether sprouts originate from dormant lateral or adventitious buds is difficult to determine, but true dormant buds have a bud trace all the way to the pith, whereas adventitious buds do not. Stump sprouts originating from parts of the stem may arise from either dormant lateral or adventitious buds (Zimmerman and Brown 1971). In contrast, root sprouts ("root suckers") always arise from adventitious buds (Kozlowski 1971; Hartman et al. 1990).

Grasses and most other herbaceous plants have no secondary tissues, so the aerial portion is relatively short-lived and the tissues comparatively soft. Consequently, the meristematic potential for roots or shoots at any node is either activated or lost within a few weeks after derivation. An "old" piece of crown and an "old" piece of rhizome will not grow, although the aging process varies somewhat among species (Dahl and Hyder 1977).

Shoot Development

Shoot is a collective term applied to the stem and leaves, or any young growing branch or twig. The stem is typically divided into nodes and internodes. The node is the part of the stem where one or more leaves are attached. Internodes are lengths of stem between two successive nodes (Kramer and Kozlowski 1979). When leaves are first formed, they occur close together, and nodes and internodes are, or appear, absent. Subsequently, as meristematic activity occurs between the leaf insertions, internodes become recognizable (Esau 1960). Hence, the shoots of most plants are composed of repeated structural units, formed in a serial sequence by an apical meristem (White 1979). These repeated structural units then are the basic unit of a shoot and in grasses they have traditionally been called *phytomers* (Beard 1973). Dahl and Hyder (1977) decided that the term "phytomer" had similar recognizable (homologous) units among dicotyledons and they suggested use of the term for these non-grass plants as well (Fig. 2). However, according to Eames (1961) and Brede (1984) phytomers do not exist and the shoot is the basic plant unit. To this argument White (1985), an Irish botanist, has responded that phytomers *do* exist in many plants and that the term has utility among field botanists and should be maintained in morphology terminology. In fact, phytomers have been recognized as a basic shoot segment for 150 years. White (1979) reported that a French botanist Gaudichaud in 1841 considered a "phyton" to consist of an internode and its upper node with the attached leaf. Asa Gray used the term "phytomer" which is now commonly defined as "a leaf, an internode, an axillary bud, or potential thereof, and a node" (Hyder 1974). It should be noted that nodes also have potential for roots as well as bud(s). The term "metamer," widely used among zoologists, has also gained acceptance among contemporary botanists for this basic shoot unit (White 1979).

In agreement with White (1985), since the concepts which these terms embody have been usefully employed by agricultural scientists, it would be better to continue their use for the benefit of plant science as a whole, rather than to discard them. Admittedly, the terms have more utility for grasses where nodes are more distinct.

Shoots may originate directly from the embryo in a seed, i.e., the primary shoot, as a branch from an axillary (lateral) bud, or adventitiously (Beard 1973). The terminology applied to the various shoots of plants has varied depending on whether woody or herbaceous plant shoots were described. However, where possible, similarities will be emphasized here, e.g., the culmless vegetative shoots of grasses such as timothy (*Phleum pratense* L.) or sideoats grama [*Bouteloua curtipendula* (Michx.) Torr.] are comparable to the short shoots of forbs such as dandelion (*Taraxacum officinale* Wiggers) and the short shoots of woody plants. The culmed vegetative shoots of grasses, such as switchgrass (*Panicum virgatum* L.), are comparable to the long shoots of alfalfa and of woody plants. The point is that mechanisms of growth control and genetics often prevent much internode elongation and the terms short and long shoots aptly describe shoots with and without significant internode elongation, respectively.

The Individual Plant

In addition to disagreeing on basic units of a plant, botanists have found it difficult to simply describe a plant. White (1979) wrote that "many biologists without any explicit theoretical justification have simply used the seemingly obvious subunitary structures of plants (however interpreted) to describe the dynamics of pasture grasses, the geometry of canopy development in trees, or some features of insect population behavior. And some plant morphologists still find the idea fruitful in new guises." Because even laymen can tell what constitutes a tree, or a grass, or a forb, we are inclined to be among those that find the use of simple plant descriptions fruitful. Nevertheless, contemporary botanists use a variety of terms pertaining to an individual plant that field botanists, i.e., range scientists and

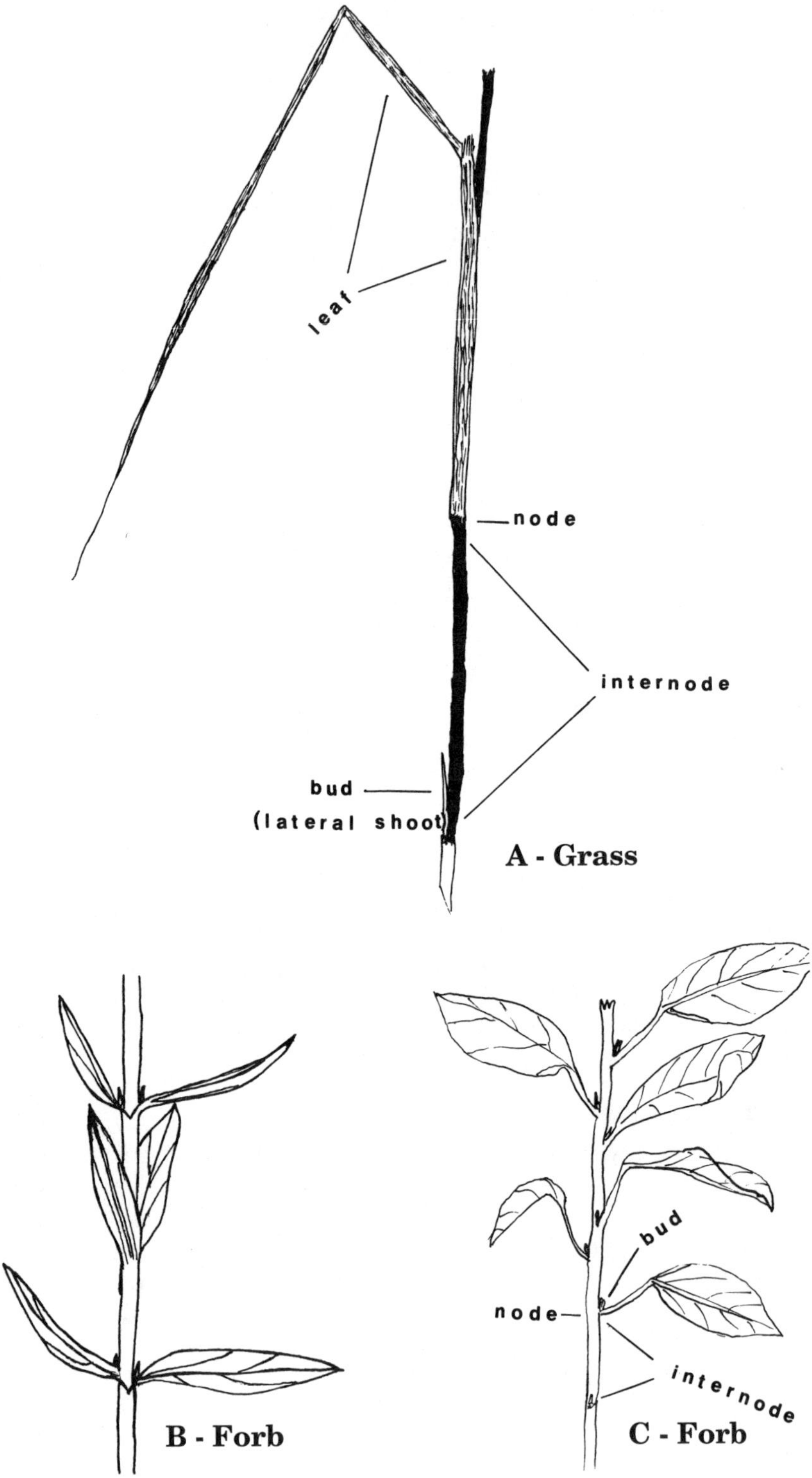
leaf
node
internode
bud
(lateral shoot)
A - Grass
B - Forb
bud
node
internode
C - Forb

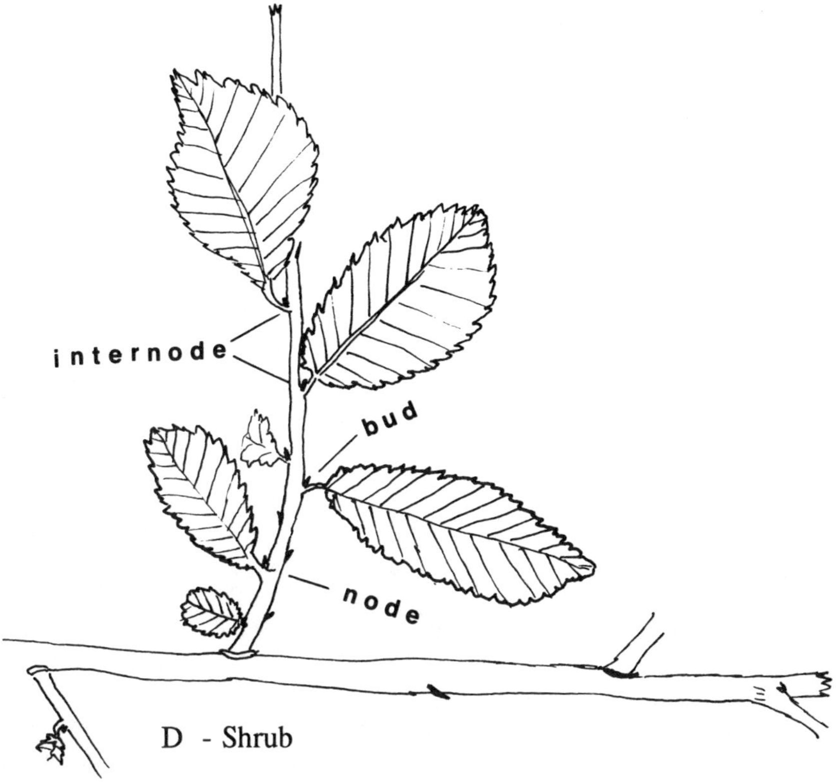

Figure 2. The phytomer (basic unit of the shoot) consists of a leaf, an internode, an axillary bud or potential bud, and a node. A—grass; B—forb (opposite leaves); C—(forb alternate leaves); D—shrub.

agronomists, should understand if they are to comprehend current botanical literature.

An *individual plant* is a plant that can be counted as physically distinct from another, regardless of whether it is genetically distinct. A *genet* is also an individual plant, of whatever size or however subdivided, derived from a single seed. Or, stated another way, a *genet* is the sum of all ramets derived from a single zygote—a genetic individual or clone. An equivalent term (common in forestry) is *ortet*. A plant has also been described as a *metapopulation* which is the shared elements that make up the morphological structure of a genetic individual, by whatever means these elements are defined (White 1979). A *clone* is defined as any organism that grows and propagates by self-replication of genetically identical units (*ramets*) that can function and survive on their own if separated from one another. Thus, plant synonomy for an individual currently includes the terms *genet*, *ortet*, *clone*, and *metapopulation* (Jackson et al. 1986; White 1979).

Relative to parts of a plant (shoots), the terms *tiller*, *module*, and *ramet* are now commonly used. A *tiller* is a shoot derived from the vertical growth of an axillary bud, especially in grasses. Similarly, a *ramet* is actually or potentially an independent member of a genet (clone). A *module* is defined as a fundamental functional unit of construction of an organism that is repeated in sequential iterative fashion in the development of a colony or clone. It is also defined as a monopodial shoot terminated by an inflorescence, by a spine or tendril, or by parenchymatization of the apical meristem. Hence, plant synonomy for members of an individual plant (a genet or clone) are *tiller*, *ramet*, or *module* (Jackson et al. 1986; White 1979). Terms more specifically related only to a kind of plant, e.g. trees, are described in the appropriate section.

Shoots of Grasses

Vegetative Shoots

In grasses, young vegetative lateral shoots growing upward within the enveloping leaf sheath (intravaginally) are called *tillers*. Such plants tend to have a tufted, caespitose, or bunch-type growth habitat. Vegetative lateral shoots that penetrate through the enveloping leaf sheath (extravaginally plagiotropic shoots of Hartman et al. 1990) result in spreading or creeping growth habits (Fig. 3 and 4). Lateral shoots that arise extravaginally with considerable aboveground, horizontal stem elongation are called stolons; shoots that elongate underground are called rhizomes (Evans 1958; Beard 1973). Plants with stolons or rhizomes or both are considered "sod forming" plants, particularly if they are grasses.

Shoots of herbaceous plants terminating in a reproductive organ or organs are vegetatively determinate. Stolons have continuous growth, and form roots and shoots at the nodes as they grow. Rhizomes may be either continuous, producing aerial shoots progressively as in quack grass [*Agropyron repens* (L.) Beauv.], or terminal, when their apices turn up and emerge as phototropic shoots as in Kentucky bluegrass (*Poa pratensis* L.) (Oakley and Evans 1921). Axillary buds of terminal rhizomes usually remain dormant unless the rhizome is severed from the plant or its apical meristem is damaged (Youngner 1969). Stolons, by successive tillering, contribute assimilates for their own growth more readily than rhizomes. However, any branch depends on the parent shoot until it has mature leaves and adventitious roots (Youngner 1969). Another phenomenon of rhizome and stolons should be mentioned as a unique manifestation of growth control; that is, they exhibit an alternating sequence of long and short internodes. An elongated internode is followed by one or more short internodes then another long one. The short internodes provide a diversion of energy supply into adventitious roots, and sometimes into shoots as well. Such a sequence has never been found in phototropic shoots. This phenomenon of rhizomes and stolons has not been reported in the literature.

Phototropic shoots are conveniently grouped into short, long, and reproductive shoots. All annual and many perennial grasses [e.g., ryegrass (*Lolium perenne* L.), orchard grass (*Dactylis glomerata* L.)], produce short shoots where the stem internodes elongate very little during vegetative growth (Jewiss 1966). While such shoots are also commonly referred to as culmless (Hyder 1972) or sometimes "nonjointing" (Heppner 1961), all subclasses of plants have long and short shoot forms and it would seem preferable to use the long-shoot, short-shoot terms to describe grass shoots. In short-shoot plants the shoot apex remains below cutting or grazing height and continues to produce new leaves (Fig. 5). In grasses, these are typically turf or sod forming plants and they are well adapted to frequent mowing and grazing. Leaf

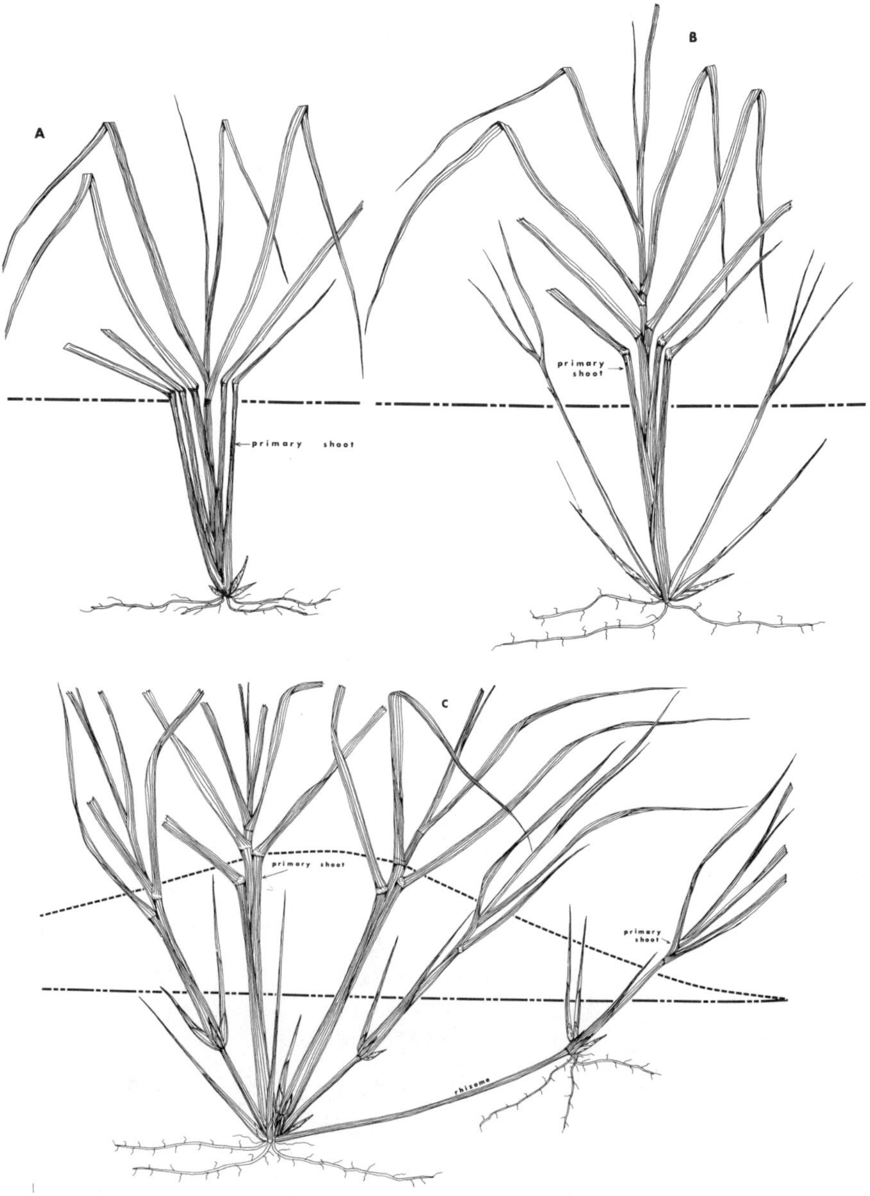

Figure 3. Lateral shoot development of sea oats (*Uniola paniculata* L.), a grass with short shoots and a spreading growth habit. (A) a single shoot at time of planting; (B) lateral shoot development on the same shoot after a few weeks; (C) lateral shoot development at the end of the first growing season. Dashed lines indicate initial planting depth (A) (B) (C) and accumulation of wind-deposited sand (C) (Dahl et al. 1975).

growth stops during unfavorable conditions. Cell division in both apical and intercalary meristems may resume with favorable conditions (Hyder 1972; Esau 1960). The leaf tips usually "die back" during unfavorable weather (e.g. winter condi-

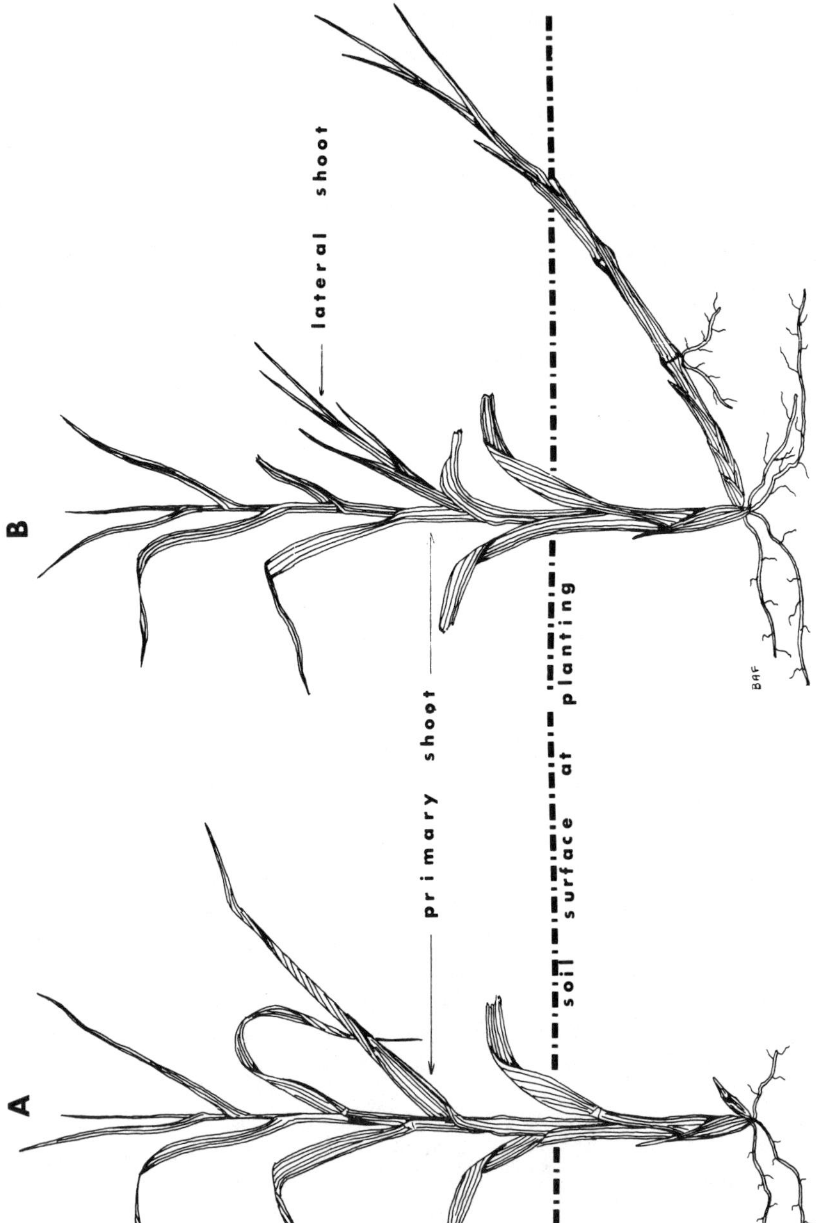

Figure 4. Lateral shoot development of bitter panicum (*Panicum amarum* Ell.), a grass with long shoots and a spreading growth habit. (A) a single shoot at time of planting; (B) lateral shoot development on the same shoot after a few weeks; (C) lateral shoot development at the end of the first growing season. Dashed lines indicate initial planting depth (A) (B) (C) and accumulation of wind-deposited sand (C) (Dahl et al. 1975).

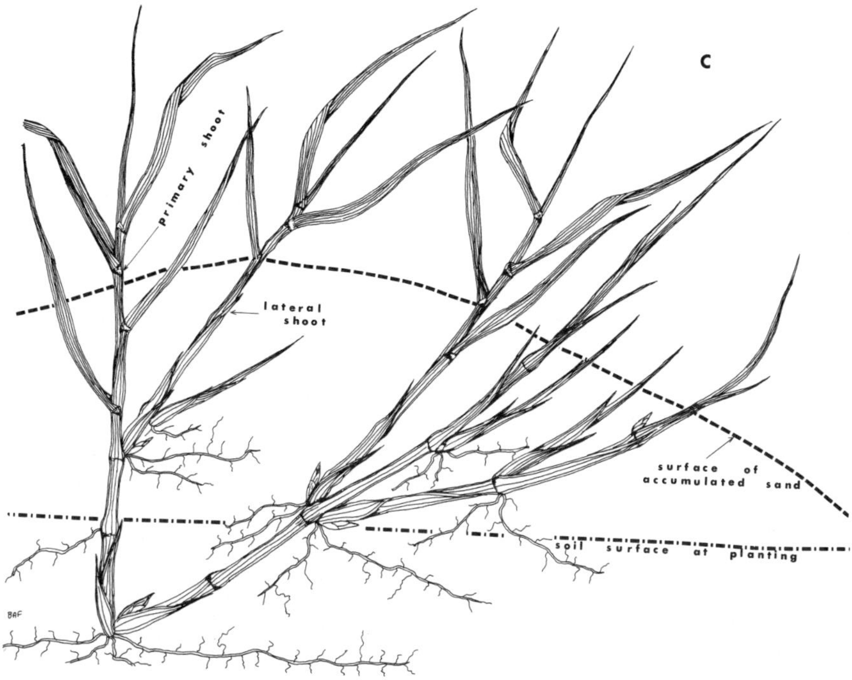

tions). Thus, following winter conditions or other situations causing leaf death, new growth surges reveal immature leaves with dead tips. This resumption of growth in residual intercalary meristems accounts for rapid "green up" following dormancy. The dead leaf tips do not come back to life (green up) as often implied. Succession of the shoot apex to reproductive status, or death by any cause, terminates these short shoots. The short shoots of perennials cannot be called annual or biennial structures because they remain active and indeterminate in producing leaf tissue until death or change to reproductive status (Hyder 1972; Beard 1973).

In long shoots (Fig. 5), internodes elongate while the apex remains vegetative (e.g. [tall oatgrass *Arrhenatherum elatius* (L.) Presl.)], switchgrass). Even these shoots are at first short because internode elongation is much delayed compared to leaf growth. Subsequent internode elongation elevates the shoot apex and leaf intercalary meristems as well. Consequently, cutting or grazing below the shoot apex stops leaf expansion and prevents leaf replacement except by the slower process of tillering from axillary buds of the crown. Long shoots are characteristically annual structures in temperate climates. Plants with such shoots are best adapted to periodic harvest, scheduled to coincide with advanced stages of growth when the axillary buds are better prepared to initiate new tillers (Evans 1958; Jewiss 1966; Youngner 1969; Hyder 1972; Beard 1973).

At this point it is well to re-emphasize the order of growth in grass shoots. Growth proceeds first in the leaf blade, the leaf sheath, and then the internode. Before elongation in the oldest phytomer, i.e., the lowermost phytomer of the shoot, is complete, the next one above begins elongation, etc. In grasses, it is common to refer to the lower portion of the shoot in various ways. *Crown*, *root crown*, *stem base*, *proaxis*,

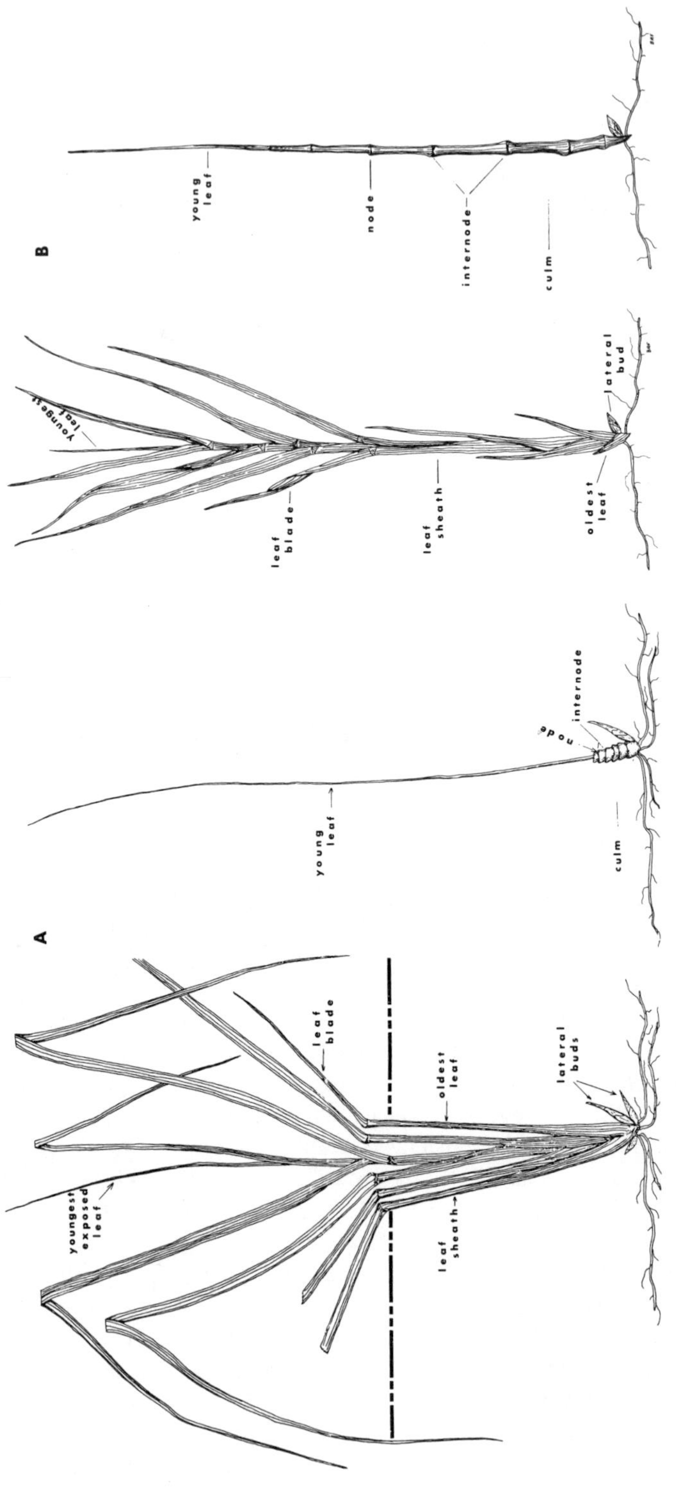

Figure 5. Vegetative shoots of (A) sea oats producing primarily short shoots and (B) bitter panicum producing primarily long shoots.

and *short basal internodes* all are terms that have been used to refer to the lower portion of the grass shoot (Branson 1953; Rechenthin 1956; Evans 1958). The term *crown* is preferred as it applies well to forbs as well as to grasses. It refers to any section of stem base where two or more nodes remain close together (Fig. 6). It includes the vertical section of a terminal rhizome, the proaxis of timothy, and in fact, the basal part of essentially all grass shoots (Hyder 1974). The number of nodes in the crown indicates the number of basal leaves produced and the potential number of tillers and adventitious roots. One also should be aware that some grasses have vertically oriented crowns (*Agropyron* sp. Gaertn. is a good example) and others have horizontal crowns that lie on the soil surface (such as blue grama) (Hyder et al. 1975). Those with vertical crowns are (more or less) subject to growing themselves out of the ground in the process of sequential tillering. Hence, the tendency of some grass clumps to develop dead centers. Internodes must elongate to elevate the growing point above the crown. Obvious advantages of plants having crowns with several nodes are: (1) a number of leaves can elongate for grazing without endangering the growing point, and (2) they provide several sites for root initiation. However, only nodes on or below ground can effectively produce roots.

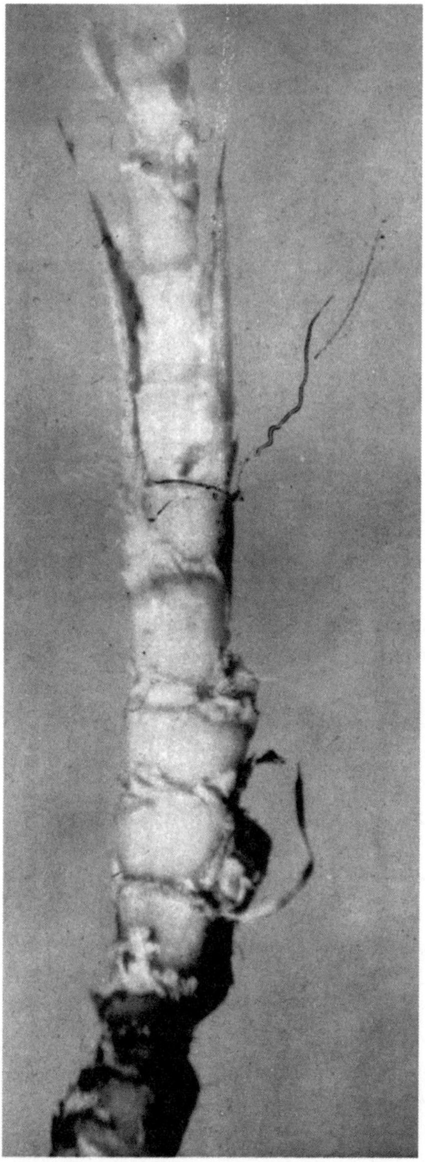

Figure 6. The crown of blue grama consists of numerous nodes and internodes that seldom exceed 2 mm in length (x 10). (Burzlaff et al. undated).

Reproductive Shoots

After attainment of a minimum amount of vegetative development, a grass shoot is subject to the stimuli that cause the process of floral induction. Thus, reproductive shoots of grasses always begin as vegetative shoots (long or short shoots). They are often divided into normal and aborted reproductive shoots (Hyder 1972, 1974). In normal reproductive shoots, mowing or grazing of leaves above rudimentary inflorescences keeps the culm obliged to grow even though the developing shoot may remain entirely leafless. Harvesting the shoot such that the inflorescence is removed stops further growth of that shoot and induces tillering at the base (Rechenthin 1956; Hyder 1972).

When disease, insects, or adverse climate kills the rudimentary inflorescence while the reproductive shoot is in the "boot" stage (inflorescence still enveloped in enclosing sheaths), it is referred to as an aborted reproductive shoot. Such shoots soon reach maximum development. They usually lose apical dominance, so tillering is stimulated earlier than in normal reproductive shoots (Hyder 1974).

In grasses, the onset of flowering is of considerable biological and economic significance, for it heralds a change from the continued production of vegetative organs to the ultimate production of seed (Calder 1966). Depending on where in shoot development one wishes to start, three or four stages can be described, although not every species exhibits all stages: (1) juvenile, (2) induction, (3) initiation, and (4) development and differentiation (Calder 1966; Youngner 1969).

During the juvenile stage, plants are insensitive to environmental conditions that later promote flowering. Depending on species, the juvenile stage may last for a few weeks as in orchardgrass and hardinggrass (*Phalaris tuberosa* L.) or up to several years as in some bamboos (*Bambusa*). Calder (1966) argues that this is a physiological stage and that it is rarely possible to relate it to morphologically recognizable stages, such as the common belief that the end of this stage can be recognized with a requisite number of leaves, i.e., a minimum leaf number. It is true that the *end* of the juvenile stage cannot be defined in terms of number of leaves, because the transition to reproductive status may depend on temperature, day length, moisture, nutrient supply, and growth rate of the shoot. However, the inverse also is true: the transition to reproductive status cannot occur *prior to* a certain minimum amount of vegetative development, which often can be defined reasonably well by number of leaves. The *end* of the juvenile stage can be defined only by the development of spikelet buds.

This argument includes the case of annual grasses in which induction may occur shortly after emergence. Unless spikelet differentiation occurs in the embryo, one can state without error that a certain minimum of vegetative growth must precede receptivity to induction.

A plant enters the inductive stage following the juvenile stage (Calder 1966). Although no morphological changes may be evident, certain physiological changes occur in response to specific environmental conditions bringing the grass plant, i.e., the apical or axillary meristems, to a state enabling it to respond to floral initiation stimuli (Youngner 1969). In other words, it has achieved the "ripe-to-flower" condition.

Actual differentiation to reproductive status is preceded by a fast elongation of the shoot apex (Cooper 1954) and spikelet primordia often appear first at the middle of this elongated apex (Evans and Grover 1940). Further differentiation proceeds upward and downward.

Floral induction is interpreted as a biochemical process that may require a cold pretreatment (vernalization), a certain day length, a certain sequence of day lengths, and favorable growing conditions. Also, some plants are genetically induced, i.e., vernalization is not required. Examples are: (1) summer annuals, (2) perennials evolved in the tropics, and (3) a few winter annuals, particularly some of Mediterranean origin. Other plants require vernalization, i.e., winter condition, and these usually fall into one of four categories: (1) those that require a cold period, (2) those that require the short days of winter, (3) those that require both, or (4) those that simply respond quicker after receiving winter conditions. In *Secale* sp. L. it is possible to vernalize embryos of developing caryopses, and many grasses can be vernalized as imbibed seeds (Naylor 1953; Calder 1966). Seedlings, as well as mature plants, can be vernalized, but older plants usually require shorter exposure.

Each tiller must be vernalized independently because there is no translocation of the vernalized state. Exposure to high temperatures immediately after vernalization may reverse the vernalized state (Calder 1966).

Floral initiation is the actual plant response to conditions that promote inflorescence initiation, i.e., the actual differentiation into a rudimentary inflorescence. For some time after a rudimentary phytomer is formed, it remains capable of developing a vegetative or a reproductive part, according to the conditions for growth. This involves chiefly photoperiod, but moisture, fertility, and shoot maturity are often indirectly involved. The leaf appears to be the organ of perception of the photoperiodic stimulus, and a substance formed there is translocated to the developing apex (Calder 1966). Please note that this shows that some photosynthetic tissue is "required." The question is "How much?" Photosynthetic tissue must provide the primary source of energy as well as of hormone. It is doubtful that induction can occur in any species until at least one leaf attains full length.

Date of floral differentiation usually will vary with several conditions. It may vary with species, variety, ecotype, and favorableness of growing conditions. If the grass shoot has not attained the proper age, attained adequate photosynthetic tissue, been vernalized, or in some way completed the juvenile stage by the time the proper day-length has arrived, the growing point cannot make the change from vegetative to reproductive status. It follows too that the better the growing conditions, the higher the proportion of stems that change from vegetative to reproductive status because rate of growth and plant development is accelerated. Commercial seed growers carefully fertilize, cultivate, and irrigate to take advantage of this principle. Length of daily photoperiod is universally considered the most important requirement for floral initiation. Lastly, latitude can drastically alter date of floral initiation. Latitude effects are largely interrelated with daylength and temperature requirements.

Plants are placed in three groups according to their photoperiodic response: long-day, short-day, and day-neutral (Salisbury 1961). In addition plants with long-short-day and short-long-day requirements have been recognized (Youngner 1969). Long-day plants flower only after exposure to day lengths greater than a critical number of hours. Such plants are more commonly found in high latitudes. Short-day plants flower only under day-lengths shorter than a critical number of hours. With increasing latitude, light and temperature conditions become less favorable for flowering in short-day plants. Day-neutral plants flower at any naturally occurring day length if other flowering requirements are satisfied (Youngner 1969).

Flowering of most cool-season (C_3) turfgrasses is promoted by long days. This is an obligate requirement of ryegrass, red fescue (*Festuca rubra* L.), creeping bentgrass (*Agrostis palustris* Huds.), and meadow fescue (*Festuca elatior* L.). Of the warm-season (C_4) grasses, all *Zoysia* Willd. species are short-day plants, but certain bermuda grass [*Cynodon dactylon* (L.) Pers.] varieties have a short-day requirement, while others may be day-neutral with some acceleration of flowering by long days (Youngner 1969).

The fourth stage in the flowering process involves the actual development and differentiation in the initiated shoot. The first visible evidence of change is an elongation of the shoot apex. Soon the double ridges indicative of an initiated shoot apex are formed by lateral bud primordia (spikelet or branch) arising in the axils of leaf primorida. Long days and high temperatures usually increase the rate of development. High temperatures may cause abortion of florets or entire spikelets. In some species, reversal to the vegetative condition (vivipary or vegetative proliferation) may occur under certain day lengths and temperatures. Florets or spikelets are then converted to "plantlets" or "bulbils," which can be propagated (Youngner 1969). This commonly occurs in certain bluegrass (*Poa* L.) species. Also, many bluegrasses produce a high proportion of their seed apomictically (i.e., production of seed without fertilization). Common dallisgrass (*Paspalum dilatatum* Poir.) is an obligate apomictic. Both cross- and self-pollination occur in grasses. Cleistogamy, pollination prior to opening of the florets, occurs in several species (Youngner 1969). Green sprangletop [*Leptochloa dubia* (H.B.K.) Nees] and Texas wintergrass (*Stipa leucotricha* Trin. & Rupr.) commonly produce viable cleistogenes.

From the literature and his own work, Calder (1966) describes the following six response patterns for inflorescence induction and initiation in grasses.

A. Juvenile stage, inductive stage and photoperiodic requirement for initiation.
B. As pattern A but with no juvenile stage. (A wide variety of grasses of temperate origin are in this category, including the winter cereals).
C. No juvenile or inductive stage, but with photoperiodic requirement for initiation. Many temperate grasses, including the spring cereals, show this type of response.
D. No environmental requirement for flowering (response pattern could be pattern E: the two cannot be distinguished).
E. Juvenile stage present but no other environmental requirements for flowering.
F. Juvenile stage present, followed by a photoperiod requirement for initiation.

One point should be emphasized regarding these groupings. As already pointed out, it is doubtful that induction can occur in any species until at least one leaf attains full length, because the leaf is the primary source of energy and of hormone (Calder 1966) for floral induction. To have no juvenile stage, a plant would have to be induced to flower at the time of germination or before. Bunting and Drennan (1966) point out that in the cereal embryo several miniature foliage leaves are present and further leaves start upon germination. Thus, it is apparently possible for an embryonic shoot to receive the floral induction stimulus (Naylor 1953). However, whether plants have a juvenile stage or not is of no real consequence, because Calder (1966) himself indicates that his group D with no juvenile stage and group E with a juvenile stage cannot be distinguished. Representative species placed in the categories suggested by Calder (1966) are: A—orchardgrass and hardinggrass; B—sheep fescue (*Festuca ovina* L.), Kentucky bluegrass, perennial ryegrass, and rye (*Secale cereale* L.); C—western wheatgrass (*Agropyron smithii* Rydb.), timothy, and corn (*Zea mays* L.); D—weeping lovegrass [*Eragrostis curvula* (Schrad.) Nees] and *Poa annua* L.; E—not distinguished from D; and F—rice (*Oryza sativa* L.) and sugarcane (*Saccharum officinarum* L.).

Shoots of Woody Plants

Long, Short, and Reproductive Shoots

Despite considerable variation in emphasis on the kind of shoots sufficiently important to be discussed, all publications reviewed recognized differences between long-shoot and short-shoot buds (Figs. 7 and 8). Because these kinds of shoots have parallels in grasses and forbs we will emphasize the long and short shoot terminology for woody plants. Long shoots have leaves separated by distinct internodes and by lateral buds, some of which develop into more long shoots and others into short shoots. Internode elongation is much reduced in short shoots, and their leaves cover the shoot tip. Their lateral buds are either lacking or develop into flowers (Wilson 1970; Kramer and Kozlowski 1979). Short shoots are also referred to as dwarf shoots on conifers and as spur shoots on deciduous trees (Zimmermann and Brown 1971).

Long-shoot buds generally are terminal and more readily removed by herbivores. Such utilization can seriously reduce production the next year because production is derived largely from long-shoot buds. Some species, such as willow (*Salix* sp. L.), readily transform some short shoot buds to long shoot status when the long-shoot buds are removed. Other species, those with preformed buds, respond more slowly to removal of long-shoot buds, and they may require a full season of growth and new bud formation to achieve the transformation from short shoot to long shoot condition. Such species are greatly affected by winter browsing. This interpretation, however, is largely theoretical and requires much experimentation.

On the other hand, control of hardwoods, e.g. honey mesquite (*Prosopis glandulosa* Torr. var. *glandulosa*) with growth-regulating herbicides seems more effective if long shoots are rare or absent. This is because carbohydrates as well as translocatable herbicides are for the most part moved to the growing young leaves at the tip of long-shoots instead of to the roots and other storage tissue. Thus for best herbicidal control, woody plants should have a higher proportion of short shoots (Wilson et al. 1975). Wilson (1984) classified branches in a way that shows why long shoots are more prevalent in young trees than in older trees. The main stem is the first order, laterals to the main stem are second order, etc., and as the order of the long shoots increases there is a general decrease in vigor. Higher order long shoots produce a smaller proportion of long shoots. Usually by the fourth or fifth order, long shoots bear no lateral long shoots—only short shoots. These short shoots bear no laterals at all, so branches higher than fifth or sixth order usually do not exist in the tree crown (Wilson 1970). Consequently, young plants with greater proportions of long shoots are more difficult to kill with translocated herbicides than older plants (Wilson et al. 1975; Sosebee and Dahl 1991).

In herbaceous species, e.g., grasses, shoots are designated *determinate* when terminated by an inflorescence, and while still vegetative shoots are regarded as *indeterminate*. In woody plants, inflorescences commonly are borne in leaf axils or on

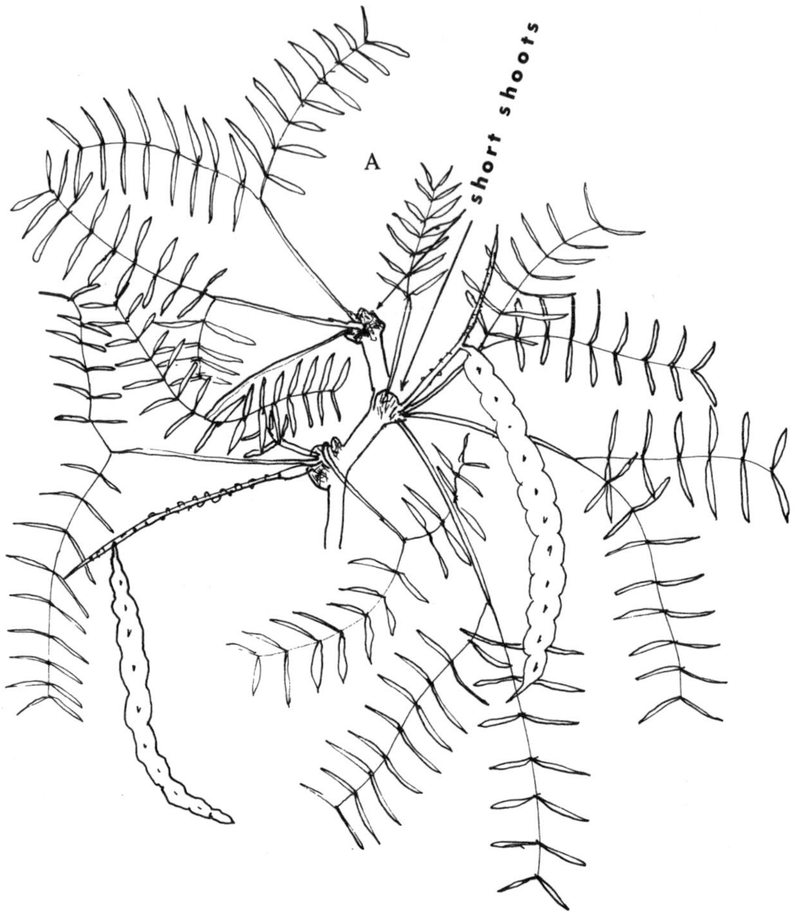

short shoots as well as at the terminal end of longer shoots. But, according to Kozlowski (1971), indeterminate growth in woody plants results when a reproductive structure terminates a branch or when a shoot tip aborts, a characteristic of most hardwood species. Determinate shoots result from opening and expansion of the contents of terminal buds on the main axis and branches. Some species of pines, spruces (*Picea* sp.), oaks, and hickories (*Carya* sp. Nutt.) characteristically have determinate shoots. Thus, these terms have a different connotation when describing grass shoots as opposed to woody plant shoots.

Preformed, Heterophyllous, and Late Season Shoots

Preformed shoots are characteristic of species whose winter buds contain a telescoped, fully formed shoot containing primorida of all the leaves that will expand during the following season. Shoot formation involves differentiation in the bud during the first year and extension of the preformed parts into a shoot the second year. Heterophyllous shoots are not fully preformed in the dormant bud and both early and late leaves are produced. The early or preformed leaves emerge at, or

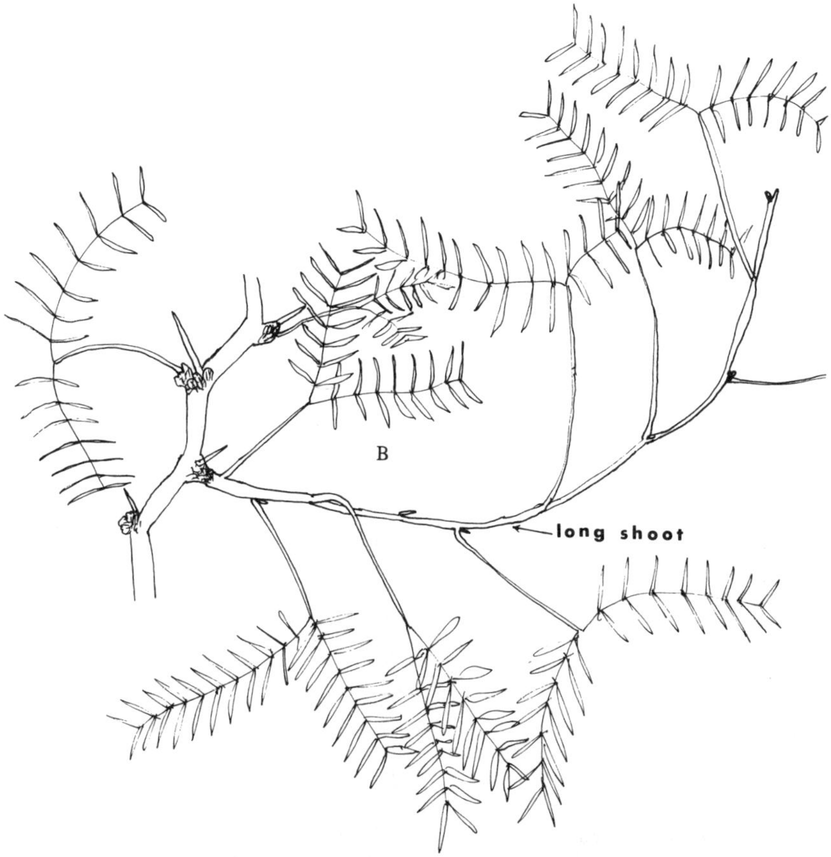

Figure 7. (A) Mesquite stem segment with predominantly short shoots (note almost total absence of internode elongation); (B) mesquite branch with a long shoot showing typical internode elongation.

shortly after, bud break. The distal or late leaves appear later in the growing season after the first leaves are well expanded. Such shoots grow for a much longer time than fully preformed shoots (Kozlowski 1971). Although not definitely described as such, preformed shoots can apparently be either short or long shoots, but heterophyllous shoots would be only long shoots.

There are a few other shoots sufficiently common to be recognized by most authors, but differ somewhat from those already named. Sudden and excessive exposure of trees often results in stimulation of dormant buds on the main stem or branches to produce epicormic shoots or "water sprouts." Most epicormic shoots develop in the upper part of the stem and only few within the crown (Kozlowski 1971). The abnormal late season flushes of growth from bursting of a current year bud have been variously named. Many late shoots are commonly called "lammas shoots" because they appear around Lammas day or August 1 (Wilson 1970; Kozlowski 1971; Zimmermann and Brown 1971; Morey 1973). Lammas shoots

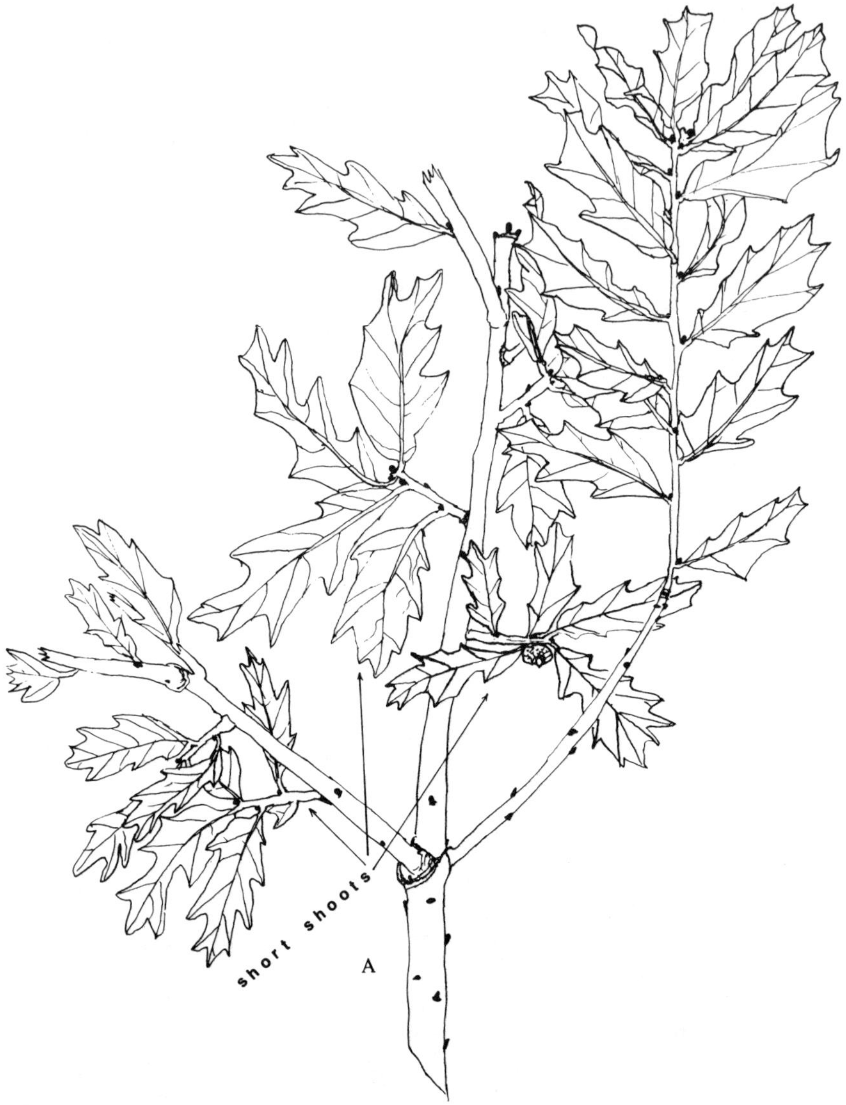

result from bursting and elongation of a current year terminal bud on the main shoot of a branch. Shoots emerging from current year lateral buds at the base of the terminal bud are occasionally recognized as distinct kinds of shoots and referred to as *proleptic* shoots. Both lammas and proleptic shoots often render trees susceptible to "winter injury" because late shoots do not always harden adequately. Both kinds of shoots cause profuse branching resulting in knotty lumber (Kozlowski 1971). These latter shoots are mostly long shoots.

Morphology of Height Growth in Woody Plants

Tree form differs depending on the way shoots develop, i.e., whether primarily from growth of the main axis or whether primarily from laterals. Trees with a

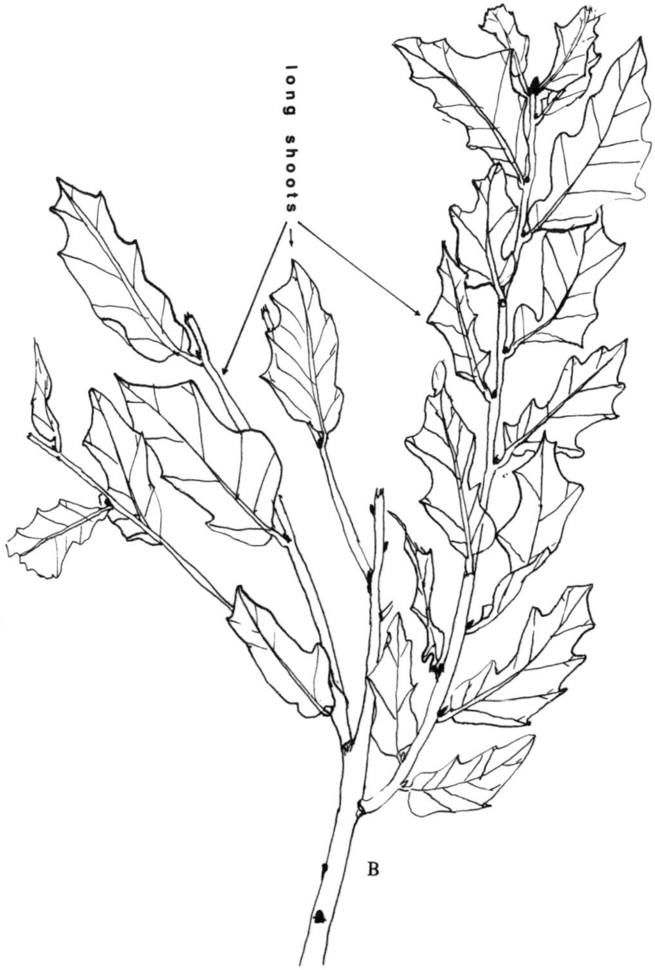

Figure 8. (A) Sand shinnery oak stem segment with predominantly short shoots (note some internode elongation on short shoots but that shoot growth ceases with 5 or 6 leaves and no lateral bud development except for flowers (an acorn in this case)); (B) stem segment with predominantly long shoots.

spiral-like form, i.e., with a continuous, unbranched main axis are said to have an *excurrent* growth pattern. Spreading trees (those with the main axis much branched above or where the growth of the main axis declines with age and the lateral branches grow more vigorously) have a *deliquescent* growth pattern (Wilson and Loomis 1962).

Woody plants also vary in the way their shoots grow throughout the year. They may be: (1) evergrowing, (2) intermittent, or (3) manifold. In evergrowing types, the apical meristem of the twig grows constantly forming new leaves throughout the year. True resting buds enclosed in bud scales are formed at the end of each period of shoot extension in intermittently growing trees. The manifold type growth is nonsynchronous among individual branches during a given period, i.e., some branches are growing while others are resting. All trees of the temperate zones show intermittent growth, but growth is variable enough that four general patterns are recognized (Zimmermann and Brown 1971).

In the first intermittent type, height growth occurs in a single flush as in maple (*Acer* sp. L.), oaks, hickories, ashes (*Fraxinus* sp. L.), fir (*Abies* sp.), and some pines. The overwintering terminal bud usually contains all the leaf primordia preformed, although the resting bud does not contain all of the cells of the next season's growth as was once believed. Numerous cell divisions occur in the subapical meristems of the elongating shoot. Normally, height growth of this type is restricted to a single period early in the growing season. In the second intermittent category are those trees that make recurrent flushes of growth each season by repeated formation and release of currently formed terminal buds. This habit is common among the pine of the southeastern United States and the most prevalent habit of growth among evergreen and deciduous trees of the subtropics.

Thirdly, in lilac (*Syringa vulgaris*), locust (*Robinia* sp. L.), elm, willow, etc., the upper portion of the shoot tip, after bud break and a period of extension growth, is regularly aborted. An upper axillary bud then replaces the aborted leader, and the shoot axis develops in a typical zig-zag or sympodial growth pattern. The fourth category of shoot extension (sometimes called continuous growth) occurs in trees where maturation of the preformed telescoped shoot may be followed immediately by further extension growth including both the initiation and development of new foliage leaves (late leaves). Such shoots end the season with formation of a distinct terminal bud. Tulip poplar (*Liriodendron tulipifera* L.), sweetgum (*Liquidambar styraciflua* L.), and Ginkgo (*Ginkgo biloba*) exhibit this growth pattern (Zimmermann and Brown 1971; Morey 1973).

Shoot Types Resulting from Management Practices

Shoot types of woody plants resulting from grazing, pruning, burning, etc., are referred to only obliquely in the major reference works in woody plant growth and development. Yet careful attention to the resultant shoots arising from such effects is necessary to properly understand consequences of management decisions. Indications of overbrowsing, trends in range condition, relative success of chemical brush control practices, nutritional value of shrubs and trees to browsing animals, the kind of growth response from a prescribed burn, and the mechanical techniques required to increase palatable browse for big game and livestock are but a few of the management problems that cannot be easily solved or recognized without a basic understanding of woody plant shoot growth and development. For example, note the basic difference required to complete shoot elongation between long and short shoots. Kozlowski (1971) reported that shoot elongation of mature, bearing apple (*Malus pumila*) or pear (*Pyrus communis* L.) trees with predominantly short shoots usually requires 6 to 8 weeks. However, shoots of young trees, with higher proportions of long shoots, often continue elongation for 10 to 12 weeks.

Which shoot growth pattern characterizes such shoots as root suckers, water shoots, and stem sprouts resulting from grazing burning, cutting, chaining, herbicides, and stand thinning has never definitely been explained. However, their growth patterns seem to fit the continuous growth (fourth type) pattern in that they usually continue to produce new foliage leaves and internodes throughout the growing season. Hence, they are long shoots. An excellent example comes from a study conducted by Bedell and Heady (1959) on chamise (*Adenostoma fasciculatum* H. & A.) in the California chaparral in which mature plants, hedged plants, and root crown sprouts were studied. Initiation of twig elongation occurred about the first day of spring for mature and hedged plants and slightly later for crown sprouts. Twig elongation ceased by the first day of summer on mature plants, by mid-summer on hedged plants, and not until late autumn with crown sprouts. Two obvious uses for recognizing shoots that may grow continuously during the summer are: (1) increased browse production, and (2) reduced effectiveness of growth-regulating herbicides. Chemical control of many woody plants depends on translocation of herbicides to the roots and root crown in sufficient quantity to kill roots and dormant stem buds. Most such herbicides move to these lower plant regions through the phloem and in conjunction with translocated photosynthate (Crafts 1953). However, Zimmermann and Brown (1971) showed that translocation from the older leaves on most woody plant shoots is primarily upward to the shoot tip and younger leaves on immature and actively elongating shoots. Only when shoot elongation is complete is photosynthate translocation primarily downward. Thus, herbicide researchers now believe that root kill from translocated herbicides is rarely possible on trees with a high proportion of long shoots where active shoot extension is occurring (Wilson et al. 1975).

Growth of forbs

Range management students were taught for many years that grasses were more resistant to heavy grazing than forbs because grasses had intercalary meristems which are lacking in forbs. While it is true that most botanists emphasize that intercalary meristems are most distinct in grass shoots (Esau 1960; Kaufman et al. 1989), and Northington and Goodin (1984) emphasize that this characteristic imparts survival potential to grasses, it is not the fact that grasses have well developed intercalary meristems that make them more grazing resistant. Rather, any herbaceous plant having shoots with the shoot apex protected from grazing is relatively grazing resistant—as in grasses and forbs having short vegetative shoots. By definition, a *node* is a part of a stem where leaves are attached. The portion of a stem between two nodes or leaves is the *internode* (Kramer and Kozlowski 1979). Since internodes do elongate, at least some, in all kinds of plants, regardless of whether the elongation is due to an intercalary meristem or from some other mechanism, the end result is exactly the same. Plants having their stem apices regularly elevated to a height making their growing points vulnerable to removal by grazing, mowing, browsing, etc., are those plants that are more easily removed from a stand by overuse regardless of whether they are a grass or a forb.

Academically, it is of interest to know how internode elongation *does* occur in forbs if not from an intercalary meristem. None of the references available directly answer the question. All of them emphasized the existence of intercalary meristems in grasses and *Equisetum* sp. L. and in the formation of leaf petioles but no direct answer for elongation of internodes was provided for herbaceous dicotyledons. In fact, most references consulted ignored the issue. The most direct answer was obtained from Esau (1960) and the following treatise is largely extracted from that 30-year old source and from a recent botanical text by Kaufman et al. (1989).

According to Kaufman et al. (1989), the site of origin of all the tissues of the stem and the appendages produced by the stem—leaves, axillary buds, and floral buds—is the shoot apical meristem. However, they recognize that the degree of development of a leaf petiole depends upon the extent of intercalary growth; prolonged intercalary growth results in the strap shaped leaves of many monocotyledons and in the needle leaves of conifers. Also, they describe the bolting response in biennials as due to enhanced rates and numbers of cell divisions in the subapical region of each shoot. This implies an intercalary meristem in biennial dicotyledons. They further emphasize that cell enlargement does occur in the region of cell division. Likewise, cell divisions occur in the regions of cell elongation and even in the region of maturation. Some cell types, such as early vascular tissues and trichomes, often mature in the regions of cell division and elongation. This would seem to imply that internode growth, including cell division, can be occurring throughout the internode rather than being limited to the base of the internode as occurs in plants with intercalary meristems.

Esau (1960) points out that cell divisions occur at levels where differentiation of specific cells and tissues occur near but not at the tip of the root or shoot. Hence, growth, in the sense of cell division, is not limited to the very tip of root or shoot but extends to levels considerably removed from the area usually called the apical meristem. In the shoot, a more intensive meristematic activity is observed at levels where new leaves are initiated than at the tip. During stem elongation, cell division extends *several internodes below the apical meristem.* Esau (1960) further indicates that in view of the gradual change from the apical meristems to the adult tissues and the integrating of the phenomena of cell division, cell enlargement, and cell differentiation, one cannot restrict the term meristem to the apex of shoot or root. The shoot and root parts, where the future tissues and organs are already partly determined, but where cell division and cell enlargement are still in progress, are also meristematic. Probably the most enlightening of all statements found for the growth of internodes is the following statement by Esau (1960): "The meristematic activity causing the elongation of an internode *may be rather uniform throughout the growing internode*, but more frequently it is more intense at the base of the arising internode than elsewhere. If internodes have a prolonged period of growth in length, the meristem at the base of the internode is rather pronounced and is called an intercalary meristem—intercolated between regions of more mature tissues."

Apparently, for internode elongation in shoots without distinct intercalary meristems, cell division, cell enlargement, and cell maturation occurs uniformly (simultaneously?) throughout the growing internode. Again, cell division is just not concentrated at the internode base as in grasses (and probably in biennial dicotyledons). The end result is an elongated internode—so whether it occurred from an intercalary internode or through a more dispersed mechanism would seem immaterial from a management perspective.

Some forbs do have stolons and rhizomes much the same as do grasses. In addition, forbs have many other forms, e.g., plants with corms and bulbs, that are not commonly important in range management. Hence, no attempt is made to describe them in this chapter.

Effects of Defoliation in Function, Development, and Yield of Range Plants

Plant Characteristics Providing Resistance to Grazing

The basic rhythm of plant growth presented to this point has been designed to acquaint readers with plant developmental processes and terminology. This final section will attempt to relate the basic information to management with special emphasis given to grazing influences.

Grazing may be timed to primarily produce a strong tillering response, harvest a maximum quantity of nutrients per unit area, improve forage quality for subsequent grazing, manipulate botanical composition, protect and improve adjacent range, or create some other desirable effect (Hyder 1974). However, the desired response from a given grazing treatment or the potential to use the principles of plant growth and development to achieve any of these desired goals will depend largely on local opportunity and the degree of control over land and livestock.

Many range plants have morphological adaptations that permit them to endure repeated partial defoliation. The apical meristems of long shoots, the source of new leaves, is elevated by stem elongation. Removal of the apical meristem stops stem elongation and leaf expansion from that shoot. Most varieties of alfalfa (*Medicago sativa* L.) and the majority of grasses do not exhibit free branching so new shoots must arise from dormant basal buds, i.e., from the crown. Therefore, grazing is best scheduled to coincide with declining vegetative activity and the initiation of a new set of stems. Otherwise plants with this type of terminal growth are susceptible to extreme reductions in herbage production (Heppner 1961; Hyder 1972). On the other hand, one should be aware that this type management with grasses often forces livestock to utilize lower quality forage as nutritional quality declines rapidly with later stages of shoot development. Consequently, in the development of grazing systems one must compromise some yield potential to utilize enhanced quality and vice-versa. The skill comes in choosing how far to go in maintaining high quality before the recuperative potential of the key management species is irreversibly penalized. The management of this kind of growth pattern can be especially challenging with deciduous shrubs that grow in this manner. Shrubs such as fourwing saltbush [*Atriplex canescens* (Pursh.) Nutt.] and bitterbrush [*Purshia tridentata* (Pursh) DC.] are vulnerable to yield reduction by excessive grazing, as a tiny nibble can remove an apical meristem and stop shoot growth (Hyder 1972). Thus, to permit such plants to maintain themselves in a diverse plant community, grazing must be adjusted to retain most of the apical meristems throughout a season or a rhythm of growth. This usually means periodic nonuse.

Plants that tend to branch prolifically from elevated leaf axils provide better adaptation to endure grazing, particularly among the nongrass forages. Russian thistle (*Salsola kali* L.), for example, produces numerous branches with enough of them decumbent to assure that some apical meristems escape removal by grazing animals (Stevens 1943). Many shrubs can branch profusely, particularly after apical meristems are removed by grazing. This involves both new shoots from axillary buds and the transition of short shoots to long shoot status when long shoot apices are removed. The literature does not contain information on such transformation, but we have seen three or four short shoots go "long" after removal of the long shoot apex in bitterbrush. Willow does it better and faster. Perhaps a better illustration is the shrubs commonly used for hedges in home landscaping, e.g. *Spirea* sp. However, even in this type of growth, shoot replacement by lateral branching is slow. Because leaves and leaf primordia are easily available to grazing animals, this type of growth, too, is better adapted to periodic harvest by machine or animal than by continuous grazing (Hyder 1972).

Short Shoots

Range plants most resistant to excessive grazing grow in a manner such that their apical meristems are inaccessible to grazing animals. Spines protect the apical meristem of some shrubs, but the apical meristems on some herbaceous plants are inaccessible by virtue of having short vegetative shoots (Branson 1953; Hyder 1972). All sod-forming grasses used for lawns, such as bermuda grass and Kentucky bluegrass and many other wildland forage grasses, some of them bunch grasses, have short vegetative shoots. *Stipa* sp. L., squirreltail grass [*Sitanion hystrix* (Nutt). J. G. Smith], most *Bouteloua* sp. Lag., and prairie sandreed [*Calamovilfa longifolia* (Hook.) Scribn.] are wildland examples. These plants have the appearance of being culmless because of their extremely short basal internodes (crowns). Fortunately, in all grasses, the leaf intercalary meristems remain in basal position, so that the elevated part of a leaf blade can be removed without stopping growth. Furthermore, most grazing animals cannot graze so closely as to remove all leaf tissue. Hence, plants with short shoots can maintain a much higher proportion of photosynthetic tissue relative to species with long shoots, whose entire leaf complement is elevated and vulnerable to removal by grazing (Peterson 1962). The higher proportion of leaves and the absence of significant internode elongation of short shoot plants also provide for much more rapid replacement of carbohydrates to stem bases, crowns, and roots than is possible with long shoot plants.

A complementary trait of most species with short vegetative shoots is that they produce a high proportion of vegetative *versus* reproductive shoots. Branson (1953) pointed out that the ratio of vegetative to reproductive shoots is an easy way to assess grazing resistance among range plants.

Hyder (1972) reported that for the shoots that remain vegetative, the apical meristem may remain vegetatively active for several years. Growth stops in quiescent periods, but cell division and expansion resume in both apical and intercalary meristems when growing conditions again become favorable. Dead leaf tips are common on spring growth when any new surge of growth exerts portions of immature leaves. Portions of immature leaves also will "die back" during drought as well as during winter cold. Intercalary meristems of leaf blades cease activity by the time of leaf collar exertion (Sharman 1947) so no further growth of that blade is possible. Forbs less commonly have the same degree of protection for leaves and shoot apexes as grasses. However, some, such as dandelion, also have basal leaf primorida that are never elevated, which protects the potential for leaf replacement.

Long Shoots

Long shoots are those that, early in the season, elevate the apical meristem. Such plants nearly always decrease in continuously heavily grazed pastures (Branson 1953). The tall grass prairie dominants mostly have this type growth as well as many important forage species both endemic and exotic in other sections of the United States. Some authors distinguish among those culmed shoots that remain vegetative and those that are reproductive (Hyder 1972) and others simply emphasize the importance of early elevation of the growing points and rudimentary inflorescences through long basal internodes (Branson 1953; Rechenthin 1956). Although it is easy to equate those plants that elevate inflorescences at an early time with those plants that have culmed vegetative shoots, they are not necessarily the same phenomenon. As pointed out by Hyder (1972), culm elongation in short shoots proceeds only after floral induction. When floral induction is not a prerequisite of internode elongation, culm growth may be linked enzymatically to day length (i.e., a certain day length is needed for internode elongation) and energetically to leaf maturation (i.e., the mature leaf contributes photosynthates to other plant parts). He found that crested wheatgrass [*Agropyron cristatum* (L.) Gaertn.] grown in the greenhouse in winter failed to exhibit culm elongation or floral differentiation until the light period was lengthened (Hyder 1972).

Plants with early elevation of inflorescences and plants with long or culmed vegetative shoots are not necessarily synonymous, as illustrated by crested wheatgrass, a plant that is known for having culmed vegetative shoots. The growth form of this plant is at first equivalent to that of plants with short shoots, in that it has a number of short basal internodes and a number of leaves arise and reach maturity before internode elongation begins (Hyder 1972). On the other hand, such plants as Indiangrass [*Sorghastrum nutans* (L.) Nash.], switchgrass, Johnsongrass [*Sorghum halepense* (L.) Pers.], and western wheatgrass have few if any short basal internodes; thus, with early lengthy internode elongation their apical meristems are quickly vulnerable to removal by herbivores (Branson 1953; Rechenthin 1956). These early elevated growing points may be destined to remain vegetative or become reproductive depending on growing conditions, stage of development, day length, etc.

Ratio of Vegetative to Reproductive Culms

Almost as important as whether shoots with early elevation of growing points are vegetative or reproductive is the ratio of vegetative to reproductive shoots. Branson (1953) found that those species producing a high proportion of reproductive shoots were less resistant to continuous heavy grazing than those species where a high proportion of the shoots remained vegetative. Thus, little bluestem [*Schizachyrium scoparium* (Michx.) Nash], which had 12 to 15 short basal internodes and a late elevation of the apical meristems to grazing height, was still easily removed by excessive grazing because over three-fourths of its shoots were reproductive. Also, Hyder (1972) emphasized that reproductive shoots are adapted for seed production rather than for tolerance to defoliation.

It is also important to remember that the percentage of vegetative or reproductive shoots within a species varies greatly from year to year and from place to place as well as among species. Just as excessive numbers of reproductive grass shoots reduce palatability and nutritional value to grazing animals, high numbers of such shoots are necessary for seed growers to make a profit or for bunchgrasses to re-establish themselves on native rangeland. Consequently, the proportion of reproductive shoots is important to grass management. Most grasses will produce a higher proportion of reproductive shoots if growing in a thin stand. Also, nitrogen fertilization increases the proportion of shoots becoming reproductive (Hyder 1972). Obviously, if one wants optimum seed production, ideally he would utilize wide spacing (plant in widely spaced rows), reduced competition (i.e., cultivate), fertilize, and irrigate. For most nutritious forage (i.e., leafy, culmless forage) one may desire to have a thicker stand to increase the proportion of vegetative shoots. This means using a narrower row spacing and/or more seed per acre if establishing a plant sward.

Relative Palatability

One can usually make the assumption that plants with 6 to 10 or more very short basal internodes and that have mostly vegetative shoots are resistant to removal by grazing from a plant stand (i.e., they are increasers). Also, we assume that plants with less than four short basal internodes will be easily removed by grazing if they have mostly reproductive shoots; however, there are some exceptions. Big bluestem (*Andropogon gerardi* Vitman.) has over six short basal internodes and only about a third of its shoots become reproductive, yet it is a well-known decreaser. Where it is adapted, it is the most palatable grass in the area; thus it is more avidly eaten than associated plants and is subjected to abnormal grazing pressure hastening its early demise despite these characteristics indicating grazing resistance. On the other hand, the abundance of old stems in little bluestems bunches discourages grazing somewhat, causing it to persist longer than other tall and mid grasses (Branson 1953). Sideoats grama and Junegrass [*Koeleria cristata* (L.) Pers.] have culmless vegetative shoots, yet they are usually decreaser species largely due to seasonal high palatability.

Also, shrubs that are about equally grazing resistant from thorns and with similar chemical defenses against herbivory show differential ability to survive due to selective browsing. DuToit et al. (1990) studied *Acacia nigrescens* and *A. tortilis* and reported that severe pruning of *A. nigrescens* by browsing ungulates in Africa reduced intershoot competition for nutrients, promoting rapid shoot regrowth. Carbohydrate demands of the regrowth reduced secondary metabolite synthesis (e.g., condensed tannins). This resulted in patches of highly palatable browse that attracted further browsing, generating a browsing—regrowth feedback loop. This selective browsing ultimately led to the replacement of *A. nigrescens* with less palatable species. *A. tortilis* is more thorny and has finer leaves and so is less preferred. This illustrates the point that relatively small differences in preference can lead to elimination of a slightly preferred plant over one even slightly less preferred.

Role of Tillering

The ability to "tiller" or branch from axillary buds (Fig. 9) without the apical meristem being removed is considered an indication of an efficient forage producer. Richards et al. (1988) found that rapid production of a new photosynthetic canopy immediately following defoliation is critical for response of species which are tol-

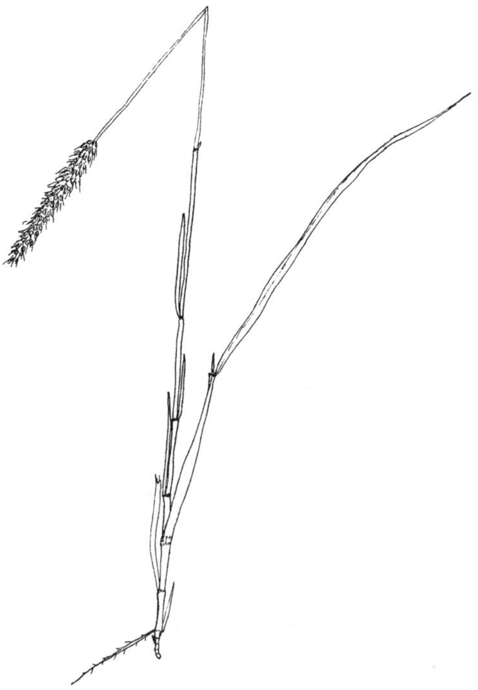

Figure 9

erant of herbivory, especially where defoliated plants must compete with undefoliated neighbors. Crested wheatgrass is an outstanding example of a species that tillers profusely following defoliation, yet the response is not due to release from apical dominance. Other factors known to influence tillering after defoliation are temperature, light quantity and quality, resource availability within and between tillers, and leaf or root produced growth regulators. Studies in Utah show that more current photosynthate was allocated to shoot growth following defoliation in crested wheatgrass, a grazing tolerant species, than in bluebunch wheatgrass [*A. spicatum* (Pursh) Scribn. & Smith], a species intolerant of close grazing, and which latter species allocated more photosynthate to root growth. This difference in resource allocation following defoliation is considered a key factor in the differential grazing tolerance of these two species (Richards 1984; Richards and Caldwell 1985; Richards et al. 1988).

Plants that produce stolons and/or rhizomes are usually considered resistant to grazing. Buffalograss [*Buchloe dactyloides* (Nutt.) Engelm.], curly mesquite [*Hilaria belangeri* (Steud.) Nash], Kentucky bluegrass, and bermuda grass are examples of resistant species. Johnsongrass, western wheatgrass, and switchgrass have vigorous rhizomes, yet they quickly elevate the shoot apical meristems above ground and have a high reproductive to vegetative shoot ratio. These characteristics make them less tolerant of grazing than the first mentioned species (Rechenthin 1956).

If intensive management is possible, such as with planted pastures, one may time grazing to obtain a particular plant response; e.g., a harvest that cuts below the inflorescence stops further growth of that shoot. Thus, regrowth requires the initiation of new tillers. Although a new set of tillers will not occur promptly, they will, given time, grow and produce leaves beneficial to plants and herbivores. However, grazing or harvesting leaves just above or through rudimentary inflorescences leaves the culm morphologically obliged to grow, even though it must remain leafless. Leafless reproductive shoots have been considered of little value either to the plant or foraging animals (Cook and Stoddart 1953; Rechenthin 1956; Heppner 1961; Hyder 1972). However, Caldwell et al. (1981) report that as much as 75% of the total carbon uptake of crested wheatgrass and bluebunch wheatgrass came from stems and sheaths for much of the growing season. Nevertheless, stems and sheaths are not preferred animal forage. Unfortunately, the degree of management required to obtain a desired tillering response is seldom possible under arid or semiarid range conditions.

Hyder (1972) reported that aborted reproductive shoots, those with peduncles and rudimentary inflorescences killed, soon reach maximum development. They usually lose apical dominance over axillary buds and new tillers emerge at an earlier time than would occur from normal reproductive shoots. This feature is also hard to capitalize on in management of native ranges.

Defoliation

Defoliation effects on developmental morphology of range plants are concerned with season (i.e., stage of growth), frequency, and severity of grazing. Usually defoliation is studied in the way it influences: (1) the growth of roots and uptake of minerals and water, (2) the amount and utilization of carbohydrates, and (3) the development of leaf area and light interception (Alcock 1964). Although most studies can concentrate on only one facet, and are thus reported, it seems unlikely that defoliation could affect only one item without simultaneously influencing the others.

Milthorpe and Davidson (1966) pointed out that while grazing may induce a transient reduction in the rate of tiller production, its most profound effect is on root growth and activity.

Root growth ceases or is markedly reduced in almost all plants following defoliation (Crider 1955). Also Oswalt et al. (1959) showed a marked reduction in nutrient uptake following clipping. Davidson and Milthorpe (1965) found severe defoliation of orchardgrass caused almost complete cessation of root extension; it reduced root respiration by two-thirds and phosphorous uptake by four-fifths. Root activity did not resume until the new leaf surface had attained considerable size. Crider's (1955) work with effects of defoliation on root growth is highly regarded and often quoted. The principles he illustrates answer many questions regarding grazing influences on root growth and should be regarded as highly informative in this subject matter area. He observed (as well as others, i.e., Zimmermann and Brown 1971) that in the yearly growth cycle of plants (in his case citrus seedlings) the growth and rest periods of roots alternated with those of the shoots. He also found that when the tops to tobacco, tomato (*Lycopersicon esculentum*), cabbage (*Brassica oleracea*) peach (*Prunus* sp. L.), plum (*Prunus* sp. L.), privet (*Ligustrum* sp.), and pine seedlings were cut back severely, root growth of all species stopped for a time. Grass roots were particularly affected by top removal whether he used field grown or laboratory plants. The percentage of roots that stopped growth varied in proportion to the amount of foliage removed. Removal of half or more of grass foliage during the growing season whether the grasses were cool or warm season, or bunch, rhizomatous, or stoloniferous caused root growth to stop for a time after each removal. Orchardgrass was the lone exception in that some root growth occurred after the first clipping, but none at all after a second clipping. Roots continued to grow in single and repeated clipping trials of all species only when 40% or less of the foliage was removed. The drastic effect where top removal exceeded

50%, causing complete and prolonged root-growth stoppage, was reflected in poor shoot development (Crider 1955).

More recent work with bluebunch wheatgrass, which is very sensitive to overgrazing, showed root elongation to continue unabated during 90 days of regrowth following severe defoliation (Richards 1984). Richards (1984) also makes the point that reduced root growth following defoliation is an effective mechanism to aid reestablishment of the photosynthetic canopy and the root:shoot balance. As such it contributes to both herbivory tolerance and maintenance of competitive ability. While we have considered the root growth decline after defoliation to be undesirable, it is apparently a greater problem to plant health if the roots continue to grow.

It should also be emphasized that any factor that retards growth relative to photosynthesis (e.g., high light, low temperature, low availability of mineral nutrients, moisture stress, and increase in age) results in increased soluble carbohydrates. Retardation of photosynthesis relative to growth as caused by darkness or low light, optimum temperature and defoliation reduces the soluble carbohydrate content (Davidson and Milthorpe 1965).

Classic works by Caldwell et al. (1981), Caldwell (1984), Richards and Caldwell (1985), and Richards et al. (1988) convincingly demonstrate that the common use of carbohydrate concentration in specific plant organs rather than the total carbohydrate pool (concentration x biomass) has led to a significantly different view of these so-called carbohydrate "reserves." Even the carbohydrate pool concept reflects only a small portion of the daily carbon flux of the plant. For example, in winterfat [*Ceratoides lanata* (Pursh) J. T. Howell] if one calculated the labile carbon pool as the total carbohydrate concentration in the shoot tissues multiplied by the shoot system biomass, this entire carbon pool would be equivalent to only one day of photosynthetic carbon fixation during the spring. Reference to labile carbon pools as "reserves" implies a sizeable carbon reservoir in the plant. Caldwell suggests that we should think of labile carbon pools in plants as a buffer, the size of which would vary from species to species, but they would be quite small in many range species relative to the magnitudes of carbon flux in photosynthesis and the demands of respiration and root growth. They do emphasize that we should not ignore the conventional wisdom with respect to carbohydrates such as the common concept that plants may be most sensitive to severe defoliation when carbohydrate concentrations are at minimal levels during the course of the year. There must indeed be critical levels of the labile carbon pool that the plant should maintain and even a small buffer may be of considerable significance (Caldwell et al. 1984).

May (1960) suggested that lowered carbohydrate reserves represents a large sink resulting in increased translocation away from the leaves. Because rate of translocation is believed a limiting factor in photosynthesis, the conclusion is that lowered carbohydrate reserves are often responsible for increase photosynthetic activity. It follows then that some grazing, other things being equal, could stimulate plant growth. Alcock (1964) was able to demonstrate this principle with winter wheat (*Triticum vulgare* Vill.). This also helps explain why plants growing in exclosures may lack the vigor of those moderately defoliated by grazing.

From carbon balance studies with orchardgrass, Davidson and Milthorpe (1965) found that, with partial defoliation, the decrease in carbohydrate content could account for all the carbohydrate used in new growth and respiration during the following 48 hrs although appreciable translocation from stubble to roots was required to account for respiration losses from the roots. With severe defoliation, decline in soluble carbohydrates could not account for new growth and respiration. They concluded that other compounds, presumably protein, must have been involved. They also reported other studies that show an appreciable increase in the nitrate concentration in the plant with defoliation. It seems that we have been overly concerned with carbohydrate studies, when other carbon compounds may prove as important as the carbohydrate fraction.

Under range situations, the level of reserve carbohydrates has been a useful tool in explaining need for periodic rest from animal grazing particularly of the preferred key management species. Recent studies that provide a more complete explanation of the total carbon flux in plants relative to defoliation should considerably enhance our ability to better manage rangelands.

Leaf Area Effects

Besides the additional growth of leaves that are not yet mature, regrowth following defoliation is largely a function of leaf replacement through tillering. Removal of the apical meristem stops growth of that tiller and soon leads to its death. Regrowth then depends on the number, extent, and type of leaves remaining on intact tillers. In many plants elongating stems inhibit the growth of their lateral buds and removal of the stem apex stimulates bud growth; less control is exerted on buds of nonelongating tillers of the same plant. With annual species, such as the cereal grains, where most of the tillers flower at the same time, decapitation early in the stem elongation process (jointing) allows a resurgence of new tillers. On the other hand, with species having only a few tillers induced to flower at any one time, decapitation of the flowering stems has little effect on the rate of tiller production (Milthorpe and Davidson 1966). In these species with weak apical dominance, other factors following leaf removal such as temperature, light quantity and quality, resource availability within and between tillers, and leaf or root produced growth regulators are more influential in the tillering process (Richards et al. 1988).

Conclusions

The principles of developmental plant morphology are important in many aspects of range management, including grazing management and the control of undesirable species by mechanical, biological, or chemical means. Further, developmental plant morphology is important in understanding forage quality and how it varies phenologically. Many aspects of revegetation involve plant morphological and developmental considerations as well. To the extent that range management is intimately involved with manipulating vegetation for specific goals, it is clear that such goals can be more readily achieved when plant developmental morphology is understood.

For example, often it is observed that desirable forage species decrease in relative abundance regardless of the specific grazing system employed (continuous, seasonal, or short duration). Can this decrease be attributable to too many livestock? Use by the wrong kind of animal? Use during the wrong season? An analysis of the situation in light of the principles of how plants grow often leads to a diagnosis of the problem. Most commonly, on year-round, continuously grazed pastures that are stocked too heavily during the growing season, it is simply that highly preferred species are continuously sought by grazing animals. With this continuous reduction in manufacturing tissues (leaves), root growth stops for certain lengths of time, tiller production is reduced, and reserve carbohydrate levels are reduced to the point that they are unable to supply respiration needs of the plant over winter or for regrowth.

As another example of the application of developmental morphology to range management, the proportion of long to short shoots of plants influences their sensitivity to herbicides: translocation of carbohydrates and herbicides is primarily upward to shoot tips and younger leaves on immature and actively elongating shoots. Thus, root kill often is difficult on trees with a high proportion of long shoots where active shoot extension is occurring.

Future Research Needs

Many principles of developmental plant morphology have been elaborated in this chapter. With these principles in mind, and with single-species stands of forage, a grazing plan (for example) could be designed for effective management. However, with rare exceptions, most range managers must deal with mixed species. Similarly, age-class dynamics usually are such that a given stand of range plants includes individuals of many ages, complicating a generalization of age-related developmental stages. Additionally, plants of a given species within a particular plant community may be in different phenological stages because of different microenvironmental conditions.

Perhaps one of the most important needs for future research involves the extension of the application of the principles of developmental plant morphology to these more complicated situations. For example, strategically timed deferment periods may be required to maintain desired key species in grazing management. But such action should not be taken at the expense of overgrazing grazed pastures. Likewise timing of herbicidal application must take into account plant development, and so must be species-, site-, and growing condition-specific.

Each of these examples deals directly with manipulation of plant photosynthetic material, or in other words, with how plants grow. Our effectiveness as range managers will be determined in large part by how well we understand this basic phenomenon.

Literature Cited

Alcock, M. B. 1964. The physiological significance of defoliation on the subsequent regrowth of grass-clover mixtures and cereals, p. 25-41. *In*: D. J. Crisp (ed.), Grazing in terrestrial and marine environments. Blackwell Scientific Publications, Oxford.

Beard, J. B. 1973. Turfgrass: science and culture. Prentice-Hall, Inc., Englewood Cliffs, N. J. 658 p.

Bedell, T. E. and H. F. Heady. 1959. Rate of twig elongation of chamise. J. Range Manage. 12:116-121.

Branson, F. A. 1953. Two new factors affecting resistance of grasses to grazing. J. Range Manage. 6:165-171.

Brede, A. D. 1984. The turfgrass phytomer—a carryover from the past. Agron. J. 76:505-506.

Bunting, A. H. and D. S. H. Drennan. 1966. Some aspects of the morphology and physiology of cereals in the vegetative phase, p. 20-38. *In*: F. L. Milthorpe and J. D. Ivins (eds.). The growth of cereals and grasses, Butterworths, London.

Burzlaff, D. F., J. Launchbaugh, and J. Stubbendieck. Undated. The growth and development of blue grama [*Bouteloua gracilis*) (H. B. K.) Lag Ex. Steud.]. Texas Tech Univ. Range and Wildlife Inf. Service. No. 1. 11p.

Calder, C. M. 1966. Inflorescence induction and initiation in the Gramineae, p. 59-73. *In*: F. L. Milthorpe and J. D. Ivins (ed.), The growth of cereals and grasses. Butterworths, London.

Caldwell, M. M., J. H. Richards, D. A. Johnson, R. S. Nowak, and R. S. Dzurec. 1981. Coping with herbivory: photosynthetic capacity and resource allocation in two semiarid *Agropyron* bunchgrasses. Oecologia 50:14-24.

Caldwell, M. M. 1984. Plant requirements for prudent grazing. Pages 117-153. *In*: Developing strategies for Rangeland Management. National Research Council/National Academy of Sciences. Westview Press. Boulder, Colo.

Cook, C. W. and L. A. Stoddart. 1953. Some growth responses of crested wheatgrass following herbage removal. J. Range Manage. 6:267-270.

Cooper, J. P. 1954. Studies on growth and development in *Lolium*. IV. Genetic control of heading responses in local population. J. Ecol. 42:521-556.

Crafts, A. S. 1953. Herbicides, their absorption and translocation. Agr. and Food Chem. 1(1):51-55.

Crider, F. J. 1955. Root-growth stoppage resulting from defoliation of grass. U. S. Dep. Agr. Tech. Bull. 1102. 23p.

Dahl, B. E., B. A. Fall, A. Lohse, and S. G. Appan. 1975. Construction and stabilization of coastal foredunes with vegetation: Padre Island, Texas. U. S. Coastal Engineering Research Center Misc. Paper 9-75. 188p.

Dahl, B. E. and D. N. Hyder. 1977. Developmental morphology and management implications. pp. 257-290. *In*: R. E. Sosebee (ed.) Rangeland Plant Physiology. Range Sci. Series No. 4. Soc. for Range Manage. Denver, Colo.

Davidson, J. L. and F. L. Milthorpe. 1965. The effect of temperature on the growth of cocksfoot (*Dactylis glomerata* L.). Ann. Bot. (N.S.) 29:407-417.

DuToit, J. T., J. P. Bryant, and K. Frisby. 1990. Regrowth and palatability of *Acacia* shoots following pruning by African savanna browsers. Ecology 71(1):149-154.

Eames, A. J. 1961. Morphology of the angiosperms. McGraw-Hill, New York. 518p.

Esau, K. 1960. Anatomy of seed plants. Wiley & Sons, New York. 376p.

Esau, K. 1965. Plant anatomy. Wiley & Sons, New York. 767p.

Evans, M. W. 1958. Growth and development in certain economic grasses. Ohio Agr. Exp. Sta. Agron. Series 147:1-123.

Evans, M. S. and F. O. Grover. 1940. Developmental morphology of the growing point of the shoot and the inflorescence in grasses. J. Agr. Res. 61:481-520.

Hartman, H. T., D. E. Kester, and F. T. Davies, Jr. 1990. Plant propagation principles and practices. 5th Ed. Prentice Hall, Englewood Cliffs, NJ.

Heppner, M. B. 1961. No summer dormancy. The Farm Quart. 16(2):60-61, 107-108, 110, 112.

Hyder, D. N. 1972. Defoliation in relation to vegetative growth, p. 304-317. *In*: V. B. Younger, and C. M. McKell (eds.). Biology and utilization of grasses. Academic Press, New York.

Hyder, D. N. 1974. Morphogenesis and management of perennial grasses in the United States, pp. 89-98. *In*: Plant morphogenesis as the basis for scientific management of range resources. Proceeding of the workshop of the U.S./Australia rangelands panel. Berkeley, Calif. March 29–April 5, 1971. U. S. Dep. Agr. Misc. Public. 1271.

Hyder, D. N., R. E. Bement, E. E. Remmenga, and D. F. Hervey. 1975. Ecologtical responses of native plants and guidelines for management of shortgrass range. U.S. Dep. Agr. Tech. Bull. 1503. 87p.

Jackson, J. B. C., L. W. Buss, and R. E. Cook. 1986. Population biology and evolution of clonal organisms. Yale Univ. Press, New Haven, Conn.

Jewiss, O. R. 1966. Morphological and physiological aspects of growth of grasses during the vegetative phase, p. 39-54. *In*: F. L. Milthorpe and J. D. Ivins (eds.). The growth of cereals and grasses. Butterworths, London.

Kaufman, P. B., T. F. Carlson, P. Dayanandan, M. L. Evans, J. B. Fisher, C. Parks, and J. R. Wells. 1989. Plants: Their biology and importance. Harper and Row, Pub. New York. 757p.

Kozlowski, T. T. 1971. Growth and development of trees. Vol. I. Academic Press. New York. 443p.

Kozlowski, T. T. and T. Keller. 1966. Food relations of woody plants. Bot. Rev. 32:293-382.

Kramer, P. J. and T. T. Kozlowski. 1979. Physiology of woody plants. Academic Press. New York. 811p.

Langer, R. H. M. 1954. A study of leaf growth in timothy (*Phleum pratense*). Brit. Grassl. Soc. J. 9:275-284.

May, L. H. 1960. The utilization of carbohydrate reserves in pasture plants after defoliation. Herb. Abstr. 30:239-245.

Meyer, R. E., H. L. Morton, R. H. Haas, E. D. Robinson, and T. E. Riley. 1971. Morphology and anatomy of honey mesquite. U. S. Dep. Agr. Tech. Bull. 1423. 186p.

Milthorpe, F. L. and J. L. Davidson. 1966. Physiological aspects of regrowth in grasses, p. 241-255. *In*: F. L. Milthorpe and J. E. Ivins (eds.). The growth of cereals and grasses. Butterworths, London.

Morey, P. R. 1973. How trees grow. Inst. Biol. Studies in Biol. 39. Edward Arnold. 58p.

Naylor, A. W. 1953. Reactions of plants to photoperiod, p. 149-178. *In*: Walter E. Loomis (ed.). Growth and differentiation in plants. The Iowa State College Press, Ames.

Northington, D. K. and J. R. Goodin. 1984. The Botanical World. Times Mirror/Mosby College Publishing, St. Louis, MO. 647p.

Oakley, R. A. and M. W. Evans. 1921. Rooting stems in timothy. J. Agr. Res. 21:173-178.

Oswalt, D. L., A. R. Bertrand, and M. R. Teel. 1959. Influence of nitrogen fertilization and clipping on grass roots. Soil Sci. Soc. Amer. Proc. 23:228-230.

Peterson, R. A. 1962. Factors affecting resistance to heavy grazing in needle-and-thread grass. J. Range Manage. 15:183-189.

Rechenthin, C. A. 1956. Elementary morphology of grass growth and how it affects utilization. J. Range Manage. 9:167-170.

Richards, J. H. 1984. Root growth response to defoliation in two *Agropyron* bunchgrasses: field observations with an improved periscope. Oecologia 64:21-25.

Richards, J. H. and M. M. Caldwell. 1985. Soluble carbohydrates, concurrent photosynthesis and efficiency in regrowth following defoliation: a field study with *Agropyron* species. J. Applied Ecol. 22:907-920.

Richards, J. H., R. J. Mueller, and J. J. Mott. 1988. Tillering in tussock grasses in relation to defoliation and apical bud removal. Ann. Bot. 62:173-179.

Romberger, J. A. 1963. Meristems, growth, and development in woody plants. U.S. Dep. Agr. Forest Serv. Tech. Bull. 1293. 214p.

Salisbury, F. B. 1961. Photoperiodism and the flowering process. Annu. Rev. Plant Physiol. 12:293-326.

Salisbury, F. B. and C. W. Ross. 1985. Plant physiology. 3rd ed. Wadsworth Publication Co. Belmont, Calif. 540p.

Sharman, B. C. 1942. Shoot apex in grasses and cereals. Nature 149:82-83.

Sosebee, R. E. and B. E. Dahl. 1991. Timing of herbicide application for effective weed control: A plant's ability to respond, pp. 115-126. *In:* L. F. James, J. O. Evans, M. H. Ralphs, and R. D. Child (eds.), Noxious Range Weeds.

Stevens, O. A. 1943. Russian thistle life history and growth. N. Dak. Agr. Exp. Sta. Bull. 326:1-20.

Wardlaw, C. W. 1957. On the organization and reactivity of the shoot apex in vascular plants. Amer. J. Bot. 44:176-185.

White, J. 1979. The plant as a metapopulation. Annu. Rev. Ecol. Syst. 10:109-145.

White, J. 1985. The rediscovery of the phytomer by botanists. Agron. J. 77:653.

Williams, R. D. 1964. Assimilation and translocation in perennial grasses. Ann. Bot. 28:419-429.

Wilson, B. F. 1970. The growing tree. The Univ. of Mass. Press. p. 28.

Wilson, B. F. 1984. The growing tree. (revised edition). The Univ. of Mass. Press. Amherst. 138p.

Wilson, C. L. and W. E. Loomis. 1962. Botany. Third Ed. Holt, Rinehart and Winston, Inc., New York. 573p.

Wilson, R. T., B. E. Dahl, and D. R. Krieg. 1975. Carbohydrate concentrations in honey mesquite roots in relation to phenological development and reproductive condition. J. Range Manage. 28:286-289.

Youngner, V. B. 1969. Physiology of growth and development. p. 187-216. *In*: C. H. Hanson and F. V. Juska (eds.). Turfgrass science. Agron. Ser. 14. Amer. Soc. Agron., Inc. Madison, Wisc.

Zimmermann, M. H. and C. L. Brown. 1971. Trees—structure and function. Springer-Verlag. New York. 336p.

Carbon and Nitrogen Dynamics in Range Plants

P.I. Coyne, M.J. Trlica and C.E. Owensby

Introduction

Range ecosystems are complex assemblages of organisms the interactions of which are important in order to understand and predict responses to management. Physiological and morphological differences among plant species which confer survival value lead to competitive advantages in different environments which can be exploited by man. Various plant growth forms have evolved different mechanisms to cope with similar environments, compete interspecifically, or fill different niches. These coping mechanisms may vary considerably with respect to biochemical costs to the plant. The manner in which range plants gain and allocate carbon and nitrogen and how these processes are affected by environmental factors, including defoliation, is fundamental to development of basic management strategies. We will examine, on a rather broad scale, the dynamics of carbon and nitrogen in range plants and some ecological implications these have for management of rangelands.

Carbon and nitrogen have both structural and regulatory roles in plant metabolism. Carbohydrates are primary components in the cellulose-based structural framework of plants and in energy-transport and storage functions. Well-known regulatory functions of N-based compounds include enzymes, which catalyze metabolic reactions and participate in electron transport, and hormones. In addition, pigments and secondary metabolites, which confer resistance to insects and disease, are also important nitrogen compounds and serve to emphasize the pivotal role of nitrogen in the plant.

To organize this discussion, reference will be made to the schematic in Fig. 1. This diagram depicts major biochemical processes and transfers which affect the carbon and nitrogen pools and, therefore, the growth and development of the plant. Management practices and environmental stresses interact to influence both pool sizes and transfers or allocation patterns. The goal of numerous, computer-based modeling efforts is to quantify these mechanistic relationships so that system behavior can be simulated and results used to test hypotheses or predict responses to environment or management inputs. Our purpose is not to quantify mechanisms

---

P.I. Coyne, Professor and Head, Agricultural Research Center, Kansas State University, Hays, KS 67601; M.J. Trlica, Professor, Department of Range Science, Colorado State University, Fort Collins, CO 80523; C.E. Owensby, Professor, Department Agronomy, Kansas State University, Manhattan, KS 66506.

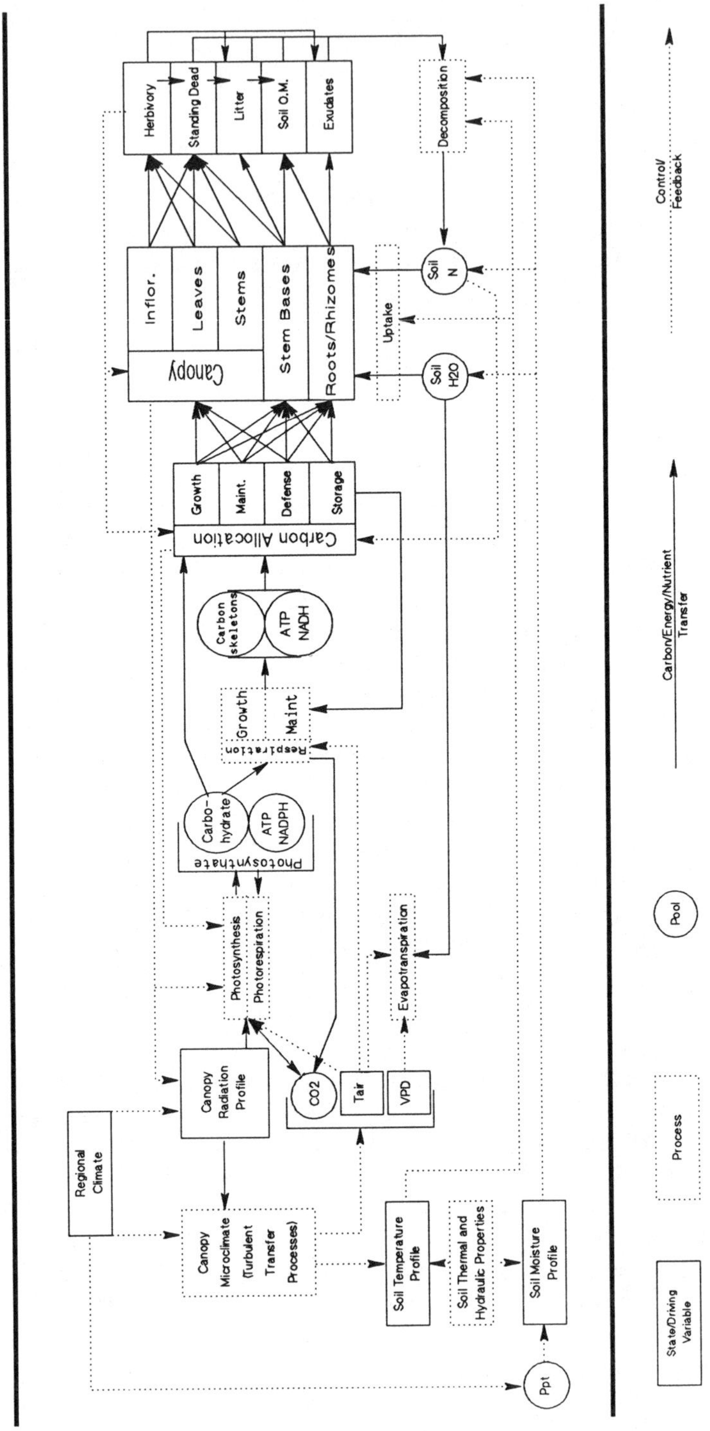

Figure 1. Schematic of carbon and nitrogen flow in a range ecosystem.

or to duplicate these modeling efforts, but rather to develop a framework in which to discuss the involvement of carbon and nitrogen in managed, forage-based, animal-production systems.

The system driving force in Fig. 1 is the regional climate which determines the input of light, thermal energy, and water to the plant canopy and soil, which in turn determines the microclimate of the plant's environment. Light energy is captured by green leaves and converted to biochemical "currency" (photosynthate) that can then be invested in growth and maintenance of tissues. Interactions between the environment and plant genotype control the allocation of carbon to major compartments or functions, and nitrogen plays an important role in influencing this entire process. Carbon sequestered in organic matter has several possible fates that are influenced greatly by the presence of herbivory. Direct removal of green tissue by grazing affects the flow of carbon to standing dead, litter, and soil organic matter, and the rate of root growth and canopy structure. This sets up a series of feedbacks that interact at numerous key points in the system to influence overall plant productivity and community stability.

The following discussion will amplify the key processes by which plants acquire, allocate, and utilize carbon and nitrogen in the context of grazed ecosystems. We will refer to the diagram in Fig. 1 to illustrate the uses and fate of carbon and nitrogen in range plants.

Carbon Dynamics

Carbon in the biosphere is in a continual state of flux as it cycles between atmospheric pools of carbon dioxide and carbon stored in living and nonliving organic matter. Fluctuations in longer-term storage forms, such as fossil fuels and sedimentary mineral deposits, affect the amount of carbon available to the biosphere. Currently, there is major concern that contemporary increases in atmospheric CO_2 resulting from the net oxidation of fossil fuels will cause undesirable and perhaps irreversible changes in the earth's heat balance and global circulation patterns.

Carbon in the form of carbon dioxide constitutes less than 0.04% of the earth's atmosphere, but makes up more than 40% of plant dry matter. The ability of green plants to utilize light energy to sequester and concentrate carbon is a truly remarkable process upon which most living organisms depend regardless of position in the food chain. Fortunately, green plants fix more carbon than is required by their own metabolism. Managers of forage-based, animal-production systems capitalize on this excess light-harvesting capability by diverting a portion of the sequestered carbon to use by ruminant animals. The skill in management comes in knowing when and how much carbon can be diverted without impinging on long-term sustainability of the system. We will in no way attempt to provide a formula or solution to this age-old problem of proper season and intensity of use. Caldwell (1984) has already addressed this issue in great detail and has vividly demonstrated the complexity associated with resolving plant requirements that must be considered in order to sustain grazing in the long term. To sustain grazing requires an understanding of plant physiology, competition ecology, plant demography, genecology, soils, and micrometeorology to supplement knowledge gained in field experiments. It demands both mechanistic and empirical approaches. Quantification of important system interactions is fundamental to development of a predictive understanding of plant response to grazing in varying environments.

In the following section, we will consider the manner in which plants gain and allocate carbon and the influence that key environmental variables have on these processes.

Carbon Acquisition

The process by which green plants acquire and allocate carbon is fundamental to all plant metabolism. Therefore, understanding carbon capture and utilization is basic to understanding management of systems which depend upon direct removal of green plant tissue.

Photosynthesis

Carbon assimilation by green plants is a photosynthetic process localized in subcellular organelles called chloroplasts. Photosynthesis can be conveniently divided into light and dark reactions for the purpose of discussion. The former traces the pathway of the electron and the latter the pathway of carbon. A general overview of these reactions that is basic to subsequent discussions of ecological implications of differences among plants relating to carbon acquisition is presented here. For detailed and systematic treatment of this subject, the reader is directed to general references such as standard plant physiology texts (e.g., Meyer et al. 1973; Devlin and Witham 1983; Salisbury and Ross 1985) as well as more specific references (e.g., Devlin and Barker 1971; Zelitch 1971; Hatch et al. 1971; Gregory 1989; Fork 1989). Books edited by Sestak et al. (1971); Coombs et al. (1985a); and Pearcy et al. (1989) are recommended for information on research methods, measurement techniques, and instrumentation.

Chloroplast structure: Devlin and Barker (1971) described higher plant chloroplasts as discoid or ellipsoidal in shape and bound by a double membrane envelope. The envelope encloses a complex system of membranes embedded in a granular matrix. The internal membranes and the surrounding matrix are called lamellae and stroma, respectively. The lamellae are paired to form thylakoids which may be stacked at intervals to form highly ordered structures called grana. Photosynthetic pigments and light reaction enzymes are associated with the lamellae and the enzymes of the dark reaction are associated with the stroma.

Light reactions (path of the electron): Photosynthetically active radiation is a wavelength band ranging from 400 to 700 nm and corresponds to the visible band of electromagnetic radiation, i.e., that to which the human eye is sensitive. Light quanta (photons) must be absorbed in order to drive a photochemical reaction. While several pigments can participate in the capture of light energy, including the various chlorophylls and carotenoids, only chlorophyll-a can directly participate in the light reactions. All other active pigments are considered accessory and function like antennae in that they transfer their absorbed photons to chlorophyll-a by resonance.

Light energy absorbed directly by or transferred to chlorophyll-a from accessory pigments causes the ejection of an electron from the chlorophyll-a molecule.

This high energy electron is used to do work in a series of electron transport chain reactions and in the process, reducing potential (NADPH, nicotinamide adenine dinucleotide phosphate) and high energy phosphate (ATP, adenosine triphosphate) are formed. These compounds in turn provide the chemical energy required to drive the dark reactions of photosynthesis. In order to prevent the chlorophyll molecule from bleaching, the lost electron must be replaced. This is accomplished by the splitting of water into protons [hydrogen ions], electrons, and oxygen. The electrons are used to replace those lost by chlorophyll-a, the protons are used in the generation of NADPH, and oxygen is evolved to the atmosphere. Thus, water is the source of the oxygen produced as a by-product in photosynthesis. Light-mediated photolysis of water and the evolution of oxygen is sometimes referred to as the Hill reaction.

The light or photochemical phase of photosynthesis actually involves two separate photosystems, generally labeled photosystems I (PSI) and II (PSII). Specific details are beyond the scope of this presentation, but briefly, PSI has a higher ratio of chlorophyll-a:chlorophyll-b than PSII and also contains carotenoids (Devlin and Witham 1983). PSII is activated by wavelengths of light less than 670 nm and is associated with noncyclic electron transport and oxygen evolution [from water]; in contrast, PSI chlorophyll absorbs wavelengths up to 700 nm and carries out both cyclic and noncyclic electron transport as well as reduction of $NADP^+$ (Lam et al. 1986). The systems are thought to be coupled serially and are commonly represented schematically in a Z-scheme first suggested by Hill and Bendall (1960). A much simplified version of that scheme including the energy trapping centers of PSI an PSII, the photolysis of water, cyclic and noncyclic photophosphorylation of ADP (adenosine diphosphate), and the formation NADPH is depicted in Fig. 2.

Dark reactions (path of carbon): The dark reactions of photosynthesis consist of a series of enzyme-catalyzed, biochemical reactions that utilize NADPH and ATP produced in the light reactions to fix and reduce atmospheric CO_2 to the level of carbohydrate. This carbohydrate pool (Fig. 1) becomes the ultimate source of all carbon skeletons utilized by the plant to synthesize other more complex carbohydrates, proteins, lipids, compounds used for defense against predators, hormones, and other products. The pool also is the ultimate source of substrate used in glycolysis and aerobic respiration to generate additional ATP, reducing potential, and carbon skeletons needed for active metabolic processes.

Alternative carbon pathways: Most forage plants of importance fall into one of two categories with respect to the initial carbohydrate produced by photosynthesis in the chloroplast stroma. One group produces a 3-carbon compound (3-phosphoglycerate, abbreviated as 3-PGA) and the other a 4-carbon dicarboxylic acid (oxaloacetate) which is rapidly reduced to form malate or transaminated (the transfer of an amino group from one compound to another) to form aspartate. This difference in the initial carboxylation product results in significant differences in the adaptation and response of plants to their environment. On this basis, plants are commonly labeled as C_3 or C_4 species depending on which carbon pathway they possess.

It is the presence or absence of PSII in the vascular bundle sheath cells that determines whether the oxaloacetate produced in the primary carboxylation reaction of

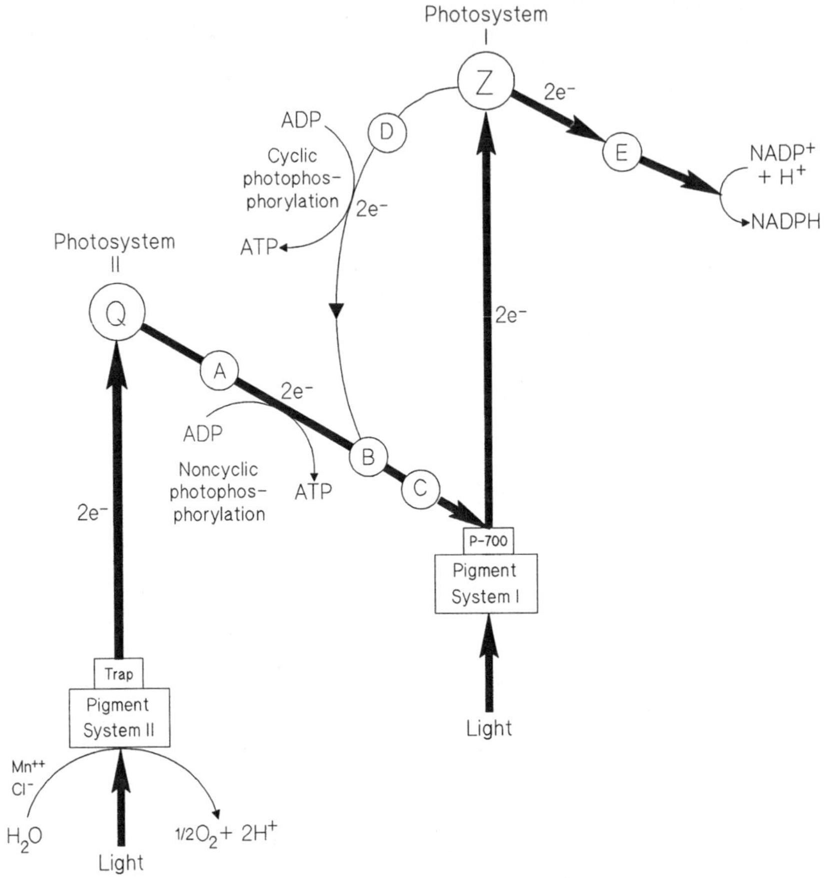

Figure 2. Simplified representation of the Hill-Bendall Z-scheme hypothesis for photosynthetic electron transport in chloroplasts. Oxidation-reduction potential increases in the vertical direction, i.e., the strongest reductants are at the top of the schematic and the strongest oxidants are at the bottom. Electrons ejected from the chlorophyll reaction centers by light energy and elevated to high redox potentials flow "downhill" to lower redox potentials via intermediate electron carriers (symbolized as A-E) and provide chemical energy to form ATP and NADPH. Definitions: A = plastoquinone a, B = cytochrome f, C = plastocyanin, D = cytochrome b_6, E = ferredoxin, Q and Z = hypothetical electron acceptors for PSII and PSI, respectively, ADP = adenosine diphosphate, ATP = adenosine triphosphate, $NADP^+$ and NADPH = oxidized and reduced forms of nicotinamide adenine dinucleotide phosphate, respectively. The trap or reaction center of PS2 is thought to be chlorophyll-a (P680). Based on information in Boardman 1971, Devlin and Barker 1971, and Lam et al. 1986.

C_4 species is primarily converted to aspartate or malate (Downton 1971a). As already described, thylakoid membranes in chloroplasts may be stacked to form grana. PSII is associated with granal-containing chloroplasts. C_4 species which are predominantly aspartate formers have granal chloroplasts in both the mesophyll and bundle sheath cells. C_4 species which are predominantly malate formers have agranal chloroplasts in the bundle sheath cells and therefore lack PSII and the ability to reduce $NADP^+$ by noncyclic electron transport. Alternative mechanisms to synthesize NADPH needed in the dark reactions have been described for these plants and will be further discussed below.

The C_3 carboxylation pathway was described by Calvin and co-workers between 1946 and 1955 and is generally referred to as the Calvin cycle, the Calvin-Benson cycle (Devlin and Barker 1971; Gregory 1989) or the photosynthetic carbon reduction cycle. Until 1965, this pathway was considered to be the primary mode of CO_2 fixation in most plants and variations in photosynthetic response among higher plants were attributed to differences in leaf physical characteristics. The primary carboxylation reaction in this cycle is catalyzed by ribulose-1,5-biphosphate carboxylase/oxygenase (RuBisCO).

Evidence for the C_4 pathway was discovered relatively recently by Kortschak et al. (1965) who noted the initial carboxylation product of sugarcane (*Saccharum officinarum* L.) was a 4-carbon acid rather than the expected 3-carbon acid. Details of this pathway were rapidly delineated and reported by Hatch and Slack (1966) and this carboxylation process is now commonly labeled the Hatch-Slack pathway in recognition of this important work. The primary carboxylation enzyme of this pathway is phosphoenolpyruvate carboxylase (PEPC).

All higher plants, including C_4 species, possess the Calvin cycle as a common denominator (Downton 1971a). Malate or aspartate produced by the Hatch-Slack pathway is eventually decarboxylated and the CO_2 released is refixed by the Calvin cycle. From this point, carbon travels the same pathway as in C_3 species. In order to keep from short circuiting the CO_2, carboxylation and decarboxylation must occur in separate cell types; C_4 carboxylation takes place in the mesophyll cells and the malate and aspartate formed are transported to the bundle sheath cells to be decarboxylated in order to supply the CO_2 used to produce 3-PGA in the Calvin cycle. An abbreviation of these reactions (intermediate metabolites and enzymes not shown) is illustrated schematically in Fig. 3. While both the aspartate and malate pathways are depicted as occurring concurrently, one or the other is usually predominant (Downton 1971a).

The Calvin cycle requires NADPH to reduce 3-PGA to glyceraldehyde-3-phosphate (3-PGAl) (Fig. 3). If bundle sheath cells are agranal and therefore lack PSII, another source of NADPH must be available to meet the demands of the Calvin cycle. It can be seen in Fig. 3 that reducing oxaloacetate (OAA) to malate in the mesophyll, transporting it to the bundle sheath, and decarboxylating it to form pyruvate produces NADPH. That NADPH can then be used by the Calvin cycle. This process generates only about half of the NADPH requirement of the Calvin cycle (Downton 1971a). To account for this deficiency in NADPH, it has been proposed that some of the 3-PGA formed in the bundle sheath by the Calvin cycle may be transported to the mesophyll to be reduced to 3-PGAl (reviewed by Downton 1971a; Randall and Tolbert 1971).

Species with the C_4 pathway have a specialized leaf anatomy ("Kranz") that permits carboxylation and decarboxylation to be separated in space (Laetsch 1971; Moser in Downton 1971a; Bolhar-Nordenkampf 1985; Gregory 1989). In the "Kranz" leaf type, vascular bundles are surrounded by two concentric chlorophyllous layers of cells consisting of an inner parenchyma bundle sheath layer surrounded by a mesophyll layer. Enzymes of the Hatch-Slack pathway are localized

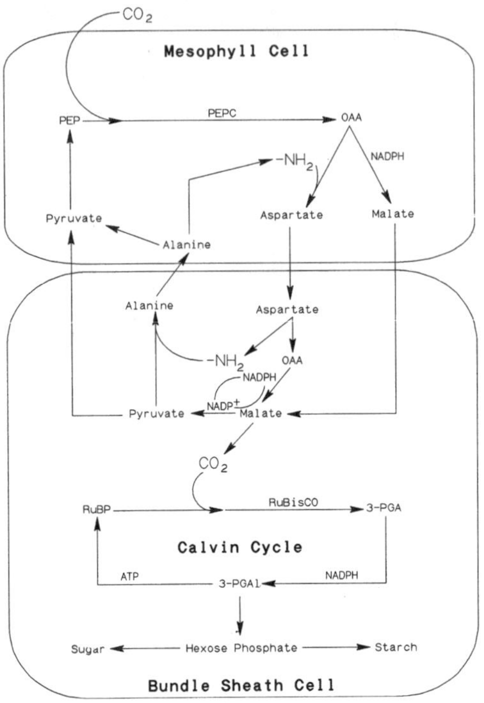

Figure 3. Schematic of the carbon fixation pathway in C_4 species. Enzymes and intermediate metabolites are not shown. Differences in cell sizes are to accommodate the included information and no significance should be implied. Initial carboxylation occurs in the mesophyll to form a 4-carbon acid. That acid is transported to the bundle sheath where it is decarboxylated. The released CO_2 is refixed by the Calvin cycle to form a 3-carbon acid. C_3 species use only the Calvin cycle to capture CO_2. The malate route of carbon transport between cell types predominates in plants having agranal chloroplasts (lack PSII and reducing potential) in their bundle sheath cells; the aspartate route is favored when granal chloroplasts are present. Abbreviations: PEP = phosphoenolpyruvate, PEPC = phosphoenolpyruvate carboxylase, OAA = oxaloacetate, RuBP = ribulose-1,5-biphosphate, RuBisCO = ribulose-1,5-biphosphate carboxylase/oxygenase, 3-PGA = 3-phosphoglycerate, 3-PGAl = 3-phosphoglyceraldehyde, ATP = adenosine triphosphate, $NADP^+$ and NADPH = oxidized and reduced forms of nicotinamide adenine dinucleotide phosphate, respectively, $-NH_2$ = amino group. Based on information in Downton (1971a), Hatch (1971), Johnson et al. (1971), Salisbury and Ross (1985), and Hay and Walker (1989).

in the mesophyll while those of the Calvin cycle are limited to the bundle sheath cells.

Although of minor importance in forage species, a third group of plants, have evolved a photosynthetic pathway that is similar to that of C_4 species, but involves compartmentalization in time rather than space. This pathway is called *Crassulacean-acid metabolism* or CAM because it is common to all members of the *Crassulaceae* family (Devlin and Barker 1971), but it is found in many *Bromeliaceace Cactaceae* and other families as well (Devlin and Barker 1971; Salisbury and Ross 1985). These are nonhalophytic plants and often succulent species that grow under semiarid conditions. The CAM pathway is a water-saving adaptation that permits stomates to open at night to take up CO_2 when evaporative demand is lower and to fix it into C_4 acids (malate) as a form of temporary storage. During daylight when the light reactions of photosynthesis can produce NADPH and ATP, the C_4 acids are decarboxylated and the CO_2 is refixed by the Calvin cycle.

Photorespiration: In order to understand the ramifications of the different photosynthetic carbon pathways, it is necessary to describe a type of respiration that is stimulated by light, and hence its name, photorespiration. Photorespiration, primarily localized in subcellular organelles called peroxisomes (Lam et al. 1986; Gregory 1989), is a series of chemical reactions that

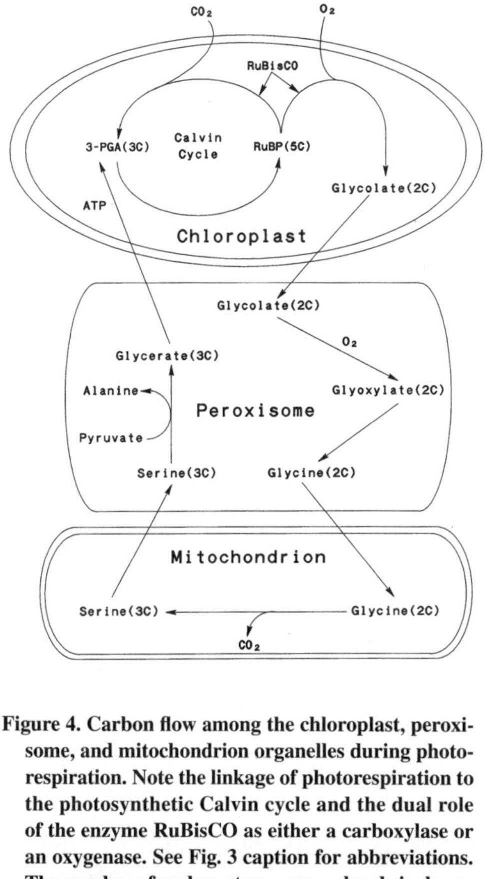

Figure 4. Carbon flow among the chloroplast, peroxisome, and mitochondrion organelles during photorespiration. Note the linkage of photorespiration to the photosynthetic Calvin cycle and the dual role of the enzyme RuBisCO as either a carboxylase or an oxygenase. See Fig. 3 caption for abbreviations. The number of carbon atoms per molecule is shown in parentheses. Based on information from Tolbert (1971) and Hay and Walker (1989).

are linked to the photosynthetic carbon cycle and results in the biosynthesis of glycolate (Fig. 4). Oxygen uptake and CO_2 production is involved and the rate increases with temperature and oxygen concentration, but decreases with increasing CO_2 concentration. Photorespiration is contrasted with dark respiration, which is confined to the mitochondria, and at one time was thought to be suppressed during photosynthesis. However, current evidence suggests respiration can proceed in the light. More details on dark respiration will be covered later.

The primary carboxylation enzyme of the Calvin cycle, RuBisCO, acts as a carboxylase in the forward direction (photosynthesis) or an oxygenase in the reverse direction (photorespiration), i.e., CO_2 and O_2 compete for the same active site on RuBisCO. Mass action determines the direction of the reaction according to the relative concentrations of CO_2 and O_2. Affinity of an enzyme for a substrate is quantified by the Michaelis constant (K_m), the substrate concentration which gives half the maximum velocity of an enzyme-catalyzed reaction. The affinity of PEPC for CO_2 in the C_4 pathway is generally thought to be much greater (K_m =7 uM) than the corresponding affinity of RuBisCO (K_m = 450 uM) in the Calvin cycle (Hatch 1971). Similarly, the CO_2 concentration required to give maximum photosynthesis with RuBisCO is 225 times higher than that required for PEPC (Hatch 1971). This difference in enzyme kinetics is particularly significant because it enables plants with the C_4 pathway to effectively concentrate CO_2 in the bundle sheath cells thereby favoring the forward reaction catalyzed by RuBisCO. As a result, C_4 species have no apparent photorespiration because the oxygenase reaction is greatly reduced by this *turbocharging* mechanism and any CO_2 that might be produced in light-stimulated respiration would be immediately refixed by PEPC in the mesophyll.

A word of caution is in inserted here. Relative differences in the affinities of PEPC and RuBisCO for CO_2 are subject to reinterpretation as more is learned about the role of the enzyme RuBisCO activase. This protein mediates activation of RuBisCO at physiological concentrations of CO_2 and ribulose-1,5-biphosphate and is dependent upon ATP (Robinson and Portis 1988). Light intensity determines the extent of RuBisCO activity *in vivo* and RuBisCO activase is thought to act as a light sensor by increasing the pH gradient across the thylakoid membranes (Portis et al. 1987). The RuBisCO CO_2 affinities mentioned above were determined *in vitro* prior to the discovery of RuBisCO activase. Apparently, in the absence of RuBisCO activase, higher substrate concentrations are required *in vitro* to attain enzyme activities corresponding to those *in vivo*. Thus, relative differences in CO_2 affinities between the two primary carboxylases may not be as great as previously thought.

Photorespiration would appear to be a wasteful process that causes a short circuit in the photochemical reduction process. However, useful functions have been proposed for photorespiration and were summarized in a review by Lam et al. (1986). Photorespiration may protect the photosynthetic and growth-regulation machinery by utilizing excess energy because it dissipates ATP and NADPH synthesized by the light reactions. In this manner, it protects against conditions of light and oxygen toxicity.

Physiological Consequences of C_3-C_4 Differences

As might be expected, the basic anatomical and biochemical differences between C_3 and C_4 species that give rise to alternate photosynthetic carbon reduction pathways are manifested in differential adaptations and responses to the environment. These are summarized in mostly qualitative fashion in Table 1. The basis for such differences are associated in large part with properties of the carboxylation enzymes.

Carboxylase affinity for CO_2: The probable higher affinity of PEPC for CO_2 results in lower CO_2 concentrations in the intercellular spaces of the leaf in C_4 compared to C_3 species. That increases the CO_2 concentration gradient between the atmosphere and the leaf interior. Movement of CO_2 into the leaf is by diffusion through the stomates; therefore, the greater the concentration gradient of CO_2, the greater the diffusion rate. Thus, for a given stomatal opening, C_4 photosynthesis rates will generally be higher than C_3 photosynthesis assuming environmental conditions are optimum for both. Also, the stomatal opening required to support a given photosynthetic rate is less for C_4 than for C_3 plants. This translates into reducing water loss by transpiration in relation to CO_2 uptake rates in C_4 plants. The result is a higher water-use efficiency (WUE), i.e., less water transpired by the plant per unit of dry matter produced.

Temperature relationships: The temperature optimum for PEPC is also higher than for RuBisCO. Temperatures in the 30-40 C range tend to inhibit C_3 photosynthesis, but promote C_4 photosynthesis (Devlin and Witham 1983). Likewise, low temperatures inhibit growth of C_4 plants and their chlorophyll is subject to photodestruction at temperatures less than 16 C. The reduced efficiency of the C_4 pathway at low temperatures compared to the C_3 pathway is attributable to lower quantum yields (Ehleringer and Bjorkman 1977; Ehleringer 1978) or higher respiration rates (Robichaux and Pearcy 1980) in C_4 species. Photosynthetic temperature

Table 1. Generalized comparison of plant characteristics associated with C_3 and C_4 photosynthetic carbon reduction pathways (developed from information in Downtown 1971a; Fitter and Hay 1981; Salisbury and Ross 1985; Farquhar et al. 1989; Hay and Walker 1989).

Characteristic	C_3	C_4
Leaf anatomy	Diffuse mesophyll; no distinct bundle sheath	Radially arranged mesophyll surrounding distinct bundle sheath ("Kranz")
Initial product of carboxylation	3-carbon acid	4-carbon acid
Primary carboxylation enzyme	Ribulose-1,5-biphosphate carboxylase/oxygenase	Phosphoenolpyruvate carboxylase
Maximum photosynthetic rates (umol $m^{-2}s^{-1}$)	6-22	19–65
Photorespiration detectable?	Yes	Bundle sheath cells only
CO_2 compensation point (uL L^{-1})	30-70	1–10
CO_2 saturation point (uL L^{-1})	500 or higher	200–250
Discrimination against ^{13}C	High	Low
Stomatal conductance	High	Low
Mesophyll conductance	Low	High
Light saturation level	Low	High
Temperature optimum	Low	High
Water-use efficiency	Low	High
Nitrogen-use efficiency	Low	High

optima for C_3 plants range from 10 to 25°C, while the range for C_4 plants (30-40°C) is narrower and higher (reviewed by Downton 1971a; Larcher 1980). These results are consistent with the observation that the C_4 pathway tends to be found in warm-season species of tropical or semi-tropical origin, while the C_3 pathway is more common among cool-season species of temperate origin.

CO_2 compensation point: The steady-state CO_2 concentration attained in a lighted, closed system containing a plant is termed the CO_2 compensation point. At this point of equilibrium, CO_2 fixed by photosynthesis is exactly balanced by CO_2 produced by all respiratory processes. The CO_2 compensation point for C_3 species ranges between 40 to 100 uL L^{-1}, but is only 0 to 10 uL L^{-1} for C_4 species (Coombs 1985). Obviously, the exact compensation point will be a function of other variables such as irradiance, temperature, and oxygen concentration. This discrepancy again reflects basic enzyme differences in affinity for CO_2 and the lack of apparent photorespiration in C_4 species.

Light saturation point: The irradiance or photon flux density (PFD) at which photosynthetic rate is maximum, assuming other environmental driving variables are held constant, is called the light saturation point. When temperature and water are optimum and CO_2 is at normal atmospheric levels (ca. 350 uL L^{-1} in 1990), C_3 species light saturate at about 50% of full sunlight (full sunlight = 2,000–2,200 umols of photons m^{-2} s^{-1}) while C_4 species do not light saturate even at full sunlight

(Coombs 1985). In C_3 species, some other factor besides light, such as RuBisCO activity or CO_2 concentration, becomes the limiting factor at high light levels.

Carbon isotope discrimination: Atmospheric CO_2 consists of two stable carbon isotopes of which about 99% is ^{12}C and the rest ^{13}C (Johnson et al. 1990). Terrestrial plant tissues contain less ^{13}C relative to ^{12}C than would be predicted from atmospheric ratios. Following the method advocated by Farquhar and Richards (1984) and Farquhar et al. (1989), carbon isotope discrimination is calculated by (R_a/R_p)-1 where R_a and R_p are the $^{13}C/^{12}C$ molar ratios in the air and plant material, respectively . Values less than zero indicate plant discrimination against ^{12}C; values greater than zero indicate discrimination against ^{13}C. Carbon isotope discrimination occurs during the initial carboxylation reaction of the dark reactions of photosynthesis as a result of mass differences in the these two stable isotopes of carbon, and RuBisCO discriminates against ^{13}C more than PEPC. However, degree of discrimination varies even within carbon-pathway groups and within a species with the same pathway. Nevertheless, this constitutes another carboxylation enzyme-based difference that can be noted between C_3 and C_4 species. Discrimination against ^{13}C by C_3 species ranges from about 20 to 35 x 10^{-3} with most species clustering around 26.5 x 10^{-3}. Discrimination in C_4 species is less variable and ranges from 7 to 15 x 10^{-3} with means of about 12.5 x 10^{-3} (Coombs et al. 1985b; Ehleringer and Osmond 1989).

Carbon isotope discrimination has practical significance. The relative difference in ^{13}C discrimination has been used to estimate relative C_3-C_4 composition of ungulate diets (Tieszen et al. 1983) and to screen for WUE in C_3 species, such as cool-season grasses (Farquhar and Richards 1984; Condon et al. 1987, 1990; Johnson et al. 1990), and cowpea [*Vigna sinensis* (L.) Endl.](Hall et al. 1990). Toft et al. (1989) reported that WUE of cold desert species measured with gas exchange techniques is relevant to long-term WUE as represented by carbon isotope composition, and that ^{13}C discrimination is a sensitive indicator of WUE in natural ecosystems. Although ^{13}C discrimination varies genetically in C_4 species, correlation of discrimination with WUE or yield in C_4 species has so far not been demonstrated (Hubick et al. 1990).

Stomatal and mesophyll CO_2 conductances: Resistances to CO_2 diffusion from the atmosphere to the chloroplast are commonly partitioned into stomatal and mesophyll. The latter may be called residual resistance because it is calculated by subtraction. It is more common to use the reciprocal of resistance, i.e., conductance, because both CO_2 uptake and water loss (transpiration) from the leaf are directly proportional to conductance. Stomatal conductances (abbreviated g_s) for C_3 species are 2–4 mm s^{-1}, whereas C_4 species have higher conductances of 5-20 mm s^{-1}. Mesophyll or residual conductances (g_r) for C_3 and C_4 species are in the ranges of 1–3 and 7–20 mm s^{-1}, respectively (Coombs 1985). What is important with respect to WUE is the ratio of $g_s/(g_s+g_r)$. C_4 plants generally have a lower ratio than C_3 plants and therefore have a higher WUE. That is why C_4 species fix more CO_2 per unit of water lost in transpiration. This does not mean that transpiration of C_4 species will always be less than that of C_3 species (commonly it is higher), rather it means that the C_4 plant has a greater capacity to fix carbon for a given transpiration rate.

Photosynthetic rates: Short-term photosynthetic rates per unit leaf area of C_4 plants (38-63 umol m^{-2} s^{-1}) are as much as two-fold or more higher compared with C_3 plants (6-22 umol m^{-2} s^{-1}) (reviewed by Downton 1971a). However, Evans

(1971) cautions that a high rate of photosynthesis is not necessarily an advantage in terms of either species survival or yield. This is because variations in relative growth rate are more a function of differences in the extent of investment in new leaf growth rather than differences in net assimilation rate. Thus, it is not necessarily correct to refer to the C_4 pathway as superior to the C_3 in all environments. In reality, optimal growing conditions for C_3 and C_4 plants are such that relative competitiveness and productivity will depend upon the environment in which the plants are growing. The advantage of C_4 photosynthesis is maximal under conditions of high photon flux density, high temperature, and limiting water (Bjorkman 1971; Waller and Lewis 1979). Pearcy and Ehleringer (1984) stated that differences in water- and nitrogen-use efficiencies, and potentially in photosynthetic capacity and growth rate, between C_3 and C_4 species are maximal under hot, high light conditions. Bull (1971) concluded that the C_4 pathway confers no advantage to plant growth which cannot be circumvented by C_3 species subjected to the required selection pressure.

Many of the advantages that appear to be associated with the C_4 pathway alluded to above may not be causal. Pearcy and Ehleringer (1984) concluded in their exhaustive review of the data comparing C_3 and C_4 plants in diverse habitats that the constraints imposed by leaf form, growth form, and the particular environment are more important determinants of photosynthetic capacity than the carbon reduction pathway. Nevertheless, range vegetation communities are commonly composed of mixtures of C_3 and C_4 forage species. Differences in their carbon reduction pathways do indeed present managers with challenges and opportunities to capitalize on associated variations in growth and forage quality. These differences give rise to seasonal changes in the mix of available forage species and make possible the complementarity of cool-season (C_3) and warm-season (C_4) components.

Origin and Occurrence of the C_4 Pathway

The C_4 pathway is found in several thousand species in at least 17 families of angiosperms, but the C_3 pathway is the dominant one throughout the plant kingdom (Pearcy et al. 1987). The taxonomic distribution suggests that the C_3 pathway is the more primitive. According to Evans (1971), the C_4 pathway has arisen polyphyletically among several of the more advanced orders of both dicots and monocots. The C_4 pathway has apparently evolved multiple times and also relatively recently because of its association with only phylogenetically advanced families (Osmond et al. 1980). The grass family (*Poaceae*) is particularly well represented among C_4 species.

Lists of plant species by family have been compiled according to photosynthetic pathway by Downton (1971b), Raghavendra and Das (1978), and Waller and Lewis (1979). The reader is encouraged to consult these for specific species classification. Discontinuities among C_3 and C_4 types occur even within a genus. For example, the genus *Atriplex* contains both C_3 and C_4 species and hybrids can be produced that are intermediate (Bjorkman et al. 1971). Several species have been shown to have a metabolism that is intermediate between the C_3 and C_4 pathways (Monson et al. 1984).

The distribution of C_3 and C_4 species is directly related to climate with the C_4 pathway consistently associated with hot, dry regions (reviewed by Potvin and Strain 1985). A comprehensive mapping of the distribution of C_3 and C_4 grasses in

75 geographical subdivisions and along environmental gradients in Australia was published by Hattersley (1983). Both C_3 and C_4 grasses increased in number with increasing precipitation within their preferred temperature range. C_4 grasses were most numerous where summers were hot and wet; C_3 where springs were cool and wet. Abundance of C_3 grasses was found to increase and that of C_4 grasses to decrease with increasing elevation in the Argentine Andes (Cavagnaro 1988) and on Mount Kenya (Tieszen et al. 1979). The geographic distribution of C_4 grasses (Teeri and Stowe 1976) and C_4 dicots (Stowe and Teeri 1978) has been correlated with climatic patterns. Results suggested that high minimum temperatures during the growing season were most correlated with the relative abundance of C_4 grasses in a regional flora. Apparently, the deleterious effects of low temperatures during growth negated any advantages of possessing the C_4 pathway in cooler environments. In comparison, distribution of C_4 dicots was mostly associated with summer pan evaporation and dryness ratio, both of which are closely related to plant water balance. Stowe and Teeri (1978) concluded that evolution of the C_4 carbon reduction pathway may confer different ecological attributes to plant taxa in different phylogenetic groups.

Factors Affecting Photosynthesis

Photosynthetic potential of an individual leaf is determined primarily by the activities of enzymes in the carbon reduction pathway (Pearcy et al. 1987) which in turn are influenced by genetic makeup and ambient environment. Individual leaves are organized hierarchically into plants and plants into communities such that canopy photosynthetic potential is strongly influenced by canopy architecture and leaf density. The extent to which potential is realized depends on a myriad of plant and environmental factors, some of which can be manipulated to influence photosynthesis and ultimately forage production and quality.

In order to understand management opportunities associated with these factors, several important ones will be discussed in this section in a univariate context (i.e., the effects of variation in a single factor will be described while other factors are assumed constant). Obviously, the norm in nature is for all factors to vary simultaneously, but isolating individual factor influences becomes difficult or impossible under a variable environment. One of the particularly useful contributions of computerized simulation models is their capacity to integrate the combined effects of complex factor interactions and to determine the relative sensitivity of response variables to individual driving variables. Factors which influence photosynthesis might be classified as abiotic driving variables (light, temperature, water, CO_2), plant factors (mineral nutrition status, insect and disease stress, leaf age, genetic), and management (timing and degree of defoliation).

Light intensity: The light dependency of leaf photosynthesis is frequently described by a rectangular hyperbola or exponential relationship assuming other environmental driving variables are within the physiological range of the plant. Response at low photon flux density (PFD) is linear, but at higher PFD, some other factor becomes limiting causing the response to degrade to curvilinear and approach a light-saturation limit asymptotically. For example, Fig. 5 depicts a generalized response of photosynthesis (as measured by CO_2 exchange rate) to light for a C_4 plant at five

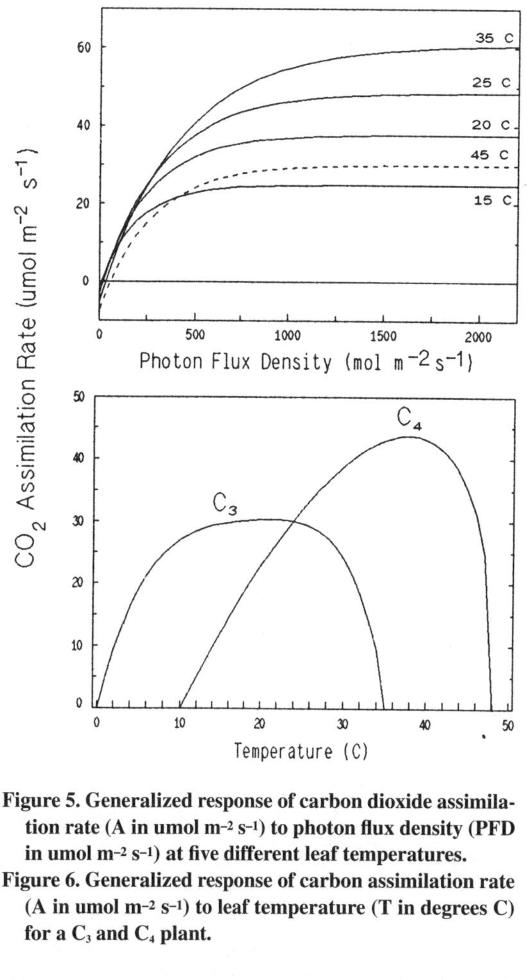

Figure 5. Generalized response of carbon dioxide assimilation rate (A in umol m^{-2} s^{-1}) to photon flux density (PFD in umol m^{-2} s^{-1}) at five different leaf temperatures.

Figure 6. Generalized response of carbon assimilation rate (A in umol m^{-2} s^{-1}) to leaf temperature (T in degrees C) for a C_3 and C_4 plant.

different temperatures. The graph clearly shows the limitation of both low and high temperatures on light-saturated photosynthesis. It is common for slopes of photosynthesis versus light in C_4 species measured at the optimum temperature to still be slightly positive even at PFDs approximating full sunlight (ca. 2000 umol m^{-2} s^{-1}). C_3 species saturate at lower PFDs so that their response curve would be flat at full sunlight and the maximum rate of CO_2 uptake would occur at temperatures closer to 25°C or lower.

The initial slope of the light response curve is a measure of photochemical conversion efficiency or apparent quantum yield (Long and Hallgren 1985) with units of mol CO_2 mol^{-1} photons absorbed. It is an apparent yield or efficiency because some light energy is reflected or transmitted and does not participate in photochemical electron transport. Estimates of true quantum yield must be corrected for these losses in energy. The maximum efficiency can only be measured at low light because the response of CO_2 uptake to light departs from linear as PFD increases as already shown. Under field conditions, canopy development may be sufficient to cause PFD to be rate limiting throughout all but the highest levels of the canopy. Hence, unlike individual leaves, canopy photosynthesis may depend more on true quantum yield than on light-saturated CO_2 assimilation rates. Quantum yield will be discussed more fully under the section heading *Efficiency of the light-harvesting process*.

The light compensation point is evident in Fig. 5 as the PFD at which CO_2 fluxes from photosynthesis and respiration are in balance (i.e., net CO_2 uptake is zero). This point is a function of species, growth environment, and conditions under which the measurements are made. However, it is usually less than 2% of full sunlight (Salisbury and Ross 1985). If the canopy of a grazed community is allowed to attain high leaf area, then much of the green tissue at lower depths may be at light levels at or below the compensation point and contribute little to net production. Such leaves may even have a net negative carbon balance during part of their life cycle. If such leaves cannot meet their own carbon demands, they may be shed by the plant.

Lower light compensation points, quantum yields, and light-saturated photosynthesis are characteristics of leaves grown in the shade such as those at the bottom of a canopy or herbaceous species within a stand of trees. These characteristics are associated with certain biochemical and morphological changes that vary with species and degree of shading. Some traits of shade-tolerant plants (summarized by Bjorkman 1989) include leaf angles that maximize light interception per unit leaf area; allocation of resources in a manner that favors leaf growth over root and other non-photosynthetic tissues; higher specific leaf area (ratio of leaf area to mass), i.e., leaves become thinner; higher chlorophyll content per chloroplast, which is related to increased grana stacking; higher ratio of chlorophyll to RuBisCO, other carbon metabolism enzymes, and electron carriers; larger antenna size (accessory pigments) in PSII; and reduced leaf respiration rates.

It could be assumed from these whole plant responses that many of the same characteristics would also be found in leaves of plants grown in the sun, but located in the shade of leaves higher in the canopy. Forage quality consists of both nutrient availability and digestibility factors. It is reasonable to assume that nutrient concentration (especially nitrogen and energy), as well as forage digestibility, would decrease with depth into well-developed canopies as a result of increasing shade effects. The same effects could also result from increasing leaf age.

Because of shading effects, the light response of canopy photosynthesis will tend to be linear to higher light levels than for individual leaves. C_4 canopies may be linear up to full sunlight, but C_3 canopies will likely still saturate before PFD at the top of the canopy is equal to full sunlight (Salisbury and Ross 1985).

Except for nutrient-deficient, insect-damaged, or diseased plants, chlorophyll density seldom limits photosynthesis. Charles-Edwards et al. (1986) stated in their review that 78% of incident light is absorbed when chlorophyll density is only 5 ug m^{-2} (Shul'gin and Kleshnin 1959) and that this relationship holds fairly constant over a wide range of leaf thicknesses, diversity of anatomical structure, species, and growth forms. Apparently, light-saturated photosynthesis is more dependent upon RuBisCO than chlorophyll density (Farquhar et al. 1980).

Temperature: Photosynthesis has been shown to involve both light and dark reactions. The temperature dependency of these separate phases is different. The rate of photochemical reactions, such as those in PSI and PSII, is essentially independent of temperature (Heath 1969). These reactions proceed at a rate proportional to the dose (intensity x time) of light received. On the other hand, rates of the biochemical reactions of photosynthesis, as well as other plant metabolic processes, are highly dependent upon the amount of heat present in the system as measured by temperature. The kinetic energy of reacting molecules increases with temperature, and this in turn increases the number of molecular collisions and hence the rate of reaction. Thus, enzyme-catalyzed reactions have a required energy of activation defined as the minimum energy required for the reaction to occur. They also have an upper temperature limit for thermal stability above which enzyme systems break down as a result of protein denaturation (Bjorkman 1981). Weis and Berry (1988) stated that the effect of high temperature on higher plants is primarily on photosynthetic functions. The heat tolerance limit of leaves in higher plant coincides with the thermal sensitivity of primary photochemical reactions occurring in the thylakoid membrane system.

The temperature coefficient of a biological process, sometimes referred to as Q_{10}, is expressed as a ratio of the velocity of a process or chemical reaction at a given temperature to that at 10 C lower (Devlin and Witham 1983). It can be calculated as

$$\ln (Q_{10}) = ((10/(T_2-T_1)) \ln (k_2/k_1)$$

where ln is the natural logarithm, k_1 and k_2 are the reaction rates measured at temperatures T_1 and T_2, respectively.

Photochemical reactions have a Q_{10} of 1.0, consistent with their lack of temperature dependence. Physical processes (e.g., diffusion) have a slight temperature dependency and a Q_{10} of about 1.1. Chemical reactions, however, are quite sensitive to added heat and have Q_{10} values of 2 to 3. Because photosynthesis consists of both photochemical and chemical reactions, it does show a strong temperature response and the Q_{10} is low (close to 1) if light is limiting and high (2 or greater) if the process is light saturated (Heath 1969).

A process such as photosynthesis, or a more complex series of integrated processes such as growth, can be ascribed cardinal temperatures. These are defined as the temperature below which a reaction does not occur (minimum), the temperature at which the rate is maximum (optimum), and the temperature above which the rate falls to zero (maximum). Cardinal temperatures vary with biochemical process, species, growth environment, and season. Within a particular physiological temperature range, reaction rates increase as temperature rises above the minimum until a maximum rate is attained, and then decline as the thermal stability of enzymes is exceeded.

Typical temperature response curves for leaf photosynthesis at full sunlight are depicted in Fig. 6 for a C_3 and C_4 plant. Note how the curve for C_4 photosynthesis, as measured by CO_2 exchange, rises fairly steeply to a relatively narrow optimum temperature peak rate and then falls rather rapidly as the upper cardinal temperature is approached. Contrast this trend to the C_3 response which has a much broader peak rate and a more gradual decline as temperature approaches the physiological maximum. The temperature optimum for C_3 species is toward the middle of the range between the lower and upper cardinal values, while the optimum for C_4 species tends to be skewed toward the upper end of this range.

While the manager can do little to influence temperature, except through controlling canopy and litter cover, response differences in photosynthesis of C_3 and C_4 forage species can be exploited. Mixtures of these species, or adjacent cool-season and warm-season communities, lengthen the period of availability of high quality forage. Forage communities dominated by C_3 species can be used in spring or fall and those dominated by C_4 species can be used during the summer to maximize the benefits from temperature-related response differences among species.

Water: Rangelands are located predominantly in semi-arid and arid regions of the world. For this reason, water is the environmental factor which most often limits forage productivity. The photosynthetic process itself utilizes infinitesimal quantities of water compared with quantities required to sustain the plant. Therefore, the influence of water deficits on primary production is indirect through direct effects on cell elongation, which retards leaf expansion. Water deficits may also reduce stomatal opening, which may restrict CO_2 uptake. Stomates must be at least partially open in order for CO_2 to diffuse from the atmosphere to carboxylation sites in the chloroplasts. Mesophyll cell surfaces are continually wet so that air in the intercellular spaces is assumed to be water saturated at leaf temperature. The loss of water from the leaf as a vapor, or transpiration, is the inevitable result whenever stomata are open and the ambient water vapor pressure is less than that of the substomatal cavities. If the rate of transpiration exceeds the rate of water uptake from the soil, the plant water deficit may increase to the point that leaf hydration can no longer be maintained and stomata close partially or completely.

It has been commonly accepted that partial stomatal closure, as measured by reduced stomatal conductance or increased resistance, resulted in a corresponding reduction in CO_2 uptake. While there is a strong relationship between photosynthesis and conductance, synthesis of the literature by Farquhar and Sharkey (1982) suggests that the opportunity costs to the plant of reduced carbon fixation by reductions in stomatal conductance are relatively small. They concluded that apart from some transients and from the effects of added abscisic acid [which stimulates stomatal closure] or reduced atmospheric humidity, reduced stomatal conductance is rarely the main cause of reductions in photosynthesis. Apparently, stomatal conductance decreases to match the sustainable rate of photosynthesis. This hypothesized coupling would operate to maximize WUE (photosynthesis/transpiration = transpiration ratio) without restricting CO_2 diffusion below the rate capable of being processed by the chloroplast.

There are however, measurable differences in the degree of coupling between CO_2 uptake and stomatal conductance among plants including those of similar genotypes. This is important in a forage species improvement program in which the goal is to match specific genotypes with a particular environment. The variation in coupling can be illustrated by calculating an index of carbon dioxide assimilation efficiency in relation to water loss as proposed by Ting et al. (1972) and Johnson (1975). This efficiency represents the ratio of observed CO_2 assimilation rate (as limited by all diffusional and biochemical resistances) to the maximum CO_2 assimilation rate that would occur if limited only by the stomatal diffusion resistance or, symbolically, A/A_0. This ratio can also be reduced to resistances (r_s/r_t) or conductances (g_r/g_t) where r_s and r_t are the stomatal and total (stomatal+residual) resistances to CO_2, respectively, and g_r and g_t are the residual and total (stomatal+residual) conductances for CO_2, respectively. The value of A/A_0 is particularly useful as an index of WUE because its value ranges from 0-1 whenever irradiances exceed the light compensation point, whereas, transpiration ratio has no easily defined theoretical limits.

Data graphed in Fig. 7 (from Coyne et al. 1982) are for four different accessions of C_4 grasses from the *Bothriochloa* genus grown in a greenhouse. They serve to illustrate inherent differences in stomatal control of photosynthesis measured under near optimal and constant conditions of light, temperature, vapor pressure gradient, and CO_2 concentration. The ordinate value is the fraction of the potential CO_2 assim-

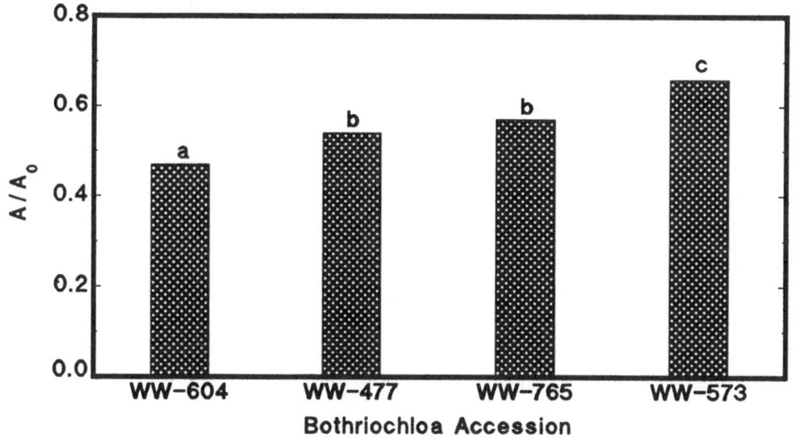

	WW-604	WW-477	WW-765	WW-573
A	25c	35a	31b	26c
E	2.5c	3.2c	2.7b	1.9c
P	648c	1,361c	2,650b	3,492a

Figure 7. The ratio of actual carbon dioxide assimilation (A) to potential assimilation (A_o) if limited only by stomatal conductance in four Old World bluestem (Bothriochloa) accessions (data from Coyne et al. 1982). Data below the bars are corresponding absolute values for assimilation (A in umol m^{-2} s^{-1}), transpiration (E in mmol m^{-2} s^{-1}), and forage production (P in kg ha^{-1}) on a dry matter basis. Bars or values marked with a common letter are not significantly different ($P<0.05$).

ilation rate, assuming photosynthesis was limited only by stomatal diffusion. The numbers below the abscissa provide additional response data on actual or measured assimilation (A) and transpiration (E) rates plus some field forage production data (P) for comparison. WW-604 and WW-573 had similar values for A ($P<0.05$)), but WW-573 was assimilating CO_2 at 66% of its potential diffusion rate compared to only 47% for WW-604. Conversely, WW-477 had a slightly higher value for A than WW-765, but was realizing about the same fraction of its potential. This would suggest stomatal conductance was higher in WW-604 and WW-477 than required to sustain the observed assimilation rates. These data also show that there is natural variation in the degree to which stomates operate to optimize water loss, i.e., open just enough to not restrict CO_2 uptake and no more. The higher the A/A_0 ratio, the higher the degree of optimization.

Data on A/A_0 obtained under controlled conditions may be useful for screening germplasm to identify genotypes with superior drought performance. Forage production, measured under drought conditions in the field (see "P" below Fig. 7), showed a strong positive relationship with degree of stomatal coupling to CO_2 uptake measured in the greenhouse.

CO_2: Carbon dioxide made up only about 0.035% (350 uL L^{-1}) of ambient air as of 1990. This concentration is insufficient to saturate photosynthesis of C_3 species under optimal conditions of light, temperature, and moisture. The response

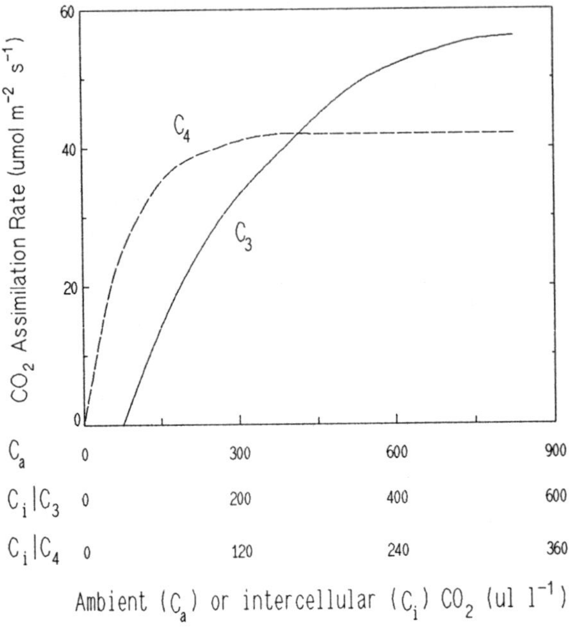

Figure 8. Generalized response of carbon assimilation rate (A in umol m-2 s-1) to ambient CO_2 concentration (C_a and C_i in ul l-1) for a C_3 and C_4 plant. Corresponding intercellular CO_2 concentrations (C_i) that would be typical for C_3 and C_4 plants are shown as parallel scales on the abscissa.

of photosynthesis to increasing CO_2 concentration (Fig. 8) is hyperbolic and therefore is similar in shape to the photosynthesis-light response curves (Fig. 5). As the CO_2 to which plants are exposed during measurement of photosynthesis is increased above the compensation concentration, i.e., where CO_2 fluxes from photosynthesis and respiration are equal, photosynthesis increases linearly at first and then curvilinearly as some other factor becomes more limiting than CO_2.

Photosynthetic response of C_3 species to increasing CO_2 is more sensitive than that of C_4 species. Photosynthesis of C_4 plants may saturate between 350-400 uL L^{-1} or less while C_3 plants saturate at CO_2 levels considerably above current ambient values (Breen et al. (1986)). This difference in behavior between C_3 and of C_4 plants is easily explained by the differences associated with their respective carbon reduction pathways as described earlier. In C_3 plants, which have a high CO_2 compensation point and a primary carboxylase with low affinity for CO_2, the direction of RuBisCO-catalyzed reactions depends on the relative concentrations of CO_2 and O_2. Increasing CO_2 relative to O_2 (as in Fig. 8) favors the carboxylation reaction over the oxygenase reaction and hence increases the rate of photosynthesis (Bjorkman 1971).

According to Long and Hutchin (1991), the drop in CO_2 concentration across the stomata is typically 100 uL L^{-1} in C_3 and 200 uL L^{-1} in C_4 species. Long and Hutchin also note that many observations show that stomatal opening is controlled to maintain a constant ratio of intercellular to external or ambient CO_2 concentration (C_i/C_a) of about 0.6 to 0.8 in C_3 and 0.3 to 0.5 in C_4 species. Maintenance of a constant C_i/C_a ratio as C_a increases suggests a corresponding decrease in stomatal conductance which should reduce transpiration and increase water-use efficiency. Because PEPC has a greater affinity for CO_2 than RuBisCO, C_4 plants reduce C_i to a lower level than do C_3 plants. This means that stomatal conductance of C_4 plants can be less than that of C_3 plants (i.e., less water loss via transpiration), when photosynthesis rates of both plant types are equal. Thus, the value of C_i needed to saturate photosynthesis in C_4 species is on the order of 140 uL L^{-1} compared to values of 345 uL L^{-1} and higher for C_3 species. This difference is depicted in the idealized C_i scale in Fig. 8. Of particular significance here is that increases in stomatal conductance beyond that which produces a saturation C_i merely increase transpiration and decrease water-use efficiency.

While the manager has no control over CO_2 concentrations in the field, the C_3-C_4 differential response has ecological implications for mixed C_3-C_4 communities growing in CO_2-enriched atmospheres that are likely to result from the burning of fossil fuels. We will return to this theme later in this discussion.

Mineral nutrition: The response of photosynthesis to deficiencies in the various essential mineral elements required by plants depends on the specific element and can be either direct or indirect, and the effects can be severe. Detailed accounts of the function and deficiency symptoms of the essential elements are provided in plant physiology texts (Devlin and Witham 1983; Salisbury and Ross 1985) and will not be reviewed here. Devlin and Barker (1971) stated that effects of nutritional deficiencies are indirect and reduce the vigor of leaves and interfere with dark reactions (i.e., reactions that do not require light energy). The loss of chlorophyll causes degeneration of chloroplast ultrastructure and this is associated with a substantial loss of photochemical activity. Deficiency symptoms for a number of macro- (N, S) and microelements (Fe, Mg, Mn, Zn) are manifested as a chlorosis related to these elements either being a component of the chlorophyll molecule or else involved in its synthesis pathway. While the manager may exert control over the mineral nutrition of tame pastures, it is seldom economical or ecologically beneficial to add fertilizer to native range communities. In addition, the effects of mineral deficiencies on animal performance may be more pronounced and visual than the effects on the plants. Managers can correct nutritional deficiencies in the animal diets resulting from low fertility soils by supplementation.

Insects and disease: Any stress, such as insects or disease, that interferes with normal metabolic function or directly damages or removes chlorophyllous tissue may cause reductions in photosynthesis. In addition, the trauma associated with insect and disease damage will likely stimulate maintenance respiration thereby further reducing net photosynthesis. Native range communities have evolved in the presence of these stresses and rarely is it feasible to implement chemical control measures to counteract the effects. Exceptions do occur, such as when populations of grasshoppers periodically explode to epidemic levels and threaten range productivity on a regional scale. The usual reaction to these stresses, however, is to accept them as part of the natural variation in annual forage production and to rely on natural predators and genetic-based resistance to keep insects and disease in check.

Leaf age: It has long been common knowledge that photosynthetic capacity is not constant over the life of a leaf. Newly initiated leaves may have a net negative carbon balance and require import of carbon from older leaves to sustain maintenance and growth. Once carbon balance of a leaf is positive, it will continue to increase until about the time the leaf is fully expanded, and then it will decline as senescence proceeds. The aging process is a function of internal factors under the control of hormones and not the intracellular environment. The class of hormones called cytokinins, which are produced in the roots and transported to the leaves, retard leaf senescence (Salisbury and Ross 1985). Research has demonstrated that it is possible to apply hormone-like chemicals or growth regulators exogenously to forage plants and delay senescence or else influence plant morphology (White 1989, 1990). As a result, forage quality is enhanced or maintained longer in a growing cycle. Towne and Owensby (1983) applied exogenous cytokinin to native tallgrass prairie and reported that treated big bluestem (*Andropogon gerardii* Vitman) leaves had higher nitrogen concentration later in the season compared with untreated leaves. Application of cytokinin in mid- to late-July increased herbage yields the following season by 22%.

Maintaining a plant canopy in a more juvenile stage of development, as by grazing, can encourage a higher canopy photosynthetic rate and forage production, and also result in a higher nutritional plane from the standpoint of the herbivore. The management implications of this trend will be discussed more fully below.

Defoliation: The most obvious effect of partial defoliation, such as results from herbivory, is to reduce photosynthesis because it directly removes a portion of the photosynthetic machinery. However, the effect on canopy photosynthesis is not in proportion to the foliage removed because of mitigating factors. A well-developed canopy causes significant shading of leaves in the lower levels. Removal of a portion of the upper layers allows deeper light penetration and perhaps more turbulent mixing of air in the canopy such that some or all of the photosynthetic capacity is recovered. In addition, deeper light penetration may also stimulate tillering of forage grasses through an increase in the ratio of red/far red radiation (Deregibus et al. 1985). As the schematic in Fig. 1 shows, herbivory triggers numerous feedback mechanisms that affect canopy microclimate, carbon allocation, and ultimately canopy architecture.

A common physiological reaction to defoliation is the immediate reduction in root growth (Crider 1955; Jameson and Huss 1959; Jameson 1963; Davidson and Milthorpe 1966; Richards 1984). While reduction of root growth following defoliation might be perceived as detrimental to the plant, results published by Richards (1984) indicated this could be an adaptive mechanism to momentarily reduce belowground carbon demand in defoliated plants. This in turn would allow more of the net carbon input to be used aboveground to promote rapid reestablishment of the canopy and hence to restore photosynthetic capacity as rapidly as possible.

Fortunately from the standpoint of forage-based, animal-production systems, plants generally have excess photosynthetic capacity that can compensate for partial defoliation and perhaps even stimulate production beyond that of ungrazed controls. Hilbert et al. (1981) reviewed the results of numerous field and laboratory studies that supported a *grazing optimization hypothesis*, i.e., that net primary productivity (NPP) of an ecosystem may be stimulated by grazing at some optimum level. The basic concept is that NPP traces a curve that increases as grazing intensity increases from 0 through some optimum level, and then declines as foliage removal exceeds the ability of the plants to compensate for loss of photosynthetic tissue. They summarized the literature that suggested possible explanatory mechanisms and listed the following possibilities for how grazing might stimulate production: increased photosynthetic rates in remaining tissue, increased proportion of current photosynthate allocated to the production of new leaf area (see Fig. 1), increased tillering or lateral bud growth from either removal of apical dominance or opening the canopy to increased light penetration, conservation of soil moisture by reducing transpirational leaf area, and a variety of other indirect mechanisms. They concluded that although the magnitude of these mechanisms is probably insufficient to completely compensate for decreased production resulting from tissue damage or reduced photosynthetic leaf surface following grazing, their existence indicates a strong potential in plants for some compensatory growth following grazing.

Photosynthetic responses of individual leaves on defoliated plants are variable depending on treatment. Painter and Detling (1981) measured net photosynthesis per unit leaf area and regrowth of 60-day old western wheatgrass (*Agropyron smithii* Rydb.) plants that were grown hydroponically and defoliated to simulate grazing. Photosynthesis of unclipped tillers declined 6–7% in both moderately (1/2 tillers clipped at 50 mm) and heavily (3/4 tillers clipped at 50 mm) defoliated plants within 30 minutes following clipping. By day 2, photosynthesis of these same tillers had increased 5–10% in both treatments compared to rates preceding defoliation, while rates of control plants had decreased 6%. From day 2 through day 10, photosynthesis rates averaged 90%, 106%, and 114% of preclipping rates in the control, moderately, and heavily defoliated treatments, respectively. Tillering did not respond to clipping during the 10-day period.

Photosynthesis of leaf tissue remaining following defoliation of that leaf may decrease or increase. Detling et al. (1979) simulated grasshopper feeding by damaging leaves of western wheatgrass at several places along their length. Photosynthesis in the remaining damaged leaf blades decreased in all cases. In one case, removal of approximately 25% of the leaf area by cutting 40 mm long notches from the base to the tip of the blade reduced photosynthesis per unit remaining tissue by one third almost immediately and photosynthesis remained at that level for over a week. In comparison, Painter and Detling (1981) and Detling and Painter (1983) reported that removing 50 to 75% of the tillers of western wheatgrass by clipping caused net photosynthesis rates of remaining undamaged leaves to average 25-35% greater than leaves of the same age on undamaged control plants. This suggests that compensatory photosynthesis may reduce the overall effect on canopy photosynthesis resulting from partial defoliation (Detling 1987).

Wallace et al. (1984) reported compensatory photosynthetic responses to clipping in three environmental chamber-grown African graminoids that were previously found to exhibit compensatory growth in the field following grazing (McNaughton 1979). The increased photosynthesis was attributed to increased stomatal opening following clipping. Nowak and Caldwell (1984) documented compensatory photosynthesis following partial defoliation in two field-grown *Agropyron* bunchgrasses that differed in their evolutionary history of grazing pressure. In both grasses, compensatory photosynthesis was attributable to delayed leaf senescence and increased soluble protein concentration. There was no measurable improvement in the water status of clipped plants. They concluded that compensatory photosynthesis was not an important ecological component of herbivory tolerance in these two species because the clipped leaves were a small fraction of total aboveground biomass, and therefore their contribution to the total photosynthate pool would be relatively small.

Hilbert et al. (1981) modeled the grazing optimization hypothesis using traditional plant growth analysis techniques (Hunt 1982). They sought to predict the extent to which plant relative growth rate (RGR, the instantaneous change in aboveground plant biomass per unit time per unit biomass present: [(1/W)(dW/dt)], where W = biomass, t = time) must change in order to generate an optimization curve. Their analysis showed the RGR at the time of grazing was an important determinant in ability of plants to maximize production following grazing. Plants growing slowly in relation to their potential RGR at the time of grazing (e.g., plants under stress) were most likely to increase production following grazing. In some circumstances, only small increases in mean RGR were required to increase production, while large increases in mean RGR were required in other cases. RGR represents a measure of production per unit of producing tissue. Because RGR integrates a large number of physiological processes, the authors encouraged its use in defoliation studies.

In addition to physiological differences, morphological characteristics, as exemplified in growth form, also determine defoliation tolerance. The effect of defoliation on photosynthesis varies among plant growth forms, as do the biochemical costs associated with mechanisms used to maintain an autotrophic canopy. Archer and Tieszen (1980) sought to determine if carbon allocation patterns among several Arctic tundra growth forms [graminoid, forb, deciduous shrub, evergreen shrub] dictated a response to defoliation that served to minimize effects on production and reproduction. They concluded that carbon allocation patterns were a function of growth form and these forms did differ in response to defoliation. Maximum photosynthetic rates were a function of growth form and inversely related to leaf longevity [evergreen leaf rates less than deciduous rates]. The impact of grazing was reduced with rapid leaf turnover and low amounts of supportive tissue. Defoliation resulted in immediate adjustment in carbon allocation patterns within the defoliated plant. Carbon allocation to maintenance tissue or long-lived productive tissue [evergreen leaves] was accompanied by mechanisms that reduced grazing. Leaf replacement rates were lowest in the evergreen shrub, highest in the perennial graminoids, and intermediate in the deciduous shrub. The synchronous leaf growth habit of shrubs enabled them to more fully exploit the most favorable portion of the Arctic growing season, but exposed the season's total complement of photosynthetic tissue to direct removal by grazing.

Growth forms such as evergreen shrubs with low photosynthetic rates apparently require the retention of leaves for more than one season to achieve a net positive carbon balance and this, in turn, requires investment in compounds to protect them from herbivores (Archer and Tieszen 1980; Chapin et al. 1987). The evergreen habit carries a high growth and maintenance cost that likely results in fewer resources available to invest in replacing tissue lost to herbivory.

Efficiency of the Light-harvesting Process

Because light energy is the ultimate source of all chemical energy and carbon skeletons stored and used in a range ecosystem, implicit among the management goals of forage-based, animal-production systems is the optimization of light energy capture. In order to understand the extent to which this goal is attained, it is productive to examine the potential and actual efficiencies of the photosynthetic process. This is because some 90% of the dry matter of higher plants is derived from light-driven assimilation of CO_2 (Zelitch 1982), yet photosynthetic efficiency of a plant canopy expressed per unit of absorbed light may only be 1-2% and the efficiency of an individual leaf is perhaps only as high as 10-20% (Bidwell 1974).

Quantum yield of photosynthesis: Photosynthetic efficiency can be expressed as a ratio of carbohydrate calories stored to calories of light energy absorbed or as a quantum yield, which is the moles of CO_2 converted per mole absorbed quanta (Devlin and Barker 1971). A quantum is the amount of energy required to activate one molecule of reactant in a chemical reaction. In the case of photochemical reactions, such as photosynthesis, the energy source is photosynthetically active radiation (PAR) or light in the 400-700 nm band of the electromagnetic spectrum. A quantum of light is called a photon and the preferred measurement of irradiance in photosynthesis studies is the photon flux density (PFD) in units of mol [photons] m^{-2} s^{-1}. From Planck's law, the energy (in joules; abbreviated J) of a mole of photons can be calculated from the equation E = Nhc/lambda, where E is energy in J mol^{-1}, N is Avogadro's number (6.02 x 10^{23} molecules mol^{-1}), h is Planck's constant (6.624 x 10^{-34} J sec^{-1}), c is the velocity of light (2.998 x 10^{8} m s^{-1}), and lambda is the wavelength of light (m).

This equation shows that E is inversely proportional to wavelength, so a blue photon (lambda = 400 nm; 4.97 x 10^{-19} J) would be expected to provide more energy to photosynthesis than a red photon (lambda = 680 nm; 2.92 x 10^{-19} J). Although the higher energy of the blue photon does eject an electron from chlorophyll and raise it to a higher level of excitation than a red photon, the excess energy of that electron, compared to one excited by a red photon, is rapidly dissipated as heat (Hay and Walker 1989). Thus red and blue photons are photochemically equivalent with respect to their ability to drive photosynthesis and the energy of a mole of red photons (176 kJ) can be used to calculate efficiencies.

The following discussion, taken from Hay and Walker (1989), shows the basis for the relatively low efficiencies commonly cited for photosynthesis. Each molecule of CO_2 reduced by photosynthesis requires 3 molecules of ATP (5 for C_4 species; Hatch and Osmond 1976) and 2 of NADPH. On a mole basis, this converts to an equivalent 525 kJ. If all absorbed photons were used with 100% efficiency, then a minimum of 3 moles of photons (525/176) would be the theoretical minimum requirement to reduce 1 mole of CO_2. When the light absorption process is corrected for energy dissipated through chlorophyll fluorescence and heat, the theoretical minimum requirement increases from 3 to 8 moles of photons. Actual measurements of the light absorption, CO_2 uptake, and respiratory losses raise the requirement still further to 15-20 moles of photons, the exact value being a function of species and temperature. The reciprocal of 15 gives a quantum yield of 0.067 mol CO_2 mol^{-1} photons.

On an energy basis, 6 moles of CO_2 are reduced to form 1 mole of glucose ($C_6H_{12}O_6$) which yields 2809 kJ when burned (oxidized) in a calorimeter. It follows that 1 mole of CO_2 is equivalent to 468 kJ (2809/6) and 15 moles of photons are equivalent to 2640 kJ (15 x 176). Therefore, the calculated maximum efficiency with which absorbed PAR can be converted into carbohydrate by single leaves in a laboratory is 17.7% (468/2640).

To move this estimate from the laboratory to the field requires additional downward adjustments. Hay and Walker (1989) calculated the conversion of incident PAR into carbohydrate by a C_3 crop. Beginning with 100% incident PAR, they subtracted 15% for losses due to reflection and transmission to get 85% absorbed PAR. The above calculations show only 17.7% (or 15% of the original 100% incident PAR) of this would end up in carbohydrate. There would be additional estimated losses for leaves in the upper canopy being light saturated (30%; less for C_4 species) and for respiration losses in the dark (40%) leaving only about 6% of the original incident PAR being converted to carbohydrate. This also is high because it assumes a canopy sufficiently developed to intercept all incident PAR and environmental conditions conducive to optimizing photosynthetic rates. Conditions are seldom optimum and various stresses do operate to reduce this efficiency still further to a range of perhaps 2-4% or even less.

The quantum yield of C_3 and C_4 species is different. The C_4 pathway itself apparently requires more light energy [2 additional ATP] to regenerate PEPC than the C_3 pathway does to regenerate RuBisCO. However, C_4 plants lose less fixed carbon by photorespiration, a process with a high temperature dependency. Hay and Walker (1989) concluded in their review of the quantum yield literature that when the temperature response of quantum yield is compared in C_3 and C_4 plants, there is a crossover temperature (quantum yields equal) at which the loss of efficiency caused by photorespiration in C_3 species is equivalent to the higher intrinsic energy costs associated with the C_4 pathway. Below this crossover temperature, the efficiency of net photosynthesis is higher in C_3 species, and above this temperature, the efficiency advantage lies with the C_4 species. The crossover temperature is highly species dependent, but according to Ehleringer and Pearcy (1983), the disadvantage in quantum yield for most C_4 plants occurs only at temperatures less than 18-23 C. While this information may be more academic than practical from the standpoint of the range manager, it does serve to enhance our understanding of relative competitive advantage among C_3 and C_4 species and how this might change with environment and season.

Canopy structure: In addition to the optical properties and photosynthetic pigment concentration of leaves, the manner in which individual leaves are arrayed in three-dimensional space determines the light-harvesting efficiency of the range plant community. The three variables used to describe canopy structure or architecture (Nobel and Long 1985) are leaf area index (LAI) and its distribution with height, leaf inclination [angle measured from the horizontal], and leaf orientation [azimuth angle measured from north]. LAI is the amount of leaf area [single side] per unit area of soil surface and is an index of the size of the photosynthetic apparatus. In grasses, LAI may include both lamina and sheaths.

Hay and Walker (1989) discussed the concept of optimum and critical values of LAI. Optimum LAI theory predicts that light interception, and therefore growth rate, increases as the canopy develops up to some optimum LAI at which all incident light is intercepted. Further increases in LAI result in reduced growth because light levels in the lower canopy fall below the light compensation point during much of the photoperiod and leaves in this layer may become parasitic. An LAI of 3 is generally required for interception of 90–95% of incident radiation. Optimal responses are seldom seen in the field. Instead, growth rate generally increases with LAI up to a critical value of 3–5 or higher, and then is maintained at a fairly constant, maximal growth rate as LAI continues to increase above the critical value. Apparently, shaded leaves do not constitute an important respiratory sink for current assimilate because overall rates of canopy respiration are more closely related to the rate of gross photosynthesis than to biomass or leaf area. Light interception is complete at both the optimum and critical LAI, but optimum implies growth rate increases with LAI below the optimum and decreases with LAI above that optimum. Critical LAI, on the other hand, implies increasing growth with LAI below the critical value, but a leveling off of growth rate as LAI increases above the critical value.

Critical LAIs vary considerably among different plant species and genotypes. This likely results from variations in leaf inclination, orientation, and optical properties. For example, Loomis and Williams (1969) used the simulation model of Duncan et al. (1967) to determine the effects of LAI and leaf inclination on crop growth rates of the C_4 crop maize (*Zea mays* L.) and the C_3 crop clover (*Trifolium* L.). When LAI was small, horizontal leaves promoted growth. At large values of LAI, more erect leaves resulted in greater growth. In keeping with the critical LAI theory, the simulations indicated no pronounced optimal LAI, providing the lower leaves adapted to the shade environment.

Managing for optimal canopy development: One of the more definitive analyses of the effects of continuous grazing on grass canopy structure and the interrelationships between LAI, shoot production, dry matter losses to respiration and death, and animal intake has been done by Parsons and colleagues (Parsons et al. 1983a, b; Parsons 1988). Their experimental system consisted of perennial ryegrass (*Lolium perenne* L., cv. S23) grazed by sheep to maintain two distinctly different leaf area (lamina only) indices over a three-year period from 1977 through 1979. The heavily-grazed sward was maintained at an LAI near 1.0; the lightly-grazed sward maintained an LAI of about 3.0. Their results, summarized in Fig. 9, show the flow of dry matter or the partitioning of photosynthate among competing sinks. The more intense grazing pressure reduced both the rates of gross photosynthesis and shoot production, as might be expected. However, there was a pronounced difference in the proportion of dry matter partitioned to death and animal intake. Relative losses to death were reduced from 42 to 25% of gross photosynthesis by reducing the LAI from 3 to 1, while absolute losses were cut by well over half. Concurrently, relative intake doubled and absolute intake increased by 1.4 (i.e., only 13% of gross photosynthate was continually harvested under light grazing compared to 25% under heavy grazing). It was clear that a greater proportion of gross photosynthate was lost prior to being harvested [grazed] in the light- compared to the heavy-grazed sward, as both swards were close to an equilibrium, and this loss was not due to greater partitioning to respiration or root growth.

Because perennial ryegrass, like many other grasses, exhibits rapid turnover of leaves, leaf tissue that is not grazed will soon die (reviewed by Parsons et al. 1983b;

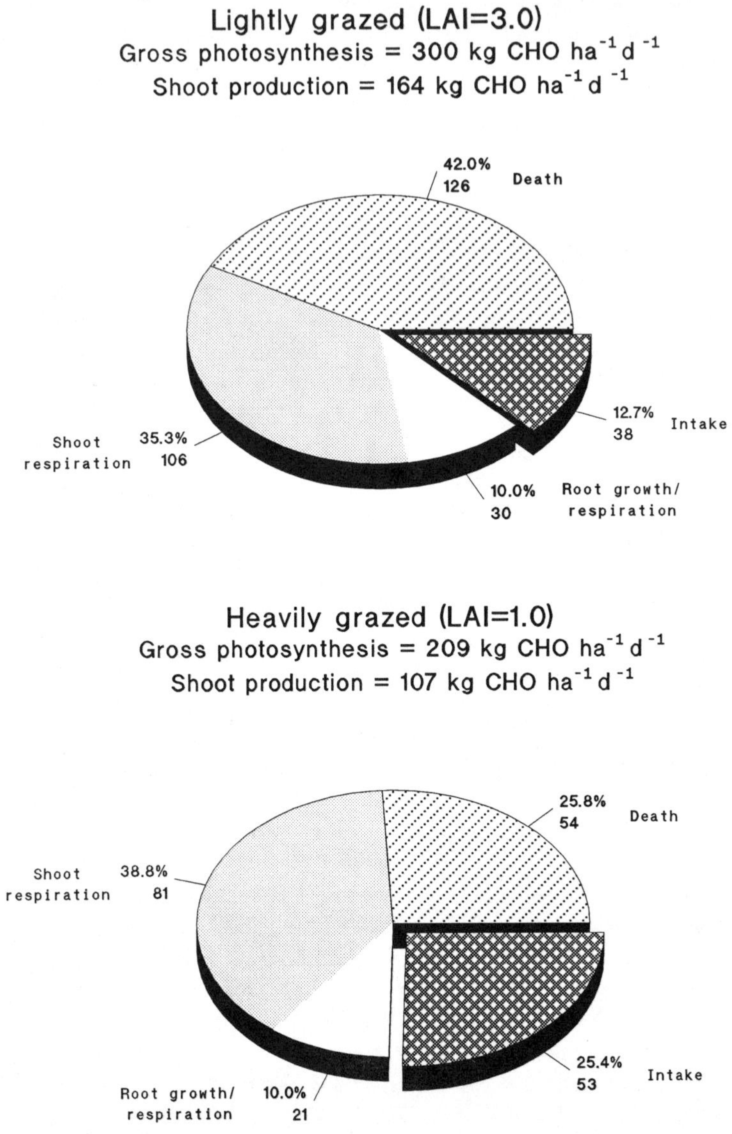

Figure 9. Partitioning of gross photosynthate in a perennial ryegrass (*Lolium perenne*) sward grazed continuously by sheep to maintain two contrasting leaf area indices (LAI). Shoot production = gross photosynthesis less shoot respiration and root growth and respiration. All units are kg CH_2O ha^{-1} d^{-1}. Data taken from Parsons et al. (1983b).

Parsons and Johnson 1986). A new leaf appears, on the average, every 11 days, but only three leaves are maintained per tiller. Average leaf longevity is just 33 days. Therefore, leaves which are not grazed are rapidly lost to death and a heavier grazing regime ensures that more of the shoot production is consumed before it has the opportunity to die. Variations in grazing pressure affect both the proportion of leaves harvested and, through leaf area reduction, photosynthesis, which affects the rate of gross tissue production.

Rates of photosynthesis and gross tissue production, according to Parsons and Johnson (1986), are close to maximum in swards maintained at a high LAI. In order to sustain this level of production, a large proportion of the leaves produced must remain in the sward (i.e., not be grazed) to contribute to photosynthesis. The inevitable result is a high rate of loss of dry matter to death, which translates into low utilization. Conversely, in swards maintained at a lower LAI, a greater proportion of leaf tissue is removed, which reduces photosynthesis and gross tissue production substantially. Nevertheless, the increase in utilization efficiency offsets the decrease in the amount of growth so that forage intake is actually increased. This clearly shows that maximum yield per unit area is not achieved in swards maintained at a high LAI for the purpose of maximizing light interception, but rather at a lower LAI that provides the best compromise among gross tissue production, forage intake, and tissue death. This concept is further illustrated in the following example.

A sward that is allowed to grow so that its canopy develops from low to high LAI (as in a hay meadow) has a different physiology of production compared to that of a continuously grazed sward (Parsons et al. 1983b). Management can influence the relative partitioning to tissue death relative to tissue production. Parsons (1980) (in Parsons et al. 1983b) used a simple crop model to analyze the optimum time to harvest swards growing from low to high LAI (Fig. 10). The concept of optimum versus critical LAI was introduced earlier. Clearly, this simulation adhered to the critical LAI hypothesis in that gross photosynthesis and gross tissue production leveled off and did not decline above the critical LAI. Prior to the time critical LAI was attained (95% interception of incident light), gross photosynthate was largely partitioned to the production and maintenance of new tissue.

Figure 10. Simulation of the relationships among rates of gross photosynthesis, loss of dry matter in respiration and death, and net accumulation of dry matter in a sward growing from low to high LAI based on a crop growth model by Parsons (1980). Adapted from Parsons et al. (1983b).

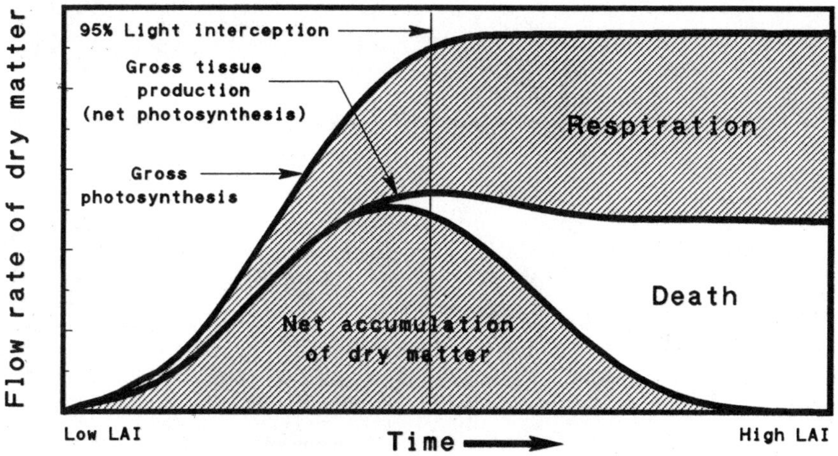

However, although changes in the rate of death lag changes in the rate of gross tissue production (Parsons and Johnson 1986), losses to death, at the expense of net accumulation of dry matter, rapidly became significant once LAI was allowed to increase beyond the critical value. Even though the rate of gross photosynthesis was maximized as LAI increased to high values, the rate at which tissue died approached the rate of tissue production and so the net accumulation of aboveground dry matter trended toward zero.

In their simulation and analysis of continuous and intermittent (rotational) grazing of grass, Parsons et al. (1988) concluded that maximum yield (forage harvested or consumed) under intermittent defoliation was achieved whenever fluctuations in LAI during regrowth and defoliation led to the same low average LAI that maximized yield under continuous grazing. This suggested that production under both continuous and intermittent grazing might best be characterized, and management rationalized, on the basis of the average sward state (average LAI) achieved.

The synthesis of Parsons et al. (1983b) provides a physiological basis for the paradox associated with the management of a continuously-grazed sward (Fig. 11).

Figure 11. Demonstration of the physiological limitation to production under continuous grazing. The top diagram shows the relationship between the uptake and loss of dry matter, and the animal intake that can be achieved in swards maintained over a wide range of LAI. Units for flow rate of dry matter are kg CH_2O ha^{-1} d^{-1}. The bottom diagram illustrates the relationship between shoot production, intake, and death and shows the proportion of shoot production that is harvested rather than lost to death. Adapted from Parsons et al. (1983b).

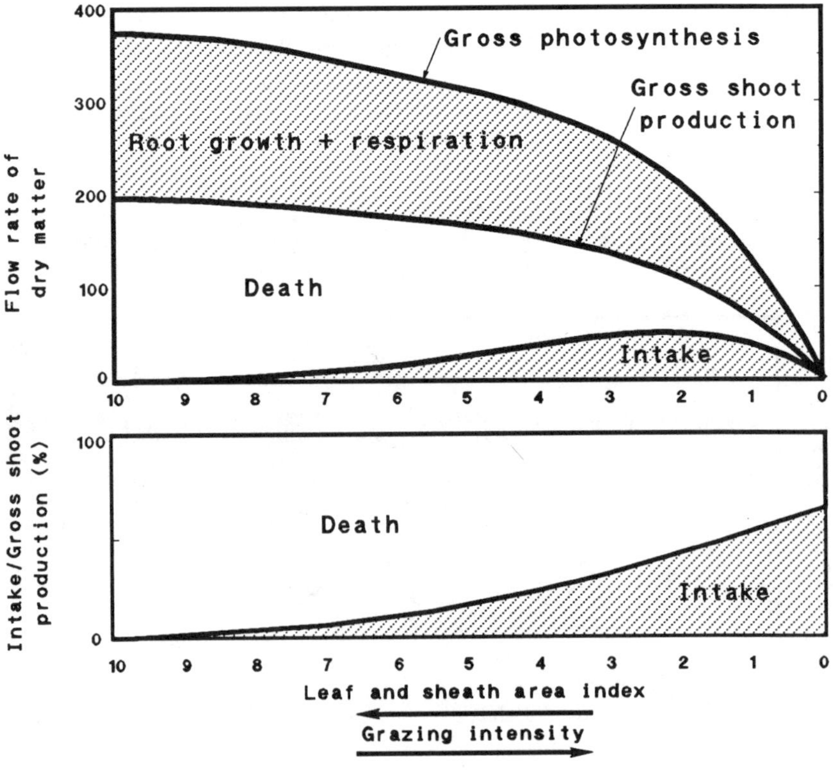

As grazing intensity increases causing a reduction in LAI, gross photosynthesis decreases more than gross shoot production because the tissue is retained at a more juvenile level which reduces tissue death rate. This allows the proportion of gross shoot production ingested to increase relative to the portion dying. Their analysis highlighted a major limitation to production under continuous grazing, namely, that high gross photosynthesis and high shoot production rates cannot be associated with a high harvest efficiency and a high harvested yield. Increases in harvest efficiency come at the expense of gross photosynthesis and shoot production. This vividly illustrates the very basic principal of optimization that range managers must use to achieve a profitable rate of forage intake while maintaining sufficient leaf area to sustain the productivity of the system.

The salable end product of forage-based, animal-production systems is animal products, not grass. The above discussion is not included to advocate overgrazing, because profitability depends on optimizing gain per unit area with gain per animal and maintaining the forage resource through time. However, it does serve, as the authors point out, to highlight the inevitable loss of production [and forage quality] that occurs through leaf death in swards maintained at even moderate LAI and the considerable increase in production that would be realized if this limitation could be overcome. The development of the common practice of intensive-early stocking (IES) of tallgrass prairie ranges in the Flint Hills of eastern Kansas (Smith and Owensby 1978) is an example of a successful attempt to deal with this issue. Under this system, stocker cattle are grazed for only the first half of the growing season (1 May–15 July), but at 2x [or higher] the normal season-long stocking (SLS) rate. This period coincides with the period of exponential growth of dominant forage species and maximum forage quality. As a result, gain per unit area may increase substantially. Gain per head during the first half of the season, when both IES and SLS treatments are stocked, is generally higher in the IES treatment, i.e., no reduction in animal performance results from doubling stocking rate during this period. Resting the range during the second half of the growing season assures that plant vigor is maintained. By capitalizing on the physiological behavior and requirements of both the plants and the animals, the forage harvested and converted to animal protein, as well as the profits made under this system, have increased significantly compared to traditional systems of management.

Management Implications of Differences in Carbon Acquisition

Differences among range plant species related to their specific carbon reduction pathway (C_3, C_4) have ecological implications that provide management opportunities as well as challenges. Some of these have already been alluded to such as WUE and seasonal complementarity of cool- and warm-season forage species. Others will be mentioned here.

Interspecific competition: Mixtures of species with differing CO_2-fixation pathways can use temperate climates more efficiently. Williams and Markley (1973) reported on the photosynthetic pathways of different species in the shortgrass area of Colorado. They hypothesized that because they found both C_3 and C_4 type plants in that grassland, there would be a minimum of interspecific competition among species which grew in different seasons. The C_3 plants grew in cool, moist periods, and the C_4 in warm, dry periods. In temperate climates, species diversity within a plant community, which included both C_3 and C_4 plants, would tend to increase primary productivity, provided other components were not limiting in the system.

Forage quality: Waller and Lewis (1979) concluded that, in general, C_4 plants have less nutritive value to grazing animals than C_3 plants at the same stage of maturity because of a lower digestibility. They cited work by Akin and Burdick (1977) which indicated that the walls of bundle sheath cells in C_4 plants were resistant to rumen microbial attack. Brazel et al. (1979) indicated that epidermal cuticle reduces digestibility. Because C_4 plants are more efficient in water use and prosper in warmer climates, cuticle development and thickness would likely be higher for C_4 than C_3 plants. That would also reduce digestibility of C_4 plants in relation to C_3.

Burning: Fire is a major ecological factor and periodic burning is essential to the existence of many vegetation types. Prescribed burning can be an effective management tool to control undesirable weed and brush species, promote earlier growth in spring, promote uniform utilization, and enhance the performance of livestock. In mixed C_3-C_4 communities at temperate latitudes, it may be possible to use fire to the differential advantage of either the C_3 or the C_4 component by burning when the component to be favored is dormant and the less desired component is actively growing. Differences in the phenology, carbohydrate reserve cycles, and meristem activity most often associated with cool- versus warm-season species, allow the range manager to control species composition of range ecosystems. Anderson et al. (1970) reported that C_3 grasses were essentially eliminated from Kansas tallgrass prairie when burning occurred during the spring (whether early or late), while early spring burning increased perennial C_3 forbs, and late spring burning reduced their populations. Wright and Bailey (1982) reported similar responses to fire in other grasslands.

Control of undesirable species: It has long been known that the success of various plant control techniques (mechanical removal, herbicides, fire) are correlated with the rate of plant growth and the level of food reserves. The phasing of growth and reserve depletion-repletion cycles are generally different for C_3 and C_4 plants and managers may capitalize on this to control C_3 brush species in ranges dominated by C_4 grasses. The topic of carbohydrate and nitrogen reserves will be considered in greater depth later in this chapter. As was the case for fire, contact and translocatable herbicides that depend upon active metabolism of the plant in order to be effective can be used to selectively remove C_3 or C_4 species if one group is growing during a portion of the dormant season of the other. Differences in the period when plants translocate soluble carbohydrates to storage organs and meristematic regions is extremely important when using foliar-applied, translocated, selective herbicides. Because the herbicide is passively swept along with the flow of food reserves to storage organs, application must occur coincident with initial active translocation of those reserves. That principle is illustrated by the results of Hyder and Sneva (1962) who found that 2,4-D effectively controlled big sagebrush (*Artemesia tridentata* Nutt.), but did not injure rubber rabbitbrush (*Chrysothamnus nauseosus* (Pall.) Britt. ssp. *nauseosus*) provided application time came before mid May.

Fertilization: Waller and Lewis (1979) implied that C_4 plants had an adaptive advantage over C_3 plants in areas with low soil fertility, because nitrogen-use efficiency was greater for C_4 than C_3 plants. Conversely, C_3 species may attain a competitive advantage over C_4 species in range ecosystems when fertilized with nitrogen because nitrogen in mesic range ecosystems often limits C_4 productivity. Numerous studies have shown dramatic shifts in species composition towards C_3 species in typically C_4-dominated plant communities when nitrogen fertilizer was applied (Launchbaugh and Owensby 1978; Mader 1956; Owensby and Smith 1979).

CO_2 enrichment: Global carbon dioxide concentration is increasing at an unprecedented rate (Keeling et al. 1982) and the consequences of that increase are the subject of much speculation (Dahlman 1984; Bazzaz 1990). Because rangelands occupy 47% of the earth's surface (Williams et al. 1968) and 54% of the contiguous United States (USDA 1972), changes in rangelands due to environmental change could be substantial. Even though organism-level responses to CO_2 enrichment have been documented, range ecosystem level responses remain unresolved (Strain and Cure 1985).

Response to CO_2 enrichment at the single plant level with respect to productivity indicates differential responses depending on photosynthetic pathway. Plants with the C_3 pathway generally show large increases in CO_2 fixation rates when CO_2 levels are increased, while C_4 plants do not (Kimball 1983; Wray and Strain 1986; Nijs et al. 1988; Reichers and Strain 1988). Those variations among plants with different carbon reduction pathways in response to enriching the atmospheric CO_2 might be expected because CO_2 and O_2 compete for the active site on RuBisCO as indicated earlier. Therefore, doubling the ambient CO_2 will effectively halve the rate of photorespiration (Long and Hutchin 1991) and thereby allow for increases in net photosynthesis.

Enhanced WUE, caused by partial stomatal closure in CO_2-enriched environments, results in enhanced productivity as well (Tolley and Strain 1984). Among the responses to CO_2 enrichment, decreased photorespiration may be responsible for a portion of the increased productivity in C_3 plants (Gifford et al. 1985). Regardless of the responses noted in controlled-environment experiments and with single plant species, the response of entire natural rangeland systems to CO_2 enrichment has yet to be documented in detail.

Response of different crop, tree, and weed species to elevated CO_2 has been documented in growth chamber and field studies (Rogers et al. 1983; Jones et al. 1985; Sionit et al. 1987; Fetcher et al. 1988). Natural ecosystem response to CO_2 enrichment has been less studied. Naturally occurring ecosystems differ from the managed cropland systems primarily in resource supplies. Essentially all nutrient resources in natural systems are supplied through nutrient cycling. In rangelands, the primary and secondary productivities of organisms in the ecosystem depend on the processes which determine the rates at which organic matter is decomposed, mineralized, and recycled through the system (Watt 1968). Most nutrients in mature, natural ecosystems are contained in organic forms. Potential response to CO_2 enrichment in naturally-occurring ecosystems most assuredly differs from the responses noted in managed cropland systems and controlled-environment studies with native plants. Limits to ecosystem productivity in natural systems often are from scarce nutrients, particularly nitrogen. Owensby et al. (1970) reported that, in the short term, nitrogen limited productivity in tallgrass prairie more than did water. Numerous authors have hypothesized that because of the differential response of C_3 and C_4 plants to CO_2 enrichment, competitive relationships may be altered. Wray and Strain (1986), working with the C_3 plant, aster (*Aster pilosus* Willd.) and the C_4 plant, broomsedge (*Andropogon virgninicus* L.), concluded that aster may out compete broomsedge in natural systems. Competition may be mitigated by differential rooting patterns. Weaver and Albertson (1943) revealed that in natural prairies, root concentrations of forbs and grasses seldom occupied the same soil layer. Also, canopy layering in prairies with an overstory of forbs provides for improved growing conditions for grasses (Launchbaugh and Owensby 1978). Therefore, shifts in botanical composition of prairies will likely occur among grass species of differing photosynthetic pathways and not between the forb and grass components. Another difficulty in predicting prairie ecosystem response to CO_2 enrichment is associated with reproductive strategies. Typically, prairies are asexually propagated (Hyder et al. 1975), which will likely lead to an increase in the sexually-reproduced, ephemeral species.

Because nutrient resources in natural systems are fixed to a great degree, the impact of increased carbon fixation in range ecosystems, resulting from higher CO_2 concentrations, may lead to dietary deficiencies of essential nutrients for ruminant herbivores. Sionit (pers. comm.) found that C:N ratios of wheat were increased with CO_2 enrichment. This trend was also suggested by the modeling exercise of Hunt et al. 1991. The C:N ratio increased in response to doubling CO_2 in the modeled shoots of crested wheatgrass [*Agropyron desertorum* (Fischer ex Link) Schultes] and blue grama [*Bouteloua gracilis* (H.B.K.) Benth. ex S. Wats]. Crested wheatgrass showed the greatest increase in C:N because net primary productivity of this C_3 plant was programmed to be more responsive to high CO_2 than the C_4 blue grama.

Reduced nitrogen concentration in forages results in reduced digestibility and efficiency of converting ingested forage to animal growth or reproductive functions. Evidence for reduced forage quality for ruminants can be extrapolated from work by Melillo et al. (1982). They reported reduced litter quality (increased lignin and fiber, decreased nitrogen) with CO_2 enrichment. Reduced litter quality slows decomposition. Because ruminant digestion is mediated by microorganisms, reduced forage quality would reduce digestibility. Lincoln et al. (1986) reported reduced forage quality for insects with CO_2 enrichment, and indicated that the soybean looper larvae (*Pseudoplusia includens* Walker) increased intake to compensate for the reduced forage quality. Ruminants cannot increase intake to compensate for a less digestible diet. The rate of passage of forage through the rumen is dependent on rate of digestibility. Before more forage may be ingested, the masticate in the rumen must be digested. As forage quality declines, intake declines as a result of reduced rate of passage of the material from the rumen.

Carbon Losses

The photosynthate pool of a plant, consisting of carbohydrates, reducing potential, and high energy phosphorus compounds, can be thought of as the biochemical currency of the plant. As shown in Fig. 1, this currency can be spent or invested in the growth of new plant tissues or in maintenance of existing tissues, put into savings, i.e., transferred to long-term storage organs (roots, rhizomes, stem bases) for use during periods when the rate of photosynthesis is insufficient to meet demands of the plant (following dormancy, severe environmental stress or defoliation), or used to produce compounds that serve to protect the plant from insects, disease, predators or environmental stress. Eventually though, all these organic compounds are oxidized and the carbon returned to the atmosphere as CO_2.

Dark Respiration

Function of respiration: The process by which green plants utilize a portion of the carbohydrate pool to generate additional chemical energy and carbon skeletons to support anabolic processes (e.g., protein, lipid, and hormone synthesis) is dark respiration, hereafter referred to simply as respiration. Thus, the function of respiration is two-fold and the overall chemical reaction is essentially the reverse of photosynthesis in that carbohydrate is oxidized to produce CO_2 and H_2O. Net photosynthesis is commonly defined as the difference between gross photosynthetic CO_2 uptake and the CO_2 produced by respiration. Respiration involves both biochemical conversions and electron transport. Central to respiration is a cyclical set of reactions known collectively as the tricarboxylic acid cycle (TCA), citric acid cycle, or the Kreb's cycle. Electrons extracted from TCA cycle intermediates are delivered to oxygen via electron transport, and in the process ATP and reducing potential are produced and oxygen is consumed. Details of the biochemistry and energetics of respiration are beyond the scope of this presentation and may be reviewed in standard biochemistry and plant physiology texts (Devlin and Witham 1983; Salisbury and Ross 1985) and in a recent synthesis by Amthor (1989).

Location of respiration: Unlike photorespiration, which takes place predominantly in the peroxisomes, the TCA cycle and electron transport chain of respiration are localized in the cell mitochondria. While photosynthesis is limited to chlorophyllous cells, respiration occurs in all living cells. Photorespiration is stimulated by light. Respiration can proceed in the dark as well as the light, but is not stimulated by light. The extent to which respiration proceeds in the light has been the subject of much debate and experimentation. However, current evidence suggests respiration does continue in the light. Gerbaud and Andre (1980) used oxygen isotopes to demonstrate mitochondrial respiration occurred in the light. In their summary of literature pertaining to this subject, Hay and Walker (1989) stated that respiration is probably not required for production of high energy compounds in the light because these can be produced in sufficient quantity by the light reactions of photosynthesis. However, respiration is required as a source of carbon skeletons for synthetic reactions. From this, they inferred that respiration rates are likely to be lower in the light and that the difference between rates in the light and dark will depend on the age of the tissue. Younger growing tissue would have higher rates of respiration in the light to support greater synthetic activities. Thus, a need for respiration to continue in the light can be identified, but considerable variation in the rates can be expected.

Partitioning Respiration into Growth and Maintenance Components

McCree (1970) searched for why the rate of plant respiration was not proportional to its size. He published evidence for white clover (*Trifolium repens* L.) that showed respiration was conceptually composed of two components. Subsequently, this concept has been confirmed in additional species (Hay and Walker 1989). According to McCree, one component was proportional to the rate of gross photosynthesis (termed *growth respiration*); the other was proportional to the total living dry weight of the plant (termed *maintenance respiration*). Hay and Walker (1989) stated that growth respiration supports the synthesis of new biomass and represents the metabolic cost of converting photosynthate into the various compounds required for growth. Maintenance respiration supports the turnover and replacement of existing cell components. With this insight and returning to the debate on whether respiration continues in the light introduced above, McCree (1986) stated that studies which showed inhibition in light were made on mature leaves and the results therefore apply only to the maintenance component of respiration, which is generally a minor component of whole plant respiration. Growth respiration, the more significant component in young growing tissue, would occur throughout a 24-h day.

McCree (1970) reduced his results to a 3-variable equation: $R = k_1P + cW$, where R, P, and W are the dark evolution rate of CO_2, the gross photosynthetic uptake of CO_2, and the total dry weight of living plant material (CO_2 equivalent basis), respectively, with units of g CO_2 m^{-2} d^{-1}; k_1 is a dimensionless constant; and c is the maintenance coefficient with units d^{-1}. Best fits of the data for white clover were obtained with $k_1 = 0.25$ and $c = 0.015$ d^{-1}. Thus, for a given day, these plants lost 25% of their gross photosynthate and an additional 1.5% of their current dry weight through respiration.

McCree's equation suggests maintenance respiration is a simple function of plant dry weight. Yet it may be better related to protein content (Hay and Walker 1989). This follows because the protein or enzyme turnover, repair and replacement of unstable cell components, and the maintenance of gradients of ions and metabolites are the main sinks for high energy compounds and carbon skeletons generated by maintenance respiration. Thus, maintenance respiration does involve some synthetic aspects, but these might be viewed as a resynthesis (e.g., protein turnover) rather than a net synthesis.

Controlled environment studies suggest young plants use 25–35% of their daily gross photosynthate to support growth and 1.5–3.0% of their dry weight (CO_2 equivalent) for maintenance processes (Hay and Walker 1989). The relative magnitude of growth and maintenance respiration in field-grown plants can be expected to vary with the species, stage of development, and environmental stress (including defoliation by grazing). However, values developed in controlled environments are often similar to those determined in the field and have potential for field application (reviewed by Amthor 1989; Hay and Walker 1989).

Factors Affecting Respiration

Two primary external or environmental factors which can significantly influence the rate of respiration are temperature and oxygen.

Temperature: The relationship between respiration and temperature is often considered to be fairly straight forward. In the temperature range of 5 to 25 C, Salisbury and Ross (1985) stated that most plants have a Q_{10} in the range of 2.0 to 2.5.

At higher temperatures (30-35 C), respiration rate increases further, but not as rapidly so that Q_{10} begins to decrease. One explanation for this decline is that oxygen penetration into the cells through the cuticle or periderm may limit respiration at these temperatures which otherwise would be expected to promote biochemical reactions. Recall that the Q_{10} for physical processes, such as the diffusion of O_2 and CO_2, is only 1.1. At still higher temperatures (40°C and higher), respiration decreases because of enzyme (protein) denaturation. As will be shown below, this view of temperature effects on respiration is perhaps an over simplification.

The mathematical analysis of growth and maintenance respiration, as introduced by McCree (1970), was further developed by Thornley (1970, 1976) and these references should be consulted for more in-depth treatment. However, comments based on Thornley's work from the synthesis of Hay and Walker (1989) are included here to help differentiate between the temperature dependency of these two respiration components. They stated that growth and maintenance respiration differed in two important aspects: the former appears independent of temperature, but is affected by limitations in [current] substrate supply; the latter is temperature dependent and not particularly sensitive to [current] substrate supply. The lack of dependence of maintenance respiration on current substrate supply is related to its capability to utilize reserve carbohydrates when the supply of recent assimilate is exhausted, and beyond that to degrade proteins [and possibly hemicellulose] for a source of substrate.

Hay and Walker (1989) point out that the maintenance coefficient, c, in McCree's (1970) equation, is a true rate constant ($g\ g^{-1}\ d^{-1}$) and is therefore dependent on temperature and its Q_{10} ranges between 1.8–2.0. The dimensionless k_1, however, is the ratio of two rate constants, each of which is affected more or less equally by temperature, thus canceling the effect of temperature on k_1 itself. The overall effect of temperature on plant respiration is, therefore, a function of the relative contributions of the growth and maintenance components. The relative mix of these will vary throughout the life-history of the plant so the Q_{10} will also vary. Maintenance of existing biomass takes precedence over growth of new tissue. Growth respiration is affected not only by substrate supply, but also by the composition of new biomass being synthesized. For example, a higher investment of substrate is required to synthesize lipid than to synthesize carbohydrate.

Oxygen: Oxygen is a direct reactant in the electron transport process. The atmosphere is about 21% oxygen and the aerial portions of plants would not be expected to be oxygen limited unless the tissue is fairly thick or dense, or temperatures are relatively high (see below). However, the rooting environment of plants may be subject to low availability of oxygen as a result of soil compaction or waterlogged conditions. Oxygen deficiency can inhibit root respiration and, therefore, interfere with the active uptake of mineral nutrients (Devlin and Witham 1983).

Respiration and Canopy Structure

The documented dependency of growth respiration on the rate of gross photosynthesis and maintenance respiration on plant dry weight or protein content is consistent with the critical LAI hypothesis presented earlier. Older leaves that grow in low light at the bottom of a canopy would have relatively low rates of growth respiration and protein concentration. As a result, their maintenance requirement would be reduced and these leaves would not be large sinks for current photosynthate. Thus, gross canopy photosynthesis would level off, as we have seen, rather than decline at high values of LAI even though net accumulation of dry matter might be strongly reduced.

Exudation and Sloughing

Root exudates represent losses of carbon and nitrogen from the plant that might otherwise be used to support production of forage. The loss of photosynthate by root exudation is ubiquitous among plants and estimates of the magnitude of this loss range from about negligible in sterile media to over 40% of net photosynthate in non-sterile soil. As potentially important as this process is to the carbon balance of a plant, remarkably little is known regarding its function. A mechanistic understanding of the exudation process under natural conditions, the degree to which it is essential to plant-soil relationships, and development of techniques to control or manipulate this potentially significant loss of carbon and nitrogen has obvious application for optimizing energy and capital inputs in forage-based, animal-production enterprises.

Exudates defined: Exudates are compounds of low molecular weight which leak from all cells into either the intercellular spaces and then to the soil via cell junctions or directly through epidermal cell walls (Rovira et al. 1979). The process is not metabolically mediated. Technically, not all materials released from actively growing roots are classified as exudates (Martin 1977). Other classes of compounds that are not considered as exudates include mucigel, which is actively secreted, and root lysates, which are released by autolysis through degeneration of epidermal cells, root hairs, and cortical cells.

For the purpose of this discussion, we will subscribe to a broader connotation for root exudates (Rovira 1969) and include in this category all those substances which are released into the surrounding medium by healthy, intact plant roots. These include sugars, amino acids, organic acids, peptides, enzymes, vitamins, nucleotides, fungal stimulators, inhibitors, and attractants of which the first three are the main energy sources released (Bowen and Rovira 1973).

Gopal (1972) stated in his review that release of the metabolites by roots is a part of their normal metabolism and is intimately associated with root physiology. He concluded that the significance of these material losses by roots had not been thoroughly investigated. Hale et al. (1971) reviewed 97 published articles and similarly concluded that the role played by root exudation in the physiology and life cycle of the plant itself and the mechanisms of exudation had been neglected. The knowledge base has not increased appreciably since these reviews were published.

Magnitude of exudation: The quality and quantity of organic substances lost by roots into the growth medium is highly variable and dependent upon numerous factors including plant species, plant age, temperature, light, plant nutrition, microorganisms, root support medium, moisture stress, anaerobiosis, and foliar sprays (pesticides, nutrients). Published studies dealing with these interactions have been reviewed by numerous authors including Rovira (1969), Hale et al. (1971), Barber and Gunn (1974), Rovira and Davey (1974), Barber and Martin (1976), Russell

(1977), Whipps (1984), and Biondini et al. (1988). One can conclude from these studies that there is potential for manipulating quality and quantity of the organic material losses from roots by either management or selection.

Russell (1977) stated that more is known of the quality of root exudates than of the quantity produced. In spite of this, it is well established that a significant, though variable, portion of the products of photosynthesis is ultimately released by the roots. The fate of root exudates is primarily to support rhizosphere microorganisms (Warembourg and Billes 1979) which, in turn stimulate further root exudation (Barber and Martin 1976). Quantity of exudates has been determined almost exclusively in artificial media by growing plants under sterile and non-sterile conditions to prevent or determine microbial alteration of the exudates. Normal soil microbial populations usually have not been included in these experiments. One of the difficulties associated with characterizing quantity of exudates is the necessity to make chemical measurements under sterile conditions to prevent microbial alteration of the exudates.

Alternatively, experiments can be conducted in normal soil by using $^{14}CO_2$ to label the photosynthate followed by monitoring of the quantity of ^{14}C recovered from the soil as CO_2 and in organic compounds. Partitioning of respiratory losses between roots and microorganisms depends on the assumption that root respiration is similar with and without the presence of microorganisms in the growth media. In spite of this arbitrary assumption, Russell (1977) stated that this is probably the most direct method of estimating the portion of photosynthate which is released by roots. This technique has been used by Barber and Martin (1976) for barley (*Hordeum vulgare* L.) and wheat (*Triticum aestivum* L.).

The use of ^{14}C labeled photosynthate to study root exudates offers several advantages (Rovira and McDougall 1967): It ensures all major components in the exudates are detected, and it is sufficiently sensitive for the study of exudate from individual plants collected over short periods of time. If labeling periods are sufficiently long to allow components of the least soluble portion to be of equal specific activity, data on the comparative composition of exudate can be obtained.

Isotope tracers also enable estimation of the time required for carbohydrate to be transported from the site of photosynthesis to the rhizosphere. Minchin and McNaughton (1984) used wheat seedlings grown hydroponically in non-sterile media and $^{11}CO_2$ to measure the rate at which current photosynthate appeared in root exudate. Thirty percent of the tracer transported to the root appeared in the media within 200 min after the leaves were labeled. Using microbial inhibitors, they estimated that 30% of the total tracer released to the media was likely root exudate. When not inhibited, the microbial population utilized all the root exudate and none accumulated in the media. Lee and Gaskins (1982) pulse-labeled sorghum (*Sorghum bicolor* L.) seedlings for 2 h with $^{14}CO_2$ and found 3.5% of the total ^{14}C recovered from shoots, roots, and the growth media 24 h following exposure, was root exudate.

One of the problems in reviewing available information on the magnitude of root exudation is a lack of uniformity in methods of recording and reporting results. Nevertheless, it is constructive to list some reported values to place the proposed research in perspective.

At the low end of the scale, two Russian studies reviewed by Samtsevich (1965) showed that roots of plants grown in sterile solution culture exuded only 0.25% and 0.7%, respectively, of total plant dry weight. Release of labeled compounds from the roots of white pine (*Pinus strobus* L.) amounted to less than 1% of the total radioactivity supplied to the shoot (Slankis et al. 1964). Ponderosa pine (*Pinus ponderosa* Laws.) roots released up to 5% of their total activity on a dry weight basis (Reid 1974), while lodgepole pine (*Pinus contorta* Doug.) (Ahrens, cited by Reid and Mexal 1977) released slightly more (up to 7%). In a later study, Reid and Mexal (1977) found that some lodgepole pine roots exuded over 80% of the total ^{14}C activity translocated to them over a period of 10 days. These results however, were based on "active" root biomass rather than total root biomass. Mycorrhizal infection did not appear to affect exudation although increasing water stress did increase exudation. That latter effect may have been in response to more labeled material being translocated to the roots in an attempt by the plants to alleviate water stress. Conversely, Reid (1974) found that water stress in solution cultures of young ponderosa pine seedlings inhibited both ^{14}C assimilation and its subsequent translocation to the roots. Root exudation increased from 0 to –0.19 MPa of substrate water potential, then decreased between –0.19 and –0.55 MPa, and finally increased again from –0.55 to –1.19 MPa. In addition, the quality of exudates was affected by decreasing soil water potential. Sugars increased, organic acids decreased, and amino acids showed only a slight decrease, with increasing water stress.

Bokhari and Singh (1974) grew western wheatgrass in hydroponic culture for an 80-day period. They found that 1 g dry weight of roots may exude from 4.5 to 6.5 mg of reducing sugars. As the plant matures, exudation per plant could be expected to increase in proportion to increase in root biomass. They compared their results to those of Rovira (1969) who measured reducing sugar losses per plant in maize of about 8.3 mg per 20-day period.

At the upper end of the scale, several authors have reported that carbon released from soil-grown plants was approximately 20% of the total dry matter (Samtsevich 1965; Barber and Martin 1976; Martin 1977; Sauerbeck and Johnen 1977). Haller and Stolp (1985) reported that 25% of the organic matter flowing to maize roots was excreted into the rhizosphere. This was equivalent to 7% of net photosynthesis. These values reflect the effects of a solid growth medium and the influence of microorganisms.

The effect of solid media on the quantity of exudates is clearly significant leading Bowen and Rovira (1976) to conclude that many exudate studies with solution-grown plants have little relevance to plants growing in the soil. In addition to the abrasive action of soil, the microbial component is also significant and decreases the relevance of studies which were restricted to sterile media. Barber and Martin (1976) found that wheat and barley roots released 5 to 10% of the photosynthetically fixed carbon when grown in sterile soil and 12 to 18% in non-sterile soil. The latter figure represented 18 to 25% of the dry matter increment of the plants. Subsequently, Martin (1977) conducted a similar study using wheat and found that an average of 39% (range 29.2 to 44.4%) of the carbon translocated to the roots was released into the soil. Martin attributed the major portion of this loss of root carbon to autolysis of the root cortex (conclusion supported by Johnen

1974). Root lysis was increased by soil microorganisms apparently without penetration of the plant cell walls.

Function of exudation: Hale et al. (1971) stated that exudates may serve as a microbial substrate or they may play a direct role in nutrient release from soil. The availability of mineral nutrients has been shown to affect root exudation both quantitatively and qualitatively. Nutrient supply in the soil and the ability of plants to utilize them can be profoundly influenced by microorganisms (Russell 1977). In addition, microbes may affect root exudation (Rovira and Davey 1974) by affecting the permeability of root cells, by affecting the metabolism of roots, and by absorbing certain compounds in root exudates and excreting several others. Kipe-Nolt et al. (1985) were not able to demonstrate a clear dependence of bacterial growth in sorghum (*Sorghum bicolor* L.) on quantity or composition of root exudates.

The predominant influence of microorganisms on the exudation process may simply be one of gradient mediation. Samtsevich (1965) (citing the work of Tueva 1926, Kostychev 1933, Geller 1954) stated that the main stimulus of exosmosis of root excretions is the difference in the concentration of dissolved substances in the root tissues and in the medium surrounding them. The leading factor in the creation of a concentration gradient under natural conditions (citing Kostychev 1933, Chesnokov and Bazyriva 1934) is the continuous consumption of organic substance released by the roots by the microorganisms which may actively stimulate the production of various organic substances from the roots. Lee and Gaskins (1982) studied root exudation of sorghum seedlings in the presence of N_2-fixing bacteria. Bacterial inoculation increased exudation. Adding an external carbon source (sucrose) to the growing media inoculated with bacteria as a means of carbon gradient manipulation caused a reduction in exudation.

Rovira and Davey (1974) concluded that partial or complete sterilization of the soil increases plant growth and that this is likely the result of modification of soil microflora. Work by Launchbaugh (1974, unpublished) has shown large increases in growth of new seedlings of range grasses when methyl bromide, a soil sterilant, was applied to the soil prior to seeding. He compared that with hand weeding and no weed control. Growth was much greater on the methyl bromide treated areas than on those having hand weeding or no weed control. Similarly, Thien et al. (1979, unpublished) added an external carbon source (sucrose) to field-grown corn and obtained a significant 23 bushel per acre increase over untreated plots. These results support the hypothesis that release of carbon from roots is a gradient mediated function.

Of particular relevance to range species is the effect of defoliation on the exudation process. Alfalfa (*Medicago sativa* L.) plants allowed to flower in sterile sand culture had significantly more carbohydrates in root exudates than did plants of the same age that were clipped to prevent flowering (Hamlen et al. 1972). Conversely, heavy clipping of western wheatgrass grown in sterile hydroponic culture stimulated root exudation (Bokhari and Singh 1974). Greater carbon loss by clipped plants was possibly offset by increased photosynthesis of young tillers that were initiated in response to the clipping process. However, the authors stated that in a total system carbon-balance study, it is important to be aware that greater carbon loss may occur through root exudation in grazed swards.

Also of importance in managed range systems is the extent to which root exudation can be manipulated. Exudates and microorganisms interact in a way that each is influenced by the other. As already pointed out, microorganisms may affect quality and quantity of exudates by differential selectivity and metabolism of exudate compounds and gradient mediation. Similarly, Gopal (1972) stated that the release of metabolites by roots is an important factor for determining the qualitative and quantitative composition of the rhizosphere microflora.

Mitigation of exudation: The quality and quantity of exudates, and in turn the microflora, can be manipulated by various techniques. Use of radioactively-labeled materials applied to foliage have revealed that many compounds are absorbed, translocated to the roots, and exudated into the surrounding soil (Rovira and Davey 1974). The implications of that knowledge to plant-microbe and plant interactions may be profound. Streptomycin applied to the leaves of coleus (*Coleus blumei* L.) suppressed gram-negative bacteria for 8–12 days (Davey and Papavizas 1961). Foliar applications of 0.1% solutions of $NaNO_3$ increased exudation of amino acids and reduced total sugars in sorghum (*Sorghum vulgare* Pers.), sunnhemp (*Crotalaria juncea* Linn.), ragi or finger millet (*Eleusine coracana* Gaertn.), and tomato (*Lycopersicon esculentum* Mill.) while Na_2HPO_4 had the opposite effect (Balasubramanian and Rangaswami 1969). Application of chloramphenicol to wheat leaves greatly increased exudation of amino acids and sugars and resulted in fewer bacteria and more fungi in the rhizosphere (Vrany et al. 1962). Jalali (1976) compared the effect of several foliar sprays [urea $(NH_4)_2SO_4$, Na_2HPO_4, KCl, chloramphenicol, 2,4-D] on wheat root exudates. Chloramphenicol, KCl, and urea were most effective in controlling the release of ribose, maltose, and raffinose. KCl reduced ribose exudates by 30.6% compared to the control and completely eliminated maltose and raffinose from the exudates.

Modification of root exudates by foliar sprays may offer a practical method of modifying the rhizosphere to protect roots from pathogens and to improve plant nutrition, but more work is needed to develop this potential (Rovira and Davey 1974). Rovira (1969), and Jalali (1976) both concluded that preferential suppression of root pathogens by manipulating root exudates with foliar sprays was possible.

Much remains to be learned about the costs and benefits of root exudation in range systems. However, it is easy to make some approximate calculations that demonstrate a significant potential to increase carbon allocation to forage production by learning to manipulate this process. Temperate grasslands, for example, allocate from 50% (tallgrass prairie; Kucera et al. 1967) to 85% (shortgrass prairie; Sims and Singh 1971) of their annual net primary productivity (NPP) to belowground growth, storage, and exudation. Let us assume for this purpose that the aboveground:belowground partition of photosynthate or NPP is 50:50 as in the tallgrass prairie; that 25% of the aboveground production is consumed by the managed grazing animal; and that the forage to gain conversion efficiency is 10:1. The amount of animal biomass produced from 100 kg of NPP would be 1.25 kg. Let us further assume exudation is a liberal process so that no negative effects on soil-plant relationships would result from reallocation of a portion of the lost carbon. Diverting just 1% (51:49) or 5% (55:45) of the NPP from the root exudate component to aboveground forage production would translate into 2% and 10% increases, respectively, in the production of animal biomass. Corresponding results for the shortgrass prairie (7% and 33%, respectively) are even more dramatic. Clearly, a relatively small amount of carbon diverted from exudates to forage production has the potential to greatly amplify the effect on animal production because of the large belowground component of grass communities.

Sloughing: During root growth in length and diameter, some of the outer tissues are sloughed and decomposed either by autolysis or through the activities of microorganisms (Hale and Moore 1979). Replacing sloughed tissue by the plant obviously represents a carbon sink that may be significant relative to total carbon loss from roots. Although such factors as plant population density and watering frequency have been found to affect sloughing in maize (Clowes and Woolston 1978), there appears to be little on a practical scale that management can do to mitigate this loss in range ecosystems.

Death

Plant organs, such as leaves and roots, are limited in their growth and longevity and plants must continually replace parts lost through natural attrition or herbivory. Caldwell (1984) pointed out that long-term stability of rangeland communities is dependent upon reproductive capacity (sexual or asexual) and recruitment of new individuals into the community plus the longevity of desirable species. The pattern of death and replacement varies with growth form and habit (e.g., perennial versus annual, woody versus herbaceous stems, and evergreen versus deciduous leaves). Deciduous growth forms may initiate and shed their entire leaf canopy synchronously in a relatively short periods of time while evergreen forms are continually turning over a fraction of their total leaf area. The tops of herbaceous plants, like the perennial grasses and other graminoids, may totally die back in response to the onset of cold temperatures or a pronounced dry period, but also are continually losing and replacing leaves asynchronously during the growing season. The rate at which a plant replaces it leaves and roots is species and environment dependent, but organ turnover represents a significant loss of the carbon gained by the photosynthetic process.

As an organ senesces, plants will normally export labile nutrients, such as nitrogen and nonstructural carbohydrates, to the growing points or else to perennating organs for storage. The carbon remaining after death initially becomes part of the standing dead compartment (Fig. 1), and eventually is transferred to litter and ultimately, soil organic matter. Grazing may enhance the rates of these transfers.

One of the most exhaustive and extensive studies of carbon and energy flow among major compartments, above- and belowground primary production, and organ turnover rates in grassland ecosystems was done under the auspices of the U.S. International Biological Program's Grassland Biome. Results for ten North American grassland sites encompassing desert grasslands, mountain grasslands, and short-, mixed-, tall-, and bunchgrass types and ranging from dry to moist and cool-season dominated to warm-season dominated were summarized in a series of four synthesis papers (Sims et al. 1978; Sims and Singh 1978a, b, c). Some of these results, as summarized by Sims and Singh (1978c), will serve to illustrate the relative magnitude of carbon pool sizes, transfers, and the losses to death in grazed and ungrazed ecosystems.

The average total net primary productivity (TNP) rate for the ten North American grassland sites was 779 g biomass m^{-2} y^{-1} for ungrazed and 806 g m^{-2} y^{-1} for previously grazed sites (Sims and Singh 1978b). The partition of TNP [in g m^{-2} y^{-1}] into aboveground (ANP) and belowground (BNP=roots plus crowns) net primary productivity was 236 (ungrazed) and 212 (grazed) for ANP and 543 (ungrazed) and 594 (grazed) for BNP. The ratios of BNP/TNP were about 0.7 for both ungrazed and grazed grasslands, which emphasizes the importance of the belowground component in these ecosystems.

For the entire growing season, the amount of live material transferred to the standing dead compartment averaged 182 and 162 g m^{-2} for ungrazed and previously-grazed grassland, respectively (Sims and Singh 1978c). These results agree with those of Parsons et al. (1983b) presented earlier in that grazing reduced the amount of current production lost to death. In addition, a very substantial amount of carbon was transferred to standing dead regardless of whether these grasslands were grazed or not. The average turnover of root material ranged from 0.10 to 0.74 and was estimated to be 0.18, 0.30, and 0.49 for the mixed-, tall-, and shortgrass prairies, respectively (Sims and Singh 1978b). Grazing did not affect root turnover significantly. For comparison, Dahlman and Kucera (1965) and Kucera et al. (1967) estimated root biomass turnover was 0.25 in a Missouri tallgrass prairie site, which was similar to the 0.30 turnover rate calculated by Sims and Singh (1978b) for a tallgrass site in Oklahoma.

Death in woody plants is more difficult to determine as a fairly large portion of roots, stems, and trunks are non-living at any point in time. As woody plants continue to grow and age, more non-living tissue is added each year so that the proportion of live/dead tissue continues to decrease throughout much of the plant's life. This non-living tissue does, however, have a functional purpose as it gives the plant mechanical strength for support and adds protection for living cambial, conductive and storage tissue.

Carbon Utilization

A plant's productivity, survival, and success are a function of its patterns of carbon utilization, as well as its capacity for carbon assimilation (Oechel and Lawrence 1981). A small change in carbon allocation from leaves to stems or roots may have a dramatic impact on nutrient uptake, water relations, and total carbon assimilation (Monsi and Murata 1970). Differential utilization of carbon then directly affects competition, recovery from disturbances such as fire or heavy grazing, and resource-use efficiency. Growth, maintenance respiration, and carbohydrate storage are major carbon sinks in plants and must be described to understand the carbon budget of plants.

One approach to studying allocation and growth involves dividing carbon and energy requirements of plant tissue into the functional categories of growth (or construction) costs and maintenance costs. These terms were defined in an earlier section. However, a somewhat different formula than that presented earlier

(McCree 1970) has been derived for partitioning total plant respiration into components associated with growth and maintenance. The maintenance respiration rate is considered to be proportional to the dry weight of the plant or plant part as in the McCree equation, whereas growth respiration rate is proportional to the relative growth rate rather than the gross photosynthetic rate (Chiariello et al. 1989). Hesketh et al. (1971) defined total respiration (R) as:

$$R = R_g(dW/dt) + R_m W,$$

where Rg is the coefficient for growth respiration, W is the dry weight, and R_m is the coefficient for maintenance respiration. Rearrangement of this equation by dividing through by W yields:

$$R/W = R_g[(dW/dt)/W] + R_m,$$

where (dW/dt)/W is the relative growth rate of biomass. Plotting the specific respiration rate of a plant (or plant part) against its relative growth rate gives a straight line relationship with a slope of R_g and intercept of R_m (Fig. 12).

This approach has been used for both whole plants and plant parts in laboratory studies. However, in field studies it is much more difficult to measure or estimate respiration rates for intact plants. So, analysis is often limited to aboveground plant parts. Belowground microbial respiration represents a significant contribution of CO_2 that complicates determination of root respiration rates. Experimentally, simultaneous measurements of specific respiration rate and growth rate are

Figure 12. Determination of maintenance respiration rate and growth coefficient of respiration by plotting specific respiration rate against relative growth rate. The data are for Helianthus tuberosus from Kimura et al. (1978), as taken from Chiariello et al. (1989).

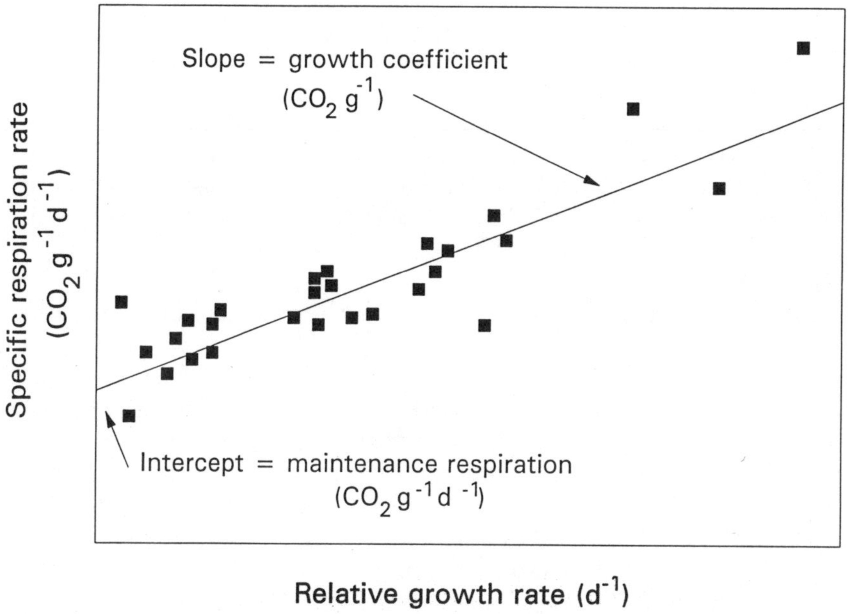

required. This requires a cohort of plant organs whose phenological stage, genetic background, and microenvironment are similar enough to insure a fairly uniform rate of growth. Respiration rates must be measured for these plant organs. Harvesting of subsamples of each cohort at time intervals of a few days can be used to determine growth rates although this is a labor-intensive process in natural populations of range plants.

Using this approach, maintenance carbon costs are simply the maintenance respiration rate as determined from the intercept of the graph of specific respiration rate versus relative growth rate (Fig. 12). Construction costs are the sum of growth respiration rate (determined from the slope of the line in Fig. 12) and the carbon included in the new biomass produced. The quantity of new carbon produced during a time period is calculated from the growth rate and a measurement of the carbon content of the biomass.

Another method for determining construction and maintenance costs from the biochemical composition of plant tissue was developed by Penning de Vries and his co-workers (Penning de Vries et al. 1974; Penning de Vries 1975). Glucose is used as the standard substrate for synthesis of plant tissues. The amount of glucose required for carbon skeletons, reducing equivalents, and ATP for production of biochemical compounds through biosynthesis is then estimated. Costs associated with transporting organic substrates across membranes and mineral accumulation are often included, as well. The mass of biochemical products produced by a unit mass of glucose is defined as the production value (PV) (Chiariello et al. 1989). Not only can the amount of glucose required be calculated, but the amount of oxygen, reductant, and ATP required can also be determined, as well as the CO_2 produced during construction of the new tissue.

This method requires an estimate of compounds present in the plant tissue and production values calculated for these compounds (Table 2). Proximate analysis may be used to determine plant constituents. Knowledge of the most probable biochemical pathway used to synthesize these compounds is also needed. In calculating biochemical pathway costs, 1 mol of glucose is assumed to yield 36 mol of ATP, and 12 mol of NADH. Any excess ATP, NADH, or NADPH produced in the process should be subtracted from the total amount of glucose required (Chiariello et al. 1989). As is evident in Table 2, the cost to produce lipids or lignin in plant tissues is much greater than the cost to produce carbohydrates.

Maintenance

The most important metabolic processes that contribute to maintenance respiration are protein turnover and active transport to maintain ion gradients (Penning de Vries 1975). Protein turnover costs about 25-55 mg of glucose per gram of protein per day (or about 7-13 mg glucose g^{-1} dw d^{-1}) in green leaves (dry weight basis), and maintenance of ion gradients may take 5-10 mg glucose $g^{-1\,dw}$ d^{-1}. Penning de Vries (1975) calculated total maintenance respiration values for a variety of tissues at 25°C of 8-60 mg glucose g^{-1} dw d^{-1}. Some actual measurements of maintenance respiration are 48 mg glucose g^{-1} dw d^{-1} for subterranean clover (McCree and Silsbury 1978) and 70 mg glucose g^{-1} dw d^{-1} for *Helianthus tuberosus* L. (Kimura et al. 1978). Maintenance respiration accounts for about 3% of a

Table 2. Glucose requirements (1/PV) for the synthesis of a few classes of organic compounds. Selected values are from (a) Penning de Vries et al. (1974); (b) Williams et al. (1987), values adjusted to include an estimated membrane transport cost of 1 mol of ATP per of mol glucose; (c) de Witt (1978); and (d) Chung and Barnes (1977) as taken from Chiariello et al. (1989).

Class of compound	Glucose requirement
	(g glucose g^{-1} compound)
Lipid	3.03 (a)
Nitrogenous compounds	
NO_3-N	2.48 (a)
NH_4-N	1.62 (a)
Carbohydrate	1.21 (a)
Sucrose	1.13 (b)
Cellulose or starch	1.21 (b)
Hemicellulose	1.24 (b)
Lignin	2.58 (b)
Organic acid	0.70 (c)
Phenolics	1.92 (d)

plant's carbon gain per day. About 30% of the carbon gain per day is utilized in construction respiration (Ryle et al. 1976; Mooney and Gulmon 1982).

In herbaceous plants, maintenance respiration has been estimated to be approximately 1% to 4% of the carbon present in the biomass (or dry weight) per day (McCree 1974). Oechel and Lawrence (1981) estimated that maintenance respiration was 47%, whereas growth respiration was 17% of the carbon allocated for growth and respiration in the California chaparral shrub *Arctostaphylos glauca* Lindl. Higher maintenance costs might be expected to support woody plants than herbaceous plants. Penning de Vries (1975) indicated that respiratory cost of maintenance of plant tissues was not as predictable as the cost of growth. Maintenance respiration varied with species, age, and carbohydrate level. Abiotic factors of temperature, irradiance, water stress, and nutrient levels also affected maintenance respiration. Larcher (1980) estimated that growth respiration was between 3 and 10 times that of maintenance respiration for actively growing plant parts.

Carbon loss from respiration may be 28% of net photosynthesis in maize and wheat (Penning de Vries 1975). Maintenance respiration decreases with water stress (Brown and Trlica 1977) and with increasing shade (Schulze 1970). It appears that differences in CO_2 release between various species and plant organs is greater than differences between plant lifeforms and environmental conditions (Schulze 1982). Plant stress will usually result in a diversion of energy from growth and production to maintenance; thus, maintenance respiration will increase (Brown and Trlica 1977; Odum 1985). However, little quantitative information exists to assess relationships between plant stress and losses of carbon through maintenance respiration. Too little is known about the dynamics of most physiological processes concerned to enable accurate estimation of maintenance respiration from basic data (Forward 1983; Salisbury and Ross 1985; Amthor 1989).

Osmoticum: Osmotic potential within the cell is maintained by concentrations of various soluble carbohydrates (sugars, organic acids, etc.) and salts (primarily Na^+, K^+, Ca^{++}, Cl^-. and $SO_4^=$). These compounds affect the ability of the cell to either take up or release water to adjacent cells or the environment. Plants can take up soil water only if the osmotic potential in root cells is lower (more negative) than that of the soil solution. Saline soils of steppes and deserts often contain high concentrations of sulfates and carbonates of Na, Mg, and Ca, which tend to make them more alkaline (Larcher 1980). These salts can affect range plants through osmotic retention of water and specific ionic effects on cell protoplasm.

Plants adapted to life in saline soils (halophytes) usually achieve this through accumulation of salt in the cell sap. This allows the plant to compensate for the low osmotic potentials characteristic of the saline soil. Monocots, and grasses in particular, usually contain less salt than dicots. Therefore, soluble carbohydrates may be used by these halophytic monocots to maintain very low osmotic potentials necessary to live in a saline soil (Larcher 1980).

Salt ions may be non-uniformly compartmented into a cell vacuole or the cytoplasm. Salt stress, to which the cytoplasmic enzyme systems are directly exposed, is less when salts are selectively stored in the vacuole (Larcher 1980). Therefore, most salts are stored in the vacuole, leaving the cytoplasm relatively low in salt. However, movement into and out of the vacuole is under control of the cell. Salt compartmentalization is brought about and maintained by ion pumps in boundary layers of the cytoplasm. The osmotic disequilibrium between cytoplasm and vacuole thus produced can be compensated for by accumulation of organic osmoticants in the cytoplasm. Energy is thus required in maintenance of the ion pumps and production of the organic compounds needed to maintain cell hydration and osmotic potential.

Starch and vacuolar sucrose are stored during the day in plant leaves. Starch is broken down at night and exported from the leaf to other sinks. Partitioning of carbon between starch and sugar is under enzymatic control, and their accumulation can feed back to cause a reduction in photosynthesis (Stitt and Quick 1989; Chapin et al. 1990). Plants export recent photosynthate before starch is mobilized. This suggests an opportunity cost for storage of starch (Chapin et al. 1990). However, this cost may not be great as increased sugar concentrations in the cytoplasm of leaves and can result in higher leaf respiration rates. Starch concentrations vary hourly and daily as fluctuations in carbon supply and sink demand affect starch levels.

Tissue repair or replacement: Plant tissue can be injured by a variety of agents such as disease, grazing, herbicides, or other environmental stresses. When tissue is injured, wound respiration begins. Respiration rate of this injured tissue may be three- to fivefold greater in the first 24 h after injury (Noggle and Fritz 1983). Thereafter, respiration rate usually declines. Increased respiration is necessary to provide the energy and carbon skeletons necessary for the repair of the tissue (callus formation) and construction of new cells. The respiration rate of plant tissue is often a sensitive indicator of health or vigor. For example, when fungal infection occurs, respiration rate of the tissue rises above normal, even before visible signs of disease are apparent (Noggle and Fritz 1983).

The physiology of the reactions to wounding and infections are complex and not well understood. There are extensive changes in protein metabolism, increases in mitochondrial respiration, increases in enzymes, and in ethylene production (Kramer and Kozlowski 1979). Evidently, substances such as ethylene or growth regulators move out of injured tissue and affect genes in adjacent cells that control production of specific enzymes. These "wound hormones" affect metabolism of adjacent tissue, resulting in compartmentalization of the wound and callus formation associated with healing. Healing involves closure of the wound, as well as walling off infected or invaded tissue associated with the wound (Kramer and Kozlowski 1979).

Healing of wounds in woody plants involves the sequential production of callus tissue and formation of new vascular cambium through conversion of callus cells to cambial cells. The amount and rate of callus formation depends on the size and location of the wound, species, and phenological stage. Rate of wound closure is positively correlated with the rate of cambial growth (Kramer and Kozlowski 1979).

Growth

Growth respiration must provide both carbon skeletons and energy for the synthesis of new plant materials. The approach taken by Penning de Vries and associates (Penning de Vries 1975; Penning de Vries et al. 1974, 1979) is of interest because it provides a theoretical basis for estimating the respiration associated with the conversion of carbohydrate substrate to structural matter, mainly proteins, lipids, cellulose, and lignin. If the composition of the tissue and the substrate are known, the amount of carbon transferred to structural compounds can be calculated, as well as the energy required for the processes involved. The consumption of O_2 and production of CO_2 in the required energy-releasing processes can then be determined. From this analysis, growth respiration can be predicted from the increase in dry matter produced.

Estimation of growth respiration by the above calculations are only approximate, as some assumptions regarding the reactions involved and some simplifications are unavoidable (Forward 1983; Amthor 1989). However, the approach does provide a rational basis for estimates of growth respiration derived from experimental data, and reasonable conformity with theoretical values have been obtained. This approach has, therefore, been widely used in simulation models of plant production.

Plant growth can be defined in several ways, including a change in size, mass, form, or number. We will restrict our discussion to growth being an increase of dry mass of a plant or plant organ through time. Growth involves various degrees of cell division and expansion. A variety of processes at different spatial, temporal and structural scales all may ultimately affect plant growth (Caldwell et al. 1981). For example, carbon allocation patterns ultimately affect shoot:root ratio, which may largely control the assimilatory capacity of the plant to take up CO_2. The importance of understanding whole plant growth has recently been emphasized (Evans 1972; Mooney 1972; Kramer and Kozlowski 1979; Schulze 1982; Tillman 1988; Pearcy et al. 1989). Mooney (1972) has outlined the distribution of carbon for use in construction of plants (Fig. 13).

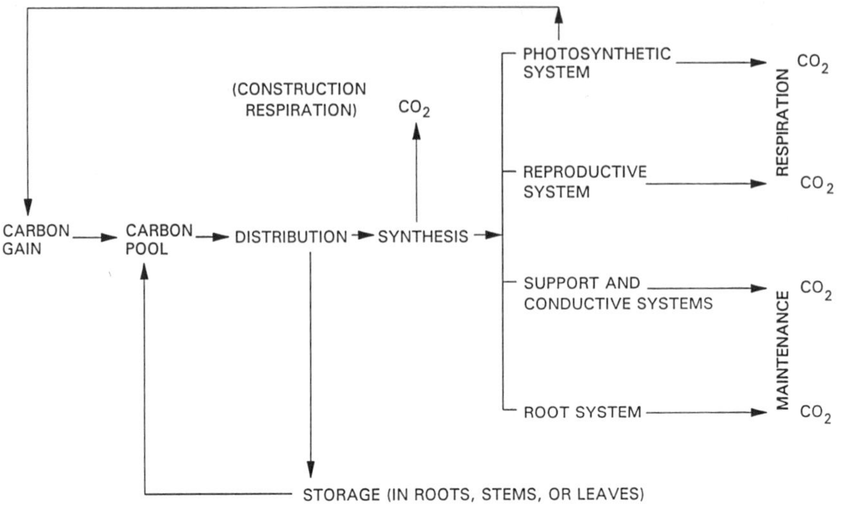

Figure 13. The use of carbon in construction of a plant. Adapted from Mooney (1972).

Natural selection probably favors high growth rate, i.e., near the maximum rate that can be maintained in a specific environment (Chapin et al. 1987). Plants that are commonly found in resource-rich environments tend to have high potential growth rates; whereas, plants from resource-poor environments (e.g., deserts, tundras, shaded understory, infertile soil) tend to have low potential growth rates (Grime 1977; Chapin 1980). Plants characteristic of resource-poor environments, when placed in resource-rich environments, still do not grow as rapidly as plants native to the resource-rich environment. The basis for this inherent limitation is not well understood, and will require studies of the control and cost of growth under varying conditions. The development of different sized hierarchies in populations has important evolutionary implications, but what controls these hierarchies is also poorly understood (Turner and Rabinowitz 1983).

Plants respond to resource limitation with growth and allocation responses that tend to increase the efficiency with which they use limiting resources (Chapin et al. 1987). Plants also appear to respond to a variety of different stresses in basically the same way. Chapin (1991) proposed that there is a basic physiological framework that regulates plant growth in response to environmental stress. This framework, although complex, involves changes in hormonal balance, water relations, carbon balance, and nutrient use. The general response of plants to stress is a reduction in the rates of growth and acquisition of all resources. Because growth rate determines plant biomass at any point in time, and many ecological phenomena are size dependent (e.g., competition and reproductive output), studies of plant growth can provide an environmental basis for plant demography (Chiariello et al. 1989).

Few studies have been conducted to simultaneously determine carbohydrate allocation and growth of different plant parts through time, and as influenced by stress. One such study was conducted by Harper and Ogden (1970) to determine allocation patterns in the annual weed, *Senecio vulgaris* L., as affected by water stress. They found that as water stress increased, greater allocation to biomass of roots occurred with less production of flowers and fruits (Fig. 14). Not only was the allocation pattern affected, but total dry matter production under the high stress treatment was only about 2% of that for well-watered plants.

Certainly rangelands provide a multitude of environmental stresses that plants must survive if they are to remain as a component in the vegetation. Field studies frequently indicate that two or more resources (e.g., water, temperature, and nitrogen) may simultaneously limit plant growth. These results suggest that either plants in these communities are simultaneously limited by several resources, or limitation switches among different resources in a short time span so that it is not possible to identify a single limiting factor over the time scale of the measurements. In either case, plants are capable of compensating for some resource imbalances, even in extreme environments (Chapin et al. 1987). Plants that are limited by multiple resources may respond synergistically to the addition of multiple resources (Lauenroth et al. 1978; Chapin and Shaver 1985).

One of the major determinants of total plant growth is the relative growth of structures that differ in function, such as photosynthetically active leaves versus stems or roots. Most CO_2 is fixed in leaves, but stems and reproductive organs of some plants may also be photosynthetically active. This fixed carbon is unique as it represents the major form in which energy is stored and transported within the plant and dominates the composition of all plant tissue. Carbon uptake, movement, storage, utilization, and loss from the plant can be traced utilizing a radioisotope of carbon, principally ^{14}C. Use of radioisotopes have allowed rates of movement to various plant organs, as well as strengths of various sinks in these organs, to be measured. Plants can modify their photosynthetic gain (growth) by several morphological and behavioral modifications in response to environmental cues. Both internal plant factors and environment influence the distribution of carbon to roots and shoots (Fig. 15).

We will now consider growth of individual plant organs and how this growth is influenced by environment. Because plant growth usually begins with the root, we will consider growth of roots and rhizomes first. These organs are usually entirely dependent on the shoot system to supply the carbon and energy necessary for growth.

Roots and rhizomes: Root growth is an essential process for water and nutrient uptake, especially at low soil water potentials and in the case of immobile soil nutrients such as phosphorus and some metals (Caldwell 1976, 1979). However, studies of root growth under natural conditions are extremely difficult. Most techniques for their study are either destructive, preventing study of the same root or root system through time, or introduce largely artificial conditions or constraints on roots (Bohm 1979; Richards 1984).

Two types of root systems exist; the taproot and fibrous root systems. Taproots, which develop from a primary root, are found in gymnosperms and dicotyledon plants. Taproots may have a number of branches and lateral roots. Another factor that contributes to overall growth and variation of the taproot system is the degree of secondary growth exhibited within various roots. Fibrous root systems are usually primarily adventitious roots that can truly be perennial, even though they seldom show

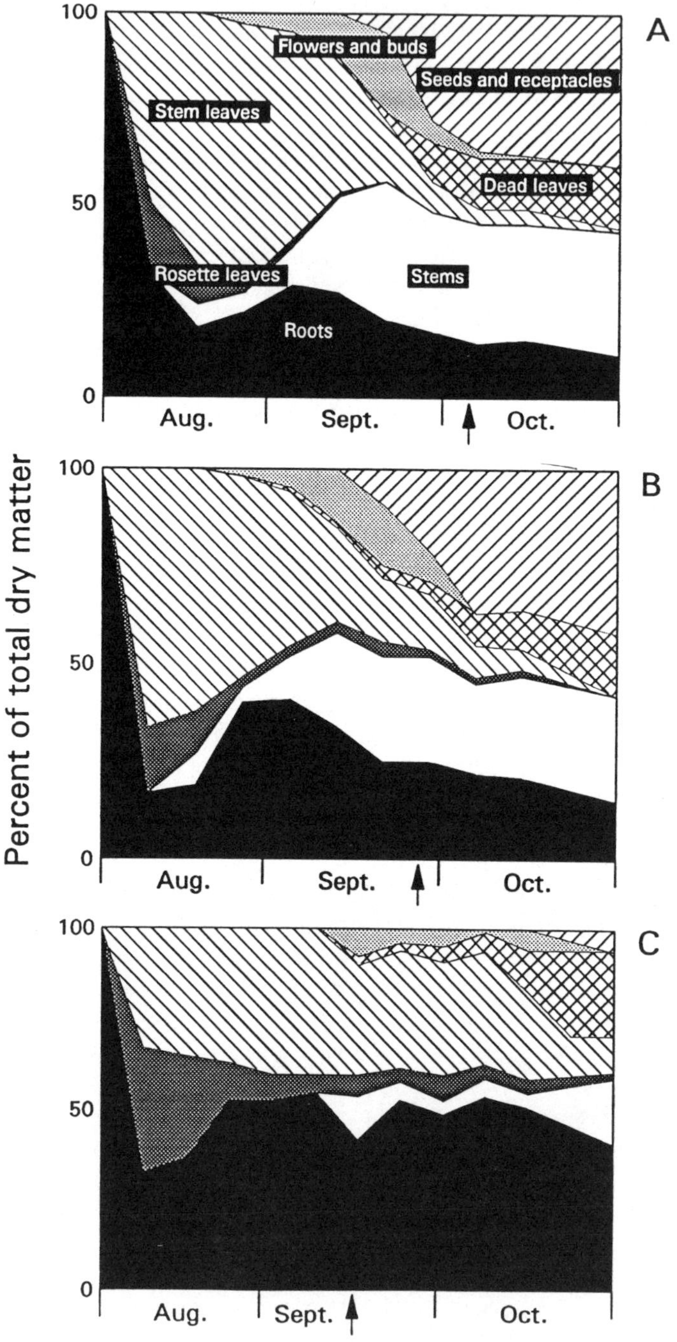

Figure 14. Allocation of total calories of biomass throughout the life cycle of *Senecio vulgaris*; A) low water stress treatment, B) moderate water stress treatment, and C) high water stress treatment. Vertical arrows mark date when maximum caloric values were obtained. Adapted from Schulze (1982) after Harper and Ogden (1970).

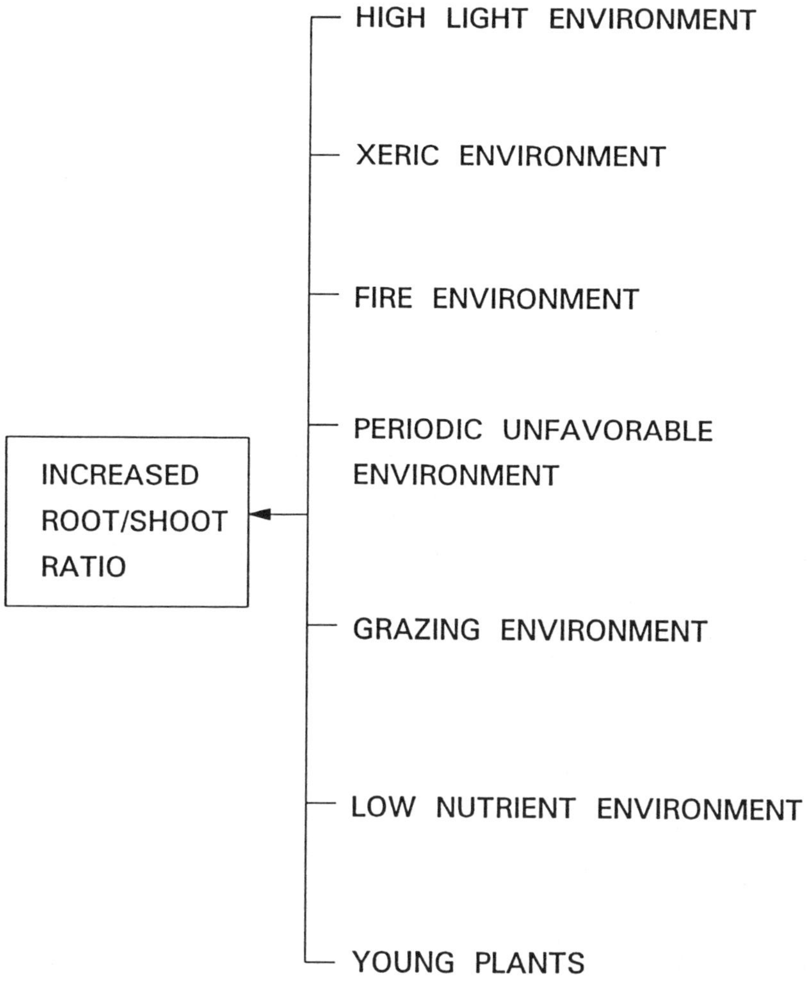

Figure 15. Factors that influence the distribution of carbon to roots and shoots. Adapted from Mooney (1972).

secondary growth. Fibrous root systems are often highly branched, have high root densities, and generally penetrate to shallower depths than taproot systems.

Root growth is often more indeterminate than is aboveground growth. Roots exist in a more favorable medium where temperature and moisture extremes are not as great as those aboveground. For these reasons, root growth may occur at various times of the year when soil temperature and moisture are favorable, and when carbohydrates and nutrients are readily available. Root growth may be suppressed during periods of rapid aboveground growth, even though other environmental conditions may be favorable (Reich et al. 1980). Photoassimilates may not be available for root growth during this time, as strong aboveground sinks exist. Root growth activity may occur over a much longer period than the defined growing season for aboveground plant organs (Fernandez and Caldwell 1975; Reich et al. 1980).

It is not unusual for root growth to begin early in the spring before significant aboveground growth has occurred (Ares 1976). At that time, root or rhizome growth will depend entirely on reserves previously stored in roots and crowns, respectively, as photosynthetic tissues have not yet developed. Root growth may decline during rapid aboveground plant growth, but accelerate again during summer (plant maturity) when photosynthetic capacity remains and less internal competition for carbohydrates and nutrients exists. Root growth may continue into fall and winter as long as abiotic conditions remain favorable and resources exist. Root growth may also be delayed for a significant period following shoot emergence (McKendrick et al. 1975). That indicates the need for a nutrient reserve because active root growth is required for nutrient uptake.

Fernandez and Caldwell (1975) examined root growth of *Atriplex confertifolia* (Torr. & Frem.) Wats. and other cold desert shrubs at varying depths throughout the growing season. Their study showed that root growth followed a wave pattern; as upper soil layers dried, root growth at lower levels increased. Still, the majority of most plant roots occur in the top meter of soil.

Soil factors other than moisture can affect root growth. Soil texture certainly affects water holding capacity, nutrient availability, and composition of soil air. Reduced oxygen, as a result of water-logging or heavy clay, can cause a significant decrease in root growth. Soil mechanical resistance can also inhibit root growth. Caliche hardpans can reduce root penetration, thus preventing roots from growing into a greater volume of soil. Roots are stimulated to grow into regions of high resource availability, including water and macronutrients (Richards 1986).

In most stands of perennial vegetation from arid and semiarid ecosystems, root system production is usually considerably greater than aboveground production (Caldwell 1987). Roots account for 50-80% of annual dry matter production for a wide variety of plants (Chapin et al. 1987). Because of difficulties involved in measuring root growth, comparatively little attention has been given them (Richards 1984). Determinations of root system biomass at different times may not be a good method to measure root production, and certainly may be inadequate for fine roots that grow, senesce and die almost continuously (Caldwell and Virginia 1989).

Rhizomes are underground stems specialized for asexual propagation of the plant. Growth of rhizomes into bare, disturbed, or uncolonized areas can be a rapid means for range plant establishment as the new plant may continue to rely on the mother plant for a supply of carbon, water, and nutrients for some time (Caraco and Kelly 1991). Buds at many of the nodes may become active, resulting in production of leaves and adventitious roots, and, eventually, the possible establishment of a new plant. Growth rates of rhizomes can be quite rapid, reaching several cm per day. Allessio and Tieszen (1978) found that about 13% of the total carbon gain of *Dupontia fisheri* R. Br. was allocated to rhizomes.

Leaves: The major carbon-gaining capacity for most plants is found in leaves. Leaves are constructed in such a way that most cells contain chlorophyll and CO_2 can diffuse readily through stomates and between cells. Energy from sunlight is used in the reduction of CO_2 to simple sugars.

Porter (1966) reviewed the information on the fate of carbon that is fixed in leaves. Developing leaves import carbon from older leaves, as well as fixing some of their own carbon in place. During this development period, the carbon is used primarily in the synthesis of polysaccharides and proteins. When the leaf attains about half its final size, it may, for a short time, both import and export carbon. Subsequently, it will supply all of its own carbon requirements, as well as exporting considerable carbon to other parts of the plant. Carbon is translocated almost entirely in the form of sucrose, a disaccharide made of glucose and fructose. The pathway for export is from chlorophyllous cells to leaf vein endings, then to phloem, and then on to various points of storage or utilization. Some of the carbon that is fixed may be exported directly as sucrose without temporary conversion to starch or inulin. Leaves near the plant apex usually transport carbon upward; whereas, those lower in the canopy transport more to the root.

Monsi (1968) presented growth equations that accounted for photosynthetic gain, maintenance and construction respiration, and distribution ratios. He vividly demonstrated how two plants with equal photosynthesis, but dissimilar distribution ratios, would have vastly different growth. Two plants, one with a distribution ratio of nonphotosynthetic system (support and root system) to photosynthetic system of 0.7/0.3 and the other with a ratio of 0.3/0.7, would accumulate 7.5 and 115 grams of dry matter, respectively, after three growth cycles. Clearly, plants with the highest carbon gain capacity and highest redistribution of carbon gain into construction of more photosynthetic system would be the most productive. However, excessive development of the photosynthetic system relative to the root systems leaves the plant quite vulnerable to environmental stresses.

Deciduous and herbaceous perennial plants lose considerable growth "interest" each year in the cost of annual replacement of their photosynthetic systems. Therefore, an evergreen growth form can be added to the attributes of the most productive system (Mooney 1972). However, evergreen needles and leaves are more expensive to construct per unit area than are deciduous leaves, as these leaves are often thicker and contain more energy-expensive organic compounds (secondary metabolites). Evergreen conifers usually have lower maximal rates of photosynthesis (4-6 mg CO_2 $g^{-1}h^{-1}$) than do broad-leaved deciduous trees (10-14 mg CO_2 $g^{-1}h^{-1}$) (Schulze 1982). In addition, costs for maintaining leaves through the winter in temperate climates must be considered, as little photosynthate would be produced during this period because of low temperatures and frozen soils.

Many evergreen species do not resprout following top removal. This may be related to lack of significant belowground carbon reserves that can be mobilized to support regrowth. That is likely an evolutionary response to having photosynthesis capacity throughout the year as a result of their evergreen nature.

The approximate carbon costs of leaf construction has been illustrated by Mooney (1972) (Fig. 16). One gram of CO_2 will yield about 0.62 grams of carbohydrate through photosynthetic reduction (Sestak et al. 1971). If this photosynthate is fixed in a mature leaf and then supplied directly to a pool utilized to construct a new leaf, about one-quarter gram of dry matter of a new leaf would result. About one-half of the original carbon gain would be lost in construction respiration. No accounting for maintenance respiration was made in Fig. 16, but this might be estimated to be 8% of gross photosynthesis (de Witt et al. 1970; Penning de Vries 1975).

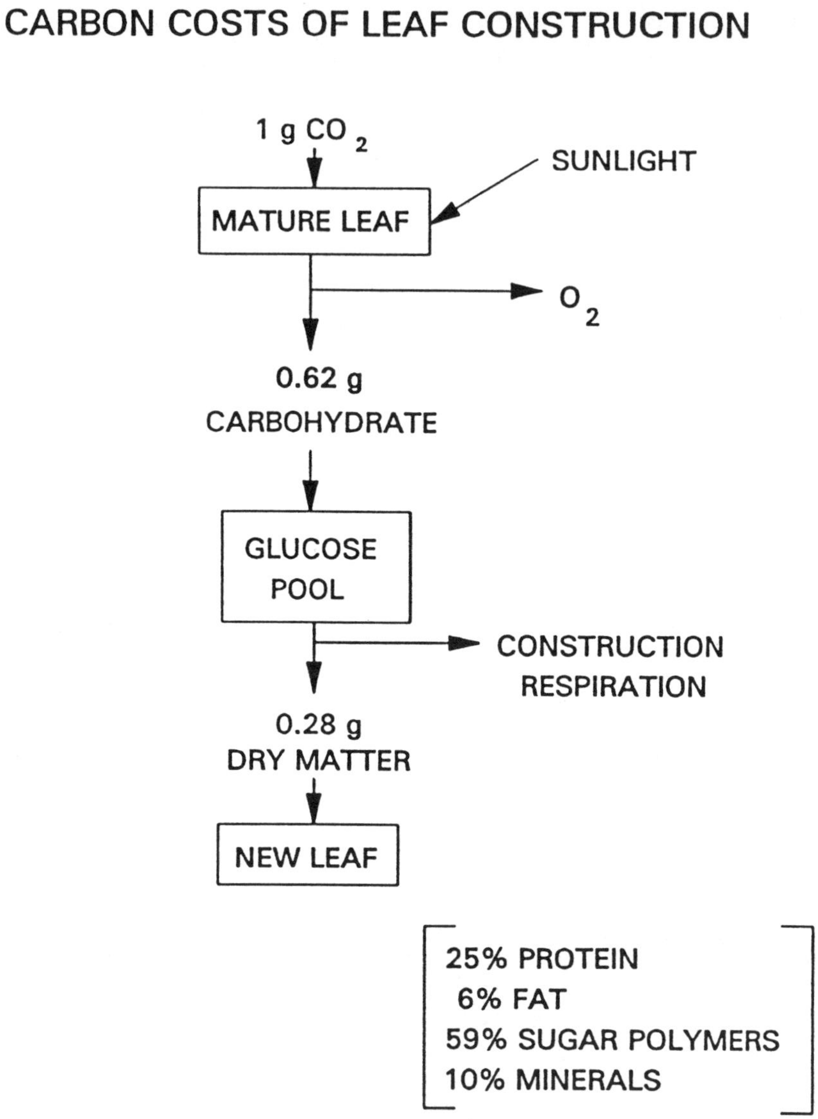

CONSTRUCTION COSTS OF INDIVIDUAL COMPONENTS
GRAMS (g) OF GLUCOSE NEEDED FOR ONE GRAM OF:

PROTEIN	= 2.35
FAT	= 2.94
SUGAR POLYMERS	= 1.15
MINERAL UPTAKE	= 0.10

Figure 16. The carbon costs for leaf construction. Adapted from Mooney (1972).

Stems: Construction costs for growth of stems, branches, and tree trunks might be expected to be greater than those of leaves, as woody structures usually contain more lignin. The glucose requirement for lignin synthesis is about twice that for synthesis of common carbohydrates, such as cellulose or starch (Table 2). Therefore, these support organs will take a great deal of carbon and energy for growth. Of course, stems give leaves the necessary horizontal and vertical projection necessary to allow for light penetration, making the plant canopy more efficient in light harvesting. The need for development of optimal systems for leaf display is an important competitive feature of plants. Therefore, leaf productivity not only must balance the construction and maintenance costs of the photosynthetic system, but also the entire shoot and root system (Mooney 1972).

Trees and shrubs differ mainly with respect to size and height, which are primarily determined by stem branching patterns and the extent of apical dominance (Schulze 1982). Limits to growth in woody plants are reached when the respiratory rates of the living portion of stems and roots approach the rate of carbon supply of the photosynthetic system (Kramer and Kozlowski 1979). Woody plants are characterized by secondary growth of stems and roots and by the continuous conversion of structural tissue into non-living, and, therefore, non-respiring biomass that is essential for physical support.

Stems of herbaceous dicots and grasses often contain chlorophyllous cells near the epidermis. These specialized cells may also be found in some trees (e.g., *Populus tremuloides* Michx.) (Strain and Johnson 1963). Therefore, photosynthesis is possible in these cells, but may be more limited by CO_2 diffusion than is photosynthesis in leaves because of low stomatal density. However, photosynthate produced in stems may at least supply all or part of maintenance and construction costs of stem growth. Stems of some grasses may even supply carbohydrates for export to other organs.

Crowns (stem bases): Plant crowns are often simply a concentration of stem bases at the junction of stems and roots. Because they may exist in close quarters and may be at or below the soil surface, chlorophyllous cells may be lacking or not well developed. Therefore, the capacity for photosynthesis is usually low. Carbon produced by the photosynthetic system is often required for crown growth and maintenance.

There is usually a concentration of axillary buds in compressed nodes of stem bases of grasses. These stem bases then represent the perennating organ for overwintering of the photosynthetic system. Large stores of carbohydrates, primarily starch or fructosans, are often found in these stem bases and utilized for early growth and initiation of new tillers (Deregibus et al. 1982).

Inflorescences: Timing of growth of plant inflorescences is often either prior to significant growth of the photosynthetic system (leaves and stems) or at the time when most active growth has ceased and leaves are nearly mature. If growth occurs early, the carbohydrate reserves stored in stems or roots will be required for growth and development of the inflorescence. If growth occurs later in the annual phenological development of the plant, then mature leaves may supply carbohydrates for inflorescence growth. The inflorescence itself often contains chlorophyllous cells and may be capable of supplying much of the carbon needed for its own growth.

Plants that reproduce once in a lifetime (e.g., annuals and monocarpic perennials) should have greater allocation of carbon to growth of reproductive structures than plants that reproduce multiple times (Bazzaz et al. 1987). Additional information on carbon allocation to sexual and asexual reproduction will be given in the following section.

Reproduction

The quantities of carbon, nitrogen, and other mineral resources invested in plant structures and activities associated with reproduction account for the direct costs of reproduction. However, resources invested in reproduction may reduce vegetative growth as vegetative and reproductive growth may compete for a common pool of resources, and an increase in one activity may result in a proportional decrease in the other activity (Bazzaz et al. 1987; Reekie and Bazzaz 1987). Most plants have fairly similar resource requirements for vegetative growth and reproduction (Chapin et al. 1987). However, differences among species, habitats, and environmental conditions result in significant variations in resource allocation. Large variability in resource allocation results from differences in the relative mass of different plant organs and the chemical composition of these structures.

Sexual reproduction: Most studies of reproductive allocation have been concerned with biomass allocated to reproductive organs (flowers, fruits, and seeds) as compared with biomass allocation to vegetative organs (leaves, stems, and roots). In many cases, measurements are made at plant maturity, so floral parts may have already been shed and costs associated with maintenance may be impossible to determine. Some reproductive structures are also capable of photosynthesis, so some of the carbon required in production of the fruit and seed is supplied by the reproductive structure itself. Photosynthesis in reproductive structures is often sufficient to meet respiratory demands (Bazzaz et al. 1987). A number of cereal grains (wheat, barley, etc.) derive much of the carbon for seed production from photosynthate produced by the glumes, awns, and flag leaf. For example, Biscoe et al. (1975) estimated that 47% of the carbon required for seed production of barley was obtained from these structures.

Reekie and Bazzaz (1987) found that sexual reproduction of *Agropyron repens* (L.) Beauv. had little detrimental effect on total growth, and, in some cases, even enhanced vegetative growth. This might happen when the reproductive organs assimilate more carbon than is required for reproduction or if strong demands of a reproductive sink results in increased photosynthesis of other vegetative structures (e.g., leaves and stems). Watson and Caspar (1984) suggested that reproduction may not be resource-limited at all.

A trend toward higher reproductive allocation in annuals, followed by herbaceous perennials, and then woody perennials, has been noted when an array of species are compared (Harper et al. 1970; Mooney 1972; Abrahamson 1979). Wild annuals may devote 15–30% of annual net carbon gain to reproduction; whereas, herbaceous perennials devote about half this amount to reproduction (Mooney 1972). This is consistent with the idea that perennials should forego some reproductive expenditure to maintain sufficient resources for perennation (Bazzaz et al. 1987). If comparisons of congeneric broad-leaved species are made, the prediction that annuals should have higher reproductive allocation than perennials is usually supported. However, there are some perennials that have very high reproductive allocations, and some annuals that have low reproductive allocation.

Construction and maintenance costs for reproductive structures are often greater than for other plant tissue. There are costs of manufacturing numerous compounds and structures (attractants, nectar, and pollen) that are used in reproduction, but then lost. For example, pollen is high in protein content (15-30%) in addition to sugar, starch, and fat (Mooney 1972), and most is not utilized by the plant.

Certain fruits may devote considerable carbon to predator protection, either in woody scales or highly toxic secondary compounds. Seeds are also costly to construct, as they often contain large amounts of reserve materials in the form of carbohydrates, lipids, or proteins. For example, seeds of *Pinus* spp. may have 20–35% fat (Kramer and Kozlowski 1979). Seeds differ from other plant tissues in that they have a high concentration of energy-rich lipids (Mooney 1972).

Environmental stresses can also have large effects on carbon allocation to reproduction. Stress usually results in a delay of flowering and a reduction in reproductive output. But extreme stress may promote flowering as though a low probability for survival triggers the perennial into a final burst of reproduction (Bazzaz et al. 1987). Also, plants on disturbed sites may allocate more resources to reproduction, but this may result from greater availability of resources or less competition at the disturbed site.

Vegetative reproduction: Clonal growth is found in a large number of plants and is an important alternative means by which plants can propagate themselves. A unique feature of clonal growth is rooting at nodes of stems, stolons, or rhizomes and may result in the eventual development of plants with the same genotype having independent fates. This trait has important ecological as well as physiological implications. Clonal growth has certain advantages over sexual reproduction in environments where grazing is severe, seed and seedling mortality is high, fires are frequent, or patchy disturbances occur (Abrahamson 1979; Bazzaz et al. 1987). However, clonal growth may represent a large sink for resources, as many of these structures have little or no photosynthetic capability. Clonal growth and sexual reproduction may also compete directly within the plant if resources are limiting.

Clonal growth does serve many of the same functions as sexual reproduction, such as providing new offspring and dispersal (Bazzaz et al. 1987). However, functional connections between mother plant and offspring allows the sharing of resources (Alpert 1991; Caraco and Kelly 1991). These vascular connections can affect competitive ability (Cook 1983), integrate patchiness in resource availability (Hartnett and Bazzaz 1985), or determine individual shoots or modules ability to survive stress or injury (Kays and Harper 1974). Until vascular connections between mother and offspring senesce or are severed, resource sharing makes it very difficult to determine construction and maintenance costs of each. More will be said about recruitment and demographics of clonal growth forms in a following chapter by Briske and Richards in this text.

Defense

Defense mechanisms in plants usually require comparatively high investments of energy and nutrient in chemical pathways (Foulds and Grime 1972; Coley 1986) or morphological structures, especially in those plant parts that are particularly vulnerable to herbivory (Meijden et al. 1988). It is doubtful, though, whether there is any species that is completely protected by its defenses against herbivory (McNaughton 1983). Therefore, a plant might utilize rapid regrowth as an alternative to chemical or structural defense mechanisms. Grasses may exhibit this behavior, as they usually have low concentrations of defensive chemicals and rapid regrowth rates (Trlica and Orodho 1989). Forbs and shrubs may produce large concentrations of some defensive compounds and have slow growth after defoliation. In this section, we will consider costs associated with the construction and maintenance of plant structures and chemistry that afford the plant some protection from various classes of herbivores.

Morphological structures, such as thorns, spines, and prickles, may give plants some protection against large grazing animals. However, these same structures may afford some degree of protection to small plant-eating insects from their winged predators. Thus, these structures may offer protection from one class of herbivore, but allow for greater use by yet another class of herbivore. These structures usually do contain high concentrations of lignin, which requires a great deal more energy for biosynthesis than other less complex carbohydrates. However, little work has been done to determine total costs of construction for these structures versus the benefits that the plant might derive for producing them. Once produced, maintenance costs should be quite low, as most of these structures consist of nonliving biomass.

Species vary greatly in the type and levels of chemical defense against herbivores and pathogens. Chemical defenses draw from an enormous group of compounds that can differ in both concentration and distribution (Bazzaz et al. 1987). Construction costs differ significantly among various defensive compounds. For example, synthesis of alkaloids or terpenes is almost twice as costly as that of tannins or lignins (McDermitt and Loomis 1981; Mooney and Gulmon 1982). Alkaloids and terpenes also have short half-lives, so maintenance costs are greater than for tannins and lignins that have very slow turnover rates (Coley et al. 1985). Alkaloids typically represent less than 1% of dry leaf mass, and glycosides and terpenes less than 5% (Rhoades and Cates 1976). Condensed tannins in leaves can make up 5 to 20% of the dry mass, and phenolic resins can attain 40% or more in concentration (McKey et al. 1978; Lincoln 1980). Lignin concentration in leaves can exceed 20% (Waring et al. 1985). There is no evidence that presence of these secondary compounds detrimentally affects photosynthetic ability of the plants that produce them (Mooney and Gulmon 1982). However, some of these compounds contain nitrogen, which may have been diverted from synthesis of carboxylating enzymes and might, thus, reduce photosynthetic gain indirectly.

Secondary compounds: Phosphoenolpyruvic acid (PEP) is formed through glycolysis. PEP is subsequently transformed to pyruvic acid and that can be converted into a number of organic compounds, including proteins and lipids, but also alkaloids and terpenes, which can act as defense compounds. PEP may also become directly involved in the shikimic acid pathway, which can lead to a variety of secondary compounds, including various phenolics, lignins, and tannins (Mooney 1972). Phenolics may play an important role in defense against both microorganisms and animals (Levins 1971). Parasitic fungal infection of a plant can result in the generation of phenolics as a defense against the infection. In terms of carbon allocation for protection, this response is very efficient.

Large variations in kinds and concentrations of secondary compounds occur among species, phenological stages, plant size, and age. These compounds are also influenced by habitat and environmental factors. When an imbalance of resources occurs, plants may divert excess resources to production of defense compounds (Bryant et al. 1983). For example, when plants are shaded and carbon limited, high leaf nitrogen levels can lead to increases in nitrogen-based defensive compounds such as alkaloids. The nitrogen in these alkaloids may later be remobilized and used in synthesis of some other compound. Plants growing in a high light environment often have higher concentrations of carbon-based defenses, such as phenolic compounds or terpenes, then do plants of the same species growing in the shade (Bazzaz et al. 1987). Both carbon- and nitrogen-based defense compounds may decrease if plants receive additional nitrogen and high light, which supports increased growth.

Trends observed for individual species are sometimes opposite the general trends among species (Bazzaz et al. 1987). Those species adapted to low-nutrient availability or low-light environments, which usually have low growth rate potential, generally have higher defense allocations than species of resource-rich habitats. Plants from resource-poor environments often have leaves that live longer; hence, the value for protecting them against loss is much greater.

Just as reproductive structures may have partially nonreproductive functions (e.g., photosynthesis), chemicals used for defense may have other functions (Bazzaz et al. 1987). Phenolics may serve as defense against predation and also filter out ultraviolet radiation (Lee and Lowry 1980). Phenolic resins on leaves may deter herbivory and reduce transpiration.

Structural compounds: Just as phenolics may have multiple functions, so do some of the structural defensive compounds. For example, lignin (a phenolic polymer derived from the shikimic acid pathway) is a major constituent in most plant cells. The prime function of lignin is to provide support for the plant through its interrelationship with the cellulose framework. However, lignin also provides resistance to cellulose against degradation and utilization by microorganisms (Mooney 1972). Lignin also makes plant tissue tough and not easily digested, so herbivore preference for tissue high in lignin often declines. High cellulose and fiber, and low concentration of cell constituents, may also help to deter herbivory and give some protection to the plant.

Large concentrations of silica in cell walls may also make tissue less attractive to herbivores, as these structures have an abrasive nature and may wear down teeth rapidly. Whether or not silica is a natural defense mechanism against herbivory is questionable. Studies by McNaughton (1984) and Brizuela et al. (1986) show that while plant leaves from areas with a history of heavy grazing may contain higher silica concentrations than leaves from lightly grazed areas, the negative effects of higher silica may not deter grazing. This follows because repeated grazing results in plants of superior nutrient quality (higher N, higher digestibility) making them more attractive to herbivores.

These structural compounds are metabolically inactive, as are condensed tannins, against the herbivore. Some of the previously considered secondary compounds, such as alkaloids and cardiac glycosides, are metabolically active and may be toxic to the animal that ingests them. These compounds may also be effective bactericidal or fungicidal agents against some microorganisms.

Storage

Storage of carbon for later use is a major function of the plant, along with acquisition, transport, growth, defense, and reproduction (Fig. 1). The term "storage" may be confusing, as it is seldom defined explicitly, and has been used differently in several disciplines. Chapin et al. (1990) defined storage quite broadly as resources that build up in the plant and are capable of being mobilized in the future to support biosynthesis for growth or other plant functions. They recognized three classes of storage: accumulation, reserve formation, and recycling.

Accumulation occurs when acquisition exceeds flows to growth, maintenance, defense, and reproduction. The accumulated products can be lost from the plant (exudates or leachates) or can contribute to future growth (Chapin et al. 1990). Reserve formation occurs when photosynthates or accumulated products are actually partitioned into storage within plant cells. Partitioning of reserves into storage may, therefore, compete with growth, defense, or reproduction. Recycling involves the breakdown of components of growth to form a pool of recycled materials that support new growth. For example, breakdown of lipids of cell membranes into simpler compounds that might then be used in growth or other cell functions would involve recycling.

The role of stored carbon should be evaluated in terms of its value to the whole plant and in light of opportunity costs which accrue because carbon was not allocated to some other function. Allocation of carbon to storage can be considered in terms of specific organs involved, source-sink relations, or biochemical allocation to specific compounds (i.e., starch, inulin, sucrose, etc.). In many range studies, we may not be too concerned with the biochemical nature of compounds and simply lump them together under total nonstructural carbohydrates (TNC), which might include a number of simple sugars, starch, fructans, and others. We should, however, be concerned with source-sink relations and organs involved, as these factors determine what will regrow and at what rate. Opportunity costs of storage are reduced if the stored compound serves other physiological functions. For example, storage of some carbohydrates imparts frost hardiness to some plants. Thus, the storage product serves dual purposes.

Carbohydrate storage and use: In a seasonal climate, storage of carbohydrate reserves is fundamental to the extension of the life cycle and initiation of growth in the following season when favorable conditions return (Schulze 1982). Storage of reserves seems to be a principle physiological feature of herbaceous perennials and a basis for their success in harsh environments. A number of excellent reviews have been written on the role of reserves in plants as related to growth and regrowth when plants are affected by weather, grazing, fire, and other perturbations (Cook 1966; Mooney 1972; White 1973; Trlica 1977; Trlica and Singh 1979; Deregibus et al. 1982; Chapin et al. 1990). Therefore, this information will not be repeated here.

In perennial plants, storage of nonstructural carbohydrates usually represents the most significant stores available to the plant. Lipids, fats, and other storage products usually are at low concentrations and may not represent a reliable source of energy and construction material. Carbohydrates are stored in cells throughout the plant, but greatest concentrations are often found in perennating organs such as stems, stem bases, roots, rhizomes, and bulbs. Carbohydrate reserves decrease in these organs during rapid growth of the plant when construction and respiration demands are great. Reserves begin to concentrate again in these organs when growth rate declines or when most active growth has been completed. Often we only consider aboveground growth as related to carbohydrate reserve levels. However, belowground processes have greater carbon demand than aboveground processes and this may influence reserve levels. Thus, root growth may require use of considerable stored products, as well as current assimilates if they are available. Current acquisition is used preferentially before stored carbohydrates when plants have access to both, which suggests an opportunity cost for storage (Deregibus et al. 1982; Chapin et al. 1990). Most analytical methods will not, however, distinguish among carbohydrates that have been accumulated versus those that were in storage.

Carbohydrate (CHO) reserve cycles are tied closely to growth periods of range plants. For warm-season perennial grasses in temperate latitudes, there is a large decline in stored CHO associated with spring growth followed by a period of replenishment. Owensby et al. (1970) reported an annual CHO reserve cycle for big bluestem which included the drawdown and replenishment periods, plus a drawdown associated with growth of roots on new rhizomes in late summer (Fig. 17). A cool-season species may have two drawdown periods associated with spring and fall growth, but the second depletion is often dependent upon fall precipitation in arid and semi-arid environments. Coyne and Cook (1970) reported spring and fall growth periods for squirreltail [*Sitanion hystrix* (Nutt.) J.G. Smith] in Utah (Fig. 18). The spring growth period used 30% of the stored reserves and the fall growth, 40%.

Milthorpe and Davidson (1966) showed that sugars that accumulated before cutting accelerated regrowth of grasses during the first few days after clipping and this initial advantage was maintained during the rapid vegetative growth stage. Davies (1965) showed that the future yield of a grass sward was dependent, to some extent, on the level of carbohydrates present in stubble at the time of defoliation. A study by Smith and Silva (1969) showed that about 68% of reserves in alfalfa roots were used in the production of new top and root growth. Blue grama, with low reserve storage, utilizes carbon in storage for the production of new leaf tissue following defoliation (Chung and Trlica 1980). Therefore, carbohydrate reserves are definitely needed in the production of regrowth, but their importance appears to be quite different for various species, and relationships can be altered considerably by other environmental conditions (Richards and Caldwell 1985).

Seasonal storage of carbohydrate is generally viewed as a carefully regulated process of reserve accumulation, distribution, and storage. However, the pattern could also be interpreted as accumulation in response to supply and demand (Zimmerman 1971; Trlica and Singh 1979). As supply continues to increase after most active growth has been completed and demand declines, reserve products accumulate and are not needed for additional construction. Thus, storage or defense are likely recipients of these products. There have, however, been few studies to distinguish between these two possibilities.

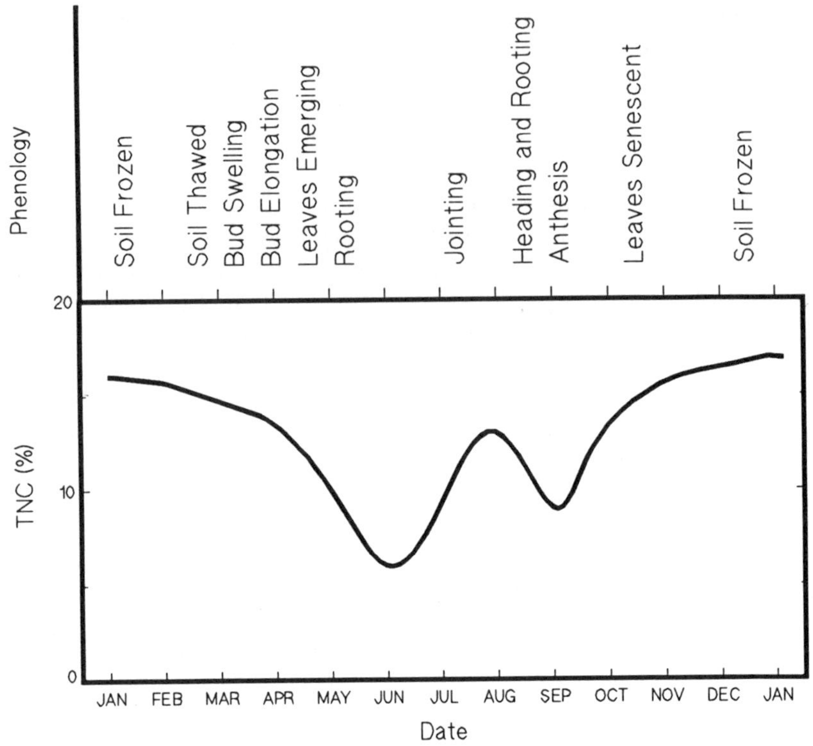

Figure 17. Total nonstructural carbohydrates (TNC) in rhizomes and stem bases on different dates for big bluestem *(Andropogon gerardii)*. Adapted from Owensby et al. (1970).

Recycling of stored carbohydrates from senescing plant tissues is not great. Most measurement of mass in recycling can be accounted for by recycled nitrogenous compounds. This supports the idea that non-nutrient containing compounds are not recycled to any large extent (Chapin et al. 1990).

Management implications: There is no clear relationship for native perennial plant species between the amount of stored carbohydrate and capacity for regrowth after defoliation (Trlica and Cook 1971; Buwai and Trlica 1977; Menke and Trlica 1983; Chapin et al. 1990). Moreover, most plants do not use a large proportion of their stored carbohydrates in response to clipping (Deregibus et al. 1982). This raises questions as to the actual accessibility of these reserves for plant use. Several studies have emphasized the magnitude of decline in carbohydrate reserves after clipping, but do not point out that substantial nonstructural carbohydrates remain in storage organs at levels of clipping that greatly repress regrowth. In fact, if plants are clipped repeatedly and kept in the dark to eliminate photosynthesis, regrowth usually ceases before carbohydrate reserves are exhausted (Richards and Caldwell 1985). Some carbohydrates may become inaccessible to the plant because they are located in dead cells and cannot be retrieved.

Because removal of photosynthetic tissue largely determines the amount of energy available for the reserve system, the time interval between removal and dor-

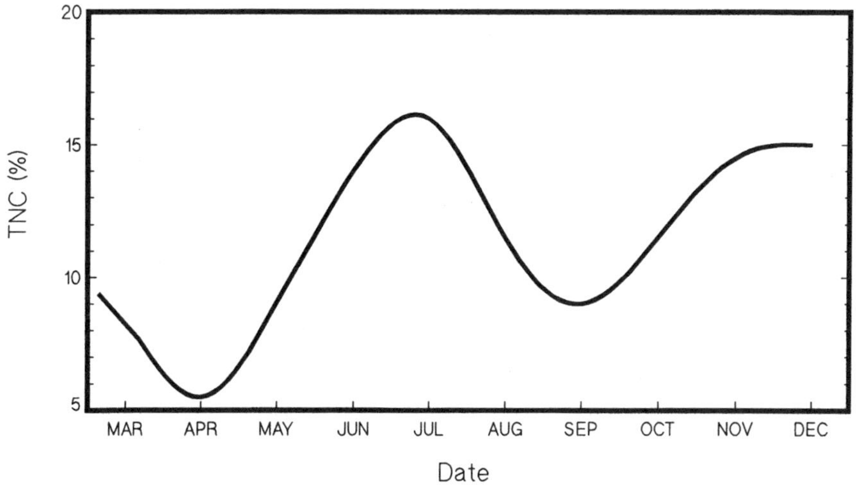

Figure 18. Total nonstructural carbohydrates (TNC) in storage organs of squirreltail (*Sitanion hystrix*) at different dates in Utah. Adapted from Coyne and Cook (1970).

mancy is critical. If removal is continuous at a high rate throughout the growing season, there is never sufficient leaf area to manufacture the photosynthate needed for reserves. If removal is moderate, leaving sufficient leaf area for photosynthate production in excess of growth, reserves will be replenished. On the other hand, if severe herbage removal comes in a relatively short time as with haying or high stocking density grazing systems, time to dormancy becomes important. In the case of heavy grazing for a short period, the plant must have time to recover reserves prior to dormancy. Owensby et al. (1977), working in the Kansas Flint Hills, compared big bluestem CHO reserves on an area stocked at twice the normal growing season stocking rate (0.68 ha/yearling steer) from May 1 to July 15 (intensive-early use) and on an area stocked at the normal stocking rate (1.36 ha/yearling steer) from May to October 1 (season-long). Because stocking rate under intensive-early use was twice that of season-long use from May l to July 15, big bluestem CHO reserves were much lower under intensive-early use compared to season-long use (Fig. 19). Following removal of steers from the intensive-early use pastures on July 15, big bluestem CHO reserves remained low for a time, but then were rapidly replenished, such that they were equal to those of plants managed under season-long use by the end of the growing season. If grazing had continued at the 0.68 ha/steer rate throughout the season, big bluestem CHO reserves would likely have been significantly reduced by the end of the growing season compared to reserves of plants stocked at the lower, season-long rate.

The length of time from severe top removal to dormancy that is necessary for adequate replenishment of CHO reserves will vary according to plant species and environment. On areas hayed at different times, accumulation of reserve CHO following mowing depends primarily on the frequency of mowing and the time interval between mowing and fall dormancy. If removal comes too late for sufficient CHO storage following the use of reserve CHO for regrowth, next year's herbage

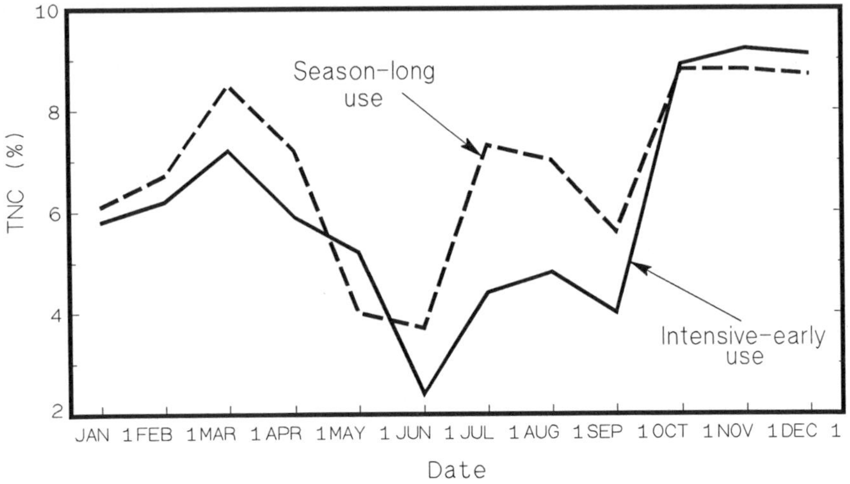

Figure 19. Total nonstructural carbohydrates (TNC) of big bluestem (*Andropogon gerardii*) on Kansas Flint Hills range stocked at twice the normal growing season stocking rate (0.68 ha/yearling steer) from May 1 to July 15 (intensive-early use) and on an area stocked at the normal stocking rate (1.36 ha/yearling steer) from May to October 1 (season-long). Adapted from Owensby et al. (1977).

yield will be reduced. Owensby et al. (1970) measured CHO reserve levels following herbage removal at different dates (Fig. 20). Removal too late lowered CHO reserve levels at dormancy and herbage yield the following season. Plant census indicated reductions in warm-season perennial grasses with August 1 and Sep-

Figure 20. Total nonstructural carbohydrates (TNC) of big bluestem (*Andropogon gerardii*) on Kansas Flint Hills range following mowing to a 50 mm cutting height on the indicated dates. Adapted from Owensby et al. (1970).

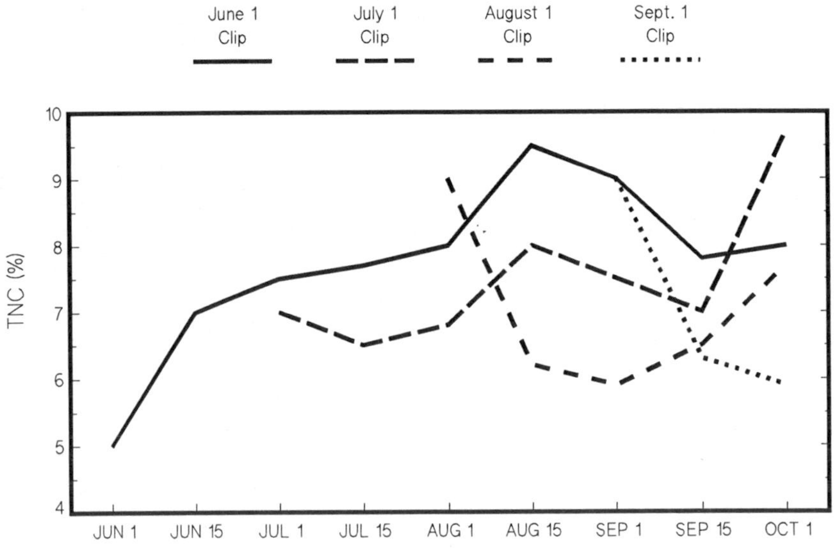

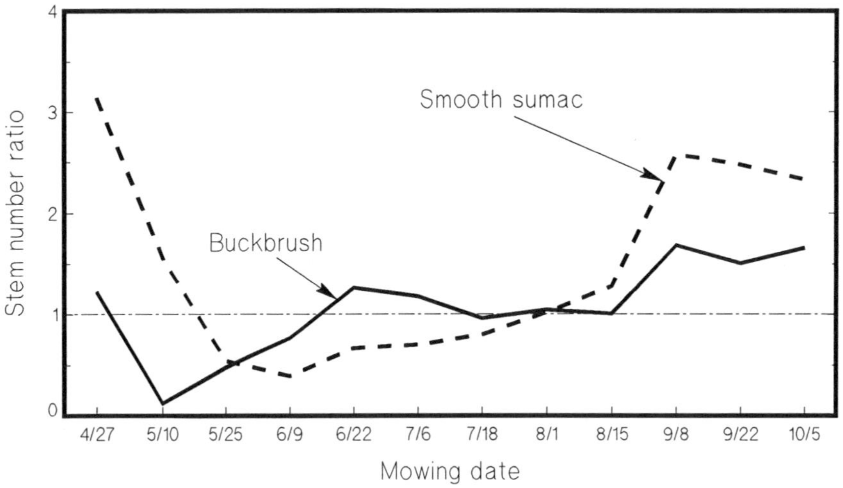

Figure 21. Average change in stem number ratio (after mowing/before mowing) of buckbrush (*Symphoricarpos orbiculatus*) and smooth sumac (*Rhus glabra*) following five years (1928-1932) of mowing in the Kansas Flint Hills. Ratios less than one indicate a decrease and numbers greater than one an increase in stem number per unit area as a result of mowing. Adapted from Launchbaugh and Owensby (1978).

tember 1 herbage removal dates, and an increase in Kentucky bluegrass (*Poa pratensis* L.).

Effective control by mechanical methods or fire of plant species which resprout is related to their food reserve cycles. With resprouting species, control is accomplished by starving the plant through removal of food producing material at a time when there is low reserve food storage, or by stimulating a drawdown in food reserves such that the plant goes into dormancy with low reserves. Usually, the best time for control occurs when the plant is at its low point in food reserves. Mowing date in a Kansas Flint Hills study reported by Launchbaugh and Owensby (1978) was extremely important in achieving control of buckbrush (*Symphoricarpos orbiculatus* Moench) and smooth sumac (*Rhus glabra* L.) as shown by changes in stem number (Fig. 21). Best control of those two species was at a date near the low point in their food reserves (Fig. 22).

Knowledge of carbohydrate pools (weight of organ x %TNC) may be much more important in some species for regrowth after defoliation or some other disturbance than knowledge of carbohydrate concentration in storage organs (Santos and Trlica 1978; Caldwell 1984; Orodho and Trlica 1990). It is the total amount of reserves that are actually available for construction and respiration that are, after all, most important. Reserve substances may only be needed by the plant for a few days to support regrowth (Richards and Caldwell 1985). Thereafter, additional growth may come from current photosynthates. Nevertheless, Coyne and Bradford (1987) found that concentrations of both carbohydrate and nitrogen reserves tracked pool sizes quite well in the roots of two Old World bluestems (*Bothriochloa*), but the relationship was not as strong in the crowns. Therefore, for these grasses it appeared that management decisions could be keyed to reserve concentrations thereby elim-

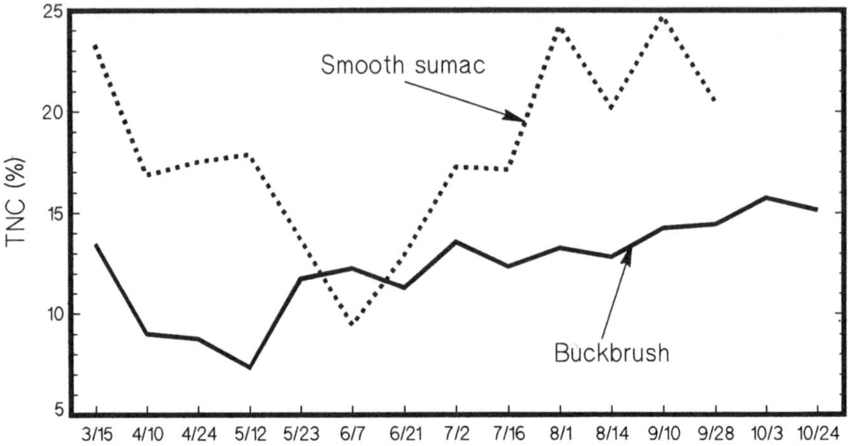

Figure 22. Five-year average (1928-1932) seasonal variation in total nonstructural carbohydrates (TNC) of buckbrush (*Symphoricarpos orbiculatus*) and smooth sumac (*Rhus glabra*) in the Kansas Flint Hills. Adapted from Launchbaugh and Owensby (1978).

inating the more laborious task of determining reserve pool sizes.

Tillering rates, one of the most important aspects of grassland yield and perennation, have been shown to be partially related to a high level of reserves (Deregibus et al. 1982). Although other factors exist that influence tiller appearance, high levels of carbohydrate reserves before defoliation have been shown to increase the number of shoots in regrowth. However, tillering rates are also influenced by red/far red light ratio and activity of axillary buds (Deregibus et al. 1985; Richards and Caldwell 1985). The beneficial effects that low clipping or heavy grazing may have on plant density through diminishing apical dominance and improving illumination conditions at the base of the plants will require the maintenance of optimum levels of carbohydrate reserves.

Root growth after defoliation has not been maintained in plants, even with high levels of reserves (Deregibus et al. 1982). However, the existence of adequate carbohydrate reserves that may allow more rapid regrowth should allow roots to renew growth shortly after cutting. Root respiration following defoliation has been shown to utilize metabolites stored prior to defoliation, which illustrates the importance of reserves for preventing root starvation and tip death. Therefore, there is a need to manage range ecosystems to maintain optimum (not maximum) reserve levels in desirable plant species to assure perpetuation of the community.

Herbage plants have evolved efficient storage as another defense against detrimental effects of defoliation or catastrophe. The range manager should utilize this characteristic to prevent deterioration of rangeland vegetation (Deregibus et al. 1982). Studies are needed to determine clipping responses of individual plants that differ only in the magnitude of stored reserves to determine recovery as related to quantity of reserves present before treatment. This could provide information on opportunity costs of storage for recovery from defoliation (Chapin et al. 1990). We also need studies to determine the period of rest required to regain normal vigor, herbage productivity, and adequate carbohydrate reserves following defoliation.

Nitrogen Dynamics

In most stable range ecosystems, nitrogen fluxes are balanced, with losses equaling acquisition. Nitrogen transformations within the plant are affected by the availability and form of N available for uptake by the plant. Range herbivores, through their ingestion and excretion of nitrogen, affect availability, distribution, and chemical form of N in the ecosystem. Nitrogen in an ecosystem is cycled in an open, biogeochemical cycle with a direct interchange between the ecosystem and the atmosphere. A schematic of the nitrogen cycle is presented in Fig. 23. For an in-depth discussion of nitrogen dynamics in the soil-plant system, the reader is referred to a synthesis book edited by Haynes (1986a).

Nitrogen Acquisition

Nitrogen in natural ecosystems enters from the atmosphere primarily through wet and dry deposition (Woodmansee et al. 1978). While Woodmansee et al. (1978) place relatively low importance on biological fixation as an input into grassland ecosystems, others (Kormondy 1976; Odum 1982; Haynes 1986b) indicate that inputs from biological fixation are nearly as important as wet and dry deposition. Nonetheless, the amount of nitrogen entering or leaving natural ecosystems is exceedingly small compared with that held in organic forms in plants or soil organic matter. Rosswall (1976) estimated that only 8% of N in the soil-plant system was lost and gained yearly. Jenkinson (1990) has reported estimated global nitrogen pools (Table 3). Most nitrogen in terrestrial ecosystems resides in the soil organic matter pool. Woodmansee et al. (1981) confirmed that the overwhelming majority of the N in grassland ecosystems is in the soil organic matter pool (Table 4).

Sources of Plant Nitrogen

Wet and dry deposition: Atmospheric nitrogen, in the form of NH_3, NO_x, and particulates, is transferred to the soil system by precipitation (Haynes 1986b). Wet deposition in non-industrial regions is largely composed of N forms from animal wastes and gaseous losses from the soil-plant system. Industrialized regions have much higher wet deposition rates for N as a result of N additions from combustion. Dry deposition results from particulate deposition in response to gravity and the direct absorption of N in the particles by the soil-plant system. Plants and soils act as both a source and sink for N. Much of the NH_3 loss from the soil system is quickly absorbed by the plant-soil system (Potter et al. 1987). NO_x and N_2O gasses have also been shown to be absorbed from the atmosphere by the plant-soil system (Haynes 1986a).

Biological N-fixation: In natural systems, biological fixation of N_2 is accomplished by free-living bacteria, symbiotic cyanobacteria, rhizocoenoses, actinomycete nodule symbiosis, and *Rhizobium* symbiosis (Haynes 1986b). The energetic costs of reducing atmospheric nitrogen in nodules is similar to the costs of converting NO_3^- to NH_4^+ (Amthor 1989). That energy, however, comes entirely from respiration,

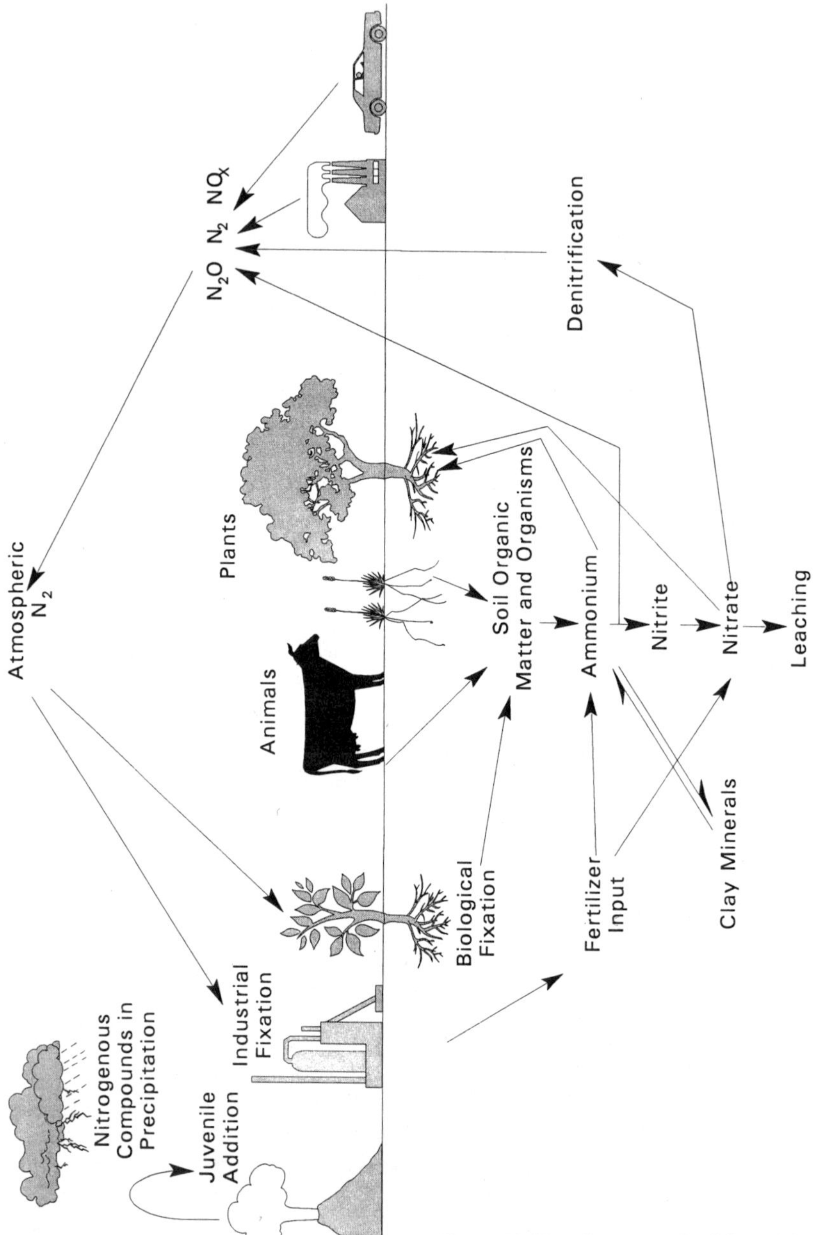

Figure 23. The nitrogen cycle. Adapted from Haynes (1986b).

whereas NO_3^- reduction may use energy from photosynthesis during the daylight. Nodulated roots have higher respiration rates than non-nodulated roots. *Rhizobium* symbiosis is estimated at 25% of the global biological N-fixation. Nitrogen fixed by microbes becomes available to plants upon microbial death. Input into a range ecosystem from N-fixation can be substantial (Burns and Hardy 1975).

Table 3. Estimates of the active global pools of nitrogen (adapted from Jenkinson, 1990).

Pool	Millions of Metric Tons
Atmospheric	
N_2	3,900,000,000
N_2O	1,400
Terrestrial	
Plants	15,000
Animals	200
Soil organic matter	150,000
Oceanic	
Plants	300
Animals	200
Inorganic	
NO_3–N	1,200,000
NH_4–N	570,000
Dissolved N_2	22,000,000

Table 4. Nitrogen partitioning in a shortgrass prairie ecosystem (data from Woodmansee et al. 1981).

	Nitrogen Quality	
Nitrogen Pool	Absolute	Relative
	(kg N ha^{-1})	percent
Total vegetation	66	1.9
Aboveground	15	0.4
Roots	51	1.5
Litter	76	2.1
Soil (0-300 mm)	3,374	96.0
Total	3,516	100.0

While most think of N-fixation in terms of *Rhizobium* symbiosis, other organisms can fix relatively large quantities of nitrogen. Eisele et al. (1990) reported that on burned, tallgrass prairie, cyanobacteria associated with *Nostoc*, a blue-green algae, fixed substantial amounts of N. Bacteria associated with the roots of many grasses fix N, a process referred to as rhizocoenoses. However, N fixed in that way represents a relatively small part of total biological N-fixation.

The magnitude of biological N-fixation is tied to the nutrient status of the ecosystem. Ecosystems that fix large amounts of carbon and also retain large amounts of nutrients in organic form have high biological N-fixation rates. Of particular importance is the *Rhizobium* association, with the relative energy status of the plant directly affecting the amount of N fixed by the symbiotic bacteria. There appears to be a direct coupling of N-fixing symbiotic bacteria to the photosynthetic output of higher plants. Therefore, constraints on photosynthesis will likely lead to

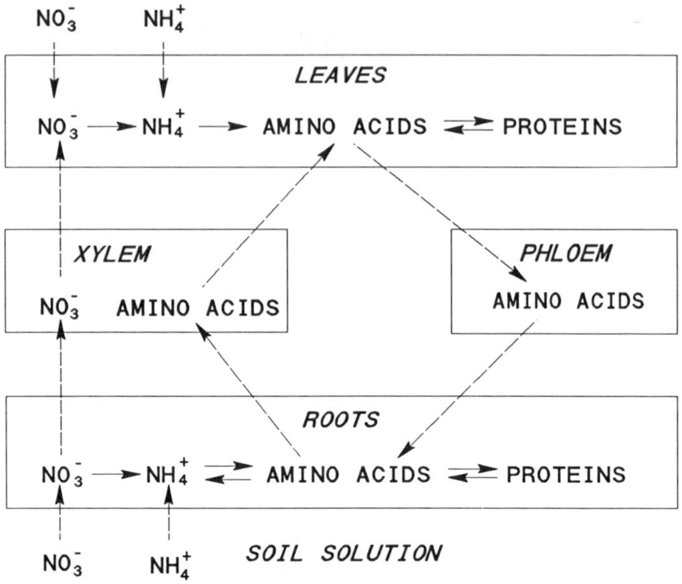

Figure 24. Basic pathway for nitrogen uptake, reduction, and synthesis into protein in plants. Adapted from Haynes (1986c).

reduced N-fixation.

Synthetic fertilizers: Normally, synthetic fertilizers play a minor role as nitrogen sources for range plants, with only a portion of the more mesic range ecosystems receiving commercial fertilizers. The greatest influence of fertilizers in rangeland ecosystems arises indirectly from increased wet deposition associated with fertilizer-induced gaseous losses from cultivated systems.

Nitrogen Uptake

Utilization of nitrogen by higher plants involves uptake, storage, translocation, and incorporation of N into organic forms (Fig. 24). Nitrogen in plants is normally found in a reduced state. Although the primary means of N uptake is by roots, foliar uptake may also occur.

Roots: Nitrogen is available in the soil system for plant uptake primarily as NH_4^+ and NO_3^-. Most N uptake by plants in terrestrial ecosystems is in the NO_3^- form (Pate 1980). Uptake of both NH_4^+ and NO_3^- require energy and exhibit a diurnal pattern with the greatest absorption taking place at midday (Pearson and Steer 1977). Apparently, nitrogen uptake is slowed at night by lack of photosynthesis.

Nitrogen uptake by plant roots can occur by passive diffusion through the epidermis of the root, or by active uptake at the outer cell membrane or plasmalemma (Haynes 1980). The ions may then be transported across the cortex, endodermis, and pericycle through a system of intercellular bridges (plasmadesmata) linking the protoplasts of adjacent cells (Fig. 25). That system is referred to as the symplast (Larcher 1980). Nitrogen can move passively into the continuum of non-living cell wall material (apoplast) of the root cortex cells and then be absorbed across the plasmalemma of cortical and endodermal cells, thereby entering the symplast. Suberin

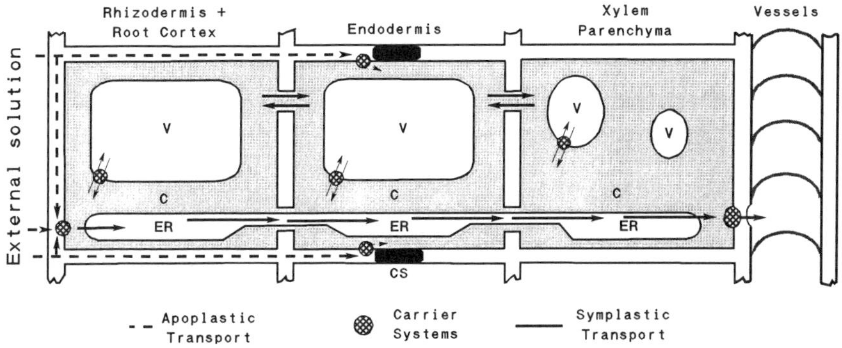

Figure 25. Schematic of nitrogen transport from the soil solution to the sieve tubes of the xylem. N, which enters the cytoplasm (C) in epidermal or cortical cells, is translocated symplastically over the endomembrane system (ER) and through plasmodesmata. N, together with water, may move through the root extracellularly until it reaches the Casparian strip (CS) where it then must move symplastically to the vascular elements. Vacuoles (V) may serve as source or sink for N, but are not an active part of the transport system. Adapted from Larcher (1980).

deposited in the radial walls of the endodermal cells (Casparian strips) restricts apoplastic movement of nitrogen from the free space of the root. Therefore, N passage through this area must be through the symplast. Absorption of ions across the plasmalemma of root cells is generally thought to be an active process that requires expenditure of energy to move ions against an electrochemical gradient (Haynes 1986c). Active uptake is likely accomplished through carriers which appear to be protein units in the plasmalemma that associate with ions in the apoplast and then discharge them into the symplast.

Most nitrogen uptake by higher plant roots is in the NO_3^- form, which may either be reduced to NH_3 or NH_4^+ and converted to amides in the root. Uptake of NO_3^- has been shown to be restricted by inhibitors of RNA and protein synthesis and inhibition of respiratory and oxidative phosphorylation (Jackson et al. 1973; Rao and Rains 1976; Tomkins et al. 1978) confirming N uptake is an active process requiring energy. Because NH_4^+ and NO_3^- uptake requires energy, any condition which reduces the available supply of photosynthate to the root system may limit ion uptake. Alternatively, NO_3^- may be transported up the xylem and reduced in the leaves. This reduction may be driven by photochemical energy captured through photosynthesis. Plants vary widely relative to where reduction of NO_3^- occurs. Pate (1973) reported that 90% of the NO_3^- reduction and assimilation occurred in the roots of lupine (*Lupinus* L.), while none occurred in the roots of cocklebur (*Xanthium strumarium* L.). Mycorrhizal symbiotic associations also enhance nitrogen uptake, particularly in soils where NH_4^+ is the major nitrogen source (Bowen and Smith 1981).

The ammonium ion is apparently toxic to plant cells and carbon skeletons must be available for immediate assimilation and detoxification. If carbon skeletons are not available, NH_4^+ uptake may be limited. Nitrate uptake is less dependent upon immediate assimilation than is NH_4^+ because NO_3^- is not nearly so toxic and may be stored in vacuoles or transported to shoots. Low temperature is a greater limitation to NO_3^- uptake than to NH_4^+ uptake.

Foliar: Nitrogen may be absorbed directly by leaves either as a gas or from solution (Haynes 1986c). Foliar absorption of soluble nitrogen is restricted primarily to solutions of urea. The gasses, NO, N_2O, NH_3, and to a lesser extent NO_2, have been shown to be absorbed by leaves (Haynes 1986b). NH_3 is likely the most absorbed nitrogen-containing gas. NH_3 is produced in large amounts near the soil surface (Denmead et al. 1976). In rangeland ecosystems, the NH_3 near the soil surface comes primarily from urine and dung patches. If sufficient active leaf canopy is available, almost all of the NH_3 liberated is absorbed by the foliage. Much of the NH_3 that is volatilized from livestock excreta is absorbed by the plant canopy rather quickly. Lemon and Van Houte (1980) found NH_3 was evolved from the top and bottom leaves in the leaf canopy of several crop species and was absorbed by leaves in the mid canopy. Therefore, it appears that much of the NH_3 lost by plants is reabsorbed within the plant canopy. Foliar uptake appears to occur through diffusion into stomata and subsequently into the intercellular spaces of leaves, where active uptake occurs. There is likely much more involved in stomatal uptake, but current knowledge is lacking in that area. While the capacity to absorb NH_3 through the foliage has been shown, the importance of this process in range plants growing in arid or semi-arid environments remains to be quantified. Certainly, foliar uptake of N would enhance conservation of N in grassland ecosystems.

Nitrogen Assimilation

Nitrate assimilation: After NO_3^- entry into the root, or following transport to the shoot, NO_3^- is assimilated by both reductive and non-reductive processes (Schrader and Thomas 1981). Nitrate may be reduced to NH_3 or NH_4^+ and converted to amides in the root. Alternatively, it may be transported to the leaves and reduced there. The overall pathway for NO_3^- assimilation into glutamate is shown in Fig. 26.

Ammonia assimilation: The primary assimilation of ammonia in plants is through the glutamate synthase cycle (Schrader and Thomas 1981). Initial incorporation of NH_4^+ into glutamine is catalyzed by glutamine synthetase. In the presence of a reducing agent, glutamate synthase catalyzes the transfer of the amide group of glutamine to alpha-oxoglutarate which results in the formation of glutamate. Alpha-keto acids are converted to amino acids through reactions catalyzed by transferases (Fig. 27). Amino acids thus formed provide building blocks for complex nitrogenous compounds that include proteins, chlorophyll, growth regulators, alkaloids, nucleosides, nucleotides, and nucleic acids. Amino acids may be found in a free state in plants as well as in nitrogen storage or defense compounds.

Photorespiration and nitrogen metabolism: Several reactions in plants release NH_3, but the primary one occurs in conjunction with photorespiration. Photorespiration involves the oxidation of glycolate (Tolbert 1971). Glycolate produced in the chloroplast is transported to the peroxisomes where it is oxidized to glyoxylate (Fig. 4). The glyoxylate is transaminated to form the amino acid glycine. Glycine is transported to the mitochondria where it is converted to serine, CO_2, and NH_3. During photorespiration, 0.18 mole of NH_3 is released for each primary glycine decarboxylation according to the following equation adapted from Canvin (1981):

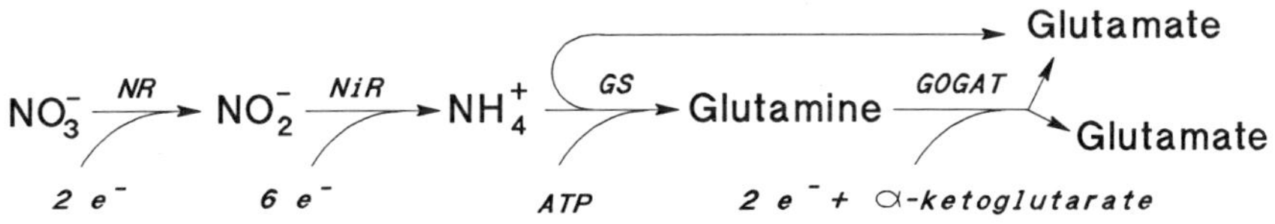

Figure 26. Assimilation of nitrate to glutamate. Abbreviations: NR = nitrate reductase, NiR = Nitrite reductase, GS = glutamine synthetase, GOGAT = glutamate synthase. Adapted from Schrader and Thomas (1981).

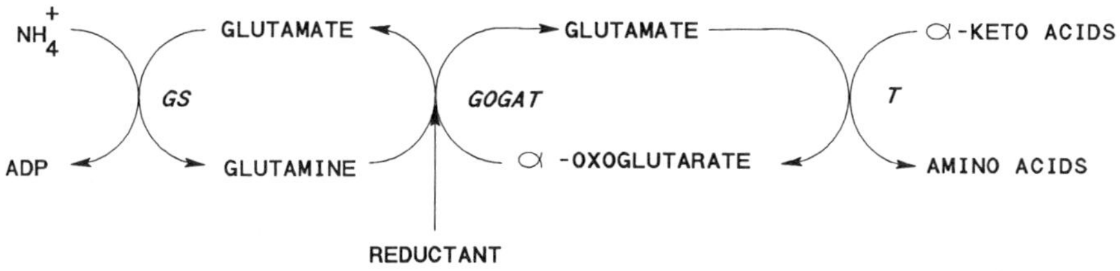

Figure 27. Assimilation of ammonium into amino acids. Abbreviations: GS = glutamine synthetase; GOGAT = glutamate synthase; T = transferases. Adapted from Schrader and Thomas (1981).

$$2\ \text{Glycine} + H_2O \longleftrightarrow \text{Serine} + CO_2 + NH_3 + 2\ H^+ + 2\ e^-$$

NH_3 produced in this manner must be reassimilated into amino acids. The NH_3 released from glycine is refixed in the glutamate synthase cycle.

Nitrogen Losses

Ammonia volatilization

When free NH_3 is near the soil surface, there is a potential for its loss (Fenn and Kissel 1976). The source of NH_3 in soils is from NH_4^+ which is supplied from organic sources such as urine or fecal material, plant residues, and native soil organic matter. Urea contributed by these sources is converted to NH_4^+ through hydrolysis as follows:

$$(NH_2)_2CO + 2\ H_2O \xrightarrow{\text{urease}} (NH_4)_2CO_3.$$

The soil complex can act as a source or sink for NH_3 (Freney et al. 1983) (Fig. 28). Whether NH_3 is lost through volatilization or not depends on the concentration gradient between the soil and the atmosphere. The usual condition is for NH_3 concentration in the atmosphere to be very low, therefore the potential for ammonia volatilization is usually extremely high. Ammonia concentration in solution of 0.5 ppm or greater usually promotes volatilization (Vlek and Craswell 1981). For that reason, ammoniacal-N added to the soil solution from any source is subject to loss. As stated earlier, atmospheric NH_3 may be reabsorbed by the plant canopy, thereby mitigating the loss.

Ammonia volatilization is affected by soil water status, pH, temperature, soil characteristics, and ammonia concentration (Farquhar et al. 1980). Because NH_3 is the most soluble gas known, there is little loss from flooded soils. Indeed, as the soil dries, volatilization losses increase. The drying process causes upward movement of

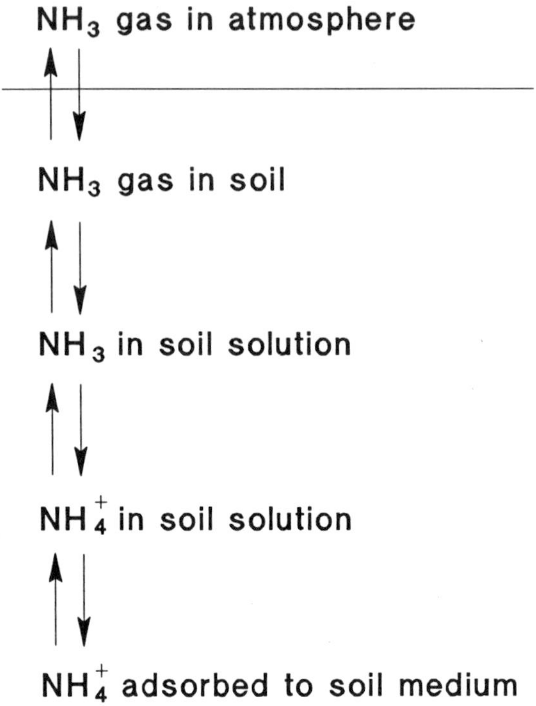

Figure 28. Source-sink relationships for ammonia in the soil-plant system. Adapted from Freney et al. (1983).

NH_3 in solution to the soil surface. As temperatures increase, so do NH_3 losses. A combination of high temperature and a dry soil can result in substantial losses. Ball and Keeney (1983) reported that NH_3 losses from urine patches averaged 5, 16, and 66% of added urine-N from cool-moist, warm-moist, and warm-dry conditions, respectively. A major factor affecting NH_3 losses is the cation exchange capacity of the soil (Freney et al. 1983). As adsorption of NH_4^+ increases, loss potential is reduced. Because NH_3 volatilization losses increase as pH increases, calcareous soils stimulate NH_3 losses. Soils with high base saturation (well-buffered) are subject to high NH_3 losses (Avnimelech and Laher 1977). Because the conversion of NH_4^+ to NH_3 increases acidity, the well-buffered soil neutralizes the acidity promoting further loss. Of course, ammonia volatilization is ultimately dependent on NH_3 concentration, with losses proportional to increased concentration. Therefore, any reduction in NH_4^+ concentration through plant uptake, nitrification, denitrification, immobilization, or clay mineral fixation will decrease NH_3 losses.

Grazing Animals

Of particular importance in rangeland ecosystems are the gaseous losses associated with excretion by grazing animals (Simpson and Steele 1983). Henzell and Ross (1973) estimated that 85-90% of ingested N was excreted in urine or feces. The localized nature of urine spots or fecal material results in a relatively high concentration of N in a small area. The average deposition on the surface in the local spot is equivalent to 500 kg N ha^{-1} (Haynes and Sherlock 1986). Urea in the urine and feces is rapidly hydrolyzed to NH_4^+ resulting in a localized patch with extremely high NH_4^+-N and a high pH, both conducive to gaseous losses of NH_3. Much of the urine may be intercepted by vegetation having high urease concentrations in the leaves, thus promoting NH_3 volatilization losses. Numerous authors have reported on losses of N from urine (Denmead et al. 1974; Ball et al. 1979;

Sherlock and Goh 1984). Their estimates of loss vary from 20 to 60%. Research reported by Sherlock and Goh (1983) indicates an immediate release of N_2O from urine patches. The mechanism for that phenomenon is unknown.

Direct Loss from Plants

Ammonia may be lost directly from plants through stomates. Usually, these losses in natural ecosystems are small, but may become important sources in senescing leaves (Farquhar et al. 1983). NH_3 lost in this way may be rapidly reclaimed by active leaves in the canopy. High temperature, low atmospheric NH_3 concentration, and high stomatal conductance all favor direct losses of NH_3 from the plant.

Denitrification

The process of reducing oxides of nitrogen to either N_2 or N_2O is denitrification. Major losses of nitrogen from the range ecosystem can be attributed to denitrification, which proceeds according to the following equation adapted from Haynes and Sherlock (1986):

$$NO_3^- \xrightarrow{NR} NO_2^- \xrightarrow{NiR} NO \xrightarrow{NoR} N_2O \xrightarrow{N_2OR} N_2$$

where NR stands for nitrate reductase, NiR for nitrite reductase, NoR for nitric oxide reductase, and N_2OR for nitrous oxide reductase.

Denitrifying bacteria have the capacity to use N oxides as electron acceptors in place of O_2 resulting in the evolution of N_2O or N_2 (Haynes and Sherlock 1986). At least 20 genera of bacteria appear to have this capacity (Payne 1973). Denitrifying bacteria are ubiquitous in surface soils. A few nitrogen-fixing bacteria also have the ability to denitrify NO_X compounds under anaerobic conditions as well. Some bacteria lack nitrate reductase, thus they are NO_2-dependent; others lack nitrous oxide reductase and yield N_2O as a terminal product. N_2O can also be produced by non-denitrifying bacteria. When grown under conditions that promote fermentation, they can produce either NH_4^+ or N_2O. The NH_4^+-oxidizing bacteria, *Nitrosomonas*, *Nitrosospira*, and *Nitrosolobus*, can produce N_2O from NH_4^+. Both biological and chemical denitrification can occur (Haynes and Sherlock 1986).

Apparently, during nitrification (oxidation of nitrogenous compounds), there is a relatively constant ratio of N_2O produced along with NO_3^- (Freney et al. 1978). Therefore, any factor promoting nitrification will promote dissimilatory denitrification. N_2O production is suppressed in the presence of oxygen, so water-saturated soils have high denitrification losses. Anaerobic microsites may occur in aerobic soils which promote denitrification even under apparent aerobic conditions. These microsites may result from localized water accumulation or as a result of intense respiratory processes which reduce oxygen supplies. In well-aerated soils, N_2O emissions are likely to originate from nitrification. High organic matter levels in soils will likely increase denitrification because they provide a substrate for heterotrophic bacteria which increases nitrification.

When nitrification is limited by some process and nitrates accumulate, then denitrification is independent of nitrification and overall denitrification is reduced

(Blackmer and Bremner 1978). The optimum pH for denitrification is from 7.0 to 8.0 (Muller et al. 1980). Lower rates of denitrification are usually found in acid soils. Low temperatures may also limit denitrification (Haynes and Sherlock 1986). Denitrification rates are low at temperatures below 10°C, but above 10°C the rates increase linearly up to 65–70°C, then rapidly decline with further temperature increases.

The rhizosphere has relatively high concentrations of carbohydrates and amino acids, and that translates into high microbial biomass. The rhizosphere is inherently low in O_2, which contributes to increased denitrification. However, active uptake of NO_3^- in the rhizosphere reduces denitrification. Therefore, active plant growth would tend to reduce denitrification losses.

Chemodenitrification appears to proceed most rapidly when there is an accumulation of NO_2 in the soil (Haynes and Sherlock 1986). In natural systems that rarely occurs. When high amounts of nitrogen fertilizer are applied, there can be accumulations of NO_2 and chemodenitrification may occur.

Other losses: Combustion losses of nitrogen from fire in natural systems depends on the completeness of the combustion process. Incomplete combustion can result in unburned litter remaining or particulate material falling back to the soil surface with nitrogen in these materials not being volatilized. Ojima et al. (1990) reported on nitrogen lost from burning tallgrass prairie in the Kansas Flint Hills. Combustion losses varied from 12.5 to 30.0 kg N ha^{-1} (63 and 89%, respectively, of aboveground biomass N) on areas burned at different sites and in different years. Sharrow and Wright (1977a) reported that 90% or more of the nitrogen in standing grass fuels was volatilized with 30 to 33 kg N ha^{-1} lost (Elwell et al. 1941; Sharrow and Wright 1977b). Immediately after fire, conditions at the soil surface likely favor ammonia volatilization because the soil surface may be hot and dry (Raison 1979; Woodmansee and Wallace 1981), but these losses have not been estimated (Seastedt and Ramundo 1990).

Leaching losses from plants and soils appear to be small in natural rangeland systems (Kilmer 1974). Seastedt and Ramundo (1990) indicated that microbes on standing dead plant material immobilize much of the inorganic-N in precipitation and provided an organic-N source as throughfall. They provided data on N concentrations in soil water and streams which indicated that the tallgrass prairie ecosystem is extremely conservative of N. Tallgrass prairie is one of the wetter rangeland ecosystems, and because no substantial losses occur in that system by leaching, we conclude that more arid rangeland ecosystems have negligible N leaching losses.

Animal tissues have higher N concentrations than plant tissues, but export of N in animal tissues from the ecosystem is relatively low in relation to other losses. Woodmansee et al. (1978) reported export losses in animal tissues from several U.S. grasslands ranged from 0.10 to 0.40 g $m^{-2}y^{-1}$, which represents less than 1% of the N in the aboveground biomass. Native hay removal represents a much greater export loss (Dyksterhuis 1961; Owensby and Anderson 1969).

Root exudation losses from plants to the soil may be substantial for some species and relatively minor for others (Klein et al. 1988), though much N lost in that manner may later be reabsorbed by the plant. The immediate effect is likely microbial immobilization of N, with nitrogen status of the system determining the extent and length of time that immobilization occurs. Biondini et al. (1988) reported that as much as 11 g N m^{-2} was exuded by blue grama roots during a 90-day simulated growth season. The form of exuded nitrogen runs the gamut of N compounds in the plant, but a high percentage of the loss comes as amino acids. Hale et al. (1971) have summarized root exudation losses and indicate that N exudation may be important in belowground plant defense, as well as enhance the ability of the plant to obtain nutrients from the soil system (Clarholm 1985; Ingham et al. 1985).

Reallocation and Storage

Because nitrogen is an element that frequently limits plant productivity in many rangeland ecosystems, its conservation and reuse within the plant becomes very important. Recycling of nitrogen stores within plants are critical; whereas, recycling of carbon stores is not nearly so important, as carbon is usually less limiting (Chapin et al. 1990). High nitrogen-use efficiency within the plant through recycling of internal nitrogenous compounds allows perennial plants to maintain high levels of productivity that could not be obtained if all nitrogen used in new growth had to be taken up by roots from the soil. Nitrogen storage acts as a buffer between asynchrony of resource supply and the demand for nitrogen in growth.

Much of the nitrogen in proteins and other organic compounds is hydrolyzed to soluble amino acids as aboveground tissues age and senesce. Amino acids may be translocated out of the senescing tissue and either reused in growth of other tissues or stored in some other part of the plant. It has been estimated that 30-70% of N in leaves is thus saved from becoming a component of litter through mobilization and translocation out of the leaf before litterfall. Estimates of retranslocation range from 0 to 90% among wild species (Goodman and Perkins 1959; Williams et al. 1978; Chapin et al. 1980, 1990). Chapin et al. (1986) estimated that from 19 to 70% of N was retranslocated from leaves of a tundra sedge (*Eriophorum vaginatum* L.) (Table 5). Other elements (Ca, Mg, etc.), and carbon compounds located in cell walls, are not nearly so mobile. Therefore, most of these elements and compounds will end up in plant litter for decomposition and mineralization processes. Few non-nitrogenous organic compounds are recycled from senescing leaves, except for some organic acids (Titus and Kang 1982; Chapin et al. 1990).

Table 5. Concentration, cost of synthesis, and retranslocation of various organic nitrogen components for leaves of the tundra sedge *Eriophorum vaginatum* (data from Chapin et al. 1986).

		Cost of Synthesis			
Component	Concentration	C	N	P	Retranslocation
	(% dry wt.)	(mg g^{-1} product)			(% of Initial pool)
Protein	10.0	650	121	0	70
Amino acid	0.7	650	121	0	19
Nucleic acid	1.0	650	155	94	68
Chlorophyll/lipoprotein	0.9	1210	63	0	63

Storage location and forms

Nitrogen can be stored in virtually all organs of the plant. Storage is only temporary in such organs as leaves and root hairs, as these structures usually have a short life span. Nitrogen in these organs must be translocated out prior to abscision if the nitrogen is to remain as part of the total plant nitrogen budget. Most longer-term storage is located in cell cytoplasm and membranes of perennating organs of the plant.

Common storage forms of N include specialized proteins, RuBisCO, amino acids (especially arginine, glutamine, asparagine), nucleic acids, and nitrate (Clark 1977; Chapin and Kedrowski 1983; Pate 1983; Chapin and Shaver 1988; Chapin et al. 1990). If these nitrogenous compounds become complexed with cell wall materials, they may no longer be mobile. For example, proteins may complex with tannins, which in turn may bind to cellulose or hemicellulose in the cell wall, making them less available.

Chapin and Shaver (1988) reported that protein was the predominant form of N storage in lichens, mosses, and stems of shrubs in the Arctic. Among most graminoids and forbs, protein was also a major form of N storage in stems, but amino acid storage was very important in several species. In leaves, 70-80% of total N may be in protein, lesser amounts in amino and nucleic acids, and small amounts in lipids.

The predominance of protein as the major N storage form in tundra shrubs (Chapin and Shaver 1988) is similar to storage patterns in evergreen and deciduous taiga trees (Chapin and Kedrowski 1983). The relative importance of amino acids and proteins as N storage forms varies among temperate herbaceous species (Pate 1983). Alkaloids, that have high N contents (e.g., caffeine, 29% N; nicotine, 17% N), may be stored in leaves. RuBisCO may function not only as a photosynthetic enzyme in evergreen leaves, but may also serve as a source of reserve N (Chiariello and Roughgarden 1984; Bloom et al. 1985). Nitrates may also be temporarily stored in leaves and often show diurnal variations in concentration. Vacuoles can serve to isolate some of these osmotically active substances. Nitrogen-rich amides (e.g., asparagine, 21% N; glutamine, 19% N) are often the transport forms of N in the plant (Bloom et al. 1985).

Plants use storage compounds that can serve multiple roles and that reduces the direct cost of storage. For example, the amino acid, proline, may serve as a principle storage molecule even though it is not a particularly nitrogen-rich molecule (12% N), but it can also serve as an osmoticum and has been linked with drought hardiness (Hsiao 1973).

Storage compounds represent high concentrations of resources that are valuable for herbivores as well as for future plant growth (Bloom et al. 1985). By their very nature, stores are usually maintained in nongrowing or slowly growing tissue. They can be protected somewhat from herbivory by location within the plant, such as belowground storage in herbaceous and grass plants. They can also be incorporated into toxic or deterrent compounds to help defend the plant against herbivory.

Seasonal Cycles

Nitrogen stores vary diurnally as well as seasonally, depending upon supply and demand within the plant. Nitrate concentrations in leaves show a diurnal pattern in many species, increasing during the night and decreasing during the day (Gebauer et al. 1984; Melzer et al. 1984; Schulze et al. 1985). During the day, there may be excess reducing power available so nitrates may be reduced with little carbon reduction. Reduced nitrogen may then be used in production of enzymes such as RuBisCO. The alternative is nitrate reduction in roots, which would be energetically more expensive (Bloom et al. 1985).

For perennial plants living in a seasonal environment, stores of nitrogen and carbon decline when growth is most rapid, and recover when growth slows or when leaf nutrients are transported to storage organs in association with the onset of senescence (Tromp 1970; Chapin et al. 1980, 1986). Current acquisitions are usually used before stores when plants have access to both, suggesting an opportunity cost for storage (Chapin et al. 1990). Seasonal storage is generally viewed as a regulated process, but the process can also be interpreted as accumulation in response to supply and demand (Clark 1977; Trlica and Singh 1979; Chapin et al. 1990). Few studies have distinguished between these two possibilities.

Nitrogen reserves may be stored in competition with growth in some plants. The biennial plant, *Arctium tomentosum* Mill., has proportional allocation of N to both growth and storage under various environmental conditions (Heilmeier et al. 1986). An increase in asparagine, arginine, and proline in the hypocotyl competed with growth as predetermined by plant genetics. Both carbon and nitrogen were allocated to the hypocotyl over a range of conditions where either carbon or nitrogen was limiting, thus indicating that storage competed directly with growth.

Chapin and Shaver (1988) found that leaf chemistry was quite similar among a number of growth forms of arctic species. The chemistry of storage organs differed among species, but the nature of stored products showed no consistent relationship among growth forms. Their study suggested that there may be less evolutionary flexibility in leaf chemistry than in storage organ chemistry. The concentration and degree of seasonal fluctuation of stored reserves generally increased from evergreen shrubs to deciduous shrubs to graminoids to forbs, which reflected the degree of woodiness or cell wall material present. Major N pools showed greatest seasonal fluctuation in rhizomes of graminoids and forbs, and less fluctuation in stems and old leaves of evergreen shrubs.

Although the major N-containing fractions tend to occur in constant proportion to one another, concentrations of specific proteins differ markedly among species and plant parts (Bloom et al. 1985). For example, leaves have a large proportion of proteins in photosynthetic enzymes; whereas, roots have proteins associated primarily with absorptive and metabolic functions.

Large quantities of N in storage organs may be required when plants begin active growth after a dormant period. This often results from high plant demand and low availability of N in the soil. With warmer soil temperatures and adequate water, microbial mineralization of organic matter results in a flush of N that may be available for plant uptake. Higher concentrations of N in young immature tissue are often observed, but this concentration is diluted with continued growth and construction of cell wall material. Compared to C_3 grasses, C_4 grasses may exhibit greater seasonal decline in N content of leaves (Coppock et al. 1983) as depicted in Fig. 29. Thus, N pools show a seasonal cycle with high levels during inactive periods or in the winter, and rapid draw-down in storage tissues to supply demands of new growth. Rapid growth is often supported more by stored nutrients than by

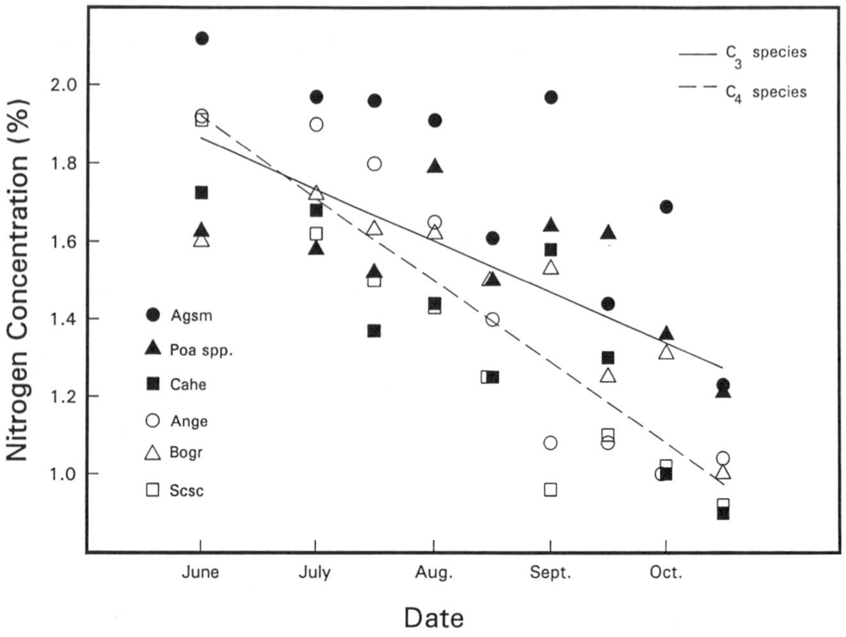

Figure 29. Seasonal course of leaf nitrogen concentration in some C_3 and C_4 grasses. Adapted from Coppock et al. 1983.

concurrent absorption (Tromp 1970; Clark 1977; Chapin 1980). The spring increase in leaf nutrient capital coincides with decreasing pool sizes of N in stems and large roots. These storage organs are gradually replenished during the summer by absorption from the soil and in autumn by translocation from senescing leaves (Greenwood 1976; Chapin 1980).

Evergreen species retain leaf nutrients *in situ* during winter rather than translocating them to stems and roots. In contrast to deciduous species, seasonal decline in leaf nitrogen in evergreen species results entirely from dilution by additional cell wall construction (Chapin 1980). In the following years, these evergreen leaves may have stable concentrations and pool sizes of N, or may have temporary gains as a result of luxury consumption during periods of inactive growth (Mooney and Rundel 1979). Leaf and shoot growth of evergreens begins earlier and is usually slower than that of deciduous perennial species. This growth might then be supported by direct absorption from the soil or from short-term storage in leaves. Thus, leaves appear to be at least as important as roots or stems for N storage in evergreen species (Chapin 1980).

Effects of Nitrogen Availability

Reserves are often allocated to root growth at the expense of shoot growth if soil nitrogen availability is low (Chapin 1980). A large decline in availability of a nutrient can cause a 1.5- to 12-fold decrease in shoot:root ratio. A low shoot:root ratio commonly found in range plants from infertile habitats is, in part, a phenotypic response to reduced nitrogen availability. However, the mechanism by which shoot:root ratio decreases in response to reduced nutrient availability remains unclear.

Plants from infertile habitats commonly have lower annual nutrient requirement as a result of increased root and leaf longevity and low relative growth rates (Chapin 1980). Increased leaf longevity of graminoids and evergreen trees and shrubs under moderate nutrient stress has been documented. However, severe nutrient stress induces leaf senescence in all plants and may cause death of meristems.

Production of chlorophyll and RuBisCO may be reduced if nitrogen is deficient in leaves, resulting in a decreased capacity for CO_2 fixation. Respiration may also decline, as well. A larger proportion of leaf protein is associated with maintenance and structure, which may explain why the efficiency of N use for photosynthesis decreases at low-leaf nitrogen concentrations (Bloom et al. 1985). Because nitrogen is also involved in other metabolic processes of growth, cell number and size may decline when N stores are not sufficient to meet growth demands.

As nitrogen availability increases, either from absorption or from stores, rapid growing species from fertile sites, particularly C_4 species, respond with greatly increased growth rates. However, species from infertile sites usually show less growth response to increased N availability, but tissue concentrations often increase (Chapin 1980). These increased tissue concentrations are frequently in the form of protein and amino acid reserves. Chapin et al. (1986) found that amino acid N (specifically arginine) increased in stems of *Eriophorum vaginatum* when excess N was available. Thus, slowly growing species that absorb nitrogen in excess of immediate growth requirements (luxury consumption) during flushes in nutrient availability may later use these reserves to support growth after soil nutrients in the root environment have been depleted (Grime 1979). Luxury consumption could be disadvantageous if it leads to accumulations of compounds or osmoticants that might become toxic to the plant or interfere with growth. Luxury consumption may also make plants toxic to grazing animals.

Effects of Environmental Factors

Environmental conditions that lead to rapid plant growth (i.e., adequate soil water, nutrients, and warm temperatures) should cause an initial large decline in nitrogen reserves. However, these same conditions should also provide for rapid photosynthesis, production of aboveground tissue, and eventual repayment of N reserves once demands have been met. When leaves senesce prior to dormancy, the N reserves should accumulate again in storage tissue.

Low ambient temperatures relative to growth optima or droughty conditions will usually result in reduced aboveground growth. With a decrease in aboveground demand, N reserves in aboveground tissues, roots, and rhizomes may increase as N uptake may exceed demand. This increase in N reserves may result in a plant that is also more cold or drought tolerant. High temperatures can lead to protein denaturization and, thus, a decline in N stores.

Defoliation may result in a rapid increase in shoot nitrogen concentration a few days following defoliation (Chapin et al. 1986; Walker et al. 1987; Jaramillo and Detling 1988). That nitrogen may come from storage, uptake, or a combination of both. Chapin et al. (1986) reported that clipping of *Eriophorum vaginatum* at 10-

day intervals resulted in increases in all stem N fractions (particularly arginine) and increases in all leaf-blade N fractions, except protein N. These responses were similar to those when plants had received added N fertilizer. Jaramillo and Detling (1988) found that blue grama had increased N uptake per unit of root biomass when plants were defoliated. They proposed that reduced retranslocation of nutrients from shoots to roots following defoliation may constitute a "signal" for increasing the rate of N uptake by roots. Leaf blades of defoliated plants had more N allocated to them than other plant parts. Decreased N immobilization and increased net N mineralization, as a result of decreased root carbon inputs to decomposer organisms in grazed grasslands, may explain why grazing induces an increase in shoot N concentration (Holland and Detling 1990).

Carbon and Nitrogen Interactions

Carbon and nitrogen interact in ways that influence plant morphology and resource-use efficiency. The abundance of N in the soil affects the manner in which carbon is allocated to competing sinks depicted in Fig. 1. This, in turn, may have significant effects on water use, nitrogen use, forage production, and forage quality. Examples of such interactions will be presented in this section.

Shoot:Root Ratio

Shoot:root ratio is defined for a particular period of growth and is determined by dividing the dry weight of shoots by the dry weight of roots produced per plant (or unit area) during that period. Fitter and Hay (1981) concluded that this ratio is very plastic and decreases with decreasing soil water, nutrients, oxygen, and temperature. In general, plants tend to put more of their resources into root production in stressful environments. Species also differ in shoot:root ratio and the ratio for a particular species undergoes changes related to physiological stage of development (Boote 1977).

Nitrogen, in particular, has profound effects on carbon allocation patterns through its effect on the internal energy and reserve carbohydrate relations of the plant (Meyer et al. 1973). If the nitrate concentration of the rooting media is low, most of the nitrates absorbed are utilized in synthesis of amino acids in the roots. These amino acids are then converted into protoplasmic proteins. Required carbohydrates are translocated downward from the leaves and only a small proportion of the available nitrogenous compounds may be translocated upward to the aerial portions of the plant. As a result, the aboveground portion of the plant may be relatively deficient in protein and growth rate of shoots will be slow in relation to that of roots. That causes the shoot:root ratio to decline. The model of Levin et al. (1989) showed that the fraction of total plant biomass that is shoot (shoot biomass/total plant biomass) tends asymptotically to a linear function of N and that this relationship holds until N-supply becomes very large.

Plants are apparently efficient scavengers of available N when it is limiting. Results of forest studies (Vitousek et al. 1982) indicated that more N was translocated from senescing leaves on sites of low fertility compared with those grown on fertile soils. Therefore, the C:N ratio of the leaf litter increased on the infertile site.

A higher C:N ratio in litter was associated with reduced N-mineralization rates and thus limited N availability to plants. If forage plants behave similarly, protein content of cured or dormant forage might be expected to be relatively low as a result of low soil fertility and additional retranslocation of N from leaves before death.

Defoliation can affect nitrogen nutrition, carbon allocation, and hence shoot:root ratio. The direct removal of shoot tissue causes an obvious and immediate effect on shoot:root ratio and a change in carbon allocation favoring replacement of the photosynthetic canopy at the expense of root growth. Temporary cessation of root growth results as the plant mobilizes reserves to replace lost photosynthetic tissue (Crider 1955; Oswalt et al. 1959).

Polley and Detling (1989) studied the effects of defoliation, nitrogen, and competition on growth and nitrogen nutrition of western wheatgrass in controlled environments. Defoliation reduced biomass yields and increased aboveground relative to belowground investment of total biomass and N. This occurred whether western wheatgrass was grown in competition with other western wheatgrass neighbors or with *Artemisia frigida* Willd. Defoliation increased N accumulation per unit root biomass of western wheatgrass when grown under intraspecific-, but not under interspecific-competition. Defoliation and high nitrogen addition rates reduced nitrogen-accumulation efficiency (N uptake/N available) under both types of competition.

Coughenour et al. (1990) found total aboveground production of the dwarf shrub *Indigofera spinosa* Forsk. was not affected by clipping, but final root biomass was reduced 17% by clipping 2/3 of the new leaves, stems, and apical meristems every month. In spite of the negative effect on root growth, total N uptake by individual plants was stimulated by defoliation as shown by higher total N in the leaves and stems compared with unclipped plants. Stimulation of N uptake by defoliation resulted from increased uptake per unit of root. Similar results were reported for blue grama (Jaramillo and Detling 1988).

Water-Use Efficiency and Drought Performance

Water-use efficiency (WUE) may be defined on the basis of leaf gas exchange (CER/E), where CER is CO_2 exchange rate and E is transpiration. WUE may also be defined as dry matter produced per unit of water used. Generally, water used includes combined losses attributable to both transpiration and evaporation from the soil and dry matter production may include only the aboveground, harvestable portion of the plant. A physiological basis for differences in WUE has already been discussed relative to the carbon pathways of photosynthesis.

Effects of N on WUE of forages have been reviewed by Rogler and Lorenz (1974). Applications of 45 to 90 kg N ha^{-1}, with or without P, increased WUE of crested wheatgrass and native range in northeastern Montana by a factor of 1.5 to 2.5 (Black 1968). In addition, total soil water extraction was increased by N fertilization. Increases in the WUE of native rangelands were also reported by Johnston et al. (1968) at five locations across southern Alberta in response to fertilization with N and P. WUE responded positively to N (57% increase) and P (11% increase) alone, but the response was greater when N and P were applied together (120% increase). Other workers have found similar relationships between WUE and N fertilization in several range plant communities (Smika et al. 1965; Owensby et al. 1970). While forage production is strongly controlled by available water and nitrogen, Frank and Ries (1990) found that morphological development of crested and western wheatgrass was a function of air temperature as measured by accumulated heat units or growing degree days. Development was similar across water and nitrogen treatments. This finding is significant because it makes it easier to model and predict the rate of forage plant development from air temperature alone. Prediction of the stage of plant development is often an important input for management decisions.

Enhancement of WUE by N may be realized even without fertilization. Owensby et al. (1970) measured WUE on loamy upland bluestem range fertilized with 56 kg N ha^{-1} annually for 4 years (Table 6). The high WUE in 1967 on unfertilized plots (equal to that of fertilized plots) was attributable to nitrogen carried over from 1966 as a result of relatively low precipitation during the 1966 growing season. Greater root exploration of the soil mass, stimulated by added nitrogen, may account for much of the increased WUE (McKell et al. 1962). Lorenz and Rogler (1966) found increased root weights with nitrogen fertilization of native rangeland in the northern Great Plains. They hypothesized that increased root exploration enhanced both WUE and nitrogen recovery.

Coughenour et al. (1990) examined the effect of the interaction between defoliation and water stress on production of *Indigofera spinosa*, a dwarf shrub. Total N yields (concentration x biomass) were reduced less than biomass yields by clipping or infrequent watering primarily because N accumulation per unit of root mass was elevated under both greater clipping intensities and less frequent watering. The least water-stressed plants were affected most negatively and the unclipped plants responded most negatively to greater water limitation. They stated that this could be interpreted as an ameliorating effect of one stress upon the response to the other. Under drought conditions, negative consequences of grazing may be reduced. The obvious effect of grazing is to immediately reduce the amount of transpiring surface area relative to root absorbing area. This may enable the plant to maintain a more favorable water status when soil water is limiting.

While adequate soil fertility, especially with regard to N, may promote efficient use of water, it may also result in greater total water use (Owensby et al. 1970). Therefore, plants may still need to limit water loss in order to maintain growth longer into a drying period. Even though plants may reduce their rate of metabolism during drought in order to survive, the grazing animal's immediate needs depend on forage production, not plant survival. Because a large portion of the

Table 6. Effect of nitrogen fertilizer on the water-use efficiency of plants from a loamy upland bluestem range over a four-year period (data from Owensby and Anderson 1969). Within year means are significantly difference ($P<0.05$) if followed by a different letter.

	Water-use efficiency (kg dry matter mm^{-1} water)			
Treatment	1965	1966	1967	1968
Unfertilized control	1.8a	2.1a	3.3a	2.0a
Fertilized (56 N kg ha^{-1})	2.8b	2.3b	3.3a	2.9b

world's rangelands are in arid or semi-arid climates, a degree of drought resistance in range plant species is generally essential to success in these environments.

Drought resistance may consist of both avoidance and tolerance mechanisms (Levitt 1972). In this terminology, avoidance implies the ability to resist coming into thermodynamic equilibrium with the water stress (i.e., the plant cells do not dehydrate). Examples of mechanisms to avoid drought include reduction of transpiration by partial stomatal closure or increased water uptake by exploiting a greater soil volume as a result of enhanced root growth. Tolerance implies the ability to come into thermodynamic equilibrium with the water stress without being killed. In drought-tolerant plants, the plants cells dehydrate, but the proteins, cell membranes, and organelles retain their functional ability in the dehydrated state or else regain it upon rehydration.

Drought resistance implies an ability to survive, endure, compensate for, or escape damage from wilting (Reitz 1974). Greater drought resistance does not necessarily translate into greater WUE (Sullivan and Eastin 1974). This is because high WUE depends not only on reduction of transpiration, but also on maintenance of photosynthesis.

Nitrogen-Use Efficiency

As we have seen, environmental conditions (through their effect on nutrient cycling), soil type, and fertilizer can all influence nitrogen supply to the plant. The efficiency with which plants absorb and use available nitrogen affects forage production, forage quality, and water use. This efficiency may vary with species, environmental stress, and management. Birk and Vitousek (1986) compared various definitions of nitrogen-use efficiency (NUE) and their applicability to forests stands. For example, NUE may relate carbon uptake or biomass production to the amount of N taken up (Gerloff 1976) or to the available N (Shaver and Melillo 1984). Production per unit of N uptake in short-lived plants is the inverse of N concentration in the plant tissue (Chapin 1980). However, in perennial plants, much of the annual N requirement is met by N resorption from senescing tissue (Switzer and Nelson 1972; Cole 1981). Consequently, Birk and Vitousek concluded that N concentrations alone are an inadequate or even misleading index of NUE in forest stands. Because of their similar growth habit, the same reasoning may apply to rangeland shrubs.

Birk and Vitousek (1986) studied the effect of nitrogen availability on N cycling and NUE in natural and fertilized stands of loblolly pine (*Pinus taeda* L.) on the upper coastal plain of South Carolina. Increased N availability increased the mass of litterfall and concentrations of N in litterfall. This was associated with reduced NUE in the high N sites. Net production per unit N of pine seedlings was significantly higher in N-limited plants. Their results showed that changes in NUE could occur within individual species in limited geographical areas and across broad-scale ranges of species and site conditions. The negative relationship between nitrogen availability and NUE was also reported by Pastor et al. (1984) for several hardwood tree species.

While adding fertilizer N has been shown to increase WUE, it may decrease NUE. Nichols et al. (1990) reported that NUE of subirrigated meadow vegetation in the Nebraska Sandhills decreased as N fertilization rates were increased, but that NUE increased when P and S were applied alone or in combination. Their results, regarding N effects on NUE, were qualitatively similar to others reported for native meadows in Oregon (Gomm 1979), the northern Rocky Mountains (Ludwick 1979), and the intermountain region of the western U.S. (Taylor et al. 1985)

NUE can be influenced by genetic factors. Alagarswamy et al. (1988) examined NUE and photosynthetic nitrogen efficiency (PNE in umol CO_2 g^{-1} N s^{-1}) under highly-controlled conditions in four pearl millet [*Pennisetum glaucum* (L.) R. Br.] genotypes known to differ in NUE. Genotypes with high NUE had thicker leaves [higher specific leaf weight (SLW), i.e., biomass investment per unit leaf area] and were generally less responsive to available N compared to N-inefficient genotypes as measured by leaf N concentrations and leaf CO_2 exchange rates. There were small differences in PNE, but those offered little explanation for overall differences in NUE. They hypothesized that differences in NUE were related to the manner in which the plants partitioned available N resources into additional leaf area.

The C_4 pathway of photosynthesis has been hypothesized to allow for more efficient use of N in CO_2 assimilation and dry matter production than the C_3 pathway. However, in their comparison of six *Panicum* species with differing CO_2 fixation pathways, Brown and Wilson (1983) concluded that NUE was not necessarily higher in the C_4 species and that any advantage conferred by C_4 metabolism may be negated by morphological or other leaf characteristics, such as SLW, leaf area expansion, and partitioning of dry weight to stems.

Management Implications

Low soil N not only has the potential to reduce forage production by shifting carbon allocation to favor roots, but can also reduce forage quality and WUE. It is seldom economically feasible to fertilize native rangeland. Even if it were, fertilizer may produce undesirable changes in botanical composition and would likely reduce NUE. Therefore, low forage protein on native rangelands is generally mitigated by supplementing the animal. More options exist in tame pastures for managing N and water and, therefore, carbon allocation and forage quality. Soil nitrogen carryover following a dry year has been documented in Kansas rangelands (Owensby et al. 1970; Launchbaugh and Owensby 1978). As a result of reduced N use during drought, forage production the year following drought was increased relative to that of other years of similar precipitation. That suggests if N fertilizer is a management input, applications rates may need to be reduced or eliminated in years following drought in order to take advantage of the N carryover.

Adding water and N are generally not options available to managers of rangeland ecosystems for the purpose of promoting forage production and quality. Knowledge of carbon-nitrogen interactions and their effect on water availability and use is, however, relevant to decisions associated with season and intensity of grazing and animal supplementation. While we can demonstrate a qualitative understanding of how carbon, nitrogen, and water may interact to influence management decisions, our quantitative understanding and ability to generalize to field conditions, plus our control over these variables as natural inputs, is quite limited.

Conclusions

The Known

A large quantity of data and information has been obtained in the past 20 to 30 years on how plants assimilate carbon dioxide. We presently know how the three photosynthetic pathways operate, the various advantages and disadvantages of these pathways in different environments, and factors that influence carbon gains and losses. Much is known about light absorption, energy transfer, and carbon biosynthesis in plants.

We have come to recognize the importance of carbon allocation within the plant and the consequences of that. A relatively minor change in allocation to either above- or belowground growth can have long-term consequences for overall plant productivity and survival. If a plant allocates too much carbon to aboveground structure and function, it is at greater risk from soil water stress. If a plant allocates too much carbon to roots, then leaf area may be insufficient to support photosynthesis at a level that will make it competitive in plant communities. These trade-offs are now recognized, and plants must be good environmental integrators to optimize carbon allocation to various above- and belowground structures and functions.

Secondary metabolites were once thought to be waste products produced by plants at considerable energetic cost. We now recognize that most of these secondary compounds can be differentially produced and may act as deterrents to herbivory. Plants that produce these chemicals may be afforded some degree of protection from defoliation by insects, microarthropods, or ungulates. Hence, there is some survival value for production of these energy-rich compounds that makes the plant more fit to its environment.

Defoliation by grazing animals is only one in a huge array of abiotic and biotic factors that can place range plants under stress. We have learned that defoliation can act antagonistically or synergistically with other environmental stress factors, resulting in plant responses that may not have been predicted if only one factor (defoliation) were considered alone. Aboveground defoliation removes leaf area, alters hormone production and distribution, affects CO_2 assimilation and respiration, influences carbon and nitrogen allocation patterns, and even affects exudation and microbial interactions. Therefore, defoliation elicits a very complex set of physical and biochemical responses in plants that has only recently become better understood. Compensatory growth after defoliation can occur, but rarely will this growth, plus what was removed, equal what is produced by the non-defoliated range plant. Because the impact of some grazing is under the control of man, we must use our knowledge of defoliation responses of plants in such a way that we do not jeopardize long-term productivity and stability of range ecosystems. Grazing animals must be managed, as man has little control over most other environmental stress factors.

The Unknown

Although we have learned a great deal about growth of range plants during the past two decades, a number of questions remain. In fact, most of the questions asked by Mooney (1972) remain unanswered today. We still know very little about root growth, structure, and function. For example, why should plants lose a considerable amount of energy to root exudation? Are the losses of these energy-rich compounds to the microflora repaid through enhanced mineralization and uptake of nutrients that are in short supply?

Other interesting questions remain unanswered concerning root growth. Why do roots stop growing almost immediately after aboveground defoliation? What are the mechanisms involved in stoppage and promotion of root growth? What causes plants of arid environments to invest more carbon in roots than do those of more mesic habitats? Why are roots of defoliated plants more capable of nitrogen uptake than are those of undefoliated plants? As is evident, a number of significant questions remain to be answered concerning the belowground arena. Most studies, so far, have been of a descriptive nature, but the time has come to address the interesting, but more difficult questions, concerning mechanisms involved in root growth, development, and death.

A number of interesting questions also await us concerning the "what" and "why" of aboveground plant growth. Why do defoliated plants remain in a vegetative stage rather than investing in reproductive structures? What causes carbon allocation to be greater in aboveground plant parts than in roots following defoliation? What causes production of secondary compounds in response to herbivory? What level of reserves is needed to insure plant survival? Is resource sharing in clonal plants desirable for long-term sustainability in plant production? And so it goes. As Mooney (1972) stated, "It is likely that through a quantitative understanding of how different plants gain and allocate their resources, it will be possible to make predictions as to their success in any given physical environment in combination with any competitor and predator." And after all, isn't it time for range managers to become more "predictive" than "descriptive" in their management of range resources?

Future Research Needs

Many of the questions that should now be addressed require sophisticated and expensive equipment and facilities. Answers to these more basic questions will require future investments that far exceed those of the past where simple procedures and equipment could be used to great advantage. This will require that research organizations and institutions recognize this need and then put greater emphasis and funds into capital equipment and research laboratories. The payoff should be in the rapid advancement in understanding of range plant physiology.

Another concern is that we address complex environmental interactions that affect plant growth and development. We should concentrate efforts to better understand synergistic and antagonistic effects that various environmental variables have on range plants. For example, what results might be predicted when range plants are heavily utilized during a drought? Will this increase or decrease the plant's competitive abilities and survival in the post-drought period? Another question is how far can we manipulate a plant to increase aboveground production at the expense of belowground root growth? This obviously would impact both above- and belowground herbivory.

Future range researchers should also search for plant materials that can efficiently utilize limited water and nitrogen resources so common on rangelands. Or, can genetic engineering produce microorganisms that can fix nitrogen and increase mineralization in range ecosystems to alleviate nitrogen shortages and is this even desirable? We should also determine the timing and extent of foliage utilization rangeland species can tolerate under varying and changing climatic conditions.

What effect will climatic change, such as increasing CO_2 concentrations, have on productivity, persistence, competition, and herbivory interactions of rangeland plants? Will C_3 species be favored over C_4 species because of the lower affinity of RuBisCO for CO_2 compared to PEPC? Or, will increasing CO_2 accompanied by decreasing precipitation actually favor the more water-use efficient, warm-season, C_4 species (Hunt et al. 1991)? Interesting questions such as these await critical experiments designed and executed by range physiological ecologists.

We know a great deal about the qualitative implications of the interactions among environmental, physiological, and morphological factors on management options for rangelands. What is sorely needed is a quantitative and more mechanistic understanding of these relationships that will allow us to predict responses to management alternatives *a priori*. We also need to transfer technology to different environments and management systems without having to repeat all the research. The sciences of systems simulation modeling and expert systems have advanced rapidly and perhaps offer the greatest promise for significant advancement of range management based on a mechanistic understanding of plant processes.

Rangeland systems are complex and so, therefore, is their management. The easy research has been done and often by individual scientists working alone or in small groups. This is not to imply that all research that has been done has been easy. Nevertheless, future advances will most likely come from integration of team research from multiple disciplines in a modeling context. Stated another way, we know a great deal about the main effects of specific factors in rangeland systems, but it is the interactions that drive these systems and hence their management. Understanding those will demand application of state-of-the-art technology.

Acknowledgments

Contribution No. 92-92B, Agricultural Research Center–Hays and Department of Agronomy, Kansas Agricultural Experiment Station, Kansas State University. Critique of this chapter by Dr. A.B. Frank, USDA/ARS, Mandan, ND, and Dr. J.K. Detling, Colorado State University, is gratefully acknowledged.

Literature Cited

Abrahamson, W.G. 1979. Patterns of resource allocation in wildflower populations of fields and woods. Amer. J. Bot. 66:71-79.

Akin, D.E., and D. Burdick. 1977. Rumen microbial degradation of starch-containing bundle sheath cells in warm-season grasses. Crop Sci. 17:529-523.

Alagarswamy, G., J.C. Gardner, J.W. Maranville, and R.B. Clark. 1988. Measurement of instantaneous nitrogen use efficiency among pearl millet genotypes. Crop Sci. 28:681-685.

Allessio, M.L., and L.L. Tieszen. 1978. Translocation and allocation of ^{14}C-photoassimilate by *Dupontia fisheri*, p. 393-413. *In* L.L. Tieszen [ed.] *Vegetation and production ecology of an Alaskan arctic tundra*. Ecol. Stud. Vol. 29. Springer-Verlag, New York.

Alpert, P. 1991. Nitrogen sharing among ramets increases clonal growth in *Fragaria chiloensis*. Ecology 72:69-80.

Amthor, J.S. 1989. Respiration and crop production. Springer-Verlag, New York. 215 p.

Anderson, K.L., C.E. Owensby, and E.F. Smith. 1970. Burning bluestem range. J. Range Manage. 23:81-92.

Archer, S., and L.L. Tieszen. 1980. Growth and physiological response of tundra plants to defoliation. Arctic and Alpine Res. 12:532-552.

Ares, J. 1976. Dynamics of the root system of blue grama. J. Range Manage. 29:208-213.

Avnimelech, Y., and M. Laher. 1977. Ammonia volatilization from soils: equilibrium considerations. Soil Sci. Soc. Am. J. 41:1080-1084.

Balasubramanian, A., and G. Rangaswami. 1969. Studies on the influence of foliar nutrient sprays on the root exudation pattern in four crop plants. Plant Soil 30:210-220.

Ball, P.R., and D.R. Keeney. 1983. Nitrogen losses from urine-affected areas of New Zealand pasture under contrasting seasonal conditions. Proc. Int. Grassl. Congr. 14th. 1981. pp. 342-344.

Ball, P.R., D.R. Keeney, P. W. Theobald, and P. Nes. 1979. Nitrogen balance in urine-affected areas of a New Zealand pasture. Agron. J. 71:309-314.

Barber, D.A., and K.B. Gunn. 1974. The effect of mechanical forces on the exudation of organic substances by the roots of cereal plants grown under sterile conditions. New Phytol. 73:39-45.

Barber, D.A., and J.K. Martin. 1976. The release of organic substances by cereal roots into soil. New Phytol. 76:69-80.

Bazzaz, F.A. 1990. The response of natural ecosystems to the rising global CO_2 levels. Ann. Rev. Ecol. Syst. 21:167-196.

Bazzaz, F.A., N.R. Chiariello, P.D. Coley, and L.F. Pitelka. 1987. Allocating resources to reproduction and defense. BioScience 37:58-67.

Bidwell, R.G.S. 1974. Plant physiology. MacMillan Publishing Co., New York. 643 p.

Biondini, M., D.A. Klein, and E.F. Redente. 1988. Carbon and nitrogen losses through root exudation by *Agropyron cristatum*, *A. smithii*, and *Bouteloua gracilis* and their potential contributions to N cycling. Soil Biol. Biochem. 20:477-482.

Birk, W.M., and P.M. Vitousek. 1986. Nitrogen availability and nitrogen use efficiency in loblolly pine stands. Ecology 67:69-79.

Biscoe, T.V., J.N. Gallagher, E.J. Littleton, J.L. Monteith, and R.K. Scott. 1975. Barley and its environment. IV. Sources of assimilate for the grain. J. Appl. Ecol. 12:295-318.

Bjorkman, O. 1971. Comparative photosynthetic CO2 exchange in higher plants, p. 18-32. *In* M.D. Hatch, C.B. Osmond, and R.O. Slatyer [ed.] *Photosynthesis and photorespiration*. Wiley-Interscience, New York.

Bjorkman, O. 1981. The response of photosynthesis to temperature, p. 273-301. *In* J. Grace, E.D. Ford, and P.G. Jarvis. [ed.] *Plants and their atmospheric environment*. Blackwell Scientific Publications, London.

Bjorkman, O. 1989. Some view points on photosynthetic response and adaptation to environmental stress, p. 45-58. *In* W.R. Briggs [ed.] *Photosynthesis*. Allen R. Liss, New York.

Bjorkman, O., M. Nobs, R. Pearcy, J. Boynton, and J. Berry. 1971. Characteristics of hybrids between C_3 and C_4 species of *Atriplex*, p. 105-119. *In* M.D. Hatch, C.B. Osmond, and R.O. Slatyer [ed.] *Photosynthesis and photorespiration*. Wiley-Interscience, New York.

Black, A.L. 1968. Nitrogen and phosphorus fertilization for production of crested wheatgrass and native grass in northeastern Montana. Agron. J. 60:213-216.

Blackmer, A.M., and J.M. Bremner. 1978. Inhibitory effect of nitrate on reduction of N_20 to N_2 by soil microorganisms. Soil Biol. Biochem. 10:187-191.

Bloom, A.J., F.S. Chapin, III, and H.A. Mooney. 1985. Resource limitation in plants—an economic analogy. Ann. Rev. Ecol. Syst. 16:363-392.

Boardman, N.K. 1971. The photochemical systems in C_3 and C_4 plants, p. 309-322. *In* M.D. Hatch, C.B. Osmond, and R.O. Slatyer [ed.] *Photosynthesis and photorespiration*. Wiley-Interscience, New York.

Bohm, W. 1979. Methods of studying root systems. Springer-Verlag, Berlin.

Bokhari, U.G., and J.S. Singh. 1974. Effects of temperature and clipping on growth, carbohydrate reserves, and root exudation of western wheatgrass in hydroponic culture. Crop Science 14:790-794.

Bolhar-Nordenkampf, H.R. 1985. Shoot morphology and leave anatomy in relation to photosynthesis, p. 107-117. *In* J. Coombs, D.O. Hall, S.P. Long, and J.M.O. Scurlock [ed.] *Techniques in bioproductivity and photosynthesis*, 2nd ed. Pergamon Press, New York.

Boote, K.J. 1977. Root:shoot relationships. Soil and Crop Sci. Soc. Florida Proc. 36:15-23.

Bowen, G.D., and A.D. Rovira. 1973. Are modelling approaches useful in rhizosphere biology? Bull. Ecol. Res. Comm. 17:443-450.

Bowen, G.D., and A.D. Rovira. 1976. Microbial colonization of plant roots. Ann. Rev. Phytopath. 14:121-144.

Bowen, G.D., and S.E. Smith. 1981. The effects of mycorrhizas on nitrogen uptake by plants, p. 405-426. *In* F.E. Clark and T. Rosswall [ed.] *Terrestrial nitrogen cycles: processes, ecosystem strategies and management impacts*. Ecolog. Bull., Stockholm.

Brazel, F. K., L. H. Harbers, and C. E. Owensby. 1979. Structural inhibitors of big and little bluestem digestion observed by scanning electron microscopy. J. Am. Sci. 48:1457-1463.

Breen, P.J., J.D. Hesketh, and D.B. Peters. 1986. Field measurements of leaf photosynthesis of C_3 and C_4 species under high irradiances and enriched CO_2. Photosynthetica 20:281-285.

Brizuela, M.A., J.K. Detling, and M.S. Cid. 1986. Silicon concentration of grasses growing in sites with different grazing histories. Ecology 67:1098-1101.

Brown, L.F., and M.J. Trlica. 1977. Interacting effects of soil water, temperature, and irradiance on CO_2 exchange rates of two dominant grasses on the shortgrass prairie. J. Appl. Ecol. 14:197-204.

Brown, R.H., and J.R. Wilson. 1983. Nitrogen response of *Panicum* species differing in CO_2 fixation pathways. II. CO_2 exchange characteristics. Crop Sci. 23:1154-1159.

Bryant, J.P., F.S. Chapin, III, and D.R. Klein. 1983. Carbon/nutrient balance of boreal plants in relation to herbivory. Oikos 40:357-368.

Bull, T.A. 1971. The C_4 pathway related to growth rate in sugarcane, p. 68-75. *In* M.D. Hatch, C.B. Osmond, and R.O. Slatyer [ed.] *Photosynthesis and photorespiration*. Wiley-Interscience, New York.

Burns, R.C., and R.W.F. Hardy. 1975. Nitrogen fixation in bacteria and higher plants. Springer-Verlag, New York. 189 p.

Buwai, M., and M.J. Trlica. 1977. Defoliation effects on root weights and total nonstructural carbohydrates of blue grama and western wheatgrass. Crop Sci. 17:15-17.

Caldwell, M.M. 1976. Root extension and water absorption, p. 63-85. *In* O.L. Lange, L. Kappen, and E.-D. Schulze [ed.] *Water and plant life*. Springer-Verlag, Berlin.

Caldwell, M.M. 1979. Root structure: the considerable cost of the belowground function, p. 408-427. *In* O.T. Solbrig, S. Jain, G.B. Johnson, and P.H. Raven [ed.] *Topics in plant population biology*. Columbia Univ. Press, New York.

Caldwell, M.M. 1984. Plant requirements for prudent grazing, p. 117-152. *In Developing strategies for rangelend management*. National Research Council/National Academy of Sciences. Westview Press, Boulder, Colorado.

Caldwell, M.M. 1987. Competition between root systems in natural communities, p. 167-185. *In* B.J. Gregory, J.V. Lake, and D.A. Rose [ed.] *Root development and function*. Cambridge Univ. Press, Cambridge.

Caldwell, M.M., J.H. Richards, D.A. Johnson, R.S. Nowak, and R.S. Dzurek. 1981. Coping with herbivory: photosynthetic capacity and resource allocation in two semiarid *Agropyron* bunchgrasses. Oecologia 50:14-24.

Caldwell, M.M., and R.A. Virginia. 1989. Root systems, p. 367-398. *In* R.W. Pearcy, J. Ehleringer, H.A. Mooney, and P.W. Rundell [ed.] *Plant physiological ecology*. Chapman and Hall, New York.

Canvin, D.T. 1981. Photorespiration and nitrogen metabolism, p. 178-194. *In* J.D. Bewley [ed.] *Nitrogen and carbon metabolism*. Martinus Nijhoff/Dr. W. Junk Publishers, The Hague.

Caraco, T., and C.K. Kelly. 1991. On the adaptive value of physiological integration in clonal plants. Ecology 72:81-93.

Cavagnaro, J.B. 1988. Distribution of C_3 and C_4 grasses at different altitudes in a temperate arid region of Argentina. Oecologia 76:273-277.

Chapin, F.S., III. 1980. The mineral nutrition of wild plants. Ann. Rev. Ecol. Syst. 11:233-260.

Chapin, F.S., III. 1991. Integrated responses of plants to stress: a centralized system of physiological responses. BioScience 41:29-36.

Chapin, F.S., III, A.J. Bloom, C.B. Field, and R.H. Waring. 1987. Plant responses to multiple environmental factors. BioScience 37:49-57.

Chapin, F.S. III, D.A. Johnson, and J.D. McKendrick. 1980. Seasonal movement of nutrients in plants of differing growth forms in an Alaskan tundra ecosystem: implications for herbivory. J. Ecol. 68:189-209.

Chapin, F.S., III, and R.A. Kedrowski. 1983. Seasonal changes in nitrogen and phosphorus fractions and autumn retranslocation in evergreen and deciduous taiga trees. Ecology 64:373-391.

Chapin, F.S., III, E.-D. Schulze, and H.A. Mooney. 1990. The ecology and economics of storage in plants. Ann. Rev. Ecol. Syst. 21:423-447.

Chapin, F.S., III, and G.R. Shaver. 1985. Individualistic growth responses of tundra plant species to environmental manipulations in the field. Ecology 66:564-576.

Chapin, F.S., III, and G.R. Shaver. 1988. Differences in carbon and nutrient fractions among arctic growth forms. Oecologia 77:506-514.

Chapin, F.S., III, G.R. Shaver, and R.A. Kedrowski. 1986. Environmental controls over carbon, nitrogen, and phosphorus chemical fractions in *Eriophorum vaginatum* L. in Alaskan tussock tundra. J. Ecol. 74:167-195.

Charles-Edwards, D.A., D. Doley, and G.M. Rimmington. 1986. Modelling plant growth and development. Academic Press, London. 235 p.

Chiariello, N., and J. Roughgarden. 1984. Storage allocation in seasonal races of an annual plant—optimal versus actual allocation. Ecology 65:1290-1301.

Chiariello, N.R., H.A. Mooney, and K. Williams. 1989. Growth, carbon allocation, and cost of plant tissues, p. 327-365. *In* R.W. Pearcy, J. Ehleringer, H.A. Mooney, and P.W. Rundell [ed.] *Plant physiological ecology*. Chapman and Hall, New York.

Chung, H.H., and R.L. Barnes. 1977. Photosynthate allocation in *Pinus taeda*. I. Substrate requirements of synthesis of shoot biomass. Can. J. For. Res. 7:106-111.

Chung, H.H., and M.J. Trlica. 1980. ^{14}C distribution and utilization in blue grama as affected by temperature, water potential, and defoliation regimes. Oecologia 47:190-195.

Clarholm, M. 1985. Interactions of bacteria, protozoa, and plants leading to mineralization of soil nitrogen. Soil Biol. Biochem. 17:181-187.

Clark, F.E. 1977. Internal cycling of ^{15}N in shortgrass prairie. Ecology 58:1322-1333.

Clowes, F.A.L., and R.E. Woolston. 1978. Sloughing of root cap cells. Ann. Bot. 42:83-89.

Cole, D.W. 1981. Nitrogen uptake and translocation in forest ecosystems, p. 219-232. *In* F.E. Clark and T.H. Rosswall [ed.] *Nitrogen cycling in terrestrial ecosystems: processes, ecosystem strategies, and management strategies*. Ecological Bulletins-NFR, Vol. 33.

Coley, P.D. 1986. Costs and benefits of defense by tannins in a neotropical tree. Oecologia 70:238-241.

Coley, P.D., J.P. Bryant, and F.S. Chapin, III. 1985. Resource availability and plant antiherbivore defense. Science 230:895-899.

Condon, A.G., G.D. Farquhar, and R.A. Richards. 1990. Genotypic variation in carbon isotope discrimination and transpiration efficiency in wheat. Leaf gas exchange and whole plant studies. Aust. J. Plant Physiol. 17:9-22.

Condon, A.G., R.A. Richards, and G.D. Farquhar. 1987. Carbon isotope discrimination is positively correlated with grain yield and dry matter production in field-grown wheat. Crop Sci. 27:996-1001.

Cook, C.W. 1966. Carbohydrate reserves in plants. Utah Agric. Exp. Sta., Utah Resource Ser. 31:1-47.

Cook, R.E. 1983. Clonal plant populations. Am. Sci. 71:244-253.

Coombs, J. 1985. Carbon metabolism, p. 139-157. *In* J. Coombs, D.O. Hall, S.P. Long, and J.M.O. Scurlock [ed.] *Techniques in bioproductivity and photosynthesis*, 2nd ed. Pergamon Press, New York.

Coombs, J., D.O. Hall, S.P. Long, and J.M.O. Scurlock [ed.]. 1985a. Techniques in bioproductivity and photosynthesis, 2 Ed. Pergamon Press, New York. 298 p.

Coombs, J., G. Hind, R.C. Leegood, L.L. Tieszen, and A. Vonshak. 1985b. Analytical techniques, p. 219-228. *In* J. Coombs, D.O. Hall, S.P. Long, and J.M.O. Scurlock [ed.] *Techniques in bioproductivity and photosynthesis*, 2nd ed. Pergamon Press, New York.

Coppock, D.L., J.K. Detling, J.E. Ellis, and M.I. Dyer. 1983. Plant-herbivore interactions in North American mixed-grass prairies. I. effects of black-tailed prairie dogs on intraseasonal aboveground plant biomass and nutrient dynamics and plant species diversity. Oecologia 56:1-9.

Coughenour, M.B., J.K. Detling, I.E. Bamberg, and M.M. Mugambi. 1990. Production and nitrogen responses of the African dwarf shrub *Indigofera spinosa* to defoliation and water limitation. Oecologia 83:546-552.

Coyne, P.I., J.A. Bradford, and C.L. Dewald. 1982. Leaf water relations and gas exchange in relation to forage production in four Asiatic bluestems. Crop Sci. 22:1036-1040.

Coyne, P.I., and J.A. Bradford. 1987. Nitrogen and carbohydrate partitioning in 'Caucasian' and 'WW-Spar' Old World bluestems. J. Range Manage. 40:353-360.

Coyne, P.I., and C.W. Cook. 1970. Seasonal carbohydrate reserve cycles in eight desert range species. J. Range Manage. 23:438-444.

Crider, F. 1955. Root-growth stoppage resulting from defoliation of grass. USDA Tech. Bul. No. 1102. 23 p.

Dahlman, R.C. 1984. Vegetation responses to carbon dioxide research plan. DOE/ER-0187. Dist. Category UC-11. U.S. Deparment of Energy, Carbon Dioxide Res. Div., Office of Basic Sci., Washington, D.C. 32 p.

Dahlman, R.C., and C.L. Kucera. 1965. Root productivity and turnover in native prairie. Ecology 46:84-89.

Davey, C.B., and G.C. Papavizas. 1961. Translocation of streptomycin from coleus leaves and its effect on rhizosphere bacteria. Science 134:1368-1369.

Davidson, J.L., and F.L. Milthorpe. 1966. The effect of defoliation on the carbon balance in *Dactylis glomerata*. Ann. Bot. 30:185-198.

Davies, A. 1965. Carbohydrate levels in regrowth in perennial ryegrass. J. Agric. Sci. 65:185-221.

de Wit, C.T. 1978. Simulation of assimilation, respiration, and transpiration of crops. John Wiley and Sons, New York. 140 p.

de Wit, C.T., R. Brouwer, and F.W.T. Penning de Vries. 1970. The simulation of photosynthetic systems, p. 47-70. *In Prediction and measurement of photosynthetic productivity*. Proc. Int. Biol. Prog. Tech. Meeting, Trebon, PUDOC, Wageningen, The Netherlands.

Denmead, O.T., J.R. Freney, and J.R. Simpson. 1976. A closed ammonia cycle within a plant canopy. Soil Biol. Biochem. 8:161-164.

Denmead, O.T., J.R. Simpson, and J.R. Freney. 1974. Ammonia flux into the atmosphere from a grazed pasture. Science 185:609-610.

Deregibus, V.A., M.J. Trlica, and D.A. Jameson. 1982. Organic reserves in herbage plants: their relationship to grassland management, p. 315-344. *In* M. Rechcigl, Jr. [ed.] *Handbook of agriculture productivity*. Vol. I. Plant Productivity. CRC Press, Boca Raton, FL.

Deregibus, V.A., R.S. Sanchez, J.J. Casal, and M.J. Trlica. 1985. Tillering responses to enrichment of red light beneath the canopy in a humid natural grassland. J. Appl. Ecol. 22:199-206.

Detling, J.K. 1987. Grass response to herbivory, p. 56-68. *In* J.L. Capinera [ed.] *Integrated pest management on rangeland: a shortgrass prairie perspective*. Westview Press, Boulder, CO.

Detling, J.K., and E.L. Painter. 1983. Defoliation responses of western wheatgrass populations with diverse histories of prairie dog grazing. Oecologia 57:65-71.

Detling, J.K., M.I. Dyer, and D.T. Winn. 1979. Effect of simulated grasshopper grazing on CO_2 exchange rates of western wheatgrass leaves. J. Econ. Entomol. 72:403-406.

Devlin, R.M., and A.B. Barker. 1971. Photosynthesis. Van Nostrand Reinhold Co., New York. 304 p.

Devlin, R.M., and F.H. Witham. 1983. Plant physiology 4th Edition. PWS Publishers, Boston. 577 p.

Downton, W.J.S. 1971a. Adaptive and evolutionary aspects of C_4 photosynthesis, p. 3-17. *In* M.D. Hatch, C.B. Osmond, and R.O. Slatyer [ed.] *Photosynthesis and photorespiration*. Wiley-Interscience, New York.

Downton, W.J.S. 1971b. Check list of C_4 species, p. 554-558. *In* M.D. Hatch, C.B. Osmond, and R.O. Slatyer [ed.] *Photosynthesis and photorespiration*. Wiley-Interscience, New York.

Duncan, W.G., R.S. Loomis, W.A. Williams, and R. Hanau. 1967. A model for simulating photosynthesis in plant communities. Hilgardia 38:181-205.

Dyksterhuis, E.G. 1961. Writing in Grassland Res. Notes #20. 2 p.

Ehleringer, J.R. 1978. Implications of quantum yield differences on the distribution C_3 and C_4 grasses. Oecologia 31:255-267.

Ehleringer, J.R., and O. Bjorkman. 1977. Quantum yields for CO2 uptake in C_3 and C_4 plants: CO_2 and O_2 concentration. Plant Physiol. 59:86-90.

Ehleringer, J.R., and C.B. Osmond. 1989. Stable isotopes, p. 281-300. *In* R.W. Pearcy, J. Ehleringer, H.A. Mooney, and P.W. Rundell [ed.] *Plant physiological ecology*. Chapman and Hall, New York.

Ehleringer, J., and R.W. Pearcy. 1983. Variations in quantum yield for CO2 uptake among C_3 and C_4 plants. Plant Physiol. 73:555-559.

Eisele, K., D.S., Schimel, L.A. Kapuska, and W.J. Parton. 1990. Fire-stimulated N_2 fixation as a result of altered P:N ratios. Oecologia (in press).

Elwell, H.M., H.A. Daniel, and F.A. Fenton. 1941. The effects of burning pasture and woodland vegetation. Oklahoma Agric. Expt. Sta. Bul. No. B-247. 14 p.

Evans, G.C. 1972. The quantitative analysis of plant growth. Blackwell Scientific Publ., Oxford, UK. 734 p.

Evans, L.T. 1971. Evoluationary, adaptive, and environmental aspects of the photosynthetic pathway:assessment., p. 130-136. *In* M.D. Hatch, C.B. Osmond, and R.O. Slatyer [ed.] *Photosynthesis and photorespiration*. Wiley-Interscience, New York.

Farquhar, G.D., J.R. Ehleringer, and K.T. Hubick. 1989. Carbon isotope discrimination and photosynthesis. Ann. Rev. Plant Physiol. and Plant Molec. Biol. 11:539-537.

Farquhar, G.D., and R.A. Richards. 1984. Isotopic composition of plant carbon correlates with water-use efficiency of wheat genotypes. Aust. J. Plant Physiol. 11:539-552.

Farquhar, G.D., S. von Caemmerer, and J.A. Berry. 1980. A biochemical model of photosynthetic CO_2 assimilation in leaves of C_3 species. Planta 149:78-90.

Farquhar, G.D., and T.D. Sharkey. 1982. Stomatal conductance and photosynthesis. Ann. Rev. Plant Physiol. 33:317-345.

Farquhar, G.D., R. Wetselaar, and B. Wier. 1983. Gaseous nitrogen losses from plants, p. 159-180. *In* J.R. Freney and J.R. Simpson [ed.] *Gaseous loss of nitrogen from plant-soil systems*. Martinus Nijhoff/Dr. W. Junk Publishers, The Hague.

Fenn, L.B., and D.E. Kissel. 1976. The influence of cation exchange capacity and depth of incorporation on ammonia volatilization from ammonium compounds applied to calcareous soils. Soil Sci. Soc. Am. J. 40:394-398.

Fernandez, O.V., and M.M. Caldwell. 1975. Phenology and dynamics of root growth of three cool semi-desert shrubs under field conditions. J. Ecol. 63:703-714.

Fetcher, N., C.H. Jaeger, B.R. Strain, and N. Sionit. 1988. Effects of long-term exposure to elevated atmospheric CO_2 concentrations on carbon exchange rates of seedlings of *Pinus taeda* L. and *Liquidambar styraciflua* L. Tree Physiol. 4:255-263.

Fitter, A.H., and R.K.M. Hay. 1981. Environmental physiology of plants. Academic Press, London. 355 p.

Fork, D.C. 1989. Photosynthesis, p. 347-390. *In* K.C. Smith [ed.] *The science of photobiology*. Plenam Press, Rosetta, New York.

Forward, D.F. 1983. Respiration: a holistic approach to metabolism, p. 459-509. *In* F.C. Stewart and R.G.S. Bidwell [ed.] *Plant physiology*, Vol. VII: Energy and Carbon Metabolism. Academic Press, London.

Foulds, W., and J.P. Grime. 1972. The response of cyanogenic and acyanogenic phenolics of *Trifolius repens* to soil moisture supply. Heredity 28:181-187.

Frank, A.B., and R.E. Ries. 1990. Effect of soil water, nitrogen, and growing degree-days on morphological development of crested and western wheatgrass. J. Range Manage. 43:257-260.

Freney, J.R., O.T. Denmead, and J.R. Simpson. 1978. Soil as a source or sink for atmospheric nitrous oxide. Nature 273:530-532.

Freney, J.R., J.R. Simpson, and O.T. Denmead. 1983. Volatilization of ammonia, p. 1-33. *In* J.R. Freney and J.R. Simpson [ed.] *Gaseous loss of nitrogen from plant-soil systems.* Martinus Nijhoff/Dr. W. Junk Publishers, The Hague.

Gebauer, G., A. Melzer, and H. Rehder. 1984. Nitrogen content and nitrate reductase activity in *Rumex obtusifolius* L. I. Differences in organs and diurnal changes. Oecologia 63:136-142.

Gerbaud, A., and M. Andre. 1980. Effect of CO_2, O_2, and light on photosynthesis and photorespiration in wheat. Plant Physiol. 66:1032-1036.

Gerloff, G.L. 1976. Plant efficiencies in the use of nitrogen, phosphorus, and potassium, p. 161-169. *In* M.J. Wright [ed.] *Plant adaptation to mineral stress in problem soil.* Cornell Univ. Agric. Exp. Sta., Ithaca, NY.

Gifford, R.M., H. Lambers, and J.I.L. Morrison. 1985. Respiration of crop species under CO_2 enrichment. Physiol. Plant. 63:351-356.

Gomm, F.B. 1979. Management alternatives for native meadowlands. Oregon State Univ. Agric. Exp. Sta. and USDA Agric. Res. Spec. Rep. 532. Corvallis, OR.

Goodman, G.T., and D.F. Perkins. 1959. Mineral uptake and retention in cottongrass (*Eriophorum vaginatum L.*). Nature 184:467-468.

Gopal, Dr. N.H. 1972. Root exudates. Indian Coffee 36(1):21-26.

Greenwood, E.A.N. 1976. Nitrogen stress in plants. Adv. Agron. 28:1-35.

Gregory, R.P.F. 1989. Biochemistry of photosynthesis, 3rd Edition. John Wiley and Sons, New York. 257 p.

Grime, J.P. 1977. Evidence for the existence of three primary strategies in plants and its relevance to ecological and evolutionary theory. Amer. Natur. 111:1169-1194.

Grime, J.P. 1979. Plant strategies and vegetation processes. John Wiley, New York. 222 p.

Hale, M.G., and L.D. Moore. 1979. Factors affecting root exudation II: 1970-1978. Advan. in Agron. 31:93-124.

Hale, M.G., C.L. Foy, and F.J. Shay. 1971. Factors affecting root exudation. Advan. Agron. 23:89-109.

Hall, A.E., R.G. Mutters, K.T. Hubick, and G.D. Farquhar. 1990. Genotypic differences in carbon isotope discrimination by cowpea under wet and dry field conditions. Crop Sci. 30:300-305.

Haller, T., and H. Stolp 1985. Quantitative estimation of root exudation of maize plants. Plant and Soil 86:207:216.

Hamlen, R.A., F.L. Lukezic, and J.R. Bloom. 1972. Influence of clipping height on the neutral carbohydrate levels of root exudates of alfalfa plants grown under gnotobiotic conditions. Can. J. Plant Sci. 52:643-649.

Harper, J.L., and J. Ogden. 1970. The reproductive strategy of higher plants. I. The concept of strategy with special reference to *Senecio vulgaris* L. J. Ecol. 58:681-698.

Harper, J.L., P.H. Lovell, and K.G. Moore. 1970. The shapes and sizes of seeds. Ann. Rev. Ecol. Syst. 1:327-356.

Hartnett, D.C., and F.A. Bazzaz. 1985. The integration of neighborhood effects by clonal agents of *Solidago canadensis*. J. Ecol. 73:415-427.

Hatch, M.D. 1971. Mechanism and function of the C_4 pathway of photosynthesis, p. 139-152. *In* M.D. Hatch, C.B. Osmond, and R.O. Slatyer [ed.] *Photosynthesis and photorespiration*. Wiley-Interscience, New York.

Hatch, M.D., and C.B. Osmond. 1976. Compartmentation and transport in C_4 photosynthesis, p. 144-184. *In* U. Heber and C.R. Stocking [ed.] *Encyclopedia of plant physiology*, Vol. 3. Springer-Verlag, New York.

Hatch, M.D., C.B. Osmond, and R.O. Slatyer. [ed.] 1971. Photosynthesis and photorespiration. Wiley-Interscience, New York. 565 p.

Hatch, M.D., and C.R. Slack. 1966. Photosynthesis by sugar cane leaves. A new carboxylation reaction and the pathway of sugar formation. Biochem. J. 101:103-111.

Hattersley, P.W. 1983. The distribution of C_3 and C_4 grasses in Australia in relation to climate. Oecologia 57:113-128.

Hay, R.K.M., and A.J. Walker. 1989. An introduction to the physiology of crop yield. Longman Scientific and Technical, Essex, England. 292 p.

Haynes, R.J. 1980. Ion exchange properties of roots and ionic interactions within the root apoplasm: their role in ion accumulation by plants. Bot. Rev. 46:75-99.

Haynes, R.J. 1986a. Mineral nitrogen in the plant-soil system. Academic Press, London. 483 p.

Haynes, R.J. 1986b. Origin, distribution, and cycling of nitrogen in terrestrial ecosystems, p. 1-44. *In* R. J. Haynes [ed.] *Mineral nitrogen in the plant-soil system*. Academic Press, London.

Haynes, R.J. 1986c. Uptake and assimilation of mineral nitrogen by plants, p. 303-358. *In* R. J. Haynes [ed.] *Mineral nitrogen in the plant-soil system*. Academic Press, London.

Haynes, R.J., and R.R. Sherlock. 1986. Gaseous losses of nitrogen, p. 242-285. *In* R. J. Haynes [ed.] *Mineral nitrogen in the plant-soil system*. Academic Press, London.

Heath, O.V.S. 1969. Physiological aspects of photosynthesis. Stanford Univer. Press, Stanford, CA. 310 p.

Heilmeier, H., E.-D. Schulze, and D.M. Whale. 1986. Carbon and nitrogen partitioning in the biennial monocarp *Arctium tomentosum* Mill. Oecologia 70:466-467.

Henzell, E.F., and P.J. Ross. 1973. The nitrogen cycle of pasture ecosystems, p. 227-245. *In* G. W. Butler and R. W. Bailey [ed.] *Chemistry and biochemistry of herbage*, Vol. 2. Academic Press, London.

Hesketh, J.D., D.N. Baker, and W.G. Duncan. 1971. Simulation of growth and yield in cotton: respiration and carbon balance. Crop Sci. 11:394-398.

Hilbert, D.W., D.M. Swift, J.K. Detling, and M.I. Dyer. 1981. Relative growth rates and the grazing optimization hypothesis. Oecologia 51:14-18.

Hill, R., and F. Bendall. 1960. Function of two cytochrome components in chloroplasts: a working hypothesis. Nature 186:136-137.

Holland, E.A., and J.K. Detling. 1990. Plant response to herbivory and belowground nitrogen cycling. Ecology 71:1040-1049.

Hsiao, T.C. 1973. Plant response to water stress. Ann Rev. Plant Physiol. 24:519-570.

Hubick, K.T., G.L. Hammer, G.D. Farquhar, L.J. Wade, S. von Caemmerer, and S.A. Henderson. 1990. Carbon isotope discrimination varies genetically in C_4 species. Plant Physiol. 91:534-537.

Hunt, R. 1982. Plant growth curves: a functional approach to plant growth analysis. University Park Press, Baltimore. 248 p.

Hunt, H.W., M.J. Trlica, E.F. Redente, J.C. Moore, J.K. Detling, T.G.F. Kittel, D.E. Walter, M.C. Fowler, D.A. Klein, and E.T. Elliot. 1991. Simulation model for the effects of climate change on temperate grassland ecosystems. Ecol. Modelling 53:205-246.

Hyder, D.N., and F.A. Sneva. 1962. Selective control of big sagebrush associated with bitterbrush. J. Range Manage. 15:211-215.

Hyder, D.N., R.E. Bement, E.E. Remmenga, and D.F. Hervey. 1975. Ecological responses of native plants and guidelines for management of shortgrass range. USDA Tech. Bul. No. 1503. 87 p.

Ingham, R.E., J.A. Trofymow, E.R. Ingham, and D.C. Coleman. 1985. Interactions of bacteria, fungi and their nematode graziers: effects of nutrient cycling and plant growth. Ecol. Monogr. 55:119-140.

Jackson, W.A., D. Flesher, and R.H. Hageman. 1973. Nitrate uptake by dark-grown corn seedlings: some characteristics of apparent induction. Plant Physiol. 51:120-127.

Jalali, B.L. 1976. Biochemical nature of root exudates in relation to root rot of wheat. Carbohydrate shifts in response to foliar treatments. Soil Biol. Biochem. 8:127-129.

Jameson, D.A. 1963. Responses of individual plants to harvesting. Bot. Rev. 29:532-594.

Jameson D.A., and D.L. Huss. 1959. The effect of clipping leaves and stems on number of tillers, herbage weights, root weights, and food reserves of little bluestem. J. Range Manage. 12:122-126.

Jaramillo, V.J., and J.K. Detling. 1988. Grazing history, defoliation, and competition: effects on shortgrass production and nitrogen accumulation. Ecology 69:1599-1608.

Jenkinson, D. S. 1990. An introduction to the global nitrogen cycle. Soil Use and Manage. 6:56-60.

Johnen, B.G. 1974. Formation, amount, and decomposition of plant roots in the soil. Ph.D. Dissertation. U. of Bonn. West Germany.

Johnson, D.A., K.H. Asay, L.L. Tieszen, J.R. Ehleringer, and P.G. Jefferson. 1990. Carbon isotope discrimination: potential in screening cool-season grasses for water-limited environments. Crop Sci. 30:338-343.

Johnson, H.B. 1975. Gas-exchange strategies in desert plants, p. 105-120. *In* D.M. Gates and R.B. Schmerl [ed.] *Perspectives of biophysical ecology*. Springer-Verlag, New York.

Johnson, H.S., C.R. Slack, M.D. Hatch, and T.J. Andrews. 1971. The CO_2 carrier between mesophyll and bundle sheath chloroplasts in C_4 pathway species., p. 189-195. *In* M.D. Hatch, C.B. Osmond, and R.O. Slatyer [ed.] *Photosynthesis and photorespiration*. Wiley-Interscience, New York.

Johnston, A., S. Smoliak, A.D. Smith, and E. Lutwick. 1968. Seasonal precipitation, evaporation, soil moisture, and yield of fertilized range vegetation. Can. J. Plant Sci. 49:123-128.

Jones, P., L.H. Allen, Jr., J.W. Jones, and R. Valle. 1985. Photosynthesis and transpiration responses of soybean canopies to short-and long-term CO_2 treatments. Agron. J. 77:119-124.

Kays, S., and J.L. Harper. 1974. The regulation of plant and tiller density in a grass sward. J. Ecol. 62:97-106.

Keeling, C.D., Bacastow, R.B., and T. Whorf. 1982. Measurement of the concentration of CO2 at Mauna Loa Observatory, Hawaii, p. 337-385. *In* W.C. Clark [ed.] *Carbon dioxide review*. Clarendon Press, New York.

Kilmer, V.J. 1974. Nutrient losses from grasslands through leaching and runoff, p. 341-362. *In* D. A. Mays [ed.] *Forest fertilization*. Amer. Soc. Agron., Madison, Wisconsin.

Kimball, B.A. 1983. Carbon dioxide and agricultural yield: an assemblage and analysis of 430 prior observations. Agron. J. 75:779-788.

Kimura, M., Y. Yokoi, and K. Hogetzu. 1978. Quantitative relationships between growth and respiration. II. Evaluation of constructive and maintenance respiration in growing *Helianthus tuberosus*. Bot. Mag. Tokyo 91:43-56.

Kipe-Nolt, J.A., U.K. Avalakki, and P.J. Dart. 1985. Root exudation of sorghum and utilization of exudates by nitrogen-fixing bacteria. Soil Biol. Biochem. 17:859-863.

Klein, D.A., B.A. Frederick, M. Biondini, and M.J. Trlica. 1988. Rhizosphere microorganism effects on soluble amino acids, sugars, and organic acids in the root zone of *Agropyron cristatum, A. smithii* and *Bouteloua gracilis*. Plant and Soil 110:19-25.

Kormondy, E.J. 1976. Concepts of ecology. Prentice-Hall, Englewood Cliffs, New Jersey. 238 p.

Kortschak, H.P., C.E. Hartt, and G.O. Burr. 1965. Carbon dioxide fixation in sugarcane leaves. Plant Physiol. 40:209-213.

Kramer, P.J., and T.T. Kozlowski. 1979. Physiology of woody plants. Academic Press, London. 811 p.

Kucera, C.L., R.C. Dahlman, and M.R. Koelling. 1967. Total net productivity and turnover on an energy basis for tallgrass prairie. Ecology 48:536-541.

Laetsch, W.M. 1971. Chloroplast structural relationships in leaves with C_4 plants, p. 323-349. *In* M.D. Hatch, C.B. Osmond, and R.O. Slatyer [ed.] *Photosynthesis and photorespiration*. Wiley-Interscience, New York.

Lam, H.L.Y., H.R. Bunjay, and L.G. Culotta. 1986. Review: an engineer looks at photosynthesis. Appl. Biochem. Biotech. 13:37-73.

Larcher, W. 1980. Physiological plant ecology, 2nd Ed. Springer-Verlag, New York. 303 p.

Lauenroth, W.K., J.L. Dodd, and P.L. Sims. 1978. The effects of water-nitrogen-induced stresses on the plant community structure in a semiarid grassland. Oecologia 36:211-222.

Launchbaugh, J.L. 1974. Unpublished research data. Ft. Hays Branch Expt. Sta., KAES, Hays, KS 67601.

Launchbaugh, J.L., and C.E. Owensby. 1978. Kansas rangelands: their management, based on a half century of research. Kansas Agric. Exper. Sta. Bul. No. 622. 56 p.

Lee, D.W., and J.B. Lowry. 1980. Young-leaf anthocyanin and solar ultraviolet. Biotropica 12:75-76.

Lee, K.J., and M.H. Gaskins. 1982. Increased root exudation of ^{14}C-compounds by sorghum seedlings inoculated with nitrogen-fixing bacteria. Plant and Soil 69:391-399.

Lemon, E., and R. van Houte. 1980. Ammonia exchange at the land surface. Agron. J. 72:876-883.

Levin, S.A., H.A. Mooney, and C. Field. 1989. The dependence of plant root:shoot ratios on internal nitrogen concentration. Ann. Bot. 64:71-75.

Levins, D.A. 1971. Plant phenolics: an ecological perspective. Am. Natur. 105:157-181.

Levitt, J. 1972. Responses of plants to environmental stresses. Academic Press, London. 697 p.

Lincoln, D.E. 1980. Leaf resin flavonoids of *Diplacus aurantiacus*. Biochem. Syst. Ecol. 8:397-400.

Lincoln, D.E., D. Couvet, and N. Sionit. 1986. Response of an insect herbivore to host plants grown in carbon dioxide enriched atmospheres. Oecologia 69:556-560.

Long, S.P., J-E. Hallgren. 1985. Measurement of CO_2 assimilation by plant in the field and the laboratory, p. 62-94. *In* J. Coombs, D.O. Hall, S.P. Long, and J.M.O. Scurlock [ed.] *Techniques in bioproductivity and photosynthesis*, 2nd ed. Pergamon Press, New York.

Long, S.P., and P.R. Hutchin. 1991. Primary production in grasslands and coniferous forests with climate change: an overview. Ecolog. Applic. 1:139-156.

Loomis, R.S., and W.A. Williams, 1969. Productivity and the morphology of crop stands: patterns with leaves, p. 27-47. *In* J.D. Eastin, F.A. Haskins, C.Y. Sullivan, and C.H.M. Van Bavel [ed.] *Physiological aspects of crop yield.* American Society of Agronomy, Madison, Wisconsis.

Lorenz, R.J., and G.A. Rogler. 1966. Root growth of Northern Plains grasses under various fertilizer and management practices. Proc. Amer. Forage and Grassland Council. p. 1-11.

Ludwick, A.E. 1979. Meadow hay production as influenced by nitrogen and phosphorus fertilization, p. 71-80. *In* R.H. Delaney and J. Borrelli [ed.] *Management of intermountain meadows.* Symp. Proc. RJ 141. Univ. Wyom., Laramie and Colo. State Univ., Ft. Collins.

Mader, E.L. 1956. The influence of certain fertilizer treatments on native vegetation of Kansas prairie. P.H. Diss., Univ. Nebraska, Lincoln. 116 p.

Martin, J.K. 1977. Factors influencing the loss of organic carbon from wheat roots. Soil. Biol. Biochem. 9:1-7.

McCree, K.J. 1970. An equation for the rate of respiration of white clover plants under controlled conditions, p. 221-229. *In* I. Setlik [ed.] *Prediction and measurement of photosynthetic productivity.* Proc. IBP/PP Technical Meeting, Centre for Agricultural Publishing and Documentation, Wageningen.

McCree, K.J. 1974. Equations for rate of dark respiration of white clover and grain sorghum, as functions of dry weight, photosynthetic rate, and temperature. Crop Sci. 14:509-514.

McCree, K.J. 1986. Measuring the whole-plant daily carbon balance. Photosynthetica 20:82-93.

McCree, K.J., and J.H. Silsbury. 1978. Growth and maintenance requirements of subterranean clover. Crop Sci. 18:13-18.

McDermitt, D.K., and R.S. Loomis. 1981. Elemental composition of biomass and its relation to energy content, growth efficiency, and growth yield. Ann. Bot. 48:275-290.

McKell, C.M., M.B. Jones, and E.R. Perrier. 1962. Root production and accumulation of root material on fertilized range. Agron. J. 59:459-461.

McKendrick, J.D., C.E. Owensby, and R.M. Hyde. 1975. Big bluestem and indiangrass vegetative reproduction and annual reserve carbohydrate and nitrogen cycles. Agro-Ecosystems 2:75-93.

McKey, D.B., P.G. Waterman, C.N. Mbi, J.S. Gartlan, and T.T. Strusaker. 1978. Phenolic content of vegetation in two African rain forests: ecological implications. Science 202:61-64.

McNaughton, S.J. 1979. Grazing as an optimization process: grass-ungulate relationships in the serengeti. Amer. Natur. 113:691-703.

McNaughton, S.J. 1983. Physiological and ecological implications of herbivory, p. 657-677. *In* O.L. Lange, P.S. Nobel, C.B. Osmond, and H. Ziegler [ed.] *Physiological plant ecology III. Response to the chemical and biological environment*, N.S. 12C, Springer-Verlag, Berlin.

McNaughton, S.J. 1984. Grazing lawns: animals in herds, plant form, and coevolution. Amer. Nat. 124:863-886.

Meijden, E. van der, M. Wijn, and H.J. Verkaar. 1988. Defense and regrowth, alternative plant strategies in the struggle against herbivores. Oikos 51:355-363.

Melillo, J.N., J.D. Aber, and J.F. Muratore. 1982. Nitrogen and lignin control of hardwood leaf litter decomposition dynamics. Ecology 62:621-633.

Melzer, A., G. Gebauer, and H. Rehder. 1984. Nitrogen content and nitrate reductase activity in *Rumex obtusifolius* L. II. Responses to nitrate starvation and nitrogen fertilization. Oecologia 63:380-385.

Menke, J.W., and M.J. Trlica. 1983. Effects of single and sequential defoliations on the carbohydrate reserve status of four range species. J. Range Manage. 36:70-74.

Meyer, B.S., D.B. Anderson, R.H. Bohning, and D.G. Fratiianne. 1973. Introduction to plant physiology. D. Van Nostrand Co., New York. 565 p.

Milthorpe, F.L., and J.L. Davidson. 1966. Physiological aspects of regrowth following defoliation, p. 241-254. *In* F.L. Milthorpe and J.D. Ivans [ed.] *The growth of cereals and grasses*. Butterworths, London.

Minchin, P.E.H., and G.S. McNaughton. 1984. Exudation of recently fixed carbon by nonsterile roots. J. exp. Bot. 35:74-82.

Monsi, M. 1968. Mathematical models of plant communities, p. 131-149. *In* F. Eckardt [ed.] *Functioning of terrestrial ecosystems at the primary production level*. UNESCO, Paris.

Monsi, M., and Y. Murata. 1970. Development of photosynthetic systems as influenced by distribution of matter, p. 115-130. *In prediction and measurement of photosynthetic productivity*. PUDOC, Wageningen, The Netherlands.

Monson, R.K., G.E. Edwards, and M.S.B. Ku. 1984. C_3-C_4 Intermediate photosynthesis in plants. Bioscience 34:563-574.

Mooney, H.A. 1972. The carbon balance of plants. Ann. Rev. Ecol. Syst. 3:315-346.

Mooney, H.A., and S.L. Gulmon. 1982. Constraints of leaf structure and function in reference to herbivory. BioScience 32:198-206.

Mooney, H.A., and P.W. Rundel. 1979. Nutrient relations of the evergreen shrub, *Adenostoma fasciculatum*, in the California chaparral. Bot. Gaz. 140:109-113.

Muller, M.M., V. Sundman, and J. Skujins. 1980. Denitrification in low pH Spodosols and peats determined by the acetylene inhibition method. Appl. Environ. Microbiol. 40:235-239.

Nichols, J.T., P.E. Reece, G.W. Hergert, and L.E. Moser. 1990. Yield and quality response of subirrigated meadow vegetation to nitrogen, phosphorus, and sulfur fertilizer. Agron. J. 82:47-52.

Nijs, I., I. Impens, and T. Behaeghe. 1988. Effects of rising atmospheric carbon dioxide concentration on gas exchange and growth of perennial ryegrass. Photosynthetica 22:44-50.

Nobel, P.S., and S.P. Long. 1985. Canopy structure and light interception, p. 41-49. *In* J. Coombs, D.O. Hall, S.P. Long, and J.M.O. Scurlock [ed.] *Techniques in bioproductivity and photosynthesis*, 2nd ed. Pergamon Press, New York.

Noggle, G.R., and G.J. Fritz. 1983. Introductory plant physiology. Prentice-Hall, Englewood Cliffs, NJ. 627 p.

Nowak, R.S., and M.M Caldwell. 1984. A test of compensatory photosynthesis in the field: implications for herbivory tolerance. Oecologia 61:311-318.

Odum, H.T. 1982. Systems ecology: an introduction. Wiley-Interscience, New York. 644 p.

Odum, E.P. 1985. Trends expected in stressed ecosystems. BioScience 35:419-422.

Oechel, W.C., and W. Lawrence. 1981. Carbon allocation and utilization, p. 185-235. *In* P.C. Miller [ed.] *Resource use by chaparral and matorral*. Springer-Verlag, Berlin.

Ojima, D.S., W.J. Parton, D.S. Schimel, and C.E. Owensby. 1990. Simulated impacts of annual burning on prairie ecosystems, p. 118-132. *In* S. L. Collins and L. L. Wallace [ed.] *Fire in the North American tallgrass prairies*. Univ. Oklahoma Press, Norman.

Orodho, A.B., and M.J. Trlica. 1990. Clipping and long-term grazing effects on biomass and carbohydrate reserves of Indian ricegrass. J. Range Manage. 43:52-57.

Osmond, C.B., O. Bjorkman, and D.J. Anderson. 1980. Physiological processes in plant ecology: toward a synthesis with *Atriplex*. Springer-Verlag, New York. 468 p.

Oswalt, D.L., A.R. Bertrand, and M.R. Teel. 1959. Influence of Nitrogen fertilization and clipping on grass roots. Soil Sci. Soc. Amer. Proc. 23:228-230.

Owensby, C.E., and K.L. Anderson. 1969. Effects of clipping date on loamy upland bluestem range. J. Range Manage. 22:351-354.

Owensby, C.E., R.M. Hyde, and K.L. Anderson. 1970. Effects of clipping and supplemental nitrogen and water on loamy upland bluestem range. J. Range Manage. 23:341-346.

Owensby, C.E., and E.F. Smith. 1979. Fertilizing and burning Flint Hills bluestem. J. Range Manage. 32:254-258.

Owensby, C.E., E.F. Smith, and J.R. Rains. 1977. Carbohydrate and nitrogen reserve cycles for continuous, season-long and intensive-early stocked Flint Hills bluestem range. J. Range Manage. 30:258-260.

Painter, E.L., and J.K. Detling. 1981. Effects of defoliation on net photosynthesis and regrowth of western wheatgrass. J. Range Manage. 34:68-71.

Parsons, A.J. 1980. The physiological basis of seasonal differences in the growth of perennial ryegrass. Ph.D. Thesis. University of Reading.

Parsons, A.J. 1988. The effect of season and management on the growth of grass swards, p. 129-175. *In* M.B. Jones and A. Lazenby [ed.] *The grass crop*. Chapman and Hall, New York.

Parsons, A.J., and I.R. Johnson. 1986. The physiology of grass growth under grazing, p. 3-13. *In* J. Frame [ed.] *Grazing*. Brit. Grassl. Soc. Occasional Symp. No. 19.

Parsons, A.J., E.L. Leafe, B. Collett, and W. Stiles. 1983a. The physiology of grazing production under grazing. I. Characteristics of leaf and canopy photosynthesis of continuously-grazed swards. J. Appl. Ecol. 20:117-126.

Parsons, A.J., E.L. Leafe, B. Collett, P.D. Penning, and J. Lewis. 1983b. The physiology of grazing production under grazing. II. Photosynthesis, crop growth, and animal intake of continuously-grazed swards. J. Appl. Ecol. 20:127-139.

Parsons, A.J., I.R. Johnson, and A. Harvey. 1988. Use of a model to optimize the interaction between frequency and severity of intermitten defoliation and to provide a fundamental comparison of the continuous and intermitten defoliation of grass. Grass and Forage Sci. 43:49-59.

Pastor, J., J.D. Aber, C.A. McClaughterty, and J.M. Melillo. 1984. Aboveground production and N and P cycling along a mineralization gradient on Blackhawk Island, Wisconsin. Ecology 65:256-268.

Pate, J.S. 1973. Uptake, assimilation and transport of nitrogen compounds by plants. Soil Biol. Biochem. 5:109-119.

Pate, J.S. 1980. Transport and partitioning of nitrogenous solutes. Ann. Rev. Plant Physiol. 31:313-340.

Pate, J.S. 1983. Patterns of nitrogen metabolism in higher plants and their ecological significance, p. 225-255. *In* J.A. Lee, S. McNeill, and I.H. Rorison [ed.] *Nitrogen as an ecology factor*. Blackwell Scientific Publ., Oxford, UK.

Payne, W.J. 1973. Reduction of nitrogenous oxides by microorganisms. Bacteriol. Rev. 37:409-452.

Pearcy, R.W., O.P. Bjorkman, M.M. Caldwell, J.E. Keeley, R.K. Monson, and B.R. Strain. 1987. Carbon gained by plants in natural environments. Bioscience 37:21-29.

Pearcy, R.W., and J. Ehleringer. 1984. Comparative ecophysiology of C_3 and C_4 plants. Plant Cell Environment 7:1-13.

Pearcy, R.W., J. Ehleringer, H.A. Mooney, and P.W. Rundell [ed.]. 1989. Plant physiological ecology. Chapman and Hall, New York. 457 p.

Pearson, C.T., and B.T. Steer. 1977. Daily changes in nitrate uptake and metabolism in Capsicum annuum. Planta 137:107-112.

Penning de Vries, F.W.T. 1975. The cost of maintenance processes in plant cells. Ann. Bot. 39:77-92.

Penning de Vries, F.W.T., A.H.M. Brunsting, and H.H. van Laar. 1974. Products, requirements, and efficiency of biosynthesis: a quantitative approach. J. Theor. Biol. 45:339-377.

Penning de Vries, F.W.T., J.M. Witlage, and D. Kramer. 1979. Rates of respiration and of increase in structural dry matter in young wheat, ryegrass, and maize plants in relation to temperature, to water stress, and to their sugar content. Ann. Bot. 44:595-609.

Polley, H.W., and J.K. Detling. 1989. Defoliation, nitrogen, and competition: effects on plant growth and nitrogen nutrition. Ecology 70:721-727.

Porter, H.K. 1966. Leaves as collecting and distributing agents of carbon. Aust. J. Sci. 29:31-40.

Portis, A.R., Jr., M.E. Salvucci, W.L. Ogren, and J. Werneke. 1987. RUBISCO activase: a new enzyme in the regulation of photosynthesis, p. 371-378. *In* J. Biggens [ed.] *Progress in photosynthesis research*, Vol. III. Martinus Nijhoff Publishers, The Netherlands.

Potter, C.S., H.L. Ragsdale, and C.W. Berish. 1987. Resorption of foliar nutrients in a regenerating southern Appalachian forest. Oecologia 73:268-271.

Potvin, C., and B.R. Strain. 1985. Photosynthetic response to growth temperature and CO2 enrichment in two species of C_4 grasses. Can. J. Bot. 63:483-487.

Raghavendra, A.S., and V.S.R. Das. 1978. The occurrence of C_4-photosynthesis: a supplementary list of C_4 plants reported during late 1974-mid 1977. Photosynthetica 12:200-208.

Raison, R.J. 1979. Modification of the soil environment by vegetation fires, with particular reference to nitrogen transformations: a Review. Plant and Soil 51:73-108.

Randall, D.D., and N.E. Tolbert. 1971. Two phosphatases associated with photosynthesis and the glycolate pathway, p. 259-266. *In* M.D. Hatch, C.B. Osmond, and R.O. Slatyer [ed.] *Photosynthesis and photorespiration*. Wiley-Interscience, New York.

Rao, K.P., and D.W. Rains. 1976. Nitrate absorption by barley. I. Kinetics. Plant Physiol. 57:55-58.

Reekie, E.G., and F.A. Bazzaz. 1987. Reproductive effort in plants 3. Effect of reproduction on vegetative activity. Amer. Nat. 129:907-919.

Reich, P.B., R.O. Teskey, P.S. Johnson, and T.M. Hinckley. 1980. Periodic root and shoot growth in oak. Forest Sci. 26:590-598.

Reichers, G.D., and B.R. Strain. 1988. Growth of blue grama (*Bouteloua gracilis*) in response to atmospheric carbon dioxide enrichment. Can. J. Bot. 66:1570-1573.

Reid, C.P.P. 1974. Assimilation, distribution and root exudation of ^{14}C by ponderosa pine seedlings under induced water stress. Plant Physiol. 54:44-49.

Reid, C.P.P., and J.G. Mexal. 1977. Water stress effects on root exudation by lodgepole pine. Soil Biol. Biochem. 9:417-421.

Reitz, L.P. 1974. Breeding for more efficient water use—is it real or a mirage? Agric. Meteor. 14:3-12.

Rhoades, D.F., and R.G. Cates. 1976. Toward a general theory of plant antiherbivore chemistry, p. 168-213. *In* J. Wallace and R.L. Mansel [ed.] *Biochemical interactions between plants and insects. Recent advances in phytochemistry*. Vol. 10. Plenum Press, New York.

Richards, J.H. 1984. Root growth responses to defoliation in two *Agropyron* bunchgrasses: field observations with an improved root periscope. Oecologia 64:21-25.

Richards, J.H. 1986. Root form and depth distribution in several biomes, p. 83-97. *In* D. Carlisle, W.L. Berry, I.R. Kaplan, and J.R. Watterson [ed.] *Mineral exploration: biological systems and organic matter*. Rubey Vol. 5. Prentice Hall, Englewood Cliffs, NJ.

Richards, J.H., and M.M. Caldwell. 1985. Soluable carbohydrates, concurrent photosynthesis, and efficiency in regrowth following defoliation: a field study with *Agropyron* species. J. Appl. Ecol. 22:907-920.

Robichaux, R.H., and R.W. Pearcy. 1980. Photosynthetic responses of C_3 and C_4 species from cool, shaded habitat in Hawaii. Oecologia 47:106-109.

Robinson, S.P., and A.R. Portis, Jr. 1988. Involvement of stromal ATP in the light activation of ribulose-1,5-biphosphate carboxylase/oxygenase in intact isolated chloroplasts. Plant Physiol. 86:293-298.

Rogers, H.H., G.E. Bingham, J.D. Cure, J. Smith, and K.A. Surano. 1983. Responses of selected plant species to elevated carbon dioxide in the field. J. Environ. Qual. 12:569-574.

Rogler, G.A., and R.J. Lorenz. 1974. Fertilization of mid-continent range plants, p. 231-254. *In* D.A. Mays [ed.] *Forage fertilization*. Amer. Soc. Agron., Madison, WI.

Rosswall, T. 1976. The internal nitrogen cycle between micro-organisms, vegetation and soil, p. 157-167. *In* B. H. Svensson and R. Soderlund [ed.] *Nitrogen, phosphorus and sulfur-global cycles*. Ecolog. Bull., Stockholm.

Rovira, A.D. 1969. Plant root exudates. Bot. Rev. 35:35-57.

Rovira, A.D., and C.B. Davey. 1974. Biology of the rhizosphere. p. 153-204. *In* E.W. Carson [ed.] *The plant root and its environment*. Univ. Press, Charlottesville, Virginia.

Rovira, A.D., R.C. Foster, and J.K. Martin. 1979. Note on terminology: origin, nature, and nomenclature of the organic materials in the rhizosphere, p. 1-4. *In* J.L. Harley and R. S. Russell [ed.] *The soil-root interface*. Academic Press, London.

Rovira, A.D., and B.M. McDougall. 1967. Microbiological and biochemical aspects of the rhizosphere. pp. 417-463. *In* A.D. McLaren and G.H. Petersen [ed.]. *Soil biochemistry*. M. Dekker, Inc. New York.

Russell, R. S. 1977. Plant root systems: their function and interaction with soil. McGraw-Hill Book Co., Ltd. London.

Ryle, G.J., J.M. Cobby, and C.E. Powell. 1976. Synthetic and maintenance respiratory losses of $^{14}CO_2$ in uniculum barley and maize. Ann. Bot. 40:571-586.

Salisbury, F.B., and C.W. Ross. 1985. Plant physiology, 3rd ed. Wadsworth Publ. Co., Belmont, CA. 540 p.

Samtsevich, S.A. 1965. Active excretions of plant roots and their significance. Soviet Plant Physiol. 12:731-740.

Santos, G.L., and M.J. Trlica. 1978. Clipping effects on production on carbohydrate reserves of blue grama and western wheatgrass, p. 384-386. *In* D.N. Hyder [ed.] *Proc. 1st Int. Rangeland Congr.*, Denver, CO.

Sauerback, D.R., and B.G. Johnen. 1977. Root formation and decomposition during plant growth. *In Soil organic matter studies*. Int. Symposium on Soil Organic Matter Studies. Brunswick, FDA. Sept. 1976.

Schrader, L.E., and R.J. Thomas. 1981. Nitrate uptake, reduction and transport in the whole plant, p. 49-93. *In* J.D. Bewley [ed.] *Nitrogen and carbon metabolism*. Martinus Nijhoff/Dr. W. Junk Publishers, The Hague.

Schulze, E.-D. 1970. Der CO_2-Gaswechsel der Buche (*Fagus silvatica* L.) in Abhängigkeit von den Klimafaktoren im Freilend. Flora 159:177-232.

Schulze, E.-D. 1982. Plant lifeforms and their carbon, water, and nutrient relations, p. 615-676. *In* O.L. Lange, P.S. Noble, C.B. Osmond, and H. Ziegler [ed.] *Encyclopedia of plant physiology*, NS, Vol. 12B. Springer-Verlag, Berlin.

Schulze, E.-D., G.W. Koch, F. Percival, H.A. Mooney, and C. Chu. 1985. The nitrogen balance of *Raphanus sativus* x *raphanistrum* plants. I. Daily nitrogen use under high nitrate supply. Plant Cell Environ. 8:713-720.

Seastedt, T.R., and R.A. Ramundo. 1990. Influence of fire on belowground processes of Tallgrass Prairie, p. 99-117. *In* S. L. Collins and L. L. Wallace [ed.] *Fire in the North American tallgrass prairies*. Univ. Oklahoma Press, Norman.

Sestak, Z., J. Catsky, and P.G. Jarvis [ed.]. 1971. Plant photosynthetic production: a manual of methods. Dr. W. Junk N.V. Publishers, The Hague. 818 p.

Sharrow, S.H., and H. Wright. 1977a. Effects of fire, ash, and litter on soil nitrate, temperature, moisture, and tobosagrass production in the Rolling Hills. J. Range Manage. 30:266-270.

Sharrow, S.H., and H. Wright. 1977b. Proper burning intervals for tobosagrass in west Texas based on nitrogen dynamics. J. Range Manage. 30:343-346.

Shaver, G.R., and J.M. Melillo. 1984. Nutrient budgets of marsh plants: efficiency concepts and relation to availability. Ecology 65:1491-1510.

Sherlock, R.R., and K.M. Goh. 1983. Initial emission of nitrous oxide from sheep urine applied to pasture. Soil Biol. Biochem. 15:615-617.

Sherlock, R.R., and K.M. Goh. 1984. Dynamics of ammonia volatilization from simulated urine patches and aqueous urea applied to pasture. I. Field experiments. Fert. Res. 5:181-195.

Shul'gin, A., and A.F. Kaleshnin. 1959. Correlation between optical properties of plant leaves and their chlorophyll content. Doklady Akad. S.S.R. 125:119-121.

Simpson, J.R., and K.W. Steele. 1983. Gaseous nitrogen exchanges in grazed pastures, p. 215-236. *In* J.R. Freney and J.R. Simpson [ed.] *Gaseous loss of nitrogen from plant-soil systems*. Martinus Nijhoff/Dr. W. Junk Publishers, The Hague.

Sims, P.L., and J.S. Singh. 1971. Herbage dynamics and net primary production in certain ungrazed and grazed grasslands in North America, p. 59-124. *In* N.R. French [ed.] *Preliminary analysis of structure and function of grasslands*. Range Sci. Dept. Sci. Ser. No. 10, Colorado State Univ., Fort Collins, Colorado.

Sims, P.L., and J.S. Singh. 1978a. The structure and function of ten western North American grasslands. II. Intraseasonal dynamics in primary producer compartments. J. Ecol. 66:547-572.

Sims, P.L., and J.S. Singh. 1978b. The structure and function of ten western North American grasslands. III. Net primary production, turnover, and efficiencies of energy capture and water use. J. Ecol. 66:573-597.

Sims, P.L., and J.S. Singh. 1978c. The structure and function of ten western North American grasslands. IV. Compartmental transfers and energy flow within the ecosystem. J. Ecol. 66:983-1009.

Sims, P.L., J.S. Singh, and W.K. Lauenroth. 1978. The structure and function of ten western North American grasslands. I. Abiotic and vegetational characteristics. J. Ecol. 66:251-285.

Sionit, N., B.R. Strain, and E.P. Flint. 1987. Interactions of temperature and CO_2 enrichment on soybean: photosynthesis an seed yield. Can. J. Plant Sci. 67:629-636.

Slankis, V., V.C. Runeckles, and G. Krotkov. 1964. Metabolites liberated by roots of white pine (*Pinus strobus* L.) seedlings. Physiologia Plantarium 17:301-303.

Smika, D.E., H.J. Haas, and J.F. Power. 1965. Effects of moisture and nitrogen fertilizer on growth and water use by native grass. Agron. J. 57:483-486.

Smith, D., and J.P. Silva. 1969. Use of carbohydrates and nitrogen root reserves in regrowth of alfalfa from greenhouse experiments under light and dark conditions. Crop Sci. 9:464-467.

Smith, E.F., and C.E. Owensby. 1978. Intensive-early stocking and season-long stocking of Kansas Flint Hills range. J. Range Manage. 31:14-17.

Stitt, M., and M.P. Quick. 1989. Photosynthetic carbon partitioning: its regulation and possibilities for manipulation. Physiol. Plant. 77:633-641.

Stowe, L.G., and J.A. Teeri. 1978. The geographic distribution of C_4 species of the Dicotyledonae in relation to climate. Amer. Nat. 112:609-623.

Strain, B.R., and J.D. Cure [ed.]. 1985. Direct effect of increasing carbon dioxide on vegetation. Carbon Dioxide Research, State of the Art. U.S. Dept. of Energy, Washington, D.C.

Strain, B.R., and P.L. Johnson. 1963. Corticular photosynthesis and ecology in *Populus tremuloides*. Ecology 44:581-584.

Sullivan, C.Y., and J.D. Eastin. 1974. Plant physiological responses to water stress. Agric. Meteor. 14:113-127.

Switzer, G.L., and L.E. Nelson. 1972. Nutrient accumulation and cycling in loblolly pine (*Pinus taeda* L.) plantation ecosystems: the first twenty years. Soil Sci. Soc. Amer. Proc. 36:143-147.

Taylor, D.T., W.J. Seamands, D.J. Menkhaus, J.J. Jacobs, and R.H. Delaney. 1985. An economic analysis of N fertilization of intermountain hay meadows. Agron. J. 77:17-20.

Teeri, J.A., and L.G. Stowe. 1976. Climatic patterns and distribution of C_4 grasses in North America. Oecologia 23:1-12.

Thien, S.J., D. Kissell, and D. Sorden. 1979. Unpublished research data. Department of Agronomy, Kansas State University, Manhattan, KS 66506.

Thornley, J.H.M. 1970. Respiration, growth, and maintenance in plants. Nature 227:304-305.

Thornley, J.H.M. 1976. Mathematical models in plant physiology. Academic Press, London. 318 p.

Tieszen, L.L., M.M. Senyimba, S.K. Imbamba, and J.H. Troughton. 1979. The distribution of C_3 and C_4 grasses and carbon isotope discrimination along an altitudinal and moisture gradient in Kenya. Oecologia 37:337-350.

Tieszen, L.L., T.W. Boutton, K.G. Tesdahl, and N.A. Slade. 1983. Fractionation and turnover of stable carbon isotopes in animal tissues: implications for delta^{13}C analysis of diet. Oecologia 57:32-37.

Tillman, D. 1988. Plant strategies and the dynamics and structure of plant communities. Princeton Univ. Press, Princeton, NJ. 360 p.

Ting, I. P., H.B. Johnson, and S.R. Szarek. 1972. Net CO_2 in Crassulacean acid metabolism plants, p. 26-53. *In* C.C. Black, Jr. [ed.] *Net carbon dioxide assimilation in higher plants*. The Souther section of the Amer. Soc. Plant Physiologists and Cotton, Raleigh, North Carolina.

Titus, J.S., and S.-M. Kang. 1982. Nitrogen metabolism, translocation, and recycling in apple trees. Hort. Rev. 4:204-246.

Toft, N.L., J.E. Anderson, and R.S. Nowak. 1989. Water use efficiency and carbon isotope composition of plants in a cold desert environment. Oecologia 80:11-18.

Tolbert, N.E. 1971. Leaf peroxisomes and photorespiration, p. 458-471. *In* M.D. Hatch, C.B. Osmond, and R.O. Slatyer [ed.] *Photosynthesis and photorespiration*. Wiley-Interscience, New York.

Tolley, L.C., and B.R. Strain. 1984. Effects of atmospheric CO2 enrichment and water stress on growth of *Liquidambar styraciflua* L. and *Pinus taeda* L. seedlings. Can. J. Bot. 10:2135-2139.

Tomkins, G.A., W.A. Jackson, and R. J. Volk. 1978. Accelerated nitrate uptake in wheat seedlings: effects of ammonium and nitrite pretreatments and of 6-methylpurine and puromycin. Physiol. Plant. 43:166-171.

Towne, G., and C. E. Owensby. 1983. Cytokinins effect on protein and chlorophyll content of big bluestem leaves. J. Range Manage. 36:75-77.

Trlica, M.J. 1977. Distribution and utilization of carbohydrate reserves in range plants, p. 73-97. *In* R.E. Sosebee [ed.] *Range plant physiology* (Range Sci. Series #4), Society for Range Management, Denver, CO.

Trlica, M.J., and C.W. Cook. 1971. Defoliation effects on carbohydrate reserves of desert species. J. Range Manage. 24:418-425.

Trlica, M.J., and A.B. Orodho. 1989. Effects of protection from grazing on morphological and chemical characteristics of Indian ricegrass, *Oryzopsis hymenoides*. Oikos 56:299-308.

Trlica, M.J., and J.S. Singh. 1979. Translocation of assimilates, and creation, distribution, and utilization of reserves, p. 537-571. *In* R.A. Perry and D.W. Goodall [ed.] *IBP 16, arid-land ecosystems*. Vol. 1. Cambridge Univ. Press, Cambridge, U.K.

Tromp, J. 1970. Storage and mobilization of nitrogenous compounds in apple trees with special reference to arginine, p. 143-159. *In* L.C. Luckwill and C.V. Cutting [ed.] *Physiology of tree crops*. Academic Press, London.

Turner, M.D., and D. Rabinowitz. 1983. Factors affecting frequency distributions of plant mass: the absence of dominance and suppression in competing monocultures of *Festuca paradoxa*. Ecology 64:469-475.

USDA, Forest Service. 1972. The nation's range resources -a forest-range environment study. Forest Resources Report No. 19.

Vitousek, P.M., J.R. Gosz, C.C. Grier, J.M. Melillo, and W.A. Reiners. 1982. A comparative analysis of potential nitrification and nitrate mobility in forest ecosystems. Ecol. Monogr. 52:155-177.

Vlek, P.L.G., and E.T. Craswell. 1981. Ammonia volatilization from flooded soils. Fert. Res. 2:227-245.

Vrany, J., W. Varcura, and J. Macura. 1962. The effects of foliar application of some readily metabolized substances, growth regulators and antibiotics on rhizosphere microflora. Folia Microbial. 7:61-70.

Walker, J.M., D.D. Briske, and R.W. Weaver. 1987. Nitrogen-15 partitioning within a three-generation tiller sequence of the bunchgrass *Schizachyrium scoparium*: response to selective defoliation. Oecologia 74:330-334.

Wallace, L.L., S.J. McNaughton, and M.B. Coughenour. 1984. Compensatory photosynthetic responses of three African graminoids to different fertilization, watering, and clipping regimes. Bot. Gaz. 145:151-156.

Waller, S.S., and J. K. Lewis. 1979. Occurrence of C_3 and C_4 photosynthetic pathways in North American grasses. J. Range Manage. 32:12-28.

Warembourg, F.R., and G. Billes. 1979. *In* J.L. Harley and R. S. Russell [ed.]. *The soil-root interface*. Academic Press, London.

Waring, R.H., A.J.S. McDonald, S. Laarson, T. Ericsson, A. Wiren, E. Arwidsson, A. Ericsson, and T. Lohammer. 1985. Differences in chemical composition of plants grown at constant relevant growth rates with stable mineral nutrition. Oecologia 66:157-160.

Watson, M.A., and D. Caspar. 1984. Morphogenetic constraints on pattern of carbon allocation in plants. Ann. Rev. Ecol. Syst. 15:233-258.

Watt, K.E.F. 1968. Ecology and resource management. A Quantitative Approach. McGraw-Hill Book Co., New York. 450 p.

Weaver, J.E., and F.W. Albertson. 1943. Resurvey of grasses, forbs, and underground plant parts at the end of the great drought. Ecol. Monogr. 13:63-118.

Weis, E., and J.A. Berry. 1988. Plants and high temperature stress, p. 329-346. *In* S.P. Long and F.I. Woodward [ed.] *Plants and temperature*. The Company of Biologists, Ltd., Cambridge, England.

Whipps, J.M. 1984. Environmental factors affecting the loss of carbon from roots of wheat and barley seedlings. J. Exp. Bot. 35:767-773.

White, L.M. 1973. Carbohydrate reserves of grasses: a review. J. Range Manage. 26:13-18.

White, L.M. 1989. Growth regulators effect on crested wheatgrass forage yield and quality. J. Range Manage. 42:46-50.

White, L.M. 1990. Mefluidide effect on Caucasian bluestem leaves, stems, forage yield, and quality. J. Range Manage. 43:190-194.

Williams, G.J. III, and J.L. Markley. 1973. The photosynthetic pathway type of North American shortgrass prairie species and some ecological implications. Photosynthetica 7:262-270.

Williams, K., F. Percival, J. Merino, and H.A. Mooney. 1987. Estimation of tissue construction cost from heat of combustion and organic nitrogen content. Plant Cell Environ. 10:725-734.

Williams, P.A., S. Mugambi, P. Nes, and K.F. O'Connor. 1978. Macro-element composition of tall tussocks (*Chionochloa*) in the South Island, New Zealand, and their relationship with soil chemical properties. N. Z. J. Bot. 16:479-498.

Williams, R.E., B.W. Allred, R.M. Denio, and H.E. Paulsen, Jr. 1968. Conservation, development, and use of the world's rangelands J. Range Manage. 21:355-360.

Woodmansee, R.G., J.L. Dodd, R.A. Bowman, F.E. Clark, and C.E. Dickinson. 1978. Nitrogen budget of a shortgrass prairie ecosystem. Oecologia 34:363-376.

Woodmansee, R.G., I. Vallis, and J.J. Mott. 1981. Grassland nitrogen, p. 443-462. *In* F.E. Clark and T. Rosswall [ed.] *Terrestrial nitrogen cycles: processes, ecosystem stratgies, and management impacts.* Ecological Bulletins, Stockholm.

Woodmansee, R.G., and L.S. Wallace. 1981. Effects of fire regimes on biogeochemical cycles, p.649-669. *In* F.E. Clark and T. Rosswall [ed.] *Terrestrial nitrogen cycles: processes, ecosystem strategies and management impacts.* Ecolog. Bull., Stockholm.

Wray, S.M., and B.R. Strain. 1986. Response of two old field perennials to interactions of CO_2 enrichment and drought stress. Amer. J. Bot. 73:1486-1491.

Wright, H.A., and A.W. Bailey. 1982. Fire ecology: United States and Southern Canada. Wiley-Interscience. New York. 501p.

Zelitch, I. 1971. Photosynthesis, photorespiration, and plant productivity. Academic Press, London. 347 p.

Zelitch, I. 1982. The close relationship net photosynthesis and crop yield. BioScience 32:796-802.

Zimmerman, M.H. 1971. Storage, mobilization, and circulation of assimilates, p. 307-322. *In* M.H. Zimmerman and C.L. Brown [ed.] *Trees: structure and function.* Springer-Verlag, New York.

Seed Physiology and Germination

David B. Wester

Introduction

A seed is a fertilized, mature ovule possessing an embryo, endosperm (and sometimes perisperm), and testa. The biological significance of a seed may be appreciated by considering these basic components. First, most seeds are the product of sexual reproduction: the embryo results from the union of male (pollen) and female (ovum) gametes. [The term "seed" may also be applied to nonsexual reproduction (apomixis) whereby seeds often nearly indistinguishable from those produced by sexual reproduction are formed from a diploid nucleus in the ovule.] The embryo itself consists of a rudimentary root (radicle), and one or more seed leaves (cotyledons) separated from the radicle by a stem portion (hypocotyl); the part of the embryo above the cotyledons containing the major leaf bud of the seed or seedling is the epicotyl (or plumule) (Copeland and McDonald 1985). The embryo is, therefore, an embryonic plant, or "plant in miniature."

However, the embryonic plant contained in a seed is separated from its immediate environment by seed coats (testa), and held in a state of quiescence by dehydration, the potential for development into an adult organism is arrested. Surrounding the embryo and providing it a source of nutrition is endosperm tissue. Endosperm may persist at seed maturity (as in cereals), or may be depleted by the developing embryo and reorganized in the cotyledons which become the major storage organ for the seed (as in many dicots). In some species (e.g., *Yucca* spp. L.), endosperm is absent or underdeveloped, and perisperm (which, unlike endosperm, is derived wholly from maternal nucellar tissue) serves as the seed's major food source.

It thus may be said that the "seed is both the culmination of … one plant generation and the start of a new generation" (Bryant 1985:3). The phenomenon of germination links these two generations. Germination is the process whereby a quiescent embryo undergoes a series of morphogenetic changes that results in its transformation into a seedling. Whereas it is certain that germination begins with imbibition, the delineation between completion of germination and commencement of seedling growth is less clear (Mayer and Poljakoff-Mayber 1989). For most researchers, a seed is considered germinated when some part of the embryo (usually the radicle) protrudes from the seed coat (Come and Thevenot 1982; Mayer and Poljakoff-Mayber 1989). Between these two phases in the life cycle of a plant, a well-orchestrated symphony of biochemical changes involving imbibition, protein hydration, synthesis and activation, subcellular structural changes, respiration, macromolecular syntheses, and cell division and elongation transforms a dehydrated seed with barely detectable metabolism into a growing, dynamic autotrophic organism. Many

Author is Associate Professor, Department of Range and Wildlife Management, Texas Tech University.

aspects of this metamorphosis are only vaguely understood. Indeed, Murray (1984a:1) described it as the "still mysterious process of seed germination."

Seed germination, seedling emergence, and early seedling growth are crucial stages in the life history of a plant in at least three respects. First, the transition from one plant generation to the next is mediated by the seed. Thus, an understanding of plant population dynamics must include considerations related to seed dispersal, germination, emergence, and recruitment into the population (Harper 1977). Second, regeneration is a critical component of the niche of a species. Differences in germination strategies are implicated in the phenomenon of species coexistence (Grubb 1977). Third, it is during germination and early seedling growth that a plant organism is most vulnerable to, and least equipped to tolerate, unfavorable environmental conditions (Angevine and Chatob 1979; Cook 1979; Silverton and Dickie 1980; Cavers 1983; Fenner 1987a).

Germination transforms a seed into an actively growing plant, and many aspects of this metamorphosis can be better appreciated with an understanding of seed development, structure, and composition. The first two sections of this review deal with these topics. With this background, the germination process is described. Factors relating to seed dormancy are then summarized, followed by a discussion of factors affecting germination. Some topics related to seed germination ecology are then presented. Finally, some aspects of germination research published in the *Journal of Range Management* are discussed.

The literature addressing seed physiology and germination is voluminous. This chapter is intended to be a summary of some topics in this area, with references to more in-depth sources. There are many excellent texts on germination, notably Khan (1982), Bewley and Black (1983, 1985), Murray (1984b, c), Copeland and McDonald (1985), and Mayer and Poljakoff-Mayber (1989) that are extensively cited.

Seed Development, Structure and Maturation

Seed development and structure will be discussed primarily as they pertain to angiosperms. Comprehensive treatments of this topic are found in Bhatnagar and Johri (1972), Johri (1982), and Raghavan (1976, 1986). Gymnosperm seed development and structure are discussed by Singh and Johri (1972), Raghavan (1976), Norstog (1982), and Steeves (1983).

Land plants evolved from an algal group by means of several vegetative adaptations including protection from desiccation (by development of a cuticle) and morphological differentiation of above- and below-ground structures (Tiffney 1981). The development of reproductive mechanisms suitable in terrestrial environments was a subsequent evolutionary step. The earliest seed-bearing plants appeared about 350 million years ago (Banks 1970; Gillespie et al. 1981). The appearance of the seed habit may be the single most important event in the evolution of land plants (Tiffney 1981; Gensel and Andrews 1984; Haig and Westoby 1989). By virtue of the fact that the seed habit is characterized by *in situ* fertilization, without the need for an external fluid medium for the transfer of male gametes, it represents a truly terrestrial mode of sexual reproduction (Steeves 1983). In fact, the evolution of the seed habit may be viewed as a means for facilitating sexual reproduction on dry land

(Tiffney 1981; Steeves 1983), whereby new and partially developed organisms survive cyclic periods of environmental conditions unfavorable for seedling growth in a state of development temporarily arrested by dehydration (Osborne 1982). Thus, true ovules and seeds derived from integumented female sporangia enjoy both protection and a source of nutrition that spores lack. These features, coupled with development of an embryo within the seed while the latter remained attached to the mother plant, enabled seed-bearing plants to supersede spore-bearing plants during the Cretaceous period (Kozlowski and Gunn 1972; Steeves 1983). In fact, the success and diversity of seed-bearing plants in nearly every terrestrial ecosystem has been attributed to their reproductive system (Galtier and Rowe 1989; DiMichele et al. 1989).

From an evolutionary viewpoint, it is widely accepted that the seed habit is a derivative of free-sporing heterospory, a conclusion supported by the facts that (1) heterospory precedes the seed habit in the fossil record, and (2) the oldest seeds display clear structural relationships to typical megaspores and megasporangia (Steeves 1983). Conceptually, the transition from heterospory to the seed habit involves (1) a structural/functional change wherein the megaspore (see below) contained in the megasporangium (seed or ovule) is not released and is in turn surrounded by an integument, and (2) the development of a delivery system of the male gametophyte to the female gametophyte (Steeves 1983; Steward 1983; DiMichele et al. 1989). Detailed discussions of related processes are found in Smith (1964; evolution of the ovule), Sussex (1966; evolution of heterospory); and Pettitt (1970), Steward (1983), and DiMichele et al. (1989) (each discusses evolution of heterospory and the seed habit).

The Female Gametophyte and Fertilization

Appearance of the archesporial cell in the nucellus of a nascent ovule marks the beginning of development of the female gametophyte (Raghavan 1986). The first division of the archesporial cell produces a parietal cell, which usually deteriorates, and a mother spore cell, or megasporocyte (Fig. 1). The megasporocyte absorbs nucellar cells during development (Evenari 1984). Four haploid cells (megaspores) are produced by meiosis, three of which generally abort. The remaining cell enlarges to become the embryo sac; its nucleus undergoes three mitotic divisions to produce eight haploid nuclei. The most common developmental course in angiosperms (Willemse 1981) at this point is termed the *Polygonum* type, wherein the eight nuclei organize into two sets of four cells. The quartet of cells at the micropylar pole forms the so-called "egg apparatus" comprised of an egg, two synergids and the upper polar nucleus. The quartet formed opposite the micropylar pole (the chalazal pole) gives rise to three antipodal cells and the lower polar nucleus. The two polar nuclei usually fuse into one cell, producing a seven-celled embryo sac. Bhatnagar and Johri (1972) detail eleven other types of embryo sac development.

The part of the embryo sac remaining after the antipodals and the egg apparatus form is called the central cell, whose primary role is to sustain the biochemical growth of the egg until the latter has achieved development necessary for fertilization. The synergids, short-lived cells which often die soon after fertilization, are important in the transportation of metabolites to the central cell (Evenari 1984). The

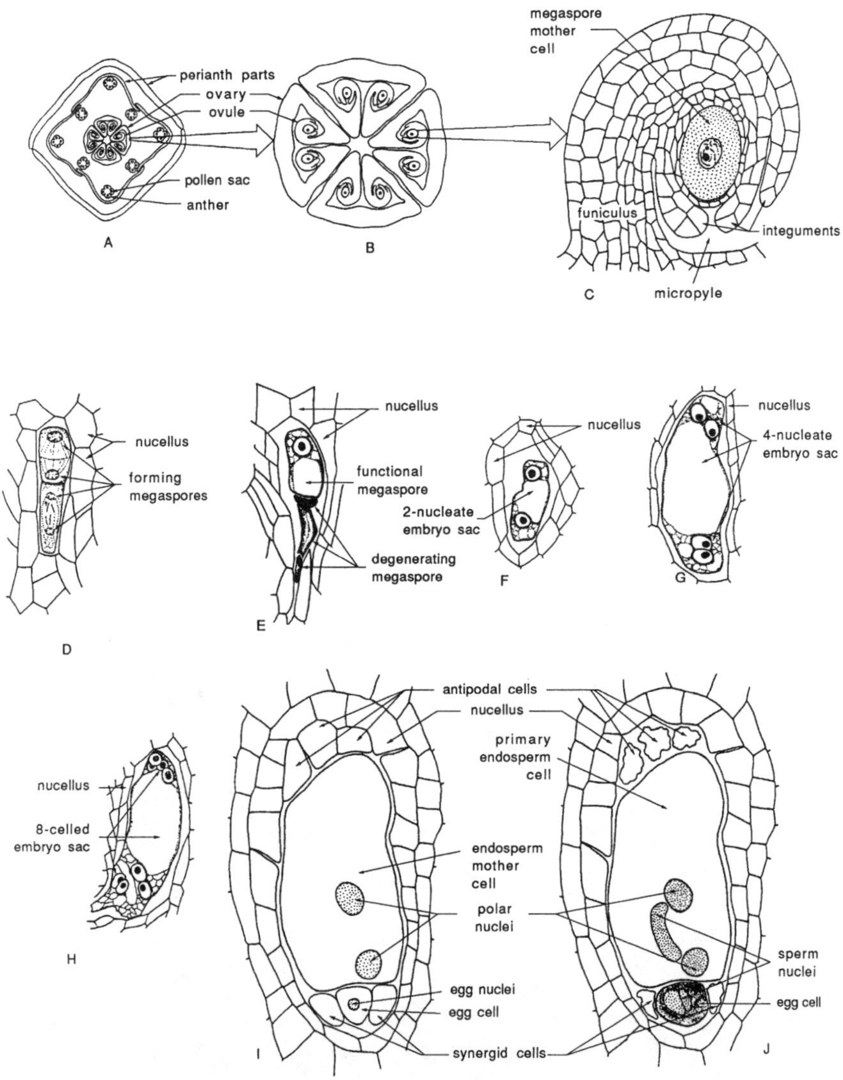

Figure 1. Angiosperm embryo sac development. A: cross section of flower bud, showing four whorls, four-carpelate ovary in center; B: enlarged view of ovary showing four carpels and ovules; C: an ovule, megaspore mother cell in prophase of meiosis; D: late telophase of second meiotic division; E: four megaspores, three degenerating, one functional; F: two-nucleate embryo sac; G: four-nucleate embryo sac; H: eight-celled embryo sac; I: mature embryo sac; J: fertilization, synergids degenerating. See text for discussion. Reproduced by permission from Weier et al. (1974).

antipodal cells are usually obliterated prior to fertilization; only rarely are they persistent. When these cells persist, they act as haustoria for the developing embryo. In some grasses they multiply to numbers suggesting they act in a nutritive role for the embryo sac (Raghavan 1976). Evenari (1984) indicated that the role of the antipodals is yet unclear.

Pollen grains germinate on the female stigma, put forth pollen tubes that move over stigmatic papillae, enter and grow the length of the style, and reach the embryo sac through the micropyle, a minute opening in the integuments of the ovule (Fig. 2). Synergids are involved in fertilization by affecting the directional growth of the pollen tube into the embryo sac; they also play a role in pollen tube discharge (Vijayaraghavan and Bhat 1983; Evenari 1984). Two sperm nuclei enter the embryo sac, the first of which fuses with the egg nucleus forming an embryo. The second sperm nucleus fuses with the two polar nuclei, the result of which is triploid endosperm. [Although double fertilization is usually considered unique to angiosperms, recently it has been reported in *Ephedra nevadensis* Wats. and *E. trifurca* Torr. where, however, endosperm does not develop from the second fertilization event (Friedman 1990, 1991). Friedman (1992) has recently shown that the second fertilization produces four free nuclei that each undergo cellularization (to produce embryos). This

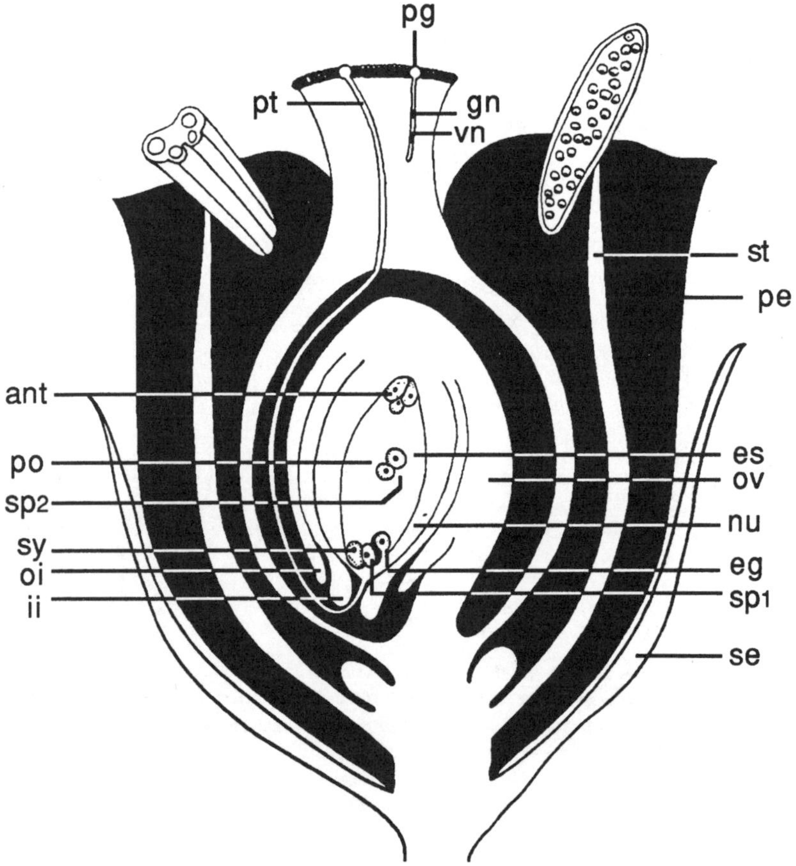

Figure 2. Generalized longitudinal section of a flower. Pollen grain (pg) germinates on stigma and produces a generative nucleus (gn) and a vegetative nucleus (vn); the pollen tube (pt) penetrates the style and enters the embryo sac (es). Other features shown are antipodal cells (ant); egg (eg); inner integument (ii); outer integument (oi); nucellus (nu); ovule (ov); petal (pe); polar nuclei (po); sepal (se); sperm cells involved in double fertilization (sp1, sp2); stamen (st); and synergid (sy). Reproduced from Bhatnaga and Johri (1972).

second fertilization event is developmentally similar to endosperm development in primitive angiosperm groups ("nuclear" endosperm development; see below]. In temporal sequence, however, fertilization of the two polar nuclei is completed prior to fertilization of the egg (Cutter 1971; Bhatnagar and Johri 1972; Evenari 1984), and hence early development of the endosperm precedes that of the embryo (Evenari 1984), lending support to the idea that the endosperm provides a source of nutrition for the developing embryo (Raghavan 1986). However, in the very early stages of embryo development endosperm tissue may be rudimentary, and the embryo derives its nutrition from other sources [e.g., from nutrients available in the embryo sac, possibly supplemented by synergids (Steeves 1983; Dute et al. 1989); see above as well as discussion of the suspensor below].

It should be appreciated that prior to fertilization, development of the mature (but unfertilized) embryo sac is under the genetic control of the maternal plant (and its *2n* complement of genetic information). However, after fertilization, genetic control of seed development becomes more complex because of double-fertilization. Structures such as the seed coat and perisperm remain under the genetic control (*2n*) of the mother plant, whereas the embryo is genetically-controlled by a *2n* genetic complement of male and female derivation, and the endosperm is under genetic control (usually *3n*) of both the mother plant (*2n*) and the male (*1n*) (Evenari 1984).

The Embryo

In embryo formation, the first cell division, yielding an apical cell and a basal cell, establishes polarity for the developing embryo (Evenari 1984; Bryant 1985). Even before this division, however, polarity is established in the fertilized egg by differential distribution of cell contents (Evenari 1984; Raghavan 1976). After the first cell division, monocot and dicot embryos undergo strikingly different development, and even within these subclasses, a number of developmental patterns have been described. Bewley and Black (1985) provide the following general description (Figs. 3, 4). In dicots, the basal cell undergoes several divisions whereas in monocots the basal cell rarely divides. For both groups, the basal cell (or its derivatives) gives rise to the suspensor. Raghavan (1986) suggested three major roles for the suspensor: (1) wall invaginations may absorb nutrients from surrounding tissues and transmit them to the growing embryo, (2) some gene products essential for embryo growth may be synthesized in the suspensor, and (3) hormones [e.g., gibberellic acid (Bewley and Black 1985)] may accumulate in the suspensor and be transported to the embryo. In dicots, the embryo is derived largely from the apical cell, although the basal cell also may contribute to embryo development. On the other hand, in monocots the single basal cell forms the terminal suspensor cell; the embryo and other few suspensor cells are formed from the apical cell (Bewley and Black 1985).

A mature dicot embryo has two cotyledons whereas a monocot has only one (see below). For both groups, the cotyledon(s) are attached to the embryonic axis.

Embryogeny in Poaceae [see Brown (1960) and Negbi and Koller (1962)] differs from that of other monocots and deserves special mention. Many texts (e.g., Bewley and Black 1985) suggest that the single cotyledon in grasses (Fig. 5) is reduced to a haustorial scutellum found adjacent to the endosperm in a mature embryo; its basal sheath is elongated to form a coleoptile which covers the first leaves. However,

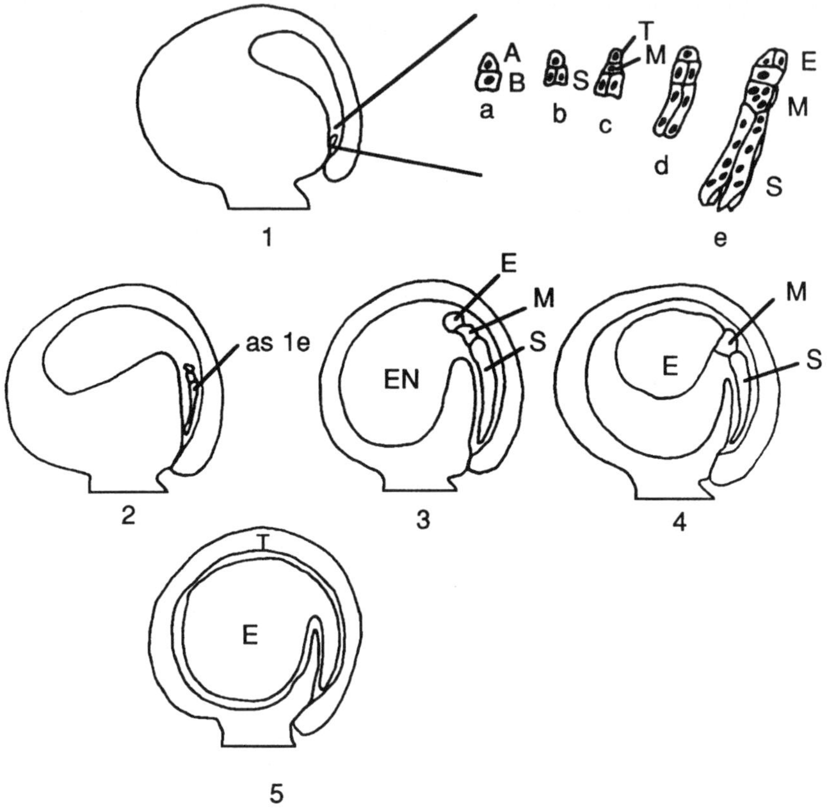

Figure 3. Embryo development in *Pisum sativum*. Early divisions take place in (1); the first division (a) of the zygote produces an apical (A) and a basal (B) cell; the basal cell divides (b) to form two suspensor (S) cells; the apical cells divides (c) to form a middle (M) and a terminal (T) cell. Further divisions (d, e) produce embryonic (E) cells. In (2–5), the suspensor elongates and embryo growth takes place into the embryo sac to utilize reserves and to occlude endosperm (EN). In (5), testa (T) surround the mature embryo. Reproduced from Bewley and Black (1985), with permission from Cooper (1938) and Marinos (1970).

Raghavan (1986) states that morphology of the scutellum and coleoptile are not well understood, and that it is not universally accepted that these organs are foliar in origin. Even if the origin of the scutellum is unclear, it is generally accepted that it has a haustorial function (Mayer and Poljakoff-Mayber 1989). In grasses, the radicle is covered by a coleorhiza on one side of which is a flap-like growth called the epiblast (Fig. 5). The epiblast is often considered a second cotyledon or part thereof (Cutter 1971). The coleorhiza is identified as part of the hypocotyl or suspensor, or as a suppressed primary root (Raghavan 1986).

Endosperm and Perisperm

After fertilization, embryo growth requires more nutrients than those provided in the embryo sac (Raghavan 1976). Nucellar and endosperm tissue serve this need. In fact, embryo development depends completely on simultaneous development of

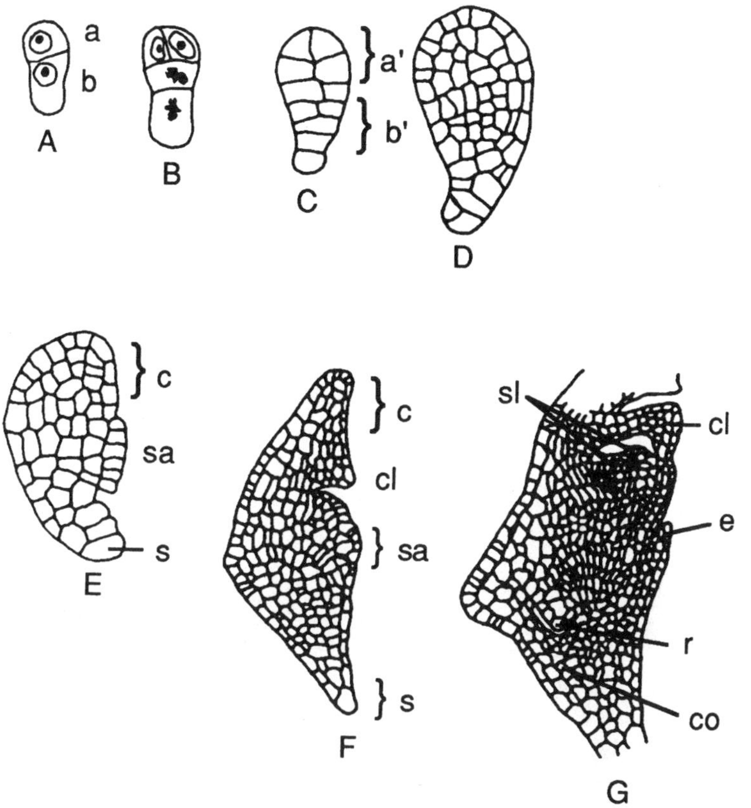

Figure 4. Embryo development in Poaceae. The first division (A) produces an apical (a) and a basal (b) cell, which form derivatives (a′) and (b′), respectively, in further divisions (B, C). After further divisions (D–F), regions that give rise to the short suspensor (s), cotyledon (c), stem apex (sa) and coleoptile (cl) can be seen (F). Near maturity (G), the coleorhiza (co), epiblast (e), first leaf (sl) and root initials (r) are delimited. [Johansen (1950),reproduced from Bewley and Black (1985)]

endosperm (Smith 1982). The nucellus, which contains the embryo sac, is usually absorbed by the developing embryo and endosperm. When the nucellus persists and acts as a storage organ, it is termed *perisperm* after the seed matures. In *Yucca* spp. L., perisperm provides the major source of nutrition for the seed, and endosperm is absent.

In most cases, perisperm fails to develop, and endosperm tissue provides an important source of nutrition for the developing embryo. Although three types of endosperm development have been described, the physiological mechanisms they involve are unknown (Evenari 1984). *Nuclear* and *cellular* endosperm development are found in both monocots and dicots; *helobial* endosperm development is less common and found only in monocots. In nuclear endosperm development, the endosperm nucleus undergoes several free-nuclear divisions prior to cell wall formation (i.e., cytokinesis does not accompany these divisions), the result of which is

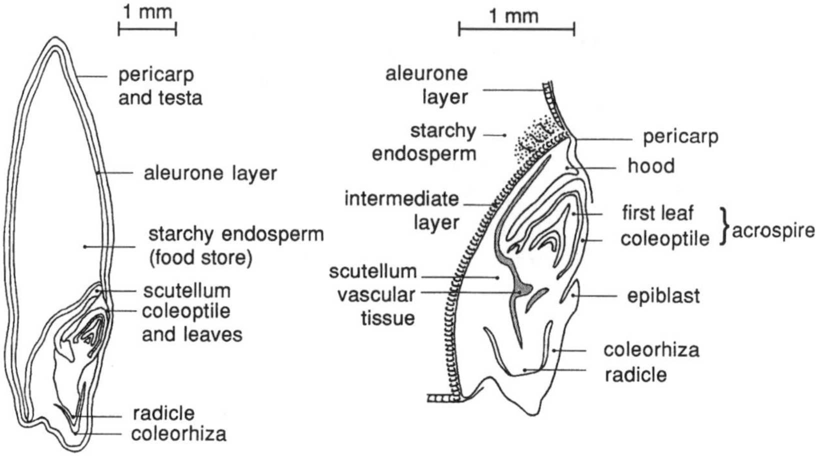

Figure 5. Longitudinal section of a *Triticum aestivum* caryopsis (left) and embryo (right). Reproduced with permission from Bewley and Black (1983).

several thousand free nuclei. These nuclei then arrange themselves peripherally in the central cell, followed by cellularization. This type of endosperm development is characteristic of gymnosperms (Steeves 1983). The cellular type of endosperm development, characteristic of angiosperms (Steeves 1983), lacks the free-nuclear phase; i.e., the regular cycle of mitosis and cytokinesis accompanies primary endosperm development. Early embryo segmentation is highly regular in this type of endosperm development, and Stebbins (1974, cited in Steeves 1983) suggests that this accelerates the process of seed maturation [in comparison to that displayed by gymnosperms; see Steeves (1983)].

In some species, endosperm is absorbed by the developing embryo, so that the former is reduced or absent at maturity; such seeds are termed *nonendospermic* (or exalbuminous). In other species (notably grasses), mature seeds have endosperm and hence are referred to as *endospermic* (or albuminous). Most endosperm cells in Poaceae are nonliving at seed maturity. These cells contain storage materials and are surrounded by living cells forming an aleurone layer whose primary function may be to release enzymes for the mobilization of nutrients in storage cells (Black et al. 1983; Ashford and Gubler 1984; Dalling and Bhalla 1984). In barley (*Hordeum vulgare* L.), for example, α-amylase (important in endospermic starch digestion) is synthesized in the aleurone layer in response to gibberellic acid produced by the embryo (Raghavan 1976). Most aleurone cells consist of oils and proteins, distinguishing them from endospermic cells (Murray 1984a). In addition to nourishing the developing embryo, the endosperm also is important as a source of energy for seed germination (Raghavan 1986).

Seed Coats (Testa)

The hard coats surrounding the embryo and endosperm and separating them from their immediate "outside" environment are called testa. Testa develop from integuments surrounding the ovule (Fig. 2) mostly by reduction in thickness. Important functions of testa include protection from mechanical injury, microorganisms, and insects; testa also may be important in seed dispersal (Bhatnagar and Johri 1972; Fahn and Werker 1972).

From a physiological viewpoint, perhaps the most interesting features of seed coats are the often fatty or waxy inner and outer cuticle layers, and one to several layers of thickened protective cells. These features render testa to some degree impermeable to water and gases [although the precise mechanisms involved are not yet fully understood (Mayer and Poljakoff-Mayber 1989)], and thus may exert some regulatory control over metabolism and growth of enclosed tissues (Brown 1972; Bewley and Black 1985). Testa also may influence seed dormancy for the same reason (Cutter 1971; see below). Mucilages that develop from the epidermis of testa of some species may be implicated in water uptake (see below).

Seed Maturation

Desiccation is the normal, final stage of seed maturation in most species (Bewley and Black 1985; Kermode and Bewley 1986; Kermode et al. 1986). In fact, a characterizing feature of most seeds is their dehydrated state. During desiccation, normally metabolizing seed tissue with a moisture content of about 90% is transformed into quiescent tissue often with a moisture content less than 10% (Brown 1972). Seeds that can be stored in a state of low moisture content are called orthodox seeds. *Recalcitrant* seeds must maintain relatively high water content for viability. Large-seeded hardwoods (e.g., *Castanea* spp. Mill. and *Quercus* spp. L.) and most aquatic species [e.g., wild rice (*Zizania aquatica* L.)] are examples of recalcitrant seeds (Bewley and Black 1985). The phenomenon of seed desiccation is related to seed storage and longevity, and thus can have an important bearing on seed germination. For example, one general rule (Bewley and Black 1985) states that seed storage life doubles for each 1% decrease in seed moisture content. Because of their dehydrated state, orthodox seeds can survive harsh conditions imposed both by temperature extremes and lack of oxygen.

Most "dry" seeds have a moisture content ranging between 4 and 16% (on a fresh weight basis) (Priestley 1986). An important finding on the state of water in "dry" seeds is that most unimbibed stored seeds possess little or no "free" water (i.e., most water is "bound" water, or water associated with large molecules). Movement of water out of hydrated, turgid cells is initially an active process (Bryant 1985); in later stages, equilibrium moisture content in open storage depends on ambient relative humidity. This relationship is conveniently summarized in a sorption isotherm (e.g., Priestley 1986: Fig. 3). Differences in equilibrium moisture content among species are largely attributable to differences in quantity of stored lipids: seeds with large amounts of lipid (e.g., Table 1) tend to have relatively lower equilibrium moisture content than seeds with large amounts of stored starch (Bewley and Black 1985; Priestley 1986).

During desiccation, storage protein synthesis ceases and other syntheses associated exclusively with the germination process begin; these processes, however, are not clearly understood (Bewley and Black 1985; Kermode et al. 1986). For example, enzymes (e.g., α-amylase) exclusively associated with post-germination events

Table 1. Chemical composition and major storage organ of some monocot and dicot seeds (from Bewley and Black [1985], Winton and Winton [1932], and Crocker and Barton [1957]).

	Average Composition %			Major storage organ
	Protein	Fat	Carbohydrate	
Grasses				
Zea mays	11	5	75	endosperm
Avena sativa	13	8	66	endosperm
Triticum aestivum	12	2	75	endosperm
Secale cereale	12	2	76	endosperm
Hordeum vulgare	12	3	76	endosperm
Legumes				
Vicia faba	23	1	56	cotyledon
Pisum sativum	25	6	52	cotyledon
Arachis hypogaea	31	48	12	cotyledon
Glycine max	37	17	26	cotyledon
Other				
Ricinus communis	18	64	trace	endosperm
Pinus ponderosa	35	48	6	megagametophyte

(e.g., storage starch hydrolysis) are not produced in response to gibberellic acid unless the seed has been desiccated (Kermode et al. 1986). Thus, it may be said that desiccation plays an important role in the transition from seed developmental processes to processes associated with germination (Osborne 1983; Bewley and Black 1985; Kermode et al. 1986; Bewley et al. 1989; Rosenberg and Rinne 1989). However, desiccation is not necessary for this transition to take place in some species. Kermode and Bewley (1989) showed that immature castor bean (*Ricinus communis* L.) seeds successfully completed the transition from a developmental to a germinative mode if they were maintained in an atmosphere of high relative humidity after detachment from the mother plant. In other words, whereas desiccation is part of the "normal" maturation process of a seed, it is not required if seeds are maintained under conditions of high relative humidity.

Chemical Composition of Seeds

About 70% of all food for human consumption comes directly from seeds, with a large portion of the remainder derived from animals fed grain (Bewley and Black 1985). Therefore, the literature on seed chemical composition is extensive, and generalizations are difficult because of wide variation in composition among species.

The chemical composition of seed storage material (Table 1) provides a basis for distinguishing between seeds that store mainly lipids and seeds that store mainly carbohydrates; the former group is the larger of the two. Both groups include seeds that store protein as well, although few species (e.g., soybean [*Glycine max* (L.) Merr.]) are predominantly protein-storers (Mayer and Poljakoff-Mayber 1989). Seed chemical composition is largely under genetic control (Mayer and Poljakoff-Mayber

1989). Environment and agronomic practices exert an influence as well, but these effects are generally minimal (Bewley and Black 1985).

Storage Locations

Seeds store nutrients in embryonic and/or non-embryonic (e.g., perisperm, cotyledons and/or endosperm, and rarely the embryonic axis) locations (Table 1). For example, dwarf bean (*Phaseolus vulgaris* L.), cucumber (*Cucumis sativus* L.), and squash (*Cucurbita pepo* L.) are nonendospermic species that store proteins and carbohydrates in cotyledons that are raised above the soil surface by an expanding hypocotyl during germination (epigeal germination). Cotyledons also are the major storage location for these nutrients in pea (*Pisum sativum* L.) and broad bean (*Vicia faba* L.); in these species, however, the hypocotyl remains short and compact, and as a consequence cotyledons are not raised above the soil surface during germination (hypogeal germination). Grasses generally are endospermic species whose primary storage material is carbohydrate (starch); wheat (*Triticum aestivum* L.), corn (*Zea mays* L.), and barley are examples of hypogeal germinators. Among the cereal grains, most of the seed's protein and starch are found in the endosperm whereas most of the oil is found in the scutellum; oil is also found in the aleurone layer. Endosperm is the storage location for oils in castor bean, an epigeal germinator. Perisperm may be the primary storage location in some species (e.g., *Yucca* spp. L.) (Bewley and Black 1983).

Seed carbohydrates may be disaccharides (sucrose), oligosaccharides (e.g., raffinose and stachyose), or polysaccharides (commonly starch). Starch may be stored in starch grains (formed in leucoplasts) in endospermic tissue in cereals or in the cotyledons of dicots, or as cell wall constituents (Mayer and Poljakoff-Mayber 1989). Although it is uncommon for sugars to be the major storage carbohydrate (Bewley and Black 1983), oligosaccharides (as well as galactomannans, a polysaccharide) play a very important role in seed germination (Mayer and Marbach 1981; Bewley and Black 1985; see below). Most cereals cultivated for food consumption [e.g., wheat, rice (*Oryza sativa* L.), corn, and sorghum (*Sorghum* spp. Moench)] store starch (occurring as granules) as the most common carbohydate (Mayer and Poljakoff-Mayber 1989). Starch serves as a major source of glucose for seed germination (Mayer and Poljakoff-Mayber 1989). Different parts of the seed show different patterns of starch accumulation and utilization. Galactomannans (often referred to as hemicelluloses; Mayer and Poljakoff-Mayber 1989) are common in certain endospermic legumes [e.g., *Gleditsia ferox* and white clover (*Trifolium repens* L.) (Bewley and Black 1983); Mayer and Poljakoff-Mayber (1989)]; for these species, starch is usually absent (Bewley and Black 1985). Non-storage carbohydrates (e.g., polyuronides) frequently are found in seeds coats, where they may be implicated with seed dispersal and water uptake (see below). Additional information on carbohydrates in seeds can be found in Matheson (1984), Duffus and Cochrane (1982), Bewley and Black (1983, 1985), Mayer and Marbach (1981), Nozzollilo et al. (1983), and Mayer and Poljakoff-Mayber (1989).

Lipids

Lipids are the most widespread storage material in seeds (Mayer and Marbach 1981). In fact, there are more species that store oils in their seeds than there are species that store proteins or carbohydrates (Bewley and Black 1983; also see Vaughan

1970). Plastids are the only sites for lipid synthesis in seeds, and such synthesis utilizes sucrose (produced elsewhere and translocated to the seed) as a carbon source (Slack and Browse 1984; Mayer and Poljakoff-Mayber 1989). Most storage lipids are unsaturated neutral fats or oils that are esters of glycerol and long-chain monocarboxylic acids occurring as triacylglycerol or triglycerides. Other storage lipids include glycolipids, phospholipids, sterols, and waxes (Mayer and Marbach 1981). Since the carbons in lipids are more highly reduced than those in carbohydrates, lipid oxidation yields more energy than carbohydrate oxidation; thus, lipid-storers maximize the quantity of stored energy in a small volume of tissue. [Of course, more energy must be expended to convert sucrose to lipid than to carbohydrate, so seed yields of oil-rich species are less than seed yields of species that store predominantly carbohydrates (Slack and Browse 1984)]. Although there is much information on the chemical composition of lipid-storers [e.g., oil and protein contents of some 4000 species have been studied by Earle and Jones (1962), Jones and Earle (1966) and Barclay and Earle (1974); also see Vaughan (1970) and references therein], there is less understanding about how storage lipids are produced (Slack and Browse 1984). Slack and Browse (1984) provide an excellent review of lipid synthesis in developing seeds (also see Mayer and Marbach 1981; Bewley and Black 1983, 1985). Galliard (1983) discusses starch-lipid complexes in cereal grains. Priestley (1986) discusses seed aging and lipid content.

Proteins

On a world-wide basis, cereals, and secondly, legumes are the most important protein sources for man and animals (Bewley and Black 1983). Therefore, there is considerable information on seed proteins for cereals and legumes (Brown et al. 1982; Daussant et al. 1983; Bewley and Black 1983, 1985; Murray 1984d; and Mayer and Poljakoff-Mayber 1989).

Proteins in seeds may be either enzymatic (metabolically active) or storage proteins. Seed proteins have been classified (Osborne 1924) into four groups based on their solubility in water, weak acids or alkalis: (1) albumins (largely enzymatic proteins) are soluble in water and salt solution, (2) glutelins are soluble only in acids and alkalis, (3) globulins are soluble in salt solution but only slightly water-soluble, and (4) prolamins are soluble in 70–80% aqueous ethanol but are not water-soluble. Globulins and storage proteins often are synonymous, especially for legumes (Mayer and Poljakoff-Mayber 1989), and prolamins and storage proteins often are synonymous in monocots.

The major storage proteins in monocots are prolamins or glutelins whereas in dicots the major reserve proteins are globulins (Mayer and Marbach 1981). In grasses, protein composition in endosperm may differ dramatically from protein composition in the aleurone layer (Bewley and Black 1983:Table 2.6), and although the endosperm may contain a major proportion of the total protein in the seed, the aleurone layer is richer in protein (Dalling and Bhalla 1984). Only small amounts of protein are found in the axis and the scutellum (Bewley and Black 1985).

Storage proteins are synthesized mainly in association with rough endoplasmic reticulum (Murray 1984d; Mayer and Poljakoff-Mayber 1989) by cells in endospermic or cotyledonary storage tissues (Osborne 1981) relatively late in seed development (Brown et al. 1982). Reserve proteins generally are deposited in specialized organelles called protein bodies. Protein accumulation usually accompanies cell expansion and often continues throughout seed desiccation (Murray 1984d). However, the pattern of protein synthesis changes during desiccation in that seed drying leads to the suppression of the synthesis of proteins associated with seed development, and induces the synthesis of proteins associated with germination and growth (Bewley et al. 1989).

The Germination Process

Germination is the process whereby a seed is transformed into an actively growing organism. The first step in germination is imbibition, or water uptake, and the completion of germination is evidenced by radicle protrusion from testa. However, radicle protrusion can occur by cell expansion alone (in the absence of cell division) (Haber and Luippold 1960a,b; Bewley and Black 1985; Mayer and Poljakoff-Mayber 1989), and the process of cell-wall plasticization is not fully understood (Bewley and Black 1985). Consequently, Bewley and Black (1983:124) concluded:

> "No satisfactory answer can therefore be given to the key questions—what biochemical events are specific to germination and how are they controlled?"

After a detailed discussion of the transition from the resting to the germinating state in seeds, Mayer and Marbach (1981:129) stated:

> "Many details [of this transition] are lacking. In particularly [sic] many of the mechanisms operating and their regulation and control processes are unknown. Interaction between the seed and its environment and recognition of changes in the environment are well established but their biochemical and physiological basis is not understood. Although it is becoming increasingly clear that the transition from the resting to the germinating state in the seed is characterized by special physiological and biochemical processes which differentiate it from growth, their nature is still essentially unclear."

Mayer and Poljakoff-Mayber (1989) provide a recent discussion of the metabolism of germinating seeds, and although this review seems less emphatic about the "missing details", it is still clear that much remains to be learned about this complex aspect of seed germination.

Some salient features of the germination process are discussed below. This discussion is taken from numerous sources. Details of the metabolism of seed germination often are species-specific, and no attempt is made here to be comprehensive; instead, this discussion is intended to present some general concepts.

Imbibition

Germination commences with water uptake and seed tissue rehydration. In fact, germination can not occur as long as water entry into the seed is prevented (Mayer and Poljakoff-Mayber 1989). Imbibition is affected by seed characteristics (e.g., seed composition and seed coat properties) and water relations between seed and substrate. Seed characteristics are emphasized here; water relations between seed and substrate are discussed in a later section.

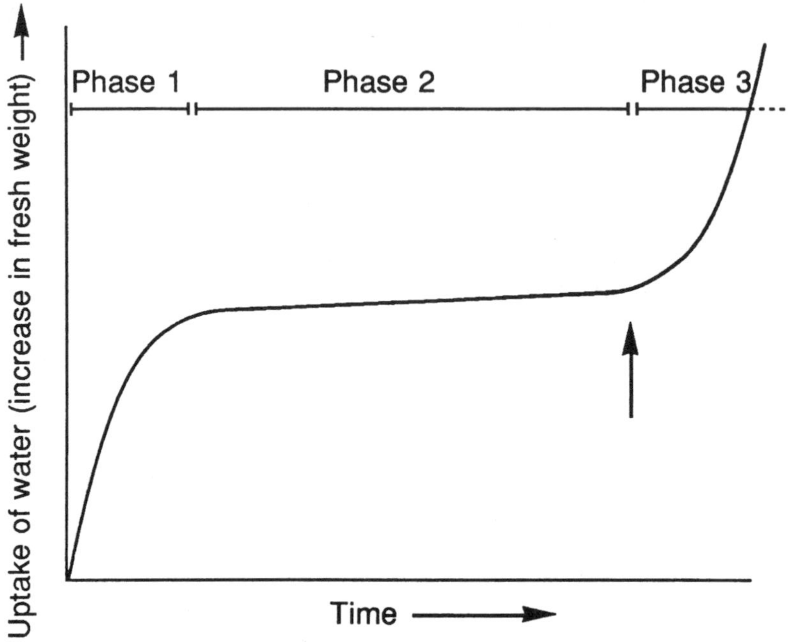

Figure 6. Pattern of water uptake by germinating seeds over time. Radicle protusion through testa (indicated by arrow) marks the completion of germination. Reproduced with permission from Bewley and Black (1985).

Imbibition takes place in three basic phases (Fig. 6), the end result of which is water uptake equal in amount to about 2 to 3 times the dry weight of the seed (Bewley and Black 1985) generally resulting in a tissue water content of 22 and 51%, depending on species (Hunter and Erikson 1952). Small seeds of a species gain proportionately more weight during early imbibition than large seeds (Knipe 1971). Imbibition initially is a passive process controlled largely by high (negative) matric potential of the seed. In fact, the water potential of desiccated seeds may be –400 MPa (Shaykewich 1973). In the second (or lag) phase of imbibition, water uptake is less rapid and is controlled mostly by osmotic and pressure potentials of the seed. Total cell water potential of the seed during this phase may not exceed –1 to –1.5 MPa. It is during this phase that the major metabolic events occur in preparation for radicle emergence. The third phase, coincident with radicle emergence, is marked by an increase in water uptake that is initially related to changes in radicle cells as they enlarge. This description is idealized: imbibition is a complex process, and although protein is the major seed component which imbibes water (Mayer and Poljakoff-Mayber 1989), different seed components hydrate at different rates (Spurny 1972; Mayer and Marbach 1981; Simon and Mills 1983; Simon 1984; Mayer 1986; McDonald et al. 1988a; Mayer and Poljakoff-Mayber 1989). For example, whereas embryonic axes of soybean show the reverse sigmoidal-shaped water absorption pattern (Fig. 6) in that water uptake in this structure increases at germination, cotyledons do not show this increase at germination and therefore these structures fail to absorb more water even after germination (McDonald et al. 1988a).

Seed coats are important in the early stages of imbibition because of their influence on permeability. Most discussions of this topic relate to seed coat impermeability, the resulting interference with water uptake, and seed dormancy (see below). However, in a detailed study of soybean seed imbibition, McDonald et al. (1988b) found that the seed coat is important in regulating rate and evenness of seed hydration. They showed that during the first 8 hours of imbibition the seed coat slowed water uptake in the embryonic axis, cotyledons, and the whole seed. Heavily cutinized palisade cells of the epidermis or the presence of phenolics are probably responsible for this initial slowing of water uptake. After 8 hours of soaking, the seed coat had no effect, or it enhanced water uptake, the latter effect due to the large water holding capacity of the seed coat. Seed coats also facilitated water movement around the seed so that both cotyledons absorbed the same amount of water regardless of their respective proximity to the water source.

Germination was described in the Introduction as "a well-orchestrated symphony of biochemical changes" involving all aspects of growth. This perhaps can be best appreciated by considering seed imbibition and resumption of active metabolism. It is well established that both rate and extent of germination are reduced as water availibility decreases. It is also widely accepted that seeds must attain a certain minimum (species-specific) critical, or threshold, hydration level for germination (Hegarty 1978). Early research related germination to whole-seed hydration. For example, Hunter and Erickson (1952) showed that soybean seeds must reach a minimum whole-seed moisture content of about 51% for germination. Much recent research also focuses on whole-seed hydration levels (e.g., Roberts and Ellis 1989). However, it also has been established that different seed components hydrate at different rates and reach different levels of hydration (McDonald et al. 1988a). These differences in seed tissue hydration are correlated with physiological activity involving structure and organization of cell constituents (Hadas and Stibbe 1973; Mayer 1986; Vertucci and Leopold 1986; Mayer and Mayber-Poljakoff 1989). Thus, some recent research has focused on tissue-specific hydration. For example, McDonald et al. (1988a) refined Hunter and Erickson's (1952) conclusion by establishing that imbibing soybean seeds reach about 55% moisture content (on a whole-seed basis) whether they germinate or not, and that for germination to occur, the embryonic axis must attain about 70% moisture content. Additionally, water content of various seed components must be compatible to achieve metabolic synchrony (Vertucci and Leopold 1986). Mayer (1986) suggested that metabolic coordination is maintained by means of growth regulators in that initiation of metabolic activity in a given seed component in determined by transport of a growth regulator from another component.

Two consequences follow very soon after imbibition. First, it is frequently observed that seeds, upon imbibition, are "leaky": organic solutes such as sugars, proteins, amino acids, and organic acids, symplastic in origin (rather than apoplastic; Simon 1984), are released along with colloids of adsorbed gases (Simon and Mills 1983; Simon 1984; Bewley and Black 1985). Leakage generally decreases as imbibition progresses. Several phenomena may be involved with leakage. Leakiness may suggest that components of the cells are damaged by the processes of desiccation and rehydration (Bryant 1985). "Selectively permeable" membranes (such as the plasmalemma and the tonoplast) that normally retain cellular solutes evidently lose their integrity during desiccation and are not effective membranes upon rehydration (Bewley and Black 1985). Cell membranes also may be damaged by the influx of water driven by the extremely sharp water potential gradient (Simon and Mills 1983). A final consideration relates to cell membrane structure. Simon [1974; also see Simon and Mills (1983)] suggested that membranes of "dry" seeds can not be described by the widely-accepted model of lamellar or bilayer membranes, but instead are "disorganized". The strong influx of water experienced by outer cells disrupts their membranes; however, under the slow imbibition experienced by inner seed cells, the lamellar or bilayer membrane structure is reestablished (Simon and Mills 1983; Simon 1984). McKersie and Senaratna (1983) found no experimental evidence for Simon's suggestion, and there is growing evidence that membranes in desiccated plant systems remain intact (Opik 1980; Leopold and Vertucci 1986).

A second event following soon after imbibition is an increase in seed respiration. Pradet (1982) indicated that ATP synthesis begins as soon as water penetrates lettuce (*Lactuca sativa* L.) seeds. Roberts and Ellis (1989) reported that lettuce and soybean seeds respiration began when seed water potential reached –14 MPa; further, respiration rate increased linearly with increasing water content until seeds attained full hydration at about –1.5 MPa. One of the major events occurring during the second phase of imbibition is activation of enzyme systems, allowing for breakdown and mobilization of reserve materials in cotyledons and/or endosperm to growing organs. For example, protein and RNA synthesis begin almost as soon as cells are rehydrated. The immediate resumption of metabolic activity indicates that components necessary for metabolism (e.g., membrane and enzyme systems) evidently "survive" not only desiccation associated with seed maturation but also rehydration coincident with imbibition (Simon 1984; Bryant 1985).

As mentioned above, leaky membranes and rapid increases in seed metabolism are observed soon after imbibition. The fact that "leakiness" decreases and eventually stops suggests that a crucial early metabolic function in seeds relates to membrane synthesis and repair (Mayer and Marbach 1981; Bryant 1985). In fact, this in one hypothesis concerning the beneficial effects resulting from osmoconditioning (see below).

Early Metabolism

Energy-Related Processes

Quiescent seeds respire only slightly. The immediate release of gas (or "wet burst") often following rehydration results from gas that is released from colloidal adsorption and not respiration. However, seed respiration soon increases dramatically upon water uptake. Three or four phases of oxygen consumption are recognizable (Fig. 7) (Bewley and Black 1985) which appear to parallel the phases of imbibition (Fig. 6) (Mayer and Poljakoff-Mayber 1989); growing tissues (e.g., the embryo) show three phases whereas non-growing (storage) tissues (e.g., cotyledons) show four phases (Pradet 1982). The first phase is due in part to activation and hydration of

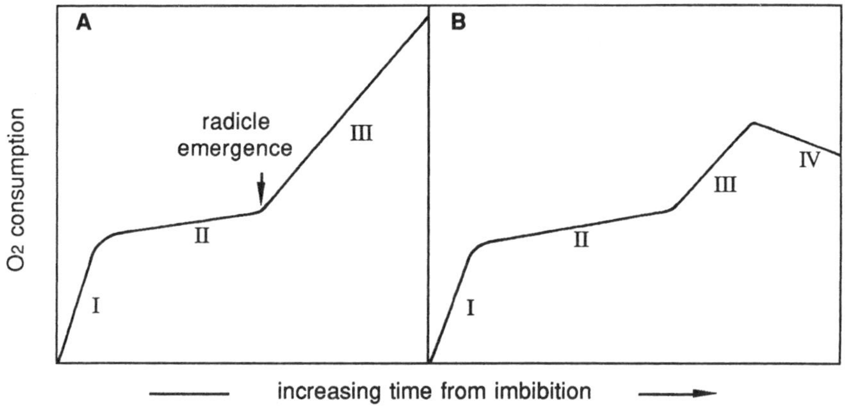

Figure 7. Embryo (A) and storage tissue (B) oxygen consumption over time. Phases of oxygen consumption (indicated by roman numerals in figure) are discussed in text. Reproduced with permission from Bewley and Black (1985).

mitochondrial enzymes. For example, ATP synthesis linked to an oxygen-dependent mechanism (probably the cytochrome electron system) begins as soon as water enters lettuce seeds (Pradet 1982). Seed tissue ATP content may increase 10-fold during the first hour of imbibition (Obendorf and Marcus 1974; Simon 1984). The lag phase in oxygen consumption is associated with full seed hydration and reactivation of pre-existing enzymes. Rapid oxygen release in phase 3 is attributable to newly formed embryonic mitochondria and respiration in newly formed cells in the growing axis. Radicle emergence from testa occurs between phases 2 and 3. It is noteworthy that the respiratory substrate during early seed metabolism is not bulk storage materials (these are usually hydrolyzed following seed germination) but sucrose and raffinose-series oligosaccharides in the embryo (Simon 1984; Bewley and Black 1985). Seed respiration generally increases with increased temperature and oxygen concentration (Mayer and Poljakoff-Mayber 1989).

Mitochondria in unimbibed seeds display poor internal differentiation and are metabolically inefficient (Vertucci and Leopold 1986). However, with water uptake, mitochondrial development has two aspects: (1) repair and activation of pre-existing organelles, and (2) production of new mitochondria (Bewley and Black 1985). Mayer and Marbach (1981) suggested that the dramatic increases in ATP observed at the onset of germination can not be accounted for by oxidative phosphorylation; they discussed alternative mechanisms of ATP production and the energy-yielding processes in germinating seeds in general. In contrast, others (e.g., Hourmant and Pradat 1981; Pradet 1982; Bewley and Black 1985) stated that oxidative phosphorylation via the normal cytochrome pathway within the mitochondria begins operation soon after the start of imbibition, and that this mechanism may be sufficient to account for ATP production during early metabolism. In a more recent literature review, Mayer and Poljakoff-Mayber (1989) stated that glycolysis linked to the tricarboxylic acid (TCA) cycle and the "usual" cytochrome pathway is the main respiratory pathway responsible for ATP production in germinating seeds, and that such production can begin within minutes of imbibition. Additionally, they suggest that an alternative respiratory pathway, the pentose phosphate pathway, produces carbon skeletons and reducing power for anabolism.

Protein, RNA and DNA Syntheses

The stage of nuclear development in quiescent seeds is species-dependent, and is best appreciated with a brief review of some basic concepts (Postlethwait and Hopson 1989). Eukaryotic cells have a four-phase cell cycle, three of which result in growth (interphase) and one in division (mitosis). Interphase is divided into three stages. The phase referred to as "G1" (or "gap 1") initiates interphase; during this phase there is general, active growth during which cellular activity is directed toward synthesis of new proteins, ribosomes, mitochondria, and endoplasmic reticulum in preparation for cell division. During this stage most of the cell's nuclei contain double-stranded DNA. Cells enter the "S" (or DNA synthesis) stage when the G1 phase is completed. During this phase, double-stranded DNA is duplicated and becomes four-stranded. After DNA copying is completed, cells enter another growth period, "G2" (or "gap 2"). The cell continues to synthesize new proteins in preparation for division during G2. Mitosis (the "M" stage) follows G2, and during this phase there is actual cell division; this stage includes cytokinesis.

With this background in mind, some brief statements about nuclear development in seeds follows (Deltour 1985). In dry seed embryos, cells blocked at mitosis have never been clearly observed. In spite of great variability between and within species, the general tendency is for most cells in quiescent embryonic axes to be arrested in the G1 phase with double-stranded DNA. The fact that most embryonic cells are in the G1 phase of nuclear development may be of considerable importance with respect to cell activation as well as early germination ecology because (1) G1 cells possess high viability for extended time periods, and (2) embryos show maximum drought resistance between the end of the G1 phase of the last mitotic cycle during maturation and the end of the G1 phase of the first mitotic cycle during germination (see below). For some species, however, cells may be arrested either in the G1 phase (with double-stranded DNA) or the G2 phase (with four-stranded DNA). When most cells are in the G1 phase, a high degree of synchrony in the first mitotic cycle may be expected when division occurs. However, regardless of their state of nuclear development, the quiescent cells of seeds develop toward mitosis, cell division, and growth following imbibition (Mayer and Poljakoff-Mayber 1989).

It has long been recognized that DNA is damaged during seed storage. Also, RNA and protein syntheses begin earlier than DNA synthesis in imbibing embryos. For example, protein and RNA syntheses in germinating rye (*Secale cereale* L.) seeds can be detected 5-10 minutes after the start of imbibition whereas DNA replication is not detected until four hours after the beginning of imbibition (Osborne et al. 1980/81). However, one of the early and important events in seed germination concerns repair of damaged DNA, which takes place prior to DNA replication (Osborne et al. 1980/81; Osborne 1983). With respect to osmoconditioning (see below), it should be noted that germinating embryos that are imbibed under osmotic stress actively metabolize but do not enlarge; while there may be slight DNA synthesis during this period of arrested development (Bray et al. 1989), DNA repair proceeds (Osborne 1982). The delay period until imbibing seeds begin DNA synthesis lengthens the longer seeds remain in a desiccated state (Osborne 1982). Osborne et al. (1980/81), Osborne (1982), and Mayer and Poljakoff-Mayber (1989) review important aspects of this topic.

Storage proteins are synthesized and stored relatively late in seed development (Brown et al. 1982) (see above). The process of desiccation plays a role, as yet not completely understood, in signalling the transition from reserve protein synthesis and storage to production of enzymes necessary for germination and subsequent reserve protein mobilization (see above; Simon 1984; Bewley and Black 1985; Kermode et al. 1986; Mayer and Poljakoff-Mayber 1989). Desiccated seeds do not contain polysomes (sites for protein synthesis); however, dry seeds contain the components necessary for protein synthesis (Simon 1984; Kermode et al. 1986; Mayer and Poljakoff-Mayber 1989). Thus, polysome production begins soon after imbibition, and germinating seeds quickly attain protein-synthesizing capabilities (Mayer and Marbach 1981; Simon 1984). Some messenger RNA evidently is stored in dry seeds and used for initial protein synthesis (Delseny et al. 1980/81; Simon 1984). However, most of this stored RNA is degraded early in the germination process (Delseny et al. 1980/81; Kermode et al. 1986; Mayer and Poljakoff-Mayber 1989). Additional RNA is required to synthesize proteins necessary for completion of germination, and this RNA is synthesized very early in the germination process (Mayer and Poljakoff-Mayber 1989). Further details about RNA synthesis in germinating seeds is provided by Dure (1979), Bewley and Black (1985), Kermode et al. 1986, and Mayer and Poljakoff-Mayber (1989).

Mayer and Poljakoff-Mayber (1989) succinctly summarize information relative to activation of enzymes associated with early protein synthesis. Such enzymic activity involves in part activation of pre-existing inactive enzymes, by which means seeds can store enzymes in an energetically-efficient way. Changes in enzyme activity, protein, RNA, and DNA syntheses during germination are summarized in

Metabolism of Reserve Materials

Proteins, lipids, and carbohydrates comprise the major storage reserves in seeds. These reserves are hydrolyzed during germination first in the seedling, then in the cotyledon or scutellum, and finally in the endosperm (Oaks 1983). That is, utilization of endosperm reserves is mostly a post-germination (i.e., post-radicle protrusion) phenomenon providing nutritional support to an enlarging embryo but is not responsible for initiating its enlargement (Jacobsen et al. 1979; Karssen et al. 1989). Carbohydrates from storage reserves or resulting from lipid metabolism provide energy and constituents for new cell formation (Oaks 1983). Also, proteins may be deanimated and then used as respiratory substrates or for synthesis of new compounds (Mayer and Poljakoff-Mayber 1989).

Proteins

As mentioned earlier, proteins are stored in protein bodies. Protein bodies also are sites for proteolysis (Mayer and Marbach 1981), a process which is detectable during the first day of germination (Laidman 1982). Synthesis of enzymes responsible for protein hydrolysis apparently are under hormonal control. Quiescent seeds con-

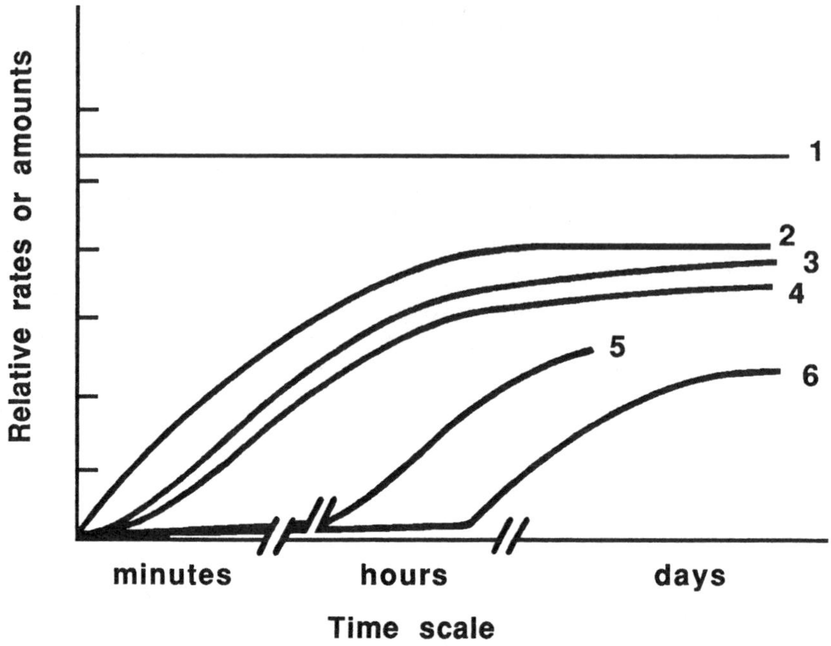

Figure 8. Generalized pattern of metabolic activity in germinating seeds. Enzymes instantly active, and rate of activity remains constant (1); enzymes activated during germination (2); enzymes expressed due to pre-existing RNA (3); all classes of RNA (4); unscheduled DNA repair (5); DNA replication (6). Reprinted from Mayer and Poljakoff-Mayber (1989, p. 150), with permission from Pergamon Press Ltd., Headington Hill Hall, Oxford OX3 0BW, UK.

tain few amino acids (Mayer and Poljakoff-Mayer 1989). Reserve proteins are hydrolyzed by proteinases during germination to produce amino acids that are utilized by the developing embryo for new protein synthesis or as a respiratory substrate (Mayer and Marbach 1981; Bewley and Black 1985; Mayer and Poljakoff-Mayber 1989). However, not all storage proteins are hydrolyzed at the same time or at the same rate during germination (Bewley and Black 1985). Mikola (1983) described three stages of proteolysis. In stage 1, hydrolysis in protein bodies of the aleurone layer yields amino acids that will be used in synthesis of enzymes required for conversion of insoluble endosperm reserves to forms suitable for transport. Major reserve proteins are hydrolyzed during the second stage, yielding amino acids for seedling growth. In stage 3, the depleted reserve tissue undergoes senescence, and cellular protein as well as various parts of the mobilization system itself are broken down to produce amino acids.

Protein breakdown in dicots generally differs from that in monocots. In dicots, protein bodies are degraded "from the inside out," at the completion of which is an empty protein body with an intact membrane. In many monocots, however, there are proteins located at the surface of the protein body which are degraded before proteins located inside the protein body (Mayer and Poljakoff-Mayber 1989).

The major reserve proteins in dicots are globulins that are stored in insoluble form (Bewley and Black 1985). Proteolyis of such proteins in peas often proceeds first by partial degradation from globulins into soluble polypeptides prior to complete hydrolysis yielding amino acids and small peptides. In many dicots, hormones that control proteolytic enzyme synthesis in cotyledons are produced in the embryonic axis; the seed coat also has been implicated (Mayer and Poljakoff-Mayber 1989).

Proteolysis in cereals occurs at three major sites: (1) enzyme synthesis and secretion in the aleurone layer, (2) storage and hydrolysis in the endosperm, and (3) uptake and mobilization of the products of this breakdown in the scutellum (Mikola 1983; Bewley and Black 1985). In the aleurone layer, regulation of protein mobilization is not well understood (Laidman 1982). Endospermic reserves are hydrolyzed by proteinases that may originate from the aleurone layer or the endosperm itself. Finally, peptides that the scutellum absorbs from the endosperm are hydrolyzed by peptide hydrolase in the scutellum. The scutellum, therefore, not only acts in an absorbing capacity; it also has a role in secondary processing (Dalling and Bhalla 1984).

Carbohydrates

The major carbohydrates of many species consist of oligosaccharides and polysaccharides such as starch, hemicellulose, and galactomannans. Degradation of smaller oligosaccharides provides an early source of energy for the germinating seed, and often precedes breakdown of bulk reserve material (starches and lipids) (Mayer and Poljakoff-Mayber 1989). Galactomannans stored in cell walls of many species are broken down next, yielding galactose and mannose which are then taken up by the cotyledons and further metabolized. The control of galactomannan breakdown may be in the aleurone layer or the embryo, depending on species. In some cereals, endospermic cell walls are degraded prior to mobilization of endospermic or cotyledonary starches and proteins (Mayer and Marbach 1981). Hydrolytic enzymes released from the aleurone layer (e.g., in barley) or the scutellum (e.g., in sorghum [*Sorghum bicolor* (L.) Moench] and maize) (Jacobsen and Chandler 1987), whose activation seems to be under hormonal control of the embryo, exert major control over these mobilization patterns. The scutellum absorbs end-products of starch metabolism (mostly glucose and maltose) and converts them to sucrose, which is then translocated to the growing roots and shoots.

Lipids

Lipid metabolism in germinating seeds is relatively well understood (Mayer and Marbach 1981). Lipid breakdown and conversion begins in the spherosomes, special organelles in which lipids are stored. Fatty acids and glycerol are the products of hydrolysis. Acetyl CoA is then produced by ß-oxidation of fatty acids in organelles called glyoxysomes [special organelles that generally are absent in dry seeds, form rapidly during germination, and disappear when lipid reserves are depleted (Mayer and Marbach 1981)]. Acetyl CoA may enter either the Krebs cycle for complete oxidation or the glyoxylate cycle for conversion to sucrose (Copeland and McDonald 1985). The latter cycle is important in the germination of seeds that store fats or oils because it provides a mechanism whereby fats, which can not be transported from storage tissue to the embryo (because of their large size and solubility), are converted to sucrose which can be transported to the embryo (Salisbury and Ross 1969).

Changes in Seed Composition During Germination

The foregoing description of some of the metabolic aspects of seed germination suggests that profound changes take place in seeds during the transition from a desiccated, quiescent state to one of active metabolism. This transition is reflected in changes in seed composition as reserve material is metabolized by the growing organism. Data of Oota et al. (1953) provide an excellent example of seed compositional changes during germination [see Mayer and Poljakoff-Mayber (1989) for additional examples].

Oota et al. (1953) studied compositional changes in cowpea (*Vigna sesquipedalis*), an epigeal germinator. Throughout germination, cotyledons decreased in dry weight as well as in content of pentose nucleic acid, protein nitrogen, and soluble sugars, polysaccharides, and nitrogen (Fig. 9). There was an increase in these compounds in the hypocotyl, epicotyl, plumule, and radicle. Growth of the latter tissues was sigmoidal whereas decreases in cotyledons was exponential. Cotyledons continued to lose constituents after hypocotyl slowed as growth in other organs increased. The hypocotyl acted as the main sink for transportation of materials out of the cotyledons until after the latter emerged from the soil surface, at which point the radicle and the plumule began to function as strong sinks for cotyledonary reserve material.

Seed Dormancy

A viable seed is termed *dormant* when it fails to germinate even when exposed to environmental conditions (see next section) appropriate for germination. Seed dormancy has received much attention in the literature because of its importance both to the ecologist and to the seed technologist. In agriculture, seed dormancy is important relative to seed testing for germination: dormancy in seeds must be "broken" before planting to insure timely emergence. To the ecologist, seed dormancy performs a crucial role in contributing to spatial and temporal distribution in germination [one aspect of the regeneration niche of Grubb (1977)], thereby improving the likelihood of successful propagation of the species.

Primary dormancy pertains to dormancy of seeds as they are dispersed from the mother plant. However, dormancy may develop in mature, dispersed seeds following exposure to unfavorable germination conditions; this type of dormancy is referred to as *secondary dormancy*. There are several schemes for categorizing conditions of primary dormancy (see Khan and Samimy 1982; Tran and Cavanagh 1984). Copeland and McDonald (1985) discuss two main conditions. In exogenous dormancy light, water, and/or temperature conditions necessary for germination are not experienced by the embryo, most often because of seed coat properties. On the other hand, endogenous dormancy is caused by inherent characteristics of the seed itself (e.g., presence of inhibitors, an underdeveloped embryo, or lack of a germination stimulator). Other authors (e.g., Bewley and Black, 1985) discuss coat-imposed dormancy (similar to exogenous dormancy) and embryo dormancy (similar to endogenous dormancy). Regardless of the general approach to discussing dormancy, it is clear that the topic is a complex one. Often, several factors interact to induce and/or break dormancy, and a complete understanding of dormancy must consider

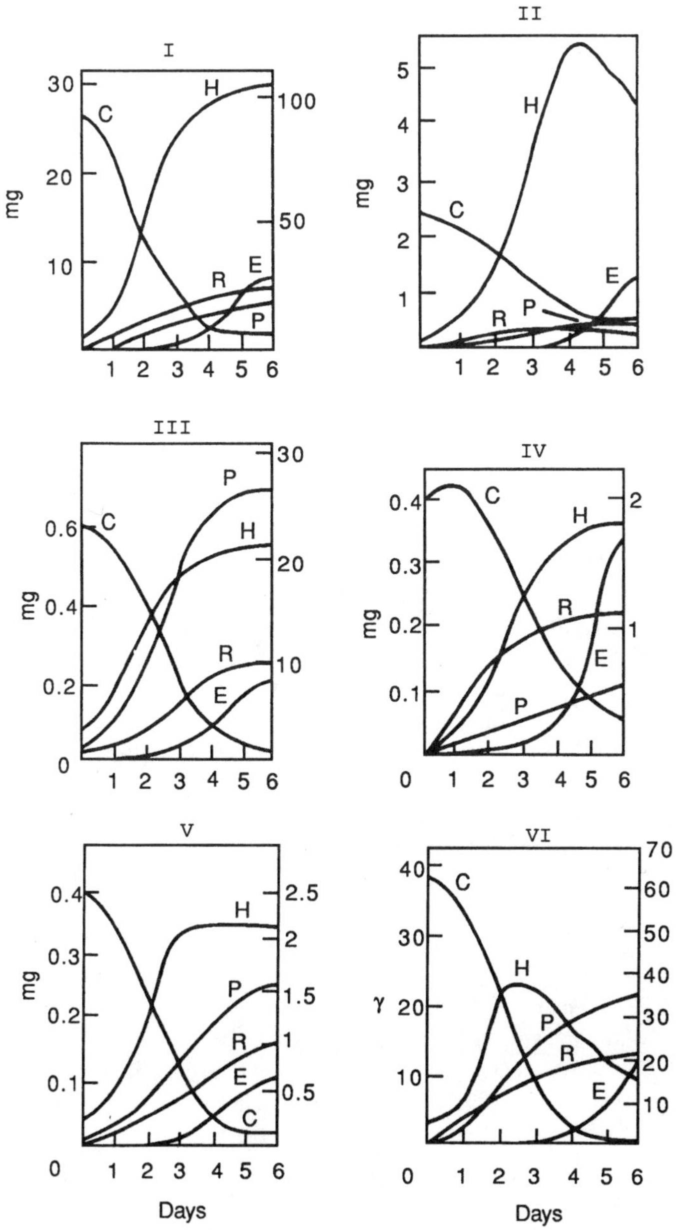

Figure 9. Changes in dry weight and amounts of constituents in various organs of *Vigna sesquipedalis* seeds during germination. Organs indicated by P: plumule; E: epicotyl; R: radicle; C: a pair of cotyledons; H: hypocotyl. I: dry weight (P, E, H, and R on the left ordinate, C on right ordinate; II: soluble sugars as glucose equivalents; III: insoluble polysaccharides as glucose equivalents (P, E, H, and R on the left ordinate, C on the right ordinate); IV: Soluble nitrogen (P, E, R, and C on the left ordinate, H on the right ordinate); V: Protein nitrogen (P, E, H, and R on the left ordinate, C on the right ordinate); VI: Pentose nucleic acid phosphorus (P, E, H, and R on the left ordinate, C on the right ordinate). [Oota et al. (1953) , reproduced from Mayer and Poljakoff-Mayber (1989)].

not only seed physiology but also environmental conditions experienced by the maternal plant before and during seed development as well as by the seed after dispersal. Many aspects of seed dormancy are only incompletely understood. A brief summary of some important concepts of seed dormancy follows; more detailed information can be found in Bewley and Black (1983, 1985), Tran and Cavanagh (1984); Ross (1984); and Mayer and Poljakoff-Mayber (1989).

Coat-Imposed Dormancy

Mechanisms implicated in coat-imposed dormancy include (1) impermeability to water, (2) impermeability to oxygen, (3) mechanical resistance to embryo growth, (4) prevention of the leakage of germination inhibitors away from the embryo, (5) impermeability to carbon dioxide diffusion, (6) supply of germination inhibitors, (7) osmotic inhibition associated with high hydrostatic pressure resulting from the large osmotic potential of the seed coat, and (8) seed coats acting as a light filter (see Werker 1980/81 and citations therein). These factors are not mutually exclusive.

Interference with Water Uptake

Seed coat impermeability develops during the final stages of seed maturation, and the extent of impermeability is, in part, related to seed moisture content (Tran and Cavanagh 1984). Additionally, oxygen concentrations experienced during maturation influence seed coat permeability to water (Marbach and Mayer 1974). Seed coat thickness may also be influenced by photoperiodic effects during seed maturation (Bewley and Black 1985). For some seeds, coats are thicker when they mature under long days than under short days. Werker (1980/81) and Tran and Cavanagh (1984) provide detailed discussions of structural aspects of seed dormancy.

Seed coat impermeability is not fully understood (Mayer and Poljakoff-Mayber 1989). Seeds of many species in the families Fabaceae, Chenopodiaceae, Convolvulaceae and Liliaceae often are characterized as "hard seeds" because their coats are impermeable, or nearly so, to water. Features of testa related to water impermeability involve a waxy cuticle, suberin (in the subcuticle), and thick-celled palisade and osteosclereid cells. Also, all seed coat openings (e.g., the micropyle) must be blocked if water uptake is to be prevented. Bewley and Black (1985) state that the osteosclereid layer of the testa of several species (which lies below the palisade layer containing macrosclereid cells) is generally the main barrier to water uptake. In contrast, Werker (1980/81) and Tran and Cavanagh (1984) indicate that it is the palisade layer [as in *Prosopis farcata* (Banks & Solander) Eig; Werker et al. (1973)] (containing macrosclereids) that is most often implicated in seed coat impermeability. Tran and Cavanagh (1984) discuss evidence bearing on this issue and conclude that information and our understanding is incomplete.

Relief of water impermeability of testa apparently is a physical event, e.g., through abrasion or cracking of the seed coat, microorganism attack, or passage through digestive tracts (Mayer and Marbach 1981). In other words, the seed itself evidently does not participate in a biochemical way to reduce or remove seed coat impermeability.

Mechanical Resistance

It is often claimed that seeds coats, as well as other tissues (e.g., endosperm) may offer considerable resistance to the emerging embryo during germination. In some cases, seed coat structures are involved in preventing radicle emergence. For example, the glumes of Johnson grass (*Sorghum halepense* (L.) Pers.) attach to the seed after dispersal, and prevent radicle emergence unless they are fractured or abraded (Marbach and Mayer 1979). Because germination is observed in some species following seed coat removal, this resistance has sometimes been offered as a cause of dormancy. However, this interpretation may fail to consider other complicating factors (Bewley and Black 1985). For example, the seed coat may contain germination inhibitors, so that coat removal also removes these inhibitors; the ensuing germination is made possible not by relief of mechanical resistance, but by inhibitor removal. Likewise, the seed coat may delay or prevent leaching of a germination inhibitor away from the embryo, so that, with coat removal, germination takes place; again, it is not the mechanical resistance of the coat that is delaying or reducing germination. Also, the strength of a seed coat is greatly reduced by imbibition (Tran and Cavanagh 1984). Tran and Cavanagh (1984) recommend that before it is concluded that mechanical seed coat resistance is responsible for preventing germination, strength and mechanical properties of seed coats must be assessed both prior to and during imbibition.

Interference with Gas Exchange

Seed coats and associated membranes can reduce the amount of oxygen reaching the embryo. This may happen because the water in the seed coat itself, or between the seed coat and the embryo, acts as an oxygen diffusion barrier [as in *Aristida contorta* F. Muell. (Mott and Tynan 1974)]. Also, as mentioned earlier, one of the first events following imbibition is a swelling of the seed, resulting in a longer distance for oxygen diffusion (Werker 1980/81). There may also be oxygen consumption in the seed coat (e.g., by oxidases that compete with the embryo for oxygen), thereby reducing its availability to the embryo (Werker 1980/81).

However, oxygen requirments for germination are so low that it is unlikely that seed coats reduce availability below a threshold amount necessary for respiration. Thus, Bewley and Black (1985) suggest that seed coats do not seem to impose dormancy by limiting the amount of oxygen to the embryo. Nevertheless, germination is often improved by mechanical disturbance of the seed coat (e.g., by scarification or puncture) or incubation in high oxygen atmospheres. Bewley and Black (1985) suggest that this phenomenon may be due to oxidation of a germination inhibitor: presence of a seed coat not only prevents the leaching of the inhibitor away from the embryo, but also prevents adequate oxygen uptake for oxidation of the inhibitor.

In summary, it appears that (1) if there are permeability barriers to oxygen, they may more likely be associated with seed membranes, mucilages, or other diffusion barriers than with seed coats, (2) oxygen requirements for germination are so low that it is unlikely that seed coats reduce oxygen availability below what is required for germination, and (3) the observed increase in germination following seed coat abrasion or incubation in high oxygen concentrations may be related to enhanced leaching and/or oxidation of a germination inhibitor.

Embryo Dormancy

Dormancy imposed by seed coats requires a physical alteration of the coat for relief. However, dormancy not specifically related to effects of the seed coat (that is, dormancy under endogenous control) is more prevalent than coat-imposed dormancy (Copeland and McDonald 1985). This so-called "embryo dormancy" is sometimes attributed to presence of inhibitors (see below) located in cotyledons (Come 1980/81). This interpretation is based largely on the fact that cotyledon removal may relieve dormancy. However, conclusive evidence is scant (Bewley and Black 1985).

Another type of embryo dormancy may be attributed to immature embryos. For example, seeds of *Plantago* spp. L., *Ilex* spp. L., and *Pinus* spp. L. (among other genera) are immature as they are dispersed, and germination is possible only after embryo maturation following dispersal (Copeland and McDonald 1985).

Another consideration related to embryo dormancy suggests that dormant embryos are in some way metabolically deficient and thus unable to germinate. Bewley and Black (1985) provide detailed discussion of this topic and conclude that as yet no metabolic or biochemical aspect of dormant seeds has been discovered to account for embryo dormancy.

Germination Inhibitors

A number of germination inhibitors (e.g., abscisic acid, coumarin, various phenolic acids; see Bewley and Black 1985: Table 5.3; Mayer and Poljakoff-Mayber 1989: Table 4.7) are located in seed tissues. The chemical nature and physiological action of some germination inhibitors is not precisely known. Inhibitors act by interfering with seed metabolism. Also, it is important to appreciate that germination inhibition and dormancy are not necessarily synonymous. Thus, respiratory inhibitors (such as fluoride and hydroxylamine) prevent germination, whereas the effects of osmotic inhibition (e.g., by sodium chloride or polyethylene glycol) or abscisic acid (see below) are reversible (Mayer and Poljakoff-Mayber 1989).

As discussed earlier, the presence of seed coats may act to restrict the leaching of inhibitors away from the embryonic axis, thus preventing germination. However, it is often difficult to determine whether the presence of these inhibitors functions directly in seed dormancy (Bewley and Black 1985; Mayer and Poljakoff-Mayber 1989).

Hormonal Considerations

Hormones have been implicated in seed dormancy for many years (e.g., Black 1980/81; Wareing 1982; Khan and Samimy 1982) and the topic is a complex one, in part because the response of seeds to growth regulators often is influenced by whether the seed is in primary or secondary dormancy (Khan and Samimy 1982). Also, it is well known that the actions of various growth regulators are often correlated. Mayer and Poljakoff-Mayber (1989) provide an important interpretational warning relative to hormones, dormancy, and germination with the reminder that observed correlations between growth hormones (e.g, gibberellic acid) and relief of dormancy may signify either a cause-and-effect relationship, or may simply be phenomena that accompany dormancy breakage. Also, it is important to bear in mind that germination responses to exogenously-applied hormones do not necessarily provide proof of a similar role for endogenously-produced hormones (Walton 1980/81; Karssen et al. 1989). Comprehensive reviews of this topic are found in Black (1980/81), Khan (1982), Ross (1984), Karssen et al. (1989), and Mayer and Poljakoff-Mayber (1989).

Abscisic Acid

Considerable research has focused on the role of abscisic acid (ABA) in germination inhibition and in seed dormancy (Walton 1980/81). The inhibitory effect of seed coats in germination of some species has been attributed to endogenous ABA in these tissues. Also, exogenously-applied ABA can inhibit germination. For example, ABA inhibits germination of dormant and non-dormant apple (*Pyrus malus* L. cv. 'Golden Delicious') embryos, and this effect is independent of oxygen (Thevenot et al. 1987). Exogenously-applied ABA may inhibit germination by reducing cell wall extensibility (Schopfer and Plachy 1985). Abscisic acid also has been observed to antagonize the action of gibberellic acid, the latter of which promotes amylase production (Ho and Varner 1976; Bewley and Black 1985; Jacobsen and Chandler 1987).

A complicating factor in this area relates to time-specific effects of ABA on germination inhibition. For example, hazel nut (*Corylus avellana* L.) embryos are initially sensitive to ABA during early seed development, and at this time, ABA is involved in imposing dormancy of freshly-harvested seed. Evidence for this interpretation comes from studies of seed coat removal: nearly all of the ABA in seeds of this species is found in the pericarp (fruit coat derived from the ovary wall) and seed coat, the removal of which allows germination. As the seed dries, however, "true" embryo dormancy is established, and at this stage the seed is relatively insensitive to ABA (Ross 1984); at this stage, stratification or gibberellic acid application to unchilled seeds can induce germination.

For many species, however, there is no correlation between depth of dormancy and retention of ABA in seeds (Bewley and Black 1985). Additionally, whereas ABA may induce dormancy in some species (e.g., Karssen et al. 1983; Mayer and Poljakoff-Mayber 1989), it is not necessarily involved in maintaining dormancy or determining it depth (Black 1980/81; Walton 1980/81; Bewley and Black 1985). Thus, a general finding is that there is no correlation between ABA content and depth of dormancy (Bewley and Black 1985). Although exogenously-applied ABA inhibits germination of non-dormant seeds, it is not clear that endogenous ABA acts in the same way (Walton 1980/81). Finally, the inhibitory effect of ABA on germination can be relieved by gibberellic acid and sometimes also cytokinins, indicating a correlation in action between growth regulators (Mayer and Poljakoff-Mayber 1989).

Cytokinins

Cytokinins, whose action interacts both with light and temperature, generally promote germination (Mayer and Poljakoff-Mayber 1989). Khan (1971; see below) suggested that cytokinins act to antagonize germination inhibitors when the latter are active; when such inhibitors are not active, however, he suggested that cytokinins have no effect on dormancy breakage. However, Ross (1984:56) suggested that the "significance, if any, of the cytokinin levels in dormant seeds remains elusive."

Gibberellic Acid

Exogenously-applied gibberellins (GA) not only break dormancy but also hasten germination of non-dormant seeds of many species (Karssen et al. 1989). However, the role of GA in seed dormancy is complex in part because other factors such as chilling and light are involved (e.g., Mayer and Poljakoff-Mayber 1989). Chilling or treatment with GA can break seed dormancy; light and GA also are very similar in their effect on the control of germination. There is some evidence that seed coats also are involved in the response of seeds to GA and light (Khan and Samimy 1982).

Gibberellic acid content is high in developing seeds, but is generally low in mature, desiccated seeds (Karssen et al. 1989). During stratification GA may remain low (as in hazel nut) or may increase [as in European beech (*Fagus sylvatica* L.)] (Ross 1984). When it increases, however, there is no apparent correlation with germination potential. In hazel nut (which requires stratification for dormancy breakage), there are large increases in GA after the chilled seeds are incubated at higher temperatures, suggesting that whereas chilling may relieve an unknown block to GA synthesis, GA synthesis takes place at higher temperatures more suitable for germination and growth (Ross 1984). Based on a review of a number of studies on GA and chilling, Ross (1984:56) suggested that "it is doubtful whether GA plays a part in the dormancy breaking process and it is more likely to be involved with metabolism associated with germination and seedling growth."

Karssen et al. (1989) found that stratification enhanced the sensitivity of *Arabidopsis thaliana* (L.) Heynh. seeds to applied GA. They suggested that the GA requirement some seeds have for germination is controlled by ABA levels during seed development via an intermediate effect of ABA-induced dormancy possibly involving suppression of cell wall extensibilty.

A General Model

Khan (1971; see Khan and Samimy 1982) proposed a model relating the involvement of hormones in seed dormancy and germination. This model suggests:

"(a) that GA is the primary stimulus for germination and the roles of cytokinin and inhibitors are secondary;
(b) that hormones have two or more sites of action;
(c) that dormancy and germination could be determined by more than one combination of hormones;
(d) that germination may occur even when there is an excess of an inhibitor and dormancy may result even though there is an excess of GA or a cytokinin;
(e) that a GA alone (in absence of an inhibitor) or a cytokinin alone (in presence of GA) might in some cases promote germination; and
(f) that a combination of a cytokinin and a GA would be expected to give an equal or better response than when they are used singly" (Khan and Samimy 1982:211).

Supporting evidence for these postulations is given by Khan and Samimy (1982). Copeland and McDonald (1985) suggest that much research is needed before the general mechanism of regulation of seed dormancy and germination proposed by Khan (1971) can be accepted. Khan and Samimy (1982:209) summarized: "… the potential for embryo growth appears to be determined by an interplay of hormonal and other endogenous factors between embryo coverings and the embryo and the nature or the intensity of these interactions is influenced markedly by the environmental factors, such as light and temperature."

Dormancy Breakage

Afterripening

The term "afterripening" refers to the process by which dry seeds slowly lose their dormancy. This process is not fully understood (Bewley and Black 1985; Mayer and Poljakoff-Mayber 1989). Because the process occurs in dry seeds, it can be delayed by exposure to excess moisture as well as by excess moisture loss. Additionally, exposure to high temperatures or very high oxygen concentrations may delay afterripening. It is noteworthy that treatment with gibberellic acid can break dormancy in seeds which have not gone through the process of afterripening, suggesting that the process of afterripening is involved in allowing the seed to synthesize gibberellic acid necessary for germination (Copeland and McDonald 1985).

Stratification

Dormancy of many species can be relieved by imbibing seeds at temperatures generally ranging from 1 to 10°C, a treatment effective whether dormancy is primary, secondary, coat-imposed, or embryonic (Bewley and Black 1985). Bewley and Black (1985) and Mayer and Poljakoff-Mayber (1989) discuss the participation of temperature-induced membrane changes in relief of dormancy.

It is obvious that many aspects of seed metabolism are temperature-sensitive. For example, the ABA content of many seeds falls during stratification. However, the cause for this decrease in ABA is not the low temperature itself, and hence when stratification breaks dormancy, it can not be concluded that such effects are attributable solely to the effect of ABA (Black 1980/81).

Stratification also increases total nucleic acid content, especially rRNA, in *tomato (Lycopersicon esculentum* Mill.) (Coolbear et al. 1990) and Norway maple (*Acer platanoides* L.) (Slater and Bryant 1987) seeds. However, at least for tomato seeds, the hastening of germination that follows stratification is not highly correlated with the total nucleic acid accumulated during the chilling period. Coolbear et al. (1990) suggested that whereas stratification-enhanced replacement of damaged rRNA in dry seeds may be an important prerequisite for successful germination, other factors (e.g., protein synthesis) may be more important in limiting the rate of radicle emergence. Another effect of stratification involves an enhanced sensitivity to exogenously-applied GA in *Arabidopsis thaliana* (L.) Heynh. seeds (Karssen et al. 1989).

As discussed above, temperature interacts with light as well as hormones in dormancy, and the situation is a complex one. Increased oxygen uptake by the embryonic axis, increased nitrogen content, and changes in enzymatic activity, protein metabolism, and various hormone levels each have been associated with lower seed temperatures (Mayer and Poljakoff-Mayber 1989).

Light

Light is an important dormancy-releasing factor in many species. The effect of light in relieving dormancy is also temperature-sensitive. Nearly all species that require light for germination (see next section) have coat-imposed dormancy (Bewley and Black 1985). Additional discussion of light and germination is found in the following section.

Germination Stimulators

Germination is stimulated by many compounds (e.g., Bewley and Black 1985: Table 5.11; Mayer and Poljakoff-Mayber 1989). The effects of selected hormones have already been discussed. The effects of thiourea and nitrates are briefly discussed here because of their wide use in seed pre-treatment (see Wester 1991).

Thiourea, a nitrogenous compound, stimulates germination of many species. Thiourea can overcome a light requirement for germination of lettuce seeds and also the inhibitory effect of high temperatures on germination of this species (Mayer and Poljakoff-Mayber 1989). The effect of thiourea on germination depends on concentration and duration of treatment and also interacts with temperature. Mayer and Poljakoff-Mayber (1989) discuss the involvement of thiourea in metabolism of germinating seeds.

Nitrates, often applied in the form of potassium nitrate, promote germination of many species. The effect of nitrates is concentration-dependent (e.g., Goudey et al. 1988) and interacts with ethylene (e.g., Saini et al. 1986), light, and temperature (Hilhorst et al. 1986; Probert et al. 1987; Bouwmeester and Karssen 1989; Hilhorst and Karssen 1989; Mayer and Poljakoff-Mayber 1989). Nitrates in *Sisymbrium officinale* (L.) Scop. are linked with phytochrome action and may be involved with gibberellin synthesis (Hilhorst et al. 1986; Probert et al. 1987; Hilhorst and Karssen 1989). However, the mechanism of nitrate-induced stimulation of germination is not well understood (Goudey et al. 1988; Hilhorst and Karssen 1989). It has been suggested that nitrates may act as electron acceptors in oxidation reactions in the pentose phosphate pathway (Roberts 1972; Bewley and Black 1985), a hypothesis proposed in connection with the suggestion that the pentose phosphate pathway was needed in the early stages of dormancy relief and germination (Roberts 1972; Mayer and Poljakoff-Mayber 1989). However, Hilhorst and Karssen (1989) indicated that this mechanism did not apply for *Sisymbrium officinale* (L.) Scop.

A Case Study in Seed Dormancy

It is evident that many factors can be involved in seed dormancy. Some seeds have a combination of dormancy-imposing mechanisms. Indian ricegrass (*Oryzopsis hymenoides* (R. & S.) Ricker.) provides an excellent example (McDonald and Khan 1977). Indian ricegrass seed coats impose a substantial barrier to water uptake that can be successfully broken by scarification with 67% sulfuric acid for 40 min. Freezing and thawing, boiling, treatment with hydrogen peroxide, various organic solvents, gibberellic acid or kinetin were ineffective in enhancing germination. Abscisic acid was present in the seed coat and associated structures, and was reduced by sulfuric acid scarification. Germination also was inhibited by abscisic acid treatment of sulfuric acid scarified seeds. In addition to these seed coat related phenomena,

Indian ricegrass seed germination also was influenced by post-harvest embryo dormancy. Four-year-old seeds required less acid scarification than freshly-harvested seeds, and both gibberellic acid and kinetin resulted in marked increases in germination of 1-year old sulfuric acid scarified seeds, but only small increases in germination of scarified 2-year-old seeds. The enhancement of 1-year-old scarified seed germination by gibberellic acid may be due to an inherent deficiency in endogenous gibberellins, and the lack of a similar response to gibberellic acid in older seeds may be attributable to increased gibberellin synthesis during the afterripening period. Thus, it is evident that dormancy factors influencing Indian ricegrass germination involve both seed coat characteristics and embryo dormancy.

A Summary Comment

It is clear that many factors interact in their effects on seed dormancy in ways that are not yet fully understood. Bewley and Black (1985) emphasize two important facts. First, they indicate that there is no universal dormancy-breaking mechanism. Second, they provide a firm reminder that it is the embryo which is ultimately and directly involved in dormancy: regardless of whether a seed is in primary or secondary dormancy, regardless of whether the dormancy is coat-imposed or due embyronic inadequacy, it is the embryo, and specifically, its axis, which achieves growth and effects germination.

Environmental Factors Affecting Germination

Seed germination is influenced by many environmental factors. Soil water, temperature, composition of air, and light quality and quantity each can play a major role in affecting germination. Additionally, these factors often interact with each other in their influence on germination. As mentioned earlier, environmental effects on seed germination are relatively well understood even though their biochemical and physiological bases are less clear (Mayer and Marbach 1981). Nevertheless, the foregoing brief introduction to the physiology and biochemistry of germinating seeds may help to direct the following summary concerning environmental effects towards a physiological perspective. It is hoped that this, in turn, will provide a basis for ecophysiological considerations in the following section related to topics in seed ecology.

Temperature

Temperature can affect seed germination in several ways. For example, dormancy of some species can be relieved by exposing seeds to some specific temperature (usually a cool or cold temperature) for a duration of time before exposing seeds to temperatures more suitable for germination (see above discussion on stratification). In these cases, an alternating temperature regime includes a chilling temperature that is involved in dormancy relief. However, alternating temperatures also may affect germination of non-dormant seeds. It is not always clear whether the effect of a temperature alternation involves dormancy relief or some other aspect(s) of the metabolism of non-dormant seeds (Bewley and Black 1983). Further, it is likely that temperature-sensing mechanisms related to stratification and dormancy relief are different from temperature-sensing mechanisms of non-dormant seeds (Mayer 1986).

The temperatures within which seeds can germinate vary widely, and depend on many factors including harvest conditions, species, seed source, genetics, seed age, and other environmental variables (Hegarty 1972; Mayer and Poljakoff-Mayber 1989). As the preceding discussion has shown, germination is a complex phenomenon whose successful completion is mediated by many and varied biochemical reactions, many of which involve enyzmes whose function is temperature-sensitive. Mayer (1986) reviewed mechanisms by which non-dormant seeds perceive temperature and concluded that membranes are probably the site at which temperature, or changes in temperature, are detected by seeds. A likely mechanism involves temperature effects on the fatty acid composition of membranes.

Temperature also affects oxygen solubility and hence influences the rate at which oxygen diffuses to the embryo (Come and Tissaoui 1972). However, neither germination rate nor final percentage necessarily increase with increasing temperature (Mayer and Poljakoff-Mayber 1989). Consequently, it is difficult to describe germination by a simple temperature coefficient (Lovato 1981; Mayer and Poljakoff-Mayber 1989).

"Optimum" temperature for germination is usually defined (e.g., Copeland and McDonald 1985; Mayer and Poljakoff-Mayber 1989) as the temperature at which there is maximum germination in the shortest period of time. Optimum germination temperature for most species is between 15 and 30°C (Copeland and McDonald 1985). Russian thistle (*Salsola kali* L.) (Dwyer and Wolde-Yohannis 1972) and cucumber (Knapp 1967) have been shown to germinate at temperatures as high as 48°C. Sosebee and Herbel (1969) reported germination of a number of species [e.g., black grama (*Bouteloua eriopoda* (Torr.) Torr.), alkali sacaton (*Sporobolus airoides* (Torr.) Torr.) and Boer lovegrass (*Eragrostis chloromelas* Steud.)] under alternating temperature regimes whose maximum soil surface temperature was 67°C (53°C at a depth of 13 mm). The minimum temperature for germination is near 0°C (Copeland and McDonald 1985). In general, (1) seeds from temperate regions require lower temperatures to germinate than do seeds from tropical areas, (2) wild species often have lower temperature requirements than domesticated plants (Copeland and McDonald 1985), and (3) within a community, seeds of annual species tend to germinate over a broader range than seeds of perennials (Ellern and Tadmor 1966, 1967; Young et al. 1973).

Optimum germination temperature was defined above as the temperature at which there is maximum germination in the shortest time period. However, temperature affects both rate and extent of germination, and germination rate is time-dependent as well. An important principle of seed germination at constant temperatures involves these considerations: as duration of incubation lengthens, the constant temperature for optimum germination often tends to be lower and to extend over a broader range (Fig. 10). For example, after four days of incubation, germination of *Danthonia* spp. Lam. & DC. is greatest at 19°C (Fig. 10); however, after 10 days of incubation, germination is greatest at 13–17°C. That is, optimum germination temperatures tended to be cooler and to occur over a broader range as incubation period lengthened.

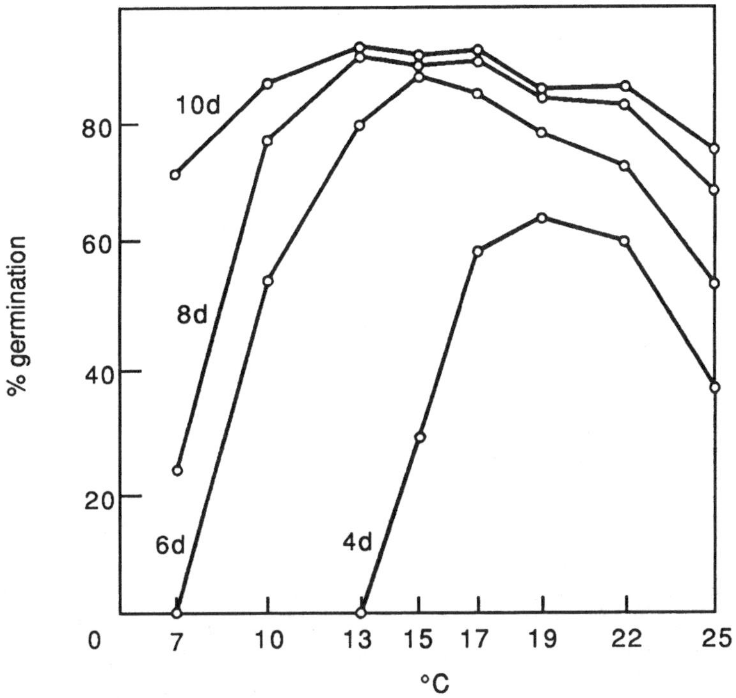

Figure 10. Percent germination of *Avena* seeds at 4-, 6-, 8-, and 10-day intervals as a function of incubation temperature. [Edwards (1932), reproduced from Mayer and Poljakoff-Mayber (1989).

Because rapid germination can have important bearing on seedling establishment under drying soil conditions, there has been considerable research on the effects of temperature, especially constant temperatures, on rate of germination. For example, Jordan and Haferkamp (1989) estimated base temperature (the minimum temperature at which germination rate is zero), heat units, and a germination index (that quantifies germination rate) for 54 species of grasses, forbs, and shrubs. Their approach, however, uses data from temperature ranges restricted to about 15–20°C above the temperature corresponding to the minimum germination rate. Covell et al. (1986) and Ellis et al. (1986, 1987) developed quantitative models describing germination rate at sub- and supra-optimal constant temperatures that require information at only four temperatures (two temperatures below, and two temperatures above, optimum temperature). Their approach appears to successfully predict germination rates even at temperatures where percentage germination is low.

Alternating temperatures also affect seed germination. In general, the amplitude of the cycle in the temperature alternation, the actual temperatures included, and the duration of exposure to each temperature have significant effects on germination (Bewley and Black 1983; Mayer and Poljakoff-Mayber 1989). However, it is important that, depending on species, effective duration of one portion of the cycle need not be long, and can be saturated (Koller 1972). Also, Cohen (1958) reported no correlation between germination stimulation and rate of temperature change.

It is not known why alternating temperatures are required by some species for germination (Mayer and Poljakoff-Mayber 1989). Several explanations have been proposed (Koller 1972) all of which seem to incorporate, in some sense, the idea that alternating temperatures influence the sequential stages of germination differently; Copeland and McDonald (1985) provide a succinct summary. Toole et al. (1956) suggested that respiratory products produced at the higher temperature of the cycle may inhibit germination at that temperature; however, these products may promote germination at the lower temperature of the cycle (which is less favorable for respiration). Thus, the alternating temperatures help maintain a balance between intermediate metabolites necessary for successful completion of germination Copeland and McDonald 1985). Cohen (1958) hypothesized the existence of a macromolecular compound which is stored in an original form that prevents germination; however, this compound is temperature-sensitive, and its prohibitive effects on germination are modified by alternating temperatures (Copeland and McDonald 1985). Thompson (1969) proposed that the successive changes in a temperature alternation are metabolically related to each other in that a product produced at one temperature is involved in reactions at the following temperature (Koller 1972). Copeland and McDonald (1985) suggest that the most likely explanation for the beneficial effect of alternating temperatures on seed germination may be that it causes a shift in a balance between a germination inhibitor and a germination promoter.

Murdoch et al. (1989) developed a relatively comprehensive quantitative model based on linear regression techniques to explain germination of *Chenopodium album* L. and *Panicum maximum* Jacq. in response to alternating temperatures. This model, which appears promising for other species as well, identifies the features of alternating temperature regimes and their interactions that are important for germination. Results for *Chenopodium album* L. are discussed here. When percent germination data were transformed to normal deviates, germination showed a positive linear relationship to increases in constant temperature up to about 25°C (which was therefore defined as optimum constant temperature), beyond which germination showed a linear decrease to increasing temperature. When seeds were exposed to an alternating temperature regime for which the mean temperature exceeded the optimum constant temperature, germination decreased as mean temperature increased; further, about 75% of the variation in germination was explained by mean temperature. However, at alternating temperature regimes for which the mean temperature was cooler than the optimum constant temperature, mean temperature, amplitude of the fluctuation, and thermoperiod (time spent at the higher temperature in the fluctuation) were important variables explaining variation in germination response. For example, when both minimum and maximum temperatures in the alternation were suboptimal, germination increased with both mean temperature and amplitude. However, there was a negative interaction between mean temperature and amplitude as they affected germination response. Additionally, germination response to amplitude was greater if seeds were exposed to the higher temperature in the alternation for 16 hours rather than for 8 hours. Additional results relating to alternating temperature regimes for which the optimum constant temperature was between the minimum and maximum temperatures in the fluctuation are discussed by Murdoch et al. (1989). Perhaps one of the most promising features of this model is its prediction of the alternating temperature regime that will maximize germination (or loss of dormancy) for a given thermoperiod. According to this model, the alternating temperature regime that will maximize germination will be one in which the maximum temperature in the alternation equals the optimum constant temperature, and the minimum temperature in the alternation will be as low as possible in order to maximize the amplitude of the alternation. This model prediction will depend on thermoperiod, and obviously must recognize possible deleterious effects of low temperatures (e.g., less than 5°C) on seed germination. By identifying the effects of temperature alternation that are important in explaining variation in germination response to fluctuating temperatures, this model should prove useful in studying germination.

Light

The effect of light on seed germination is a complex topic that has been the subject of numerous detailed reviews (Koller 1972; Taylorson 1982; Kendrick and Kronenberg 1986; Cone and Kendrick 1986; Mayer 1986). Only a brief summary of some important concepts is given here, together with references to more detailed literature.

There are two types of seed germination response to light (Cone and Kendrick 1986). The first type of response (the low fluence response, or LFR) involves the phytochrome system by influencing the proportion of Pfr (the form of the phytochrome pigment after exposure to red light) and Pr (the form of the phytochrome pigment after exposure to far red light, heat, or dark). This type of response may or may not be influenced by irradiation time (Taylorson 1982; cf Cone and Kendrick 1986) and shows photoreversibility (see below). The second type of response (the high irradiance response, or HIR) involves an inhibition of germination by long periods of far red irradiance (Mancinelli and Rabino 1979). This response is both time- and irradiance-dependent and does not show photoreversibility (Taylorson 1982). Prolonged exposure of seeds to white light (sun light) may also inhibit germination (Bewley and Black 1985); this inhibition is usually due to the far red radiation in the white light (Mayer and Poljakoff-Mayber 1989).

Low Fluence Response (LFR)

The response of seed germination to light varies among species. Mayer and Poljakoff-Mayber (1989) recognized four categories: (1) species that require light for germination, (2) species that germinate only in the dark, (3) species that require at least an exposure to light, and (4) species whose germination response to light is indifferent. Over half of the species tested for light sensitivity have been shown to require light for germination. For most of these species, the greatest promotive effects occur at wavelengths between 560 nm and 700 nm; wavelengths greater than 700 nm and less than 290 nm usually inhibit germination (Mayer and Poljakoff-Mayber 1989). Both light intensity and duration of illumination are involved in light effects on germination (Mayer and Poljakoff-Mayber 1989).

It is clear that the response of seeds to light involves the phytochrome system (Cone and Kendrick 1986). However, the mechanisms involving the active form of phytochrome with germination are not fully understood (Mayer and Poljakoff-

Mayber 1989). Light sensitivity in seeds appears to be induced mainly during seed dehydration and maturation (e.g., Gutterman 1982; Bryant 1985; Cone and Kendrick 1986). While on the mother plant, maternal tissue encompassing the seed is high in chlorophyll that acts as a filter to screen out red light. Light reaching the seed is therefore relatively rich in far red wavelengths, establishing a low Pfr (i.e., phytochrome in the form following exposure to red wavelengths) level (Cresswell and Grime 1981).

Phytochrome is trapped in the form of "intermediates" (*sensu* Rudiger 1986) during seed desiccation that "relax" upon rehydration during imbibition forming predominantly Pfr and resulting in the promotion of germination. It is the percentage of Pfr, rather than its absolute amount, that is generally important to germination (Mayer and Poljakoff-Mayber 1989). Characteristic of the phytochrome system, prolonged far red radiation inhibits germination by causing a cycling of phytochrome. Thus, seed germination may show photoreversibility: germination occurs when the last exposure is in the red wavelengths (Cone and Kendrick 1986). The effect of blue light on germination also involves the phytochrome system, and may be stimulatory or inhibitory, depending on its exposure time relative to stage of imbibition (Mayer and Poljakoff-Mayber 1989). Seeds that germinate in the dark may do so because of endogenously-produced Pfr or production of some other factor, e.g., gibberellic acid, that promotes germination (Cone and Kendrick 1986). Additionally, it should be noted that light may inhibit germination of some species. In these cases, it appears that far red radiation acts to maintain a low percentage of Pfr (Mayer and Poljakoff-Mayber 1989).

Photoconversion of phytochrome between Pr and Pfr takes place through several intermediates. Some of the steps in this photoconversion process take place only in hydrated seeds, and other steps occur in desiccated seeds. Thus, whereas dry seeds show some light sensitivity, full photoconversion from Pr to Pfr can not take place in dry seeds (Bewley and Black 1985). Therefore, seeds of light-sensitive species that are desiccated generally do not respond to red light. Upon imbibition and irradiation with red light, Pfr is formed. Germination will occur with subsequent dehydration and imbibition even in the dark because of the presence of Pfr formed during the initial irradition of the moist seeds (Cone and Kendrick 1986).

Some species also show germination photoperiodicity, or sensitivity to alternations between light and dark periods (Koller 1972; Gutterman 1982; Bewley and Black 1985; Mayer and Poljakoff-Mayber 1989). For example, seeds of *Chenopodium album* L. show higher germination when exposed during development to short days than when exposed to long days (Karssen 1970, 1980/81a). The phytochrome system evidently is involved with photoperiodic effects (Koller 1972; Mayer and Poljakoff-Mayber 1989; *cf* Bewley and Black 1985).

The specific involvement of the phytochrome system with seed germination is not fully understood. Current hypotheses suggest that the phytochrome system may influence germination through its effects on (1) gene activation, (2) gibberellin production, (3) membrane permeability, or (4) kinase activity (Copeland and McDonald 1985; Mayer 1986; Mayer and Poljakoff-Mayber 1989). Copeland and McDonald (1985) summarized some evidence on these hypotheses, and Mayer (1986) reviewed the topic of phytochrome and membrane permeability. For example, several hydrolytic enzymes [e.g., endo-ß-mannanase and α-galactosidase, which are important in the mobilization of endosperm cell wall hemicellulose (Bewley et al. 1983)] increase following activation of the phytochrome system; the production of these enzymes may be stimulated by phytochrome activation (also see Bewley and Halmer 1980/81). The dormancy of many light-sensitive species can be broken by gibberellin instead of light, suggesting that Pfr promotes gibberellin production. Because phytochrome is located in cell membranes it may influence membrane permeability. Finally, the increase in kinase activity due to phytochrome activation may cause a change in the balance between the pentose phosphate shunt and the glycolytic pathway thereby modifying the capacity of seeds to germinate (Copeland and McDonald 1985).

Light sensitivity interacts in complex ways with other environmental factors as well as with physiological stress (Koller 1972; Taylorson 1982). For example, a light requirement may depend on seed age: only freshly harvested seeds of *Salvia pratensis* L. and *Epilobium angustifolia* L. require light for germination (Mayer and Poljakoff-Mayber 1989). Light sensitivity of seeds also generally increases with imbibition (Mayer and Poljakoff-Mayber 1989). The protective role of maternal tissues in establishing initial concentrations of Pfr was mentioned above. This consideration suggests that the light environment of the mother plant may influence the light sensitivity of seeds it produces. For example, seeds of *Arabidopsis thaliana* (L.) Heynh. that mature on plants irradiated with flourescent light (which contains little far red wavelengths) germinate readily in the dark because of high concentrations of Pfr; in contrast, seeds maturing on plants irradiated with incandescent light (which is rich in far red wavelengths) show poor dark germination because of low Pfr (McCullough and Shropshire 1970). Seed germination also may be influenced not only by the day length regime of the maternal plant, but also by this day length regime relative to stage of seed maturation (Gutterman 1982).

The interaction between light and temperature is complex and varied, and it is difficult to state general rules. For example, whereas lettuce seeds readily germinate in the dark at 20°C, dark germination was inhibited at temperatures greater than 20°C (thermoinhibition). A single 5-minute exposure to red light completely overcame this inhibition at temperatures up to 29°C, but was ineffective at 33°C. However, repeated exposures to light were effective in promoting germination at 34°C (Saini et al. 1989). Germination of peppergrass (*Lepidium* spp. L.) is increased when red light follows a period of low temperature but precedes a period of higher temperature (Copeland and McDonald 1985). A three-way interaction between light, temperature, and osmotic stress has been shown for lettuce seed germination (Khan 1960). In this case, a light requirement for germination developed under osmotic stress; further, osmotic inhibition of germination was reversed by exposure to light at temperatures under which light was not required for germination (Koller 1972). Mayer and Poljakoff-Mayber (1989) give additional details and examples.

High Irradiance Response (HIR)

As discussed above, short term exposure to red light effects a conversion of the phytochrome pigment, forming Pfr. If short term exposure to far red light follows, the phytochrome pigment converts to Pr. Germination is promoted by Pfr following red light exposure (in those seeds that are light sensitive), and, following the completion of phytochrome action (which may take from a few hours to 2 days), far red light no longer inhibits germination (Cone and Kendrick 1986). The concept of "escape time" refers to the completion of phytochrome action, after which time far red light no longer inhibits germination (i.e., the Pfr has "escaped" from far red light reversibility). At this stage, however, prolonged exposure to far red light may inhibit germination of some species. This response, termed the high irradiance response (or HIR), is less well-understood than the LFR and apparently is not widespread (Thompson 1989). After prolonged exposure to far red light, germination may be induced by red light, establishing high levels of Pfr (Frankland 1986). In addition, prolonged exposure to white light (sun light) may inhibit germination. This response is often temperature-dependent (Bewley and Black 1985).

Gases (Oxygen and Carbon Dioxide)

Because of their low metabolism, dehydrated seeds have a very low oxygen requirement. However, with the rapid increase in metabolism following imbibition, the oxygen demand of a germinating seed rises. Most species show delayed and reduced germination if oxygen falls below the atmospheric level of 20%. The effect of carbon dioxide is generally opposite the effect of oxygen: most species fail to germinate if carbon dioxide tension is greatly increased (reducing carbon dioxide below atmospheric levels (about 0.03%) generally has no effect on germination). Atmospheric pressure appears to have little effect on seed germination (Mayer and Poljakoff-Mayber 1989).

Oxygen requirements are temperature-dependent. As temperature of incubation increases, so do seed metabolic rates and hence oxygen demand. However, at higher temperatures oxygen availability is decreased because of reduced oxygen solubility in the water of the seed coat (Come and Tissaoui 1972).

There are many exceptions to the foregoing generalities. For example, *Typha latifolia* L. *and Cynodon dactylon* (L.) Pers. germinate better with 8% oxygen than with 20% oxygen (Morinaga 1926a,b cited in Mayer and Poljakoff-Mayber 1989). A number of species [e.g., rice, barnyard grass (*Echinochloa crus-galli* (L.) Beauv. var. *oryzicola*), *Nelumbo nucifer* Gaertn. and *Peltandra virginica* (L.) Kunth.)] are capable of germinating in an oxygen-free environment (although seedling growth under these conditions is abnormal) (Koller 1972; Mayer and Poljakoff-Mayber 1989).

Light, atmospheric gas, and temperature interact in their effects on seed germination. It was mentioned earlier that lettuce is light-sensitive at temperatures exceeding 20°C. This light requirement can be overcome by high concentrations of carbon dioxide if oxygen is maintained at atmospheric concentrations (Koller 1972). Light and temperature interact in their effects on germination of *Phacelia tanacetafolia* Benth. Continuous light inhibits germination of this species; this inhibition can be overcome by high oxygen pressure (Koller 1972).

Water

Factors affecting seed water uptake fall into two general categories: (1) characteristics of the seed, and (2) water relations between seed and substrate (Hadas 1982;

Bewley and Black 1983). The description of the imbibition process presented earlier emphasized seed characteristics. Some considerations relative to seed/substrate water relations are discussed below. Detailed discussions are found in Hadas (1982), Bewley and Black (1983, 1985), and Mayer and Poljakoff-Mayber (1989).

Seed water uptake is controlled by the water potential gradient between the seed and its surrounding medium, and by water diffusivities involved in this seed/substrate system (Hadas 1982). Water potential in desiccated seeds may be –400 MPa (Shaykewich 1973). Thus, exposure of seeds to less negative water potentials creates a water potential gradient that may result in water uptake. The rate and extent of water uptake is affected by factors that impact water transport (e.g., diffusivity of water from substrate to seed and permeability of seed coat and seed tissue). Impermeable seed coats have already been discussed. In order to describe seed water uptake, the discussion below will assume seed coat permeablility.

Seeds make contact with soil water in two ways: seeds may be in contact with the surface of a soil particle, or seeds may be "in contact" with a pore between soil particles. An important principle follows from this consideration that should be kept in mind in the discussion below. Since less energy is required to extract water from a pore (especially a large one) than from the surface of a particle, soil water is depleted first from soil pores. Therefore, any effect that impacts seed/soil contact will have critical bearing on water uptake (Hadas 1982; Bewley and Black 1983).

The two most important components contributing to total soil water potential as it relates to water movement from soil to seed and seed hydration are soil matric and soil osmotic potential (Hadas 1982). However, seeds perceive changes in these two components differently: seed water uptake generally is retarded more by a given decrease in soil matric potential than by a similar decrease in soil osmotic potential, because the former involves a recession of water film from larger to smaller pores and hence a decrease in both seed/substrate contact and water conductivity (Hadas 1982).

As a seed imbibes water, its water potential becomes less negative, and because the surrounding medium is "losing" water to the seed, its water potential becomes more negative (Hadas 1982). Of course, this phenomenon takes place only in the area immediately surrounding the seed: it is noteworthy that water in soils may often flow only over about a 10 mm distance (Dasberg 1971). Thus, as imbibition proceeds, the water potential gradient initially present between seed and substrate decreases; this, in turn, results in less rapid water uptake. This decline in water uptake is more rapid in soils with low water-holding capacity (e.g., sands in comparison with clays) (Bewley and Black 1985). Another factor involved is seed swelling, which is one of the first morphological changes that takes place in imbibing seeds. Seed swelling generally results in increased compaction at the seed/soil interface and increased total seed/soil contact area; however, there is also a decrease in soil water content and reduced pore space immediately surrounding the seed, the net effect of which is reduced hydraulic conductivity and less water uptake (Hadas 1982).

Thus, soil water potential is only one aspect of the seed/substrate system. The hydraulic conductivity of a soil may be reduced if its matric potential is low, and this effect reduces the amount of water arriving at the seed surface (Bewley and Black 1983). However, additional factors that include surface contact phenomena between seed and soil are more important. These considerations lead to a general principle related to soil texture: an unsaturated sandy soil has both a lower hydraulic conductivity than a clay soil as well as a smaller area of seed/water contact (because of larger particle sizes). Therefore, at a specified soil water potential, germination of a given species will be higher in a clay soil than a sandy soil (Hadas 1982; Bewley and Black 1985).

The foregoing discussion was related to seed/substrate relations between seed and soil. A different situation occurs in petri dishes. Seeds on blotter paper in a petri dish generally have excellent contact with the surrounding aqueous medium. Consequently, the most significant factors affecting water uptake are the difference between the water potential of the seed and its environment (e.g., the blotter paper), and the properties of the seed itself (e.g., composition and seed coat characteristics). This situation is much simpler than that faced by a seed in or on the soil, where substrate matric potential and seed/substrate contact phenomena must be considered as well (Bewley and Black 1983).

Another factor that must be considered with respect to seed/soil water relations involves mucilage associated with the seed coat. Mucilages are polysaccharides that are or contain polyuronides (Bewley and Black 1983). The occurrence of mucilaginous seeds is common in the families Brassicaceae, Euphorbiaceae, Lamiaceae, Onagraceae, and Plantaginaceae but apparently absent in Poaceae and Asteraceae (Young and Evans 1973; Hadas 1982). In general, mucilage acts to extend the pathways for water flow from soil to seed, thereby enhancing water uptake (Hadas 1982; Bewley and Black 1983). However, Hadas (1982) suggests that mucilaginous seeds have little advantage over non-mucilaginous seeds under severely limiting soil water conditions. Also, oxygen uptake may be impaired if too much water is taken up by mucilage (e.g., Negbi et al. 1966; Heydecker and Orphanos 1968; Haferkamp and Jordan 1977; Hadas 1982). Mucilage is also important as a seed dispersal mechanism (e.g., Fahn and Werker 1972).

It is generally recognized that seedling growth immediately following germination (radicle emergence) can take place at water potentials more negative than those which permit germination (Hegarty and Ross 1980/81). Hegarty and Ross (1980/81) suggested that the process of germination most sensitive to water stress is initiation of cell elongation (rather than cell division). That is, initiation of growth ("germination") is more sensitive to water stress than subsequent seedling growth.

Evidence bearing on this conclusion involves the effect of hormones on germination. The stimulatory effect of light and hormones (e.g., cytokinin and gibberellic acid) on germination of dormant seeds has already been discussed. These factors often have a similar effect in enhancing germination of water-stressed seeds (Hegarty and Ross 1980/81). For example, water potentials at which there is a 50% reduction in germination are more negative for seeds treated with kinetin and gibberellic acid than for untreated seeds (Hegarty and Ross 1980/81: Tables 2, 3). Hegarty and Ross (1980/81) have shown, however, that the stimulatory effect of these factors in water-stressed seeds may also be accompanied by reduction in growth, the latter possibly due to reduction in mitotic activity. The conclusion is that processes involved with initiation of cell elongation respond differently to hormones than processes involved with growth itself, suggesting that metabolic control of the two types of processes is different. These differences are manifested in water relationships such that initiation of cell elongation associated with early germination is more sensitive to water stress than the process of early seedling growth.

Much research has focused on the effects of reduced water availability on seed germination. It is well established that both rate and extent of germination are reduced as water availability decreases. This is a direct function of seed tissue hydration: imbibition is the essential, first step in germination of non-dormant seeds, and when imbibition is impaired, for whatever reason, seed tissues undergo less hydration and subsequently experience slower or reduced germination. It is clear that soil water potential becomes less negative with the addition of water. However, seed imbibition is a complex process that involves not only seed characteristics but also seed/soil water relations as well. Whereas considerable research has been devoted to understanding the effects of water potential on seed germination, much of this effort has involved petri dish studies, and extrapolation from these results to the field often is difficult because processes involved with water uptake in a petri dish are different from those involved in the field. Some additional topics related to effects of water on seed germination are discussed in the next section.

Some Topics in the Ecology of Germination

Much of the foregoing discussion has focused on seed germination from a physiological viewpoint, with an emphasis on single factor effects considered largely in isolation and from a laboratory perspective. However, seed germination "in the field" is a much more complex phenomenon than in the laboratory because of interaction among factors (e.g., light, temperature, and water) that vary both abiotically and biotically. Selected topics in the ecology of germination are discussed in this section in an attempt to relate physiological principles presented earlier to seed germination in the field.

Topics related to seed demography and seed banks are not discussed here. However, it is appropriate that these subjects are receiving increased attention from plant ecologists. In light of the report that at least 95% of all plant mortality takes place at the level of the individual seed (Hickman 1979), it is clear that a complete understanding of plant population dynamics must include the fate of seeds prior to germination as well as the dynamics of seed banks. Some references include Harper (1977), Solbrig et al. (1979), Grime (1979), Cavers (1983), Fenner (1985), Thompson (1987), Hendrix (1988), and Leck et al. (1989).

Gap Processes

One of the fundamental resources that plants compete for is space. In fact, many of the niche differences usually offered to explain coexistence among plants may be "refinements of a basic explanation in terms of competition for space" (Yodzis 1986:482). In order to establish itself in existing vegetation, a plant requires a certain amount of belowground space from which to extract necessary water and nutrients, and a certain amount of aboveground space to capture light. With this perspective in mind, gap processes become an integral component in an understanding of the continued coexistence of species in a stand of vegetation.

Gaps occur in existing vegetation when occupied space is made available for a plant by mortality or removal of an earlier occupant of a site. Gaps provide one of the principal means by which species richness is maintained in vegetation. Many tree species, for example, maintain their representation in closed-canopy forests by gap colonization (Hubbell and Foster 1986); gap processes are also important in grasslands (Loucks et al. 1985). Additionally, many phenomena related to succession can be understood in the context of gap processes. The ability to detect a gap may therefore be critical to the establishment of a plant in a site. Some examples of gap-detecting mechanisms in seeds are discussed below.

Light and Gap Detection

The light environment in vegetation is a complex and ever-changing phenomenon. The spectral composition of light changes diurnally, with far red enrichment at sunrise and sunset. Similarly, leaf canopy acts as a red light filter: red light is absorbed by the chlorophyll of leaves, while far red light passes through. Additionally, the light environment of gaps is different from that beneath a canopy of vegetation. In gaps, there is higher light intensity and longer periods of illumination, as well as different spectral composition, than under a vegetation canopy (see Denslow 1987). Finally, light quality changes as it penetrates the soil. For example, only about 2% of the incident light passes through 2 mm of sand, and this light has wavelengths longer than 700 nm (Woolley and Stoller 1978). The phytochrome system in plants and seeds may act as an environmental sensor of surrounding light conditions.

The germination of light-requiring seeds may be effectively inhibited by soil burial because of the quality of light penetrating soil. Because light that penetrates soil is relatively rich in far red wavelengths, phytochrome in buried seeds may be relatively rich in Pr, effectively inhibiting germination. Thus, germination of these seeds may take place only when they are very near the soil surface (Bewley and Black 1985). Another factor may be involved here. A light requirement for seed germination is often associated with small seeds (Grime 1979; Thompson and Grime 1983; Denslow 1987; Foster 1987; Mayer and Poljakoff-Mayber 1989). For these seeds, germination near the soil surface may be advantageous if limited reserves are inadequate to support the needs of the seedling as it penetrates soil from a deeper depth (Grime 1979; Thompson and Grime 1983; Bewley and Black 1985; Mayer and Poljakoff-Mayber 1989). Additionally, small seeds of many species that require light for germination not only have the ability to remain dormant for long time periods, but also are stimulated to germinate by increases in incident light, temperature, and the ratio of red to far-red radiation (Denslow 1987). Each of these factors may increase in gap openings.

It is well known that many pioneer (or early successional) species have a light requirement for germination (e.g., Koller 1972; Grime et al. 1981; Fenner 1987b). Russian thistle is a notable exception (Wallace et al. 1968). Also, there is considerable evidence that the far red light which filters through a leaf canopy may inhibit germination of ruderal species (Taylorson and Borthwick 1969; Grime 1979; Fenner 1980a; Fenner 1987b). For example, Fenner (1980a) germinated *Bidens pilosa* L. seeds under a variety of shaded situations. A control treatment of "neutral light" was devised by germinating seeds under the shade of a piece of white paper. Germination under "neutral shade" was 85%. In contrast, germination of seeds under a shade of one layer of banana leaf was 0.7%. There was no germination of seeds planted under a canopy of grass 40–90 cm tall or under a canopy of dense weeds, and only about 25% germination of seeds planted among scattered weeds with about 20–30% cover.

Vasquez-Yanes and Smith (1982) provide another excellent example of phytochrome control of seed germination in the field. *Cecropia obtusifolia* Bertol and *Piper auritum* (H.B.K.) are tropical forest pioneer species whose field germination is apparently under the strong control of light. Seeds of these species can remain dormant in darkness or in soil for at least one year. Seeds planted in gap centers attained 70–80% germination in 10 days; seeds planted at the periphery of a gap reached 50% germination in 14 days; and seeds planted under a leaf canopy of the forest showed no germination after a month. In addition to light wavelength, duration of light exposure was important for seed germination. Maximum germination required repeated exposure to red wavelengths on consecutive days, and long exposures (e.g., 2 hours) were more effective than short exposures (e.g., 10 minutes). However, a single 6-hour exposure to red light was insufficient to stimulate initiation of germination. Vasquez-Yanes and Smith (1982) suggested that the requirement for long or intermittent exposure to red light for maximum germination may be a mechanism by which seeds of these species "determine" gap size and "distinguish" the light experienced in a gap from sun flecks in a canopy.

It is reasonably clear that a light requirement for germination of pioneer species may help insure that seeds germinate only in canopy gaps. In addition, however, many of these early pioneer species have photosynthetic and growth rate characteristics that involve most optimum performance under non-shade conditions (Hubbell and Foster 1986). A germination strategy that insures germination in gaps also places the seedling in well-lighted conditions that are most conducive for subsequent vegetative growth (Cone and Kendrick 1986). For example, *Cecropia obtusifolia* Bertol seedlings had a relative growth rate of about 0.018 g/g/day under light conditions (7 hours of direct insolation, with between 39 and 53% of the daily photosynthetic photon flux of that above a forest canopy) similar to that experienced in a large (500 m^2) gap. In contrast, under light conditions (1 hour of direct insolation, with between 2 and 6% of the daily photosynthetic photon flux of that above a forest canopy) similar to that of a small (50 m^2) gap, seedlings of this species had a relative growth rate of about 0.004 g/g/day (Popma and Bongers 1988).

Another phenomenon of considerable ecological importance relates to the inducement of a light requirement in seeds by exposure to light filtered through a leaf canopy. It is generally accepted that germination inhibition by light filtering through a canopy can be a mechanism that prevents germination in conditions where the seedling would face competition from neighboring vegetation. However, this mechanism would apply only to seeds on the soil surface. If such seeds were buried, they would no longer experience the light that is filtered through the leaf canopy, and they may be expected to germinate. The germination of these buried seeds would lead to seedling emergence under leaf canopies where competition from neighbors may be intense. Fenner (1980b) suggested that exposure to canopy-transmitted light may *induce* a light requirement for germination in some species. By this mechanism, canopy-sensitive species have a means of avoiding germination in heavily-vegetated areas even if they become buried. Fenner (1980b) showed that *Bidens pilosa* L. seeds that remained on the soil surface for 1 or 2 hours (thereby experiencing the light conditions transmitted through the leaf canopy) before burial acquired a light requirement for germination. In this manner, seeds avoided germination if buried. Fenner (1980b:106) suggested that "it is as if the seeds 'remember' the circumstances of their burial. They will remain dormant until a further disturbance brings them to the surface, into a situation unshaded by neighboring plants."

A final topic related to light concerns the influence of parental tissue surrounding seeds as they mature. Earlier it was discussed that chlorophyll in maternal tissues that envelop maturing seeds may act to filter out red light. It may also be recalled that because phytochrome is a protein that can not fully alter its configuration when desiccated, photoconversion can only take place in seeds prior to desiccation, or during imbibition (see above). Thus, it may be expected that seeds maturing in surrounding tissue that remains green will be exposed to light rich in the far red wavelengths. Most of the phytochrome in these seeds will be in the Pr form, and thus these seeds may be expected to have a light requirement for germination. On the other hand, if tissues surrounding seeds lose much of their chlorophyll (or are removed) prior to seed desiccation, then the enclosed seeds will be exposed to light unfiltered by chlorophyll. Much of the phytochrome in these seeds may be in the Pfr form, and so these seeds will germinate under light or dark conditions (Cresswell and Grime 1981).

With this background in mind, Cresswell and Grime (1981) investigated relationships among seed maturation, patterns of chlorophyll retention, and loss in parental tissues surrounding seeds, and germination requirements of a variety of species. They found a strong correlation between chlorophyll retention in surrounding parental tissues and a light requirement for germination. Additionally, they showed that light requirements for germination may change during seed maturation. For example, germination of *Hordeum murinum* L. was similar under light and dark conditions when seed coats were removed in full daylight prior to seed desiccation. However, when dried in the dark before removal of surrounding tissues, immature seeds failed to germinate in the dark. That is, surrounding parental tissue imposed a light requirement for germination of immature seeds. However, this light requirement was lost during the usual course of seed maturation because chlorophyll in parental tissue decreased sharply prior to desiccation.

Cresswell and Grime (1981) discussed the ecological significance of their results. They stated that plants that retain chlorophyll in parental tissues surrounding seeds during seed maturation also tend to be small-seeded species that develop more or less persistent soil seed banks. These species thus have the ability to remain dormant while buried because of a light requirement induced by their surrounding tissues during seed maturation. Other species that retain chlorophyll during seed maturation that do not tend to develop persistent seed banks (e.g., large-seeded composites) may be unusually susceptible to canopy-induced dormancy. That is, these species may have a light requirement for germination that might enable them to exploit vegetation canopy gaps. Seeds surrounded by tissue that loses chlorophyll early in the maturation process tend to belong to species that germinate in the fall in response to water conditions rather than a light requirement, and to legumes whose germination is controlled by hard seed coats.

Temperature and Gap Detection

In the study of the effect of alternating temperatures on seed germination, it is important to distinguish the effects of the temperature alternation itself on germination from the effects on germination of specific constant temperatures that may be included in the alternation (Thompson and Grime 1983). Many species, particularly those of wetland habitats, show increased germination when exposed to high temperatures and light. *Chenopodium rubrum* L., for example, does not require alternating temperatures for germination when the temperature is constant and exceeds 30°C; however, when the temperature is 22°C, fluctuating temperatures are required for germination (Thompson and Grime 1983). Therefore, to focus on the effects of alternating temperatures on germination, it may be important to use various depressions below a certain baseline temperature in order to avoid the high-temperature stimulation of germination that may not be due to the temperature fluctuation itself but instead may be attributable to high temperatures (Thompson et al. 1977; Thompson and Grime 1983). This consideration is important when field germination is studied as well as when alternating temperatures are implemented in laboratory conditions.

Soil and air temperatures are generally higher and fluctuate more widely in gaps (especially larger gaps) than under canopy or litter cover (e.g., Thompson et al. 1977; Vasquez-Yanes and Orozco-Segovia 1982; cf Collins and Pickett 1987). The temperature fluctuation sensitivity some seeds show for germination may act as an environmental sensor of surrounding vegetation and/or soil depth (Fenner 1985). For example, *Heliocarpus donnell-smithii* Rose is a tropical tree found only in light gaps and young secondary vegetation. Relatively constant soil temperatures under forest canopies appear to induce dormancy of seeds of this species that can be broken when gaps appear and temperature fluctuations increase. *Heliocarpus donnell-smithii* Rose seeds showed 67% germination when planted at gap centers where soil temperatures in the upper 2 cm fluctuated 15°C diurnally. However, seeds planted under a forest canopy where soil temperatures fluctuated less that 5°C showed only 25% germination (Vasques-Yanes and Orozco-Segovia 1982). Additionally, the few seedlings that emerged under a forest canopy did not successfully establish because of low light intensity. Temperature fluctuation sensitivity thus appears to be a mechanism whereby seeds of this heliophyte can detect gaps in surrounding vegetation where light conditions are more conducive for successful establishment.

Temperature fluctuation sensitivity with respect to germination may also provide seeds with a means of "determining" depth of burial (e.g., Dubey and Mall 1972; Thompson 1974; Thompson and Grime 1983). Thompson and Grime (1983) found an association between small seed size and a requirement for a large temperature fluctuation for germination. They suggested that since seed size and ability to emerge from soil depth are related (Bewley and Black 1985; cf Mayer and Poljakoff-Mayber 1989), a larger temperature fluctuation requirement for germination of some small-seeded species may enable these seeds to "sense" when they are near the soil surface. A similar mechanism applies for some wetland species. Thompson et al. (1977) found that of 68 wetland species examined, 43% responded to fluctuating temperatures. For example, seeds of *Fimbristylis littoralis* Gaudich., a common weed in lowland rice fields whose germination is reduced under water, require fluctuating temperatures for germination, and this sensitivity is apparently a means whereby depth of submergence under water can be "detected" (Pons and Schroder 1986).

Requirements for light and for fluctuating temperatures are to some extent interchangeable for germination of some species. For example, *Holcus lanatus* L. showed 50% germination in darkness at a constant temperature of 22°C, and germination increased to about 85% with 12°C temperature fluctuations in darkness (Thompson and Grime 1983). However, germination was nearly 100% in the light at a constant temperature of 22°C. It appears that the temperature fluctuation requirement for high germination of this species can be replaced by light. For other species (e.g., *Chenopodium rubrum* L.), light reduces the amplitude of temperature fluctuation required for high germination (Thompson and Grime 1983).

Nitrates and Germination

The ecological significance of nitrate stimulation on germination lies in the observation that soil nitrate concentration often rises sharply in connection with soil warming in the spring (Russell 1966, cited in Harper 1977), soil microorganism activity (Fenner 1985) and short-term fluctuations in soil moisture (Goudey et al. 1988), and may thus influence emergence patterns of many species. For example, Popay and Roberts (1970) found that some seasonal flushes of *Senecio vulgaris* L. emergence immediately followed periods of maximum soil nitrate concentrations combined with maximum diurnal soil temperature fluctuation at the soil surface. However, seed nitrate concentration may be passively controlled by soil nitrate concentration, with the effect that high external nitrate concentrations can inhibit germination (Goudey et al. 1988). Bouwmeester and Karssen (1989) discuss seasonal patterns of seed dormancy and seedling emergence in relation to light, temperature, moisture and seed desiccation, and Pons (1989) discussed nitrate concentration in vegetation gaps and its role in germination stimulation.

Water and Germination

An understanding of water and germination must begin with a clear view of the processes of imbibition. As discussed earlier (Fig. 6), imbibition takes place in three phases, the first of which is largely a passive process controlled by the high negative matric potential of the seed. During the second phase of imbibition, water uptake is less rapid and is controlled mostly by osmotic and pressure potentials in the seed. It is during this second phase of imbibition that the major metabolic events take place that precede radicle emergence. For example, very soon after imbibition begins there are dramatic increases in seed respiration; polysomes are produced and protein-synthesizing capabilities are quickly attained with the early production of new messenger RNA; and damaged DNA is repaired (Fig. 8). All of these events are crucial for successful radicle emergence. An important consideration is, therefore, the effect of water stress during this second phase of imbibition. In particular, the effects of temporal pattern of water supply to the seed must be understood. An understanding of this issue relates to practical considerations in seeding such as effects of rainfall pattern on seed germination and effects of alternate wetting and drying cycles on germination.

Osmoconditioning

It is useful to begin with the topic of osmoconditioning, or osmotic priming. This process involves exposing seeds to a low osmotic potential for a period of time prior to exposure to less negative water potentials at which germination occurs (Mayer and Poljakoff-Mayber 1989). The ultimate results of osmoconditioning may include more rapid, and often more complete, germination; more synchronous germination; earlier seedling development (possibly conferring a competitive advantage); and increased yield (e.g., Hegarty 1978; Khan et al. 1980/81; Bewley and Black 1983). Osmoconditioning may also reduce deterioration of stored seeds (Thanos et al. 1989) and restore vigor (as indicated by germination rate) of aged seeds (Dell'Aquila and Tritto 1990). All of these effects are important in understanding seed germination, both to the plant ecologist and to the seed technologist.

During osmoconditioning, the initial low water potential is one at which there is sufficient imbibition for essential metabolic processes (e.g., RNA and protein syntheses as well as cell division) to begin prior to germination, but radicle emergence is prevented (Hegarty 1978; Bewley and Black 1985). The physiological basis of osmoconditioning is not fully understood (Hegarty 1978; Bewley and Black 1985; Bray et al. 1989). Khan et al. (1980/81) summarized pertinent literature on biochemical and physiological considerations related to osmoconditioning. They reported that DNA synthesis was probably not involved in the advancement of germination associated with osmoconditioning. Bray et al. (1989) showed that DNA synthesis occurs, albeit at low rates, in leek (*Allium porrum* L.) seeds during osmoconditioning. However, during osmoconditioning, when seeds are metabolically active but prevented from germinating, DNA repair may take place (Osborne et al. 1980/81; Osborne 1982). It may be recalled that DNA synthesis and replication take place relatively late in the germination process (Fig. 8). The period of osmoconditioning effectively lengthens the period between imbibition and DNA replication, during which time the repair of DNA may ultimately result in fewer chromosomal aberrations when replication does take place (Osborne 1982). Additionally, increased amounts of RNA in osmoconditioned seeds, as well as the enhanced ability of osmoconditioned seeds to synthesize RNA during subsequent germination, is likely due to activation and/or synthesis of enzymes related to RNA metabolism as a result of the prolonged imbibition and hydration that takes place during osmoconditioning. Finally, osmoconditioned seeds have higher rates of protein synthesis than untreated seeds, and this apparently is involved in both control of germination and advancement of germination of osmoconditioned seeds (Khan et al. 1980/81; Bray et al. 1989).

Alternate Wetting and Drying Cycles

A topic that is related to osmoconditioning is alternate cycles of seed wetting and drying. Following early imbibition and up to the commencement of cell division, seeds may be dried without usually adversely affecting ultimate germination and growth (Bewley and Black 1983). Embryos are drought-resistant when their cells are between the end of the G1 phase of the last mitotic cyle during maturation and the end of the G1 phase of the first mitotic cycle during germination (Deltour 1985; see above). Many seeds show more rapid, and sometimes more complete, germination if initial imbibition is followed by a drying and then a wetting period (e.g., Keller and Bleak 1968, 1969; Bleak and Keller 1969, 1970, 1972; Keller et al. 1970; Hauser 1981; Bewley and Black 1983), and subsequently emerging seedlings may also show improved drought and frost tolerance (Bewley and Black 1983; Deltour 1985) and increased yield (Bleak and Keller 1970; Bewley and Black 1983; see Hegarty (1978) for a review). Metabolic activity (e.g., protein and RNA syntheses) during the initial hydration period may result in the accumulation of substrates and/or the enhancement of enzymes which are stable following dehydration (Hegarty 1978; Simon 1984); seeds at this time are poised for subsequent germination when exposed to rehydration (Bewley and Black 1985). However, as indicated above, timing of dehydration is critical.

Many of these results are highly variable and species-specific (Bewley and Black 1983). With respect to speed of germination, it may be a mis-statement to say that such wetting and drying cycles hasten germination. Bewley and Black (1983) suggested that the total time of imbibition, or hydration, is the same for "pre-soaked" and untreated seeds, and that hastening of germination is apparent only if the initial imbibition time is disregarded. Additionally, improved drought or frost tolerance, and improved yields, may be related to timing of germination: pre-soaked seeds may emerge earlier in the field than untreated seeds, and this may confer a competitive and/or developmental advantage to the seedling that is produced.

Repeated cycles of wetting and drying may eventually reduce the germinability of some seeds (Plummer 1943; Maynard and Gates 1963; Cook et al. 1967; Hillel 1972). Seeding failures have been attributed to excessive periods of alternate seed wetting and drying that ultimately may be explained by deleterious interruption of the germination process (Robertson 1947 cited in Cook et al. 1967; Cook et al. 1967) or to the accumulation of a germination inhibitor during dehydration that induces secondary dormancy (Hegarty 1978). For example, it may be recalled that seed ATP content increases dramatically during the first hours of imbibition (e.g., Simon 1984; also, Fig. 7). Wilson et al. (1970) measured seed phosphate ester content to assess metabolic responses of seeds to environmental changes. They planted crested wheatgrass (*Agropyron desertorum* (Fisch.) Schult.) seeds in the fall both at a 2.5-cm depth and broadcast on the soil surface and measured synthesis of phosphate esters at various intervals after planting. Net synthesis of phosphate esters was not detected in the absence of precipitation. Rainfall of 0.4 cm during a 5-day period stimulated seed metabolism and marked increases in phosphate esters were detected. There was no precipitation for the following seven days, and while there was little change in phosphate esters in the drilled seeds, these compounds were largely broken down in broadcast seeds. With additional rainfall, seeds resumed biochemical activity. Net synthesis of ATP suggested that intermittent storms and dry periods resulted in frequent starting and stopping of the germination process in broadcast seeds. Similarly, Morohashi and Bewley (1980) found that rate of oxygen uptake in mitochondria of pea cotyledons subjected to repeated cycles of wetting and drying was only 59% of the rate obtained from cotyledons that were not desiccated following initial hydration.

Pattern of Precipitation

Pattern of rainfall is another aspect of temporal water supply to the seed that has important implications for successful germination and seedling emergence. Many seeding failures have been attributed to unfavorable rainfall. Often there may be sufficient amounts of rainfall for seed germination and/or seedling emergence, but its pattern of occurrence is not conducive to successful germination and/or emergence. For example, a given amount of rainfall, say, 2 cm, occurring on two consecutive rainy days may present a seed with different imbibition and hydration conditions than the same amount of rainfall also occurring on two days, but with each rainy day separated by one or several dry days. [Of course, successful seedling establishment depends in part on soil moisture conditions following emergence, during which time crucial initiation as well as growth of roots takes place (e.g., Briske and Wilson 1978; Wilson and Briske 1979)]. These considerations of pattern of rainfall may be related to osmoconditioning as well as effects of alternate wetting and drying cycles, discussed above.

In general, both rate and extent of emergence as well as seedling survival are affected by (1) total amount of water seeds are exposed to, (2) how many days of water seeds are exposed to, (3) whether initial moisture is followed by additional moisture, and if so, how may dry days elapse between waterings, (4) whether rainfall occurs in single-day applications or on consecutive days, (5) soil temperature and texture, and (6) species in question (Wester and Dahl 1977, 1983; Wester 1979; Wester et al. 1986; also see Frasier et al. 1984, 1985, 1987; Frasier 1989). Some of these principles will be illustrated with data from weeping lovegrass (*Eragrostis curvula* (Schrad.) Nees.) and sideoats grama (*Bouteloua curtipendula* (Michx.) Torr.), two grasses with contrasting germination and emergence behavior in relation to pattern and amount of rainfall. Seeds of both species were planted at a depth of 6 mm in sand maintained at 30°C in a growth chamber, and supplied with 16 patterns of watering (see legend in Fig. 11); each day of water provided 5 mm.

Figure 11 shows that at a 30°C soil temperature, weeping lovegrass seeds receiving two initial consecutive days of water produced seedlings that emerged two days after the beginning of watering. Seeds receiving one day of water, followed by one to five dry days before the second day of watering, produced seedlings which emerged three to eight days after the beginning of watering; emergence rate decreased linearly as the number of dry days increased between applications of initial and additional moisture. These results suggest that weeping lovegrass required at least 10 mm of water to emerge: when only 5 mm on one day occurred, emergence was delayed until a second 5 mm was provided. Emergence was greater in treatments that supplied 15 mm of water (three days of watering) regardless of pattern of application (Fig. 12); emergence was also relatively high when 10 mm of water occurred within two to three days (treatments 1 and 2), but declined as the number of dry days increased between single applications of 5 mm. After emergence in response to treatments was completed, all seeds were watered at a rate of 5 mm per day for three consecutive days to investigate the germinability of seeds which had not previously germinated ("follow-up" emergence). Follow-up emergence for weeping lovegrass was approximately inversely related to treatment emergence): seeds that did not produce seedlings after initial treatment watering retained viability

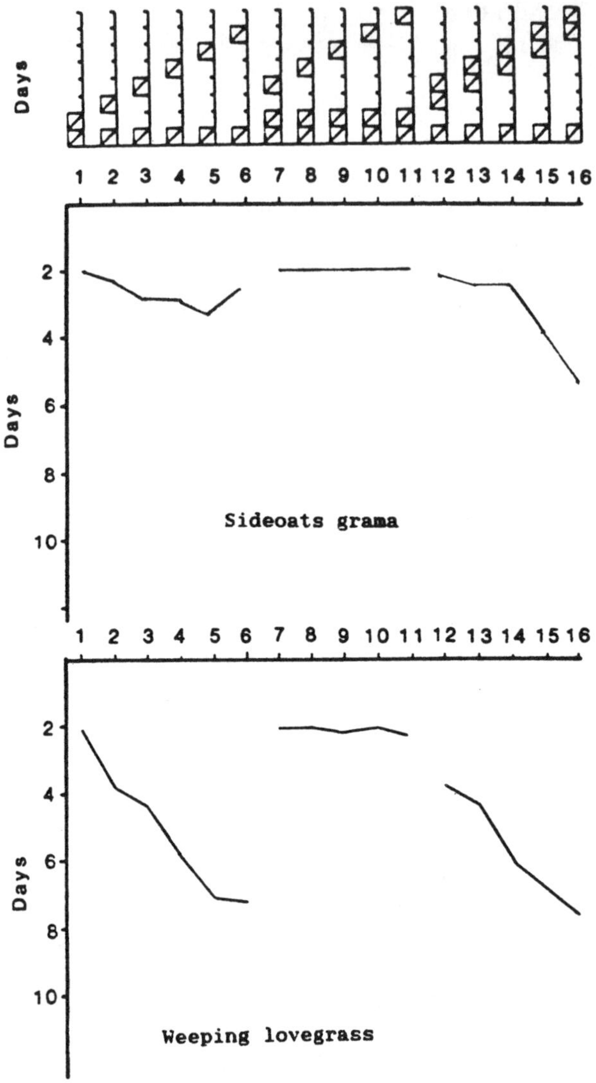

Figure 11. Days after the start of watering until beginning of sideoats grama (top) and weeping lovegrass (bottom) seedling emergence at 30°C soil surface temperature. Each hatched square in figure legend represents one day of water (5mm) [modified from Wester (1979) and Wester et al. (1986)].

for subsequent germination (Fig. 12). Field experiments (Wester et al. 1986) supported these findings.

Sideoats grama showed important differences in its germination and emergence behaviour. Although seedlings emerged soonest with two consecutive days of water, there was nevertheless some emergence with only one day of moisture (5 mm) (Fig. 11). Additionally, extent of emergence was highest with either two

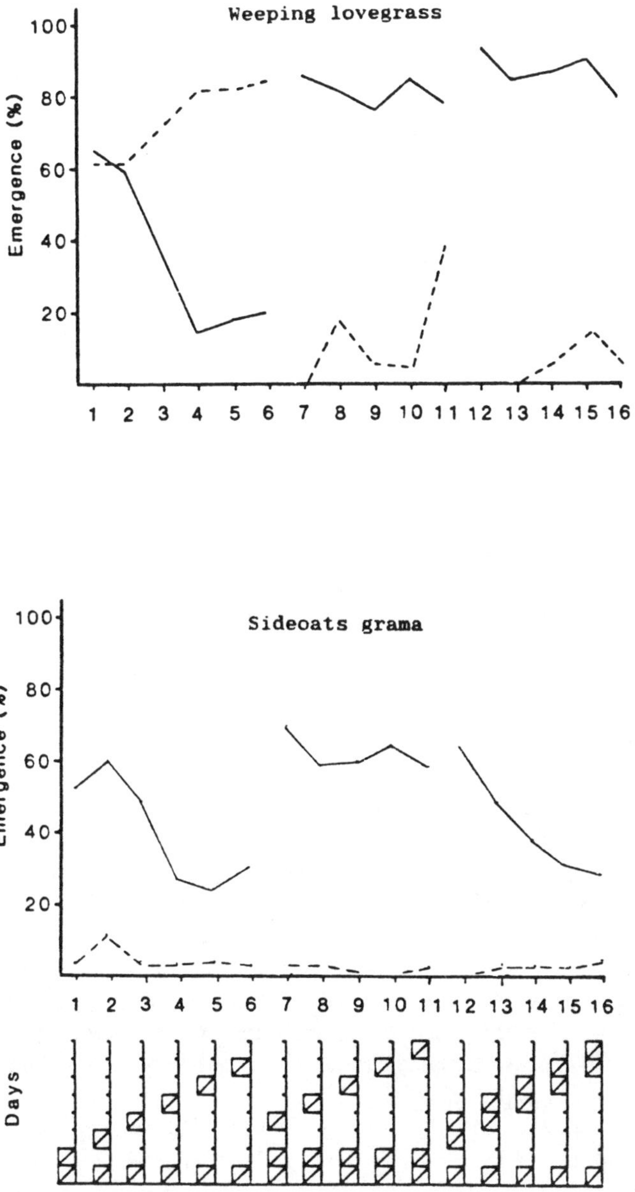

Figure 12. Treatment (solid line) and follow-up (dashed line) emergence (%) of weeping lovegrass (top) and sideoats grama (bottom) at 30°C soil surface temperature. Each hatched square in figure legend represents one day of water (5mm) [modified from Wester 1979) and Wester et al. (1986)].

initial consecutive days of water (regardless of total amount of water) or with two days of water supplied over a 4-day period (Fig. 12). Follow-up emergence was low regardless of treatment emergence: initial exposure to water, even if it was insufficient to stimulate high emergence, reduced later germinability and emergence of seeds.

These studies show that both pattern and amount of moisture can be important in determining field emergence. From a practical standpoint, revegetation projects will have an increased chance of success if seeding is done when there is a high probability of encountering rainfall patterns and amounts conducive to successful seedling emergence. Heermann et al. (1971) tabulated the probabilities of occurrence of specific patterns of wet and dry days for 11 western states and Texas. With information about emergence response to rainfall patterns, these data may improve chances for seeding success.

Seed-Soil Contact

As discussed earlier, there are important differences in imbibition of seeds on filter paper surfaces in petri dishes and seeds in soil. In the former setting, seeds generally have excellent contact with the imbibing medium, and consequently the most significant factors affecting water uptake are the differences in water potential of the seed and the filter paper, and properties of the seed. For seeds germinating in soil, however, soil matric potential as well as seed-soil contact are important; two further considerations related to the latter may be mentioned. In contrast to the "simpler" imbibitional settings encountered by seeds on petri dish filter paper, it is important to appreciate that soil water content differs, even at the same tension, in soils that are drying as compared to soils that are being wetted, and likewise, at the same water content, drying soils have different water tensions than soils that are being wetted (Hillel 1971; Harper 1977). This may be an important consideration relative to germination and pattern of water supply, discussed earlier.

Germination also is influenced by seed size and morphology, soil microtopography, and the position of the seed on or in the soil. Harper (1977) reviews several excellent examples. For example, small smooth seeds often have a tendency to fall into crevices between soil particles where conditions are more conducive for germination compared to positions adjacent to soil particles, where larger or rough-surfaced seeds may lodge (Harper et al. 1965). Seed hairs of *Blepharis persica* (Burm.) Kuntze. swell on contact with moist soil, reorienting the seed so that its micropyle makes contact with the soil and imbibition is improved (Gutterman et al. 1967; also see Harper 1977). As discussed earlier, many seeds produce mucilage upon wetting, and this may impact water uptake.

Seeds positioned on the soil surface encounter a situation inimical to seed germination in at least two respects. It is clear that both levels and fluctuations of irradiance, temperature and soil moisture on the soil surface may be hazardous to seed germination and seedling establishment (e.g., Hadas 1982). In addition to these considerations, however, the imbibitional setting for seeds on the soil surface differs from that encountered by buried seeds in several important respects. Compared to buried seeds, seeds on the soil surface make only partial contact with soil particles, reflected in fewer contact points and less total contact area between seed and soil. Both conditions result in impaired water uptake (Hadas 1982) and reduced germination even at high soil water potentials (Collis-George and Hector 1966, cited in Hadas 1982). Additionally, when an imbibing seed on the soil surface swells, it does not compact surrounding or underlying soil. Under these conditions, contact impedance (related to water flow across the seed-soil interface) increases and water uptake decreases (Hadas 1982).

Seasonal Aspects of Dormancy and Seed Germination

The induction of seed dormancy, its relief, and its involvement in seasonal patterns of seedling emergence is a complex topic beyond the scope of this chapter. It is clear that many factors including light, temperature, moisture, soil nutrient status, maternal effects on seed development, time of seed shattering, time and depth of seed burial, and disturbance are all involved, and interactions among these effects are complicated. Some discussions of this subject can be found in Karssen (1980/81a,b), Fenner (1985), and Baskin and Baskin (1985, 1988, 1989).

Seed Germination Research Published in the *Journal of Range Management*

Rangeland includes environmental settings that range from wetlands to xeric deserts to mountain grasslands. A broad diversity of plants can be found in many rangelands. Germination information for many of these species is limited. More familiar compilations of seed germination information (e.g., Rules for Testing Seeds, AOSA 1988) often do not include information on range plants. The *Journal of Range of Range Managment* includes a wealth of information on seed germination research.

This research can be classified under two general headings. Field studies are carried out under conditions where some environmental factor or factors are uncontrolled. In these studies, the variable most often recorded is seedling emergence rather than seed germination. Also, in many of these field studies emergence data is collected at time intervals after which some seedlings may have emerged but wilted or died because of water stress, disease, or insect damage. In comparison to these field studies, research carried out in growth chambers and petri dishes may be more specifically related to seed germination. Even in greenhouse studies where seeds are planted in soil, it may be argued that emergence is similar to germination when seeds are not planted too deeply and data are recorded early enough that quickly emerging seedlings are noted prior to possible wilting or death.

Wester (1991) reviewed and summarized research of the latter type that has been published in the *Journal of Range Management* since its inception in 1948. This review focuses on methodological issues. Papers that were reviewed were summarized with respect to (1) species studied (arranged alphabetically), (2) author, (3) experimental scope as reflected by type of data collected, (4) germination medium (e.g., soil, filter paper, etc), (5) water stress imposed, (6) constant temperatures used, (7) alternating temperatures used, (8) light conditions, (9) seed age, (10) seed size, (11) seed source, (12) pH, (13) specific ions and concentrations, (14) seed depth, and (15) seed pre-treatment. This review includes 790 entries pertaining to 430 species taken from 212 papers.

An extract from this tabular summary is given in Table 2. For example, Stidham et al. (1980) studied germination [Exp(eriment) =1] of *Fallugia paradoxa* (D. Don)

Table 2. An extract from a summary of seed germination research published in the *Journal of Range Management*, from Wester (1991). See text and Wester (1991) for an explanation of notation.

Species	Source	Exp.	Med.	Water Stress	Temperature Con.	Temperature Alt.	Light	Seed Age	Seed Size	Source	pH	Conc.	Ion Depth	Seed Pre-treatment
Eurotia lanata	Booth 1989	2	GP	0.1– –0.8	15	–	L/D	–	–	–	–	N	–	S:DIW, N A SC:D
Eurotia lanata	Booth 1990	2	GP	–	–	5/15	L/D	–	–	–	–	–	–	S:W,0C,2d see footnote
Eysenhardtia texana	Whisenant and Ueckert 1982b	2	P	0– 1 –1.2	0-45	–	L,D; L	0–42m	–	–	5–9	–	–	SC:M
Fallugia paradoxa	Stidham et al. 1980	1	V	–	–	12/20	L/D	–	–	–	–	–	–	C,S:K,0.2%,W,2C,0–10w; T,0.3%;0–60mi
Fallugia paradoxa	Jordan and Haferkamp 1989	1	P	–	5-30	–	L/D	–	–	–	–	–	–	
Festuca altaica ssp. *hallii*	Romo et al. 1991	2	P	0– –1.64	0-40	0/2- 35/40	D+	–	–	+	–	–	–	
Festuca arundinacea	Hull 1973	1	S	–	–	16/28	L/D	19-24y	–	–	–	–	–	
Festuca arundinacea	Senter et al. 1975	1	P	–	–	20/30	L/D	–	–	–	–	–	–	ST:with AN,(d),humid C:5-10C(m),5d see footnote
Festuca arundinacea	Miller and Chapman 1978	1	P	–	–	15/25	D+	–	–	–	–	+/ 4-16 mmhos	–	
Festuca arundinacea 'K-31'	Undersander 1986	1	P	–	25	–	L	–	–	–	–	+/ .034-.102M	–	S:AM
Festuca idahoensis	Doescher et al. 1985	1	P	0– –0.9	5-35	–	D?	–	–	+	–	–	–	SC:D
Festuca idahoensis	Chambers 1989	1W	–	30	D	+	–	–	–	–	–	–	–	

Endl. seeds in vermiculite [Med(ium) =V]; water stress was not studied, and seeds were exposed to a single temperature alternation [Alt(ernating temperature) =12/20°(C)]; seeds were exposed both to light and dark conditions, but this exposure was not a treatment factor of interest [Light =L/D]; seeds were given a pretreatment that included chilling and soaking in a 0.2% solution of potassium nitrate, as well as water, at 2°C for periods ranging from 0 to 10 weeks, and a 0.3% solution of thiourea for periods ranging from 0 to 60 minutes [C, S:K, .2%,W, 2C, 0–10w, T, 0.3%, 0–60mi]. Romo et al. (1991) studied both germination and some measure of seedling performance [Exp(eriment) =2] of *Festuca altaica* Trin. ssp. *hallii* (Vasey) Harms seeds in a petri dish [Med(ium) =P]; water stress imposed ranged from 0 to –1.64 MPa; seeds were exposed to constant temperatures ranging from 0 to 40°C, as well as several sets of alternating temperatures, the lowest of which included an alternation of 0/2 C, and the warmest of which included an alternation of 35/40°C; although the authors indicated that the research involved germination under dark conditions, the data presented suggested that seeds were exposed to light [Light =D+]; seeds from several sources also were studied [Source =+]. Hull (1973) studied germination of *Festuca arundinacea* Schreb. of several ages ranging from 19 to 24 years. Whisenant and Ueckert (1982) exposed seeds of *Eysenhardtia texana* Scheele. to light and dark conditions, and this was a treatment factor of interest [Light =L,D]; seeds were studied in constant light as well [Light =L]; seeds were mechanically scarified [Pre-treatment = SC:M]. A complete description of the notation used in this table is given in Wester (1991).

This table can be used in several ways. Suppose one is interested in germinating a particular species; this table provides a methodological summary to the kinds of conditions under which a given species has been studied and published in the *Journal of Range Management*. Additionally, this table can be examined to see what factors have not been studied. Also, by locating a species in this table, one can possibly find additional papers about the particular species in question published in other journals through the literature cited of the paper appearing in the *Journal of Range Management.*

This table also can be used to assess the kinds of seed germination research published in *the Journal of Range Management*. For example, of the 212 papers included in the review, only 31 papers (15%) studied light conditions as a specific treatment factor of interest. In 41 papers (19%), the authors stated that germination was conducted under dark conditions, but presented data (e.g., daily or weekly germination counts) that suggested that seeds may have been exposed to light. As discussed earlier, the response of seeds to light is a complex phenomenon that interacts with seed age, temperature, and moisture conditions. It is critically important that when the intention is to study germination under dark conditions, seeds not be exposed to light in wavelengths that can interact with the phytochrome system.

Constant temperature conditions were studied in 118 papers (56%), alternating temperatures were studied in 105 papers (50%), and both constant and alternating temperatures were examined in 48 papers (23%). It is clear that a seed physiologist studying imbibition or enzyme dynamics may impose constant temperatures because these processes are temperature-dependent. However, a range scientist studying revegetation must consider germination under conditions of alternating temperature, because these are the conditions seeds experience in the field. When seed germination research is intended to be predictive of field responses, it should be conducted under conditions that approximate the environment seeds will experience in nature.

Only 51 papers (24%) varied moisture supply to the seed, and most of these did so by varying the concentration of an osmoticum to impose water stress on the seed. It is likely that many seedings fail because of unfavorable moisture conditions during or following seeding (because of improper depth or date of seeding) than because of unfavorable temperature or light conditions. An area of seed germination research that needs more effort is seed germination in response to water stress, and this research should be carried out under conditions that approximate the water stresses encountered by seeds in the field. For example, of the 51 papers studying moisture stress, 18 papers used soil as a germination medium, 42 papers used something other than soil (e.g., filter paper in a petri dish), and 10 papers used both soil and non-soil (e.g., filter paper). As discussed earlier, filter paper in a petri dish offers an imbibitional setting to a seed that is very different from that encountered in or on soil. Soil moisture conditions in the field are in an almost constant flux. Soils in a drying phase and soils in a wetting phase have different water contents at the same water tension (and vice versa) (Hillel 1971). Beyond the general conclusion that both rate and extent of germination decrease as moisture stess increases, it may therefore be difficult to extrapolate results of water stress studies from a petri dish to the field. The following discussion of directions in seed germination research is taken from Wester (1991).

Some Directions in Seed Germination Research

Laboratory-oriented seed germination research published in the *Journal of Range Management* (Wester 1991) represents an impressive body of information, and there can be little doubt that much has been learned through this effort. Research ranges in scope from data on enzyme dymanics as related to environmental conditions, to the influence of specific ions on imbibition, to the effect of seed age on germinability. Basic principles of germination response to various moisture and temperature conditions have been studied and applied to hundreds of species. Nearly every conceivable kind of seed pre-treatment can be found in the papers reviewed by Wester (1991). This review can be used to assess our past directions of work, and to suggest future avenues of research.

Two general areas or types of research are discussed below, labelled "seed-oriented" research and "environment-oriented" research. Of course, these labels are somewhat vague, and they are far from exclusive. However, with these labels, I hope to distinguish research that is directed more by purely physiological considerations, on the one hand, from research that deals directly with the sorts of environment under which germination occurs.

Seed-Oriented Research

Many types of seed germination research are best done under the carefully controlled conditions of a petri dish or a growth chamber. For example, much of the research investigating methods of seed pre-treatment to relieve dormancy may be best done in a petri dish environment: in this research, the major interest is on the seed pre-treatment itself, and it is certainly efficient from a methodological viewpoint to expose seeds to the "best possible" germination conditions after pre-treatment. For example, Rasmussen and Wright (1988) first established that under constant temperature conditions, flameleaf sumac [*Rhus lanceolata* Gray (Britt.)] seeds germinated best when exposed to a 20°C temperature with 8 hours of light and 16 hours of dark. With this information in hand, they then studied 28 conditions of scarification (various combinations of wet and dry heat, and durations of exposure to acid).

Of course, factor interaction between, say, type of scarification and germination conditions (e.g., light and temperature) is fundamentally important. However, it is often best to begin by studying one factor at a time, and after some basic understanding is gained, then it is usually advisable to begin studying effects of several factors simultaneously. Thus, even with the wide variety of statistical methods that allow for simultaneous inclusion of several factors into an experimental design, an excellent case may be made for the general methodological approach wherein environmental conditions are held constant in order to focus on other factors.

Environment-Oriented Research

Another general direction of research involves studying effects of environment on seed germination. In many of these studies, the principal environmental factors under study are moisture and/or temperature, and other factors such as seed age or pre-treatment are of secondary interest. This research is often used as a basis to predict field germination responses. However, it is obvious that germination in response to moisture stress imposed in a petri dish can be markedly different from germination in response to moisture stress in the field. Similar considerations apply to germination temperature research. Soil temperature maxima at a 13-mm depth can average 57°C (and commonly reach 66°C) in southwestern deserts (Sosebee and Herbel 1969), a temperature that is seldom studied in growth chambers. Even when average daily air temperatures may not exceed 24°C during early May seeding periods (Wester et al. 1986), soil surface (0–5mm depth) temperatures of a bare seedbed may commonly reach 50°C (Wester et al., unpublished data). Most research dealing with germination response to temperature shows that seed germination often may be "optimum" over a broad range of temperatures, and that major differences occur only at temperature extremes. And yet extreme temperatures in the field, and not average temperatures, are probably the most significant factors influencing germination. From this perspective, it may be questioned whether the methodological approach of studying varying environmental conditions in a petri dish yields information that is directly and efficiently transferrable to the practitioner in the field.

Some Unanswered Questions

There are many physiological processes in seed germination that are only vaguely understood, and much work in this area is required. Interactions between the seed itself and its environment throughout the germination process are imminently important. Below, some general questions are offered.

The area of seed pre-conditioning, or osmoconditioning, has long been studied, but many questions remain. Numerous applied studies have shown that pre-soaking seeds may improve chances for seeding success, and whereas it is well-known that pre-soaked seeds often emerge sooner in the field that non-pre-soaked seeds, specific processes involved are not fully understood. It is certainly clear that during pre-soaking, the seed begins many of the initial processes that precede radicle protrusion. But how does the process of seed pre-conditioning actually enable the seed to "poise" itself for germination? A general answer may be given by observing the importance of early RNA and protein syntheses, and DNA repair that take place in the first stages of imbibition. These phenomena now need to be studied for specific rangeland species that are commonly used in revegetation studies.

A related question concerns why it is that for some species (e.g., sideoats grama), exposure of seeds to water in amounts insufficient to cause germination drastically reduces the potential for future germination when adequate moisture is given, whereas other species (e.g., weeping lovegrass) under similar conditions retain the potential for germination after exposure to earlier moisture that is inadequate to stimulate high germination (Wester 1979; Wester et al. 1986)? It is likely that the area of seed pre-conditioning holds an answer to this question. With this sort of specific information, concrete recommendations can be made to the revegetationist who wishes to know, for example, whether an initial but perhaps light rainfall has essentially eliminated the germination potential of seeds he has planted, or whether these seeds still retain the potential to germinate with future adequate rainfall.

The ultimate goal of rangeland revegetation is the establishment of a productive stand of grass. The probability of success in this endeavor depends on many factors including appropriate species, proper seedbed preparation, planting depth and planting date, and suitable environmental conditions for germination, emergence and eventual establishment. Laboratory or growth chamber seed germination research whose results are intended to be predictive of field responses should be done under conditions that approximate field conditions so that resulting information can be more readily transferred to the practitioner. A comprehensive methodological approach that studies the environment (e.g., moisture, temperature and light conditions) as it is experienced by seeds in the soil will ultimately yield the most useful information. For example, in a soil of a given texture and initial water content, how does water availability change with a given addition of water, how quickly is water lost to evaporation, and how does water availability decrease as a result? Answers to these questions will of course be very "situation-specific", but they will set the stage for understanding the environment as encountered by the seed. Given this information, how deeply should a seed be planted to maximize its chances for experiencing adequate soil moisture for the longest period of time without risking planting too deeply? Then it may be asked, what are the specific patterns and amounts of rainfall necessary for seed germination? How soon is follow-up rainfall required, and in what patterns and amounts, to ensure development of adventitious roots and eventual seedling survival? Answers to these questions will require a comprehensive understanding of the germination process as it relates to the seed in the soil, as well as a full appreciation of the factors that enhance seedling establishment.

Acknowledgments

The author wishes to thank C.M. Britton, N.W. Hopper, and B.E. Dahl (deceased) (Texas Tech University), M.B. McDonald, Jr. (The Ohio State University), and J.A. Young (USDA-ARS, Reno, NV.) for reviewing an earlier draft of this paper. This is contribution number T-9-733, College of Agricultural Sciences and Natural Resources, Texas Tech University, Lubbock, Texas.

Literature Cited

Angevine, M.W. and B.F. Chabot. 1979. Seed germination syndromes in higher plants, pp. 188-206. In: O.T. Solbrig, S. Jain, G.B. Johnson and P.H. Raven (eds.), Topics in Plant Population Biology. Columbia Univ. Press, New York, 589 pp.

AOSA. 1988. Rules for Testing Seeds. Assoc. Official Seed Analysts. J. Seed Technol. 12:1-109.

Ashford, A.E. and F. Gubler. 1984. Mobilization of polysaccharide reserves from endosperm, pp. 117-162. In: D.R. Murray (ed.), Seed Physiology, vol. 2. Academic Press, New York, 295 pp.

Banks, H.P. 1970. Major evolutionary events and the geological record of plants. Biol. Rev. 47:451-454.

Barclay, A.S. and F.R. Earle. 1974. Chemical analyses of seeds, III: Oil and protein content of 1253 species. Economic Bot. 28:178-236.

Baskin, J.M. and C.C. Baskin. 1985. The annual dormancy cycle in buried weed seeds: A continuum. Bioscience 35:492-498.

Baskin, C.C. and J.M. Baskin. 1988. Germination ecophysiology of herbaceous plant species in a temperature region. Amer. J. Bot. 75:286-305.

Baskin, J.M and C.C. Baskin. 1989. Physiology of dormancy and germination in relation to seed back ecology, pp. 53-66. In: M.A. Leck, V.T. Parker and R.L. Simpson (eds.), Ecology of Soil Seed Banks. Academic Press, New York, 462 pp.

Bewley, J.D. and P. Halmer. 1980/81. Embryo-endosperm interactions in the hydrolysis of lettuce seed reserves. Israel J. Bot. 29:118-132.

Bewley, J.D. and M. Black. 1983. Physiology and Biochemistry of Seeds, 2 volumes. Springer-Verlag, New York.

Bewley, J.D., D.W.M. Leung and F.B. Ouellette. 1983. The cooperative role of endo-ß-mannase, ß-mannosidase and a-galactosidase in the mobilization of endosperm cell wall hemicelluloses of germinating lettuce seed, pp. 137- 152. In: C. Nozzolillo, P.J. Lea and F.A. Loewus (eds.), Mobilization of Reserves in Germination, Recent Adv. in Phytochem., Vol. 17, 312 pp.

Bewley, J.D. and M. Black. 1985. Seeds, Physiology of Development and Germination. Plenum Press, New York, 367 pp.

Bewley, J.D., A.R. Kermode and S. Misra. 1989. Desiccation and minimal drying treatments of seeds of castor bean and *Phaseolus vulgaris* which terminate development and promote germination cause changes in protein and messenger RNA synthesis. Ann. Bot. 63:3-17.

Bhatnagar, S.P. and B.M. Johri. 1972. Development in angiosperm seeds, pp. 77-149. In: T.T. Kozlowski (ed.), Seed Biology, vol. 1. Academic Press, New York, 416 pp.

Black, M.J. 1980/81. The role of endogenous hormones in germination and dormancy. Israel. J. Bot. 29:181-192.

Black, M., J. Chapman and H. Norman. 1983. The ability of wheat aleurone tissue to participate in endosperm mobilization, pp. 193-211. In: C. Nozzolillo, P.J. Lea and F.A.

Loewus (eds.), Mobilization of Reserves in Germination, Recent Adv. in Phytochem., vol. 17, 312 pp.

Bleak, A.T. and W. Keller. 1969. Effects of seed age and pre-planting seed treatment on seedling response of crested wheatgrass. Crop Sci. 9:296-299.

Bleak, A.T. and W. Keller. 1970. Field emergence and growth of crested wheatgrass from pre-planted vs non-treated seeds. Crop Sci. 10:85-87.

Bleak, A.T. and W. Keller. 1972. Germination and emergence of selected forage species following pre-planting seed treatment. Crop Sci. 12:9-13.

Bouwmeester, H.J. and C.M. Karssen. 1989. Environmental factors influencing the expression of dormancy patterns in weed seeds. Ann. Bot. 63:113-120.

Bray, C.M., P.A. Davison, A. Ashraf and R.M. Taylor. 1989. Biochemical changes during osmopriming of leek seeds. Ann. Bot. 63:185-193.

Briske, D.D. and A.M. Wilson. 1978. Moisture and temperature requirements for adventitious root development in blue grama seeedings. J. Range Manage. 31:174-178.

Brown, J.W.S., D.R. Ersland and T.C. Hall. 1982. Molecular aspects of storage protein synthesis during seed development, pp. 3-42. In: A.A. Khan (ed.), The Physiology and Biochemistry of Seed Development, Dormancy and Germination. Elsevier Biomedical Press, New York, 547 pp.

Brown, R. 1972. Germination, pp. 3-48. In: F.C Steward (ed.), Plant Physiology, vol. VIC. Academic Press, New York, 450 pp.

Brown, W.V. 1960. The morphology of the grass embryo. Phytomorphology 10:215-223.

Bryant, J.A. 1985. Seed Physiology. Edward Arnold, London, 76 pp.

Cavers, P.B. 1983. Seed demography. Can. J. Bot. 61:3578-3590.

Cohen, D. 1958. The mechanism of germination stimulation by alternating temperatures. Bull. Res. Counc. Israel., Sect. D 6:111-117.

Collins, B.S. and S.T.A. Pickett. 1987. Influences of canopy opening on the environment and herb layer in a northern hardwood forst. Vegetatio 70:3-10.

Collis-George, N. and J.B. Hector. 1966. Aust. J. Soil Res. 4:145-164, cited in Hadas, A. 1982. Seed-soil contact and germination, pp. 507-527. In: A.A. Khan (ed.), The Physiology and Biochemistry of Seed Development, Dormancy and Germination. Elsevier Biomedical Press, New York, 547 pp.

Come, D. 1980/81. Problems of embryonal dormancy as exemplified by apple embryo. Israel J. Bot. 29:145-156.

Come, D. and T. Tissauoi. 1972. Interrelated effects of imbibition, temperature and oxygen on seed germination, pp. 157-168. In: W. Heydecker (ed.), Seed Ecology, Butterworths, London, 578 pp.

Come, D. and C. Thevenot. 1982. Environmental control of embryo dormancy and germination, pp. 271-298. In: A.A. Khan (ed.), The Physiology and Biochemistry of Seed Development, Dormancy and Germination. Elsevier Biomedical Press, Amsterdam, 547 pp.

Cone, J.W. and R.E. Kendrick. 1986. Photocontrol of seed germination, pp. 443-465. In: R.E. Kendrick and G.H.M. Kronenberg (eds.), Photomorphogenesis in Plants, Martinus Nijhoff Publ., Dordrecht, 580 pp.

Cook, C.W., L.A. Stoddart and P.L. Sims. 1967. Effects of season, spacing and intensity of seeding on the development of foothill range grass stands. Utah Agr. Exp. Stat. Bull. 467, 73 p.

Cook, R.E. 1979. Patterns of juvenile mortality and recruitment in plants, pp. 207-231. In: O.T. Solbrig, S. Jain, G.B. Johnson and P.H. Raven (eds.), Topics in Plant Population Biology. Columbia Univ. Press, New York, 589 pp.

Coolbear, P., R.J. Slater and J.A. Bryant. 1990. Changes in nucleic acid levels associated with improved performance of tomato seeds after low temperature presowing treatment. Ann. Bot. 65:187-195.

Cooper, D.C. 1938. Embryology of *Pisum sativa*. Bot. Gaz. 100:123-132.

Copeland, L.O. and M.B. McDonald. 1985. Principles of Seed Science and Technology, 2nd edition. Burgess Publ. Co,. Minneapolis, 321 pp.

Covell, S., R.H. Ellis, E.H. Roberts and R.J. Summerfield. 1986. The influence of temperature on seed germination rate of grain legumes. I. A comparison of chickpea, lentil, soyabean and cowpea at constant temperatures. J. Exp. Bot. 37:705-715.

Cresswell, E.G. and J.P. Grime. 1981. Induction of a light requirement during seed development and its ecological consequences. Nature 291:583-585.

Crocker, W. and L.V. Barton. 1957. Physiology of Seeds. Chronica Botanica, Waltham, Ma.

Cutter, E.G. 1971. Plant Anatomy: Experiment and Interpretation, 2 volumes, Addison-Wesley Publ. Co., Reading.

Dalling, M.J. and P.L. Bhalla. 1984. Mobilization of nitrogen and phosphorus from endosperm, pp. 163-199. In: D.R. Murray, (ed.), Seed Physiology, vol. 2. Academic Press, New York, 295 pp.

Dasberg, S. 1971. Soil water movement to germinating seeds. J. Exp. Bot. 22:999-1008.

Daussant, J., J. Mosse and J. Vaughan (eds.) 1983. Seed Proteins. Ann. Proc. Phytochem. Soc. of Europe, No. 20. Academic Press, New York, 335 pp.

Dell'Aquila, A. and V. Tritto. 1990. Ageing and osmotic priming in wheat seeds: Effects upon certain components of seed quality. Ann. Bot. 21-26.

Delseny, M., L. Aspart and R. Cooke. 1980/81. Studies on ribonucleic acid during radish seed germination: A search for potential regulatory mechanisms. Israel J. Bot. 29:246-258.

Deltour, R. 1985. Nuclear activation during early germination of the high plant embryo. J. Cell. Sci. 75:43-83.

Denslow, J.S. 1987. Tropical rainforest gaps and tree species diversity. Annu. Rev. Ecol. Syst. 18:431-451.

DiMichele, W.A., J.I. Daves and R.G. Olmstead. 1989. Origins of heterospory and the seed habit: The role of heterochrony. Taxon 38:1-11.

Dubey, P.S. and L.P. Mall. 1972. Ecology of germination of weed seeds. 1. Role of temperature and depth of burial in soil. Oecologia 10:105-110.

Duffus, C. M. and M. P. Cochrane. 1982. Carbohydrate metabolism during cereal grain development, pp. 43-66. In: A.A. Khan (ed.), The Physiology and Biochemistry of Seed Development, Dormancy and Germination. Elsevier Biomedical Press, Amsterdam, 547 pp.

Dure, L.S. 1979. Role of stored messenger RNA in late embryo development and germination, pp. 113-128. In: I. Rubenstein, R.L. Phillips, C.E. Green and B.G. Gengenbach (eds.), The Plant Seed: Development, Preservation, and Germination. Academic Press, New York, 266 pp.

Dute, R.R., C.M. Peterson and A.E. Rushing. 1989. Ultrastructural changes of the egg apparatus associated with fertilization and proembryo development of soybean, *Glycine max* (Fabaceae). Ann. Bot. 64:123-135.

Dwyer, D.D. and K. Wolde-Yohannis. 1972. Germination, emergence, water use and production of Russian thistle (*Salsola kali*). Agron. J. 64:52-55.

Earle, F.R. and Q. Jones. 1962. Analysis of seed samples from 113 plant families. Economic Bot. 16:221-250.

Edwards, T.I. 1932. Temperature relations of seed germination. Q. Rev. Biol. 7:428-443.

Ellern, S.J. and N.H. Tadmor. 1966. Germination of range plant seeds at fixed temperatures. J. Range Manage. 19:341-345.

Ellern, S.J. and N.H. Tadmor. 1967. Germination of range plant seeds at alternating temperatures. J. Range Manage. 20:72-77.

Ellis, R.H., S. Covell, E.H. Roberts and R.J. Summerfield. 1986. The influence of temperature on seed germination rate in grain legumes. II. Intraspecific variation in chickpea

(*Cicer arietinum* L.) at constant temperatures. J. Exp. Bot. 37:1503-1515.

Ellis, R.H., G. Simon and S. Covell. 1987. The influence of temperature on seed germination rate in grain legumes. J. Exp. Bot. 38:1033-1043.

Evenari, M. 1984. Seed physiology: Its history from antiquity to the beginning of the 20th century. Biol. Rev. 50:119-142.

Fahn, A. and E. Werker. 1972. Anatomical mechanisms of seed dispersal, pp. 151-221. In: T.T. Kozlowski (ed.), Seed Biology, vol. 1. Academic Press, New York, 416 pp.

Fenner, M. 1980a. The inhibition of germination of *Bidens pilosa* seeds by leaf canopy shade in some natural vegetation. New Phytol. 84:95-101.

Fenner, M. 1980b. The induction of a light requirement in *Bidens pilosa* seeds by leaf canopy shade. New Phytol. 84:103-106.

Fenner, M. 1985. Seed Ecology. Chapman and Hall, London, 151 pp.

Fenner, M. 1987a. Seedlings. New Phytol. 106(Suppl.):35-47.

Fenner, M. 1987b. Seed characteristics in relation to succession, pp. 103-114. In: A.J. Gray, M.J. Crawley and P.J. Edwards (eds.), Colonization, Succession and Stability. Blackwell Sci. Publ., Oxford, 482 pp.

Foster, S.A. 1987. On the adaptive value of large seeds for tropical moist forest species: A review and synthesis. Bot. Rev. 52:260-299.

Frankland, B. 1986. Perception of light quantity, pp. 219-235. In: R.E. Kendrick and G.H.M. Kronenberg (eds.), Photomorphogenesis in Plants. Martinus Nijhoff Publ., Dorcrecht, 580 pp.

Frasier, G.W., D.A. Woolhiser and J.R. Cox. 1984. Emergence and seedling survival of two warm-season grasses as influenced by the time of precipitation: A greenhouse study. J. Range Manage. 37:7-11.

Frasier, G.W., J.R. Cox and D.A. Woolhiser. 1985. Emergence and survival response of seven grasses for six wet-dry sequences. J. Range Manage. 38:372-377.

Frasier, G.W., J.R. Cox and D.A. Woolhiser. 1987. Wet-dry cycle effects on warm-season grass seedling emergence. J. Range Manage. 40:2-6.

Frasier, G.W. 1989. Characterization of seed germination and seedling survival during the initial wet-dry periods following planting. J. Range Manage. 42:299-303.

Friedman, W.E. 1990. Double fertilization in *Ephedra,* a nonflowering seed plant: Its bearing on the origin of angiosperms. Science 247:951-954.

Friedman, W.E. 1991. Double fertilization in *Ephedra trifurca*, a non-flowering seed plant: the relationships between fertilization events and the cell cycle. Protoplasma 165:106-120.

Friedman, W.E. 1992. Evidence of a pre-angiosperm origin of endosperm: implications for the evolution of flowering plants. Science 255:336-339.

Galliard, T. 1983. Starch-lipid complexes and other non-starch components of starch granules in cereal grains, pp. 111-136. In: C. Nozzolillo, P.J. Lea and F.A. Loewus (eds.), Mobilization of Reserves in Germination, Recent Adv. in Phytochem., Vol. 17, 312 pp.

Galtier, J. and N.P. Rowe. 1989. A primitive seed-like structure and its implications for early gymnosperm evolution. Nature 340:225-227.

Gensel, P.G. and H.N. Andrews. 1984. Plant Life in the Devonian. Praeger, New York, 380 pp.

Gillespie, W.H., G.W. Rothwell and S.E. Scheckler. 1981. The earliest seeds. Nature 293:462-464.

Goudey, J.S., H.S. Saini and M.S. Spencer. 1988. Role of nitrate in regulating germination of *Sinapis arvensis* L. (wild mustard). Plant, Cell and Environment 11:9-12.

Grime, J.P. 1979. Plant Strategies and Vegetation Processes. John Wiley and Sons, New York, 222 pp.

Grime, J.P., G. Mason, A.V. Curtis, J. Rodman, S.R. Band, M.A.G. Mowforth, A.M. Neal and S. Shaw. 1981. A comparative study of germination characteristics in a local flora. J. Ecol. 69:1017-1059.

Grubb, P.J. 1977. The maintenance of species-richness in plant communities: The importance of the regeneration niche. Biol. Rev. 52:107-145.

Gutterman, Y. 1982. Phenotypic maternal effect of photoperiod on seed germination. In: A.A. Khan (ed.), The Physiology and Biochemistry of Seed Development, Dormancy and Germination. Elsevier Biomedical Press, Amsterdam, 547 pp.

Gutterman, Y., A. Wiztum and M. Evenari. 1967. Seed dispersal and germination in *Blepharis persica*. Israel J. Bot. 16:213-234.

Haber, A.H. and H.J. Luippold. 1960a. Separation of mechanisms initiating cell division and cell expansion in lettuce seed germination. Plant Physiol. 35:168-173.

Haber, A.H. and H.J. Luippold. 1960b. Effects of gibberellin, kinetin, thiourea, and photomorphogenic radiation on mitotic activity in dormant lettuce seed. Plant Physiol. 35:486-494.

Hadas, A. 1982. Seed-soil contact and germination, pp. 507-527. In: A.A. Khan (ed.), The Physiology and Biochemistry of Seed Development, Dormancy and Germination. Elsevier Biomedical Press, New York, 547 pp.

Hadas, A. and E. Stibbe. 1973. An analysis of soil water movement towards seedlings prior to emergence, pp. 97-106. In: A. Hadas, D. Swartzendruber. P.E. Rijtema, M. Fuchs, and B. Yaron (eds.), Physical aspects of Soil Water and Solutes in Ecosystems. Chapman and Hall, Ltd., London.

Haferkamp, M.R. and G.L. Jordan. 1977. The effect of selected presowing seed treatments on germination of Lehmann lovegrass. J. Range Manage. 30:151- 153.

Haig, D. and M. Westoby. 1989. Selective forces in the emergence of the seed habit. Biol. J. Linn. Soc. 38:215-238.

Harper, J.L., J.T. Williams and G.R. Sagar. 1965. The role of seeds in soils. Part 1. The heterogeneity of soil surfaces and its role in determining the establishment of plants from seed. J. Ecol. 53:273-286.

Harper, J.L. 1977. Population Biology of Plants. Academic Press, New York, 892 pp.

Hauser, V.C. 1981. Planting germinated grass seed. Trans. of the ASAE. 24:1127-1130.

Heerman, D.F., M.D. Finkner and E.A. Hiler. 1971. Probability of sequences of wet and dry days for 11 western states and Texas. Colorado State Univ. Exp. Stat. Tech. Bull. 117.

Hegarty, T.W. 1972. Temperature relations of germination in the field, pp. 411-432. In: W. Heydecker (ed.), Seed Ecology, Butterworths, London, 578 pp.

Hegarty, T.W. 1978. The physiology of seed hydration and dehydration, and the relation between water stress and the control of germination: A review. Plant, Cell and Envir. 1:101-119.

Hegarty, T.W. and H.A. Ross. 1980/81. Investigations of control mechanisms of germination under water stress. Israel J. Bot. 29:83-92.

Heydecker, W. and P.I. Orphanos. 1968. The effects of excess moisture in the germination of *Spinacia oleracea* L. Planta 83:237-247.

Hendrix, S.D. 1988. Herbivory and its impact on plant reproduction, pp. 246-263. In: J.L. Doust and L.L. Doust (eds.), Plant Reproductive Ecology, Patterns and Strategies. Oxford University Press, Oxford, 344 pp.

Hickman, J.C. 1979. The basic biology of plant numbers. pp. 232-263. In: O.T. Solbrig, S. Jain, G.B. Johnson and P.H. Raven, (eds.), Topics in Plant Population Biology. Columbia Univ. Press, New York, 589 pp.

Hilhorst, H.W.M., A.I. Smitt and C.M. Karssen. 1986. Gibberellin-biosynthesis and -sensitivity mediated stimulation of seed germination of *Sisymbrium officinale* by red light and nitrate. Physiol. Plant. 67:285-290.

Hilhorst, H.W.M. and C.M. Karssen. 1989. Nitrate reductase independent stimulation of seed germination in *Sisymbrium officinale* L. (hedge mustard) by light and nitrate. Ann. Bot. 63:131-137.

Hillel, D. 1971. Soil and Water. Academic Press, New York, 288 pp.

Hillel, D. 1972. Soil moisture and seed germination, pp. 65-89. In: T.T. Kozlowski (ed.), Water Deficits and Plant Growth, vol. 3, Academic Press, New York.

Ho, D.T. and J.E. Varner. 1976. Response of barley aleurone layers to abscisic acid. Plant Physiol. 57:175-178.

Hourmant, A. and A. Pradet. 1981. Oxidative phosphorylation in germinating lettuce seeds (*Lactuca sativa*) during the first hours of imbibition. Plant Physiol. 68:631-635.

Hubbell, S.P. and R.B. Foster. 1986. Canopy gaps and the dynamics of a neotropical forst, pp. 77-96. In: M.J. Crawley (ed.), Plant Ecology. Blackwell Sci. Publ., Oxford, 496 pp.

Hull, A.C., Jr. 1973. Germination of range plant seeds after long periods of uncontrolled storage. J. Range Manage. 26:198-200.

Hunter, J.R. and A.E. Erickson. 1952. Relation of seed germination to soil moisture tension. Agron. J. 44:107-109.

Jacobsen, J.V., T.J.V. Higgins and J.A. Zwar. 1979. Hormonal control of endosperm function during germination, pp. 241-262. In: Rubenstein I., R.L. Phillips, C.E. Green and B.G. Gengenbach (eds.), The Plant Seed, Development, Preservation, and Germination. Academic Press, New York, 266 pp.

Jacobsen, J.V. and P.M. Chandler. 1987. Gibberellin and abscisic acid in germinating cereals, pp. 164-193. In: P.J. Davies (ed.), Plant Hormones and Their Role in Plant Growth and Development. Martinus Nijhoff Publ., Dorcrecht, 881 pp.

Johansen, D.A. 1950. Plant Embryology. Embryology of the Spermatophyta. Chronica Botanica, Waltham, Ma.

Johri, B.M. (ed.). 1982. Experimental Embryology of Vascular Plants. Springer-Verlag, Berlin, 273 pp.

Jones, Q. and F.R. Earle. 1966. Chemical analyses of seeds, II: Oil and protein content of 759 species. Economic Bot. 20:127-155.

Jordan, G.L. and M.R. Haferkamp. 1989. Temperature responses and calculated heat units for germination of several range grasses and shrubs. J. Range Manage. 42:41-45.

Karssen, C.M. 1970. The light promoted germination of the seeds of *Chenopodium album* L.: Effects of the photoperiod during growth and development of the plants on the dormanncy of the produced seeds. Acta. Bot. Neerl. 19:81-94.

Karssen, C.M. 1980/81a. Environmental conditions and endogenous mechanisms involved in secondary dormancy of seeds. Israel J. Bot. 29:45-64.

Karssen, C.M. 1980/81b. Patterns of change in dormancy during burial of seeds in soil. Israel J. Bot. 29:65-73.

Karssen, C.M., D.L.C. Brinkhorst-van der Swan, A.E. Breekland and M. Koorneef. 1983. Induction of dormancy during seed development by endogenous abscisic acid deficient genotypes of *Arabidopsis thaliana* (L.) Heyhn. Planta 157:158-165.

Karssen, C.M., S. Zagorski, J. Kepczynski and S.P.C. Groot. 1989. Key role for endogenous gibberellins in the control of seed germination. Ann. Bot. 63:71-80.

Keller, W. and A.T. Bleak. 1968. Preplanting treatment to hasten germination and emergence of grass seed. J. Range Manage. 21:213-216.

Keller, W. and A.T. Bleak. 1969. Root and shoot growth following pre-planting treatment of grass seed. J. Range Manage. 22:43-46.

Keller, W., A.T. Bleak and A.A. Hanson. 1970. Pre-planting seed treatment may reduce failures in range seeding, pp. 118-119. In: Proc. XI Int. Grassland Congr., Univ. of Queensland Press, Queensland, Australia, 956 pp.

Kendrick, R.E. and G.H.M. Kronenberg (eds.). 1986. Photomorphogenesis in Plants. Martinus Nijhoff Publ., Dordrecht. 580 pp.

Kermode, A.R. and J.D. Bewley. 1986. Alteration of genetically regulated syntheses in seeds by desiccation, pp. 59-84. In: A.C. Leopold (ed.), Membranes, Metabolism, and Dry

Organisms. Comstock Publ. Assoc., Cornell Univ. Press, Ithaca, New York, 374 pp.

Kermode, A.R., J.D. Bewley, J. Dasgupta and S. Misra. 1986. The transition from seed development to germination: A key role for desiccation? Hortscience 21:1113-1118.

Kermode, A.R. and J.D. Bewley. 1989. Developing seeds of *Ricinus communis* L., when detached and maintained in an atmosphere of high relative humidity, switch to a germinative mode without the requirement for complete desiccation. Plant Physiol. 90:702-707.

Khan, A.A. 1960. An analysis of "dark-osmotic inhibition" of germination of lettuce seed. Plant Physiol. 35:1-7.

Khan A.A. 1971. Cytokinins: Permissive role in seed germination. Science 171:853-859.

Khan, A.A., N.H. Peck and C. Samimy. 1980/81. Seed osmoconditioning: physiological and biological changes. Israel J. Bot. 29:133-144.

Khan, A.A. (ed.). 1982. The Physiology and Biochemistry of Seed Development, Dormancy and Germination. Elsevier Biomedical Press, New York, 547 pp.

Khan, A.A. and C. Samimy. 1982. Hormones in relation to primary and secondary seed dormancy, pp. 203-241. In: A.A. Khan (ed.), The Physiology and Biochemistry of Seed Development, Dormancy and Germination. Elsevier Biomedical Press, New York, 547 pp.

Knapp, R. 1967. Über möglichkeiten der Keimung bei sehr hohen Temperaturen. Flora 157b:31-35.

Knipe, O.D. 1971. Imbibition by alkali sacaton seeds. J. Range Manage. 24:71-73.

Koller, D. 1972. Environmental control of seed germination, pp. 2-102. In: T.T Kozlowski (ed.), Seed Biology, vol. 2, Academic Press, New York, 447 p.

Kozlowski, T.T. and C.R. Gunn. 1972. Importance and characteristics of seeds, pp. 1-20. In: T.T. Kozlowski (ed.), Seed Biology, vol. 1. Academic Press, New York, 416 pp.

Laidman, D.L. 1982. Control mechanisms in the mobilization of stored nutrients in germinating cereals, pp. 371-405. In: A.A. Khan (ed.), The Physiology and Biochemistry of Seed Development, Dormancy and Germination. Elsevier Biomedical Press, Amsterdam, 547 pp.

Leck, M.A., V.T. Parker and R.L. Simpson (eds.). 1989. Ecology of Soil Seed Banks. Academic Press, New York, 462 pp.

Leopold, A.C. and C.W. Vertucci. 1986. Physiological attributes of desiccated seeds, pp. 22-34. In: A.C. Leopold (ed.), Membranes, Metabolism and Dry Organisms. Comstock Publ. Assoc., Cornell Univ. Press, Ithaca, New York, 374 pp.

Loucks, O.L., M.L. Plumb-Mentjes and D. Rogers. 1985. Gap processes and large-scale disturbances in sand prairies, pp. 71-83. In: S.T.A. Pickett and P.S. White (eds.), The Ecology of Natural Disturbance and Patch Dynamics. Academic Press, New York, 472 pp.

Lovato, A. 1981. Germination of seeds, pp. 86-120. In: J.R. Thomson (ed.), Advances in Research and Technology of Seeds, pt. 6, Pudoc, Wageningen, 124 pp.

Mancinelli, A.L. and J. Rabino. 1979. The "high irradiance responses" of plant photomorphogenesis. Bot. Rev. 44:129-180.

Matheson, N.K. 1984. The synthesis of reserve oligosaccharides and polysaccharides in seeds, pp. 167-208. In: D.R. Murray (ed.), Seed Physiology, vol. 1. Academic Press, New York, 279 pp.

Marbach, I. and A.M. Mayer. 1974. Permeability of seed coats to water as related to drying conditions and metabolism of phenolics. Plant Physiol. 54:817-820.

Marbach, I. and A.M. Mayer. 1979. Germination, utilization of storage materials and potential for cyanide release in cultivated and wild sorghum. Physiol. Plant. 47:100-104.

Marinos, N.G. 1970. Embryogenesis of the pea (*Pisum sativum*). I. The cytological environment of the developing embryo. Protoplasma 70:261-279.

Mayer, A.M. 1986. How do seeds sense their environment? Some biochemical aspects of the sensing of water potential, light and temperature. Isr. J. Bot. 35:3-16.

Mayer, A.M. and I. Marbach. 1981. Biochemistry of the transition from resting to germinating state in seeds, pp. 95-136. In: L. Reinhold, J.B. Harborne and T. Swain (eds.), Progress in Phytochemistry, vol. 7. Pergamon Press, Oxford, 344 pp.

Mayer, A.M. and A. Poljakoff-Mayber. 1989. The Germination of Seeds, 4th edition. Pergamon Press, Oxford, 270 pp.

Maynard, M.L. and D.H. Gates. 1963. Effects of wetting and drying on germination of crested wheatgrass seed. J. Range Manage. 16:119-121.

McCullough J.M. and W. Shropshire. 1970. Physiological predetermination of germination responses in Arabidopsis thaliana (L.) Heynh. Plant Cell. Physiol. 11:139-148.

McDonald, M.B., Jr., and A.A. Khan. 1977. Factors determining germination of Indian ricegrass seeds. Agron. J. 69:558-563.

McDonald, M.B., Jr., C.W. Vertucci and E.E. Roos. 1988a. Soybean seed imbibition: Water absorption by seed parts. Crop Sci. 28:993-997.

McDonald, M.B., Jr., C.W. Vertucci and E.E. Roos. 1988b. Seed coat regulation of soybean seed imbibition. Crop Sci. 28:987-992.

McKersie, B.D. and T. Senaratna. 1983. Membrane structure in germinating seeds, pp. 29-52. In: C. Nozzolillo, P.J. Lea and F.A. Loewus (eds.), Mobilization of Reserves in Germination. Plenum Press, New York, 312 pp.

Mikola, J. 1983. Proteinases, peptidases, and inhibitors of endogeneous proteinases in germinating seeds, pp. 35-52. In: J. Daussant, J. Mosse and J. Vaughan (eds.), Seed Proteins, Academic Press, New York. 335 p.

Morinaga, T. 1926a. Germination of seeds under water. Amer. J. Bot. 13:126-140.

Morinaga, T. 1926b. The favorable effect of reduced oxygen supply upon the germination of certain seeds. Amer. J. Bot. 13:159-166.

Morohashi, Y. and J.D. Bewley. 1980. Development of mitochondrial activities in pea cotyledons. Plant Physiol. 66:637-640.

Mott, J.J. and P.W. Tynan. 1974. Mechanisms controlling dormancy in the arid zone grass *Aristida contorta*, II. Anatomy of the hull. Aust. J. Bot. 22:647-653.

Murdoch, A.J., E.H. Roberts and C.O. Goedert. 1989. A model for germination responses to alternating temperatures. Ann. Bot. 63:97-111.

Murray, D.R. 1984a. The seed and survival, pp. 1-40. In: D.R. Murray (ed.), Seed Physiology, vol. 1. Academic Press, New York, 279 pp.

Murray, D.R. (ed.) 1984b. Seed Physiology, vol. 1. Academic Press, New York, 279 pp.

Murray, D.R. (ed.) 1984c. Seed Physiology, vol. 2. Academic Press, New York, 295 p.

Murray, D.R. 1984d. Accumulation of seed reserves of nitrogen, pp. 83-137. In: D.R. Murray (ed.), Seed Physiology, vol. 1. Academic Press, New York, 279 pp.

Negbi, M. and D. Koller. 1962. Homologies in the grass embryo — a re-evaluation. Phytomorphology 12:289-295.

Negbi, M., E. Rushkin and D. Koller. 1966. Dynamic aspects of water-relations in germination of *Hirschfeldia incana* seed. Plant Cell Physiol. 7:363- 376.

Norstog, K. 1982. Experimental embryology of gymnosperms, pp. 25-51. In: B.M. Johri (ed.), Experimental Embryology of Vascular Plants. Springer-Verlag, Berlin, 273 pp.

Nozzolillo, C., P.J. Lea and F.A. Loewus, (eds.). 1983. Mobilization of Reserves in Germination, Recent Adv. in Phytochem., Vol. 17, 312 pp.

Oaks, A. 1983. Regulation of nitrogen metabolism during early seedling growth, pp. 53-75. In: C. Nozzolillo, P.J. Lea and F.A. Loewus (eds.), Mobilization of Reserves in Germination. Plenum Press, New York, 312 pp.

Obendorf, R.L. and A. Marcus. 1974. Rapid increase in adenosine 5'-triphosphate during early wheat embryo germination. Plant Physiol. 53:779-781.

Oota, Y., R. Fujii and S. Osawa. 1953. Changes in chemical constituents during the germination state of a bean, *Vigna sesquipedalis*. J. Biochem. 40:649- 661.

Opik, H. 1980. The ultrastructure of coleoptile cells in dry rice grains. New Phytol. 85:521-529.

Osborne, D.J. 1981. Physiological and biochemical events in seed development, pp. 43-85. In: J.R. Thomson (ed.), Advances in Research and Technology of Seeds, pt. 6, Pudoc, Wageningen, 124 pp.

Osborne, D.J. 1982. Deoxyribonucleic acid integrity and repair in seed germination: the importance in viability and survival, pp. 435-463. In: A.A. Khan (ed.), The Physiology and Biochemistry of Seed Development, Dormancy and Germination. Elsevier Biomedical Press, New York, 547 pp.

Osborne, D.J. 1983. Biochemical control systems operating in the early hours of germination. Can. J. Bot. 61:3568-3577.

Osborne, D.J., R. Sharon and R. Ben-Ishai. 1980/81. Studies on DNA integrity and DNA repair in germinating embryos of rye (*Secale cereale*). Israel J. Bot. 29:259-272.

Osborne, T.B. 1924. The Vegetable Proteins, 2nd edition, London: Longmans, Green.

Pettitt, J. 1970. Heterospory and the origin of the seed habit. Biol. Rev. 45:401-415.

Plummer, A. P. 1943. The germination and early seedling development of twelve range seeds. Agron. J. 53:19-34.

Pons, T.L. 1989. Breaking of seed dormancy by nitrate as a gap detection mechanism. Ann. Bot. 63:139-143.

Pons, T.L. and H.F.J.M. Schroder. 1986. Significance of temperature fluctuation and oxygen concentration for germination of the rice field weeds *Fimbristylis littoralis* and *Scirpus juncoides*. Oecologia 68:315-319.

Popay, A.I. and E.H. Roberts. 1970. Ecology of *Capsella bursa-pastoris* (L.) Medik and Senecio vulgaris L. in relation to germination behaviour. J. Ecol. 58:123-139.

Popma, J. and F. Bongers. 1988. The effect of canopy gaps in growth and morphology of seedlings of rain forest species. Oecologia 75:625-632.

Postlethwait, J.H. and J.L. Hopson. 1989. The Nature of Life. McGraw-Hill Book Co., New York, 820 pp. + appendices.

Pradet, A. 1982. Oxidative phosphorylation in seeds during initial phases of germination, pp. 347-367. In: A.A. Khan (ed.), The Physiology and Biochemistry of Seed Development, Dormancy and Germination. Elsevier Biomedical Press, New York, 547 pp.

Priestley, D.A. 1986. Seed Aging, Implications for Seed Storage and Persistence in the Soil. Comstock Publ. Assoc., Cornell Univ. Press, Ithaca, New York, 304 pp.

Probert, R.J., K.H. Gajjar and I.K. Haslam. 1987. The interactive effects of phytochrome, nitrate and thiourea on the germination response to alternating temperatures in seeds of *Ranunculus sceleratus* L.: A quantal approach. J. Exp. Bot. 38:1012-1025.

Raghavan, V. 1976. Experimental Embryogenesis in Vascular Plants. Academic Press, London, 603 pp.

Raghavan, V. 1986. Embryogenesis in Angiosperms, a Developmental and Experimental Study. Cambridge University Press, Cambridge, 303 pp.

Rasmussen, G.A. and H.A. Wright. 1988. Germination requirements of flameleaf sumac. J. Range Manage. 41:48-52.

Roberts, E.H. 1972. Oxidative processes and the control of seed germination, pp. 189-218. In: W. Heydecker (ed.), Seed Ecology, Butterworths, London, 578 pp.

Roberts, E.H. and R.H. Ellis. 1989. Water and seed survival. Ann. Bot. 63:39-52.

Robertson, J.H. 1947. Season, date and depth of planting crested wheatgrass and other grasses at lower elevations in northern Nevada. Intermountain For. and Range Exp. Stat. Res. Pap. no. 14.

Romo, J.T., P.L. Grilz, C.J. Bubar and J.A. Young. 1991. Influences of temperature and water stress on germination of plains rough fescue. J. Range Manage. 44:75-81.

Rosenberg, L.A. and R.W. Rinne. 1989. Protein synthesis during rehydration, germination and seedling growth of naturally and precociously matured soybean seeds (*Glycine max*). Ann. Bot. 64:77-86.

Ross, J.D. 1984. Metabolic aspects of dormancy, pp. 45-75. In: D.R. Murray (ed.), Seed Physiology. vol. 2, Academic Press, New York, 295 pp.

Rudiger, W. 1986. Phytochrome, pp. 17-33. In: R.E. Kendrick and G.H.M. Kronenberg (eds.), Photomorphogenesis in Plants, Martinus Nijhoff Publ., Dorcrecht, 580 pp.

Russell, E.W. 1966. Soil Conditions and Plant Growth. Longmans, London.

Saini, H.S., P.K. Bassi and M.S. Spencer. 1986. Use of ethylene and nitrate to break seed dormancy of common lambsquarters (*Chenopodium album*). Weed Sci. 34:502-506.

Saini, H.S., E.D. Consolacion, P.K. Bassi and M.S. Spencer. 1989. Control processes in the induction and relief of thermoinhibition of lettuce seed germination. Plant Physiol. 90:311-315.

Salisbury, F.B. and C. Ross. 1969. Plant Physiology. Wadsworth Publ. Co., Inc., Belmont, 747 pp.

Schopfer, P. and C. Plachy. 1985. Control of seed germination by abscisic acid. III. Effect on embryo growth potential (minimum turgor pressure) and growth coefficient (cell wall extensibility) in *Brassica napus* L. Plant Physiol. 77:676-686.

Shaykewich, C.F. 1973. Proposed method of measuring swelling pressure of seeds prior to germination. J. Exp. Bot. 24:1056-1061.

Silverton, J.W. and J.B. Dickie. 1980. Seedling survivorship in natural populations of nine perennial chalk grassland plants. New Phytol. 88:555- 558.

Simon, E.W. 1974. Phospholipids and plant membrane permeability. New Phytol. 73:377-420.

Simon, E.W. 1984. Early events in germination, pp. 77-115. In: D.R. Murray (ed.) Seed Physiology, vol. 2. Academic Press, New York, 295 p.

Simon, E.W. and L.K. Mills. 1983. Imbibition, leakage and membranes, pp. 9-27. In: C. Nozzolillo, P.J. Lea and F.A. Loewus (eds.), Mobilization of Reserves in Germination. Plenum Press, New York, 312 pp.

Singh, H. and B.M. Johri. 1972. Development of gymnosperm seeds, pp. 21-75. In: T.T. Kozlowski (ed.) Seed Biology, vol. 1. Academic Press, New York, 416 pp.

Slack, C.R. and J.A. Browse. 1984. Synthesis of storage lipids in developing seeds, pp. 209-244. In: D.R. Murray (ed.), Seed Physiology, vol. 1. Academic Press, New York, 279 pp.

Slater, R.J. and J.A. Bryant. 1987. RNA polymerase activity during breakage of seed dormancy by low temperature treatment of fruits of *Acer platanoides* (Norway maple). J. Exp. Bot. 38:1026-1032.

Smith, D.L. 1964. The evolution of the ovule. Biol. Rev. 39:137-159.

Smith, D.L. 1982. Seed development and structure, pp. 62-86. In: J.R. Thomson (ed.) Advances in Research and Technology of Seeds, pt. 7, Pudoc, Wageningen, 140 pp.

Solbrig, O.T., S. Jain, G.B. Johnson, and P.H Raven, (eds.). 1979. Topics in Plant Population Biology. Columbia Univ. Press, New York, 589 pp.

Sosebee, R.E. and C.H. Herbel. 1969. Effects of high temperatures on emergence and initial growth of range plants. Agron. J. 61:621-624.

Spurny, M. 1972. The imbibition process, pp. 367-390. In: W. Heydecker (ed.), Seed Ecology, Butterworths, London, 578 pp.

Stebbins, G.L. 1974. Flowering Plants: Evolution Above the Species Level. Belknap Press of the Harvard Univ. Press, Cambridge, Ma.

Steeves, T.A. 1983. The evolution and biological significance of seeds. Can. J. Bot. 61:3550-3560.

Steward, W.N. 1983. Paleobotany and the Evolution of Plants. Cambridge Univ. Press, Cambridge, 405 pp.

Stidham, N.D., R.M. Ahring, J. Powell, and P.L. Claypool. 1980. Chemical scarification, moist prechilling, and thiourea effects on germination of 18 shrub species. J. Range Manage. 33:115-118.

Sussex, I.M. 1966. The origin and development of heterospory in vascular plants, pp. 140-152. In: E.G. Cutter (ed.) Trends in Plant Morphogenesis. John Wiley, New York. 329 pp.

Taylorson, R.B. and H.A. Borthwick. 1969. Light filtration by foliar canopies; significance for light-controlled weed seed germination. Weed Sci. 17:48- 51.

Taylorson, R.B. 1982. Interaction of phytochrome and other factors in seed germination, pp. 323-346. In: A.A. Khan (ed.), The Physiology and Biochemistry of Seed Development, Dormancy and Germination. Elsevier Biomedical Press, Amsterdam, 547 p.

Thanos, C.A., K. Georghiou and H.C. Passam. 1989. Osmoconditioning and ageing of pepper seeds during storage. Ann. Bot. 63:65-69.

Thevenot, C., J. Ralambosoa and E. Simond-Cote. 1987. Influence of abscisic acid and oxygen supply on germination of more or less dormant apple embryos. Israel J. Bot. 36:101-112.

Thompson, K. 1987. Seeds and seed banks. New Phytol. 106(Suppl):23-34.

Thompson, K. 1989. A comparative study of germination responses to high irradiance light. Ann. Bot. 63:159-163.

Thompson, K., J.P. Grime and G. Mason. 1977. Seed germination in response to diurnal fluctuations of temperature. Nature 267:147-149.

Thompson, K. and J.P. Grime. 1983. A comparative study of germination responses to diurnally-fluctuating temperatures. J. Appl. Ecol. 20:141- 156.

Thompson, P.A. 1969. Germination of *Lycopus europaeous* L. in response to fluctuating temperatures and light. J. Exp. Bot. 20:1-11.

Thompson, P.A. 1974. Effects of fluctuating temperature on germination. J. Exp. Bot. 25:164-175.

Tiffney, B.H. 1981. Diversity and major events in the evolution of land plants, pp. 193-230. In: K.J. Niklas (ed.), Paleobotany, Paleoecology and Evolution, vol. 2. Praeger, New York, 279 pp.

Toole, E.H., S.B. Hendricks, H.A. Borthwick and V.K. Toole. 1956. Physiology of seed germination. Annu. Rev. Plant Physiol. 7:229-324.

Tran, V.N. and A.K. Cavanagh. 1984. Structural aspects of dormancy, pp. 1-44. In: D.R. Murray (ed.), Seed Physiology, vol. 2. Academic Press, New York, 295 p.

Vasquez-Yanes, C. and H. Smith. 1982. Phytochrome control of seed germination in the tropical rain forest pioneer trees *Cecropia obtusifolia* and *Piper auritum* and its ecological significance. New Phytol. 92:477-485.

Vasquez-Yanes and A. Orozco-Segovia. 1982. Seed germination of a tropical rain forest tree (*Heliocarpus donnell-smithii*) in response to diurnal fluctuation of temperature. Physiol. Plant. 56:295-298.

Vaughan, J.G. 1970. The Structure and Utilization of Oil Seeds. Chapman and Hall, London, 279 pp.

Vertucci, C.W. and A.C. Leopold. 1986. Physiological activities associated with hydration levels in seeds, pp. 35-49. In: A.C. Leopold (ed.), Membranes, Metabolism and Dry Organisms. Comstock Publ. Assoc., Cornell Univ. Press, Ithaca, New York, 374 pp.

Vijayaraghavan, M.R. and U. Bhat. 1983. Synergids before and after fertilization. Phytomorphology 33:74-84.

Wallace, A., W.A. Rhodes and E.F. Frolich. 1968. Germination behavior of *Salsola* as influenced by temperatures, moisture, depth of planting, and gamma radiation. Agron. J. 60:76-78.

Walton, D.C. 1980/81. Does ABA play a role in seed germination? Israel. J. Bot. 29:168-180.

Wareing, P.F. 1982. Hormonal regulation of seed dormancy — past, present and future, pp. 185-202. In: A.A. Khan (ed.), The Physiology and Biochemistry of Seed Development, Dormancy and Germination. Elsevier Biomedical Press, New York, 574 pp.

Weier, T.E., C.R. Stocking and M.G. Barbour. 1974. Botany: An Introduction to Plant Biology, 5th ed., John Wiley and Sons, New York, 693 pp.

Werker, E., A. Dafni and M. Negbi. 1973. Variability in *Prosopis farcata* in Israel: Anatomical features of the seed. Bot. J. Linn. 66:223-232.

Werker, E. 1980/81. Seed dormancy as explained by the anatomy of embryo envelopes. Israel J. Bot. 29:22-44.

Wester, D.B. and B.E. Dahl. 1977. Water and temperature influences on germination of range seedlings, p. 42. In: R.E. Sosebee and H.A. Wright (eds.) Research Highlights: Noxious Brush and Weed Control, Dept. Range and Wildlife Manage., Texas Tech Univ., Lubbock, Tx., 54 pp.

Wester, D.B. 1979. Moisture and temperature effects on seed germination and seedling survival. M.S. Thesis, Texas Tech Univ., Lubbock, Tx., 161 pp.

Wester, D.B. and B.E. Dahl. 1983. Rainfall/germination interface, pp. 46-54. In: H.T. Wiedemann and J.F. Cradenhead (eds.), Proc. Range and Pasture Seedings in the So. Great Plains. Texas A&M Agric. Res. and Ext. Center, Vernon, Tx., 116 pp.

Wester, D.B., B.E. Dahl and P.F. Cotter. 1986. Effects of pattern and amount of simulated rainfall on seedling dynamics of weeping lovegrass and Kleingrass. Agron. J. 78:851-855.

Wester, D.B. 1991. A Summary of Range Plant Seed Germination. International Center for Arid and Semiarid Plant Studies, Publ. No. 91-2, 112 pp.

Whisenant, S.G. and D.N. Ueckert. 1982. Germination responses of *Eysenhardtia texana* and *Leucaena retusa*. J. Range Manage. 35:748-750.

Willemse, M.T.M. 1981. Polarity during megasporogenesis and megagametogenesis. Phytomorphology 31:124-134.

Wilson, A.M., J.R. Nelson and C.J. Goebel. 1970. Effects of environment on the metabolism and germination of crested wheatgrass seeds. J. Range Manage. 23:283-288.

Wilson, A.M. and D.D. Briske. 1979. Seminal and adventitious root growth of blue grama seedlings on the Central Plains. J. Range Manage. 32:209-213.

Winton, A.L. and K.B. Winton. 1932. The Structure and Composition of Foods. John Wiley and Sons, New York.

Woolley, J.T. and E.W. Stoller. 1978. Light penetration and light-induced seed germination in soil. Plant Physiol. 61:597-600.

Yodzis, P. 1986. Competition, mortality, and community structure, pp. 480-491. In: J. Diamond and T.J. Case (eds.), Community Ecology. Harper and Row, Publ., New York, 665 pp.

Young, J.A., R.A. Evans, and B.L. Kay. 1973. Temperature requirements for seed germination in an annual-type rangeland community. Agron. J. 65:656-659.

Young, J.A. and R.A. Evans. 1973. Mucilaginous seed coats. Weed Sci. 21:52-54.

Morphology and Seedling Establishment

D. T. Booth and M. R. Haferkamp

Introduction

Morphology is the study of structure and form. The morphologies of plants, as diaspores, embryos, and seedlings, influence their interaction with the environment and therefore the probability for their establishment and survival. Seedlings are "established" when they have expanded a photosynthetic surface and are theoretically capable of an existence independent of their seed reserves (Harper 1977). Thus defined, seedling establishment is a specific and observable event in the morphological development of seed plants. Seedling establishment studies are particularly important for those investigating the processes of seedbed ecology and physiology. Land managers are concerned with stands of plants. Stand establishment refers to the density and stability of a plant population. An "established" stand implies that the period of high seedling mortality, characteristic of new populations, is over and that an acceptable number of the plants have adequate root and canopy development to maintain their place in the community. The focus of our discussion is morphology as it affects individual seedling establishment and survival during the period of high seedling mortality.

Diaspore Morphology and Function[1]

"Diaspore" is from the Greek *diaspora*, which means scattering, sowing; the root is *sporos*, meaning seed or the act of sowing. In the botanical sense a diaspore is a seed-containing dispersal unit. Therefore diaspores can be seeds, fruits, fruits with attached structures, or even whole aerial portions of plants (e.g., *Salsola*). Much has been written about the influence of diaspore morphology on seed dispersal. Ridley (1930) discussed diaspore morphology and modes of dispersal—wind, water, gravity, birds and animals, and self dispersal. The mechanics (Burrows 1973, Augspurger and Franson 1987) and ecological implications (Harper 1977, van der Pijl 1982) of diaspore dispersal have also been well reviewed.

Booth is a rangeland scientist, USDA, Agricultural Research Service, High Plains Grasslands Research Station, Cheyenne, WY 82009; and Haferkamp is a rangeland scientist, USDA, Agricultural Research Service, Fort Keogh Livestock and Range Research Laboratory, Miles City, MT 59301.

[1] Adapted from Booth 1990.

We now recognize that diaspores are more than units for seed dispersal. The structures surrounding or attached to seeds often function in one or more nondispersal phenomenon that contribute to seedling recruitment. The nature of these functions may be protective, facilitating, regulatory, or any combination; and the ways and means by which nondispersal functions are accomplished is a most interesting aspect of seedbed ecology. In the following sections we review the morphology and mechanisms evolved by plants to promote reproduction from dispersed seed.

Protection

Hairiness, thickness, and hardness of seed coats, and small seed size are all factors that contribute to seed survival (Janzen 1971, Thompson 1987, Fried et al. 1988). Likewise, awns and spines protect seeds from herbivory (Smith 1968). Injury to livestock by diaspores from three-awn (*Aristida*), bromegrass (*Bromus*), and barley (*Hordeum*) is well documented (U.S. Forest Service 1937, Stoddart and Smith 1955). A study of the botanical composition of cattle diets in Wyoming provides more recent evidence that cattle avoid awns like those of needle-and-thread [*Stipa comata* Trin. and Rupr. (Samuel and Howard 1982)].

Not all decimating factors are biological, and seed coverings of various kinds protect seeds from abiotic injury and limit precocious germination (Sankary and Barbour 1972, Stevens et al. 1977, Quinn 1987). Anderson (1985) reported that the thick, indurate cupule surrounding the caryopsis of eastern gamagrass (*Tripsacum dactyloides* L.) appears to be important in conditions that limit germination (and has a limited role in preventing seed predation). Certain diaspore characteristics improve the microclimate of seeds or seedlings during adverse environmental conditions. For example, the fluffy hairs on winterfat [*Eurotia lanata* (Pursh) Moq.; *Ceratoides lanata* (Pursh) J. T. Howell] diaspores (Fig. 1) can cause them to collect into groups; Murbeck called the grouping process synaptospermy (Murbeck 1919/20 as cited by van der Pijl 1982) and noted that the diaspore complex absorbed and retained more moisture than isolated seeds. The advantages of synaptospermy have been discussed for buffalo grass [*Buchloe dactyloides* (Nutt.) Engelm.] (Quinn 1987), and Woodmansee and Potter (1971) have described the accumulation of winterfat diaspores in soil depressions and the importance of sheltering materials to seedling survival.

Perhaps the ultimate microclimate modification by a diaspore occurs in species with cryptogeal germination (Jackson 1974). Diaspores of these species develop a positively geotropic pseudoradicle, formed by elongating cotyledon bases, which carries the embryo axis 8 cm or more below the soil surface prior to development of the true root and shoot (Schlising 1969, Jackson 1974, Clarkson and Clifford 1987). Cryptogeal germination is regarded as an adaptation to protect seedlings from fire and drought.

Seed Positioning and Orientation

The position and orientation of seeds in the soil influence seed fixation and imbibition, contact of the seminal root with the soil, water uptake, and seedling growth rate. Therefore seed placement influences establishment and survival. Stevens et al.

Figure 1. A natural aggregation of winterfat [*Eurotia lanata* (Pursh) Moq.] diaspores and seedlings on the lee side of a mother-plant. The two arrows indicate germinants from diaspores not anchored to the soil surface. The roots of these plants are unlikely to penetrate the soil (compare figure 2), but the hairy bracts provide a mulch for underlying diaspores and seedlings.

(1986) placed seed of white rubber rabbitbrush *[Chrysothamnus nauseosus* (Pallas) Britt. ssp. *albicaulis* Nutt.) Rydb.] in 12 positions in and on the soil. They found that the number of seedlings surviving 30 days after planting varied by position and orientation. The greatest number of seedlings survived when seeds were placed upright with the basal attachment scar contacting the soil. They concluded that the seed pappus was important in proper seed positioning and in aiding seedling establishment. Similar conclusions have been drawn from studies of related genera (Sheldon and Burrows 1973, Sheldon 1974) and of oatgrass (*Danthonia*; Simpson 1952, panicgrass *Panicum*; Koller and Roth 1963) and *Blepharis* (Gutterman et al. 1967).

The influences appendages have on seed position has been used to explain the relative success of different species on different soils. Seed position affects water uptake, and soil penetration by emerging radicles; these factors are cited as contributing to differential establishment between different types of soils (Harper et al. 1965, Harper et al. 1970, Harper 1977, Peart 1979, Peart and Clifford 1987).

Fixation (Anchoring)

Fixation holds seeds in place so that dispersal agents can not move them about. Fixation also counteracts soil resistance to radicle penetration and thus enhances or enables radicle entry. Diaspores are often fixed by anchoring connections formed between appendages and soil. This anchoring has been described for the long diaspore hairs of winterfat (Fig. 2) (Booth and Schuman 1983), the diaspore wings of *Haloxylon articulatum* (Cav.) Bge. (*Hammada scoparia*) Iljin (Sankary and Bar-

Figure 2. Winterfat seedling from anchored diaspore. The cotyledons are partially covered by the hairy bracts and soil particles cling to the hairs which anchored the diaspore to the soil surface. Note good soil penetration by the seminal root.

bour 1972), the pappus of rabbitbrush (Stevens et al. 1986), and the bur of buffalo grass (Quinn 1987). Backwardly directed bristles, barbs, and teeth also fix diaspores within soil cracks (Sheldon 1974), Peart 1979, 1981, 1984). These reports all note the importance of fixation to radicle penetration. Others have demonstrated that fixation is necessary for maximum transmission of growth pressure to the penetrating root (Taylor and Gardner 1960, Stolzy and Barley 1968).

Fixation by burial is common. Self-burial and orientation may occur through the hygroscopic action of seed appendages like those of oats (*Avena*), mountain mahogany (*Cercocarpus*), filaree (*Erodium*), needle grass (*Stipa*) (Fig. 3), and others. The action causes the diaspore to bore into soft soil (Zohary 1973 as cited by van der Pijl 1982, Young et al. 1975) or to creep across soil surfaces until it is lodged in cracks or other safe sites (Harper et al. 1965, Harper 1977, Peart 1979).

Moisture Absorption

Hydraulic conductivity at the diaspore/soil interface is closely related to seed fixation and positioning. Adhesion to soil by hairs, bur spines, pappi, awns, or mucilage improves hydraulic conductivity by increasing the seed's surface area available for water absorption and by enhancing the immediacy of soil particles and absorbing surfaces (Harper and Benton 1966, Young et al. 1970, Hadas and Russo 1974 a and b). The effect of diaspore characteristics (size, shape, and mucilage) on water absorption by seeds has been studied under various degrees of water tension (Harper 1977, Harper and Benton 1966, Young et al. 1970). Enhanced hydraulic conductivity by seed appendages is particularly beneficial to species where water absorption is not restricted to one portion of the diaspore or to the micropyle of the seed.

Figure 3. Diaspore of needle-and-thread grass (*Stipa comata* Trin. and Rupr.) that was broadcast onto bare soil. Fluctuating relative humidities have provided the motive force for the twisted, hygroscopic awn to move the seed. The diaspore has lodged at a small crack in the soil surface, has oriented the seed, and is gradually boring into the soil with the aid of its hard and sharp point (callus). The callus has short barb-like hairs that hold the seed in the soil.

Diaspores may also restrict or prevent moisture uptake. The diaspores of *Rhus javanica* L. remain viable in moist forest soils for long periods of time and germinate when dormancy is broken by heat (Washitani and Takenaka 1986). The dormancy of this species is "coat-imposed"—a well-known germination-preventing mechanism (Bewley and Black 1982).

Light Filters

Seed coverings affect the wavelength and quantity of light reaching the embryo before and after dispersal (Cresswell and Grime 1981; Widell and Vogelmann 1985, 1988; Cardoso and Felippe 1988). Transmission of light into diaspores is affected by the morphology, pigmentation (especially chlorophyll), and moisture content of seed-surrounding structures. During seed maturation these structures often set the state of phytochrome and therefore control the light requirement for germination (Cresswell and Grime 1981). After the seeds are dispersed, the seed coverings influence the wave length and the intensity of light reaching the diaspore interior (Widell and Vogelmann 1985, 1988). Light sensors are thought to be located in the hypocotyl, and the light-filtering function appears to interact with other germination-regulating mechanisms in both negative and positive photoblastic seeds. The sequence of events may be a mechanism of monitoring the environment of the mother plant and setting germination conditions for gap detection (Cresswell and Grime 1981).

Regulation of Respiration

Seed coverings may impose dormancy by preventing or significantly limiting oxygen from reaching the embryo (Thornton 1945, Kozlowski and Gentile 1959, Zietsman and Botha 1987). Removing or abrading these coverings, as is done for Indian ricegrass [*Oryzopsis hymenoides* (Roiem. and Schult.) Ricker] (Fendall 1965, Zemetra et al. 1983, Griffith and Booth 1988) or *Halimium halimifolium* L. (Pena et al. 1988), appears to end dormancy by increasing seed respiration.

The seed coverings of some diaspores may also improve germination and enhance seedling vigor by decreasing the partial pressure of embryo oxygen (Come et al. 1985). Oxidation of seed substrates can occur prior to mitochondrial respiration in dry seed and during early imbibition (Vertucci and Leopold 1986, 1987; Wilson and Bonner 1971). The apparent result of this oxidation is decreased seedling vigor because of substrate loss and the harmful effects of oxidation products. Seed coverings influence the oxidation process by affecting seed moisture and the rate of seed-gas exchange. Thus, removing or disrupting seed coverings may contribute to an accelerated loss of seed mass or to aging during storage (Thalen 1979, Griffith and Booth 1988).

Significance of Diaspore Morphology to Plant Recruitment

Most any plant community, regardless of its successional status or life forms, has a diverse array of diaspore morphologies. The diversity results from niche variety within communities, the different genetic histories of community members, and the inherent randomness of evolutionary pressures and species response. The common factor in the diversity is reproductive success. The significance of a specific morphology is evident when the natural form is altered. For example, threshing *Haloxylon* and winterfat improved germination but reduced seedling establishment (Sankary and Barbour 1972, Booth and Schuman 1983). Likewise, removing diaspore appendages reduced establishment of the grasses studied by Simpson (1952) and Peart (1979). Threshing prostrate kochia [*Kochia prostrata* (L.) Schrad.] increased abnormal germination (Romo and Haferkamp 1987), and threshing buffalo grass reduced seed germination and seedling survival (Quinn 1987).

Injury to seed embryos during threshing was only one of several reasons for reduced establishment from threshed winterfat (Booth 1984, 1987) and was a minor element in reducing oatgrass establishment (Simpson 1952). Embryo injury was not a factor in the reduced establishment of threshed *Haloxylon*, kochia, buffalo grass (Sankary and Barbour 1972, Romo and Haferkamp 1987, Quinn 1987), or of the threshed species studied by Peart (1979); instead, reduced establishment was attributed to loss of assisting functions by the parts removed in threshing.

Heydecker (1973) has suggested that one obstacle to understanding seed ecology is that "... ingenious mechanisms often outwit our teleological imagination." His statement stresses the folly of assuming, conclusively, the function, nonfunction, or sole function of diaspore attributes. In any given environment, dominant selective foreces have influenced and continue to influence the gross morphology of the diaspore to favor the most critical roles for seedling recruitment in that environment. But gross morphology is also influenced by minor functions or combination of functions that further an enlargement of the reproducing population. The result is multifunctional morphology. Understanding seedbed ecology and the signficance of diaspore morphology to plant recruitment requires the student to remember that diaspore structure and function reflect many facets in the several processes that allow a plant species to establish on and occupy a site.

Embryo Morphology

Morphological Descriptions

Three early-stage morphologies in plant development interact with seedbed microenvironments to influence establishment and first-year survival. The embryo, germinant, and seedling stages may be viewed as first, second, and third order morphology, respectively. The plant in the seed is an embryo and becomes a germinant when either the radicle or the hypocotyl ruptures the testa. The germinant becomes a seedling when both a root and a shoot have developed outside the seed.

A seed is a mature or ripened ovule consisting of an embryo, and perhaps stored food (gametophyte tissue in gymnosperms and endosperm or perisperm in angiosperms), enclosed within a seed coat or coats. Embryos consist of cotyledons or seed leaves, an axis or stalk, and depending upon the degree of differentiation and development, primordia for various vegetative organs (Martin 1946, Esau 1965). At the lower end of the axis is the radicle or its meristem, and at the top is the apical meristem or shoot apex. Some species develop an epicotyl or plumule before the embryo matures.

Atwater (1980) used works by Martin (1946) and others, and data from her own studies to correlate patterns of dormancy and germination in seeds with morphologically similar embryos. She has described eight morphological divisions, including five separations using embryo morphology (rudimentary, linear, miniature, peripheral linear, and embryos in nonendospermic seeds). Her separations of flowering herbs and the associated germination characteristics can also be applied to most gymnosperms and to many woody dicots (Martin 1946, Schopmeyer 1974). In this section we have used Atwater's five separations of embryo morphology and Martin's (1946) classification of grass embryos (lateral) as a basis to discuss embryo morphology and potential relationships of embryo morphology to seedling establishment.

Seeds classified as endospermic have embryos dominated by stored food. The embryos are immature and dependent, meaning that cell division may need to occur before cell extension produces radicle emergence. Embryo growth is dependent upon the endosperm as an external food source. The cotyledons in these seeds are often thin and delicate and may serve primarily as an organ to absorb and mobilize stored food (Bewley and Black 1978). Generally, delayed development of the embryo is a primary reason for seed dormancy.

Rudimentary embryos of endospermic seeds have a basal location—the cotyledons are obscure and may be limited to a few cells—and the endosperm occupies most of the seed. Seed imbibition is prompt, but germination is delayed while the embryo develops. Embryo growth is governed by temperature and by any inhibitors that might be present (Atwater 1980, Baskin and Baskin 1986, Dehgan and Schutzman 1989). Rudimentary embryos occur in a variety of monocots and dicots (Martin 1946, Atwater 1980).

Linear and miniature embryos lie in a central axillary position, are larger than rudimentary, but may require some additional development. Stored food occupies half or less of the seed and surrounds the embryo. Environmental controls on germination, such as a light requirement, are often encountered. [All gymnosperms are classed as linear except ephedra (Martin 1946).]

Peripheral-linear seeds have mature embryos that lie in a peripheral position surrounding centrally located perisperm or endosperm. The cotyledons are thin, narrow, and equal to the axis in length. Many of the seeds in this group germinate rapidly. If dormancy is present, it is often due to inhibitors in seed coverings. These inhibitors can be removed by leaching.

Lateral embryos (poaceae) are located basally and tend to extend laterally along the periphery of the endosperm. With a few exceptions, grass embryos account for a small fraction of their caryopsis weight [1% for (*Lolium perenne* L.) S24 (Anslow 1966 as cited by Hayes 1966)]. The highly differentiated grass embryos contain many parts, including the primordia for adventitious roots (Esau 1965, Robbin et al. 1965), and can be subdivided into two basic types, panicoid and festucoid. Four characteristics of taxonomic significance divide these two types into the following six subdivisions (Reeder 1957): the true festucoids, true panicoids, chloridoid-eragrostoid, bambusoid, oryzoid-olyroid, and arundinoid-danthonioid (Reeder 1957). The characteristics utilized include (1) course of the vascular system [whether the trace to the scutellum (cotyledon) and embryonic foliage leaves diverge at about the same point or are separated by a more or less elongated internode], (2) presence or absence of an epiblast, (3) presence or absence of a coleorhiza on the lower part of the scutellum, and (4) quantity of bundles in the embryonic leaf and the degree of overlap of these bundles (observed in cross-section of the embryonic leaf).

Nonendospermic seeds have residual or no endosperm and have mature, independent embryos that occupy most of the seed. The cotyledons are large and thickened, and dominate the axis. Seed dormancy is due to impermeable seed coats or to germination inhibitors contained within the seed.

Example Seedbed Ecologies

The following five examples cover four embryo types, illustrating morphology and development of diverse plants in their habitats. We have included two examples of seeds without endosperm to present an interesting contrast of plants with comparable embryo morphology.

Isopyrum biternatum (Raf.) T. & G. (= *Enemion biternatum* Raf.) is a herbaceous perennial of mesic deciduous forests from southern Ontario to Florida. The seedbed ecology was described by Baskin and Baskin (1986). In central Kentucky the plant flowers from mid-March through early May and seeds are dispersed in mid to late May. The embryo is rudimentary, the seeds are not dormant. Post-dispersal embryo development is delayed by high temperatures so that seeds are quiescent until fall. During the fall the embryos mature and elongate. This is followed by germination and seedling establishment. Winter is passed in the seedling stage, and the young plant develops rapidly in the spring.

Ponderosa pine (*Pinus ponderosa* P. & C. Laws.) a large, slow-growing, drought-resistant tree of the western United States, is typical of the Pinaceae family. Ponderosa flowers in late spring to early summer, but fertilization takes place about 13 months after pollination (Schopmeyer 1974). The embryo, which is linear with multiple cotyledons, develops during the second summer following fertilization. Lipid, starch, and protein food reserves are deposited in the gametophyte tissue surrounding the embryo (Bewley and Black 1978). Seed dispersal begins in August or September (Schopmeyer 1974). The seeds are somewhat dormant. A 28-day moist-prechill at 3 to 5°C, followed by incubation at alternating 20 and 30°C is recommended (AOSA 1981) for germination tests. Animal caches with multiple seeds account for 14 to 21% of natural regeneration (Geier-Hayes 1987). Germination and seed-supported growth is slow compared to that of sagebrush (*Artemisia tridenta* Nutt.) and bitterbrush [*Purshia tridentata* (Pursh) D.c.] which are both noted cover plants for ponderosa seedlings (Booth and Morgan 1992, Geier-Hayes 1987). As with other conifers, survival of new seedlings is enhanced by shade from antecedent vegetation (Tappeiner and Helms 1971, Conard and Radosevich 1982, Geier-Hayes 1987).

Winterfat is a low, often suffrutescent, shrub of western North America. Its range extends from Alaska to Mexico in habitats characterized by drought and temperature extremes. The plants flower June through August, depending upon the location. The seeds mature by late fall or early winter and are shed as diaspores consisting of a utricle and enclosing bracts (Booth 1988). Like many members of Chenopodiceae, winterfat seeds have typical peripheral-linear embryos surrounding centrally located perisperm. The seeds are not dormant and germinate with the diaspore on the soil surface, often under wet snow. Seedling survival depends upon early spring growth sufficient to withstand summer heat and drought.

Russian-thistle (*Salsola kali* L.) is an annual herb, native to Eurasia, but common in cultivated fields and waste places of the arid U.S. west. Unlike the seeds of most chenopods, Russian-thistle is without perisperm. The embryo has been described as a spirally coiled (like a clock spring), fully differentiated, "seedling" that will break through the testa in less than ½ hour, given optimum temperature and moisture (Fig. 4) (Wallace et al. 1968). The coiled nature of the embryo and the tension developed during imbibition aids in rapid germination and in driving the radicle and part of the stem into the soil (Wallace et al. 1968, Young and Evans 1972).

Willow (*Salix*) is representative of the cosmopolitan Salicacea family. Most species flower in the spring and short-lived, non-dormant, cotton-covered seeds are dispersed in the summer (Schopmeyer 1974, Densmore and Zasada 1983). Simak (1980) described the germination and establishment of *Salix caprea*. The seeds are about 1 mm long (1000 seeds weigh 0.06 to 0.2 g) and have thin, transparent seed coats. Endosperm consists of only a few cells. The embryo has two dominant cotyledons, a hypocotyl-ring, and a radicle. Viable seeds turn green a few hours after imbibing water. Densmore and Zasada (1983) noted that some willow seeds are green when they are dispersed and will begin photosynthesis as soon as they are moistened. During germination the hypocotyl increases in length and diameter, becomes positively geotropic, and bends so that the hypocotyl ring and the base of the seed is placed against the moist substrate (Fig. 4). A fringe of radial hairs grow from the ring and anchor the seed. The hypocotyl then raises the cotyledons to a vertical position and elongates, rupturing the seed coat. On the second day, the seed

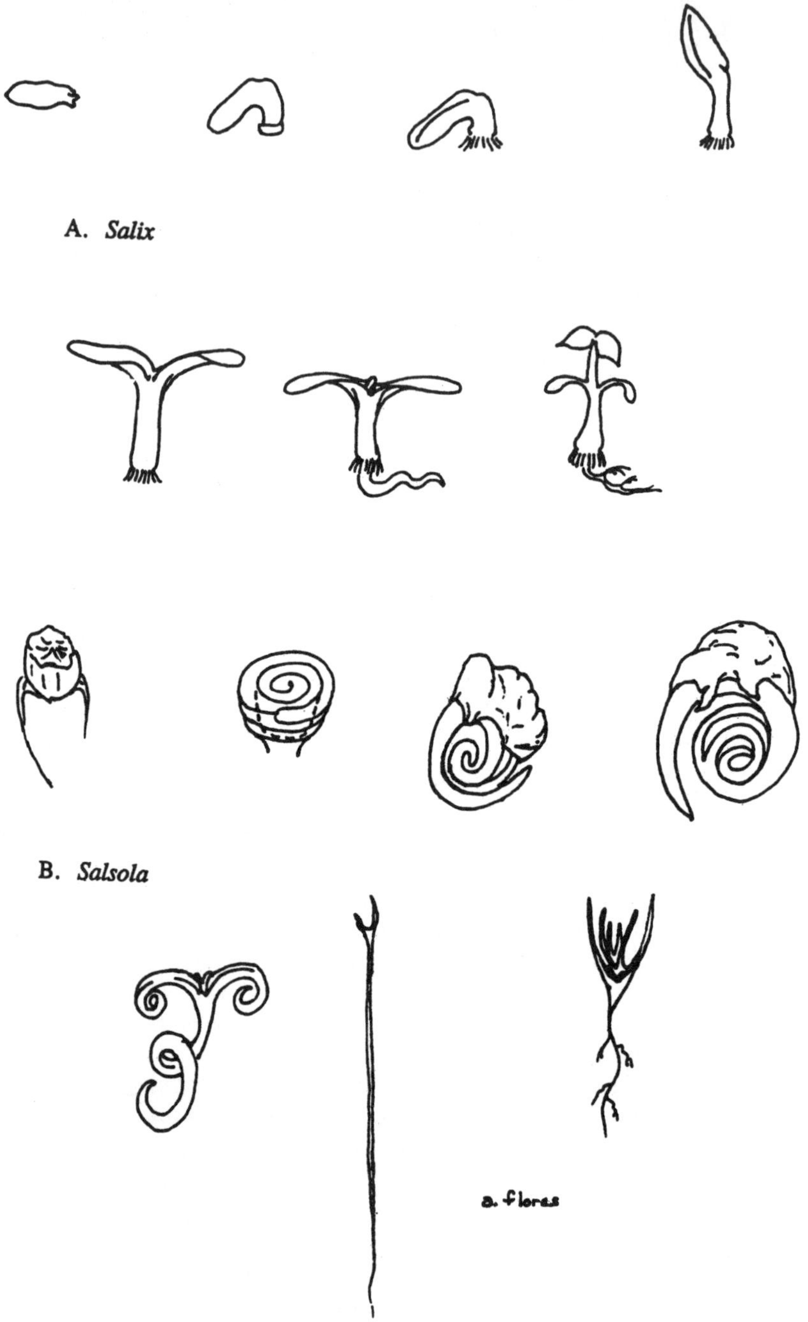

Figure 4. Early morphologies of willow (*Salix* (A) and tumbleweed (*Salsola)* (B) from seed/utricle to the true-leaf stage. The willow seed is about 1 mm long. The tumbleweed utricle is 3 to 6 mm in diameter enclosing a seed 1.5 to 2 mm in diameter.

coat is stripped off by the emerging cotyledons while the radicle begins growing through the hypocotyl ring and into the soil. On the third day the main root, which develops from the radicle, is clearly visible.

Notice the symmetry of embryo morphology with the ecological requirements of the niche. The rudimentary embryo of *Isopyrum biternatum* fits a scheme of seedling establishment when many competing neighbors are dormant. The large seed and slow growth (seed reserves are depleted slowly), associated with the linear morphology of pine embryos, is advantageous for establishing the species into existing vegetation of a preceding seral stage. Winterfat's peripheral-linear seeds facilitate early-season establishment—a pattern that may be more correlated with precipitation and temperature than to existing vegetation. Both Russian-thistle and willow are pioneers with advanced embryos. Russian-thistle will establish on disturbed areas where surface moisture exists briefly. Willows will establish on the exposed and drying mud banks left by receding waters of ephemeral floods. In all species the embryo morphology appears related to other adaptive characteristics. This is true also of the grasses, and the structural influences on seedling establishment of that highly differentiated embryo are discussed in detail later in this chapter.

Seed Size, Germination, and Seedling Growth

Production Factors Influencing Seed Size

Variation in size among seeds from the same mother plant results from different rates of development and different stages of development at harvest or natural dispersal (Copeland 1976, Zasada et al. 1978, Silvertown 1984). Often the first seeds produced by annual species are larger than those produced later in the season (Cavers and Steel 1984). Differences in seed size are due to the environment (Walter and Jensen 1970, Zasada 1988, Booth 1990), the location on the mother plant (Gutterman 1980, Zasada 1988) and the number of seeds in a fruit. Stanton (1985) noted that seeds produced in few-seeded fruits of radish (*Raphanus raphanistrum* L.) tend to be heavier than seeds from many-seeded fruits. Finally, variability in seed or embryo size may be inherited as occurs in heterocarpy (Koller and Roth 1964, Werker and Many 1974).

Embryo Size as Related to Seed Size

Larger seeds may contain larger embryos, have a greater amount of stored food, or have thicker seed coats. Among domesticated species, increased seed size may result from an increase in embryo size for seeds without endosperm (Black 1956, Harper and Obeid 1967) or from an increase in both endosperm and embryo for seeds with endosperm (Wood et al. 1977). Among other species, larger seeds are likely to indicate larger embryos and greater endosperm when the true seed is considered (Cooper and MacDonald 1970, Rogler 1954, Tadmor and Cohen 1968).

Fenner (1983) ranked seeds of 24 herbaceous species of Asteraceae and found that the percent of total seed weight contributed by the embryo decreased as seed weight increased and that the percentage contributed by the seed coat increased. In Poaceae larger seed sizes reduce the ratio of surface-area to volume, therefore the proportion of the caryopsis weight represented by bracts might be expected to be smaller in larger seeds (Anslow 1966, as cited by Hayes 1976).

Rate of Germination

Rapid germination is often cited as a factor in seedling establishment and survival (Knapp and Smith 1982, Van der Toorn and Pons 1988). Seed size, however, has not been shown to consistently affect germination rate for species lacking endosperm (Harper and Obeid 1967, Dalianis 1980, Mobayen 1980).

In seeds having endosperm or perisperm, especially those seeds from arid environments (Baker 1972), large seeds from dicots (Springfield 1973, Putievsky 1980, Philipupillai and Ungar 1984) and grasses (Rogler 1954, Vogel 1963, Green and Hansen 1969, Wright 1977) appear to germinate faster than medium or small seeds. In blue panicgrass (*Panicum antidotale* Retz.) for example, Wright (1977) reported that the genotypes having heavy seed and having a greater quantity of endosperm performed better during all phases of growth from germination through reproduction than did genotypes with light seed. Two reports indicate no affect of seed size on the speed of germination (Kneebone and Cremer 1955, Green and Hansen 1969).

Efforts have been made to measure the factors contributing to faster germination of large seed. Abernethy et al. (1977) reported a positive correlation between seed weight and adenylate energy charge in blue panicgrass. During the first 72 hours of imbibition, respiration rates were higher for heavy seed than for light seed. Heavy seeds contained more free sugars for respiratory substrate and more adenosine phosphates for phosphorylation. Enhanced respiratory metabolism in heavy seeds resulted in increased rates of formation of adenosine triphosphate. They concluded that differences in germination speed were partly due to differences in amounts of carbohydrates and of mitochondria per dry seed. Austenfeld (1988) compared nutrient reserves of the more rapidly germinating large seed produced on median flowers of *Salicornia europaea* L. with the nutrient reserves of the slower germinating, smaller seeds produced by lateral flowers. Like Abernethy et al. (1977), Austenfeld found that differences in germinability were related to total nutrients per seed.

Grime et al. (1981) ranked, by seed weight, over 400 herbaceous species occurring near Sheffield, England. They reported that the representation of rapid-germinating species declined as seed weight increased.

So, when seeds with endosperm are compared within species, larger seeds germinate in less time. Otherwise, it seems that larger seed size may have a weak general influence toward decreased germination rate.

Seedling Size

Within a cultivar and often within species, seed size influences the size and weight of a seedling up to the time when its leaves begin photosynthesis. Many studies have reported correlations between seed size and seedling vigor in dicots (Table 1) and grasses (Table 2). In dicots, larger seed contributes to larger hypocotyls, including a greater hypocotyl radius, larger cotyledons, and a greater potential for axis elongation. The larger size may increase survival. For example, in

Table 1. Reports correlating large seed size with dicot seedling vigor.

Authors and date	Species	Lines tested
Beveridge and Wilsie 1959 *(dry weight)	*Medicago sativa* L.	3
Black 1956 *(emergence, hypocotyl length, cotyledon area, and leaf area)	*Trifolium subterraneum* L.	1
Carleton and Cooper 1972 *(dry weight)	*Lotus corniculatus* L.	6
	Medicago sativa L.	8
	Onobrychis viciaefolia Scop.	7
*Choe et al. 1988 *(dry weight)	*Raphanus raphanistrum* x *R. sativa*	1
Clor et al. 1976 *(seedling length)	*Haloxylon salicornicum*	1
Curtis and McKersie 1984 *(seedling length, dry weight)	*Lotus corniculatus* L.	1
Dalianis 1980 *(seedling length)	*Trifolium alexandrinum*	1
	Trifolium resupinatum	1
Evers 1982 *(dry weight, leaf number)	*Trifolium subterraneum* L.	3
	Trifolium vesiculosum Savi.	3
Fransen and Cooper 1976 *(emergence, Cotylendon area, leaf area, seedling weight)	*Onobrychis* spp	10
Harper and Obeid 1967 *(emergence, hypocotyl length, seedling length, dry weight, and leaf area)	*Linum usitatissum*	6
Haskins and Gor 1975 *(emergence, seedling length)	*Melilotus officinalis*	2
Koller and Roth 1964 *(mortality, fresh weight)	*Gymnarrhena micrantha*	1
Stanton 1985 *(emergence, mortality, cotyledon width)	*Raphanus raphanistrum*	1
Taylor and Baltensperger 1987 *(dry weight)	*Alysicarpus vaginalus*	10
Townsend 1972	*Astragalus cicer* L.	1
	Astragalus falcatus Lan.	
*(emergence, emergence not well correlated; respectively)		
Townsend 1979 *(emergence, hypocotyl length & diameter, seedling length, dry weight, radicle length)	*Astragalus cicer* L.	2
Twamley 1967 *(fresh weight)	*Lotus corniculatus* L.	1

*Seedling characteristics studied that correlated with seed size.

a natural stand of radish the probability of a seedling dying before flowering decreased from 0.52 in the smallest cotyledon size class to 0.03 in the largest size class (Stanton 1985). In the grasses, larger seed size has been positively related to

Table 2. Reports correlating large seed size with grass seedling vigor

Authors and date	Species tested	lines
Assay and Johnson 1980 *(emergence during and after drought, emergence and seedling vigor from a 7.6-cm depth)	*Elymus junceus* Fisch.	134
Assay and Johnson 1983 *(emergence and height, shoot dry weight, percentage stand)	*Agropyron cristatum* (L.) Gaertn. *Agropyron desertorum* (Fisch ex Link) Shult	175
Carren et al. 1987 a and b *(emergence, weight per unit length of subcoelopltile internode, coleoptile length, shoot weight, adventitious root weight)	*Bouteloua gracilis* (H.B.K.) Lag. ex Steud.	3
Davies 1927 *(percentage establishment)	several grasses and legumes	
Green and Hanson 1969 *(germination rate)	*Bouteloua curtipendula* *Bouteloua gracilis* *Panicum virgatum* L. *Sorghastrum nutans* (L.) Nash	1 1 2 1
Hunt and Miller 1968 *(emergence, height, coleoptile length)	*Agropyron intermedium* (Host) Beauv.	20
Kalton et al. 1959 *(seedling vigor, rate of emergence)	Bromegrass	12
Kneebone 1956 *(seedling vigor)	*Andropogon hallii* Hack. *Bouteloua curtipendula*	22 4
Kneebone and Cremer 1955 *(seedling emergence and rate of growth)	*Andropogon hallii* *Buchloe dactyloides* (Nutt.) Engel. *Bouteloua curtipendula* *Panicum virgatum* 3 *Sorghastrum nutans* 1	2 1 5
Lawerence 1963 *(seedling vigor-emergence from 38-mm depth)	*Elymus junceus* Fisch.	12
Nason et al. 1987 *(shoot weight)	*Bouteloua gracilis*	4
Robinson and Thomas 1963 *(seedling weight)	*Bromus inermis* Leyss	47
Rogler 1954 *(seedling emergence and establishment from 5-cm depth, seedling weight)	*Agropyron cristatum* *Agropyron desertorum*	8
Tossell 1960 *(seedling vigor)	*Bromus inermis*	10
Trupp and Carlson 1971 *(seedling weight, seedling vigor-emergence from 5-cm depth)	*Bromus inermis*	8
Whalley et al. 1966a *(seedling length)	*Phalaris tuberosa* var. *stenoptery* (Hack.) Hitch *Phalaris tuberosa* var. hirtiglamis (Batt. and Trab.) Batt. and Trab.	2 2

	Phalaris coerulescens (Desf.)	1
	Oryzopsis miliacea (L) Benth and Hook	1
	Schismus arabicus Nees	1
Wright 1977 *(germination through reproductive stage)	*Panicum antidotale* Ret.	10

*Seedling characteristics studied that correlated with seed size

increased seedling weight, increased weight per unit of subcoleoptile internode, increased coleoptile length, increased shoot weight, increased root weight, increased weight of adventitious roots, and to general seedling vigor. Large seeds of certain species also appear to produce seedlings that emerge from deeper planting than small seeds (Kneebone and Cremer 1955, Knipe 1970, Wilson et al. 1981). When seeds of perennial ryegrass were planted at similar depths, the seedlings arising from heavy seed were heavier and developed more leaves and tillers than those from light seeds (Arnott 1969). In fairway crested wheatgrass (*Agropyron cristatum* L. Gaertn.) and standard crested wheatgrass [*Agropyron desertorum* (Fisch.) Schult.] Rogler (1954) found a very close relationship between seed size and seedling vigor and found a high positive correlation between seed size and emergence of seedlings from deeper plantings. Vogel (1963) showed planting depth markedly influenced emergence of beardless wheatgrass [*Agropyron inerme* (Scribn. and Smith) Rydb.] from seed of three sizes, and seedlings emerged better from large than from small seeds.

The positive influence of seed size on seedling vigor was also evident in studies of heterocarpic species from three plant families. In all cases, the larger seeds produced larger, more vigorous seedlings than the smaller form (Koller and Roth 1964, Cavers and Harper 1966, Baker 1972). These reports and that of Hill et al. (1985) contrast with the report by Marshall and Naylor (1985) in which no significant correlations were found between seed weights of nine lots of Italian ryegrass (*Lolium multiflorum* Lam.) and field emergence at two locations in the U. K.

Generally, within a species, small seeds that have small embryos, limited reserves, and limited cotyledon area, produce small seedlings. Conversely, large seeds produce seedlings with greater cotyledon area, greater embryo weight, and greater photosynthetic activity immediately after emergence (Cooper and Fransen 1974, Wood et al. 1977).

Among species, larger seeds can produce larger seedlings and greater growth, but not always. Among species of legumes, a highly positive correlation exists between stem area and seed mass, and between stem area and maximum emergence force (Inouye et al. 1979). Yet, growth rates for seedlings of Engelman spruce [*Picea englemannii* (Parry) Engelm.] was 0.041 g/g/day from seeds averaging 2.9 mg/seed while subalpine fir [*Abies lasicarpa* (Hook) Nutt.] averaged 0.028 g/g/day from 7.9 mg seed (Knapp and Smith 1982). Arndt (1965b) compared axial morphology of seedlings of peanut (*Arachis hypogaea* L.), cotton (*Gossypium* spp.), and guar (*Cyamopsis* spp.), including diameters of various parts of the seedlings, and observed that the relatively small seeds of guar produced stouter seedlings than the larger cotton seeds.

Within wide limits, grasses having the heaviest seed send up shoots from greater depths, although a strict relationship of this kind is not clearly apparent (Plummer 1943, Watt and Whalley 1982a). Kneebone and Cremer (1955) and Whalley et al. (1966a) found that seed weight, seedling vigor, and seedling length were not related for several rangeland grasses. Also, Kittock and Patterson (1959) believed that seed sizes among the wheatgrasses, bromes, and fescues were not correlated significantly with seedling root penetration or survival (Kittock and Patterson 1959).

McWilliam et al. (1970) suggested that despite the obvious importance of seed reserves during germination and early seedling growth, reserves were in excess of a seedling's requirements under normal growing conditions for Koleagrass (*Phalaris tuberosa* L.), perennial ryegrass, and orchardgrass (*Dactylis glomerata* L.), species commonly seeded in pastures in temperate Australia. Their conclusion was based on evidence that nutrients are taken up and the compensation point is attained in young seedlings long before exhaustion of endogenous reserves. Reserves may only be fully exploited when adverse conditions, such as those that retard emergence or cause shading of seedlings, or tend to delay the onset of autotrophic growth.

The greatest effect of seed size on seedling growth appears to occur soon after germination. The maximum beneficial effect of heavier grass seeds, as measured by seedling height, was noted 3 weeks after seeding and declined after 5 weeks (Kittock and Patterson 1959). The rate of root penetration continued to receive additional benefit from heavier seeds for the full 7 weeks of the trial. Growth rates of perennial ryegrass seedlings from heavy and light seed were similar when sown at depths that produced the highest proportion of emerged shoots (Arnott 1969). A weight advantage was gained by seedlings developing from heavy seed. This was maintained, but was not increased after photosynthesis began. Seed weight affected the yield of seedling dry matter from tall fescue (*Festuca arundinacea* Schreb.) in the very early stages of growth (Lewis and Garcia 1979), but development of a tiller in the axil of the coleoptile, which varied significantly among populations, had a far greater and more permanent effect on dry matter yield and tiller number than seed weight. Seed weight of perennial ryegrass was positively correlated with all characteristics of seedling vigor during initial seedling growth, but the influence decreased as plants matured (Thomas 1966).

Although effects of seed weight are generally confined to early stages of growth in grasses, these effects were noted up to 3 months after sowing smooth brome (*Bromus inermis* Leyss., Trupp and Carlson 1971). Hill et al. (1985) reported that growth rates of Italian ryegrass, rescuegrass (*Bromus catharticus* Vahl), and tall fescue during the first 20 days after sowing were closely related ($r = 0.87$ to 0.99) to weights of seed and caryopses when perenniality (associated with relatively slow growth) and polyploidy (associated with faster growth) were considered. They showed a multivariate relationship between growth rate, temperature, and the weights of seed and caryopsis ($R^2 = 0.88$).

Curtis and McKersie (1984) found that the axes of seedlings from large seeds of birdsfoot trefoil (*Lotus corniculatus* L.) accumulated dry matter at a faster rate than those from small seeds. Conversely, when dry weights for the entire seedling were compared, growth rates of subterranean clover (*Trifolium subterraneum* L.; Black 1956) and radish (Choe et al. 1988), were not influenced by initial seed weight. Kittock and Law (1968) found that seed weight was correlated with seedling vigor and tetrazolium chloride reduction by germinating seeds. However, the tetrazolium chloride test was a measure of the total reduction, not the rate per unit of seed weight. McDaniel (1969) found that heavier seed had more mitochondrial protein and higher mitochondrial respiration. However, respiration per unit of mitochondrial protein did not differ among seed of various weights. He suggested that seed weight was less important to growth rate than genetic factors influencing respiratory control. Genetic differences in respiratory control seem to explain the lack of a consistent pattern among species for the relationship between seed size and germination rate or seedling growth rate. These genetic differences also explain the more consistent pattern found within species and cultivars.

Other considerations besides differences in respiratory control affect the overall growth rate of germinants and seedlings. Schmalstig and Cosgrove (1988) noted that stem elongation in etiolated pea (*Pisum sativum* L.) was inhibited by 60% within 5 hours of cotyledon excision. The growth inhibition was found to be linked to an increase in the threshold for cell-wall yield. Bennett et al. (1988) demonstrated that root acidification was severely depressed by excising cotyledons of soybean (*Glycine max* L.) seedlings. Root acidification was also depressed in radish seedlings that had low cotyledonary reserves. The cell-wall-yield threshold and root acidification are related to H^+ physiology. Root acidification is especially important in seedlings, which compared to other life stages, have the highest nutrient requirements per unit of root length (Mengel and Kirkby 1982, page 80). Bennett et al. (1988) suggested that depletion of seedling cotyledonary reserves before the developing foliage begins exporting photosynthate may result in poor establishment of stressed seedlings.

The literature available on the effect of seed size leads us to conclude that large seeds do not give rise to seedlings with intrinsically higher growth rates. Large seeds, relative to other seeds in a lot, produce larger post-germination seedlings that often have a greater capacity for extracting moisture from the soil and for photosynthesis. Large seed reserves protect against growth-inhibiting stress from depleted respiratory substrate and allow growth to continue at a rate that is closer to an inherent potential rate determined by the genetic program for respiratory control. The amount of seed reserves and the post-germination seedling size can affect the axis growth rate and thus seedling morphology by producing a slightly taller stem or a slightly longer seminal root. The resulting morphological variations may or may not significantly influence early seedling dry matter accumulation but could influence a seedling's ability to compete for resources during a critical period. Over a growing season, this influence may result in a greater accumulation of dry matter as was demonstrated for yellow sweet clover [*Melilotus officinalis* (L.); Haskins and Gorz 1975], or it may simply result in increased seedling survival as suggested by Arnott (1969). So while a direct influence of large seed on seedling morphology may appear to exist for only a very short time, the result has long term advantages.

The Significance of Seed Size in Range Plant Adaptation

In the arid environments that typify most rangelands, the lack of water is a constant factor selecting for large seed size and early germination in native species (Baker 1972, Stebbins 1974, Schirman 1981) and imposing similar criteria for species seeded by man (Young et al. 1983, Jordan 1983). In many environments, selection for larger seed is balanced by increased predatory pressure on larger compared to smaller seeds, as indicated by studies finding large seed more attractive to predators (Fried at al. 1988). [This may explain why large seeds might tend to have thicker seed coats (Fenner 1983).]

Baker's (1972) conclusion that aridity encouraged evolution to larger seeds was based upon an analysis of nearly 2,500 taxa of the California flora. He found that seed weights for herbaceous plants and trees tended to be greater for species growing in climates where seedlings are exposed to drought soon after establishment. He deduced that larger seed resulted from this natural selection and that seedlings from large seed develop roots faster. [For shrubs, seed weights tended to increase where shading and competitive stress appeared to be influencing factors. Thompson (1987) has noted that large seed provided an advantage for seedling establishment where plant competition was a dominant factor. He concluded that establishment in closed vegetation was facilitated by a low rate of exhaustion of seed reserves. The low exhaustion rate was accomplished by a large seed, by a low growth rate, or both.] Baker (1972) also found that seed weights within a species and within community types decreased with increasing altitude. He cited evidence that seed weight might be genetically controlled and observed that shorter growing seasons at higher altitudes may result in less photosynthate being available to seeds. Generally, growing conditions are more mesic at higher altitudes which may explain why lighter weight seeds succeeded.

In a somewhat analogous situation, winterfat growing along the eastern gradient of the Rocky Mountains was suggested to have adapted to windier sites through changes in the balance between the capacity to produce photosynthate, and the capacity to limit water loss by transpiration (Booth 1990 and unpublished data). Ecotypes with the greatest capacity to limit transpiration also had the least diaspore production (grams/plant), even under ideal conditions.

As with seed size and germination rate/seedling growth, the significance of seed size in adaptation is fairly discernable within a species but is muddled when species are compared. Why, for instance, do the sagebrushes (*Artemisia* spp.) with relatively tiny seeds and advanced embryo morphology, share habitats around the world with cheatgrass (*Bromus tectorum* L.), which has relatively large seeds and more dependent embryos? Obviously seed size is only one of many factors determining whether a species can colonize and occupy arid-land habitats. Comparisons of the seeds and habitats of species emphasizes that speciation is a much more complicated process than ecotypic change facilitating range extension.

Morphology of Dicot and Gymnosperm Seedlings

Seedlings can be divided into two types according to the fate of the cotyledons. In an epigeal seedling, cotyledons are raised out of, or spread above, the soil; they generally become green, photosynthesize, and can contribute significantly to plant growth (Machado et al. 1974, Lovell and Moore 1970, Wood et al. 1977). This advantage may be particularly important for small seeded species such as willow or sagebrush. In hypogeal seedlings, one or both cotyledons remain below the soil surface. Hypogeal growth is by elongation of the epicotyl, which raises the first true leaves out of the soil. Chinkapin (*Castanopsis*) and jojoba (*Simmondsia*) are examples of rangeland dicots with the hypogeal characteristic (Schopmeyer 1974). Epigeal and hypogeal seedlings arise from endospermic and nonendospermic seeds, but hypogeal germination from endospermic seeds is rare among dicots (Bewley and Black 1978).

In species adapted to germination below the soil surface, epigeal morphology hinders emergence by imposing on the seedling a large surface area that must be pulled or pushed to the surface. Plant geneticists have suggested the possibility of replacing the epigeal characteristic with hypogeality as a means of improving plant establishment (Machado et al. 1974, Wood et al. 1977, Inouye et al. 1979).

Many epigeal species germinate on the soil surface. Adaptations to surface germination are usually reflected in diaspore morphology but may also be seen in seedling morphology. Root hairs just below the stem/root junction supersede bract hairs as the anchoring mechanism during establishment of winterfat (Booth 1982). Young and Martens (1988) noted that early root hairs of sagebrush seedlings appear to be an adaptation facilitating surface germination, since the root hairs do not elongate or branch and they gradually disappear as the seedlings grow older. The root hairs in the hypocotyl ring of Salicaceae seedlings (Simak 1980) also disappear after the primary root has developed (Zasada, personal communication). Compared to subsurface germination, surface germination requires less energy for seedling establishment but places seeds and seedlings in more jeopardy to desiccation, fire and predation.

Epigeal species may also be adapted to group emergence. The emergence of five to 10 bitterbrush seedlings clumped together will improve establishment and seedling survival (Ferguson and Basile 1967, Booth 1980). *Floerkea proserpinacoides* Willd., a forest floor annual, also has a better chance of emergence if seedlings are clumped so that they can emerge simultaneously to push aside surface litter (Fenner 1987). In group emergence the benefits of epigeality (early photosynthesis) and seed burial can be obtained while the energy required for emergence of individual seedlings is moderated by group action. The disadvantage is competition among seedlings.

Morphology of Grass Seedlings

Growth of a grass seedling occurs in three stages (Whalley et al. 1966a): the heterotrophic stage, which begins when water is imbibed by the seed and ends when photosynthesis commences; the transition stage, which begins at the commencement of photosynthesis and exhaustion of seed reserves; and the autotrophic stage which begins after endosperm reserves are exhausted and is the period when complex organic compounds are being provided by photosynthesis.

Grass seedlings are hypogeal—the cotyledon [scutellum (Avery 1936)] remains below ground during germination. Emergence of the embryonic shoot occurs by several mechanisms. In warm-season grasses the coleoptile, a tubular structure surrounding the embryonic plumule bud, is short and contributes little if any to the emergence process. However, the subcoleoptile internode, the first internode located below the divergence of the coleoptile, may contribute greatly to emergence. In cool-season grasses the subcoleoptile internode is short; and the coleoptile is long, and important for emergence.

Subcoleoptile internodes and coleoptiles are especially adapted for emergence because meristematic activity is located in a narrow band of cells near the top of these structures. True leaf and stem internodes above the coleoptilar node are not adapted to pushing through soil because meristematic cells are located at the base of these internodes. If the coleoptile cannot reach the soil surface, the first true leaf might be moved a short distance from the coleoptile tip. However, the soft meristem at the base of the true-leaf internode will kink and twist, preventing the blade from pushing through the soil.

The seedling's first internode located below the divergence of the coleoptile has been described by many terms (Table 3). Discussions and arguments for using the terms appear in Avery (1930), Boyd and Avery (1936), Brown (1960), Newman (1986), Tischler and Voit (1987) and Ries and Hofmann (1991). We have used subcoleoptile internode as proposed by Hyder et al. (1971).

The second internode, located above the divergence of the coleoptile but developed inside the coleoptile, was more recently referred to as the intracoleoptile internode by Hyder et al. (1971) and Hyder (1974). The second and third internode have collectively been called the sub-crown internode (Taylor and McCall 1936, Webb and Stephens 1936). If elongated, the first internode above the coleoptilar node is truly an intracoleoptile internode because there is no true leaf from the coleoptilar node. If the coleoptilar internode does not elongate, the first shoot node will be established very close to the coleoptilar node. If the subsequent internode elongates, it is truly an intraleaf (or intrasheath) internode even though it may be located inside the coleoptile, and thus this subsequent internode could be defined as intracoleoptile as well. Each node, beginning with the coleoptilar node, initially has the meristematic potential for developing tiller and root.

Ries and Hofmann (1991) have recently proposed the following terminology for describing structures of grass seedlings. Nodes are defined as scuttelar node, coleoptilar node, and leaf nodes named in sequence from first to last. They propose mesocotyl for the internode between the scutellar and coleoptilar nodes and first leaf internode for the internode developing inside the colcoptile between the coleoptilar and first leaf node. Subsequent internodes were named for the leaf node immediately above; e.g. second leaf internode.

The root system of grasses typically consists of seminal roots and adventitious roots (Hyder 1974; Fig. 5). The seminal root system consists of one to five roots—the primary root (radicle) and two pairs of lateral roots arising in the same plane from the scutellar note (Hyder 1974). A single root may also arise opposite the epiblast (Boyd and Avery 1936; Fig. 6). Seminal roots are slender and branched and can be short (Olmsted 1941, Weaver and Zink 1945) or long lived (Olmsted 1941, Van der Sluis and Hyder 1974, Harris 1967).

Adventitious roots arising from nodes in the crown are called adventitious,

Table 3. Terminology used in describing morphology of grasses.

PLANT PART	TERM USED	REFERENCE
First internode of seedling located below divergence of the coleoptile	First internode	Maun and Riach 1981 Rogler 1954
	Tissue between coleoptile and scutellum node	Boyd and Avery 1936
	Mesocotyl	Hoshikawa 1968 Berlyn 1972 Ries and Hoffmann 1989 Ries and Hoffmann 1991
	Subterranean internode	Lawerence and Ashford 1964
	Subcoleoptile internode	Hyder et al. 1971 Hyder 1974 Van der Sluijs and Hyder 1974 Wilson et al. 1976 Briske and Wilson 1977 Wilson and Briske 1978 Briske and Wilson 1980 Tischler and Monk 1980 Cornish 1982 Tischler and Voigt 1983 Cornish et al. 1984 Fulbright et al. 1985 Newman 1986 Nason et al. 1987 Newman and Moser 1988a Newman and Moser 1988b
Second and third internode	Subcrown internode	Taylor and McCall 1936 Webb and Stephens 1936
Second internode located above the divergence and inside the coleoptile	Intracoleoptile internode	Hyder et al. 1971 Hyder 1974
Seminal lateral roots	Transitory node roots	Hoshikawa 1969
Adventitious roots developing from mesocotyl or sub-coleoptile internode	Mesocotylar roots	Hoshikawa 1969
	Subcoleoptile roots	Newman and Moser 1988a Newman and Moser 1988b
	Permanent roots Hsu and Nelson 1986 a and b	Hsu et al. 1985b
Adventitious roots developing from nodes	Cotyledonary node roots	Hoshikawa 1969
	Adventitious roots	Hyder 1974 Newman and Moser 1988a Newman and Moser 1988b
	Secondary roots	Cornish 1982 Cornish et al. 1984 Watt 1981 Watt and Whalley 1982 a and b
	Nodal adventitious roots	Tischler and Voigt 1987
	Permanent roots	Hsu et al. 1985b Hsu and Nelson 1986 a and b

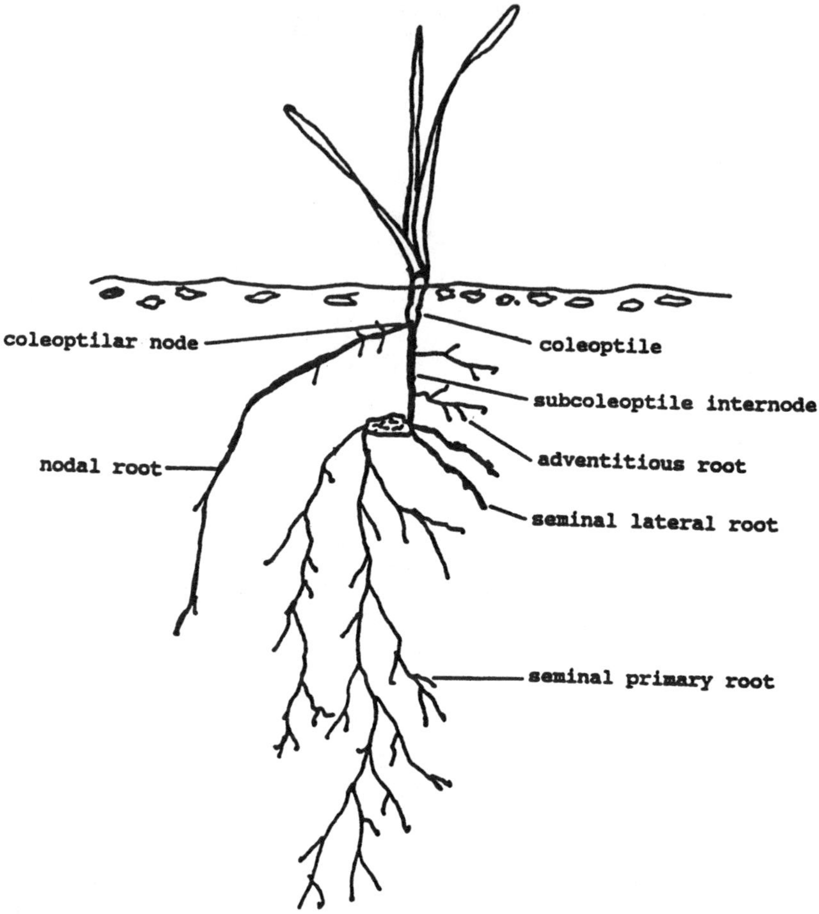

Figure 5. Grass seedling.

nodal, crown, or secondary roots; adventitious roots arising from internodes are identified by the internode from which they grew (Troughton 1957, Hyder et al. 1971, Tischler and Voigt 1987). Development of nodal roots is necessary for establishment of grass seedlings, and collectively, adventitious roots are generally considered the root systems of mature plants (Esau 1965, Tischler and Voigt 1987). Nomenclature for adventitious roots also varies in current literature (Table 3).

Hoshikawa (1969) reported that although all grass species initially produce a seminal root in germination, they have different features in subsequent development of underground organs. He described seedling morphology of 219 grass species in 88 genera and classified the species into six types based on presence or absence of seminal lateral roots, elongation or nonelongation of the subcoleoptile internode, and presence or absence of adventitious roots on the subcoleoptile internode. He also classified the species into seven types based on mode of seedling establishment. Ries and Hofmann (1991) suggested that primary differences

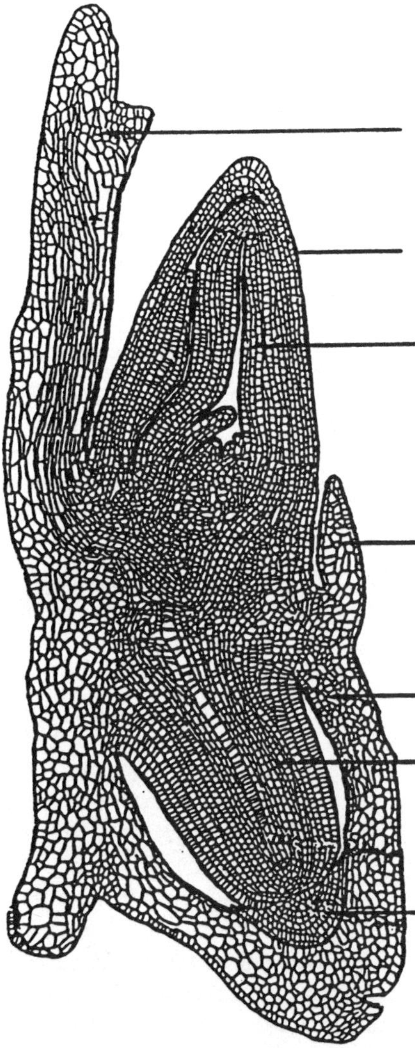

Figure 6. Grass Embryo (From Robbins et al. 1965).

between the types of seedlings are in the limited development or nondevelopment of the subcoleoptile internode, and the nonelongation and/or variable elongation of all internodes (subcoleoptile, intracoleoptile, and intraleaf).

Most warm-season, panicoid-type grasses with an elongated subcoleoptile internode develop roots on the subcoleoptile internode. Most cool-season, festucoid-type grasses lack or have little subcoleoptile internode elongation and have seminal lateral root development. However, this classification is not exact, and exceptions exist (Tischler and Voigt 1987).

Seedling Emergence and Development

Depth of Planting

Planting depth and the environmental factors of moisture, temperature, and light are known to affect the developmental morphology of seedlings. Most species have a range of planting depths from which the probability for seedling establishment is good. Seedlings from surface-germinating seeds readily develop a photosynthetic surface, so their establishment depends on soil penetration; whereas seedlings from buried seeds must push through the soil before beginning photosynthesis. Growth through the soil is a fundamental requirement for seedling establishment. Differences in the way this growth is accomplished are largely a matter of timing, stored energy, and mechanical differences in the way diaspores and germinants are anchored. The capability of a plant to grow through the soil has been described by the formula: EA = (EF) (Ch) (Sp); where EA is the emergence (or penetration) ability, EF is the elongation force asserted by the plant in darkness, Ch is the cross-sectional area and physiological characteristics of the plant, and Sp is a measure of the speed of elongation of the plant in darkness (Inouye et al. 1979). Elongation force has been positively correlated with seed weight in several studies (Jensen et al 1972, Williams 1956, Williams 1963, Taylor and Ten Broeck 1988, Inouye et al. 1979). Because our focus is on morphology, we will examine the influence of cross-sectional area.

Plants have adapted to the need for growing through the soil with two somewhat opposite morphologies. Monocots are noted to have great emergence ability because they have a small cross-sectional area (Taylor and Ten Broeck 1988). This is needed in the grasses because intercalary meristems, located at the tops of the coleoptile and subcoleoptile internodes, tend to push the coleoptile through the soil. In contrast, dicots have large cross-sectional areas which decreases flexibility of the stem and increases lifting or pushing power (Arndt 1965a).

The ability of a cylinder (including the embryo axis) to support a load, increases by the fourth power of the radius (Knittle and Burris 1979). And as cylinder length is increased, the buckling tendency of a loaded column increases independently of the cross-sectional area (Arndt 1964b). Of the two factors, length and radius, length is manageable by controlling seeding depth, but also has the least effect on buckling per unit of change. The most effective change, that of increasing the axis diameter, is a natural seedling response to loads (Knittle and Burris 1979, Taylor and Ten Broeck 1988, Gerard 1980, Wanjura and Minton 1981, Liptay and Geier 1983). Soil around a seedling axis provides a degree of radial support (Arndt 1965b) that is somewhat analogous to an increase in cylinder radius. The strength of this support depends on soil conditions and varies with soil moisture, but the interaction of these factors on seedling emergence has not been studied.

Soil crusts vary in their influence on seedling success and on the elongation force required to penetrate them. For example, crusts form surface domes when butts of crust plates, forced up by germinating seeds, jam against adjacent plates (Arndt 1965b). Such conditions increase cylinder load and remove radial soil support in the space beneath the dome. Here, flexible plants with small cross-sectional surface areas may need to exert as much force as large-stemmed plants in order to emerge.

Since they do not have the cross-sectional area to exert that force, they buckle and fail. Seeds germinating as a group, however, may exert enough force to break through the crust. Therefore, seeding rates should be increased on soils subject to crusting.

Specific effects of planting depth on morphology have been investigated for several species. Although planting any species too deeply reduces emergence, the chief morphological effects are reduced root development and reduced seedling height [Black 1956 for subterranean clover, Fulbright et al. 1985 for green needlegrass (*Stipa viridula* Trin.)]. Similar observations have been made when species adapted to surface germination were planted (Everitt 1983a and b, Everitt et al. 1983, Stevens and Van Epps 1984, Booth 1987, Young and Martens 1988). Conversely, if a species adapted to burial is sown on the surface, establishment or survival will be reduced. Nelson et al. (1970) found that bunchgrass seedlings from broadcast seed never appeared well anchored until after growth resumed in the fall of their second growing season. Evans et al. (1977) found that germination of Sandberg bluegrass (*Poa sandbergii* Vasey) was limited on the soil surface and at depths greater than 30 mm.

The influence of planting depth on grass seedling morphology has been studied extensively. Coleoptile length and crown depth have been positively correlated with actual planting depth in smooth brome, and subcoleoptile internode length has been positively correlated with actual planting depth in big bluestem (*Andropogon gerardii* var *gerardii* Vitman), Indiangrass [*Sorghastrum nutans* (L.) Nash], and switchgrass (*Panicum virgatum* L.) (Newman and Moser 1988b). Deeper planting did not enhance nodal root development in the latter three species, since crown depth was not correlated with planting depth in these species. When emerging from an intermediate depth (25 mm), smooth brome had intermediate-length subcoleoptile internodes (1 to 6 mm) and coleoptiles (23 to 27 mm) and had elongated intracoleoptile and intraleaf internodes (Ries and Hofman 1989). When sideoats grama [*Bouteloua curtipendula* (Michx.) Torr.] seedlings emerged from a 25-mm depth, lengths of subcoleoptile internodes ranged from 23 to 25 mm, and coleoptiles were 4.5 to 8 mm long (Ries and Hoffman 1989). Subcoleoptile internode elongation was relative to the planting depth, and the coleoptilar node occurred near the soil surface. Western wheatgrass (*Agropyron smithii* Rydb.) seedlings emerged from 6- to 102-mm depths with elongation of the coleoptile and leaves (Ries and Hoffman 1989). When sideoats grama seeds were planted at 51 mm, crown placement was at the coleoptilar node at the top of the subcoleoptile internode, at an average depth of 6 mm below the soil surface (Ries and Hofmann 1991). Crown placement for western wheatgrass was usually at the coleoptilar node located near planting depth, about 49 mm below the soil surface. Smooth brome crown placement did not occur at the coleoptilar node, instead, it occurred at the first leaf node, second leaf node, or fourth leaf node. Depth of crown placement averaged 22.5, 34.0, and 25.5 mm, respectively, for crowns placed at the first, second, and fourth leaf nodes on the six seedlings measured by Ries and Hoffman. Coleoptile extension was less than 12 mm for sand bluestem (*Andropogon hallii*, Hack.), but switchgrass seedlings appeared capable of emerging from a 30-mm depth (Ohlenbush 1976). Both species were able to extend the first leaf beyond the coleoptile tip and emerge through a limited soil depth. When planting depth was increased, the coleoptile length of green needlegrass seedlings increased, but the length of seminal primary and nodal roots and the number of nodal roots decreased (Fulbright et al. 1985). The subcoleoptile internode length ranged from 4 to 16 mm and generally was 2 to 5 mm longer for 'Lodorm' than for SD-93 (Fulbright et al. 1985). Emergence of prairie sandreed [*Calamovilfa longifolia* (Hook) Scribn.] seedlings was negatively correlated with depth of planting (Maun and Riach 1981). The coleoptile lengths averaged 8 mm for seedlings emerging from below the soil surface and 2 mm for seedlings emerging from the soil surface. Length of the subcoleoptile internode was positively correlated with depth of planting. Maximum length of the subcoleoptile internode was 60 mm; the most frequent length was between 20 and 30 mm. When seeds were planted at 20 mm, elongation of the subcoleoptile internode and coleoptile ceased when the coleoptilar node was situated at or very near the soil surface. The maximum potential depth from which a prairie sandreed seedling can emerge is about 74 mm (6-mm plumule, 60-mm subcoleoptile internode, and 8-mm coleoptile). Seedlings of Russian wildrye (*Elymus junceus* Fisch.) emerging from 25- and 38-mm depths were consistently taller than seedlings emerging from a 13-mm depth (McGinnies 1974).

Tischler and Voigt (1983a) used subcoleoptile internode length to determine actual planting depth (10, 20, 40, 60, and 80 mm) for kleingrass (*Panicum coloratum* L.), weeping lovegrass [*Eragrostis curvula* (Schrad.) Nees], and Wilman lovegrass (*Eragrostis superba* Peyr.). Shoot and root weight and number of leaves decreased and days to emergence increased with increasing subcoleoptile internode lengths. Nodal root length and root number decreased with increasing subcoleoptile internode length for kleingrass and Wilman lovegrass. The lovegrass seedlings produced the most growth at shallow depths, and kleingrass seedlings produced a greater fresh weight at 60 mm. Tischler and Voigt (1983b and 1984) hypothesized that the faster growth of the lovegrass at shallow depths, and poorer growth at deeper depths, may have been related to faster utilization of seed reserves by lovegrass compared to kleingrass.

Cornish (1982) found that surface drying encountered with surface placement of perennial ryegrass and *Phalaris aquatica* L. seed prevented nodal root primordia from extending even when plants were otherwise supplied with water. Since initial root growth of a deep planted seed explores no deeper than a shallow planted seed, the only advantage of deeper planting is better protection from surface drying (Tischler and Voigt 1987). Deeper plantings require greater precipitation to wet the soil at planting depth and initiate germination. The increased precipitation sufficient to wet the soil profile improves the probability for successful germination and establishment. Planting seeds below the soil surface reduces losses to rodents and birds, to fluctuating moisture conditions which cause germination to start and stop (Nelson et al. 1970), and to extreme temperatures and fire (Harper 1977, Peart 1984).

Temperature

When moisture is adequate, temperature determines the rate of development in all organisms. Seedling growth of several warm-season grasses in growth chambers was reduced dramatically at 20°C compared to seedling growth of the grasses at

25 or 30°C (Hsu et al. 1985b). Speed of emergence of several warm-season grass species was increased as the planting date was delayed, suggesting that soil temperature was a major factor associated with emergence (Hsu et al. 1986a and b). Base soil temperatures required for seedling emergence ranged between 9.4 and 11.4°C for big bluestem, Indiangrass, switchgrass, and caucasian bluestem [*Bothriochloa caucasica* (Trin) C.E. Hubb] (Hsu and Nelson 1986a). Soil moisture, however, probably played an overriding role on emergence of the four species as the soil became drier during later plantings. Emergence of the species was better in the field when soil water potential at the 0 to 3-cm depth was above –0.4 Mpa. This was more easily attained at cool temperatures. The cool soil temperatures of early spring also partially met the chilling requirement for these seeds (Hsu et al. 1985a). Though emergence is slower when seeds are planted in late April and mid May, soil moisture supply is better with cooler temperatures, reduced evapotranspiration, and spring rainfall.

Robocker et al. (1953) studied seedling emergence of indiangrass, big bluestem, switchgrass, Canada wildrye (*Elymus canadensis* L.), Virginia wildrye (*Elymus virginicus* L.), little bluestem (*Andropogon scoparius* Michx.), sand dropseed [*Sporobolus cryptandrus* (Torr.) A. Gray], sand lovegrass [*Eragrostis trichodes* (Nutt.) Wood], sideoats grama, and green needlegrass. For each species the time required for emergence was remarkably uniform under conditions of relatively uniform temperature, depth of planting, type of soil, and soil moisture. The number and length of roots of both warm-and cool-season grasses increased with increasing temperature. Emergence of the coleoptile of 'Blackwell' switchgrass and sand bluestem from a 1.2-cm planting depth was influenced by soil temperature and the soil temperature-soil water potential interaction (Ohlenbush 1976). Greater emergence occurred at a soil temperature of 45°C for the bluestem and 25 and 30°C for switchgrass. Soil water potentials of –0.3 MPa and higher were best for both species.

Laude and Chaugule (1953) reported that the heat tolerance of newly emerged brome seedlings was good compared to that of older seedlings. When seedlings were kept in the dark before heat treatments occurred, older seedlings had reduced heat tolerance, but younger seedlings did not. They related the unchanged tolerance in younger seedlings to the presence and availability of unused food reserves in the seed of the young seedling. High temperatures observed under range conditions in New Mexico appeared to be detrimental to all species even when moisture was adequate (Sosebee and Herbel 1966). Seedlings of most species had stopped growing or were growing very slowly in the high temperature regime (18–67°C).

Laude (1956) found a greater reduction in seedling emergence as short-term freezes occurred later in the preemergence period for several species of cool-season grasses. Percentage survival of fairway crested wheatgrass, smooth brome, and slender wheatgrass [*Agropyron trachycaulum* (Link) Malte] seedlings was positively associated with stage of development at freeze-up (White and Horner 1943). Over 90% of the seedlings survived when 3 or more leaves were developed. Fall seedling size of standard crested wheatgrass, Russian wildrye and intermediate wheatgrass [*Agropyron intermedium* var. *trichophorum* (Link.) Reichb. ex Hegi], greatly affected plant development the following spring (White 1984). White hypothesized that winter damage would be less and subsequent spring and fall growth would be maximized if seedlings developed to the 3-leaf stage or beyond prior to winter dormancy (White 1984, White and Currie 1980). Frischknecht (1951) suggested that fall-planted seeds of certain grasses germinate earlier and are stimulated to grow and develop faster than from spring plantings.

Light

Photomorphogenesis in plants includes virtually all aspects of plant development: germination, etiolation, growth via photosynthesis, phototropism, phototaxis of chloroplasts, opening and closing of leaflets, entrainment of circadian rhythms, pigment formation, plant form, light damage (including ultraviolet), flower formation, bud dormancy, and many other responses (Salisbury 1974). Dark-grown plants differ in several important and easily recognizable ways from plants grown in the light. Dark-grown seedlings tend to have elongated internodes and reduced size of leaves and roots compared to those of light-grown seedlings.

The effects of shading on seedling survival have been evaluated for several species. Shade substantially reduced maximum soil temperatures but did not increase survival of grass seedlings; shaded plants were elongated and spindly whereas unshaded plants were short and had a rosette form of vegetative growth (Benedict 1941, McGinnies 1966). Dry weight accumulation in alfalfa (*Medicago sativa* L.) seedlings was essentially proportional to light intensity when the plants were grown under 8073, 16146, and 32292 lx (Bula et al. 1959). Light saturation of photosynthesis appeared to be in excess of 32292 lx. Cooper (1966) reported that less dry matter was partitioned into legume seedling roots as light intensity decreased. Thus, alfalfa seedlings grown in shade may become more susceptible to drought because of restricted root development. Further, reduced light may cause a proportional decrease in root nodulation (Pritchett and Nelson 1951). Low light intensities also increased mortality of seedlings of bare-ground species, *Satureja acinos* (L.) Scheele, and germander (*Teucrium botrys* L.) compared to mortality of the closed microhabitat species carrot (*Daucus carota* L.) and the perennial hawkbit (*Leontodon hispidus* L.) (Verkaar and Schenkeveld 1984). Under the light conditions found in dense stands, seedling mortality of species from open microhabitats, carlina (*Carlina vulgaris* L.) and scaboisa (*Scaboisa columbaria* L.), also was rather high. Mortality resulting from light intensities below the compensation points may be more related to the amount of seed reserves than to negative growth rates.

Shading is, however, a definite benefit to many tree seedlings, especially forest species in the post-germination stages of growth. Maximum light saturation values for both conifers and hardwoods are reported to range from about 1500 to 3200 fc (16,146 to 34,445 lx), and excessively high intensities reduce growth and lead to injury or death of seedlings (Larson 1974). Data collected on conifer seedling establishment in the field shows the best establishment and survival occurs in shaded microsites (Tappeiner and Helms 1971, Conard and Radosevich 1982, Geier-Hayes 1987). The benefits of shading to seedling survival have also been documented for the shrubs *Artemisia herba-alba* (Friedman and Orshan 1975) and winterfat (Woodmansee and Potter 1971).

Morphology and Survival

The relationship between seedling morphology, environmental conditions, and ease of seedling establishment/survival has received considerable attention, especially among the grasses. Reviews on the topic have been published by Evans (1946), Hyder (1974), Silcock (1980), Johnson (1986), Newman (1986), and Tischler and Voight (1987). Morphological development occurs in stages, and progression through the stages is affected by both genetic potential and environmental conditions. Any factor inherent in the plant or the environment that effectively impedes germination and emergence, or that delays the development of roots and shoots in a way that impairs the competitive capabilities of the seedling, will diminish establishment or survival.

The development of the root system, particularly the nodal root system of grasses, which supplies large amounts of water to a growing seedling, is critical to establishment of seedlings on ranges and pastures (Ries and Svejcar 1991). High root:shoot ratios, very rapid germination, and rapid seminal root elongation have all been reported as characteristics important for seedling establishment in semi-arid or arid environments (Simanton and Jordan 1986). These characteristics, however, are not always related to successful establishment of seedlings, as evidenced during a 7-day planting trial in Arizona. Seedlings of sideoats grama had comparatively high root:shoot ratios, very rapid germination, and rapid seminal root elongation and did not establish well in arid environments. Seedlings of 'Cochise' lovegrass (*Eragrostis lehmanniana* Nees x *Eragrostis trichophora* Coss & Dur.) had a low root:shoot ratio, slow germination rate, and the slowest seminal root development of the species tested but were successfully established on arid ranges. Similar results were found by Frasier et al. (1984). High percentages (50 to 70%) of sideoats grama seedlings emerged after a 1- to 2-day wetting period, but over 50% of the seedlings died in a 5-day drying period. Seeds of 'Cochise' germinated more slowly, and seedlings were less susceptible to the 5-day drying period. In another study, seedling survival varied between species and between varieties of the same species during short-term drought (Frasier 1985). Although similar responses occurred in the field, seeds generally did not germinate after the 1- to 2-day wet period, and nongerminated seed survived the dry period as viable seed (Frasier 1987). When seeds germinated in a 5-day wet period, they usually survived dry periods of 5 to 7 days. Wester et al. (1988) also reported that moisture in specific patterns and amounts is important for seedling success of kleingrass and weeping lovegrass.

Morphology of seminal primary roots that develop immediately after germination vary by species. The primary root system of sideoats grama seedlings is finely branched, whereas the root system of creeping foxtail (*Alopercurus arundinaceous* Poir.) seedlings is poorly developed (Newman and Moser 1988a). A small percentage of seedlings of kleingrass 'Selection 75' and weeping lovegrass 'Ermelo' produced no primary root at emergence (Tischler and Voigt 1981). A higher percentage of the no-root seedlings was produced at 35°C than at 20°C, and more no-root seedlings emerged from lighter (less mature) seed than from heavier (more mature) seed. Exposure of coleoptiles to light produced normal root growth in no-root seedlings of kleingrass but not in those of the lovegrass.

Seminal root development appears to be related to leaf and physiological development of seedlings. During growth and development of the first foliage leaf, seminal roots of switchgrass and sand bluestem seedlings extended deeper into the soil, but did not branch (Ohlenbush 1976). Root length of seedlings for several perennial grasses reached a maximum length of 40 to 60-mm length and no more before the onset of photosynthesis (Tadmor and Cohen 1968). Elongation of the primary root was a biphasic process in kleingrass seedlings (Tischler and Monk 1988). A limited 60-mm elongation occurred, and then growth stopped until the coleoptile perceived light. The terminal preemergence length of the primary root was 10 mm and varied from 0 to 30 mm until the coleoptile reached the soil surface. Resumption of primary root growth upon emergence of the coleoptile did not appear to be a response to photosynthate, since 5 minutes of light was as promotive as continuous light. They suggested that light initially altered partitioning of seed reserves by reducing elongation of the subcoleoptile internode, thus making seed reserves available to other parts of the seedling. Radicle emergence in several species of canarygrass (*Phalaris*) was closely associated with a rapid rise in sugar content (Whalley et al. 1967). Seedling growth rate of some canarygrass species was limited by the ability of the seedlings to use the sugars supplied by the endosperm, and in others growth was limited by the supply from the endosperm. McWilliam et al. (1970) reported that the early onset of autotrophic growth in young seedlings, as judged by the utilization of external nutrients and the attainment of positive net photosynthesis, occurred within 5 days of imbibition under favorable environmental conditions.

Seminal lateral roots have been found on seedlings of *Agropyron*, *Elymus*, *Lolium*, *Phalaris*, and *Stipa* (Hassanyar and Wilson 1978, Cornish 1982, Fulbright et al. 1984, Newman and Moser 1988a). Hoshikawa (1969) reports that these roots also occur on seedlings of *Oryzopsis*, *Stebrochaeta*, *Brachypodium*, *Bromus*, *Lolium*, *Briza*, *Deschampsia*, *Avena*, *Phalaris*, and *Milium*, and most genera of the Triticeae tribe. Several authors have stressed the importance of these roots for seedling survival when the primary seminal root is destroyed or damaged by drought (Hassanyar and Wilson 1978, Fulbright et al. 1984, Bassiri et al. 1988a and b). Cornish (1982) found, however, that seminal lateral roots on perennial ryegrass and *Phalaris aquatica* had small xylem vessels that did not significantly reduce axial resistance to flow of water into the seedling.

Adventitious roots commonly develop on internodes of seedlings in the Panicoideae, Eragrostoideae, and Arundinoideae subfamilies (Hoshikawa 1969). Adventitious roots developing from internodes are not always easy to distinguish from nodal roots in seedlings that have slight internode elongation (Olmsted 1941). Newman and Moser (1988a) found that adventitious roots occurred on subcoleoptile internodes of 71 to 100% of seedlings in the Eragrostideae, Andropogoneae, and Paniceae tribes. They found that adventitious roots were more numerous on Caucasian bluestem and sand lovegrass than on other warm-season grasses. They also found that these roots were more numerous on reed canarygrass, which has a long subcoleoptile internode, than on other cool-season species. In contrast, only a few long roots grew at the acropetal end of the long subcoleoptile internode of creeping foxtail seedlings. They hypothesized that these long roots may have an important function in the juvenile root phase of this species but not in other species. Longer subcoleoptile internodes had more adventitious roots than shorter ones on seedlings of kleingrass (Tischler and Monk 1988). Tischler and Monk (1988) suggested that the adventitious roots on the subcoleoptile internode may be valuable when seeds are planted deeply and the surface soil is too dry to allow initiation of nodal roots; the moisture present at the deeper depths can be extracted by these adventitious roots. Roots also grew on subcoleoptile internodes of *Phalaris aquatica* seedlings grown in limited light (Cornish 1982). These roots contained xylem vessels that appeared to be too small to allow adequate water supply to developing seedlings. Thus, we lack convincing evidence that the adventitious roots from either the subcoleoptile internodes or the shoot internodes contribute significantly to survival.

Nodal roots or adventitious roots developing at nodes near the crown or base of seedlings form the major root system of mature plants. Initiation of these roots is critical to the survival of grass seedlings. Cornish et al. (1984) found that delaying development of nodal roots reduced transpiration rates in ryegrass and canarygrass within 15 days of emergence and reduced tillering and leaf area in ryegrass within 22 days; in canarygrass leaf area was reduced within 28 days, and tiller number was reduced within 42 days. They suggested that these reductions were caused by high axial resistance to water flow through the plant. Nodal roots were essential for survival of the canarygrass and ryegrass seedlings. About 20 days after the seed was sown, seedlings with no seminal roots survived on nodal roots. Seedlings that had seminal roots but no secondary roots survived less than 4 months when growing in moist soil. Nodal roots were initiated after the second leaf began developing in seedlings of switchgrass and sand bluestem but no branching occurred during this stage (Ohlenbush 1976). Nodal roots began to develop on curly mesquite [*Hilaria belangeri* (Steud.) Nash.] seedlings between 21 and 28 days after planting, and 2 roots were present at 35 days (Merrill and Young 1962). The nodal root system was very important to the seedling after 56 days, and the seedling was totally dependent on this system 98 days after planting. Of all of the species and tribes studied by Newman and Moser (1988a), except for fairway crested wheatgrass and intermediate wheatgrass and members of the Andropogoneae (bluestem) tribe, nodal roots were present on 70 to 100% of seedlings at the third-leaf stage. A strong positive relationship was found between nodal root development and time from planting. The leaf developmental stage of seedling shoots, apparently, was not associated with nodal root development as much as seedling age or genetic differences among species. A rapid increase in nodal root number occurred in 4-week-old seedlings of big bluestem, indiangrass, and switchgrass after three or more consecutive days of rainfall. Nodal root dry weight was greater for cool-season grasses than for warm-season grasses when plants were grown at 28°C day and 23°C night temperatures.

Tischler and Voigt (1987) suggested that range grasses, particularly warm-season grasses, were difficult to establish because of their morphological development and the hostile environment in which they grow. Poor seedling establishment of blue grama is due to the elongation of the subcoleoptile internode and to the placement of the coleoptilar node and point of origin of nodal roots near the soil surface (Hyder et al. 1971). The depth of the coleoptilar node dictates whether the lowest nodal roots will develop at the planting depth or at the surface of the soil (Tischler and Voigt 1987). Nodal roots are generally developed at the planting depth for the festucoid grasses and in the upper 5 mm of soil for the panicoid grasses. Tischler and Voigt also described the process whereby seedlings that have subcoleoptile internode elongation, may have a coleoptilar node emerge above the soil surface. Emergence of the coleoptilar node generally occurs when seedlings break the soil surface after dark. Without a light stimulus, the subcoleoptile internode continues to elongate, thus moving the node to the surface. They based their description on cereal crops, since data concerning the mechanism by which light regulates elongation of the coleoptile or subcoleoptile internode were lacking for forage grasses.

The morphological limitations of panicoid grasses are also aggravated by the warm-season characteristic (C4 photosynthetic pathway). Panicoid grasses cannot be planted in the fall because they will not grow to an adequate size to overwinter. Yet spring plantings face the prospect of summer drought. The top 20 mm of the soil dries much more rapidly in warmer periods (when panicoid grasses are emerging) than in cool periods (when festucoid grasses are emerging). However, cool temperatures can also limit festucoid grass emergence at times of optimum soil moisture. In areas such as the Great Basin, where most of the precipitation occurs in fall, winter, and spring, temperatures are often too cool for germination and seedling growth until the soils begin to dry (Young and Evans 1986). Germination and seedling development at cooler temperatures would enhance establishment of grass stands before the onset of dry summers.

Generally, moisture availability increases as soil depth increases. Tischler and Voigt (1987) stated that the potential advantages of festucoid morphology are realized only in very specific situations. They suggested that grass seedlings of panicoid or festucoid morphology should become established with equal success when soil moisture is adequate and temperatures are in reasonable ranges. When seeds are planted at shallow depths (less than 5 mm), festucoid morphology is not an advantage because nodal roots must then initiate growth near the soil surface, where nodal roots of panicoid grasses develop. Panicoid seedlings emerge better from deep plantings, since the subcoleoptile internode appears to have potential for greater elongation than the coleoptile.

Panicoid – Blue grama

A series of studies, beginning with Hyder et al. (1971), described the relationship of seedling morphology of blue grama and the difficulty of establishing seedlings of that species in the harsh environment of the Central Great Plains. They reported that blue grama was poorly adapted for natural regeneration from seed on rangelands where it is the ecological dominant. When planted in warm-moist soil, seedlings emerged quickly and produced thick stands on most plots. However, the seedlings died in 6 to 10 weeks, presumably because they had not developed nodal roots and could not survive increasing temperatures supported only by seminal roots. During emergence of plants with the panicoid-type morphology, the coleoptile and shoot were elevated to the soil surface by elongation of the subcoleoptile internode. The coleoptilar node and the origin of the nodal roots were within 6 mm of the soil surface. Although most plants reached the two-tiller stage, development occurred with only seminal roots, as the nodal roots were usually dead stubs.

Without nodal roots, the seedlings could not survive. Even at optimum growing temperatures, blue grama seedlings died after 9 weeks when grown only on the seminal primary root (Weaver and Zink 1945). These observations led to the assumption that there might be an inherent limit to the longevity of blue grama's seminal primary root. Van der Sluijs and Hyder (1974) tested the assumption in a greenhouse experiment by excising all adventitious roots as they developed. After 22 weeks they concluded that there was no inherent limit to seminal root longevity. Tillering began at 3 weeks and continued at a linear rate through 112 days. Maximum length of green leaves on primary shoots was attained at about 40 days and decreased slowly thereafter as the older leaves died. Nevertheless, with the continuation of tillering, the total length of green leaves per seedling increased in a linear progression through 90 days. Van der Sluijs and Hyder hypothesized that termination of an increase in green leaf tissue should reveal the maximum leaf tissue that could be supported by water flow from the seminal root through the subcoleoptile internode, and that the maximum leaf area that could be supported by a seminal root in moist soil would decrease as atmospheric drought increased. Wilson et al. (1976) defined the "growth drought resistance" of blue grama seedlings, which had only seminal roots, as the environmental drought just sufficient to halt the expansion of leaf area. At the seminal-root stage of development, the constant for growth drought resistance was calculated from measurements of evaporation, water uptake, and maximum leaf area. They found that seedlings avoided drought by increasing water uptake and/or by adjusting leaf area to reduce the transpiration rate. Their findings and equations were expanded in subsequent publications (Wilson and Sarles 1978, Wilson and Briske 1978, Wilson and Briske 1979, Wilson 1981, Wilson 1984). The essential conclusion is this: blue grama seedlings, and surely most warm-season grass seedlings with a seminal root system in moist soil, will expand leaf area to the maximum that can be supported at the minimum transpiration rate, which rate is determined by water flow through the subcoleoptile internode. When maximum sustainable leaf area is maintained, a slow and gradual increase in atmospheric drought causes leaf area adjustment (increasing death rate of older leaves), whereas a sudden increase in atmospheric drought causes death.

Nodal roots are essential for increasing water uptake, relieving water stress, and assuring survival. Blue grama achieved the capacity for nodal roots when they were 2 to 3 weeks of age; but their nodal roots could not enter the soil and continue growth, since soil moisture was inadequate (nodal root growth requires the soil to be moist for 2 to 3 days). The following characteristics should be used in selecting and breeding warm-season grass varieties for improved establishment: (1) increased coleoptile length, (b) increased diameter of the subcoleoptile internode and concomitantly increased cross-sectional area of the xylem vessel, (c) possibly a slower rate of leaf area expansion, (d) increased maximum sustainable water stress, (3) better capacity for growing nodal roots at a younger age, and perhaps other characteristics not yet defined.

Festucoid – Crested wheatgrass

The festucoid type morphology imparts an advantage to seedling establishment via an elongated coleoptile and lack of an elongated subcoleoptile internode (Hyder et al. 1971, Hyder 1974, Johnson 1986). Elongation of the intracoleoptile internode may also occur but does not affect emergence. Seminal primary, seminal lateral, and nodal roots develop at or near the planting depth of the seed and escape, to some degree, the harsh environment near the soil surface. Rogler (1954) noted that when seeds of fairway crested wheatgrass and standard crested wheatgrass were planted deeper than 25 mm, elongation of both the coleoptile and the subcoleoptile internode occurred. The length of the subcoleoptile internode ranged from 12 to 50 mm at the 75-mm depth of planting. Crested wheatgrass seedlings survived better than blue grama seedlings for two major reasons: the wheatgrass seedlings have a longer lived seminal primary root and have nodal roots that have a deeper origin (Hyder et al. 1971).

Hassanyar and Wilson (1978) stressed the importance of seminal lateral roots for seedling survival of crested wheatgrass. When temporary droughts of –37, –91, and –158 Mpa, were imposed on seedlings of standard crested wheatgrass, 75, 58, and 24% of the seedlings developed seminal lateral roots, respectively. Bassiri et al. (1988b) reported that survival of wheatgrass seedlings depended on the development of seminal lateral roots from the scutellar node when seminal primary roots were killed by dehydration. When seminal primary and seminal lateral roots of seedlings were dependent upon a similar one-time water application for the first 60 days of growth, both types of roots penetrated to the same depth and were similarly effective in extracting water from the upper 500 mm of the soil profile (Bassiri et al. 1988a).

Morphology and Establishment of Other Grasses

Watt (1981) examined several grasses to determine if a consistent relationship could be found between root morphology and seedling establishment. Species listed in order of decreasing establishment capability included *Dichantium sericeum* L., *Danthonia linkii* Kunth, *Bothriochloa macra* (Steud), S. T. Blake, *Panicum decompositum* R.Br., *Setaria pophyrantha* Stapf, kleingrass, buffelgrass (*Cenchrus ciliaris* L.), and rhodesgrass (*Chloris gayana* Kunth). Nodal root growth was restricted in the native grasses due to dry surface soils. However, the native grasses compensated by having a better developed seminal primary root system and a less well-developed nodal root system. Also the native grasses appeared to have a lower ratio of top growth to root growth, and thus their roots should have been able to supply the plant better during periods of moisture stress.

Morphology and Establishment of Dicots

Studies of the relationship between seedling morphology and survivability have been extended to the dicots, alfalfa and cicer milkvetch (*Astragalus cicer*), both of which are important forage legumes for rangelands (Bassiri et al. 1988a and b). The alfalfa seedlings emerged faster under dehydration treatments and gained more phytomass than milkvetch. Both species had similar patterns of root development, although cicer roots generally formed more branches. In tests of plant ability to produce a replacement taproot which would be needed if the primary root was killed by drought, the investigators found that replacement roots from alfalfa grew as long, and were as effective in water uptake, as the first root. Cicer replacement roots were less effective in water uptake; although they were greater in number, they did not penetrate to the same depth as the first roots.

The observed tendency for alfalfa roots to branch less often than those of the milkvetch indicates that resources were apportioned toward fewer, but deeper roots—a factor which may partially account for alfalfa's superior seedling survival. However, deep rooting morphology is not an assurance of success. Frazer and Davis (1988) compared seedling survival of *Ceanothus megacarpus* Nutt. and *Rhus laurina* Nutt. and found that survival rates were 54% for the former but only 0.1% for the latter. The higher mortality of *Rhus* was attributed to greater sensitivity of its tissues to water stress, since seedlings of the two species appeared to have similar rooting depths.

Morphology, Synecology, and Survival

Seedling survival requires that resources for plant growth be available and obtainable by the plant on a more or less sustained basis during the growing season. Competition occurs when the demands of neighboring plants exceed the resource supply. Competition induces stress and results in the eventual elimination of the less effective neighbors. Competitive stress [also called density stress (Harper 1977)] affects morphology. Gorham (1979) reviewed data on 19 stands of 15 species and analyzed these for effects of plant density on both shoot height and shoot dry weight. He found that plant density explained 81% of the variance in shoot height and 98% of the varience in shoot weight. Besides showing effects of competition on morphology, Gormham's work is evidence that in crowded, monospecific stands, the self-thinning rule developed by Yoda et al (1963). (also called the -3/2 power rule of self-thinning) has general application to the plant kingdom (Weller 1989). The rule has also been confirmed for mixed stands of forest trees (White 1981) indicating it may have application to competition among different species where plant density exceeds what the environment can support.

High seedling density can have a positive influence on seedling survival. The probability of a bitterbrush seedling surviving the first growing season increases when it is part of a group of seedlings, as occurs when cached seeds germinate (Ferguson and Basile 1967). The increased survival rate may be related to group emergence (Booth 1980) and resulting morphology, to the microsites chosen by caching rodents, or to unrecognized factors.

Competitive effectiveness and survivability of a seedling may result from any number of endogenous and exogenous factors (Samuel and DePuit 1987). These factors may include a larger embryo, advantageous microsite (perhaps secured by unique diaspore morphology), opportune timing of dispersal and of germination, faster seedling development, or other superior adaptations of a physiological, morphological, or behavioral nature. Early and rapid germination increases the probability that plant roots will grow and expand in the soil profile ahead of soil drying. Within a species, larger seed size and the relatively greater capacity to capture limiting resources, is a preponderant aspect of survival morphology.

Among neighboring plants of different species, the competitive importance of seed size and growth rate remains but is less distinct because it interacts with other adaptive characteristics. Early germination and taller leaves and cotyledons allowed *Plantago lanceolata* L. seedlings to establish and survive in stands of grass where *Plantago major* L. seedlings could not (Van der Toorn & Pons 1988). Tallness was also a factor in survival of bigleaf maple (*Acer Macrophyllum* Pursh) seedlings (Fried et al. 1988). However, tall seedlings are often more susceptible to grazing, and grazing is known to reduce the competitive pressure exerted by dominant plants (Svejcar et al. 1988).

Below ground, rates of primary root elongation—particularly at low temperature—has a major influence on competition and survivability. Harris and coworkers (Harris 1967, Harris and Wilson 1970, DeWitt 1969, Harris 1977, Harris and Goebel 1976) have studied the competitive relationships among annuals and perennial wheatgrasses. Bluebunch wheatgrass [*Agropyron spicatum* (Pursh) Scribn. & Smith] roots grow slowly in the fall and cease growth in the winter. Adventitious (nodal) roots develop later and grow slower than roots of competing annuals. In contrast, cheatgrass and medusahead (*Taeniatherum asperum* Nevski) begin post-germination growth at a rapid rate and continue growing all winter. At 2°C the primary roots of standard crested wheatgrass grow at a rate similar to that of annual grasses, making it a more successful competitor than blue bunch. Svejcar (1990) also found cheatgrass seedlings produced leaf area and root length more efficiently than 'Nordan' crested wheatgrass seedlings. In another study, root and shoot development were generally similar for seedlings of cheatgrass, Hycrest (*Agropyron desertorum* x *Agropyron cristatum*), Whitmar (a cultivar of bluebunch wheatgrass), and Secar—a cultivar of Snake River wheatgrass [*Elymus lanceolatus* (Scribner & J.G. Smith) Gould] but, they differed significantly in their cumulative growing degree day requirement for initiating seminal and nodal root elongation and root branching (Aguirre 1989, Aguirre and Johnson 1991a). Cheatgrass elongated adventitious roots earlier and at colder temperatures than Hycrest, and Whitmar seedlings did not produce adventitious roots at low temperatures (Aguirre and Johnson 1991b). Order of branching of roots associated with the first foliar node was also greater in cheatgrass than in Hycrest or Whitmar at all water applications and dates (Johnson and Aguirre 1991). Seedlings of cheatgrass may be better suited for growth with limited moisture compared to Hycrest and Whitmar seedlings because of their greater: (1) order of branching of seminar roots, (2) branching density on the main axis, and (3) length of lateral roots and external–internal links (external links = the segment between a meristem and a branch junction; internal links = the segment between two junctions). Hycrest seedlings with greater root branching densities, lateral root lengths, and external-external link lengths) were able to grow better than Whitmar seedlings with limited moisture. Different root morphologies is also a factor affecting seedling growth of the panicoid, guinea grass (*Panicum maximum* Jacq.), in a native *Heteropogon* grassland community that was given supplementary irrigation (Cook and Ratcliff 1984).

Factors that prevent seedling establishment or decrease seedling survival vary greatly from place to place and year to year (Fenner 1987, Durrant et al. 1988). Occurrence of these factors result in natural selection for different genetic qualities and tend to maintain a high degree of genetic diversity within populations relying upon seeds for plant recruitment.

Watt and Whalley (1982a and b) found that no single seed or seedling feature could be identified as essential for the establishment of grasses on the self-mulching black soils of the northwestern slopes of New South Wales. They suggested that the sequence of events from seed imbibition to final establishment is a continuum and that different seed and seedling features are important at different stages. Advantageous features at one stage compensate for weaknesses at another stage. Fast root and shoot growth and the ability to emerge from deep plantings may compensate for slow germination. Poor seedling survival in one case might be explained by slow emergence caused by deep planting and the associated short primary root system. Good seedling survival in another case might result from increased moisture and protection occurring with deep planting.

The requirements of each species for germination and establishment must be considered. Moreover they may be thought of as having temporal as well as spatial aspects; the temporal aspect of the regeneration may be the more important to niche occupation and the relative abundances of species and their coexistence (Fowler 1986). The precise timing of the decline in water potential determines whether seeds will remain imbibed and ungerminated or whether radicle protrusion will commence (Marshall and Naylor 1985). The resulting effect may be only a temporary cessation in germination and seedling emergence but may cause a decline in the proportion of seeds able to continue or resume germination. Thus, successful establishment and survival requires the occurrence of germination and emergence and the development of critical morphological and physiological components necessary for existence within genetic and environmental constraints.

For successful seedling recruitment the required events must occur within a temporal framework consistent with opportunity "windows" allowed by the climate and by the associated species in the community. When environmental conditions are adequate, morphological differences are not critical. However, environmental stress is normal on rangelands. Therefore diaspore morphology and function, germination modes, and seedling growth and morphological development, with and without stress, need to be better understood. This information can then be used by plant breeders, land managers, and others who seek to understand and wisely use our renewable rangeland resources.

Glossary

Acropetal—Development produced in succession from the base toward the apex or top.

Adenosine Phosphates—Adenosine diphosphate (ADP), monophosphate (AMP), and triphosphate (ATP) are compounds composed of adenine, d-ribose, and the respective phosphate. The phosphate bonds provide energy to drive important chemical reactions of cell metabolism.

Adenylate Energy Charge—The amount of metabolically available energy in the adenosine phosphates of a living cell.

Autotrophic—Capable of producing food from inorganic materials, as is done by chlorophyll-containing plants.

Caryopsis—A dry one-seeded indehiscent fruit in which the seed coat and fruit wall (pericarp) are fused, as in the grasses.

Cell-wall Yield—The plastic expansion of a cell wall during cell elongation.

Circadian Rhythms—Movement with a regular repetition occurring at intervals of about 24 hours.

Coleoptile—A sheath enclosing the grass shoot.

Cryptogeal Germination—Germination well below the soil surface and within the protecting confines of elongated cotyledon bases and pseudo-radicle. Regarded as an adaptation to protect seedlings from fire and drought.

Diaspore—The seed-containing unit of a plant that functions in seed dispersal, embryo protection, and in promoting the establishment and survival of the seedling.

Differentiation—The physiological and morphological changes that occur as meristematic tissue becomes specialized for specific functions.

Dimorphic—The occurrence of two types of structures, as two types of seeds or two types of leaves, within one genotype.

Embryo—The rudimentary plant within a seed.

Embryo Axis—The central structure of the embryo extending lengthwise from the tip of the embryonic shoot to the radicle (root) cap.

Endosperm—The nutritive tissue formed within the embryo sac of seed plants. Compare with perisperm.

Epiblast—A small appendage of grass embryos. See scutellum and Figure 6, page 261.

Epicotyl—The portion of a plant embryo or seedling above the cotyledons that produces the shoot. See plumule.

Etiolation—The result of plant development with insufficient light.

Heterocarpy—The occurrence of more than one type of fruit within a genotype. Comparie with dimorphic.

Heterotrophic—Requiring preformed organic compounds as energy sources. Embryonic plants are heterotrophic.

Hydraulic conductivity—The rate at which water moves through a matrix (soil), or between matrices (soil to seed).

Mesic—Characterized by a moderate amount of moisture.

Mitochondria—Cell organelles where oxidative phosphorylation occurs.

Niche—A suitable place or position. In ecology it refers to the place occupied by an organism relative to other organisms in a particular environment.

Nodal—A node is a joint in a stem. Nodal refers to such a region.

Nucellus—The central part of the ovule in which the embryo sac originates. The nucellus does not develop as a result of fertilization of the oosphere as is the case for the embryo and the endosperm.

Ovule—The part of a plant that develops into a seed.

Oxidation—The loss of electrons from a substance. Oxidative phosphorylation is the process in which ATP is formed as electrons are transferred from NADH or FADH to oxygen by a series of reactions known as the electron transport chain.

Perisperm—Food storage tissue derived from the nucellus.

Phosphorylation—A reaction that generates a high-energy phosphate compound such as ATP.

Photoblastic—The condition in which seed germination is controlled by light. Positive photoblastic seeds require light to germinate.

Photomorphogenesis—Light-stimulated alterations in plant development.

Photosynthate—The product of photosynthesis.

Phototaxis—The orientation of various chloroplasts to light.

Phototropism—Bending or turning in response to light.

Plumule—A bud at the top of the embryo axis. The term is used interchangeably with epicotyl.

Polymorphic—Existing in more than one form.

Polyploidy—Having a chromosome number that is in excess of the normal diploid.

Primordia—Organized cells in their earliest stage of differentiation.

Radicle—The embryonic root.

Root Acidification—The excretion of hydrogen ions (organic acids) by rootlets. This increases the solubility of soil nutrients making them more available for plant use.

Scutellum—The cotyledon of grass embryos. Some interpret the epiblast to be the other cotyledon (Esau 1965). See Figure 6, page 261.

Seed—A mature ovule consisting of an embryo enclosed within a seed coat or coats. Stored food in the form of endosperm or perisperm may be present.

Self-thinning Rule—The rule relates plant density (p) (declining over time) to the mean biomass (w) (increasing over time) of surviving plants in monospecific populations undergoing density-dependent mortality: log w –log K –1.5 log p (White 1981).

Seminal—Pertaining to seed.

Stress—A condition imposed by environment that acts against optimum plant growth.

Suffrutescent—Having a woody base.

Synaptospermy—Clustering or adhering of diaspores to form groups (Ridley 1930, van der Pijl 1982) or more than one closely associated seeds in a diaspore (Quinn 1987).

Testa—The integument(s) surrounding the embryo and any nutritive tissue. The seed coat.

Tetrazolium Test—A viability test used for dormant seeds. Active dehydrogenase found in live tissue will reduce tetrazolium to a bright red form. The degree of red stain in a sectioned seed is an indication of seed viability.

Xeric—Characterized by relatively little moisture.

Literature Cited

Abernethy, R.H., L.N. Wright, and K. Matsuda. 1977. Association of seedling respiratory metabolism and adenylate energy charge with seed weight of *Panicum antidotale* Retz. Crop Sci. 17:563-566.

Aguirre, L. 1989. Early seedling development in three range grasses as influenced by temperature, drought, and competition. Ph.D. Diss., Utah State Univ. Logan, UT.

Aguirre, L. and D.A. Johnson. 1991a. Root morphological development in relation to shoot growth in seedlings of four range grasses. J. Range Manage. 44:341-346.

Aguirre, L. and D.A. Johnson. 1991b. Influence of temperature and cheatgrass competition on a seedling development of two bunchgrasses. J. Range Manage. 44:347-354.

Anderson, R.C. 1985. Aspects of the germination ecology and biomass production of eastern gamagrass (*Tripsacum dactyloides* L.). Bot. Gaz. 146:353-364.

Anslow, R.C. 1966. Report of the crop physiology department. Ann. Rep. for the Grassland Res. Inst., Hurley, 1966.

Arndt, W. 1965a. The nature of the mechanical impedance to seedlings by soil surface seals. Aust. J. Soil Res. 3:45-54.

Arndt, W. 1965b. The impedance of soil surface seals and the forces of emerging seedlings. Aust. J. Soil Res. 3:55-68.

Arnott, R.A. 1969. The effect of seed weight and depth of sowing on the emergence and early seedling growth of perennial ryegrass (*Lolium perenne*). J. Brit. Grassl. Soc. 24:104-110.

Asay, K.H., and D.A. Johnson. 1980. Screening for improved stand establishment in Russian wild ryegrass. Can. J. Plant Sci. 60:1171-1177.

Asay, K.H., and D.A. Johnson. 1983. Genetic variability for characters affecting stand establishment in crested wheatgrass. J. Range Manage. 36:703-706.

Atwater, B.R. 1980. Germination, dormancy and morphology of the seeds of herbaceous ornamental plants. Seed Sci. & Technol. 8:523-573.

Augspurger, C.K., and S.E. Franson. 1987. Wind dispersal of artificial fruits varying in mass, area and morphology. Ecology 68:27-42.

Austenfeld, F. 1988. Seed dimorphism in *Salicornia europeae*: Nutrient reserves. Physiol. Plant. 73:502-504.

Avery, G.S., Jr. 1930. Comparative anatomy and morphology of embryos and seedlings of maize, oats, and wheat. Bot. Gaz. 89:1-39.

Baker, H.G. 1972. Seed weight in relation to environmental conditions in California. Ecology 53:997-1010.

Bassiri, M., A.M. Wilson, and B. Grami. 1988a. Root excision and dehydration effects on water uptake in four range species. J. Range Manage. 41:378-382.

Bassiri, M., A.M. Wilson, and B. Grami. 1988b. Dehydration effects on seedling development of four range species. J. Range Manage. 41:383-386.

Baskin, J.M. and C.C. Baskin. 1986. Germination ecophysiology of the mesic deciduous forest herb *Isopyrum biternatum*. Bot. Gaz. 147:152-155.

Benedict, H.M. 1941. Growth of some range grasses in reduced light. Bot. Gaz. 102:582-589.

Bennett, J.H, N.J. Chatterton, and P.A. Harrison. 1988. Rhizosphere physiology of crested wheatgrass and legume seedlings: root-shoot carbohydrate interactions. J. Plant Nut. 11:1099-1116.

Berlyn, G.P. 1972. Seed germination and morphogenesis. p. 223-312. *In*: T.T. Kozlowski (ed.) Seed Biology, Vol. I. Academic Press, New York, NY.

Beveridge, J.L. and C.P. Wilsie. 1959. Influence of depth of planting, seed size, and variety on emergence and seedling vigor in alfalfa. Agron. J. 51:731-734.

Bewley, J.D. and M. Black. 1978. Physiology and Biochemistry of Seeds, Vol. I. Springer-Verlag, New York.

Bewley, J.D., and M. Black. 1982. Physiology and Biochemistry of Seeds. Vol. II Springer-Verlag, New York.

Black, J.N. 1956. The influence of seed size and depth of sowing on pre-emergence and early vegetative growth of subterranean clover (*Trifolium subterraneum* L.). Aust. J. Agr. Res. 7:98-109.

Booth, D.T. 1980. Emergence of bitterbrush seedlings on land disturbed by phosphate mining. J. Range Manage. 33:439-441.

Booth, D.T. 1982. The ecological, physiological and morphological effects of threshing winterfat fruits. p. 133-137. *In*: E.F. Aldon and W.R. Oaks (eds.) Proc. Reclamation of Mined Lands in the Southwest. Soil Con. Soc. Amer.—New Mexico Capter. P.O. Box 2142, Albuquerque, NM 87102.

Booth, D.T. 1984. Threshing damage to radicle apex affects geotropic response of winterfat. J. Range Manage. 37:222-225.

Booth, D.T. 1987. Contributions to the reproductive autecology of winterfat [*Eurotia lanata* (Pursh) Moq.] with notes on direct seedling methods. Ph.D. Diss. Univ. of Wyoming, Laramie, WY. Diss. Abstr. 48:1568-B.

Booth, D.T. 1988. Winterfat diaspore morphology. J. Range Manage. 41:351-353.

Booth, D.T. 1990. Seedbed ecology of winterfat: effects of mother-plant transpiration, wind stress and nutrition on seedling vigor. J. Range Manage. 43:20-24.

Booth, D.T. 1990. Plant diaspore functions. J. Seed Tech. 14:61-73.

Booth, D.T., and G.E. Schuman. 1983. Seedbed ecology of winterfat: fruits versus threshed seeds. J. Range Manage. 36:387-390.

Booth, D.T., and D.R. Morgen. 1992. Post-germination growth related to time-to-germination for four woody plants. J. Seed Tech. 16:30-38.

Boyd, L., and G.S. Avery, Jr. 1936. Grass seedling anatomy: The first internode of *Avena* and *Triticum*. Bot. Gaz. 97:765-779.

Briske, D.D. and A.M. Wilson. 1977. Temperature effects on adventitious root development in blue grama seedlings. J. Range Manage. 30:276-280.

Briske, D.D., and A.M. Wilson. 1978. Moisture and temperature requirements for adventitious root development in blue grama seedlings. J. Range Manage. 31:174-178.

Briske, D.D., and A.M. Wilson. 1980. Drought effects on adventitious root development in blue grama seedlings. J. Range Manage. 33:323-327.

Brown, W.V. 1960. The morphology of the grass embryo. Phytomorphology 10:215-223.

Brown, W.V. 1965. The grass embryo—a rebuttal. Phytomorphology 15:274-284.

Bula, R.J., C.L. Rhykerd, and R.G. Langston. 1959. Growth responses of alfalfa seedlings under various light regimes. Agron. J. 51:84-86.

Burrows, F.M. 1973. Calculation of the primary trajectories of plumed seeds in steady winds with variable convection. New Phytol. 72:647-664.

Cardoso, V.J.M., and G.M. Felippe. 1988. The action of the testa upon the germination of seeds of *Cucumis anguria* L. Biologia Plantarum. 30:48-52.

Carleton, A.E. and C.S. Cooper. 1972. Seed size effects upon seedling vigor of three forage legumes. Crop Sci. 12:183-186.

Carren C.J., A.M. Wilson, R.L. Cuany, and G.L. Thor. 1987a. Caryopsis weight and planting depth of blue grama I. Morphology, emergence, and seedling growth. J. Range Manage. 40:207-211.

Carren, C.J., A.M. Wilson, and R.L. Cuany. 1987b. Caryopsis weight and planting depth of blue grama II. Emergency in marginal soil moisture. J. Range Manage. 40:212-216.

Cavers, P.B. and J.L. Harper. 1966. Germination polymorphism in *Rumex crispus* and *Rumex obtusifolius*. J. Ecology 54:367-382.

Cavers, P.B. and M.G. Steel. 1984. Patterns of change in seed weight over time on individual plants. The Am. Nat. 124:324-335.

Choe, H.S., C. Chu, G. Koch, J. Gorham, and H.A. Mooney. 1988. Seed weight and seed resources in relation to plant growth rate. Oecologia 76:158-159.

Clarkson, J.R. and H.T. Clifford. 1987. Germination of *Jedda multicaulis* J.R. Clarkson (Thymelaeaceae). An example of cryptogeal germination in the Australian flora. Aust. J. Bot. 35:715-720.

Clor, M.A., T.A. Al-ani, and F. Charchafchy. 1976. Germinability and seedling vigor of *Haloxylon salicornicum* as affected by storage and seed size. J. Range Manage. 29:60-62.

Come, D., C. Perino, and J. Ralambosoa. 1985. Oxygen sensitivity of apple (*Pyrus malus* L.) embryos in relation to dormancy. Israel J. Bot. 34:17-23.

Conard, S.G., and S.R. Radosevich. 1982. Growth responses of white fir to decreased shading and root competition by montane chaparral shrubs. Forest Sci. 28:309-320.

Cook, S.J., and D. Ratcliff. 1984. A study of the effects of root and shoot competition on the growth of green panic (*Panicum maximum* var. trichoglume) seedlings in an existing grassland using root exclusion tubes. J. Appl. Ecol. 21:971-982.

Cooper, C.S. 1966. Response of birdfoot trefoil and alfalfa to various levels of shade. Crop Sci. 6:63-66

Cooper, C.S., and S.C. Fransen. 1974. Contribution of cotyledons to growth of the sainfoin seedling. Crop Sci. 14:732-735.

Cooper, C.S., and P.W. MacDonald. 1970. Energetics of early seedling growth in corn (*Zea mays* L.). Crop Sci. 10:136-139.

Copeland, L.O. 1976. Principles of seed science and technology. Burgess Publishing Company, Minneapolis, MN.

Cornish, P.S. 1982. Root development in seedlings of ryegrass (*Lolium perenne* L.) and phalaris (*Phalaris aquatica* L.) sown into the soil surface. Aust. J. Agric. Res. 33:665-677.

Cornish, P.S., J.R. McWilliam, and H.B. So. 1984. Root morphology, water uptake, growth and survival of seedlings of ryegrass and phalaris. Aust. J. Agric. Res. 35:479-492.

Cresswell, E.G. and J.P. Grime. 1981. Induction of a light requirement during seed development and its ecological consequences. Nature 291:583-585.

Curtis, K. and B.D. McKersie. 1984. Growth potential of the axis as a determinant of seedling vigor in birdsfoot trefoil. Crop Science 24:47-50.

Dalianis, C.D. 1980. Effect of temperature and seed size on speed of germination, seedling elongation and emergence of berseem and Persian clovers (*Trifolium alexandrinum* and *T. resupinatum)* Seed Sci. & Technol. 8:323-331.

Davies, W. 1927. Seed mixture problems. Soil germination, seeding and plant establishment with particular reference to the effects of environmental and agronomic factors. Bulletin. Welsh Plant Breeding Station, Univ. College of Wales Ser. H. No. 6:39-60.

Dehgan, B. and B. Schutzman. 1989. Embryo development and germination of Cycas seeds. J. Amer. Soc. Hort. Sci. 114:125-129.

Densmore, R. and J. Zasada. 1983. Seed dispersal and dormancy patterns in northern willows: ecological and evolutionary significance. Can. J. Bot. 61:3207-3216.

DeWitt, F.P. 1969. Early developmental response to low temperature in *Agropyron spicatum* collected from native stands in western North America. M.S. Thesis. Washington State Univ. Pullman, WA.

Durrant, M.J., R.A. Dunning, K.W. Jaggard, R.B. Bugg, and R.K. Scott. 1988. A census of seedling establishment in sugar-beet crops. Ann. Appl. Biol. 113:327-345.

Esau, K. 1965. Plant anatomy, second edition. John Wiley and Sons, Inc., New York.

Evans, M.W. 1946. The grasses: their growth and development. Agronomy Mimeograph No. 105. Ohio Agr. Exp. Sta., Wooster, Ohio.

Evans, R.A., J.A. Young, and B.A. Roundy. 1977. Seedbed requirements for germination of Sandberg bluegrass. Agron. J. 69:817-820.

Everitt, J.H. 1983a. Seed germination characteristics of two woody legumes (Retama and Twisted Acacia) from South Texas. J. Range Manage. 36:411-414.

Everitt, J.H. 1983b. Seed germination characteristics of three woody plant species from South Texas. J. Range Manage. 36:246-249.

Everitt, J.H., M.A. Alaniz, and J.B. Lee. 1983. Seed germination characteristics of *Kochia scoparia*. J. Range Manage. 36:646-648.

Evers, G.W. 1982. Seedling growth and nodulation of arrowleaf, crimson, and subterranean clovers. Agron. J. 74:629-632.

Fendall, R.K. 1964. An investigation into the site and cause of seed dormancy of *Stipa viridula* and *Oryzopsis hymenoides*. Ph.D. Diss. North Dakota State Univ. Fargo, ND.

Fenner, M. 1983. Ash content and seedling growth in twenty-four species of Compositae. New Phytol. 95:697-706.

Fenner, M. 1987. Seedlings. New Phytol. 106 (suppl.):35-47.

Ferguson, R.B. and J.V. Basile. 1967. Effect of seedling numbers on bitterbrush survival. J. of Range. Manage. 20:380-382.

Fowler, N.L. 1986. Microsite requirements for germination and establishment of three grass species. Amer. Midland Nat. 115:131-145.

Fransen, S.C., and C.S. Cooper. 1976. Seed weight effects upon emergence, leaf development, and growth of the sainfoin (*Onobrychis* ssp.) seedling. Crop Sci. 16:434-437.

Frasier, G.W., J.R. Cox, and D.A. Woolhiser. 1985. Emergence and survival response of seven grasses for six wet-dry sequences. J. Range Manage. 38:372-377.

Frasier, G.W., J.R. Cox, and D.A. Woolhiser. 1987. Wet-dry cycle effects on warm-season grass seedling establishment. J. Range Manage. 40:2-6.

Frasier, G.W., D.A. Woolhiser, and J.R. Cox. 1984. Emergence and seedling survival of two warm-season grasses as influenced by the timing of precipitation: A greenhouse study. J. Range Manage. 37:7-11.

Frazer, J.M. and S.D. Davis. 1988. Differential survival of chaparral seedlings during the first summer drought after wildfire. Oecologia 76:215-221.

Fried, J.S., J C. Tappeiner, II, and D.E. Hibbs. 1988. Bigleaf maple seedling establishment and early growth in Douglas-fir forests. Can. J. For. Res. 18:1226-1233.

Friedman, J. and G. Orshan. 1975. The distribution, emergence and survival of seedlings of *Artemisia herba-alba* Asso in the Negev desert of Israel in relation to distance from the adult plants. J. Ecol. 63:627-632.

Frischknecht, N.C. 1951. Seedling emergence and survival of range grasses in central Utah. Agron. J. 43:177-182.

Fulbright, T.E., A.M. Wilson, and E.F. Redente. 1984. Effects of temporary dehydration on growth of green needlegrass (*Stipa viridula* Trin.) seedlings. J. Range Manage. 37:462-464.

Fulbright, T.E., A.M. Wilson, and E.F. Redente. 1985. Green needlegrass seedling morphology in relation to planting depth. J. Range Manage. 38:266-270.

Geier-Hays, K. 1987. Occurrence of conifer seedlings and their microenvironments on disturbed sites in central Idaho. Research Paper INT-383. Ogden, UT: U.S. Dept. Agric., Forest Service, Intermountain Research Station.

Gerard, C.J. 1980. Emergence force by cotton seedlings. Agronomy J. 72:473-476.

Gorham, E. 1979. Shoot height, weight and standing crop in relation to density of monospecific plant stands. Nature 279:148-150.

Green, N.E., and R.M. Hansen. 1969. Relationship of seed weight to germination of six grasses. J. Range Manage. 22:133-134.

Griffith, L.W., and D.T. Booth. 1988. Indian ricegrass seed damage and germination responses to mechanical treatments. J. Range Manage. 41:335-337.

Grime, J.P., G. Mason, A.V. Curtis, J. Rodman, S.R. Band, M.A.G. Mowforth, A.M. Neal and S. Shaw. 1981. Germination characteristics in a local flora. J. Ecology 69:1017-1059.

Gutterman, Y., 1980. Influences on seed germinability: Phenotypic maternal effects during seed maturation. Israel J. Bot. 29:105-117.

Gutterman, Y., A. Witztum, and M. Evenari. 1967. Seed dispersal and germination in *Blepharis persica* (Burm.) Kunte. Israel J. Bot. 16:213-234.

Hadas, A., and D. Russo. 1974a. Water uptake by seeds as affected by water stress, capillary conductivity and seed-soil water contact. I. Experimental Study. Agron. J. 66:643-647.

Hadas, A. and D. Russo. 1974b. Water uptake by seeds as affected by water stress, capillary conductivity and seed-soil water contact. II. Analysis of experimental data. Agron. J. 66:647-652.

Harper, J.L. 1977. Population biology of plants. Academic press. New York.

Harper, J.L., and R.A. Benton. 1966. The behavior of seeds in soil II. the germination of seeds on the surface of a water supplying substrate. J. Ecol. 54:151-166.

Harper, J.L. and M. Obeid. 1967. Influence of seed size and depth of sowing on the establishment and growth of varieties of fiber and oil seed flax. Crop Sci. 7:527-532.

Harper, J.L., P.H. Lovell, and K.G. Moore. 1970. The shapes and sizes of seeds,. pp. 327-356. *In*: Johnston, Richard F. et al. (eds.) Annual review of Ecology and Systematics. Vol. I. Annual Reviews Inc. Palo Alto, CA,

Harper, J.L., J. T. Williams, and G. T. Sagar. 1965. The behavior of seeds in soil. The heterogeneity of soil surfaces and its role in determining the establishment of plants from seed. J. Ecol. 53:273-286.

Harris, G.A. 1967. Some competitive relationships between *Agropyron spicatum* and *Bromus tectorum*. Ecol. Mongr. 37:89-111.

Harris, G.A. 1977. Root phenology as a factor of competition among grass seedlings. J. Range Manage. 30:172-177.

Harris, G.A., and C.J. Goebel. 1976. Factors of plant competition in seeding Pacific Northwest bunchgrass ranges. College of Agriculture Research Center Bull. 820. Washington State University. Pullman, WA.

Harris, G.A., and A.M. Wilson. 1970. Competition for moisture among seedlings of annual and perennial grasses as influenced by root elongation at low temperature. Ecology 51:530-534.

Haskins, F.A. and H.J. Gorz. 1975. Influence of seed size, planting depth, and companion crop on emergence and vigor of seedlings in sweetclover. Agron. J. 67:652-654.

Hassanyar, A.S., and A.M. Wilson. 1978. Drought tolerance of seminal lateral root apices in crested wheatgrass and Russian wildrye. J. Range Manage. 31:254-258.

Hayes, P. 1976. Seedling growth of four grasses. J. Brit. Grassl. Soc. 31:59-64.

Heydecker, W. 1973. Seed Ecology, pp 1-3 In: Heydecker, W. (ed.), Seed Ecol. Proc. Nineteenth Easter School in Agricultural Sciences. Univ. of Nottingham, Penn. State Univ. Press, London.

Hill, M.J., C.J. Pearson, and A.C. Kirby. 1985. Germination and seedling growth of prairie grass, tall fescue and Italian reygrass at different temperatures. Aust. J. Agric. Res. 36:13-24.

Hoshikawa, K. 1969. Underground organs of the seedlings and the systematics of gramineae. Bot Gaz. 130:192-203.

Hsu F.H., and C.J. Nelson. 1986a. Planting date effects on seedling development of perennial warm-season forage grasses. I. Field emergence. Agron. J. 78:33-38.

Hsu F.H., and C.J. Nelson. 1986b. Planting date effects on seedling development of perennial warm-season forage grasses. II. Seedling growth. Agron. J. 78:38-42.

Hsu, F.H., C.J. Nelson, and A.G. Matches. 1985a. Temperature effects on germination of perennial warm-season forage grasses. Crop Sci. 25:215-220.

Hsu, F.H., C.J. Nelson, and A.G. Matches. 1985b. Temperature effects on seedling development of perennnial warm-season forage grasses. Crop Sci. 25:249-255.

Hunt, O.J., and D.G. Miller. 1965. Coleoptile length, seed size, and emergence in intermediate wheatgrass [*Agropyron intermedium* (Host) Beauv.]. Agron. J. 57:192-195.

Hyder, D.N. 1974. Morphogenesis and management of perennial grasses in the United States, p. 89-98. *In*: K.W. Kreitlow and R.H. Hart (eds.), Plant morphogenesis as the basis for scientific management of range resources. Proc. The Workshop of the United States-Australia Rangelands Panel. USDA Misc. Publ. No. 1271.

Hyder, D.N., A.C. Everson, and R.E. Bement. 1971. Seedling morphology and seeding failures with blue grama. J. Range Manage. 24:287-292.

Inouye, J., S. Tanakamaru, and K. Hibi. 1979. Elongation force of seedlings of leguminous crops. Crop Science 19:599-602.

Jackson, G. 1974. Cryptogeal germination and other seedling adaptations to the burning of vegetation in savanna regions: The origin of the pyrophytic habit. New Phytol. 73:771-780.

Janen, G.H. 1971. Seed predation by animals, pp. 465-492. *In*: R. F. Johnson (ed.), Annual Review of Ecology and Systematics, Annual Reviews Inc. Palo Alto, CA.

Jensen, E.H., J.R. Frelich, and R.O. Gifford. 1972. Emergence force of forage seedlings. Agron. J. 64:635-639.

Johnson, D.A. 1986. Seed and seedling relations of crested whatgrass: A review, pp. 65-90. *In*: K.L. Johnson (ed.), Crested wheatgrass: its values, problems and myths; symposium proceedings. Utah State Univ., Logan, UT.

Jordan, G.L. 1983. Planting limitations for arid, semiarid, and salt-desert shrublands, pp.11-16. *In*: S.B. Monsen and N. Shaw (compilers), Proc. Managing intermountain rangelands—improvement of range and wildlife habitats. USDA-Forest Serv. Ogden, UT, Gen. Tech. Rep. INT-157.

Kalton, R.R., R.A. de Long, and D.S. McLeod. 1959. Cultural factors in seedling vigor of smooth bromegrass and other forage species. Iowa State J. Sci. 34:47-80.

Kittock, D.L. and A.G. Law. 1968. Relationship of seedling vigor to respiration and tetrazolium chloride reduction by germinating wheat seeds. Agron. J. 60:286-288.

Kittock, D.L. and J.K. Patterson. 1959. Measurement of relative root penetration of grass seedlings. Agron. J. 51:512.

Knapp, A.K. and W.K. Smith. 1982. Factors influencing understory seedling establishment of Engelmann spruce (*Picea engelmannii*) and subalpine fir (*Abies lasiocarpa*) in southeast Wyoming. Canadian J. Bot. 60:2753-2761.

Kneebone, W.R. 1956. Breeding for seedling vigor in sand bluestem (*Andropogon hallii* Hack.) and other native grasses. Agron. J. 48:37-40.

Kneebone, W.R., and C.L. Cremer. 1955. The relationship of seed size to seedling vigor in some native grass species. Agron. J. 17:472-477.

Knipe, O.D. 1970. Large seeds produce more, better alkali sacaton plants. J. Range Manage. 23:369-371.

Knittle, K.H. and J.S. Burris. 1979. Effect of downward force on soybean hypocotyl growth. Crop Sci. 19:47-51.

Koller, D., and N. Roth. 1963. Germination-regulating mechanism in some desert seeds VII. *Panicum turgicum* (Gramineae). Israel J. Bot. 12:64-73.

Koller, D., and N. Roth. 1964. Studies on the ecological and physiological significance of amphicary in *Gymnarrhena micrantha* (Compositae). Amer. J. Bot. 51:26-35.

Kolowski, T.T., and A.C. Gentile. 1959. Influence of the seed coat on germination, water absorption, and oxygen uptake of eastern white pine seed. Forest Sci. 5:389-395.

Larson, Philip R. 1974. The upper limit of seedling growth, pp. 62-76. *In*: Tinus, R.W. et al. (eds.), Proceedings of the North American containerized forest tree seedling syposium. Great Plains Agricultural Council Publication No. 68.

Laude, H.M. 1956. The seedling emergence of grasses as affected by low temperature. Agron. J. 48:558-560.

Laude, H.M., and B.A. Chaugule. 1953. Effect of stage of seedling development upon heat tolerance in bromegrasses. J. Range Manage. 6:320-324.

Lawrence, T. 1963. A comparison of methods of evaluating Russian wild ryegrass for seedling vigor. Can. J. Plant Sci. 43:307-312.

Lawrence, T., and R. Ashford. 1964. Mesocotyl development in Russian wild ryegrass and its effect on survival of shoot apices. Nature 201:727-728.

Lewis, E.J., and J.A. Garcia. 1979. The effect of seed weight and coleoptile tiller development on seedling vigor in tall fescue, *Festuca arundinacea* Schreb. Euphytica 28:393-402.

Liptay, A. and T. Geier. 1983. Mechanism of emergence of tomato (*Lycopersicon esculentum* L.) seedlings through surface-crusted or compressed soil. Ann. Bot. 51:409-412.

Lovell, P.H. and K.G. Moore. 1970. A comparative study of cotyledons as assimilatory organs. J. Exper. Botany 21:1017-1030.

Machado, A.D., W.A. Williams, and C.L. Tucker. 1974. Dry matter contribution by cotyledons of lima beans and other epigeal legumes. Crop Sci. 14:90-93.

Marshall, A.H., and R.E.L. Naylor. 1985. Ryegrass seed-water relationships. Seed Sci. & Technol. 13:829-845.

Martin, A.C. 1946. The comparative internal morphology of seeds. The Amer. Midland Naturalist. 36:513-660.

Maun, M.A., and S. Riach. 1981. Morphology of caryopses, seedlings and seedling emergence of the grass *Calamovilfa longifolia* from various depths in sand. Oecologia (Berl) 49:-137-142.

McDaniel, R.G. 1969. Relationships of seed weight seedling vigor and mitochondrial metabolism in barley. Crop Sci. 9:823-827.

McGinnies, W.J. 1966. Effects of shade on the survival of crested wheatgrass seedlings. Crop Sci. 6:482-484.

McGinnies, W.J. 1973. Effects of date and depth of planting on the establishment of three range grasses. Agron. J. 65:120-123.

McGinnies, W.J. 1974. Effect of planting depth on seedling growth of Russian wildrye (*Elymus junceus* Fisch.). J. Range Manage. 27:305-307.

McWilliam, J.R., R.J. Clements, and P.M. Dowling. 1970. Some factors influencing the germination and early seedling development of pasture plants. Aust. J. Agri. Res. 21:19-32.

Mengel, K. and E.A. Kirkby. 1982. Principles of plant nutrition. International Potash Institute, Worblaufen-Bern, Switzerland.

Merrill, L.B., and V.A. Young. 1962. Germination and root establishment in seedlings of curly mesquitegrass. Texas Agric. Exp. Sta. MP-615.

Mobayen, R.G., 1980. Germination and emergence of citrus and tomato seeds in relation to temperature. J. Hort. Sci. 55:291-297.

Murbeck, S. 1919/20. Beitrage zur biologie der Wustenpflanzen I, II. Lunds University Arsskr 15(10), 17(1).

Nason, D.A., R.L. Cuany, and A.M. Wilson. 1987. Recurrent selection in blue grama. I. Seedling water uptake and shoot weight. Crop Sci. 27:847-851.

Nelson, J.R., A.M. Wilson, and C.J. Goebel. 1970. Factors influencing broadcast seeding in bunchgrass range. J. Range Manage. 23:163-170.

Newman, P.R. 1986. Seedling root development of forage grasses. M.S. Thesis. Univ. Nebraska, Lincoln, Neb.

Newman, P.R., and L.E. Moser. 1988a. Seedling root development and morphology of cool-season and warm-season forage grasses. Crop Sci. 28:148-151.

Newman, P.R., and L.E. Moser. 1988b. Grass seedling emergence, morphology, and establishment as affected by planting depth. Agron. J. 80:383-387.

Ohlenbush, P.D. 1976. Emergence and early growth of two warm-season grasses. Ph.D. Diss. Texas A & M Univ., College Station, TX.

Olmsted, C.E. 1941. Growth and development in range grasses. I. Early development of *Bouteloua curtipendula* in relation to water supply. Bot. Gaz. 102:499-519.

Peart, M.H. 1979. Experiments on the biological significance of the morphology of seed-dispersal units in grasses. J. Ecol. 67:843-864.

Peart, M.H. 1981. Further experiments of the biological significance of the morphology of seed-dispersal units in grasses. J. Ecol. 69:425-436.

Peart, M.H. 1984. The effects of morphology, orientation and position of grass diaspores on seedling survival. J. Ecol. 72:437-453.

Peart, M.H., and H.T. Clifford. 1987. The influence of diaspore morphology and soil-surface properties on the distribution of grasses. J. Ecol. 75:569-576.

Pena, J., P. Aparicio-Tejo, and M. Sanche-Dia. 1988. Dormancy mechanism and the effect of scarification in the germination of *Halimium halimifolium* seeds. J. Plant Physiol. 132:54-58.

Philipupillai, J. and I.A. Ungar. 1984. The effect of seed dimorphism on the germination and survival of *Salicornia europaea* L. populations. Amer. J. Bot. 71:542-549.

Plummer, A.P. 1943. The germination and early seedling development of twelve range grasses. J. Amer. Soc. Agron. 35:19-34.

Pritchett, W.L., and L.B. Nelson. 1951. The effect of light intensity on the growth characteristics of alfalfa and bromegrass. Agron. J. 43:172-177.

Putievsky, E. 1980. Germination studies with seed of caraway, coriander and dill. Seed Sci. & Technol. 8:245-254.

Quinn, J.A. 1987. Relationship between synaptospermy and dioecy in the life history strategies of *Buchloe dactyloides* (Gramineae). Amer. J. Bot. 74:1167-1172.

Reeder, J.R. 1957. The embryo in grass systematics. Amer. J. Bot. 44:756-768.

Rhodes, I. 1968a. The growth and development of some grass species under competitive stress 1. Competition between seedlings, and between seedlings and established plants. J. Brit. Grassl. Soc. 23:129-136.

Rhodes, I. 1968b. The growth and development of some grass species under competitive stress 3. The nature of competitive stress, and characters associated with competitive ability during seedling growth. J. Brit. Grassl. Soc. 23:330-335.

Ridley, H.N. 1930. The dispersal of plants throughout the world. Reeve and Co. Ltd. Ashford. U.K.

Ries, R.E., and L. Hofmann. 1989. Mechanism of emergence and crown placement of three perennial grasses. Abstract No. 056, Abstracts 42nd annu. meeting Soc. for Range Manage. 19-24 February 1989. Billings, MT.

Ries, R.E., and L. Hofmann. 1991. Mechanism of emergence and crown placement of three perennial grasses. J. Range Manage. 44:404-407.

Robbins, W.W., T.E. Weier, and C.R. Stocking. 1965. Botany. John Wiley and Sons Inc. New York.

Robinson, L.R., and H.L. Thomas. 1963. Combining ability for seedling vigour in *Bromus inermis*. Crop Sci. 3:358-359.

Robocker, W.C., J.T. Curtis, and H.L. Ahlgren. 1953. Some factors affecting emergence and establishment of native grass seedlings in Wisconsin. Ecology 34:194-199.

Rogler, G.A. 1954. Seed size and seedling vigor in crested wheatgrass. Agron. J. 46:216-220.

Romo, J.T., and M.R. Haferkamp. 1987. Forage kochia germination response to temperature, water stress, and specific ions. Agron. J. 79:27-30.

Salisbury, F.B. 1974. Effects of light and photoperiod on morphogenesis, pp. 219-232. *In*: K.W. Kreitlow and R.H. Hard (eds.), Plant morphogenesis as the basis for scientific management of range resources. Proc. The Workshop of the United States-Australia Rangelands Panel. USDA Misc. Publ. No. 1271.

Samuel, M.J. and E.J. DePuit. 1987. Competition and plant establishment, pp.138-148. *In*: G.W. Frasier and R.A. Evans (eds.), Proc. seed and seedbed ecology of rangeland plants. USDA-ARS. Dist. by Natl. Tech. Info. Serv., Springfield, VA 22161.

Samuel, M.J., and G.S. Howard. 1982. Botanical Composition of Summer Cattle Diets on the Wyoming High Plains. J. Range Manage. 35:305-308.

Sankary, M.N., and M.G. Barbour. 1972. Autecology of *Haloxylon articularum* in Syria. J. Ecol. 60:697-711.

Schirman, R. 1981. Seed production and spring seedling establishment of diffuse and spotted knapweed. J. Range Manage. 34:45-47.

Schlising, R.A. 1969. Seedling morphology in Marah (Cucurbitaceae) related to the Californian Mediterranean climate. Amer. J. Bot. 56:552-561.

Schmalstig, J.G. and D.J. Cosgrove. 1988. Growth inhibition, turgor maintenance, and changes in yield threshold after cessation of solute import in pea epicotyls. Plant Physiol. 88:1240-1245.

Schopmeyer, C.S. (Tech. Coord.) 1974. Seeds of woody plants in the United States. Agri. Handbook No. 450. Forest Service, USDA. Washington, D.C.

Sheldon, J.C. 1974. The behavior of seeds in soil III. The influence of seed morphology and the behavior of seedlings on the establishment of plants form surface-lying seeds. J. Ecol. 62:47-66.

Sheldon, J.C., and F.M. Burrows. 1973. The dispersal effectiveness of the achene-pappus units of selected Compositae in steady winds with convection. New Phytol. 72:665-675.

Silcock, R.G. 1980. Seedling characteristics of tropical pasture species and their implication for ease of establishment. Trop. Grasslands 14:174-180.

Silvertown, J.W. 1984. Phenotypic variety in seed germination behavior: the ontogeny and evolution of somatic polymorphism in seeds. The Amer. Naturalist 124:1-16.

Simak, M. 1980. Germination and storage of *Salix caprea* L. and *Populus tremula* L. seeds, pp. 142-160. *In*: B.S.P. Wang and J.A. Pitel (eds.), Seed problems, proceedings of the International Symnposium on Forest Tree Seed Storage. Canadian Forestry Service, Ottowa.

Simanton, J.R. and G.L. Jordan. 1986. Early root and shoot elongation of selected warm-season perennial grasses. J. Range Manage. 39:63-67.

Simpson, M. 1952. Value of the awn in establishing seed of *Danthonia penicillata* (Labill.) Palisot. *New Zealand* J. Science and Technology. 34:360-365.

Smith, D.F. 1968. The growth of barley grass (*Hordeum leporinum*) in annual pasture 1. Germination and establishment in comparison with other annual pasture species. Australian J. Exp. Agr. and Anim. Husb. 8:478-483.

Sosebee, R.E., and C.H. Herbel. 1969. Effects of high temperatures on emergence and intial growth of range plants. Agron. J. 61:621-624.

Springfield, H.W. 1973. Larger seeds of winterfat germinate better. J. Range Manage. 26:153-154.

Stanton, M.L. 1985. Seed size and emergence time within a stand of wild radish (*Raphanus raphanistrum* L.): the establishment of a fitness hierarchy. Oecologia 67:524-531.

Stebbins, G. 1974. Adaptations for seed development and dispersal and for seedling establishment, pp. 68-101. *In*: G. Stebbins, Flowering plants-evolution above the species level. Belknap Press of Harvard Univ. Press, Cambridge, Mass.

Steen, E.B. 1971. Dictionary of biology. Barnes and Noble Books, New York.

Stevens, R., and G.A. Van Epps. 1984. Seeding techniques to improve establishment of forage kochia (*Kochia prostrata* [L.] Schrad.) and fourwing saltbush (*Atriplex canescens* [Pursh] Nutt.), pp. 269-272. *In*: A.R. Tiedemann et al. (compilers), Proc. — symposium on the biology of Atriplex and related chenopods. Ogden, UT. General Tech. Rep. INT-172 USDA, Forest Service Intermountain Forest and Range Experiment Station.

Stevens, R., B.C. Giunta, K.R. Jorgensen, and A.P. Plummer. 1977. Winterfat. Utah State Division of Wildlife Resources. Publication No. 77-2.

Stevens, R., K.R. Jorgensen, J.N. Davis, and S.B. Monsen. 1986. Seed pappus and placement influences on white rubber rabbitbrush establishment. *In* : E.D. McArthur, and B.L. Welch (compilers), Proc. —symposium on the biology of *Artemisia* and *Chrysothamnus*: Ogden, UT: Gen. Tech. Rep. INT-200. USDA, Forest Service Intermountain Forest and Range Experiment Station.

Stoddart, L.A., and A.D. Smith. 1955. Range Management. McGraw-Hill Book Co., NY.

Stoly, L.J., and K.P. Barley. 1968. Mechanical resistance encountered by roots entering compact soils. Soil Sci. 105-297-301.

Svejcar, T.J., T.W. Boutton, and S. Christiansen. 1988. Rooting dynamics of *Medicago sativa* seedlings growing in association with *Bothriochloa caucasica*. Oecologia 77:453-456.

Tadmor, N.H., and Y. Cohen. 1968. Root elongation in the preemergence state of Mediterranean grasses and legumes. Crop Sci. 8:416-419.

Tappeiner, J.C. and J.A. Helms. 1971. Natural regeneration of Douglas fir and white fir on exposed sties in the Sierra Nevada of California. Am. Midl. Nat. 86:358-370.

Taylor, A.G. and C.W. Ten Broeck. 1988. Seedling emergence forces of vegetable crops. Hort. Science 23:367-369.

Taylor, H.M., and H.R. Gardner. 1960. Use of wax substrates in root penetration studies. Soil Science Society Amer. Proc. 24:79-81.

Taylor, J.W., and M.A. McCall. 1936. Influence of temperature and other factors on the morphology of the wheat seedling. J. Agr. Res. 52:557-568.

Taylor, S.G. and D.D. Baltensperger. 1987. Seedling vigor of selected *Alysicarpus* accessions. Agron. J. 79:101-103.

Thalen, D.C.P. 1979. Ecology and utilization of desert shrub rangelands in Iraq. Dr. W. Junk Pub., The Hague.

Thomas, R.L. 1966. The influence of seed weight on seedling vigour in *Lolium perenne*. Ann. Bot. 30:111-121.

Thompson, K. 1987. Seeds and seed banks. New Phytol. 106 (suppl.):251-263.

Thornton, N.D. 1945. Importance of oxygen supply on secondary dormancy and its relation to the inhibiting mechanisms regulating dormancy. Contributions From Boyce Thompson Institute. 13:487-500.

Tischler, C.R., and R.L. Monk. 1980. Variability in root system characteristics of kleingrass seedlings. Crop Sci. 20:384-386.

Tischler, C.R., and P.W. Voigt. 1981. Non-genetic factors affecting primary root absence in lovegrass and kleingrass. Crop sci. 21:427-430.

Tischler, C.R., and P.W. Voigt. 1983a. Effects of planting depth on vegetative characteristics of three forage grasses at 14 days post emergence. Crop Sci. 23:481-484.

Tischler, C.R., and P.W. Voigt. 1983b. Seedling characteristics and rates of seed reserve utilization of Wilman lovegrass and kleingrass. Crop Sci. 23:953-955.

Tischler, C.R., and P.W. Voigt. 1984. Screening and selection to improve establishment of warm-season forage grasses in arid regions, pp. 115–119. *In*: Proc. 1984 Forage and Grassland Congr. 23-26 January 1984. Forage systems leading U.S. agriculture into the future. American Forage and Grassland Council.

Tischler, C.R. and P.W. Voigt. 1987. Seedling morphology and anatomy of rangeland plant species, pp. 5-13. *In*: G.W. Frasier and R.A. Evans (eds.), Proc. of symposium. "Seed and Seedbed Ecology of Rangeland Plants. USDA-ARS. Distrib. by Nat. Tech. Info Serv. Springfield, VA 22161.

Tossell, W.E. 1960. Early seedling vigour and seed weight in relation to breeding in smooth bromegrass, *Bromus inermis* Leyss. Can. J. Plant Sci. 40:268-280.

Townsend, C.E. 1972. Influence of seed size and depth of planting on seedling emergence of two milkvetch species. Agron. J. 64:627-630.

Townsend, C.E. 1979. Associations among seed weight, seedling emergence, and planting depth in cicer milkvetch. Agron. J. 71:410-414.

Troughton, A. 1957. Underground organs of herbage grasses. Commonwealth Bureau of Pastures and Field Crops Bull. No. 44, Hurley, Berkshire.

Trupp, C.R., and I.T. Carlson. 1971. Improvement of seedling vigor of smooth brome (*Bromus inermis* Leyss.) by recurrent selection for high seed weight. Crop Sci. 11:225-228.

Twamley, B.E. 1967. Seed size and seedling vigor in birdsfoot trefoil. Can. J. Plant Sci. 47:603-609.

U.S. Forest Service. 1937. Range Plant Handbook. U.S. Government Printing Office. Washington, D.C.

van der Pijl, L. 1982. Principles of dispersal in higher plants. Springer-Verlag, New York.

Van der Sluijs, D.H., and D.N. Hyder. 1974. Growth and longevity of blue grama seedlings restricted to seminal roots. J. Range Manage. 27:117-119.

van der Toorn, J. and T.L. Pons. 1988. Establishment of *Plantago lanceolata* L. and *Plantago major* L. among grass. Oecologia 76:341-347.

Verkaar, H.J., and A.J. Schenkeveld. 1984. On the ecology of short-lived forbs in chalk grasslands: seedling development under low photon flux density conditions. Flora 175:135-141.

Vertucci, C.W., and A.C. Leopold. 1986. Physiological activities associated with hydration level in seeds, pp. 35-49. *In*: A.C. Leopold (ed.), Membranes, metabolism and dry organisms. Cornell University Press. Ithaca, NY.

Vertucci, C.W., and A.C. Leopold 1987. Oxidative processes in soybean and pea seeds. Plant Physiol. 84:1038-1043.

Vogel, W.G. 1963. Planting depth and seed size influence emergence of beardless wheatgrass seedlings. J. Range Manage. 16:273-274.

Wallace, A., W.A. Rhods, and E.F. Frolich. 1968. Germination behavior of *Salsola* as influenced by temperature, moisture, depth of planting, and gamma irradiation. Agron. J. 60:76-78.

Walter, L.E. and E.H. Jensen. 1970. Effect of environment during seed production on seedling vigor of two alfalfa varieties. Crop Science 10:635-638.

Wanjura, D.F. and E.B. Minton. 1981. Delayed emergence and temperature influences on cotton seedling vigor. Agron. J. 73:594-597.

Washitani, I., and A. Takenaka. 1986. 'Safe sites' for the seed germination of *Rhus javanica*: a characterization by responses to temperature and light. Ecol. Res. 1:71-82.

Watt, L.A. 1981. Establishment of grasses on cracking clay soils: seedling morphology characteristics, p. 451-453. *In*: J.A. Smith and V.w. Hays (eds.), Proc. XIV Int. Grassl. Congr. Westview Press, Boulder, CO.

Watt, L.A., and R.D.B. Whalley. 1982a. Effect of sowing depth and seedling morphology on establishment of grass seedlings on cracking black earths. Aust. Rangel. J. 4:52-60.

Watt, L.A., and R.D.B. Whalley. 1982b. Establishment of small-seeded perennial grasses on black clay soils in north-western New South Wales. Aust. J. Bot. 30:611-623.

Weaver, J.E., and E. Ink. 1945. Extent and longevity of the seminal roots of certain grasses. Plant Physiol. 20:359-379.

Webb, R.B., and B.E. Stephens. 1936. Crown and root development of wheat varieties. J. Agric. Res. 52:569-583.

Weller, D.E. 1989. The interspecific size-density relationship among crowded plant stands and its implications for the -3/2 power rule of self-thinning. The Am. Nat. 133:20-41.

Werker, E. and T. Many. 1974. Heterocarpy and its otogeny on *Aellenia autrani* (Post) Zoh. light- and electron-microscope study. Israel J. Botany 23:132-144.

Wester, D.B., B.E. Dahl, and P.F. Cotter. 1986. Effects of pattern and amount of simulated rainfall on seedling dynamics of weeping lovegrass and kleingrass. Agron. J. 78:851-855.

Whalley, R.D.B., and C.M. McKell. 1967. Interrelation of carbohydrate metabolism, seedling development, and seedling growth rate of several species of Phalaris. Agron. J. 59:223-226.

Whalley, R.D.B., C.M. McKell and L.R. Green. 1966a. Seedling vigor and early non-photosynthetic stage of seedling growth in grasses. Crop Sci. 6:147-150.

Whalley, R.D.B., C.M. McKell and L.R. Green. 1966b. Effect of environmental conditions during the parent generation on seedling vigor of subsequent seedlings of *Oryopsis milliacea* (L.) Benth. and Hook. Crop Sci. 6:510-512.

White, J. 1981. The allometric interpretation of the self-thinning rule. J. Theor. Biol. 89:475-500.

White, R.S. 1984. Stand establishment: The role of seedling size and winter injury in early growth of three perennial grass species. J. Range Manage. 37:206-211.

White, R.S., and P.O. Currie. 1980. Morphological factors related to seedling winter injury in three perennial grasses. Can. J. Plant Sci. 60:1411-1418.

White, W.J., and W.H. Horner. 1943. The winter survival of grass and legume plants in fall sown plots. Sci. Agr. 23:399-408.

Widell, K., and T.C. Vogelmann. 1985. Optical properties of *Lactuca* and *Taraxacum* seed and fruit coats: their role as light filters. Physiol. Plant 64:34-40.

Widell, K., and T.C. Vogelmann. 1988. Fiber optic studies of light gradients and spectral regime within *Lactuca sativa* achenes. Physiol. Plant 72:706-712.

Williams, W.A. 1956. Evaluation of the emergence force exerted by seedlings of small seeded legumes using probit analysis. Agron. J. 48:273-274.

Williams, W.A. 1963. The emergence force of forage legume seedlings and their response to temperature. Crop Science 3:472-474.

Wilson, A.M. 1973. Responses of crested wheatgrass seeds to environment. J. Range Manage. 26:43-46.

Wilson, A.M. 1981. Air and soil temperature effects on elongation of adventitious roots in blue grama seedlings. Agron. J. 73:693-697.

Wilson, A.M. 1984. Nonstructural carbohydrates and root development in blue grama seedlings. J. Range Manage. 37:28-30.

Wilson, A.M., and D.D. Briske. 1978. Drought and temperature effects on the establishment of blue grama seedlings, p. 359-361. *In*: D.N. Hyder (ed.) Proc. First Int. Rangeland Congress. Society for Range Manage. Denver, CO.

Wilson, A.M., and D.D. Briske. 1979. Seminal and adventitious root growth of blue grama seedlings on the Central Plains. J. Range Manage. 32:209-213.

Wilson, A.M., and J.A. Sarles. 1978. Quantification of growth drought tolerance and avoidance of blue grama seedlings. Agron. J. 70:231-237.

Wilson, A.M., D.N. Hyder, and D.D. Briske. 1976. Drought resistance characteristics of blue grama seedlings. Agron. J. 68:479-484.

Wilson, A.M., R.L. Cuany, J.G. Fraser, and W.R. Oaks. 1981. Relationships among components of seed yield in blue grama. Agron. J. 73:1058-1062.

Wilson, S.B., and W.D. Bonner, Jr. 1971. Studies of electron transport in dry and imbibed peanut embryos. Plant Physiol. 48:340-344.

Wood, D.W., P.C. Longden, and R.K. Scott. 1977. Seed size variation: its extent, source and significance in field crops. Seed Sci. & Technol. 5:337-352.

Woodmansee, R.G., and L.D. Potter. 1971. Natural reproduction of winterfat (*Eurotia lanata*) in New Mexico. J. Range Manage. 24:24-31.

Wright, L.N. 1977. Germination and growth response of seed weight genotypes of *Panicum antidotale* Ret. Crop Sci. 17:176-178.

Young, J.A., R.A. Evans, R.O. Gifford, and R.E. Eckert Jr. 1970. Germination characteristics of three species of Crudiferae. J. Weed Science Society Ameri. 18:41-48

Young, J.A. and R.A. Evans. 1972. Germination and establishment of *Salsola* in relation to seedbed environment. I. Temperature, after ripening, and moisture relations of *Salsola* seeds as determined by laboratory studies. Agron. J. 64:214-218.

Young, J.A., and R.A. Evans. 1986. Seed and seedbed ecology of crested wheatgrass, p. 61-64. *In*: K.L. Johnson (ed.), Crested wheatgrass: its values, problems and myths; symposium proceedings. Utah State Univ. Logan, UT.

Young, J.A., and E. Martens. 1988. Juvenile root adaptations for germination on the surface of seedbeds. Weed Sci. Sco. Amer. Abstr. p. 266.

Young, J.A., R.A. Evans, and B. L. Kay. 1975. Dispersal and germination dynamics of broadleaf filaree, *Erodium botrys* (Ca.) Bertol. Agron. J. 67:54-57.

Young, J.A., R.A. Evans, R.O. Gifford, and R.E. Eckert Jr. 1970. Germination characteristics of three species of Cruciferae. J. Weed Sci. 18:41-48.

Young, J.A., R.A. Evans, B. Roundy, and G. Cluff. 1983. Moisture stress and seed germination. Agri. Res. Serv. (Western Region), USDA. ARM-W-36, Oakland, Calif.

Zemetra, R.S., C. Havstad, and R.L. Cuany. 1983. Reducing seed dormancy in Indian ricegrass (*Oryopsis hymenoides*). J. Range Manage. 36:239-241.

Zietsman, P.C., and F.C. Botha. 1987. Seed germination of *Ziziphus mucronata* ssp *mucronata*. South African J. Bot. 53:341-34.

The Water Relations of Range Plants: Adaptations to Water Deficits[†]

Ray W. Brown[1]

Introduction

Ideally, rangelands are managed to optimize the productivity of plants for livestock and wildlife grazing consistent with sustained soil and watershed values. For the most part, rangelands are uncultivated and unirrigated natural lands supporting vegetation suitable for grazing. Vast regions of native rangelands throughout the World occur in climatic regimes where water may be limiting to plant growth during at least portions of the growing season. Thus, the practice of range management is dramatically affected by often severe and limiting climatic variables.

The availability of water is the most important environmental factor affecting the distribution, growth, and survival of rangeland plants throughout the World. Precipitation largely determines the type of vegetation and relative species composition of plant communities. Temporal variations in the amount and distribution of precipitation dramatically affect plant productivity, density, and cover on rangelands. In addition, precipitation is the primary variable affecting the storage and availability of water in soils. When extended periods of little or no precipitation coincide with the growing season of plants, the availability of water is largely limited to that stored in the soil. During such periods it is this water alone that determines plant productivity, tugidity, nutrition, and palatability of vegetation on rangelands. Thus, it is little wonder that the importance of water and its availability for plant growth and survival on rangelands has been recognized for decades (Craddock and Forsling 1938; Humphrey 1962; Keller 1971; Nelson 1934; Weaver and Clements 1938).

Rangelands commonly occur in regions that experience variable periods of suboptimum precipitation or droughts. In a meteorological and environmental sense, drought is a prolonged period of less than normal precipitation during which soil water is depleted and, as a result, water deficits result in plant damage (Kramer 1980; Swindale and Bidinger 1981). The frequency, duration, and intensity of a drought are important in determining its severity. The variations from normal expected precipitation that cause drought vary significantly with region and

---

† Chapter in: *Rangeland Plant Physiology and Morphology*, Society for Range Management, D. Bedunah and R. E. Sosebee (eds.)

[1] Plant Physiologist, USDA Forest Service, Intermountain Research Station, Forestry Sciences Laboratory, 860 North 1200 East, Logan, Utah 84321.

ecosystem (Heathcote 1973). In arid regions drought may arguably be considered a nearly permanent condition, or alternatively a rare event in view of normally low precipitation. In semi-arid and mesic regions, especially in mid-continental land masses (such as western North America), mild drought is common and normally seasonal in nature; however, severe drought that extends across seasons for several consecutive years may reoccur regularly about every decade (Heathcote 1973; Humphrey 1962; Swindale and Bidinger 1981). In sub-humid and humid climates drought is less common, but more predictable (Kramer 1980).

Although native vegetation normally maintains relatively stable cover and species composition during extended periods of optimum precipitation, drought conditions can lead to severe disruptions in plant communities and ecosystems (Chamrad and Box 1965; Glantz 1987; Schulze 1986b; Whalley 1973). The distribution ranges of most plant species are conspicuously dynamic, fluctuating spatially and geographically with the periodic occurrence of drought. It is well known that severe droughts may result in wide-spread mortality of plants, and in the worst cases may result in the complete loss of some species in an area. Some plant species are highly sensitive to even mild drought, whereas others appear to be highly drought tolerant. Hence, the onset and termination of a drought for one plant species may not coincide with that of another if they have different tolerance ranges to stress.

Although somewhat controversial, it appears that grazing can accentuate the effects of drought for many plant species (Craddock and Forsling 1938; Humphrey 1962; Hurtt 1951; Nelson 1934; Orodho et al. 1990; Sampson 1952). The mechanisms of drought and grazing tolerances in many plant species are normally quite different, and often there is little, if any, correlation between them. Under grazing pressure, mild drought periods that result in plant water deficits may result in sufficient declines in plant productivity of both drought tolerant and intolerant species alike to warrant light to moderate reductions in numbers of grazing animals on ranges. However, under severe drought conditions major reductions may be required to protect long-term values of rangelands (Hurtt 1951; Humphrey 1962). Range management practices are thus implemented largely to regulate the rate of plant use by livestock and wildlife animals (Jameson 1963).

Water directly affects range plant palatability as well as nutritive and energy values. In arid and semiarid environments, water availability largely controls range condition and trend (Ellison et al. 1951) as well as plant regrowth and development following grazing. Because range management is so intimately and dynamically dependent upon plant growth and survival, which are so impacted by climatic variables such as drought, it is essential that range managers have a firm knowledge of the principles of plant water relations.

The study of plant water relations is concerned with plant responses to variations in the availability of water; plant adaptations to water deficits; and the movement, function, and distribution of water in the environment (Kramer 1983). In the discussion of plant water relations that follows, literature pertaining to specific rangeland species is used to illustrate water relations principles where possible. However, the water relations of rangeland species have generally not been studied as extensively as many crops, cultivated plants, forest tree species, unique desert species, and ornamentals. Thus, in some instances the literature of plant water relations has been cited irrespective of plant type with the understanding that adaptations and the degree of response to water deficits and other environmental factors may differ somewhat. It is emphasized that generally there are no data to suggest that the water relations of rangeland plant species are any more unique than those of other groups of plants, aside from perhaps specific adaptations to and interactive effects of grazing. The objective of this chapter is to summarize the principles of plant water relations that have significance to range managers, and to describe plant adaptations to water deficits that may be useful in making land management decisions.

Importance of Water to Plants

Water is physiologically essential for all plants, as it is for all organisms. Water makes up 80 to 90% of the fresh weight of herbaceous plant tissues and about 50% of the fresh weight of woody plants (Kramer 1983; Turner and Burch 1983). Most of the water in plants occurs within the individual cells of the organism, but significant amounts also occur within the transport tissues, cell walls, and apoplastic regions. Water is the primary solvent within which all physiological processes occur, and in which gases, minerals, and solutes enter into the plant and are translocated. It is also the primary substance responsible for the high degree of permeability of cell membranes to diffusion and osmotic processes. Water is essential for the maintenance of turgidity in plants, which controls stomatal opening, gas exchange, cell enlargement, and growth.

Despite its relative abundance in the environment, water is a highly unique substance with numerous unusual chemical and physical properties (Stillinger 1980). These properties affect the nature of how water occurs in plants and soils, how it moves within the environment, and how it interacts with the various components of soil, plants, and the atmosphere. It is important to understand these properties and how they affect the status of water to understand plant water relations.

Water (H_2O) molecules are composed of two atoms of hydrogen and one of oxygen for a combined molecular weight of about 18. The water molecule is formed by covalent bonds between the two hydrogen protons and the oxygen atom, and carries a dipolar charge (the two H protons are positive, and the O is negative). The dipolar nature of water makes it an excellent solvent for electrolytes (Hammel 1976) and nonelectrolytes as well. Because water forms very strong hydrogen bonds between itself and other substances, it tends to be readily adsorbed on surfaces such as soil particles, cellulose, proteins, and other matter, which plays a major role in water relations of plants and soils. Hydrogen bonds between individual water molecules gives water an unusually high boiling temperature compared with other substances of similar molecular weight. In addition, its significantly high latent heat of fusion requires that great amounts of heat must be removed for water to freeze. When freezing does occur, the density of the solid phase, unlike most substances, decreases (Runnels 1966). These qualities, together with its unusually high specific heat, give water the unique attribute of being remarkably stable to changes in heat. Yet, evaporation of water has a dramatic cooling effect and condensation has an equally striking warming effect; these latter two apparent anomalies are caused by water's exceptionally elevated heat of vaporization (the highest known for any common substance!). The high cohesive forces between water molecules in the liquid phase gives it an unusually high surface tension and is the primary reason why water has such a high tensile strength, which is employed in water transport in the conductive tissues of plants (Scholander 1972). Water is virtually transparent to visible light, thus allowing photosynthesis within the aqueous environment of cells, and yet is nearly opaque to longer wavelengths allowing water to absorb high amounts of heat.

Status of Water in Soils and Plants

Status of Water

The water relations of range plants is necessarily concerned with the distribution, movement, function, and status of water in soils, plants, and the atmosphere. The distribution and movement of water among these various constituents of the environment occur largely in the liquid and gaseous phases. The driving force for this movement are gradients in free energy of water imposed by such factors as differences in temperature, the occurrence of salts and other solutes, the amount of water available, gravity, and other forces. Thus, soil, plants, and the atmosphere can be considered as constituents of an elaborate system intimately linked together by a common energy bond (Cowan 1965; Gates 1962; Philip 1964, 1966; Slatyer and Denmead 1964). Philip (1966) proposed the term "soil-plant-atmosphere continuum" (SPAC) to describe this system which conceptually illustrates the idea of an energy continuum (Fig. 1).

Water movement in the SPAC occurs along gradients of free energy, moving from regions of higher free energy to regions of lower free energy (Kramer 1983; Slatyer 1967; van den Honert 1948). Under normal conditions, gradients in free energy decrease and progressively steepen from the soil to the plant, through the vascular system of plants to leaf surfaces, and from the leaves to the surrounding atmosphere (Fig. 1). Water is normally available to plants primarily from the soil in the liquid phase, but water occurs in both the liquid and gaseous phases in soil. As it is absorbed by the plant, water forms a liquid continuum throughout the vascular system. Within the substomatal cavities of leaves and other organs of the plant, water undergoes a phase-change from liquid to a vapor, and escapes in response to the evaporative demand of the atmosphere. Thus, the free energy and mass of water is seldom, if ever, in equilibrium within the environment.

The status of water occurring in soils and plants can be considered as either a capacity term (such as amount, expressed as weight or volume), or as an intensity term (expressed as free energy). Although the weight or volume of water in plants and soils is a useful measure for some applications, research indicates that the energy status of water is generally more useful for expressing the dynamics of water movement within the SPAC and the physiological significance of water to plants (Crafts 1968; Hillel 1971; Philip 1966).

By convention the fundamental unit of expression for the free energy of water in soils, plants, and the atmosphere is **water potential** (Slatyer and Taylor 1960; Taylor 1968). Water potential terminology is based on thermodynamic principles (Owen 1952), and replaces confusing terms such as "diffusion pressure deficit,"

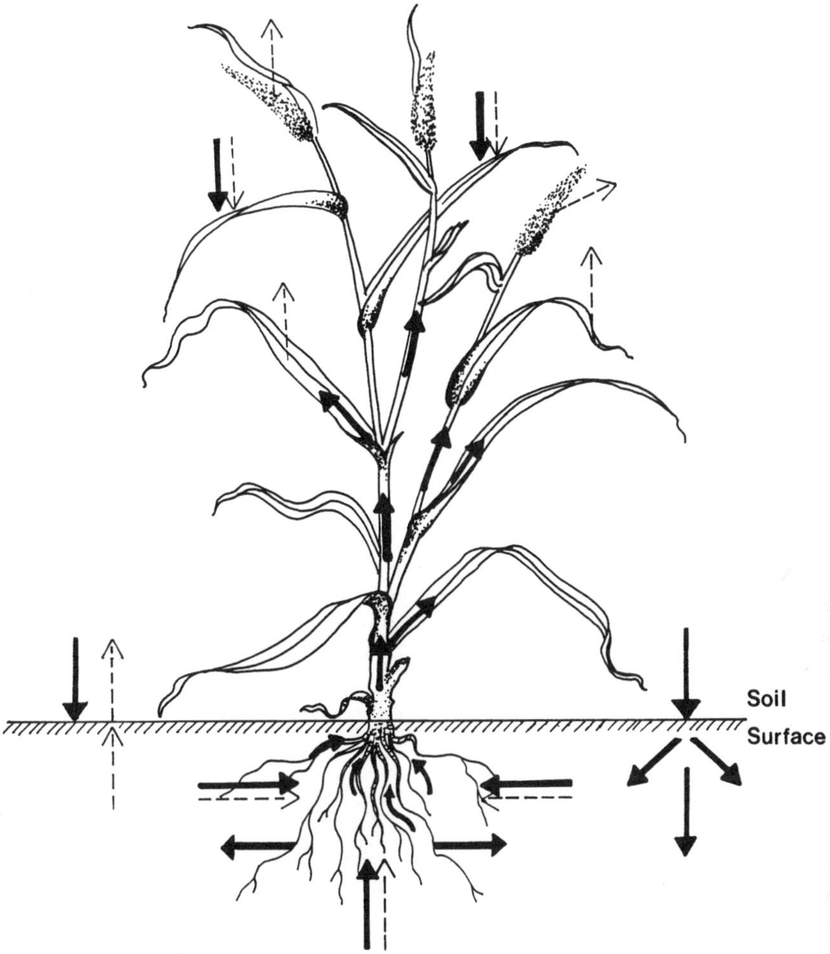

Figure 1. Diagrammatic sketch of the soil-plant-atmosphere continuum, showing the movement of liquid water (solid lines) and vapor (broken lines).

"water suction," "moisture stress," "hydrature," and other terms. The water potential expresses the difference in free energy between pure water at a reference temperature and pressure and water in soil, plants, or the atmosphere under the same conditions. The units of water potential are expressed as megapascals (MPa, where 1 MPa = 10 bars = 10^7 dynes cm^{-2}). Pure free water at standard temperature and pressure is, by convention, arbitrarily assigned a water potential of 0 MPa. However, water in soils, plants, or the atmosphere is affected by various forces that result in either a lower or higher free energy than that of pure free water (e.g., a water potential of –1.3 MPa describes a free energy that is 1.3 MPa lower than that of pure water at the same temperature and pressure).

The **total water potential** (ψ) is partitioned into a number of components, each of which is dependent upon the specific forces acting on water. The more important components of water potential include:

1. **Osmotic potential** (ψ_π) arises from forces due to the presence of dissolved solutes in solution with water (replaces such terms as solute potential or osmotic pressure). The presence of solutes in water lowers the water potential by increasing solution entropy and by the formation of various associations between water molecules and the solute. The dipolar nature of water causes the water molecules to orient themselves around the electric field of each solute constituent in solution, and depending upon the nature of the solute, forms either hydrogen bonds, ion-dipole bonds, or other associations that reduce the kinetic energy of water. The water potential of the solution is lowered below that of pure free water, and hence osmotic potential is considered a negative quantity. In plant cells the osmotic potential is one of the primary components of total water potential within vacuoles and cytoplasm, but is a minor component within apoplastic regions. The osmotic potential is primarily responsible for controlling biochemical processes in cells (Wiebe 1972). In soils this component arises from the presence of salts and other solutes in association with soil particles.
2. **Matric potential** (ψ_τ) arises from capillarity, adsorption, and electrostatic forces resulting from the contact of water with a matrix, or surface. According to Spanner (1972), these are Newtonian forces that seize water molecules and "pull" them onto the matrix surface. The matric potential lowers the water potential below that of pure free water, and thus is a negative quantity. In plants, the matric potential is primarily a factor of total water potential within intercellular spaces, around the surfaces of cell walls, within the vascular system, and may be active around colloidal surfaces within the cytoplasm (Miller 1972; Noy-Mier and Ginzburg 1967; Passioura 1980; Warren Wilson 1967a). Matric potential is the primary component of total water potential in soil and arises from adsorptive, capillary, and electrostatic forces between water molecules and the surface of soil particles.
3. **Pressure potential** (ψ_p) arises from forces of hydrostatic pressure, and can be either positive or negative. In plants this component is commonly referred to as **turgor potential** and arises from pressure exerted against the cell wall and internal membranes of protoplasts. In growing plants the pressure potential is normally positive and provides rigidity to plant tissues, is involved in cell enlargement, and is the primary control of stomatal movement (Turner and Burch 1983; Wiebe 1972). During water deficits the pressure potential decreases gradually, and when it approaches atmospheric pressure or 0 MPa, wilting of tissues may occur. Under severe wilting, negative pressure potentials have been reported (Brown 1974, 1977; Noy-Meir and Ginzburg 1967; Warren Wilson 1967b). Pressure potential is rarely a component of soil water status, although under saturated conditions it may be a force at depth. A positive pressure results in an increase in water potential whereas a decrease in pressure lowers water potential.
4. **Gravitational potential** (ψ_g) arises from forces due to gravity. These forces are usually quite small if the elevation between the reference state and the water in the system are similar, hence are not considered to be as important as other forces (Kramer 1983; Slatyer 1967).

Interrelations of Water Potential Components

Osmotic, matric, and pressure potentials represent the three principle forces affecting the total water potential in soils and plants (Kramer 1983; Noy-Meir and Ginzburg 1967, 1969; Turner and Burch 1983; Warren Wilson 1967a, 1967b, 1967c). At equilibrium, the energy status of water can be described as the algebraic sum of the various components such that;

$$\psi = \psi_{\pi} + \psi_{\tau} + \psi_{p} \qquad (1)$$

Other terms such as the gravitational component, and an interactive term (Noy-Meir and Ginzburg 1967) are omitted for simplicity (see Brown 1977 for a more detailed explanation).

Water in Soil

Water is retained in soil primarily by matric and osmotic forces (Hillel 1971; Slatyer 1967; Taylor 1965; Taylor and Ashcroft 1972). Interfacial tension and adsorptive forces between the soil particles and water, or matric forces, are the major factors affecting total soil water potential. These forces reduce the total free energy of soil water below that of pure free water and are affected by the size and distribution of soil particles in the soil matrix together with the amount of water in the soil. Fine-textured particles (clay and silt), because of their greater surface area per unit mass, exert greater forces of retention on water than larger particles (sand). Consequently, fine-textured soils retain larger amounts of water at a given water potential than coarse-textured soils. Under saturated conditions, the water in soil fills the air spaces within the soil matrix (Fig. 2) such that matric forces are minimized, and the total soil water potential is approximately 0 MPa. Hydraulic pressure greater than atmospheric pressure may then result in a transitory positive pressure potential component lasting only until the free water drains from the system due to gravity. When the free water (that water not held by matric forces, and which is subject to forces of gravity) drains from the soil, the remaining water is held by a matric force of about –0.033 MPa, which by convention is termed the "field capacity" (Fig. 2). As the soil dries further, water within the soil matrix is

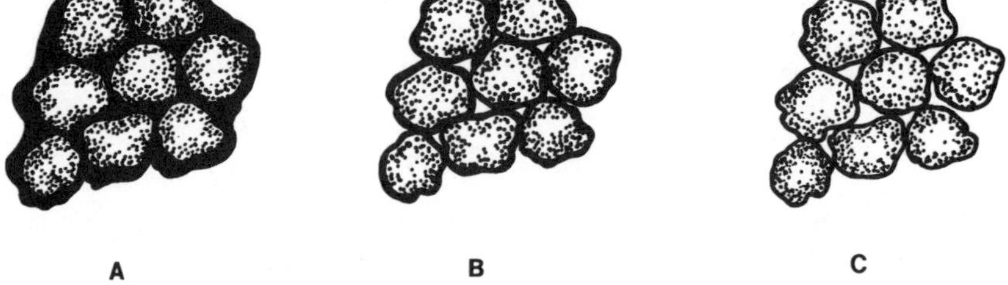

Figure 2. Representation of water films around individual soil particles at different levels of soil water status: A., at saturation water fills the air spaces within the soil; B., at field capacity the free water has drained leaving some air spaces; and C., as the soil dries further, the films of water surrounding soil particles becomes thinner, and is held with greater energy.

retained as a film (composed of multiple layers of water molecules) surrounding the individual soil particles (Fig. 2). With continued drying this film becomes thinner, and the remaining water is held by increasingly stronger matric forces.

The historic research on plant wilting in drying soils by Hendrickson and Veihmeyer (1945), Richards and Weaver (1943), and Veihmeyer and Henrickson (1928, 1950) established the so-called "permanent wilting point" at –1.5 MPa soil water potential. They felt that the soil was a reservoir that supplied water to plants between the field capacity and the permanent wilting point, and that water was equally available throughout this range. Further, they believed that the upper and lower limits of water availability were fixed by these two soil water conditions ("In fact, the authors believe they are the only ones of any practical value for consideration in connection with plant growth.", Veihmeyer and Hendrickson 1950, p. 285). More recently it has been shown that the field capacity and the permanent wilting point are not really constants at all relative to plant growth, although they have been retained in soil physics as relative points of comparison for different soils. In plant water relations, however, the range of available water cannot be defined so simply because most plant species have wide ranging adaptations that do not correspond to any particular soil water constant (Gardner and Nieman 1964). Gardner (1972, p. 237) probably clarified the confusion about the significance of soil water constants best ("It seems quite clear that water content at the wilting point does not represent the lower limit of anything. Plant growth is seriously retarded well before this point is reached. The region beyond –15 bars [–1.5 MPa] is sometimes viewed as analogous to beyond the edge of the earth—an unidentified nonexistence. However, plants do exist, transpire, and some even grow a bit beyond this point. This is particularly true of prairie and desert plants.").

In addition to matric forces, different soils also contain varying amounts of dissolved salts that affect total water potential due to osmotic forces. In arid and semiarid rangeland soils, dissolved salts may comprise a significant contribution to the total soil water potential. Although the total soil water potential is the sum of both matric and osmotic forces, these components do not act completely independent of each other (Hillel 1971; Taylor and Ashcroft 1972). At high water contents when matric potential is also high (near 0 MPa), the total soil water potential is influenced more by solute concentrations, or the osmotic potential, than by other forces. At low water contents both matric and osmotic forces influence total soil water potential, and the relative magnitude of each component is dependent upon particle size, water content, solute concentration, and soil temperature (Nielsen et al. 1972; Taylor and Ashcroft 1972).

Water in Plants

The three most important water potential components in plants are osmotic, pressure, and matric potentials (Kramer 1983; Turner and Burch 1983; Wiebe 1972). The status of water in plant tissues is intimately related to cell water relations because most plant water occurs within cells (Kramer 1983; Paleg and Aspinall 1981; Turner and Kramer 1980). Brown (1977) reviewed the interactions of plant water potential components in some detail, therefore only a brief summary will be presented here.

Plant tissues are far more complex systems than soils relative to the distribution, interactions, and function of water potential components. Generally, water moves throughout the plant (as throughout the SPAC; Fig. 1) along water potential gradients from regions of high water potential to regions of low water potential. Although the total water potential described in equation (1) is the sum of its various components, the interactions and dynamic functions of these components are not readily apparent.

Within living tissues the total water potential and its components are in a constant state of flux, responding to water absorption, transpiration losses, translocation among plant organs, temperature and osmotic gradients within the plant, and other factors. The tendency is for water potential to achieve equilibrium between and among tissues, but under natural conditions it is doubtful that true equilibrium really occurs. As such, a state of continual adjustment of water potential and components occurs in different parts of the plant as water is absorbed, translocated, stored, and lost through transpiration.

Under equilibrium conditions, however, where fluctuations in conditions are minimized (as might occur in the early morning before sunrise, or on humid cloudy days when transpiration is minimized), the total water potential within a given tissue or organ would be expected to be nearly uniform. Under these conditions the components of water potential may vary widely. Within a single turgid cell the total water potential would be identical with that of the cell wall, the apoplastic intercellular regions, and all surrounding cells. Also, within a single cell the total water potential of the cytoplasm and vacuoles, as well as the intracellular bodies (e.g., chloroplasts, etc.) would be the same (Oertili 1976a, 1976b).

The components of water potential within each of the different parts of the tissue may vary greatly. For example, within the cellulose fibers of the cell wall external to the plasmalemma membrane, the total water potential is composed primarily of matric forces as the result of adsorptive attraction between water molecules and cellulose (Fig. 3). A minor osmotic component may be present within the cell wall and within the intercellular spaces (apoplastic region), but probably can be considered as insignificant.

The symplastic region of the cell is composed of both the cytoplasm and vacuole, located within the plasmalemma membrane (Fig. 3). Within the cytoplasm of the cell, enclosed between the plasmalemma and tonoplast membranes, are found solutes, colloids, and organelle bodies. The total water potential in this region is affected primarily by osmotic and pressure components, with some matric forces potentially important (Fig. 3). The osmotic component is probably the most important within the cytoplasm because of the concentration of solutes in that region of plant cells. The matric term is probably of lesser importance, but it may also be important because of the attractive forces that colloids and organelle bodies may have for water. The pressure potential, or the positive hydrostatic pressure of water against the plasmalemma and tonoplast membranes, partially counter-balances the combined negative forces of osmotic and matric components such that the net water potential in the cytoplasm equals that of the cell wall.

Within vacuoles enclosed within the tonoplast membrane, the total water potential is primarily affected by osmotic and pressure forces (Fig. 3). The water within the vacuole contains solutes of salts and organic acids, but has few colloids

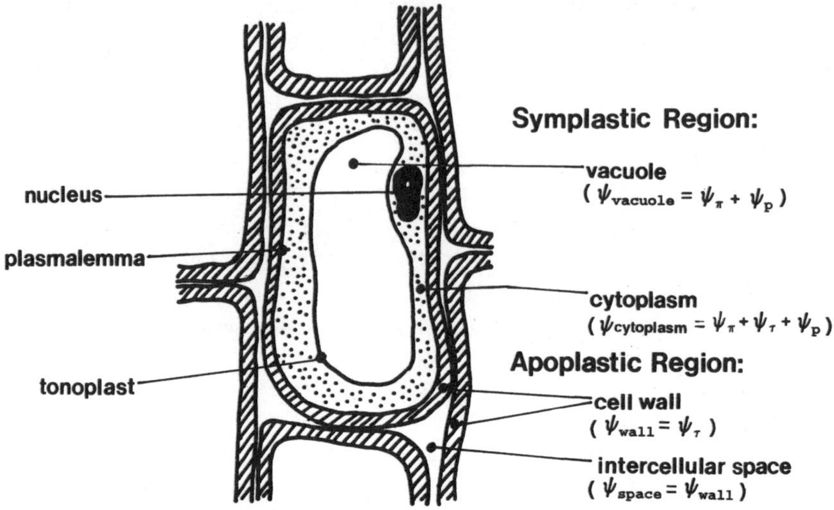

Figure 3. Diagram of a theoretical plant cell showing structural features and the distribution of water potential components in both the symplastic region (comprising the vacuole and cytoplasm) and the apoplastic region (comprising the cell wall and intercellular spaces).

or other substances. Hence, matric forces are probably insignificant in the vacuoles of plant cells. The positive pressure term here also partially counter-balances the osmotic term such that the total vacuole water potential equals that of the cytoplasm and cell walls.

The total water potential within tissues at equilibrium can be summarized as:

$$\psi_{tissue} = \psi_{cell} = \psi_{wall} = \psi_{cytoplasm} = \psi_{vacuole} \quad (2)$$

where the total water potential of the tissue equals that of the cell, cell wall, cytoplasm, and vacuoles.

Water Deficits in Plants

In both seedlings and mature plants transitory periods of water deficits are common occurrences throughout the growing season, and have their greatest and most immediate effect on temporary reductions in plant biomass and growth. This has obvious significance to land management because biomass production and cover are essential to providing forage for grazing animals and site protection from soil erosion. Regardless of the frequency or severity of water deficit periods, all plants probably endure some degree of stress due to dehydration during their life-cycle (Kramer 1983; Oertli 1983; Osmond et al. 1987; Schulze et al. 1987). Water deficits in plants obviously do not occur just during drought periods, although they can be expected to occur more frequently then. Terrestrial plants may be subjected to water deficits periodically during growing seasons of normal or even greater than normal precipitation because of such factors as warm temperatures and low humidity (Campbell and Harris 1981; MacMahon and Schimpf 1981; Schulze 1986a, 1986b), drying winds (Mansfield and Davies 1981), cool soil temperatures (Levitt 1972), waterlogging (Bradford and Yang 1981; Wallace and Nelson 1990), damage to root systems, or other conditions. The severity of water deficits can even be lethal during non-drought periods if transitory conditions are critical, although this is usually a localized effect.

Development of Plant Water Deficits

Water deficits occur when the availability of water decreases sufficiently to cause a decrease in plant physiological activities such as growth and reproduction or photosynthesis below the potential for that plant, species, or genotype (Kramer 1983; Osmond et al. 1987). In the case of plant communities, water deficits become significant when most of the floristic components of the community are negatively affected.

In individual plants, water deficits begin to develop when the rate of water loss due to transpiration exceeds the rate of water absorption by the roots (Kramer 1983; Kaufmann 1981; Turner and Burch 1983). In the majority of terrestrial plants water loss occurs through the leaf blades as transpiration, hence the leaves are the primary organs in which water deficits first develop. Initially, when leaves are fully hydrated, leaf water content is near 100%. At this point, the osmotic concentration within the cells is diluted at about the saturated level because of the high tissue water content, and the pressure (turgor) exerted against the plasmalemma membrane is at an equivalent force, but opposite in sign (Fig. 4). From equation (1), the net total water potential is near 0 MPa. However, as the leaf loses water to the surrounding environment through transpiration, a water deficit develops in the cells because the rate of water loss exceeds the rate of water absorption and replacement (the so-called absorption "lag"; see Kramer 1983). As the volume of water loss continues, total leaf water potential decreases as the osmotic concentration of cell solutes increases (causing a decrease in osmotic potential), and as the positive pressure (against the cell wall) of the pressure potential declines (Fig. 4). In most species, leaf water potential declines gradually as water content decreases, but the rate of decline in water potential and its components increases with continued loss of water (Fig. 4).

As water deficits continue to develop and the cells lose more water, the total water potential and osmotic potential eventually approach equality as the pressure potential approaches 0 MPa (Fig. 4). Wilting, leaf folding or rolling, and other visible responses to water deficits become obvious in many species at or near this point. However, this point is by no means a constant because many species display no visible signs of wilting. Continued loss of water from the tissues may result in negative pressure potentials (when the total water potential is higher than osmotic potential) creating a condition known as "plasmolysis" or physical separation of the plasmalemma membrane and the cell wall (Kramer 1983). Many workers feel that observations of negative pressure potentials in most herbaceous and many woody plants are artifacts of the techniques used to measure water potential and its components (Tyree 1976).

Water deficits probably occur in plants nearly every day of the growing season in terrestrial environments as the result of transpiration when direct solar radiation results in increasing plant and air temperatures and lower humidity. When soil water is freely available and plant water deficits are relatively mild, diurnal oscillations of leaf water potential and its components are primarily affected by a

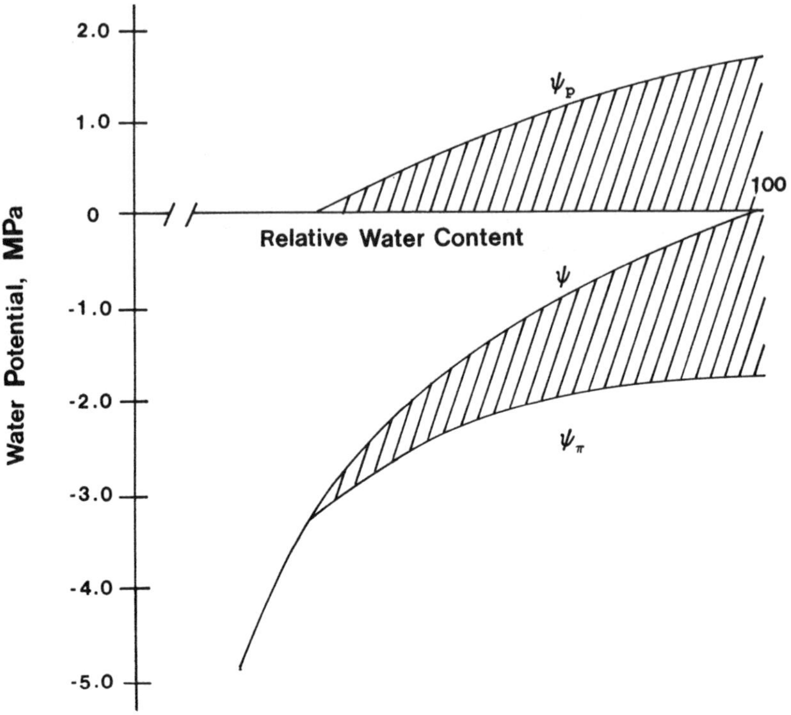

Figure 4. Example of the relationships among total leaf water potential (ψ), osmotic potential (ψ_π), pressure potential (ψ_p), and leaf relative water content (the X-axis is not labeled quantitatively here because actual data vary widely with different species). As a leaf loses water during periods of water deficits, total leaf water potential, osmotic potential and pressure potential also decline. The relative rate of decline in water potential components increases with each increment of water content loss. Note that ψ_p is the difference between ψ and ψ_π (cross-hatched areas); ψ_p normally falls to 0 MPa as most plants wilt, but may fall below 0 MPa (see text).

transitory lag in water absorption in response to transpiration during daylight hours (Fig. 5). Resistance to water absorption by roots is normally higher than that due to transpiration loss (Boyer 1985; Kramer 1983; Passioura 1981, 1988; Taylor 1980; Turner 1986), hence absorption tends to lag behind the rate of transpiration, and thus plants commonly experience some midday stress even when soil water is readily available. At night, full recovery of leaf water potential and its components are normally complete. Prior to sunrise the stomata are closed, and leaf water potential and its components, together with the soil water potential adjacent to the root absorbing surfaces, are at their daily maximum. At this point, leaf and soil water potentials rarely achieve equilibrium because resistances to water movement within the tissues of the plant remain greater than those between the soil and roots.

As the sun rises on clear summer days, the stomata open in response to increasing light, and transpiration and photosynthesis begin (Fig. 5). The water potential components within the guard cells and tissues surrounding the stomata almost immediately begin to respond as the accumulated bulk water within these cells is lost to the transpiration stream. Water in more distant cells and tissues is initially buffered from immediate loss. However, transpiration quickly exhausts the reser-

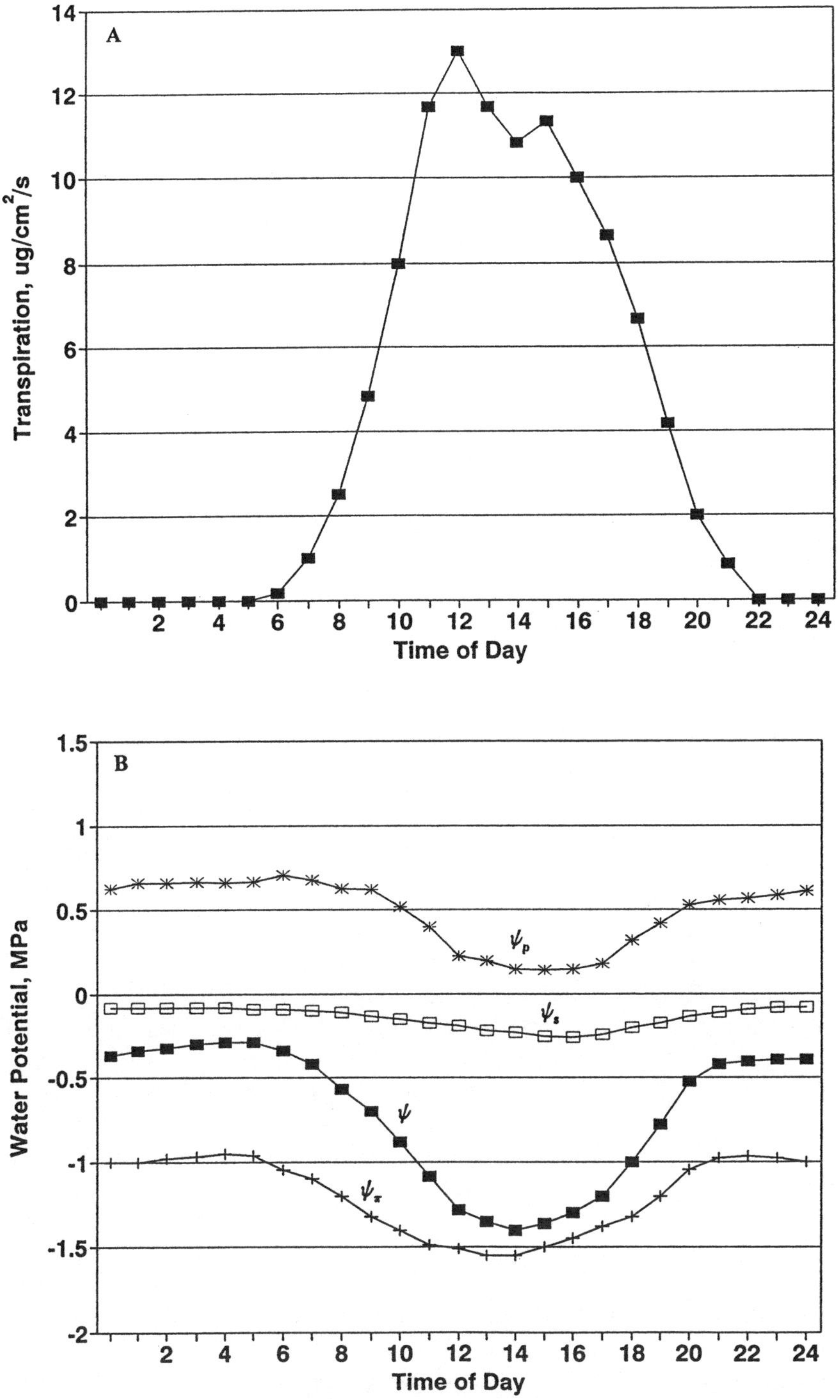

Figure 5. (A) Diurnal course of transpiration (ug cm^{-2} s^{-1}), and (B) total leaf water potential (ψ), leaf osmotic potential (ψ_π), leaf pressure potential (ψ_p), and soil water potential (ψ_s). Data collected (Brown 1992, unpublished) from *Oryzopsis hymenoides* in Logan, Utah.

voir of stored water in most plants, and water potential components begin to decline almost immediately. As water is lost from the tissues, a water deficits gradually develop as total leaf water potential, osmotic potential, and pressure potential decline. The water potential gradient within the vascular system between the leaves and roots steepens slightly, and absorption begins to increase from the soil. The soil water potential adjacent to the roots also begins a gradual decline as the rate of water absorption increases in response to the steepening gradient. The soil water potential in this region continues to decline throughout the day in response to continued transpiration losses from the plant shoot (Fig. 5). Gradients in soil water potential from the bulk soil to the root are steepest near the root absorbing surfaces, and gradually decline with distance away from the root. The rate of mass movement of soil water toward the roots in response to these gradients is limited by soil hydraulic conductivity and water content such that it is commonly exceeded by the rate of absorption.

If sufficient water is available to maintain open stomata, transpiration rate continues to increase during the morning, and perhaps early afternoon hours (Fig. 5). However, near midday water loss within the tissues of most leaves exceeds that replaced by absorption from the soil, the gradient between soil and leaf water potential is maximized, and pressure potential within the guard cells of the stomata may approach 0 MPa. In response, the guard cells may then begin a transient closure, and transpiration and photosynthesis may thus decline (Fig. 5). Midday reductions in transpiration and photosynthesis are common, even when soil water is not limiting, because resistance to water movement within the cells of leaves and in guard cells is higher than that to absorption of water by roots (Kramer 1983; Turner 1986). Midday reductions in transpiration are particularly common when high air temperatures, low humidities, and possibly mild winds create a steepening gradient in the vapor pressure deficit between plant leaves and the atmosphere. Later in the afternoon as the solar angle and the vapor pressure gradient decline, slight recovery in guard cell turgidity may occur as absorbed water from the roots replaces some of that lost by the leaves earlier in the day. Transpiration and photosynthesis may then increase again for a short time. With the approaching sunset late in the afternoon, stomata begin to close for the day when the dwindling light becomes limiting (Fig. 5).

As transpiration declines in the late afternoon or evening, leaf water potential and its components begin a slow recovery as soil water absorption replaces water lost within the tissues during the day. The lag in rate of water absorption behind rate of water loss is overcome as stomatal closure progresses, and ultimately absorption exceeds the rate of loss. Assuming that soil water is freely available, the water potential in the soil immediately adjacent to the absorbing surfaces of the roots may also experience a slight increase as the water potential gradient between the root and bulk soil decreases (Fig. 5). By night-fall the stomata are closed, and leaf water potential and its components gradually increase throughout the evening and night, eventually reaching a maximum sometime during the early morning hours before the next sunrise. At that point the gradient between soil and leaf water potential is once again at a minimum, just prior to when the diurnal cycle begins again.

In most rangeland locations extended periods of clear dry weather often prevail during the growing season. As long as water is available to the roots, the loss of water during the day can be replaced at night, and hence osmotic and pressure potentials recover most (but not all) of the deficit. Although there is some loss in positive pressure (turgor) potential and possibly lower photosynthesis when midday stomatal closure occurs, the net effect of this stress is only temporary and of little consequence.

Longer-term plant water deficits occur if water is removed from the soil over a period of days or weeks, but is not replaced by precipitation or irrigation (Fig. 6). Under such conditions, a daily cycle of water absorption begins with mild transitory periods of stress during the day in which stomata may temporarily close when guard cell turgidity falls, as described above. However, with each successive day that the plant removes additional water from the soil, the resistance to soil water movement to the roots increases (water movement becomes slower), and eventually exceeds the ability of the plant to replace that lost to continued transpiration (Fig. 6). Both total leaf and soil water potentials decrease a little more each day, slowly at first but becoming more rapid as the degree of water deficit increases. Gradually, the stomata remain closed for longer periods each day as the pressure potential of the guard cells continues near 0 MPa. Also, leaves may begin to wilt due to a loss in pressure potential, and absorption at night replaces progressively less of the water deficit in the shoots. Wilting of the plant leaves may occur (day 5 in Fig. 6) when the soil water potential declines (or when soil water hydraulic conductivity is too low to maintain liquid water flow) below the level at which the

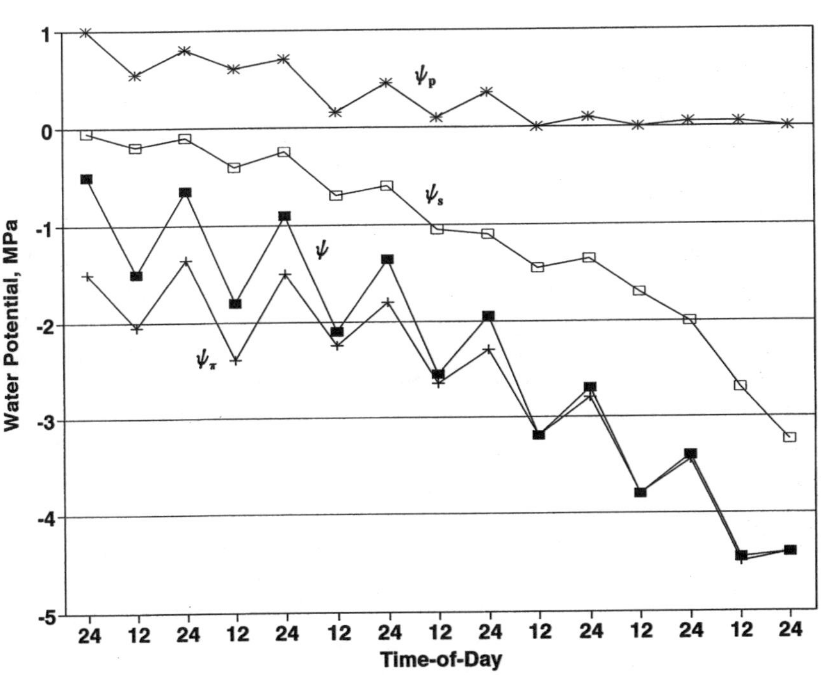

Figure 6. Graphic representation of the trend in a developing water deficit for a plant showing total leaf water potential ( ), osmotic potential (ψ), pressure potential (ψ_p), and soil water potential (ψ_s) at noon and midnight over a 7-day period.

plant can absorb additional water. At this point, leaf pressure potential remains at or near 0 MPa, and total leaf water potential is equivalent to leaf osmotic potential (e.g., see Figures 4–6).

Generally, mortality of tissues or plants is not significant at this point during a water deficit cycle. If soil water is replaced during, or soon after wilting occurs or when pressure potential remains at 0 MPa, damage to the plant is largely restricted to a temporary loss in photosynthesis, but recovery of the plant can be rather rapid. However, if the soil water deficit persists, the plant will continue to dry down, the leaves will remain wilted and eventually die. Mortality of leaves and some shoot material due to water deficits is not necessarily equated with mortality of the plant in most range plant species. Some shoot tissue mortality due to transitory water deficits may be common in arid or semiarid regions, but complete plant mortality may not occur at all if water deficits are relieved. However, if water deficits persist, or occur frequently and severely enough that regrowth and periodic recovery are precluded, lethal levels of dehydration to meristems and critical physiological processes may ultimately result in plant death (see discussion below).

Effects of Plant Water Deficits

The effects of water deficits are of considerable concern to range managers because their occurrences have dramatic effects on seed germination, seedling development and survival, plant growth, productivity, vigor, competitive ability, reproduction, and ultimately on individual plant and plant community survival. Although water deficits that occur during drought rarely result in the total mortality of a flora, they do have profound effects on short-term management options and stocking rates of grazing animals on ranges. Obviously, loss of biomass and plant development are immediate consequences of water deficits. With loss of vigor and competitive ability of desirable plants there is an increasing probability that less desirable species will invade and become established. Continued grazing pressure under such conditions can ultimately lead to deteriorated range condition resulting in lower stocking rates, deteriorated vegetation, and soil erosion (Meeuwig 1960, 1965).

Plants are susceptible to water deficits throughout their life-cycle, but are particularly vulnerable during seed germination, seedling emergence and early growth, and during periods of active growth and reproduction as mature organisms. Rangeland plants in semiarid and arid environments are especially subject to the effects of water deficits throughout the growing season at all stages of plant development. The discussion below summarizes some of the effects of water deficits on plants at critical stages of growth and development, and attempts to illustrate some of the management options available to minimize the effects of water deficits on range plants. Although overall effects of water deficits appear to be negative, deficits can also have some positive effects on plants. An attempt is made here to emphasize both the beneficial effects of water deficits, such as acclimation, as well as the obvious negative ones.

Seed Germination

Seed germination is the initial process in a complex sequence of steps leading to the development of seedlings. Germination involves imbibition of water, development of the seed embryo, rupture of the seed coat, and emergence of the radicle or root (Osmond et al. 1980a). The imbibition of water by a dry seed involves the absorption of water by the dormant living embryo and the nonliving endosperm reserves, terminates the dormancy state, and initiates metabolic activity leading to germination. Prior to imbibition the seed is least sensitive to environmental extremes, but after imbibition environmental variations have considerable control over the germination process.

The rate of seed germination can be a critical factor in successful establishment of plants, especially in arid and semiarid environments where favorable conditions of water availability and temperature regimes may only occur over a short period of time (Johnson 1980; MacMahon and Schimpf 1981; Owens and Call 1985). Generally the seeds of most plants can maintain physiological activity at much lower water potentials than can other developmental stages in the life cycle of the plant. Seeds of many aridland species commonly remain viable for years under extremely dry conditions and germinate only when conditions are optimum. Many species of range plants avoid the consequences of drought in the seed stage of development rather than in vegetative stages.

Seeds within the soil under natural conditions rely on a highly variable and sometimes hostile environment to meet the requirements for germination (Chambers 1989; MacMahon and Schimpf 1981; Meyer and Monsen 1992). Soil water status, temperature, insects, rodents, pathogens, salinity, light, and seed viability are but a few of the variables that affect germination following imbibition. Variability in precipitation amount and duration, together with temperature and rate of soil drying with depth also affect the germination environment of seeds (Amen 1966; MacMahon and Schimpf 1981). Also, the seeds of many species are dormant when released from the plant and may remain dormant for several years before environmental conditions are favorable for initiating germination (Amen 1966, 1968; Baskin and Baskin 1985; McDonald 1976; Young and Evans 1984). In some species (Young and Evans 1984) dormancy occurs because of an impermeable seed coat to water and/or oxygen, whereas in other species metabolic arrest of the seed embryo appears to cause dormancy (Amen 1966; McDonald 1976; Jones 1990). In some species (e.g.,Indian ricegrass [*Oryzopsis hymenoides*], Jones, 1990) both mechanical and physiological, or embryo, dormancy occurs. Mechanical dormancy may be broken by physical scarification through contact between the seed and soil particles during freezing and thawing events (Young et al. 1983) or as a result of ingestion and later expulsion by animals (e.g., McAdoo et al. 1983) to remove or penetrate hardened seed coats. Physiological dormancy is normally overcome by stratification under cool, moist conditions, or by use of certain hormones such as gibberellic acid or kinetin (Osmond et al. 1980a; Jones 1990). Many species require long periods of leaching by water to remove endogenous inhibitors (MacMahon and Schimpf 1981), and in arid and semi-arid environments numerous precipitation events may be required before leaching is complete. Some species stagger the periods of dormancy among different seeds, which appears to favor germination success when environmental conditions are optimum (Baskin and Baskin 1985; Meyer and Monsen 1992). Taylorson (1986) suggests that some species (e.g., giant foxtail, *Setaria faberi*) may even require a water deficit cycle to break dormancy and that exposure to water deficits may enhance germination (afterripening).

The seeds of some species require light for germination in addition to water, temperature, and other requirements (Amen 1966; Baskin and Baskin 1985; Haggas et al. 1987; Osmond et al. 1980a). Generally, light penetration into soil is restricted to the upper few mm (Osmond et al. 1980a), hence seeds in this zone may be subjected to a far harsher environment than those deeper in the soil (Haggas et al. 1987). Periodic drying, together with large temperature extremes near the soil surface, probably result in high mortality of seeds within a few mm of the soil surface.

The effect of water availability on seed imbibition and germination has been studied extensively in crops (e.g., Hadas 1970; Koller and Hadas 1982; Osmond et al. 1980a), but a large number of range species have also been studied (e.g., El-Sharkawi et al. 1989; Frasier 1989; Herbel and Sosebee 1969; Knipe 1968; McDonough 1975a, 1975b; Mueller and Bowman 1989; Owens and Call 1985; Romo and Eddleman 1985, 1988; Roundy 1985a, 1985b; Roundy et al. 1985; see MacMahon and Schimpf 1981 for a summary of critical germination water potentials of desert species). The effects of low water potentials on biochemical processes within seeds has been discussed by Wilson (1971) and Wilson and Harris (1966, 1968). Seed germination of most species is inhibited dramatically by decreases in seed or substrate water status, but because of broad variability among (and within) species, generalizations are difficult to make.

MacMahon and Schmipf (1981) feel that seed germination may be the life cycle stage in plants requiring the highest water potential. A major source of variation in defining the critical water status at which seed germination is inhibited is temperature, which appears to mediate the optimum and minimum water potential for germination of many species (El-Sharkawi et al. 1989; Romo and Eddleman 1985, 1988; MacMahon and Schimf 1981; Osmond et al. 1980a; Owens and Call 1985; Romo et al. 1991; Sharma 1976). Generally, temperatures at the high end of the favorable range shorten the time of germination and allow seeds to germinate under only transiently favorable water conditions (McDonough and Harniss 1974; Owens and Call 1985). In addition, the optimum water and temperature levels for germination may shift in different seasons and for seeds collected in different years (e.g., Romo and Eddleman 1988), presumably due to environmental conditions during seed development. Romo et al. (1991) found that exposure to descending temperatures and water deficits reduces germination in plains rough fescue (*Festuca altaica* subsp. *hallii*). Decreasing temperatures and water deficits together are characteristic of summer or fall conditions, and these may induce seed dormancy that may result in the preservation of seedbanks for later germination. However, water deficits under increasing temperatures, characteristic of spring conditions, were less restrictive on germination of this species. Chambers (1989) found high variability in alpine seed viability in different years, and attributed this to the harsh and unpredictable environmental conditions under which the seeds are produced.

In general, little apparent correlation has been observed between the minimum critical water potential at which germination occurs and the origin of the species (MacMahon and Schimpf 1981); many desert and rangeland species appear to require similar water potentials for germination as do many crops and other drought intolerant species. However, Doescher et al. (1985) found that collections of Idaho fescue (*Festuca idahohensis*) from sagebrush-grass habitats in Oregon have higher germination rates over wider temperature and water deficit regimes than collections from conifer-grass habitats. Despite large variation among species and environmental conditions, the minimum water potentials at which seeds of species from widely different habitats germinate is surprising. For example, Koller and Hadas (1982) quote critical soil matric potentials for germination of crop species such as corn (*Zea mays*), sorghum (*Sorgham bicolor*), clover (*Trifolium alexandrium*), cotton (*Gossypium* sp.), and chickpea (*Cicer arietinum*) of –1.4, –2.0, –0.45, –1.1, and –1.5 MPa, respectively. Lindstrom et al. (1976) observed winter wheat (*Triticum aestivum* L.) germination at soil water potentials as low as –2.0 MPa, although complete seedling emergence did not occur below –1.0 MPa.

The limited available data of critically low water potentials for seed germination of rangeland species show similar results as those reported for crops. As with most plants, when seeds of rangeland species are subjected to increasing levels of water deficit, total germination is reduced, and the amount of time required for germination is increased (e.g., McGinnies 1960). Roundy et al. (1985) found that tall wheatgrass (*Agropyron elongatum*) and Great Basin wildrye (*Elymus cinereus*), common grasses of the Great Basin in North America, both germinated at soil matric potentials of about –2.0 MPa, although germination declined rapidly below –0.5 MPa. Sharma (1976) found similar results for three Australian semi-arid species; wallaby grass(*Danthonia caespitosa*), oldman salt bush (*Atriplex nummularia*), and bladder saltbush (*A. vesicaria*). All three species displayed some, although different, germination levels at the lowest tested matric potential of –1.5 MPa. As with the Great Basin grasses (Roundy et al. 1985), the three Australian species showed marked decreases in germination as matric potential was reduced below about –0.8 MPa. In both studies, maximum germination was observed at 0 MPa, although sharp declines were not observed until matric potentials fell below about –0.4 to –0.5 MPa. McDonough (1975b) found that Russian thistle (*Salsola kali*) seed germination is progressively inhibited at water potentials below –0.2 MPa, but that perceptible imbibition and swelling occur down to –3.1 MPa. McDonough (1975a) studied the changes in water potential that occur in seeds during the germination process of 11 native species and found that generally seed water potential increases sharply during initial imbibition, then declines dramatically as water is redistributed within the osmotic and matric environment of the seed. Although seeds were placed in a water potential environment near 0 MPa, the actual water potential of the germinating seeds fell to as low as –1.7 MPa. This suggests that actual germination may occur at lower water potentials than experimental conditions might indicate. Johnson and Asay (1978) developed a technique for studying seed germination and seedling emergence as affected by soil matric potential. They found that seeds of various genotypes of crested wheatgrass (*Agropyron desertorum*) varied widely in their response.

Graded osmotic solutions have been used more extensively than matric potential in studies to determine the effect of water potential on seed germination of different range species, primarily because the water potential of osmotic solutions is easier to control than soil matric potential. For example, Knipe (1968) found that the effect of water status on germination of alkali sacaton (*Sporobolus airoides*), galleta (*Hilaria jamesii*), and blue grama (*Bouteloua gracilis*) not only varied with species, but that germination occurs down to osmotic potentials of -1.6 MPa. Hardegree and Emmerich (1990) and Emmerich and Hardegree (1991) observed germination of sideoats grama (*Bouteloua curtipendula*), kleingrass (*Panicum coloratum*), buffelgrass (*Cenchrus ciliaris*), and Lehmann lovegrass (*Eragrostis lehmanniana*) at osmotic potentials as low as –1.2 MPa in various solutions of salts and polyethylene glycol. They observed both lower and slower germination rates with decreasing water potential. In addition, they found that $CaCl_2$ had the least detrimental effect, and sometimes had a positive effect, on germination when compared with NaCl, K_2SO_4, and polyethylene glycol. Generally they observed that K_2SO_4 had a more negative effect on germination than NaCl. Romo and Eddleman (1988) show evidence of germination in green rabbitbrush (*Chrysothamnus nauseosus* (Pallas) Britt. subsp. *graveolens* (Nutt.) Piper.) and gray rubber rabbitbrush (*C. nauseosus* (Pallas) Britt. subsp. *nauseosus* (Nutt.) Piper.) at osmotic potentials between about –0.8 and –1.5 MPa, depending on temperature. Also using osmotic solutions, Romo and Eddleman (1985) observed germination in greasewood (*Sarcobatus vermiculatus*) at –3.6 MPa at 10 and 20°C, but only at –3.0 MPa at 30°C temperature. Generally, data from these studies show that seed germination appears to occur at lower osmotic potentials than that reported for matric potentials of soil.

Unfortunately, the use of osmotic solutions as a medium for definition of critical water potentials may be misleading because seeds tend to germinate more readily in solutions than in soil (matric) environments (Koller and Hadas 1982; Wright 1975). Also, in some cases the results may be difficult to interpret because of changes in solution water potential during the course of germination (Emmerich and Hardegree 1991). As pointed out by MacMahon and Schimpf (1981), an equal water potential of an osmotic solution and a soil matrix are only equivalent in terms of the total water potential gradient between the substrate and the seed. A given matric potential in soil has a lower resistance to water movement and exchange area between the seed and the soil particles than an equivalent osmotic solution (Roundy et al. 1985). Although the solute itself in osmotic solutions may stimulate or retard germination, and thus make interpretation of data difficult (e.g., Roundy et al. 1985; Hardegree and Emmerich 1990), other workers (e.g., Emmerich and Hardegree 1990) have not observed any adverse affects of polyethylene glycol contact on seed germination.

Our understanding of the interactions of seed germination and soil water status is not adequate for most rangeland species and more comprehensive research is needed. However, based on what is currently known about the water relations of seed germination, the land manager does have some important options available that may be useful during range improvement and revegetation efforts. For example, during revegetation of arid or semiarid ranges, seeds of adapted species should be planted during the proper season and at specific depths in the soil to ensure optimum germination (Hull 1974; Hull and Holmgren 1964; Dowling et al. 1971). Also, the soil surface should be stabilized against the drying effects of wind and protected from raindrop impact and surface erosion, and from other forces that can dislodge or redistribute seeds prior to germination (Chambers et al. 1990). In arid and semiarid regions surface mulch may be necessary to reduce the effects of adverse soil drying in the zone where critical moisture is required for germination (Chambers 1989; Chambers et al. 1990; Packer and Aldon 1978; Sidle and Brown 1988). Seeds requiring light for germination (photoblastic) must be applied on the soil surface, or only slightly below the surface, and will require surface mulching to minimize wind redistribution and rapid evaporative drying (Haggas et al. 1987). Such species may need to be seeded separately from other species that can be incorporated into the soil. Using incorporated organic matter in poorly developed or coarse-textured soils appears to improve both water and nutrient holding characteristics of soil during germination (Brown and Chambers 1989, 1990; Anderson and Coleman 1985).

Seedling Growth and Survival

The most profound transition in the life of a vascular plant occurs during development of the germinating seed to an emerging seedling (Osmond et al. 1980a). During seedling establishment, the plant emerges from the dark moist soil to the drying effects of a terrestrial environment as its roots and shoots begin to coordinate the processes of absorption, photosynthesis, and other physiological processes that will carry it through to maturity and reproduction. The life of an emerging seedling is particularly perilous because of uncertainties that adequate quantities of both water and nutrients will be supplied to the plant.

Most of the water relations research on seedlings of rangeland plant species has been conducted with grasses rather than forbs or shrubs, presumably because the Gramineae family is viewed as having a higher economic value for grazing and other range management concerns than other lifeforms. Also, grasses are generally regarded as easier to work with and are usually more available commercially for revegetation applications than other types of plants. However, virtually all lifeforms of plants are important in range management whether for grazing, erosion control, slope stability, or other concerns. Many species have unique characteristics that link their physiological and morphological features to specific water relations concerns even though a high degree of variability exists among individual plants of a taxon. Thus, although the discussion here is dominated by principles derived from research on grasses, the basic tenets presented apply equally to all lifeforms.

The basic development and growth characteristics of seedlings of both monocotyledonous (grasses and similar lifeforms) and dicotyledonous (most forbs, shrubs, and trees) plants are reviewed and contrasted in Figures 7 and 8, respectively. Figure 7 represents the various major stages of development from germination to self-sustainability in many common grasses, and Figure 8 represents these stages for many forb and shrub species common to rangelands. The stages and conditions of establishment of grass seedlings have been discussed in detail by Hyder (1974) and Ries and Svejcar (1991). Ries and Svejcar (1991) summarize three primary stages during establishment: (1) the heterotrophic stage, when the young seedling is dependent upon reserves of the germinating seed (Fig. 7A); (2) a transition stage (Fig. 7B); and (3) the autotrophic stage when the seedling is no

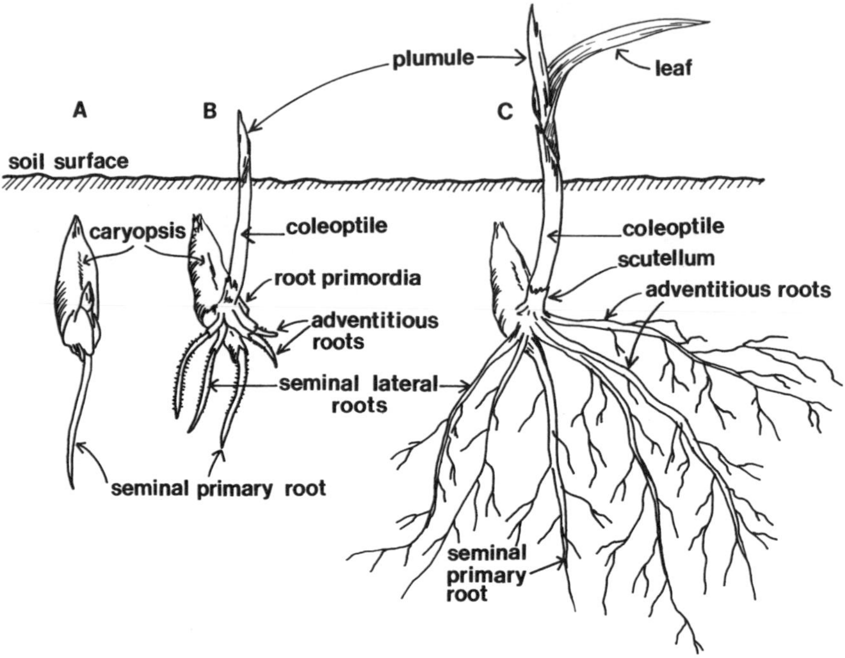

Figure 7. Diagram of an emerging grass (monocotyledon) seedling showing the major structures during development. A) during germination the seminal primary root (radicle) first emerges from the caryopsis (or seed), and the coleoptile begins to elongate from the scutellum toward the soil surface: B) seminal lateral roots and adventitious roots begin to emerge from root primordia, and the coleoptile approaches the soil surface: C) adventitious roots dominate rooting development, and the shoot begins to emerge from the coleoptile, from which emerges the plumule, becoming the culm or stem of a grass plant.

longer reliant on seed food reserves (Fig. 7C). Seedlings are highly susceptible to water deficits primarily because their tissues are not usually "hardened" or acclimated to stress conditions, and because root development and penetration may be inadequate to exploit large enough volumes of soil during periods of water deficits. The primary defense against water deficits during seedling development is the rapid proliferation of roots in both depth and surface area to outpace the rate of soil drying, and to tap larger volumes of available soil water (Hassanyar and Wilson 1978; Johnson et al. 1981; Osmond et al. 1980a; Plummer 1943). The entire range of growth and development represented in Figures 7 and 8 commonly occur in a matter of days in most range species under optimum conditions (Aguirre and Johnson 1991a; Hyder 1974; Johnson 1986).

In most arid or semiarid rangeland regions, such as western North America, the main periods of precipitation are normally seasonal and highly unpredictable. Because the wettest season does not necessarily occur during the growing season of plants in many rangeland regions, plant establishment is often affected by severe water deficits. For example, in many mid-continental regions precipitation occurs primarily during winter and spring months, hence plant growth and development during the summer and fall must rely on stored soil water (Johnson et al.

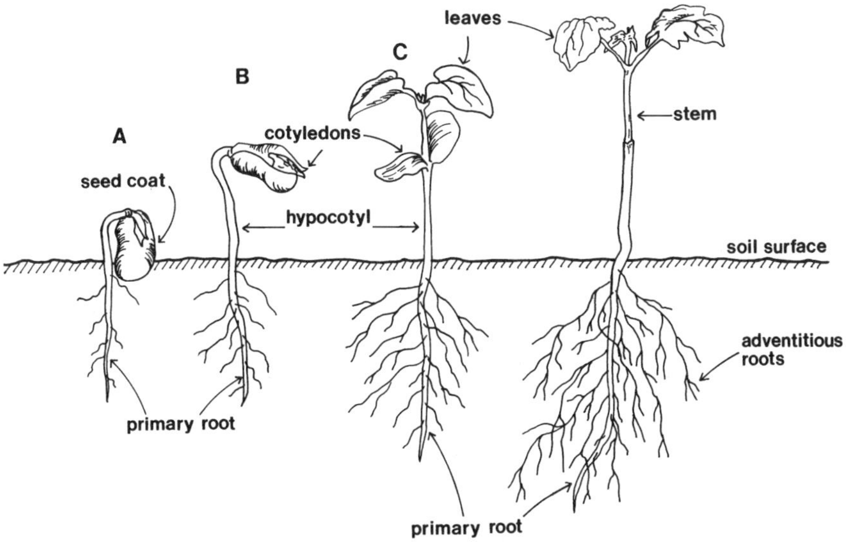

Figure 8. Diagram of an emerging dicotyledon seedling (forbs and shrubs) showing the development of the seedling from the germinating seed. A) the seed and endosperm, bearing the cotyledons, are often lifted above the soil surface during emergence, and the radical becomes the primary root: B) the seed coat is shed freeing the cotyledons, and the hypocotyle elongates: C) the true stem emerges above the cotyledons and bears the first leaves; the primary root elongates and the first adventitious roots originate from primordia within the lower hypocotyle: D) adventitious roots begin to predominate the root system, the cotyledons are shed, and the seedling is supported by photosynthesis of the emerging leaves on the developing stem.

1981). Under these conditions, seed germination and seedling emergence must occur during the spring when water and temperature conditions are most favorable. However, conditions are not always optimal during normally moist periods; expected amounts, duration, and distribution of precipitation can often be far below or above normal, and temperatures can often be much colder or warmer than normal. Thus, periodic water deficits are common as are periods of waterlogging (Bradbury and Yang 1981; Wallace and Nelson 1990), devastating freezes, and other conditions during periods of plant establishment when plants are most susceptible to severe environmental stresses. Because of such conditions, Cox and Jordan (1983) estimated that successful revegetation in southern Arizona may only be possible in one of every 10 years. The importance and severity of such occurrences during range plant establishment has led to numerous efforts to improve methods of seedling establishment (Chambers et al. 1990), and also to improve the predictability of successful seedling establishment on rangelands (e.g., Frasier 1989).

Among the more important factors affecting seedling survival under water deficits are: severity of water deficits, recoverability after water deficits, vigor during emergence, and rate of root development and other morphological characteristics (Aguirre and Johnson 1991a, 1991b; Asay and Johnson 1983a; Johnson 1980, 1986; Johnson and Aquirre 1991; Herbel and Sosebee 1969; Lindstrom et al. 1976; McDonough and Mathews 1970; Newman and Moser 1988; Ries and Svejcar 1991). Generally, as the severity of a water deficit increases the survival of developing seedlings decreases (Fullbright et al. 1984; Herbel and Sosebee 1969; Johnson et al. 1981; Lindstrom et al. 1976). Data on critical water potentials at which mortality occurs in seedlings is largely lacking for most range plants. However, variability among species, and among genetic strains within species (Asay and Johnson 1983a, 1990), appears to be high and is strongly correlated with temperature (Herbel and Sosebee 1969). Herbel and Sosebee (1969) found that both biomass and length of roots and shoots decreased with increasing water deficits in black grama (*Bouteloua eriopeoda*) and boer lovegrass (*Eragrostis chloromelas*). Lower soil temperatures were generally more favorable for survival and growth under water deficits than higher temperatures in both species. They concluded that more than three-times as much water would be required for seedling survival under high soil temperatures than under cooler soil temperatures. Lindstrom et al. (1976) found similar results for two winter wheat lines in which the minimum water potential for emergence increased (became wetter) with increasing temperature. Bouaziz and Bruckler (1989a, 1989b) feel that under dry conditions temperature linearly affects seedling growth rates of both corn and wheat, that time from sowing to emergence increases with decreasing soil water potential, and that O_2 diffusion rate rarely limits seedling growth. They showed that full seedling emergence of wheat was delayed from 5 to 13 days when soil water potential decreased from –0.02 to –1.31 MPa. Their analyses of the various parameters affecting wheat seedling emergence emphasize the importance soil water potential, and their subsequent model (Bouaziz and Bruckler 1989a, 1989b) was specifically designed for semiarid conditions to estimate seed losses during germination and seedling emergence.

Grass seedlings are generally considered successfully established when adventitious roots are prolific enough to supply adequate water and nutrients for continued growth and survival (Ries and Hofmann 1991). The crown of a grass seedling is defined by the placement of 2 or more nodes, and is particularly important for water relations because the crown defines the location from which adventitious roots originate. The depth at which seeds are planted (Lawrence et al. 1991), the morphological attributes of the developing seedling (Johnson and Aquirre 1991; Ries and Hofmann 1991), and the characteristic growth rates and patterns of both seminal and adventitious root can have a significant effect on seedling survival and growth in arid and semiarid environments (Hassanyar and Wilson 1978; Johnson and Aguirre 1991; Ries and Hofmann 1991). The primary seminal root is the first to develop from the germinating seed of many range grasses, and is analogous to the radicle in most other lifeforms of plants (Esau 1967). Seminal roots, or radicles, die shortly after adventitious roots begin to develop in most species, but play a significant role in water and nutrient absorption early in the life of the seedling. In many arid and semiarid regions the primary seminal root may die if exposed to early water deficits, but many range grasses such as crested wheatgrass and Russian wildrye (*Elymus junceus*) have lateral root primordia that can become active and replace the primary seminal root with lateral seminal roots (Hassanyar and Wilson 1978). The lateral seminal roots then become the principal sources of water and nutrient absorption.

Seedling vigor during emergence and early growth is widely used as an indicator of survivability during water deficits. Variables that have been used to indicate relative vigor include ability of seedlings to compete, ability to emerge from deep soil depths, rate of shoot and root growth during emergence, rate of respiration, and rate of germination under high temperature stress (e.g., Wright 1975). Generally, measures of seedling vigor are relative comparisons of plant responses to specific environmental conditions that affect the genotypic expression of that plant. Asay and Johnson (1980) used seedling emergence with depth to screen individual breeding lines of Russian wildrye (*Psathyrostachys juncea*, formerly *Elymus junceus*) and found significant differences among lines for emergence and plant height. In arid or semiarid environments seeding depth is an important consideration of seedling vigor because deeper planted seeds, and subsequent seedlings, generally are exposed to more moderate water deficit conditions than those nearer the soil surface (Asay and Johnson 1980; Johnson 1980, 1986). Plummer (1943) found that seeding depths of about 0.64 cm (0.25 in) resulted in the greatest seedling emergence of 12 different range grasses. Deeper depths (about 3.8 cm) resulted in a marked reduction in shoot emergence of most except the larger-seeded species, but did not totally prevent emergence of even small seeded species. Larger, denser seeded species apparently have an advantage for emerging from greater depths. Asay and Johnson (1983a) found that seed weight of crested wheatgrass and Russian wildrye may be a useful criterion to judge relative seedling vigor for breeding and selection within a species.

One of the most critical attributes in selecting adapted plants for range improvement and revegetation in semiarid environments is that of early root initiation and elongation, followed by rapid root extension (Aguirre and Johnson 1991b; Johnson 1980, 1986; Johnson and Aguirre 1991; Johnson and Asay 1987; Plummer 1943; Ries and Svejcar 1991). The ability of seedlings to survive water deficits is strongly related to their ability to develop deep and spreading root systems that penetrate the soil more quickly than the rate of soil drying (Johnson and Aguirre 1991; Plummer 1943). The rate of root growth and total root length formed by developing seedlings following germination varies widely with plant species, time after germination, and environmental conditions. Plummer (1943) found that species with slower root development are normally more difficult to establish in range seedings, whereas species with rapid root development generally can be established more successfully. He feels that the universal success of crested wheatgrass (*Agropyron cristatum*) in western U.S. range seedings is in part related to the greater total root length observed in this species as a seedling than that of most other species.

Manipulation of seeding depth is often suggested as a means to avoid, or minimize, unfavorable soil water conditions near the soil surface during seedling emergence (Hyder 1974; Lawrence et al. 1991; Redmann and Qi 1992). However, the morphological characteristics of emerging seedlings of different species can critically limit the depth at which seeds can be planted, especially in arid or semiarid environments (Hyder 1974; Johnson 1986; Johnson and Aguirre 1991; Redmann and Qi 1992). Two of the more common morphological characteristics of grass seedlings that may significantly affect plant water relations are illustrated in and follow the classification suggested by Hyder (1974). Type A seedlings (Hyder 1974) include those grass species characterized by an elongated sub-coleoptile

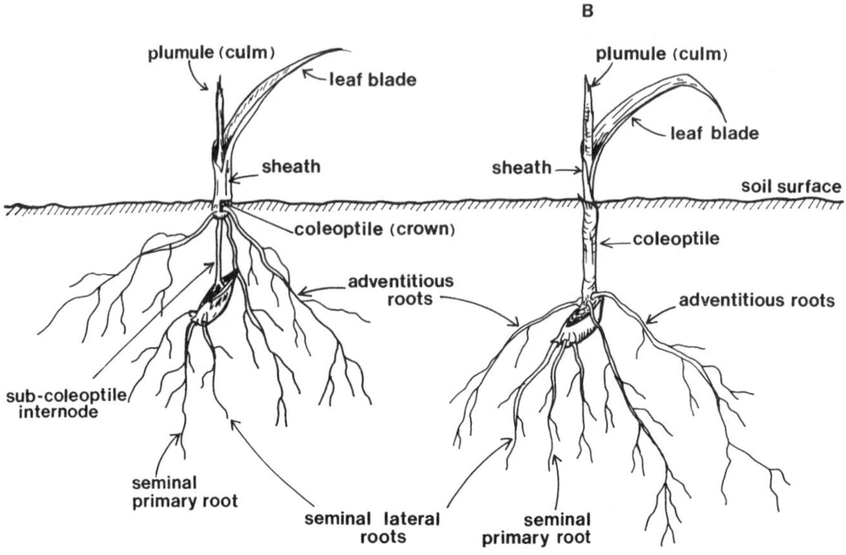

Figure 9. Comparison of two different types of seedling morphological characteristics that may influence seedling water relations. A) elongated sub-coleoptile internode and short coleoptile, with adventitious roots originating near the soil surface. B) Long coleoptile lacking elongated sub-coleoptile internode, showing origin of adventitious roots at deeper soil depths (after Johnson 1986).

internode (the internode between the cotyledonary and coleoptilar nodes) that develops from the caryopsis during germination (Fig. 9A). The elongating sub-coleoptile internode supports a short coleoptile from which the primordia for culms and adventitious roots originate. Type B seedlings include those species without an elongated sub-coleoptile internode (Fig. 9B), but which have the potential for developing an elongated coleoptile. The growth form of type A seedlings places the crown (or base of the coleoptile, hence the primordia that develop culms and adventitious roots) near the soil surface where the greatest variability in soil water and temperature occur. In such seedlings, early seedling development is entirely dependent upon the seminal roots for water and nutrient absorption. The growth patterns in type B seedlings, however, establish the crown of the seedling at the coleoptilar node adjacent to the caryopsis. Type A plants include many C_4 and warm season species such as blue grama, sideoats grama, and little bluestem (*Andropogon scoparius*), many of which are common to the North American Great Plains (Jones 1985; Newman and Moser 1988; Redmann and Qi 1992). Type B plants include many cool season and C_3 species such as thickspike wheatgrass (*Agropyron dasystachyum*), crested wheatgrass, western wheatgrass (*A. smithii*), alti wildrye (*Elymus angustus*), Russian wildrye, and others (Johnson 1986; Jones 1985; Newman and Moser 1988; Redmann and Qi 1992).

Seedling developmental characteristics are important to plant water relations because growth pattern affects the placement of the plant crown relative to the soil surface. Obviously, seed depth in the soil is dependent upon many variables following natural seed dispersal, and includes seed size, seed shape, soil characteristics, and other factors (Winkel et al. 1991b). Seeding during revegetation (see section on Rangeland Revegetation and Plant Water Relations, below) may provide an opportunity to control this variable to avoid water deficits during seedling development for some species, but it may not always be a reliable mechanism for all species (Hyder et al. 1971; Lawrence et al. 1991; Tapia and Schmultz 1971; Wilson and Briske 1979; Wilson et al. 1976). Type A species tend to elevate the plant crown to within a few mm or closer of the soil surface regardless of seed depth, thus exposing the adventitious root and shoot primordia to potentially lethal water deficits. In fact, type A species may require light exposure at the node between the sub-coleoptile internode and coleoptile before adventitious roots develop (Roohi et al. 1991). In blue grama, Roohi et al. (1991) found that light is required at the 3-leaf or later stages during shoot development before adventitious roots can form at the coleoptile, thus insuring exposure of the crown to potentially desiccating conditions. Type B species, however, tend to initiate adventitious roots and shoots at greater depths (if seed placement is appropriate), thus avoiding such large variations in soil water availability.

The interactions of plant morphological characteristics and soil-plant water relations have far-reaching significance in arid and semiarid regions, especially when seedling development is dependent upon stored soil water (Johnson 1986). The seminal primary root is normally short-lived and may not always penetrate the soil at a rate sufficient to equal or exceed the rate of soil drying (Hassanyar and Wilson 1978; Wilson and Briske 1979), especially under severe drought conditions or when temperatures are sub-optimal for growth. In some species seminal lateral roots may emerge and sustain water absorption long enough for the initiation of adventitious roots, but in some type A species (e.g., blue grama), seminal lateral roots are lacking. In addition, there is a limit to the amount of water seminal roots can absorb, hence a limit to the leaf area that can be supported by them. When conditions for transpiration are maximized (hot dry winds, low humidity, and high air temperatures), seedlings become particularly vulnerable to desiccation. Redmann and Qi (1992) showed that type A seedlings have smaller diameter xylem vessels than type B seedlings, and that the number of xylem vessels decrease within the sub-coleoptile internode with depth in the soil. These characteristics tended to result in lower hydraulic conductances in type A plants than type B plants. Even when adventitious roots begin to develop, there is always the danger that the rate of soil drying will exceed the rate of adventitious root initiation and growth, especially if the coleoptilar node is near or at the soil surface.

Ries and Svejcar (1991) evaluated the development of adventitious roots and leaf growth to determine when blue grama and crested wheatgrass become established under field conditions. They found that for the conditions of their study, 21 days were required for establishment following seedling emergence. At emergence, crested wheatgrass had 4 leaves, 2 adventitious roots to a soil depth of 80 mm, and one tiller, whereas blue grama had about 6 leaves, 2 adventitious roots to a depth of 100 mm, and 2 tillers per plant. At this level of development, most seedlings of both species were capable of absorbing soil water, surviving winter conditions, and initiating growth the following spring. Lawrence et al. (1991)

suggested that tetraploid cultivars of some Triticeae species may be less affected by variable seed depths than many other more widely used species.

The morphology of root development during germination and emergence at least partially determines tolerances to water deficits in developing seedlings. Ries and Hofmann (1991) and Johnson and Aguirre (1991) have stressed the importance of root development morphology and adaptations of various range grasses for competition and water relations. Fullbright et al. (1984) found that the development of both seminal lateral roots and adventitious roots in seedlings of green needlegrass (*Stipa viridula*) was stimulated by the initiation of water deficits during seedling growth. They also noted that the number and length of seminal roots in seedlings decreased with increasing severity of water deficits, but adventitious root length was not affected. Hassanyar and Wilson (1978) observed root development in crested wheatgrass and Russian wildrye following exposure to various levels of water deficits. They observed that seedling survival under temporary water deficits, which may often kill primary seminal roots, depends on subsequent development of seminal lateral roots. Under adequate soil water conditions, no differences were observed between the two species in the percentage of seedlings that developed seminal lateral roots or in root number. However, following various levels of temporary water deficits, crested wheatgrass developed more and longer lateral seminal roots than Russian wildrye. They concluded that species and cultivars whose root apices possess low tolerance to water deficits, that exhibit a low inherent capacity for seminal lateral root development, or that entirely lack seminal lateral roots, are more susceptible to injury by water deficits. Differences among species in seminal lateral root development following a temporary water deficit apparently are due to variability in latent potential for growth of seminal lateral roots and to differences in tolerance to water deficits in root apices (Hassanyar and Wilson 1978). Species whose root apices are not tolerant of water deficits may not be able to form lateral seminal roots.

The ability of a seedling to recover following a water deficit is an important characteristic of potential plant survival on arid and semiarid rangelands (Bassiri et al. 1988; Hassanyar and Wilson 1978; McAlister 1944). Exposure of developing seedlings to dehydration and determinations of subsequent recovery have shown that generally grasses are more tolerant than legumes (Bassiri et al. 1988). Bassiri et al. (1988) exposed seedlings of crested wheatgrass, Russian wildrye, alfalfa (*Medicago sativa*), and Cicer milkvetch (*Astragalus Cicer*) to water potentials ranging from –10 to –220 MPa. Survival of the grasses depended upon the formation of seminal lateral roots (Figures 7 and 9), but legume survival depended on development of new meristems along the hypocotyl-root axis (see Fig. 8). Following exposure at –220 MPa for 4 days, percent emergence for crested wheatgrass, Russian wildrye, alfalfa, and Cicer milkvetch was 59, 35, 6, and 1, respectively, and percent rooting was 58, 12, 3, and 1, respectively. Asay and Johnson (1983b) found significant differences in seedling recovery following water deficits among crested wheatgrass [(*Agropyron desertorum*) and (*A. cristatum*)] progeny, but not among those for Russian wildrye. A time interval of just a few hours between the initiation of a water deficit and when water is made available was critical to seedling survival, being shorter for Russian wildrye than for crested wheatgrasses. Asay and Johnson (1980) and Johnson and Asay (1978) also screened various genetic lines of both crested wheatgrass and Russian wildrye for seedling emergence under drying conditions and found significant differences in response to water deficits. Generally, seedling emergence of crested wheatgrass was lower than for Russian wildrye with decreasing soil water potential.

Highly competitive species may replace desirable species on rangelands, and their success is often related to seedling vigor and growth characteristics (Aguirre and Johnson 1991a, 1991b). Cheatgrass (*Bromus tectorum*) is a well known aggressive annual invader on western rangelands in the U.S., which is replacing desirable native species such as bluebunch wheatgrass (*Pseudoroegneria spicata*, formerly *Agropyron spicatum*) and Snake River wheatgrass (*Elymus lanceolatus*). Because bluebunch wheatgrass and Snake River wheatgrass are poor competitors with cheatgrass, use of other more aggressive perennial species such as the hybrid "Hycrest" (*Agropyron desertorum* x *A. cristatum*) may have several advantages for use in rangeland revegetation. In a comparative greenhouse study, Aguirre and Johnson (1991b) found that seedlings of cheatgrass had greater plant height, leaf area, shoot dry weight, primary root length, number and order of root branching of the second group of seminal roots, order of branching of the first group of adventitious roots, and total root dry weight than Hycrest, bluebunch wheatgrass, or Snake River wheatgrass (Whitmar and Secar). Aguirre and Johnson (1991a) found that cheatgrass elongated adventitious roots earlier and at cooler temperatures than the perennial species. Hycrest was more competitive with cheatgrass than bluebunch wheatgrass, although competition reduced seedling growth of both perennials. Soil water was reduced more by cheatgrass-Hycrest mixtures than in cheatgrass-bluebunch wheatgrass mixtures, suggesting that Hycrest seedlings were more effective competitors with cheatgrass than bluebunch wheatgrass. Johnson and Aquirre (1991) compared the root branching order of cheatgrass, Hycrest, and bluebunch wheatgrass under different levels of soil water availability. They found that cheatgrass had a greater number and order of branching of seminal roots, and initiated adventitious roots earlier than the other grasses under all water treatments. Under the lowest levels of watering, cheatgrass had longer seminal roots on all sampling days than either Hycrest or bluebunch wheatgrass. However, they found that Hycrest had greater root branching densities, lateral root length, and external-external link lengths than bluebunch wheatgrass, and hence is a more successful competitor with cheatgrass under water deficit conditions than bluebunch wheatgrass. These studies illustrate the types of data needed to quantify characteristics of different range species for improved seedling establishment of perennial range grasses and other lifeforms on rangelands.

The soils of many rangelands are affected by salinity, and under arid or semi-arid conditions these soils can be limiting to successful seed germination, seedling development, and survival (Mueller and Bowman 1989; Roundy 1985a, 1985b). High salt concentrations not only lower the osmotic potential of soil water, but also may be toxic to germinating seeds and seedlings, and frequently delay germination and emergence of most range plants. If plants are to become established in such areas, not only must they be adapted to salinity, but the soil must remain moist for longer periods into the spring and summer. Mueller and Bowman (1989) found that seeds of Russian wildrye and western wheatgrass require longer germination periods than crested wheatgrass, and that during this time interval, soil water potentials declined sufficiently to reduce germination and emergence. However, they found that pregerminated seeds established more rapidly and were less affected by salts and water deficits during emergence than untreated seeds in salt-affected soils subject to rapid drying. Also, salinity apparently slows the rate of root growth in many species, leading to increased chances of water deficits later in the growing season. Mueller and Bowman (1989) suggested that using pregerminated seeds of plants in range reseedings of salt-affected soils could result in not only more rapid germination, but also in greater seedling emergence, root biomass, and root length. However, they cautioned that cultural practices for successfully planting pregerminated seeds in large-scale seedings that minimize desiccation of seedling radicles have not been developed.

Roundy (1985a) studied seedling establishment of Great Basin wildrye and tall wheatgrass in saline arid soils of Nevada. He found that Great Basin wildrye was not as tolerant of salts or water deficits as tall wheatgrass, and concluded that arid saline soils in the Great Basin Region should not be seeded without supplemental irrigation. Although tall wheatgrass apparently has greater tolerance to water deficits and salinity as a seedling than Great Basin wildrye, he found it was less tolerant and persistent on dryland saline soils when mature than Great Basin wildrye. If Great Basin wildrye, a native grass in the Great Basin, is to establish itself on saline soils, high and frequent precipitation is required in the seedling stage.

Roundy (1983) studied the effects of various salts on growth and development of basin wildrye and tall wheatgrass seedlings. He found that on an absolute basis, tall wheatgrass produced much higher shoot growth than basin wildrye under all salinity conditions, and only slightly greater root biomass. Root/shoot ratios of basin wildrye doubled when salinized to a soil osmotic potential of –1.5 MPa, but then decreased with further decreasing osmotic potential. However, tall wheatgrass root/shoot ratios decreased by half when salinized to only –0.5 MPa soil osmotic potential, and then increased slightly as osmotic potentials decreased further. He feels the increase in root/shoot ratios of basin wildrye down to –1.5 MPa may be an adaptation to arid soils where reduced transpiration may result in longer periods of higher soil water status and extended growth period. He also found that salinized plants of both species are capable of maintaining higher turgor potentials at low total leaf water potentials than unsalinized plants.

McCree and Richardson (1987a) observed greater water-use efficiency (grams of water transpired per gram of biomass produced) in salinized crop plants (sugarbeets and cowpea) than in unsalinized plants. Power (1980) found that nitrogen fertilization of perennial grasses on the Great Plains also resulted in increased water-use efficiency for smooth brome (*Bromus inermis*). Data on water use efficiency of range plants under arid saline conditions is generally not well documented, but studies in this area would be highly beneficial.

In addition to salinity effects, some rangelands are also affected by high concentrations of various elements that may be toxic to plant establishment. For example, Roundy (1985b) found that B (boron) uptake by Great Basin wildrye and tall wheatgrass seedlings is affected by both boron concentration and soil water status, pH, organic matter content, and other factors. He found that the lower tolerance of Great Basin wildrye to B may be due indirectly to reduced seedling survival by decreased root growth, which prevents root extension below the zone of soil drying. Tall wheatgrass is apparently more tolerant of B, and may be able to support more extensive root development and therefore avoid severe soil drying during the growing season. As soil water content decreased due to drying, B concentrations increased substantially.

Mature Plant Growth and Survival

Water deficits affect virtually every aspect of plant growth, development, and reproduction. The consequences of water deficits are far too numerous to discuss here in exhaustive detail, and the interested reader is referred to other sources for a more complete treatment (e.g., Bradford and Hsiao 1982; Gates 1968; Hanson and Hitz 1982; Hsiao 1973; Hsiao et al. 1976a, 1976b; Kramer 1983; Slatyer 1967; Turner and Burch 1983).

Water deficits occur in plants when the internal water potential and pressure potential of tissues decline sufficiently to interfere with normal physiological activity. The most important consequences of water deficits in plants during the growing season occur during periods of active vegetative growth and reproduction. In mature range plants the most serious consequences of water deficits are reduced vegetative growth, smaller plant size, and hence lower total plant biomass (Barker et al. 1989; Bittman and Simpson 1987, 1989a, 1989b; Ganskopp and Bedell 1979; Knapp 1984; McIntyre 1987; Miller et al. 1986a, 1986b; Sala et al. 1982; Sampson 1952; Stoddart et al. 1975). During vegetative growth of mature plants, water deficits have widely varying effects depending on the duration and intensity of dehydration, and upon the age, stage of development, and recent history of the affected plant. Generally, the longer the duration and intensity of water deficits, the more injurious the effects will be (Bradford and Hsiao 1982; Kramer 1983). In the long-term, water deficits probably reduce range plant growth and development more than all other influences combined, primarily because they are so pervasive. Many of the responses to water deficits discussed above for range plant seedlings are equally relevant to mature plants.

The sequence of events that occur within a mature plant in response to water deficits and their physiological consequences differ with species, type of plant, current environmental conditions, and the past environmental conditions to which the plant was subjected. However, generally as water deficit intensity and duration persists, a slowing in leaf growth and shoot development occurs first. Leaf expansion and rigidity are affected first by a loss of pressure (turgor) potential followed by visible signs of wilting, folding, discoloration, or other distortions. Oosterhuis et al. (1987) found that visible color change in peas in response to increasing water deficits was an easily identifiable indicator of the onset of water deficits. Tillering in grasses and the development of new shoots in other plants are reduced. Mild stress appears to affect shoots more than roots, and may even stimulate deep root production at the expense of shoots (Sharp and Davies 1979). Tissue age and the stage of development of plants when a deficit occurs are important in determining the effect of water deficits. Older tissues are generally more sensitive to water deficits in terms of the initiation of growth reductions than younger tissues (Catsky 1962; Knipling 1967; Thomas and Stoddart 1980; Wenkert et al. 1978).

The effects of increasing water deficits on the metabolic activity of plants have been summarized for various plant species (Hsiao 1973; Hsiao et al. 1976a, 1976b; Kaufmann 1981; Kramer 1983; Turner and Burch 1983), and are only generalized here. Although the exact sequence of events may vary with species or environmental effects, the same physiological processes affected are generally observed in most terrestrial vascular plants. Some of the consequences of progressively increasing levels of water deficits on the physiological activities of plants are summarized in Table 1.

As leaf growth rate decreases as a result of loss in cell pressure potential, there is also a reduction in cell wall formation, cell division, and protein synthesis. Enzyme activity and the formation of protochlorophyll soon decline or cease (Virgin 1965). As the stomata close, not only is gas exchange (photosynthesis and respiration) reduced, but abscisic acid (ABA) concentrations increase (Davies et al. 1990), sometimes dramatically, and in some plants osmotic potential may decrease as solute concentration increases (Turner and Jones 1980). When water deficits are severe, proline often accumulates and other amino acids and xylem tissues may cavitate and become blocked by vapor spaces. Cell membranes become disorganized and cease normal function, and Brown (1974) speculated that protoplast collapse may occur in some plants in response to plasmolysis.

Table 1: Generalized sequential responses of plant physiological processes and morphology to progressive increases in levels of water deficits ranging from mild to severe.

Effect:	Intensity of Water Deficit:
Cell growth and division decrease	Mild
Cell wall synthesis declines	
Protein synthesis decreases	
Protochlorophyll formation ceases	
Activity of nitrate reductase and other enzymes decline	
ABA concentration increases	
Leaf expansion declines or ceases	
Stomatal closure occurs	
Photosynthesis slows or ceases	Mildly severe
Leaf surface temperatures increase	
Respiration decreases	
Xylem conductance decreases	
Proline and other amino acids accumulate	
Sugar levels increase, starch declines	
Leaf wilting, folding, rolling, occurs	
Membrane disorganization begins	
Tissue and organ mortality	Severe
Plant death	

The ultimate effect of prolonged water deficits in plants is mortality. Mortality due to water deficits can occur in every stage of plant development (Bewley 1979; Jones et al. 1981; Kramer 1983; Lange et al. 1976; Levitt 1972; MacMahon and Schimpf 1981). However, recognizing mortality due to water deficits in the field is difficult, and except during extreme conditions, is probably rare. Seedlings are generally more susceptible to water deficit mortality than mature plants primarily because of their smaller less extensive root systems (Johnson 1980; Zahner 1968). Also, seedlings have less protective tissues around meristems to mediate the effects of dehydration due to their generally smaller size. Additionally, seedlings generally are not capable of acclimating, or hardening, to dehydration because of their shorter exposure times. There is no evidence that seedlings are less capable of acclimating to water deficits than mature plants given the same opportunity to slowly acclimate to dehydration (Bewley 1979).

Mortality of mature plants due to water deficits has been observed in the field (Box 1978; Chamrad and Box 1965; Gaff 1971; Glantz 1987; Whalley 1973). However, quantitative data on the extreme water potentials at which plant mortality occurs are limited (some exceptions are discussed by Gaff 1980, and MacMahon and Schimpf 1981). It is well known that some perennial plants are adapted to extremely low water potentials (e.g., Caldwell et al. 1977; Love and West 1972; MacMahon and Schimpf 1981), but many perennials experience periods of dormancy induced by water deficits that are confused with mortality (Whalley and Davidson 1969). Mortality of some, or all, leaves due to water deficits rarely means the plant has died, especially in perennials. Until recently, methods of studying water relations have not been available to measure the extreme water potentials at which plant mortality occurs. However, even if techniques are available, the water potential at which mortality occurs is difficult to measure because plant death is more a process than an absolute event. Leaves are usually the first organs to die in response to water deficits because they are exposed to environmental extremes more than most other organs. Older tissues are usually subject to mortality sooner than younger tissues because water and nutrients are often diverted to younger tissues, hence they tend to be less capable of acclimating to water deficits than younger tissues (Catsky 1962; Knipling 1967; Thomas and Stoddart 1980). Soon after leaf death occurs, roots may begin to experience mortality as the soil dries, with death progressing from the root tips toward the root crown (Caldwell 1976; Kaufmann 1981; Kramer 1983). Ultimate plant death occurs when the meristems become dehydrated below the limits required to maintain cell turgidity and biochemical activity.

Although mortality represents the ultimate consequence of water deficits, it is but one of many that occur in plants (Kramer 1983). Generally, of greater importance are the consequences affecting growth, development, and reproduction. Drought rarely results in the total elimination of a flora, but the associated reduction in biomass productivity, loss of physiological vigor, and effects on plant competition and species composition are well known to land managers. Most native species in arid or semiarid regions have very effective adaptations and abilities to acclimate to drought, hence mortality is either postponed or avoided during most short-term drought periods. Obviously, long-term drought may have more severe effects in causing shifts in species composition, biomass, and productivity at both the species and community or ecosystem level.

Periodic water deficits during the growth cycle of some species may ultimately lead to major alterations in the floristic components of plant communities. If dormancy is induced in response to water deficits frequently enough in some species such that competitive abilities are impaired, other more tolerant species may become established that ultimately exclude the more sensitive species from the community. Thus, plant mortality may not always occur in direct response to climatic drought and subsequent water deficits, but rather may be indirectly attributed to poorer competitive abilities (Rahman and Rutter 1980). Some major changes noted in floristic composition of communities during periods of climatic drought may not have occurred exclusively because of mortality due to water deficits, but rather some species may have declined due to reduced competitiveness as a result of water deficits.

Reproduction

Reproductive tissues and meristems are particularly sensitive, and water deficits during anthesis and flower development can be devastating to seed production of plants resulting in poor seed germinability and viability. Water deficits have particularly severe consequences during flowering and reproductive stages of plant development and result in reduced flower development, fewer inflorescences, and lowered seed production (e.g., Ekanayake et al. 1990; Mahalakshmi et al. 1988; Ouattar et al. 1987; Westgate and Grant 1989). The duration and intensity of the water deficit, together with the timing of its occurrence during plant development, are most critical to flowering and seed production in plants (Ekanayake et al. 1990; Kramer 1983; Mahalakskmi et al. 1988; Morgensen 1980; Richardson et al. 1989). For instance, Sandhu and Horton (1977) found that water deficits in oats (*Avena sativa*) during anthesis caused significantly greater floret sterility and fewer number of heads than water deficits during the boot stage. Similar results have been observed with corn (Westgate and Boyer 1986), soybean (*Glycine max*, Westgate and Grant 1989), millet (*Pennisetum americanum*, Mahalakshmi et al. 1988), rice (*Oryza sativa*, Ekanayake et al. 1990), wheat (Fischer and Kohn 1966), and alfalfa (Halim et al. 1989). Although none of these species are range plants, most plant species probably respond similarly with some exceptions (e.g., Southwick and Davenport [1986] report that water deficits are essential to induce flowering in citrus [*Citrus latifolia*]).

Data on the effects of water deficits on flowering and reproduction are limited. Although the adverse effects of water deficits and drought on flower and seed production of range plants has been widely observed (Knapp 1984; Miller et al. 1986a), unfortunately these effects have not been studied in detail. In downy brome (*Bromus tectorum*; referred to above as cheatgrass), Richardson et al. (1989) noted that seed production was highly dependent on available soil water during the reproductive stage of development with a strongly curvilinear decrease in seed number as water availability decreased. They found that water deficits reduced panicle production, or prevented it altogether in some cases. Number of seeds per panicle was reduced under severe water deficits (–1.1 MPa) during seed fill stages, seed number was reduced during anthesis or during seed fill under moderate stress (–0.5 MPa), and seed number was reduced during anthesis only under mild stress (–0.1 MPa). Plants severely stressed (–1.1 MPa) for 7 days during anthesis developed glumes, lemmas, and paleas, but no seeds. However, of the seeds produced under water deficits, germination percentages and seed weights were not affected. Water deficits also may delay the onset of flowering in plants (Marc and Palmer 1976), and appear to significantly reduce the period of active flowering in some species (Clarkson and Russell 1976; Andrews et al. 1977). In some species, water deficit history may influence the frequency of flower formation. For example, Nobel (1987b) found that inflorescence emergence in desert agave (*Agave deserti*) was correlated with the number of wet days during the growing season 2 years previous to flowering, indicating in this species at least, that floral primordia were differentiated 2 years before inflorescences appeared.

Measurement of Plant and Soil Water Status

It is often fashionable, and certainly easier, to simply dismiss discussions of water relations methods and techniques as being somewhat mundane and less challenging than the far more complex issues of basic concepts and principles. This is unfortunate because methods and techniques provide the data and information from which principles and concepts are formulated. The very nature of methods and techniques dictate the quality and complexity with which principles in science are generated. If the methods are faulty, misunderstood, or misapplied, it is likely that the concepts and principles derived will probably be fundamentally flawed as well. Some methods currently in use were developed decades ago, whereas others originated more recently; but more important than how contemporary a technique may be is the absolute requisite that they continue to be scrutinized with care. The most useful methods and techniques continue to evolve and change as new technology and innovation contribute to their improvement. Thus, we always must view methods and techniques with extreme care, objective caution, and a bit of skepticism.

Techniques commonly used in water relations research and related fields are far too broad and complex for detailed coverage in this chapter; only methods of measuring water status are considered here. However, other useful techniques in water relations, such as those for measuring transpiration, photosynthesis, plant growth, root development and other characteristics related to water relations have been thoroughly discussed elsewhere (e.g., Pearcy et al. 1989; Sestak et al. 1971; Taylor 1987). Methods of measuring plant and soil water status have been studied extensively, and the large number of techniques available have been discussed comprehensively in the literature (Brown and Bartos 1982; Brown and Oosterhuis 1992; Brown and Van Haveren 1972; Campbell 1988; Campbell and Gee 1986; Gardner 1986, 1988; Hillel 1971; Johnson 1962; Koide et al. 1989; Kramer 1988; Pearcy et al. 1989; Richards 1949, 1954; Slavik 1974; Spomer 1985; Taylor and Ashcroft 1972; Turner 1981, 1988). Only a brief summary of the more important techniques is included here (see Table 2).

Water status in plant tissues and soils is usually expressed in terms of either water content or free energy. Water content can be thought of as a capacity term in which the amount of water is described by its weight or volume relative to that

Table 2a. Summary of some important methods and techniques of measuring soil and plant water content. Symbols: $_m$ = gravimetric water content; $_v$ = volummetric water content; m_w = wet wt.; m_d = dry wt; $_b$ = soil bulk density; fwt = fresh wt.; twt = turgid wt.; dwt = dry wt.

Technique	Range of Sensitivity	Applications	Advantages	Disadvantages
		WATER CONTENT		
		SOILS		
Gravimetric water content	Saturation to complete dryness	soil water content as mass of water per unit mass of soil $\theta_m = [(m_w - m_d)/m_d] * 100$ Ref. Gardner 1986; Bloom 1989	-inexpensive. -simple and direct. -precise with care.	–destructive. –many steps can lead to errors –time consuming, laborious. –soil drying is arbitrary at 100–110°C –organic matter oxidation –may lead to errors.
Volummetric water content	0 - 100% by volume	soil water volume per unit volume dry soil $\theta_v = (\rho_b) * \theta_m$ Ref. Gardner 1986; Bloom 1989	-inexpensive. -precise (with care). -expresses water as unit volume.	–same as gravimetric. –bulk density difficult to measure especially in rocky or high organic matter soil.
Neutron probe	0 - 50% water content	soil water content at different depths; Ref. Bloom 1989; Gardner 1986	-non-destructive after access tubes installed. -averages over large volume. -repeated readings at same location and depths.	–expensive initial cost. –calibration required for different soil types. –potential radioactive hazard –cannot measure water content at surface. –errors due to air voids around access tubes. –destructive during installation.
		PLANT TISSUES		
Relative water content ("turgidity")	0 - 100% water content	plant tissue water content relative to water content at full turgor. RWC=[(fwt-dwt)/(twt-dwt)]*100 Ref. Barrs 1968; Kramer 1988; Weatherley 1950, 1951.	-direct measure of water in tissues. -inexpensive. -tissue dry wt. may change during hydration.	–significance not comparable between species. –time consuming (1-2 days). –many steps may lead to error. –full turgidity is artificial.

Water saturation deficit (Saturation deficit)	0 - 100% deficit	measures percent water deficit - from full turgidity.	WSD=(100 - RWC), or =[(twt-fwt)/(twt-dwt)]*100 Ref. Barrs 1968; Kramer 1988	same as RWC above.	-same as RWC above.

of plant tissues or soils. Most commonly, the amount of water in plants or soils is referenced to either the maximum amount that could be held, or to the weight or volume of water retained at a given time following its removal (usually by drying). Free energy or the water potential, on the other hand, describes the forces acting on water that affect its movement and retention.

Methods of Measuring Water Content

The amount of water in soils has traditionally been referred to as "soil moisture" or "soil moisture content", and methods of measuring this quantity have probably been studied more than any other aspect of the physical properties of soils (e.g., Alessi and Prunty 1986; Briggs and Shantz 1912; Gardner 1986; Hendrickson and Veihmeyer 1945; Johnson 1962; Lull and Reinhart 1955; Olson and Hoover 1954; Veihmeyer and Hendrickson 1949). Most often, two different quantities are measured: water content expressed as a mass, and volummetric water expressed as the volume of water per unit volume of soil (Table 2a and 2b).

The gravimetric method, or soil water content based on mass, is probably the most widely used method of expressing soil water content. A soil sample is collected, weighed, and oven dried at between 100°C–110°C until a constant weight is achieved, and then re-weighed (e.g., see Gardner 1986 for concerns about precautions, accuracy, and precision of the gravimetric method). Note that the equation in Table 2a references mass water content to soil dry weight. Although this method is simple and direct, inexpensive, and can be very precise if extreme care is given to measurements, some additional disadvantages not listed in Table 2 can be significant. The most important of these is that soil water content has little or no relationship to plant water status (either water content or water potential); for any given soil it varies with individual plant species, whereas for a given plant species it varies with differences in soil characteristics. In addition, because water content is so strongly affected by rockiness and textural properties, there is little comparability of water contents among different soil types. Among the disadvantages listed in Table 2a, perhaps the most significant is that of arbitrary dryness achieved at temperatures of 100 to 110°C; the degree of dryness achieved certainly varies with textural differences, actual soil temperature achieved in the oven, sample size, length of time of drying, and with methods of weighing the dry sample after it is removed from the oven (see Gardner 1986).

The volummetric expression of soil water content (Table 2a) describes the volume of water in a soil sample per

Table 2b. Summary of some important methods and techniques of measuring soil and plant water potential.

Technique	Range of Sensitivity	Applications	Advantages	Disadvantages
		WATER POTENTIAL		
		SOILS		
Tensiometric	0 to -0.08 MPa	Soil matric potential at fixed locations and depths Ref. Cassell and Klute 1986.	–Direct measure of water potential. –repeated measurements. –no calibration needed. –insensitive to temp. –portable. –inexpensive. –automatic data acquisition possible.	–limited range of sensitivity. –water reservoir requires refilling frequently. –fragile ceramic cup; may crack or clog with clay. –ceramic may attract roots. –not suited for small volumes or near soil surface.
Soil Moisture Blocks	0 to -1.5 MPa	Soil matric potential as a function of electrical resistance. Ref. Campbell and Gee 1986.	–relatively inexpensive. –commercially available or made locally. –no complicated electrical circuity required. –used *in situ*.	–deteriorate with age. –calibration not stable nor linear. –slow response times to changing water potential. –temperature sensitive. –requires separate circuit to measure soil temperature. –subject to hysteresis effects during wetting-drying cycles.
Pressure plate apparatus /0 to -1.5 MPa		Water retention relations between water content and soil matric potential under pressure. Ref. Klute 1986; Richards 1949.	–desired matric potential pre-set with pressure gauge. –yields water retention characteristic curves. -straight forward.	–expensive. –equipment is heavy, large. –potential hazards at high pressures. –slow; 1 to 3 days. –data collection cannot be automated.
		PLANTS		
Pressure Chamber ("pressure bomb")	0 to -10 MPa	xylem potential, or apoplast matric potential; osmotic and turgor are inferred from pressure Ref. Ritchie and Hinckley 1975; Turner 1988; Waring and Cleary 1967.	–simple to operate. –rapid measurements of total water potential. –volume measurements –useful for leaves, stems, roots. –can derive osmotic and	–potential hazard at high pressures. –data acquisition not –portable in field or lab. automated. –pressure-volume data is labor intensive, slow.

Table 2b (cont.). Summary of some important methods and techniques of measuring soil and plant water potential (cont.).

Technique	Range of Sensitivity	Applications	Advantages	Disadvantages
			pressure potentials; symplastic and apoplastic volumes; other pressure volume data.	–end-point exudation also not always evident. –instrument cumbersome, – heavy for remote locations. –apparent water potential may not correlate with psych.
		PLANTS AND SOILS		
Thermocouple psychrometer Peltier type (or Spanner type)	–0.1 to -8.0 MPa	Plant and soil total water potential, either *in situ* of samples in chambers; also components of plant tissue in chambers. Ref. (Barrs 1968; Brown and Van Haveren 1972; Brown and Oosterhuis 1992; Rawlins and Campbell 1986; Spanner 1951)	–versatile for both *in situ* and lab. water potential. –water potential components for plant tissues. –can be very precise. –commercially available, or can be constructed locally. –stable calibration. –all equipment commercially available. –data acquisition can be automated. –sensitive in active range of physiological processes. –also sense ambient temp.	–subject to errors caused by: —temperature gradients —corrosion of junctions —leaky chambers —defective electronics —external electronic noise —poor calibration —poorly trained users –only precise under true isothermal conditions. –delicate, easily broken. –require elaborate careful calibration and periodic re-calibration. –require equilibrium of both vapor pressure and temp. –require extensive training. –expensive, including meters and peripheral equipment.
Wet-loop	0 to -15 MPa	soil and plant water potential; components of plant tissues under laboratory conditions. Richards and Ogata 1958; Boyer 1969	–wider range of sensitivity than Peltier types. –very precise measurements possible. –less expensive electronics required.	–limited to laboratory conditions; waterbath. –thermocouple must be removed from chamber to be wetted; disturbs vapor pressure equilibrium. –automated outputs more difficult than Peltier. –not as portable as Peltier.

unit volume of soil. Soil bulk density expresses the ratio of the mass of dry soil particles to the total volume of the soil (including particles and pores). Bulk density of soil is determined by sampling the soil with a standard cylindrical tube sampler yielding a known volume of soil. The sample is oven dried at 100 to 110°C until a constant weight is reached, followed by determination of mass water content. Although the same advantages and disadvantages apply to volummetric measurements as those for gravimetric methods, additional disadvantages include potential errors in estimating soil bulk density, particularly when rocks and organic matter are substantial components of the sample (e.g., see Gardner 1986).

Numerous other techniques are available to determine soil water content, but the most commonly used method, other than gravimetric and volummetric techniques, is the neutron probe (Bloom 1989; Gardner 1986). The principle of this method involves the use of fast neutrons emitted by a probe source lowered into a tube buried in the soil. When fast-moving neutrons encounter a body of similar mass, such as hydrogen nuclei, some of the energy is transferred from the neutron to the proton, and the neutron rebounds at a much slower rate. The rebounding slower neutrons are counted by a detector in the probe, and the total is proportional to the number of collisions between neutrons and hydrogen nuclei, which in turn reflects the soil water content. This principle is particularly useful in measurements of soil water content because the primary source of H in soils is associated with water, although significant amounts of organic matter in the soil may cause erroneous interpretation of the data. The technique is convenient and is widely used because it permits repeated measurements of water content with depth over time once the access tube is installed in the soil. Numerous problems with the technique have been noted, however, including the requirements for calibration (Bruce and Luxmoore 1986), the effects of air voids around access tubes (Richardson and Boroughs 1972), effects of soil texture variation with depth (Bloom 1989), and numerous others (e.g., see Bruce and Luxmoore 1986; Gardner 1986).

For plant tissues, numerous methods have been devised to determine water content referenced either to fresh or dry weight (e.g., see Kramer 1988). Because of numerous difficulties with interpretation of plant tissue water content based on a fresh weight basis (e.g., large diurnal and seasonal variations in dry weight), the most widely used technique continues to be the relative water content or relative "turgidity" method (Table 2a) which is referenced to maximum turgid weight (Barrs 1968; Hellmuth 1970; Weatherley 1950, 1951). A related calculation, the water deficit (Kramer 1988) or water saturation deficit (WSD), measures the percent deficit from full turgor. Both are related terms, wherein (RWC = 100 –WSD).

Both measures of plant water content represent inexpensive and convenient methods of measuring tissue water status, but they have several significant disadvantages. One of the major complications with these techniques involves interpretation of the turgid weight of the tissue. Several methods have been proposed to estimate full turgidity, including floating on water, standing the tissue in a shallow pool of water for various lengths of time, and allowing vapor transfer to achieve full turgidity in a closed system. Transfer of bulk water by infiltration into the tissue through cut edges of detached leaves and other tissues can cause major errors in estimates of RWC and WSD (Barrs 1968; Hellmuth 1970). Additionally, blotting the tissue dry after hydration is arbitrary, the many steps and time involved make the technique subject to several sources of compounding error, leaf fresh weight may change during long periods of hydration, and temperature variation and light contamination during water or vapor transfer have also been found to be sources of error (Barrs 1968; Millar 1966; Roberts 1969; Slavik 1974). Another important disadvantage of these techniques is that they are not related in any way to soil water status (Kramer 1988), nor are they comparable among different plant species.

Methods of Measuring Water Potential

Water potential is now more widely used as the primary unit of water status than water content, and offers the distinct advantage of expressing water in the same terms for both soils and plant tissues. Numerous methods are available for measuring water potential and its components in both soils and plant tissues, the most common of which are summarized in Table 2b. For soils, some of these include tensiometers (Cassell and Klute 1986), soil "moisture blocks" (Campbell and Gee 1986), and the pressure plate apparatus (Klute 1986; Richards 1949). For plant tissues, the pressure chamber (Turner 1988) has been a major development for the measurement of both total plant water potential and pressure-volume relations that permit derivation of water potential components and other plant water relations characteristics.

However, it was the development of the thermocouple psychrometer (Barrs 1968; Brown and Van Haveren 1972; Rawlins and Campbell 1986; Spanner 1951) that made it possible to measure both soil and plant water potentials and their components in identical terms, using the same instrument. This technique has revolutionized the field of water relations research in terms of quantity of data collected, and has become the standard against which most other techniques are gauged. These instruments utilize vapor equilibrium between the soil or plant tissue and the air surrounding a sensing thermocouple junction to infer the water potential. Basically, the vapor pressure of the air at constant temperature and pressure will come into equilibrium with the water potential of the soil or tissue, such that:

$$\psi = [(RT)/V] * \ln (p/p^{\circ}) \qquad (3)$$

where the water potential (ψ) is determined by the universal gas constant (R = 8.31432 joules $mole^{-1}$), the Kelvin temperature (T), the partial molal volume of water (V = 18.06875 x 10^{-6} m^3 mol^{-1}), and the natural log of the relative vapor pressure, where p is the vapor pressure of the sample, and p° is the vapor pressure of pure water at reference temperature and pressure. The thermocouple psychrometer is used to determine the relative vapor pressure term in the Kelvin equation, from which the water potential is inferred

Details on theory of operation, construction, calibration, applications, and precautions in use of thermocouple psychrometers are contained in Barrs 1968, Brown and Bartos 1982, Brown and Oosterhuis 1992, Brown and Van Haveren 1972, Rawlins and Campbell 1986, Spanner 1951, and Wiebe et al. 1971.

Two primary thermocouple psychrometer designs are currently used; Peltier, or Spanner, and "wet-loop" types. Of these, the Peltier or Spanner type is by far the most widely used because it can be used in a wide variety of applications and is commercially available (Table 2b). The psychrometer sensor is sealed either in a closed chamber with the soil or tissue sample, or is sealed in contact with the *in situ* soil or plant tissue, such that the vapor pressure around the thermocouple comes into equilibrium with the water potential of the sample or medium being studied. After isothermal and vapor pressure equilibrium conditions are achieved within the psychrometer-sample system, the water potential of the air surrounding the system is measured. A small electrical current (usually about 5 mA) is passed through the circuit for a short duration (usually 15–30 sec.), which cools the sensing thermocouple junction about 0.6°C below the ambient temperature (the "Peltier effect"). This causes vapor in the air around the thermocouple, which originated from the soil or plant tissue, to condense as a liquid onto the junction. After the Peltier current is terminated, the accumulated water on the junction evaporates back into the surrounding air at a rate proportional to the relative vapor pressure of the air in equilibrium with the soil or plant tissue water potential. The junction is thus cooled as a function of the evaporation rate, or water potential of the sample, and the amount of cooling is measured as a microvoltage output.

The primary difference between Peltier and "wet-loop" psychrometers is the manner by which water is placed on the thermocouple junction prior to evaporative cooling. Instead of Peltier cooling initiated electronically, the "wet-loop" psychrometer is wetted physically. Following thermal and vapor equilibrium in a chamber with the sample of soil or tissue, the thermocouple is quickly withdrawn and a small droplet of water is placed by hand on a small ring attached to the sensing thermocouple, usually with an eye dropper or similar device (Boyer 1969; Richards and Ogata 1958). The thermocouple assembly is then quickly re-sealed in the chamber adjacent to the sample, and as evaporation of the water droplet from the ring cools the thermocouple junction as a function of the relative vapor pressure or water potential in the chamber, the microvoltage output of the circuit is monitored until it stabilizes. The equilibrium microvoltage output of the thermocouple is interpreted as water potential, similar to Peltier instruments.

Both psychrometer instruments are calibrated under carefully controlled temperature conditions using solutions of known water potential (Brown and Bartos 1982). Only Peltier models, however, can be used for *in situ* measurements of both soil and plant water potential in the field (e.g., Moore and Caldwell 1972; Richards and Caldwell 1987; Wiebe et al. 1971), as well as under controlled conditions with small samples in the laboratory. Peltier psychrometers are sensitive to a water potential range extending from about –0.1 to about –8.0 MPa, corresponding to the same range in which most plant physiological activity occurs. Wet-loop instruments have a broader range of sensitivity, although this is rarely discussed in the literature. The water potentials of extracted samples of soil or plant tissues are easily measured by sealing the sample in a chamber with either type of psychrometer, and allowing vapor pressure equilibrium to become achieved under isothermal conditions. With soil samples, the total soil water potential can thus be measured. For plant tissue samples, however, the total water potential is determined first, then the entire sample and psychrometer assembly is thoroughly frozen in liquid N or dry ice to disrupt cellular membranes and other structures. Following thorough freezing the sample and psychrometer assembly is thawed, then allowed to re-establish vapor equilibrium, and then the osmotic potential of the tissue can be measured. The turgor potential is then calculated as the difference between total water potential and osmotic potential, or

$$\psi - \psi_\pi = \psi_p \tag{4}$$

Various instrument designs for different applications are available commercially, but they may also may be constructed by the user with patience and extreme care (Brown 1970, Wiebe et al. 1971). In addition, thermocouple psychrometers can provide very precise estimates of water potential (e.g., ±0.01 MPa) under controlled temperature and pressure conditions (Brown and Bartos 1982). The relative advantages and disadvantages of types of psychrometers are summarized in Table 2b and in the literature cited.

Plant Water-Use Efficiency

Although not a method of measuring water status, interest in measuring plant water-use efficiency is growing among range plant physiologists and ecologists in recent years. Water-use efficiency is variably defined either as the amount of water consumed by a plant in transpiration per unit gain in growth or biomass productivity, or alternatively as gain in biomass per unit of water transpired (Ehleringer et al. 1993; Kramer 1983). Historically, plant species were believed to have unique "water requirements" for growth, wherein plants were thought to require specific quantities of water for specific gains in biomass. It was further believed that these requirements could be used to define relative "drought tolerances" for different species. However, later research showed that species, ecotypes, and cultivars have highly variable rates of water use relative to biomass production, and that atmospheric conditions such as precipitation, the diffusion pressure deficit between the plant and the bulk air, and wind influence water-use rates far more than demands for plant growth. Water-use rates also vary with stage of plant development and soil physical and chemical properties (Stanhill 1986). Therefore, water-use efficiency is not a fixed constant characteristic of each species, nor is it a unique index of "drought tolerance" (Kramer 1983). However, water-use efficiency is of enormous interest and importance to plant physiologists, plant breeders, and range managers because this quantity is used to define and understand the interactions of water use and carbon gain as they affect plant growth, survival, responses to stress, and in selecting for adaptations to water deficits.

The most meaningful expression of water-use efficiency is defined by the relationship of net photosynthesis as CO_2 uptake (A) and transpiration expressed as water loss (E), or A/E (Ehleringer et al. 1993). Traditional techniques of measuring water-use efficiency involve measuring gains in plant tissue mass over time as a function of transpiration loss. These methods are largely restricted to potted plants and have proven to be cumbersome, slow, often inaccurate, and result in data that rarely reflect actual conditions appropriate to field-grown plants (Radin 1992). Instantaneous measurements of transpiration and photosynthesis can be made easily enough with modern equipment, but these observations provide only spot-measures of water-use efficiency that may not indicate long-term growth responses under field conditions over time (Johnson et al. 1993; Sandquist et al. 1993; Wright et al. 1994).

However, a more time-integrated estimate of water-use efficiency was made possible by the discovery that Calvin cycle (C_3) plants discriminate against the incorporation of the heavy isotope carbon-13 (^{13}C) relative to the lighter isotope carbon-12 (^{12}C) during photosynthesis (Farquhar et al. 1989; Ehleringer et al. 1993). The basis for discrimination against the heavier ^{13}C isotope is associated with the primary photosynthetic carboxylation reaction and by differences in diffusion rates between ^{13}C and ^{12}C. A quantitative relationship has been established between intercellular CO_2 concentration and water-use efficiency thus linking the two quantities for C_3 plants (Ehleringer et al. 1993; Farquhar et al. 1989). Carbon isotope ratios ($^{13}C/^{12}C$) become more negative as intercellular CO_2 concentrations increase. Therefore, the carbon isotope composition of plant tissues is now used to provide a time-integrated estimate of A/E, or water-use efficiency and water-use patterns in general as a means of understanding ecological systems (Ehleringer et al. 1993; Hall et al. 1990; Sandquist et al. 1993). In addition, carbon isotope discrimination has become a widely used technique in breeding programs for selecting plants adapted to growth in water-limiting environments (e.g., Hall et al, 1990; Johnson et al. 1993; Wright et al. 1994).

Plant Adaptations to Water Deficits

Adaptation and Acclimation

Adaptations are any heritable qualities, or modifications in structure or process that increase the probability of surviving a given stressful environmental effect (Conrad 1983; Kramer 1983; Oertli 1983; Whalley 1973). These qualities originate during evolution in response to random mutations and recombinations of genetic material, and are maintained in a population by natural selection (Harper 1977). Contrary to teleological belief, however, adaptations in plants do not materialize in response to stress as a strategy to improve survival (Kramer 1983). There is no evidence to support the idea that a "strategic plan" is manifested by plants to alter physiological function or anatomical structure in an attempt to improve chances of survival in response to environmental stress.

Acclimation, in contrast to adaptation (Kramer 1980; 1983) is a process whereby nonheritable phenotypic modifications are produced in response to variations in environmental factors. An equivalent term, "hardening," is frequently used, and refers to the process of preconditioning plants by gradually, and perhaps repeatedly, exposing plants to mild forms of stress. It is generally found that plants subjected to several cycles of mild water deficits are less affected by severe stress than "unhardened" plants.

The differences between adaptation and acclimation are not always readily apparent, and although it may seem unimportant on a practical level, it can be quite significant when selecting plants apparently "adapted" to particular environments. Quite often we hear reference made to a plant being "adapted" to a specific environmental factor or level of stress, when what was really intended was a reference to the ability of that plant or species to acclimate. As will hopefully become clear later in this chapter, perhaps the ability of plants or species to acclimate may be of greater advantage in some instances than is the possession of some highly specialized adaptation that does not have any plasticity to acclimate to variable environmental conditions (Hall 1981). For example, Ehleringer and Mooney (1978) and Ehleringer (1982) discuss development of leaf pubescence in the aridland shrub *Encelia farinosa* as an adaptation which reduces leaf absorptance, and thus reduces the surface heat load on the plant during periods of water deficits. However, during periods of severe water deficit, decreasing leaf water potentials and increasing temperatures both independently increase the amount of leaf pubescence (Ehleringer 1982). The ability to alter pubescence in response to changing environmental conditions appears to be a favorable acclimation feature, whereas the presence of pubescence appears to be an adaptation. In another example, Cox and Hutchinson (1980) note that the bunchgrass tufted hairgrass (*Deschampsia caespitosa*) has a wide range of tolerance to available metals in mine spoils, but generally the species is not restricted to metal contaminated soils. However, one particular population of this species was found that appears to be restricted exclusively to mine spoils with extremely high concentrations of Cu and Ni. Since subsequent generations of this population retain the specificity for contaminated spoils, the authors feel it evolved an adaptation specific for the spoils with little or no ability to acclimate to uncontaminated soils. Thus, when selecting plants for range improvement, a knowledge of the differences in adaptation and acclimation in plant species could be crucial to successful establishment of plants.

Water Deficit Adaptations

The demonstration of adaptive significance implies that a specific adaptation empowers an organism with relatively more fitness for an environment than its competitors (Harper 1977; Osmond et al. 1980b). Harper (1977) suggests that it is the nature of adaptation to the environment that the ultimate degree of adaptiveness by plants is possible if the environment does not change. However, environments are not static, and in fact appear to be in a constant state of change; competition, parasitism, predation, climate change, and in the modern World pollution and human intervention and disturbance can be dominant forces of change. Organisms, non-static environmental factors, and perhaps even seemingly imperceptible global events form part of every plants' environment. Hence, changes in any or all of these also change the selective forces affecting plants and their adaptations, which in-turn activate feedback leading to more modification (Harper 1977). Thus, "proving" the significance of (or in some cases, even the existence of) an apparent adaptation is extremely difficult. Demonstration of differences in performance or response between individuals or species, although suggestive, is hardly proof that adaptive significance exists. As Harper (1977) explains, the fact that organisms are different is easy to explain, but the question of how so many kinds of organisms are able to persist in the same environment at the same time is much more difficult and interesting. It is this process "... in which evolution is an existential game in which success is measured by continuing to play the game rather than by winning." (Harper 1977, p. 750).

Plant adaptations to long-term water deficits, or drought, have been classified into two major categories: (1) avoidance or (2) tolerance (according to the classification suggested by Kramer 1980 and 1983). Drought (or water deficit) avoidance mechanisms allow plants a certain degree of flexibility to alter phenological development and ontogeny according to climatic and environmental conditions. Under such circumstances, plants may "avoid" periods of severe water and temperature stresses in a stage of development (e.g., as seeds) that insures maximum chances for survival. Some authors prefer the term "drought escape" to describe this type of adaptation (e.g., Levitt 1972), but it seems that "drought avoidance" is more useful and precise because it more accurately describes what actually happens. "Drought tolerance" adaptations, also often referred to as "drought resistance adaptations" by some authors (Levitt 1972; Paleg and Aspinall 1981), allow plants to function under varying degrees of stress ranging from mild to severe, and include morphological and physiological mechanisms that either postpone the effects of water deficits, or permit endurance of them during growth and development.

Although widely accepted, use of the term "drought" in reference to the condition of water deficits in plants is unfortunate and misleading, and should be discouraged. In the discussion here I have taken the liberty of distinguishing the word "drought" as a climatic term that does not necessarily imply water deficits in plants; water deficits certainly occur during periods of drought, but they also occur during periods of normal or high precipitation. Despite suggestive terminology to the contrary, plants do not literally escape, avoid, or postpone drought although they may endure, or even tolerate it (even these terms may arguably be interpreted as teleological, and are probably best left out of any discussion on plant adaptations; e.g., see Kramer 1980; Oertli 1983; and Jones et al. 1981 for alternative views concerning terminology). When water deficits occur, plants have few options other than to; (1) alter their phenological state by either passing the period as seeds or going into a state of dormancy, and thus "avoid" the stress period, (2) postpone the stress by altering physiological processes and/or anatomical-morphological structures to reduce water loss, or (3) alter physiological conditions that permit maintenance of turgor and protoplasmic endurance of the stress.

In the context used here a plant may be thought of as being adapted to water deficits if it is capable of completing its entire life cycle during most years under periodic drying conditions that may occur in the environment in which it is growing. The types of adaptations to water deficits manifested in plants are not usually dictated by phylogeny or specific taxa, growth form, geographic origin, or habitat. There is a danger of confusion here; most terrestrial plant species can potentially possess similar adaptations for water deficits as other species. Some grasses share many similar adaptations as trees or shrubs, and many mesophytes have similar traits under some conditions as classically defined xerophytes. In addition, many species display more than one adaptation to water deficits; it is not uncommon to observe different species growing adjacent to each other that have similar adaptations, but which might manifest themselves by means that are quite different. Even annual plants (discussed below) may have various adaptations similar to some perennial species even though they typically are credited with only altering phenology as a means of coping with water deficits. Some adaptations are referred to as being morphological, but these may be affected by various interacting physiological adaptations as well. Therefore, in order to circumvent some of the confusion associated with "pigeonholing" specific adaptations into yet another classification scheme, I have chosen not to categorize water deficit adaptations into traditional "escape," "avoidance," "postponement," and "tolerance" divisions. Rather, various broadly defined adaptations (or apparent adaptations) that have been discussed in the literature are summarized here individually, and reference is made to traditional classification schemes where needed for purposes of clarification. Of course, the mere use of broadly defined categories of adaptations can be criticized justifiably for the same reasons that "drought tolerance" classifications are, but it seems that the lumping done here leads to less confusion.

Most range plants display one or more adaptations that increase their tolerance to water deficits. Within the genetic limits of adaptation in each plant and the range of environmental variability that occurs on a site, most range plants are adapted to regulate their own water relations to some extent. Adaptations to water deficits in plants include such features as:

1). Phenological modifications in response to periods of water deficits.
2). Stomatal control of water loss, including early or rapid closure in response to water deficits, and yet that permit net carbon gain.
3). Morphological features such as leaf surface structures that restrict water loss and reduce heat loads, leaf movements and rolling, folding, leaf curling in response to water deficits, and leaf shedding.
4). Root systems that are able to respond to physiological demands and environmental conditions, and that may either be deep and extensive, or shallow and competitive.
5). Alternative photosynthetic carbon pathways.
6). Osmotic adjustment.
7). Halophytic adaptations.
8). Dehydration tolerance mechanisms, often referred to as xerophytism.

Ontogenetic or Phenological Adaptations

Typically, adaptations referred to as "avoidance" or "evading" mechanisms (e.g., Kramer 1980, 1983; Levitt 1972) for the most part (although not exclusively) apply to the annual and ephemeral growth form in plants. In environments in which water availability and temperature extremes are favorable for only short durations during the year, such as in deserts, it is important that plants be adapted to: (1) take advantage of required resources for growth, development, reproduction, and survival when available; and (2) survive the least favorable periods when environmental conditions exceed their physiological tolerances. One group of plants that appears to be particularly well adapted to such severe shifts in seasonal climatic conditions are emphemerals, or those plants that complete their entire life cycles during favorable periods. Ephemeral annual plants occur in many climatic zones, but are best known in desert and Mediterranean climates where they germinate, grow, flower, and produce seeds within relatively short periods when water is readily abundant and temperatures are moderate. In mesic environments, the annual growth form permits "avoidance" of both seasonal water deficits and winter conditions during the vegetative stage of development. During periods when water and/or temperature are limiting these plants generally exist as seeds, and do not germinate and resume normal vegetative growth until favorable conditions return the following season. One of the best examples of this group of plants on North American rangelands is Russian thistle (Young 1991).

Ephemerals in arid climates have been classified according to the seasonality of rainfall and the time of resumption of growth (MacMahon 1985). Winter emphemerals normally germinate in response to precipitation received during the late fall, winter, or early spring, and complete their life cycles during a relatively short period afterward when water is available and temperatures are moderate (MacMahon 1985; Mulroy and Rundel 1977). Most winter ephemerals in North America tend to be exclusively of the C_3 photosynthetic pathway (Mulroy and Rundel 1977; Nobel 1987a). Summer annuals, on the other hand, tend to be C_4 plants, and germinate following precipitation that occurs in the summer or early fall (Gutierrez and Whitford 1987; MacMahon 1985). The hot deserts of North America vary in their precipitation patterns, hence so do the types and concentrations of annual plants. The Sonoran Desert, with biseasonal precipitation periods, has both winter and summer annual flora; the Mojave Desert is primarily a winter annual zone; the Chihuahuan Desert is primarily a summer annual region; and the Great Basin Desert, with a scarcity of annual flora, is somewhat more complex because of its characteristic climatic patterns, but supports mostly winter annuals when any occur at all (MacMahon 1985). Smith and Nowak (1990) suggest that perennial forbs in the Great Basin belong in the winter ephemeral category because they respond primarily to snowmelt or winter-spring precipitation, and then senesce by early summer. It may even be argued (perhaps not convincingly) that many perennial plants in the Great Basin can similarly be classified as winter ephemerals since most respond to snowmelt and winter-spring precipitation, and then undergo at least partial dormancy during the summer (this is another example of the problems associated with "pigeonholing" adaptations).

The water relations of summer annuals appear to be mostly mesophytic, and have relatively few apparent morphological or physiological adaptations to water deficits other than life cycle timing as would be expected in aridland species (Bennert and Mooney 1979; Mulroy and Rundel 1977; Nobel 1987a). Summer ephemerals tend to maintain relatively high transpiration rates, leaf water potentials, and photosynthesis rates during active vegetative growth, indicative of their rapid luxuriant development during compressed periods of available water (Bennert and Mooney 1979). Most summer emphemerals are large showy plants with an upright, often robust, stature, and may invest heavily in large showy flowers and a high productivity of seeds. Because of their largely C_4 carbon pathway adaptations, these plants often have very high temperature and solar photon flux density optima for carbon fixation, with high net productivity and water use efficiency (Mulroy and Rundel 1977).

Winter annuals, however, appear to have far different adaptations to water deficits. These plants tend to use the Calvin cycle or C_3 photosynthesis pathway, which is effective for carbon gain during periods of moderate temperature (Mulroy and Rundel 1977). The majority of, but not all, winter annuals begin vegetative growth as rosettes, which may be an adaptation that moderates soil surface temperature extremes (Gutierrez and Whitford 1987; Mulroy and Rundel 1977). Winter annuals appear to be more sensitive to water deficits during vegetative stages of growth than summer annuals, wherein growth, flowering and seed production may be interrupted or aborted in response to moderate levels of stress (Gutierrez and Whitford 1987; Klikoff 1966; Mulroy and Rundel 1977). Because of lower evaporation demands during the winter months, these plants tend to have deeper root systems than summer annuals (Gutierrez and Whitford 1987), and thus may be able to tap soil nutrients and water supplies normally not available to summer annuals.

Stomatal Adaptations

Transpiration is the loss of water from plants in the form of vapor (Kramer 1983; Meidner and Mansfield 1968). Kramer (1983) feels that transpiration is the dominant process in plant water relations because it produces the energy gradient responsible for water movement in plants, thus for absorption by roots and the ascent of sap in the conductive tissues of virtually all terrestrial plants. Stomata (Fig. 10) are the primary regulatory structures in plants affecting the amount and rate of transpirational water loss from the plant, and through which CO_2 flows for carbon gain during photosynthesis (Allaway and Milthorpe 1976; Meidner and Mansfield 1968). Levitt (1976) suggests that stomata fulfill three major functional adaptations in plants: (1) they permit the entrance of CO_2 at a rate sufficient to support photosynthesis for net carbon gain; (2) they permit the exchange of O_2 to support respiration; and (3) they are adapted to open and close in response to gains and losses of water in plants, hence are responsible in-part for the regulation of heat exchange with the environment. Thus, stomata play a key role in controlling the balance between growth and transpiration (Ludlow 1980). The interested reader is referred to numerous and far more complete sources of in-depth discus-

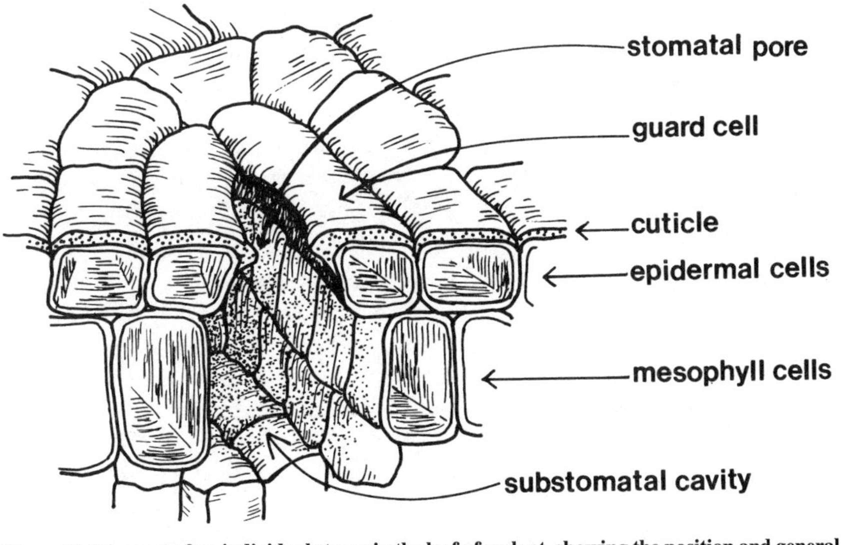

Figure 10. Diagram of an individual stoma in the leaf of a plant, showing the position and general shape of guard cells, mesophyll cells, epidermis, cuticle, the stomatal pore, and the substomatal cavity.

sion of stomatal physiology and control of transpiration (e.g., Allaway and Milthorpe 1976; Bradford and Hsiao 1982; Delwiche and Cooke 1977; Kramer 1980; Levitt 1976; Mansfield 1976; Mansfield and Davies 1981, 1985; Mansfield 1976; Meidner and Mansfield 1968; Raschke 1976; Schulze 1986a; Turner and Burch 1983; Waggoner and Zelitch 1965; Zelitch 1969). Of primary concern here will be a general broad overview of those adaptations that are thought to be influential in affecting water deficits in plants.

Transpiration from plants is dependent upon; (1) an energy source to satisfy the latent heat required for vaporization, (2) a water vapor concentration gradient from inside the stomata to the open atmosphere, and (3) the diffusive resistances in the vapor pathway (Kramer 1983; Slatyer 1967). Mechanisms of heat transfer in plants are well known (e.g., Drake et al. 1970; Gates 1965; Gates et al. 1968; Idso and Baker 1967; Raschke 1960; 1975). Basically, plants absorb heat from their environment and dissipate it by three basic methods: re-radiation, convection, and latent heat transfer (transpiration). The main source of heat energy to plants is solar radiation, and the primary heat transfer mechanism in plants is reradiation back to the environment (Gates 1965; Idso and Baker 1967). In addition, a substantial amount of heat can also be exchanged by convection, or exchange due to the disruption of the boundary layer by the movement of air (wind) over the plant surface. However, it is the latent heat transfer mechanism, or transpiration, over which plants have some degree of control, hence that may have some adaptive advantage for affecting water deficits.

The overall major driving forces for transpiration are gradients in water vapor pressure between the evaporating cell surfaces within the stomatal cavities and the open atmosphere. As water diffuses out of plant leaves it encounters multiple resistances to movement due to a tortuous path via anatomical structures of cells and diffusion of water vapor in air, termed the total leaf resistance (r_l; Fig. 11). The leaf resistance term is composed of two resistances, both linked in parallel as cuticular resistance (r_c) and the stomatal resistance (r_s). Thus,

$$1/r_l = 1/r_c + 1/r_s \tag{5}$$

The cuticular resistance arises from the effective barrier created by the outer cutinized surfaces of leaves and other organs, and tends to be relatively high compared with stomatal resistance. Although the cuticular resistance is high, it is not totally impermeable in most species (e.g., Schonherr 1982; Schulze 1986a; Sowell et al. 1982). Some water vapor may escape from plants through the cuticle of leaves (Fig. 11) and stems when the stomata are closed, but generally the relative amount of water loss via this pathway is extremely small. In most range plants virtually all transpirational loss of water occurs through stomata. Stomatal resistance is normally much lower when the stomata are open, hence is a far more direct pathway for vapor movement (Kramer 1983; Slatyer 1967). The stomatal resistance is comprised of several linked resistances, and can be represented as:

$$r_s = r_m + r_i + r_p \tag{6}$$

where r_m is the mesophyll resistance to water movement through the mesophyll cells within the leaf, r_i is the resistance to vapor movement within the intercellular

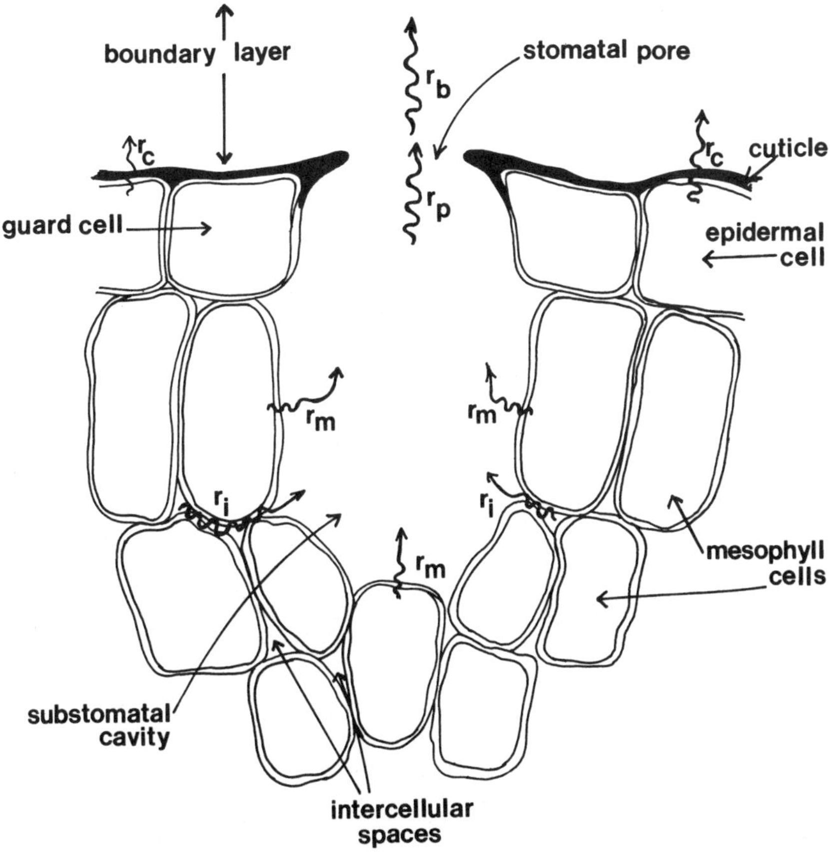

Figure 11. Diagram of a stoma showing the relative resistances to water vapor movement and structures, including guard cells, stomatal pore, substomatal cavity, cuticle, intercellular spaces, and boundary layer. Resistances include: r_c cuticular resistance; r_m mesophyll resistance; r_i intercellular space resistance; r_p pore resistance; and r_b boundary layer resistance.

spaces, and r_p is the resistance due to the stomatal pore which varies with degree of stomatal opening and anatomical structure.

In addition to total leaf resistance, or internal resistances, an external resistance (sometimes termed the boundary layer resistance; Fig. 11) imposes further opposition to vapor movement. The external resistance varies with wind speed and leaf shape and size. Generally, mild increases in wind speed disturb the boundary layer of still air adjacent to the leaf, thus increasing transpiration and reducing the external resistance to vapor movement. Although highly complex (see section below on wind effects) and somewhat debatable, in many species higher wind speeds may actually cause a reversal and result in stomatal closure (Kramer 1983). Equally complex is the effect of leaf shape and size (Drake et al. 1970; Raschke 1960; see section below on leaf adaptations).

Stomata are affected by numerous environmental factors that influence their aperture and physiology. Stomata are surprisingly responsive to changes in environmental conditions, often adjusting aperture opening within several minutes of sensed changes (Mansfield and Davies 1985). The principal environmental factors affecting stomata are guard cell turgidity, light, CO_2 concentration, humidity, temperature, plant age, and growth regulators. Unfortunately, it is extremely difficult to isolate the specific effects of one factor on stomatal behavior without complicating it with the effects of other factors.

As water deficits develop in plants from soil water depletion and/or as a result of a high atmospheric evaporative demand (e.g., low humidity), the water status within the plant tissues declines. As the guard cells around stomata lose water and turgidity the individual stoma begin to close, thus ultimately restricting water vapor loss and CO_2 uptake. It is well established that increases in guard cell turgor relative to adjacent cells in the leaf result in stomatal opening, and loss of turgor during water deficits results in closure (Kramer 1983; Schulze 1986a; Slatyer 1967). As soil water depletion occurs, causing increased resistance to water absorption by roots, or as the humidity of the atmosphere declines as solar radiation heats the air and leaves of plants, mild water deficits may result in either total or partial stomatal closure. The vapor pressure gradient between the leaf and air around it steepens, creating a "transpiration demand" that results in greater water loss from the plant. During warm dry periods, high rates of transpiration may deplete water in the guard cells at a rate exceeding that of replacement by root absorption that results in a loss of guard cell turgor, and ultimately leading to mid-day stomatal closure (see Fig. 9) in response to steepening vapor pressure gradients between the leaf and air (e.g., Oosterhuis and Walker 1987; Zelitch 1969). Except under severe drying conditions, periods of a partial lag in water transport during the warmest times of the day (Boyer 1985) are most often relieved toward evening when the solar angle declines, the transpiration demand is moderated, and hence root absorption once again sustains transpiration (see Development of Plant Water Deficits, above).

Different plant species have unique ranges of stomatal sensitivity to declining leaf water potentials, in some cases their stomata remaining open over a considerable range of water potentials before closure occurs. However, in most species a critical threshold water potential triggers stomatal closure as deficits increase (e.g., Ehlig and Gardner 1964; Lopushinsky 1969; Oosterhuis and Walker 1987; Wendt et al. 1968). Threshold water potentials commonly observed at which stomatal closure occurs among different species vary widely (e.g., Ceulemans et al. 1988; Oosterhuis and Walker 1987). For instance, Lopushinsky (1969) reported threshold leaf water potentials for stomatal closure of –1.4 MPa for lodgepole pine (*Pinus contorta*), –1.7 MPa for ponderosa pine (*P. ponderosa*), –2.0 MPa for Douglas fir (*Pseudotsuga menziesii*), –1.6 MPa for Englemann spruce (*Picea engelmannii*), and –2.5 MPa for grand fir (*Abies grandis*). Similarly, Running (1976) studied threshold water potentials in forest trees of the Pacific Northwest (USA), and concluded that humidity-induced stomatal closure is generally smaller in magnitude than plant water deficit-induced closure by about 50%. Leaf water potential-induced closure may reduce stomatal conductance by a factor of 10. Hsiao (1973) reported threshold values for tomato (*Lycopersicon esulentum*) ranging from –0.6 to -0.9 MPa. Oosterhuis and Walker (1987) found threshold values for wheat that varied from –1.0 to about –1.7 MPa, the variability ranging according to such factors as soil water status, plant canopy size, and previous water deficit history. Ceulemans et al. (1988) cited leaf water potentials for stomatal closure for six hybrids of poplar (*Populus* spp.), and noted a very high degree of variability and population plasticity corresponding to hybrid origin. Threshold values ranged from about –2 MPa to lower than –4 MPa, with some hybrids never really showing stomatal closure at any water potential. Ludlow (1980) feels that the threshold leaf water potential at stomatal closure may vary in a species with rate of water deficit development; rapidly developing deficits may result in higher water potentials of closure than slowly developing deficits. Turner and Burch (1983) and Kramer (1983) suggest that plant age, temperature, and other environmental conditions affect the responsiveness of stomata. Unfortunately, very little critical data of this type are available for native range plants.

Data such as the critical or threshold leaf water potential at which stomatal closure occurs appear to be of some significance in evaluating the relative adaptability of plants to water deficits. One could reasonably expect that high leaf water potentials for stomatal closure would imply a high degree of stomatal responsiveness to developing water deficits, with consequent reduction or termination of transpiration. Alternatively, however, early stomatal closure might be interpreted as hardly a useful adaptation at all if photosynthesis and net carbon gain are terminated long before critically important levels of water deficits that might threaten survival are reached. Thus, despite the apparent significance of threshold water potentials (Ludlow 1980), there is considerable confusion and doubt about their real importance. For instance, Running (1976) found Sitka spruce (*Picea sitchensis*) and noble fir (*Abies procera*) to have relatively high leaf water potential thresholds compared with Douglas fir. He attributes this to the fact that the two former species are restricted to foggy coastal areas where evaporation is low and precipitation is high. He concludes that they must therefore have the ability to close their stomata at high water potentials because they inhabit humid environments where water deficit adaptations are presumably not needed. Observations of stomatal control in both apparently intolerant species were not evident, yet he observed adjacent Douglas fir trees, which are supposedly more tolerant of water deficits, that exhibited significant stomatal control at the same time.

Humidity, or the vapor pressure concentration between the leaf and the open atmosphere, and temperature affect stomatal opening and closure in most plants (Mansfield and Davies 1985). Running (1976) suggests that the effects of steep vapor pressure gradients act similarly to those of lowered leaf water potentials, but to a moderated degree. It is not uncommon to observe lower stomatal conductances, or higher resistances, on drier sites or during warm dry periods of the growing season (Fetcher 1976; Kaufmann 1982a, 1982b). With coniferous forest trees, Kaufmann (1982b) found that leaf conductance during the growing season declined in response to previous night temperatures lower than about 4°C. He feels, however, that stomatal conductance of conifers is affected more by photosynthetic photon flux density and absolute humidity differences between the leaf and air than by temperature and other factors. He emphasizes the importance of studying and evaluating the significance of stomatal control mechanisms under natural habitat conditions of a species rather than under artificial greenhouse or growth chamber conditions, where results can be meaningfully misrepresented and seriously misinterpreted. Wendt et al. (1968) found that transpiration in mesquite (*Prosopis glandulosa*) increased with increasing soil temperatures, but increasing vapor pressure deficits had a diminishing effect, indicating that mesquite does not continue to transpire proportionally with vapor pressure deficits.

Ludlow (1980) suggests that the rapid feed-forward response of stomata to humidity is an adaptive mechanism in arid and semiarid environments because it minimizes sudden increases in transpiration demand, it is readily reversible with few consequences, and because it maximizes daily water-use efficiency. He feels this mechanism may be equally significant for survival of unhardened plants to sudden water deficit conditions, to plants lacking dehydration tolerance, and for those at critical stages of development such as during flowering. Bjorkman et al. (1980) show that stomatal conductance and photosynthesis responses to temperature are dependent on the environmental conditions under which the plants were grown. In many species, stomatal conductance increases with increasing temperature up to some critical maxima, and then declines rapidly as temperature continues to increase. However, some native desert shrubs, such as creosote bush (*Larrea divaricata*) and big saltbush (*Atriplex lentiformis*) are capable of temperature-acclimation over an extremely wide range of temperatures. In these species, and probably in many others as well, photosynthesis capacity under cool temperatures can be high if the plants have been growing under low temperatures; photosynthesis under high temperatures is also high if the plants are acclimated to elevated temperatures.

Stomatal aperture is also influenced by CO_2 concentration in the intercellular spaces, opening when a minimum concentration is reached within the leaf, and closing when a certain maximum concentration is achieved (Kramer 1983; Slatyer 1967). Mansfield and Davies (1985) feel that the main function of CO_2 sensitivity is to bring about partial stomatal closure in response to increasing wind speed. Wind transports water away from the leaf while bringing CO_2 towards it; hence, adaptations that result in stomatal closure exert some degree of control over transpiration. This theory is complicated by changes in leaf temperature because increasing wind may cool the leaf, and thus reduce the vapor pressure in the intercellular spaces. As CO_2 concentration increases above a threshold value, partial stomatal closure may result in reduced transpiration, thus conserving water while permitting simultaneous photosynthesis. Osmond et al. (1980b) document instances of water deficits causing stomatal closure that result in declining photosynthesis leading to shifts in photosynthetic pathways from C_3 to CAM ([Crassulacean Acid Metabolism] see Alternative Photosynthetic Pathway Adaptations, below).

Growth regulators such as abscisic acid (ABA) and cytokinins have been implicated in stomatal control. The possible role of ABA is particularly attractive as a control mechanism for stomatal action in response to soil water drying in the root zone (Davies and Zhang 1991). Current opinion of some (e.g., Davis and Zhang 1991; Davies et al. 1990; Steuer et al. 1988; Zhang and Davies 1989) suggests that ABA may be a "messenger" or "signal" that alerts stomata when the soil begins to dry. They point out that declining soil and root water potentials are not always accompanied by declining stem water potentials, and yet stomatal closure may occur. This suggests that a water deficit signal is sent to the stomata by the drying roots, thus by-passing the shoot tissues until further drying occurs. Henson et al. (1989) found that stomata on the abaxial (lower) surface of leaves of lupine (*Lupinus*) and wheat responded more readily than adaxial (upper) stomata to both induced water deficits and artificially applied ABA. Kramer (1983) remains skeptical and suggests that ABA alone does not explain the mid-day decline in transpiration often observed in many plants, and believes that too much emphasis has been attached to ABA and other growth regulators as adaptive mechanisms in plant water relations.

Stomata can occur on any aerial portion of terrestrial plants, including flowers and rhizomes, but most are located on the leaves (Esau 1967). The location and distribution of stomata may play a role in adaptiveness to water deficits (Sundberg 1985), although the evidence for this is not convincing. It appears that opening and closing responsiveness to environmental conditions is far more important as adaptive mechanisms to water deficits than number or distribution of stomata. Amphistomaty, or the presence of stomata on both leaf surfaces, appears to be widely distributed throughout the plant kingdom, but does not appear to be correlated with plant lifeform or habitat (Mott et al. 1982). However, most monocots, succulents, nearly all desert plants, virtually all C_4 plants, and many dicots have stomata on both surfaces. Amphistomaty appears to occur in most crop plants (e.g., cotton, corn, sorgham) as well, and Miller and Shultz (1987) observed it in western juniper (*Juniperus occidentalis*). Mott et al. (1982) suggest that amphistomaty has major adaptive advantage for plants with a high photosynthetic capacity, for growth in full-sun environments, and that experience rapidly fluctuating soil water depletion. However, many broadleaf forest trees appear to have stomata on only one surface: hypostomatic leaves have stomata only on the lower, or abaxial surface (e.g.,oaks [*Quercus*]), and epistomatic leaves with stomata only on the upper, or adaxial surface (e.g., some species of poplar and willow [*Salix* spp.]). Stomatal density varies widely among plant species, as well as within species, and appears to be linked to environmental conditions under which plants grow. Density also varies widely on different leaf surfaces of the same plant (e.g., Wullschleger and Oosterhuis 1989). Sundberg (1985) feels that desert plants have evolved two major stomatal distribution "strategies"; CAM plants have large stomates at low density, whereas non-CAM desert plants may have smaller stomata but in much higher densities. The supposed adaptive features of these distributions are rationalized for CAM plants as providing a mechanism for rapid gas exchange at night while minimizing water loss, whereas non-CAM plants have the capability for high daytime transpiration and thus cooler tissue temperatures.

At least part of the adaptive significance of stomatal control over transpiration and water deficits are complicated by the interaction of all the factors in the natural environment. It is not enough to simply partition the environment into separate components (water, temperature, light, humidity, wind, CO_2, etc.) and expect to understand how stomata behave. Adaptive significance to water deficits implies that stomatal responsiveness to developing water deficits minimizes water loss before serious damage to physiological processes can result, yet also permits a net carbon gain during periods of mild stress.

Leaf and Other Morphological Adaptations

Probably few other characteristics of plants and how they cope with water deficits have received more attention, observation, discussion, and debate than apparent morphological adaptations to dehydration. Observations of plants from arid regions, or plants from other areas that appear to be tolerant of water deficits, have revealed what are often claimed to be unusual structures, shapes, sizes, and colors of plant parts, and have led to numerous teleological conclusions and explanations of water deficit adaptations (Stocker 1960). Inferences about water deficit tolerance and adaptations that are based on observation alone are tenuous at best, but many observations have led to quantitative research that has either resulted in definitive evidence, or as often happens, unanswered questions requiring further study. For the interested reader the literature on this topic is extensive (e.g., Allaway and Milthorpe 1976; Begg 1980; Gindel 1973; Harris and Campbell 1981; Kozlowski 1976; Levitt 1972; MacMahon and Schimpf 1981; Miller et al. 1987; Oppenheimer 1960; Smith and Nowak 1990; Solbrig and Orians 1977; Stocker 1960). The following is a list of just some of the leaf and other morphological features of plants that have variously been shown, or alternatively argued, to represent adaptations to aridity and water deficit tolerance (note; literature citations listed are not exhaustive, but are merely to indicate where the interested reader may find more in-depth information):

1. Reduced cell size and increased cell wall thickness have been suggested as mechanisms that may maintain positive turgor to lower water potentials, and that promote greater water retention in cell walls (Cutler et al. 1977). Nobel (1980a) cites evidence that smaller cell size changes the amount of tissue available for photosynthesis. He shows that an increase in the ratio of mesophyll cell surface area per unit leaf area commonly occurs in plants subjected to water deficits, although it appears to be highly species-dependent. Water-use efficiency appears to be enhanced greatly as the ratio increases. Many so-called "sun" leaves, or leaves grown in mostly sunlighted environments, appear to have much higher ratios of internal to external surfaces than the "shade" leaves of plants. Many desert species display similarly higher ratios of internal to external surface area than mesic-site species, and this is often used to explain the much higher rates of transpiration observed in many desert plants compared with mesic plants when water is available (e.g., Kramer 1983). However, when water becomes limiting and the stomata close, dry-site species tend to have lower cuticular transpiration presumably due to their thicker cuticles.
2. Stomatal size, stomatal density on leaves, sunkeness in the epidermis, and surface distribution (see discussion on Stomatal Adaptations, above) are suggestive of greater conductance to enhance photosynthesis and/or leaf cooling, that sunken stomata (Fig. 12a) may encourage thicker boundary layers to impede water loss while maintaining carbon gain (Allaway and Milthorpe 1976; Mott et al. 1982; Oppenheimer 1960; Stocker 1960).
3. Small leaf size, thick fleshy or succulent-like, or schlerophylus-like leaves, and leaf shape (Fig. 12b) indicate potential advantages for heat exchange in the case of small leaves, or enhanced water storage in succulent leaves,

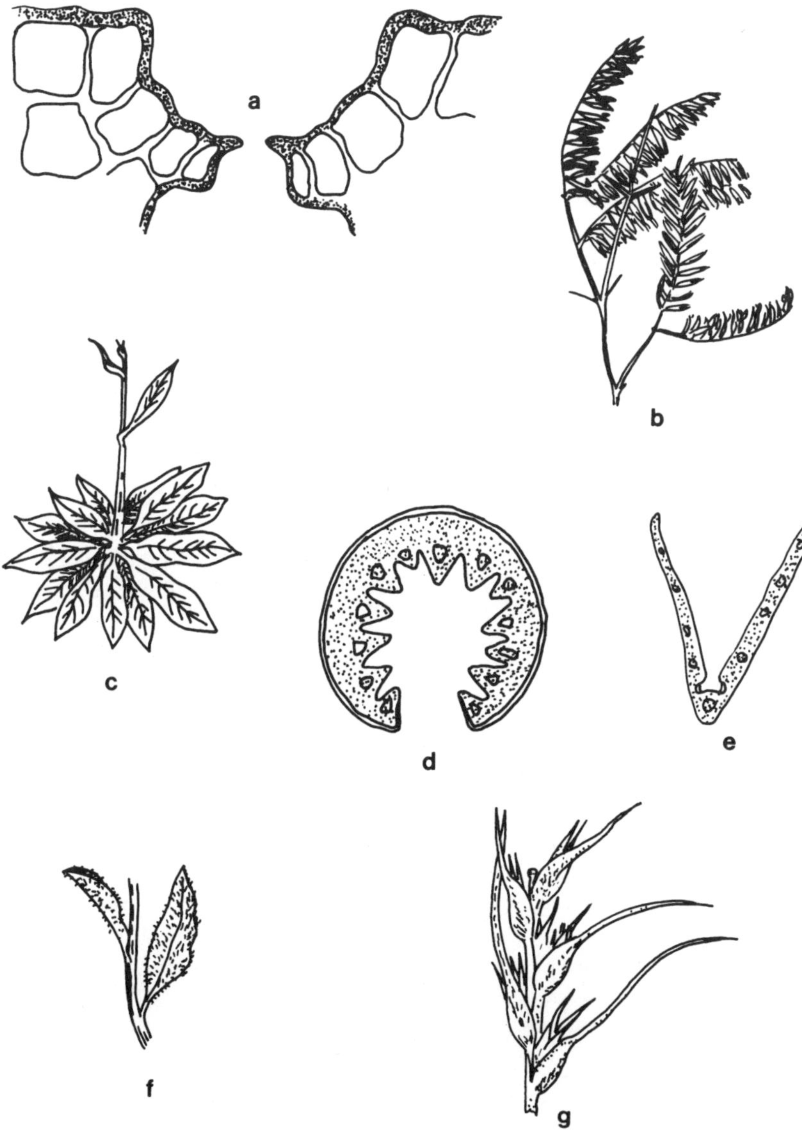

Figure 12. Illustrations of some leaf adaptations to water deficits; (a) sunken stomata; (b) small leaflets of *Prosopis* sp. change orientation during water deficits [stressed above, unstressed below]; (c) rosette lifeform; (d) cross-section of leaf rolling common in grasses [*Deschampsia caespitosa* shown]; (e) cross-section of leaf folding, common in grasses such as *Poa* sp.; (f) leaf pubescence; and (g) modified leaves and bracts of *Salsola kali*.

and a tendency to retard transpiration in schlerophylly (Abrams 1988; Blue and Jensen 1988; Farris 1984; Gindel 1973; Meyer and Walker 1981; Miller and Schultz 1987; Mulroy and Rundel 1977; Smith 1978; Solbrig and Orians 1977). Deep leaf dissection or the development of small leaflets (Fig. 12b) in place of single larger leaves in some desert plants is thought to enhance rapid CO_2 and heat exchange (Mulroy and Rundel 1977). Low leaf surface/volume ratios are reported to be special morphological adaptations of plants in desert regions (Solbrig and Orians 1977). Prostate rosette life-forms (Fig. 12c) of some plants are believed to conserve water by creating favorable microhabitats near the soil surface and which mediate temperature extremes of the air (Burgess 1985; Mulroy and Rundel 1977).

4. Thick leaf cuticles, internal cuticles, and leaf cuticular waxes possibly lead to reduced cuticular loss of water when stomata are closed, and generally are felt to increase water conservation during water deficit periods (Blaker et al. 1989; Fischer and Wood 1979; Hadley and Smith 1990; Jacoby et al. 1990; Oosterhuis et al. 1988; Sowell et al. 1982; Tulloch and Bergter 1981; Wullschleger and Oosterhuis 1989). Glaucousness (dull-white to bluish-white coloration, or "bloom" on leaves of some grasses) in wheat appears to result in cooler tissue temperatures, reduced photosynthesis and transpiration, but increased water use efficiency (Johnson et al. 1983; Richards et al. 1986).
5. Leaf shedding, dormancy, and senescence is generally viewed as a possible mechanism for reducing exposed leaf surfaces to water loss and heat gain during periods of water deficits. Severe deficits may induce dormancy (Whalley and Davidson 1969); shedding may result from a build-up of ABA due to water deficits (Hake et al. 1989), but this feature can be viewed as a disadvantage because it is irreversible and may be catastrophic should favorable climatic conditions prevail following a period of severe water deficit (Turner and Begg 1981). However, dormancy appears to be a major mechanism for coping with seasonal drought and extended water deficits in mid-continental regions such as the North American Great Basin (Harris and Campbell 1981; Smith and Nowak 1990). Shedding and senescence appear to be tissue age-related; older tissues and leaves appear to be shed sooner than younger leaves during periods of stress (Thomas and Stoddart 1980).
6. High root/shoot ratios are believed to favor tolerance to water deficits by increasing root surface area in the soil for water absorption while decreasing leaf exposure to potential water loss (Kummerow 1980; Turner and Begg 1981). Some species, such as creosote bush and big sagebrush (*Artemisia tridentata*) appear to shift their root/shoot ratio seasonally by shedding larger winter leaves during the spring or summer and replacing them with smaller leaves (Harris and Campbell 1981). Kummerow (1980) cautions that the biomass ratio of roots/shoots has no known adaptive advantage, but that some adaptive advantage might be observed for high ratios of root surface area/leaf surface area.
7. Changes in leaf orientation, leaf rolling, folding, and curling (Fig. 12d and 12e) responses to water deficits are viewed as adaptations because they reduce incident solar radiation, protect stomata from forces of evaporation, reduce exposure to wind, and increase heat transfer by other methods than latent heat. They are generally viewed as being turgor regulated, and result in few if any long-term adverse effects because of generally rapid recovery to full leaf expansion and orientation when stress is relieved (Begg 1980;

Bittman and Simpson 1989a; Clarke 1986; Oosterhuis et al. 1985; O'Toole and Cruz 1980; Turner and Begg 1981). Such adaptations have been observed widely in forage grasses, forbs, and shrubs, and are probably widely manifested in range plants throughout arid and semiarid regions.

8. Leaf pubescence (Fig. 12f) is viewed as an adaptation that results in water conservation primarily through reduction of the heat load on leaves, hence lowering leaf temperatures and transpiration, of many plants (Ehleringer 1982; Ehleringer and Mooney 1978; Meinzer and Goldstein 1985; Smith and Nobel 1977). However, an apparent disadvantage to plants is that usually pubescent leaves result in lower photosynthesis rates because pubescence reflects solar energy that might otherwise be available. Temperature optima for photosynthesis in the desert shrub *Encelia farinosa* are much cooler than normal air temperatures, hence maximum photosynthesis can only be achieved if leaf temperature is reduced by the presence of pubescence; a lack of pubescence in these plants results in leaf temperatures that exceed optima for photosynthesis (Ehleringer and Mooney 1978; Ehleringer and Bjorkman 1976).
9. Modified leaves (awns, etc.), appendages, bracts, and other modification structures (Fig. 12g) appear to have somewhat more confusing roles in plant water relations. Wullschleger et al. (1990) are uncertain about the role of bracts in cotton, but speculate that photosynthates produced by bracts may be exported directly and entirely to the developing boll, and that this may increase during water deficit periods. Young (1991) suggests that the bracts in tumbleweed (referred to as Russian thistle here, *Salsola kali*) may absorb water from the soil during seed germination. Geniculated awns in some grasses are essential for self-burial in the soil where water is available before seed germination and seedling emergence occurs (Somody et al 1985). Modified leaves in cactus have been reduced to spines, and may have been an evolutionary trend toward reducing transpiration (Woodell 1973).

Numerous other potential morphological adaptations to water deficits in plants have also been observed, but most have not been studied quantitatively to definitively establish significance. The relative internal air space and total internal surface area to external surface area ratio of leaves (e.g., Nobel 1980a) requires further study before definitive conclusions can be made relative to their adaptive significance for moderating water deficits. Also, the role of water storage in some tissues (e.g., pith volumes in some plants, Meinzer et al. 1985) is still unclear and needs further investigation. The apparent role of such leaf surface structures as trichomes other than pubescence is unclear (Fahn 1969). Similarly, the ratio of leaf area to total sapwood area in conifers is reported to be relatively high on mesic sites and low in more arid sites, and yet some anomalies have been observed as in the semiarid species western juniper, which appears to have much lower stomatal conductance than other conifers (Miller et al. 1987). As Miller et al. (1987) caution, there is a danger in interpreting the significance of a single apparent adaptive structure without considering other possible compensatory adaptations or acclimation capabilities. The apparent lack of an expected morphological adaptation should not be surprising; it is possible that none exists in a given species, that its absence may either be to greater advantage for unforeseen reasons, counterbalanced by other morphological, anatomical, or physiological adaptations, or by the ability to acclimate sufficiently to periods of water deficits. We must be on guard against teleological expectations and extrapolations about plant adaptations to water deficits; generally they are wrong.

Morphological adaptations in plants to water deficits are usually classified as "drought postponement adaptations" in traditional classification schemes (e.g., Kramer 1980, 1983).

Root System Adaptations

In rangeland ecosystems virtually all water and nutrient uptake by plants occurs through their root systems. Root system development represents a large investment in resources by the plant, and annually can constitute over half of its total net primary production (Caldwell 1976; Caldwell and Richards 1986). Caldwell (1987) emphasizes that even in such diverse ecosystems as deserts, tundra, and deciduous forests, the primary productivity directed to roots substantially exceeds that devoted to shoots. The heavy investment in root production made by plants is essential, especially in environments where such resources as water and nutrients are limiting and where competition from other plants is prevalent (Caldwell 1987). Because root adaptations appear to postpone the ultimate effects of water deficits, they would probably be lumped under the traditional headings of "drought postponing adaptations" (Kramer 1980, 1983).

Most dicotyledonous plants (e.g., most forbs, shrubs, and trees) tend to have a taprooted architecture that spreads laterally by adventitious roots that originate from the taproot or meristems at the root crown later in their development (Fig. 13a and b). Fig. 13a illustrates the typically multi-layered root system of big sagebrush, a widespread highly adapted cool-season shrub found on arid to semiarid rangelands throughout the western U.S. Fig. 13b illustrates the common architecture of dicot forbs showing extensive elongation of primary adventitious roots (scarlet globemallow [*Sphaeralcea coccinea*], a common forb found in arid to semiarid regions of the western U.S.). Monocotyledonous plants (e.g., grasses) produce a seminal primary root at germination which is soon replaced by seminal lateral roots, and eventually adventitious, or nodal, roots originating at the root crown base. The common root architectural pattern in all grasses is primarily fibrous (Fig. 13c) for all growth forms, including species with the bunch (tussock) habit of growth, rhizomatous (underground stems), or stoloniferous (above surface stems).

Most water and nutrient absorption appears to occur in the younger portions of roots, particularly where the primary xylem is mature behind the growing tip (Esau 1967). It is in these regions where most root hairs develop (Fig. 14a). Root hairs occur in virtually all terrestrial plants regardless of over-all root structure, originate as elongations or tubular extensions of individual epidermal cells (Fig. 14b), and have large vacuoles with thin walls that favor water and nutrient absorption (Peterson 1992). Root hair formation results in a substantial increase in surface area contact with the soil, hence in an increase in absorbing surface. Usu-

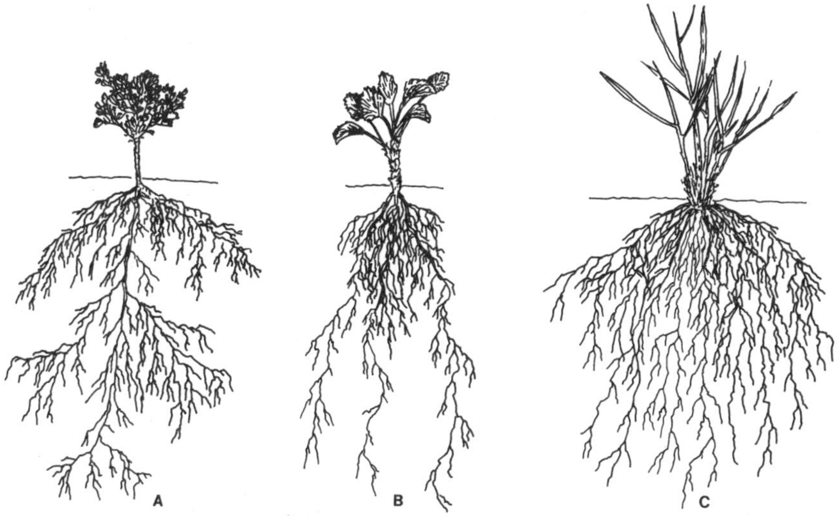

Figure 13. Illustration of some of the common architectural patterns of range plant root systems. (A) multiple layered root form of big sagebrush (*Artemisia tridentata*); (B) common root form of a forb, scarlet globemallow (*Sphaeralcea coccinea*); and (C) fibrous root form of monocots (*Sitanion hystrix*).

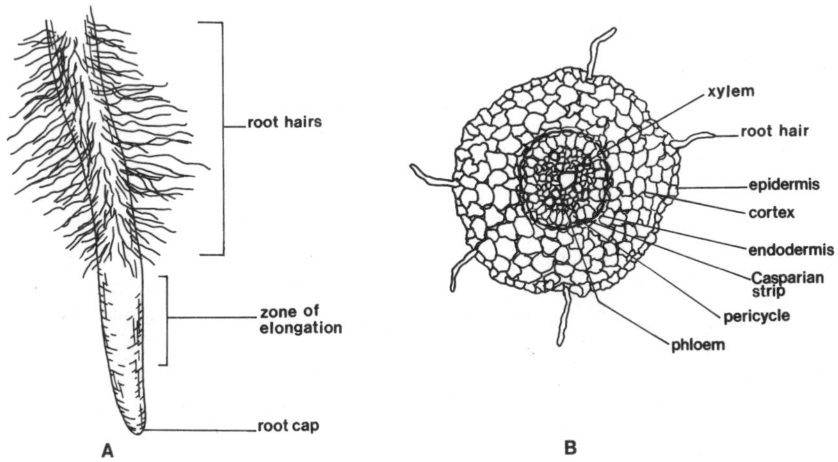

Figure 14. Typical root anatomical features of most vascular plants. (A) illustrates location of root hairs behind the region of elongation in most roots; (B) cross section of typical root of most vascular plants showing epidermis, root hairs, cortex, endodermis, Casparian strip, pericycle, and vascular tissues.

ally root hairs form near the region of rapid growth and elongation, but in some species they may occur nearly throughout the entire root length. Development of root hairs is inhibited by very dry or very wet soils, and their numbers vary widely among species and plant growth form. Normally these structures are short-lived, being destroyed as the root elongates, and new ones are formed by new epidermal cells as the root tip elongates. It is speculated that root hairs are particularly well adapted for the absorption of P, and that their occurrence is correlated with dependence on symbiotic VAM (vesicular-arbuscular mycorrhizae). Peterson (1992) cites research that suggests species with few root hairs are more dependent on VAM than species with abundant root hairs, but that root hair occurrence does not preclude colonization of roots by VAM.

It is likely that the pathway of least resistance to water movement into roots of virtually all vascular plants passes through the epidermis and root hairs and into the cortex (Fig. 14b). There are three possible avenues for water movement through the cortex; (1) via the vacuoles; (2) via the cytoplasm of the cortex cells; and (3) via the apoplast, or cell walls (Passioura 1981). Generally, the pathway of least resistance appears to be through the apoplast (Kramer 1983; Passioura 1981). From the cortex, water must pass through the endodermis and associated Casparian strip, or zone of suberin and sometimes lignin deposition on the radial walls, which renders the cells in this region relatively impermeable to water and ion passage from the cortex to the stele. Numerous pits and plasmodesmata in many dicots and branch roots in monocots (or "passage cells"; Esau 1967), together with cytoplasmic transport through endodermal cells, permit transverse movement of water through the endodermis to the pericycle and ultimately the xylem vessels for transport within the vascular system of the plant.

The total quantity of water and nutrients available to any plant are determined by the volume of soil occupied by its root system (Caldwell 1976, 1987; Caldwell and Richards 1986; Taylor and Klepper 1978). In many plants the lateral spread of roots often (but not always) exceeds the crown cover canopy of the plant, a trait often observed in aridland species. Only under ideal conditions will roots of plants achieve their full genetic potential (Richards 1986). Interestingly, the root systems of most studied range plant species and plant communities tend to be concentrated near the surface of the soil, even in arid and semiarid environments where soils are well drained (Bartos and Jameson 1974; Bartos and Sims 1974; Holechek 1982; Manning et al. 1989; Richards 1986; Tabler 1964; Turner and Burch 1983; Weaver 1977, 1982; Weaver and Clements 1938). In the northern Rocky Mountains, for example, Weaver (1977) found that 55–65% of the total root biomass of plants occurs within the 0–20 cm layer of soil, 20 to 27% in the 20–40 cm layer, and 6–25% in the 40–60 cm layer. Weaver (1982) speculates that root growth near the surface 2 cm of soil declines rapidly because of environmental harshness. In a shortgrass prairie in northeastern Colorado, Bartos and Sims (1974) found that 55% of the total root biomass occurs between 0–10 cm soil depth, and 69% in the upper 20 cm of soil. However, characterizing the rooting form and behavior of established plants in the field is extremely difficult and widely variable (e.g., Frank and Bauer 1991; Manning et al. 1989; Proffitt et al. 1985; Rickard 1985; Svejcar 1990; Weaver 1982). Individual observations of rooting depth and lateral spread, while instructive, may be misleading unless numerous observations are made.

Deep extensive rooting is perhaps the most effective means for plants to postpone the effects of water deficits and to maintain optimum shoot water potentials during periods of limited water supply. Markhart (1985) found that deeper rooted varieties of pea (*Phaseolus vulgaris* and *P. actifolius*) maintained higher leaf water potentials throughout water deficit cycles than shallow rooted varieties. Also, he found that deeper rooted varieties tended to maintain higher stomatal resistances as leaf water potential declined during stress, and had more sensitive stomata that tended to close earlier in water deficit cycles than shallow rooted plants. However, there appears to be a cost associated with deeper rooting behavior, primarily in reduced shoot growth and biomass that may eventually translate into lower leaf surface area and seed production. Maximum rooting depths of numerous species have been reported (e.g., Foxx et al. 1984a, 1984b; Richards 1986; Wyatt et al. 1980), some extending to many meters. Phillips (1963) reports finding roots of mesquite to depths of 53 m below the surface in southern California. Solbrig and Orians (1977) report that roots of Argentine mesquite (*Prosopis nigra*) in Argentina have been found to depths of 60 m, but that 10 to 30 m are more normal. Sturges (1977) studied the root system of big sagebrush on several sites in Wyoming, and found that maximum depth of rooting depended on local site conditions, but generally ranged from 152 to 213 cm. However, maximum lateral spread normally occurred in the upper 30.5 cm of soil, and extended outward from the plant base to distances of about 122 to 213 cm depending on site conditions. In rocky areas, lateral root spread was confined to deeper depths between 61 and 122 cm. Over 50% of the total root biomass in big sagebrush was found in the upper 30 cm of soil, and an additional 25-30% was found between 30 and 60 cm deep. Tabler (1964) found 62% of big sagebrush roots in the upper 60 cm of soil. A large portion of the root system of big sagebrush is devoted to lateral spread near the soil surface which appears to take advantage of summer precipitation, whereas the deeper roots appear to supply water primarily after surface soil water is exhausted (Campbell and Harris 1977; Daddy et al. 1988; Sturges 1977).

Although the root systems of most temperate grassland plant communities appear to be concentrated near the soil surface, the total root length and soil volume exploited by plants can be staggering (e.g., Svejcar 1990; Manning et al. 1989). For example, Taylor and Klepper (1978) cite evidence that individual corn plants rooted to a depth of only about 2 m may have a total root length of as much as 10 km. Kummerow (1980) cites total root lengths for several chaparral species ranging from 2.11 to 3.52 km/m^2 soil surface area, and 7.0 km/m^2 for Douglas fir. I have observed total root lengths of 6 month old Indian ricegrass plants under controlled conditions in large "bins" (1 m deep x 1.5 m wide x 1.5 m long) of over 680 m per plant (Brown 1992, unpublished).

Root form, including maximum depth and lateral spread characteristics, appears to be genetically controlled in each species, but is determined more by environmental constraints in the soil (Feldman 1988; Richards 1986; Taylor and Klepper 1978). Thus, it appears that most plants are potentially capable of tapping great volumes of soil if conditions are favorable, but individual rooting behavior is highly site specific. The most important primary factors determining, or restricting, the distribution of roots in soil include the availability of water and nutrients. However, significantly important physical factors include soil compactness or mechanical resistance to root extension (such as the presence of a caliche layer), soil aeration, pore volume (bulk density), and soil texture (Drew 1987; Richards 1986). The presence of toxic substances such as salts (Poljakoff-Mayber 1975), pyrites and heavy metals (as in mine spoils; e.g., Mays and Bengtson 1978), permafrost (Billings 1974), waterlogging (Bradford and Yang 1981), or very low temperatures (Eissenstat and Caldwell 1987) also impede root extension and exploration.

It appears that there are few, if any, true root system adaptations instrumental in postponing the effects of water deficits (Kummerow 1980). Rather, it is generally agreed that deep, wide spreading, root systems are more effective for maintaining plant water status than shallow confined root systems, and that root form is subject to considerable modification by environmental constraints. Obviously roots are the primary organs through which range plants obtain water and nutrients, but root form alone is not the only variable in plants responsible for maintaining plant internal water status or imparting water deficit tolerance. However, some general inferences can be made about plant roots on rangeland areas that might be useful in predicting soil depths at which most water and nutrient absorption may occur. Rooting depths will normally be rather shallow in areas where water is either severely limiting during the growing season (e.g., many desert areas), or in areas where precipitation is abundant throughout the season such that water deficits are unlikely. Also, regions with abundant precipitation with deep well developed soils where compaction, salinity, and aeration are not limiting may also be characterized by shallow rooting because plant primary productivity costs are unessential (Caldwell and Richards 1986) Hardpan or caliche layers in the soil may also dictate shallow rooting. Deep rooting depths would likely be expected in areas receiving abundant precipitation in the winter and spring and thus allowing for slow infiltration in areas where normal mid-season water deficit periods are likely (Richards 1986), particularly if soil conditions are not limiting. Phreatophytes are normally characterized as having deep wide spreading root systems that insure an adequate water supply from ground water sources throughout the growing season (Nilsen et al. 1983).

Peterson (1992) feels that root branching is an important adaptive feature of root systems, primarily as it affects surface area and total occupied soil volume. He suggests that the evolutionary trends in angiosperms is toward the development of finely branched root systems, like grasses, with a concomitant reduction in dependence on fungal associations. Fitter and Strickland (1991) and Fitter et al. (1991) propose that root system architecture is suggestive of adaptive value, and that high exploitation efficiency of roots is associated with a herringbone topology (branching is predominately on the main axis) and with long interior and exterior links, especially for plants growing in low nutrient conditions. Simulation modeling of root system growth confirmed the harringbone pattern for efficiency of exploitation in dicots, but in grasses the model suggests that link lengths among branches are more important than topology alone.

Kummerow (1980) feels that plant rooting depth and lateral spread, although important features in water deficit tolerance, are not truly adaptive features but rather are expressions of morphological plasticity. A high degree of morphological plasticity is seen as a highly favorable characteristic for range plants growing in arid or semiarid environments because it allows for greater flexibility in exploitation of large soil volumes given suitable conditions. However, maximum depth and lateral spread of rooting alone appear to be unreliable indicators of the effectiveness of root systems to protect plants against water deficits. Although considerable attention has been paid to root:shoot or root:whole-plant biomass ratios as an indicator of water deficit tolerance in plants, Kummerow (1980) doubts biomass is the significant quantity to which plants respond. Rather, he feels that the ratio of the absorbing surface of roots to the photosynthetically active leaf area is more meaningfully descriptive of the adaptive nature of roots to water deficits. For example, he cites data showing ratios of root surface area to leaf surface area for several chaparral species from southern California that had been exposed to various water regime treatments. In most species the ratio of root surface area to leaf surface area reflected the level of water regime treatment applied (being highest for dehydrated plants), whereas simple comparisons of biomass ratios were inconsistent.

In addition to taproot and fibrous root structures, many range plants also have rhizomatous root systems which are really a combination of modified horizontal stems with fibrous roots. Rhizomes are vegetative reproductive organs that normally grow at shallow depths close to the soil surface, although those of some species may grow quite deep in specific soil conditions. These organs provide a method of rapid vegetative propagation that, under optimal conditions, appear to instill an aggressive competitive advantage for site occupation and consumption of water and nutrients. Alpert (1990) found that in the rhizomatous desert grass, saltgrass (*Distichlis spicata*), water sharing among ramets that are connected along rhizomes may be an important survival mechanism in arid areas. When rhizomes were severed, water deficits and plant mortality of the isolated ramets increased significantly within about a day, reaching tissue water potentials below –6 MPa. Lau and Young (1988) found similar results in the herb *Lycopodium flabelliforme*. Severing rhizomes resulted in significant decreases in the xylem potentials of affected ramets compared with unsevered individuals within minutes following illumination. Their studies also verified a bidirectional movement of water in rhizomes, thus allowing resource sharing that may buffer the effects of variably favorable and unfavorable environments encountered at different times during the growing season. According to Lau and Young (1988), the xylem integration pattern observed in this species suggests adaptive advantages for colonizing largely unfavorable environments.

Root growth and extension generally decline as soil water status decreases, and alternatively, during flooding (Kramer 1983). However, studies of the effect of drying soil on root growth and extension suggest that roots are intrinsically less sensitive to drying conditions than aerial plant tissues (Sharp et al. 1988; Westgate and Boyer 1985). Obviously root elongation in drying soil would be a significant advantage to young developing seedlings and mature plants alike, especially if a large proportion of the plant roots are confined to surface soils where drying can be rapid during the growing season. Drew (1987) feels that it is unlikely that roots can continue to absorb water at soil water potentials below about –1.5 MPa, not necessarily due to physiological reasons, but more because of a loss of soil-root contact (also see Tinker 1976) and the greatly restricted mobility of ions and water molecules in drying soil. Cellular dysfunction does occur in roots of both desiccation tolerant and intolerant species if dehydration results in disorganization of cellular constituents (Drew 1987). Drew (1987) believes that an important property of desiccation tolerance in roots is the apparent ability to repair and resynthesize structures following desiccation provided the roots are not dehydrated below some critical value (70% water content is cited as the lethal limit for some species). Roots exposed to sub-lethal levels of water deficits gradually over extended periods show reduced growth with enhanced endodermal suberization tissue development forming close to the root tip, especially in woody species and some dicotyledonous herbaceous species. In this condition roots may remain dormant for extended periods, and although growth and extension are stopped, lethal limits of water loss are minimized. Presumably, desiccation tolerant species will later resume normal growth following rehydration. In herbaceous species, however, (e.g., particularly monocots) extensive secondary growth and suberization of roots in response to near-lethal desiccation and/or age does not occur (Esau 1967). In these lifeforms, dormancy or root death result from lethal levels of desiccation. However, development of new adventitious roots in grasses can often be stimulated by water deficits (Bassiri et al. 1988; Drew 1987; Hassanyar and Wilson 1978; Wilson and Briske 1979), and generally root initiation and replacement from the root crown of grasses is highly vigorous under optimum conditions of soil water availability. Under severe desiccation, grass roots show signs of extensive cortex and epidermal collapse, but continued axial movement of water is frequently observed, perhaps because of the almost universal lignification that occurs external to the vascular system (Esau 1967) which may protect inner tissues from dehydration (Drew 1987).

As the soil dries during the growing season, root growth activity in most plant species appears to shift to progressively deeper depths as water deficits in shallower layers intensifies. For example, Fernandez and Caldwell (1975) observed root growth at various soil depths in the desert shrubs shadscale (*Atriplex confertifolia*), winterfat (*Ceratoides lanata*), and big sagebrush. They found that not only is the growing season for roots significantly extended over that for the shoot portion of the plant (DePuit and Caldwell 1973), but root activity was phased over a long period of time with increasing depth. They feel this progression in root growth may be linked to soil temperature increases as the growing season progresses, but that ultimately it appears to prolong the ability of these shrubs to absorb water. Sturges (1977) also observed water use zones under sagebrush to shift outward from the plant base and downward into the soil as the growing season progresses.

Research has shown that roots have a substantial capability for osmotic adjustment during drying conditions (Oosterhuis and Wullschleger 1987; Sharp and Davies 1979; Westgate and Boyer 1985), which may be important for maintaining turgor in root growing regions. There is even some evidence that water deficits may enhance root development, especially the ratio of roots/shoots (Begg 1980). Markhart (1985) found that water deficits resulted in an increase in root/shoot ratios of pea plants. It appears that declining soil water potentials cause suberization of older larger roots and shedding of younger fine roots, and in some species root contraction may also occur in response to water deficits (Kummerow 1980). Bouaziz and Bruckler (1989a) found that both total root length and root growth rate in emerging wheat seedlings decline with decreasing soil water potential. They found that seminal root length decreased from 200 to 120 mm as soil water potential was decreased from –0.02 to –1.31 MPa, respectively. Root growth terminated at about –1.4 to -1.7 MPa, depending on genetic variety. Water deficits in roots may develop not only directly as soil water potential declines, but also as a result of minimal contact between soil particles and the root surface (Drew 1987; Tinker 1976). It is probable that root contact is affected most often in clay soils that dry and shrink away from the root, although new root growth may follow larger old root channels. Tinker (1976) cites evidence that up to 40% of the root surface in some plants may not be in direct contact with the soil at any one time.

Competition significantly complicates rooting behavior, and may greatly influence root distribution and function (Caldwell and Richards 1986; Caldwell et al. 1985; Samuel and Hart 1992). Deep or extensive lateral root spread may be modified or inhibited by competition for water or nutrients by adjacent plants. Caldwell et al. (1985) have demonstrated the relative competitiveness of two wheatgrass (*Agropyron*) species with big sagebrush. They showed that big sagebrush is able to extract more than 6 times as much phosphorous in shared volumes of soil with bluebunch wheatgrass than from soil volumes shared with crested wheatgrass, indicating that bluebunch wheatgrass is less competitive than crested wheatgrass. Bluebunch wheatgrass is less tolerant of low soil temperatures than desert wheatgrass, although this is apparently not attributable to differences in relative growth rate (Eissenstat and Caldwell 1987). Rather, it appears that the greater competitiveness of both crested wheatgrass and big sagebrush is related to their ability to more rapidly extract water from the soil than bluebunch wheatgrass (Eissenstat and Caldwell 1988).

Related to the effects of competition on root form and function is the phenomenon of "hydraulic lift" observed in some range plant communities. Richards and Caldwell (1987) observed nocturnal increases in soil water potential of dry surface soils by water "lifted" by big sagebrush roots from wetter soil at greater depths. At night water escapes from the upper roots of big sagebrush into surrounding drier soil. Thus, they suggest that roots form a bridge between soil layers of different water status, and that the drier surface soils serve as a temporary storage reservoir during the night for water that is later extracted by shallow root layers during the day. Caldwell and Richards (1989) showed that neighboring roots of grasses growing with big sagebrush can parasitize the water released in the upper layers of soil. Thus, grasses and other plants growing in immediate association with the shrub derive benefits from a source of water that may have been restricted otherwise. Pyke and Archer (1991) emphasize the importance of apparently mutualistic interactions, and feel they may be important in successful revegetation. Caldwell and Richards (1989) suggest that despite the loss of some water to adjacent plants, hydraulic lift is of advantage to big sagebrush because without the temporary storage of water in the upper soil at night, water uptake during mild water deficit periods when the surface is dry would be restricted to only deeper roots.

In addition to root system form, other possible adaptive features have been observed relevant to water deficit tolerance in plants. Despite Kummerow's (1980) concerns that root systems probably have few adaptations to water deficits, Fahn

(1969) discusses what he considers numerous adaptations to xeric conditions. These include such characteristics as root system form, root succulence, development of a thick root bark, sclerification of cortical cells, and the isolation of the vascular cylinder by periderm formation or by the necrosis of the cortical parenchyma. He cites evidence that secondary thickening of the endodermal cells occurs in monocots found growing in unfavorable (presumably arid) environments. Well developed xylem tissues that help rapid conduction of water may be considered an adaptive feature (Fahn 1969). Fahn (1969) also cites evidence suggesting that aridland species have reduced cortical layers in their roots that shorten the distance between the soil and the internal root stele. Many xerophytes with wider Casparian strips than mesophytes have been observed, implying (perhaps teleologically?) a more efficient barrier to the passage of solutes into the central cylinder of the root. Peterson (1992) suggests the Casparian strip is an evolutionary protective mechanism that prevents internal collapse of the stele during drying conditions, and is a barrier to pathogen entry into the vascular system of terrestrial plants.

The presence of rhizosheaths, or sheaths of sand grains adhering to the outer surfaces of roots, have been implicated in enhancing water absorption and reducing resistance to water flow within the rhizosphere (Bristow et al. 1985; Duell and Peacock 1985). Rhizosheaths have been observed on the roots of many aridland grasses such as Indian ricegrass, needle-and-thread (*Stipa comata*), thickspike wheatgrass, and purple three-awn (*Aristida purpurea*), but also in some mesophytic grasses such as tall fescue (*Festuca arundinacea*), orchardgrass (*Dactylis glomerata*), timothy (*Phleum pratense*), reed canarygrass (*Phalaris arundinacea*), quackgrass (*Agropyron repens*), bermudagrass (*Cynodon dactylon*), perennial ryegrass (*Lolium perenne*), and cereal rye (*Secale cereale*) (Duell and Peacock 1985). Associations of VAM with the roots of many arid and semiarid plants have been shown to enhance their water relations characteristics (Allen and Allen 1986; Peterson 1992). VA mycorrihizae are soil fungi with hyphae that penetrate into the root of the host plant to form vesicles and arbuscles within the root cortex. The hyphae of the fungus extend out into the soil mass, and increase the volume of soil exploited by the plant, thus increasing accessibility to water and nutrient pools in the rooting environment. In studies with western wheatgrass, Allen and Allen (1986) showed that leaf water potential increased and stomatal resistance decreased when the plants were inoculated with VA mycorrhizas. Further, they showed that the competitive effects of annual plants with inoculated western wheatgrass for water were greatly reduced.

Alternative Photosynthetic Pathway Adaptations

Because plants must lose water to the air in order to gain CO_2 for photosynthesis through open stomata, they are in constant danger of dehydration during periods of active growth. Carbon balance is viewed as a useful concept for studying and understanding how plants function in different environments. Carbon is treated as the currency of costs and benefits of a plant having particular physiological or morphological adaptations that permit it to cope with environmental limitations (Pearcy et al. 1987). Generally, plants in resource-rich environments have much higher rates of photosynthesis than plants in regions where water, nutrients, or other essential factors are deficient for optimum growth. Pearcy et al. (1987) note that species adapted to high light environments tend to have higher photosynthetic capacities than plants from shade environments, and agricultural plants adapted to highly fertile soils have higher rates than plants from relatively infertile areas. Further, plants from resource-poor environments tend to respond less to increases in resource availability than do species from environments where resources are readily available.

It appears that the pathway by which CO_2 is fixed as carbon compounds within the tissues of plants may have significant effects on various physiological and ecological characteristics of plants, and may be influential in affecting their water relations. There are three primary photosynthetic pathways in terrestrial plants that may influence their water relations characteristics: (1) the Calvin cycle (or C_3 pathway); (2) the Kranz type (or C_4 pathway); and (3) the CAM (or Crassulacean Acid Metabolism pathway).

Most species originating in cool season temperate regions fix CO_2 by the Calvin cycle, or the C_3 pathway, wherein CO_2 is incorporated to produce the 3-carbon compound, 3-phosphoglyceric acid. However, an alternative pathway that fixes CO_2 into 4-carbon compounds (and thus referred to as the C_4 pathway) has been observed in plants of tropical origin and in some species from warm arid regions (Jones 1985; Waller and Lewis 1979). Initial carboxylation of CO_2 within the mesophyll cells of C_4 species occurs by the enzyme phosphoenolpyruvate carboxylase, which has a particularly high affinity for CO_2 (Kriedemann and Downton 1981). Decarboxylation occurs within the particularly thick bundle sheath cell walls of these species, which tend to act as CO_2 concentrating sites for carbon fixation by ribulosebiphosphate carboxylase with a minimum of carbon loss due to photorespiration.

Plants with these two photosynthesis pathways differ significantly both anatomically and physiologically. Anatomically, C_4 plants have closely packed mesophyll cells with few air spaces that are radially arranged around prominent bundle sheaths with prominent chloroplasts and starch, whereas C_3 plants lack well-defined bundle sheaths and normally have large intercellular spaces within the mesophyll. The mesophyll surrounding the bundle sheaths in C_4 plants captures any photorespiratory CO_2 loss and redirects it back into the photosynthesis pathway; no such mechanism is available to C_3 plants. However, physiologically the two pathways are even more distinct primarily due to the double carboxylation and CO_2-concentrating mechanisms characteristic of C_4 plants. C_4 plants lack significant photorespiration and have very low CO_2 compensation points, whereas C_3 plants normally experience significant light-induced loss of CO_2 with relatively high compensation points. Also, C_4 plants tend to have higher growth rates, higher photosynthesis rates (sometimes three to four-times as great, although not always), higher light saturation, and higher optimum temperatures for growth than C_3 plants (Waller and Lewis 1979). C_3 plants tend to have long-day sensitivity for flowering, whereas C_4 plants are day-neutral or short-day plants (Jones 1985).

The primary advantage of the C_4 photosynthesis pathway to plants in arid and semiarid environments is that it permits efficient photosynthesis while simultaneously restricting water loss due to partially closed stomata (Ludlow 1976, 1980; Jones 1985; Osmond et al. 1980b; Pearcy et al. 1987; Waller and Lewis 1979).

This is due in part to a favorable cellular environment in which carbon can be concentrated even during periods when leaf stomata are partially closed during mild water deficit conditions (Caldwell et al. 1977). The primary features of the C_4 photosynthesis pathway that may contribute to water deficit adaptations include; (1) optimal rates of carbon fixation in the C_4 pathway occur at higher temperatures than in C_3 plants, (2) optimal carbon fixation occurs at higher solar flux densities, (3) net productivity is higher, and (4) relative water use efficiency is higher than that in C_3 plants. C_4 plants have been demonstrated to have greater water use efficiencies than C_3 plants (grams of tissue produced per gram of water absorbed). Nobel (1980b) found that big galleta (*Hilaria rigida*), a hardy colonizing C_4 desert grass that grows on harsh dry sites in both the Mojave and Sonoran deserts, has a water use efficiency about three times that of adjacent C_3 plants. He feels this, combined with the highest photosynthesis rates of other plants, to be a definite competitive advantage. Generally, C_4 plants have a lower mesophyll resistance and a higher stomatal resistance than C_3 plants (Jones 1985; Ludlow 1976; Osmond et al. 1980b; Turner and Burch 1980). Also, the lack of photorespiration in C_4 plants implies that there is no, or less, loss of CO_2 during periods when stomata are open than in C_3 plants, hence a higher efficiency per unit gain in carbohydrates per unit of water absorbed. McCree and Richardson (1987a, 1987b) suggest that C_4 plants are better equipped to exercise stomatal control over water loss without appreciably reducing carbon gain during deficit periods because their photosynthesis system operates at lower internal CO_2 concentrations than C_3 plants. This, combined with potential influences of other adaptations that some species may have for water deficits (such as osmotic adjustment), could substantially improve net carbon gain during extended periods of dehydration. However, Osmond et al. (1980b) and Ludlow (1976) argue that it may be dangerous to assign too much significance to the C_4 carbon pathway in terms of water deficit tolerance because the potential photosynthetic advantages are often exceeded by the intensity or duration of water deficits experienced under field conditions.

Based on the apparent attributes of the C_4 carbon pathway, Caldwell et al. (1977) hypothesized that C_4 plants should display several more favorable adaptations to desert environments than C_3 plants. They posed the following hypotheses concerning the C_4 pathway as it might play a role in adaptations to desert plant survival and growth: (1) they speculated that C_4 plants should display higher photosynthesis rates per unit foliage mass than C_3 plants; (2) since C_4 plants have higher temperature optima than C_3 plants, they might delay growth and photosynthetic activity until the warmest times of the growing season, while C_3 plants initiate growth earlier in the season; (3) based on these apparent attributes, C_4 plants would be expected to carry on photosynthesis in water-limiting environments with stomata partially closed in order to restrict the rate of water loss during periods of potential water deficits; (4) C_4 plants would thus be expected to exhibit higher net photosynthesis/transpiration ratios than that of C_3 plants, that they thus would be enabled to allocate less energy to root production and more to above-ground biomass, and (5) that on a community level C_4 plant communities should have higher biomass production in relation to water use (greater water-use-efficiency) than C_3 plant communities.

These hypotheses were tested by comparing the responses of the C_4 plant shadscale with those of the C_3 species winterfat, both prominent components of arid saline ecosystems of the Great Basin in the western U. S. Caldwell et al. (1977) found, contrary to expectations, that the high net photosynthesis rates anticipated by the C_4 plants were never observed, and that often the C_3 species had higher rates. Late in the season when temperatures were elevated, photosynthesis ceased in C_3 plants, but some was measured in C_4 plants. The C_4 plants did not delay photosynthetic activity until later in the season than C_3 plants, and the period of maximum photosynthesis was about coincident for both. There was no detectable difference in leaf diffusion resistance, indicating that C_4 plants do not have a significantly lower transpiration rate than C_3 plants under comparable environments. Also, they found little to support the hypothesis that C_4 plant communities (composed of shadscale) have either higher photosynthesis or lower transpiration rates than C_3 plant communities (composed of winterfat). The water-use-efficiency of the C_4 species was higher for plant growth, and these plants were able to incorporate about 37% more carbon into productivity than the C_3 species. However, both communities appeared to transpire equivalent amounts of water annually. Therefore, they concluded that the supposed benefits from C_4 photosynthesis were not readily evident, and that great caution is warranted in generalizing about the ecological significance of the C_4 photosynthesis pathway. Wiebe et al. (1974) noted that woody tissues, leaves, bark, and inflorescence bracts of shadscale all are capable of photosynthesis, and may contribute to total food supply at different times seasonally.

Some concern has been generated regarding global warming as a result of increasing CO_2 concentrations in the atmosphere, and the possible implications this may have for range plants (e.g., Kirkham et al. 1991). In midcontinental regions, like the prairie regions of North America, concern is that soil water status may decline as atmospheric temperatures and CO_2 increase, and that this may have severe effects on range plants. Kirkham et al. (1991) studied the effects of elevated CO_2 on photosynthesis of big bluestem (*Andropogon gerardii*), a warm season C_4 grass dominant in the tallgrass prairie. They found that transpiration resistance was increased by elevated CO_2, but that photosynthesis was unaffected. They also found that average plant canopy temperatures and soil water contents were both increased as a result of lower transpiration. They concluded that for this species at least, elevated CO_2 may reduce the amount of water required to produce the same amount of biomass by about 35% of that produced under current levels of CO_2, and that soil water levels may increase.

An additional form of carbon metabolism in plants thought to have considerable significance as an adaptive feature to water deficits is the CAM pathway. Originally observed only in plant taxa within the family Crassulaceae, and generally associated with succulent life-forms, CAM has more recently been observed in numerous other plant families as well (e.g., Aizoaceae, Agavaceae, Bromeliaceae, Cactaceae, etc. [Fu and Hew 1982; Kluge 1976; Nobel 1976; Nobel and Quero 1986; Winter and Luttge 1976]).

The salient feature of CAM as an adaptive mechanism for water deficits is linked to the peculiar nature of CAM plant stomatal movement and physiology. In contrast to most other terrestrial plants, the stomata of CAM plants open at night, when most atmospheric CO_2 is fixed, and tend to remain partially or totally closed during the day (Kluge 1976). The CO_2 taken up at night is fixed by carboxylation of phosphoenolpyruvate (PEP), and then catalyzed by PEP carboxylase to form malic acid. During the day the malic acid is consumed by decarboxylation enzymes and converted into photosynthesis end-products such as starch and sugars. The advantages of the CAM carbon pathway are apparent in arid environments where periods of water deficits are common. The inverted pattern of stomatal activity in CAM plants insures that transpiration losses are minimized by opening primarily at night when vapor pressure deficits between the leaf and open atmosphere are minimized. During the day, when vapor pressure deficits are large between the plant and the atmosphere, the stomata remain either totally or partially closed. Some CAM plants remain photosynthetically active both day and night during favorable periods when water is readily available, shifting to night activity only during periods of water deficits, and other species appear to be obligated to night CO_2 fixation. In either case, carbon gain is maintained even during periods of extended drought when water deficits may be limiting to other plants (Kluge 1976; Winter and Luttge 1976).

CAM plants are generally considered to have low growth rates and low primary productivities in natural environments, and although exceptions are known (Nobel and Quero 1986), in the larger context of rangeland management few are likely candidates to provide extensive forage for grazing animals. However, aside from the obvious adaptive features of CAM for survival in arid environments, evidence is clear that CAM plants often shift their photosynthetic pathway between CAM and C_3 in response to environmental conditions (e.g., Brulfert et al. 1988; Winter and Luttge 1976; Hanscom and Ting 1978; Salisbury and Ross 1992). The CAM pathway is favored in hot, dry environments where water deficits are common, but shifts from CAM to C_3 occur in response to favorable conditions such as following precipitation or moderated temperatures. Salisbury and Ross (1992) describe CAM plants as facultative C_3 plants that thus take advantage of higher photosynthesis rates during favorable environmental periods. The adaptive advantages of such shifts are readily apparent when the trade-offs between carbon gain and water deficits are considered. C_3 photosynthesis tends to be expensive in terms of water loss because of obligate stomatal activity during the day when water deficits are normally most severe. However, during short periods of readily available water following precipitation events, C_3 carbon gain normally results in higher growth rates and potentially greater investments in root exploration and other adaptive mechanisms than CAM (Kluge 1976). That some species are "induced" to switch carbon fixation pathways from C_3 to CAM when water deficits threaten to shut-down photosynthesis activities altogether under normal C_3 cycling, although teleologically attractive, still requires further research. Although recent studies have identified many species that display CAM–C_3 shifts, it has been observed mostly in succulents such as stonecrop (*Sedum spectabile*, Brulfert et al. 1988), *Portulacaria afra*, *Peperomia jerreyrae*, and *Zygocactus truncatus* (Hanscom and Ting 1978), and *Diamorpha cymosa* (Teeri 1985).

Because alternative photosynthesis pathways delay or postpone the effects of water deficits, these adaptations would be categorized as "drought postponing" under the traditional classification scheme of water deficit adaptations (Kramer 1983; Turner and Burch 1983).

Osmotic Adjustment Adaptations

Maintenance of plant turgor is generally recognized as one of the most important means by which the effects of water deficits on metabolic processes can be moderated (Jones et al. 1981; Kramer 1983; Turner and Jones 1980). Osmotic adjustment refers to the lowering of osmotic potential of cell contents by the active accumulation of solutes in response to water deficits or salinity, with the maintenance of positive pressure potential (or turgor). A clear distinction is to be made between normal increases in osmotic concentration as water is lost from a plant cell during dehydration, and accumulation of, or a net gain in solutes, within the cells in response to water deficits. The former condition is not to be confused with the mechanism of osmotic adjustment; during dehydration the loss of water within cells results in an increased concentration of solutes, hence a lower osmotic potential. Osmotic adjustment also results in lowered osmotic potential, but is achieved by an active gain in solute concentration within cells quite separate from changes due only to water loss. Many species of plants, primarily halophytes and some xerophytes, store high concentrations of solutes within their tissues during normal growth, even when water deficits are absent. These plants have quite low osmotic potentials even when fully hydrated, and may or may not have the ability to adjust concentration levels of solutes within their tissues in response to water deficits. Most mesophytes, however, normally store much lower concentrations of solutes within their tissues, and in the absence of osmotic adjustment may experience large declines in turgor during water deficits even though the overall osmotic concentration increases due to a loss of water. The definition provided by Turner and Jones (1980) and Jones et al. (1981) clarifies ambiguities between the two conditions.

The ability of a plant to adjust osmotically implies that solutes are accumulated from outside the water deficit-affected cells, and as a consequence lower their internal osmotic potential. The net effect, illustrated in Fig. 15, is to maintain either partial or full turgor in tissues as water deficits increase (Turner and Jones 1980). The loss of turgor is thus extended to lower plant water potentials than would have occurred without osmotic adjustment. The primary benefits of turgor maintenance at lower water potentials, hence of osmotic adjustment, include: (1) maintenance of stomatal opening and net carbon gain by photosynthesis; (2) greater cell volumes with continued cell expansion; (3) greater tolerance to dehydration; (4) deferring leaf rolling and death to lower leaf water potentials; and (5) extending periods of root growth and exploration of larger soil volumes and at higher mechanical impedances (Munns 1988; Turner 1986; Turner and Jones 1980). Thus, osmotic adjustment could be classified as a dehydration tolerance (or escape; Turner 1986) mechanism in the traditional classification schemes of Kramer (1983).

The degree of osmotic adjustment that plants are capable of achieving varies with species, rate at which water deficits occur, tissue age, history of growth conditions experienced by the plant, and current environmental conditions. The

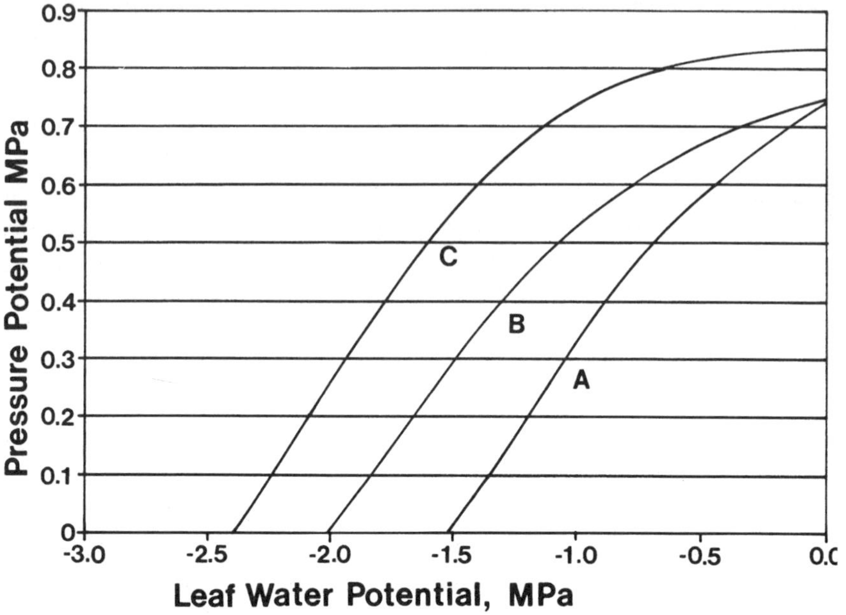

Figure 15. Illustration of osmotic adjustment in plants. (A) illustrates typical decrease in pressure potential (turgor) with rapidly declining leaf water potential with no osmotic adjustment. (B) illustrates the same plant when osmotic adjustment occurs during slow development of water deficit. (C) illustrates response of the same plant during a subsequent water deficit if turgor maintenance occurs during osmotic adjustment. Note shift in leaf water potential at 0 turgor point (B and C), and increased maximum turgor at maximum leaf water potential (C). All data are hypothetical.

amount of osmotic adjustment achieved is usually determined by comparing the tissue osmotic potential at full turgor before water deficits with that after a water deficit and rehydration to full turgor (Fig. 15, b and c; Turner and Jones 1980). Also, a comparison of the osmotic potentials at zero turgor (when $\psi_{leaf} = \psi_{\pi leaf}$) has also been used (Fig. 15, a and b or c; Turner and Jones 1980).

Although not all terrestrial plants are capable of osmotic adjustment, a surprisingly large number of species have been shown to possess this adaptation. Osmotic adjustment appears to be a fairly common adaptation in many plants, and has been noted in numerous crop species (Cutler et al. 1977; Henson 1982; Oosterhuis and Wullschleger 1987; Turner and Jones 1980), in woody shrubs and trees (Auge et al. 1990; Auge and Stodola 1989; Hinckley et al. 1980; Myers and Neales 1986; Tschaplinski and Blake 1989), in desert halophytes (Jefferies 1981; Richardson and McKell 1980), in mesophytic grasses and forbs (Drake and Gallagher 1984; Thomas 1986, 1990; West et al. 1990), and in numerous important rangeland species (Knapp 1985; Monson and Smith 1982). Normally, osmotic adjustment occurs when water deficits develop slowly in plants (Thomas 1986; Turner and Jones 1980). However, Oosterhuis and Wullschleger (1987) observed substantial osmotic adjustment in cotton with water deficit durations as short as four days. The amount of osmotic adjustment is highly variable among different species and tissues. For example, Oosterhuis and Wullschleger (1987) observed osmotic potentials at full turgor had adjusted in cotton leaves by about –0.41 MPa, whereas in roots osmotic adjustment was only about –0.19 MPa. West et al. (1990) noted in tall fescue (*Festuca arundinaceae*) that osmotic adjustment was far greater both in absolute osmotic potential and as a percentage of change, in tissues closest to the growing point, and least in mature tissues further removed from the basal meristems. They speculated that osmotic adjustment in meristematic tissues may be most critical in favoring survival of grass tillers during severe deficits.

Like most adaptations to water deficits, osmotic adjustment is probably not the only mechanism operating in plants during periods of dehydration. For example, McCree and Richardson (1987b) found that stomatal closure in combination with active osmotic adjustment may result in larger net gains of carbon over a longer time during deficit periods than either mechanism operating alone.

Plant adaptations frequently display a seasonality (e.g., Mooney 1980), and osmotic adjustment appears to be no exception. Bowman and Roberts (1985) showed that osmotic adjustment and the point at zero turgor potential in chaparral shrubs vary diurnally and seasonally as temperatures increase and water potentials decline later in the growing season. Johnson (1978) showed that environmental effects contribute significantly to the maintenance of turgor in range grasses. Plants at low temperatures appear to maintain turgor over a wider range of leaf water potentials than plants at high temperatures, and the degree of maintenance varies widely in different species. Monson and Smith (1982) noted osmotic adjustment in four out of seven desert species studied in the Sonoran Desert of Arizona. The desert species in which osmotic adjustment was felt to be a major component of water deficit adaptation included the evergreen tree *Olneya tesota*, the drought-deciduous shrub *Encelia farinosa*, evergreen creosote bush (*Larrea tridentata*), and the halophytic shrub *Atriplex polycarpa*. For evergreen creosote bush and *A. polycarpa* they observed a potential range of osmotic adjustment of only about –0.5 MPa, but both of these species had the highest turgor potentials and lowest osmotic potentials of all the species studied throughout the entire growing season. Knapp (1985) also observed seasonal changes in osmotic adjustment in the prairie grasses big bluestem and *Panicum virgatum* that had been burned and unburned. The total ranges of adjustment varied with species and burning treatment, with significantly lower osmotic potentials at 0 turgor for *A. gerardii* than *P. virgatum*. Burning appeared to result in lower osmotic potentials (greater adjustment) at both 0 turgor and full turgor for both species by as much as -0.4 MPa. Total osmotic adjustment in both species ranged from about –1 MPa to about –2.04 MPa throughout the season. Dibble and Spomer (1987) observed osmotic adjustments as great as –1.7 MPa in bluebunch wheatgrass, one of the most important rangeland grasses in the Intermountain area of the western U.S., during the growing season. They felt that osmotic adjustment in this species may be greater than that observed in any crop species, and is among the greatest degree of adjustment observed in any plant. They attribute the high degree of osmotic adjustment in this species to the very high elasticity of its tissues.

Elasticity of cell walls has been implicated as an important controlling factor in osmotic adjustment (Jones et al. 1981; Turner and Jones 1980), wherein species with inelastic cell walls may be unable to adjust osmotically to developing water deficits. A decrease in cell volume occurs as water is withdrawn during a water deficit, and the amount of volume change is related to the elasticity of the cell wall. The relationship between volume changes and changes in turgor is expressed as;

$$\epsilon = (\Delta p/\Delta v)*v \quad (7)$$

where ϵ is the volummetric modulus of elasticity, p is turgor, and v is cell volume. The higher the elasticity, the lower ϵ. Very often relative water content (RWC; see Table 2a) is used in place of v for routine calculations because determinations of cell volume are tedious and often more inaccurate than apparent changes in volume estimated by RWC. Plants with rigid cell walls would thus have relatively low elasticities, or high ϵ, whereas plants with thin elastic cell walls would have relatively lower ϵ characteristics. The thicker and more rigid (high ϵ) the cell wall, the less volume change would occur per unit water withdrawn from the cells, and conversely, the thinner and more elastic (low ϵ) the cell wall, the greater the volume change that would occur per unit water withdrawn.

Maxwell and Redmann (1978) observed elasticity of the cell walls of the grass thickspike wheatgrass, a component of the xeric grasslands of Canada. They found the cell walls of this species to be inelastic with highly rigid cell walls, and although they did not measure osmotic adjustment directly, they concluded that adjustments to water deficits in this species are probably as much related to changes in tissue structure and turgor as they are to simple adjustments in osmotic potential. Monson and Smith (1982) observed that cell elasticity of desert plants changes during the season, and noted that lower elasticity correlated with greater decreases in turgor per unit loss in tissue relative water content.

Osmotic adjustment tends to be a transient adaptation in most plants following the relief of water deficits. Normally the effects of osmotic adjustment are retained for only several days after water deficits are relieved (Turner and Jones 1980). Thus, when osmotic adjustment develops in a plant during periods of water deficits, it can be expected to disappear shortly after precipitation rehydrates the plant. Also, there appears to no beneficial carry-over effect during water deficit cycles from past osmotic adjustments that may have occurred in the recent history of the plant (Turner and Jones 1980), although this appears to be somewhat contradictory to other experiences in water relations research. Generally, a history of water deficits in plants is related to increased plant "tolerance" to future water deficits (Kramer 1983). For example, Cutler and Rains (1978) found in cotton that preconditioning plants to water deficits resulted in lower osmotic potentials and the maintenance of turgor to lower leaf water potentials than in plants having no previous water deficit conditioning. They found that the solutes involved in osmotic potential were at the same concentrations in both groups of plants, and therefore concluded that solute accumulation was not responsible for the observed differences. However, preconditioned plants had less water per unit dry weight of tissue than unconditioned plants, leading the authors to conclude that water deficits result in increased soluble solids and insoluble constituents such as cell walls and cuticular material within leaf tissues.

The solutes responsible for osmotic adjustment appear to vary with different species. In many mesophytes (Jones et al. 1981) the primary solutes involved in osmotic adjustment appear to be sugars and amino acids, but various salts and ionic constituents have also been implicated (e.g., K^+ and Cl^-). Under severe levels of dehydration other solutes such as carboxylic acid, sugar phosphates, and amino acids appear to be mainly responsible for lowering of osmotic potential. In halophytes salts of Na, K, and Cl^- ions, and oxalate appear to be responsible for most osmotic adjustment (Bennert and Schmidt 1984; Jefferies 1981). Munns (1988) reviews the literature on solute build-up in plants and discusses the concepts of solute increases due to reduced growth under decreasing temperatures, but feels that organic solutes may build up when demand by tissues falls below supply. During osmotic adjustment growth rate usually declines, and Munns (1988) believes the excess supply is the primary source of solutes observed in osmotic adjustment. Hence, he feels osmotic adjustment cannot occur concurrently during periods of active growth, and thus concludes that this adaptation is primarily important for survival during water deficits rather than as a mechanism to enhance growth during water deficits.

Osmotic adjustment appears to be finite in its ability to maintain turgor during water deficits. Generally, when water deficits exceed a critical minimum leaf water potential, osmotic adjustment is no longer capable of maintaining turgor sufficiently to sustain physiological activity (Jones et al. 1981; Turner and Jones 1980). Thus, this adaptation has definable limits in specific species for maintaining tissue hydration under conditions of water deficits.

Halophytic Adaptations

Halophytes are plants adapted to highly saline soils in such habitats as salt marshes, salt pans, and salt desert regions. These plants are important rangeland components, occupying millions of hectares of aridland regions such as the deserts of North America, and have unique adaptations that affect their water relations. Most rangeland halophytes of interest are found in the family Chenopodiaceae, and include shrubs and forbs of such genera as *Atriplex*, *Sarcobatus*, *Suaeda*, *Ceratoides*, *Salsola*, *Kochia*, *Halogeton*, *Salicornia*, and *Chenopodium*. However, some halophytic rangeland grass genera are also well known, such as *Distichlis* and *Spartina*. Halophytes accumulate large quantities of salts and ions within the vacuoles of their cells, principally Na^+ and Cl^- (Clipson et al. 1985; Flowers 1985; Jefferies 1981). Water potential equilibrium of the cytoplasm with the high ionic concentrations in the vacuoles is achieved by an accumulation of sugars and organic acids (Flowers 1985). Some halophytes accumulate salts continuously throughout their life cycle, and most regulate salt uptake by mechanisms such as active salt glands that secrete concentrated solutes, and by continuous enlargement of epidermal bladder cells that accumulate salts. These latter plants can thus maintain nearly constant osmotic potentials, aside from daily or seasonal variations due to water availability, throughout their lives (Hellebust 1976). Because halophytes accumulate such high concentrations of solutes, their osmotic potentials can be many times lower than those observed in normal glycophytes (non-halophytes). As such, halophytes are generally considered to be capable of absorbing soil water at much lower water potentials than most other plants, and thus are able to maintain positive turgor potentials needed to support photosynthesis and reproduction under extremely dry conditions (Flowers 1985; Jones et al 1981; Monson and Smith 1982; Jefferies 1981).

Some clarification may be helpful regarding osmotic adjustment adaptations and halophyte physiology and water relations. Halophytes accumulate salts whether under water deficits or not; most glycophytes do not (Jones et al. 1981). Although halophytes are subject to the same hazards of water deficits as glycophytes, the ranges of water potentials normally involved in halophyte physiology are much greater than those for glycophytes because of the high concentrations of salts in their tissues. Also, many halophytes are just as capable of osmotic adjustment in response to water deficits as many glycophytes, but their accumulation of solutes is not dependent on water deficits. Thus, salt accumulation is the dominant force in the water relations of halophytes.

The total water and osmotic potentials of halophytic plants have fascinated plant physiologists and ecologists for decades, and hence have been the subject of much study. Jefferies (1981) indicates that osmotic potentials of expressed sap from halophytes typically range between –3 and –4 MPa. Moore et al. (1972) observed osmotic potentials below –20 MPa in shadscale from the Great Basin. Moore and Caldwell (1972) feel that shadscale and winterfat are capable of extracting water and carrying on active photosynthesis when soil water potentials reach minima between –6.5 to –8 MPa. MacMahon and Schimpf (1981) list numerous examples of minimum water potentials measured in desert plants, and water potentials of halophytes can exceed –10 MPa. Love and West (1972) reported water potentials of winterfat and shadscale lower than –8.0 MPa during the middle of the growing season in the Great Basin, and cite literature of similar, or even drier, observations for other halophytes. Similarly, Detling and Klikoff (1973) observed seasonally low water potentials for the four halophytes annual seepweed (*Suaeda depressa*), perennial seepweed (*S. fruiticosa*), greasewood, and the halophytic grass, saltgrass (*Distichlis spicata*; note the common name for both *D. stricta* and *D. spicata* is saltgrass, reproduced here as used by the respective authors). Everett et al. (1977) studied the seasonal patterns of water potentials in southwestern Great Basin species, and observed that predawn leaf water potentials of shadscale were consistently lower than those of the associated non-halophytic species galleta grass, rabbitbrush (*Chrysothamnus viscidiflorus*), and big sagebrush. They noted that shadscale leaf water potentials were highest (about –4 MPa) in the spring and fall, but during the mid-summer period they fell below –7 MPa. Monson and Smith (1982) observed seasonal variations in dawn leaf osmotic potentials for *Atriplex polycarpa* at the point when turgor is maximum and at 0 turgor potential. Generally, at full turgor potential, osmotic potentials were nearly constant throughout the season, and varied from about –2.5 to –3 MPa. However, osmotic potentials when turgor was at 0 MPa varied from about –3 MPa in the early season when water was available to about –4 MPa during the middle of the season when water deficits were maximized. Associated non-halophytic species responded similarly with seasonal shifts in precipitation, but their osmotic potentials tended to be higher. Unfortunately, not many water relations studies of halophytes include seasonal and diurnal measurements of all the various water potential components.

Dehydration Adaptations

During prolonged and persistent periods of water deficits, adaptations such as phenological timing, stomatal control, rooting form, morphological and anatomical structural features, alternative photosynthesis mechanisms, osmotic adjustment, and halophyte characteristics contribute more to the deferral of damage rather than to the dehydration tolerance of plants. Eventually, when deficits become severe enough under persistently dry conditions, these adaptations are no longer able to sustain continued physiological activity. Continued water deficits beyond that point in most plants may result in irreversible injury, and ultimately in death. The only alternative for survival under such conditions in most terrestrial rangeland plants involves a near cessation of normal physiological processes, with the initiation of the condition of dormancy.

MacMahon and Schimpf (1981) consider true xerophytes to be non-succulent non-halophytic plants of arid regions that maintain living above-ground organs throughout the year, including periods of severe water deficits, and that obtain their water supply from local precipitation or atmospheric water only. Such plants would be classified as poikilhydric, suggesting that they come into equilibrium with the water status of the environment (Levitt 1972). Gaff (1980) has reviewed the characteristics of a group of dehydration tolerant plants that he refers to as "resurrection plants." Generally these include a small number of known genera and species with limited distribution in rangeland areas with few flowering plants, and mostly include dryland ferns, lichens, and microorganisms. Gaff (1977) lists 35 known species of angiosperms to display resurrection plant characteristics, including species from the families Scrophulariaceae, Gramineae, Cyperaceae, Polypodiaceae, Schizeaceae, and Actiniopteridaceae. However, only a few grass genera are familiar to range scientists, and include some species of *Sporobolus*, *Eragrostis*, and *Tragopogon* which appear to have a "predisposition" for the development of resurrection-plant-like characteristics (Gaff et al. 1985). Generally, it has been determined that resurrection plant tissues can tolerate dehydration down to about air dryness, and then be "resurrected" when water becomes available with restoration of physiological activity. For instance, Gaff (1971) found that numerous resurrection plants, both from South Africa and North America, could survive equilibrium water potentials to less than –100 MPa, as determined by equivalent relative humidities down to about 15–30%. More recently, Gaff and Bole (1986) found 9 grass species in India that appear to be desiccation tolerant down to 0% relative humidity (about –200 MPa at 28°C); they feel these plants may have some usefulness for grazing purposes. This is at least an order of magnitude drier than most terrestrial plants can survive, including those from desert regions.

The metabolic responses of plants to extreme dehydration have been reviewed by numerous authorities (Levitt 1972; Kramer 1983; Gaff 1980), and will not be pursued here. However, Kramer (1983) explains that even after several decades of research in this area, no adequate explanation has been advanced for why the protoplasm of some kinds of plants can tolerate extreme dehydration, but most cannot. Turner (1986) concurs, but suggests that the answer may lay at the molecular level, and depends on membrane structure and enzyme activity (e.g., see Gaff 1980; Levitt 1972).

Range Management Practices and Plant Water Relations

The very nucleus of the range ecosystem is composed of the interrelations among plants, soils, and climatic variables around which management practices are designed. Virtually every decision made by the range manager affects these interrelations in some manner. Management practices are designed and followed to provide and protect forage productivity, and by implication if not by direct necessity, are intended to also protect the soil from erosion and contamination, to preserve watershed values, and to achieve long-term sustainability of all range resources within the limits of the climatic conditions of the area. As such, few management decisions are made that do not in some way affect, or interact with, the water relations of plants and soils on rangelands.

Such activities as grazing, revegetation, water developments, plant and animal control, burning, range fertilization, fencing, and other operations affect on-site and adjacent plant and soil water relations on rangelands. With just a basic understanding of the principles of plant and soil water relations, it is not particularly challenging to imagine what some of the localized effects these practices might have on water relations of plants and soils. For example, the range improvement practice of building water developments can involve transporting and storing water in reservoirs or tanks either for use by grazing animals or for irrigation of pasture. The direct effects of water spreading and irrigation in arid or semiarid rangelands are intuitively obvious: increasing the availability of water for plant growth may enhance forage production and insure sustained range condition over longer periods of time, and may minimize reliance on the vagaries of climate. Whether the results of such practices are judged beneficial or detrimental depend not only on the objectives of the manager, but also on a knowledge of the long-term consequences of such management techniques. For example, significant increases in available soil water during the growing season above natural levels and variations may encourage the invasion and establishment of other, less desirable, plant species. It may also encourage heavier livestock and wildlife concentrations that can lead to overgrazing, severe trampling, accumulation and concentration of organic matter and chemicals from fecal matter and urine, increased sediment production, loss of infiltration, and other associated consequences. Use of salt blocks and fencing to control or encourage diverse livestock distribution on the range, plant control through burning, chaining, root ripping, or herbicides (e.g., see Valentine 1989 for a more complete discussion of range improvement methods) and other practices all have many unique, yet similar, benefits and consequences.

While it is not possible or appropriate to discuss the implications of every range management practice on water relations here, some examples of the consequences of these practices on plant and soil water relations will be discussed. Grazing and revegetation are two widely different, yet important activities practiced on rangelands that influence the water relations of both plants and soils. The effects of these management practices on basic plant and soil water relations will be discussed here to illustrate some of the ways management decisions can affect basic physiological and physical processes vital to range plants and soils.

Grazing and Plant Water Relations

Sensitivity of plants to herbivory, and particularly regrowth and replacement of both vegetative and reproductive tissues, has been studied extensively (e.g.,

Blaisdell and Pechanec 1949; Busso et al. 1989; 1990; Caldwell et al. 1981; Cook and Stoddart 1963; Cook et al. 1958; Craddock and Forsling 1938; Ellison 1960; Gold and Caldwell 1989, 1990; Jameson 1963; Johnson 1953; McNaughton 1992; Mohammad et al. 1982; Mueggler 1970, 1972, 1975; Nelson 1934; Nowak and Caldwell 1984; Oesterheld and McNaughton 1991; Olson and Richards 1988a, 1988b; Reed and Peterson 1961; Richards 1984; Richards and Caldwell 1985; Richards et al. 1988; Schuster 1964; Simoes and Baruch 1991; Wraith et al. 1987). Although numerous factors influence plant responses to herbivory (e.g., frequency and intensity of grazing, stage of plant development, plant growth-form, competition, morphology, recent physiological history, levels of nutrient and water deficits, etc.), the influences of grazing pressure on plant water relations and the ability to cope with periods of water deficits subsequent to grazing deserve concern. It is evident that removal of vegetative and reproductive tissues during the growing season by either grazing or clipping (which presumably mimics herbivory) can have significant effects, either direct or indirect, on physiological processes and morphological development that affect plant water relations, and ultimately plant survival in harsh rangeland environments. The main concerns addressed here relate to the influences of defoliation on plant water relations. Although rangeland herbivores are usually thought of as large livestock and wildlife grazing animals (e.g., cattle, sheep, horses, elk, deer, bison, etc.), it is to be noted that the effects of herbivory on plant water relations inflicted by insects, birds, rodents, small mammals, and other defoliators are considered to be equally important (Kelrick et al. 1986; Kemp et al. 1991; Mattson and Haack 1987; Naiman 1988; Parker 1985; Rissing 1988).

Tolerances to Grazing

The sensitivity of plants to grazing damage is highly dependent upon the physiological and morphological characteristics of individual plant species, and varies widely with climatic conditions, grazing habits of animals, and the intensity and duration of grazing (Coughenour 1991). Archer and Pyke (1991) feel that adaptations to herbivory can be classified as either grazing tolerance or grazing avoidance. In their discussion, grazing tolerance relates to characteristics that facilitate the re-establishment of foliage following defoliation, whereas avoidance relates to morphological or physiological characteristics that minimize the probability of herbivory. Further, they feel that traits conferring grazing tolerance are often characteristic of competitive plants, whereas avoiders are plants with traits similar to stress tolerators. Unfortunately, the physiological bases for determining adaptations to herbivory, let alone proper use and season of grazing, are rarely known for most range species, and hence are most widely based on experience and observation. Although it is arguably misleading to be broadly inferential about the entire array of range plants subjected to grazing, detailed physiological data are accumulating for a number of important range species to permit a few cautious generalizations about the sensitivity of plants to grazing. It is to be emphasized that responses of species to grazing are highly variable and dependent upon a wide spectrum of physiological and morphological characteristics (Dahl and Hyder 1977), and that generalizations are often contradictory when applied to specific individuals (Coughenour 1991).

On arid and semiarid rangelands at least (characteristic of the climatic zones of most rangelands), the periods of lowest sensitivity to grazing appear to occur:

1) during early vegetative development before flowering begins or during early floral formation, usually in the spring of the year (e.g., Caldwell et al. 1981; Olson and Richards 1988a, 1988b).
2) during later stages of maturity following flower and seedhead development (Cook and Stoddart 1963; Dahl and Hyder 1977; Reed and Peterson 1961).
3) when soil water is either abundant (e.g., spring) or during periods when demand for water is minimal, such as late summer, fall, or winter (e.g., Blaisdell and Pechanec 1949; Busso et al. 1990; Cook et al. 1958).

Periods of maximum plant sensitivity to damage appear to coincide with:

1) advancing phenological development ranging from late vegetative and early floral initiation through flower maturity (Blaisdell and Pechanec 1949; Cook et al. 1958; Jameson 1963).
2) during periods of declining seasonal soil water availability (Cook and Stoddart 1963; Craddock and Forsling 1938; Jameson 1963).

Grazing appears to be particularly detrimental during periods of limited water availability regardless of levels of intensity or duration of grazing or phenological stage of plant development. Herbage removal during periods of severe plant water deficits appears to limit shoot regrowth and root extension more than when adequate water is available (Cook et al. 1958; Mohammad et al. 1982). Even under heavy grazing, however, the availability of water appears to have a greater effect on productivity and regrowth than defoliation alone (Blaisdell and Pechanec 1949; Craddock and Forsling 1938; Nelson 1934; Mohammad et al. 1982).

Water Relations Under Grazing Pressure

The immediate short-term effects of herbivory may appear to improve plant and soil water relations (e.g., implying and increase in water potential, or a prolonging of higher water potentials) as the result of a reduction in transpiring leaf surface area and consequent diminished soil water depletion. In some cases grazing has been shown to have no effect on soil water status with depth (Orodho et al. 1990; Wraith et al. 1987), or to have a positive effect on available water (Svejcar and Christiansen 1987). Wraith et al. (1987) found that grazing did not significantly affect the overall seasonal depletion of soil water in a pure crested wheatgrass stand in Utah. However, during the early part of the season soil water depletion rate was significantly higher in ungrazed stands than in grazed stands, whereas during the latter part of the season grazed plots ranged from having significantly greater rates of depletion to being no different from ungrazed plots. There is little doubt that soil water status may temporarily increase if the transpiring surface area of plants is reduced during periods of active growth.

It can also be demonstrated that grazing results in higher tissue water status in the short-term, primarily in response to a lowered transpiring surface area. For example, Svejcar and Christiansen (1987) found that grazing can apparently improve the water relations of leaf tissue primarily as the result of increasing the ratio of root absorbing surface to transpiring leaf area. Studies of Caucasian bluestem (*Bothriochloa caucasica*) showed that heavy grazing consistently resulted in higher leaf xylem potentials and generally higher stomatal conductances than lightly grazed plants. Wraith et al. (1987) found no significant effects of grazing on predawn leaf water potentials of crested wheatgrass in the early part of the season, but later in the season ungrazed plants had significantly lower leaf water potentials. Midday leaf water potentials over the entire season did not differ significantly between grazed and ungrazed plants, although midday water potentials were significantly drier in ungrazed plots early in the season.

Despite any apparent short-term advantages of herbivory on plant water relations, the longer-term effects of defoliation are far more important for plant survival, growth, and overall prolonged range condition and trend (Ellison et al. 1951). Perhaps the most significant extended effect of grazing on plant water relations is the impact of repeated defoliation on reductions in root growth and survival. Obviously, a reduction in root growth can severely affect the water relations of plants since both nutrient and water absorption would be diminished during optimum periods of rapid regrowth. Generally, removal of shoot tissue by grazing results in the termination or reduction of root growth, reduced root extension both laterally and with depth, and may affect the growing points of roots more at depth than those near the soil surface (Crider 1955; Jameson 1963). In addition, it appears that reductions in root growth are proportional to the amount and intensity of defoliation (Crider 1955). Archer and Pyke (1991) emphasize that belowground herbivores, such as nematodes, grubs, and others may directly consume more vegetation in some ecosystems than aboveground grazers. They point out the direct effects of root consumption on regrowth potential, survival, and the interaction effects of aboveground herbivory that may increase susceptibility to nematode consumption of roots.

Thus, the overwhelming evidence suggests that any short-term gains in either plant or soil water status (e.g., Wraith et al. 1987) resulting from grazing are no conciliation in view of the known long-term responses of plant root systems to defoliation (Crider 1955; Dahl and Hyder 1977; Richards 1984; Stoddart et al 1975; Simoes and Baruch 1991). Schuster (1964) showed that repeated long-term moderate and heavy grazing resulted in major reductions in total root biomass, rooting depth, and lateral spread of grass roots. However, shrubs and forbs were not as severely affected by grazing. Mohammad et al. (1982) found that root biomass decreased as grazing intensity increased in both crested wheatgrass and Russian wildrye. The highest root biomass in both species was produced in undefoliated plants, and the least was produced in those defoliated by 80%. Also, they found that reductions in root biomass were significantly affected more by water deficits than by defoliation alone, and that crested wheatgrass showed greater resistance to water deficits and defoliation than Russian wildrye. Perhaps contrary to expectations, however, they found that light levels of grazing (e.g., 40% defoliation) had no effect on root production, even under moderate levels of water deficits.

The role of root responses in defining grazing tolerance was clearly demonstrated by Richards (1984) in which he found that clipping resulted in significant reductions in root length growth in the relatively grazing tolerant species crested wheatgrass, whereas root length growth continued unabated in the less tolerant bluebunch wheatgrass. He found that crested wheatgrass allocates relatively more carbon resources to aboveground growth following shoot removal than does bluebunch wheatgrass, and that this regrowth aids in the re-establishment of optimum root:shoot ratios. Although bluebunch wheatgrass root growth continued following clipping, significant root mortality resulted during the following winter and growing season. The immediate reduction in root growth observed in crested wheatgrass following clipping, although usually interpreted to be detrimental, was considered in this study to be a mechanism by which the reestablishment of the photosynthetic canopy and balance in roots and shoots could be maintained. In turn, this is interpreted to enhance overall competitiveness and tolerance to herbivory. Busso et al. (1990) also found that regrowth of aboveground shoots of unclipped crested wheatgrass was greater than that of bluebunch wheatgrass. The relative grazing tolerances, regrowth characteristics, resource allocation, and potential significance of these characteristics in crested wheatgrass and bluebunch wheatgrass following herbivory have been studied extensively (e.g., Busso et al. 1989, 1990; Caldwell 1984; Caldwell et al. 1981, 1983; Eissenstat and Caldwell 1987, 1988; Gold and Caldwell 1989, 1990; Nowak and Caldwell 1984; Olson and Richards 1988a, 1988b; Richards 1984; Richards and Caldwell 1985; Richards et al. 1988).

In addition to the immediate impacts on water relations, reductions in root and shoot productivity from herbivory often lead to marked reductions in carbohydrates in both above- and belowground tissues (Dahl and Hyder 1977; Jameson 1963; Richards and Caldwell 1985). Although a review of the effects of herbivory on plant carbohydrates is inappropriate here, it is to be emphasized that any force that affects overall plant vigor and condition may impact plant water relations either directly or indirectly, and may be critical to plant survival in arid or semiarid environments. For example, Busso et al. (1990) found that total nonstructural carbohydrates in the roots and crowns of crested wheatgrass and bluebunch wheatgrass in the spring tend to be affected less by clipping than by the availability of soil water the previous year. They found that total nonstructural carbohydrates in the spring were usually higher following drought conditions the previous year than in plants that had grown under natural precipitation or irrigation. In crested wheatgrass and bluebunch wheatgrass, Richards and Caldwell (1985) found that meristematic limitations, or limited flexibility for reallocation of resources, is more important for regrowth following defoliation than stored carbohydrates. They concluded that photosynthetic replacement of tissues in crested wheatgrass is quantitatively more important than stored carbon resources. Recent research (e.g., Caldwell et al. 1981, 1983; Eissenstat and Caldwell 1987, 1988; Gold and Caldwell 1989, 1990; Nowak and Caldwell 1984; Olson and Richards 1988a, 1988b; Richards 1984; Richards and Caldwell 1985; Richards et al. 1988) suggests that carbohydrate reserves alone are poor indicators of potential grazing tolerance, but rather is better related to regrowth potential and shoot characteristics. However, evidence indicates that carbohydrate constituents of plants vary not only among species, but also with season, and that these differences may play a major role in defining grazing preference among herbivores (e.g., Chapin et al. 1986).

There is considerable speculation that plant biomass productivity under moderate defoliation of shoot tissues may equal or exceed that of undefoliated plants (especially grasses) in a number of different species through increased tiller production or other mechanisms (e.g., Dahl and Hyder 1977; Taylor and Aarssen 1988). It may be tempting to suspect that growth and production of shoot tissues in thick, dense, ungrazed tussock grasses is light inhibited, and that moderate grazing could more efficiently permit entry of solar energy that would stimulate additional growth. Caldwell et al. (1981) speculated that the differences in architecture between the relatively dense canopy of created wheatgrass and the more open splayed canopy of bluebunch wheatgrass may be significant in production of regrowth following defoliation. Measurements of photosynthetically active radiation within the canopies of crested wheatgrass suggested that foliage production may be light limited, and that there may be a possibility that partial canopy opening by defoliation may stimulate the production of more foliage than the canopy can efficiently display under undefoliated conditions. However, later studies (Caldwell 1984; Caldwell et al. 1983) showed that undefoliated tussocks of crested wheatgrass were just as efficient at capturing light energy as more open rhizomatous grasses, and that partial defoliation resulted in a lower water-use efficiency, primarily as the result of poor photosynthetic light harvesting by the remaining tissues. They feel that for defoliation to improve water-use efficiency, tissue removal would need to be executed selectively with respect to age and position of individual foliage elements within the tussock, a procedure not likely performed consistently by grazing animals. Removal of dead standing litter from grasses during dormant periods can result in significant reductions in growth and development when growth resumes the following growing season. For example, Willms (1988) found that clipping dead litter during dormancy in rough fescue (*Festuca scabrella*) resulted in reduced plant height and increased tillering compared with unclipped plants. Growth and development responses to the loss of standing litter is interpreted to result from modified microenvironmental conditions since it does not affect plant energy relations directly. Thus, the role of light (and perhaps other microenvironmental factors) as exclusive casual agents contributing to the production of shoot tissue following defoliation appears unresolved.

Some workers have speculated that light to moderate levels of defoliation in combination with water deficits may improve plant survival and productivity. Mohammed et al. (1982) feel that light to moderate defoliation of plants (presumably implying grazing) under low to moderate levels of water deficits may enhance production and survival of both crested wheatgrass and Russian wildrye plants. They speculated that light to moderate defoliation reduced the transpiring surface sufficiently to diminish the negative impacts of water deficits. Light defoliation of both species resulted in higher aboveground production than undefoliated control plants even under severe levels of soil water deficit ("wilting point"). Similarly, Daddy et al. (1988) observed that herbaceous above-ground production and cover, and soil water content, were higher under moderate levels of grazing than under heavy grazing or complete protection in sagebrush-grass communities. They found about 80% of the total root biomass occurred in the 0–40 cm depth range under grazing, whereas only about 40% were observed in this depth range in the protected sites. Grass root distribution appeared not to be affected by grazing history.

Although controversial, Simoes and Baruch (1991) concluded that defoliation and water deficits interact strongly to reduce the effects of defoliation alone in the C_4 grasses *Hyparrhenia rufa* (introduced) and *Trachypogon plumosus* (native) in the savannas of Venezuela. They speculated that water deficits lead to: (1) less growth, and thus result in lower biomass loss during defoliation; (2) water deficits develop more slowly in defoliated plants because of lower leaf areas, thus leading to diminished effects of water deficits compared with undefoliated plants; and (3) a suspension of leaf aging due to water deficits ameliorates responses to defoliation alone. On the other hand, flooding and defoliation appear to have opposite interactive effects. Oesterheld and McNaughton (1991) found that flooding and defoliation suppress growth rates more than either effect alone. Species adapted to flooding appear to be less tolerant of grazing than plants not so adapted, whereas plants unadapted to flooding appear to be more grazing tolerant. Two potential mechanisms were suggested to explain this response: (1) flooding-adapted plants have a greater investment in stature and support tissues than flood-unadapted plants, hence defoliation of the limited photosynthetic tissues is critical; and (2) flooding temporarily allocates greater amounts of carbon resources to adventitious root production away from shoot regrowth. Similar results were observed by Etherington (1984) for orchardgrass. It was concluded that clones of this species from drought-prone areas tend to be more grazing tolerant than clones from moist soil areas. Two mechanisms were suggested to explain this conclusion: (1) moist site clones tend to have larger above-ground gas-exchanging surfaces that defoliation removes, and thus reduces oxygen supply to the roots; and (2) dry-site clones tend to have deeper more extensive root systems which may provide a greater competitive advantage for regrowth.

Competition and Grazing

Competitiveness among individual plants and species in environments where water is limiting during key periods of the growing season is a major factor determining plant vigor and survival, but competition also affects plant responses to herbivory (Eissenstat and Caldwell 1987, 1988; Mueggler 1975; Reader 1992). Mueggler (1970, 1972, 1975) effectively demonstrated the interactive effects of competition and defoliation on bluebunch wheatgrass and Idaho fescue. For both species, the negative effects of defoliation on herbage and flower stalk production decreased significantly as the level of competition decreased among individual plants. His data (Mueggler 1975) clearly show the differences in response to grazing and competition among species. Bluebunch wheatgrass is more sensitive to defoliation than Idaho fescue, and may require from 5–8 years for recovery of full vigor, whereas Idaho fescue may only require 3–6 years. Recovery periods following defoliation vary markedly with initial plant vigor; plants of low initial vigor require almost twice as long to recover following defoliation as plants with moderately low initial vigor. Eissenstat and Caldwell (1987, 1988) found that crested wheatgrass is more competitive than bluebunch wheatgrass, which is primarily described in terms of the rate of soil water extraction. Crested wheatgrass is able to extract water from the soil at a higher rate than bluebunch wheatgrass mainly because of its greater total root length and root growth in winter and early spring. Thus, in water limiting environments where seasonal competition for available water is expected to be high among individuals, permissible levels of grazing pressure designed to maintain optimum levels of vegetative and reproductive tissue production may vary from year to year as climatic history and past grazing varies. It appears that the long term sustainability of plant vigor is at least in-part determined by the interactive effect of climatic variables, grazing pressure, and competition, hence on the sensitivity of the land manager in adjusting allowable grazing pressure to insure the maintenance of vigor and condition and trend (Ellison et al. 1951).

Although quantitative data are lacking, it is tempting to at least partially attribute long-term trends in range condition, especially in arid and semiarid regions, to the effects of grazing on plant water relations. Although numerous factors interact to define trends in range condition, continued grazing pressure contributes significantly to a weakened regrowth potential and competitiveness in many species such that long-term grazing effects on water relations is certainly involved in defining the nature of "decreaser" and "increaser" species on rangelands (e.g. Stoddart et al. 1975). Certainly for tree species, there are clear indications of the interactive effects of grazing and competition by herbaceous species of tree seedling survival (Doescher et al. 1989). Trees seedling survival, water relations characteristics, and growth appear to be unquestionably enhanced under grazing where competition by herbaceous vegetation is reduced by grazing.

Other Effects of Grazing on Water Relations

Range plant and soil water relations are also affected by other activities of grazing animals in addition to just defoliation. Such associated by-products of grazing as fecal and urine deposition, and trampling have been found to significantly affect the localized water relations of plants and soils. The interactive effects of animals and vegetation suggest that dung and urine deposition can have significant effects on biomass and nutrient cycling for localized patches of vegetation, although definitive responses appear to be highly species specific (Day and Detling 1990; McNaughton et al. 1988). For example, Day and Detling (1990) discuss the effects of bison urine on shoot and root biomass production and nitrogen content of Kentucky bluegrass (*Poa pratensis*), (a C_3 grass), and little bluestem (*Schizachyrium scoparius*, formerly *Andropogon scoparius*, a C_4 grass) in the northern mixed prairie, and its subsequent influence on leaf and soil water relations. They found that Kentucky bluegrass tended to be influenced more by urine deposition than little bluestem in that its seasonal and diurnal leaf conductances tended to be significantly higher on urine patches than on unaffected patches, whereas leaf and soil water potentials were generally lower on urine patches than untreated areas. Consistent with observations of other C_3 plants, they feel these data suggest that Kentucky bluegrass may be able to maintain higher photosynthesis rates at lower water potentials on urine patches than the C_4 grass, and that this may explain the higher observed shoot and root biomass and nitrogen content of its tissues. The C_4 grass, little bluestem, was less affected by urine deposition; its stomatal conductances, leaf water potentials, and biomass were generally similar on both treated and untreated areas.

Trampling by grazing animals can significantly affect soil properties through compaction and disturbance. In general, factors such as soil bulk density, water infiltration rates, erodability, and water holding capacity are affected most by moderate to heavy sustained grazing (Branson et al. 1981; Warren et al. 1986). In most cases, as trampling intensity and duration increase, sediment production and overland flow also increase and infiltration rates decrease (Branson et al. 1981; Packer 1953; Warren et al. 1986). Soil bulk density, infiltration rates, and sediment production loads are generally affected more as soil water content increases. The effects of trampling appear to be less severe on vegetated ranges than on poor condition range or bare soil (Packer 1953; Warren et al. 1986).

Livestock trampling and range imprinting have been recommended as methods to bury seeds and enhance seedling emergence during range revegetation (Roundy et al. 1992; Winkel and Roundy 1991; Winkel et al. 1991a). It is suggested that trampling enhances germination and seedling emergence by increasing soil-seed contact and water movement from soil particles to the seed (Winkel et al. 1991a). Winkel and Roundy (1991) found that greater soil disturbance due to trampling, imprinting, root plowing, or ripping resulted in increased seedling emergence under moderately moist periods than light trampling or no disturbance. In drier periods, the degree of disturbance had less effect on seedling emergence. Roundy et al. (1992) feel that trampling and imprinting may enhance available soil water for only one to several days longer than undisturbed soils, but that this may be critical for the establishment of seedlings in arid sites. Trampling and imprinting appear to shorten the period of time that seeds and seedlings are exposed to potentially lethal levels of soil water potential between precipitation events, and this may be critical to plant establishment in arid and semiarid regions (Roundy et al. 1992). Thus, although there are some negative effects of trampling on soil properties under severe intensities, moderate trampling may have some positive effects on seedling establishment and seedling survival.

Rangeland Revegetation and Plant Water Relations

Rangeland revegetation is usually implemented to restore or enhance forage production and increase grazing capacity (Laycock 1982; Vallentine 1989). The concept of stable states and thresholds (e.g., Laycock 1991) suggests that grazing management alone may not be sufficient to alter existing states of range condition, and that revegetation may play an essential role in improving the state of range condition. In addition, incentives to revegetate rangelands may also include the reduction of erosion and sedimentation resulting from severe disturbances such as overgrazing, road construction, fire, mining activities, recreation, natural seismic activities, and other causes. Concern about disturbances on rangelands stem more from their consequences for forage production, grazing capacity, and watershed values than from the cause of disturbance. The importance of rangeland revegetation is heightened with the realization that the effects of disturbance are often exacerbated when two or more causes of disturbance occur simultaneously, wherein the occurrence of one often intensifies the effects of the other (e.g., overgrazing on steep unstable slopes prone to landslides; Sidle and Amacher 1990). The interaction of these activities can be cumulative in both space and time as they relate to the availability of limiting resources such as water and nutrients, and ultimately to grazing capacity and biomass productivity of the range.

The range manager faced with the decision to implement revegetation is challenged with the need to select not only cultural techniques (e.g., methods of seeding and planting, fertilizing, mulching, etc.) that will maximize success, but also must confront the dilemma of species selection that will optimize long-term range condition. The spectrum of techniques available to enhance successful revegetation narrows considerably with the severity of environmental conditions. In regions where water is limiting during most of the growing season and where the soil is completely lost as the result of disturbance, barriers to successful revegetation are far more formidable than those where water and/or native soil are abundant. Obviously environmental conditions rarely provide ideal revegetation opportunities. In arid and semiarid regions water is normally the major limiting resource to successful revegetation during the growing season. But even in more mesic areas such as subalpine and alpine ecosystems where water is not normally as limiting (e.g., Chambers et al. 1990), severe disturbances may completely destroy the natural soil and uncover geologic material of poor quality as a growing medium for plants (Brown and Chambers 1990; Brown et al. 1978). Following severe disturbance it is not uncommon for residual surface materials to contain little or no organic matter, have high concentrations of rock and hence to have low water and nutrient holding capacities, and/or to contain toxic concentrations of metals or other chemicals limiting to both plants and animals (Brown and Chambers 1989, 1990).

Importance of Species Selection

One of the most important steps to successful revegetation is selection of adapted plant species (Brown and Johnston 1979; Brown et al. 1978; Laycock 1982; Thornburg 1982). For the most part, plant species are selected for revegetation based on desired expectations and experience by the range manager rather than on basic quantitative physiological data. Although this has proven satisfactory under many circumstances, experience alone can be misleading. Experience gained in one region or climatic zone can lead to misjudgments in another when making decisions about plant species adaptability. The overall environmental characteristics of an area or region, including annual and seasonal climatic variability, soil properties, topography, local populations of insect pollinators, predators, diseases, and other conditions in combination over centuries of natural selection define the genetic range of physiological tolerances of each species or population of plants. Those species capable of germination, emergence, growth, development, competition for resources, and reproduction are generally considered adapted to a particular region or area. However, severe disturbances can result in new soil and microclimatic conditions that exceed the tolerances of many otherwise adapted species by either broad or, in some cases, subtle margins that may not be immediately evident based on visual observation and experience. And, recent concerns about global climate change emphasize that conditions are rarely static, and may be in a state of more rapid transformation than previously realized. Unfortunately, knowledge of the physiological tolerances of most range plants suitable for rangeland revegetation is lacking or largely unavailable to the manager in a useful format for application purposes. Hopefully, future research on the basic physiology of water relations, revegetation, and in plant breeding (e.g., Asay 1986; Asay and Johnson 1983a, 1983b, 1990) will address these issues.

As an example, crested wheatgrass is one of the most widely selected and used range revegetation species in the western U.S. (e.g., Johnson 1986). The species is well-known for its wide ecological adaptability (Thornburg 1982), tolerance to grazing (Caldwell et al. 1981; Nowak and Caldwell 1984), competitiveness (Caldwell and Richards 1986), regrowth potential and vigor (Johnson 1986; Richards 1984; Richards and Caldwell 1985), palatability (Reynolds and Springfield 1953), genetic attributes and breeding potential (Asay 1986; Asay et al. 1986), and many other desirable attributes (Johnson 1986). Vast areas of western rangeland have been seeded with this species, many stands of which have become virtual monocultures displaying rather profound ecological stability and permanence (e.g., Trlica and Biondini 1990; West et al. 1984). One of the more important concerns about the wide-spread use of any species in arid or semiarid rangeland revegetation is the relative demand for water resources that species places on the environment in which it is planted. Trlica and Biondini (1990) compared the soil water dynamics, transpiration, and water-use-efficiency of a pure stand of crested wheatgrass with that of an adjacent native shortgrass stand consisting of blue grama and western wheatgrass. The native grasses had greater root biomass and greater soil water depletion with increasing soil depth than the crested wheatgrass stand, suggesting that the native stand may be more competitive in a water-limited environment. Seasonal transpiration of the native grasses was higher and their leaf water potentials were lower than for crested wheatgrass. However, crested wheatgrass displayed a greater degree of leaf rolling in response to water deficits than the native grasses, which may indicate a water conservation adaptation that contributes to its overall competitiveness and ability to allocate greater carbon resources to aboveground production. These relationship may explain why crested wheatgrass used less soil water from greater soil depths than the native grasses. They feel that the greater soil water reserves with depth under crested wheatgrass stands may explain why this species is better able to regrow in the late summer and fall than native grasses. Although the water use patterns of different species are highly different, such research (Trilca and Biondini 1990) provides basic quantitative water relations data that help explain the physiological bases for species selection and adaptability, and should be encouraged more.

Soil Surface Treatments

The principal objective of cultural practices in revegetation is to ameliorate or modify local site conditions that are limiting to plant germination, emergence, growth, and survival, and/or to alter site conditions to reduce erosion and enhance other land management goals. Practices such as contouring and shaping are implemented to improve slope angles, aspects, and surface characteristics to optimize water relations and other conditions for plant growth and survival, and to control water movement and distribution. Emphasis should be given to minimizing steep windward-facing slopes that encourage wind scour and desiccation. Shaping, pitting, ridging, and furrowing offer opportunities to construct topography that favors the capture of precipitation, primarily in the form of snow (Schuman et al. 1987). These practices are especially favored in arid or semiarid regions where available soil water may become limiting during seedling establishment. Similarly, ripping, disking, and other physical soil modification techniques can be used to improve infiltration of water, incorporation of nutrients, to enhance aeration of roots, and to reduce competition for soil water and nutrients by undesirable species (Ferguson and Frischnecht 1985; Laycock 1982).

Nutrient Enhancement

Fertilizers are frequently used to enhance nutrient availability, which can have a significant effect on biomass productivity of both natural range and reseeded disturbances (DePuit and Coenenberg 1979; Laycock 1982). In addition, fertilizers may improve root production and distribution, and provide desirable species with competitive advantages over annual weeds or other undesirable species (Fisher et al. 1988; West et al. 1984). Nutrient enhancement can also be used to improve palatability, extend periods of growth, and expedite recovery of overgrazed ranges. On disturbances, fertilizer may be essential for seedling emergence and early growth (Brown and Johnston 1979; Laycock 1982), but is probably not important for germination (Chambers et al. 1990; McDole and Booth 1979). In high elevation areas where growing seasons are short, it is usually recommended that fertilizer be applied in the fall after seeding and planting are accomplished, instead of the spring when seedling growth begins following snowmelt. Although nutrient losses will certainly occur throughout the winter due to leaching and volatilization of some nutrients (especially nitrogen), access to high elevation areas in the spring at the proper time for seedling emergence, initiation of growth, and development cannot be guaranteed due to unpredictable weather conditions. On other range areas at lower elevations fertilization is recommended in the spring when growth begins (McDole and Booth 1979).

The nutrients contained in most commercial fertilizers are manufactured primarily in the form of salts or other soluble forms which require sufficient water to dissolve and become available for absorption by plants. Thus, supplemental additions of nutrients are only beneficial when adequate soil water is available. In arid regions fertilizers applied at heavy rates or at inappropriate times can have adverse effects on plant water relations because limited water availability may decrease the osmotic potential of the soil to lethal limits within the root zone of young seedlings, or may increase concentrations of otherwise essential nutrients to toxic levels (California Fertilizer Association 1985). In order to minimize the chances of creating nutrient toxicity conditions or other disastrous effects, it is essential that complete soil analyses be conducted for a site before any supplemental nutrients are added. Supplemental irrigation (not discussed here) may be required in arid environments where nutrient enhancement is considered essential for plant establishment (Powell et al. 1990).

The water relations of range plants are notably affected by the availability of nutrients. Fertilization of rangelands will likely favor certain species more than others (Power 1980; Schuman et al. 1991), and may affect root:shoot ratios and other growth characteristics depending upon the component nutrients added (Chambers et al. 1987; Ram et al. 1991; Tiedemann and Lopez 1982). The effects of nutrient enhancement also appear to vary widely with growth form and seral status of many species (Chambers et al. 1987; West et al. 1984). Nitrogen is normally the most deficient nutrient on rangelands, and is most commonly added to enhance shoot biomass productivity. However, enhancement of shoot growth with N alone may significantly increase the transpiring surface area of plants without compensatory increases in root development and volume of soil occupied (Ram et al. 1991). This almost certainly will increase the potential for, and perhaps duration of, water deficits in arid and semiarid regions during the growing season. Although the interactive effects of water deficits and N availability appear to be additive (Bennett et al. 1986; Morgan 1986), the effects of N fertilization are dependent on minimal thresholds of available water that vary with soil type, species, and other site conditions (Chambers et al. 1987; Fisher et al. 1988; Power 1980; Schuman et al. 1991). The evidence is clear that N enhancement can increase water-use-efficiency in some grass species (Power 1980), although this appears to be specific by species, growth form, and location (e.g., Bennett et al. 1986; Morgan 1986).

Although a complete quantitative soil analysis should be completed before adding any nutrients to the soil, a balance of both N and P appears to be more beneficial than supplemental N alone (Laycock 1982; Packer and Aldon 1978). Generally, data from rangeland fertilization studies show that applications of both N and P together are more effective for maintaining sustained long-term increases in vegetative biomass than N applications alone (e.g., Laycock 1982). In addition to the obvious benefits of supplemental N for vegetative growth, P is essential for root and reproductive development, and hence for establishing more favorable water relations characteristics (Tiedemann and Lopez 1982; West et al. 1984). Sustained root production and the subsequent benefits for reproduction suggest that supplemental P would enhance rangeland plant water relations more than N fertilization alone, especially in areas where deficiencies in P are common. West et al. (1984) found that a crested wheatgrass stand absorbed more P from the soil, and recycled it faster, than an adjacent native semidesert shrub community. A deficiency of any essential nutrient, including micronutrients such as Fe, Cu, Zn, and B, have been shown to exacerbate the effects of water deficits on plants (e.g., Sharma and Sharma 1987).

In addition to the well-known effects of nutrient deficiencies on plant water relations, toxic levels of nutrients and other chemicals also impede the ability of plants to respond to water deficits. Similar to the interactive effects of N with plant water relations, Al (and presumably other elements) toxicity is exacerbated under water deficit conditions for a number of species (Brown and Amacher 1990; Krizek and Foy 1988; Krizek et al. 1988). Brown and Amacher (1990) studied the effects of acid mine drainage (AMD) containing toxic concentrations of available Cu on the water relations characteristics of a dominant alpine grass, sedge, and forb (tufted hairgrass, Payson's sedge [*Carex paysonis*], and sibbaldia [*Sibbaldia procumbens*], respectively) under both field and laboratory conditions. AMD-affected plants all had lower leaf water potentials, lower transpiration rates, exhibited wilting sooner, and had lower growth rates than control plants unaffected by AMD.

Mulching Treatments

Surface mulches are commonly used during revegetation to improve site conditions and ameliorate such factors as temperature extremes, to conserve soil water, reduce frost activity during seedling growth and emergence, to reduce wind and water erosion, and to minimize the redistribution of soil fines and seeds (Fraser and Wolfe 1982; Kay 1982; Slick and Curtis 1985; Unger 1978). Mulches usually consist of organic residues such as manure, straw, hay, native plant litter, jute netting, wood fibers and chips, sewage sludge, and other materials. However, industrial textiles, fabrics, netting, papers, excelsior, and other artificial products are also considered mulches (Kay 1982). Humic matter incorporation (Skujins and Richardson 1985) can be considered mulch also, but generally only substances applied on the surface for stabilization or modification of limiting conditions are generally regarded as true mulches (Slick and Curtis 1985).

Mulching has been shown to result in significantly higher levels of plant cover, density, and biomass production than adjacent unmulched areas (Day and Ludeke 1987; Kay 1982). In addition, mulching is an effective means of reducing erosion by wind, water, and gravity (Parr et al. 1989). In arid regions, mulch appears to significantly moderate air temperatures adjacent to the soil surface and soil immediately below the surface. Especially on darker-colored materials, mulching results in lower soil surface temperatures due to the interception of direct solar radiation. In cold alpine regions, mulch appears to act as a heat-trap by causing a reverse in average mean maxima and minima soil surface temperatures (Unger 1978). This effect appears to be crucial in minimizing temperature extremes conducive to frost activity, leading to seedling mortality during revegetation (Brown and Chambers 1989, 1990). Mulching in cool regions, such as alpine rangelands, may require some unclear compromises; reduced soil surface temperatures during the day (which can lead to slower plant growth an development), vs. higher surface temperatures at night that reduce the incidence of frost action (Unger 1978). Generally, mulching results in higher levels of soil water status, especially in arid and semiarid environments during the early stages of seedling emergence and growth (Brown et al. 1991). However, the longer term effects of mulch on soil water availability are not as optimistic; Brown et al. (1991) found that straw mulch continues to have a positive effect on soil water availability in the surface soil for only about 10 months after application, although it has a major effect on snow trapping (Fig. 16) and seedling establishment during the first growing season in arid environments (Sidle and Brown 1988). The use of crimped straw in arid environments can lead to the so-called phenomenon of "wicking," wherein soil surface water is absorbed and evaporated at a greater rate through the straw fibers than from unmulched bare soil. However, crimping does appear to result in significant snow trapping at the surface during spring storms (Sidle and Brown 1988). It is likely that

Figure 16. Photograph illustrating the snow trapping efficiency and accumulation on the surface of straw mulch erosion blankets compared with that on bare ground (Logan, Utah).

newer mulching products such as erosion blankets and similar developments will replace the need for distributing and crimping straw and similar organic residues, at least on smaller critical areas where mulching is essential and where economics justify their use.

Conclusions

The universal importance of water is one of the more deeply ingrained truths in range management. The very nature of rangeland climates, soils, and resulting vegetation defines the implicit character of water as the controlling substance dominating the successful practice of range management. The majority of rangelands throughout the World occur in arid and semiarid climatic zones, hence water is commonly a limiting environmental factor during at least portions of most growing seasons. In many rangeland areas, extended periods of water deficits varying in amplitude from moderate to severe are occasionally tempered by transitory intervals of either drought or all-out flooding. The so-called "happy medium" or "average year" is probably more an anomaly than a reality in rangeland regions; anticipating broad shifts and wide variations in climatic factors is almost always more realistic than expectations of a fleeting dream that rarely comes true. Under such conditions, it seems, good range management is not an easy thing to do. Juggling the vagaries of an uncompromising climate with grazing and other pressures of land management can greatly muddle our understanding of the effects of individual factors on plant water relations. The interactive effects of water deficits with grazing, nutrient deficiencies, or other factors attest to the complexity that multiple impacts have on plant responses. Thus, the fact that range plants must often cope with periods of water deficits is not surprising; it seems to go with the territory. What is astonishing, however, is that so many plant species are so successful under the broad spectrum of harsh conditions found on typical rangelands.

The overwhelming image of arid and semiarid climatic regions is that of persistent drought and searing heat. However, the almost whimsical nature of climate also occasionally goes to the other extreme. For example, much of the normally arid to semiarid western United States experienced successive years of above normal precipitation from about 1977–1986 (Wallace and Nelson 1990). Almost simultaneously vast expanses of native shrub die-off were observed in the Great Basin and surrounding areas, primarily in 1983–1985 (Nelson et al. 1989, 1990). Although the search for a causal relationship is still progressing, numerous studies have proposed plausible hypotheses to explain this massive mortality. It is obvious that even arid and semiarid plants have upper limits of tolerance to available water just as they, and all plants, have lower limits of tolerance to water availability.

An awareness of the physiological tolerances and adaptations of range plants to water deficits, grazing, and other environmental variables is having a gradual and evolving positive impact on range management. Although the significance of basic knowledge about range plants is not always readily apparent at the land management level, better decisions about grazing, revegetation, and other management concerns are being made because of it. The fact that an unfortunate few plant species have attracted so much more basic research attention than others only highlights the need for more accelerated research on others. Species such as crested wheatgrass, smooth brome, Lehmann lovegrass, and others have been studied intensively in terms of grazing tolerance, water relations, morphology, seed germination, and other factors. The observation frequently made in the literature that "introduced" plant species appear to respond more favorably to grazing or revegetation treatments than many "native" species is probably partially true. It should be no surprise that some species, many of which happen to be introduced, display more favorable characteristics than others. Those species that evolved under conditions of aridity and grazing pressure can be expected to respond more favorably to herbivory in water limiting environments than those that did not. Many species were identified decades ago as having superior traits, and since then have undergone generations of selection, breeding, and physiological study to the extent that many of them are no longer the same genetic entities they were when first identified. As a result, many of these species are more widely used and preferred in range management because of various factors such as commercial availability, familiarity, and adaptation. Perhaps more native species with promising drought tolerance and other characteristics deserve the same degree of attention. Maybe then management policies and decisions would be based more on species adaptability than on species origin. The research currently underway by range plant breeders is some of the most exciting and promising science being conducted in range management, and it should be encouraged and expanded.

Great strides are also being made in plant and soil water relations research with the development of new techniques and sophisticated instruments. Automated data collection systems and electronic sensors are being used in both the laboratory and at remote field settings to collect orders of magnitude more data than ever. Within definitive limits, it is possible to measure the quantity and energy status of water in multiple numbers of plants, at different soil depths, and various locations simultaneously. Despite this, however, the central problem of identifying why some plants survive severe water deficits while others are irreparably injured, or identifying the limits of tolerance to water deficits in species and ecotypes, and how these interact with other processes or factors at different stages in the life cycle of plants remain illusive. We can measure with great precision how dry or wet a plant and soil become under different situations, but how this translates into better management is not always so precisely known. It would seem that our ability to achieve precision in instrumentation and techniques is approaching some degree of maturity; unfortunately, wise interpretation of our data is often still an infant.

Linking basic knowledge gained through research to successful application on the ground hinges on the essential realization that plants are far more enigmatic organisms than can possibly be described in a single chapter of a book. The adaptations to water deficits and other characteristics of plants discussed here have been dissected and isolated into artificially discrete units for the sake of convenience, but it is important to realize that this is probably a gross over-simplification. In reality, plants are composed of a highly complicated network of interacting adaptations and physiological mechanisms that all operate interactively. It is probably rare that single features or adaptations are ever individually and separately influential in a significant way during growth and development. Plants that are capable of osmotic adjustment, for example, frequently also have responsive stomata to dehydration, and many also have various modifications in leaf or stem structure, root form, or other characteristics that *collectively* and *in combination* enhance their chances of surviving dehydration. There are no known instances where osmotic adjustment (or any other adaptive feature) is the sole determinant of survival. Many plant species are totally incapable of osmotic adjustment, and yet are well suited for life in extremely arid environments; also, many plant species have shallow roots, relatively unresponsive stomata, the C_3 photosynthesis pathway, or a thin glabrous epidermis, and yet are well adapted and very successful in arid climates. The individual adaptations of an organism also probably function quite differently in different situations such that the relative contribution made by any one adaptation varies with the nature and extent of the environmental stress.

Adaptations were developed throughout the entire evolutionary history of an organism and encompass a broad array of heritable attributes in combination that function to increase the probability of survival and reproduction. They certainly do not develop in response to environmental pressures, nor do plants initiate "strategies" designed to overcome environmental stresses. Further, it is also probably inaccurate (and maybe teleological) to visualize the plant environment as being divided into periods of "stress" alternating with periods of non-stress, and that adaptations "kick-in" only when stressful periods occur. Presumably, it is more accurate to envision that some degree of stress, including water deficits or water-logging, nutrient deficiencies or toxicities, high or low temperatures, and other factors, universally and persistently affect the growth and development of virtually every plant to some degree. In fact, all of a plants' adaptations are operational and functioning every minute of its life, or the plant dies.

It is more than a little humbling to realize that we probably know far less about why some plants can tolerate water deficits while others can not than we like to believe. Although it may not be surprising that a research scientist would call for additional research on this and related fields, the facts are clear. Without this knowledge, the World's rangelands will not be managed as efficiently as they should be. Certainly the costs are high, but the potential gain is even higher; the track-record of benefits of past research to range management speak for themselves. Therefore, a little pessimism, although healthy, should not deter the promises of new knowledge. Outstanding advances have been made in understanding how plants respond to environmental pressures and the interactive effects of range management practices on them. The fact that the water relations of many important range species have never been studied, or that we still do not understand everything about water deficit tolerances, or that we frequently are unsure of how to apply the basic knowledge we do have, although unfortunate, is not reason for despair. The realization that species vary widely in their responses to water deficits and other environmental pressures, and that particular morphological or physiological adaptations are often instrumental in defining how species react to, and survive, stress periods can be viewed as a major advance toward better land use.

As an almost last-line of defense against the whims of climate and the ravages of human use and demand, the land manager must rely on regulatory practices and physical improvements as a buffer between the harsh physical and biological realities of the range and those of economic gain. The plants, after all, make the range. Although the study of plant water relations may appear to be a much more narrowly focused endeavor than that of range management, its implications are pervasive throughout. Perhaps without realizing it, the manager manipulates practices that are designed to improve the condition and sustainability of the range, and in so doing is really reaching down and touching the very plant cells and tissues that make it all possible in the first place.

Acknowledgements

Deepest appreciation is extended to the Society for Range Management, and especially Drs. Don Bedunah (University of Montana) and Ron Sosebee (Texas Tech University), for offering the invitation to participate in the preparation of this book. I also want to express sincere gratitude to the following people who reviewed the manuscript, either in part or in total, and for their many excellent suggestions and criticisms: Drs. Dale L. Bartos (USDA Forest Service), Martyn M. Caldwell (Utah State University), Jeanne C. Chambers (USDA Forest Service), Douglas A. Johnson (USDA Agricultural Research Service), Walter F. Mueggler

(USDA Forest Service, retired), Derrick M. Oosterhuis (University of Arkansas), and M. J. Trlica (Colorado State University). I accept full responsibility for any errors or mistakes in content, scientific interpretation, or judgement used in preparing this chapter. Thanks also to Bryan D. Williams, Angie Willis, and Garth Goodrich (USDA Forest Service) for assistance in preparing some of the figures and for help with numerous details throughout preparation of the manuscript.

Literature Cited

Abrams, M. D. 1988. Genetic variation in leaf morphology and plant and tissue water relations during drought in *Cercis canadensis* L. Forest Sci. 34:200-207.

Aguirre, L. and D. A. Johnson. 1991a. Influence of temperature and cheatgrass competition on seedling development of two bunchgrasses. J. Range Manage. 44:347-354.

Aguirre, L. and D. A. Johnson. 1991b. Root morphological development in relation to shoot growth in seedlings of four range grasses. J. Range Manage. 44:341-346.

Alessi, R. S. and L. Prunty. 1986. Soil-water determination using fiber optics. Soil Sci. Soc. Am. J. 50:860-863.

Allaway, W. G. and F. L. Milthorpe. 1976. Structure and functioning of stomata. *In*: Kozlowski, T. T. (ed.). Water Deficits and Plant Growth. Vol. IV. Soil Water Measurement, Plant Responses, and Breeding for Drought Resistance. Academic Press. New York. p. 57-102.

Allen, E. B. and M. F. Allen. 1986. Water relations of xeric grasses in the field: interactions of mycorrhizas and competition. New Phytol. 104:559-571.

Alpert, P. 1990. Water sharing among ramets in a desert population of *Distichlis spicata* (Poaceae). Amer. J. Bot. 77:1648-1651.

Amen, R. D. 1966. The extent and role of seed dormancy in alpine plants. Quart. Rev. Biol. 41:271-281.

Amen, R. D. 1968. A model of seed dormancy. Bot. Gaz. 34:1-31.

Anderson, D. W. and D. C. Coleman. 1985. The dynamics of organic matter in grassland soils. J. Soil Water Conserv. 40:211-216.

Andrews, P., W. J. Collins, and W. R. Stern. 1977. The effect of withholding water during flowering on seed production in *Trifolium subterraneum* L. Aust. J. Agric. Res. 28:301-307.

Archer, S. and D. A. Pyke. 1991. Plant-animal interactions affecting plant establishment and persistence on revegetated rangeland. J. Range Manage. 44:558-565.

Asay, K. H. 1986. Breeding strategies in crested wheatgrass. p. 53-57. *In*: K. L. Johnson (ed.). Crested Wheatgrass: its Values, Problems, and Myths. Sympos. Proc. Utah Sate Univ.

Asay, K. H., D. R. Dewey, F. B. Gomm, W. H. Horton, and K. B. Jensen. 1986. Genetic progress through hybridization of induced and natural tetraploids in crested wheatgrass. J. Range Manage. 39:261-263.

Asay, K. H. and D. A. Johnson. 1980. Screening for improved stand establishment in Russian wild ryegrass. Can. J. Plant Sci. 60:1171-1177.

Asay, K. H. and D. A. Johnson. 1983a. Breeding for drought resistance in range grasses. Iowa State Journal of Res. 57(4):441-455.

Asay, K. H. and D. A. Johnson. 1983b. Screening for improved stand establishment in Russian wild ryegrass. Can. J. Plant Sci. 60:1171-1177.

Asay, K. H. and D. A. Johnson. 1990. Genetic variances for forage yield in crested wheatgrass at six levels of irrigation. Crop Sci. 30(1):79-82.

Auge, R. M. and A. J. W. Stodola. 1989. Analysis of water potential isotherms in two ornamental shade tree species entering winter dormancy. J. Amer. Soc. HortSci. 114:666-673.

Auge, R. M., A. J. W. Stodola, and B. D. Pennell. 1990. Osmotic and turgor adjustment in *Rosa* foliage drought stressed under varying irradiance. J. Amer. Soc. Hort. Sci. 115:661-667.

Barker, R. E., A. B. Frank, and J. D. Berdahl 1989. Cultivar and clonal differences for water-use efficiency and yield in four forage grasses. Crop Sci. 29:58-61.

Barrs, H. D. 1968. Determination of water deficits in plant tissues. p. 235-368. *In*: Kozlowski, T. T. (ed.). Water Deficits and Plant Growth. I. Development, Control, and Measurement. Academic Press. New York.

Bartos, D. L. and D. A. Jameson. 1974. A dynamic root model. Amer. Midland Nat. 91:499-504.

Bartos, D L. and P. L. Sims. 1974. Root dynamics of a shortgrass ecosystem. J. Range Manage. 27:33-36.

Baskin, J. M. and C. C. Baskin. 1985. The annual dormancy cycle in buried weed seeds: a continuum. BioScience 35(8):492-498.

Bassiri, M., A. M. Wilson, and B. Grami. 1988. Dehydration effects on seedling development of four range species. J. Range Manage. 41:383-386.

Begg, J. E. 1980. Morphological adaptations of leaves to water stress. *In*: Turner, N. C. and P. J. Kramer (eds.). Adaptation of Plants to Water and High Temperature Stress. John Wiley and Sons. New York. p. 33-42.

Bennert, H. W. and H. A. Mooney. 1979. The water relations of some desert plants in Death Valley, California. Flora 168:405-427.

Bennert, H. W. and B. Schmidt. 1984. On the osmoregulation in *Atriplex hymenelytra* (Torr.) Wats. (Chenopodeacea). Oecologia 62:80-84.

Bennett, J. M., J. W. Jones, B. Zur, and L. C. Hammond. 1986. Interactive effects of nitrogen and water stresses on water relations of field-grown corn leaves. Agron. J. 78:273-280.

Bewley, J. D. 1979. Physiological aspects of desiccation tolerance. Ann. Rev. Plant Physiol. 30:195-238.

Billings, W. D. 1974. Arctic and alpine vegetation: plant adaptations to cold summer climates. *In*: Ives, J. D. and R. G. Barry (eds.). Arctic and Alpine Environments. Methuen and Co., LTD. London. p. 401-443.

Bittman, S. and G. M. Simpson. 1987. Soil water deficit effect on yield, leaf area, and net assimilation rate of three forage grasses: crested wheatgrass, smooth brome, and alta wildrye. Agron. J. 79:768-774.

Bittman, S. and G. M. Simpson. 1989a. Drought effect on leaf conductance and leaf rolling in forage grasses. Crop Sci. 29:338-344.

Bittman, S. and G. M. Simpson. 1989b. Drought effects on water relations of three cultivated grasses. Crop Sci. 29:992-999.

Bjorkman, O., M. R. Badger, and P. A. Armond. 1980. Response and adaptation of photosynthesis to high temperatures. *In*: Turner, N. C. and P. J. Kramer (eds.). Adaptation of Plants to Water and High Temperature Stress. John Wiley and Sons. New York. p. 233-249.

Blaisdell, J. P. and J. F. Pechnaec. 1949. Effects of herbage removal at various dates on vigor of bluebunch wheatgrass and arrowleaf balsamroot. Ecology 30:298-305.

Blaker, T. W., R. I. Greyson, and D. B. Walden. 1989. Variation among inbred lines of maize for leaf surface wax composition. Crop Sci. 29:28-32.

Bloom, A. J. 1989. Principles of instrumentation for physiological ecology. p. 1–56. *In*: Pearcy, R. W., J. Ehleringer, H. A. Mooney, and P. W. Rundel (eds.). Plant Physiological Ecology: Field Methods and Instrumentation. Chapman and Hall. New York.

Blue, M. P. and R. J. Jensen. 1988. Positional and seasonal variation in oak (*Quercus*; Fagaceae) leaf morphology. Amer. J. Bot. 75:939-947.

Bouaziz, A. and L. Bruckler. 1989a. Modeling wheat seedling growth and emergence. I. Seedling growth affected by soil water potential. Soil Sci. Soc. Amer. J. 53:1832-1838.

Bouaziz, A. and L. Bruckler. 1989b. Modeling wheat seedling growth and emergence. II. Comparison with field experiments. Soil Sci. Soc. Amer. J. 53:1838-1846.

Bowman, W. D. and S. W. Roberts. 1985. Seasonal and diurnal water relations adjustments in three evergreen chaparral shrubs. Ecology 66:738-742.

Boyer, J. S. 1969. Measurement of the water status of plants. Ann. Rev. Plant Physiol. 20:351-364.

Boyer, J. S. 1985. Water transport. Ann. Rev. Plant Physiol. 36:473-516.

Box, T. W. 1978. The arid lands revisited: 100 years after John Wesley Powell. 57th Ann. Faculty Honor Lecture, Utah State Univ., Logan, UT. 30 p.

Bradbury, K. J. and S. F. Yang. 1981. Physiological responses of plants to waterlogging. HortSci. 16(1):25-30.

Bradford, K. J. and T. C. Hsiao. 1982. Physiological responses to moderate water stress. *In*: Lange, O. L., P. S. Nobel, C. B. Osmond, and H. Ziegler (eds.). Encyclopedia of Plant Physiology, New Series, Vol. 12B. p. 263-324. Springer-Verlag, Berlin.

Bradford, K. J. and S. F. Yang. 1981. Physiological responses of plants to waterlogging. HortSci. 16:25-30.

Branson, F. A., G. F. Gifford, K. G. Renard, and R. F. Hadley. 1981. Rangeland Hydrology. Second Edition. Range Science Series. No. 1. Soc. Range Manage. Kendall/Hunt Publ. Co., Dubuque, Iowa. 340 p.

Briggs, L. J. and H. L. Shantz. 1912. The wilting coefficient for different plants and its indirect determination. USDA Bureau of Plant Industry. Bulletin No. 230. Wash. D.C. 83 p.

Bristow, C. E., G. S. Campbell, L. H. Wullstein, and R. Neilson. 1985. Water uptake and storage by rhizosheaths of *Oryzopsis hymenoides*: a numerical simulation. Physiol. Plant. 65:228-232.

Brown, R. W. 1970. Measurement of water potential with thermocouple psychrometers: construction and applications. USDA Forest Service. Intermountain Forest and Range Exp. Sta. Res. Paper INT-80. 27 p.

Brown, R. W. 1974. Some wilting phenomena in grasses. Plant Physiol. (Supplement). Amer. Soc. Plant Physiol. Ann. Meeting. p. 56.

Brown, R. W. 1977. Water relations of range plants. *In*: R. E. Sosebee (ed.). Rangeland Plant Physiology. Soc. Range Manage., Range Science Series 4. pg. 97-140.

Brown, R. W. and D. L. Bartos. 1982. A calibration model for screen-caged Peltier thermocouple psychrometers. USDA Forest Service, Intermountain Research Sta. Res. Paper INT-293. 155 p.

Brown, R. W. and M. C. Amacher. 1990. Some effects of acid mine drainage on plant water relations. p. 34. *In*: Agronomy Abstracts. Amer. Soc. Agron. Ann. Meeting, Oct. 21–26. San Antonio, TX.

Brown, R. W. and J. C. Chambers. 1989. Reclamation of severely disturbed alpine ecosystems: new perspectives. p. 59-68. *In*: Reclamation, a Global Perspective. Can. Land Reclam. Assoc., Amer. Soc. Surface Mining and Reclam. Calgary.

Brown, R. W. and J. C. Chambers. 1990. Reclamation practices in high-mountain ecosystems. p. 329-334. *In*: Schmidt, W. C. and C. J. McDonald (eds.). Symposium on Whitebark Pine Ecosystems: Ecology and Management of a High-Mountain Resource. USDA Forest Service, Intermountain Research Sta. Gen. Tech. Report INT-270.

Brown, R. W. and R. S. Johnston. 1979. Revegetation of disturbed alpine rangelands. p. 76-94. *In*: Johnson, D. A. (ed.). Special Management Needs of Alpine Ecosystems. Soc. Range Manage. Denver, CO.

Brown, R. W., R. S. Johnston, and K. Van Cleve. 1978. Rehabilitation problems in alpine and arctic regions. p. 23-44. *In*: Reclamation of Drastically Disturbed Lands. Amer. Soc. Agron. Madison, WI.

Brown, R. W. and D. M. Oosterhuis. 1992. Measuring plant and soil water potentials with thermocouple psychrometers: some concerns. Agron. J. 84:78-86.

Brown, R. W. and B. P. Van Haveren (eds.). 1972. Psychrometry in water relations research. 342 p. Utah Agric. Exp. Stat., Utah State Univ., Logan, Utah.

Brown, R. W., R. C. Sidle, M. C. Amacher, and J. Kotuby-Amacher. 1991. Reclaiming aridland mine spoils in the Great Basin: Research results on the Borealis Mine 1988-1991. p. 34-36. *In*: Zamudio, D. C. (ed.). Arid Mined Land Revegetation Symposium. Soil and Water Conserv. Soc., NV Chapter. Mar. 12-14, 1991, Tonopah, NV.

Bruce, R. R. and R. J. Luxmoore. 1986. Water retention: field methods. p. 663-686. *In*: Methods of Soil Analysis, Part 1: Physical and Mineralogical Methods. Agron. Monog. no. 9 (2nd ed.). Amer. Soc. Agron. and Soil Sci. Soc. Amer., Madison, WI.

Brulfert, J., M. Kluge, S. Guclu, and O. Qeiroz. 1988. Combined effects of drought, daylength and photoperiod on rapid shifts in the photosynthetic pathways of *Sedum spectabile*, a CAM species. Plant Physiol. Biochem. 26:7-16.

Burgess, T. L. 1985. *Agave* adaptation to aridity. Desert Plants 7:39-50.

Busso, C. A., R. J. Mueller, and J. H. Richards. 1989. Effects of drought and defoliation on bud viability in two caespitose grasses. Ann. Bot. 63:477-485.

Busso, C. A., J. H. Richards, and N. J. Chatterton. 1990. Nonstructural carbohydrates and spring regrowth of two cool-season grasses: interaction of drought and clipping. J. Range Manage. 43:336-343.

Caldwell, M. M. 1976. Root extension and water absorption. *In*: Lange, O. L., Kappen, L., and Schulze, E.-D. (eds.). Water and Plant Life. Springer-Verlag, Berlin. p. 63-85.

Caldwell, M. M. 1984. Sunlight and plants: some pursuits in physiological ecology. 69th Faculty Honor Lecture. Utah State Univ., Logan, Utah. 25 p.

Caldwell, M. M. 1987. Competition between root systems in natural communities. p. 167-185. *In*: P. J. Gregory, J. V. Lake, and D. A. Rose (eds.). Root Development and Function. Cambridge University Press. Cambridge.

Caldwell, M. M., T. J. Dean, R. S. Nowak, R. S. Dzurec, and J. H. Richards. 1983. Bunchgrass architecture, light interception, and water-use efficiency: assessment by fiber optic point quadrats and gas exchange. Oecologia 59:178-184.

Caldwell, M. M., D. M. Eissenstat, J. H. Richards, and M. F. Allen. 1985. Competition for phosphorus: differential uptake from dual-isotope-labeled soil interspaces between shrub and grass. Science 229:384-386.

Caldwell, M. M. and J. H. Richards. 1986. Competing root systems: morphology and models of absorption. *In*: Givnish, T. J. (ed.). On the Economy of Plant Form and Function. Cambridge Univ. Press, Cambridge. p. 251-273.

Caldwell, M. M. and J. H. Richards. 1989. Hydraulic lift: water efflux from upper roots improves effectiveness of water uptake by deep roots. Oecologia 79:1-5.

Caldwell, M. M., J. H. Richards, D. A. Johnson, R. S. Nowak, and R. S. Dzurec. 1981. Coping with herbivory: photosynthetic capacity and resource allocation in two semiarid *Agropyron* bunchgrasses. Oecologia 50:14-24.

Caldwell, M. M., R. S. White, R. T. Moore, and L. B. Camp. 1977. Carbon balance, productivity, and water use of cold-winter desert shrub communities dominated by C_3 and C_4 species. Oecologia 29:275-300.

California Fertilizer Association. 1985. Western Fertilizer Handbook. The Interstate Printers and Publ., Inc. Danville, IL. 288 p.

Campbell, G. S. 1988. Soil water potential measurement: an overview. Irrig. Sci. 9:265-273.

Campbell, G. S. and G. W. Gee. 1986. Water potential: miscellaneous methods. p. 619-633. *In*: Methods of Soil Analysis, Part 1: Physical and Mineralogical Methods. Agron. Monog. no. 9 (2nd ed.). Amer. Soc. Agron. and Soil Sci. Soc. Amer. Madison, WI.

Campbell, G. S. and G. A. Harris. 1977. Water relations and water use patterns for *Artemisia tridentata* Nutt. in wet and dry years. Ecology 58:652-659.

Campbell, G. S. and G. A. Harris. 1981. Modeling soil-water-plant-atmosphere systems of deserts. *In*: Evans, D. D. and Thames, J. L. (eds.). Water in Desert Ecosystems. USIBP Synthesis Series II. Dowden, Hutchinson, and Ross, Inc. Strousburg. p. 75-91.

Cassell, D. K. and A. Klute. 1986. Water potential: tensiometry. p. 563-596. *In*: Methods of Soil Analysis, Part 1. Physical and Mineralogical Methods. Agron. Monog. no. 9 (2nd. ed.). Amer. Soc. Agron. and Soil Sci. Soc. Amer. Madison, WI.

Catsky, J. 1962. Water saturation deficit in the wilting plant. The preference of young leaves and the translocation of water from old into young leaves. Biol. Plant. 4:306-314.

Ceulemans, R., I. Impens, and R. Impler. 1988. Stomatal conductance and stomatal behavior in *Populus* clones and hybrids. Can. J. Bot. 66:1404-1414.

Chambers, J. C., J. A. MacMahon, and R. W. Brown. 1987. Response of an early seral dominant alpine grass and a late seral dominant alpine forb to N and P availability. Reclam. Reveg. Res. 6:219-234.

Chambers, J. C. 1989. Seed viability of alpine species: variability within and among years. J. Range Manage. 42:304-308.

Chambers, J. C., J. A. MacMahon, and R. W. Brown. 1990. Alpine seedling establishment: the influence of disturbance type. Ecology 71(4):1323-1341.

Chamrad, A. D. and T. W. Box. 1965. Drought-associated mortality of range grasses in south Texas. Ecology 46(6):780-785.

Chapin, F. S. III., J. D. McKenrick, and D. A. Johnson. 1986. Seasonal changes in carbon fractions in Alaskan tundra plants of differing growth form: implications for herbivory. J. Ecol. 74:707-731.

Clarke, J. M. 1986. Effect of leaf rolling on leaf water loss in *Triticum* spp. Can. J. Plant Sci. 66:885-891.

Clarkson, N. M. and J. S. Russell. 1976. Effect of water stress on the phasic development of annual *Medicago* species. Aust. J. Agric. Res. 27:227-234.

Clipson, N. J. W., A. D. Tomos, T. J. Flowers, and R. G. Wyn Jones. 1985. Salt tolerance in the halophyte *Suaeda maritima* L. Dum. Planta 165:392-396.

Conrad, M. 1983. Adaptability: the Significance of Variability from Molecule to Ecosystem. Plenum Press. New York. 383 p.

Cook, C. W. and L. A. Stoddart. 1963. The effect of intensity and season of use on the vigor of desert range plants. J. Range Manage. 16:315-317.

Cook, C. W., L. A. Stoddart, and F. E. Kinsinger. 1958. Responses of crested wheatgrass to various clipping treatments. Ecol. Monogr. 28:237-272.

Coughenour, M. B. 1991. Spatial components of plant-herbivore interactions in pastoral, ranching, and native ungulate ecosystems. J. Range Manage. 44:530-542.

Cowan, I. R. 1965. Transport of water in the soil-plant-atmosphere system. J. Appl. Ecol. 2:221-239.

Cox, J. R. and G. L. Jordan. 1983. Density and production of seeded range grasses in southern Arizona (1970-1982). J. Range Manage. 36:649-652.

Cox, R. M. and T. C. Hutchinson. 1980. Multiple metal tolerances in the grass *Deschampsia cespitosa* (L.) Beauv. from the Sudbury smelting area. New Phytol. 84:631-647.

Craddock, G. W. and C. L. Forsling. 1938. The influence of climate and grazing on spring-fall sheep range in southern Idaho. U.S.D.A. Technical Bulletin No. 600. 42 p.

Crafts, A.S. 1968. Water deficits and physiological processes. *In*: Kozlowski, T. T. (ed.).

Water Deficits and Plant Growth. Vol II. Plant Water Consumption and Response. Academic Press. New York. p. 85-133.

Crider, F. J. 1955. Root-growth stoppage resulting from defoliation of grass. USDA Tech. Bulletin 1102. 23 p.

Cutler, J. M. and D. W. Rains. 1978. Effects of water stress hardening on the internal water relations and osmotic constituents of cotton leaves. Physiol. Plant. 42:261-268.

Cutler, J. M., D. W. Rains, and R. S. Loomis. 1977. The importance of cell size in the water relations of plants. Physiol. Plant. 40:255-260.

Daddy, F., M. J. Trlica, and C. D. Bonham. 1988. Vegetation and soil water differences among big sagebrush communities with different grazing histories. Southwestern Nat. 33:413-424.

Dahl, B. E. and D. N. Hyder. 1977. Developmental morphology and management implications. p. 257-290. *In*: R. E. Sosebee (ed.). Rangeland Plant Physiology. Range Sci. Series No. 4. Soc. Range Manage. Denver, CO.

Davies, W. J., T. A. Mansfield, and A. M. Hetherington. 1990. Sensing of soil water status and the regulation of plant growth and development. Plant Cell Environ. 13:709-719.

Davies, W. J. and J. Zhang. 1991. Root signals and the regulation of growth and development of plants in drying soil. Ann. Rev. Plant Physiol. 42:55-76.

Day, A. D. and K. L. Ludeke. 1987. Effects of soil materials, mulching treatments, and soil moisture on the growth and yield of western wheatgrass for coal mine reclamation. Desert Plants 8:136-139.

Day, T. A. and J. K. Detling. 1990. Changes in grass leaf water relations following bison urine deposition. Amer. Midland Nat. 123:171-178.

Delwiche, M. J. and R. Cooke. 1977. An analytical model of the hydraulic aspects of stomatal dynamics. J. Theor. Biol. 69:113-141.

DePuit, E. J. and M. M. Caldwell. 1973. Seasonal pattern of net photosynthesis of *Artemisia tridentata*. Amer. J. Bot. 60:426-435.

DePuit, E. J. and J. G. Coenenberg. 1979. Response of revegetated Coal Strip mine spoils to variable fertilization rates, longevity of fertilization program and season of seeding. Research Report 150. Montana Agric. Exp. Stat., Montana State Univ. Bozeman. 81 p.

Detling, J. K. and L. G. Klikoff. 1973. Physiological response to moisture stress as a factor in halophyte distribution. Amer. Midland Nat. 90:307-318.

Dibble, M. S. and G. G. Spomer. 1987. Cell water potential components in the adaptation of *Pseudoroegneria spicata* (Pursh.) A. Love to Various habitat moisture conditions. Bot. Gaz. 148:73-78.

Doescher, P., R. Miller, and A. Winward. 1985. Effects of moisture and temperature on germination of Idaho fescue. J. Range Manage. 38(4):317-320.

Doescher, P. S., S. D. Tesch, and W. E. Drewien. 1989. Water relations and growth of conifer seedlings during three years of cattle grazing on a southwest Oregon plantation. Northwest Sci. 63:232-240.

Dowling, P. M., R. J. Clements, and J. R. McWilliam. 1971. Establishment and survival of pasture species from seeds sown on the soil surface. Aust. J. Agric. Res. 22:61-74.

Drake, B. G. and J. L. Gallagher. 1984. Osmotic potential and turgor maintenance in *Spartina alterniflora* Loisel. Oecologia 62:368-375.

Drake, B. G., K. Raschke, and F. B. Salisbury. 1970. Temperature and transpiration resistances of *Xanthium* leaves as affected by air temperatures, humidity, and wind speed. Plant Physiol. 46:324-330.

Drew, M. C. 1987. Function of root tissues in nutrient and water transport. p. 71-101. *In*: P. J. Gregory, J. V. Lake, and D. A. Rose (eds.). Root Development and Function. Cambridge University Press. Cambridge.

Duell, R. W. and G. R. Peacock. 1985. Rhizosheaths on mesophytic grasses. Crop Sci. 25:880-883.

Ehleringer, J. 1982. The influence of water stress and temperature on leaf pubescence development in *Encelia farinosa*. Amer. J. Bot. 69:670-675.

Ehleringer, J. R. and O. Bjorkman. 1976.Leaf pubescence: effects on absorptance and photosynthesis in a desert shrub. Science 192:376-377.

Ehleringer, J. R., A. E. Hall, and G. D. Farquhar (eds.). 1993. Stable Isotopes and Plant Carbon-Water Relations. Academic Press, Inc. San Diego. 555 p.

Ehleringer, J. R. and H. A. Mooney. 1978. Leaf hairs: effects on physiological activity and adaptive value to a desert shrub. Oecologia 37:183-200.

Ehlig, C. F. and W. R. Gardner. 1964. Relationship between transpiration and the internal water relations of plants. Agron. J. 56:127-130.

Eissenstat, D. M. and M. M. Caldwell. 1987. Characteristics of successful competitors: an evaluation of potential growth rate in two cold desert tussock grasses. Oecologia 71:167-173.

Eissenstat, D. M. and M. M. Caldwell. 1988. Competitive ability is linked to rates of water extraction: a field study of two aridland tussock grasses. Oecologia 75:1-7.

Ekanayake, I. J., P. L. Steponkus, and S. K. De Datta. 1990. Sensitivity of pollination to water deficits at anthesis in upland rice. Crop Sci. 30:310-315.

Ellison, L. 1960. Influence of grazing on plant succession of rangelands. Bot. Rev. 26:1-78.

Ellison, L., A. R. Croft, and R. W. Bailey. 1951. Indicators of condition and trend on high range-watersheds of the Intermountain Region. USDA Agric. Handbook No. 19. 66 p.

El-Sharkawi, H. M., K. A. Farghali, and S. A. Sayed. 1989. Interactive effects of water stress, temperature and nutrients in the seed germination of three desert plants. J. Arid Environ. 17:307-317.

Emmerich, W. E. and S. P. Hardegree. 1990. Polyethylene glycol solution contact effects on seed germination. Agron. J. 82:1103-1107.

Emmerich, W. E. and S. P. Hardegree. 1991. Seed germination in polyethylene glycol solution: effects of filter paper exclusion and water vapor loss. Crop Sci. 31:454-458.

Esau, K. 1967. Plant Anatomy. John Wiley and Sons. New York. 767 p.

Etherington, J. R. 1984. Relationship between morphological adaptation to grazing, carbon balance and waterlogging tolerance in clones of *Dactylis glomerata* L. New Phytol. 98:647-658.

Everett, R. L., R. O. Meewig, P. T. Tueller, and R. A. Evans. 1977. Water potential in sagebrush and shadscale communities. Northwest Sci. 51:271-281.

Fahn, A. 1969. Plant Anatomy. Pergamon Press, Oxford. 534 p.

Farquhar, G. D., J. R. Ehleringer, and K. T. Hubrick. 1989. Carbon isotope discrimination and photosynthesis. Ann. Rev. Plant Physiol. Mol. Biol. 40:503-537.

Farris, M. A. 1984. Leaf shape variation associated with drought stress in *Rumex acetosella* L. (Polygonaceae). Amer. Midland Nat. 111:358-363.

Feldman, L. J. 1988. The habits of roots. BioScience 38:612-618.

Ferguson, R. B. and N. C. Frischknecht. 1985. Reclamation on Utah's Emery and Alton coal fields: techniques and plant materials. USDA Forest Service. Intermountain Forest and Range Exp. Sta., Res. Paper INT-335. 78 p.

Fernandez, O. A. and M. M. Caldwell. 1975. Phenology and dynamics of root growth of three cool semi-desert shrubs under field conditions. J. Ecol. 63:703-714.

Fetcher. 1976. Patterns of leaf resistance to lodgepole pine transpiration in Wyoming. Ecology 57:339-345.

Fischer, R. A. and G. D. Kohn. 1966. The relationship of grain yield to vegetative growth and post-flowering leaf area in the wheat crop under conditions of limited soil moisture. Aust. J. Agric. Res. 17:281-295.

Fischer, R. A. and J. T. Wood. 1979. Drought resistance in spring wheat cultivars. III. Yield associations with morpho-physiological traits. Aust. J. Agric. Res. 30:1001-1020.

Fisher, F. M., J. C. Zak, G. L. Cunningham, and W. G. Whitford. 1988. Water and nitrogen effects on growth and allocation patterns of creosotebush in the northern Chihuahuan desert. J. Range Manage. 41:387-391.

Fitter, A. H. and T. R. Strickland. 1991. Architectural analysis of plant root systems 2. Influence of nutrient supply on architecture in contrasting plant species. New Phytol. 118:383-389.

Fitter, A. H., T. R. Strickland, M. L. Harvey, and G. W. Wilson. 1991. Architectural analysis of plant root systems 1. Architectural correlates of exploitation efficiency. New Phytol. 118:375-382.

Flowers, T. J. 1985. Physiology of halophytes. Plant Soil 89:41-56.

Foxx, T. S., G. D. Tierney, and J. M. Williams. 1984a. Rooting depths of plants on low-level waste disposal sites. Los Alamos National Lab. LA-10253-MS, Los Alamos, NM. 23 p.

Foxx, T. S., G. D. Tierney, and J. M. Williams. 1984b. Rooting depths of plants relative to biological and environmental factors. Los Alamos National Lab. LA-10254-MS, Los Alamos, NM. 26 p.

Frank, A. B. and A. Bauer. 1991. Rooting activity and water use during vegetative development of crested and western wheatgrass. Agron. J. 83:906-910.

Fraser, J. G. and H. G. Wolfe. 1982. Effects of mulches and mulch tackifiers on the establishment of native species in New Mexico. p. 151-156. *In*: Aldon, E. F. and W. R. Oaks (eds.). Reclamation of Mined Lands in the Southwest: A Symposium. Soil Conserv. Soc., NM Chapter. Oct. 20-22, 1982. Albuquerque, NM.

Frasier, G. W. 1989. Characterization of seed germination and seedling survival during initial wet-dry periods following planting. J. Range Manage. 42(4):299-303.

Fu, C. F. and C. S. Hew. 1982. Crassulacean acid metabolism in orchids under water stress. Bot. Gaz. 143:294-297.

Fullbright, T. E., A. M. Wilson, and E. F. Redente. 1984. Effects of temporary dehydration on growth of green needlegrass (*Stipa viridula* Trin.) seedlings. J. Range Manage. 37(5):462-464.

Gaff, D. F. 1971. Desiccation-tolerant flowering plants in southern Africa. Science 174:1033-1034.

Gaff, D. F. 1977. Desiccation tolerant vascular plants of southern Africa. Oecologia 31:95-109.

Gaff, D. F. 1980. Protoplasmic tolerance to extreme water stress. *In*: Turner, N. C. and P. J. Kramer (eds.). Adaptation of Plants to Water and High Temperature Stress. John Wiley and Sons. New York. p. 207-230.

Gaff, D. F. and P. V. Bole. 1986. Resurrection grasses in India. Oecologia 71:159-160.

Gaff, D. F., H. Ziegler, and U. Zimmermann. 1985. Electrofusion of protoplasts from desiccation tolerant grass species with desiccation sensitive grass protoplasts. J. Plant Physiol. 120:375-380.

Ganskopp, D. C. and T. E. Bedell. 1979. Response of bluebunch wheatgrass to drought and climatic fluctuations: a review. Agric. Exp. Stat., Oregon State Univ., Corvallis. Cir. Infor. 680. 15 p.

Gardner, W. H. 1986. Water content. p. 493-544. *In*: Methods of Soil Analysis, Part 1. Physical and Mineralogical Methods. Agron. Monog. No. 9 (2nd ed.). Amer. Soc. Agron-Soil Sci. Soc. Amer. Madison, WI.

Gardner, W. H. 1988. A historical perspective on measurement of soil and plant water status. Irrig. Sci. 9:255-264.

Gardner, W. R. 1972. Comments on the "field capacity" and "permanent wilting point" concepts. *In*: Brown, R. W. and Van Haveren, B. P. (eds.). Psychrometry in Water Relations Research. Utah State Univ. Agric. Exp. Station. p. 237.

Gardner, W. R. and R. H. Nieman. 1964. Lower limit of water availability to plants. Science 143:1460-1462.

Gates, C. T. 1968. Water deficits and growth of herbaceous plants. *In*: Kozlowski, T. T. (ed.). Water Deficits and Plant Growth. Vol.II Plant Water Consumption and Response. p.135-190. Academic Press, New York.

Gates, D. M. 1962. Energy Exchange in the Biosphere. Harper and Row, Publ. New York. 151 p.

Gates, D. M. 1965. Heat transfer in plants. Sci. Amer. 213:76-84.

Gates, D. M., R. Alderfer, and E. Taylor. 1968. Leaf temperatures of desert plants. Science 159:994-995.

Gindel, I. 1973. A New Ecophysiological Approach to Forest-Water Relationships in Arid Climates. Dr. W. Junk B. V., Publ. The Hague. 142 p.

Glantz, M. H. 1987. Drought in Africa. Sci. Amer. 256(6):34-40.

Gold, W. G. and M. M. Caldwell. 1989. The effects of the spatial pattern of defoliation on regrowth of a tussock grass. II. Canopy gas exchange. Oecologia 81:437-442.

Gold, W. G. and M. M. Caldwell. 1990. The effects of the spatial pattern of defoliation on regrowth of a tussock grass. III. Photosynthesis, canopy structure and light interception. Oecologia 82:12-17.

Gutierrez, J. R. and W. G. Whitford. 1987. Chihuahuan desert annuals: importance of water and nitrogen. Ecology 68:2032-2045.

Hadas, A. 1970. Factors affecting seed germination under soil moisture stress. Israel J. Agric. Res. 20(1):3-14.

Hadley, J. L. and W. K. Smith. 1990. Influence of leaf surface wax and leaf area to water content ratio on cuticular transpiration in western conifers, U.S.A. Can. J. For. Res. 20:1306-1311.

Haggas, L., R. W. Brown, and R. S. Johnston. 1987. Light requirement for seed germination of Payson sedge. J. Range Manage. 40:180-184.

Hake, K., G. Guinn, and D. M. Oosterhuis. Environmental causes of shed. Cotton Physiology Today. December. 4 p.

Halim, R. A., D. R. Buxton, M. J. Hattendorf, and R. E. Carlson. 1989. Water-deficit effects on alfalfa at various growth stages. Agron. J. 81:765-770.

Hall, A. E. 1981. Adaptation of annual plants to drought in relation to improvements in cultivars. HortSci. 16:37-38.

Hall, A. E., R. G. Mutters, K. T. Hubrick, and G. D. Farquhar. 1990. Genotypic differences in carbon isotope discrimination by cowpea under wet and dry field conditions. Crop Sci. 30:300-305.

Hammel, H. T. 1976. Colligative properties of a solution. Science 192:748-756.

Hanscom, Z. and I. P. Ting. 1978. Responses of succulents to plant water stress. Plant Physiol. 61:327-330.

Hanson, A. D. and W. D. Hitz. 1982. Metabolic response of mesophytes to plant water deficits. Ann. Rev. Plant Physiol. 33:163-203.

Hardegree, S. P. and W. E. Emmerich. 1990. Partitioning water potential and specific salt effects on seed germination of four grasses. Ann. Bot. 66:587-595.

Harper, J. L. 1977. Population Biology of Plants. Academic Press. London. 892 p.

Harris, G. A. and G. S. Campbell. 1981. Morphological and physiological characteristics of desert plants. *In*: Evans, D. D. and J. L. Thames. (eds.). Water in Desert Ecosystems. USIBP Synthesis Series 11. Dowden, Hutchinson, and Ross, Inc. Stroudsburg. p. 59-74.

Hassanyar, A. S. and A. M. Wilson. 1978. Drought tolerance of seminal lateral root apices in crested wheatgrass and Russian wildrye. J. Range Manage. 31:254-258.

Heathcote, R. L. 1973. Drought perception. p. 17-40. *In*: J. V. Lovett (ed.). The Environmental, Economic and Social Significance of Drought. Angus and Robertson, Publishers. Sydney.

Hellebust, J. A. 1976. Osmoregulation. Ann. Rev. Plant Physiol. 27:485-505.

Hellmuth, E. O. 1970. Measurement of leaf water deficit with particular reference to the whole plant method. J. Ecol. 58:409-417.

Hendrickson, A. H. and F. J. Veihmeyer. 1945. Permanent wilting percentages of soil obtained from field and laboratory trials. Plant Physiol. 20:517-539.

Henson, I. E. 1982. Osmotic adjustment to water stress in pearl millet (*Pennisetum americanum* (L.) Leeke) in a controlled environment. J. Exp. Bot. 33:78-87.

Henson, I. E., C. R. Jensen, and N. C. Turner. 1989. Leaf gas exchange and water relations of lupins and wheat. III. Abscisic acid and drought-induced stomatal closure. Aust. J. Plant Physiol. 16:429-442.

Herbel, C. H. and R. E. Sosebee. 1969. Moisture and temperature effects on emergence and initial growth of two range grasses. Agron. J. 61:628-631.

Hillel, D. 1971. Soil and Water Physical Principles and Processes. Academic Press, New York. 288 p.

Hinckley, T. M., F. Duhme, A. R. Hinckley, and H. Richter. 1980. Water relations of drought hardy shrubs: osmotic potential and stomatal reactivity. Plant Cell Environ. 3:131-140.

Holechek, J. 1982. Root biomass on native range and mine spoils in southeastern Montana. J. Range Manage. 35:185-187.

Hsiao, T. C. 1973. Plant responses to water stress. Ann. Rev. Plant Physiol. 24:519-570.

Hsiao, T. C., E. Acevedo, E. Fereres, and D. W. Henderson. 1976a. Water stress, growth, and osmotic adjustment. Philos. Trans. R. Soc. London. Series B. 273:479-500.

Hsiao, T. C., E. Fereres, E. Acevedo, and D. W. Henderson. 1976b. Water stress and dynamics of growth and yield of crop plants. *In*: Lange, O. L., L. Kappen, and E. D. Schulze (eds.). Water and Plant Life. p. 281-305. Springer-Verlag. New York.

Hull, A. C. 1974. Species for seeding arid rangeland in southern Idaho. J. Range Manage. 27:216-218.

Hull, A. C. and R. C. Holmgren. 1964. Seeding southern Idaho rangelands. USDA Forest Service. Intermountain Forest and Range Exp. Sta. Res. Paper INT-10. 31 p.

Humphrey, R. R. 1962. Range Ecology. The Ronald Press Co. New York. 234 p.

Hurtt, L. C. 1951. Managing northern Great Plains cattle ranges to minimize effects of drought. USDA Circular No. 865. 24 p.

Hyder, D. N. 1974. Morphogenesis and management of perennial grasses in the United States. pg. 89-98. *In*: K. W. Kreitlow and R. H. Hart (eds.). Plant Morphogenesis as the Basis for Scientific Management of Range Resources. USDA Misc. Publ. No. 1271.

Hyder, D. N., A. C. Everson, and R. E. Bement. 1971. Seedling morphology and seeding failures with blue grama. J. Range Manage. 24:287-292.

Idso, S. B. and D. G. Baker. 1967. Relative importance of reradiation, convection, and transpiration in heat transfer from plants. Plant Physiol. 42:631-640.

Jacoby, P. W., R. J. Ansley, C. H. Meadors, and A. H. Huffman. 1990. Epicuticular wax in honey mesquite: seasonal accumulation and intraspecific variation. J. Range Manage. 43:347-350.

Jameson, D. A. 1963. Responses of individual plants to harvesting. Bot. Rev. 29:532-594.

Jefferies, R. L. 1981. Osmotic adjustment and the response of halophytic plants to salinity. BioScience 31:42-46.

Johnson, A. I. 1962. Methods of measuring soil moisture in the field. U.S. Geological Survey, Water Supply Paper 1619-U. 25 p.

Johnson, D. A. 1978. Environmental effects on turgor pressure response in range grasses. Crop Sci. 18:945-948.

Johnson, D. A. 1980. Improvement of perennial herbaceous plants for drought-stressed western rangelands. p. 419-433. *In*: N. C. Turner and P. J. Kramer (eds.). Adaptations of Plants to Water and Temperature Stress. John Wiley, New York.

Johnson, D. A. 1986. Seed and seedling relations of crested wheatgrass: a review. p.65-90. *In*: K. L. Johnson (ed.). Crested wheatgrass: its values, problems and myths; symposium proceedings. Utah State Univ., Logan.

Johnson, D. A. and L. Aguirre. 1991. Effect of water on morphological development in seedlings of three range grasses: root branching patterns. J. Range Manage. 44:355-360.

Johnson, D. A. and K. H. Asay. 1978. A technique for assessing seedling emergence under drought stress. Crop Sci. 18:520-522.

Johnson, D. A. and K. H. Asay. 1987. Selection procedures for improving seedling establishment in cool-season range grasses. p. 131-137. *In*: G. W. Frasier and R. A. Evans (eds.). Proceedings of Sympos. Seed and Seedbed Ecology of Rangeland Plants. USDA, Agric. Res. Service.

Johnson, D. A., K. H. Asay, and J. J. Read. 1993. Genotypic and environmental variation for carbon isotope discrimination in crested wheatgrass, a perennial forage grass. p. 269-280. *In*: Ehleringer, J. R., A. E. Hall, and G. D. Farquhar (eds.). Stable Isotopes and Plant Carbon-Water Relations. Academic Press, Inc. San Diego.

Johnson, D. A., M. D. Rumbaugh, and K. H. Asay 1981. Plant improvement for semiarid rangelands: possibilities for drought resistance and nitrogen fixation. Plant Soil 58:279-303.

Johnson, D. A., R. A. Richards, and N. C. Turner. 1983. Yield, water relations, gas exchange, and surface reflectances of near-isogenic wheat lines differing in glaucousness. Crop Sci. 23:318-325.

Johnson, W. M. 1953. Effect of grazing intensity upon vegetation and cattle gains on ponderosa pine-bunchgrass ranges of the Front Range of Colorado. USDA Circular No. 929. Wash., D.C. 36 p.

Jones, C. A. 1985. C_4 Grasses and Cereals: Growth, Development, and Stress Response. John Wiley and Sons, New York. 419 p.

Jones, M. M., N. C. Turner, and C. B. Osmond. 1981. Mechanisms of drought resistance. *In*: Paleg, L. G. and Aspinall, D. (eds.). The Physiology and Biochemistry of Drought Resistance in Plants. Academic Press, New York. p. 15-37.

Jones, T. A. 1990. A viewpoint on Indian ricegrass research: Its present status and future prospects. J. Range Mange. 43:416-420.

Kay, B. L. 1982. Uses and effectiveness of mulch products. p. 159-164. *In*: Aldon, E. F. and W. R. Oaks. (eds.). Reclamation of mined Lands in the Southwest: A Symposium. Soil Conserv. Soc., NM Chapter. Oct. 20-22, 1982. Albuquerque, NM.

Kaufmann, M. R. 1981. Water relations during drought. *In*: Paleg, L. G. and D. Aspinall (eds.). The Physiology and Biochemistry of Drought Resistance in Plants. Academic Press. Sydney. p. 55-70.

Kaufmann, M. R. 1982a. Evaluation of season, temperature, and water stress effects on stomata using a leaf conductance model. Plant Physiol. 69:1023-1026.

Kaufmann, M. R. 1982b. Leaf conductance as a function of photosynthetic photon flux density and absolute humidity difference from leaf to air. Plant Physiol. 69:1018-1022.

Keller, W. 1971. Limits on western range forage production-water or man. J. Range Manage. 24:243-247.

Kelrick, M. I., J. A. MacMahon, R. R. Parmentor, and D. V. Sisson. 1986. Native seed preferences of shrub-steppe rodents, birds and ants: the relationships of seed attributes and seed use. Oecologia 68:327-337.

Kemp, W. P., J. S. Berry, and J. M. Caprio. 1991. Use of ornamental lilac and honeysuckle phenopases as indicators of rangeland grasshopper development. J. Range Manage. 44:583-587.

Kirkham, M. B., H. He, T. P. Bolger, D. J. Lawlor, and E. T. Kanemasu. 1991. Leaf photosynthesis and water use of big bluestem under elevated carbon dioxide. Crop Sci. 31:1589-1594.

Klikoff, L. G. 1966. Competitive responses to moisture stress of a winter annual of the Sonoran desert. Amer. Midland Nat. 75:383-391.

Kluge, M. 1976. Crassulacean acid metabolism (CAM): CO_2 and water economy. *In*: Lange, O. L., Kappen, L., and Schulze, E.-D. (eds.). Water and Plant Life: Problems and Modern Approaches. Springer-Verlag, Berlin. p. 313-322.

Klute, A. 1986. Water retention: laboratory methods. p. 635-662. *In*. Methods of Soil Analysis, Part 1. Physical and Mineralogical Methods. Agron. Monog. no. 9 (2nd. ed.). Amer. Soc. Agron. and Soil Sci. Soc. Amer. Madison, WI.

Knapp, A. K. 1984. Water relations and growth of three grasses during wet and drought years in a tallgrass prairie. Oecologia 65:35-43.

Knapp, A. K. 1985. Effect of fire and drought on the ecophysiology of *Andropogon gerardii* and *Panicum virgatum* in a tallgrass prairie. Ecology 66:1309-1320.

Knipe, O. D. 1968. Effects of moisture stress on germination of alkali sacaton, galleta, and blue grama. J. Range Manage. 21(1):3-4.

Knipling, E. B. 1967. Effect of leaf aging on water deficit-water potential relationships of dogwood leaves growing in two environments. Physiol. Plant. 20:65-72.

Koide, R. T., R. H. Robichaux, S. R. Morse, and C. M. Smith. 1989. Plant water status, hydraulic resistance, and capacitance. p. 161-183. *In*. Pearcy, R. W., J. R. Ehleringer, H. H., Mooney, and P. W. Rundel (eds.). Plant Physiological Ecology: Field Methods and Instrumentation. Chapman and Hall, New York.

Koller, D. and A. Hadas. 1982. Water relations in the germination of seeds. p. 401-431. *In*: O. L. Lange, P. S. Nobel, C. B. Osmond, and H. Ziegler (eds.). Encyclopedia of Plant Physiol. New Series, Vol. 12B. Physiol. Plant Ecol. II. Water Relations and Carbon Assimilation. Springer-Verlag. Berlin.

Kozlowski, T. T. 1976. Water supply and leaf shedding. *In*: Kozlowski, T. T. (ed.). Water Deficits and Plant growth. Vol. IV. Soil Water Measurement, Plant Responses, and Breeding for Drought Resistance. Academic Press. New York. p. 191-231.

Kramer, P. J. 1980. Drought, stress, and the origin of adaptations. *In*: N. C. Turner and P. J. Kramer (eds.). Adaptation of Plants to Water and High Temperature Stress. John Wiley and Sons. New York. p. 7-20.

Kramer, P. J. 1983. Water Relations of Plants. Academic Press. New York. 489 p.

Kramer, P. J. 1988. Measurement of plant water status: historical perspectives and current concerns. Irrig. Sci. 9:275-287.

Kriedemann, P. E. and W. J. S. Downton. 1981. Photosynthesis. *In*: Paleg, L. G. and Aspinall, D. (eds.). The Physiology and Biochemistry of Drought Resistance in Plants. Academic Press, Sydney. p. 283-314.

Krizek, D. T. and C. D. Foy. 1988. Role of water stress in differential aluminum tolerance of two barley cultivars grown in an acid soil. J. Plant Nutr. 11:351-367.

Krizek, D. T., C. D. Foy, and W. P. Wergin. 1988. Role of water stress in differential aluminum tolerance of six sunflower cultivars grown in acid soil. J. Plant Nutr. 11:387-408.

Kummerow, J. 1980. Adaptation of roots in water-stressed native vegetation. *In*: Turner, N. C. and Kramer, P. J. (eds.). Adaptation of Plants to Water and High Temperature Stress. John Wiley and Sons, New York. p. 57-73.

Lange, O. L., L. Kappen, and E.-D. Schulze. 1976. Water and Plant Life. Springer-Verlag. Berlin. 536 p.

Lau, R. R. and D. R. Young. 1988. Influence of physiological integration on survivorship and water relations in a clonal herb. Ecology 69:215-219.

Lawrence, T., C. D. Ratzlaff, and P. G. Jefferson. 1991. Emergence of several Triticeae range grasses influenced by depth of seed placement. J. Range Manage. 44:186-187.

Laycock, W. A. 1982. Seeding and fertilizing to improve high-elevation rangelands. USDA Forest Service. Intermountain Forest and Range Exp. Sta., Gen. Tech. Report INT-120. 19 p.

Laycock, W. A. 1991. Stable states and thresholds of range condition on North American rangelands: A viewpoint. J. Range Manage. 44:427-433.

Levitt, J. 1972. Response of Plants to Environmental Stresses. Academic Press. New York. 697 p.

Levitt, J. 1976. Physiological basis of stomatal response. *In*: Lange, O. L., Kappen, L., and Schulze, E.-D. (eds.). Water and Plant Life: Problems and Modern Approaches. Springer-Verlag, Berlin. p. 160-168.

Lindstrom, M. J., R. I. Papendick, and F. E. Koehler. 1976. A model to predict winter wheat emergence as affected by soil temperature, water potential, and depth of planting. Agron. J. 68:137-141.

Lopushinsky, W. 1969. Stomatal closure in conifer seedlings in response to leaf moisture stress. Bot. Gaz. 130:258-263.

Love, L. D. and N. E. West. 1972. Plant moisture stress patterns in *Eurotia lanata* and *Atriplex confertifolia*. Northwest Sci. 46:44-51.

Ludlow, M. M. 1976. Ecophysiology of C_4 grasses. *In*: Lange, O. L., Kappen, L., and E.-D. Schulze (eds.). Water and Plant Life: Problems and Modern Approaches. Springer-Verlag, Berlin. p.364-386.

Ludlow, M. M. 1980. Adaptive significance of stomatal responses to water stress. *In*: Turner, N. C. and Kramer, P. J. (eds.). Adaptation of Plants to Water and High Temperature Stress. John Wiley and Sons, New York. p. 123-138.

Lull, H. W. and K. G. Reinhart. 1955. Soil-moisture measurement. USDA Forest Service. Southern Forest Exp. Sta. Occasional Paper 140. 56 p.

MacMahon, J. A. 1985. Deserts: The Audubon Society Nature Guides. Alfred A. Knopf. New York.. 638 p.

MacMahon, J. A. and D. J. Schimpf. 1981. Water as a factor in the biology of North American plants. p. 114-171. *In*: D. D. Evans and J. L. Thames (eds.). Water in Desert Ecosystems. USIBP Synthesis Series II. Dowden, Hutchinson, and Ross, Inc. Stroudsburg, Pa.

Mahalakshmi, V., F. R. Bidinger, and G. D. Rao. 1988. Timing and intensity of water deficits during flowering and grain-filling in pearl millet. Agron. J. 80:130-135.

Manning, M. E., S. R. Swenson, T. Svejcar, and J. Trent. 1989. Rooting characteristics of four intermountain meadow community types. J. Range Manage. 42:309-312.

Mansfield, T. A. 1976. Stomatal behaviour: chemical control of stomatal movements. Phil. Trans. R. Soc. Lond. B. 273:544-550.

Mansfield, T. A. and W. J. Davies. 1981. Stomata and stomatal mechanisms. p. 315-346. *In*: L. G. Paleg and D. Aspinall (eds.). The Physiology and Biochemistry of Drought Resistance in Plants. Academic Press. Sydney.

Mansfield, T. A. and W. J. Davies. 1985. Mechanisms for leaf control of gas exchange. BioScience 35:158-164.

Marc, J. and J. H. Palmer. 1976. Relationship between water potential and leaf and inflorescence initiation in *Helianthus annuus*. Physiol. Plant. 36:101-104.

Markhart, A. H. 1985. Comparative water relations of *Phaseolus vulgaris* L. and *Phaseolus actifolius* Gray. Plant Physiol. 77:113-117.

Mattson, W. J. and R. A. Haack. 1987. The role of drought in outbreaks of plant-eating insects. BioScience 37:110-118.

Mays, D. A. and G. W. Bengtson. 1978. Lime and fertilizer use in land reclamation in humid regions. *In*: Reclamation of Drastically Disturbed Lands. Amer. Soc. Agron., Madison, WI. p. 307-328.

Maxwell, J. O. and R. E. Redmann. 1978. Leaf water potential, component potentials and relative water content in a xeric grass, *Agropyron dasystachyum* (Hook.) Scribn. Oecologia 35:277-284.

McAdoo, J. K., C. C. Evans, B. A. Roundy, J. A. Young, and R. A. Evans. 1983. Influence of heteromyid rodents on *Oryzopsis hymenoides* germination. J. Range Mange. 36:61-64.

McAlister, D. F. 1944. Determination of soil drought resistance in grass seedlings. J. Amer. Soc. Agron. 36:324-336.

McCree, K. J. and S. G. Richardson. 1987a. Salt increases the water use efficiency in water stressed plants. Crop Sci. 27:543-547.

McCree, K. J. and S. G. Richardson. 1987b. Stomatal closure vs. osmotic adjustment: a comparison of stress responses. Crop Sci. 27:539-543.

McDole, R. E. and D. T. Booth. 1979. Fertilization of seedlings on range and disturbed lands. Univ. Idaho. Agric. Exp. Stat. Current Info. Series No. 513.

McDonald, M. B. 1976. Improving the germination of Indian ricegrass seeds. J. Seed Tech. 1:44-54.

McDonough, W. T. 1975a. Water potential of germinating seeds. Bot. Gaz. 136(1):106-108.

McDonough, W. T. 1975b. Water potentials of *Salsola kali* L. at differing levels of substrate water availability. Plant Soil 42:387-393.

McDonough, W. T. and R. O. Harniss. 1974. Effects of temperature on germination in three subspecies of big sagebrush. J. Range Manage. 27:204-205.

McDonough, W. T. and V. B. Mathews. 1970. Germination and emergence of three mountain range grasses as influenced by controlled temperatures and depths of planting. USDA Forest Service, Intermountain Forest and Range Exp. Sta. Research Note INT-111. 6 p.

McGinnies, W. J. 1960. Effects of moisture stress and temperature on germination of six range grasses. Agron. J. 52:159-162.

McIntyre, G. I. 1987. The role of water in the regulation of plant development. Can. J. Bot. 65:1287-1298.

McNaughton, S. J. 1992. Laboratory-simulated grazing: interactive effects of defoliation and canopy closure on Serengeti grasses. Ecology 73:170-182.

McNaughton, S. J., R. W. Ruess, and S. W. Seagle. 1988. Large mammals and process dynamics in African ecosystems. BioScience 38:794-800.

Meeuwig, R. O. 1960. Watersheds A and B—a study of surface runoff and erosion in the subalpine zone of central Utah, J. Forestry 58:556-560.

Meeuwig, R. O. 1965. Effects of seeding and grazing on infiltration capacity and soil stability of a subalpine range in central Utah. J. Range Manage. 18:173-180.

Meidner, H. and T. A. Mansfield. 1968. Physiology of Stomata. McGraw-Hill Book Co. New York. 179 p.

Meinzer, F. and G. Goldstein. 1985. Some consequences of leaf pubescence in the Andean giant rosette plant *Espeletia timotensis*. Ecology 66:512-520.

Meinzer, F. C., G. H. Goldstein, and P. W. Rundel. 1985. Morphological changes along an altitude gradient and their consequences for an Andean giant rosette plant. Oecologia 65:278-283.

Meyer, S. E. and S. B. Monsen. 1992. Big sagebrush germination patterns: subspecies and population differences. J. Range Manage. 45(1):87-93.

Meyer, W. S. and S. Walker. 1981. Leaflet orientation in water-stressed soybeans. Agron. J. 73:1071-1074.

Millar, B. D. 1966. Relative turgidity of leaves: temperature effects in measurement. Science 154:512-513.

Miller, L. N. 1972. Matric potentials in plants: means of estimation and ecophysiological significance. *In*: Brown, R. W. and Van Haveren, B. P. (eds.). Psychrometry in Water Relations Research. Utah State Univ., Agric. Exp. Sta., Logan. p. 211-218.

Miller, R. F., P. S. Doescher, T. Svejcar, and M. R. Haferkamp. 1986a. Growth and internal water status of three subspecies of *Artemisia tridentata*. *In*: McArthur, E. D., Welch, B. L. (compilers). Proc. Symposium on the Biology of *Artemisia* and *Chrysothamnus*. July 9-13, Provo, Utah. Gen. Tech. Report INT-200. USDA Forest Service. p. 347-352.

Miller, R. F., J. M. Seufert, and M. R. Haferkamp. 1986b. The ecology and management of bluebunch wheatgrass (*Agropyron spicatum*): a review. Agric. Exp. Stat., Oregon State Univ., Corvallis. Sta. Bul. 669. 39 p.

Miller, R. F., L. E. Eddleman, and R. F. Angell. 1987. Relationship of western juniper stem conducting tissue and basal circumference to leaf area and biomass. Great Basin Nat. 47:349-354.

Miller, R. F. and L. M. Schultz. 1987. Water relations and leaf morphology of *Juniperus occidentalis* in the northern Great Basin. Forest Sci. 33:690-706.

Mohammad, N., D. D. Dwyer, and F. E. Busby. 1982. Responses of crested wheatgrass and Russian wildrye to water stress and defoliation. J. Range Manage. 35:227-230.

Monson, R. K. and S. D. Smith. 1982. Seasonal water potential components of Sonoran Desert plants. Ecology 63:113-123.

Mooney, H. A. 1980. Seasonality and gradients in the study of stress adaptations. *In*: Turner, N. C. and P. J. Kramer (eds.). Adaptation of Plants to Water and High Temperature Stress. John Wiley and Sons. New York. p. 279-294.

Moore, R. T., S. W. Breckle, and M. M. Caldwell. 1972. Mineral ion composition and osmotic relations of *Atriplex confertifolia* and *Eurotia lanata*. Oecologia 11:67-78.

Moore, R. T. and M. M. Caldwell. 1972. The field use of thermocouple psychrometry in desert soils. *In*: Brown, R. W. and B. P. Van Haveren (eds.). Psychrometry in Water Relations Research. Utah State Univ., Agric. Exp. Sta. p. 165-171.

Morgan, J. A. 1986. The effects of N nutrition on the water relations and gas exchange characteristics of wheat (*Triticum aestivum* L.). Plant Physiol. 80:52-58.

Morgansen, V. O. 1980. Drought sensitivity at various growth stages of barley in relation to relative evapotranspiration and water stress. Agron. J. 72:1033-1038.

Mott, K. A., A. C. Gibson, and J. W. O'Leary. 1982. The adaptive significance of amphistomatic leaves. Plant Cell Environ. 5:455-460.

Mueggler, W. F. 1970. Influence of competition on the response of Idaho fescue to clipping. USDA Forest Service, Intermountain Forest and Range Exp. Sta., Research Paper INT-73. 10 p.

Mueggler, W. F. 1972. Influence of competition on the response of bluebunch wheatgrass to clipping. J. Range Manage. 25:88-92.

Mueggler, W. F. 1975. Rate and pattern of vigor recovery in Idaho fescue and bluebunch wheatgrass. J. Range Mangmt. 28:198-204.

Mueller, D. M. and R. A. Bowman. 1989. Emergence and root growth of three pregerminated cool-season grasses under salt and water stress. J. Range Manage. 42(6):490-495.

Mulroy, T. W. and P. W. Rundel. 1977. Annual plants: adaptations to desert environments. BioScience 27:109-114.

Munns, R. 1988. Why measure osmotic adjustment? Aust. J. Plant Physiol. 15:717-726.

Myers, B. A. and T. F. Neales. 1986. Osmotic adjustment, induced by drought, in seedlings of three *Eucalyptus* species. Aust. J. Plant Physiol. 13:597-603.

Naiman, R. J. 1988. Animal influences on ecosystem dynamics. BioScience 38:750-752.

Nelson, E. W. 1934. The influence of precipitation and grazing upon black grama grass range. U.S.D.A. Technical Bulletin No. 409. 32 p.

Nelson, D. L., K. T. Harper, K. C. Boyer, D. J. Weber, B. A. Haws, and J. R. Marble. 1989. Wildland shrub dieoffs in Utah: an approach to understanding the cause. p. 119-135. *In*: A. Wallace, E. D. McArthur, and M. R. Hafenkamp (eds.). Proceedings—symposium on shrub ecophysiology and biotechnology. USDA Forest Service. Intermountain Research Sta. Gen. Tech. Report INT-256.

Nelson, D. L., D. J. Weber, and S. C. Garvin. 1990. The possible role of plant disease in the recent wildland shrub dieoff in Utah. p. 84-90. *In*: E. D. McArthur, E. M. Romney, S. D. Smith, and P. T. Tueller (eds.). Proceedings— symposium on cheatgrass invasion, shrub die-off, and other aspects of shrub biology and management. USDA Forest service. Intermountain Research Sta., Gen. Tech. Report INT-276.

Newman, P. R. and L. E. Moser. 1988. Seedling root development and morphology of cool-season and warn-season forage grasses. Crop Sci. 28:148-151.

Nielsen, D. R., R. D. Jackson, J. W. Cary, and D. D. Evans. (eds.). 1972. Soil Water. Amer. Soc. Agron., Madison, Wis.175 p.

Nilsen, E. T., M. R. Sharifi, P. W. Rundel, W. M. Jarrell, and R. A. Virginia. 1983. Diurnal and seasonal water relations of the desert phreatophyte *Prosopis glandulosa* (honey mesquite) in the Sonoran desert of California. Ecology 64:1381-1393.

Nobel, P. S. 1976. Water relations and photosynthesis of a desert CAM plant, *Agave deserti*. Plant Physiol. 58:576-582.

Nobel, P. S. 1980a. Leaf anatomy and water use efficiency. *In*: Turner, N. C. and P. J. Kramer (eds.). Adaptation of Plants to Water and High Temperature Stress. John Wiley and Sons. New York. p. 43-55.

Nobel, P. S. 1980b. Water vapor conductance and CO_2 uptake for leaves of a C_4 desert grass, *Hilaria rigida*. Ecology 61:252-258.

Nobel, P. S. 1987a. Photosynthesis and productivity of desert plants. *In*: Berkofsky, L. and M. G. Wurtele (eds.). Progress in Desert Research. Rowman and Littlefield, Publs. Totowa, N.J. p. 41-66.

Nobel, P. S. 1987b. Water relations and plant size aspects of flowering for *Agave deserti*. Bot. Gaz. 148:79-84.

Nobel, P. S. and E. Quero. 1986. Environmental productivity indices for a Chihuahuan desert CAM plant, *Agave lechuguilla*. Ecology 67:1-11.

Nowak, R. S. and M. M. Caldwell. 1984. A test of compensatory photosynthesis in the field: implications for herbivory tolerance. Oecologia 61:311-318.

Noy-Meir, I. and B. Z. Ginzburg. 1967. An analysis of the water potential isotherm in plant tissue. I. The theory. Aust. J. Biol. Sci. 20:695-721.

Noy-Meir, I. and B. Z. Ginzburg. 1969. An analysis of the water potential isotherm in plant tissue. II. Comparative studies on leaves of different types. Aust J. Biol. Sci. 22:35-52.

Oertli, J. J. 1976a. The soil-plant-atmosphere continuum. *In*: Lange, O. L., Kappen, L., and Schulze, E.-D. (eds.). Water and Plant Life: Problems and Modern Approaches. Springer-Verlag, Berlin. p. 32-41.

Oertli, J. J. 1976b. The states of water in the plant: theoretical consideration. *In*: Lange, O. L., Kappen, L., and Schulze, E.-D. (eds.). Water and Plant Life: Problems and Modern Approaches. Springer-Verlag, Berlin. p. 19-31.

Oertli, J. J. 1983. Adaptations of plants to limited water availability. Proc. 17th Coll. Int. Potash Institue, Bern. p. 49-62.

Oesterheld, M. and S. J. McNaughton. 1991. Interactive effect of flooding and grazing on the growth of Serengeti grasses. Oecologia 88:153-156.

Olson, B. E. and J. H. Richards. 1988a. Annual replacement of the tillers of *Agropyron desertorum* following grazing. Oecologia 76:1-6.

Olson, B. E. and J. H. Richards. 1988b. Tussock regrowth after grazing: intercalary meristem and axillary bud activity of tillers of *Agropyron desertorum*. Oikos 51:374-382.

Olson, D. F. and M. D. Hoover. 1954. Methods of soil moisture determination under field conditions. USDA Forest Service. Southeastern Forest Exp. Sta. Station Paper No. 38. 28 p.

Oosterhuis, D. M., F. Le Maire, and C. Le Maire. 1987. Leaf water potential and crop color change in water-stressed peas. HortSci. 22:429-431.

Oosterhuis, D. M., M. L. Parker, S. D. Wullschleger, and K. S. Kim. 1988. The citrus leaf cuticle in relation to measurement of leaf water potential using thermocouple psychrometers. Plant Cell Environ. 11:129-135.

Oosterhuis, D. M. and S. Walker. 1987. Stomatal resistance measurement as an indicator of water deficit stress in wheat and soybeans. South African J. Plant Soil 4:113-120.

Oosterhuis, D. M., S. Walker, and J. Eastham. 1985. Soybean leaflet movements as an indicator of crop water stress. Crop Sci. 25:1101-1106.

Oosterhuis, D. M. and S. D. Wullschleger. 1987. Osmotic adjustment in cotton (*Gossypium hirsutum* L.) leaves and roots in response to water stress. Plant Physiol. 84:1154-1157.

Oppenheimer, H. R. 1960. Adaptation to drought: xerophytism. *In*: Plant-Water Relationships in Arid and Semi-Arid Conditions. UNESCO. Arid Zone Research XV. p.105-138.

Orodho, A. B., M. J. Trlica, and C. D. Bonham. 1990. Long-term grazing effects on soil and vegetation in the Four Corners region. Southwestern Nat. 35(1):9-14.

Osmond, C. B., M. P. Austin, J. A. Berry, W. D. Billings, J. S. Boyer, J. W. H. Dacey, P. S. Nobel, S. D. Smith, and W. E. Winner. 1987. Stress physiology and the distribution of plants. BioScience 37(1):38-48.

Osmond, C. B., O. Bjorkman, and D. J. Anderson. 1980a. Physiological Processes in Plant Ecology. Springer-Verlag. Berlin. 468 p.

Osmond, C. B., K. Winter, and S. B. Powles. 1980b. Adaptive significance of carbon dioxide cycling during photosynthesis in water-stressed plants. *In*: Turner, N. C. and Kramer, P. J. (eds.). Adaptations of Plants to Water and High Temperature Stress. John Wiley and Sons, New York. p. 139-154.

O'Toole, J. C. and R. T. Cruz. 1980. Response of leaf water potential, stomatal resistance, and leaf rolling to water stress. Plant Physiol. 65:428-432.

Ouattar, S., R. L. Jones, R. K. Crookston, and M. Kajeiou. 1987. Effect of drought on water relations of developing maize kernels. Crop Sci. 27:730-735.

Owen, P. C. 1952. The relation of the germination of wheat to water potential. J. Exp. Bot. 3:188-203.

Owens, D. W. and C. A. Call. 1985. Germination characteristics of *Helianthus maximilianai* Schard. and *Simsia calva* (Engelm, & Gray). J. Range Manage. 38(4):336-339.

Packer, P. E. 1953. Effects of trampling disturbance on watershed condition, runoff, and erosion. J. Forestry 51:28-31.

Packer, P. E. and E. F. Aldon. 1978. Revegetation techniques for dry regions. p. 425-450. *In*: Reclamation of Drastically Disturbed Lands. Amer. Soc. Agron., Madison, WI.

Paleg, L. G. and D. Aspinall. 1981. The Physiology and Biochemistry of Drought Resistance in Plants. Academic Press. Sydney. 492 p.

Parker, M. A. 1985. Grasshopper attack and water balance of the shrub *Gutierrezia microcephala*. Amer. Midland Nat. 113:193-197.

Parr, J. F., R. I. Papendick, S. B. Hornick, and D. Colacicco. 1989. Use of organic amendments for increasing the productivity of arid lands. Arid Soil Res. Rehab. 3:149-170.

Passioura, J. B. 1980. The meaning of matric potential. J. Exp. Bot. 31(123):1161-1169.

Passioura, J. B. 1981. Water collection by roots. *In*: Paleg, L. G. and Aspinall, D. (eds.). The Physiology and Biochemistry of Drought Resistance in Plants. Academic Press, New York. p. 39-53.

Passioura, J. B. 1988. Water transport in and to roots. Ann. Rev. Plant Physiol. 39:245-265.

Pearcy, R. W., O. Bjorkman, M. M. Caldwell, J. E. Keeley, R. K. Monson, and B. R. Strain. 1987. Carbon gain by plants in natural environments. BioScience 37:21-29.

Pearcy, R. W., J. R. Ehleringer, H. A. Mooney, and P. W. Rundel (eds.). 1989. Plant Physiological Ecology: Field Methods and Instrumentation. Chapman and Hall. New York. 457 p.

Peterson, R. L. 1992. Adaptations of root structure in relation to biotic factors. Can J. Bot. 70:661-675.

Philip, J. R. 1964. Sources and transfer processes in the air layers occupied by vegetation. J. Applied Meteor. 3:390-395.

Philip, J. R. 1966. Plant water relations: some physical aspects. Ann. Rev. Plant Physiol. 17:245-268.

Phillips, W. S. 1963. Depth of roots in soil. Ecology 44:424.

Plummer, A. P. 1943. The germination and early seedling development of twelve range species. J. Amer. Soc. Agron. 35:19-34.

Poljakoff-Mayber, A. 1975. Morphological and anatomical changes in plants as a response to salinity stress. *In*: Poljakoff-Mayber, A. and Gale, J. (eds.). Plants in Saline Environments. Springer-Verlag, Berlin. p. 97-117.

Powell, K. B., R. B. Vincent, E. J. DePuit, J. L. Smith, and F. E. Parady. 1990. Role of irrigation and fertilization in revegetation of cold desert mined lands. J. Range Manage. 43:449-455.

Power, J. F. 1980. Response of semiarid grassland sites to nitrogen fertilization: I. Plant growth and water use. Soil Sci. Soc. Amer. J. 41:545-550.

Proffitt, A. P. B., P. R. Berliner, and D. M. Oosterhuis. 1985. A comparative study of root distribution and water extraction efficiency by wheat grown under high- and low-frequency irrigation. Agron. J. 77:655-662.

Pyke, D. A. and S. Archer. 1991. Plant-plant interactions affecting plant establishment and persistence on revegetated rangeland. J. Range Manage. 44:550-557.

Radin, J. W. 1992. Reconciling water-use efficiencies of cotton in field and laboratory. Crop Sci. 32:13-18.

Rahman, M. S. and A. J. Rutter. 1980. A comparison of the ecology of *Deschampsia cespitosa* and *Dactylis glomerata* in relation to the water factor. II. Controlled experiments in glasshouse conditions. J. Ecol. 68:479-491.

Ram, J., S. P. Singh, and J. S. Singh. 1991. Effect of fertilizer on plant biomass distribution and net accumulation rate in an alpine meadow in central Himalaya, India. J. Range Manage. 44:140-143.

Rasche, K. 1960. Heat transfer between the plant and the environment. Ann. Rev. Plant Physiol. 11:111-126.

Raschke, K. 1975. Stomatal action. Ann. Rev. Plant Physiol. 26:309-340.

Raschke, K. 1976. How stomata resolve the dilemma of opposing priorities. Phil. Trans. R. Soc. Lond. B. 273:551-560.

Rawlins, S. L. and G. S. Campbell. 1986. Water potential: thermocouple psychrometry. p. 597-618. *In*. Methods of Soil Analysis, Part 1. Physical and Mineralogical Methods. Agron. Monog. no. 9 (2nd. ed.). Amer. Soc. Agron.-Soil Sci. Soc. Amer. Madison, WI.

Reader, R. J. 1992. Herbivory as a confounding factor in an experiment measuring competition among plants. Ecology 73:373-376.

Reed, M. J. and R. A. Peterson. 1961. Vegetation, soil, and cattle responses to grazing on northern Great Plains range. USDA Forest Service. Technical Bulletin No. 1252. 79 p.

Redmann, R. E. and M. Q. Qi. 1992. Impacts of seeding depth on emergence and seedling structure in eight perennial grasses. Can. J. Bot. 70:133-139.

Reynolds, H. G. and H. W. Springfield. 1953. Reseeding southwestern range lands with crested wheatgrass. USDA Farmer's Bull. No. 2056. 20 p.

Richards, J. H. 1984. Root growth response to defoliation in two *Agropyron* bunchgrasses: field observations with an improved root periscope. Oecologia 64:21-25.

Richards, J. H. 1986. Root form and depth distribution in several biomes. *In*: Carlisle, D., Berry, W. L., Kaplan, I. R., and Watterson, J. R. (eds.). Mineral Exploration: Biological Systems and Organic Matter. Rubey Vol. V. Prentice-Hall, Englewood Cliffs, N.J. p. 83-97.

Richards, J. H. and M. M. Caldwell. 1985. Soluble carbohydrates, concurrent photosynthesis and efficiency in regrowth following defoliation: a field study with *Agropyron* species. J. Appl. Ecol. 22:907-920.

Richards, J. H. and M. M. Caldwell. 1987. Hydraulic lift: substantial nocturnal water transport between soil layers by *Artemisia tridentata* roots. Oecologia 73:486-489.

Richards, J. H., R. J. Mueller, and J. J. Mott. 1988. Tillering in tussock grasses in relation to defoliation and apical bud removal. Ann. Bot. 62:173-179.

Richards, L. A. 1949. Methods of measuring soil moisture tension. Soil Sci. 68:95-112.

Richards, L. A. 1954. The measurement of soil water in relation to plant requirements. Sci. Monthly 78:307-313.

Richards, L. A. and L. R. Weaver. 1943. Fifteen-atmosphere percentage as related to the permanent wilting percentage. Soil Sci. 56:331-339.

Richards, L. A. and G. Ogata. 1958. Thermocouple for vapor pressure measurements in biological and soil systems at high humidity. Science 128:1089-1090.

Richards, R. A., H. M. Rawson, and D. A. Johnson. 1986. Glaucousness in wheat: its development and effect on water-use efficiency, gas exchange and photosynthetic tissue temperatures. Aust. J. Plant Physiol. 13:465-473.

Richardson, B. Z. and E. R. Burroughs. 1972. Effects of air gaps and saturated voids on accuracy of neutron moisture measurements. USDA Forest Service, Intermountain Forest and Range Exp. Sta. Res. Paper INT-120. 20 p.

Richardson, J. M., D. R. Gealy, and L. A. Morrow. 1989. Influence of moisture deficits on the reproductive ability of downy brome (*Bromus tectorum*). Weed Sci. 37:525-530.

Richardson, S. G. and C. M. McKell. 1980. Water relations of *Atriplex canescens* as affected by the salinity and moisture percentage of processed oil shale. Agron. J. 72:946-950.

Rickard, W. H. 1985. Biomass and shoot production in an undisturbed sagebrush-bunchgrass community. Northwest Sci. 59:126-133.

Ries, R. E. and L. Hofmann. 1991. Research observations: standardized terminology for structures resulting in emergence and crown placement of 3 perennial grasses. J. Range Manage. 44:404-407.

Ries, R. E. and T. J. Svejcar. 1991. The grass seedling: when is it established? J. Range Manage. 44:574-576.

Rissing, S. W. 1988. Seed-harvester ant association with shrubs: competition for water in the Mojave desert? Ecology 69:809-813.

Ritchie, G. A. and Hinckley. 1975. The pressure chamber as an instrument for ecological research. Advan. Ecol. Research 9:165-254.

Roberts, B. R. 1969. Light as a source of error in estimates of water potential by vapor equilibration. Plant Physiol. 44:937-938.

Romo, J. T. and L. E. Eddleman. 1985. Germination responses of greasewood (*Sarcobatus vermiculatus*) to temperature, water potential and specific ions. J. Range Manage. 38(2):117-120.

Romo, J. T. and L. E. Eddleman. 1988. Germination of green and gray rubber rabbitbrush and their establishment on coal mined land. J. Range Manage. 41(6):491-495.

Romo, J. T., P. L. Grilz, C. J. Bubar, and J. A. Young. 1991. Influences of temperature and water stress on germination of plains rough fescue. J. Range Manage. 44:75-81.

Roohi, R., D. A. Jameson, and N. Nemati. 1991. The effect of light on adventitious root formation in blue grama. J. Range Manage. 44:184-185.

Roundy, B. A. 1983. Response of basin wildrye and tall wheatgrass seedlings to salination. Agron. J. 75:67-71.

Roundy, B. A. 1985a. Emergence and establishment of Basin wildrye and tall wheatgrass in relation to moisture and salinity. J. Range Manage. 38(2):126-131.

Roundy, B. A. 1985b. Germination and seedling growth of tall wheatgrass and basin wildrye in relation to boron. J. Range Manage. 38(3):270-272.

Roundy, B. A., V. K. Winkel, H. Khalifa, and A. D. Matthais. 1992. Soil water availability and temperature dynamics after one-time heavy cattle trampling and land imprinting. Arid Soil Res. Rehab. 6:53-69.

Roundy, B. A., J. A. Young, and R. A. Evans. 1985. Germination of basin wildrye and tall wheatgrass in relations to osmotic and matric potential. J. Range Manage. 77:129-135.

Runnels, L. K. 1966. Ice. Sci. Amer. 215(6):118-126.

Running, S. W. 1976. Environmental control of leaf water conductance in conifers. Can. J. For. Res. 6:104-112.

Sala, O. E., W. K. Lauenroth, and W. J. Parton. 1982. Plant recovery following prolonged drought in a shortgrass steppe. Agric. Meteor. 27:49-58.

Salisbury, F. B. and C. W. Ross. 1992. Plant Physiology: Fourth Edition. Wadsworth Publ. Co. Belmont, Calif. 682 p.

Sampson, A. W. 1952. Range Management Principles and Practices. John Wiley and Sons, Inc. New York. 570 p.

Samuel, M. J. and R. H. Hart. 1992. Survival and growth of blue grama seedlings in competition with western wheatgrass. J. Range Manage. 45:444-448.

Sandquist, D. R., W. S. F. Schuster, L. A. Donovan, S. L. Phillips, and J. R. Ehleringer. 1993. Differences in carbon isotope discrimination between seedlings and adults of southwestern desert perennial plants. Southwestern Nat. 38(3):212-217.

Sandu, B. S. and M. L. Horton. 1977. Response of oats to water deficit. II. growth and yield characteristics. Agron. J. 69:361-364.

Scholander, P. F. 1972. Tensile water. Amer. Sci. 60:584-590.

Schonherr, J. 1982. Resistance of plant surfaces to water loss: transport properties of cutin, suberin and associated lipids. p. 153-179. *In*: O. L. Lange, P. S. Nobel, C. B. Osmond, and H. Ziegler (eds.). Physiological Plant Ecology II: Water Relations and Carbon Assimilation. Encyclopedia of Plant Physiology. New Series, Vol. 12B. Springer-Verlag. Berlin.

Schulze, E.-D. 1986a. Carbon dioxide and water vapor exchange in response to drought in the atmosphere and in the soil. Ann. Rev. Plant Physiol. 37:247-274.

Schulze, E.-D. 1986b. Whole-plant responses to drought. Aust. J. Plant Physiol. 13:127-141.

Schulze, E.-D., R. H. Robichaux, J. Grace, P. W. Rundel, and J. R. Ehleringer. 1987. Plant water balance. BioScience 37(1):30-37.

Schuman, G. E., F. Rauzi, and G. S. Howard. 1987. Vegetation response to soil surface modification in mined land reclamation. Reclam. Reveg. Res. 6:49-54.

Schuman, G. E., E. M. Taylor, and F, Rauzi. 1991. Forage production of reclaimed mined lands as influenced by nitrogen fertilization and mulching practice. J. Range Manage. 44:382-384.

Schuster, J. L. 1964. Root development of native plants under three grazing intensities. Ecology 45:63-70.

Sestak, Z., J. Catsky, and P. G. Jarvis. 1971. Plant Photosynthetic Production: Manual of Methods. Dr. W. Junk N.V. Publs. The Hague. 818 p.

Sharma, C. P. and P. N. Sharma. 1987. Mineral nutrient deficiencies affect plant water relations. J. Plant Nutr. 10:1637-1643.

Sharma, M. L. 1976. Interaction of water potential and temperature effects on germination of three semi-arid plant species. Agron. J. 68:390-394.

Sharp, R. E., and W. J. Davies. 1979. Solute regulation and growth of roots and shoots of water stressed maize plants. Planta 147:43-49.

Sharp, R. E., W. Kuhn Silk, and T. C. Hsiao. 1988. Growth of the maize primary root at low water potentials I. Spatial distribution of expansive growth. Plant Physiol. 87:50-57.

Sidle, R. C. and M. C. Amacher. 1990. Effects of mining, grazing, and roads on sediment and water chemistry in Birch Creek, Nevada. p. 463-472. *In*: Watershed Planning and Analysis in Action. Sympos. Proc. Amer. Soc. Civil Eng.

Sidle, R. C. and R. W. Brown. 1988. Soil water availability for seedling establishment on arid mine spoils. p. 48. *In*: Agron. Abstracts, Amer. Soc. Agron. Ann. Meeting, Nov. 27-Dec. 3 Anaheim, CA.

Simoes, M. and Z. Baruch. 1991. Responses to simulated herbivory and water stress in two tropical C_4 grasses. Oecologia 88:173-180.

Skujins, J. and B. Z. Richardson. 1985. Humic matter enrichment in reclaimed soils under semiarid conditions. Geomicrobiology J. 4:299-311.

Slatyer, R. O. 1967. Plant Water Relationships. Academic Press. New York. 366 p.

Slatyer, R. O. and O. T. Denmead. 1964. Water movement through the soil-plant-atmosphere system. *In*: Hills, E. S. (ed.). Water Resources Use and Management. Univ. Melbourne Press. p. 276-289.

Slatyer, R. O. and S. A. Taylor. 1960. Terminology in plant-and soil-water relations. Nature 187:922-924.

Slavik, B. 1974. Methods of Studying Plant Water Relations. Springer-Verlag, Berlin. 449 p.

Slick, B. M. and W. R. Curtis. 1985. A guide for the use of organic materials as mulches in reclamation of coal minespoils in the eastern United States. USDA Forest Service. Northeastern Forest Experiment Station. Gen. Tech. Report NE-98. 14 p.

Smith, S. D. and R. S. Nowak. 1990. Ecophysiology of plants in the Intermountain lowlands. *In*: Osmond, C. B., L. F. Pitelka, and G. M. Hidy (eds.). Plant Biology of the Basin and Range. Springer-Verlag. Berlin. p. 179-241.

Smith, W. K. 1978. Temperature of desert plants: another perspective on the adaptability of leaf size. Science 201:614-616.

Smith, W. K. and P. S. Nobel. 1977. Influences of seasonal changes in leaf morphology on water-use efficiency for three desert broadleaf shrubs. Ecology 58:1033-1043.

Solbrig, O. T. and G. H. Orians. 1977. The adaptive characteristics of desert plants. Amer. Sci. 65:412-421.

Somody, C.N., J. D. Nalewaja, and S. D. Miller. 1985. Self-burial of wild oat florets. Agron. J. 77:359-362.

Southwick, S. M. and T. L. Davenport. 1986. Characterization of water stress and low temperature effects on flower induction in citrus. Plant Physiol. 81:26-29.

Sowell, J. B., D. L. Koutnik, and A. J. Lansing. 1982. Cuticular transpiration of whitebark pine (*Pinus albicaulis*) within a Sierra Nevadan timberline ecotone, U.S.A. Arctic and Alpine Res. 14:97-103.

Spanner, D. C. 1951. The Peltier effect and its use in the measurement of suction pressure. J. Exp. Bot. 2:145-168.

Spanner, D. C. 1972. Plants, water, and some other topics. p. 29-39. *In*: R. W. Brown and B. P. Van Haveren (eds.). Psychrometry in Water Relations Research. Utah Agr. Exp. Sta., Utah State Univ., Logan.

Spomer, L. A. 1985. Techniques for measuring plant water. HortSci. 20:1021-1028.

Stanhill, G. 1986. Water use efficiency. Adv. Agron. 39:53-85.

Steuer, B., T. Stuhlfauth, and H. P. Fock. 1988. The efficiency of water use in water stressed plants is increased due to ABA induced stomatal closure. Photosyn. Res. 18:327-336.

Stillinger, F. H. 1980. Water revisited. Science 209(4455):451-457.

Stocker, O. 1960. Physiological and morphological changes in plants due to water deficiency. *In*: Plant-Water Relationships in Arid and Semi-Arid Conditions. UNESCO. Arid Zone Research XV. p. 63-104.

Stoddart, L. A., A. D. Smith, and T. W. Box. 1975. Range Management: Third Edition. McGraw-Hill Book Co., New York. 532 p.

Sturges, D. L. 1977. Soil water withdrawal and root characteristics of big sagebrush. Amer. Midland Nat. 98:257-274.

Sundberg, M. D. 1985. Trends in distribution of stomata in desert plants. Desert Plants 7:154-157.

Svejcar, T. 1990. Root length, leaf area, and biomass of crested wheatgrass and cheatgrass seedlings. J. Range Manage. 43:446-448.

Svejcar, T. and S. Christiansen. 1987. Grazing effects on water relations of Caucasian bluestem. J. Range Mangmt. 40:15-18.

Swindale, L. D. and F. R. Bidinger. 1981. Introduction: the human consequences of drought and crop research priorities for their alleviation. *In*: L. G. Paleg and D. Aspinall (eds.). The Physiology and Biochemistry of Drought Resistance in Plants. Academic Press. Sydney. p 1-13.

Tabler, R. D. 1964. The root system of *Artemisia tridentata* at 9,500 feet in Wyoming. Ecology 45:633-636.

Tapia, C. R. and E. M. Schmutz. 1971. Germination responses of three desert grasses to moisture stress and light. J. Range Manage. 24:292-295.

Taylor, D. R. and L. W. Aarssen. 1988. An interpretation of phenotypic plasticity in *Agropyron repens* (Gramineae). Amer. J. Bot. 75:401-413.

Taylor, H. M. 1980. Modifying the root systems of cotton and soybean to increase absorption. *In*: Turner, N. C. and Kramer, P. J. (eds.). Adaptation of Plants to Water and High Temperature Stress. Academic Press. New York. p. 75-84.

Taylor, H. M. (ed.). 1987. Minirhizotron Observation Tubes: Methods and Applications for Measuring Rhizosphere Dynamics. Amer. Soc. Agron., Madison, WI. 143 p.

Taylor, H. M. and B. Klepper. 1978. The role of rooting characteristics in the supply of water to plants. Adv. Agron. 30:99-128.

Taylor, S. A. 1965. Measuring soil-water potential. *In*: Methodology of Plant Eco-physiology: Proc. Montpellier Sympos. UNESCO. p. 149-157.

Taylor, S. A. 1968. Terminology in plant and soil water relations. *In*: Kozlowski, T. T. (ed.). Water Deficits and Plant Growth. Vol. I. Development, Control, and Measurement. Academic Press. New York. p. 49-72.

Taylor, S. A. and G. L. Ashcroft. 1972. Physical Edaphology: The Physics of Irrigated and Nonirrigated Soils. W. H. Freeman and Co. San Francisco. 533 p.

Taylorson, R. B. 1986. Water stress-induced germination of giant foxtail (*Setaria faberi*) seeds. Weed Sci. 34:871-875.

Teeri, J. A. 1985. Crassulacean acid metabolism in the winter annual *Diamorpha cymosa* (Crassulaceae) in its native habitat. Ecology 66(1):308-311.

Thomas, H. 1986. Effect of rate of dehydration on leaf water status and osmotic adjustment in *Dactylis glomerata* L., *Lolium perenne* L., and *L. multiflorum* Lam. Ann. Bot. 57:225-235.

Thomas, H. 1990. Osmotic adjustment in *Lolium perenne*; its heritability and the nature of solute accumulation. Ann. Bot. 66:521-530.

Thomas, H. and J. L. Stoddart. 1980. Leaf senescence. Ann. Rev. Plant Physiol. 31:83-111.

Thornburg, A. A. 1982. Plant materials for use on surface-mined lands in arid and semi-arid regions. USDA Soil Conserv. Serv. SCS-TP-157. 88 p.

Tiedemann, A. R. and C. F. Lopez. 1982. Soil nutrient assessments of mine spoils. p. 66-79. *In*: E. F. Aldon and W. R. Oaks (eds.). Reclamation of Mined Lands in the Southwest: A Symposium. Soil Conserv. Soc., NM Chapter.

Tinker, R. P. 1976. Roots and Water: transport of water to plant roots in soil. Phil. Trans. R. Soc. Lond. B. 273:445-461.

Trlica, M. J. and M. E. Biondini. 1990. Soil water dynamics, transpiration, and water losses in a crested wheatgrass and native shortgrass ecosystem. Plant Soil 126:187-201.

Tschaplinski, T. J. and T. J. Blake. 1989. Water-stress tolerance and late-season organic solute accumulation in hybrid poplar. Can. J. Bot. 67:1681-1688.

Tulloch, A. P. and L. Bergter. 1981. Epicuticular wax of *Juniperus scopulorum*. Phytochem. 20:2711-2716.

Turner, N. C. 1981. Techniques and experimental approaches for the measurement of plant water status. Plant Soil 58:339-366.

Turner, N. C. 1986. Crop water deficits: a decade of progress. Adv. Agron. 39:1-51.

Turner, N. C. 1988. Measurement of plant water status by the pressure chamber technique. Irrig. Sci. 9:289-308.

Turner, N. C. and J. E. Begg. 1981. Plant-water relations and adaptation to stress. Plant Soil 58:97-131.

Turner, N. C. and G. J. Burch. 1983. The role of water in plants. p. 73-126. *In*: I. D. Teare and M. M. Peet (eds.). Crop-Water Relations. John Wiley and Sons. New York.

Turner, N. C. and M. M. Jones. 1980. Turgor maintenance by osmotic adjustment: a review and evaluation. *In*: Turner, N. C. and P. J. Kramer (eds.). Adaptation of Plants to Water and High Temperature Stress. John Wiley and Sons. New York. p. 87-103.

Turner, N. C. and P. J. Kramer. 1980. Adaptations of Plants to Water and High Temperature Stress. John Wiley and Sons, New York. 482 p.

Tyree, M. T. 1976. Negative turgor pressure in plants: fact or fallacy. Can. J. Bot. 54:2738-2746.

Unger, P. W. 1978. Staw mulch effects on soil temperature and sorgham germination and growth. Agron. J. 70:858-864.

Vallentine, J. F. 1989. Range Development and Improvement. Third Edition. Academic Press, Inc. New York. 524 p.

van den Honert, T. H. 1948. Water transport in plants as a catenary process. Discuss. Faraday Soc. 3:146-153.

Veihmeyer, F. J. and A. H. Hendrickson. 1928. Soil moisture at permanent wilting of plants. Plant Physiol. 3:355-357.

Veihmeyer, F. J. and A. H. Hendrickson. 1949. Methods of measuring field capacity and permanent wilting percentage of soils. Soil Sci. 67:75-94.

Veihmeyer, F. J. and A. H. Hendrickson. 1950. Soil moisture in relation to plant growth. Ann. Rev. Plant Physiol. 1:285-304.

Virgin, H. I. 1965. Chlorophyll formation and water deficit. Physiol. Plant. 18:994-1000.

Waggoner, P. E. and I. Zelitch. 1965. Transpiration and the stomata of leaves. Science 150:1413-1420.

Wallace, A. and D. L. Nelson. 1990. Wildland shrub dieoffs following excessively wet periods: a synthesis. p. 81-83. *In*: Proceedings-Symposium on cheatgrass invasion, shrub die-off, and other aspects of shrub biology and management. USDA Forest Service. General Technical Report INT-276.

Waller, S. S. and J. K. Lewis. 1979. Occurrence of C_3 and C_4 photosynthetic pathways in North American grasses. J. Range Manage. 32:12-28.

Waring, R. H. and B. D. Cleary. 1967. Plant moisture stress: evaluation by pressure bomb. Science 155:1248-1254.

Warren, S. D., T. L. Thurow, W. H. Blackburn, and N. E. Garza. 1986. The influence of livestock trampling under intensive rotation grazing and soil hydrologic characteristics. J. Range Manage. 39:491-495.

Warren Wilson, J. 1967a. The component of leaf water potential. I. Osmotic and matric potentials. Aust J. Biol. Sci. 20:329-347.

Warren Wilson, J. 1967b. The components of leaf water potential. II. Pressure potential and water potential. Aust. J. Biol. Sci. 20:349-357.

Warren Wilson, J. 1967c. The components of leaf water potential. III. Effects of tissue characteristics and relative water content on water potential. Aust. J. Biol. Sci. 20:359-367.

Weatherley, P. E. 1950. Studies in the water relations of the cotton plant. I. The field measurement of water deficits in leaves. New Phytol. 49:81-97.

Weatherley, P. E. 1951. Studies in the water relations of the cotton plant. II. Diurnal and seasonal variations in relative turgidity and environmental factors. New Phytol. 50:36-51.

Weaver, J. E. and F. E. Clements. 1938. Plant Ecology. McGraw Hill Book Co., Inc. New York. 601 p.

Weaver, T. 1977. Root distribution and soil water regimes in nine habitat types of the northern Rocky Mountains. *In*: Marshall, J. (ed.). The Belowground Ecosystem. Range Sci. Dept. Ser. #26. Colo. State Univ., Fort Collins. p. 239-244.

Weaver, T. 1982. Distribution of root biomass in well-drained surface soils. Amer. Midland Nat. 107:393-395.

Wendt, C. W., R. H. Haas, and J. R. Runkles. 1968. Influence of selected environmental variables on the transpiration rate of mesquite [*Prosopis glandulosa* var. *glandulosa* (torr.) Cockr.]. Agron. J. 60:382-384.

Wenkert, W., E. R. Lemon, and T. R. Sinclair. 1978. Water content-potential relationship in soya bean: changes in component potentials for mature and immature leaves under field conditions. Ann. Bot. 42:295-307.

West, C. P., D. M. Oosterhuis, and S. D. Wullschleger. 1990. Osmotic adjustment in tissues of tall fescue in response to water deficit. Environ. Exp. Bot. 30:149-156.

West, N. E., R. A. Griffin, and J. J. Jurinak. 1984. Comparison of phosphorus distribution and cycling between adjacent native semidesert shrub and cultivated grass-dominated ecosystems. Plant Soil 81:151-164.

Westgate, M. E. and J. S. Boyer. 1985. Osmotic adjustment and the inhibition of leaf, root, stem and silk growth at low water potentials in maize. Planta. 164:540-549.

Westgate, M. E. and J. S. Boyer. 1986. Reproduction at low silk and pollen water potentials in maize. Crop Sci. 26:951-956.

Westgate, M. E. and D. L. T. Grant. 1989. Water deficits and reproduction in maize: response of the reproductive tissue to water deficits at anthesis and mid-grain fill. Plant Physiol. 91:862-867.

Whalley, R. D. B. 1973. Drought and vegetation. p. 81-98. *In*: J. V. Lovett (ed.). The Environmental, Economic and Social Significance of Drought. Angus and Robertson, Publs. Sydney.

Whalley, R. D. B. and A. A. Davidson. 1969. Drought dormancy in *Astrebla lappacea*, *Chloris acicularis*, and *Stipa aristiglumis*. Aust. J. Agric. Res. 20:1035-1042.

Wiebe, H. H. 1972. The role of water potential and its components in physiological processes of plants. p. 194-197. *In*: R. W. Brown and B. P. Van Haveren (eds.). Psychrometry in Water Relations Research. Utah Agric. Exp. Station, Utah State Univ.

Wiebe, H. H., H. A. Al-Saadi, and S. L. Kimbal. 1974. Photosynthesis in the anomalous secondary wood of *Atriplex confertifolia* stems. Amer. J. Bot. 61:444-448.

Wiebe, H. H., G. S. Campbell, W. H. Gardner, S. L. Rawlins, J. W. Cary, and R. W. Brown. 1971. Measurement of plant and soil water status. Utah Agric. Exp. Stat. Bulletin 484. Utah State Univ., Logan, UT. 71 p.

Willms, W. D. 1988. Response of rough fescue (*Festuca scabrella*) to light, water, temperature, and litter removal, under controlled conditions. Can. J. Bot. 66:429-434.

Wilson, A. M. 1971. Amylase synthesis and stability in crested wheatgrass seeds at low water potentials. Plant Physiol. 48:541-546.

Wilson, A. M. and D. D. Briske. 1979. Seminal and adventitious root growth of blue grama seedlings on the Central Plains. J. Range Manage. 32:209-213.

Wilson, A. M. and G. A. Harris. 1966. Hexose-, inositol-, and nucleoside phosphate esters in germinating seeds of crested wheatgrass. Plant Physiol. 41:1416-1419.

Wilson, A. M. and G. A. Harris. 1968. Phosphorylation in crested wheatgrass seeds at low water potentials. Plant Physiol. 43:61-65.

Wilson, A. M., D. N. Hyder, and D. D. Briske. 1976. Drought resistance characteristics of blue grama seedlings. Agron. J. 68:479-484.

Winkel, V. K. and B. A. Roundy. 1991. Effects of cattle trampling and mechanical seedbed preparation on grass seedling emergence. J. Range Manage. 44:176-180.

Winkel, V. K., B. A. Roundy, and D. K. Blough. 1991a. Effects of seedbed preparation and cattle trampling on burial of grass seed. J. Range Manage. 44:171-175.

Winkel, V. K., B. A. Roundy, and J. R. Cox. 1991b. Influence of seedbed microsite characteristics on grass seedling emergence. J. Range Manage. 44:210-214.

Winter, K. and U. Luttge. 1976. Balance between C_3 and CAM pathway of photosynthesis. *In*: Lange, O. L., Kappen, L., and Schulze, E.-D. (eds.). Water and Plant Life: Problems and Modern Approaches. Springer-Verlag, Berlin. p. 323-334.

Woodell, S. R. J. 1973. Xerophytes. Oxford Univ. Press. London. 16 p.

Wraith, J. M., D. A. Johnson, R. J. Hanks, and D. V. Sisson. 1987. Soil and plant water relations in a crested wheatgrass pasture: response to spring grazing by cattle. Oecologia 73:573-578.

Wright, G. C., R. C. Nageswara Rao, and G. D. Farquhar. 1994. Water-use efficiency and carbon isotope discrimination in peanut under water deficit conditions. Crop Sci. 34:92-97.

Wright, L. N. 1975. Improving range grasses for germination and seedling establishment under stress environments. *In*: Campbell, R. S and Herbel, C. H. (eds.). Improved Range Plants. Soc. Range Manage. Range Symposium Series No. 1. p. 3-22.

Wullschleger, S. D. and D. M. Oosterhuis. 1989. The occurrence of an internal cuticle in cotton (*Gossypium hirsutum* L.) leaf stomates. Environ. Exp. Bot. 29:229-235.

Wullschleger, S. D., D. M. Oosterhuis, and S. A. Rutherford. 1990. Importance of bracts in the carbon economy of cotton. Arkansas Farm Res. July-Aug. p. 4.

Wyatt, J. W., D. J. Dollhopf, and W. M. Schafer. 1980. Root distribution in 1-to 48-year-old stripmine spoils in southeastern Montana. J. Range Manage. 33:101-104.

Young, J. A. 1991. Tumbleweed. Sci. Amer. 264:82-87.

Young, J. A. and R. A. Evans. 1984. Germination of seeds of "Paloma" and "Nezpar" Indian ricegrass. J. Range Mange. 37:19-21.

Young, J. A, R. A. Evans, and B. A. Roundy. 1983. Quantity and germinability of *Oryzopsis hymenoides* seed in Lahontan sands. J. Range Manage. 36:82-86.

Zahner, R. 1968. Water deficits and growth of trees. *In*: Kozlowski, T. T. (ed.). Water deficits and Plant Growth. Vol. II: Plant Water Consumption and Response. Academic Press, New York. p. 191-254.

Zelitch, I. 1969. Stomatal control. Ann. Rev. Plant Physiol. 20:329-350.

Zhang, J. and W. J. Davies. 1989. Abscisic acid produced in dehydrating roots may enable the plant to measure the water status of the soil. Plant Soil Environ. 12:73-81.

Effects of Nutrient Stress

J. O. Klemmedson and A. R. Tiedemann

Introduction

Basic to an understanding of the effects of nutrient stress on morphology and physiology of rangeland plants is an understanding of the status of nutrients in rangeland ecosystems. Of the 20 elements essential for plant growth (Tisdale et al. 1985), the soil supplies all the plant's needs for 14 of these (P, K, Ca, Mg, Cu, Zn, B, Fe, Mn, Mo, Co, Na, V, Si), and except for leguminous plants, all of the needs for N. The other five elements (C, H, O, S and Cl) are at least partially supplied by the atmosphere (Fried and Broeshart 1967). Our attention here is focused primarily on nutrients supplied by the soil.

Soil-Forming Processes and Factors

The amount and availability of any soil nutrient is primarily determined by the soil-forming (pedogenic) processes (Buol et al. 1989) and the soil-forming factors (Jenny 1980). All soil-forming processes operate to some degree in all soils (Pritchett 1979). The relative importance of any process for any soil is largely a function of climate. Because rangelands occupy arid to subhumid regions, soil-forming processes favored by those climates, e.g. calcification, salinization and alkalinization, are relatively important in determining amount and supply of nutrients, while processes of humid regions, e.g. podzolization and laterization, are less important. As climate changes across a region so will the intensity of eluviation and illuviation, i.e. processes involving translocation and accumulation of soil constituents in the soil mantle.

Soil-forming processes concern additions to the soil, transformations and transfers within the soil, and removals from the soil (Buol et al. 1989). Soil-forming factors regulate the magnitude, rate, and intensity of soil-forming processes, i.e. they define the state of the soil system. Our knowledge of the soil-forming factors began with Dokuchaev (1898) and Hilgard (1914), but development to its current status can be attributed largely to Jenny (1941, 1980). His soil-forming factor (state factor) equation

$$S, s = f(cl,o,r,p,t \ldots) \quad (1)$$

Klemmedson is Professor Emeritus, School of Renewable Natural Resources, University of Arizona, Tucson, AZ 85721, and Tiedemann is Chief Ecologist (Retired), Pacific Northwest Forest and Range Experiment Station, La Grande, OR 97850.

states that the soil (S), or any soil property (s), is a function of regional climate (cl), the biotic factor (o), topography (r), parent material (p) and time of soil formation (t). Jenny (1941) defined the five soil-forming factors within the parentheses of equation (1) as independent variables.

Climate (cl) is the most important factor in determining the properties of soils (Pritchett 1979, Birkeland 1984). Climate determines plant species composition, amount of biomass produced, and microbial activity, and hence governs organic matter content of soils (Campbell 1978). Moisture (precipitation) and temperature are the dominant sub-factors of climate in terms of its effect on soil properties. Their influence on nutrients and soil fertility is most dramatic for nutrients associated with organic matter (i.e. C, N, P, S). This is illustrated in a surface model (Fig. 1) depicting the combined effect of moisture (m) and temperature (T) on the accumulation of soil nitrogen (Jenny 1980). Although the equation in Figure 1 was developed for the Great Plains area of the United States, the observed N-temperature and N-moisture trends are similar to those for other parts of the world (Jenny 1980). Moreover, organic carbon-climate functions will be very similar to the N-functions (Jenny 1930). Each point on these curves is a reflection of the combined effect of the production of biomass, its consumption by decomposer organisms, and the soil's capacity for retention of humus (Jenny 1980).

The biotic factor (o) is second most important and the most misunderstood factor in Jenny's equation. It is commonly confused with vegetation (Jenny 1958). From the broadest viewpoint, the biotic factor comprises all species present or able to migrate into the ecosystem, including humans. It is the potential flora and fauna.

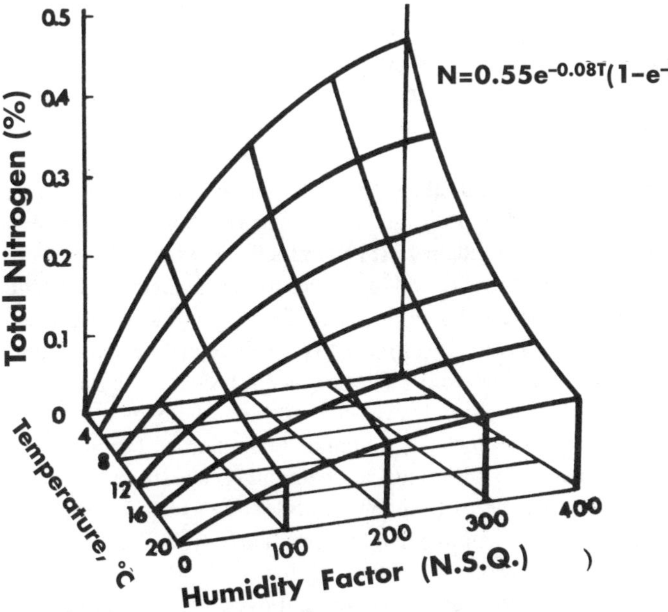

Figure 1. Nitrogen content of loamy grassland soils of the U.S. Great Plains as a function of mean annual temperature and the annual humidity factor (N.S.Q.); from Jenny 1980.

Like soil (S) and soil properties (s), vegetation (V) and vegetation properties (v) are dependent variables and are functionally related to the the biotic factor and other vegetation-conditioning variables (Major 1951), which are identical to the soil-forming factors. This is apparent in Jenny's (1961) statement of a more comprehensive state factor equation

$$l, s, v, a = f(cl, o, r, p, t ...) \quad (2)$$

where a = animal properties and l = s + v + a or any ecosystem property. Influence of the biotic factor is best expressed in the horizontal pattern of soil properties of shrub deserts (Charley and West 1975, West and Klemmedson 1978). These patterns, which develop through nutrient cycling processes and shading by plants, are graphically depicted in Figure 2. They are especially evident when species comprising a plant community differ markedly in their affinity for uptake of specific elements (Fireman and Hayward 1952, Sharma and Tonway 1973) as portrayed in Figure 3, or when N-fixing species are present, thus creating "islands of fertility" (Garcia-Moya and McKell 1970).

Humans (o_h) are often overlooked as part of the biotic factor. Their impact on nutrients of croplands has been reviewed by Jenny (1941), Stevenson (1965) and Campbell (1978). Likewise, man influences nutrients of rangelands by burning, shrub removal, livestock grazing and other practices. Klemmedson and Wienhold (1991) have recently described the effects of biotic and topographic factors on N and P status in an Arizona chaparral system. To the extent that humans have influenced the fire and erosion regime of this system, they have contributed to a level of productivity limited by low supply of P, an element that is not readily replaced once lost from the soil-plant system. In this same system, N, which is commonly the most limiting nutrient, is well supplied, partly through symbiotic N-fixation by shrubs. For continued productivity of this system, managers (o_h) may want to put high priority on management practices geared to maintain P on the site.

Topography (t) affects the amount and distribution of nutrients through its influence on moisture, temperature and radiation intensity at the soil surface, runoff, and erosion (Jenny 1980, Birkeland 1984). Differential effects of north and south aspects or ridge and valley positions on soil and vegetation, even over short distances, can be striking (Aandahl 1948, Klemmedson 1964).

Parent material (p) influences soil properties to varying degrees depending on the property and kind of parent material (Jenny 1980). Its influence is greatest in drier regions and young soils and may be masked entirely in mature soils of humid zones (Birkeland 1984). Soils derived from basic igneous rocks (e.g. basalt, andesite) and their sedimentary and metamorphic counterparts are usually more fertile than those derived from acid igneous rocks (e.g. granite, granodiorite, rhyolite) or their counterparts. Parent materials with unusual chemical composition (i.e. deficiency of nutrients and/or presence of toxic elements), such as serpentine, may lead to highly stressed nutrient conditions, low productivity, and endemic vegetation (Walker 1954, Jenny 1980).

Time (t) for full development of soil properties is highly variable. It may range from less than 100 years to about 1000 years for organic matter and N to reach a steady state (Crocker and Major 1955, Olsen 1958, Hall et al. 1982), depending on

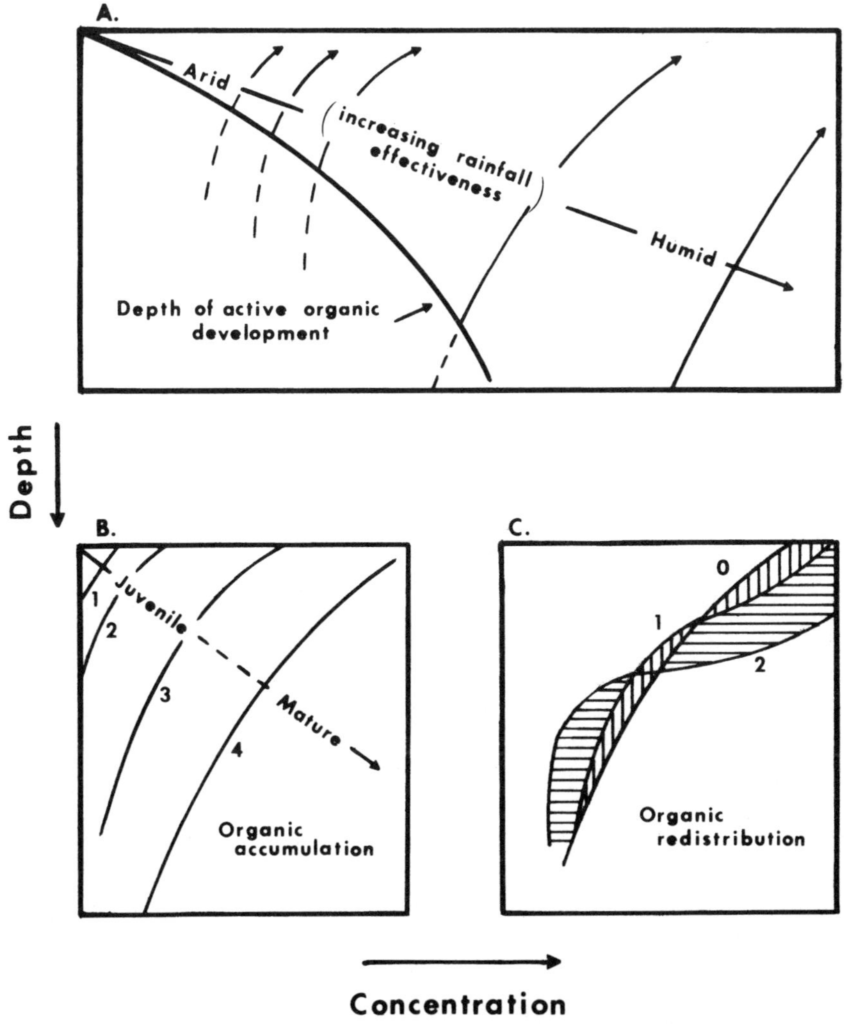

Figure 2. Idealized sequences of soil organic matter accumulation in relation to (A) rainfall effectiveness, (B) maturity, and (C) the localizing influence of perennial shrubs of increasing age or size (0 to 2); from Charley and West 1975.

climate, parent material, and other factors. Completion of trends in Fe, Al and P accumulation, and development of argillic horizons, which strongly influence nutrient status, may take up to several hundred thousand years (Birkeland 1984).

Plant Nutrition and Response to Nutrient Stress

In this paper, we describe nutrient stress as the condition of limited growth or reproduction of plants associated with external constraints limiting nutrient acquisition (Grime 1977, 1989). This definition distinguishes stress from disturbance and competition (Grime 1977), each of which also may limit plant biomass

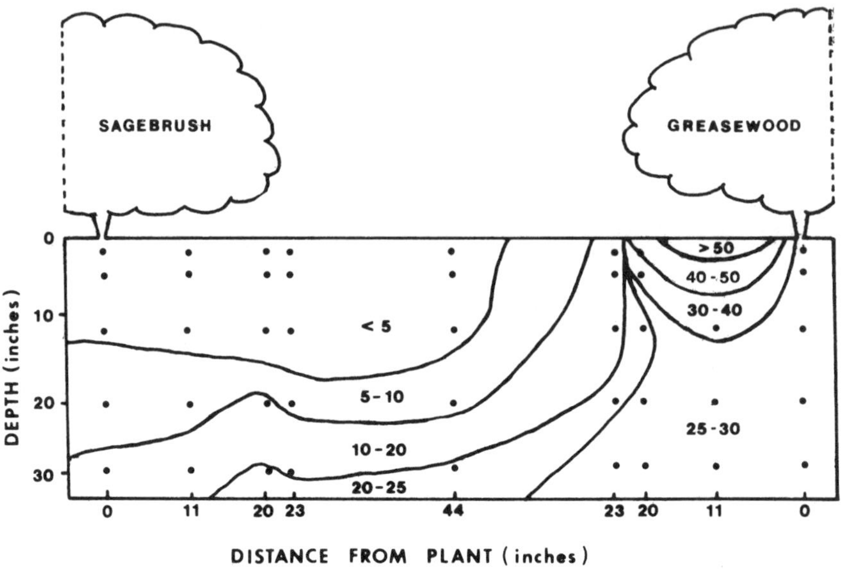

Figure 3. Sagebrush *Artemisia tridentata* Nutt.) and greasewood (*Sarcobatus vermiculatus* [Hook.] Torr.) display contrasting patterns of sodium accumulation (exchangeable sodium percentage) compared with intervening barren areas in the Utah desert; after Fireman and Hayward 1952.

accumulation and reproduction, but which are not considered here. By implication, Grime (1977) and other authors (e.g. Chapin 1980) appear to restrict nutrient stress to deficiency conditions only.

An understanding of nutrient stress in range plants requires an understanding of the nutrition of plants and how they respond and adapt to nutrient stress. Broadly speaking, plant nutrition involves the processes of absorption and allocation of nutrients by plants. The basic considerations of nutrient absorption and allocation, including factors governing concentration of the soil solution, nutrient mobility, soil-root interactions, and root and shoot characteristics, have been briefly reviewed by Chapin (1980). Greater detail can be found in Fried and Broeshart (1967), Nye and Tinker (1977) and Marschner (1986).

In response to low plant nutrient status resulting from low availability of soil nutrients, nutrients are preferentially allocated to root growth (Chapin 1980, Marschner 1986). Thus, nutrient deficiency usually leads to higher root:shoot ratios. McIntyre (1964) reported that nitrogen supply is an especially significant morphogenic factor in bud development and the balance between rhizome and shoot production. In quackgrass (*Agropyron repens* (L.) Beauv.) with low N status, 90% of the buds in the axils of the first leaf developed into rhizomes and only 5% into tillers; with high N status the proportion of rhizomes and tillers was largely reversed. Once in the plant, mobile nutrients move preferentially to regions of greatest meristematic activity (Langer 1966, McIntyre 1977). Reduced nutrient status has a greater impact on size and number of lateral shoots and tillers than on leaf area in the apical shoot. McIntyre (1977) reported that apical dominance can be precisely controlled by varying N supply; P is also effective in this role.

When nutrients in the rhizosphere become deficient, nutrients for continued growth of meristems and new leaves are withdrawn from older or basal leaves with resulting senescence (Williams 1955). This is readily observed in pot experiments where a deficient nutrient is excluded from an otherwise complete nutrient treatment; in cereals or grasses the lower leaves may senesce well before plant maturity. The deficient nutrients are withdrawn from inorganic storage pools or readily hydrolyzable compounds, usually at the expense of photosynthetic rate (Chapin 1980).

For many species, a strong correlation exists between leaf nitrogen concentration and photosynthesis (Evans 1989); nitrogen is a component of photosynthetic enzymes and chlorophyll (Chapin 1980). Loss of nutrients from senescing tissues during retranslocation varies from as much as 90% (Chapin, 1980) for highly mobile nutrients (K, N, P) to none for Ca (Van Den Driessche 1984).

If the supply of nutrients is still inadequate to maintain meristematic growth following depletion of nutrients from soil and senesced leaves, plants exhibit visible deficiency symptoms (Walker and Gessel 1990) as well as greatly reduced nutrient absorption, photosynthesis, and respiration (Chapin 1980).

Plant Adaptations to Nutrient Stress

Plant adaptations to stress have been well documented for many physical factors (moisture, temperature, light) and some soil factors, such as salinity and alkalinity, that limit plant growth (Levitt 1980). Although adaptations that enable plants to grow and survive in nutrient-limited situations are not as well studied, they are an important part of the ability of plants to survive stressful environments.

Among three adaptive plant strategies described by Grime (1977), the strategy of stress-tolerant plants, i.e. those associated with environments of high stress and low disturbance, is of interest here. Although emphasis here will be on nutrient stress that arises from the impoverished nature of habitats, it should be recognized that severe nutrient stress also can arise when nutrients are sequestered in the living and nonliving parts of biomass (Grime 1977). Not only are the sequestered nutrients, especially those in aboveground biomass, temporarily removed from the cycling pool of nutrients, but they are vulnerable to loss via catastrophic events and man's activities (Walker and Gessel 1990), e.g. fire (Klemmedson 1976) and livestock grazing.

Plant Morphology. A common feature of plants across the entire range of nutrient-deficient soils is similar plant morphology (Grime 1977). Among grasses, narrow-leaved, creeping, and rosette forms predominate. Where woody vegetation prevails, the common dominants are trees (often coniferous) and sclerophyllous shrubs. Most species of the stress-tolerant strategy are characterized by small stature (Grime and Hunt 1975). The plants of serpentine soils, noted for their extreme infertility, possess many distinguishing morphological characteristics (Walker 1954, Proctor and Woodell 1975); most notable are reduction in size of leaves and other organs, shrubbiness of growth, and stunting.

Growth Rate. An inherently low rate of growth is characteristic of species found on nutrient-deficient soils (Grime 1977, Chapin 1980), and with few exceptions, is associated with small stature in herbaceous species (Grime and Hunt 1975). Low growth rate is usually accompanied by low rates of nutrient absorption. A good illustration of this adaptive feature is provided by Clarkson (1967). He attributed the ability of *Agrostis setacea* Curt. to maintain itself on soils of low P availability to inherently slow growth rate which placed a low demand on P supply. Other examples of this adaptation of plants to nutrient stress are described by Beadle (1954) for phosphate-deficient soils in Australia and Kruckeburg (1954) and Huenneke et al. (1990) for serpentine soils.

Grime and Hunt (1975) suggested three ways in which low growth rate can be adaptive to conditions of stress: (a) slow-growing plants make modest demands and hence are less likely to exhaust available resources of their habitat, (b) low rates of incorporation of photosynthate and mineral nutrients into structure may permit build-up of plant reserves, and (c) under conditions of protracted and extreme stress in which no growth is possible, species of low relative growth rate may be better fitted to survive.

Nutrient Acquisition. Although species from soils of low fertility exhibit low nutrient absorption rates, they maximize nutrient acquisition by several means: a) maintaining a large root biomass, hence high root:shoot ratio (Chapin 1980, Marschner 1986), b) increased length (Marschner 1986) and branching (Troughton 1980) of roots, c) strongly developed mycorrhizal associations (Mosse 1973) and d) greater root longevity (Chapin 1980). Thus, plants from infertile sites survive not because of adaptations for enhanced nutrient absorption, but because of adaptations that compensate for low nutrient absorption (Chapin 1980). Low nitrogen, and to a lesser extent phosphorus supply, especially during early growth, leads to longer roots and root hairs, hence increase in the volume of soil explored. Roots also become finer (Marschner 1986).

Growth of many range plants is influenced by association of the roots with mycorrhizae. Most rangeland species have vesicular-arbuscular endomycorrhizal associations (Trappe 1981) as contrasted to trees and shrubs with predominance of ectomycorrhizal associations (Grime 1977). Mycorrhizal fungi are essential for many vascular plants for absorption of phosphate and other nutrients from the soil (Molina et al. 1978, Grime 1977). Mycorrhizae seem especially important for plants growing on nutrient-poor soils or in extremely stressful habitats, such as sand dunes. Chapin (1980) concluded that mycorrhizae provide the greatest benefit to plants where nutrient limitation is a result of slow diffusion in the soil. The order of benefit is generally phosphate > ammonium > potassium > nitrate, i.e. from least to most mobile. Chapin (1980) reported mycorrhizal associations achieve their best development on infertile sites, but Reeves et al. (1979) found that vegetation on highly disturbed sites in primary succession was dominated by non-mycorrhizal species. Grime (1977) concluded that "the main benefit of a mycorrhizal association to a host plant growing under conditions of mineral nutrient stress may be an absorptive system which, because it remains functional over a long time, allows exploitation of temporary periods of mineral nutrient availability at lower cost to the synthetic resources of the plant."

Plants can directly alter nutrient availability in the rhizosphere in several ways (Chapin 1980, Marschner et al. 1986), but there is little evidence that species from infertile soils are more effective than other species in this respect (Chapin 1980). Some of these adaptive mechanisms of plants are (a) changes in pH of the rhizosphere, (b) mobilization and immobilization of mineral elements by root exudates, and (c) symbiotic and rhizosphere N_2-fixation (Chapin 1980, Marschner et al. 1986). Although root exudates can affect rhizosphere pH, the most marked pH change is caused by the cation/anion absorption balance, especially involving nitrate and ammonium supply. Ammonium supply is usually correlated with preferential cation uptake, hence higher net excretion of H^+ compared with HCO_{3-} or OH^-, while nitrate supply causes the reverse (Marschner et al. 1986). In legumes, N_2-fixation has been shown to affect the cation/anion uptake ratio and hence rhizosphere pH (Aguilar and van Diest 1981, Marschner et al. 1986). The effect is similar to that of ammonium uptake. Each of the three adaptive mechanisms mentioned above can be expected to increase absorption of iron and phosphorus (Marschner et al. 1986), both of which are relatively insoluble at the pH expected in arid and semiarid soils.

Numerous examples exist of biological N_2-fixation, both symbiotic and nonsymbiotic, in range vegetation. By far the greatest amount of N_2-fixation is attributed to the symbiotic association of bacteria of the genus *Rhizobium* with members of the Fabaceae family (Hanes 1986). Many genera of this plant family are important range plants; *Lupinus*, *Lathyrus*, *Medicago*, *Astragalus*, *Thermopsis*, *Trifolium*, and *Vicia* are only a few of the genera known to have effective N_2-symbiosis with *Rhizobium* (Vlassak et al. 1973, D.H. McNabb, Oreg. State Univ., Corvallis, unpublished report). Actinorhizal plants, especially species of *Ceanothus*, *Cercocarpus*, and *Purshia*, are important components of shrub-dominated vegetal types of the western U.S. and undoubtedly contribute to the nitrogen economy of western ranges (Klemmedson 1979). The adaptability of species of these and other actinorhizal genera to arid, infertile soils and often to pioneer successional stages, is widely documented (Hellmers and Kelleher 1959, Beasley and Klemmedson 1976, Klemmedson 1979, Hanes 1986).

The blue-green algae (Cyanobacteria), though often overlooked, are an important component of the vegetation in nutrient-stressed environments, especially as pioneering plants on virgin landscapes (Stevenson 1986). They are widely distributed around the world; on rangelands, especially more arid regions where densities of legumes are low, cryptogamic crusts may be the major N fixers (West and Skujins 1977). Although little data exists, N_2-fixation in sand grain rhizosheaths (Fig. 4) seems to be a mechanism permitting xeric grasses to inhabit sites with very low soil N levels (Wullstein et al. 1979). Acetylene reduction ranging from 515 to 920 nmol C_2H_4/ g (dry wt.) for a 6-day period was reported for whole plant specimens of Indian ricegrass (*Oryzopsis hymenoides* (Roem. & Schult.) Ricker), thickspike wheatgrass (*Agropyron dasystachyum* (Hook.) Scrib.), needle and thread grass (*Stipa comata* Trin. & Rupr.) and purple threeawn (*Aristida purpurea* Nutt.) collected from sand dunes in west-central Utah.

Root Exchange Capacity. The cation exchange capacity (CEC) of roots influences the effectiveness of plants in the uptake of cations from the soil. Cation exchange capacity differs between monocotyledonous and dicotyledonous plants (Heintze 1961, Crooke and Knight 1962, Crooke 1964, Woodward et al. 1984). In

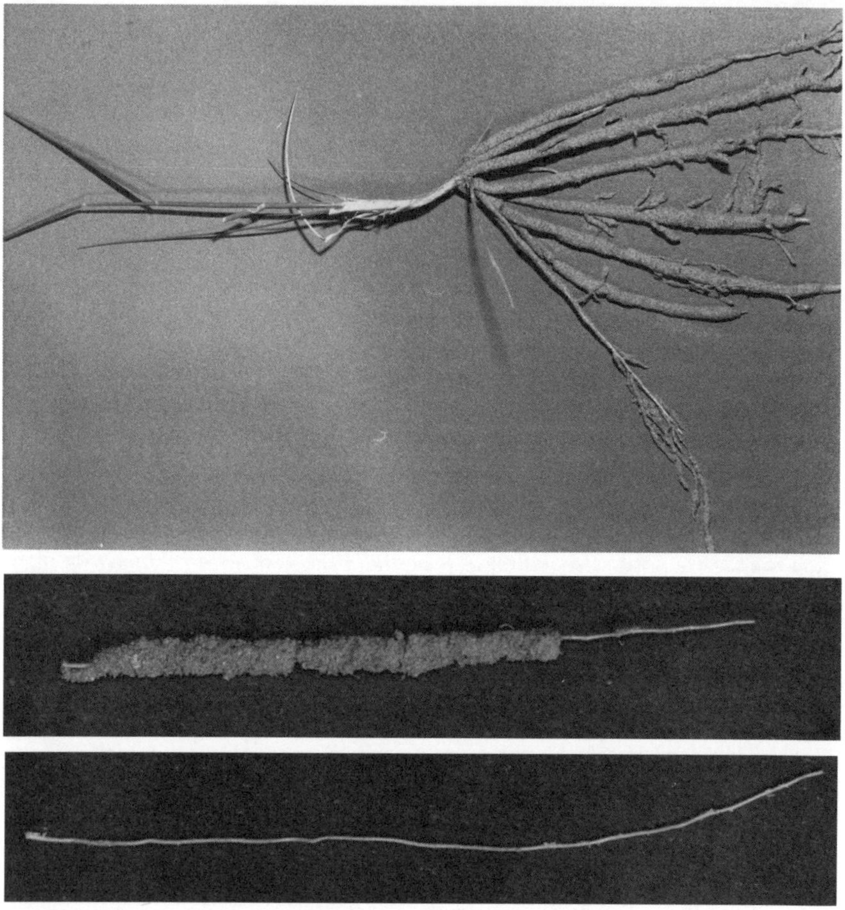

Figure 4. Root system of thickspike wheatgrass (*Agropyron dasystachum* [Hook.] Scribn.) covered with rhizosheaths and individual roots of Indian ricegrass (*Oryzopsis hymenoides* [Roem and Schult.] Ricker.) showing sand grain root sheath intact and root with sheath removed. Photographs furnished by L. H. Wullstein.

a study of Utah range plants, Woodward et al. (1984) demonstrated that root CEC values were significantly greater for dicots than for monocots. The relative amounts of monovalent and divalent cations taken up were strongly correlated with root CEC. Dicots tended to take up divalent cations more efficiently than monocots; the reverse was true for monocots. This relationship not only helps explain differential distribution of grasses and shrubs in wildlands but also may be an important consideration for management following revegetation. For example, on range sites deficient in magnesium, but having adequate potassium, plant species (dicots) with large root CEC values may have a competitive advantage. Species with low root CEC (monocots) would likely perform poorly on such sites. To achieve a balanced mixture of grasses and shrubs in this situation, supplements of magnesium may be required. Another option would be to revegetate with shrub varieties with low CEC values and adapted grasses with high CEC values.

Nutrient Storage and Retranslocation. Remobilization (retranslocation) of nutrients is important for a number of plant functions (Marschner 1986). On infertile habitats where perennial species tend to dominate (Grime and Hunt 1975), seasonal storage and retranslocation of nutrients takes on added significance (Chapin 1980, Vitousek 1982). In perennials, internal recycling of nutrients during senescence of leaves and other plant organs permits reuse of the same unit of nutrient (Clark 1977, Turner 1977), thus increasing efficiency of nutrient use. In the past, the role of retranslocation has been largely ignored as a source of N for new plant growth Clark (1977). Transfer of nutrients, especially nitrogen and phosphorus, from root and stem reserves may be more dependable than scarce soil nutrients during rapid spring growth. These reserves are gradually replenished by absorption from soil during summer and transfer from senescing leaves during fall (Chapin 1980). The concept of nutrient use efficiency would seem to have special significance to rangelands, as in cropland agriculture where the extent of remobilization of mineral nutrients is attracting much attention in the selection and breeding of genotypes for nutrient efficiency (Marschner 1986). This is consistent with a shift from high-input approaches (e.g. fertilization) in agricultural technology toward fitting the plants to the soils (Marschner et al. 1986). Despite differences in how nutrient efficiency is defined (Chapin 1980, Vitousek 1982, Graham 1984) and differences in plants native to fertile and infertile habitats, the concept has relevance for rangelands. From a study of forest vegetation, Vitousek (1982) concluded that nitrogen use efficiency varies as an inverse function of nitrogen availability (i.e. data suggested that on sites with low nitrogen availability, retranslocation during senescence increased, leading to lower nitrogen concentration of litterfall).

Photosynthetic Adaptations. Plants with C_4 photosynthesis have been shown to have higher efficiency for nitrogen use than C_3 plants (Marschner 1986), thus suggesting this may be an adaptation to nutrient stress. Studies of the relations between soil nutrient levels and C_3/C_4 plant distribution indicate that C_4 species are more abundant where soils are of moderate to low fertility (Barnes et al. 1983, White 1971). However, Barnes et al. (1983) conclude this may be fortuitous and a result of other interrelated factors such as soil moisture and internal soil drainage. Research by Sage and Pearcy (1987) showed that best growth and apparent competitive ability of *Amaranthus retroflexus* L. (C_4) compared to *Chenopodium album* L. (C_3) was at high nitrogen levels. At low N levels, *C. album* L. held the competitive advantage. At this point, relationships between C_4 and C_3 plants with respect to a nutrient stress advantage has not been firmly established. The importance of soil nutrients in determining the distribution of C_3 and C_4 plants needs to be examined more thoroughly (Barnes et al. 1983).

Reproductive Strategies. Some dioecious plant species such as fourwing saltbush (*Atriplex canescens* (Pursh) Nutt.) have been shown to differentially distribute male and female plants along environmental stress gradients (Freeman et al. 1976). Male plants, because of smaller investment of resources to reproduction than female plants, occur in greater abundance on xeric sites than do female plants. Although moisture appears to be the principal driving force in this distribution, nutrient gradients tend to follow moisture gradients—thus, xeric sites would be expected to be lower in nutrients than mesic sites.

Freeman et al. (1980) developed general relationships for the response of plant sexuality to a large array of plant physiological parameters and environmental conditions. Higher levels of soil nutrition (i.e. naturally rich or fertilized soil) generally resulted in a shift from maleness to femaleness for vascular plants (both monoecious and dioecious). Nutrient requirements for females may be inferred to be greater than those for males because they may allocate more of their resources such as biomass, calories, nitrogen, and phosphorus to reproductive tissues. Wallace and Rundel (1979) found this to be the case with *Simmondsia chinensis* (Link) Schneid. Male plants allocated from 10 to 15% of their resources to reproductive tissues, whereas female plants allocated 30 to 40% to reproduction. These findings may support a hypothesis of nutrient advantage for male plants on nutrient-limited sites because of their allocation strategy. But, to date no relationship between the relative proportions of male to female plants and nutrient stress gradients in soils has been established.

Plant Succession

That successional dynamics may be closely tied to soil nutrient status is not a new idea (Tilman 1986). Sampson's (1952) lucid description of primary succession as the simultaneous development of soil and vegetation embodies the early ideas of Cowles (1899) and Clements (1916) and implies the gradual accumulation of organic matter and mineral nutrients with seral development. Numerous studies since (Crocker and Major 1955, Olson 1958, Dickson and Crocker 1953abc) have described characteristic patterns of vegetal development in relation to nutrient status. These studies illustrate the role of the time factor in soil and vegetation development.

In secondary succession, initial colonization following disturbance is ordinarily by species which are able to survive because they have low nutrient requirements or, in the case of leguminous species, are able to offset a short supply of nitrogen (Lloyd and Pigott 1967, Tilman 1986). In young systems, where nutrient reserves are yet small, the vegetation may soon return to closely resemble its original state following disturbance (Walker et al. 1981). In old systems where disturbance has significantly depleted the nutrient capital, vegetation will likewise decline, to be reversed only by an input of nutrients. This situation is illustrated by long-term overgrazing in the Wasatch subalpine, which resulted in extensive erosion and loss of the original plant community (Sampson and Weyl 1918, Sampson 1919). Site potential appears to have been reduced significantly because of loss of most of the A-horizon. Recovery of this system has apparently stabilized with a mid-seral plant community (Ellison 1954, Klemmedson and Tiedemann, 1994). Soil phosphorus or sulfur is the likely key to restoration of soil organic pools and further development of the vegetation in the Wasatch subalpine (Walker 1965, Walker and Syers 1976).

Effects of Management

Management can influence nutrient stress in several ways. Total amounts of nutrients, quantities of available nutrients, plant/soil nutrient cycling, and transfer off-site may all be influenced by management. In this section, our intent is to focus on major management activities that influence nutrient status of rangeland ecosystems with emphasis on ways to ameliorate nutrient stress. The primary management activities on rangelands that influence the status and distribution of nutrients are grazing, fire, and habitat conversion. Fertilization and plant selection (i.e. matching plants to soils) are the primary means of ameliorating or coping with nutrient stress (Marschner 1986).

Grazing and Browsing. In natural, undisturbed wildlands, nutrients are accumulated, cycled, and lost from the soil-plant system in an orderly, relatively predictable manner. Grazing or browsing can exert a pronounced effect on nutrient cycling, distribution, and availability. The primary effect of grazing is alteration of cycling pathways and added complexity to nutrient cycles (Fig. 5). Grazing also tends to accelerate and alter the timing of nutrient transfers, and increase amounts of nutrients cycled from plants to soil (Chapin et al. 1986, Pieper 1977). Green leaves consumed by herbivores contain higher concentrations of nutrients than leaves returned to the soil as litter because the latter have suffered nutrient loss via retranslocation and leaching during senescence (Clark 1977, Chapin et al. 1986).

Freely grazing livestock are not uniform distributors of nutrients (Peterson et al. 1956). In sloping topography, grazing results in transfer of nutrients from ridges and slopes to gentler slopes and valley bottoms where animals prefer to graze and where they rest and water (Mueggler 1965, Gillingham 1982). Hence, large amounts of nutrients may be redistributed to relatively small areas. For sheep grazing in pastures, Hilder (1964) concludes that up to 30% of the fecal pellets may be deposited on only 5% of the area. Thus, grazing accentuates the already unequal distribution of nutrients that results because of topography (Jenny, 1980).

Nutrients in animal excreta are deposited in a form that is more readily available to plants than if cycling had occurred directly from plants to soil (Floate 1970). Nitrogen in urine is readily converted by an enzyme (urease) in the soil to ammonia and ammonium-N, a form readily used by vascular plants. Nutrients in animal excreta are differentially distributed between urine and feces. Most of the N, K and S are concentrated in the urine whereas P, Ca, and Mg are higher in feces (Crampton and Lloyd 1959, Maynard and Loosli 1962, Peiper 1977). For N and P, this is of reverse benefit to plants. Nitrogen is more mobile than P and urine deposits cover about 3X greater area than feces (Peterson et al. 1956). Although P in feces is highly soluble initially, it becomes immobilized and largely unavailable as feces dry (Blair et al. 1975).

Nitrogen may be lost from the soil-plant system in excess of that which occurs in the absence of grazing or browsing animals, because of the susceptibility of excreted N to volatilization and leaching (Watson and Lapins 1969). According to Ball et al. (1979), these mechanisms may account for up to 35% of N deposited as urine.

Grazing usually results in increased export of nutrients from the system. Animals retain about 20% of the nutrients in the forage they consume (Dean et al. 1975). This represents a permanent loss to the system when livestock or wildlife are removed. Woodmansee (1978) concludes that N may accumulate under light grazing; with moderate grazing a balance may occur and at higher grazing intensities, actual losses of N may occur. With prolonged N depletion via this route, the organic pool may act as a temporary buffer, but that pool is finite and will diminish quickly

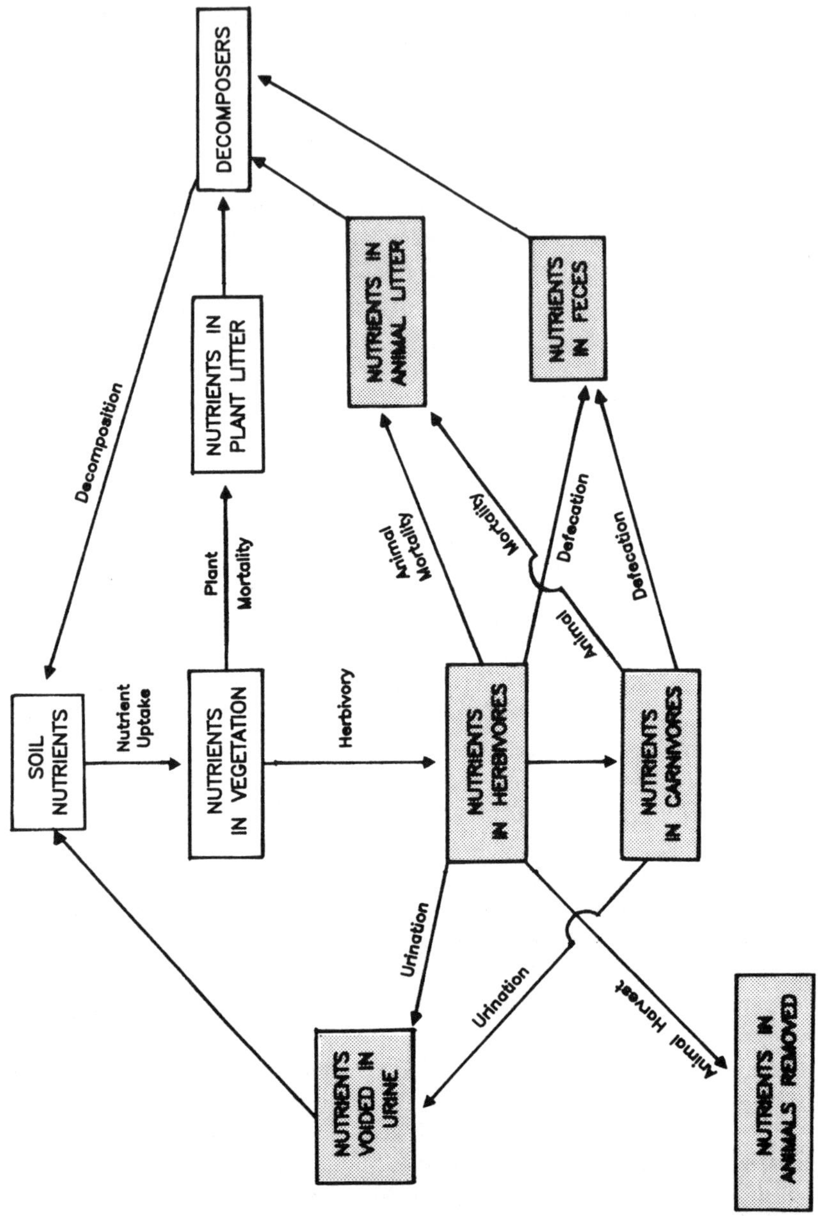

Figure 5. Grazing alters pathways and adds complexity to nutrient cycles of this idealized range ecosystem. Shaded boxes and related pathways are associated with grazing animals; after Pieper 1977.

if the system is unstable, e.g. accelerated erosion is occurring. Protracted depletion of the organic pool will eventually lead to reduced site productivity.

Erosion is potentially the most serious export loss to range ecosystems, especially for N and P (Jawson et al. 1982, Klemmedson and Wienhold 1991). On well-managed rangelands with stable soils and for nutrients with mechanisms for rapid replacement (e.g. nitrogen), losses via erosion are probably balanced by inputs to the system (via several means of N fixation) over relatively short periods of time (Stevenson 1986). However, on unstable rangelands where erosion is or has been severe and for nutrients, such as phosphorus, which are replaced very slowly, chiefly by weathering (Smeck 1985, Stevenson 1986), management should adopt strategies to stabilize the soil-plant system and preserve or rebuild critical nutrients. Present and future productivity may hinge on how limiting nutrients are managed (Klemmedson and Wienhold, 1991).

Fire. Fire, whether by prescription or wildfire, is a rapid, almost instantaneous, process of mineralizing nutrients in plant biomass and litter (Tiedemann 1981). Depending on conditions at the time of burning (i.e. temperature, moisture content of soil, plants and litter, and wind), there are several possible pathways for nutrients contained in ecosystem components that are combusted or exposed to elevated temperatures of fire (Fig. 6). Some nutrients, notably N, S, P and K, are volatilized in varying degrees and lost to the atmosphere during combustion (Evans and Allen 1971; DeBano and Conrad 1978, Wright and Bailey 1982, Raison et al. 1985, Tiedemann 1987b, DeBano and Klopatek 1988). Among the nutrients, N and S are most susceptible to volatilization (Wells et al. 1979, Boerner 1982). Nitrogen losses during combustion from common heather (*Calluna vulgaris* (L.) Hull) were 42 and 57% of the total N in unburned plant material at 310°C and 750°C, respectively (Evans and Allen 1971). Losses of N as high as 88% were reported for piled slash and forest floor burned in ponderosa pine (*Pinus ponderosa* Dougl. ex Laws.) forests (Klemmedson et al. 1962, Klemmedson 1976).

Based on several studies, Raison et al. (1985) reported the proportion of N and P lost from plant material during combustion increased linearly with percentage of fuel weight lost during combustion. For N, the correlation was high and the slope of the regression approached unity. Moreover, N losses were almost entirely nonparticulate. Recent research by Lobert et al. (1990) has shown that as much as 50% of the N lost during combustion of biomass is lost as N_2. Although loss of P can reach 60%, it depends on many factors. DeBano and Klopatek (1988) reported a 50% loss of total P in litter from burning pinyon-juniper (*Pinus edulis* E./*Juniperus osteosperma* (Torrey) Little). Nonparticulate loss of P may range between 30 and 50% of total losses and are of particular ecological significance (Raison et al. 1985). Whereas N is readily replaced from its atmospheric pool by several mechanisms of fixation, P, once lost is only slowly replaced by weathering of primary apatites from soil-forming rocks (Smeck 1985, Stevenson 1986).

Wright and Bailey (1982) cited losses of total N ranging from 30 kg/ha for grasslands to 907 kg/ha for heavy slash burns. Tiedemann (1987b) reported potential losses of S ranging from 24 to 79% at temperatures of 375 to 575°C.

Losses of metallic cations such as Ca, Mg, K and Na are usually small (5 to 20%) in burning of vegetation (Evans and Allen 1971) and occur largely as particulates in the smoke column, depending on burning and atmospheric conditions (Raison et al. 1985). Because of higher volatility, loss of K and Na may be greater and with a higher proportion as nonparticulates, especially if temperatures are elevated (Grier 1975, Tiedemann 1981, Raison et al.1985).

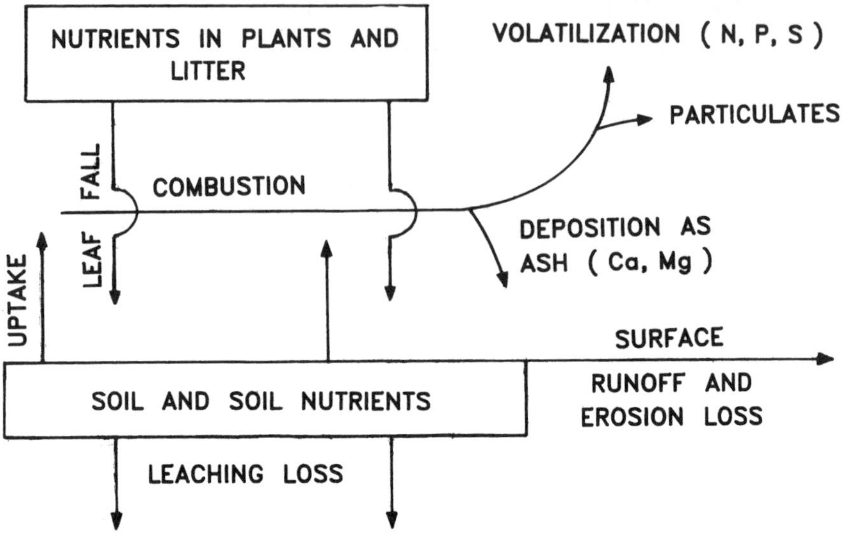

Figure 6. Possible pathways of nutrients in soil, plants and litter combusted or exposed to elevated temperature during fire; after Tiedemann 1981.

Elements that remain in the ash are largely converted to oxides and deposited at the surface as ash. Here in the presence of CO_2 and H_2O of the atmosphere, these cations are converted to bicarbonate salts (Tiedemann 1981). Once the oxides are converted to the bicarbonate form, solubility increases and nutrients are then more easily transferred into the soil mantle. In the oxide form, losses occur chiefly from wind and gravitational redistribution of ash. In steep topography, nutrients in either form are vulnerable to loss from the site if torrential rains follow soon after catastrophic fires. To the extent that burning is a planned human activity (e.g. prescribed burning), control of nutrient losses by this mechanism is within the realm of management options.

Although rangeland burning usually results in a net loss of N and S, and to a lesser extent K and P from biomass through volatilization and particulate loss in smoke, changes in availability of soil nutrients associated with burning can positively influence nutrient status. Burning generally enhances the short-term supply of soil nitrogen (St. John and Rundel 1976, Dunn et al. 1979, Jurgensen et al. 1981, Wright and Bailey 1982). The immediate effect of burning is an increase in ammonium-N in the surface soil (Dunn et al. 1979, DeBano et al. 1979, Jurgensen et al. 1981, Covington et al. 1991). Nitrate-N increases are usually manifested after several days or weeks, depending on favorable conditions for nitrification (Christensen and Muller 1975, Dunn et al. 1979, Jurgensen et al. 1981) and may persist for three years (Covington et al. 1991). Fire favorably influences nitrifiers and other microorganisms through the fertilizing effect of nutrients deposited in the ash and higher temperatures in the darkened soil surface.

Available P levels also are enhanced by fire (St. John and Rundel 1976, DeBano and Klopatek 1988) but increases appear short-lived (less than 45 days) (DeBano and Klopatek 1988). Oxidation and subsequent conversions of metallic elements such as Ca, Mg, and K to bicarbonates described earlier results in enhanced availability of these elements (Ahlgren and Ahlgren 1960, Allen 1964, St. John and Rundel 1976, Wright and Bailey 1982).

When considering fire effects on nutrients, it is important to distinguish between total and available nutrients (both of which have been discussed above), and to consider these responses in terms of the entire nutrient cycle. To do otherwise may lead to faulty conclusions. Using N as an example, we can illustrate. Most studies have shown a short-term improvement in N availability following fire. Often this is coupled with a marked increase in biomass production (Harris and Covington 1983). If these results are considered alone, they might well be construed to suggest repeated burning at frequent intervals to maintain the short-term effects on productivity. However, this viewpoint is based only on effects of burning to the available pool of N. What was the source of the increase in available N that led to increased vegetal production? Intuitively, one would conclude the source was mineralization of easily decomposable organic N compounds (Clark, 1977), i.e. conversion of total N to available N. What is the response to total N? Probably a loss, based on the literature reviewed above.

This simple example points out the importance of evaluating nutrient response to fire using a mass balance approach. With this approach, one would determine the sources and rates of loss and gain to the soil-plant system for the nutrient(s) in question, as well as the rates of transfer between various compartments of the nutrient cycle. This information provides a more realistic basis for evaluating the effects of fire—or any other management tool—on the long-term productivity of an ecosystem. With this information, a manager can then say with confidence how frequently prescribed burns can be conducted without deminishing site productivity.

Another point is worth making in regard to the evaluation of nutrient losses to fire or other management effects. The total pool for most nutrients (e.g. C, N, P, S) consists of several fractions, some readily available and some very resistant (Paul 1970, Clark 1977, Campbell 1978). Based on turnover time, Paul (1970) identified three fractions of organic matter with mean residence time ranging from a few years for decomposing fresh residues and associated biomass to 250 to 1500 years for the resistant fraction. Unfortunately, our knowledge of the distribution of nutrients among labile and resistent fractions is very poor, e.g. for C and N (Klemmedson 1989). Jansson (1971) concluded that living active biomass comprised only a minor part of soil organic N. This suggests it may be unwise to dismiss seemingly small nutrient losses as unimportant when loss is expressed as percentage total nutrients. More conservative estimates of nutrients losses, especially in environments of nutrient stress, should be based on biomass nutrients rather than the entire pool of ecosystems nutrients (Klemmedson 1976).

Type Conversions. Eliminating vegetation such as shrubs or trees and supplanting them with herbaceous vegetation may result in large losses and shifts in distribution of nutrients. Native, undisturbed, grass-forb dominated and shrub-forb-grass ecosystems may have as little as 0.2 to 2.0 %, respectively, of the total ecosystem capital of N contained in aboveground biomass (Table 1). In shrub and tree systems such as pinyon-juniper, mesquite (*Prosopis juliflora* (Swartz) DC), and palo verde (*Cercidium floridum* Benth), percentages of total system N present as aboveground biomass N range from 15 to 26% (Barth and Klemmedson 1982, Tiedemann 1987a). Type conversions that involve burning, particularly involving pile and burn methods, have the potential for losses of N, P, S, and K through volatilization, and redistribution of P and cations such as Ca and Mg. The result of type conversions that cause large losses and disproportionate redistribution (i.e. into burn piles) of nutrients may be increased nutrient stress from the perspective of long-term site productivity.

Type conversions that result in a marked change in canopy shading of the soil surface, e.g. shrub or tree cover to grass in arid regions with little ground cover, could potentially lower soil organic C, N, P and S because of a change in the temperature at the soil surface (i.e. a change in the state of the system). This follows from the classic work of Jenny (1930), where he demonstrated the influence of precipitation and temperature on organic matter. Tiedemann and Klemmedson (1986) provide data supporting this hypothesis.

Fertilization. If nutrients are limiting, and the specific nutrients and degree of limitation are known, fertilization is probably the simplest procedure to ameliorate deficiencies. This option may be infeasible in some situations because of economic limitations. However, fertilization is gaining favor in wildland management for: improving emergence and survival of seeded species (Klock et al. 1975, Whalley and McKell 1976, Clary and Tiedemann 1984); increasing soil protection by enhanced foliar cover (Tiedemann 1983); improving plant nutritional quality (Duncan and Hylton 1970); managing use patterns of livestock and wildlife (Hanson and Smith 1970, Brown and Mandery 1962); and extending the duration of green forage (Hanson and Smith 1970).

Liebig's law of the minimum and the Mitscherlich's law (Stålfelt 1972) are seldom given appropriate consideration in decisions to fertilize. The basic concept of these related "laws" is that growth of a plant is dependent upon the amount of nutrient presented to it in minimum quantity. If more than a single nutrient is limited in supply, the addition of one alone may produce only small increases in growth compared to the growth resulting from addition of all nutrients that are limiting (Dean 1957). The addition of a single nutrient such as N to soil with marginal sufficiency of other nutrients may create a deficiency (nutrient stress) of one or more of these other nutrients.

Bioassay trials by Klemmedson and Ferguson (1973) and Klock et al. (1971) demonstrated the interaction of deficient elements. Data from Klock et al.(1971) showed that supplying N alone at three levels produced no yield response until a deficiency of S was satisfied. When S was added at a single rate, yields increased with each successive increase in N (Fig. 7).

In addition to nutrient stress that may result from fertilization, interactions between fertilization and moisture stress are common. Soil moisture reserves may be depleted because of rapid growth of annual species in response to fertilization (Sneva 1978, Wilson et al., 1966, and Kay and Evans 1965). Applied N increases competition between cheatgrass (*Bromus tectorum* L.), an annual, and native perennial species causing a decline in the perennials. The competitive ability of cheatgrass over bluebunch wheatgrass (*Agropyron spicatum* (Pursh)

Table 1. Distribution of dry matter and nitrogen in components of a low shrub/grass/forb ecosystem in central Waqshington, a sagebrush/grass ecosystem in southern Idaho and a pinyon-juniper ecosystem (composited data from several studies). Adapted from A. R. Tiedemann 1987a.

Component	Low Shrub/Grass/Forb			Sagebrush/Grass			Pinyon-Juniper		
	Dry matter	N Total	N Percent of total	Dry matter	N Total	N Percent of total	Dry matter	N Total	N Percent of total
	kg/ha	kg/ha	%	kg/ha	kg/ha	%	kg/ha	kg/ha	%
Standing crop									
Trees									
Boles							39000	133	1.7
Branches				10400	46	1.0	46000	184	2.3
Leaves				1000	13	0.3	12000	108	1.3
Dead				5100	29	0.6			
Subtotal				16500	88	1.9	97000	425	5.3
Herbage	600	6	0.1						
Roots	1200	12	0.3	16800	150	3.1	36000	30	0.3
Total	1800	18	0.4	33300	238	5.0	133000	455	5.6
Litter	3000	1.5	0.1	350	10	0.2	100000	1000	12.4
Soil		4200	99.5		4527	94.8		6615	82.0
Grand total	4800	4219.5	100.0	18650	4775	100.0	233000	8078	100.0

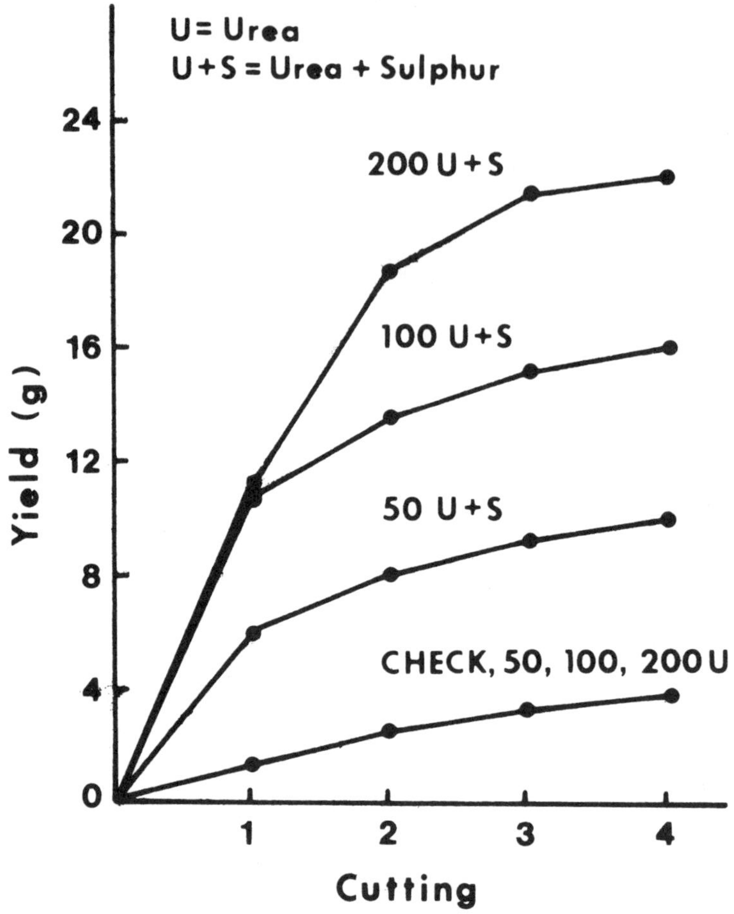

Figure 7. Cumulative pot yields of orchardgrass (*Dactylis glomerata* L.) in Tolo soil with control, urea without sulfur, and urea plus elemental sulfur treatments; after Klock et al. 1971.

Schribn. & Smith) for both moisture and nutrients is enhanced by its greater root elongation at soil temperatures as low as 2°C (Harris and Wilson 1970).

At present, decisions regarding range fertilization are focused largely on benefits expected from responses in forage production. However, economic assessment of fertilization probably should be based on multiple benefits such as soil stabilization, forage quality, improved animal distribution and wildlife habitat improvement. As a more complete understanding of the combined benefits of fertilization emerges, this practice may become a more viable option for managing nutrient stress in rangeland habitats.

Plant Selection and Development. Fertilization is referred to as a "high input" option by Marschner (1986). The "low input" option, and perhaps the most feasible means of managing nutrient stress, is to encourage the development of species that are either adapted to or can tolerate low nutrient levels, or those that can ameliorate N limitations by symbiotic N_2 fixation. Numerous genera of native plant species exhibit this trait (Klemmedson 1979). Spurred by the demands of intensive agriculture, research has progressed rapidly on genetic effects on transport, accumulation, and efficiency of nutrient use by plants and in development of species tolerant of nutrient stress (Vose 1983). Notable examples are work by Brown and Foy and their groups at the Plant Stress Laboratory in Beltsville, Maryland, and by Epstein and Lauchli at the University of California, Davis (Marschner 1986).

Plant selection and development is rapidly emerging as a viable option to achieve a wide range of wildland management goals (McArthur 1988). Finding plant species and cultivars that germinate and develop on nutrient-limited sites is one goal of this western-wide U.S. effort centered at the Shrub Sciences Laboratory in Provo, Utah. Its importance is indicated by the fact that 131 cultivars of range grasses have been released to date (McArthur 1988). Several forbs, shrubs, and half-shrubs have been released and numerous others are currently under study.

Acknowledgments

The authors are indebted to Justine McNeil for preparation of illustrations, to James L. Clayton and Leonard F. DeBano for technical review, and to Lynn Starr for editorial review of the manuscript.

Literature Cited

Aandahl, A.R. 1948. The characterization of slope positions and their influence on the total nitrogen content of a few virgin soils of western Iowa. Soil Sci. Soc. Am. Proc. 13:449-454.

Aguilar, S.A. and A. van Diest. 1981. Rock phosphate mobilization induced by the alkaline uptake pattern of legumes utilizing symbiotically fixed nitrogen. Plant Soil 61:27-42.

Ahlgren, I.F. and C.E. Ahlgren. 1960. Ecological effects of forest fires. Bot. Rev. 26:483-533.

Allen, S.E. 1964. Chemical aspects of heather burning. J. Appl. Ecol. 1:347-368.

Ball, R., D.R. Keeney, P.W. Theobold, and P. Nes. 1979. Nitrogen balance in urine affected areas of a New Zealand pasture. Agron. J. 71:309-314.

Barnes, P.W., L.L. Tieszen, and D.J. Ode. 1983. Distribution, production, and diversity of C3- and C4-dominated communities in a mixed prairie. Can. J. Bot. 61:741-751.

Barth, R.C. and J.O. Klemmedson. 1982. Amount and distribution of dry matter, nitrogen and organic carbon in soil-plant systems of mesquite and palo verde. J. Range Manage. 35:412-418.

Beadle, N.C. 1954. Soil phosphate and the delimitation of plant communities in eastern Australia. Ecology 35:370-375.

Beasley, R.S. and J.O. Klemmedson. 1976. Water stress in bristlecone pine and associated plants. Commun. Soil Sci. Plant Anal. 7:609-618.

Birkeland, P. W. 1984. Soils and geomorphology. Oxford University Press, NY.

Blair, G.C., A.R. Till, and R.C.G. Smith. 1975. The phosphorus cycle—what are the sensitive areas? Reviews in Rural Sci. III:9-19.

Boerner, R.E. 1982. Fire and nutrient cycling in temperate ecosystems. Bioscience 32:187-192.

Brown, E.K. and J.H. Mandery. 1962. Planting and fertilization as a possible means of controlling distribution of big game animals. J. For. 60:33-35.

Buol, S. W., F.D. Hole and R.J. McCracken. 1989. Soil genesis and classification. 3rd Edition. The Iowa State Univ. Press, Ames.

Campbell, C.A. 1978. Soil organic carbon, nitrogen and fertility, p.173-271. *In:* M. Schnitzer and S.U. Khan (eds.) Soil Organic Matter. Elsevier Scientific Publ. Co., Amsterdam.

Chapin, F.S. III. 1980. The mineral nutrition of wild plants. Ann. Rev. Ecol. Syst. 11:233-260.

Chapin, F.S., P.M. Vitousek, and K. Van Cleve. 1986. The nature of nutrient limitation in plant communities. The Amer. Naturalist 127:48-58.

Charley, J.L. and N.E. West. 1975. Plant-induced soil chemical patterns in some shrub-dominated semi-desert ecosystems of Utah. J. Ecology 63:945-963.

Christensen, N.L. and C.H. Muller. 1975. Effects of fire on factors controlling plant growth in *Adenostoma* chaparral. Ecol. Monogr. 45:29-55.

Clark, F.E. 1977. Internal cycling of ^{15}nitrogen in shortgrass prairie. Ecology 58:1322-1333.

Clarkson, D.T. 1967. Phosphorus supply and growth in species of *Agrostis* L. J. Ecol. 55:111-118.

Clary, W.P. and A.R. Tiedemann. 1984. Development of 'Rincon' fourwing saltbush, winterfat, and other shrubs from seed following fire, p. 273-280. *In:* A. R. Tiedemann et al. (Compilers). Proceedings—symposium on the biology of *Atriplex* and related chenopods. USDA For. Serv. Gen. Tech. Rep. INT 172. Intermountain For. and Range Exp. Sta., Ogden, UT.

Clements, F.E. 1916. Plant succession: An analysis of the development of vegetation . Carnegie Inst. Pub. 242, Wash., D.C.

Covington, W.W., L.F. DeBano and T.G. Huntsberger. 1991. Soil nitrogen changes associated with slash burning in pinyon-juniper woodlands. For. Sci. 37:347-355.

Cowles, H.C. 1899. The ecological relations of the vegetation on the sand dunes of Lake Michigan. Bot. Gaz. 27:95-117, 167-202, 281-308, 361-369.

Crampton, E.W. and L.E. Lloyd. 1959. Fundamentals of nutrition. W. Freeman and Co., San Francisco.

Crocker, R.L. and J. Major. 1955. Soil development in relation to vegetation and surface age at Glacier Bay, Alaska. J. Ecol. 43:427-448.

Crooke, W.M. and A.H. Knight. 1962. An evaluation of published data on the mineral composition of plants in the light of the cation-exchange capacities of their roots. Soil Sci. 93:365-373.

Crooke, W.M. 1964. The measurement of the cation-exchange capacity of plant roots. Plant and Soil 21:43-49.

Dean, L.A. 1957. Plant nutrition and soil fertility. In: Stefferud, A.(ed). Soil: the 1957 yearbook of agriculture. U.S. Dep. of Agric. Wash., D.C.

Dean, R., J.E. Ellis, R.W. Rice, and R.E. Bement. 1975. Nutrient removal by cattle from a shortgrass prairie. J. Appl. Ecol. 12:25-29.

DeBano, L.F. and C.W. Conrad. 1978. Effects of fire on nutrients in a chaparral ecosystem. Ecol. 59:489-497.

DeBano, L.F., G.E. Eberlein, and P.H. Dunn. 1979. Effects of burning on chaparral soils: I. Soil nitrogen. Soil Sci. Soc. Am. J. 43:504-509.

DeBano, L.F. and J.M. Klopatek. 1988. Phosphorus dynamics of pinyon-juniper soils following simulated burning. Soil Sci. Soc. Am. J. 52:272-277.

Dickson, B.A. and R.L Crocker. 1953a. A chronosequence of soils and vegetation near Mt. Shasta, California. I. Definition of the ecosystem investigated and features of the plant succession. J. Soil Sci. 4:123-141.

Dickson, B.A. and R.L. Crocker. 1953b. A chronosequence of soils and vegetation near Mt. Shasta, California. II. The development of soils and forest floors and the carbon and nitrogen profiles of the soils. J. Soil Sci. 4:142-154.

Dickson, B.A. and R.L. Crocker. 1953c. A chronosequence of soils and vegetation near Mt.

Shasta, California. III. Some properties of the mineral soils. J. Soil Sci. 4:173-190.

Dokuchaev, V.V. 1898. Writings (in Russian), Vol. 6, p. 381, 1951. Akad. Nauk, Moscow.

Duncan, D.O. and L.D. Hylton Jr. 1970. Effects of fertilization on quality of range forage, p. 57-62. *In*: Paulson, H.A., Jr. and E.H. Reid. Range and wildlife habitat evaluation-research symposium. U.S. Dep. Agric. For. Serv. Misc. Publ. 1147. Wash., D.C.

Dunn, P.H., L.F. DeBano, and G.E. Eberlein. 1979. Effects of burning on chaparral soils. II. Soil microbes and nitrogen mineralization. Soil Sci. Soc. Am. J. 43:509-514.

Ellison, L. 1954. Subalpine vegetation of the Wasatch Plateau, Utah. Ecol. Monogr. 24:89-184.

Evans, C.C. and S.E. Allen. 1971. Nutrient losses in smoke produced during heather burning. Oikos 22:149-154.

Evans, J.R. 1989. Photosynthesis and nitrogen relationships in leaves of C_3 plants. Oecologia 78:9-19.

Fireman, M. and M.E. Hayward. 1952. Indicator significance of some shrubs in the Escalante Desert, Utah. Bot. Gazette 114:143-155.

Floate, M.J.S. 1970. Decomposition of organic materials from hill soils and pastures. II. Comparative studies on the mineralization of carbon, nitrogen, and phosphorus from plant materials and sheep feces. Soil Biol. and Biochem. 2:173-185.

Freeman, D.C., L.G. Klikoff, and K.T. Harper. 1976. Differential resource utilization by the sexes of dioecious plants. Science 193:597-599.

Freeman, D.C., K.T. Harper, and E.L. Charnov. 1980. Sex change in plants: old and new observations and new hypotheses. Oecologia (Berl.) 47:222-232.

Fried, M. and H. Broeshart. 1967. The soil-plant system in relation to inorganic nutrition. Academic Press, New York. 358 p.

Garcia-Moya, E. and C.M. McKell. 1970. Contribution of shrubs to the nitrogen economy of a desert-wash plant community. Ecology 51:81-88.

Gillingham, A.G. 1982. Topographic and management effects on dung distribution by grazing sheep. Proc. New Zealand Grassland Assoc. 43:161-170.

Graham, R.D. 1984. Breeding for nutritional characteristics in cereals, p. 57-102. *In:* P.B. Tinker and A. Lauchli (eds.) Advances in Plant Nutrition, Vol. 1, Praeger, NY.

Grier, C.C. 1975. Wildfire effects on nutrient distribution and leaching in a coniferous ecosystem. Can. J. For. Res. 5:599-607.

Grime, J.P. 1977. Evidence for the existence of three primary strategies in plants and its relevance to ecological and evolutionary theory. The Amer. Naturalist 111:1199-1194.

Grime, J.P. 1989. The stress debate: symptom of impending synthesis? Biol. J. Linnean Soc. 37:3-17.

Grime, J.P. and R. Hunt. 1975. Relative growth rate: its range and adaptive significance in a local flora. J. Ecol. 63: 393-422.

Hall, G.F., R.B Daniels and J.E. Foss. 1982. Rate of soil formation and renewal in the USA. p.23-39. *In*: Determinants of soil loss tolerance. ASA Special Publ. No. 45, Amer. Soc. Agron.-Soil Sci. Soc. Amer., Madison, WI.

Hanes, R.J. 1986. Origin, distribution, and cycling of nitrogen in terrestrial ecosystems. p. 1-51. *In*: R.J. Hanes (ed.) Mineral nitrogen in the plant-soil system. Academic Press, Inc.

Hanson, W.O. and J.G. Smith. 1970. Significance of forage quality as a tool in wildlife management, P. 25-31. *In*: Paulson, H.A. Jr. and E.H. Reid. Range and wildlife habitat evaluation-research symposium. U.S. Dep. Agric. For. Serv. Misc. Publ. 1147. Wash., DC.

Harris, G.A. and A.M. Wilson. 1970. Competition for moisture among seedlings of annual and perennial grasses as influenced by root elongation at low temperature. Ecol. 51:530-534.

Harris, G.R. and W.W. Covington. 1983. The effect of a prescribed fire on nutrient concentration and standing crop of understory vegetation in ponderosa pine. Can. J. For. Res. 13:501-507.

Heintze, S.G. 1961. Studies on cation-exchange capacities of roots. Plant and Soil 13:365-383.

Hellmers, H., and J.M. Kelleher. 1959. *Ceanothus leucodermis* and soil nitrogen in southern California mountains. For. Sci. 5:275-278.

Hilder, E.J. 1964. The distribution of plant nutrients by sheep at pasture. Proc. Aust. Soc. Anim. Prod. 5:241-248.

Hilgard, E.W. 1914. Soils. The Macmillian Co., NY.

Huenneke, L.F., S.P. Hamburg, R. Koide, H.A. Mooney, and P.M. Vitousek. 1990. Effects of soil resources on plant invasion and community structure in Californian serpentine Grassland. Ecology 71:478-491.

Jansson, S.L. 1971. Use of ^{15}N in studies of soil nitrogen, p. 129-166. *In*: A.D. Mclaren and J.Skujins (eds.), Soil Chemistry, Vol. 2. Marcel Dekker, NY.

Jawson, M.D., L.F. Elliott, K.E. Saxton, and D.H. Fortier. 1982. The effect of cattle grazing on nutrient losses in a Pacific Northwest setting. J. Environ. Qual. 11:628-632.

Jenny, H. 1930. A study of the influence of climate upon the nitrogen and organic matter content of the soil. Missouri Agric. Exp. Sta. Res. Bull. 152.

Jenny, H. 1941. Factors of soil formation. McGraw-Hill Book Co., Inc.

Jenny, H. 1958. Role of the plant factor in the pedogenic functions. Ecology 39:5-16.

Jenny, H. 1961. Derivation of state factor equations of soils and ecosystems. Soil Sci. Soc. Amer. Proc. 25:385-388.

Jenny, H. 1980. The soil resource: origin and behavior. Springer-Verlag, New York.

Jurgensen, M.F., A.E. Harvey, and M.J. Larsen. 1981. Effects of prescribed fire on soil nitrogen levels in a cutover Douglas-fir/western larch forest. USDA For. Serv. Res. Pap. INT-275. Intermountain For. and Range. Exp. Sta. Ogden, Utah.

Kay, B.L. and R. Evans. 1965. Effects of fertilization on a mixed stand of cheatgrass and intermediate wheatgrass. J. Range Manage. 18:7-11.

Klemmedson, J.O. 1964. Topofunction of soils and vegetation in a range landscape. p. 176-189. *In*: Forage Plant Physiology and Soil-Range Relationships. Am. Soc. Agron. Spec. Publ. 5.

Klemmedson, J.O. 1976. Effect of thinning and slash burning on nitrogen and carbon in ecosystems of young ponderosa pine. For. Sci. 22:45-53.

Klemmedson, J.O. 1979. Ecological importance of actionomycete nodulated plants in the western United States. Bot. Gaz. 140 (Suppl.):S91-S96.

Klemmedson, J. O. 1989. Soil organic matter in arid and semiarid ecosystems: Sources, accumulation and distribution. Arid Soil Res. and Rehabil. 3:99-114.

Klemmedson, J.O. and R.B. Ferguson. 1973. Effect of sulfur deficiency on yield and nitrogen content in bitterbrush (*Purshia tridentata*) seedlings on granitic soils. Soil Sci. Soc. Amer. Proc. 37:947-951.

Klemmedson, J.O., A.M. Schultz, H. Jenny and H.H. Biswell. 1962. Effect of prescribed burning of forest litter on total soil nitrogen. Soil Sci. Soc. Amer. Proc. 26:200-202.

Klemmedson, J. O., and A.R. Tiedemann. 1994. Soil and vegetation development in an abandoned sheep corral on degraded subalpine rangeland. Great Basin Naturalist 54:301-312.

Klemmedson, J.O. and B.J. Wienhold. 1991. Aspect and species influences on nitrogen and phosphorus availability in Arizona chaparral soils. Soil Sci. Soc. Amer. J. 55:1735-1740.

Klock, G.O., J.M. Geist, and A.R. Tiedemann. 1971. Erosion control fertilization—from pot study to field testing. Sulphur Inst. J. 7(3):7-10.

Klock, G.O., A.R. Tiedemann, and W. Lopushinsky. 1975. Seeding recommendations for disturbed mountain slopes in north central Washington. USDA For. Serv. Res. Note PNW-244. Pacific Northwest For. and Range. Exp. Stn., Portland, Oreg.

Kruckeberg, A.R. 1954. The ecology of serpentine soils. III.Plant species in response to serpentine soils. Ecology 35:267-274.

Langer, R.H.M. 1966. Mineral nutrition of grasses and cereals. p. 213-226. *In*: F.L. Milthrope and D.J. Irvins (eds.) The growth of cereals and grasses. Butterworths, London.

Levitt, J. 1980. Responses of plants to environmental stresses. Vol. 1. Acad. Press, New York. 497 p.

Lloyd, P.S. and C.D. Pigott. 1967. The influence of soil conditions on the course of succession on the chalk of southern England. J. Ecol. 55:137-146.

Lobert, J.M., D.H. Scharffe, W.M. Hao, and P.J. Crutzen. 1990. Importance of biomass burning in the atmospheric budgets of nitrogen-containing gases. Nature 346:552-554.

Major, J. 1951. A functional, factorial approach to plant ecology. Ecology 32:392-412.

Marschner, H. 1986. Mineral nutrition of higher plants. Academic Press, London.

Marschner, H., V. Romheld, W.J. Horst and P. Martin. 1986. Root-induced changes in the rhizosphere: Importance for the mineral nutrition of plants. Z. Pflanzenernaehr. Bodenk. 49:441-456.

Maynard, L.A. and J.K. Loosli. 1962. Animal Nutrition. 5th. ed., McGraw-Hill Co. Inc., New York.

McArthur, E. D. 1988. New plant development in range management, p. 81-112. *In*: Tueller, P.T. (ed.) Vegetation science applications for rangeland analysis and management. Kluwer Academic Publishers, Boston.

McArthur, E.D., S.E. Stranathan and G.L. Noller. 1984. 'Rincon' fourwing saltbush—proven for better forage and reclamation. Rangelands 6:62-64.

McIntyre, G.I. 1964. Influence of nitrogen nutrition on bud and rhizome development in *Agropyron repens* L. Beauv. Nature, (Lond.) 203:1084-1085.

McIntyre, G.I. 1977. The role of nutrition in apical dominance, p. 251-273. *In*: D.H. Jennings (ed.) Integration of activity in the higher plant. Symp. of the Soc. for Expt. Biol., Cambridge Univ. Press, Cambridge.

Molina, R.J., J.M. Trappe, and G.S. Strickler. 1978. Mycorrhizal fungi associated with Festuca in the western United States and Canada. Can. J. Bot. 56:1691-1695.

Mosse, B. 1973. Advances in the study of vesicular-arbuscular mycorrhiza. Ann. Rev. Phytopathol. 11:171-196.

Mueggler, W. F. 1965. Cattle distribution on steep slopes. J. Range Manage. 18:255-257.

Nye, P.H. and P.B. Tinker. 1977. Solute movement in the plant-root system. Univ. of California Press, Berkeley.

Olsen, J. 1958. Rates of succession and soil changes on southern Lake Michigan sand dunes. Bot. Gaz. 119:125-170.

Paul, E.A. 1970. Characterization and turnover rate of soil humic constituents, p. 63-76. *In*: S. Pawluk (ed.) Pedology and Quaternary Research. Univ. of Alberta Press, Edmonton.

Pieper, R.D. 1977. Effects of herbivores on nutrient cycling and distribution, p. 249-275. *In*: Proc. 2nd. United States/Australian Rangeland Panel, Adelaide, 1972. Aust. Rangeland Soc., Perth, W. Aust.

Peterson, R.G., W.W. Woodhouse, Jr., and H.L. Lucas. 1956. The distribution of excreta by freely grazing cattle and its effect on pasture fertility: II. Effect of refined excreta on the residual concentration of some fertilizer elements. Agron. J. 48:444-449.

Pritchett, W.L. 1979. Properties and management of forest soils. John Wiley & Sons.

Proctor, J. and S.R.J. Woodell. 1975. The ecology of serpentine soils. Adv. Ecol. Res. 9:255-366.

Raison, R.J., P.K. Khana, and P.V. Woods. 1985. Mechanisms of element transfer to the atmosphere during vegetation fires. Can. J. For. Res. 15:132-140.

Reeves, F.B., D. Wagner, T. Moorman, and J. Kiel. 1979. The role of endomycorrhizae in revegetation practices in the semiarid west. 1. A comparison of incidence of mycorrhizae in severely disturbed vs. natural environments. Amer. J. Bot. 66:6-13.

Sage, R.F. and R.W. Pearcy. 1987. The nitrogen use efficiency of C3 and C4 plants. Plant Physiol. 84:954-958.

Sampson, A.W. and L.H. Weyl. 1918. Range preservation and its relation to erosion control on western grazing lands. U.S. Dept. Agric. Bull. No.675.

Sampson, A.W. 1919. Plant succession in relation to range management. U.S. Dept. Agric. Bull No. 791.

Sampson, A.W. 1952. Range management, principles and practices. John Wiley & Sons, Inc. NY

Sharma, M.L. and D.J. Tongway. 1973. Plant induced soil salinity patterns in two saltbush (*Atriplex* spp.) communities. J. Range Manage. 26:121-125.

Smeck, N.E. 1985. Phosphorus dynamics in soils and landscapes. Geoderma 36:185-199.

Sneva, F. 1978. Nitrogen and sulfur impacts on the cold desert biome, p. 678-680. *In*: Hyder, D.M.(ed.) Proc. First Intl. Rangeland Congr., Soc. Range Manage., Denver, Colo.

St. John, T.V. and P.W. Rundel. 1976. The role of fire as a mineralizing agent in a Sierran conferous forest. Oecologia (Berl.) 25:35-45.

Stålfelt, M.G. 1972. Stålfelt's plant ecology. (English translation Longman Group Limited), John Wiley & Sons (Halsted Press), NY.

Stevenson, F.J. 1965. Origin and distribution of nitrogen in soil. p. 1-42. *In*: W.V. Bartholomew and F.E. Clark (eds.) Soil Nitrogen, Amer. Soc. Agron. Inc., Madison, WI.

Stevenson, F.J. 1986. Cycles of soil: carbon, nitrogen phosphorus, sulfur, micronutrients. John Wiley & Sons. NY.

Tiedemann, A.R. 1981. Stream chemistry, nutrient economy, and site productivity consequences of wildland management and wildfire, p. 183-201. *In*: D.M. Baumgartner (Compiler and Editor), Interior West watershed management. Washington State Univ. Coop. Ext., Pullman, WA.

Tiedemann, A.R. 1983. Response of bitterbrush and associated plant species to broadcast nitrogen, phosphorus, and sulfur fertilization, p. 240-254. *In*: Tiedemann, A.R. and K.L. Johnson (Compilers). Proceedings—research and management of bitterbrush and cliffrose in western North America. USDA For. Serv. Gen. Tech. Rep. INT 152. Intermountain For. and Range Exp. Sta., Ogden, Ut.

Tiedemann, A.R. 1987a. Nutrient accumulations in pinyon-juniper ecosystems—managing for future site productivity, p.352-359. *In*: Everett, R.L. (Compiler) Proceedings—pinyon-juniper conference. USDA For. Serv. Gen. Tech. Rep. INT-215. Intermountain For. & Range Exp. Sta., Ogden, Ut.

Tiedemann, A.R. 1987b. Combustion losses of sulfur from forest foliage and litter. For. Sci. 33:216-223.

Tiedemann, A.R. and J.O. Klemmedson. 1986. Long-term effects of mesquite removal on soil characteristics. I. Nutrients and bulk density. Soil Sci. Soc. Am. J. 50:472-475.

Tilman, D. 1986. Nitrogen-limited growth in plants from different successional stages. Ecology 67:555-563.

Tisdale, S.L., W.L. Nelson and J.D. Beaton. 1985. Soil fertility and fertilizers. Macmillian Publ. Co., NY.

Trappe, J.M. 1981. Mycorrhizae and productivity of arid and semiarid rangelands, p. 581-599. *In*: Advances in food producing systems for arid and semiarid lands. Acad. Press Inc., NY.

Troughton, A. 1980. Environmental effects upon root-shoot relationships, p.25-41. *In*: D.N. Sen (ed.) Environment and root behavior. Geobios International, India.

Turner, J. 1977. Effect of nitrogen availability on nitrogen cycling in a Douglas-fir stand. For. Sci. 23:307-316.

Van Den Driessche, R. 1984. Nutrient storage, retranslocation and relationship of stress to nutrition, p.181-209. *In*: G.D. Bowen and E.K.S. Nambiar (eds.) Nutrition of Plantation Forests. Academic Press, London.

Vitousek, P. 1982. Nutrient cycling and nutrient use efficiency. The Amer. Naturalist 119:553-572.

Vlassak, K., E.A. Paul, and R.F. Harris. 1973. Nitrogen fixation in grasslands and associated sites. Plant and Soil 38:637-649.

Vose, P.B. 1983. Rationale of selection for specific nutritional characters in crop improvement with *Phaseolus vulgaris* L. as a case study. Plant and Soil 72:351-364.

Walker, J., C.H. Thompson, I.F. Fergus and B.R.Tunstall. 1981. Plant succession and soil development in coastal sand dunes of subtropical eastern Australia, p.107-131. *In:* D.C. West et al. (eds.) Forest Succession: Concepts and Application. Springer-Verlag, NY.

Walker, R.B. 1954. The ecology of serpentine soils. II. Factors affecting plant growth on serpentine soils. Ecology 35:259-266.

Walker, R.B. and S. P. Gessel. 1990. Mineral deficiency symptoms in Pacific Northwest conifers. Western J. Applied For. 5: 96-98.

Walker, T.W. 1965. The significance of phosphorus in pedogenesis, p. 295-315. *In*: E.G. Hallsworth and D.V.Crawford (eds.) Experimental Pedology. Butterworths, London.

Walker, T.W. and J.W. Syers. 1976. The fate of phosphorus during pedogenesis. Geoderma 15:1-19.

Wallace, C.S. and P.W. Rundel. 1979. Sexual dimorphism and resource allocation in male and female shrubs of *Simmondsia chinensia*. Oecologia (Berl.) 44:34-39.

Watson, E.R. and P. Lapins. 1969. Losses of nitrogen from urine on soils from southwestern Australia. Aust. J. Exp. Agric. and Anim. Husb. 9:85-91.

Wells, C.G., R.E. Campbell, C.E. Lewis, and others. 1979. Effects of fire on soil—a state-of-knowledge review. National Fire Effects Workshop, Denver, Colo., Apr. 10-14, 1978. USDA For. Serv. Gen. Tech. Rep. WO-7. USDA Forest Service, Wash., DC. 34 p.

West, N.E. and J.O. Klemmedson. 1978. Structural distribution of nitrogen in desert ecosystems, p.1-15. *In*: N.E. West and J. Skujins (eds.) Nitrogen Processes in Desert Ecosystems. Dodd, Hutchinson and Ross, Stroudsbury, Pa.

West, N.E. and J. Skujins. 1977. The nitrogen cycle in North American cold-winter semi-desert ecosystems. Ecol. Plant. 12:45-53.

Whalley, R.D.B., and C.M. McKell. 1976. Seedling vigor of some important range species as affected by nitrogen and phosphorus. Agron. J. 68:565-569.

White, E.M. 1971. Some soil age-range vegetation relationships. J. Range Manage. 24:360-365.

Williams, R.F. 1955. Redistribution of mineral elements during development. Ann. Rev. Plant Physiol. 6:25-42.

Wilson, A.L., G.A. Harris, and D.H. Gates. 1966. Fertilization of mixed cheatgrass-bluebunch wheatgrass stands. J. Range Manage. 19:134-137.

Woodmansee, R.G. 1978. Additions and losses of nitrogen in grassland ecosystems. Bioscience 28:448-453.

Woodward, R.A., K.T. Harper, and A.R. Tiedemann. 1984. An ecological consideration of the significance of cation-exchange capacity of roots of some Utah range plants. Plant and Soil 79:169-180.

Wright, H.A. and A.W. Bailey. 1982. Fire ecology, United States and southern Canada. John Wiley and Sons, Inc.

Wullstein, L.H., M.L. Bruening and W.B. Bollen. 1979. Nitrogen fixation associated with sand grain root sheaths (rhizosheaths) of certain xeric grasses. Physiol. Plant. 46:1-4.

Plant Adaptations to Saline Environments

Richard F. Miller and Paul S. Doescher

Introduction

The land area affected by saline soils has been estimated at 2.2 to 6.3% of the world's land surface, an area equivalent to 44 to 122% of the United States (Flowers et al. 1986). A significant proportion of these saline environments, both natural and human induced, are rangelands. Saline environments include salt deserts, wet maritime marshes, mangrove swamps and interior salt marshes. Human activities through irrigation, dryland farming and mining activities have increased the amount of land surface affected by salts. Human activities, beginning as early as 2400 B.C., have increased salinity on as much as 4×10^7 ha, or approximately one-third of the world's irrigated land (Peck 1975). With increased demands being placed on food production, worldwide salinization of soil and water resources is expected to increase in the near future (Bettenay 1986).

The ability of a plant species to survive and successfully compete in these saline environments is dependent upon mechanisms which allow it to avoid or tolerate water stress, ion toxicity and nutrient deficiency. Due primarily to the concerns of salt accumulations on agricultural lands, a large body of literature has been written on the effects of salts on plants and the mechanisms allowing plants to survive in these saline environments. However, with the exception of species in the saltbush genus *Atriplex*, little work has been done on range plants.

This chapter is separated into four sections which include; (1) a description of saline soils and environments, (2) a general description of the kinds of plants adapted to these environments, (3) the direct and indirect effects of salts on plants, and (4) mechanisms which enable a plant to complete its life cycle in a saline environment. To conserve space, review articles by Caldwell (1974), Flowers et al. (1977), Greenway and Munns (1980), Wyn Jones (1981), Yeo (1983), Gorham et al. (1985), Flowers (1985) and Flowers et al. 1986, and books by Waisel (1972), Larcher (1980), Levitt (1980), and Marschner (1986) are frequently cited. For a historical perspective on salt tolerance in plants the reader should refer to Uphof (1941).

Saline Environments

Soils

Soluble ions which commonly accumulate in soils in varying proportions are sodium, calcium, magnesium, chloride and sulfate. Potassium, bicarbonate, carbon-

---

Miller is Professor of Rangeland Resources, Eastern Oregon Agricultural Research Center, Oregon State University; and Doescher is Associate Professor of Rangeland Resources, Oregon State University.

ate and nitrate occur in smaller amounts. The most common found in marine saline environments is NaCl (Flowers et al. 1977). In arid regions, saline soils are more frequently alkaline, containing the less soluble sulfates and carbonates of Na, Mg, and Ca, in addition to chlorides (Larcher 1980). In the salt deserts of the Great Basin, however, saline and saline-sodic soils are more commonly composed of the soluble salts NaCl and Na_2SO_4, rather than the less soluble salts $CaSO_4$ and $CaCO_3$ (Roundy 1984). In Australia, soluble salt levels are generally low in soils of the northern deserts, but high levels (e.g. chlorides and sulfates, 1.6% or more) are common in the southern deserts (Williams and Calaby 1985). In mine oil shale deposits in the western United States, the two most common salts are Na_2SO_4 and $MgSO_4$ (Schmehl and McCaslin 1973).

Salt affected soils are classified into three categories; saline, non-saline alkali (sodic), and saline alkali (Shainberg 1975) (Table 1). Saline soils contain levels of soluble salts sufficient to inhibit growth of most crop plants, but not enough exchangeable sodium to alter soil characteristics. Non-saline alkali soils contain limited levels of soluble sodium with sufficient levels of exchangeable sodium to inhibit plant growth and alter soil physical and chemical properties. In these soils, exchangeable sodium influences infiltration, permeability, soil structure, and stability of aggregates. When high levels of exchangeable sodium occur, soil particles disperse which blocks soil pores thus decreasing hydraulic conductivity and aeration. These soils are characterized by a pH of 8.5 to 10. The third category, saline alkali soils are similar to saline soils as long as soluble salt content is high. As soluble salt content declines, soil properties become similar to alkali soils.

Solonchak (saline) and Solonetz (alkali) soils are commonly found in salt deserts throughout the world. These soils are a result of; (1) climate, (2) vegetation, (3) parent material, (4) relief and (5) time (Naphan 1966). Solonchak soils are formed by high water tables rising through capillary action to the soil surface and evaporation causing salts to accumulate on the surface. These soils are frequently found around saline lakes or where drainage has been impeded forming a shallow water table (Millar et al. 1966, Brady 1974). Solonetz soils form where soluble salts are leached through the soil profile leaving 15% or more exchangeable sodium on clay particles. Saline soils can be reclaimed by leaching soluble salts to the water table. Reclama-

Table 1. Three categories of salt affected soils.

SOILS	SALTS	SOIL CHARACTERISTICS	FORMATION
Saline	soluble salts	not altered	rising and falling water table depositing soluble salts near surface
Sodic	exchangeable sodium	decreased infiltration, soil structure and aggregate stability	soluble salts are leached from the upper soil profile, >15% exchangeable sodium deposited on clay particles
Saline alkali	soluble salt content > exchangeable sodium	not altered	

tion of alkali soils, however, is more difficult due to low permeability and a requirement for Ca-salts to replace exchangeable sodium. Throughout this chapter, saline soils or environments will be used for all salt affected soils or environments unless special reference is made to sodic conditions.

Soils can also be affected by the plant species which occupy them. Salt-tolerant plants which absorb high levels of salts can induce significant changes in the soil surface layer through decomposition of salt-rich leaves (Fireman and Hayward 1952, Rickard 1965a, Rickard and Keough 1968, Sharma and Tongway 1973). Decomposition of leaves slowly releases cations and anions (e.g. Na^+, K^+, Ca^{++}, Mg^{++}) (Fireman and Hayward 1952) which increases salinity levels. Leaves of sodium rich greasewood (*Sarcobatus vermiculatus* (Hook.)Torr.) and K^+ rich spiny hopsage have been shown to increase the Na^+ and K^+ content in the soils beneath their canopy (Rickard 1965b). Eckert and Kinsinger (1960) reported that plots occupied by halogeton (*Halogeton glomeratus* [Bieb.] C.A.Mey) for 7 years increased in Ph, electrical conductivity, exchangeable sodium, and soluble sodium. Increased Ph levels have also been found beneath the canopies of bladder saltbush (*Atriplex vesicaria* Heward and Benth) and old man saltbush (*A. nummularia* Lindley and T.Mitch) in the 0–7.5 cm soil layer (Sharma 1973) and greasewood and shadscale (*Atriplex confertifolia* (Tort. and Grem.) Wats.) (Fireman and Hayward 1952). Increased salt content from decomposing leaves can cause a deterioration of surface soil structure (Eckert and Kinsinger 1960, Sharma 1973).

Plant Distribution

Some of the primary edaphic factors influencing plant species distribution across salt desert landscapes are matric and osmotic potentials of the soil solution during the growing season, water table depth, salinity and ionic composition of the ground water, quantity and quality of soil salts, distribution of soil salt concentrations, and soil fertility, texture, structure, surface morphology and depth (Branson et al. 1967, Detling and Klikoff 1973, Roundy and Young 1985).

Ecotypic variation also adds to the complexity of plant distribution in saline environments (Goodman and Caldwell 1971). It is apparent that no single chemical (Gates et al. 1956), physical factor (Mitchell et al. 1966) or biological factor can solely explain the distribution of vegetation in saline environments.

West and Ibrahm (1968) found landforms, soil texture and depth to the saline horizon closely related to vegetation distribution in the shadscale zone of southeastern Utah. In the Great Basin, vegetation patterns are frequently influenced by a salinity gradient, increasing in ion concentrations from the upper terraces of ancient Pleistocene Lakes to the lake bottoms (Flowers and Evans 1966). Breckle (1976) illustrated the relationship of plant species distribution with increasing salinity and aridity in the Intermountain Salt Desert (Fig. 1). In soils containing shadscale, salinity was significantly greater at the 30–40 cm depth as compared to soils occupied by winterfat (*Ceratoides lanata* (Pursh.) J.T. Howell). Early ecological work generally showed shadscale was indicative of soils being saline below the first foot (Weaver and Clements 1938). Billings (1949), however, found shadscale communities in Nevada and California growing in nonsaline soils generally too dry for big sagebrush. In the Sonoran desert, creosote bush (*Larrea divericata* Cav.) failed

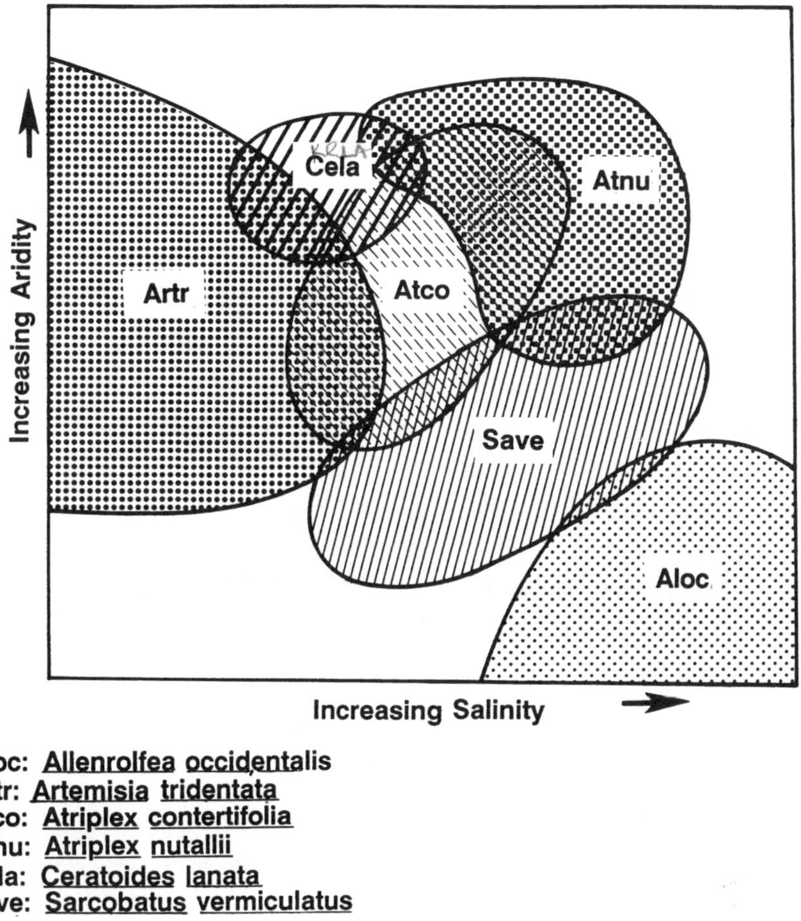

Figure 1. Position of salt desert shrubs in northern Utah in relation to aridity and salinity (from Breckle 1976).

to grow in soils where the C horizon was highly saline (Wallace et al. 1973a).

Water table depth has also been shown to influence the distribution of saline tolerant species such as greasewood. The presence of greasewood in a shadscale dominated community indicated a water table shallow enough for greasewood roots to reach it during at least part of the year (Roundy and Young 1985). As the water table approached to within 1-m of the surface, shrub dominance declined and salt tolerant graminoides such as rush (*Juncus* spp.), alkali grass (*Puccinellia* spp. Parl.), saltgrass (*Distichlis spicata* (L.) Greene, basin wildrye (*Elymus cinereus* Scribn and Merr.) and alkali cordgrass (*Spartina gracilis* Trin.) increased. As soil aeration becomes limited, such as in the ancient lake bottoms, greasewood is confined to dunes and mounds (Fig. 2).

Salinity levels can become extreme on ephemeral desert lake and valley bottoms, reducing soil osmotic potentials to very low levels, even at saturation. Under these conditions only extremely salt tolerant species can survive such as pickleweed (also

Figure 2. Greasewood (*Sarcobatus vermiculatus*) growing on mounds in an ancient pleistocene lake bottom.

called iodine bush) (*Allenrolfea occidentalis* [Wats.] Kuntze), seablite (*Suaeda* spp. Forsk) and the most tolerant of the three, samphire (also called glasswort or saltwort) (*Salicornia* spp. [L.] Dumort). European samphire (*Salicornia europeaea* L.) and *Halocnemum strobilaceum* (Pall.) Bieb., two extremely salt tolerant species, are the first to colonize salt crusted solonchak soils in the Karakum Desert of Russia (Walter and Box 1983). Both plants are leafless succulents capable of storing large amounts of sodium chloride in their cells. In the Great Salt Lake basin, samphires *Salicornia rubra* and *S. utahensis* are the first species to pioneer areas where lake waters have recently receded and soils have a high salt content compared to nearby plant communities (Flowers 1934). Second only to the samphires in salt tolerance in this basin is pickleweed.

In salt and brackish marshes, distribution of vegetation is also very complicated due to gradients of salinity, climate and varying water-levels (Dijkema 1990). In the salt marshes of western Utah, vegetation distribution is significantly influenced by both salt concentrations and plant competition (Bolen 1964). For further reviews of salt marsh ecology see Penfound and Hataway (1938), Purer (1942) and Miller and Egler (1950).

Salt Deserts

Salt deserts are one of the most extensive terrestrial saline environments with examples found throughout the arid and semi-arid regions of Eurasia, Africa, Australia, South America, and North America (West 1983, Evenari et al. 1985). In the western United States, salt desert shrub communities occupy 15 to 17 million ha in the Great Basin, Colorado Plateau, Mojave, Sonoran and Chihuahuan Deserts (Kuchler 1964, Branson et al. 1967). The Great Basin accounts for the largest proportion of these salt desert shrublands.

In arid and semi-arid regions of the world, salt deserts most frequently develop when ions weathered from soil minerals become concentrated in the upper soil layers. This is most often caused by inadequate leaching and high evaporation rates (Shainberg 1975), or rising water tables, causing the upward movement of salts (Richards 1954, Orshan 1986). Saline soils may also be the result of internal drainage into closed basins, such as those in the Great Basin where saline soils characterize the bottoms of ancient pleistocene lakes and enclosed valleys (Papke 1976, Blaisdell and Holmgren 1984). Salt desert shrub communities in areas such as the Snake River Plains in southwestern Idaho, the Big Horn Basin in Wyoming, the Painted Desert in the Colorado Plateau, the Badlands of South Dakota and the breaks along the Missouri River of Montana are primarily caused by marine shale outcroppings deposited by ancient shallow seas (West 1983, Blaisdell and Holmgren 1984). Along the Arabian Gulf Coast, salt desert communities are also found on soils of predominately marine origin (Abd el Rahman 1986). In some areas of the Negev Desert in Israel and Sinai, soil salinity is induced by the precipitation of seawater spray carried by wind over long distances (Yaalon 1964).

The combination of saline soils and high evaporative demands in salt deserts generally places severe physiological demands on plants. Species adapted to these deserts must survive high salt concentrations in the soil in addition to low humidity, limited precipitation, large temperature extremes and frequently other soil conditions such as low fertility and excessive levels of toxic elements (Caldwell 1974). Extreme environmental conditions makes these ecosystems frail and difficult to restore.

Due to extreme conditions in most salt deserts, the seasonal growing period is very short. Environmental conditions in the salt desert contribute to the sensitivity of shrubs and grasses to defoliation during the growing season (Plummer and Chapman 1976, Cook 1977, Norton 1978). For example, by 1900, improper grazing had caused extensive reductions of palatable salt desert shrubs and grasses in the Great Basin of North America (Griffiths 1902). As a result of overgrazing, winterfat (Griffiths 1902) and black sage (*Artemisia nova* (A. Nels.) Ward.) declined or were eliminated in many salt desert communities while less palatable species, such as shadscale, Russian thistle (*Salsola kali* L.) and halogeton increased (Hancock 1966). Well into the 20th century, improper grazing continued to cause large reductions in shrubs and grasses in addition to accelerating wind erosion in the salt deserts (Stewart et al. 1940). Livestock carrying capacities in the salt deserts of the western United States had been reduced 70% by the 1930s (Clapp 1936). Carrying capacities averaging 5 ac/AUM had declined to 18 ac/AUM (McArdel et al. 1936). Proper grazing management of this resource usually entails predominately winter use (Roundy and Young 1985).

During the last 10 years, cheatgrass (also known as downy brome) (*Bromus tectorum* L.) has been observed invading into the margins of salt desert communities in the Great Basin (Young and Tipton 1990). Although fire has played an important role in shaping landscapes throughout the sagebrush steppe, it occurred very rarely within the salt desert. An increase in fire frequencies due to cheatgrass invasion would have a large affect on plant composition in these salt desert communities.

Saline Seeps

Saline seeps are becoming an increasing problem in the Great Plains and Texas, particularly in areas where native rangeland has been plowed and crop fallow systems installed (Brown et al. 1983, Berg et al. 1987). Brown et al. (1983) defines a saline seep as an "Intermittent or continuous saline water discharge, at or near the soil surface downslope from recharge areas under dryland conditions, that reduces or eliminates crop growth in the affected area because of increased soluble salt concentration in the root zone." Saline seeps have been caused by a combination of geologic, climatic, and cultural conditions (Halvorson 1988). They form in areas where native vegetation has been removed and replaced by crops with lower water requirements, and on dryland farms where land lays fallow for 21 months over a two year period (Brown et al. 1983, Bettenay 1986, Berg et al. 1987, Halvorson 1988). Increased water storage, due to decreased evapotranspiration, causes excess water to percolate through a permeable salt rich strata below the rooting zone underlain by an impermeable layer creating a perched water table. The saline water moves downslope along the impermeable layer until it is discharged, forming the seep, frequently on sidehills and toe-slopes. Salt concentrations increase as discharged water evaporates. As soil salinity increases, native vegetation or crops decline and are replaced by more salt tolerant species such as kochia (*Kochia scoparia* [L.] Schrad.) (Berg et al. 1987).

Saline seeps were first noticed about 40 years ago and have rapidly increased within the last 20 years in the Midwest (Berg et al. 1987) and in many countries around the world (Olson 1978, Vander Pluym 1978, Bettenay 1986). In Montana, saline seeps increased from 20,500 to 81,000 ha in the 1970s (Brown et al. 1983). They have frequently formed in the deepest and most productive soils, decreasing site productivity and increasing salt concentrations in adjacent drainages (Berg et al. 1987). Saline seeps have appeared only within the last 40 years due to increased water storage caused by more efficient farming practices. A method for controlling saline seeps has been the planting of deep rooted perennials such as alfalfa (*Medicago sativa* L.), Russian wildrye (*Elymus junceus* Fischer) and tall wheatgrass (*Agropyron elongatum* (Host) Beauv.) in the recharge areas (Halvorson 1988). For a more detailed description of control and reclamation of saline seeps see Brown et al. (1983) and Halvorson (1988).

Plants Adapted to Saline Environments

Plant Taxa

Plant species which are adapted to saline environments are found in over 100 families of flowering plants and in approximately 500 genera (Heywood 1978, Flowers et al. 1986). Chenopodiaceae has the highest number of salt-tolerant species, containing 312, and Poaceae second with 109 salt-tolerant species currently reported. Common genera found in saline areas of the western United States are *Atriplex, Prosopis, Allenrolfea, Distichlis, Sporobolus* and *Suaeda* (West 1983, MacMahon and Wagner 1985).

Halophytes and Glycophytes

Plants are generally classified as being either halophytes or glycophytes, based on their ability to grow in saline environments. Halophytes have been defined as native flora of saline environments (Jennings 1976, Greenway and Munns 1980, 1983). For terrestrial plants, this means growing on sites with a minimum salt concentration of 100 mM in the soil solution (Flowers et al. 1986). In contrast, glycophytes are plant species characterized by relatively low salt-tolerance. Flowers et al. (1986) suggested the most useful criteria for separating the two groups is the ability of halophytes to complete their life cycle at concentrations in excess of 100 to 200 mM NaCl. Strogonov (1964) reported that some halophytes are capable of growing in soils containing up to 20% salt in the internal solution, although the majority of saline environments are characterized by soil solutions containing 2–6% salts. In contrast, some glycophytes may exhibit limited growth in soil solutions containing as little as 0.3% salt.

Defining salt tolerance, however, in terms of percentages of salts, can be misleading since plants respond to ions in solution as a function of such factors as ion pair formation, solubility coefficients and soil water content (Tanji 1969, Roundy 1984). Defining absolute salt tolerance of a species is also difficult due to many interactions between the plant, soil, water and environmental factors which affect the ability of a plant to tolerate salt (Maas and Hoffman 1977) and the experimental conditions (Caldwell 1974). For example, halophytes can tolerate higher salt concentrations when soil water depletion is slow as compared to rapid (Flowers 1934). In the Lake Bonneville basin, samphires thrive in areas where salt content exceeds 4% on a dry weight basis and the water table is slow to withdraw. However, on sites where soil water depletion rates are rapid, samphires are found on soils containing 2% salt on a dry weight basis.

A clear separation between halophytes and glycophytes does not exist since there is a continuum from plants highly tolerant to salinity to plants highly sensitive to salinity (Fig. 3) (Greenway and Munns 1980, Wyn Jones 1981). Salt-tolerance also varies not only between species but within populations of a single species (McArthur et al. 1978).

The ability of plants to tolerate high external concentrations of salts is achieved in two general ways: (1) by uptake of salts, which requires avoidance of ion toxicity in plant tissue; or (2) through exclusion of salts which requires avoidance of internal water deficits (Flowers et al. 1977, Marschner 1986). Halophytes typically achieve high salt-tolerance by inclusion of salts. Glycophytes exhibit relatively limited salt-tolerance by excluding salt from the leaves, although some may accumulate Na^+ in their roots and stems. Glycophytes may osmotically adjust by producing biochemically-compatible organic solutes within the cytoplasm, such as glycinebetaine (see section on cytosolutes). Although both halophytes and glycophytes are frequently classified as includers and excluders, respectively, Cheeseman (1988) cautions that this terminology is imprecise since most salt-sensitive species will accumulate salt to a certain degree when available. It is the amount of salts accumulated which generally separates halophytes from glycophytes (Caldwell 1974, Wyn Jones 1981). The rate of absorption and translocation of Na^+ is considerably greater in halophytes than in glycophytes.

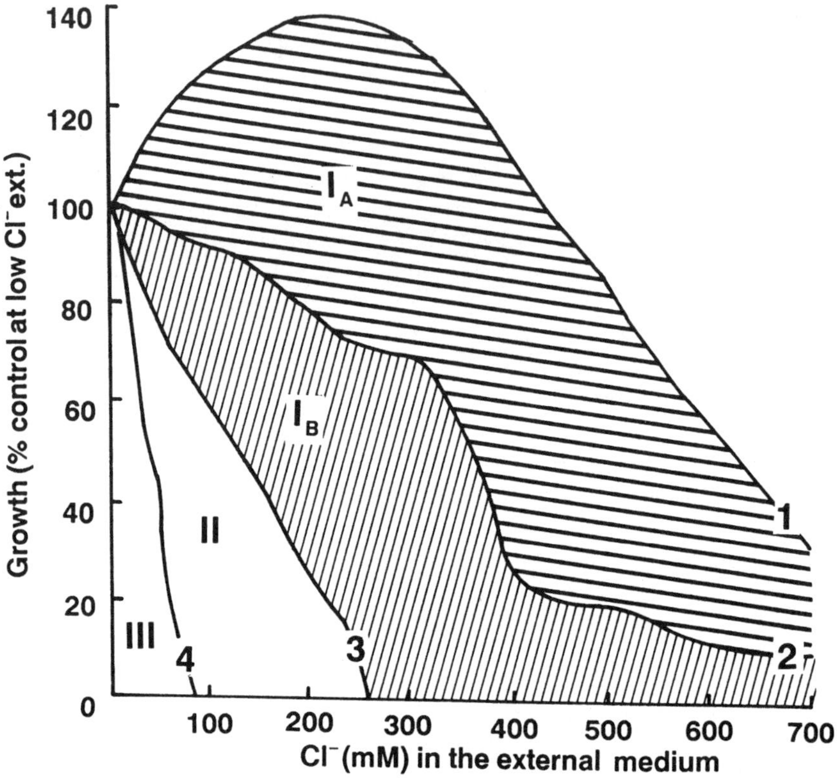

Figure 3. Growth response of different species to salinity; growth after 1-6 months at high external Cl^-. Curve 1: herbaceous seablite (*Suaeda maritima*); Curve 2: sugar beet and cordgrass (*Spartina townsendii*); Curve 3: cotton; Curve 4: beans. Group I - halophytes; Group II - halophytes and nonhalophytes; Group III - very salt sensitive nonhalophytes (from Greenway and Munns 1980).

Halophytes and glycophytes also differ in their regulation of Na^+ and K^+ content. Storey and Wyn Jones (1979) reported that halophytes have a higher affinity for Na^+ and lower affinity for K^+ than glycophytes. Ion ratios of K^+/Na^+ for shadscale, an ion includer, and winterfat, an ion excluder, were 0.25 and 2.0, respectively (Breckle 1976). Caldwell et al. (1971) reported shadscale shoot tissue contained ten times more Na^+ than winterfat tissue. In the Mojave Desert, dry weight content of sodium measured in the leaves of the halophytes shadscale, desert saltbush (*Atriplex polycarpa* (Torr.) Wats.), pickleweed, and greasewood were 6.83, 11.11, 8.45 and 5.91%, respectively (Wallace et al. 1973b). Sodium content in glycophyte leaves was considerably less, with big sagebrush (*Artemisia tridentata* Nutt.), creosote and Indian ricegrass (*Oryzopsis hymenoides* (R.&S.) Ricker) respectively containing 0.01, 0.06 and 0.004%.

Ion concentrations in the leaves also varies on a seasonal basis. Figure 4 shows yearly changes in Na^+, K^+, and Cl^- concentrations on a dry weight basis for old and new leaves of the glycophyte winterfat and the halophyte shadscale (Breckle 1976). Although winterfat is adapted to saline environments shadscale is better able to survive extreme drought and saline conditions (Caldwell 1971).

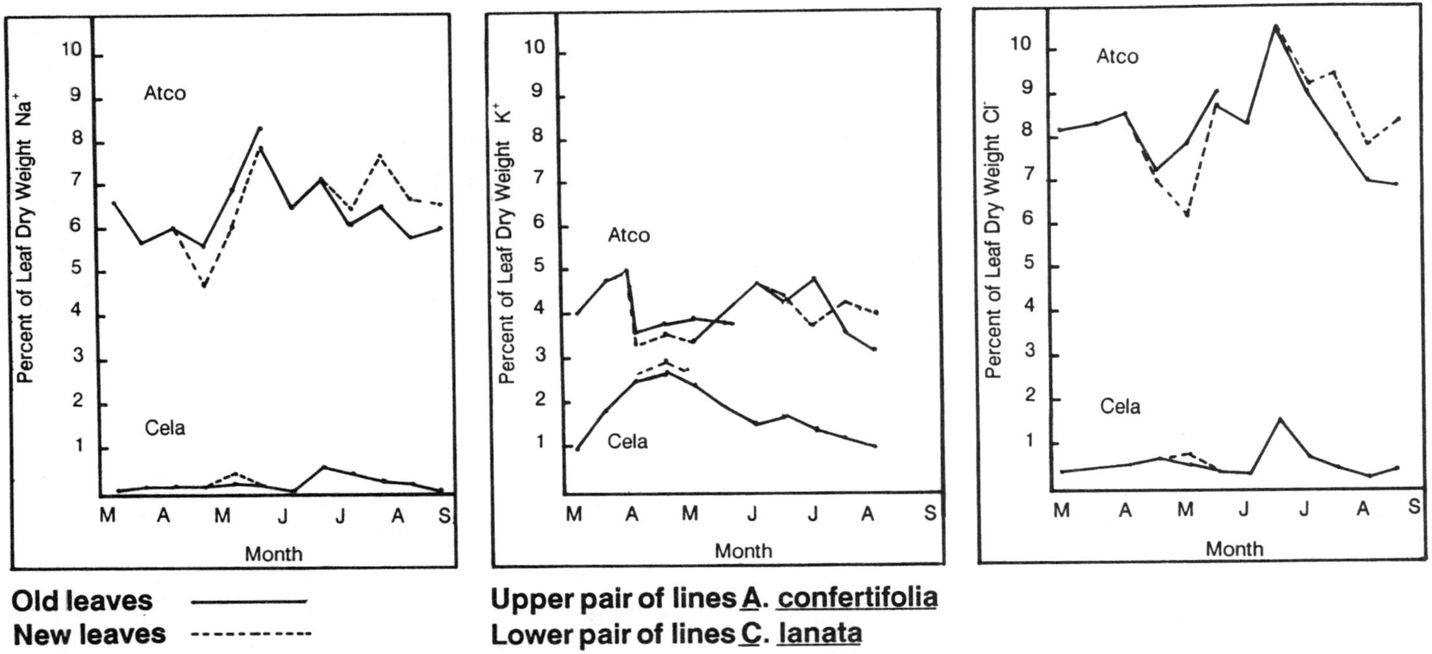

Figure 4. Seasonal leaf content (March - Sept.) of Na^+, K^+, and Cl^- fore shadscale (*Atriplex confertifolia*) and winterfat (*Ceratoides lanata*) (from Breckle 1976).

One of the distinguishing characteristics of many halophytes in contrast to glycophytes, is increased growth in the presence of moderate concentrations (50 to 250 mM) of NaCl (Nieman 1962, Flowers et al. 1977, Flowers et al. 1986, Marschner 1986). Kleinkopf et al. (1975) reported maximum leaf production in shadscale was 2.2 times greater for plants grown in 50 me/l of NaCl than plants grown in its absence (Table 2). Increased growth with the addition of NaCl to the external solution has also been reported for shadscale, desert-holly (*Atriplex hymenelytra* (Torr.) Wats.), and to a lesser degree fourwing saltbush (*A. canescens* (Pursh.) Nutt.) (Wallace et al. 1973b) and *A. triangularis* Willdenow (Drake and Ungar 1989). In the Great Salt Lake Region, halophytes become smaller and more spindling if soil alkali content becomes less than optimum for growth (Flowers 1934). Increased growth, produced by NaCl, however, in many halophytes may be a physical rather than a metabolic response.

Although halophytes are well adapted to saline environments, seed germination of most halophytic species increases as saline conditions decline (Ungar 1978, 1982). The ability to germinate under increasing salinity also varies within species (Workman and West 1967). Ungar (1978) reported that halophyte seeds can survive long periods of exposure to saline conditions, but successful germination is likely dependent on prior precipitation, particularly in arid environments (Flowers et al. 1986). Seed germination in salt deserts in the Great Basin, usually occur following the winter period of high precipitation and minimum evaporation, when total water potentials are least negative due to high water content and salts are leached from the upper soil profile (Roundy et al. 1984). In the absence of spring rains, total water potential of the surface soil decreases rapidly as soil matric and osmotic potentials decrease due to water loss, upward movement of salts and deposition of salts on the soil surface. Roundy (1987) and Roundy et al. (1984) characterized surface soil and seedbed ecology in salt deserts. It is likely that germination and establishment of plants in salt deserts is restricted to years of unusually high precipitation (Went 1955, West 1979). Flowering in halophytes was either reduced or not effected with increasing salt concentrations in the external media, but seedling survival remained high (Okusanya and Ungar 1983). Saline conditions increased flowering in glycophytes but severely reduced seedling survival.

Competition from associated vegetation also plays a major role in determining the success of seedling establishment. Bakker et al. (1985) reported that although glycophyte seeds were capable of germinating under saline conditions, seedlings were unable to compete with halophytic species. In an Alaskan salt marsh, plant

Table 2. Some leaf characteristics of *Atriplex confertifolia* grown at different salt levels (from Kleinkopf et al. 1975).

NaCl me/l	Leaf yield dry wt. g/plant*	Dry weight per leaf mg	% dry matter
0	6.52	6.6	24.2
10	3.22	11.2	22.0
50	14.31	14.3	26.9
250	10.14	13.3	19.2
400	5.34	16.1	18.6

*LSD .05 3.00. LSD .01 4.26

species composition along a salt gradient were sorted out by limitation of physiological tolerance towards one end of the gradient, and by competitive ability towards the other (Snow and Vince 1984).

Effects of Salts on Plants

Specific responses of plants to salinity are often difficult to isolate (Caldwell 1974, Greenway and Munns 1980) because effects resulting from ion toxicity and/or water deficit cannot always be separated. In addition, phenology may affect the level of sensitivity to salt ions (Bernstein and Haywood 1958). Effects of salt ions on plants are also influenced by environmental conditions, type of ions present in the solution, duration of exposure to ions and concentration of salts (Caldwell 1974, Marschner 1986). For example, salt stress in most species increases with increasing vapor pressure deficit, causing transpiration rates to increase mass flow of ions to the root surface (Greenway and Munns 1980, Wyn Jones 1981, Flowers et al. 1986) and movement from root to shoot (Salim 1989). However, for some species such as the salt-tolerant silvery orache (*Atriplex halimus* L.), presence of NaCl in the soil solution appears to be necessary for plants to maintain optimum growth with increasing vapor pressure deficit (Gale et al. 1970).

Past research, which examines salinity effects on plants may be difficult to interpret due to (1) inadequate characterization or control of the root environment, (2) changes in salt concentrations with time in the root zone and (3) Na^+ being the most frequently used ion to determine salt effects on plants with few experiments evaluating the effects of other ions such as Ca^{++} and Mg^{++}. For instance, recent work reported several species of gum tree (*Eucalyptus* spp. L'Her.) responded differently to Na^+, Mg^{2+} and Cl^- (Marcar and Termaat 1990).

Salt injury can be classified as either direct (primary) or indirect (secondary) (Levitt 1980). High soil salinity primarily affects plants in four ways; directly by (1) influencing metabolic processes and (2) membrane permeability, and indirectly by (3) reducing soil water availability and (4) inducing nutrient deficiency (Bernstein and Francois 1964, Wyn Jones 1981, Marschner 1986). Reduction of plant growth through direct effects of high salinity on metabolic processes not only includes inhibition of metabolic processes but also increases in energy demands (Levitt 1980, Yeo 1983, Marschner 1986). Increased energy is required for maintenance of ion balance, moving ions against large thermodynamic gradients, osmotic adjustment, energy needed for excess ion secretion, repairing cellular damage, and movement of ions into cellular compartments.

Salinity stress can also induce morphological and anatomical changes in plants which include increased succulence and thickening of the cuticle, decreased stomata density, extensive development of callosity (hardening), earlier occurrence of lignification, inhibition of differentiation, and changes in number and diameter of xylem vessels (Strogonov 1964, Waisel 1972, Poljakoff-Mayber 1975). Some of the effects of increasing salinity in shadscale leaves are summarized in Table 3 (Kleinkopf et al. 1975).

In salt-sensitive species, the earliest response to salinity is a reduction in leaf growth (Munns and Termaat 1986). Shoot growth is generally affected more than root growth and the effect may occur within minutes of exposure to salinity (Matsuda and Riazi 1981). Munns and Termaat (1986) concluded that the immediate (or short term) reduction in leaf expansion is probably induced by low soil water potential, communicated from the root to shoot via growth regulators (e.g., cytokinins). Long term reduction in leaf expansion is caused by excessive ion accumulation in the plant causing leaf mortality which may exceed leaf production.

However, growth increases in many halophytes in the presence of moderate concentrations of NaCl. Increased dry weight may simply be caused by an increase in NaCl content of the plant. Also, NaCl is an important osmotica necessary for osmotic adjustment, in saline environments (Williams 1960, Caldwell 1974). Salt may also increase carbon assimilation in some halophytes due to faster leaf expansion induced by succulence or increased turgor pressure (Caldwell 1974). Sodium induced growth, however, may also be due to an increase in ATP production (Waisel 1972). In succulent halophytes, Na^+ crosses the plasmalemma without an expenditure of energy. Movement of Na^+ across the tonoplast and into the vacuole is thought to result in ATP production due to the functioning of H^+-translocating ATPase. In some halophytes, Na^+ is also thought to replace K^+ in some metabolic processes (Levitt 1980, Flowers 1985).

Plants which exist in saline environments have developed unique mechanisms to deal with these problems. It is important to note that plants adapted to these environments must cope with all four factors (metabolic processes, membrane permeability, and soil water and nutrient deficiency) collectively. For example, avoidance of salt stress through exclusion of salt uptake reduces the danger of ion toxicity but may decrease water availability in the external solution due to inadequate osmotic adjustment by the plant. Absorption of salts may improve plant water availability, but at the same time increase potential ion toxicity.

Metabolic Processes

The primary effect of ion toxicity is inhibition of enzyme activity, resulting in disruption of protein synthesis, the accumulation of non-protein nitrogen compounds, altered nucleic acid metabolism, stimulation of respiration, and inadequate energy production by photophosphorylation (Slayter 1967, Marschner 1986). Disruption in protein metabolism is attributed to a decrease in protein synthesis and an increase in protein hydrolysis (Uprety and Sarin 1976). Prikhod'ko and Klyshev (1963) suggested the decline in protein synthesis may be due to inhibition in amino

Table 3. Some physiological effects* of salt applied to soil on *Atriplex confertifolia* leaves (from Kleinkopf et al. 1975).

NaCl me/l	Chlorophyll	Soluble protein	Nitrate	Nitrate reductase	Water potential	Leaf Thickness
	- - - - - - - - -	mg/g fw	- - - - - - - - -	mols/g fw	bars	mm
0	0.75	12.8	204	4.4	19.3	0.2
10	0.75	9.7	167	4.9	21.4	0.3
50	0.58	9.0	102	3.3	34.8	0.4
250	0.41	5.4	93	0.6	52.8	0.6
400	0.33	5.1	93	0.7	50.3	0.8

*Chlorophyll and soluble protein, mg per g fresh weight. Nitrate, mg per g fresh weight. Nitrate reductase, moles nitrate per hr per g fresh weight. Water potential. - bars.

acid metabolism. The Na^+/K^+ ratio is also important in determining the sensitivity of protein synthesis to salt. High Na^+/K^+ ratios inhibit protein synthesis through the decline in translational efficiency (Wyn Jones et al. 1979). However, increased protein synthesis has also been reported to occur in some plants under salt stress (Levitt 1980). Abd el Rahman et al. (1972) reported salinity increased nitrogen content and decreased carbohydrate concentrations in several desert forage plants.

Surprisingly, enzymes in salt-tolerant species are as sensitive to salinity as nonadapted species (Greenway and Osmond 1972, Flowers et al. 1977, Greenway and Munns 1980). Greenway and Osmond (1972) reported that the enzyme PEP carboxylase (primary carboxylation enzyme in C_4 pathway) was affected more by salt in salt-tolerant species than salt-sensitive species. However, internal salt concentrations in salt-tolerant species do not exert the same disruptions in protein metabolism as that which occurs in salt-sensitive species (Caldwell 1974). Mechanisms which allow salt-tolerant species to protect enzymes from excess ions will be discussed in later sections.

The study of enzyme sensitivity to ions has proven to be rather difficult (Cheeseman 1988). Generally, enzymes are removed from plant tissue which may affect sensitivity of an enzyme to salt ions (Flowers et al. 1986). It is also possible that production of additional enzymes may compensate for enzyme inhibition under saline conditions. Billard and Boucaud (1982) reported that activity of a salt-sensitive enzyme increased when plants were grown in the presence of NaCl in the external solution. Relatively high concentrations of Cl^- (100–200 mM) stimulated NADP malic enzyme activity in two C_4 umbrella-sedge species (*Cyperus* spp. L.)(Murata et al. 1989). Maintenance of a low ratio of Cl/(Na+K) ions in the cytoplasm of most salt-tolerant species is also important since Cl^- is very disruptive to enzyme activity (Gimmler et al. 1984).

Gas exchange has been shown to also be affected by salinity (Levitt 1980, Marschner 1986), although to a lesser extent in salt-tolerant than salt-sensitive species (Caldwell 1974). In several studies, salinity influenced photosynthesis by decreasing chlorophyll content, leaf area, stomata conductance, and mesophyll resistance (Poljakoff-Mayber 1975, Levitt 1980, Marschner 1986, Chavan and Karadge 1986). Warren and Brockelman (1989) found reduced stem growth in saltgrass with increasing salinity which was a result of reduced photosynthesis and increased leaf respiration. Reduced rates of CO_2 fixation with increased levels of salinity in saltgrass were partially attributed to increased stomatal resistance (Kemp and Cunningham 1981).

Salinity also has been shown to limit some of the biochemical pathways of photosynthesis. Bowman (1987) concluded that reduced net CO_2 uptake in a salt-sensitive population of bushy beardgrass (*Andropogon glomeratus* Walt.) was due to changes in carboxylation efficiency. He suggested that salinity decreased the carboxylation/oxidation ratio of ribulosebiphosphate carboxylase-oxygenase (RuBisCo, the primary carboxylation enzyme in the Calvin cycle–C_3). Heuer and Plaut (1989) also concluded that salt-induced effects on photosynthesis were caused by inhibition of RuBisCo activity. However, Flanagan and Jefferies (1988) reported that the decline in photosynthesis in saline soils was not caused by a reduction in biochemical capacity but reduced leaf conductance and intercellular CO_2 concentrations. Although photosynthetic rate generally decreases with an increase in salinity (Marschner 1986), several plant species reportedly increase photosynthesis rate with the addition of salt to the external media (Chavan and Karadge 1986, Heuer and Plaut 1989). In other plant species, increases in salinity had minimal effect on photosynthesis until cell damage occurred (Plaut et al. 1989).

Salinity affects both mesophyll and stomatal resistance. Mesophyll resistance may decrease due to an increase in mesophyll area/leaf area ratios induced by an increase in leaf succulence and thickness (Longstreth and Nobel 1979). However, these researchers found as external salt concentrations increased, mesophyll resistance changed little in salt-tolerant species and increased substantially in salt-sensitive species. Transpiration rates were affected by salinity via increased stomata resistance caused by a reduction in stomata opening and density (Flowers 1985, Flanagan and Jefferies 1988). Flowers (1985) reported the decrease in stomata density was a consequence of succulence in saltgrass and samphire. The effects of salinity on stomatal resistance have also been shown to be influenced by temperature and vapor pressure deficit. In shadscale, elevated salt concentrations did not affect transpiration rates at 20°C, but did decrease transpiration at 30°C (Kleinkopf et al. 1975). Sen and Chawan (1975) reported that although salt stress may cause stomatal closure, transpiration may increase if additional salts enter the guard cells, inducing stomates to open.

Membrane Permeability

Increased salinity can also increase permeability of the plasmalemma, thereby, decreasing ion selectivity, disrupting osmoregulation and leading to ion toxicity (Wyn Jones 1981). Calcium is an essential ion in maintaining ion selectivity and low cell permeability (Epstein 1972). High Na^{+} in the external media increases the potential for this ion to exchange for Ca^{++} at the cell membrane in the root. If the ratio of monovalent to divalent ions increases, membrane permeability increases leading to ion toxicity in saline soils. Hannon and Barber (1972) report that the highest concentrations of Ca^{++} were found in salt-tolerant grasses as compared to salt sensitive grasses. Calcium maintains ion selectivity through the ion transport mechanism (Epstein 1972). If Ca^{++} availability is limited in the external medium, uptake of other cations will occur. In sodic soils, availability of Ca^{++} in the soil solution becomes limited due to precipitation of $CaCO_3$ and $CaSO_4$. Lipid composition in root membranes which also affects ion selectivity (Radda 1975), may be altered when exposed to salinity (Marschner 1986).

Soil Water Availability

In order for salt-tolerant species to exist in highly saline environments, they must be able to acquire soil moisture against very steep water potential gradients. Water availability is directly related to the water potential gradient between the plant root and the soil (Wyn Jones 1981, Marschner 1986). Water uptake by plants is not an active process but must move along a water potential gradient from less negative (soil) to more negative (plant root) (Levitt 1980). As salt concentrations increase in the soil, osmotic potentials become more negative, lowering soil water potentials. Water flow into the plant will decrease unless the plant is able to continue to adjust its internal water potential below that of the soil.

Nutrient Deficiency

Salt ions such as Na^+, Cl^-, Mg^{++} and SO_4^{--} can induce nutrient deficiencies by limiting uptake of essential nutrients, e.g., K^+, Ca^{++}, HPO_4^-, NH_4^+ and NO_3^- (Waisel 1972, Levitt 1980, Wyn Jones 1981, Marschner 1986, Bradley and Morris 1991). For instance, inhibition of NO_3^- uptake by NaCl reduces the nitrogen supply in the plant (Cram 1973, Smith 1973, Deane-Drummond and Glass 1982). Reasons for reduced uptake of essential nutrients are several and include competition for the uptake site with large numbers of salt ions in solution (Caldwell 1974, Levitt 1980) and competition with ions of similar valence (Epstein 1972).

In barley (*Hordeum vulgare* L.) and several species of saltbush, uptake of Na^+ in place of K^+ caused potassium deficiency (Greenway and Rogers 1963, Osmond 1968). Kleinkopf et al. (1975) reported elevated concentrations of NaCl reduced internal levels of nitrate and uptake of Mg, K, and P in shadscale. Decreased nitrogen uptake may also occur due to low rates of ammonification and nitrification in saline ecosystems (Wollenweber and Zechmeister-Boltenstern 1989) or inhibition of nitrogen fixation in legumes (Tu 1981).

In addition to the above factors, salt-induced water stress may reduce a plant's capacity to absorb nutrients. This may be caused by salinity limiting root growth (Levitt 1980) and impairing long distance transport in the plant of important nutrients (Drier and Goering 1974, Flowers 1985). For instance transport of Ca^{++} from root to shoot and its release into the root xylem is very sensitive to increased salinity (Lynch and Lauchli 1985). Lynch et al. (1988) suggests that the initial decrease in leaf growth of barley in saline soils may be associated with reduced flow of Ca^{++} to regions of expanding leaves. However, in terrestrial saline environments, high Ca^{++} levels may be toxic as well as Na^+ (Maas and Neiman 1978).

Although Na^+ can cause nutrient deficiency in plants (see section on nutrient deficiency), Na^+ appears to be a required micronutrient in C_4 species (Nable and Brownell 1984). Bladder saltbush, growing in cultures with no Na^+, became chlorotic, developed necrotic patches on the leaf tips, made little growth, and over time some individuals eventually died (Brownell and Wood 1957, Brownell 1964, Brownell and Crossland 1972). Brownell and Crossland (1972) also reported that a species of purslane (*Portulaca* spp. L., a C_4 plant) failed to flower in the absence of NaCl. C_3 species studied did not require Na^+. Brownell (1967) found a similar distinction between other C_3 and C_4 species. NaCl stress has been shown to induce C_3 metabolism to shift to CAM photosynthesis in ice plant (*Mesembryanthemum crystallinum* L.) (Winter and Von Willet 1972). Bloom (1979) suggested that Na^+ accumulation is necessary to induce CAM metabolism in this succulent. Sodium may be a required nutrient in the PEP-carboxylation process found in both C_4 and CAM plants (Brownell 1967). Although evidence indicates a difference in requirement for Na^+ between photosynthetic pathways, both C_3 and C_4 species include successful halophytes (Caldwell 1974). In general, mechanisms for salt accumulation, excretion and synthesis of organic acids do not appear to be different between C_3 and C_4 halophytes.

Adaptations to Saline Environments

Enzymes or metabolic processes resistant to high salt concentrations within the plant does not appear to be a mechanism allowing plants to grow in saline environments. Primary mechanisms leading to plant adaptations to saline environments are; (1) maintenance of cell turgor, (2) formation of cytosolutes, (3) compartmentation, (4) succulence, (5) elimination and (6) ion selectivity.

Maintenance of Cell Turgidity

Maintenance of cell turgor is essential for plant growth, continued root exploration and CO_2 fixation (Hsiao 1973). The ability to adjust osmotic potential to varying and very low soil water potentials is an important mechanism allowing salt tolerant species to survive in saline environments. Osmotic adjustment enables plants to maintain turgor while continuing to acquire soil water at very low soil water potentials. The capacity to accumulate solutes (lowering osmotic potential), highly elastic cells, and small cells all aid in maintaining positive turgor pressure as water content in plant tissues decrease (Turner and Jones 1980).

To acquire moisture, plant roots must maintain a more negative water potential than that of the soil. Osmotic adjustment (the lowering of osmotic potential arising from the net accumulation of solutes in the vacuole) is a mechanism by which plants may increase the water potential gradient from roots to soil. Osmotic adjustment may also occur through a decrease in tissue water content which concentrates salts in the cell sap, typical of many succulent and Poaceae species (Wyn Jones 1981, Glenn and O'Leary 1984). Succulent species have been observed to reversibly tolerate water losses of up to 50% in their shoot (Glenn and O'Leary 1984).

Osmotic adjustment in dicotyledonous halophytes is primarily via accumulation of salts in the vacuole (Larcher 1980, Marschner 1986). It has been reported that salt concentrations in some halophytes exceed that of sea water (Flowers et al. 1977). However, as noted earlier, evidence indicates enzymes in halophytes are equally as sensitive to high salt concentrations as glycophytes (Caldwell 1974, Flowers et. al. 1977). Salt accumulating halophytes must be able to protect enzymes in the cytoplasm and chloroplast from high concentrations of toxic ions such as Na^+ and Cl^- (Marschner 1986). Although Na^+ is the most common ion used by halophytes for osmotica, a diversity of ions and organic solutes may be accumulated. Monocotyledonous halophytes, particularly Poaceae, accumulate sugars and potassium in the vacuole to attain osmotic adjustment (Flowers et al. 1986). Flowers 1985 and Marschner 1986 reported that internal concentrations of Na^+ and Cl^- in Poaceae do not exceed external concentrations. In glycophytes osmotic adjustment depends upon the synthesis of compatible solutes such as soluble sugars, amino acids or uptake of nutrient ions such as K^+, Ca^{++}, or NO_3^- (Yeo 1983, Marschner 1986)(also see section on cytosolutes).

The ability of a dicotyledonous halophyte to osmotically adjust is related to the quantity of ions delivered to the shoot per unit of root mass and time (Flowers et. al. 1977). Rate of ion absorption and translocation is considerably greater in halophytes than glycophytes (Levitt 1980, Wyn Jones 1981). Clipson (1987) reported tolerance of dicotyledonous halophytes to large changes of salinity in the external solution is attributed to rapid uptake and movement of Na^+ and Cl^- for quick osmotic adjustment and maintenance of shoot water potentials substantially more negative than the external solution.

Close regulation of osmotic adjustment has been shown to be an important characteristic in determining salt-tolerance in both halophytes and glycophytes. Glenn and O'Leary (1984) reported that halophytes from more saline habitats maintained an osmotic potential within the shoot proportional to levels in the external media better than did halophytes from less saline environments. In another study, Gorham et al. (1985) reported similar results of tightly controlled osmotic adjustment for cereals that survived and continued to grow under saline conditions (exposed to 250 mol m^{-3} NaCl). Change in sap osmotic potential in basin wildrye and Russian wildrye closely followed the change in external osmotic potential (% of control, in Table 4), whereas less tolerant species, wheat (*Triticum aestivum* L.) and wild wheat (*Triticum dicoccoides* Koern.), increased sap osmotic pressure to a much greater degree than the external media. They concluded that the degree of salt-tolerance increased with strict control of ion uptake, rather than ion exclusion.

The ability of plant species to regulate tissue water content and internal osmotic potential may also be important in determining their competitiveness on saline sites. Moore et al. (1972a) reported that shadscale, a NaCl includer, had a broader range of tissue moisture contents and osmotic potentials during the growing season than winterfat, an excluder. High ion uptake by shadscale was one of the factors attributed to prolonging the growing period late into the summer. Breckle (1976) reported that the salt-excluder big sagebrush, although moderately salt-tolerant, was not capable of competing on saline sites with more salt-tolerant species that accumulate high concentrations of salts. Osmotic adjustment through salt uptake may allow halophytes a longer period of resource acquisition on saline sites.

Development and maintenance of water potential gradients between the soil and plant root, and within the plant, requires an expenditure of energy. Plants must accumulate and maintain a greater concentration of solutes in root cells than exists in the soil solution. Halophytes commonly maintain concentrations of Na^+ several times higher than the external solution. Osmotic potential gradients within the plant may be as large as 6.1 to 8.1 MPa (Caldwell 1974). There are two possible ways

Table 4. Osmotic adjustment (mOsmol kg^{-1}) in leaves of *Leymus sabulosus*, *Elytrigia juncea*, *Triticum aestivum* cv. Ciano 79 and *T. dicoccoides* (from Gorham et al. 1985

Parameter	*Leymus sabulosus*		*Elytrigia juncea*		*Triticum aestivum*		*Triticum dicoccoides*	
External NaC1(molml)	0	250	0	250	0	100	0	100
Days at full stress		54		66		21		21
Measured sap osmotic pressure	502	1003	468	911	443	883	392	1184
Sap osmotic pressure corrected for sucrose hydrolysis	497	990	458	862				
Contribution of Na + K salts (% of total)	410 (82%)	810 (82%)	311 (68%)	650 (75%)	245 (55%)	637 (72%)	283 (72%)	851 (72%)
Sap O.P. - external O.P. (% of control)	487	530 (110%)	448	402 (98%)	433	693 (160%)	382	94 (260%)

Na^+ may be accumulated in the root vacuole for osmotic adjustment (Yeo 1983). First, Na^+ entry may be limited by membrane permeability, requiring an ion carrier to accumulate Na^+. Second, the membrane is permeable to Na^+, leading to excess ion accumulation which the plant must exude actively to prevent toxicity. In either case, energy will be required to control Na^+ concentrations. However, osmotic adjustment through exclusion of salts and synthesis of organic solutes could require considerably more energy costs than the active uptake of salts (to be discussed in more detail in a later section) (Greenway and Munns 1980, Yeo 1983).

Cell elasticity (the ability of the cell membrane to stretch or contract) is one mechanism which allows a plant to maintain turgor pressure with a decrease in water content (Bowman 1988). Tissue elasticity has been shown to increase with salinity. The development of smaller cells, induced by water stress, also tends to increase cell elasticity (Turner and Jones 1980). The stretching and contracting of the cell membrane for maintenance of turgor pressure will improve a plant's ability to grow and compete in a saline environment.

Cytosolutes

The accumulation of metabolically compatible organic solutes (cytosolutes) is necessary for osmotic adjustment, maintaining osmotic equilibrium under saline conditions in both halophytes and glycophytes and maintaining cation-anion balance (Greenway and Munns 1980, Wyn Jones 1981, Marschner 1986, Haynes 1990). Cytosolutes are used for osmotic adjustment in salt-excluders and maintenance of osmotic equilibrium within the cells of salt-includers.

In halophytes, large electrolyte gradients across the tonoplast are developed due to accumulations of Na^+ and Cl^- in the vacuole for osmoregulation (Wyn Jones 1981). In many chenopods, this gradient is modified by an accumulation of K^+ and glycinebetaine in the cytoplasm. Phloem, which also generally contains low levels of Na^+ and Cl^-, requires cytosolutes to maintain its water balance (Wyn Jones 1981). Although most halophytes accumulate high concentrations of Na^+ and Cl^- in the shoots, some species contain relatively low levels (Waisel 1972). For instance, Albert and Popp (1977) reported high concentrations of soluble sugars in several halophytes, capable of contributing to leaf osmotic pressure. In nonhalophytic graminoides, differences in salt-tolerance among ecotypes has been partially attributed to production of organic solutes for osmotic adjustment (Hudson et al. 1985). It has also been proposed that cytosolutes may protect enzymes from inhibitory effects of salts, although little data exist to support this hypothesis (Pollard and Wyn Jones 1979, Greenway and Munns 1980).

The accumulation and degradation of organic acids and their translocations within the plant may be an important mechanism in maintenance of cation-anion balance in plants (Haynes 1990). Plants usually absorb unequal quantities of cations (NH_4^+, Ca^{2+}, Mg^{2+}, K^+, Na^+) and anions (NO_3^-, Cl^-, SO_4^{2-}, $H_2PO_4^-$). Organic salts (malate, oxalate, citric etc.) which make up a major portion of electrolytes in many halophytes (Caldwell 1974) may also maintain cation-anion balance in the plant. In one study, Williams (1960) reports that halogeton contained over 30% oxalates on a dry weight basis. He found the major portion of sodium absorbed is metabolized into soluble sodium oxalate. The role of oxalate production in the plant is unknown (Caldwell 1974). However, Epstein (1972) and Flowers et al. (1977) speculate that organic acids may balance unequal rates of cation and anion uptake. Oxalate being negatively charged may balance internal cell electrostatic gradients when cation uptake from the soil solution is greater than anion uptake. Charge balance is also maintained by excreting H^+ and OH^- into the rhizophere. The mechanism of excretion, however, is poorly understood.

Cytosolutes may also be important in the initial phases of osmoregulation following exposure to osmotic shock (McNulty 1985). Osmotic shock, the sudden exposure to elevated salinity, can cause an increase in solutes other than Na^+ and Cl^- during the initial phases of osmotic adjustment. Doddema et al. (1986) concluded that carbohydrates were used in the initial phases of osmotic adjustment when a plant was exposed to osmotic shocks. However, other compatible compounds were necessary to maintain long-term osmotic equilibrium. McNulty (1985) found an immediate osmotic response with European samphire when exposed to a sudden increase in NaCl (100 mM). Magnesium concentrations increased within the first 2 hours, while sugars increased 5-fold within 6 hours which provided adequate amounts of solutes in the cytoplasm for osmotic balance. However, within 24 hours osmotica had decreased with the exception of Na^+, Mg^{++}, Cl^- and proline. Although elevated levels of metabolically-compatible osmotica were not maintained at sufficient levels to balance the increase in salts in the vacuole 72 hours after increasing NaCl in the external solution, osmotic balance within the cell may have occurred by redistribution of osmotica in the cell. Storey and Wyn Jones (1977) reported that halophytes in the family Chenopodiaceae maintained high concentrations of betaine (an important solute compatible with metabolic processes) even under non-saline conditions. These solutes may be reapportioned from the vacuole to the cytoplasm to maintain osmotic equilibrium during conditions of increasing salinity (Flowers et al. 1986).

Cytosolutes include sugars, sugar alcohols, amino acids, glycinebetaine, sorbitol and proline. The specific cytosolutes accumulated in a plant appear to be taxonomically related (Wyn Jones 1981). Glycinebetaine is primarily found in Chenopodiaceae, beta-alaninebetaine in Plumbaginaceae, sorbitol in Plantaginaceae, and prolinebetaine in Fabaceae. Soluble sugars are dominant cytosolutes in many salt excluding monocot species, predominately in the Juncaceae, Cyperaceae and Poaceae families.

Glycinebetaine is a primary cytosolute found in many salt-tolerant species (Wyn Jones 1981) and generally increases with salt-tolerance (Storey and Wyn Jones 1979). It is absent in salt-sensitive species (Greenway and Munns 1980). Histochemical studies of herbaceous seablite (*Suaeda maritima* [L.] Dumort) showed glycinebetaine to be present in the cytoplasm and absent in the vacuole (Hall et al. 1978).

The role of proline as a cytosolute is less clear. Proline increases in many plants with water stress (Wyn Jones 1981) and is found in salt-tolerant ecotypes of nonhalophytic graminoides (Rozema et al. 1978, Ahmad et al. 1981, Hudson et al. 1985). However, in several salt-sensitive and tolerant turfgrasses, increases in proline content were too low in relation to Na^+ accumulation to have any osmoregulatory role (Torello and Rice 1986). In pop saltbush (*Atriplex spongiosa* R. Muell.)

proline only increased when NaCl was high enough to inhibit growth and leaf succulence, whereas glycinebetaine increased under only small inhibitions of growth (Greenway and Munns 1980). In several other species, glycinebetaine increased with no concurrent increases in proline over an increasing range of salinity in the external media. Greenway and Munns (1980) suggested the adaptive role of proline is related to survival rather than to maintenance of growth.

Organic solutes are an expensive source of osmotica, competing for energy with other plant processes (Greenway and Munns 1980, Yeo 1983). Although ions such as Na^+ are potentially toxic osmotica, they are an energetically cheap source of solute for osmotic adjustment (Marschner 1986). Reduced stem growth of saltgrass with increasing salinity has been partially attributed to the diversion of fixed carbon to the production of osmotica (Daines and Gould 1985). Monocots which rely on organic solutes typically grow slower under saline conditions than dicots which are ion accumulators (Yeo 1983). Osmotic adjustment in mature cells, where vacuoles make up 90 to 95% of the cell volume, requires large amounts of solutes, which may account for 20 to 30% of the solute dry weight (Greenway and Munns 1980). Yeo (1983) suggests that a mature vacuolated cell osmotically adjusting to 200 mol m^{-3} of NaCl in the external solution with sugar would require 0.5-g glucose in solution for each existing gram of dry weight.

Compartmentation

The ability of halophytes to osmotically adjust requires substantial movement of ions into leaf cells (Flowers et al. 1977, Wyn Jones 1981, Marschner 1986). Accumulation of high concentrations of ions in the plant requires either metabolic tolerance in the presence of high ion concentrations, or separation of non-compatible ions from metabolic processes (Yeo 1983). Metabolic tolerance is unlikely since enzyme activity is inhibited by ion concentrations commonly found in halophytes (Caldwell 1974, Flowers et al. 1977, Yeo 1983).

One means to reduce potential toxicity is through the process of compartmentation. Yeo (1983) defines compartmentation as the separation of inorganic ions in the vacuole so that they do not interfere with metabolic processes in the cytoplasm. Compartmentation is considered to be one of the primary mechanisms enabling halophytes to tolerate high internal salt concentrations. Sensitivity of some species to salts has been attributed to inadequate compartmentation (Greenway and Munns 1980) and leakage of ions from the vacuole back to the cytoplasm (Seemann and Critchley 1985). However, little evidence directly relates compartmentation to salt-tolerance (Cheeseman 1988).

The separation of salts into different parts of the cell strongly supports compartmentation as a mechanism allowing halophytes to absorb large amounts of ions into leaf cells. Under saline conditions, when tissue concentrations exceed 200 mol m^{-3}, salts absorbed into the leaves are mainly accumulated in the vacuole (Gorham et al. 1985). Marschner (1986) reported ion ratios of Na^+/K^+ in pop saltbush were 0/29 in the cytoplasm and 12/4 in the vacuole. The majority of Na^+ and Cl^- ions in herbaceous seablite leaf mesophyll cells, both in terms of total amount and concentrations were located in the vacuole (Harvey et al. 1981). Flowers (1985) estimated that 90% of the salt ions present in this species were present in the shoot and 95%

of those were located in the vacuole. The majority of the mesophyll cells contained an ion ratio for Na^+/K^+ of 23 in the vacuole and a Na^+ concentration 5.2 times greater in the vacuole than in the cytoplasm. This demonstrates the sequestering of Na^+ in the vacuole to reduce toxicity. To maintain osmotic equilibrium across the tonoplast, organic solutes compatible with metabolic processes are present in the cytoplasm.

Distribution of solutes within the cell is controlled by active and passive properties of the plasmalemma and tonoplast (Epstein 1972, Gorham et al. 1985, Marschner 1986). In most halophytes, the tonoplast is highly impermeable to salt, preventing passive diffusion between the protoplasm and vacuole. The tonoplast in glycophytes, however, appears to be less effective in sustaining ion gradients and solute compartmentation (Jeschke 1979). Accumulation of various ions into the vacuole requires transport from the protoplasm into the vacuole (Levitt 1980). Ion pumps have been proposed as a means of transporting ions across the tonoplast membrane. Ion pumps are ATPases (ATP hydrolyzing enzymes) that actively transport ions using energy from ATP hydrolysis (Salisbury and Ross 1985). A K^+/Na^+ ion pump moves 2 K^+ ions into the cytoplasm and 3 Na^+ out of the cytoplasm and into the cell wall or vacuole. Jeschke (1983) concluded that ion pumps move Na^+ and Ca^{++} from the cytoplasm into the vacuole. Chloride also activates a pump moving Cl^- into the vacuole. These ion pumps operate to move ions into and out of the vacuole against large electrolyte gradients. The ability of ATPase enzymes to function under increasing salinity appears to be limited by the energy available to the plasma membrane, not through direct inhibition of the enzyme (Burggemann and Janiesch 1989).

Rates of ion uptake and translocation are critical factors for the process of compartmentation to effectively work (Gorham et al. 1985, Cheeseman 1988). Small imbalances between rate of ion supply to the shoot and the rate of ion accumulation into cells would have a detrimental effect to plant cells (Yeo and Flowers 1986). Ion uptake and translocation rates exceeding ion movement into the vacuole, would result in excessive ion concentrations in the cytoplasm or apoplast (interconnecting cell walls and intercellular spaces). However, limited ion uptake and transport into the cell would cause inadequate osmotic adjustment inducing cell dehydration.

Succulence

Internally, plants respond to concentrations of salts within the cell rather than absolute quantities of salt (Larcher 1980). The progressive uptake of salts which may lead to increased salt concentrations can be controlled by succulence (Caldwell 1974, Larcher 1980, Levitt 1980, Marschner 1986). Succulence (mg water cm^{-2} of leaf surface area) is the result of increasing internal plant water content which causes cells to distend due to salt uptake (Jennings 1976, Longstreth and Nobel 1979). This mechanism enables plants to dilute salt concentrations in the cell. Jennings (1968) noted water content increased in plants growing in salt solutions. He proposed the hypothesis that water uptake increases in order to dilute sodium concentrations, thus allowing for continued plant growth. Apparent resistance of the protoplasm and yield of plants growing in saline environments correlates well with succulence (Repp et al. 1959). An extended period of leaf development and metabolic activity in saline environments have also been partially attributed to succulence

(Black 1958, Moore et al. 1972a).

Increased succulence as a response to salinity is primarily found in dicots (Greenway and Munns 1980), although it has been reported in saltgrass (Kemp and Cunningham 1981). In some halophytes, regulation of ion concentrations with succulence is found only in mature leaves (Greenway et al. 1966, Karimi and Ungar 1983). Succulence induces an increase in leaf thickness and cell size, and a decrease in leaf number and stomatal density (Flowers et al. 1986). Also associated with succulence is an increase in thickness of both the epidermal cell walls and the cuticle.

It appears that high concentrations of Na^+ and Cl^- in the cell contribute to leaf succulence (Moore et al. 1972b). The capacity for Na^+ and Cl^- uptake increases with cell size and vacuole/cytoplasm volume ratio (Gorham et al. 1985). Both cell size and vacuole/cytoplasm ratio increase the H_2O content per unit area of cell (Mendoza 1971). Binet (1985) reported that salinity affected the chemical composition of cell walls which in turn increased cell succulence. Salt may increase cell wall plasticity through an increase in arabinose, maintenance of a high degree of methylation of pectic compounds and high cellulase activity. He also speculated that salt maintained cell plasticity, allowing for dilution of absorbed salts within the cells.

Growth is another factor that influences ion concentrations in the plant (Caldwell 1974, Flowers et al. 1977, Yeo 1983). A balance between rates of ion uptake and growth is critical in determining internal ion concentrations and osmotic potential in the plant. If growth declines and net ion uptake continues, a rapid increase in ion concentrations can lead to ion toxicity (Yeo 1983). If ion uptake exceeds growth, plants may maintain an acceptable balance through secretion, retranslocation or discontinuous distribution.

Elimination

Elimination of high internal salt concentrations becomes necessary when plants take-up large amounts of ions and are unable to adequately sequester them into internal tissue (Caldwell 1974, Yeo 1983). This is particularly true in halophytes when control of ion uptake and transport is limited. Excess salt within a plant may be eliminated by: downward transport through the phloem and secretion from the root; exudation and recretion at the shoot surface; shedding of plant parts heavily loaded with salts, such as abscission of older leaves; or excretion into or through gland cells such as vesicular hairs (Larcher 1980).

Very little evidence supports the theory that excess salts are transported in the phloem and secreted from the roots (Flowers 1985). Evidence indicates that Na^+ and Cl^- are relatively immobile in the phloem of halophytes (Larcher 1980, Yeo 1981, Flowers 1985, Gorham et al. 1985), and that once xylem transport has deposited Na^+ and Cl^- in a leaf, it is then relatively immobile.

Karimi and Ungar (1983) report that essential nutrients were withdrawn from mature leaves prior to leaf abscission while Na^+ and Cl^- remained. Yeo (1983) also reported little movement of Na^+ from mature leaves to young developing leaves. The immobility of Na^+ and Cl^- ions through the phloem suggested that removal of excess ions from the plant by leaf fall is a primary mechanism for reducing excess salts. Several desert shrubs (Marschner 1986) and salt marsh grasses (Kostina 1974) have been shown to remove excess salt ions through leaf fall. Albert (1975) reported that although water content increased substantially in the older leaves of succulent halophytes, the increase was not sufficient to compensate for the increase in ion concentrations. He concluded that abscission of old salt saturated leaves was an important mechanism in regulating salt concentrations.

Gland cells are one of the most studied mechanisms for salt-excretion (Luttge 1971). These salt excreting cells vary in structure and methods of salt release (Esau 1977) and are capable of excreting large amounts of salt (Marschner 1986). Breckle (1976) estimates that on a community level, shadscale excreted 0.4 ton/ha of salt in a single year through bladder hairs. Bladder hairs are a glandular trichome consisting of a large vacuolated bladder cell resting on a stalk cell attached to the epidermis (Fig. 5). Bladder hairs (also called vesicular hairs) are commonly found in the genus *Atriplex* (Schrimer and Breckle 1982). Osmotic potentials in the bladder cell have been reported as low as –51 to –72 MPa (Mozafar and Goodin 1970). Although bladder hairs are an important excretion mechanism, they do not contribute to the osmotic gradient in the leaf since they are external to the leaf surface (Caldwell 1974). In the past, it was suggested that bladder hairs may absorb water from the atmosphere, due to their very low osmotic potential (Volkens 1884, Wood 1923). Mozafar and Goodin (1970), however, reported water was not absorbed by the bladder hairs due to a waxy coating on the outer surface, forming a hydrophobic covering.

Bladder hairs have been shown to be an effective salt-excretion mechanism in species of saltbush (Pallaghy 1970). Bladder hairs function by storing excess salt and eventually burst, depositing salt on the leaf surface. In shadscale, 18% of the dry weight of bladders hairs was salt (Breckle 1976). In several other saltbush species, 25% of the leaf dry weight was inorganic ions, half of which were estimated to be contained in the bladder hairs (Karimi and Ungar 1983). Mozafar and Goodin (1970) reported that as salt concentration increased in the external solution, salt concentration in bladder hairs dramatically increased while xylem exudate increased only slightly.

Saltbush leaves have been shown to potentially produce numerous crops of bladder hairs on a single leaf (Osmond et al. 1969). Schirmer and Breckle (1982)

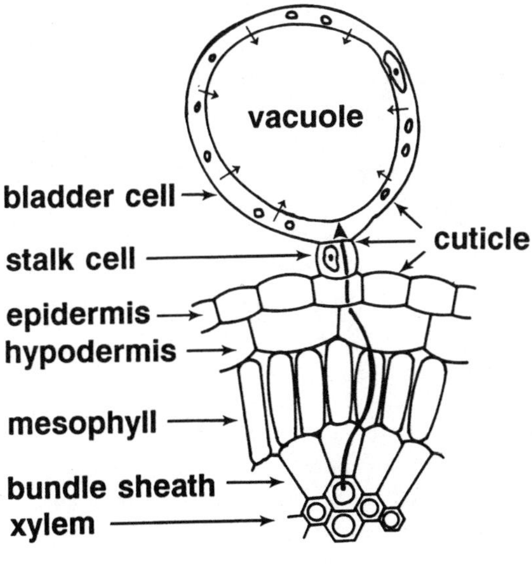

Figure 5. Diagram of a salt-- excreting trichome shown attached to part of an *Atriplex* leaf. The long arrow indicates the path of movement of ions from the xylem of a vein to the bladder cell of the trichome. Short arrows indicate release of ions into the vacuole (from Esau 1977).

found this to be true for shadscale but not for garden orache (*Atriplex hortensis* L.) where bladder hairs only occurred on young leaves. Bladder cells also only occur on young *A. triangularis* leaves, and once burst were not replaced (Karimi and Ungar 1983). For this species, salt concentration in the older leaves was controlled by succulence. Loss of salt hairs in mature leaves has also been reported for other saltbush species (Frankton and Bassett 1968, Troughton and Card 1974). The number of bladder hairs on shadscale leaves was not influenced by external salt concentrations but was genetically fixed (Breckle 1976). However, increased salinity decreased salt hair density and increased bladder cell size in *A. triangularis* (Karimi and Ungar 1989). Decreased light intensity also reduced salt hair density.

Salt ions have been reported to be delivered to bladder cells by transport through the cytoplasm and plasmodesmata in the transpiration stream (Osmond et al. 1969, Esau 1977). Campbell et al. (1974), however, reported apoplastic transport of ions to the glands are also important. As salt ions enter the bladder cell they were carried across the tonoplast into a relatively large vacuole for storage. Transport of ions into bladder hairs against a large ion gradient has been reported to be a metabolically active process. Uptake of Cl^- into bladder hairs was strongly stimulated by light, indicating an energy requirement for transport (Osmond et al. 1969, Luttge 1971). However, limited photosynthesis occurred in the bladder and stalk cytoplasm, energy produced was too small to account for the energy required for active ion uptake by bladder hairs. Luttge and Osmond (1970) reported that the active driving energy source for the chloride pump was photosynthetic electron flow produced in the mesophyll. This energy produced in mesophyll chloroplasts was delivered to the bladder cells through the symplast (interconnected protoplasm not including the vacuoles).

Another type of salt-excreting gland is found on salt cedar (*Tamarix aphylla* (L.) Karst.) (Fig.6) (Esau 1977). Rather than bursting, this gland secretes salts directly to the outside through pores in the surface layer. The gland is made up of six secretory cells, enclosed by a cuticular layer, and two basal collecting cells. Salts are transported from mesophyll cells through the collecting cells to the secretory cells via the plasmodesmata. Five stamen tamarisk (*T. pentandra* Pall. Fl. Ross) leaves contain similar excretion glands, predominately located on the abaxial leaf surface with a rosette of epidermal cells surrounding each salt gland (Decker 1961, Campbell and Strong 1964).

Liphschitz and Waisel (1974) and Liphschitz et al. (1974) reported that excretion glands occur in the genera *Spartina, Aeluropus, Distichlis* and *Chloris*. These glands consisted of two cells, a basal cell and cap cell, and lack a central vacuole. In California cordgrass (*Spartina foliosa* Trin.) the basal cell contained long extracellular channels (Levering and Thomson 1971). Ions moved through the extracellular spaces along an osmotic gradient with the passive flow of water. The ions were moved into the cap cell and excreted out through pores in the cuticle.

Selectivity and Transport

Salt-tolerance begins in both halophytes and glycophytes by regulating uptake and movement of salt ions into and through the plant (Khan and Marshall 1981, Yeo 1983). Ion uptake and transport rate in the plant is essential in maintaining several important plant processes, including (1) osmotic balance, (2) preferential

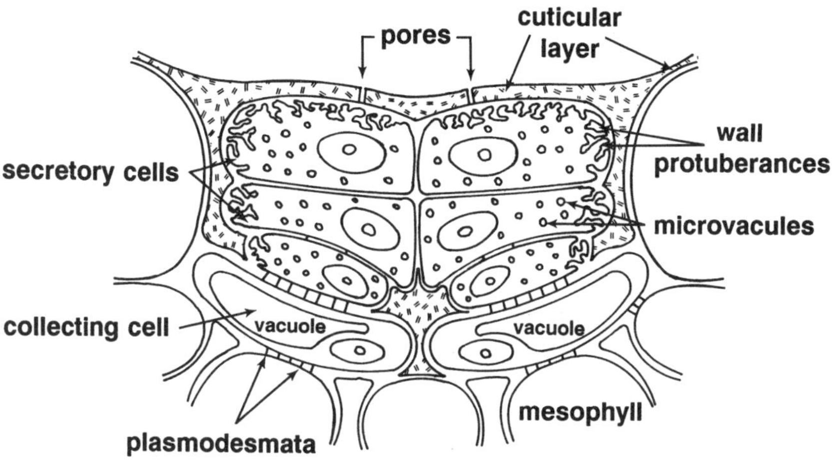

Figure 6. Diagram of a salt-excreting gland of salt cedar (*Tamarix aphylla*). The complex of eight cells, six of which are secretory and two collecting cells, is embedded in the epidermis and is in contact with the mesophyll. Cuticle and cutinized wall are indicated jointly by cross hatching (from Esau 1977).

uptake and movement of ions of nutritional importance that occur at relatively low levels as compared to salt ions, and (3) the internal distribution of these ions in the plant (Flowers et al. 1977, Yeo and Flowers 1986).

To maintain a proper balance of ions in saline environments, uptake must limit entry of specific ions to avoid toxicity while at the same time accumulating other ions typically found in low amounts (e.g. K^+)(Marschner 1986, Bowman 1988, Mahmood and Malik 1987). For example, in seedlings of mesquite (*Prosopis* spp. L.), Na^+/K^+ ratios were approximately 1 in the roots and 5 in the external solution with K^+ selectivity increasing with salt stress (Winter 1988). Regulation of ion uptake at the root have also been reported for *Atriplex falcata* (Breckle 1976), four-wing saltbush (Richardson 1982), river saltbush (*A. rhagodiides* F. Muell.) (Mahmood and Malik 1987), and big sagebrush (Karimi and Ungar 1983). Low Na^+/K^+ ratios in the plant with respect to the saline media are mainly due to selectivity against Na^+ and for K^+ by root uptake (Jeschke 1984, Winter 1988). Ion selectivity has been shown to occur in the root epidermis and outer root cortex (Fig. 7) for *Atriplex hastata* (L.) Hall & Clem. (Kramer et al. 1978). However, in *Puccinellia peisonis* (Stelzer and Lauchli 1978) and barley, the inner cortical cells and endodermis were reported to reduce Na^+/K^+ ratios (Yeo et al. 1977).

Effectiveness of compartmentation as a means to reduce toxic levels of ions in leaf tissues also depends upon ion uptake and transport rates (Yeo 1983). Rapid uptake will overwhelm the plant's ability to sequester toxic ions in the vacuole. Control of salt ion uptake and transport rates, however, is likely less critical in plants capable of excreting excess ions.

Uptake of ions into plant roots depends upon a two stage process: (1) movement of ions to the cell membrane; and (2) transport of ions across the membrane. Flow of ions to the cell membrane occurs as solutes move either by diffusion or mass flow into the water-filled intercellular spaces of the root cortex (Marschner 1986).

Ions are prevented from passively entering the plant by the plasmalemma of the endodermis and cortical cells, or into the xylem stream by the suberized barrier referred to as the Casparian strip (Fig. 7). These barriers prevent uncontrolled ion flow through the apoplast which would quickly lead to excess ion accumulation and toxicity (Flowers et al. 1986).

Special anatomical features in roots of halophytes have been shown to enhance the control of Na^+ and Cl^- uptake (Marschner 1986). Poljakoff-Mayber (1975) reported that the Casparian strip width was 2 to 3 times greater in halophytes as compared to glycophytes. Saline conditions may also influence the development of the Casparian strip. In herbaceous seablite, the Casparian strip was developed within 3 mm of the root tip under saline conditions as compared to 15 mm from the root tip in nonsaline conditions (Flowers 1985). Length of the Casparian strip was also longer (0.93 um) in saline than in nonsaline (0.46 um) conditions. The earlier development of the Casparian strip would decrease uncontrolled ion flow through the apoplast. Control of ion uptake in some halophytes may also be increased by differentiation of the inner cortex layer into a second endodermis, increasing impermeability (Stelzer and Lauchli 1977). Permeability may also be affected by the frequency and size of transient pores in the membrane (Hall and Baker 1977). In *Prosopis farcta* (Banks & Sol) Macbride, transfer cells (cell wall protuberances in the root hypodermis) were formed when roots were exposed to salinity (Winter 1988). An increase in K^+ selectivity occurred with the onset of transfer cell development in the roots.

Movement of ions into plant cells appears to occur both actively and passively, although exact mechanisms are still not fully understood (Marschner 1986). In the active process, carriers move ions across an impermeable membrane. Ion carriers are subunits of the membrane that bind with the ion at the external surface of the membrane, forming a carrier-ion complex (Epstein 1972). This complex then tra-

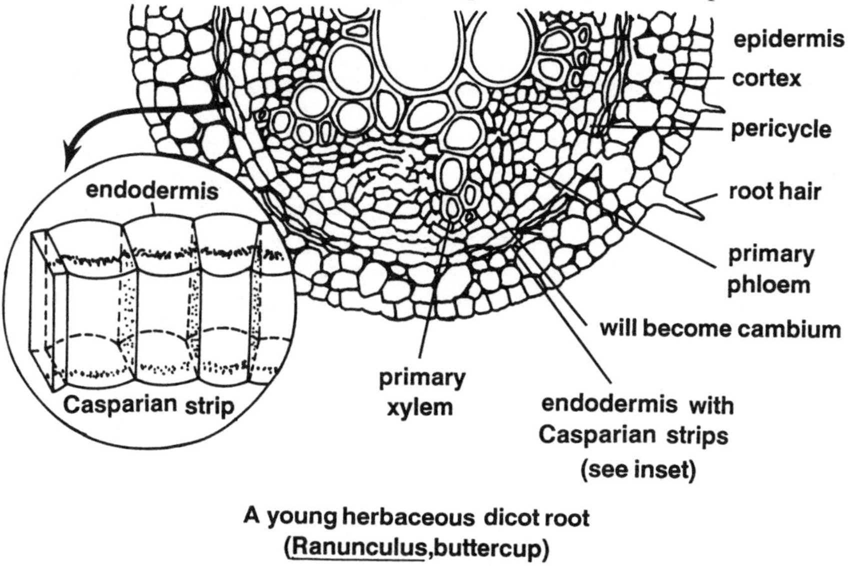

Figure 7. Anatomy of a young primary root (from Salisbury and Ross 1985).

verses the membrane, rotates and releases the ion to the inner space. Proteins probably serve as carrier molecules and transport is an active process requiring the expenditure of energy. Ions are commonly thought to move against concentration gradients in this manner.

Evidence indicates that carriers are ion-specific, with selectivity being maintained by Ca^{++}. Accumulation of K^+ against a concentration gradient in winterfat was attributed to a highly specific active transport carrier (Moore et al. 1972b). In contrast, shadscale, an ion accumulator, contained a higher percentage of Na^+ carriers.

Active ion transport competes with other processes for energy (Yeo 1983). As salinity increases, increasing ion transport rates, the demands for energy will also increase.

Another transport mechanism involves movement in response to an electropotential gradient (Marschner 1986). Ions at high concentrations outside the plasmalemma may diffuse across the cell membrane with the aid of lipid-associated carrier molecules, or through aqueous pores in the cell membrane. This is a passive process in response to electrical gradients, although it is believed that when concentrations of ions within the cell exceed desired amounts, an energy-consuming pump excretes them. There is evidence to suggest that for Na^+, such an efflux pump operates to reduce internal cell concentrations (Jeschke and Jambor 1981). Additional information on active and passive ion transport is provided by Marschner (1986).

Ions are transported from the root through the plant in the transpiration stream. To avoid ion toxicity in the root, ion uptake must decline when transpiration decreases (Yeo and Flowers 1986). These authors suggests that ion uptake may be limited by one or more of the following: (1) saturation of influx rates at the plasmalemma of the epidermis and cortex; (2) diffusion of ions to the cortical plasmamembranes; and (3) saturation of the capacity for cyclosis to transfer ions across the cell at an acceptable cytoplasmic ion concentration.

In many species ion movement is limited to certain plant parts. In *Prosopis farcta*, salt ions are largely retained in the roots (Levitt 1980). Partitioning is primarily determined by transmembrane events, with little movement through the apoplast (Lazof and Cheeseman 1988). The mechanisms involved in partitioning salt ions between roots and stems, however, are poorly understood. Yeo et al. (1977) suggested salts may be absorbed during translocation by specialized xylem parenchyma cells. In two grass species varying in salt-tolerance, shoot concentrations of Na^+ were lower in the glycophyte than the halophyte even though uptake of Na^+ was similar (Taleisnik 1989). Lower Na^+ content in the shoot of the glycophyte was attributed to recirculation of the ion from the shoot back to the root and greater loss of recently absorbed Na^+. Kramer et al. (1977) reported Na^+/K^+ ratios in xylem parenchyma >300 mm from the root tip were 5.4 compared to ratios equaling 0.01 in xylem parenchyma 100 mm from the root tip. Although this evidence supports the existence of an ion filtration system in the xylem (via xylem parenchyma cells), the mechanisms involved are poorly understood.

Conclusions

Terrestrial saline environments, including salt deserts, abandoned agricultural lands, saline seeps and mine spoils, occupy a significant portion of the earths surface. Although often low in productivity, saline environments provide forage, wildlife habitat, genetic resource of many unique plants, recreation and open space allowing for dilution of air pollutants. Plants growing in these environments are exposed to multiple stresses such as limited water availability, ion toxicity, nutrient deficiency, and extreme climatic conditions. Due to these conditions, many saline environments are very fragile. Many salt deserts throughout the world, particularly those with palatable shrubs, have been degraded through overgrazing causing erosion, a change in composition and a loss of productivity. Increased recreation use and the invasion of exotic plant species have also contributed to the degradation of these environments. Extreme environmental conditions and low productivity of these deserts have made reclamation very difficult and most often unsuccessful.

Plants have evolved numerous mechanisms enabling them to exist in saline environments. Among these are the ability to (1) use Na^+ in metabolic processes, (2) osmotically adjust, (3) sequester toxic ions in vacuoles of leaf cells, (4) increase succulence in order to dilute high ion concentrations, (5) regulate ion uptake and (6) excrete excess ions. However, the way many of these mechanisms actually work are frequently not well understood.

Both the study of effects of salt ions on plants and mechanisms allowing them to survive in saline environments is difficult. Interactions between salt ions, water availability, soils, and other environmental factors influencing plant response in saline environments are difficult to separate out. Experimental conditions also make it difficult to draw definite conclusions on the effects of ions on plants and mechanisms of adaptation.

It is generally accepted that plants either avoid or tolerate salt ions. Plants which tolerate salt ions accumulate greater quantities of salts than avoiders and are usually better adapted to saline environments. However, metabolic processes in both salt tolerant and salt sensitive species appear to be equally sensitive to salt ions. This has lead to a great deal of work describing mechanisms allowing salt tolerant species to accumulate salt ions but avoid ion toxicity. However, research evaluating the sensitivity of enzymes to ions is difficult. Enzymes are removed from plant tissue and studied in an artificial environment, causing some researchers to challenge the conclusion that enzymes in halophytes are sensitive to ions.

Gland cells which allow plants to excrete excess salts are probably the best understood mechanism characteristic of some salt tolerant species. However, other mechanisms such as movement of ions in the plant, partitioning of salt ions and ion filtration are poorly understood. Knowledge of the ecology and management of terrestrial saline environments are also not well understood.

In the future, many agricultural and arid areas of the world face the problem of increased soil salinization. An understanding of salt-tolerance mechanisms may improve vegetation manipulation and management. Development of salt-tolerant plants through breeding or biotechnology must be considered if salt affected areas are to be revegetated. Mechanisms which aid in salt-tolerance, but require the least amount of energy to function may, impart the greatest advances in a breeding program (Marschner 1986). Desirable characteristics include compartmentation, use of salt ions for osmotic adjustment, maintenance of high K^+/Na^+ ratios and excretion of excess salts. An increased understanding of the physiology of salt-tolerance will aid in efforts to speed up the process of gene transfer for the development of salt-tolerant plants (Flowers et al. 1986).

Acknowledgments

Authors wish to thank Tony Svejcar, Bruce Roundy, Wesley Jerrel and William Krueger for their critical review of this manuscript and Susan Kramer for the translation of Breckle 1976. I also wish to thank Martin Vavra for allowing the first author the flexibility in taking on these projects.

Literature Cited

Abd el Rahman, A.A., M.N. El-Shourbagy, A.F. Shalaby and M.O. El-Monayeri. 1972. Salinity effects on growth and water relations of some desert range plants. Flora 161:495-508.

Abd el Rahman, A.A. 1986. The deserts of the Arabian Peninsula, p. 29-54. *In*: Evenari, M. I. Noy-Meir and D.W. Goodall (eds.), Ecosystems of the world: hot deserts and arid shrublands. vol. 12A. Elsevier, New York.

Ahmad, I., S.J. Wainwright, and G.R. Stewart. 1981. The solute and water relations of *Agrostis stolonifera* ecotypes differing in their salt tolerance. New Phytol. 87:615-629.

Albert, R. and M. Popp. 1977. Chemical composition of halophytes from the Neusiedler lake region in Austria. Oecol. 27:449-464.

Albert, R. 1975. Salt regulation in halophytes. Oecol. 21:57-71.

Bakker, J.P., M. Dijkstra and P.T. Russchen. 1985. Dispersal, germination and early establishment of halophytes and glycophytes on a grazed and abandoned salt-marsh gradient. New Phytol. 101:291-308.

Berg, W.A., C.R. Cail, D.M. Hungerford, J.W. Naney and G.A. Sample. 1987. Saline seep on wheatland in northwest Oklahoma, p. 265-272. *In*: D.M. Fairchild (ed.), Ground water quality and agricultural practices. Lewis Publ.

Bernstein, L. and L.E. Francois. 1964. Salt tolerance of plants. USDA Agric. Inform. Bull. 283.

Bernstein, L. and H.E. Haywood. 1958. Physiology of salt tolerance. Ann. Rev. Plant Physiol. 9:25-46.

Bettenay, E. 1986. Salt affected soils in Australia. Reclam. Reveg. Res. 5:167-179.

Billard, J.P. and J. Boucaud. 1982. Effect of sodium chloride on the nitrate reductase of *Suaeda maritima* var. *macrocarpa*. Phytochem. 21:1225-1228.

Billings, W.D. 1949. The shadscale vegetation zone of Nevada and eastern California in relation to climate and soils. Amer. Midl. Natur. 42:87-109.

Binet, P. 1985. Salt resistance and the environment of the cell wall of some halophytes. Vegetatio 61:241-246.

Black, R.F. 1958. Effect of sodium chloride on leaf succulence and area of *Atriplex hasata* L. Aust. J. Bot. 11:306-321.

Blaisdell, J.P. and R.C. Holmgren. 1984. Managing intermountain rangelands—salt-desert shrub ranges. USDA Intermountain For. Range Exp. Sta. Gen. Tech. Rep. INT-163.

Bloom, A.J. 1979. Salt requirement for crassulacean acid metabolism in the annual succulent, *Mesembryanthemum crystallinum*. Plant Physiol. 63:749-753.

Bolen, E.G. 1964. Plant ecology spring-fed salt marshes in western Utah. Ecol. Monog.

34:143-166.

Bowman, W.D. 1987. Effect of salinity on leaf gas exchange in two populations of a C_4 non-halophyte. Plant Physiol. 85:1055-1058.

Bowman, W.D. 1988. Ionic and water relations responses of two populations of a non-halophyte to salinity. J. Exp. Bot. 39:97-105

Bradley, P.M. and J.T. Morris. 1991. The influence of salinity on the kinetics of NH_4^+ uptake in *Spartina alterniflora*. Oecol. 82:375-380.

Brady, N.C. 1974. The nature and properties of soils. MacMillan Publ. Co. New York.

Branson, F.A., R.F. Miller, and I.S. McQueen. 1967. Geographic distribution and factors affecting the distribution of salt desert shrubs in the United States. J. Range Manage. 20:287-296.

Breckle, S.W. 1976. Zur Okologie und zu den Mineralstoffverhaltnissen absalzender und nichtabsalzender Xerohalophyten, Habilitationsschrift, Universitat Bonn.

Brown, P.L., A.D. Halvorson, F.H. Siddoway, H.F. Mayland and M.R. Miller. 1983. Saline-seep diagnosis, control and reclamation. USDA Conser. Res. Rep. 30.

Brownell, P.F. and J.G. Wood. 1957. Sodium as an essential micronutrient element for *Atriplex vesicaria*, Heward. Nature 179:635-636.

Brownell, P.F. 1964. Sodium as an essential micronutrient element for a higher plant (*Atriplex vesicaria*). Plant Physiol. 39:460-468.

Brownell, P.F. and C.J. Crossland. 1972. The requirements of sodium as a micronutrient by species having the C_4 Dicarboxylic photosynthesis pathway. Plant Physiol. 49:794-797.

Brownell, P.F. 1967. Sodium as an essential micronutrient for some high plants. Plant Soil 28:161-164.

Burggemann, W. and P. Janiesch. 1989. Comparison of plasma membrane ATPase from salt-treated and salt-free grown *Plantago maritima* L. J. Plant Physiol. 134:20-25.

Caldwell, M.M., N.E. West and P.J.Goodman. 1970. Autecological studies of *Atriplex confertifolia* and *Eurotia lanata*. IBP Desert Biome Prog. Rep.

Caldwell, M.M. 1974. Physiology of desert halophytes, p. 355-378. *In*: R.J. Reimold and W.H.Queen (eds.) Ecology of halophytes. Academic Press. New York.

Campbell, C.J. and J.E. Strong. 1964. Salt gland anatomy in *Tamarix pentandra* (Tamaricaceae). S.W. Natur. 9:323-238.

Campbell, N., W.W. Thomson and K. Platt. 1974. The apoplastic pathway of transport to salt glands. J. Exp. Bot. 84:61-69.

Chavan, P.D. and B.A. Karadge. 1986. Growth, mineral nutrition, organic constituents and rate of photosynthesis in *Sesbania grandiflora* L. grown under saline conditions. Plant Soil 93:395-404.

Cheeseman, J.M. 1988. Mechanisms of salinity tolerance in plants. Plant Physiol. 87:547-550.

Clapp, E.G. 1936. The major problems and their solution. p. 1-70. *In*: W.A. Wallace, Letter from the Secretary of Agriculture of the U.S.. Senate-Report on western Rangelands - A great but neglected resource. Senate Doc. 199. Wash. D.C.

Clipson, N.J.W. 1987. Salt tolerance in the halophyte *Suaeda maritima* L. Dum. Growth, ion and gas exchange in response to altered salinity. J. Exp. Bot. 38:1996-2004.

Cook, C.W. 1977. Effects of season and intensity of use on desert vegetation. Utah Agr. Exp. Sta. Bull. 483. Utah State Univ., Logan.

Cram, W.J. 1973. Internal factors regulating nitrate and chloride influx in plant cells. J. Exp. Bot. 24:328-341.

Daines, R.J. and A.R. Gould. 1985. The cellular basis of salt tolerance studied with tissue cultures of the halophytic grass *Distichlis spicata*. J. Plant Physiol. 119:268-280.

Deane-Drummond, C.E., and A.D.M. Glass. 1982. Studies of nitrate influx into barley roots by the use of $^{36}ClO_3^-$ as a tracer for nitrate. I. Interactions with chloride and other ions. Can. J. Bot. 60:2147-2153.

Decker, J.P. 1961. Salt secretion by *Tamarix pentandra* Pall. For. Sci. 7:214-217.

Detling, J.K. and L.G. Klikoff. 1973. Physiological response to moisture stress, as a factor in halophyte distribution. Am. Midl. Nat. 90:307-318.

Dijkema, K.S. 1990. Salt and brackish marshes around the Baltic Sea and adjacent parts of the North Sea: Their vegetation and management. Biol. Conser. 51:191-209.

Doddema, H., R.S. Eddin and A. Mahasneh. 1986. Effects of seasonal changes of soil salinity and soil nitrogen of the N-metabolism of the halophyte *Arthrocnemum fruticosum* (L.) Moq. Plant Soil 92:279-293.

Drake, D.R. and I.A. Ungar. 1989. Effects of salinity, nitrogen and population density on the survival, growth and reproduction of *Atriplex triangularis* (Chenopodiaceae). Am. J. Bot. 76:1125-1135.

Drier, W. and H. Goering. 1974. The influence of high salt concentrations on various physiologic parameters of maize roots. Wiss. Z. Humboldt-Univ. Berl. Math. Naturwiss. Reihe. 23:641-644.

Eckert, R.E. Jr. and F.E. Kinsinger. 1960. Effects of *Halogeton glomeratus* leachate on chemical and physical characteristics of soils. Ecol. 41:764-772.

Epstein, E. 1972. Mineral Nutrition of Plants: Principles and Perspectives. John Wiley and Sons, Inc., New York, NY.

Esau, K. 1977. Anatomy of seed plants. John Wiley and Sons, New York, NY.

Evenari, M., Imanuel Noy-Meir and D.W. Boodall (eds). 1985. Ecosystems of the World: Hot desert and arid shrublands. Elsevier, New York.

Fireman, M. and H.E. Haywood. 1952. Indicator significance of some shrubs in the Escalante Desert, Utah. Bot. Gaz. 114:143-155.

Flanagan, L.B. and R.L. Jefferies. 1988. Stomatal limitation of photosynthesis and reduced growth of the halophyte, *Plantago maritima* L., at high salinity. Plant Cell Environ. 11:239-245.

Flowers, S. 1934. Vegetation of the Great Salt Lake region. Bot. Gaz. 45:353-418.

Flowers, S. and F.R. Evans. 1966. The flora and fauna of the Great Salt Lake region, Utah. *In*: H. Boyko (ed.), Salinity and aridity. The Hague: Junk publ.

Flowers, T.J., P.F. Troke and A.R. Yeo. 1977. The mechanism of salt tolerance in halophytes. Ann. Rev. Plant Physiol. 28:89-121.

Flowers, T.J. 1985. Physiology of halophytes. Plant Soil 89:41-56.

Flowers, T.J., M.A. Hajibagheri, and N.J.W. Clipson. 1986. Halophytes. The Quart. Rev. Biol. 61:313-337.

Frankton, C. and T.J. Bassett. 1969. The genus *Atriplex* (Chenopodiaceae) in Canada. I. Four native annuals, *A. heterosperma, A. oblongifolia, A. hortensis*. Can. J. Bot. 46:1309-1313.

Gale, J., R. Naaman, and A. Poljakoff-Mayber. 1970. Growth of *Atriplex halimus* L. in sodium chloride salinated culture solutions as affected by the relative humidity of the air. Aust. J. Biol. Sci. 23:947-952.

Gates, D.H., L.A. Stoddart and C.W. Cook. 1956. Soil as a factor influencing plant distribution on salt-deserts of Utah. Ecol. Monogr. 26:155-175.

Gimmler, H., R. Kaaden, U. Kirchner, and A. Weyand. 1984. The chloride sensitivity of *Dunaliella parva* enzymes. Z. Pflanzenphysiol. 114:131-150.

Glenn, E.P and J.W. O'Leary. 1984. Relationship between salt accumulation and water content of dicotyledonous halophytes. Plant Cell Environ. 7:253-261.

Goodman, P.J. and M.M. Caldwell. 1971. Shrub ecotypes in a salt desert. Nature 232:571-572.

Gorham, J., R.G.W. Jones, and E. McDonnell. 1985. Some mechanisms of salt tolerance in crop plants. Plant Soil 89:15-40.

Greenway, H. and A. Rogers. 1963. Plant responses to saline substrates. III. Effect of nutrient concentration on the growth and ion uptake of *Hordeum hulgare* during a sodium chloride stress. Aust. J. Biol. Sci. 16:616-628.

Greenway, H. and C.B. Osmond. 1972. Salt responses of enzymes from species differing in salt tolerance. Plant Physiol. 49:256-259.

Greenway, H. and R. Munns. 1980. Mechanisms of salt tolerance in nonhalophytes. Ann. Rev. Plant Physiol. 31:149-190.

Greenway, H., A. Gunn, and D.A. Thomas. 1966. Plant response to saline substrates. VIII. Regulation of ion concentrations in salt-sensitive and halophytic species. Aust. J. Biol. Sci. 19:741-756.

Greenway, H. and R. Munns. 1983. Interaction between growth, uptake of Cl^- and Na^+, and water relations of plants in saline environments. II. Highly vacuolated cells. Plant Cell Environ. 6:575-589.

Griffiths, D. 1902. Forage conditions on the northern border of the Great Basin. USDA Bull. 15. Bureau of Plant Industry.

Hall, J.L. and D.A. Baker. 1977. Cell membranes and ion transport. Longman, London, pp. 23-24. ISBN 0-582-44192-7.

Hall, J.L., D.M.R. Harvey, and T.J. Flowers. 1978. Evidence for the cytoplasmic localization of betaine in leaf cells of *Suaeda maritima*. Planta 140:59-62.

Halvorson, A.D. 1988. Role of cropping systems in environmental quality: saline seep control. p. 179-191. *In*: Cropping strategies for efficient use of water and nitrogen. ASA-CSSA-SSSA, Spec. Publ. no. 51.

Hannon, N.J. and H.N. Barber. 1972. The mechanisms of salt tolerance in naturally selected populations of grasses. Search 3:259-260.

Hancock, N.V. 1966. Wildlife use of the salt desert shrub areas of the Great Basin, p. 101-112. *In*: Salt desert symposium, Cedar City Utah. U.S.D.I. Bureau of Land Management.

Harvey, D.M.R., J.L. Hall, T.J. Flowers, and B. Kent. 1981. Quantitative ion localization within *Suaeda maritima* leaf mesophyll cells. Planta 151:555-560.

Haynes, R.J. 1990. Active ion uptake and maintenance of cation-anion balance: A critical examination of their role in regulating rhizophere pH. Plant Soil 126:247-264.

Heuer, B. and Z. Plaut. 1989. Photosynthesis and osmotic adjustment of two sugarbeet cultivars grown under saline conditions. J. of Exp. Bot. 40:437-440.

Heywood, V.H. (ed). 1978. Flowering plants of the world. University Press, Oxford.

Hudson, M.J., H. Opik, and S.J. Wainwright. 1985. Changes in ion and water content of individual shoot organs in a salt-tolerant and a salt-sensitive clone of *Agrostis stolonifera* L. during and subsequent to treatment with sodium chloride. Plant Cell Environ. 8:657-658.

Hsiao, T.C. 1973. Plant responses to water stress. Ann. Rev. Plant Physiol. 24:519-570.

Jennings, D.H. 1976. The effect of sodium chloride on higher plants. Biol. Rev. Cambridge Philos. Soc. 51:453-486.

Jennings, D.H. 1968. Halophytes, succulence and sodium in plants - a unified theory. New Phytol. 67:899-911.

Jeschke, W.D. 1979. Univalent cation selectivity and compartmentation in cereals, p. 37-61. *In*: D.L. Laidman, R.G. Wyn Jones (eds), Recent advances in the Biochemistry of cereals. Academic Press, San Francisco.

Jeschke, W.D. and W. Jambor. 1981. Determination of unidirectional sodium fluxes in roots of intact sunflower seedlings. J. Exp. Bot. 2:1257-1272.

Jeschke, W.D. 1983. K^+ - Na^+ exchange at cellular membranes, intracellular compartmentation of cations and salt tolerance, p. 71-86. *In*: R.C. Staples and G.A. Toenniessen (eds.), Salinity Tolerance in Plants - Strategies for Crop Improvement, Wiley International, New York.

Jeschke, W.D. 1984. K^+/Na^+ exchange at cellular membranes, intracellular compartmentation of cations, and salt tolerance, p. 37-67. *In*: R.C. Staples and G.H. Toenniessen (eds.), Salinity tolerance in plants - strategies for crop improvement. Wiley and Sons, New York.

Karimi, S.H. and I.A. Ungar. 1983. The effect of salinity on the ionic content and water relation of *Atriplex triangularis*, p. 124-132. *In*: A.R. Tiedemann, E.D. McArthur, H.C. Stutz, R. Stevens and K.L. Johnson (eds.), Proc. - Symposium on the biology of *Atriplex* and related chenopods. USDA For. Ser. Gen. Tech Rep. INT-172.

Karimi, S.H. and I.A. Ungar. 1989. Development of epidermal salt hairs in *Atriplex triangularis* Willd. in response to salinity, light intensity, and aeration. Bot. Gaz. 150:68-71.

Kemp, P.R. and G.L. Cunningham. 1981. Light, temperature and salinity effects on growth, leaf anatomy and photosynthesis of *Distichlis spicata* (L.) Greene. Am. J. Bot. 68:507-516.

Khan, A.H. and C. Marshall. 1981. Salt tolerance within populations of chewing fescue (*Festuca rubra*, L.). Commun. Soil Sci. Plant Anal. 12:1271-1281.

Kleinkopf, G.E., A. Wallace, and J.W. Cha. 1975. Sodium relations in desert plants: 4. Some physiological responses of *Atriplex confertifolia* to different levels of sodium chloride. Soil Sci. 120:45-48.

Kostina, H.O. 1974. Gradient distribution of chlorine and ash in plants and its economical significance. Ukr. Bot. Zh. 31:322-326.

Kramer, D., A. Lauchli, and A.R. Yeo. 1977. Transfer cells in roots of *Phaseolus coocineus*: ultrastructure and possible function in exclusion of sodium from the shoot. Ann. Bot. 41:1031-1040.

Kramer, D., W.P. Anderson, and J. Preston. 1978. Transfer cells in the root epidermis of *Atriplex hastata* L. as a response to salinity. A comparative cytological and X-ray microprobe investigation. Aust. J. Plant Physiol. 5:734-747.

Kuchler, A.W. 1964. Potential natural vegetation of the conterminous Untied States. Spec. Publ. 36. Amer. Geogr. Soc.

Kuchler, A.W. 1970. Potential natural vegetation, p. 90-91. *In*: U.S. Geol. Survey, The National Atlas of the U.S.A. U.S. Government Printing Office, Washington D.C.

Larcher, W. 1980. Physiological plant ecology. Springer-Verlag, New York.

Lazof, D. and J.M. Cheeseman. 1988. Sodium and potassium compartmentation and transport across the roots of intact *Spergularia marina*. Plant Physiol. 88:1274-1278.

Levering, C.A. and W.W. Thomson. 1971. The ultrastructure of the salt gland of *Spartina foliosa*. Planta 97:183-196.

Levitt, J. 1980. Responses of plants to environmental stresses: water, radiation, salt, and other stresses. Academic Press, New York.

Liphschitz, N., Adiva-Shomer-Ilan, A. Eshel and Y. Waisel. 1974. Salt glands on leaves of Rhodes grass (*Chloris gayana* Kth.). Ann. Bot. 38:459-462.

Liphschitz, N. and Y. Waisel. 1974. Existence of salt glands in various genera of the Gramineae. New Phytol. 73:507-513.

Longstreth, D.J. and P.S. Nobel. 1979. Salinity effects on leaf anatomy. Consequences for photosynthesis. Plant Physiol. 63:700-703.

Luttge, U. and C.B. Osmond. 1970. Ion absorption in *Atriplex* leaf tissue. III. Sites of metabolic control of light dependent chloride secretion to epidermal bladders. Aust. J. Biol. Sci. 23:17-25.

Luttge, U. 1971. Structure and function of plant glands. Ann. Rev. Plant Physiol. 22:23-44.

Lynch, J. and A. Lauchli. 1985. Salt stress disturbs the Ca nutrition of barley. New Phytol. 99:345-354.

Lynch, J., G. Thiel and A. Lauchli. 1988. Effects of salinity on the extendibility and Ca availability in the expanding region of growing barley leaves. Bot. Acta 101:355-361.

Maas, E.V. and G.J. Hoffman. 1977. Crop salt tolerance - current assessment. J. Irrig. Drainage Div., Am. Soc. Civil Eng. 103:115-134.

Maas, E.V. and R.H. Nieman. 1978. Physiology of plant tolerance to salinity, p. 277-299. *In*: G.A. Jung (ed.), Crop tolerance to suboptimal land conditions. Volume 32. Chap 13. Am. Soc. Agron. Spec Publ.

MacMahon, J.A. and F.H. Wagner. 1985. The Mojave, Sonoran and Chihuahuan Deserts of North America, p. 105-202. *In*: Evenari, M. I. Noy-Meir and D.W. Goodall (eds.), Ecosystems of the world: hot deserts and arid shrublands. vol. 12A. Elsevier, New York.

Mahmood, K. and K.A. Malik. 1987. Salt tolerance studies on *Atriplex rhagodioides* F. Muell. Environ. and Exp. Bot. 27:119-125.

Marcar, N.E. and A. Termaat. 1990. Effects of root-zone solutes on *Eucalyptus camaldulensis* and *Eucalyptus bicostata* seedlings: Responses to Na^+, Mg^{2+} and Cl^-. Plant and Soil 125:245-254.

Marschner, H. 1986. Mineral nutrition of higher plants. Academic Press, Orlando, Florida.

Matsuda, K. and A. Riazi. 1981. Stress-induced osmotic adjustment in growing regions of barley leaves. Plant Physiol. 68:571-576.

McArdel, R.E. et al. 1936. The whiteman's toll. p. 81-116. *In*: W.A. Wallace, Letter from the Secretary of Agriculture of the U.S.. Senate-Report on western Rangelands - A great but neglected resource. Senate Doc. 199. Wash. D.C.

McArthur, D.E., A.P. Plummer and J.N. Davis. 1978. Rehabilitation of game range in the salt desert. p. 23-50. *In*: Wyoming shrublands. Univ. of Wyoming, Laramie.

McNulty, I.B. 1985. Rapid osmotic adjustment by a succulent halophyte to saline shock. Plant Physiol. 78:100-103.

Mendoza, M.M. 1971. The effects of NaCl on anatomical and physiological processes in *Atriplex hastata* L. M.S. thesis. Univ. Utah, Salt Lake City.

Millar, C.E., L.M. Turk and H.D. Foth. 1966. Fundamentals of soil science. John Wiley and Sons, New York.

Miller, W.R. and F.E. Egler. 1950. Vegetation of the Wequetequock-Pawcatuck tidal marshes. Ecol. Monog. 20:143-172.

Mitchell, J.E., N.E. West and R.W. Miller. 1966. Soil physical properties in relation to plant community patterns in the shadscale zone of northwestern Utah. Ecol. 47:627-630.

Moore, R.T., S.W. Breckle and M.M. Caldwell. 1972b. Mineral ion composition and osmotic relations of *Atriplex confertifolia* and *Eurotia lanata*. Oecol. 11:67-78.

Moore, R.T., R.S. White and M.M. Caldwell. 1972a. Transpiration of *Atriplex confertifolia* and *Eurotia lanata* in relation to soil, plant and atmospheric moisture stress. Can. J. Bot. 50:2411-2418.

Mozafar, A. and J.R. Goodin. 1970. Vesiculated hairs: a mechanism for salt tolerance in *Atriplex halimus* L. Plant Physiol. 45:62-65.

Munns, R. and A. Termaat. 1986. Whole-plant responses to salinity. Aust. J. Plant Physiol. 13:143-160.

Murata, T., O. Ueno and R. Ohsugi. 1989. Stimulatory effect of chloride ions on NADP enzyme from leaves of tow C_4 species of *Cyperus*. Plant Cell Physiol. 30:851-855.

Nable, R.O. and P.F. Brownell. 1984. Effect of sodium nutrition and light upon concentrations of alanine in leaves of C-4 plants. Aust. J. Plant Physiol. 11:319-324.

Naphan, E.A. 1966. Soils of the salt desert shrub area and their productive capabilities, p. 44-68. *In*: Salt desert symposium, Cedar City Utah. U.S.D.I. Bureau of Land Management.

Nieman, R.H. 1962. Some effects of sodium chloride on growth, photosynthesis, and respiration of twelve crop plants. Bot. Gaz. 123:279-285.

Norton, B.E. 1978. The impact of sheep grazing on long term successional trends in salt desert shrub vegetation of southwestern Utah, p. 610-613. *In*: D.N. Hyder (ed.), Proc. 1st Int. Rangeland Congress, Denver, Colo. USA.

Okusanya, O.T. and I.A. Ungar. 1983. The effects of time of seed production on the germination response of *Spergularia marina*. Physiol. Plant. 59:335-341.

Olson, G.W. 1978. Some observations on dryland saline seepage in several countries and some comments on causes, effects, and cures, p. 7-29. *In* Proc. Sub-Commission on Salt-Affected Soils. 11th International Soil Sci. Soc. Cong. Sec 2, Edmonton, Alberta.

Orshan, G. 1986. The deserts of the Middle East, p. 1-28. *In*: Evenari, M. I. Noy-Meir and D.W. Goodall (eds.), Ecosystems of the world: hot deserts and arid shrublands. vol. 12B. Elsevier, New York.

Osmond, C.B., U. Luttge, K.R. West, C.K. Pallaghy and B. Shacher-Hill. 1969. Ion absorption in *Atriplex* leaf tissue II. secretion of ions to epidermal bladders. Aust. J. Biol Sci. 22:797-814.

Osmond, C.B. 1968. Ion absorption in *Atriplex* leaf tissue. I. Absorption by mesophyll cells. Aust. J. Biol. Sci. 21:1119-1130.

Papke, K.G. 1976. Evaporites and brines in Nevada playas. Nevada Bureau of Mines and Geology Bull. 87. Univ. of Nevada, Reno.

Pallaghy, C.K. 1970. Salt relations of *Atriplex* leaves, p. 57-62. In. R. Jones (ed.), The biology of *Atriplex*. Deniliquin.

Peck, A.J. 1975. Effects of land use on salt distribution in the soil, p. 77-90. *In*: A. Poljakoff-Mayber and J. Gale, (eds.), Plants in saline environments. Springer-Verlag, New York.

Penfound, W.T. and E.S. Hathaway. 1938. Plant communities in the marshlands of southeastern Louisiana. Ecol. Monog. 8:1-56.

Plaut, Z., C.M. Grieve and E. Federman. 1989. Salinity effects on photosynthesis in isolated mesophyll cells of cowpea leaves. Plant Physiol. 91:493-499.

Plummer, A.P. and S.R. Chapman. 1976. Clipping effects on dry matter yields and survival of basin wildrye. J. Range Manage. 29:311-312.

Poljakoff-Mayber, A. 1975. Morphological and anatomical changes in plants as a response to salinity, p. 97-117. *In*: A. Poljakoff-Mayber and J. Gale, (eds.), Plants in saline environments. Springer-Verlag, New York.

Pollard, A.S, and R.G. Wyn Jones. 1979. Enzyme activities in concentrated solutions of glycinebetaine and other solutes. Planta 144:291-298.

Prikhod'ko, L.S. and L.K. Klyshev. 1963. The change in metabolism of pea seedlings caused by salinization of the substrate. Tr. Inst. Bot. Akad. Nauk. Kaz. SSR 16:81-96.

Purer, E.A. 1942. Plant ecology of coastal salt-marshes of San Diego County, California. Ecol. Monogr. 12.81-111.

Radda, G.K. 1975. Algal cells, p. 124-160. *In*: D.A. Baker and J.L. Hall (eds.), Ion transport in plant cells and tissues. North Holland, Amsterdam.

Repp, G.I, D.R. McAllister and H.H. Wiebe. 1959. Salt resistance of protoplasm as a test for the salt tolerance of agricultural plants. Agron. J. 51:311-314.

Richardson, S.G. 1982. High and low sodium biotypes of fourwing saltbush: Their responses to sodium and potassium in retorted oil shale. J. Range Manage. 35:795-797.

Rickard, W.H. 1965a. The influence of greasewood on soil-moisture penetration and soil chemistry. Northwest Sci. 39:36-42.

Rickard, W.H. 1965b. Sodium and potassium accumulation by greasewood and hopsage leaves. Bot. Gaz. 126:116-119.

Rickard, W.H. and R.F. Keough. 1968. Soil-plant relationships of two steppe desert shrubs. Plant Soil. 29:205-212.

Roundy, B.A. 1984. Estimation of water potential components of saline soils of Great Basin rangelands. Soil Sci. Soc. Am. J. 48:645-650.

Roundy, B.A., R.A. Evans and J.A. Young. 1984. Surface soil and seedbed ecology in salt-desert plant communities, p. 66-74. *In*. A.T.Tiedemann, D.E. McArthur, H.C. Stutz, R. Stevens and K. Johnson (eds.), Proceedings-symposium on the biology of *Atriplex* and related chenopods. USDA For. Ser. Gen. Tech. Rep. INT-172.

Roundy, B.A. and J.A. Young. 1985. Salt deserts of the Great Basin, p. 39-49. *In*: Symposium: Cultural, physical and biological characteristics of the range livestock industry in the Great Basin. Proc. selected papers presented at the 38th Annual Meeting of the Society for Range Management. Denver, Colorado.

Roundy, B.A. 1987. Seedbed salinity and the establishment of range plants, p. 68-81 *In*. G.W. Fraiser and R.A. Evans (eds.), Proceedings-symposium seed and seedbed ecology of rangeland. USDA Agric. Res. Ser. Nat. Tech. Info. Ser.

Rozema, J., E. Rosema-Dijst, A.H.J. Freijsen, and J.J.L. Huber. 1978. Population differentiation within *Festuca rubra* L. with regard to soil salinity and soil water. Ibid. 34:329-341.

Salim, M. 1989. Effects of salinity and relative humidity on growth and ionic relations of plants. New Phytol. 113:13-20.

Salisbury, F.B. and C.W. Ross. 1985. Plant physiology. Wadsworth Publ. Co., Belmont, CA.

Schirmer, U. and S.W. Breckle. 1982. The role of bladders for salt removal in some Chenopodiaceae (mainly *Atriplex* species), p. 215-231. *In*: D.N. Sen and K.S. Rajpurohit (eds.), Contributions to the ecology of halophytes. Junk, The Hague.

Schmehl, W.R. and B.D. McCaslin. 1973. Some properties of spent oil shale significant to plant growth, p. 7-25. *In*: M.B. Bloch and P.D. Kilburn (eds.), Processed oil shale revegetation studies, 1965-1973. Atlantic Richfield Co., Colony Development Operation.

Seemann, J.R. and C. Critchley. 1985. Effects of salt stress on the growth, ion content, stomatal behavior and photosynthetic capacity of a salt-sensitive species, *Phaseolus vulgaris* L. Planta 164:151-162.

Sen, D.N. and D.D. Chawan. 1975. Role of abcisic acid in the closure of stomata in some arid zone plant species. Biol. Plant. 17:198-201.

Shainberg, I. 1975. Salinity of soils— effects of salinity on the physics and chemistry of soils, p. 39-55. *In*: A. Poljakoff-Mayber and J. Gale, (eds.), Plants in saline environments. Springer-Verlag, New York.

Sharma, M.L. 1973. Soil physical and physico-chemical variability induced by *Atriplex nummularia*. J. Range Manage. 26:426-430.

Sharma, M.L. and D.J. Tongway. 1973. Plant induced soil salinity patterns in two saltbush (*Atriplex* spp.) communities. J. Range Manage. 26:121-125.

Slayter, R.O. 1967. Plant-water relationships. Academic Press, New York.

Smith, F.A. 1973. The internal control of nitrate uptake into excised barley roots with differing salt contents. New Phytol. 72:769-782.

Snow, A.A. and S.W. Vince. 1984. Plant zonation in an Alaskan salt marsh. II. An experimental study of the role of edaphic conditions. J. Ecol. 72:669-684.

Stelzer, R. and A. Lauchli. 1977. Salz-und Uberflutungstoleranz von *Puccinellia peisonis* II. Differenzierung der wurzel in beziehung zur funktion. Z. Pflanzenphysiol. 84:95-108.

Stelzer, R. and A. Lauchli. 1978. Salt- and flooding tolerance of *Puccinellia peisonis* III. Distribution and localization of ions in the plant. Z Pflanzenphysiol. 88:437-448.

Stewart, G., W.P. Cottam and S.S. Hutchings. 1940. Influence of unrestricted grazing on northern salt desert plant associations in western Utah. J. Agric. Res. 60:289-316.

Storey, R. and R.G. Wyn Jones. 1977. Taxonomic and ecological aspects of the distribution of glycinebetaine and related compounds in plants. Oecol. 27:319-332.

Storey, R. and R.G. Wyn Jones. 1979. Responses of *Atriplex spongiosa* and *Suaeda* *monoica* to salinity. Plant Physiol. 63:156-162.

Strogonov, B.P. 1964. Physiological basis of salt tolerance of plants. Acid. Sci. USSR, Davey and Co. New York, NY.

Taleisnik, E.L. 1989. Sodium accumulation in *Pappophorum* I. Uptake, transport and recirculation. Ann. Bot. 63:221-228.

Tanji, K.K. 1969. Predicting specific conductance from electrolytic properties and ion association in some aqueous solutions. Soil Sci. Soc. Am. Proc. 33:887-890.

Torello, W.A. and L.A. Rice. 1986. Effects of NaCl stress on proline and cation accumulation in salt sensitive and tolerant turfgrasses. Plant Soil 93:241-247.

Troughton, J.H. and K.A. Card. 1974. Leaf anatomy of *Atriplex buchananii*. New Zealand J. Bot 12:167-177.

Tu, J.C. 1981. Effect of salinity on rhizobium-root-hair interaction, nodulation and growth of soybean. Can. J. Plant Sci. 61:231-239.

Turner, N.C. and M.M. Jones. 1980. Turgor maintenance by osmotic adjustment: A review and evaluation, p. 87-103. *In*: N.C. Turner and P.J. Kramer (eds.), Adaptation of plants to water and high temperature stress. John Wiley & Sons, New York, NY.

Ungar, I.A. 1978. Halophyte seed germination. Bot. Rev. 44:233-264.

Ungar, I.A. 1982. Germination ecology of halophytes, p. 143-154. *In*: D.N. Sten and K.S. Kajpurohit (eds.), Tasks for vegetation science 2. Junk, The Haque.

Uphof, J.C.T. 1941. Halophytes. Bot. Rev. 7:1-58.

Uprety, D.C. and M.N. Sarin. 1976. Physiological studies on salt tolerance in *Pisum sativum* L: IV. Tonic composition and nitrogen metabolism. Acta. Agron. Acad. Sci. Hung. 25:455-460.

Vander Pluym, H.S.A. 1978. Extent, causes and control of dryland saline seepage in the Northern Great Plains of North America, p. 1.48-1.58. *In*: Proc. Meet. Sub-Commission on Salt-Affected Soils, 11th Int. Soil Sci. Soc. Congr., Edmonton, Alberta.

Volkens, G. 1884. Die Kalkdrusen der Plumbagineen. Ber. Dtsch. Bot. Ges. 2:334-342.

Waisel, Y. 1972. Biology of halophytes. Academic Press. New York.

Wallace, A., E.M. Romney and V.Q. Hale. 1973a. Sodium relations in desert plants: 1. Cation contents of some plant species from the Mojave and Great Basin deserts. Soil Sci. 115:284-287.

Wallace, A., R.T. Mueller and E.M Romney. 1973b. Sodium relations in desert plants. 2. Distribution of cations in plant parts of three different species of *Atriplex*. Soil Sci. 115:390-394.

Walter, H. and E.O. Box. 1983. Middle Asian deserts, p. 79-104. *In*: N.E. West. (ed.), Ecosystems of the world. V. Temperate deserts and semi-deserts. Elsevier Scientific Publ. Co., New York.

Warren, R.S. and R.M. Brockelman. 1989. Photosynthesis, respiration, and salt gland activity of *Distichlis spicata* in relation to soil salinity. Bot. Gaz. 150:346-350.

Weaver, J.E. and R.E. Clements. 1938. Plant ecology. McGraw-Hill. New York.

Went, F.W. 1955. The ecology of desert plants. Sci. Am. 192:68-75.

West, N.E. and K.I. Abraham. 1968. Soil-vegetation relationships in the shadscale zone of southeastern Utah. Ecology 49:445-455.

West, N.E. 1979. Survival patterns of major perennials in salt desert shrub communities of southwestern Utah. J. Range Manage. 32:442-445.

West, N.E. 1983. Intermountain salt-desert shrubland, p. 375-397. *In*: N.E. West (ed.), Ecosystems of the world. V. Temperate deserts and semi-deserts. Elsevier Scientific Publ. Co., New York.

Williams, M.C. 1960. Effect of sodium and potassium salts on growth and oxalate content of halogeton. Plant Physiol. 35:500-505.

Williams, O.B. and J.H. Calaby. 1985. The hot deserts of Australia, p. 269-312. *In*: Evenari, M. I. Noy-Meir and D.W. Goodall (eds.), Ecosystems of the world: hot deserts and arid shrublands. vol. 12A, Elsevier, New York.

Winter, E. 1988. Salt-induced hypodermal transfer cells in roots of *Prosopis farcta* and ion distribution within young plants. Bot. Acta 101:174-181.

Winter, K., and D.J. Von Willet. 1972. NaCl induced crassulacean acid metabolism in *Mesembryanthemum crystallinum*. Z. Pflanzenphysiol 67:166-170.

Wollenweber, B. and S. Zechmeister-Boltenstern. 1989. Nitrogen fixation and nitrogen assimilation in a temperate saline ecosystem. Bot. Acta 102:96-105.

Wood, J.G. 1923. On transpiration in the field of some plants from the arid portions of South Australia, with notes on their physiological anatomy. Trans. R. Soc. S. Aust. 47:259-278.

Workman, J.P. and N.E. West. 1967. Germination of *Eurotia lanata* in relation to temperature and salinity. Ecology 48:

Wyn Jones, R.G., C.J. Brady and J. Speirs. 1979. Ionic and osmotic relations in plant cells, p. 63-103 *In*: D.L. Laidman and R.G. Wyn Jones (eds.), Recent advances in biochemistry of cereals. Academic Press, London.

Wyn Jones, R.G. 1981. Salt tolerance, p. 271-292. *In*: C.B. Johnson (ed.), Physiological Processes Limiting Plant Productivity. Butterworth, London.

Yaalon, D.H. 1964. Airborne salts as an active agent in pedogenetic processes, p. 997-1000. *In*: Proc.8th Int. Congr. Soil Sci, Bucharest.

Yeo, A.R., A. Lauchli, and D. Kramer. 1977. Ion measurements by X-ray microanalysis in unfixed, frozen, hydrated plants cells of species differing in salt tolerance. Planta 134:35-38.

Yeo, A.R. 1983. Salinity resistance: Physiologies and prices. Physiol. Plant. 58:214-222.

Yeo, A.R. and T.J. Flowers. 1986. Ion transport in *Suaeda maritima*: Its relations to growth and implications for the pathway of radial transport of ions across the root. J. Exp. Bot. 37:143-159.

Young, J.A. and F. Tipton. 1990. Invasion of cheatgrass into arid environments of the Lahontan Basin, p. 37-40. *In*: E.D. McArthur et al. (eds.), Symposium on cheatgrass invasion, shrub die-off, and other aspects of shrub biology and management. USDA For. Ser. Inter. Res. Sta., Gen. Tech. Rep. INT-276.

Allelopathy

Rick G. Kelsey and Richard L. Everett

I. Introduction

Biological life is chemistry—dynamic reactions and equilibria, bound and mediated by the physical laws of nature. As our chemical knowledge has expanded and our instruments and methods used to study chemical processes have improved, the boundaries between chemistry and the traditional biological sciences have become less distinct. As ecologists have searched more thoroughly for factors influencing plant interactions, it has become increasingly clear that chemicals produced within plants (phytochemicals) can have ecological functions. This has led to the development of biochemical plant ecology which focuses on the study of phytochemicals that mediate plant-animal, plant-microorganism, or plant-plant interactions (Sondheimer and Simeone 1970, Whittaker and Feeny 1971, Harborne 1988). In this chapter, we focus primarily on the plant-plant interactions, and exclude from our discussion most plant disease relations that have traditionally fallen within the discipline of plant pathology.

Although the effects of allelopathy have been observed for hundreds of years, it is only in the 20th century that significant scientific evidence of allelopathy has been generated. The term allelopathy was introduced in 1937 by Molisch, and the first comprehensive text on the subject was published just 20 years ago (Rice 1974). The latter publication helped to focus hypotheses and identify areas where knowledge was lacking. As a consequence, research in this field has expanded tremendously, as evidenced by a second edition of the "Allelopathy" text in 1984 and the appearance of three other books since 1984 (Thompson 1985, Putnam and Tang 1986, and Waller 1987). In addition, there have been some excellent reviews (Einhellig 1985, Putnam 1985) and special journal issues devoted to this subject (Silverstein and Simeone 1983, Putnam 1987). We have used these sources extensively in the preparation of this chapter. Because allelopathy is an interdisciplinary subject that draws upon knowledge and research from various fields, it is difficult to incorporate all of the available literature. Also, because of the potential significance of allelopathy to agriculture, an extensive literature has begun to build around crops and weeds associated with heavily managed and manipulated ecosystems. This information was referenced as needed to provide a thorough understanding of the allelopathic process; but in our discussion of case histories, we selected examples of more natural environments and species with a greater relevance to rangelands. After reviewing the allelopathic literature and discussing relevant examples, we tried to evaluate what this phenomena means to range ecology and the management of rangelands. Finally, we speculate on what impact allelopathy will have on range

The authors are research scientist and plant ecologist, Pacific Northwest Research Station, Forestry Sciences Laboratory, Corvallis, OR and Wenatchee, WA, respectively.

management in the future as our knowledge of this phenomena and the field of biotechnology improve.

II. Definition of Allelopathy

Allelopathy as presented by Molisch (1937) was intended to include all inhibitory and stimulatory interations between plants, including microorganisms, but the term was derived from two Greek words; allelon, meaning each other, and pathos, meaning to suffer (Mandava 1985). When put together, they mean the injurous effects of one on another. Focusing on inhibition, Rice (1974) excluded stimulatory interactions and defined allelopathy as "any direct or indirect harmful effect by one plant (including microorganisms) on another through production of chemical compounds that escape into the environment." In the 1984 edition of "Allelopathy," the definition was revised to include stimulatory effects.

Many compounds stimulate plant growth at low concentrations and inhibit it at high concentrations (Stevens and Merrill 1985, Leather and Einhellig 1988), and stimulatory compounds may play more important roles in ecological interactions than previously recognized (Rice 1986). It is our contention, however, that stimulatory interactions should not be grouped with inhibitory interactions under the term allelopathy, because their physiological modes of action and ecological consequences are distinctly different. Inhibitory interactions result in negative interference between plants, and stimulatory interactions results in positive interference; they have opposite effects on the fitness of the recipient plant. Stimulatory interactions derived from chemicals placed into the environment should be given their own term to avoid confusion and inconsistencies in terminology, and allelopathy should be restricted to the inhibitory effects as defined by Rice (1974). This is the definition we used in preparing this chapter.

Whittaker and Feeny (1971) define allelochemicals as "chemicals by which organisms of one species affect the growth, health, behavior, or population biology of organisms of another species (excluding substances used only as food by the second species)." Allelochemicals may be produced by plants or animals and affect animal-animal, animal-plant, or plant-plant interactions. In plants, a single compound or class of compounds may influence both plant-plant and animal-plant relationships. If the donor species derives an advantage from a chemical it produces, the compound is classified as an allomone; if the species receiving the chemical benefits then the compound is called a kairomone (Whittaker and Feeny 1971). An allelochemical toxic to plants is a phytotoxin and it could be produced by animals or plants. An allelopathic agent or allelopathic compound, by definition of allelopathy, refers only to phytochemicals toxic to other plants. If an allelopathic compound is toxic to members of the species producing it (intraspecific), the compound is autotoxic. If the allelopathic chemical is toxic to a species other than the one producing it (interspecific), the compound is allotoxic. Many phytochemicals are both auto- and allotoxic.

Other terms have been suggested to provide a better description of the type of allelopathic interaction taking place, but most have been slow to be incorporated into common use. Grummer (1955) proposed the following terms be used: kolines, meaning phytochemicals produced by higher plants affecting other higher plants; phytoncides, meaning phytochemicals produced by higher plants affecting microorganisms; marasmins, phytochemicals produced by microorganisms affecting higher plants; and antibiotics, phytochemicals produced by microorganisms affecting microorganisms. Fuerst and Putnam (1983) have suggested substituting "phytoinhibitins" for "kolines" and "saproinhibitins" for "marasmins" because these terms more clearly describe the phenomena involved.

III. Plant Interactions: Competition and Allelopathy

Ecology is the science of interactions between organisms and their environment. Ecologists have categorized the responses of different species, sharing the same environment, relative to their living apart (Burkholder 1952, Odum 1971, Radosevich and Holt 1984). When two or more species live together they may interfere with each other. This interference can have a positive, negative, or neutral effect on the fitness of each species. Competition occurs when there is a limited resource, or resources, in an environment where species are interacting. The result is negative interference because the interacting species will be less successful in the presence of a competitor than they would have been if growing alone in the limiting environment.

Allelopathy differs from competition in its mode of action and its effect on species interactions. Allelopathy functions through the addition of phytochemicals to the environment by one of the interacting species, thereby resulting in the inhibition of its neighbor sharing the habitat simultaneously or sequentially (Muller 1969, Fuerst and Putnam 1983). In this situation, the fitness of the phytochemical donor species remains unchanged relative to its growth alone, whereas the recipient species, suffering from growth inhibition, will be less successful because of reduced fitness or negative interference. The mechanism of interference can be a direct or indirect process. Phytotoxins from allelopathic species can directly inhibit germination, establishment, or growth of itself (autoallelopathy) or associated species. An example of indirect interference would be a phytotoxin that inhibits a microorganism, such as mycorrhizae, that benefit an associated plant species.

Newman and Rovira (1975) propose two ways for allelopathy to function. First, in situations where competition is occurring between species, any advantage one species has over its competitor will increase the probability that it will survive and possibly dominate. The synergistic effect between competition and allelopathy may provide the producer of phytotoxins with a necessary advantage. The other mechanism they propose is "self-balancing negative-feedback," which suggests that stressed plants produce more allelochemicals to suppress the growth of immediate vegetation, thereby providing enough reduction in competition to remain in the community. This assumes the allelopathic compounds are from living material and are not decomposition products.

Because of the complex cause and effect relations between plants and their biotic and abiotic environment, it is extremely difficult, if not impossible, to design field experiments to unequivocally separate interference by allelopathy from interference by competition. Harper (1977) discusses inherent problems associated with studying differences between species as they struggle for existence. He notes the need to stop focusing on germination and growth to determine interference. He encourages researchers to identify and study specific symptoms recognizable in the field, repeatable in the laboratory, and intentionally reproducible in the field. To help facilitate experimental design in the study of interference, Dekker et al. (1983) discuss the merits and shortcomings of various methodologies to provide qualitative and quantitative information on plant interactions. Fuerst and Putnam (1983) further developed Harper's (1977) ideas and propose the criteria in Figure 1 they consider essential for the proof of competition and allelopathy.

At a specific toxin concentration the degree of inhibition exhibited by the recipient plants is dependent on the number of plants adsorbing a toxin that occurs in limited amounts. Increasing plant density can reduce the quantity of toxin available per plant, thus decreasing the degree of inhibition per plant. Experiments that demonstrate decreased inhibition with increased plant density can be used as an indicator of toxins in the soil (Weidenhamer et al. 1989).

A complete and thorough study of interference between two plant species that satisfies all the criteria in Figure 1 never has been accomplished. To help separate the effects of interference by competition from interference by allelopathy, more innovative field experiments and monitoring techniques are needed. Because of the experimental difficulties and the complexity of biotic and abiotic factors influencing natural interactions, not all ecologists and biologists are convinced that allelopathy is a significant ecological phenomena in natural environments (Harper 1977, Begon et al. 1986, Henn et al. 1988, Qasem and Hill 1989a).

Criteria for competition:

1. Identify symptoms of interference and measure intensity if possible.
2. Show that the reduced utilization of resources by the susceptible plant is correlated with the presence of the causal plant. This indicates but does not prove competition is the cause of interference.
3. Identify a limiting resource (or resources) depleted by the causal plant. Resource interaction may be a complicating factor.
4. Reduce levels of the suspected limiting resource (or resources) to simulate interference from the causal plant. Symptoms of interference should become apparent in the susceptible plant when the resource is reduced to levels occurring in the presence of the causal plant. Resources should be tested individually to prove which is causing the observed interference (this is again complicated by resource interactions, the efficiency of control over the resource, and the sensitivity and precision for monitoring the resource).

Criteria for allelopathy:

1. Identify symptoms of the interference and quantify the intensity of interference if possible.
2. Using bioassays, isolate and purify the toxins. Structures must be characterized by chemical and instrumental methods and proven by synthesis.
3. Apply the toxins in the field at natural rates and at the appropriate phenological stages of the susceptible plants. Symptoms of observed interference should be apparent.
4. Toxin release, movement, and uptake should be monitored. Measured concentrations must be adequate to account for the observed interference. Supplemental studies are necessary to show that the toxin is not present in the susceptible plant when that plant is grown in the absence of the causal plant.

Figure 1. Experimental criteria necessary for the proof of competition and allelopathy (Fuerst and Putnam 1983).

IV. Phytochemicals With Allelopathic Potential

Phytochemicals can be divided into a number of classes depending on the degree of structural similarity desired among compounds within a class. They are frequently organized according to biosynthetic similarities or functional group composition, or both. Rice (1984) recognized 19 categories of allelopathic compounds (Table 1); however, Einhellig (1985) stresses, many of these categories are poorly represented in the allelopathic literature. The most commonly reported allelopathic compounds are terpenes and phenolic derivatives. Nearly all classes of terpenoids have at least a few compounds known to be germination and plant growth regulators (Mandava 1979, Fischer 1991). Fischer (1986) identified 14 monoterpenes, 15 sesquiterpenes, and 16 sesquiterpene lactones with suspected allelopathic activity. He reported an additional 26 sesquiterpene lactones with alleged plant growth regulatory activity. Since 1980 approximately 60 terpenoids have been investigated for their phytotoxic and allelopathic activities (Fischer 1991).

Phenolics are the most frequently identified class of compounds responsible for soil toxicity. The types of phenolics most often encountered are in the cinnamic acid, simple phenols/benzoic acid, and coumarin categories of Table 1 (Rice 1984). Many phenolics, such as cinnamic acid, caffeic acid, p-coumaric acid, ferulic acid, and vanillic acid, are present in the soil when tissues decay because they are biosynthesized in all plants and are essential components of structural polymers (Dao 1987). Coumarins, such as scopoletin, isoscopoletin, esculin, and esculetin, also are distributed widely in plants.

The biological activity of a particular compound, or class of compounds, is dependent on the types of functional groups present (double bonds, alcohols, ketones, lactones, amines, sulphydryls, etc.) and the structural arrangement of the molecule. Acyclic, unbranched organic compounds containing 6 to 9 carbon atoms are released by plant residues. Six classes of these compounds (n-alkanes, 2-alkanols, n-alkanals, 2-alkenals, 2-alkanones, and n-alkanoic acids) were bioassayed with onion (*Allium cepa* L.), carrot (*Daucus carota* L.), and tomato (*Lycopersicon esculentum* Mill.) seeds to evaluate the relationship between structure and germination inhibition (Bradow 1991). Overall the 2-alkenals were most inhibitory, then the 2-alkanols, n-alkanals, and 2-alkanones. The degree of inhibition depended on plant species, class of compound, chain length, and concentration.

Reynolds (1977, 1978, 1987, 1989) tested the effects of molecular structure and functional group composition of numerous acyclic, cyclic, heterocyclic, and aromatic compounds on the inhibition of lettuce (*Lactuca sativa* L.) seed germination. For all of the compounds tested lipophilicity was the one factor that correlated most often with biological activity. Lipophilicity influences a compounds ability to penetrate the membranes that surround the cell and cellular organelles. The type of functional groups present and the structural arrangement of the molecule also affects the inhibitory activity.

V. Allelochemicals in Plant Tissues and Methods of Release

Allelochemicals can be found in all organs of plants: seeds, flowers, pollen, fruits, leaves, stems, roots, and rhizomes. Factors influencing the compounds

Table 1. Chemcial classes of allelopathic compounds (as listed by Rice 1984).

Chemical class	Structure
Organic acids, aliphatic aldehydes and ketones, straight-chain alcohols.	CH_3COOH CH_3CH_2COOH $CH_3CH_2CH_2COOH$
Simple unsaturated lactones.	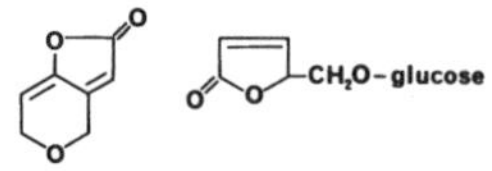
Long-chain fatty acids and polyacetylenes.	$CH_3C\equiv C-C\equiv C-C\equiv C-\overset{H}{C}=\overset{H}{C}COOCH_3$
Quinones (naphthoquinones, anthroquinones, and complex quinones).	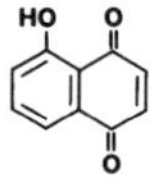
Simple phenols/benzoic acid and their derivatives.	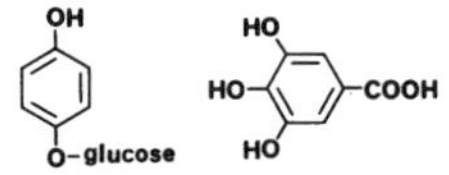
Cinnamic acid and its derivatives.	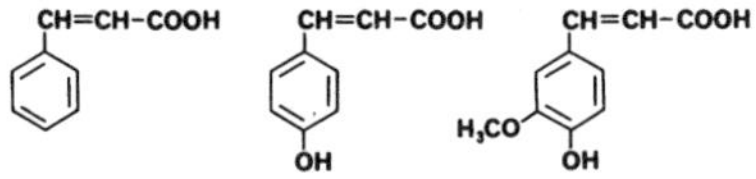
Coumarins.	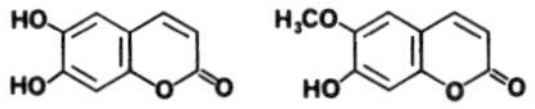
Flavonoids.	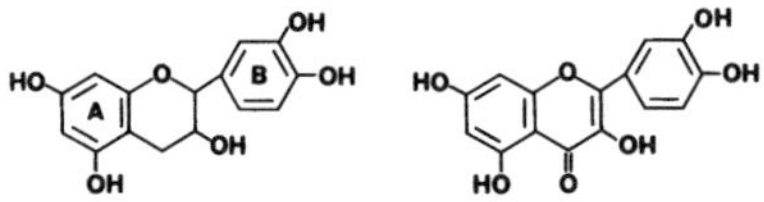
Tannins (hydrolyzable and condensed).	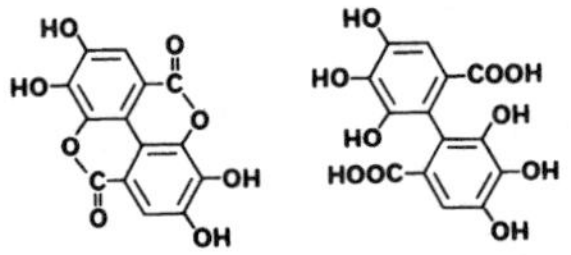
Terpenoids and steroids.	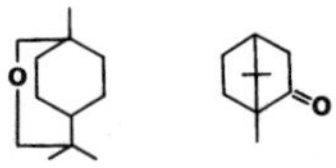
Amino acids and peptides.	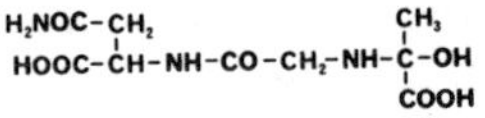
Alkaloids and cyanohydrins.	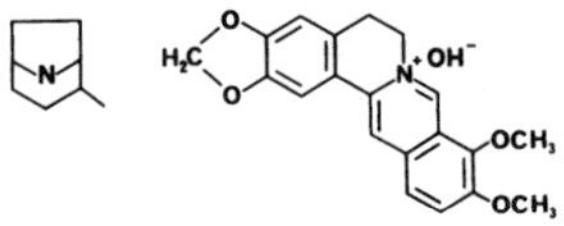

Table 1. Continued

Chemical class	Structure
Sulfides and mustard oil glycosides.	$CH_2{=}CHCH_2C({=}NOSO_3)\text{S-glucose}$ $CH_2{=}CHCH_2NCS$
Purines and nucleosides.	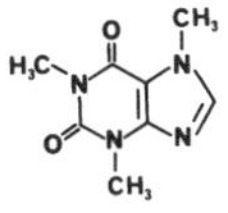
Miscellaneous.	$CH_2{=}CH_2$

present and their concentrations include genetics, organ/tissue type, environmental conditions affecting growth including stress from physical (Gershenzon 1984, Waterman and Mole 1989) and biological sources (pathogens and herbivores), physiological/phenological stage, and age. Each plant part may contain very different chemical components. When the same compound occurs in more than one organ, the concentrations can be quite different (Locken and Kelsey 1987). Leaves generally will contain the greatest number and quantities of toxins; roots rank second. If compounds are physiologically active, how do the producing plants avoid toxicity? Allelochemicals usually are compartmentalized within special tissues, cells, or organelles within the cells to isolate them from critical metabolic processes. Terpenoids, when produced in large quantities, always are associated with specialized cells responsible for biosynthesis, secretion, and storage (Kelsey et al. 1984). If not compartmentalized, the compounds may be maintained in an inactive form, such as a glycoside with a sugar molecule(s) attached (Balke et al. 1987, Hogan and Manners 1991).

Environmental factors that significantly impact biosynthesis of allelochemicals are the same as those influencing growth i.e. light, water, and nutrients. Because their effects on growth and allelochemical biosynthesis are not mutually exclusive, it is difficult to predict how a change in one of these parameters will affect tissue concentrations of a particular allelochemical or class of allelochemicals. Some generalizations for expected responses nevertheless can be made. Plants stressed by limited nutrients or water tend to produce greater concentrations of allelochemicals, depending on the class of compounds and the species (Gershenzon 1984). In these situations, a reduction in growth results in a greater availability of photosynthate for use in the biosynthesis of allelochemicals, particularly the carbon-based classes such as phenolics and terpenes. When nutrients or water are limited, the synthesis of allelochemicals containing nitrogen may or may not increase depending on the availability of nitrogen. Light intensity is important because it influences the quantity of photosynthate. It is reasonable to expect a plant grown in full light to produce more allelochemicals than a plant of the same species grown in the shade (Cooper-Driver et al. 1977). This is not always true, however, and under natural conditions the response will be mitigated by the nutrients and water available for growth (Waterman and Mole 1989). The toxicity of extracts prepared from tissue residues of two mustard (*Brassica*) species increased when plants were stressed by high temperatures, short days, or reduced nutrients. Extracts of residue from plants subjected to intermediate water stress were more inhibitory than extracts from residues of severely stressed and nonstressed plants (Mason-Sedun and Jessop 1989).

Except in the case of pathogens, where microorganisms are in direct contact with their host, allelopathic compounds must be released into the environment surrounding the donor plant. The four recognized mechanisms of release are illustrated in Figure 2. Volatilization is restricted to compounds with a vapor pressure high enough to release sufficient quantities to the atmosphere directly from live tissues or plant residues (Bradow and Connick 1987). If vapors are concentrated enough for sufficient time, they can inhibit seed germination and growth directly without being deposited to the soil (Asplund 1968, Klarich and Weaver 1973, Hoffman and Hazlett 1977, Weaver and Klarich 1977, Groves and Anderson 1981, Bradow and Connick 1988a, 1988b). In this situation, the soils merely provide the platform on which a seed rests while it accumulates volatiles dispersed from adjacent plants (Heisey and Delwiche 1984). When environmental conditions are favorable, volatile allelopathic compounds can be adsorbed onto soil particles (Muller and del Moral 1966, del Moral and Muller 1970, Halligan 1976), thereby accumulating sufficient concentrations to retard root growth of young seedlings coming in contact with these particles. This mechanism requires periods of dry warm weather to facilitate volatilization and soil adsorption followed by a wet season with sufficient rain and high enough temperatures to allow seed germination and growth. These conditions are characteristic of the Mediterranean climate found along the Pacific coast of the western United States.

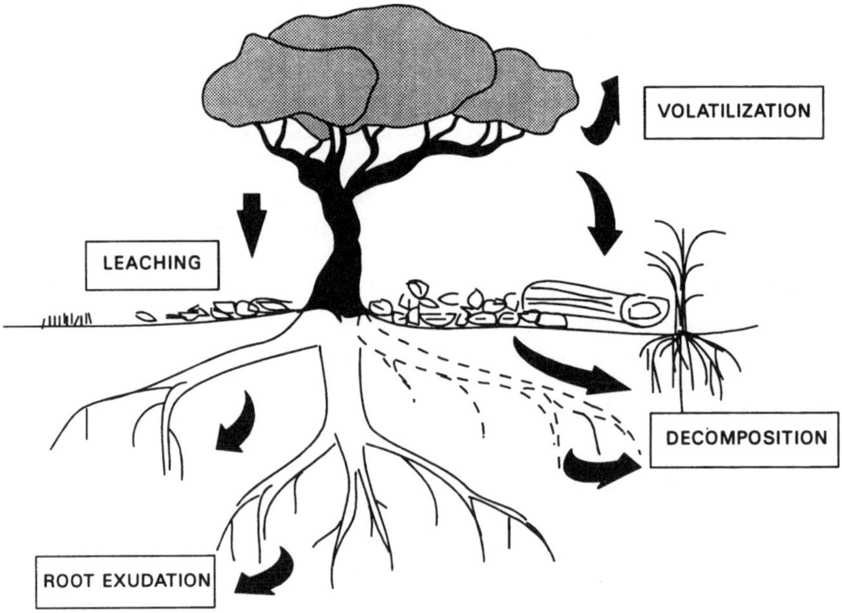

Figure 2. The four mechanisms that release phytotoxins from plant tissues.

Allelochemicals can accumulate on the epidermal surfaces of plants by leakage through membranes (Tukey 1970), active exudation, or from specialized epidermal structures such as glandular trichomes where these compounds are stored or exuded (Fahn 1979). Leaching occurs when water from rain, fog, or dew contacts the epidermal surfaces, dissolves minute quantities of these compounds, and transports them directly to the tissues of other plants, seeds on the soil, or to the soil particles below. Allelochemicals also may be leached from the litter and tissue residues accumulated on the soil surface. The amount of water, its frequency of occurrence, and the time of year will all effect the efficiency of leaching and inhibitor concentrations (Tukey 1966, 1970, Halligan 1976, Richardson and Williamson 1988).

Root exudation, the secretion of substances from healthy intact root tissues to the surrounding medium (Rovira 1969), is difficult to study under natural field conditions. All factors affecting plant growth affect the quantity or quality of exudates (Rovria and Davey 1974). Particular care must be taken to avoid alteration or additions of compounds by microorganisms and contamination from decaying tissues and leachates of foliage and litter. The biotic and abiotic conditions of the rhizosphere, the soil in the immediate vicinity of the roots, play an important role in this mechanism of release. The interface between roots and soil is a complex environment with active populations of bacteria, fungi, and soil fauna (Rovira and Davey 1974, Ingham et al. 1985, Foster 1988). Some microbes may be attached to the outer surface of the root (Hoagland and Williams 1985), and others, like ectotrophic mycorrhizae, bridge the barrier by penetrating the root cortex to establish a symbiotic association. The presence or absence of mycorrhizae may be regulated in part by allelochemicals (Perry and Choquette 1987, Vierheilig and Ocampo 1990, Wacker et al. 1990). There are quantitative and qualitative differences in the microorganisms around plant roots compared to the root free soils. Microorganisms are more abundant around plant roots in part because of the many compounds exuded from roots that can be used in their growth and development. Compounds exuded by roots include sugars, amino acids, organic acids, and various allelochemicals such as phenolics (Rovira and Davey 1974, Tang and Young 1982), alkaloids (Wink 1983), isoflavonoids (d'Arcy Lameta and Jay 1987), and benzofurans or thiophenes (Tang et al. 1987).

Of the four release mechanisms of allelochemicals shown in Figure 2, tissue decomposition has the most potential for producing high concentrations of phytotoxins. During decomposition there are several processes that can result in toxic compounds in the soil: (1) direct release of toxins previously present in the plant tissues, (2) transformation or conversion of inactive plant derived compounds into toxins, and (3) toxins biosynthesized by microbes. Direct release is initiated when cells start dying and membranes begin to break down. After death, physical and biological fragmentation of membranes and structural polymers make once inaccessible compounds available for leaching or for uptake by penetrating root hairs. As compounds are released, they may be rearranged or transformed into other forms. This may or may not involve the action of microbes. Hydroxyjuglone, for example, upon leaving black walnut (*Juglans nigra* L.) tissues is readily oxidized to juglone, a toxic quinone (Rietveld 1983). Carveol and (-)-carvone from spearmint (*Mentha spicata* L.) plants can be converted by soil microbes into numerous components including the bioactive (+)-bottrospicatol (Noma 1980, Noma et al. 1982, Nishimura et al. 1983). Included in this second category is the production of toxins from polymer degradation, such as the release of various phenolic derivatives during lignin breakdown (Dao 1987). In the third category, toxins are biosynthesized by the decomposers themselves when they use the plant tissues as a carbon and energy source. High concentrations of organic acids are commonly produced by microbes, especially under anaerobic conditions (Lynch and Penn 1980, Gussin and Lynch 1981).

Allelopathic effects from residue decomposition can be seasonal in nature. Patrick et al. (1963) report maximum toxicity early in residue decomposition of barley (*Hordeum vulgare* L.), rye (*Secale cereale* L.), wheat (*Triticum aestivum* L.), sudan grass (*Sorghum bicolor* L.), and vetch (*Vicia* sp.) with stimulatory effects occurring later. Phytotoxicity from decomposing barley residue reaches a maximum in 3 weeks and then declines after 6–7 weeks (Toussoun et al. 1968). Decomposing grass residues are inhibitory to the growth of other grasses and clover (*Trifolium repens* L.) after 10 days; less activity is observed at 20 days (Gussin and Lynch 1981). In the Pacific Northwest, allelopathic effects of litter from decomposing bracken fern (*Pteridium aquilinum* (L.) Kuhn.) coincides with breaking of spring dormancy and emergence of competing species (Gliessman 1976). Soils under California sagebrush (*Artemisia californica* Less.) are toxic at the start of the wet season and soon made innocuous as toxins leach from the soil surface (Halligan 1976).

VI. Allelopathic Compounds and Soils

Regardless of how allelopathic compounds are released, in most instances they enter the soil. Soil toxicity of allelopathic compounds can be mediated by several factors, as indicated in Figure 3. Concentration and availability will be affected by adsorption/desorption, which is mediated by soil type, particle size, pH, soil water, and organic matter content among others (Dao 1987, Dalton et al. 1989). Various phenolic acids inhibitory to the germination and growth of several crop species when tested on paper, were not inhibitory when the same solutions were added to soil, despite the fact that concentrations added to soil were greater than amounts previously reported from soil (Krogmeier and Bremner 1989). Del Moral and Muller (1970) clearly demonstrate how soil texture can influence toxicity. Sand was far less efficient at adsorbing and retaining volatile terpene and phenolic toxins than was loam. The sand, consequently was less inhibitory than the loam when exposed to the same concentrations of toxins. Bare zones at the edges of soft chaparral stands, created and maintained in part by allelopathy, develop best on heavy fine-textured Zaca clay soils (Muller 1966). Conversely, residues and extracts of mustard species are more toxic in sand than in clay soils (Mason-Sedun and Jessop 1988). Toxic effects of litter from bur buttercup (*Ranunculus testiculatus* Crantz) are less in fine-textured soils than in coarse-textured soils. Buchanan et al. (1978) ascribe this to greater microbial decomposition in the fine soils.

Both physiochemical and biological processes are involved in the recoverability of allelochemicals from soil. For example, ferulic acid in high concentrations

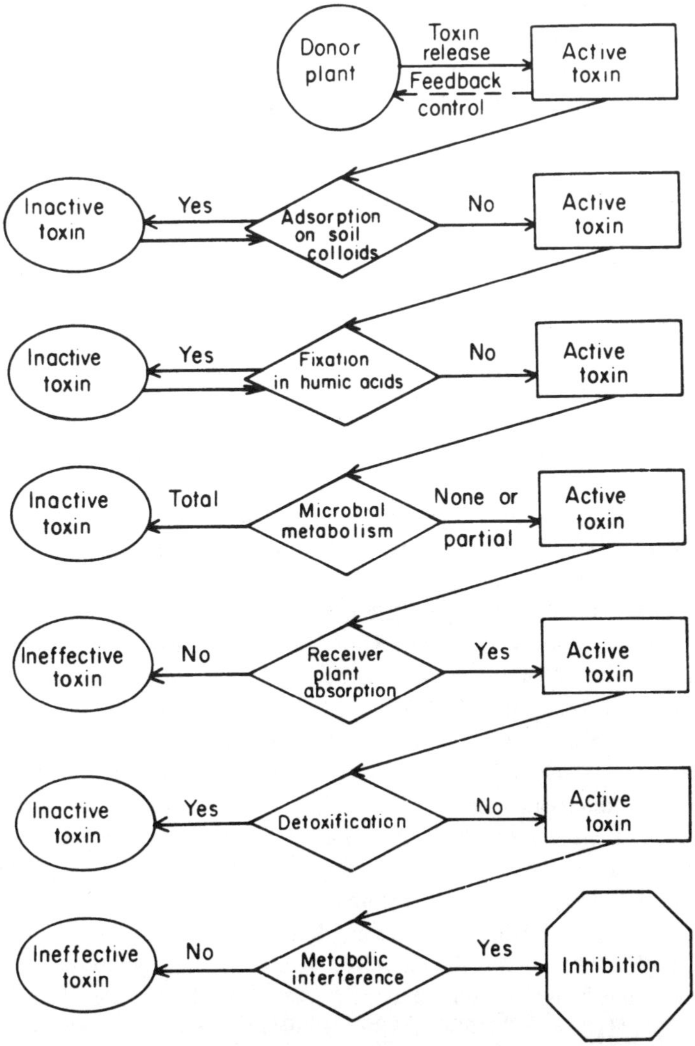

Figure 3. The potential fate of allelopathic compounds in the environment. Reprinted with permission from Fisher (1979), copyright, Academic Press.

(1000 mg/kg) decreases rapidly when added to tropical forest soils under laboratory conditions. The greatest rate of loss is in nonsterile soils and immediate recovery is inversely correlated with the percentage of organic carbon in the soil. In nearly all nonsterile soils, water-extractable ferulic acid essentially disappeared in two days, and for some soils it dissappeared in only one day (Dalton 1989). Ferulic acid bound ("trapped") to the soil and not extractable by water, could be released by extraction with sodium acetate solutions. In the bound form, ferulic acid was protected from microbial degradation. In a similar study (Sparling et al. 1981), various phenolic acids were added in high concentrations (5000 mg/kg) to nonsterile agricultural soils. Immediate recovery was very low (2.6 to 25% of total added), and by 28 days the concentrations of phenolic acids were similar to unamended controls. Alkaline extraction of soils removes all free and bound phenolics potentially available to inhibit plants (Sparling et al. 1981), but it is believed that only water soluble phenolics in free form, and possibly some in the bound form, can affect plant growth. Yet total combined concentrations of the free and bound phenolics found in field soils are usually too low to have an effect on plant growth (Whitehead et al. 1981, 1982, 1983). Concentrations of phenolic acids isolated from soils of various studies have been summarized by Einhellig (1985).

Soil microbes may be involved in allelopathy in various ways. As already mentioned they may synthesize toxins during growth or alter the structure of a plant-derived compound, converting it from an inactive to an active form. Microbes can significantly limit toxicity by metabolizing allelochemicals being released. Microorganisms capable of metabolizing a specific toxin are associated with the plants producing the toxin. For example, bacteria (Pseudomonas J1) that utilize juglone as a sole source of carbon and energy were isolated only from soils collected around black walnut trees which produce juglone (Schmidt 1988). Their metabolic activity was not restricted to juglone, however, as other phenolic compounds also could be catabolized. Microbial populations can adapt rapidly to take advantage of readily available carbon and energy sources to increase their population biomass, which in turn can result in greater rates of toxin breakdown (Stowe and Osborn 1980, Sparling et al. 1981, Vaughan et al. 1983). In the presence of phenolics the microbial population shifted so a greater proportion was capable of metabolizing phenols (Blum and Shafer 1988). Terpenes also can be used as a carbon and energy source by soil microbes (Vokou et al. 1984).

Microbes may have indirect effects on allelopathy. A plant and its rhizosphere community of microorganisms are in direct competition for nutrients and moisture (de Jong and Klinkhamer 1985). Quantitative and qualitative changes in the rhizosphere will affect growing conditions for the plant and its ability to produce exudates. Lovett (1987) suggests allelopathy occurs when environmental conditions prevent soil microbes from metabolizing phytotoxins, thus permitting toxin accumulation to active levels. If the microbes are pathogenic, then phytotoxin weakened plants may be more susceptible to infection (Lynch and Penn 1980, Penn and Lynch 1982, Hassan et al. 1989a, 1989b, 1989c).

Decomposition of allelopathic compounds in soils seems to be delayed by cold, acidic, and reducing environments (Einhellig 1985). Jameson (1970) reports the accumulation of phytotoxins from juniper leaves under conditions of poor aeration and drainage. Maximum toxicity of decomposing field crops occurs in cold, wet, and heavy soils during early stages of decomposition when organic acids accumulate under inadequate aeration (Toussoun et al. 1968, Patrick 1971, Lynch and Penn 1980, Gussin and Lynch 1981). Ballester et al. (1979) suggests that the low pH (5.1) in the heathland soils of northwest Spain prevents decomposition of some allelopathic compounds. Adding lime to increase soil pH will release some bound phenolics, thereby raising phenolic concentrations in the soil solution (Whitehead et al. 1982), but at the same time the higher pH may reduce their toxic activity (Blum et al. 1989, Chou 1987).

The availability of soil nutrients ameliorates allelopathic effects. Barley plants grown hydroponically are strongly inhibited by phenolic acids in $CaSO_4$ solution, whereas there is no effect when grown with phenolics and a complete inorganic nutrient medium (Glass 1976). In sand cultures with various concentrations of nitrogen and phosphorus, phenolic acids are inhibitory to barley only at low nutrient concentrations (Stowe and Osborn 1980). Application of nutrients directly to the soil will eliminate growth inhibition in some instances (Christensen and Muller 1975a, Rice 1976, Fisher et al. 1978, Hall et al. 1983), but not in others (Buchholtz 1971, de Jong and Klinkhamer 1985, Klein and Blum 1990a). Although rangeland species may contribute allelopathic compounds to the soil, they also may be lessening the potential impact by releasing nutrients at the same time. Nutrient enrichment under shrub and woodland trees is well documented (Tiedemann and Klemmedson 1973, Charley and West 1975, Barth 1980).

Soil microenvironments must be considered as factors in the presence and absence of allelopathy (Foster 1988), especially in natural habitats and no-till agriculture (Cast et al. 1990). Yearly deposition of litter may result in heterogeneous distributions of organic matter within communities. Concentrations of inhibitors released or derived during the decomposition of this material will differ with the amounts of debris present. Higher concentrations of inhibitors are likely to occur where litter is more abundant. The volume of soil containing toxins is important because the severity of growth inhibition increases as the proportion of a plant's root system exposed to toxin increases (Klein and Blum 1990b). In addition, the amount and composition of the organic matter will influence many other soil factors; e.g., nutrients, microbial populations, fauna, and water holding capacity. All of these will affect the concentrations and toxicity of inhibitors, as well as the overall conditions for germination and plant growth. Consequently, the degree of allelopathic influence can be substantially different for the distinct microenvironments within a community. Our understanding of allelopathy will be enhanced tremendously as our knowledge of soil chemistry and ecology improves.

VII. Bioassays, Isolation, and Identification of Allelochemicals

Stowe (1979) conducted a survey of bioassay results from 84 research papers (Table 2) and concluded there exists a high probability of discovering phytotoxins in plant tissues using laboratory bioassays. In addition, he conducted his own bioassays with 7 introduced grass species abundant in an abandoned field in Illinois. Most bioassays indicated phytotoxicity, even for species not suspected of being allelopathic or showing signs of allelopathy in the field. Autotoxicity also was a common occurrence. No correlations were found between bioassay results and species distribution patterns measured in the field. He concluded bioassay results are dependent on the procedures and may have no ecological meaning. The relevance of petri dish bioassays has also been questioned by Henn et al. (1988).

Laboratory bioassays have limited value for interpreting field observations, but they always will be an integral part of comprehensive allelopathic investigations. They are essential for the extraction, isolation, and identification of the toxic chemicals. As emphasized by Stowe (1979), the absence of standardized procedures is a major problem for allelopathic bioassays. Leather and Einhellig (1986, 1988) have reviewed the procedures and problems of bioassay techniques used in allelopathic investigations.

Table 2. Percentage of laboratory bioassay studies (84 surveyed) reporting phytotoxic activity (Stowe 1979).

1

Plant material treatment	% with significant inhibition	Material assayed	% with significant inhibition	Bioassay procedure	% with significant inhibition
Left intact	54	Shoots	63	Seed germination in petri dishes	56
Cut up	69	Roots	61	Seed germination in sand or soil	62
Dried & ground	68	Litter	73	Root elongation in petri dishes	73
Ground fresh in blender	60	Soil	50	Growth in sand or soil	51
				Growth in water culture	87

Interference with germination (inhibition or stimulation) is the most frequently used bioassay. The advantages are simplicity, speed, minimal cost, and limited volumes of test solution or purified toxin required. Disadvantages include low sensitivity compared to other assays, variation in sensitivity depending on the plant species used, and lack of knowledge of germination physiology (Leather 1987a). A phytotoxin's mode of action may be interference with a particular physiological process not functional during germination. Testing for an effect on germination therefore would not be appropriate.

Bioassays monitoring interference with growth or physiological processes are more sensitive than germination assays, but they also are more difficult to conduct, take more time, and require more solution and purified inhibitors. Growth assays usually measure the length or dry weights of radicles, hypocotyls, or entire seedlings. Measuring the dry weight of a radicle is much easier, faster, and more accurate than trying to measure its length, because of curling or tissue damage (Leather and Einhellig 1986, 1988); however, methods have been devised to speed up and simplify measurements of radicle elongation (Leather and Einhellig 1985, 1986). For example, seeds or pregerminated seedlings can be placed on top of agar containing known concentrations of the active compound. The agar can be used in test tubes (Tang and Waiss 1978), petri dishes (Stevens and Merrill 1980, Merrill 1989), or the small wells of cluster tissue culture dishes (Dornbos and Spencer 1990). This latter procedure has the additional advantage of requiring minimal quantities of compound. Seedlings are easily removed from the agar for length measurement or other physiological assays, including the the analysis of seedling tissues for the presence of toxin. For physiological bioassays, any number of metabolic processes can be monitored (photosynthesis, respiration, enzyme activity, or concentrations of specific compounds), but the ones selected for use should depend on the suspected mode of action of the toxin.

Recently a bioassay was developed (Einhellig et al. 1985a) that uses aquatic species of *Lemna* L. (duckweed). It is fast and sensitive and requires only small volumes of solution. It can be used to measure effects on growth, growth rate, reproduction, and physiological processes. With *L. obscura* (Aust.) Daubs, anthocyanin production is the most sensitive indicator of inhibition (Leather and Einhellig 1985, 1986, and 1988). The duckweed bioassay has been found to correlate with terrestrial plant toxicity but has the advantage of being from 10 to 100 times more sensitive (Leather, personal communication).

Many external factors can influence bioassays and create procedural errors. Species selected for bioassay should include those suspected of receiving allelopathic interference in the field. Nonsusceptible species can also be used as a check. It is helpful to know the germination requirements of assay species because they could change in the presence of toxins. Specific bioassay parameters requiring attention are solution pH (Blum et al. 1984), osmotic potential (Anderson and Loucks 1966, Bell 1974, Wardle et al. 1992), how the solution is applied to the seed bed (Leather and Einhellig 1986), and volume of solution or toxin per seed (Weidenhamer et al. 1987).

Bioassays can provide information about the source of toxins. They may not prove a particular release mechanism is functioning in the field, but they can help eliminate those routes that are not functioning. Different experimental procedures are associated with each source-and-release mechanism. Although laboratory and greenhouse experiments are designed to control as many variables as possible, upon close scrutiny some allelopathic designs and treatments are not as clear and easy to interpret as they first seem. There may be more than one possible explanation for the observed response from bioassay species (Qasem and Hill 1989a).

In tests for volatilization, the bioassays are set up so that the only contact between donor tissues, or plants, and the recipient test species is through the atmosphere (Muller et al. 1964, Klarich and Weaver 1973, Hoffman and Hazlett 1977, Weaver and Klarich 1977, Groves and Anderson 1981). Another approach is to demonstrate that soils in an enriched chemical atmosphere can adsorb sufficient quantities of compound to become toxic (Muller and del Moral 1966, del Moral and Muller 1970). In the field, volatiles must be trapped from the atmosphere around the donor plants (Muller, C.H. 1965, 1966) and isolated from soils when conditions for release by other routes is unlikely—during dry periods when leaching has not occurred for some time and when there is insufficient soil water for microbial growth and decay (del Moral and Muller 1970, Halligan 1975).

Toxins transported by leaching can be collected by artificially leaching fresh foliage (McPherson and Muller 1969), live plants, or litter in the laboratory or field. The best procedure is to collect natural rain leachates (Christensen and Muller 1975a, Halligan 1976). The solutions are then tested by watering seeds and seedlings in petri dishes, or in pots containing sand, greenhouse soils, or field soils. Litter leaching is simulated by placing litter on top of potted soil and then pouring or spraying water over the litter while watering the germinating seeds and seedlings (del Moral and Muller 1970). In the field, effects of litter can be monitored by comparing germination and seedling growth on paired plots with and without litter (del Moral and Muller 1970, McConnell and Smith 1971, Parker and Muller 1979) or by adding litter to plots where none previously existed (del Moral and Muller 1970, Muir et al. 1987).

Release of toxins during decomposition, or their formation by microbes during decomposition, is examined by incorporating the various fresh plant tissues (Toai and Linscott 1979, Hall et al. 1983) or their dead litter into soil subsequently planted with the assay species. Most often this is done in pots, but it can be set up to simulate the more natural conditions of the field (Leather 1987b). In the laboratory, it is possible to separate toxins of plant origin from those of microbial origin, or rearrangement, by including a series of sterile controls (Ohman and Kommedahl 1964, Solomon 1983, Weston and Putnam 1986, Weston et al. 1987). Adding organic matter to soil can stimulate microbial growth and thereby reduce available nutrients (Katz et al. 1987, de Jong and Klinkhamer 1985). To avoid erroneous interpretations resulting from microbial competition rather than inhibitors, a non-toxic organic matter such as peat moss or cellulose should be added to the control soils at concentrations equal to the treatments (Ohman and Kommedahl 1964, Wilson and Rice 1968, Lodhi and Rice 1971, Balke 1985). Microbial competition for nutrients can be further minimized by supplementing treatments and controls with nutrient solutions, preferably after the plant material or organic matter is incorporated (Qasem and Hill 1989a). Other complications in these experiments, as discussed by Qasem and Hill include changes in soil texture, water-holding capacity, and pH. The decaying tissues also may stimulate the growth of disease organisms.

Root exudation is one of the most difficult release mechanisms to investigate because of dead root tissues and rhizosphere microbes in the soil. Sampling undisturbed roots is a problem partially overcome by several interesting techniques. Donor plants can be grown in agar and the agar tested for toxic exudates (Putnam and Duke 1978). The donor and recipient species can be grown together in pots while the competition for essential resources is minimized (Rietveld 1975). More commonly, donor plants are grown in containers of soil or sand and are supplemented with nutrient solutions. The container is leached and the leachate used to water assay plants grown in petri dishes, sand, or soil (Leather 1983, Kil and Lee 1987). The stairstep design (Figure 4a) is a more automated modification of this technique (Wilson and Rice 1968, Bell and Koeppe 1972, Lovett and Jokinen 1984). Brede (1991) constructed a horizontal, electronically controlled modification of the stair step apparatus in the field. Another similar alternative is a one-on-one donor and recipient exudate recycling system (Figure 4b) (Stevens and Tang 1985, Tang 1986). Qasem and Hill (1989a) were concerned that the stairstep technique did not adequately control for differences in species' nutrient requirements and the species' abilities to accumulate nutrients. Weeds, in some instances, are more efficient at nutrient uptake than crop or forage plants. To help minimize any possibility of nutrient competition, Qasem and Hill (1989b) developed the drip channel irrigation method (Figure 4c). Limiting the opportunities for competition is critical in all these experimental designs.

To facilitate chemical studies of exudates in undisturbed root cultures, organic compounds escaping into the culture nutrient solutions are continuously trapped by XAD-4

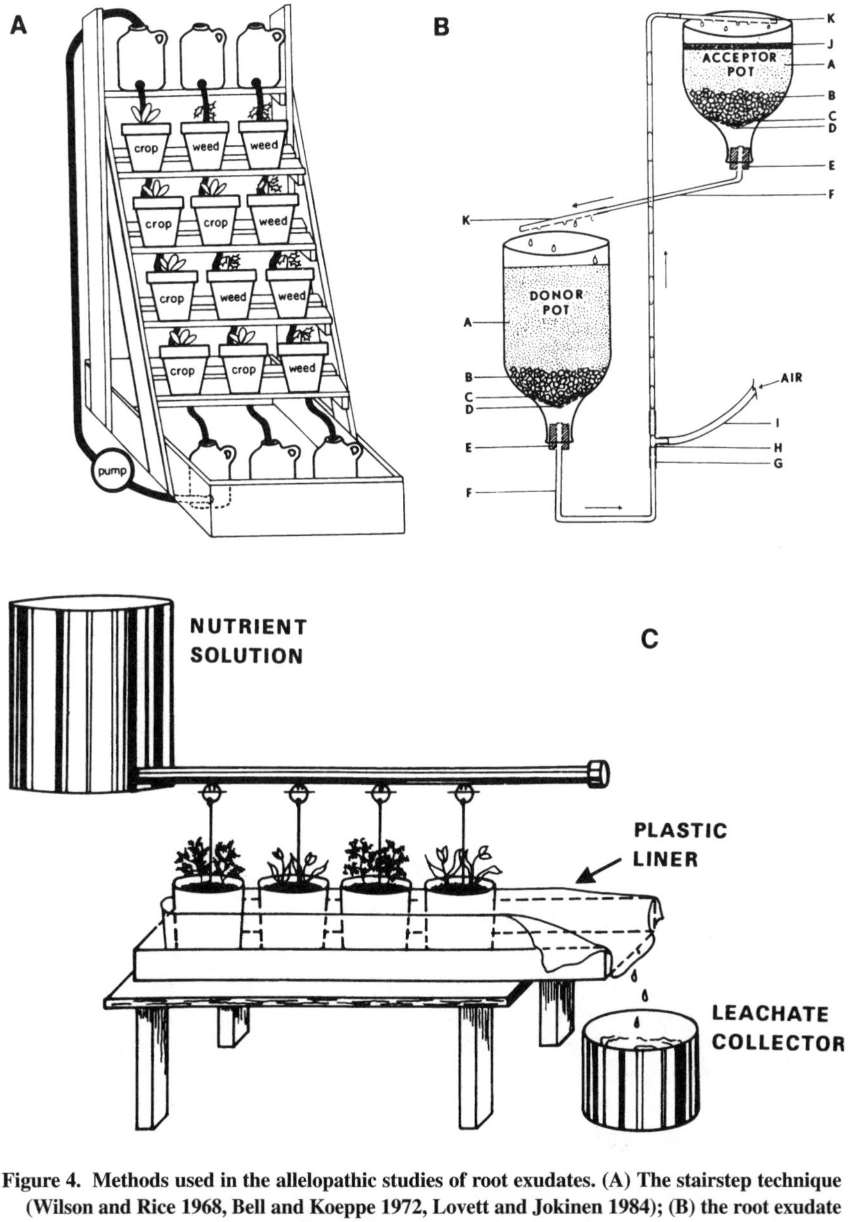

Figure 4. Methods used in the allelopathic studies of root exudates. (A) The stairstep technique (Wilson and Rice 1968, Bell and Koeppe 1972, Lovett and Jokinen 1984); (B) the root exudate recycling system (Stevens and Tang 1985); (C) the drip irrigation system (Qasem and Hill 1989b), pots containing donor and reciever plants are alternated in one channel, whereas pots containing all reciever plants are placed in a separate channel as a control. Our diagram shows only 4 pots, Qasem and Hill (1989b) used 12 per channel; (D) the hydrophobic root exudate trapping system (Tang and Young 1982).

Figures reprinted with permission: (A) Radosevich and Holt, Weed Ecology. Implications for Vegetation Management, copyright 1984, John Wiley & Sons; (B) Stevens and Tang, copyright 1985, Plenum Publishing Corporation; Figure (C) was drawn by us as described in Qasem and Hill (1989b); (D) Tang and Young, copyright 1982, American Society of Plant Physiologists.

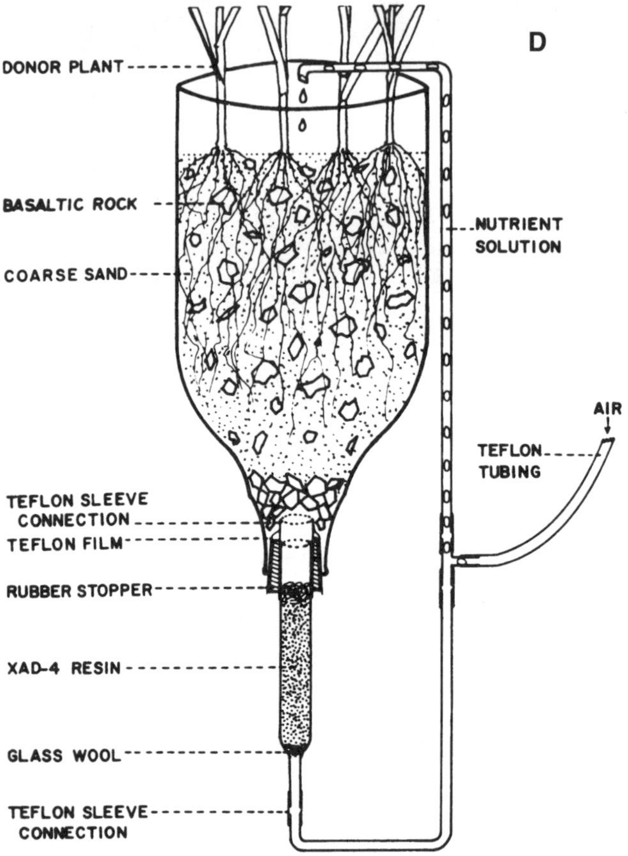

Figure 4d.

resin (Figure 4d). Adsorbed organics are eluted later from the traps for further isolation, identification, and bioassay (Tang and Young 1982, Tang 1986, Tang et al. 1987). Compounds labeled with a radioactive nuclide have been useful in conducting studies of root exudates (Rovira and Davey 1974, Van Veen et al. 1989), but they have not been used much in studies of allelopathy. As Horsley pointed out in 1977, this is a useful technique for following the release and movement of inhibitors through the environment, including uptake by the recipient plants.

Another method has been developed (Unamma and Akobundu 1989) to facilitate the separation of allelopathic effect from competition when donor and recipient plants are grown in the same container (Figure 5). A container is divided into donor and recipient soil zones with plastic sheets. Through a drain at the bottom of the donor soil, leachate is collected for addition to the recipient soil without root competition. A comparison of the recipient growth responses in this container with the container where roots are not separated provides a measure of the allelopathic influence. The containers can be placed outdoors to take advantage of natural conditions where rain leachates can be syncronized with the growth and phenological development of the species being tested.

Once allelopathic activity has been supported by experimental evidence, the task of isolating and identifying toxins begins. For plant tissues, standard phytochemical techniques are used (Harborne 1984). The same, or similar, methods (Read and Jensen 1989) can be used to extract soils; but as observed for phenolic acids, the extracting solution may significantly affect the quantities isolated (Whitehead et al. 1981, Dalton et al. 1987, Dalton 1989). Water is the most natural solvent, and it

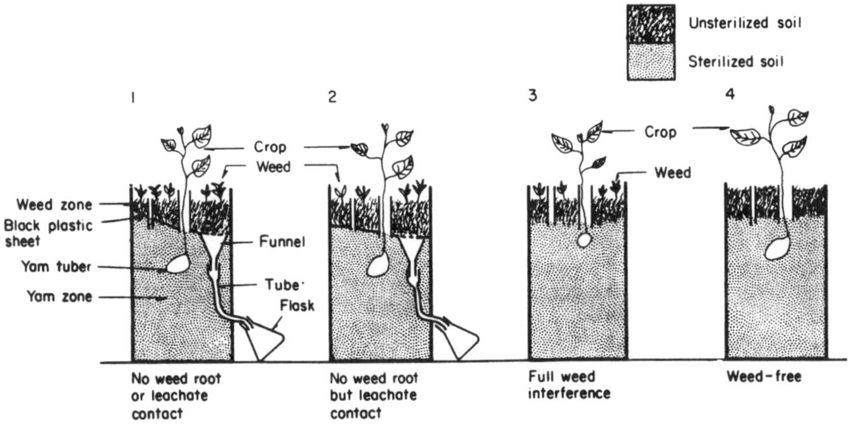

Figure 5. A method for investigating allelopathy in deep-rooted plants. Reprinted with permission from Unamma and Adobundu (1989), copyright, European Weed Research Society.

probably removes those compounds present in the soil solution available to microbes and plants (Whitehead et al. 1981, Dalton 1989). Water has been used frequently to extract toxins from soils (Halligan 1976, Bradow and Connick 1987, Lodhi et al. 1987). Water can be used to extract plant tissues (Lehle and Putnam 1983), but it usually is not the best solvent for efficiently extracting many organic compounds. If the suspected toxin is a plant constituent, larger quantities usually can be obtained by using an appropriate organic solvent. This is determined by successively extracting a tissue sample with solvents of increasing polarity. Each solvent fraction must then be bioassayed to locate the active compound. These initial fractions will still be complex mixtures. Knowledge of which solvent extracts the toxin can be useful information for further analysis and purification techniques.

Separating pure compounds requires some form of chromatography (Sewell and Clarke 1987). Many common terpene and phenolic compounds can be identified with chromatography by using authentic reference standards and without isolating and purifying each individual constituent. Concentrations also can be accurately determined by these techniques. Gas chromatography in combination with mass spectrometry is used to identify volatiles, and high-performance liquid chromatography (HPLC) is used to study nonvolatile, and often more polar, compounds. Less common and unknown compounds must be purified and their structures determined by instrumental methods. Nuclear magnetic resonance, infrared, ultraviolet, and mass spectroscopy are the techniques employed (Williams and Fleming 1980). Structures of crystalline compounds can be determined precisely by X-ray crystallography. Ideally, the toxin should be synthesized and the activity shown to be the same as the isolated natural product (Fuerst and Putnam 1983). Currently, synthesis is seldom accomplished. Pure compounds commonly are bioassayed under laboratory or greenhouse conditions (Demos et al. 1975, Rasmussen and Einhellig 1977, Einhellig and Rasmussen 1978, Williams and Hoagland 1982, Barnes and Putnam 1983) and used in experiments to study modes of action. Applications of purified toxins in the field are lacking, in part, because of the relatively large quantities needed. This is a significant obstacle for those compounds not easily synthesized or isolated. It also can be difficult to apply pure compounds to the native environment in a manner that simulates natural conditions.

Under natural conditions there seldom is just a single toxin; complex mixtures are common (del Moral and Muller 1970, Halligan 1975, Whitehead et al. 1982, Tang 1986, Chou et al. 1989, Hartung et al. 1990). Mixtures of phenolic acids are often more inhibitory to germination and seedling growth than individual compounds or that expected by adding the inhibition from two or more compounds tested separately. This synergistic enhancement of toxicity was observed for mixtures of *p*-coumaric and ferulic acids (Rasmussen and Einhellig 1977) and vanillic and *p*-hydroxybenzoic acids (Einhellig and Rasmussen 1978). Williams and Hoagland (1982) tested combinations of phenolic acids on germination of three bioassay species. They found additive effects for some combinations, but combinations most often gave nearly the same results as the most potent compound alone. They doubted the importance of synergistic effects from combinations of phenolic acids at normal concentrations in the inhibition of germination. Caffic acid in equimolar mixtures with ferulic or *p*-coumaric acid antagonized the inhibition of lettuce germination (Duke et al. 1983). Antagonistic interactions were observed for the inhibition of cucumber (*Cucumis sativus* L.) radicle growth using two compound mixtures of phenolic acids (Blum et al. 1984). Joint action analysis is a statistical procedure designed to help evaluate results from experiments involving compound mixtures (Gerig et al. 1989). This procedure was used to analyze the effects of mixed phenolic acids in soil on the leaf growth of cucumber seedlings (Blum et al. 1989, Gerig and Blum 1991). Mixtures were antagonistic or additive, and their ability to cause inhibition was pH sensitive. Inhibition was more pronounced at pH 5.2 than at 6.0, and none of the mixtures were active at pH 6.9.

Factors to consider in choosing a bioassay method include the amount of compound available for testing, degree of sensitivity desired, the suspected source of the toxin, and its possible mode of action (Leather and Einhellig 1985, 1988). Because many variables are associated with bioassay experiments, it is advisable to use more than one type while investigating a potential incident of allelopathic interference. Standardization and improvements in bioassay procedures undoubtedly will enhance reliability. Bioassays will continue to be essential for isolating toxins and studying modes of action. As our technology advances, the correlation between laboratory data and field observation should improve. Because of the complexity of the biotic and abiotic interactions in the field, there will be nothing more valuable in advancing our knowledge of allelopathy than new tools and experimental innovation used directly in the field.

VIII. Allelopathic Modes of Action

The most obvious effects of allelopathic compounds are inhibition of germination and reduction in growth. These responses are manifestations, however, of interference with normal metabolic processes at the cellular level. As emphasized by Harper (1977), little progress will be made in understanding allelopathy until we begin to study symptoms other than germination and growth. This can be done by closely monitoring the physiological responses of recipient plants.

To separate primary and secondary effects of allelochemicals, Einhellig (1985, 1986) proposed a model (Figure 6) for an action sequence based on membrane perturbations as a common denominator for the modes of action. This is a reasonable choice for a primary effect, because one of the first organelles allelochemicals encounter when they contact plant tissues is the plasma membrane. The more lipid-soluble the allelochemicals, the less resistance they may encounter in entering the cell. In root cells, the internal water status, nutrient uptake, and elongation may be significantly impacted by allelochemical molecules simply passing through the membrane, binding to membrane associated enzymes, interfering with hormone active sites, or some combination of these effects.

Once inside the cell, allelochemicals may be metabolized or structurally changed, thereby resulting in a modification of their toxic effects. The formation of glycosides (Balke et al. 1987, Hogan and Manners 1991) is a common detoxification mechanism that blocks reactive functional groups, increases molecular size and decreases lipid solubility of the compound, thus reducing its membrane penetrability. If sufficient allelochemical molecules can enter the cell, those escaping immediate detoxification are free to affect enzymes, block active sites, or disrupt

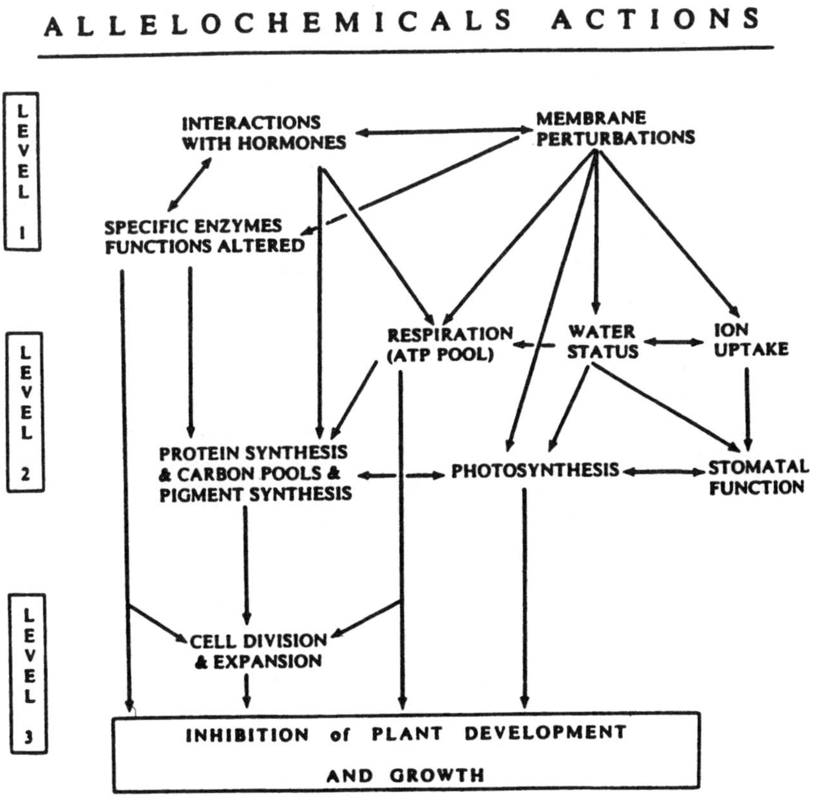

Figure 6. An action sequence showing the hypothetical effects of allelochemicals on higher plants (Einhellig et al. 1985b, Einhellig 1986). Arrows indicate negative effects on plant processes. Modified with permission from Einhellig et al. (1985b), copyright, American Chemical Society.

membranes of internal organelles (Moreland and Novitzky 1987). The consequence of these actions will be interruption of metabolic processes such as respiration, photosynthesis, protein synthesis, pigment synthesis, and carbon metabolism, all at the secondary level of effects as shown in Figure 6. The end result of these interferences appears at level three with disruption of cell division and elongation, which is readily visible as inhibited growth and development. Of course, many of these events occur almost simultaneously, thereby making the precise determination of the primary cause of toxicity difficult to determine experimentally.

Membrane structure and function emphasizing nutrient uptake

Membranes critical to the organization and regulation of metabolic processes in cells (Hall et al. 1982) can be disrupted by allelopathic compounds. Because everything moving in and out of a cell passes through the plasma membrane, its disruption is particularly significant to the normal functioning of roots. Balke (1985) reviewed the literature on allelopathic inhibition of mineral uptake and concluded that the best evidence for inhibition has come from experiments with excised roots in the presence of either phenolic acids or flavonoids. Pertinent observations from these studies include (1) lack of ion specificity—absorption of cations, or anions, can be inhibited; (2) pH has a dramatic effect on the inhibitory activity of a specific phenolic acid concentration; (3) absorption inhibition can be measured shortly after experiments begin, thus supporting the hypothesis of direct inhibition; (4) the inhibition of phosphorus and potassium absorption by phenolic acids is reversible; and (5) phenolic acids do not competitively inhibit absorption of phosphorus and potassium. The modes of action possibly responsible for these effects, based on an active mineral absorption model, include (1) changing membrane permeability to ions, (2) inhibiting ATPase enzymes, (3) reducing ATP content of the cell, and (4) changing the electrical potential across the plasma membrane and tonoplast. H^+-ATPase activity of the microsomal fraction isolated from roots is inhibited by the phytotoxic monoterpene, deacetylpiquerol. The degree of inhibition is greater for plasma membrane ATPase (67.2 %) than for tonoplast ATPase (31.4%) (Ortega et al. 1990).

Numerous nutrient studies also have been conducted with seedlings. Uptake of nitrogen, phosphorus, potassium, iron and molybdenum by cowpea seedlings (*Bigna sinensis* L.) was reduced by most concentrations (10^{-4} M, 5×10^{-4} M, and 10^{-3} M) of caffeic, protocatechuic and syringic acids. Calcium uptake was reduced by a few concentrations, whereas magnesium was not reduced by any (Alsaadawi et al. 1986). Of the 6 elements analyzed in the tissue of sorghum (*Sorghum bicolor* (L.) Moench) seedlings, the uptake of phosphorus was most strongly suppressed by ferulic acid in nutrient solution. Both root and shoot tissues had lower potassium levels, whereas magnesium concentrations were only lower in the roots (Kobza and Einhellig 1987). Leaf nitrogen in tomato seedlings decreased when the tomatoes were grown in nutrient solution containing 10^{-3} M concentrations of caffeic, *p*-coumaric, chlorogenic, ferulic, or vanillic acids. There were no decreases at 10^{-4} M concentrations. Tomato leaf phosphorus was reduced by most compounds at both 10^{-3} and 10^{-4} M. Effects on other nutrients depended on the allelopathic compound and its concentration (Mersie and Singh 1988). When allelopathic compounds interfere with plant nutrition, phosphorus is the most commonly reported element showing changes (Kobza and Einhellig 1987).

The plasma membrane is not the only membrane that can be affected. Phenolics inhibit mitochondrial electron transport by alterations of the inner mitochondrial membrane (Moreland and Novitzky 1987). When root tips of cucumber seedlings are exposed to the volatile terpenes from purple sage (*Salvia leucophylla* Greene), membranes surrounding organelles are severely disrupted and fragmented (Lorber and Muller 1976). Both the inner and outer membranes surrounding mitochondria are abnormal, the double membrane around the nucleus becomes separated in places by large vacuole like areas, and some membranous structures such as dictyosomes are absent.

Water uptake and retention

Closely associated with root function and nutrient uptake is the absorption and retention of water. Soybean seedlings (*Glycine max* (L.) Merr.) inhibited by aqueous leaf extracts of velvetleaf (*Abutilon theophrasti* Medic.) show evidence of water stress, with a reduced water content and lower leaf water potential than the controls (Colton and Einhellig 1980). Leaf diffusion resistance increases in the treated seedlings, suggesting that stomata are partially closed. Aqueous extracts of hackberry leaves (*Celtis laevigata* Willd.) are inhibitory to the shoot and root growth of three grass species. Most inhibited seedlings had significantly lower water content than controls (Lodhi and Nickell 1973). When soybean and grain sorghum (*Sorghum bicolor* (L.) Moench) are treated with 0.25 mM ferulic or *p*-coumaric acid, leaf water potential is lowered to about –10 bars compared to –5 bars in controls, with little effect on stomatal diffusive resistance. When concentrations of either compound is increased to 0.5 mM, plant growth is stunted, diffusion resistance increases significantly, and leaf water potential drops further. Both osmotic potential and turgor contribute to these changes in water potential (Einhellig 1985, Einhellig et al. 1985b). Leaves of cucumber seedlings subjected to 1 mM or greater concentrations of ferulic acid wilt rapidly, but then recovered after 24 to 48 hours (Blum and Dalton 1985). The initial water imbalance inhibits leaf cell expansion causing a delay in leaf area expansion. Removing seedlings from the ferulic acid results in rapid recovery of leaf growth rates.

Stomata on the leaves of tobacco (*Nicotiana tabacum* L.) and sunflower (*Helianthus annuus* L.) closed one day after treatment with 10^{-3} and 5×10^{-4} M concentrations of either scopoletin or chlorogenic acid (Einhellig and Kuan 1971) and remain closed for several days before reopening. Sixteen hours after starting treatments with 10^{-3} M concentrations of phenolic acids the stomatal conductance of soybean seedlings was reduced by 25 to 50% (Patterson 1981). At 40 hours after treatment initiation, the reduction in conductance was 75 to 98% of the controls.

Enzyme activity and biosynthetic products

Enzymes responsible for controlling metabolic processes are present in the cytoplasm, membranes, and organelles. Any interference with their synthesis, breakdown, or function will impact the products they biosynthesize and, thus, normal plant growth. Enzymes inhibited by various allelochemicals include cellulase, sucrase, catalase, peroxidase, phosphorylases, b-systathionase, phenylalanine ammonia-lyase, and pectolytic enzymes (Rice 1984). In tobacco tissue cultures, scopoletin, and its monoglucoside, scopolin, inhibit glucose-β-phosphate dehydrogenase activity. This enzyme contributes to the formation of NADPH and compounds initiating the shikimic acid pathway (Kajinami et al. 1971).

Sesquiterpene lactones are a class of terpenes with allelopathic effects and plant growth regulatory properties (Fischer 1986, 1991, Picman 1986). These compounds occur in introduced weeds and native species on rangelands (Kelsey and Shafizadeh 1979, 1980). Their cytotoxicity (Picman 1986) often is due to the presence of an α, β-unsaturated ketone system (O=C–C=C–) in the form of an α-methylene-γ-lactone or cyclopentenone. These functional groups are quite reactive and can readily bind with free sulphydral (SH) or amino (NH) groups common in amino acids and proteins. The SH groups often are present in or near enzyme active sites. Consequently, sesquiterpene lactones can bind with and deactivate enzymes. Although much of this enzymatic research has been conducted with animal cell systems, similar types of activity will occur in plants. The types of enzymes these lactones inhibit include deoxyribonuclease, ribonuclease, phosphofructokinase, hexokinase, acetyl-CoA, citrate-lyase, succinate dehydrogenase, acid phosphatase, and alkaline phosphatase (Lee et al. 1977, Hall et al. 1980, Narasimhan et al. 1985).

Cell division and growth

A reduction in the size of a plant organ is the result of fewer cells being produced by cellular division or smaller cells because of limited cell expansion, or some combination of the two. Cell division in root tips of cucumber seedlings is inhibited when the seedlings are exposed to the terpene vapors from macerated leaves of purple sage. Levels of inhibition increase with large quantities of leaves, and total inhibition occurs in the presence of two grams of leaf material (Muller and Hauge 1967). Mitosis in onion roots grown in soils exposed to terpene vapors of *Salvia* is inhibited after 24 hours and nearly ceases by 72 hours (Muller 1986). Treated chromosomes in metaphase are shorter and thicker than those in the controls after 24 hours of exposure. Breakage is observed in metaphase chromosomes after 48 hours of exposure. Roots of bioassay species inhibited by aqueous shoot extracts from Jericho resurrectionmustard (*Anastatica hierochuntica* L.) have a lower mitotic index [(number of dividing cells/total cells counted) × 100] than untreated controls (Hegazy et al. 1990).

Two sesquiterpene lactones from sunflower inhibit DNA and RNA synthesis but not protein synthesis (Spring et al. 1982). Parthenolide, a lactone in species from several Asteraceae (Composite) genera (Seaman 1982) inhibits DNA replication and induces single-strand breakage of cellular DNA (Woynarowski and Konopa 1981, Woynarowski et al. 1981). Other lactones, including coronopilin and parthenin, two allelopathic components of the tropical weed, guayule, *Parthenium hysterpohorus* L. (Kanchan and Jayachandra 1980, Picman and Picman 1984), clearly cause chromosomal aberrations (Abeysekera et al. 1985).

Cell division and expansion are regulated by plant hormones (Bidwell 1979) and allelochemicals can alter normal growth by interfering with the production, degradation, or function of these hormones. Cell elongation is particularly important in extension growth. It is regulated by the hormones indoleacetic acid (IAA) and gibberellins. Cells in the roots of cucumber seedlings exposed to monoterpenes expanded radially and tangentially but not longitudinally (Muller, W.H. 1965, Muller and Hauge 1967). In some cases, normal root cells are three to four times longer than exposed ones.

Numerous sesquiterpene lactones possessing one or more α, β-unsaturated ketone (O=C–C=C–) functional groups inhibit auxin-mediated cell elongation (Shibaoka et al. 1967a, 1967b, Sequeira et al. 1968, Kupchan 1975, Watanabe et al. 1982). This activity can be eliminated by chemical reactions with the carbon double bond. Two sunflower sesquiterpene lactones strongly inhibit the IAA-induced elongation growth of sunflower itself and oat (*Avena sativa* L.). Experimental data suggests the auxin attaches to a receptor site, thereby freeing an SH group that is irreversibly bound to the lactones forming a complex unable to initiate elongation growth (Spring and Hager 1982). IAA-induced cell elongation is dependent on hormone concentrations which is regulated by synthesis and IAA oxidase degradation. Phenolic acids may mediate cell elongation by their influences on IAA concentrations. These compounds can be divided into two groups: those stimulating IAA destruction (vanillic, *p*-hydroxybenzoic, syringic, and *p*-coumaric acids), and those protecting or retarding its breakdown (ferulic, chlorogenic, protocatechuic, and caffeic acids) (Lee and Skoog 1965, Stenlid 1968, Lee 1977, Einhellig 1985).

Numerous tannins and many common allelopathic phenolics found in soils are inhibitors of gibberellin-induced growth of dwarf peas, *Pisum sativum* L. (Corcoran et al. 1972, Green and Corcoran 1975). The antagonistic activity of tannins is specific for gibberellins as demonstrated by cucumber hypocotyl bioassays where gibberellin-induced growth is inhibited, and auxin growth is stimulated (Corcoran et al. 1972). Phenolics also antagonize the activity of abscisic acid, a hormone well known for its inhibition of cell elongation. Although many of the compounds tested are known inhibitors of plant growth (cinnamic acid, ferulic acid, tannic acid, coumarin, and others), they all stimulate hypocotyl growth of pigweed (*Amaranthus caudatus* L.) in the presence of inhibitory concentrations of abscisic acid. The effect is not caused by the decomposition or rearrangement of the hormone (Ray et al. 1980).

Photosynthesis and respiration

When all modes of action associated with allelochemicals are considered, it is not surprising that a common consequence of their action is a measurable change in chlorophyll content and photosynthetic activity. The chlorophyll content of soybean leaves is significantly reduced by aqueous-leaf extracts from velvetleaf (Colton and Einhellig 1980) and by several phenolic acids at 10^{-3} M concentration (Patterson 1981). In bioassays with cowpea seedlings, most concentrations (10^{-4} M, 5×10^{-4} M, and 10^{-3} M) of syringic, caffeic, and protocatechuic acids decrease seedling growth, chlorophyll a, and total chlorophyll content but do not decrease the concentrations of chlorophyll b (Alsaadawi et al. 1986). Similar results are observed with tomato seedlings treated with caffeic, vanillic, and chlorogenic acids at 10^{-3} M concentrations (Mersie and Singh 1988). Coumaric and ferulic acids at this concentration have no effect on the tomato chlorophyll.

Inhibition of photosynthesis in soybean by 10^{-3} M concentrations of various phenolic acids ranges from 0 to 80% after 16 hours of treatment. All compounds are inhibitory after 40 hours (Patterson 1981). This same concentration of scopoletin inhibits net photosynthesis in tobacco seedlings two days after treatment, whereas decreases in sunflower seedlings do not become apparent until days three and four (Einhellig et al. 1970). Treated tobacco plants fix only 51% of the CO_2 fixed by the controls. Ferulic acid at 0.25 and 0.5 mM reduces $^{14}CO_2$ fixation by 30 and 38%, respectively, in cucumber seedlings (Blum and Rebbeck 1989).

There seems to be no consistent effect of allelochemicals on repiration (Einhellig 1985). Volatiles of sagebrush (*Artemisia* sp. L.), primarily monoterpenes including camphor and cineole (Kelsey et al. 1978, 1983), inhibit the respiration of germinating seeds and young leaf tissues but stimulate respiration of mature leaf tissues (Weaver and Klarich 1977). Stimulation is observed for both mature conifer and grass leaves. But, when tested on mature wheat leaves, vapors from camphor decreased respiration, whereas vapors from cineole and a-pinene caused an increase. Respiration in ripgut brome (*Bromus rigidus* Roth) grass seedlings and in the roots of cucumber is dramatically decreased when germinated in the presence of monoterpenes released from purple sage leaves (Muller et al. 1968). Cineole, one of the sage (*Salvia* L.) monoterpenes, reduces respiration of wild oat (*Avena fatua* L.) roots to 40% of the control after 3 hours exposure. Cineole also inhibits the oxygen uptake and Krebs cycle metabolism of mitochondrial suspensions from wild oat (Muller et al. 1969, Muller 1986).

Variable effects on respiration also have been observed with phenolics (Van Sumere et al. 1972). Of the numerous aromatic and phenolic compounds tested on yeast, some have no effect on respiration, some cause inhibition, but most, at $5 \times$ 10-4 or 10^{-3} M, stimulate respiration, except for the quinones. Uncoupling of oxidative phosphorylation may be the cause (Van Sumere et al. 1972). Phenolics inhibiting growth of mung bean (*Phaseolus aureus* L.) (Demos et al. 1975) either reduce mitochondrial respiration (*p*-coumaric, tannic, and gentisic acids) or have no effect on mitochondrial processes (syringic, *p*-hydroxybenzoic, caffeic, and ferulic acids). Disruption of mitochondrial function may be the mechanism of growth inhibition for sorgoleone, a toxic quinone isolated from the roots of grain sorghum (Rasmussen et al. 1992).

Seed protection, germination inhibition, and autotoxicity

Seedling establishment is the most critical stage in a plant's development (Harper 1977). Once seeds are dispersed they must survive until the proper combination of abiotic and internal biotic conditions trigger the metabolic processes resulting in germination. Allelochemicals may function as part of the survival mechanism of seeds. Compounds in the seeds with antibiotic activity (phytoncides) may provide protection from decay by bacteria and fungi. Germination inhibitors may help to ensure environmental conditions are adequate to meet the needs of germinating seedlings. A single compound could provide both of these functions. Van Sumere et al. (1972) found that phenolics inhibitory to yeast growth also are germination regulators at a particular concentration. Horsley (1977) did not discuss endogenous seed germination inhibitors in his review of allelopathy, because the inhibitors do not have to be released to the environment to be effective and therefore do not satisfy all criteria of the allelopathy definition.

Rice (1984) speculates that the production of microbial inhibitors by seeds is one of the most consistent and important ecological roles of allelopathy for plants growing in natural environments. As he points out, though, there is little direct evidence compared to indirect evidence to support this assumption. It has been well documented that various phenolic compounds occur in seeds and fruits, and it also is known that phenolics are common inhibitors of bacterial and fungal growth. The presence of antibacterial compounds was demonstrated by submerging seeds and some fruits from 512 plant species and varieties into an agar medium innoculated with test bacteria (Ferenczy 1956). After 20 hours of bacterial growth, zones of inhibition were observed around the seeds. The bacteria used were common soil saprophytes, and some plant parasites. Approximately 1% of the species tested contained antibacterial compounds in their seeds or fruit. Some antifungal compounds were detected, although fungi were not tested thoroughly. Mold resistance of sorghum strains has been correlated with the concentration of flavan-4-ols in the seed extracts (Jambunathan et al. 1986, 1990). Apigenindin, a derivative from one of the flavan-4-ols and a natural component in sorghum seed, inhibits the growth of numerous fungi, including *Alternaria solani*, a sorghum pathogen (Schutt and Netzly 1991).

The two most obvious functions of endogenous germination inhibitors are to prevent premature germination in undesirable environmental conditions that might exist after the seeds are dispersed and to establish a seed reserve that will extend germination over a longer time. This prevents seeds from germinating simultaneously and increases the probability that some seedlings will encounter adequate growing conditions to reach maturity. Oat caryopses contain the germination inhibitor dihydrohematinic acid, a tricarboxylic acid. Other species, radish and amaranth (*Raphanus sp.* L. and *Amaranthus caudatus* L., respectively), are even more sensitive to the inhibitory activity of this compound than are oats (Rudiger and Lohaus 1987). These investigators also isolated three inhibitors from dormant seeds of dog rose (*Rosa canina* L.), including abscisic acid, an α-pyrone dicarboxylic acid and a furan carboxylic acid derivative.

Another mechanism of controlling germination is autotoxicity. Instead of inhibitors occurring within the seeds, they are produced in other tissues of the plant and inhibit its own seed germination when released to the environment. In laboratory bioassays, autotoxicity is not uncommon (Newman and Rovria 1975, Bokhari 1978, Stowe 1979, Picman and Picman 1984, Heisey and Delwiche 1985, Kelsey and Locken 1987, Chou 1989, Merrill 1989). In environments where water is limited, an autotoxic mechanism to help synchronize germination with the wettest season, or to regulate plant density to minimize intraspecific competition, would have survival value. Certainly some of the autotoxicity being reported is not functional in the field (Stowe 1979). Newman and Rovira (1975) suggest that truly autotoxic species should grow better in communities of mixed species than in pure stands. In native short-grass prairie ecosystems this seems to be true for blue grama (*Bouteloua gracilis* (H.B.K.) Lag. ex Griffiths) and western wheatgrass (*Agropyron smithii* Rydb.) (Bokhari 1978). For dropseed, (*Sporobolus pyramidatus* (Lam.) A. Hitchc.), pure stands occur for a while and then become decadent in the center, possibly after toxins have accumulated (Rasmussen and Rice 1971). The "fairy ring" pattern exhibited by prairie sunflower, *Helianthus rigidus* (Cass.) Desf. (Curtis and Cottam 1950), and some fungi is caused by autotoxicity resulting in rapid changes in community patterns. Old communities of soft chaparral with purple sage and California sagebrush show the characteristics associated with autotoxicity. Mature plants within the stand have smaller crowns with low vigor, compared to seedlings and young plants recently established at the community's edge in relatively bare border zones (Muller 1966). Bare soil is common between old plants, but few seedlings are being established to replace those that die. The autotoxic inhibition of horsenettle (*Solanum carolinense* L.) seed germination is reversible and may function to maintain a seed-reserve bank in the soil (Solomon 1983). The role of allelochemicals in seed protection, germination inhibition, and autotoxicity is poorly understood. These three mechanisms could have significant implications for the management of plants affected by them.

Soil biology and the nitrogen cycle

In ponderosa pine (*Pinus ponderosa* Dougl. ex Laws.) forests growing on alkaline soils (pH 7.25–7.75), nitrification is thought to be inhibited by chemicals in needles, bark, and surface soil extracts (Lodhi and Killingbeck 1980). In New Mexico, net N mineralization in a ponderosa pine stand was inversely correlated with total monoterpene content in the soil (White 1991) and nitrification in the mineral soil was inversely correlated with the concentration of monoterpenes containing terminal carbon double bonds in the litter horizon. In mixed species forests, we can expect variability in soil nitrification potential in response to litter differences among trees (Lodhi 1977). Tannins found in grass, oak, and pine species completely inhibit nitrification by *Nitrosomonas*, but high concentrations are required to inhibit nitrification by *Nitrobacter* (Rice and Pancholy 1973). Baldwin et al. (1983) found that removal of tannins and phenolics from the subalpine balsam fir (*Abies balsamea* (L.) Mill.) forest floor stimulates nitrification. Clearcutting of forests rapidly increases nitrifier populations and soil $N0_3$-nitrogen (Smith et al. 1968). Nitrate reductase activity of crinkled hairgrass (*Deschampsia flexuosa* (L.) Trin.) in a clearcut was two to sixteen times greater than in adjacent forests (Hogberg et al. 1986).

Plants and associated soil microorganisms are thought to form mutualistic partnerships (Perry et al. 1989). Plant-soil microbe allelopathic effects may inhibit the development of such relations or cause their disruption. Research has emphasized impacts of allelopathic materials on the form of soil nitrogen present (ammonium, NH_4 or nitrate, NO_3), nitrification, and nitrogen fixation. Rice and Pancholy (1972, 1973) proposed that climax vegetation produces compounds inhibitory to populations of *Nitrosomonas* and *Nitrobacter* and their oxidation of NH_4. The result is a decline in $N0_3$ and an increase in NH_4 during succession. Suggested allelopathic compounds that inhibit nitrification include ellagic, gallic, digallic, caffeic, chlorogenic, isochlorogenic, neochlorogenic, ferulic and *p*-coumaric acids, quercetin, condensed tannins, gallotannins, and monoterpenes (Rice 1965a, Rice and Pancholy 1973, Lodhi 1982, Lodhi and Killingbeck 1980, White 1986, 1991).

Maintenance of soil nitrogen as NH_4 reduces leaching losses of nitrogen and conserves energy transformations within the plant. Positively charged ammonium ions interacting with negatively charged soil particles resist leaching more than do negatively charged nitrate ions (Rice and Pancholy 1972). The leaching of NO3-nitrogen in early successional stages has been observed as increased nitrate levels in adjacent streams (Todd et al. 1975).

Associated with the decline in soil nitrate and population sizes of nitrifiers in the rhizosphere is a decline in plant nitrate reductase enzyme activity (NRA) (Blacquiere 1986, Lodhi and Ruess 1988). Nitrate reductase, the enzyme involved in the first step of nitrate reduction within plants, is more active in pioneer species than climax species (Franz and Haines 1977). With the decline in soil NO_3 any competitive advantage of increased NRA present in pioneer species declines as well. Ammonium is reported to play a more critical role than NO_3 in the nutrition of late successional species (Zak and Pregitzer 1988). If a climax species preferentially uses NH_4-nitrogen, rather than NO_3-nitrogen, then it conserves the energy required for NO_3 reduction back to NH_3 within the plant (Lodhi 1982). Given the successional shift in soil nitrogen toward NH_4-nitrogen, the NRA level reflects both successional position and physiological adaptation of a species to the available form of nitrogen in the soil (Smith and Rice 1983). The conservation of soil nitrogen may be a prerequisite for later successional species that have greater nitrogen requirements (Rice et al. 1960). Conversely, inhibiting the accumulation of soil nitrogen would favor the persistence of early successional species with low nitrogen requirements.

The hypothesis that climax vegetation inhibits nitrification by allelopathic compounds (phenols and tannins) is not an accepted fact. Neal (1969) reports that needle and thread grass (*Stipa comata* Trin. & Rupr.) was the only climax grass species out of six tested that inhibited nitrification, yet grass and forb species invading overgrazed areas had an inhibitory effect. Root exudates of the grass genus *Hyparrhenia* Anderss. ex Fourn. in the veldt savanna of Rhodesia were suggested to inhibit nitrification (Boughey et al. 1964, Munro 1966), but Purchase (1974) argued that nitrification in soils under plants of this genus is limited by availability of ammonium.

The ability of tannins and phenolics to inhibit nitrification was questioned by McCarty and Bremner (1986), although they show nitrification inhibition by two patented nitrification inhibitors (nitrapyrin and etridiazole). They suggest that previous work by Rice and Pancholy (1973) did not measure nitrification of NH_4, but rather NO_2 production in a soil microbe inoculated system where conversion from NO_2 to NO_3 was more rapid than oxidation of NH_4 to NO_2. Robertson (1982) argues that nitrification is controlled by ammonium availability, not inhibitors, in most successional stages of primary sand dune and secondary old-field seres. Inhibition of nitrification by suspected allelopathic compounds in incubated soils could often be ascribed to the additional carbon and hydrogen (pH changes) rather than toxicity to nitrifiers. Nitrifiers are poor competitors for inorganic nitrogen which can be immobilization by heterotrophs after the addition of high carbon organic matter (Jones and Richards 1977). Only 6 of 49 potential plant and soil sources of inhibitors significantly inhibited nitrification over the effects of added carbon and altered pH; however, inhibitory substances were not continually supplied to the soil during incubation and initial amounts could have been rapidly degraded to innocuous compounds (Robertson 1982). Vitousek et al. (1989) suggest that nitrogen form and availability during succession is related to the type of disturbance and subsequent successional stage, (primary vs secondary vs intermediate). In primary stages nitrogen steadily increases, but in intermediate and secondary successional stages a nitrification pulse occurs followed by a decline. The data linking allelopathic inhibition to reduced nitrification is more correlative than causal in nature. The decline in soil nitrate, nitrifiers, and nitrate reductase activity with secondary succession does not appear in dispute, but the role allelopathy plays in these phenomena remains unclear.

IX. Allelopathy on Rangelands

Vegetation response to allelopathy is observed on rangelands as it affects the establishment and persistence of individual plants. The cumulative adverse impacts of allelopathy on a given plant population and the competitive advantage provided to associated species populations are seen as changes in site productivity, species patterns, and plant community composition. Allelopathic interference affects rangeland productivity by reducing plant density (establishment and survival) and biomass of desirable species. Allelopathy can affect the composition and spatial distribution of species within a community during a single growing season and also mediate successional community changes through time. Case histories are provided of reported allelopathic species and their effects on rangeland productivity, composition, plant distribution, and succession.

Grasslands

Abandonment of previously cultivated fields in the grassland ecosystem of Oklahoma and southeast Kansas has provided opportunities to observe and study ecological changes in community structure involving allelopathy. Four successional stages (Booth 1941) have been identified: (1) pioneer weed, lasting 2 to 3 years; (2) annual grass, 9 to 13 years; (3) perennial bunch grass, known to last up to 40 years; and (4) true prairie climax, which never has been fully reestablished. The first year after abandonment, the community is composed of remnant species that were present before abandonment and a variety of new species, primarily weeds. In the next 1 to 2 years, rapid changes occur in community composition as some weeds disappear and prairie threeawn (*Aristida oligantha* Michx.), a small rather innocuous annual grass, the species that will dominate the second successional stage, begins to appear. As prairie threeawn increases in abundance, the successional process begins to slow and the community composition changes less rapidly. After an extended period of a decade or more, bunch grass species may enter with a gradual shift in dominance to species such as little bluestem (*Andropogon scoparius* Michx. = *Schizachyrium scoparium* (Michx.) Nash).

Other species from the last successional stage eventually will become established in the bunch grass community, thereby changing its structure to the climax prairie.

The pioneer weed stage is dominated by several vigorous plants such as Japanese brome (*Bromus japonicus* Thumb.), common sunflower (*Helianthus annuus* L.), lambsquarters (*Chenopodium album* L.), western ragweed (*Ambrosia psilostachya* D.C.), Johnsongrass (*Sorghum halepense* (L.) Pers.), horseweed (*Erigeron canadensis* L.), and crabgrass (*Digitaria sanguinalis* (L.) Scop.). These disappear rapidly, however, from the community as prairie threeawn invades. A series of studies demonstrated that root exudates, leaf leachates, and decomposing tissues from many of the dominant weeds in stage 1 were inhibitory to the germination or growth of other weeds but not to prairie threeawn (Abdul-Wahab and Rice 1967, Wilson and Rice 1968, Parenti and Rice 1969, Neill and Rice 1971, Rice 1974, Kapustka and Moleski 1976). Phenolics were identified as the toxic compounds being released by the weeds. When pure compounds (*p*-hydroxybenzaldehyde, isochlorogenic, chlorogenic, *p*-coumaric, gallic, and tannic acids) were bioassayed at various concentrations with rough pigweed (*Amaranthus retroflexus* L.), Japanese brome, and threeawn, the results reaffirmed previous studies and field observations. At least one concentration of each compound inhibited the growth of both weed species, except for *p*-coumaric acid which retarded only pigweed. Pigweed, the most readily inhibited species in these bioassays, disappears from the weed community after the first year. Prairie threeawn was not inhibited by any of the compounds and was stimulated by one concentration of chlorogenic acid. Although threeawn is inhibited by crabgrass, crabgrass is rapidly displaced by the allelopathic effects of sunflower, thus allowing threeawn to invade.

Once prairie threeawn is established, it dominates for about a decade, although nearly pure stands have been reported 30 years after abandonment (Rice 1976, Owensby and Launchbaugh 1977). As prairie threeawn begins to dominate the site in stage 2, the rate of succession slows and remains quite stable in subsequent years as a result of limited availability of nutrients, particularly nitrogen and phosphorus. This annual grass stage eventually gives way to the perennial bunch grass dominated by little bluestem. Climax is finally reached when the community contains a mixture of little bluestem, big bluestem (*Andropogon gerardi* Vitman), Indiangrass (*Sorghastrum nutans* (L.) Nash), and switchgrass (*Panicum virgatum* L.). The nitrogen and phosphorus requirement of prairie threeawn, little bluestem, and switchgrass increase in their order of appearance in the abandoned fields (Rice et al. 1960).

In the early successional stages, soil nitrogen is maintained at a low level by the release of phytotoxins from various species that inhibit the growth or metabolic processes of microorganisms associated with the nitrogen cycle. Organisms affected by the toxins include *Nitrosomonas* (convert NH_4 to NO_2), *Nitrobacter* (convert NO_2 to NO_3), *Azotobacter* (free living nitrogen-fixers), *Rhizobium* (a nitrogen-fixing symbiont of legume nodules), and nitrogen-fixing blue-green algae (Rice 1964, 1965a, 1965b, 1971, 1977, Wilson and Rice 1968, Blum and Rice 1969, Parks and Rice 1969, Rice and Pancholy 1973, 1974).

Direct measurement of nitrogen-fixation in soils collected from various successional stages provides additional evidence of inhibition. There is a much higher fixation rate in the later prairie stages (3 and 4) than in the early pioneer weed (1) and annual grass (2) stages (Kapustka and Rice 1976, 1978). The late perennial bunchgrass successional stage (3) has more NH_4 and less NO_3 nitrogen than the two earlier stages. This is accompanied by a lower number of nitrifying bacteria (*Nitrosomonas* and *Nitrobacter*) compared to the early weedy stage.

Although climax prairie species such as little bluestem and fleabane (*Erigeron strigosus* Muhl. ex Wild.) inhibit nitrification, they have no effect on nitrogen-fixing bacteria (Rice 1964). Thus, nitrogen-fixation and nitrogen-conservation can occur simultaneously. The ability of plant extracts from big and little bluestem (climax prairie species) to reduce *Nitrosomonas* numbers in a soil suspension increases with the growing season, whereas soil extracts are toxic at all times (Lodhi 1982). Apparently soil toxicity is being supplemented by the leaching of inhibitors from maturing or decaying vegetation.

Verification of the role nutrients play in old-field succession was achieved by applying fertilizer to field plots. A heavily grazed field, dominated by prairie threeawn 31 years after abandonment, was selected for study. Species frequency and biomass were examined in October, 6 months before treatment (Rice 1976). The following March, 3 plots were treated with various levels (low, medium, and high) of nitrogen as ammonium nitrate, 3 with various levels of nitrogen and phosphorus as ammonium phosphate, and 3 were left as controls. After one growing season, the frequency and biomass of prairie threeawn were lower than controls when averaged for all nitrogen and phosphorus treatments, but were most greatly reduced in the high nitrogen and phosphorus (201 + 251 kg/ha, respectively) treatment. The combined frequency and biomass of little bluestem and beardgrass (*Andropogon ternarius* Michx.), a species also common in the bunch grass and prairie climax stage, increased substantially with the high nitrogen and phosphorus. Western ragweed and heath aster (*Aster ericoides* L.), forbs common in the climax stage, increased in plots treated with nitrogen and phosphorus. In one growing season, the application of high levels of nitrogen and phosphorus had accelerated the succession from an annual grass to a late perrenial grass or early climax prairie. Rapid changes continued during the second growing season after fertilization, with the complete elimination of prairie threeawn from the high nitrogen and phosphorus treatment plots. The increased abundance of little blue stem and beardgrass resulted in the establishment of an excellent prairie sod in just two years.

To briefly summarize, allelopathy apparently functions both directly and indirectly in old-field succession. Community composition in the initial weedy stage and the subsequent rapid changes in community structure during the first few years are probably caused by direct toxicity of allelochemicals. Toxin effectiveness may be enhanced by the low nutrient status of soil which is perpetuated in part by indirect allelopathic effects toward microorganisms responsible for nitrogen-fixation and nitrification. Nutrient requirements of key species correlate with their appearance in the fields; the more nutrients they need, the later they appear. The rate of succession is accelerated tremendously by the application of nutrients. In secondary succession of old fields in Oklahoma, indirect allelopathic effects altering available nitrogen may be the most significant process.

In the open oak-savannah of southern California, wild oats are the primary grass species with lesser amounts of longarm ripgut brome (*Brome diandrus* Roth.) associated with scattered California live oak trees (*Quercus agrifolis* Nee.). Beneath the oak canopy is a tree-dependent broad-leaved dicot, pholistoma (*Pholistoma auritum* (Lindl.) Lilja), growing in nearly pure stands or in association with ripgut brome. The relatively pure stands recur year after year even though surrounded by communities of annual grasses. Dense stands of philostoma usually produce a thick layer of litter, tangled and bound together by retrorse prickles, which act as physical barrier to the soil. Grass seeds have difficulty penetrating this layer. The litter also contains a leachable toxin(s) inhibitory to grass seed germination in laboratory and field bioassays (Parker and Muller 1979). Although the relative importance of these two mechanisms in maintaining the vegetation pattern has not been determined, it is likely that allelopathy makes a significant contribution.

Although it is important to know when allelopathy may be functioning in a community, it is equally important to know when it is not. If tall grass prairie is protected from burning, mowing, or grazing, the litter accumulates and productivity decreases after just a few years. Various explanations include decreased soil temperatures from the insulating effect of litter, a slower rate of mineral cycling, and allelopathy. After conducting a series of experiments to evaluate these various possibilities, Rice and Parenti (1978) conclude that higher soil temperatures early in the growing season, in conjunction with non-Festucoid grass species with a high temperature optimum for growth, are responsible for greater productivity in burned and mowed plots. Allelopathy is not a factor.

Further examples of allelopathy affecting grass species and grasslands are discussed in other sections on shrublands, woodlands, and weeds. Table 3 lists some of the grass species examined for allelopathic potential. It should not be assumed that species in this table have been proven to be allelopathic because references reporting no toxicity have been included. The function of the table is to provide a brief but incomplete summary of work completed and a source of additional references.

Chaparral

Adjacent to the annual grasslands of coastal southern California are communities of soft chaparral dominated by purple sage and California sagebrush. Within the chaparral stands are bare areas between shrub plants with little herbaceous growth. A seemingly bare, or sparsely vegetated, zone also extends 1 to 2 m beyond shrub crowns at the edge of the community (Muller 1966, Halligan 1973). Bare looking border zones are dependent on shrub density; they occur near large stands and clusters of shrubs but not single shrubs (Halligan 1973, 1974). Three meters beyond shrub thickets is a community of annual grasses and other herbaceous plants. Of the three areas, the grassland is most densely populated, followed by the border zone and then the shrub zone. Although areas within and adjacent to shrubs appear bare, there are many plants present, mostly herbs of small stature rather than the larger more conspicuous grasses. Each zone is also characterized by the presence of some species not found in the other zones (Halligan 1973). Analysis of soils collected from the three zones did not reveal any significant differences in mineral content or physical characteristics (Muller 1966).

Table 3. Grass species investigated for allelopathic activity.

Grass species	Reference
Acroceras macrum Stapf Nile grass	Chou and Young 1975.
Agropyron glaucum Rhoem. & Shult.	Grodzinsky and Panchuck 1974, cited in Rice 1984.
Agropyron repens (L.) Beauv. Quackgrass	Stowe 1979, Toai and Linscott 1979, Gabor and Veatch 1981, Weston and Putnam 1985, Weston et al. 1987.
Agropyron smithii Rydb. Western wheatgrass	Bokhari 1978, Schmidt and Reeves 1989.
Agrostis stolonifera L. Carpet bentgrass	Gussin and Lynch 1981.
Alopecurus pratensis L. Meadow foxtail	Gussin and Lynch 1981.
Andropogon gerardi Vitman Big bluestem	Rice and Parenti 1978.
Andropogon nodosus Willem. Nash Angelton grass	Chou and Young 1975.
Schizachyrium scoparium (Michx.) Nash Little bluestem	Rice 1964, Rice and Parenti 1978.
Andropogon virginicus L. Broomsedge	Rice 1972.
Aristida oligantha Michx. Prairie threeawn grass	Rice 1964, Parks and Rice 1969.
Anthoxanthum odoratum L. Sweetvernal grass	Newman and Rovira 1975, Newman and Miller 1977, Gussin and Lynch 1981.
Aristida adscensionis Linn. Sixweeks threeawn	Sridhara Murthy and Ravindra 1975.
Avena fatua L. Wild oats	Tinnin and Muller 1972.
Bouteloua gracilis (H.B.K.) Lag. Blue grama	Jameson 1968, Bokhari 1978.
Brachiaria mutica (Forsk.) Stapf Para' grass	Chou and Young 1975, Chou 1989.
Bromus inermis Leyss. Smooth brome	Stowe 1979.
Bromus japonicus Thumb. Japanese brome	Rice 1964, Rice and Parenti 1967.
Buchloe dactyloides (Nutt.) Englm. Buffalo grass	Bokhari 1978.
Cenchrus ciliaris L. Buffel grass	Cheam 1984.
Chloris gayana Kunth Rhodes grass	Chou and Young 1975.
Cortaderia selloana (Schult.) Aschers and Graebn. Pampas grass	Chou and Young 1975.
Cynodon dactylon (L.) Pers. Common bermudagrass	Horowitz and Friedman 1971, Chou and Young 1975, Weller et al. 1985.
Cynosurus cristatus L. Crested dog's tail	Newman and Rovira 1975.

Table 3. Continued.

Grass species	Reference
Dactylis glomerata L. Orchardgrass	Stowe 1979, Wardle et al. 1991.
Digitaria decumbens Stent Pangola grass	Chou and Young 1975, Chou 1989.
Digitaria sanguinalis (L.) Scop. Crabgrass	Schreiber and Williams 1967, Bieber and Hoveland 1968, Parenti and Rice 1969.
Echinochloa crus-galli (L.) Beauv. Barnyard grass	Bhowmik and Doll 1982.
Eragrostis curvula Nees. Weeping lovegrass	Chou and Young 1975, Creek and Wade 1985.
Festuca arizonica Vasey Arizonia fescue	Jameson 1968, Rietveld 1975, 1977.
Festuca arundinacea Schreb. Tall fescue	Beiber and Hoveland 1968, Peters 1968, Walters and Gilmore 1976, Peters and Zam 1981, Luu et al. 1982, Creek and Wade 1985, Peters and Luu 1985, Pederson 1986, Wardle et al. 1992.
Festuca elatior L. Meadow fescue	Stowe 1979.
Festuca rubra L. Red fescue	Fales and Wakefield 1981, Gussin and Lynch 1981.
Hemarthria altissima (Poir.) Stapf and Hubb. Bigalta	Young and Bartholomew 1981, Tang and Young 1982.
Holcus lanatus L. Yorkshire fog	Newman and Rovira 1975, Gussin and Lynch 1981, Wardle et al. 1992.
Imperata cylindrica (L.) Beauv. Cogan grass	Chou 1989.
Leptochloa fusca (L.) Kunth. Kallar grass	Mahmood et al. 1989.
Lolium perenne L. Perennial ryegrass	Newman and Rovira 1975, Gussin and Lynch 1981, Fales and Wakefield 1981.
Miscanthus floridulus (Labill.) Warb Silvergrass	Chou and Chung 1974.
Miscanthus transmorrisonensis Hay Silvergrass	Chou and Lee 1991.
Muhlenbergia montana (Nutt.) A.Hitchc. Mountain muhly	Rietveld 1975, 1977.
Panicum dichotomiflorum Michx. Fall panicum	Bhowmik and Doll 1982.
Panicum maximum Jacq. Guinea grass	Chou and Young 1975.
Panicum repens L. Torpedo grass	Chou 1989.
Panicum virgatum L. Switch grass	Rice and Parenti 1978.
Paspalum notatum Flugge Bahaigrass	Bieber and Hoveland 1968.

continued on next page

Table 3. Continued.

Grass species	Reference
Pennisetum clandestinum Hochst. ex Chiov. Kikuyu grass	Chou et al. 1987, 1989.
Phalaris tuberosa L. Canarygrass	Wardle et al. 1992.
Phleum pratense L. Timothy	Stowe 1979.
Paspalum plicatulum Michx. Brownseed paspalum	Chou and Young 1975.
Poa annua L. Annual bluegrass	Brede 1991.
Poa compressa L. Canada bluegrass	Stowe 1979.
Poa pratensis L. Kentucky bluegrass	Fales and Wakefield 1981, Kolb 1988.
Poa trivialis L. Rough bluegrass	Gussin and Lynch 1981.
Setaria faberii Herrm. Giant foxtail	Schreiber and Williams 1967, Bell and Koeppe 1972, Bhowmik and Doll 1982.
Setaria glauca (L.) Beauv. Yellow foxtail	Schreiber and Williams 1967, Bhowmik and Doll 1982.
Setaria viridis (L.) Beauv. Green foxtail	Bhowmik and Doll 1982, Mallik and Tesfai 1988.
Setaria sphacelata (Schumach.) Staph and Hubbard Golden timothy	Chou and Young 1975.
Sitanion hystrix (Nutt.) J.G. Smith Bottlebrush squirreltail	Jameson 1968.
Sorghastrum nutans (L.) Nash Indian grass	Rice and Parenti 1978.
Sorghum halepense (L.) Pers. Johnsongrass	Abdul-Wahab and Rice 1967, Bieber and Hoveland 1968, Horowitz and Friedman 1971, Lolas and Coble 1982.
Sporobolus pyramidatus (Lans.) A. Hitchc. Dropseed	Rasmussen and Rice 1971.
Stipa pulchra A.Hitchc. Purple needlegrass	Hull and Muller 1977.
Tripsacum laxum Nash Guatemala grass	Chou and Young 1975.

In the vicinity of California sagebrush, allelopathy, herbivory by small animals, and soil moisture contribute to the vegetation patterns. The shrubs provide cover and protection for birds and small rodents that eat grassland species near the shrubs (Bartholomew 1970, Halligan 1974), leaving behind the smaller, protected or unpalatable herbs. Grass species with larger coarse leaves are more likely to be eaten than those with small thin-leaves. Seed predation is also prevalent near the shrubs. Grass species dominant in the grassland zone tend to produce large seeds that are more susceptible to discovery and predation, especially in grazed or bare areas (Halligan 1974). Consequently, fewer seeds of grassland species are available to germinate near the shrubs. This provides an opportunity for herbs that are not inhibited by shrub toxins, and cannot compete in the dense grassland zone, to become established in the bare areas near shrubs. Shrub canopies intercept and channel rain within the shrub zone. This causes areas of wet and dry soil which also contributes to areas of "lush" and depauperate herb growth (Halligan 1973).

Ripgut brome is one of the species restricted to the grassland zone. Although it is sensitive to shrub toxins in various bioassays (Muller 1966, Muller and del Moral 1966, Muller et al. 1968), allelopathy is not as important as herbivory in excluding it from the shrub and border zone communities (Halligan 1974, 1976). When protected from animals in the shrub zone it actually grows larger than in the surrounding grasslands. Foxtail fescue (*Festuca megalura* Nutt.) is a grass common to all three zones (Halligan 1974, 1976). Its presence is influenced to some degree by allelopathy and herbivory, but never to the extent it is completely excluded from a particular community.

Chile tarweed (*Madia sativa* Torr. & Gray) and smooth cat's-ear (*Hypochoeris glabra* L.) are two herb species sensitive to the toxins in California sagebrush. During the dry season volatiles and leachates from the foliage, and soils collected beneath the shrubs, are toxic to both herbs. During the wet season none of these materials are toxic. The allelopathic activity of California sagebrush is associated with the Pacific Mediterranean climate. The winter rainy season results in the production of new shrub foliage containing low levels of terpenes. As the leaves grow and mature through the spring and summer the total terpene concentration increases including the amount of camphor, the most toxic volatile. During the drought of summer and fall the terpenes are volatilized into the atmosphere and adsorbed onto dry soil particulates, eventually reaching toxic levels near the end of the dry season, by October and November. At this time the compounds can be detected in the shrub and border zone soils by chromatography (Halligan 1975) and the soils are toxic to sensitive bioassay species (Halligan 1976). Fresh litter that accumulates on the soil surface also contains the toxic compounds.

During the first days of the winter rains in December the leaf drip from California sagebrush is toxic. The rain leaches toxins from the foliage and litter to the soil adding them to the compounds previously deposited by volatilization during the dry season. Seed germination of many species coincides with the onset of the rainy season and maximal toxicity of the soil solutions. Germination and growth of sensitive species like Chile tarweed and smooth cat's ear is inhibited, but not all seedlings affected may die until soil moisture becomes limiting and they are unable to compete with plants having more fully developed root systems. Because of allelopathy Chile tarweed is restricted to the grassland. The growth of smooth cat's ear is inhibited by sagebrush toxins and by toxins leached from straw of ripgut brome in the grassland. Consequently, its growth is restricted to more open grazed areas in the grassland or under shrubs other than California sagebrush. Seedlings of purple sage and California sagebrush are seldom found in mature chaparral stands, they are more common in the border zone and edge of the grassland (Muller 1966). Autotoxicity may play a role in regulating shrub density, thus minimizing intraspecific competition for water during the dry season.

Although the initial winter rain facilitates allelopathy in and around California sagebrush by releasing toxins into solution and transporting them to germinating seedlings, additional rain helps to eliminate toxicity by diluting and leaching toxin concentrations on the soil surface and initiating microbial decomposition which may breakdown active compounds in the litter and soil (Halligan 1976). For California sagebrush the rain also initiates the replacement of old leaves with the growth of new foliage. Leachates from the new tissue are not toxic because it has a lower terpene concentration and a much smaller proportion of camphor than the older mature foliage (Halligan 1975, 1976). Although allelopathy plays a role in the vegetation patterning of soft chaparral, other parameters can mediate its severity. Soil type, weather pattern, and shading ameliorate the allelopathic effects, whereas grazing, seed predation, and drought intensify its effect (Muller 1966). Allelopathy and livestock grazing together hasten the conversion of annual California grasslands to shrublands. These studies of soft chaparral demonstrate the importance of closely examining each species in a community to unravel the complex interactions that exist in the field.

On the more xeric sites of southern California, chaparral vegetation is dominated by the shrub chamise (*Adenostoma fasciculatum* H. & A). In mature stands of chamise, annual herbaceous and suffrutescent plants are noticeably lacking beneath the canopy but may be present in canopy openings and interspaces. The first year after a fire, the same site is covered with herbs; from germinating fire-annuals, opportunistic-annuals, suffrutescents, and sprouting of dormant perennial bulbs and rhizomes. By the second growing season, fire-annuals (Keeley et al. 1985) disappear. Herbs, including opportunistic-annuals, remain abundant the second year and then begin to decline. Suffrutescents increase in size the third and fourth years and then decline. By 10 to 12 years, the resprouting shrubs begin to reach their mature size and the cycle is nearly complete, with a few perennial herbs or suffrutescents remaining under the canopy (McPherson and Muller 1969, Christensen and Muller 1975a).

Germination requirements of species from the different life-history groups revealed some interesting characteristics contributing to the observed vegetational patterns and successional changes associated with fire (Keeley and Keeley 1982, Keeley et al. 1985). For example, in order for the fire-annuals to break dormancy and germinate, there must be a fire-related cue from heat, charred wood, or the optimal combination of heat and charred wood. Species most sensitive to these stimuli germinate only in the first season after a fire. Opportunistic-annuals that gradually recede into the intershrub spaces as the canopy returns, have polymorphic seeds. Some seeds do require heat from fire or charred wood to germinate and others are sufficiently stimulated by high temperatures that occur on the surface of bare soil, or require no treatment at all. The latter two seed types can germinate and colonize gaps in the shrub canopy for extended periods. Suffrutescent seed germination after fire is stimulated by heat, charred wood, or both and some germinated without any of these treatments (Keeley et al. 1985). Animal grazing and nutrient availability are unimportant in the germination response (Christensen and Muller 1975b), but after germination, animal grazing has a significant impact on seedling survival. Other factors contributing to limited growth and survival of herbaceous species in chamise are low soil nutrients, possibly low light levels and allelopathy (Christensen and Muller 1975a, 1975b).

The significance of allelopathy in chamise, however, has not been clearly determined. Indeed, extracts and leachates of chamise inhibited growth of ripgut brome and native dicots *in vitro* (McPherson and Muller 1969), but only after they were concentrated. Most inhibition occurred with leachates 4× and 10× more concentrated than originally prepared. Seeds of herbaceous perennials and suffrutescents were not sensitive to unconcentrated chamise foliar leachate and only 2 of 22 annuals were inhibited by leachate, whereas 9 were stimulated (Keeley et al. 1985). Phenolics have been identified in leaf leachates and soils collected beneath and away from the shrubs. Compared to leachates, soils had fewer phenolics, in lower concentrations and with less toxicity (McPherson et al. 1971). However, burned soils had higher extractable phenolic concentrations than unburned soils (Christensen and Muller 1975a). Soil collected from beneath chamise, heated and then returned to its original field location, showed no inhibition of germination or growth (Christensen and Muller 1975b), thereby shedding further doubt on the toxicity of natural foliage leachates. Inhibition of lettuce seed germination in chamise soils was caused by inhibitors produced by soil microbes rather than toxins released from foliage (Kaminsky 1981). This effect was not rcpeatable, however, with five native species and eight chaparral soils (Pack 1985). The role of allelopathy in the chamise successional fire cycle requires additional study.

Sagebrush steppe

Various *Artemisia* L. species occur on rangelands of North America, but none probably is more important than the shrubby sagebrushes (subgenus *Tridentatae* (Rydb.) McArthur). Their foliage contains high concentrations of volatile monoterpenes, including camphor and cineole (Kelsey et al. 1978, 1983) and sesquiterpene lactones (Kelsey and Shafizadeh 1979, 1980). Sagebrush volatiles, aqueous extracts, and individual compounds can interfere with the germination and growth of other plants in vitro, including many native species and *Artemisia* itself, (Asplund 1969, Klarich and Weaver 1973, McCahon et al. 1973, Hoffman and Hazlett 1977, Kelsey et al. 1978, Groves and Anderson 1981). Compounds present in leaves also are present in litter and some of the foliar sesquiterpene lactones and coumarins have been detected in the soils under the shrubs (Kelsey et al. 1978). Nitrogen fixation by blue-green algae-lichen crusts in the Great Basin is much lower under canopies of big sagebrush (*Artemisia tridentata* Nutt.) than it is in the interspaces between shrubs. Lichen crusts watered with foliage extracts are completely inhibited from fixing nitrogen (Rychert and Skujins 1974). Wikeem and Pitt (1982) report high concentrations of NO_3-nitrogen in interspace soils relative to soils under the crown of big sagebrush and suggest this is due to inhibition of nitrifiers, under the shrubs. Sagebrush litter inhibits nitrification (Fulgham et al. 1975), and sagebrush utilizes NH_4-nitrogen in preference to NO_3-nitrogen (Wallace et al. 1978). Cheatgrass (*Bromus tectorum* L.), an annual grass and common associate of sagebrush, inhibits *Nitrosomonas* and *Nitrobacter* (Neal 1969).

Considering the allelopathic activities of California sagebrush and *Artemisia herba-alba* in the Negev desert (discussed below), it seems highly probable that the widely distributed North American sagebrushes are allelopathic; however, there are differences in habitat and climate that could change the frequency and intensity of allelopathy. Most sagebrush species grow under cold desert conditions, with a dry hot summer, followed by a cold winter and a cool spring. Although winters may not be as wet as in the Mediterranean climate, snow melt and spring rains can keep the soil moist for extended periods when many associated species germinate. Can volatile terpenes adsorbed on soil particles the previous summer and fall, or leached during the winter, remain at toxic levels until spring germination? Does sagebrush litter, in the spring, have sufficient concentrations of toxins to interfere with growth? What happens to the large quantities of sesquiterpene lactones in the foliage of sagebrush? Are they ever inhibitory? To fully understand the importance of allelopathy in the ecology of sagebrush shrublands, many studies will be required to cover the array of species, habitats, and weather patterns where these plants abound.

Desert

Like the coastal chaparral environments of southern California, the northern Negev desert in Israel is a Mediterranean type climate with hot dry summers (June–August) and cool wet (December–January) winters. In the southern portion of the northern Negev highlands, north slopes support communities with *Artemisia herba-alba*, a semidwarf aromatic shrub. On adjacent south slopes are communities dominated by the nonaromatic shrub beancaper (*Zygophyllum dumosum* Boiss). Both sites receive the same amount of rain, the south slope gets 23% more solar radiation in a year, and the soils are similar but not identical. Yields of annual plants growing among beancaper on the south slope are significantly greater (2.5 to 13×) than those growing with sagebrush on the north slopes (Friedman 1987). The number of annual plants growing near crowns of *Artemisia* is less than half the number found 0.7 m away. There is no difference in the number of annuals near the crown or 3 m away from beancaper (Friedman et al. 1977). Volatiles and aqueous extracts from sagebrush shoots inhibit the germination of native annuals and tomato. Camphor, cineole, α-thujone and β-thujone are major components in the foliage of this species (Benjilali and Richard 1980, Friedman 1987). Inhibition is reversible by removing seeds from the toxic source. Rate of reversion is accelerated by leaching of residual toxins from the seeds. Dispersal of sagebrush seeds near the parental canopy and the preponderance of germination and seedling survival away from the canopy suggest autotoxicity occurs. Importance of climate in the functioning of allelopathy for this species is evident, because in less arid regions of Israel the annuals aggregate and are more abundant near the *Artemisia* canopies (Friedman 1987).

Just because foliage contains potent toxins, does not insure these compounds will function as allelopathic agents. Burrow-weed (*Franseria dumosa* Gray) and incienso (*Encelia farinosa* Gray) are two shrubs distributed in southern California desert communities. They are significantly different in their associated flora. Annual shrub-dependent herbs such as chickory (*Rafinesquia neomexicana* Gray), phacelia (*Phacelia distans* Benth.), desert dandelion (*Malacothrix californica* D.C.), and others are confined to sites under the canopy of burrow-weed (Went

1942). The under canopy of incienso is noticably devoid of such herbs, although nondependent species, normally in the open areas, may be found. A phytotoxin isolated from incienso foliage and identified as 3-acetyl-6-methoxybenzaldehyde was considered an important contributor to the absence of associated vegetation (Gray and Bonner 1948a, 1948b). Further investigation led to the discovery of a toxin (unidentified) in burrow-weed foliage that is more active than the one in incienso (Muller 1953). Compounds from both shrubs are inhibitory to tomato plants and shrub dependent and independent herbs (Muller and Muller 1956). The toxins obviously are not a primary cause of the vegetational patterns. Differences in canopy morphology and soil organic matter accumulation are considered the controlling factors. Rhizomatous shoots of the long-lived burrow-weed branch at ground level and trap wind-blown organic matter, thereby forming a good seedbed for the herbs. The relatively shorter lived incienso branches at a higher level above ground and does not trap the necessary debris required for dependent herb germination and growth.

Woodlands

Conifer and deciduous woodlands are climax vegetation following grass or shrub serial stages on 30.4 million ha of rangelands in the western United States (West et al. 1975). The expansion of woodlands and increased tree density over the last several decades is thought to be the result of superior tree competitiveness enhanced by livestock grazing of understory and fire exclusion (Tausch et al. 1981, Bolsinger 1989). The role allelopathy plays in tree encroachment and site dominance is unknown, but allelopathic effects on understory species have been associated with volatiles from foliage leachates (Jameson 1968), litter (Jameson 1961, 1970; Wilt et al. 1988), and tree roots (Lavin et al. 1968).

Laboratory and greenhouse tests support the hypothesis that phytotoxins from trees can inhibit establishment of understory species in the canopy litter zone. This zone is subject to the combined effects of foliage and litter leaching, volatiles, and litter decomposition. Jameson (1968) found growth of squirreltail (*Sitanion hystrix* Raf.) and blue grama to be significantly inhibited by extracts of Utah juniper (*Juniperus osteosperma* (Torr.) Little) foliage. Lavin et al. (1968) found Utah juniper extracts inhibitory to the germination of blue grama, crested wheatgrass (*Agropyron desertorum* (Fisch. ex Link) J. A. Schultes), and sideoats grama (*Bouteloua curtipendula (Michx.) Torr.*). Weeping lovegrass (*Eragrostis curvula* (Schrad.) Nees.) was slightly inhibited by the extracts and germination of fourwing saltbrush (*Atriplex canescens* (Pursh) Nutt.) was stimulated. Using wheat radicles as a bioassay, Jameson (1961) found water extracts of pinyon (*Pinus edulis* Engelm.) and juniper (Utah, one-seeded, *Juniperus monosperma* (Engelm.) Sarg.) and alligator, *Juniperus deppeana* Steud.) foliage to reduce growth of wheat radicles by 67 to 85%. In greenhouse studies, loose litter of singleleaf pinyon (*Pinus monophylla* Torr. & Frem.) inhibited the germination and emergence of Sandberg's bluegrass (*Poa sandbergii* Vassey) and Idaho fescue (*Festuca idahoensis* Elmer) to a greater degree than it did squirreltail or cheatgrass (Everett 1987). This study supports field observations of squirreltail and cheatgrass rapidly reoccupying the canopy litter zone of singleleaf pinyon after felling, while the Sandberg's bluegrass and Idaho fescue do not (Everett and Sharrow 1985).

Allelopathic effects from woodland trees are suggested from understory zones around stems of one-seeded juniper (Arnold 1964, Johnsen 1962, Schott and Pieper 1985), Utah juniper (Jameson 1970), and singleleaf pinyon (Everett et al. 1984). The complete absence of vegetation under the canopy of Utah juniper (Jameson 1970) and singleleaf pinyon (Everett et al. 1986) could be additional evidence for allelopathic effects. This phenomenon also appears in the deciduous oak and and bay woodlands of California where understory species are excluded from under California laurel (*Umbellularia californica* Nutt.) (Barbour and Major 1988). Klopatek and Klopatek (1987) report fewer nitrifying bacteria *Nitrosomonas* in soils under pinyon and Utah juniper than in interspace soils between trees.

Grasses, however, do well or are even most vigorous at the crown drip edge where allelopathic effects are likely to occur (Jameson 1970, Everett et al. 1984, and Schott and Pieper 1985). Pinyon ricegrass (*Piptochaetium fimbriatum* (H.B.K.) A. Hitchc.) grows only under the crown of one-seeded juniper, and its basal area is proportional to tree size (Schott and Pieper 1985). When pinyon trees are removed without disturbing the litter layer, established plants in the litter zone dramatically increase in vigor (Everett and Sharrow 1985). A similar release of cool season grasses growing under alligator juniper after felling is reported by Clary (1974). Any allelopathic effects of pinyon or juniper litter on established grasses seem to be overcome by increased nutrient availability in the canopy litter zone (Barth 1980, Doescher et al. 1987). The physical effect of litter buildup may be a more significant factor than allelopathy in the loss of understory under the crown. Arnold (1964) found that blue grama basal area and production declined under the tree crown of Utah juniper and pinyon, and Jameson (1966) showed this to be more correlated with litter depth than canopy coverage.

Allelopathy may play a significant role in the maintenance of understory patterns in woodlands once they have been created. Koniak and Everett (1982) found soil seed reserves of herbaceous species much lower under the tree crowns than in interspace soils. Fewer soil seed reserves combined with fewer remnant understory species reduces the potential of tree litter microsites to respond once trees are removed. Patterns remain the same for many understory species if the litter layer is left undisturbed during removal of singleleaf pinyon (Everett 1987). Preharvest grass dominants, Sandberg's bluegrass and Idaho fescue, and the perennial forb, low phlox (*Phlox hoodii* Richards), did not rapidly reestablish in the litter zone after tree removal. The canopy litter zone can remain void of any vegetation for several years. Similar allelopathic inhibition of seedling establishment for decades after crown removal are reported for oak woodlands of California (Barbour and Major 1988).

Transitional range

Allelopathy also occurs in transitional forest-range systems. There are two distinct vegetation types along the dry sandy ridges of relict shoreline in Florida. Closed canopy sand pine (*Pinus clausa* (Chapm. ex Engelm.) Vasey ex Sarg.) dominates the sand pine scrub, with a dense evergeen oak understory and few if any herbaceous plants. In contrast there is the sandhill or high pineland. This is an open woodland with native grasses dominated by slash pine (*P. elliottii* Engelm.), longleaf pine (*P. palustris* Mill.), and deciduous oaks. There is an abrupt ecotone between these two vegetation types characterized by a bare zone. Fire plays a role in maintaining these vegetation types. The open sandhill pine is characterized by frequent fires (3-8 years) fueled by grass and young pines (Richardson and Williamson 1988). The sand pine scrub burns infrequently (40–60 years), the plants are killed by frequent fires and require long intervals between fires to regenerate. Consequently, the invasion of herbs and grasses from the sandhill pine type into the fire sensitive pine scrub would facilitate conditions favoring more frequent fires in the pine scrub. Richardson and Williamson (1988) hypothesize that allelopathy helps maintain the pine scrub type free of fire by preventing grasses from invading, thus reducing the fire hazard. Evidence that allelopathy is involved includes aqueous foliage leachates from pine scrub species, sandheath (*Ceratiola ericoides* (A. Gray) Heller), savory (*Calamintha ashei* (Weatherby) Shinner), and bluesage conradina (*Conradina canescens* Torr and Gray), that inhibit the germination and growth of native grasses (Richardson and Williamson 1988, Williamson et al. 1989, Fischer et al. 1988, 1989). Also, transplanted pine seedlings and plugs of wiregrass (*Aristida stricta* Michx.) grow more slowly at scrub sites than at sandhill sites (Richardson and Williamson 1988).

Water washes of fresh sandheath leaves contain ceratiolin as the major constituent (Tanrisever et al. 1987). Ceratiolin is not toxic to the native grass, little bluestem, but it readily undergoes a photo-induced degradation to hydrocinnamic acid (HCA) which is toxic to bluestem germination and root growth. The decomposition of ceratiolin to HCA explains why water washes of litter are more toxic than washes from fresh leaves (Fischer et al. 1989). HCA was found in soils during summer months at concentrations known to inhibit bluestem germination (Fischer et al. 1988). Water washes of savory, which are inhibitory to little bluestem germination and growth, contain the triterpene, ursolic acid, the sesquiterpene, caryophyllene oxide and several menthofuran-type monoterpenes (Tanrisever et al. 1987, Macias et al. 1989). Phytotoxic monoterpenes, including camphor and cineole, and the triterpene, ursolic acid were isolated from water used to soak fresh leaves of bluesage conradina (Fischer et al. 1988, 1989, Williamson et al. 1989). Ursolic acid, present in the leaf washes of all three scrub species above may act as a detergent increasing solubility of non-polar compound in rain water washes. It may also facilitate the transport of toxins into the membranes of recepient plants (Fischer et al. 1988, 1989).

Soils collected seasonally from beneath another scrub species, jointweed (*Polygonella myriophylla* (Small) Horton), and from scrub bare zones reduced the germination of bahiagrass (*Paspalum notatum* Fluegga), a species common along roadsides but absent from the scrub (Weidenhamer and Romeo 1989). Low nutrient availability also is a potential contributor to the formation and maintenance of bare zones around jointweed. The involvement of other factors such as herbivory, disease, and subtle mycorrhizal effects are subjects for future studies.

In North Dakota, open ponderosa pine stands are characterized by understory vegetation with limited biomass and cover. Soil pH and light do not restrict understory growth. In laboratory bioassays, germination and growth of two understory species, little bluestem and big bluestem, were inhibited by decaying pine needles, needle leachates, and soils collected from within the pine stand (Lodhi and Killingbeck 1980). Various phenolic compounds (caffeic acid, chlorogenic acid, quercetin, and condensed tannins) were extracted and identified from the pine tissues and soils.

Understory communities also can effectively exclude or retard tree encroachment. In the ponderosa pine-bunchgrass communities of northern Arizona, grass species interfere with pine regeneration. Arizona fescue (*Festuca arizonica* Vassey) and mountain muhly (*Muhlenbergia montana* (Nutt.) A. Hitchc.) are the dominant understory bunchgrass species. When openings are created in the climax pine forests by fire or logging, these grasses are quick to establish and form dense communities. Large interspaces between mature trees remain relatively free of pine seedlings for many years, even though there is a ready seed source. Species other than pine also seem to be excluded by the bunchgrass. Yellow sweetclover (*Melilotus officinalis* (L.) Lam.) will invade sites only after a disturbance but then disappears in a few years as the grass reestablishes. The grasses are strong competitors for water, nutrients, and space (Larson and Schubert 1969, Rietveld 1975), but allelopathy also may play some role. Phytotoxins occur in the live foliage of both grasses (Jameson 1968, Rietveld 1975), and the most probable routes of release are leaching or microbial decay of dead tissue. Leachates of the green tissues from both grasses inhibit the germination of yellow sweetclover seeds (Rietveld 1977). More field studies are required before the importance of allelopathy can be clearly determined.

Disturbed rangelands and weeds.

Weeds are a category of plants considered undesirable because they have limited uses and interfere with the growth and productivity of more desirable species. A major portion of allelopathy research has focused on interference between agricultural crops and weeds. Weed species with allelopathic potential have been listed by Rice (1984) and Putnam (1985). Case histories already discussed have included species considered weeds: the pioneer weed stage in abandoned fields in Oklahoma and sagebrush of western rangelands.

Quackgrass (*Agropyron repens* L. Beauv.) is a weed of the Northern Hemisphere (Holm et al. 1977) found most often on cultivated lands rather than native ranges, and it has considerable allelopathic potential. The presence of phytotoxins in the tissues of quackgrass was reported in 1948 by Osvald. Subsequent observations indicate that the productivity of numerous crops is reduced when grown simultaneously or sequentially in quackgrass-infested soils (Isleib 1960, Bandeen and Buchholtz 1967, Kommedahl et al. 1970, Buchholtz 1971, Weston and Putnam 1985). Separating the combined effects of competition and allelopathy (both direct and indirect) has been a difficult task (Welbank 1961, 1964, Ohman and Kommedahl 1964, Buchholtz 1971).

Phytotoxins have been isolated from the shoots and rhizomes of quackgrass. Water extracts of these tissues inhibit a variety of species in petri dish bioassays (Osvald 1948, Tourneau and Heggeness 1957, Gabor and Veatch 1981). Tissue extracts and leachates from pots growing quackgrass also are inhibitory when used to water bioassay species grown in soil (Kommedahl 1957, Kommedahl et al. 1959, Ohman and Kommedahl 1960). A methanol-water extract of dried rhizomes led to the isolation and purification of a growth inhibitor tentatively identified as a glycoside (Gabor and Veatch 1981). From the aqueous extract of shoots two flavonoids, tricin and a partially characterized unknown compound, were isolated. A 50% reduction in the radicle growth of cress (*Lepidium sativum* L.) occurred at 123 ug/ml of tricin or 59 ug/ml of the unknown. Both compounds are found in field-grown shoots and rhizomes with greatest quantities in the shoots (Weston et al. 1987). The higher shoot concentrations correlate with the greater toxicity of shoot extracts to a variety of bioassay species (Weston and Putnam 1986).

Hagin (1989) suspected additional toxins in quackgrass because tricin occurs in a combined form in various grass species, including alfalfa, and yet alfalfa is particularly sensitive to quackgrass toxins. His chemical analysis uncovered high concentrations of 5-hydroxyindole-3-acetic acid (5-HIAA) and 5-hydroxytryptophan (5-HTP) glucosides in the roots and shoots. Removal of the sugar molecules enhanced their toxicity. These two compounds act like plant hormones or affect auxin controlled interactions. Together with other tissue phenolics, such as tricin, they may produce supraoptimal levels of growth hormones that inhibit the growth of susceptible species.

Glycoside hydrolysis could explain changes in toxicity when dried quackgrass leaves and rhizomes are incubated with soils (Toai and Linscott 1979). Incubation at high temperatures, 30°C, increases decay and toxin production whereas low temperatures 5–10°C prevent additional toxin formation. Also, under anaerobic conditions organic acids (acetic, proprionic, and butyric) and injurious levels of ethylene can be produced from decomposing quackgrass tissues (Welbank 1963, Lynch and Penn 1980, Harvey and Linscott 1978). The production of these compounds, however, is not unique to quackgrass (Patrick et al. 1963, Harvey and Linscott 1978).

The deleterious effects of quackgrass toxins on associated plants may provide more desirable conditions for the attack and growth of pathogenic microbes (Penn and Lynch 1982). Rhizomes of quackgrass can be infected with pathogens that cause blight on small grain plants. Barley seedlings grown in the presence of the root-rotting fungus *Helminthosporium sativum* and ground rhizomes are more diseased than seedlings in the presence of fungus alone (Kommedahl 1957). Barley grown with quackgrass rhizomes innoculated with root rot fungi (*Fusarium culmorum*, isolated from quackgrass rhizomes) was substantially smaller than plants grown with sterile rhizomes (Lynch and Penn 1980). Damage from decaying rhizomes was more severe in drier soils than in very wet soils under anaerobic conditions because dry soil favors the development of *Fusarium* rot. The success and aggressive behavior of quackgrass in its interactions with other species is complex and cannot be explained adequately without a consideration for the direct and indirect effects of allelopathy.

Several species of *Centaurea* (spotted knapweed, *C. maculosa* Lam.; diffuse knapweed, *C. diffusa* Lam.; Russian knapweed, *C. repens* L.; and yellow starthistle, *C. solstitialis* L.) were introduced to western North America around the start of the 20th century (Watson and Renney 1974, Watson 1980). Yellow starthistle was present in California much earlier (Maddox et al. 1985). These species exhibit the characteristics of "typical" weeds (Zimmerman 1976, Radosevich and Holt 1984), and possess the ecological behavior associated with allelopathic plants. All produce allelochemicals in appreciable quantities (Kelsey and Bedunah 1989), and many of the compounds are toxic in laboratory bioassays. Once a few individual plants become established on a site, their density will usually increase, eventually eliminating other species in the community (Tyser and Key 1988) or drastically reducing their productivity. Many *Centaurea* populations are monoculturelike in appearance, and self-perpetuating. Normal successional changes are very slow.

To understand the factors contributing to the ecological success of these species, some allelopathic studies have been conducted. Diffuse knapweed is the most thoroughly studied species. Toxins were isolated from the extracts of leaf and root tissues (Muir and Majak 1983), but other experiments indicate that the toxins are probably not effective under natural conditions (Muir et al. 1987). For example, the germination of two grass species was unaffected when sown on soil previously infested with diffuse knapweed. Solvent extracts of knapweed soils also had no effect on germination. Chemical analysis of root exudates did not detect any apparent toxins such as phenols, terpenes, or polyketides (Muir et al. 1987). In density controlled experiments, seedlings of three grass species were grown in pots with knapweed seedlings. Growth of only one grass was retarded. Alkaline extractable phenol concentrations in knapweed soils were less than or equal to the concentrations in nonknapweed soils (Muir et al. 1987). Because alkaline extraction removes some phenols partially bound in the soil and unavailable to plants, natural soil concentrations were considered too low to be inhibitory to grasses when compared to the experimental concentrations of soluble *p*-coumaric acid required to retard grass growth. Fall application of knapweed litter to pasture and range sites did not reduce grass growth in spring. In most instances where litter was added, grass growth was stimulated over the control, probably as a result of the mulching effect. Muir et al. (1987) conclude it is unlikely that allelopathy contributes to the ecological success of diffuse knapweed.

Spotted knapweed and diffuse knapweed contain the same major foliar toxin, the sesquiterpene lactone cnicin. Its strong autotoxicity in laboratory bioassays (Kelsey and Locken 1987), slow release from the foliage, low extractable concentrations in the soil, and ineffectiveness in diffuse knapweed habitats suggest that it causes minimal interference with associated species (Locken and Kelsey 1987). Consequently, allelopathy is not considered an important contributor to spotted knapweed aggressiveness.

Foliage of Russian knapweed contains several sesquiterpene lactones (Stevens 1982); four of them stimulate radicle growth of lettuce seedlings at low concentrations and inhibit growth at higher concentrations (Stevens and Merrill 1985). Seven polyacetylenes have been isolated from the roots, and one of them is phytotoxic to several bioassay species (Stevens 1986). The active compound was isolated from soils surrounding the roots at concentrations that were inhibitory in the bioassays. This one root toxin, in conjunction with the complex interlinking root system of Russian knapweed, could provide an effective allelopathic mechanism. Although autotoxicity by this compound has not been investigated, it could be avoided by the rhizomatous asexual reproduction and perennial life history of this species. Of these four weedy *Centaurea* species, Russian knapweed is the most likely to be allelopathic (Kelsey and Bedunah 1989).

Yellow starthistle contains many of the same foliar lactones as Russian knapweed (Stevens and Merrill 1985), and in addition, it has two toxic chromenes that are also strongly autotoxic in laboratory bioassays (Merrill 1989). This is a winter annual species with a different root system than Russian knapweed. Its root chemistry is yet to be investigated. Aside from its aggressive behavior in the field, there is insufficient information to evaluate allelopathy.

X. Rangeland Management and Allelopathy

The known

The practical consequence of allelopathy, and the reason for resource managers to be concerned, is the potential effect it has on the management of rangelands. The more comprehensive our knowledge of allelopathic species interactions, the more successful we will be at managing plant communities to yield products we desire. Managing allelopathy is complicated by the many mitigating factors influencing its occurrence and intensity. Allelopathy is only part of a complex of processes affecting, growth, vigor, and reproductive potential of plants. In fact, in most instances, it is probably not the single most important cause of changes in plant patterns, succession, productivity, or plant response to management. We are just starting to identify allelopathic effects on rangelands and the environmental conditions that enhance or weaken allelopathic effects, and to develop management prescriptions that consider allelopathy in achieving the desired plant community.

Physiological impairment of rangeland species by allelopathic chemicals and subsequent decline in plant establishment or growth has been described previously in this chapter. The cumulative physiological impacts on plants can cause or at least maintain bare soil or species inhibition zones around allelopathic species. The summation of these microsite effects can change the plant community composition to one of dominance by the allelopathic species. Site dominance by allelopathic shrubs and woodland tree species reduces forage production and stocking rates on western ranges (Bedell and Bunch 1978) and delays maximum forage production in grasslands (Owensby and Launchbaugh 1977). Removal of allelopathic shrubs and trees increases forage production significantly (Aro 1971) if site quality allows (Thatcher and Hart 1974), but we do not know the amount of increase due to reduced competition compared to reduced chemical interference.

There is ample evidence that allelopathy is more likely to be functional and of greater intensity when the recipient plant is under stress from other factors: low nutrients, limited water, high temperature, and herbicides (Einhellig 1987). Herbivory by livestock, wildlife, and insects is a major cause of stress to rangeland plants and can enhance or detract from allelopathic processes depending on whether it is the allelopathic donor or recipient species being used and to what degree. Unpalatable allelopathic species would thus have a twofold competitive advantage. For example, bur buttercup, an unpalatable forb to livestock, invades overgrazed sites in the Great Basin, where it maintains a population in part because of allelopathic interference with competing species (Buchanan et al. 1978) and its low palatability. Prairie threeawn, a grass of low forage preference, has maintained itself for decades in disturbed grasslands partially through allelopathic interactions (Owensby and Launchbaugh 1977). Allelopathic volatile oils and tannins in shrubs are associated with reduced forage preference by herbivores (Mole and Waterman 1987, Reichardt et al. 1987) and livestock poisoning (Ruyle et al. 1986). Preference by mule deer for different sagebrush taxa is strongly influenced by terpenoid compounds on the leaf surface (Personius et al. 1987).

Grazing can play an important role in the establishment of allelopathic species both in successional regression to weedy communities and in hastening normal succession from grass and forbs to allelopathic shrubs and trees. Muller (1969) noted this as occurring in the chaparral. Loss of plants from under shrub or tree crowns and the inhibition of new plant establishment reduces the area producing forage and seems to force increased grazing pressure on remaining plants. The eventual reduction in herbaceous biomass from the interspace areas between allelopathic shrubs or trees makes sites "fire safe" and protects the allelopathic species (Tausch et al. 1981, Richardson and Williamson 1988). Inhibition of understory species establishment reinforces the current plant community until the stand declines from senescence or severe burn conditions, thereby initiating a new successional cycle.

Once allelopathic interferences by range plants are identified, and the consequences determined to be undesirable, the most direct solution to the problem is to eliminate the donor plant. We may also want to remove residual allelopathic chemicals in surface litter and soil as well. The method of plant control should be dependent on the biology and ecology of the target species, anticipated post-treatment ecological changes, available technology, social acceptability, previous experience, and costs. Cost is often a formidable obstacle to large-scale control of undesirable plants, especially on rangelands of low to moderate productivity. Prescribed fire has been used successfully to remove allelopathic shrubs, forbs, and grasses at reasonable cost and has the added benefit of reducing allelopathy in litter and soils as well (Blaisdell et al. 1982, Wright and Bailey 1982). Biological control of allelopathic species is also possible. For example, leafy spurge a noxious weed with allelopathic properties (Selleck 1972, Manners 1987) is difficult and costly to control. Sheep consume significant quantities particularly during the late summer, thus reducing the above ground biomass (Landgraf et al. 1984). Some sort of integrated control program may be necessary to provide the most cost-effective benefits.

Toxin management is an alternative for controlling allelopathic interactions in situations where the donor plants are a desirable component in the community. In these cases, it may be possible to lessen allelopathic influence by interfering with the mechanism of release or preventing the accumulation of chemicals to active levels. If the inhibitors accumulate in the litter (Wilt et al. 1988), then periodic litter reduction by fire or mechanical removal (Lodhi 1981) may be effective. Fisher (1976) suggests "where fire is a common element one of its functions could be the destruction or deactivation of allelochems" such that herbaceous species can once again establish on the site. Soil microbes can utilize various toxins as a carbon source for energy and growth (Sparling et al. 1981, Vokou et al. 1984, Blum and Shafer 1988, Schmidt 1988, Weyman-Kaczmarkowa and Wojcik-Wojtkowiak 1991). Inhibitory concentrations may be reduced by inoculating soils with microbes capable of utilizing or degrading the specific toxin being produced.

Nutrient availability also mediates toxicity. Fertilization of rangelands is economically feasible in some regions to increase forage production (Vallentine 1977) and also may ameliorate allelopathic effects. Rice (1976) accelerated succession from annual allelopathic grass species to perennial grasses with high nitrogen and phosphorus fertilization. Soil nutrient status explain in part range plant distributions that differ from predictions based on known allelopathic relations among species.

Seeding rangelands to reestablish desirable forage species is a common management practice. The time of seed germination and initial seedling establishment is a critical period when plant species are most susceptible to allelopathic inhibition. The presence of allelopathic chemicals in soils have been demonstrated for several rangeland communities (Jameson 1970, del Moral and Cates 1971, Ballester et al. 1979). Given the large acreages seeded and the portion of soil surface previously occupied by allelopathic grass, shrub, or tree species, attention to allelopathic effects seems warranted.

Method of seeding is likely to alter the importance of allelopathy on species establishment. In greenhouse studies, seedling emergence from buried seeds of crested wheatgrass (*Agropyron cristatum*), smooth brome (*Bromus inermis* Leyss.), blue flax (*Linum lewesii* Pursh.), and ladak alfalfa (*Medicago sativa* L.) were inhibited by both singleleaf pinyon and Utah juniper litter (Everett 1987). Small burnet (*Sanguisorba minor* Scop.) was unaffected by tree litter. When these seeds were broadcast on the soil-litter surface, seedling establishment was unaffected or improved by the litter, which suggested adverse allelopathic effects were more than offset by a favorable seedbed microenvironment.

Longevity of range seedings may be impacted by the invasion of allelopathic weedy species such as sagebrush or woodland tree species and the autotoxicity of the seeded species themselves. Water extracts of sagebrush inhibit the commonly seeded species, bluebunch wheatgrass (*Agropyron spicatum* (Pursh) Scribn. & Smith), smooth brome, crested wheatgrass, and thickspike wheatgrass (*Agropyron dasystachyum* (Hook) Scribn.) (Hoffman and Hazlett 1977). Juniper extracts inhibit germination of crested wheatgrass, sideoats grama, and blue grama but are less inhibitory to pubescent wheatgrass or weeping lovegrass (Lavin et al. 1968). Seeding longevity of smooth brome is impacted by its own autotoxicity, which naturally thins established stands (Benedict 1941). Blue grama and western wheatgrass also exhibit autotoxicity and grow better in mixed stands than they do in monoculture (Bokhari 1978).

It is relatively safe to say that the influences of allelopathy are not being addressed directly in the management of rangelands today. Furthermore, it is not likely to become a critical management criterion until it is more thoroughly understood. Allelopathy is a subject deserving of intense investigation in the next decade.

The unknown and future research needs

Future allelopathic studies will undoubtedly improve knowledge of allelopathic interactions on rangelands and allow for better estimates of the causal mechanisms driving spatial and temporal succession. Research reviewed in this chapter shows that competition and allelopathy can act in concert to rapidly degrade or alter rangelands, and grazing can either promote or reduce these processes. Opportunities exist to modify grazing systems to inhibit or promote allelopathic species. The future may find us using the knowledge of species allelopathy to provide the successional stage we desire. The current practice of interseeding shrubs into monoculture stands of seeded grass for enhanced biodiversity (Richardson et al. 1984) may use allelopathy to maintain shrubs within the created community.

The amelioration of allelopathy by soil nutrients suggests fertilization could be used in the future to manipulate allelopathic effects on rangelands. If allelopathic species can reduce nitrogen buildup in soils and perpetuate themselves (Rice 1964), then fertilization can be used to hasten succession through undesirable allelopathic species-dominated stages (i.e., threeawn) to a more productive forage producing community (Rice 1976).

Land managers already are controlling allelopathic species on many range plant communities invaded by undesirable shrubs (sagebrush), woodland trees (pinyon, juniper, oak), and weedy annuals (cheatgrass, threeawn). In the future, similar rangeland practices may be used, but with increased knowledge of allelopathy we will better understand why some practices work better than others.

By understanding the ecology of allelopathic species, we may mediate their adverse impacts on range communities. Roder et al. (1988) recommend no-till seedings of switchgrass, because sandbur (*Cenchruys longispinus* (Hack.) Fern.) develops rapidly on disturbed soils and inhibits switchgrass germination and root growth. The sensitivity of prairie threeawn to fall burning provides a way to control it and to increase more desirable forage (Owensby and Launchbaugh 1977).

Seeding grasses, forbs, and shrubs to improve the forage base is common on rangelands and pastures; unfortunately we do not know the allelopathic relations among many of our commonly seeded species and may be placing antagonistic species in the same seed mix (Nurdin and Fulbright 1990). Conversely, we know some species are autoallelopathic and should be seeded in mixes (Benedict 1941, Bokari 1978). Inclusion of nitrogen-fixing species to enhance soil nutrition may ameliorate allelopathic interactions among species as suggested by Putnam and Duke (1978). Some strains of nitrogen-fixing, root nodule *Rizobium* sp. can utilize phenolics as a carbon source for growth (Gajendran and Mahadevan 1990). This desirable characteristic may be enhanced to overcome soil phenolic toxicity to nitrogen-fixing symbiotic bacteria.

We have yet to appreciate the impacts of allelopathic compounds present in the seedbed. Vegetation patterning and associated soil toxicity present before seeding treatments have not been evaluated for impact on seeding success, seeded species distribution, and stand longevity. Fire can reduce allelopathic compounds in soils and mechanical disturbance from drilling seed may also lessen seedbed toxicity for seeded species, but residual soil microsite effects on seedings have not been reported. Considering the large acreages seeded after removal of allelopathic species and the proportion of the soil surface occupied by allelopathic shrub, tree, or weed species (Everett and Koniak 1981, Eckert et al. 1986, Smith 1989), an evaluation of soil allelopathic effects on commonly seeded species is needed.

Opportunities exist to select range species for seeding that have desirable allelopathic characteristics. We already are capitalizing on the aggressive traits of sagebrush species and are trying to selectively breed plants to improve palatability and nutrient content (Welch et al. 1986). Fay and Duke (1977) report finding plant introductions of oat that exuded more scopoletin than others and depressed weed growth of wild mustard (*Brassica kaber* (D.C.) L.C. Wheeler). Peters and Luu (1985) report leaf and root leachates of tall fescue inhibit growth of *Brassica nigra*, and the intensity of inhibition differed with the fescue genotype. The genetic transfer of toxins in crossing experiments has been demonstrated for wheatgrass (*Agropyron glaucum* Roem. & Schult.) and wheat (Grodzinsky and Panchuk 1974, cited in Rice 1984). By genetically altering desirable plant species, we may be able to reduce levels of autotoxicity or develop plant varieties with enhanced allelopathic activity against weeds, thus using the allelopathic process to our benefit (Putnam and Duke 1978, Putnam 1988b). So far enhancement of allelopathic potential in crop plants by selection and breeding has achieved only limited success (Putnam 1988b, Purvis 1990). Rather than enhancing a plants allelopathic potential it may be possible to increase its resistance to a specific phytotoxin. Plants growing near the allelopathic tree of heaven (*Ailanthus altissima* Swingle) produce offspring better suited for growth near the tree than do plants of the same species previously unexposed to the trees toxins (Lawrence et al. 1991).

The allelochemicals themselves, from higher plants or microorganisms may have utility as specific, biodegradable, and environmentally safe herbicides (Heisey et al. 1985, 1990, Duke 1986, Putnam 1988a, Anaya et al. 1990) or fungicides (Kumar and Tripathi 1991). Agrostemin, an allelopathic compound from cocklebur, has been used to decrease weedy forbs and increase grass yields in pastures (Gajic 1973, cited in Rice 1984). The development of plants with greater concentrations of allelochemicals and the production of allelopathic compounds for practical uses have high probabilities of success, especially if one considers the genetic advancements occurring in biotechnology with microorganisms (Lindow et al. 1989) and with higher plants (Gasser and Fraley 1989).

Allelopathy will be a rewarding field of research into the next century; the opportunities are unlimited, the challenges difficult, and the benefits potentially enormous. Laboratory and greenhouse studies always will be an integral part of allelopathic investigations, but in the future we need to follow through into the natural environment. Field experiments need to be designed with greater creativity and incorporate the most advanced technological procedures and equipment.

Acknowledgments

The authors wish to thank Drs. Gerald R. Leather, Raymond A. Evans, and Roger Rosentreter for their comments and review of this manuscript.

Literature Cited

Abdul-Wahab, A.S., and E.L. Rice. 1967. Plant inhibition by johnson grass and its possible significance in old-field succession. Bull. Torrey Bot. Club 94:486-497.

Abeysekera, B.F., Z. Abramowski, and G.H.N. Towers. 1985. Chromosomal abberrations caused by sesquiterpene lactones in Chinese hamster ovary cells. Biochem. Syst. Ecol. 13:365-369.

Alsaadawi, I.S., S.M. Al-Hadithy, and M.B. Arif. 1986. Effects of three phenolic acids on chlorophyll content and ions uptake in cowpea seedlings. J. Chem. Ecol 12:221-227.

Anaya, A.L., M.R. Calera, R. Mata, and R. Pereda-Miranda. 1990. Allelopathic potential of compounds isolated from *Ipomoea tricolor* Cav. (Convolvulaceae). J. Chem. Ecol 16:2145-2152.

Anderson, R.C., and O.L. Loucks. 1966. Osmotic pressure influence in germination tests for antibiosis. Science 152:771-773.

Arnold, J.F. 1964. Zonation of understory vegetation around a juniper tree. J. Range Manage. 17:41-42.

Aro, R. 1971. Evaluation of pinyon-juniper conversion to grassland. J. Range Manage. 24:188-197.

Asplund, R.O. 1968. Monoterpenes: Relationship between structure and inhibition of germination. Phytochemistry. 7:1995-1997.

Asplund, R.O. 1969. The phytotoxicity of essential oils of different species of sagebrush. Wyo. Range Manage. Issue 270:40-43.

Baldwin, I.T., R.K. Olson, and W.A. Reiners. 1983. Protein binding phenolics and the inhibition of nitrification in subalpine balsam fir soils. Soil Biol. and Biochem. 15:419-423.

Balke, N.E. 1985. Effects of allelochemicals on mineral uptake and associated physiological processes, p.163-178. *In:* A.C. Thompson (ed.), The chemistry of allelopathy: Biochemical interactions among plants. ACS Symp. Series 268. Amer. Chem. Soc., Washington, D.C.

Balke, N.E., M.P. Davis, and C.C. Lee. 1987. Conjunctions of allelochemicals by plants. Enzymatic glucosylation of salicylic acid by *Avena sativa*, p. 214-227. *In:* G.R. Waller (ed.), Allelochemicals: Role in agriculture and forestry. ACS Symp. Series 330. Amer. Chem. Soc., Washington, D.C.

Ballester, A., A.M. Vieitez, and E. Vieitez. 1979. The allelopathic potential of *Erica australis* L. and *E. arborea* L. Bot. Gaz. 140:433-436.

Bandeen, J.D., and K.P. Buchholtz. 1967. Competitive effects of quackgrass upon corn as modified by fertilization. Weeds. 15:220-224.

Barbour, M. G., and J. Major. 1988. Terrestrial vegetation of California. California Native Plant Soc. Spec. Pub. No. 9. 1020 pp.

Barnes, J.P., and A.R. Putnam. 1983. Rye residues contribute weed suppression in no-tillage cropping systems. J. Chem. Ecol. 9:1045-1057.

Barth, R.C. 1980. Influence of pinyon pine trees on soil chemical and physical properties. Soil Sci. Soc. Am. J. 44:112-114.

Bartholomew, B. 1970. Bare zone between California shrub and grassland communities: The role of animals. Science 170:1210-1212.

Bedell, T. and T. Bunch. 1978. Effects of western juniper on forage production and livestock grazing management, p. 163-167. *In:* Proceedings of the western juniper ecology and management workshop. USDA, Forest Serv. Gen. Tech. Rep. PNW-74.

Begon, M., J.L. Harper, and C.R. Townsend. 1986. Ecology: Individuals, populations, and communities, p. 261. Sinauer Associates, Inc., Sunderland, Massachusetts.

Bell, D.T. 1974. The influence of osmotic pressure in tests for allelopathy. Trans. Ill. Acad. Sci. 67:312-317.

Bell, D.T., and D.E. Koeppe. 1972. Noncompetitive effects of giant foxtail on the growth of corn. Agron. J. 64:321-325.

Benedict, H.M. 1941. The inhibitory effect of dead roots on the growth of bromegrass. J. Amer. Soc. Agron. 33:1108.

Benjilali, B., and H. Richard. 1980. Etude de quelques peuplements d'armoise blanche du Marco *Artemisia herba alba*. Rivista Italiana 62:69-74.

Bhowmik, P.C., and J.D. Doll. 1982. Corn and soybean response to allelopathic effects of weed and crop residues. Agron. J. 74:601-606.

Bidwell, R.G.S. 1979. Plant physiology. MacMillan Publishing Co., Inc., New York.

Bieber, G.L., and C.S. Hoveland. 1968. Phytotoxicity of plant materials on seed germination of crownvetch, *Coronilla varia* L. Agron. J. 60:185-188.

Blacquiere, T. 1986. Nitrate reductase in the leaves and numbers of nitrifiers in the rhizosphere of *Plantago lanceolata* in two contrasting sites. Plant Soil 50:371-385.

Blaisdell, J., R.B. Murray, and E. McArthur. 1982. Managing intermountain rangelands. USDA, Forest Serv. Gen. Tech. Rep. INT-134. 41 p.

Blum, U., and B.R. Dalton. 1985. Effects of ferulic acid, an allelopathic compound, on leaf expansion of cucumber seedlings grown in nutrient culture. J. Chem. Ecol. 11:279-301.

Blum, U., B.R. Dalton, and J.O. Rawlings. 1984. Effects of ferulic acid and some of its microbial metabolic products on radicle growth of cucumber. J. Chem. Ecol. 10:1169-1191.

Blum, U., T.M. Gerig, and S.B. Weed. 1989. Effects of mixtures of phenolic acids on leaf area expansion of cucumber seedling grown in different pH Portsmouth A, soil material. J. Chem. Ecol. 15:2413-2423.

Blum, U., and J. Rebbeck. 1989. Inhibition and recovery of cucumber roots given multiple treatments of ferulic acid in nutrient culture. J. Chem. Ecol. 15:917-928.

Blum, U., and E.L. Rice. 1969. Inhibition of symbiotic nitrogen-fixation by gallic and tannic acid, and possible roles in old-field succession. Bull. Torrey Bot. Club 96:531-544.

Blum, U., and S.R. Shafer. 1988. Microbial populations and phenolic acids in soil. Soil Biol. Biochem. 20:793-800.

Bokhari, U.G. 1978. Allelopathy among prairie grasses and its possible ecological significance. Ann. Bot. 42:127-136.

Bolsinger, C.L. 1989. California's western juniper and pinyon-juniper woodlands: Area, stand characteristics, wood volume and fenceposts. USDA, Forest Serv. Resource Bull. PNW-RB-166.

Booth, W.E. 1941. Revegetation of abandoned fields in Kansas and Oklahoma. Amer. J. Bot. 28:415-422.

Boughey, A.S., P.E. Munro, J. Meiklejohn, R.M. Strang, and M.J. Swift. 1964. Antibiotic reactions between African savanna species. Nature, Lond. 203:1302-1303.

Bradow, J.M. 1991. Relationships between chemical structure and inhibitory activity of C6 through C9 volatiles emitted by plant residues. J. Chem. Ecol 17:2193-2212.

Bradow, J.M., and W.J. Connick, Jr. 1987. Allelochemicals from palmer amaranth, *Amaranthus palmeri* S. Wats. J. Chem. Ecol. 13:185-202.

Bradow, J.M., and W.J. Connick, Jr. 1988a. Seed-germination inhibition by volatile alcohols and other compounds associated with *Amaranthus palmeri* residues. J. Chem. Ecol. 14:1633-1648.

Bradow, J.M., and W.J. Connick, Jr. 1988b. Volatile methyl ketone seed-germination inhibitors from *Amaranthus palmeri* S. Wats. residues. J. Chem. Ecol. 14:1617-1631.

Brede, A.D. 1991. Field apparatus for testing allelopathy of annual bluegrass on creeping bentgrass. Crop Sci. 31:1372-1374.

Buchanan, B.A., K.T. Harper, and N.C. Frischknecht. 1978. Allelopathic effects of bur buttercup tissue on germination and growth of various grasses and forbs *in vitro* and in soil. Great Basin Nat. 38:90-96.

Buchholtz, K.P. 1971. The influence of allelopathy on mineral nutrition, p. 86-89. *In:* Biochemical interactions among plants. National Academy of Science, Washington D.C.

Burkholder, P.R. 1952. Cooperation and conflict among primitive organisms. Amer. Sci. 40:601-631.

Cast, K.G., J.K. McPherson, A.J. Pollard, E.G. Krenzer, Jr., and G.R. Waller. 1990. Allelochemicals in soil from no-tillage versus conventional-tillage wheat (*Triticum aestivum*) fields. J. Chem. Ecol. 16:2277-2289.

Charley, J.L., and N.E. West. 1975. Plant-induced soil chemical patterns in some shrub-dominated semi-desert ecosystems of Utah. J. Ecol. 63:945-963.

Cheam, A.H. 1984. A "natural herbicide" against calotrope? J. Agric. West. Aust. 25:42-43.

Chou, C.H. 1987. Allelopathy in subtropical vegetation and soils in Taiwan. p. 102-117. *In:* G.R. Waller (ed.), Allelochemicals: Role in agriculture and forestry. ACS Symp. Ser. 330. Amer. Chem. Soc., Washington, D.C.

Chou, C.H. 1989. Allelopathic research of subtropical vegetation in Taiwan. IV. Comparative phytotoxic nature of leachate from four subtropical grasses. J. Chem. Ecol. 15:2149-2159.

Chou, C.H., S.J. Chang, C.M. Cheng, Y.C.Wang, F.H. Hsu, and W.H. Den. 1989. The selective allelopathic interaction of a pasture-forest intercropping in Taiwan. II. Interaction between kikuyu grass and three hardwood plants. Plant Soil 116:207-215.

Chou, C.H., and Y.T. Chung. 1974. The allelopathic potential of *Miscanthus floridulus*. Bot. Bull. Acad. Sinica 15:14-27.

Chou, C.H., S.Y. Hwang, C.I. Peng, Y.C. Wang, F.H. Hsu, and N.J. Chung. 1987. The selective allelopathic interaction of a pasture-forest intercropping in Taiwan. Plant Soil 98:31-41.

Chou, C.H., and Y.R. Lee. 1991. Allelopathic dominance of *Miscanthus transmorrisonensis* in an alpine grassland community in Taiwan. J. Chem. Ecol 17:2267-2281.

Chou, C.H., and C.C. Young. 1975. Phytotoxic substances in twelve subtropical grasses. J. Chem. Ecol. 1:183-193.

Christensen, N.L., and C.H. Muller. 1975a. Effects of fire on factors controlling plant growth in *Adenostoma* chaparral. Ecol. Mono. 45:29-55.

Christensen, N.L., and C.H. Muller. 1975b. Relative importance of factors controlling germination and seedling survival in *Adenostoma* chaparral. Amer. Midl. Nat. 93:71-78.

Clary, W.P. 1974. Response of herbaceous vegetation to felling of alligator juniper. J. Range Manage. 27:387-389.

Colton, C.E., and F.A. Einhellig. 1980. Allelopathic mechanisms of velvetleaf (*Abutilon theophrasti* Medic., Malvaceae) on soybean. Amer. J. Bot. 67:1407-1413.

Cooper-Driver, G., S. Finch, T. Swain, and E. Bernays. 1977. Seasonal variation in secondary plant compounds in relation to the palatability of *Pteridium aquilinum*. Biochem. Syst. Ecol. 5:177-183.

Corcoran, M.R., T.A. Geissman, and B.O. Phinney. 1972. Tannins as gibberellin antagonists. Plant Physiol. 49:323-330.

Creek, R., and G.L. Wade. 1985. Excretion of phenolic compounds from the roots *of Festuca arundinacea, Eragrostis curvula,* and *Lespedeza striata*. Trans Kentucky Acad. Sci. 46:51-55.

Curtis, J. T., and G. Cottam. 1950. Antibiotic and autobiotic effects in prairie sunflower. Bull. Torrey Bot. Club. 77:187-191.

Dalton, B.R. 1989. Physiochemical and biological processes affecting the recovery of exogenously applied ferulic acid from tropical forest soils. Plant Soil 115:13-22.

Dalton, B.R., U. Blum, and S.B. Weed. 1989. Differential sorption of exogenously applied ferulic, *p*-coumaric, *p*-hydroxybenzoic, and vanillic acids in soil. Soil Sci. Soc. Am. J. 53:757-762.

Dalton, B.R., S.B. Weed, and U. Blum. 1987. Plant phenolic acids in soils: A comparison of extraction procedures. Soil Sci. Soc. Amer. J. 51:1515-1521.

Dao, T.H. 1987. Sorption and mineralization of plant phenolic acids in soil. p. 358-370. *In:* G.R. Waller (ed.), Allelochemicals: Role in agriculture and forestry. ACS Symp. Series 330. Amer. Chem. Soc., Washington, D.C.

d'Arcy Lameta, A., and M. Jay. 1987. Study of soybean and lentil root exudates III. Influence of soybean isoflavonoids on the growth of rhizobia and some rhizospheric microorganisms. Plant Soil 101:267-272.

de Jong, T.J., and P.G.L. Klinkhamer. 1985. The negative effects of litter of parent plants of *Cirsium vulgare* on their offspring: Autotoxicity or immobilization? Oecologia 65:153-160.

Dekker, J.H., W.F. Meggitt, and A.R. Putnam. 1983. Experimental methodologies to evaluate allelopathic plant interactions: The *Abutilon theophrasti-Glycine max* model. J. Chem. Ecol. 9:945-981.

del Moral, R., and R.G. Cates. 1971. Allelopathic potential of the dominant vegetation of western Washington. Ecology 52:1030-1037.

del Moral, R., and C.H. Muller. 1970. The allelopathic effects of *Eucalyptus camaldulensis*. Amer. Midl. Nat. 83:254-282.

Demos, E.K., M. Woolwine, R.H. Wilson, and C. McMillan. 1975. The effects of ten phenolic compounds on hypocotyl growth and mitochondrial metabolism of mung bean. Amer. J. Bot. 62:97-102.

Doescher, P., L. Eddleman, and M. Vaitkus. 1987. Evaluation of soil nutrients, pH and organic matter in rangelands dominated by western juniper. Northwest Sci. 61:97-102.

Dornbos, D.L., Jr., and G.F. Spencer. 1990. Natural products phytotoxicity. A bioassay suitable for small quantities of slightly water-soluble compounds. J. Chem. Ecol. 16:339-352.

Duke, S.O. 1986. Microbially produced phytotoxins as herbicides—a perspective, p. 287-304. *In:* A.R. Putnam and C.S. Tang (eds.), The science of allelopathy. John Wiley & Sons, New York.

Duke, S.O., R.D. Williams, and A.H. Markhart III. 1983. Interaction of moisture stress and three phenolic compounds on lettuce seed germination. Ann. Bot. 52:923-926.

Eckert, R., F. Peterson, M. Meurisse, and J. Stephens. 1986. Effects of soil-surface morphology on emergence and survival of seedlings in big sagebrush communities. J. Range Manage. 39:414-420.

Einhellig, F.A. 1985. Allelopathy-a natural protection, allelochemicals. p. 161-200. *In:* N.B. Mandava (ed.), CRC handbook of natural pesticides: Methods, Vol. 1, CRC Press, Inc., Boca Raton, Florida.

Einhellig, F.A. 1986. Mechanisms and modes of action of allelochemicals. p. 171-188. *In:* A.R. Putnam and C.S. Tang (eds.), The science of allelopathy. John Wiley & Sons, New York.

Einhellig, F.A. 1987. Interactions among allelochemicals and other stress factors of the plant environment, p. 343-357. *In:* G.R. Waller (ed.), Allelochemicals: Role in agriculture and forestry. ACS Symp. Series 330. Amer. Chem. Soc., Washington, D.C.

Einhellig, F.A., and L.Y. Kuan. 1971. Effects of scopoletin and chlorogenic acid on stomatal aperture in tobacco and sunflower. Bull. Torrey Bot. Club 98:155-162.

Einhellig, F.A., G.R. Leather, and L.L. Hobbs. 1985a. Use of *Lemna minor* L. as a bioassay in allelopathy. J. Chem. Ecol. 11:65-72.

Einhellig, F.A., M.S. Muth, and M.K. Schon. 1985b. Effects of allelochemicals on plant-water relationships, p. 179-195. *In:* A.C. Thompson (ed.), The chemistry of allelopathy: Biochemical interactions among plants. ACS Symp. Series 268. Amer. Chem. Soc., Washington, D.C.

Einhellig, F.A., and J.A. Rasmussen. 1978. Synergistic inhibitory effects of vanillic and *p*-hydroxsybenzoic acids on radish and grain sorghum. J. Chem. Ecol. 4:425-436.

Einhellig, F.A., E.L. Rice, P.G. Risser, and S.H. Wender. 1970. Effects of scopoletin on growth, CO2 exchange rates, and concentration of scopoletin, scopolin, and chlorogenic acids in tobacco, sunflower, and pigweed. Bull. Torrey Bot. Club 97:22-33.

Everett, R.L. 1987. Allelopathic effects of pinyon and juniper litter on emergence and growth of herbaceous species, p. 62-67. *In:* Seed and seedbed ecology of rangeland plants: A symposium. Tucson, AZ. USDA, Agric. Res. Serv.

Everett, R., and S. Koniak. 1981. Understory vegetation in fully stocked pinyon-juniper stands. Great Basin Nat. 41:467-476.

Everett, R.L., and S.H. Sharrow. 1985. Response of grass species to tree harvesting in singleleaf pinyon-Utah juniper stands. USDA, Forest Serv. Res. Pap. INT-334.

Everett, R.L., S.H. Sharrow, and R.O. Meeuwig. 1984. Pinyon-juniper woodland understory distribution patterns and species associations. Bull. Torrey Bot. Club 110:454-463.

Everett, R., S. Sharrow, and D. Thran. 1986. Soil nutrient distribution under and adjacent to singleleaf pinyon crowns. Soil Sci. Soc. Amer. J. 50:788-792.

Fahn, A. 1979. Secretory tissues in plants. Academic Press. London.

Fales, S.L., and R.C. Wakefield. 1981. Effects of turfgrass on the establishment of woody plants. Agron. J. 73:605-610.

Fay, P. and W. Duke. 1977. An assesment of allelopathic potential in *Avena* germ plasm. Weed Sci. 25:224-228.

Ferenczy, L. 1956. Occurrence of antibacterial compounds in seeds and fruits. Acta Biol. Acad. Sci. Hung. 6:317-323.

Fischer, N.H. 1986. The function of mono and sesquiterpenes as plant germination and growth regulators, p. 203-218. *In:* A.R. Putnam and C.S. Tang (eds.), The science of allelopathy. John Wiley & Sons, New York.

Fischer, N.H. 1991. Plant terpenoids as allelopathic agents, p. 377-398. *In:* J.B. Harborne and F.A. Tomas-Barberan (eds.), Ecological chemistry and biochemistry of plant terpenoids. Clarendon Press, Oxford.

Fischer, N.H., N. Tanrisever, and G.B. Williamson. 1988. Allelopathy in the Florida scrub community as a model for natural herbicide actions, p. 233-249. *In:* J.G. Cutler (ed.), Biologically active natural products: Potential use in agriculture. ACS Symp. Series 380. Amer. Chem. Soc., Washington, D.C.

Fischer, N.H., G.B. Williamson, J.D. Weidenhamer, N. Tanrisever, A. de la Pena, E.D. Jordan, and D.R. Richardson. 1989. Allelopathic mechanisms in the Florida scrub community, p. 183-193. *In:* C.H. Chou and G.R. Waller (eds.), Phytochemical ecology: Allelochemicals, mycotoxins and insect pheromones and allomones. Inst. Bot., Academia Sinica Mono. Series No. 9. Taipei, Republic of China.

Fisher, R.F. 1976. Allelopathic interference among plants. I. Ecological significance, p. 73-92. *In:* Proceedings, fourth North American forest biology workshop. State Univ. of New York, Syracuse.

Fisher, R.F. 1979. Allelopathy, p. 313-330. *In:* J.G. Horsfall and E.B. Cowling (eds.), Plant disease. An advanced treatise, Vol. IV, Academic Press, New York.

Fisher, R.F., R.A. Woods, and M.R. Glavicic. 1978. Allelopathic effects of goldenrod and aster on young sugar maple. Can. J. For. Res. 8:1-9.

Foster, R.C. 1988. Microenvironments of soil microorganisms. Biol. Fertil. Soil 6:189-203.

Franz, E. H., and B. L. Haines. 1977. Nitrate reductase activities of vascular plants in a terrestrial sere: Relationship of nitrate to uptake and the cybernetics of biogeochemical cycles. Bull. Ecol. Soc. Amer. 58:62.

Friedman, J. 1987. Allelopathy in desert ecosystems, p. 53-68. *In:* G.R. Waller (ed.), Allelochemicals: Role in agriculture and forestry. ACS Symp. Series 330. Amer. Chem. Soc., Washington, D.C.

Friedman, J., G. Orshan, and Y. Ziger-Cfir. 1977. Suppression of annuals by *Artemisia herba-alba* in the Negev desert of Israel. J. Ecol. 65:413-426.

Fuerst, E.P., and A.R. Putnam. 1983. Separating the competitive and allelopathic components of interference: Theoretical principles. J. Chem. Ecol. 9:937-944.

Fulgham, P.T., B. Klubek, and J. Skujins. 1975. Differential inhibition of nitrification in desert shrub communities. Annual meeting of the Amer. Soc. for Microbiol. (Abstract).

Gabor, W.E., and C. Veatch. 1981. Isolation of a phytotoxin from quackgrass (*Agropyron repens*) rhizomes. Weed Sci. 29:155-159.

Gajendiran, N., and A. Mahadevan. 1990. Growth of *Rhizobium* sp. in the presence of catechol. Plant Soil 125:207-211.

Gajic, D. 1973. The effect of agrostemins as a means of improvement of the quality and quantity of grass-cover of the Zlatibor-as a preventive measure against weeds. Yugoslav symposium on weed control in hilly and mountainous areas. Sarajevo. Cited in Rice 1984.

Gasser, C.S., and R.T. Fraley. 1989. Genetically engineering plants for crop improvement. Science 244:1293-1299.

Gerig, T.M., and U. Blum. 1991. Effects of mixtures of four phenolic acids on leaf area expansion of cucumber seedlings grown in Portsmouth B_1 soil materials. J. Chem. Ecol. 17:29-40.

Gerig, T.M., U. Blum, and K. Meier. 1989. Statistical analysis of the joint inhibitory action of similar compounds. J. Chem. Ecol. 15:2403-2412.

Gershenzon, J. 1984. Changes in the levels of plant secondary metabolites under water and nutrient stress. Recent Adv. Phytochem. 18:273-320.

Glass, A.D.M. 1976. The allelopathic potential of phenolic acids associated with the rhizosphere *of Pteridium aquilinum*. Can. J. Bot. 54:2440-2444.

Gliessman, S.R. 1976. Allelopathy in a broad spectrum of environments as illustrated by bracken. Bot. J. Linnean Soc. 73:95-104.

Gray, R., and J. Bonner. 1948a. An inhibitor of plant growth from the leaves of *Encelia farinosa*. Amer. J. Bot. 35:52-57.

Gray, R., and J. Bonner. 1948b. Structure determination and synthesis of a plant growth inhibitor, 3-acetyl-6-methoxybenzaldehyde, found in the leaves of *Encelia farinosa*. J. Amer. Chem. Soc. 70:1249-1253.

Green, F.B., and M.R. Corcoran. 1975. Inhibitory action of five tannins on growth induced by several gibberellins. Plant Physiol. 56:801-806.

Grodzinsky, A.M., and M.A. Panchuk. 1974. Allelopathic properties of crop residues of wheat-wheat grass hybrids, p. 51-55. *In:* A.M. Grodzinsky (ed.), Physiological-biochemical basis of plant interactions in phytocenoses. Naukova Duma, Kiev. Cited in Rice 1984.

Groves, C.R., and J.E. Anderson. 1981. Allelopathic effects of *Artemisia tridentata* leaves on germination and growth of two grass species. Amer. Midl. Nat. 106:73-79.

Grummer, G. 1955. Die gegenseitige beeinflussung hoherer pflanzen-allelopathie. Fischer, Jena.

Gussin, E.J., and J.M. Lynch. 1981. Microbial fermentation of grass residues to organic acids as a factor in the establishment of new grass swards. New Phytol. 89:449-457.

Hagin, R.D. 1989. Isolation and identification of 5-hydroxyindole-3-acetic acid and 5-hydroxytryptophan, major allelopathic aglycons in quackgrass (*Agropyron repens* L. Beauv.). J. Agric. Food Chem. 37:1143-1149.

Hall, A.B., U. Blum, and R.C. Fites. 1983. Stress modification of allelopathy of *Helianthus annuus* L. debris on seedling biomass production of *Amaranthus retroflexus* L. J. Chem. Ecol. 9:1213-1222.

Hall, J.L., T.J. Flowers, and R.M. Roberts. 1982. Plant cell structure and metabolism, 2nd Ed. Longman, London.

Hall, I.H., K.H. Lee, C.O. Starnes, O. Muraoka, Y. Sumida, and T.G. Waddell. 1980. Antihyperlipidemic activity of sesquiterpene lactones and related compounds. J. Pharm. Sci. 69:694-697.

Halligan, J.P. 1973. Bare areas associated with shrub stands in grassland: The case of *Artemisia californica*. Bioscience 23:429-432.

Halligan. J.P. 1974. Relationship between animal activity and bare areas associated with California sagebrush in annual grassland. J. Range Manage. 27:358-362.

Halligan, J.P. 1975. Toxic terpenes from *Artemisia californica*. Ecology 56:999-1003.

Halligan, J.P. 1976. Toxicity of *Artemisia californica* to four associated herb species. Amer. Midl. Nat. 95:406-421.

Harborne, J.B. 1984. Phytochemical methods. 2nd Ed. Chapman and Hall, London.

Harborne, J.B. 1988. Introduction to ecological biochemistry, 3rd Ed. Academic Press, London.

Harper, J.L. 1977. Population biology of plants. Academic Press, London.

Hartung, A.C., M.G. Nair, and A.R. Putnam. 1990. Isolation and characterization of phytotoxic compounds from asparagus (*Asparagus officinalis* L.) roots. J. Chem. Ecol. 16:1707-1718.

Harvey, R.G., and J.J. Linscott. 1978. Ethylene production in soil containing quackgrass rhizomes and other plant materials. Soil Sci. Soc. Am. J. 42:721-724.

Hassan, M.S., I.S. Alsaadawi, and A. El-Behadli. 1989a. Citrus replant problem in Iraq. II. Possible role of allelopathy. Plant Soil 116:157-160.

Hassan, M.S., A.H. El-Behadli, and I.S. Alsaadawi. 1989b. Citrus replant problem in Iraq. I. Possible role of soil fungi and nematodes. Plant Soil 116:151-155.

Hassan, M.S., A.H. El-Behadli, and I.S. Alsaadawi. 1989c. Citrus replant problem in Iraq. III. Interactive effect of soil fungi and allelopathy. Plant Soil 116:161-166.

Hegazy, A.K., K.S. Mansour, and N.F. Abdel-Hady. 1990. Allelopathic and autotoxic effects of *Anastatica hierochuntica* L. J. Chem. Ecol 16:2183-2193.

Heisey, R.M. 1990. Allelopathic and herbicidal effects of extracts from tree of heaven (*Ailanthus altissima*). Amer. J. Bot. 77:662-670.

Heisey, R.M., J. DeFrank, and A.R. Putnam. 1985. A survey of soil microorganisms for herbicidal activity, p. 337-349. *In:* A.C. Thompson (ed.), The chemistry of allelopathy: Biochemical interactions among plants. ACS Symp. Series 268. Amer. Chem. Soc., Washington, D.C.

Heisey, R.M., and C.C. Delwiche. 1984. Phytotoxic volatiles from *Trichostema lanceolatum* (Labiatae). Amer. J. Bot. 71:821-828.

Heisey, R.M., and C.C. Delwiche. 1985. Allelopathic effects of *Trichostema lanceolatum* (Labiatae) in the California annual grassland. J. Ecol. 73:729-742.

Henn, H., D. Petit, and P. Vernet. 1988. Interference between *Hieracium pilosella* and *Arrhenatherum elatius* in colliery spoils of north France. Oecologia 76:268-272.

Hoagland, R.E., and R.D. Williams. 1985. The influence of secondary plant compounds on the associations of soil microorganisms and plant roots, p. 301-325. *In:* A.C. Thompson (ed.), The chemistry of allelopathy: Biochemical interactions among plants. ACS Symp. Series 268. Amer. Chem. Soc., Washington, D.C.

Hoffman, G.R., and D.L. Hazlett. 1977. Effects of aqueous *Artemisia* exracts and volatile substances on germination of selected species. J. Range Manage. 30:134-137.

Hogan, M.E., and G.D. Manners. 1991. Differential allelochemical detoxification mechanism in tissue cultures *of Antennaria microphylla* and *Euphorbia esula*. J. Chem. Ecol. 17:167-174.

Hogberg, P., A. Granstrom, T. Johansson, A. Lundmark-Thelin, and T. Nasholm. 1986. Plant nitrate reductase activity as an indicator of availability of nitrate in forest soils. Can. J. For. Res. 16:1165-1169.

Holm, L.G., D.L. Pluckett, J.V. Pancho, and J.P. Herberger. 1977. The worlds worst weeds: Distribution and biology, p. 153-168. University Press, Honolulu, Hawaii.

Horowitz, M., and T. Friedman. 1971. Biological activity of subterranean residues of *Cynodon dactylon* L., *Sorghum halepense* L. and *Cyperus rotundus* L. Weed Res. 11:88-93.

Horsley, S.B. 1977. Allelopathic interference among plants. II. Physiological modes of action, p. 93-135. *In:* H.E. Wilcox and A.F. Hamer (eds.), Proc. 4th North Amer. For. Biol. Workshop. College of Environmental Science and Forestry, Syracuse, New York.

Hull, J.C., and C.H. Muller. 1977. The potential for dominance by *Stipa pulchra* in a California grassland. Amer. Midl. Nat. 97:147-175.

Ingham, R.E., J.A. Trofymow, E.R. Ingham, and D.C. Coleman. 1985. Interactions of bacteria, fungi, and their nematode grazers: Effects on nutrient cycling and plant growth. Ecol. Monogr. 55:119-140.

Isleib, D.R. 1960. Quackgrass control in potato production. Weeds 8:631-635.

Jambunathan, R., L.G. Butler, R. Bandyopadhyay, and L.K. Mughogho. 1986. Polyphenol concentrations in grain, leaf, and callus tissues of mold-susceptible and mold-resistant sorghum cultivars. J. Agric. Food Chem. 34:425-429.

Jambunathan, R., M.S. Kherdekar, and R. Bandyopadhyay. 1990. Flavan-4-ols concentration in mold-susceptible and mold-resistant sorghum at different stages of grain development. J. Agric. Food Chem. 38:545-548.

Jameson, D.A. 1961. Growth inhibitors in native plants of northern Arizona. USDA, Forest Serv. Res. Note RM-61.

Jameson, D.A. 1966. Pinyon-juniper litter reduces growth of blue grama. J. Range Manage. 19:214-217.

Jameson, D.A. 1968. Species interactions of growth inhibitors in native plants of northern Arizona. USDA, Forest Serv. Res. Note RM-113.

Jameson, D.A. 1970. Degradation and accumulation of inhibitory substances from *Juniperus osteosperma* (Torr.) Little. Plant Soil 33:213-224.

Johnsen, T. 1962. One-seed juniper invasion of northern Arizona grasslands. Ecol. Monogr. 32:187-207.

Jones, J.M., and B.N. Richards. 1977. Effect of reforestation on turnover of N-15 labeled ammonium plus nitrate in relation to changes in soil microflora. Soil Biol. Biochem. 9:383-392.

Kajinami, S., S.H. Wender, and E.C. Smith. 1971. Inhibition of glucose-6-phosphate dehydrogenase from tobacco tissue culture by scopoletin and scopolin. Phytochemistry 10:1501-1503.

Kaminsky, R. 1981. The microbial origin of the allelopathic potential of *Adenostoma fasciculatum*. H & A. Ecol. Monogr. 51:365-382.

Kanchan, S.D., and Jayachandra. 1980. Allelopathic effects of *Parthenium hysterophorus* L. Part IV. Identification of inhibitors. Plant Soil 55:67-75.

Kapustka, L.A., and F.L. Moleski. 1976. Changes in community structure in Oklahoma old field succession. Bot. Gaz. 137:7-10.

Kapustka, L.A., and E.L. Rice. 1976. Acetylene reduction (N2-fixation) in soil and old field succession in central Oklahoma. Soil Biol. Biochem. 8:497-503.

Kapustka, L.A., and E.L. Rice. 1978. Acetylene reduction (N2-fixation) of glucose-amended soils from central Oklahoma old field succession plots. Southwest. Nat. 23:389-396.

Katz, D.A., B. Sneh, and J. Friedman. 1987. The allelopathic potential *of Coridothymus capitatus* L. (Labiatae). Preliminary studies on the roles of the shrub in the inhibition of annuals germination and/or to promote allelopathically active actinomycetes. Plant Soil 98:53-66.

Keeley, S.C., and J.E. Keeley. 1982. The role of allelopathy, heat, and charred wood in the germination of chaparral herbs, p. 128-134. *In:* Dynamics and management of mediterranean-type ecosystems. USDA, Forest Serv. Gen. Tech. Rep. PSW-58.

Keeley, J.E., B.A. Morton, A. Pedrosa, and P. Trotter. 1985. Role of allelopathy, heat, and charred wood in the germination of chaparral herbs and suffrutescents. J. Ecol. 73:445-458.

Kelsey, R.G., and D. Bedunah. 1989. Ecological significance of allelopathy for *Centaurea* species in the northwestern United States, p. 10-32. *In:* P.K. Fay, and J.R. Lacey (eds.), Proceedings of the 1989 knapweed symposium. Montana State Univ., Bozeman.

Kelsey, R.G., and L.J. Locken. 1987. Phytotoxic properties of cnicin, a sesquiterpene lactone from *Centaurea maculosa* (spotted knapweed). J. Chem. Ecol 13:19-33.

Kelsey, R.G., G.W. Reynolds, and E. Rodriguez. 1984. The chemistry of biologically active constituents secreted and stored in plant glandular trichomes, p. 187-241. *In:* E. Rodriguez, P.L. Healey, I. Mehta (eds.), Biology and chemistry of plant trichomes. Plenum Press, New York.

Kelsey, R.G., and F. Shafizadeh. 1979. Sesquiterpene lactones and systematics of the genus *Artemisia*. Phytochemistry 18:1591-1611.

Kelsey, R.G., and F.Shafizadeh. 1980. Glandular trichomes and sesquiterpene lactones of *Artemisia nova* (Asteraceae). Biochem. Syst. Ecol. 8:371-377.

Kelsey, R.G., T.T. Stevenson, J.P. Scholl, T.J. Watson, Jr., and F. Shafizadeh. 1978. The chemical composition of the litter and soil in a community of *Artemisia tridentata* ssp. *vaseyana*. Biochem. Syst. Ecol. 6:193-200.

Kelsey, R.G., W.E. Wright, F. Sneva, A. Winward, and C. Britton. 1983. The concentration and composition of big sagebrush essential oils from Oregon. Biochem. Syst. Ecol. 11:353-360.

Kil, B.S., and S.Y. Lee. 1987. Allelopathic effects of *Chrysanthemum morifolium* on germination and growth of several herbaceous plants. J. Chem. Ecol. 13:299-308.

Klarich, D., and T. Weaver. 1973. Effects of vapors from *Artemisia tridentata* Nutt. on seed germination. Proc. Mont. Acad. Sci. 33:31-36.

Klein, K., and U. Blum. 1990a. Effects of soil nitrogen level on ferulic acid inhibition of cucumber leaf expansion. J. Chem. Ecol. 16:1371-1383.

Klein, K., and U. Blum. 1990b. Inhibition of cucumber leaf expansion by ferulic acid in split-root experiments. J. Chem. Ecol. 16:455-463.

Klopatek, C., and J. Klopatek. 1987. Mycorrhizae, microbes and nutrient cycling processes in pinyon-juniper systems, p. 360-364. *In:* Proceedings, pinyon-juniper conference. USDA, Forest Serv. Gen. Tech. Rep. INT-215.

Kobza, J., and F.A. Einhellig. 1987. The effects of ferulic acid on the mineral nutrition of grain sorghum. Plant Soil 98:99-109.

Kolb, T.E. 1988. Allelopathic effects of Kentucky bluegrass on northern red oak and yellow-popular. J. Arboriculture 14:281-283.

Kommedahl, T. 1957. Quackgrass can be toxic to crop seedlings. Down to Earth 13:4-5.

Kommedahl, T., J.B. Kotheimer, and J.V. Bernardini. 1959. The effect of quackgrass on germination and seedling development of certain crop plants. Weeds 7:1-12.

Kommedahl, T., K.M. Old, J.H. Ohman, and E.W. Ryan. 1970. Quackgrass nitrogen effects on succeeding crops in the field. Weed Sci. 18:29-32.

Koniak, S., and R. Everett. 1982. Soil seed reserves in successional stages of pinyon woodland. Amer. Midl. Nat. 108:295-303.

Krogmeier, M.J., and J.M. Bremner. 1989. Effects of phenolic acids on seed germination and seedling growth in soil. Biol. Fertil. Soils 8:116-122.

Kumar, A., and S.C. Tripathi. 1991. Evaluation of the leaf juice of some higher plants for their toxicity against soil borne pathogens. Plant Soil 132:297-301.

Kupchan, S.M. 1975. Advances in the chemistry of tumor-inhibitory natural products. Rec. Adv. Phytochem. 9:167-188.

Landgraf, B.K., P.K. Fay, and K.M. Havstad. 1984. Utilization of leafy spurge (*Euphorbia esula*) by sheep. Weed Sci. 32:348-352.

Larson, M.M., and G.H. Schubert. 1969. Root competition between ponderosa pine seedlings and grass. USDA, Forest Serv. Res. Pap. RM-54.

Lavin, F., D. Anderson, and F. Gomm. 1968. Juniper extracts and deficient aeration effects on germination of six range species. J. Range Manage. 21:262-263.

Lawrence, J.G., A. Colwell, and O.J. Sexton. 1991. The ecological impact of allelopathy in *Ailanthus altissima* (Simaroubaceae). Amer. J. Bot. 78:948-958.

Leather, G.R. 1983. Sunflowers (*Helianthus annuus*) are allelopathic to weeds. Weed Sci. 31:37-42.

Leather, G.R. 1987a. The seed germination process, p. 1-4. *In:* Seed and seedbed ecology of rangeland plants: A symposium. Tucson, AZ. USDA, Agric. Res. Serv.

Leather, G.R. 1987b. Weed control using allelopathic sunflowers and herbicide. Plant Soil 98:17-23.

Leather, G.R., and F.A. Einhellig. 1985. Mechanisms of allelopathic action in bioassays, p. 197-205. *In:* A.C. Thompson (ed.), The chemistry of allelopathy: Biochemical interactions among plants. ACS Symp. Series 268. Amer. Chem. Soc., Washington, D.C.

Leather, G.R., and F.A. Einhellig. 1986. Bioassays in the study of allelopathy, p. 133-145. *In:* A.R. Putnam and C.S. Tang (eds.), The science of allelopathy. John Wiley & Sons, New York.

Leather, G.R., and F.A. Einhellig. 1988. Bioassay of naturally occurring allelochemicals for phytotoxicity. J. Chem. Ecol. 14:1821-1828.

Lee, T.T. 1977. Role of phenolic inhibitors in peroxidase-mediated degradation of indole-3-acetic acid. Plant Physiol. 59:372-375.

Lee, K.H., I.H. Hall, E.C. Mar, C.O. Starnes, S.A. ElGebaly, T.G. Waddell, R.I. Hadgraft, C.G. Ruffner, and I. Weidner. 1977. Sesquiterpene antitumor agents: Inhibitors of cellular metabolism. Science 196:533-536.

Lee, T.T., and F. Skoog. 1965. Effects of hydroxybenzoic acids on indoleacetic acid inactivation by tobacco callus extracts. Physiol. Plant. 18:577-585.

Lehle, F.R., and A.R. Putnam. 1983. Allelopathic potential of sorghum (*Sorghum bicolor*): Isolation of seed germination inhibitors. J. Chem Ecol. 9:1223-1234.

Lindow, S.E., N.J. Panopoulos, and B.L. McFarland. 1989. Genetic engineering of bacteria from managed and natural habitats. Science 244:1300-1307.

Locken, L.J., and R.G. Kelsey. 1987. Cnicin concentrations in *Centaurea maculosa*, spotted knapweed. Biochem. Syst. Ecol. 15:313-320.

Lodhi, M.A.K. 1977. The influence and comparison of individual forest trees on soil properties and possible inhibition of nitrification due to intact vegetation. Amer. J. Bot. 64:260-267.

Lodhi, M.A.K. 1981. Accelerated soil mineralization, nitrification, and revegetation of abandoned fields due to the removal of crop-soil phytotoxicity. J. Chem. Ecol. 7:685-694.

Lodhi, M.A.K. 1982. Additional evidence of inhibition of nitrifiers and possible cycling of inhibitors produced by selected plants in a climax community. Bull. Torrey Bot. Club. 109:199-204.

Lodhi, M.A.K., R. Bilal, and K.A. Malik. 1987. Allelopathy in agroecosystems: Wheat phytotoxicity and its possible roles in crop rotation. J. Chem. Ecol. 13:1881-1891.

Lodhi, M.A.K., and K.T. Killingbeck. 1980. Allelopathic inhibition of nitrification and nitrifying bacteria in a ponderosa pine (*Pinus ponderosa* Dougl.) community. Amer. J. Bot. 67:1423-1429.

Lodhi, M.A.K., and G.L. Nickell. 1973. Effects of leaf extracts of *Celtis laevigata* on growth, water content, and carbon dioxide exchange rates of three grass species. Bull. Torrey Bot. Club 100:159-165.

Lodhi, M.A.K., and E.L. Rice. 1971. Allelopathic effects of *Celtis laevigata* Bull. Torrey Bot. Club 98:83-89.

Lodhi, M.A.K., and R.W. Ruess. 1988. Variation in soil nitrifiers and leaf nitrate reductase activity of selected tree species in a forest community. Soil Biol. Biochem. 20:939-943.

Lolas, P.C., and H.D. Coble. 1982. Noncompetitive effects of johnsongrass (*Sorghum halepense*) on soybeans (*Glycine max*). Weed Sci. 30:589-593.

Lorber, P., and W.H. Muller. 1976. Volatile growth inhibitors produced by *Salvia leucophylla*: Effects on seedling root tip ultrastructure. Amer. J. Bot. 63: 196-200.

Lovett, J.V. 1987. Allelopathy in Australia: Bacterial mediation, p. 156-174. *In:* G.R. Waller (ed.), Allelochemicals: Role in agriculture and forestry. ACS Symp. Series 330. Amer. Chem. Soc., Washington, D.C.

Lovett, J.V., and K. Jokinen. 1984. A modified stairstep apparatus for studies of allelopathy and other phytotoxic effects. J. Agr. Sci. Finland 56:1-7.

Luu, K.T., A.G. Matches, and E.J. Peters. 1982. Allelopathic effects of tall fescue on birdfoot trefoil as influenced by N fertilization and seasonal changes. Agron. J. 74:805-808.

Lynch, J.M., and D.J. Penn. 1980. Damage to cereals caused by decaying weed residues. J. Sci. Food Agric. 31:321-324.

Macias, F.A., F.R. Fronczek, and N.H. Fischer. 1989. Menthofurans from *Calamintha ashei* and the absolute configuration of desacetylcalcminthone. Phytochemistry 28:79-82.

Maddox, D.M., A. Mayfield, and N.H. Poritz. 1985. Distribution of yellow starthistle (*Centaurea solstitialis*) and Russian knapweed (*Centaurea repens*). Weed Sci. 33:315-327.

Mahmood, K., K.A. Malik, K.H. Sheikh, and M.A.K. Lodhi. 1989. Allelopathy in saline agricultural land: Vegetation successional changes and patch dynamics. J. Chem. Ecol. 15:565-579.

Mallik, M.A.B., and K. Tesfai. 1988. Allelopathic effect of common weeds on soybean growth and soybean-*Bradyrhizobium* symbiosis. Plant Soil 112:177-182.

Mandava, N.B. 1979. Natural products in plant growth regulation, p. 135-213. *In:* N.B. Mandava (ed.), Plant growth substances. ACS Symp. Series 111, Amer. Chem. Soc., Washington, D.C.

Mandava, N.B. 1985. Chemistry and biology of allelopathic agents, p. 33-54. *In:* A.C. Thompson (ed.), The chemistry of allelopathy: Biochemical interactions among plants. ACS Symp. Series 268. Amer. Chem. Soc., Washington, D.C.

Manners, G.D. 1987. The role of phytochemistry in attacking the leafy spurge (*Euphorbia esula*) problem, p. 228-237. *In:* G.R. Waller (ed.), Allelochemicals: Role in agriculture and forestry. ACS Symp. Series 330. Amer. Chem. Soc., Washington, D.C.

Mason-Sedun, W., and R.S. Jessop. 1988. Differential phytotoxicity among species and cultivars of the genus *Brassica* to wheat. II. Activity and persistence of water-soluble phytotoxins from residues of the genus *Brassica*. Plant Soil 107:69-80.

Mason-Sedun, W., and R.S. Jessop. 1989. Differential phytotoxicity among species and cultivars of the genus *Brassica* to wheat. III. Effects of environmental factors during growth on the phytotoxicity of residue extracts. Plant Soil 117:93-101.

McCahon, C.E., R.G. Kelsey, R.P. Sheridan, and F. Shafizadeh. 1973. Physiological effects of compounds extracted from sagebrush. Bull. Torrey Bot. Club 100:23-28.

McCarty, G.W., and J.M. Bremner. 1986. Effects of phenolic compounds on nitrification in soil. Soil Sci. Soc. Amer. J. 50:920-923.

McConnell, B.R., and J.G. Smith. 1971. Effect of ponderosa pine needle litter on grass seedling survival. USDA Forest Serv. Res. Note PNW-155.

McPherson, J.K., C.H. Chou, and C.H. Muller. 1971. Allelopathic constituents of the chaparral shrub *Adenostoma fasciculatum*. Phytochemistry 10:2925-2933.

McPherson, J.K., C.H. Muller. 1969. Allelopathic effects of *Adenostoma fasciculatum*, "Chamise," in the California chaparral. Ecol. Monogr. 39:177-198.

Merrill, G.B. 1989. Eupatoriochromene and encecalin, plant growth regulators from yellow starthistle (*Centaurea solstitialis* L.) J.Chem. Ecol. 15:2073-2087.

Mersie, W., and M. Singh. 1988. Effects of phenolic acids and ragweed parthenium (*Parthenium hysterophorus*) extracts on tomato (*Lycopersicon esculentum*) growth and nutrient and chlorophyll content. Weed Sci. 36:278-281.

Mole, S., and P.G. Waterman. 1987. Tannins as antifeedants to mammalian herbivores—still an open question? p. 572-587. *In:* G.R. Waller (ed.), Allelochemicals: Role in agriculture and forestry. ACS Symp. Series 330. Amer. Chem. Soc., Washington, D.C.

Molisch, H. 1937. Der einfluss einer pflanze auf die andere-allelopathie. Fischer, Jena.

Moreland, D.E., and W.P. Novitzky. 1987. Effects of phenolic acids, coumarins, and flavonoids on isolated chloroplasts and mitochondria, p. 247-261. *In:* G.R. Waller (ed.), Allelochemicals: Role in agriculture and forestry. ACS Symp. Series 330. Amer. Chem. Soc., Washington, D.C.

Muir, A.D., and W. Majak. 1983. Allelopathic potential of diffuse knapweed (*Centaurea diffusa*) extracts. Can. J. Plant Sci. 63:989-996.

Muir, A.D., W. Majak, F. Balza, and G.H.N. Towers. 1987. A search for the allelopathic agents in diffuse knapweed, p. 238-246. *In:* G.R. Waller (ed.), Allelochemicals: Role in agriculture and forestry. ACS Symp. Series 330. Amer. Chem. Soc., Washington, D.C.

Muller, C.H. 1953. The association of desert annuals with shrubs. Amer. J. Bot. 40:53-60.

Muller, C.H. 1965. Inhibitory terpenes volatilized from *Salvia* shrubs. Bull. Torrey Bot. Club 92:38-45

Muller, C.H. 1966. The role of chemical inhibition (allelopathy) in vegetational composition. Bull. Torrey Bot. Club 93:332-351.

Muller, C.H. 1969. Allelopathy as a factor in ecological process. Vegetation 18:348-357.

Muller, C.H., and R. del Moral. 1966. Soil toxicity induced by terpenes from *Salvia leucophylla*. Bull. Torrey Bot. Club 93:130-137.

Muller, C.H., W.H. Muller, and B.L. Haines. 1964. Volatile growth inhibitors produced by aromatic shrubs. Science 143:471-473.

Muller, W.H. 1965. Volatile materials produced by *Salvia leucophylla*: Effects on seedling growth and soil bacteria. Bot. Gaz. 126:195-200.

Muller, W.H. 1986. Allelochemical mechanisms in the inhibition of herbs by chaparral shrubs, p. 189-199. *In:* A.R. Putnam and C.S. Tang (eds.), The science of allelopathy. John Wiley & Sons, New York.

Muller, W.H., and R. Hauge. 1967. Volatile growth inhibitors produced by *Salvia leucophylla*: Effect on seedling anatomy. Bull. Torrey Bot. Club 94:182-191.

Muller, W.H., P. Lorber, and B. Haley. 1968. Volatile growth inhibitors produced by *Salvia leucophylla*: Effect on seedling growth and respiration. Bull. Torrey Bot. Club. 95:415-422.

Muller, W.H., P. Lorber, B. Haley, and K. Johnson. 1969. Volatile growth inhibitors produced by *Salvia leucophylla*: Effect on oxygen uptake by mitochondrial suspensions. Bull. Torrey Bot. Club 96:89-95.

Muller, W.H., and C.H. Muller. 1956. Association patterns involving desert plants that contain toxic products. Amer. J. Bot. 43:354-361.

Munro, P.E. 1966. Inhibition of nitrite-oxidizers by roots of grass. Appl. Ecol. 3:231-238.

Narasimhan, T.R., N. Harindranath, S. Premlata, B.S. Kesava Murthy, and P.V. Subba Rao. 1985. Toxicity of the sesquiterpene lactone parthenin to cultured bovine kidney cells. Planta Med. 194-197.

Neal, J.L. 1969. Inhibition of nitrifying bacteria by grass and forb root extracts. Can. J. Microbiol. 15:633-635.

Neill, R.L., and E.L. Rice. 1971. Possible role of *Ambrosia psilostachya* on pattern and succession in old-fields. Amer. Midl. Nat. 86:344-357.

Newman, E.I., and M.H. Miller. 1977. Allelopathy among some British grassland species. II. Influence of root exudates on phosphorus uptake. J. Ecol. 65:399-411.

Newman, E.I., and A.D. Rovira. 1975. Allelopathy among some British grassland species. J. Ecol. 63:727-737.

Nishimura, H., S. Hiramoto, J. Mizutani, Y. Noma, A. Furusaki, and T. Matsumoto. 1983. Structure and biological activity of bottrospicatol, a novel monoterpene produced by microbial transformation of (-)-*cis*-carveol. Agric. Biol. Chem. 47:2697-2699.

Noma, Y. 1980. Conversion of (-)-carvone by strains of *Streptomyces*, A-5-1, and *Nocardia*, 1-3-11. Agric. Biol. Chem. 44:807-812.

Noma, Y., H. Nishimura, S. Hiramoto, M. Iwami, and C. Tatsumi. 1982. A new compound, (4R, 6R)-(+)-6,8-oxidomenth-1-en-9-ol produced by microbial conversion of (-)-cis-carveol. Agric. Biol. Chem. 46:2871-2872.

Nurdin, and T.E. Fulbright. 1990. Germination of 2 legumes in leachate from introduced grasses. J. Range. Manage. 43:466-467.

Odum, E.P. 1971. Fundamentals of ecology, 3rd Ed. W.B. Saunders Company, Philadelphia.

Ohman, J.H., and T. Kommedahl. 1960. Relative toxicity of extracts from vegetative organs of quackgrass to alfalfa. Weeds 8:666-670.

Ohman, J.H., and T. Kommedahl. 1964. Plant extracts, residues, and soil minerals in relation to competition of quackgrass with oats and alfalfa. Weeds 12:222-231.

Osvald, D.R. 1948. Toxic exudates from the roots of *Agropyron repens* (couch-grass). J. Ecol. 36:192-193.

Ortega, R.C., A.L. Anaya, M. Gavilanes-Ruiz, S.S. Nieto, and M.N. Estrada. 199O. Effect of diacetyl piquerol on H+-ATPase activity of microsomes from *Ipomoea purpurea*. J. Chem. Ecol. 16:2253-2261.

Owensby, C. and J. Launchbaugh. 1977. Controlling prairie threeawn (*Aristida oligantha* Michx) in central and eastern Kansas with fall burning. J. Range Manage. 30:337-339.

Pack, P.E. 1985. The interaction of microbial toxins and charate in the germination of chaparral herb seeds. M.A. Thesis, Occidental College, Los Angeles, California. Cited in Keeley et al. 1985.

Parenti, R.L., and E.L. Rice. 1969. Inhibitional effects of *Digitaria sanguinalis* and possible role in old-field succession. Bull. Torrey Bot. Club 96:70-78.

Parker, V.T., and C.H. Muller. 1979. Allelopathic dominance by a tree-associated herb in a California annual grassland. Oecologia (Berlin) 37:315-320.

Parks, J.M., and E.L. Rice. 1969. Effects of certain plants of old-field succession on the growth of blue-green algae. Bull. Torrey Bot. Club 96:345-360.

Patrick, Z.A. 1971. Phytotoxic substances associated with the decomposition in soil of plant residues. Soil Sci. 111:13-18.

Patrick, Z.A., T.A. Toussoun, and W.C. Snyder. 1963. Phytotoxic substances in arable soils associated with decomposition of plant residues. Phytopathology 53:152-161.

Patterson, D.T. 1981. Effects of allelopathic chemicals on growth and physiological responses of soybean (*Glycine max*). Weed Sci. 29:53-59.

Pederson, G.A. 1986. White clover seed germination in agar containing tall fescue leaf extracts. Crop Sci. 26:1248-1249.

Penn, D.J., and J.M. Lynch. 1982. The effects of bacterial fermentation of couch grass rhizomes and *Fusarium culmorum* on the growth of barley seedlings. Plant Pathol. 31:39-43.

Perry, D., M. Amaranthus, J. Borchers, S. Borchers, and R. Brainerd. 1989. Bootstrapping in ecosystems. Bioscience 39:230-237.

Perry, D.A., and C. Choquette. 1987. Allelopathic effects on mycorrhizae. Influence on structure and dynamics of forest ecosystems, p. 185-194. *In:* G.R. Waller (ed.), Allelochemicals: Role in agriculture and forestry. ACS Symp. Series 330. Amer. Chem. Soc., Washington, D.C.

Personius, T.L., C.L. Wambolt, J.R. Stephens, and R.G. Kelsey. 1987. Crude terpenoid influence on mule deer preference for sagebrush. J. Range Manage. 40:84-88.

Peters, E.J. 1968. Toxicity of tall fescue to rape and birdsfoot trefoil seeds and seedlings. Crop Sci. 8:650-653.

Peters, E.J., and K.T. Luu. 1985. Allelopathy in tall fescue, p. 273-283. *In:* A.C. Thompson (ed.), The chemistry of allelopathy: Biochemical interactions among plants. ACS Symp. Series 268. Amer. Chem. Soc., Washington, D.C.

Peters, E.J., and A.H.B. M. Zam. 1981. Allelopathic effects of tall fescue genotypes. Agron. J. 73:56-58.

Picman, A.K. 1986. Biological activities of sesquiterpene lactones. Biochem. Syst. Ecol. 14:255-281.

Picman, J., and A.K. Picman. 1984. Autotoxicity in *Parthenium hysterophorus* and its possible role in control of germination. Biochem. Syst. Ecol. 12:287-292.

Purchase, B.S. 1974. Evaluation of the claim that grass root exudates inhibit nitrification. Plant Soil 41:527-539.

Purvis, C.E. 1990. Allelopathy: A new direction in weed control. Plant Protect. Quarter. 5:55-59.

Putnam, A.R. 1985. Weed allelopathy, p. 131-155. *In:* S.O. Duke (ed.), Weed physiology, Vol.1. CRC Press, Inc., Boca Raton, Florida.

Putnam, A.R. (ed.). 1987. Special allelopathy edition. Plant Soil 98:1-159.

Putnam, A.R. 1988a. Allelochemicals from plants as herbicides. Weed. Technol. 2:510-518.

Putnam, A.R. 1988b. Allelopathy: Problems and opportunities in weed management, p. 77-88. *In:* M.A. Altieri and M. Liebman (eds.), Weed management in agroecosystems: Ecological approaches. CRC Press, Inc., Boca Raton, Florida.

Putnam, A.R., and W.B. Duke. 1978. Allelopathy in agroecosystems. Ann. Rev. Phytopathol. 16:431-451.

Putnam, A.R., and C.S. Tang (eds.). 1986. The science of allelopathy. John Wiley & Sons, New York.

Qasem, J.R., and T.A. Hill. 1989a. On difficulties with allelopathy methodology. Weed Res. 29:345-347.

Qasem, J.R., and T.A. Hill. 1989b. Possible role of allelopathy in the competition between tomato, *Senecio vulgaris* L. and *Chenopodium album* L. Weed Res. 29:349-356.

Radosevich, S.R., and J.S. Holt. 1984. Weed ecology. John Wiley & Sons. New York.

Rasmussen, J.A., and F.A. Einhellig. 1977. Synergistic inhibitory effects of *p*-coumaric and ferulic acids on germination and growth of grain sorghum. J. Chem. Ecol. 3:197-205.

Rasmussen, J.A., A.M. Hejl, F.A. Einhellig, and J.A. Thomas. 1992. Sorgoleone from root exudate inhibits mitochondrial functions. J. Chem. Ecol. 18:197-207.

Rasmussen, J.A., and E.L. Rice. 1971. Allelopathic effects of *Sporobolus pyramidatus* on vegetational patterning. Amer. Midl. Nat. 86:309-326.

Ray, S.D., K.N. Guruprasad, and M.M. Laloraya. 1980. Antagonistic action of phenolic compounds on abscisic acid-induced inhibition of hypocotyl growth. J. Exp. Bot. 31:1615-1656.

Read, J.J., and E.H. Jensen. 1989. Phytotoxicity of water soluble substances from alfalfa and barley soil extracts on four crop species. J. Chem. Ecol. 15:619-628.

Reichardt, P., T. Clausen, and J. Bryant. 1987. Plant secondary metabolites as feeding deterrents to vertebrate herbivores, p. 37-42. *In:* Proceedings, symposium on plant-herbivore interactions. USDA, Forest Serv. Gen. Tech. Rep. INT-222.

Reynolds, T. 1977. Comparative effects of aliphatic compounds on inhibition of lettuce fruit germination. Ann. Bot. 41:637-648.

Reynolds, T. 1978. Comparative effects of aromatic compounds on inhibition of lettuce fruit germination. Ann. Bot. 42:419-427.

Reynolds, T. 1987. Comparative effects of alicyclic compounds and quinones on inhibition of lettuce fruit germination. Ann. Bot. 60:215-223.

Reynolds, T. 1989. Comparative effects of heterocyclic compounds on inhibition of lettuce fruit germination. J. Exp. Bot. 40:391-404.

Rice, E.L. 1964. Inhibition of nitrogen-fixing and nitrifying bacteria by seed plants. I. Ecology 45:824-837.

Rice, E.L. 1965a. Inhibition of nitrogen-fixing and nitrifying bacteria by seed plants. II. Characterization and identification of inhibitors. Physiol. Plant. 18:255-268.

Rice, E.L. 1965b. Inhibition of nitrogen-fixing and nitrifying bacteria by seed plants. IV. The inhibitors produced by *Ambrosia elatior* and *A. psitaslachya*. Southwest. Nat. 10:248-255.

Rice, E.L. 1971. Inhibition of nodulation of inoculated legumes by leaf leachates from pioneer plant species from abandoned fields. Amer. J. Bot. 58:368-371.

Rice, E.L. 1972. Allelopathic effects of *Andropogon virginicus* and its persistence in old fields. Amer. J. Bot. 59:752-755.

Rice, E.L. 1974. Allelopathy. Academic Press, New York.

Rice, E.L. 1976. Allelopathy and grassland improvement, p. 90-111. *In:* J.R. Estes and R.J. Tyrl (eds.), The grasses and grasslands of Oklahoma. Noble Foundation, Ardmore, Oklahoma.

Rice, E.L. 1977. Some roles of allelopathic compounds in plant communities. Biochem. Syst. and Ecol. 5:201-206.

Rice, E.L. 1984. Allelopathy, 2nd Ed. Academic Press, New York.

Rice, E.L. 1986. Allelopathic growth stimulation, p. 23-42. *In:* A.R. Putnam, and C.S. Tang (eds.), The science of allelopathy. John Wiley & Sons, New York.

Rice, E.L., and S.K. Pancholy. 1972. Inhibition of nitrification by climax ecosystems. Amer. J. Bot. 59:1033-1040.

Rice, E.L., and S.K. Pancholy. 1973. Inhibition of nutrification by climax ecosystems. II. Additional evidence and possible role of tannins. Amer. J. Bot. 60:691-702.

Rice, E.L., and S.K. Pancholy. 1974. Inhibition of nitrification by climax ecosystems. III. Inhibitors other than tannins. Amer. J. Bot. 61:1095-1103.

Rice, E.L., and R.L. Parenti. 1967. Inhibition of nitrogen-fixing and nitrifying bacteria by seed plants. V. Inhibitors produced by *Bromus japonicus* Thunb. Southwest. Nat. 12:97-103.

Rice, E.L., and R.L. Parenti. 1978. Causes of decreases in productivity in undisturbed tall grass prairie. Amer.J. Bot. 65:1091-1097.

Rice, E.L., W.T. Penfound, and L.M. Rohrbaugh. 1960. Seed dispersal and mineral nutrition in succession in abandoned fields in central Oklahoma. Ecology 41:224-228.

Richardson, B., S. Monsen, and D. Bowers. 1984. Interseeding selected shrubs and herbs on mine disturbances in southeastern Idaho, p. 134-139. *In:* Proceedings—symposium on the biology of *Artemisia* and *Chrysothamnus.* USDA, Forest Serv. Gen. Tech. Rep. INT-200.

Richardson, D.R., and G.B. Williamson. 1988. Allelopathic effects of shrubs of the sand pine scrub on pines and grasses of the sandhills. For. Sci. 34:592-605.

Rietveld, W.J. 1975. Phytotoxic grass residues reduce germination and initial root growth of ponderosa pine. USDA, Forest Serv. Res. Pap. RM-153.

Rietveld, W.J. 1977. Phytotoxic effects of bunchgrass residues on germination and initial root growth of yellow sweetclover. J. Range Manage. 30:39-43.

Rietveld, W.J. 1983. Allelopathic effects of juglone on germination and growth of several herbaceous and woody species. J. Chem. Ecol. 9:295-308.

Robertson, G.P. 1982. Factors regulating nitrification in primary and secondary succession. Ecology 63:1561-1573.

Roder W., S. Waller, and J. Stubbendieck. 1988. Allelopathic effects of sandbur leachate on switchgrass germination: Observations. J. Range manage. 41:86-88.

Rovira, A.D. 1969. Plant root exudates. Bot. Rev. 35:35-57.

Rovira, A.D., and C.B. Davey. 1974. Biology of the rhizosphere, p.153-204. *In:* E.D. Carson (ed.), The plant roots and its environment. Univ. Virgina Press, Charlottesville.

Rudiger, W., and E. Lohaus. 1987. Germination and growth inhibitors as allelochemicals, p. 118-128. *In:* G.R. Waller (ed.), Allelochemicals: Role in agriculture and forestry. ACS Symp. Ser. 330. Amer. Chem. Soc., Washington, D.C.

Ruyle, G., R. Grumbles, M. Murphy, and R. Cline. 1986. Oak consumption by cattle in Arizona. Rangelands 8:124-126.

Rychert, R.C., and J. Skujins. 1974. Nitrogen fixation by blue-green algae-lichen crusts in the Great Basin desert. Soil Sci. Soc. Amer. Proc. 38:768-771.

Schmidt, S.K. 1988. Degradation of juglone by soil bacteria. J. Chem. Ecol. 14:1561-1571.

Schmidt, S.K., and F.B. Reeves. 1989. Interference between *Salsola Kali* L. seedlings: Implications for plant succession. Plant Soil 116:107-110.

Schott, M., and R. Pieper. 1985. Influence of canopy characteristics of one seed juniper on understory grasses. J. Range Manage. 38:329-331.

Schreiber, M.M., and J.L. Williams, Jr. 1967. Toxicity of root residues of weed grass species. Weeds 15:80-81.

Schutt, C., and D. Netzly. 1991. Effects of apiforol and apigeninidin on growth of selected fungi. J. Chem. Ecol. 17:2261-2266.

Seaman, F.C. 1982. Sesquiterpene lactones as taxonomic characters in the Asteraceae. Bot. Rev. 48:121-595.

Selleck, G.W. 1972. The antibiotic effects of plants in laboratory and field. Weed Sci. 20:189-194.

Sequeira, L., R.J. Hemingway, and S.M. Kupchan. 1968. Vernolep *in:* A new, reversible plant growth inhibitor. Science 161:789-790.

Sewell, P.A., and B. Clarke. 1987. Chromatographic separations. John Wiley & Sons. Chichester, England.

Shibaoka, H., M. Mitsuhashi, and M. Shimokoriyama. 1967a. Promotion of adventitious root formation by heliangine and its removal by cysteine. Plant Cell Physiol. 8:161-170.

Shibaoka, H., M. Shimokoriyama, S. Iriuchijima, and S. Tamura. 1967b. Promoting activity of terpenic lactones in *Phaseolus* rooting and their reactivity toward cysteine. Plant Cell Physiol. 8:297-305.

Silverstein, R.M., and J.B. Simeone (eds.). 1983. Special issue: North American symposium on allelopathy. J. Chem. Ecol. 9:937-1292.

Smith, A.E. 1989. The potential allelopathic characteristics of bitter sneezeweed (*Helenium amarum*). Weed Sci. 37:665-669.

Smith, W.H., F.H. Bormann, and G.E. Likens. 1968. Response of chemoautotrophic nitrifiers to forest cutting. Soil Sci. 106:471-473.

Smith, J.L., and E.L. Rice, 1983. Differences in nitrate reductase activity between species of different stages in old field succession. Oecologia 57:43-48.

Solomon, B.P. 1983. Autoallelopathy in *Solanum carolinense*: Reversible delayed germination. Amer. Midl. Nat. 110:412-418.

Sondheimer, E., and J.B. Simeone. 1970. Chemical ecology. Academic Press, New York.

Sparling, G.P., B.G. Ord, and D. Vaughan. 1981. Changes in microbial biomass and activity in soils amended with phenolic acids. Soil Biol. Biochem. 13:455-460.

Spring, O., and A. Hager. 1982. Inhibition of elongation growth by two sesquiterpene lactones isolated from *Helianthus annuus* L. Planta 156:433-440.

Spring., O., J. Kupka, B. Maier, and A. Hager. 1982. Biological activities of sesquiterpene lactones from *Helianthus annuus*: Antimicrobial and cytotoxic properties; influence on DNA, RNA, and protein synthesis. Z. Naturforsch 37c:1087-1091.

Sridhara Murthy M., and R. Ravindra. 1975. Allelopathic effects of *Aristida adscensionis*. Oecologia (Berl.) 18:243-249.

Stenlid, G. 1968. On the physiological effects of phloridzin, phloretin and some related substances upon higher plants. Physiol. Plant. 21:882-894.

Stevens, K.L. 1982. Sesquiterpene lactones from *Centaurea repens*. Phytochemistry 21:1093-1098.

Stevens, K.L. 1986. Allelopathic polyacetylenes from *Centaurea repens* (Russian knapweed). J. Chem. Ecol. 12:1205-1211.

Stevens, K.L., and G.B. Merrill. 1980. Growth inhibitors from spikerush. J. Agric. Food Chem. 28:644-646.

Stevens, K.L., and G.B. Merrill. 1985. Sesquiterpene lactones and allelochemicals from *Centaurea* species, p. 83-98. *In:* A.C. Thompson (ed.), The chemistry of allelopathy. ACS Symp. Series 268. Amer. Chem. Soc., Washington, D.C.

Stevens, G.A., Jr., and C.S. Tang. 1985. Inhibition of seedling growth of crop species by recirculating root exudates of *Bidens pilosa* L. J. Chem. Ecol. 11:1411-1425.

Stowe, L.G. 1979. Allelopathy and its influence on the distribution of plants in an Illinois old-field. J. Ecol. 67:1065-1085.

Stowe, L.G., and A. Osborn. 1980. The influence of nitrogen and phosphorus levels on the phytotoxicity of phenolic compounds. Can. J. Bot. 58:1149-1153.

Tang, C.S. 1986. Continuous trapping techniques for the study of allelochemicals from higher plants, p. 113-131. *In:* A. R. Putnam and C.S. Tang (eds.), The science of allelopathy. John Wiley & Sons, New York.

Tang, C.S., and A.C. Waiss, Jr. 1978. Short-chain fatty acids as growth inhibitors in decomposing wheat straw. J. Chem. Ecol. 4:225-232.

Tang, C.S., C.K. Wat, and G.H.N. Towers. 1987. Thiophenes and benzofurans in the undisturbed rhizosphere of *Tagetes patula* L. Plant Soil 98:93-97.

Tang, C.S., and C.C. Young. 1982. Collection and identification of allelopathic compounds from the undisturbed root system of bigalta limpograss (*Hemarthria altissima*). Plant Physiol. 69:155-160.

Tanrisever, N., F.R. Fronczek, N.H. Fischer, and G.B. Williamson. 1987. Ceratiolin and other flavonoids from *Ceratiola ericoides*. Phytochemistry 26:175-179.

Tausch, R., N. West, and A. Nabi. 1981. Tree age and dominance patterns in Great Basin pinyon-juniper woodlands. J. Range Manage. 34:259-264.

Thatcher, A., and V. Hart. 1974. Spy mesa yields better understanding of pinyon-juniper in range ecosystem. J. Range Manage. 27:354-357.

Thompson, A.C. (ed.) 1985. The chemistry of allelopathy: Biochemical interactions among plants. ACS Symp. Series 268. Amer. Chem. Soc., Washington, D.C.

Tiedemann, A.R., and J.O. Klemmedson. 1973. Effect of mesquite on physical and chemical properties of the soil. J. Range Manage. 26:27-29.

Tinnin, R.O., and C.H. Muller. 1972. The allelopathic influence of *Avena fatua*: The allelopathic mechanism. Bull. Torrey Bot. Club. 99:287-292.

Toai, T.V., and D.L. Linscott. 1979. Phytotoxic effect of decaying quackgrass (*Agropyron repens*) residues. Weed Sci. 27:595-598.

Todd, R.L., W.T. Swank, J.E. Douglas, P.C. Kerr, D.L. Brockway, and C.D. Monk. 1975. The relationship between nitrate concentration in southern Appalachian mountain streams and terrestrial nitrifiers. Agro. Ecosystems 2:127-137.

Tourneau, D.L., and H.G. Heggeness. 1957. Germination and growth inhibitors in leafy spurge foliage and quackgrass rhizomes. Weeds 5:12-19.

Toussoun, T.A., A.R. Weinhold, R.G. Linderman, and Z.A. Patrick. 1968. Nature of phytotoxic substances produced during plant residue decompsition in soil. Phytopathology 58:41-45.

Tukey, H.B., Jr. 1966. Leaching of metabolites from above-ground plant parts: Its implications. Bull. Torrey Bot. Club 93:385-401.

Tukey, H.B., Jr. 1970. The leaching of substances from plants. Ann. Rev. Plant Physiol. 21:305-324.

Tyser, R.W., and C.H. Key. 1988. Spotted knapweed in natural area fescue grasslands: An ecological assessment. Northwest Sci. 62:151-160.

Unamma, R.P.A., and I.O. Akobundu. 1989. Effects of tropical weeds on yield in white yam (*Dioscorea rotundata* Poir). Weed Res. 29:1-6.

Vallentine, J. 1977. Range development and improvements. Brigham Young University Press, Provo, Utah.

Van Sumere, C.F., J. Cottenie, J. DeGreef, and J. Kint. 1972. Biochemical studies in relation to the possible germination regulatory role of naturally occurring coumarin and phenolics. Recent Adv. Phytochem. 4:165-221.

Van Veen, J.A., R. Merckx, and S.C. Van de Geijn. 1989. Plant-and soil related controls of the flow of carbon from roots through the soil microbial biomass. Plant Soil 115:179-188.

Vaughan, D., G.P. Sparling, and B.G. Ord. 1983. Amelioration of the phytotoxicity of phenolic acids by some soil microbes. Soil Biol. Biochem. 15:613-614.

Vierheilig, H., and J.A. Ocampo. 1990. Effect of isothiocyanates on germination of spores of *G. mosseae*. Soil Biol. Biochem. 22:1161-1162.

Vitousek, P.M., P.A. Matson, and K. Van Cleve. 1989. Nitrogen availability and nitrification during succession: Primary, secondary and old-field seres. Plant Soil 115:229-239.

Vokou, D., N.S. Margaris, and J.M. Lynch. 1984. Effects of volatile oils from aromatic shrubs on soil microorganisms. Soil Biol. Biochem. 16:509-513.

Wacker, T.L., G.R. Safir, and C.T. Stephens. 1990. Effects of ferulic acid on *Glomus fasciculatum* and associated effects on phosphorus uptake and growth of asparagus (*Asparagus officinalis* L.). J. Chem. Ecol. 16:901-909.

Wallace, A. E., M. Romney, G. E. Kleinkopf, and S. M. Soufi. 1978. Uptake of mineral forms of nitrogen by desert plants, p. 130-151. *In:* N.E. West and J. Skujins (eds.), Nitrogen in desert ecosystems. Dowden, Hutchinson and Ross, Stromberg, Pennsylvania.

Waller, G.R. (ed.). 1987. Allelochemicals: Role in agriculture and forestry. ACS Symp. Series 330. Amer. Chem. Soc., Washington, D.C.

Walters, D.T., and A.R. Gilmore. 1976. Allelopathic effects of fescue on the growth of sweetgum. J. Chem. Ecol. 2:469-479.

Wardle, D.A., K.S. Nicholson, and M. Ahmed. 1992. Comparison of osmotic and allelopathic effects of grass leaf extracts on grass seed germination and radicle elongation. Plant Soil 140:315-319.

Watanabe, K., N. Ohno, H. Yoshioka, J. Gershenzon, and T.J. Mabry. 1982. Sesquiterpene lactones and diterpenoids from *Helianthus argophyllus*. Phytochemistry 21:709-713.

Waterman, P.G., and S. Mole. 1989. Extrinsic factors influencing production of secondary metabolites in plants, p. 107-134. *In:* E.A. Bernays (ed.), Insect-plant interactions, Vol.1. CRC Press. Boca Raton, Florida.

Watson, A.K. 1980. The biology of Canadian weeds. 43. *Acroptilon* (*Centaurea*) *repens* (L.) DC. Can. J. Plant Sci. 60:993-1004.

Watson, A.K., and A.J. Renney. 1974. The biology of Canadian weeds. 6. *Centaurea diffusa* and *C. maculosa*. Can. J. Plant Sci. 54:687-701.

Weaver, T.W., and D. Klarich. 1977. Allelopathic effects of volatile substances from *Artemisia tridentata* Nutt. Amer. Midl. Nat. 97:508-512.

Weidenhamer, J.D., D.C. Hartnett, and J.T. Romeo. 1989. Density-dependent phytotoxicity: Distinguishing resource competition and allelopathic interference in plants. J. Appl. Ecol. 26:613-624.

Weidenhamer, J.D., T.C. Morton, and J.T. Romeo. 1987. Solutin volume and seed number: Often overlooked factors in allelopathic bioassays. J. Chem. Ecol. 13:1481-1491.

Weidenhamer, J.D., and J.T. Romeo. 1989. Allelopathic properties of *Polygonella myriophylla*. Field evidence and bioassays. J. Chem. Ecol. 15:1957-1970.

Welbank, P.J. 1961. A study of the nitrogen and water factors in competition with [*Agropyron repens* (L.) Beauv.] Ann. Bot. N.S. 25:116-137.

Welbank, P.J. 1963. Toxin production during decay of *Agropyron repens* (couch grass) and other species. Weed Res. 3:205-214.

Welbank, P.J. 1964. Competition for nitrogen and potassium in *Agropyron repens*. Ann. Bot. N.S 28:1-16.

Welch, B.L., E.D. McArthur, D.L. Nelson, J.C. Pederson, and J.N. Davis. 1986. "Hobble Creek"—a superior selection of low-elevation mountain big sagebrush. USDA, Forest Serv. Res. Pap. INT-370.

Weller, S.C., W.A. Skroch, and T.J. Monaco. 1985. Common bermudagrass (*Cynodon dactylon*) interference in newly planted peach (*Prunus persica*) trees. Weed Sci. 33:50-56.

Went, F.W. 1942. The dependence of certain annual plants on shrubs in southern California deserts. Bull. Torrey Bot. Club. 69:100-114.

West, N.E., K.H. Rea, and R.J. Tausch. 1975. Basic synecological relationships in juniper-pinyon woodlands, p. 41-53. *In:* The pinyon-juniper ecosystem: A symposium. Logan, UT. Utah State Univ.

Weston, L.A., B.A. Burke, and A.R. Putnam. 1987. Isolation, characterization and activity of phytotoxic compounds from quackgrass [*Agropyron repens* (L.) Beav.]. J. Chem. Ecol. 13:403-421.

Weston, L.A., and A.R. Putnam. 1985. Inhibition of growth, nodulation, and nitrogen fixation of legumes by quackgrass. Crop Sci. 25:561-565.

Weston, L.A., and A.R. Putnam. 1986. Inhibition of legume seedling growth by residues and extracts of quackgrass (*Agropyron repens*). Weed Sci. 34:366-372.

Weyman-Kaczmarkowa, W., and D. Wojcik-Wojtkowiak. 1991. The dynamics of microflora and the occurrence of phytotoxic substances in different soils with corn residues and inorganic nitrogen. Plant Soil 132:11-20.

White, C.S. 1986. Volatile and water-soluble inhibitors of nitrogen mineralization and nitrification in a ponderosa pine ecosystem. Biol. Fertil. Soils 2:97-104.

White, C.S. 1991. The role of monoterpenes in soil nitrogen cycling processes in ponderosa pine. Results from laboratory bioassays and field studies. Biogeochem. 12:43-68.

Whitehead, D.C., H. Dibb, and R.D. Hartley. 1981. Extractant pH and the release of phenolic compounds from soils, plant roots and leaf litter. Soil Biol. Biochem. 13:343-348.

Whitehead, D.C., H. Dibb, and R.D. Hartley. 1982. Phenolic compounds in soil as influenced by the growth of different plant species. J. Appl. Ecol. 19:579-588.

Whitehead, D.C., H. Dibb, and R.D. Hartley. 1983. Bound phenolic compounds in water extracts of soils, plant roots and leaf litter. Soil Biol. Biochem. 15:133-136.

Whittaker, R.H., and P.P. Feeny. 1971. Allelochemics: Chemical interactions between species. Science 171:757-770.

Wikeem, S.J., and M.D. Pitt. 1982. Soil nitrogen gradients as influenced by sagebrush canopy in southern British Columbia. Northwest Sci. 56:276-286.

Williams, D.H., and I. Fleming. 1980. Spectroscopic methods in organic chemistry, 3rd Ed. McGraw-Hill, London.

Williams, R.D., and R.E. Hoagland. 1982. The effects of naturally occurring phenolic compounds on seed germination. Weed Sci. 30:206-212.

Williamson, G.B., N.H. Fischer, D.R. Richardson, and A. de la Pena. 1989. Chemical inhibition of fire-prone grasses by fire-sensitive shrub, *Conradina canescens*. J. Chem. Ecol. 15:1567-1577.

Wilson, R.E., and E.L. Rice. 1968. Allelopathy as expressed by *Helianthus annuus* and its role in old-field succession. Bull. Torrey Bot. Club 95:432-448.

Wilt, F.M., G.C. Miller, and R.L. Everett. 1988. Monoterpene concentrations in litter and soil of singleleaf pinyon woodlands of the western Great Basin. Great Basin Nat. 48:228-231.

Wink, M. 1983. Inhibition of seed germination by quinolizidine alkaloids. Planta 158:365-368.

Woynarowski, J.W., T.A. Beerman, and J. Konopa. 1981. Induction of deoxyribonucleic acid damage in HeLa S3 cells by cytotoxic and antitumor sesquiterpene lactones. Biochem. Pharmacol. 30:3005-3007.

Woynarowski, J.M., and J. Konopa. 1981. Inhibition of DNA biosynthesis in HeLa cells by cytotoxic and antitumor sesquiterpene lactones. Molec. Pharmacol. 19:97-102.

Wright, H.A., and A.W. Bailey. 1982. Fire ecology: United States and southern Canada. John Wiley & Sons, Inc., New York.

Young, C.C., and D.P. Bartholomew. 1981. Allelopathy in a grass-legume association: I. Effects of *Hemarthria altissiman* (Poir.) Stapf. and Hubb. Root residues on the growth of *Desmodium intortum* (Mill.) Urb. and *Hemarthria altissima* in a tropical soil. Crop. Sci. 21:770-774.

Zak, D.R., and K.S. Pregitzer. 1988. Nitrate assimilation by herbaceous ground flora in late successional forests. J. Ecol. 76:537-546.

Zimmerman, C.A. 1976. Growth characteristics of weediness in *Portulaca oleracea*. Ecology 57:964-974.

Fire as a Physical Stress

Allen A. Steuter and Guy R. McPherson

Introduction

The ecological role and human use of fire on native grazing lands is broadly defined by the evolutionary history of specific plant communities (e.g. Denslow 1980, Axelrod 1985). Adaptations by plant species and communities to fire evolved under a set of environmental conditions with unique temporal and space scales. Yet, fire stress is expressed as reduced productivity of individual plants or communities, and by changes in species composition. The extent of fire stress is adjusted by topoedaphic and climatic variation. Today, grazing lands may be subjected to accelerated rates of species change (extirpation, introduction), and fragmentation. The challenge to range managers is to recognize which adaptations to fire operate at their scale, can be used to achieve a desired result, and will maintain the resource in the long-term.

Fire interacts with other disturbances (Collins 1987, Vinton et al. 1993) to change the light, temperature, moisture, and nutrient regimes for plants (Raison 1979). Fire stress, therefore, ranges from direct and short-term to indirect and long-term. For example, fire may stress individual plants by consuming stored energy reserves needed for continued growth (direct), and plant communities by reducing soil moisture through increased evapotranspiration or runoff (indirect). Long fire-free intervals represent a stress to species or communities adapted to short fire-return intervals, and the reverse is also true. An appreciation of fuel conditions, and fire size, season, and return-interval are all required to organize plant and community responses. It is within this context that the mechanisms by which plants tolerate, avoid, or respond to fire are outlined.

Individual Plant Response

Many mechanisms facilitate survival and resilience of individual plants to fire (Table 1). However, traits enhancing reproductive success in fire-prone communities may also enhance success in the presence of other stressful environmental factors (Lotan et al. 1985). Many fire-adapted species evolved or achieved dominance during the Pliocene, presumably in response to the much drier climate 5–6 million years ago (Axelrod 1985). Thus, caution should be exercised when interpreting adaptive traits. Although they often did not evolve in response to fire, adaptive traits nonetheless confer a competitive advantage to species with these traits in fire-affected communities.

Effects of fire vary widely between and within species. Response is affected by physical factors (e.g., fuel condition, weather, slope and aspect) as well as biologi-

---

Allen Steuter, Director of Science and Stewardship-NE, The Nature Conservancy; Guy McPherson, School of Renewable Natural Resources, University of Arizona, Tucson, AZ.

Table 1. Mechanisms exhibited by plants which enable them to avoid, resist, or respond positively to fire. See text for discussion.

Life-stage	Strategy	Mechanisms
Seeds	Avoidance	Burial Insulative plant tissue
	Resistance	Hard seed coat Increased germination Competitive advantage
Juveniles	Avoidance	Rapid growth
	Resistance	Aboveground buds protected by insulative plant tissue Belowground buds protected by soil
	Response	Resprouting at a young age
Mature plants	Avoidance	Life cycle shorter than fire-return interval Protective tissue around buds Flowering phenology
	Resistance	Thick, platy, corky, fissured bark Aboveground buds protected by insulative plant tissue Belowground buds protected by soil
	Response	Resprouting Fire-obligate flowering Increased flowering (?)

cal factors (plant morphology and physiology), and varies during the life cycle. Therefore, few generalizations can be made regarding response of individuals to fire. Calling a species "fire tolerant" or "fire adapted" nearly always requires qualification. In this section, mechanisms by which plants resist, escape, or respond to fires in various life stages are discussed.

Seeds

Seeds are generally well-adapted to survive in environments with fire (Wright and Bailey 1982), a phenomenon at least partially attributable to the insulating effect of soil. Seeds are capable of tolerating relatively high temperatures (Daubenmire 1968), and seeds of woody plants are generally more resistant to fire than seeds of herbs (Wright and Bailey 1982). Seeds are not only capable of tolerating high temperature, but germination percentage increases with increasing heat for some species (Cushwa et al. 1968, Pratt et al. 1984, Keeley et al. 1985, Hodgkinson and Oxley 1990). Hard, water-impermeable seed coats frequently require scarification to germinate. Heat cracks or melts the seed coat, allowing water to enter the seed and initiate germination (Fig. 1). Shrubs [e.g., *Adenostoma fasciculatum* H. & A., *Ceanothus* L. spp., *Rhus ovata* Wats., *R. lanceolata* (Gray) Britton, *Salvia apiana* Jepson, *Acacia longifolia* (Andr.) Willd.] are best

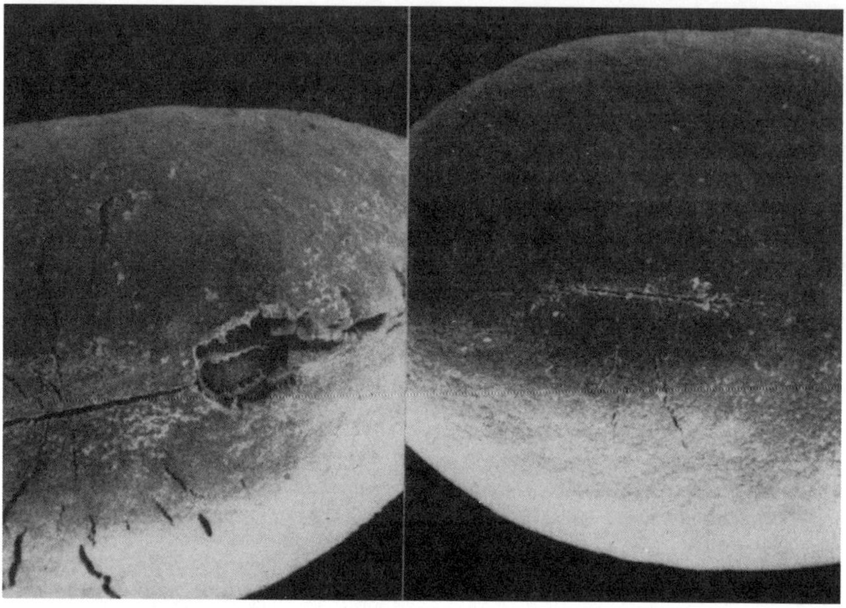

Figure 1. Seeds of flameleaf sumac (*Rhus copallina* L.) have a hard, waxy seed coat which inhibits germination. Heat melts the seed coat, allowing germination. The seed on the left is untreated; the seed on the right was heated for 97°C for 3 min (scanning electron micrographs, 45X, from Rasmussen and Wright 1988).

known for this characteristic (Stone and Juhren 1953, Quick and Quick 1961, Pieterse and Cairns 1986, Keeley 1987, Rasmussen and Wright 1988). Germination percentage of chaparral species usually increases over controls when seeds are exposed to temperatures of 115° to 120°C (Sampson 1944, Went et al. 1952, Keeley 1987). However, exposure to temperatures above 120°C for a few minutes decreases germination of most species, and temperatures greater than 150°C for a few minutes are usually lethal (Went et al. 1952, Stone and Juhren 1953, Keeley et al. 1985). Keeley's (1991) excellent review of seed germination in California chaparral made it clear that generalizations about seed response to elevated temperatures are difficult to make in light of the many other mechanisms which affect seed dormancy.

Grassland fire temperatures vary from 50° to 682°C at the soil surface, depending on fuel load (Smith and Sparling 1966, Stinson and Wright 1969, Britton and Wright 1971, Bailey and Anderson 1980, Hodgkinson and Oxley 1990). Temperatures decrease rapidly with increasing soil depth, and changes are generally negligible below 2 cm (Table 2). When dry, seeds of most herbaceous species can survive temperatures of 80–120°C with no significant reduction in germination (Sampson 1944, Keeley et al. 1985). When wet, moisture-absorbing seeds are damaged by temperatures exceeding 60°C. However, buried seeds are rarely affected by grassland fires. Steward et al. (1990) outlined a technique for predicting depth of lethal heat penetration in soils as a function of soil and fuel properties.

Seeds of some species have evolved mechanisms to facilitate seed burial. For example, hygroscopic styles or awns [e.g., mountain-mahogany (*Cercocarpus* HBK.

Table 2. Soil temperatures at various soil depths in wildland fires (data summarized from Nelson and Sims 1934, Isaac and Hopkins 1937, Heyward 1938, Beadle 1940, Bentley and Fenner 1958, Hare 1961, Whittaker 1961, Norton and McGarity 1965, Smith and Sparling 1966, Stinson and Wright 1969, Britton and Wright 1971, DeBano et al. 1977, Bailey and Anderson 1980, Auld 1986, Stott 1986, Hodgkinson and Oxley 1990).

	Temperature (°C)	
Depth (cm)	Grassland	Shrubland/Woodland
0	50-682	260-1005
2.5	ambient	90-195
5.0	ambient	50-122
7.5-10.0	ambient	57-95
>10.0	ambient	ambient

spp.), New Mexico feathergrass (*Stipa neomexicana* Thurb. Scribn.), broadleaf filaree (*Erodium botrys* Bertol.)] encourage movement and subsequent burial of seeds with alternate wet and dry periods. This mechanism is especially common among species of the Mediterranean region (Naveh 1975). Seed burial of many Australian species is facilitated by unappendaged diaspores tightly enclosed in imbricate bracts (Peart 1984). Fire-seed bank interactions have been reviewed recently for shrublands (Parker and Kelly 1989, Keeley 1991) and forests (Archibold 1989).

Seeds can also escape fires by being protected in an insulating layer of plant tissue. Cones, fleshy fruits, or mucilaginous seed coats allow seeds of some species to escape lethal temperatures. Ashton (1986) hypothesized that the packing of capsules in umbels on some Australian shrubs allows seeds in the cluster centers to avoid lethal temperatures. The moisture content of capsule walls may protect the seeds of these species.

In addition to having seeds which are resistant to (or stimulated by) high temperatures, some woody species possess serotinous cones, which are held closed by a resinous bond until heat melts the bond and releases the seeds. Lamont et al. (1991b) concluded that canopy seed storage in woody plants is most closely associated with fire-prone habitats. Some serotinous cones are consumed by high-intensity fires, but most open to release viable seeds (Wright and Bailey 1982, Lamont and Barker 1988). Serotinous species release a high density of seeds to the soil surface in a short period following fire (Gill 1975, Lotan et al. 1985, van Wilgen and Viviers 1985, Lamont et al. 1991a), which may be problematic where serotinous species have been introduced. On rangelands, serotiny is particularly common among South African species in the Protaceae family (van Wilgen and Viviers 1986) and Australian genera such as *Banksia*, *Casuarina*, *Calothamnus*, *Hakea*, and *Xylomelum* (Gill 1975 and references therein). Temperatures required to release seeds vary from 45°C in lodgepole pine (*Pinus contorta* Dougl.) (Lotan 1974) to 200°C in knobcone pine (*P. attenuata* Lemm.) (Vogl 1973). However, fire is not always necessary to open serotinous cones. In some species, serotiny can be broken by solar heating near the soil surface during the summer months (Bates et al. 1929, Lotan 1964, Lotan et al. 1985). Several serotinous species are known to have populations with non-serotinous cones.

Elevated temperature is not the only germination cue associated with fires. Smoke released by burning fynbos plant material triggers germination of *Audouinia capitata* (L.f.) Brongn. (de Lange and Boucher 1990). Ethylene and ammonia are released during wildland fires (Lewcock 1937, Russell et al. 1974), and both gases stimulate germination of some species (Esashi and Leopold 1969, Cairns and de Villiers 1986). Germination of some chaparral herbs is stimulated by heated xylan (a hemicellulose) and glucuronic acid (a common side chain of xylan and other hemicelluloses); xylan and glucuronic acid are released when wood is heated (Keeley and Pizzorno 1986). Charred wood stimulates germination of at least 45 chaparral species (Keeley 1991). Germination of *Erica hebecalyx* Beth., a South African rangeland species, appears to be stimulated by multiple cues such as elevated temperature, ethylene, and ammonia (van de Venter and Esterhuizen 1988).

Seed response in some species is strongly affected by the post-fire environment. Fire volatilizes compounds such as sulfur, phosphorus, and nitrogen (though often increasing plant-available forms of some nutrients). Diurnal temperature extremes increase, as does light striking the soil surface. Soil water relations are changed, and competition for resources usually declines (Chandler et al. 1983). Early successional species usually possess several characteristics which allow them to colonize following fire. They are typically opportunistic generalists with widely-dispersed seeds, rapid early growth rates, and a high ratio of photosynthesis to respiration. Species with these characteristics hold a competitive advantage over other species in the immediate post-fire environment. Increased germination of seeds which do not require heat treatments may be related to reduced competition, litter, or allelopathic compounds produced by established plants (Muller et al. 1968, Hanes 1971, Christensen and Muller 1975). In addition, seed predation by insects may be reduced in burned areas (Hartnett and Richardson 1989).

Juveniles

Few plant species are capable of tolerating or escaping fire as juveniles. Therefore, response to fire early in the life cycle depends primarily on the ability of individuals to recolonize. In general, early successional species are capable of establishing on burned sites more quickly than late successional species. Annuals are especially well adapted to fire; seeds which survive can rapidly exploit post-fire habitats.

Most woody species are incapable of surviving even low-intensity fires as juveniles. Few woody species resprout after top removal until several years after establishment. Honey mesquite (*Prosopis glandulosa* Torr.) is exceptional in this regard, becoming extremely fire tolerant at only 3 years of age (Wright et al. 1976). By contrast, most sprouting species require several years for development of roots, rhizomes, lignotubers, or root crowns. Fires which occur before these regenerative tissues develop kill a large proportion of young plants.

Sprouting woody species capable of surviving fires as juveniles generally exhibit one of two characteristics (or both): regenerative tissue develops quickly (e.g., honey mesquite), or the sprouting region is protected by soil at an early age (Fig. 2). Due to differences in soil movement, the latter phenomenon occurs more quickly on deep-soiled, moderately-sloping sites than on shallow-soiled, flat sites (Steuter and Britton 1983, McPherson and Wright 1989).

Figure 2. Redberry juniper (*Juniperus pinchotii* Sudw.) is one of five conifers which resprout as mature plants in North America. This picture shows the swollen stem base, from which stems arise following top-kill by fire. The sprouting region is covered with soil in 10–20 years, depending on soil depth and site productivity. Therefore, this species is well-adapted to maintaining itself in grasslands subjected to fire less than once every 10–20 years.

Longleaf pine (*Pinus palustris* Mill.) possesses unique juvenile adaptations to fire (Garren 1943, Chapman 1944, Fowells 1965). It remains in a grass-like morphological stage for 2 to 10 years after emergence. During this stage, the growing point is protected by a cluster of long needles. Periodic fires benefit the plant by removing dead leaves harboring brown-spot disease (*Septoria acicola*), which would otherwise kill the seedlings. After storing carbohydrates in the roots for several years in the grass stage, longleaf pine seedlings grow rapidly (up to a meter per year) for several years. This rapid growth elevates the terminal bud beyond the reach of surface fires.

Mature Plants

Vascular plant tissue is easily killed by heat, although lethal temperature is variable between species and depends strongly upon exposure time and moisture content (reviewed by Wright and Bailey 1982). Mature plants of some species possess morphological features which affect their probability of surviving and recovering from fires.

Bark is a good insulator for woody plants. Heat resistance increases with decreased density, increased moisture content, and increased thickness of bark (Spalt and Reifsnyder 1962). Resistance is further enhanced by fissured or corky bark (Davis et al. 1980). Insulative bark is most commonly associated with forest species such as ponderosa pine (*Pinus ponderosa* Dougl. ex Laws.) and western larch (*Larix occidentalis* Nutt.).

Resprouting is a relatively common attribute of woody plants following fire, and results from protection of buds by bark, dense leaf bases, or soil (Gill 1977). Protection of buds is probably more important than protection of cambial tissue for rangeland woody plants. Following fire-induced death of aboveground tissue, plants with protected buds can exploit the post-fire environment, which is characterized by higher soil temperature, increased availability of some nutrients, and decreased competition from neighboring plants (Rundel 1981). Sprouting vigor decreases with decreased moisture content of the soil and duff, and is generally less following fires during the growing season compared to dormant-season fires (Wright and Bailey 1982). Many woody angiosperms resprout following fire, but resprouting by mature conifers in North America is limited to 5 species: alligator juniper (*Juniperus deppeana* Steud.), redberry juniper (*J. pinchotii* Sudw.), pitch pine (*Pinus rigida* Mill.), bigcone Douglas-fir [*Pseudotsuga macrocarpa* (Vasey) Mayr], and coastal redwood [*Sequoia sempervirens* (D. Don) Endl.]. Several conifer species resprout as juveniles.

Growth form and location of growing points are important factors contributing to individual plant survival and resilience. Aboveground portions of herbaceous species rarely survive fires, although some species minimize the incidence of fire by growing in rocky areas which escape fires. In areas which do burn, growing points of rhizomatous plants usually escape lethal temperatures because they are located below the soil surface (Allen and Partridge 1988). On the other hand, stoloniferous species like black grama (*Bouteloua eriopoda* Torr.) and buffalograss [*Buchloe dactyloides* (Nutt.) Engelm.] are severely damaged by fire, especially if they have low vigor when burned or are grazed soon after the fire (Reynolds and Bohning 1956, Cable 1975). Bunchgrasses are susceptible to fire damage if they have growing points near the soil surface [e.g., threeawns (*Aristida* L. spp.)] or if they accumulate large amounts of dead plant material within the bunch [e.g., needle-and-thread (*Stipa comata* Trin. & Rupr.)]. The latter characteristic may result in high temperatures for a long time within the center of the plant after the main fire passes (Wright 1971). Thus, rhizomatous grasses and small bunchgrasses recover quickly after a fire compared to large, decadent bunchgrasses (Wright and Bailey 1982), although additional factors interact to influence response of bunchgrasses to fire (Britton et al. 1990). In temperate regions, burning Idaho fescue (*Festuca idahoensis* Elmer) with high soil moisture content is beneficial in the spring (immediately before growth begins), but detrimental in the fall. The latter treatment stimulates late-season growth shortly before the onset of freezing temperatures (Britton et al. 1983).

Considerable variability in rooting habit exists between herbaceous species, but differences in rooting habit generally do not affect direct response to fire. Soil is an excellent insulator which protects even shallow-rooted species. Increases in soil temperature are negligible below 2.5 cm during grassland and shrubland fires (DeBano et al. 1979), although soil heating may produce water repellency in forests and shrublands (DeBano 1981). The effect of fire on the post-fire environment probably has greater impact on plant roots than fire's direct effects. Burning typically increases primary productivity, including roots, by elevating soil temperatures (Raison 1979, Rundel 1981). However, increased soil temperatures, along with decreased moisture-holding capacity (resulting from combustion of organic matter), tends to make burned sites more xeric than unburned sites. Therefore, shallow-rooted species are sometimes susceptible to drought injury on burned sites (Wright and Bailey 1982).

Canopy architecture of woody plants may contribute to their ability to survive fires. Shrubs and trees with layered foliage (e.g., juniper) trap heat, increasing the likelihood of ignition or crown scorch. By contrast, open canopies (e.g., pines) allow heat to dissipate through the canopies without damaging foliage. Some pines are noted for long needles whorled around buds, an adaptation which causes needles to absorb heat, thereby protecting the buds.

Although it is rarely recognized as an important factor in ignition, chemical composition of plant tissue influences thermal behavior (Philpot 1969, Rogers et al. 1986 and references therein). Ether extract content may be particularly important for some species (reviewed by Bunting et al. 1983). Relatively large amounts of ether extracts, along with other traits, make some species predisposed to fire. Species which have become more flammable over evolutionary time may have done so because natural selection has favored the development of characteristics that make them more flammable (Mutch 1970; but see Snyder 1984, Troumbis and Trabaud 1989). These species usually respond quickly and positively to fire.

Flowering and Fruit Set

Flowers and fruits, like most other vascular tissue, are easily damaged by fire. Notable exceptions include serotinous cones and capsules which require heat to release seeds. The fact that fires occur in a particular region during a relatively short period of the year can permit flowers to escape fires. Thus, in addition to spatial refugia offered in tall trees or in needle clusters, temporal refugia exist in the form of flowering phenologies. Plants which complete the flowering process non-coincident with the fire season effectively avoid the possibility of fire-induced damage. Similarly, plants which complete their life cycle in a period shorter than the fire-return interval avoid interruption of the reproductive process by fire. Annual and biennials are obvious examples, but even long-lived plants may complete their life cycle between fire occurrences. Conversely, multiple fires in conifer stands may kill seed-producing trees (first fire), then kill newly-established trees before they reach reproductive age (second fire) (Koch 1942, Bailey and Poulton 1968).

Flowering response of rangeland plants to fire has been described with respect to flowering phenology (Curtis and Partch 1950, Gill and Ingwerson 1976, Abrahamson 1984, Lamont 1993) and number of flowering stems (Stone 1951, Kucera and Ehrenreich 1962, Old 1969, Cornett and Stewart 1986, Knapp and Hulbert 1986, Cheplick and Quinn 1988, Swamy and Ramakrishnan 1988). In grasslands, burned plots usually produced more flowering stems than unburned plots, but considerable differences in flowering response may result from fires conducted during different seasons (Gill 1981, Henderson et al. 1983, Lovell et al. 1983, Snyder 1986, Platt et al. 1988a). Furthermore, there is little experimental evidence to indicate that increased number of flowering stems translates to increased seed production or increased survival of germinants. In fact, Stur and Humphreys (1987) found that increased seed production per inflorescence partially compensated for decreased inflorescence density among burning treatments. The few studies which directly measured fruit or seed production reported highly variable responses between species (Burton 1944, Cornelius 1950, Miyanishi and Kellman 1986, Hartnett and Richardson 1989). Based on very limited data, seed production of perennial bunchgrasses in the northwestern United States appears to increase following fire, although the response may be delayed for 3 to 5 years (Patton et al. 1988).

Fire-stimulated flowering has been attributed to increased soil temperature, decreased competition, and release of ethylene, a component of wood smoke (Kimmins 1987 and references therein). Flowering of perennial species is obligately tied to burning in some habitats with high fire frequencies (Christensen 1985, Auld 1987, Zammit and Westoby 1987, Platt et al. 1988). This response has not been reported for other systems.

Plant Community Response

As fire moves across the landscape it re-casts the set of characters encountered through a combination of direct and indirect stresses. Fire consumes or damages photosynthetic material, regenerative organs, storage organs, and structural material from the fixed set of plant species in a community. The post-fire community responds with regrowth of surviving individuals, and establishment and growth of new individuals.

Individual plant species may have morphological and physiological adaptations to counter the stress of an individual fire event (small scale). However, at the managed landscape level, plant communities respond to the overall fire regime (Platt et al. 1988b, McFarland 1988, Gibson 1988). The result is a dynamic mosaic of vegetation patch sizes changed by the combined forces acting within the community (Wiens et al. 1985). When the fire regime is altered the community responds to the new set of stresses by adding and/or subtracting species and re-structuring relationships between species.

Plant communities produce fuels which determine their characteristic fire behavior (Sneeuwjagt and Frandsen 1977, Anderson 1982, Rothermel 1983), fire season (Bragg 1982, Steuter 1986), and fire frequency (Whitney 1986, Gibson 1988, Riggan et al. 1988, Clark 1990). All three factors influence the pattern of community diversity and biomass on rangelands (Armour et al. 1984, Steuter 1987, Biondini et al. 1989). Grazing stress also influences community diversity and biomass (see chapter by Briske and Richards in this book). In fact, the interaction between fire and large herbivores often defines plant community boundaries (Zimmerman and Neuenschwander 1984, Turner and Bratton 1987, Belsky 1989, Moe et al. 1990). Stress from this interaction results in a dynamic spatial and temporal pattern of seral communities in the landscape (Fig. 3). Relative to impacts on vegetation, fire frequency has a bioregional pattern of impact, while fire season has a landscape pattern, and fire intensity has a local or plant association impact.

Plant growth is reduced as fire-related stress intensifies. However, fire-adapted communities maintain their dominant species and achieve maximum productivity when fire occurs with characteristic frequency (Fig. 4). Species diversity also appears to be maximized under intermediate levels of fire stress (Fox and Fox 1986, Biondini et al. 1989). For example, in the absence of fire perennial grasses or shrubs

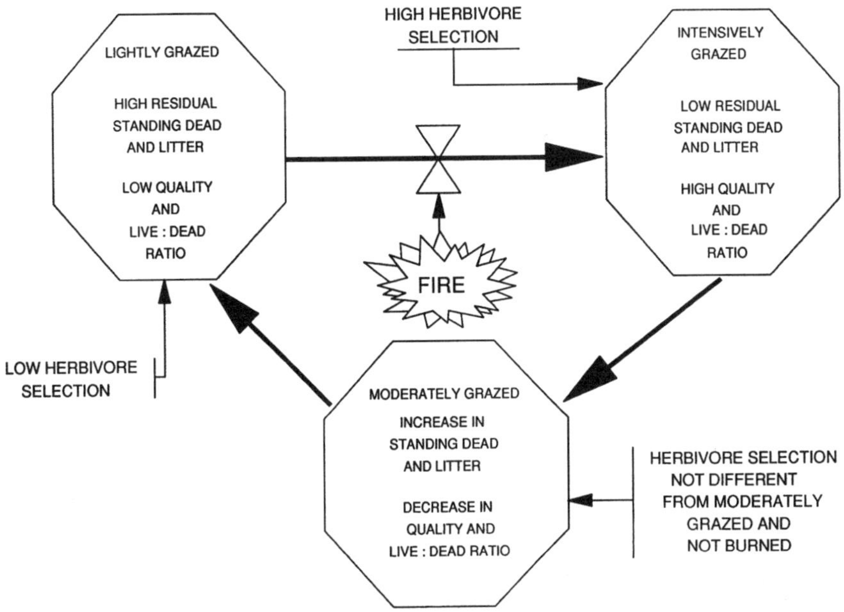

Figure 3. A conceptual model describing the interaction between fire and large herbivore grazing leading to vegetation patches driven by the fuel/forage production of the landscape. The temporal and spatial dynamics of this interaction are constrained by regional climate, weather, topoedaphic forces, and management actions.

may increasingly monopolize light and nutrients at the expense of forbs. This is consistent with the intermediate disturbance hypothesis of Connell (1978), and the concept that species diversity is related to the evolutionary disturbance regime for a community (Denslow 1980). In a fundamentally significant modification of this theory, Miller (1982) suggests that the effect of disturbance rate (fire frequency in our case) on community diversity is influenced by disturbance size (Fig. 5). This concept has not been integrated into rangeland fire management, as evidenced by the fact that most current guidelines for prescribed burning base fire size on economic or logistical criteria (e.g. Bunting et al. 1987, Rasmussen et al. 1988) rather than community spatial adaptations.

Grassland Ranges

Grasslands are best developed under the combined stresses of a semi-arid climate, large grazers, and frequent fire. At least in part, adaptations are due to a fire regime intensified by humans to promote grasslands at the expense of woodlands (Nisbet 1991). In these communities, seasonal production originates from growing points or seeds near the soil surface. Nutrients and energy move rapidly between above- and below-ground biomass. Grassland communities cycle carbon, nitrogen and phosphorus more rapidly in the presence of fire, especially when interacting with herbivores (Risser and Parton 1982, Shimel et al. 1985). Predictably, photosynthetic rate and aboveground nitrogen concentration increase during at least the early part of the post-fire growing season (Britton and Steuter

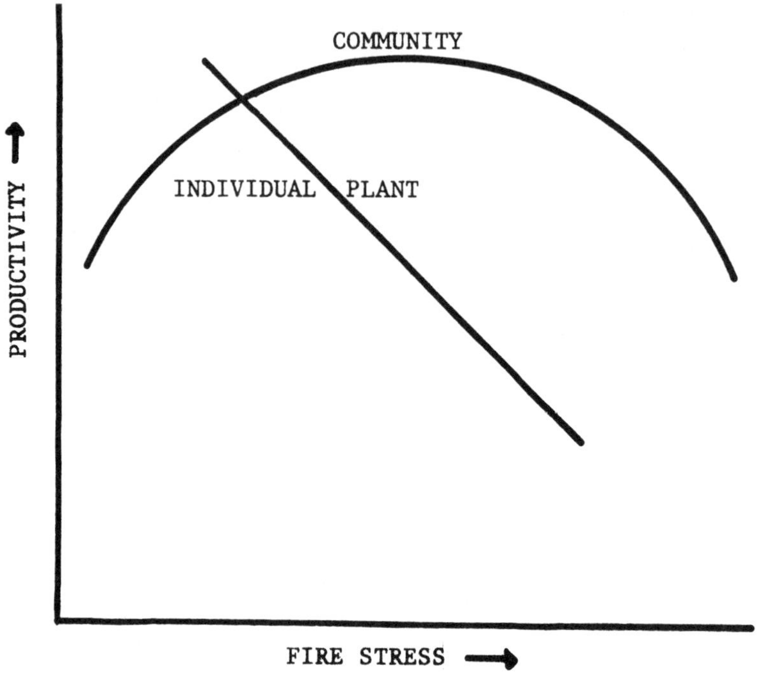

Figure 4. Generalized relationship between productivity and fire stress for individual plants and fire-adapted communities.

1983, Svejcar and Browning 1988). With adequate moisture this condition can be extended with grazing (Coppock and Detling 1986). Increased nitrogen mineralization can result from fire on some grassland ranges (Sharrow and Wright 1977, Stock and Lewis 1986). However nitrogen mineralization in North American tallgrass prairie is reduced by long-term annual burning (Ojima et al. 1990), and plant production remains nitrogen-limited (Seastedt 1988, Seastedt et al. 1991).

The phytomass of grasslands is well aerated and in the fine fuel class. These fuels are most conducive to fire spread when cured. Grassland fire regimes are characterized by declining fire frequencies with increasing aridity (lower fuel load) and an increase in caespitose growth forms (less fuel continuity). Grassland communities in large gently rolling landscapes developed under higher fire frequencies than similar fuels in small dissected landscapes (Wells 1970, Axelrod 1985).

Humid grasslands tend to have high fire frequencies and increased post-fire productivity (Kucera et al. 1967, Risser et al. 1981, Towne and Owensby 1984). Individual species may be reduced by fire-related stress (Curtis and Partch 1948, Gibson and Hulbert 1987, Ewing and Engle 1988), but fire in humid grasslands is appropriately viewed as a regenerative event. Fire improves the light, temperature, and nitrogen regime for the dominant species (Knapp 1984, Hulbert 1988). However, localized stress, phenology of associated species, fire intensity, and other disturbances maintains the heterogeneity required for community diversity (Gibson 1988). In the absence of fire and large herbivore grazing these communities accumulate large amounts of litter which suppresses the growth of characteristic

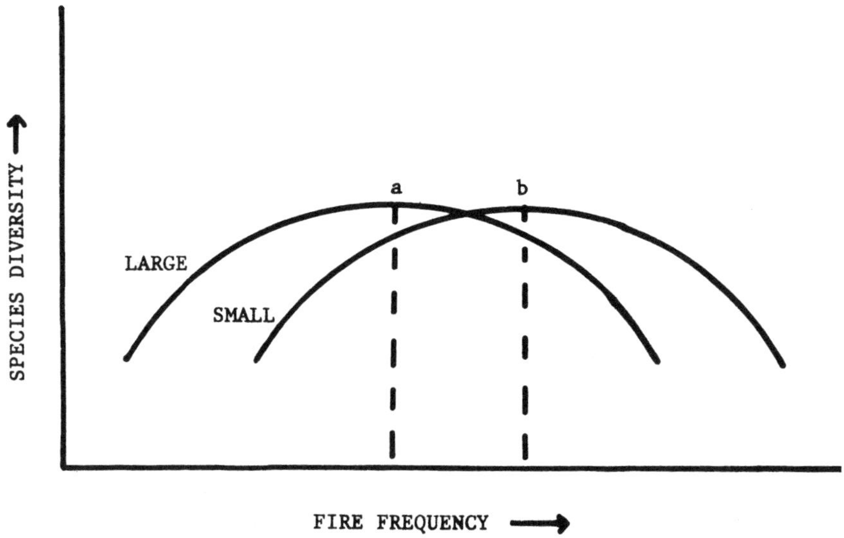

Figure 5. For fire-adapted communities, peak diversity should occur at intermediate fire frequencies (relative to the range possible for that community), but communities characterized by large fires might be expected to achieve maximum diversity at a lower fire frequency. Adapted from Miller (1982).

species—a community stress response. Long fire-free intervals lead to the relatively rapid development of woody vegetation (Bragg and Hulbert 1976) and ultimately a change in physiognomy.

Compared to humid grasslands, semi-arid grasslands have lower fire frequencies and a wider range of pre-fire conditions and post-fire response (Launchbaugh 1972, Wright and Bailey 1980). Fire intensifies plant water stress by increasing evapotranspiration (Redmann 1978, Savage and Vermeulen 1983). Mixed prairie communities dominated by tall or coarse species occur on mesic sites and maintain or increase production following fire even under moderate drought conditions (Britton and Steuter 1983, Engle and Bultsma 1984, Steuter 1987). These communities are primarily light-limited even though they occur in a semi-arid climate. Mixed prairie communities dominated by mid-height and short species occur on xeric sites and generally maintain production following fire in years with normal precipitation. However, they have reduced production following fire in years with 80% or less of normal precipitation (Steuter and Wright 1983, Engle and Bultsma 1984). These communities are limited by moisture, not light.

Apparently fire has been a consistent enough event in semi-arid grasslands for communities to develop structural adaptations to its periodic stress. For example, the forb component tends to be more diverse and have a higher number of individuals following fire than in long-term unburned communities (Bailey and Anderson 1978, Collins and Barber 1985). The magnitude of change in forb composition resulting from different fire seasons is similar to, though independent of, that resulting from year-to-year weather effects suggesting similar levels of stress (Fig. 6).

Fire season effects are most thoroughly described for semi-arid grasslands. The

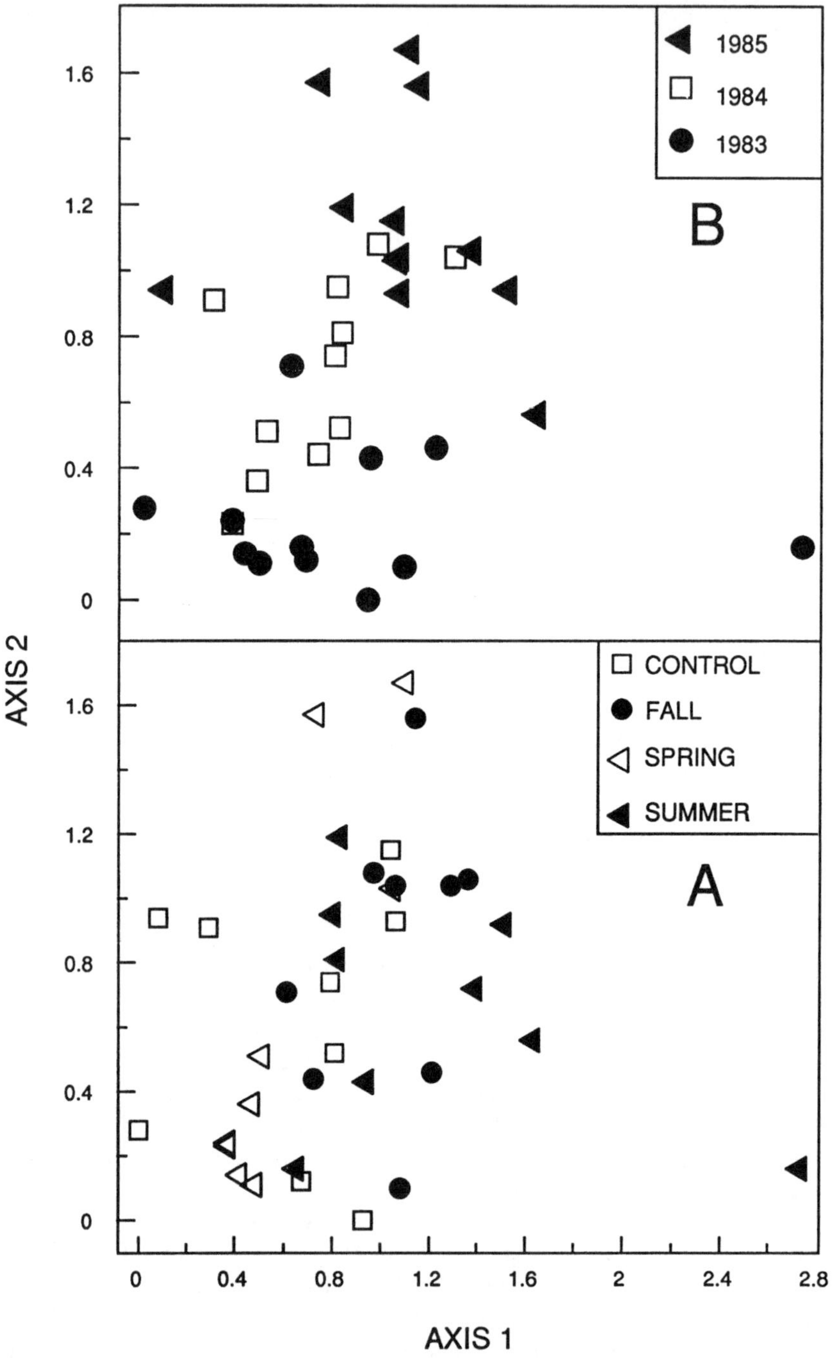

Figure 6. Ordination axis values from a detrended correspondence analysis of forb composition samples collected from unburned (control) and spring, summer, and fall burned plots during a wet (1983), normal (1984), and dry (1985) year in the northern mixed prairie, USA. In (A) plots are identified by fire season, and in (B) plots are identified by year (Biondini et al. 1989). The response range for year effect is similar to that for season of burn.

fuels of semi-arid grasslands may support high rates of fire spread when cured (Rothermel 1983), or be too discontinuous or actively growing to carry a fire (Andrews 1986), depending on plant growth form and phenology. As a result, native communities probably developed under a characteristic range of fire size and frequency for the different seasons (Steuter 1986). Thus fire stress may be very severe when a burn occurs during peak growth of the C_3 or C_4 guild in a mixed stand (e.g. Engle and Bultsma 1984, Steuter 1987), but these fires are relatively small due to high fuel moisture resulting in low rates of spread (Table 3). Alternatively, fire stress may be intermediate in intensity but large in area, as when dormant woody species are top-killed in a dormant grassland with low fuel moisture and high rates of spread (e.g. Neuenschwander et al. 1978). Fire early in the dormant season affects composition indirectly, and at a relatively large scale, by increasing evapotranspiration in the next growing season (Owensby and Anderson 1967, Redmann 1978, White and Currie 1983) and increasing the mosaic diversity of the forb community (Fig. 7). The season in which a fire occurs has the potential to generate a dynamic spatial and temporal mosaic within the landscape.

Arid grassland communities have developed under the lowest fire frequencies of the grassland types (Wright and Bailey 1980). Plant production may be maintained (Heirman and Wright 1973), but is usually reduced following fire (Dwyer and Pieper 1967, Launchbaugh 1972). Fire reduces the vigor and abundance of the dominant species. Subsequent long-term changes in community structure (Dix 1960, Cable 1967) are indicative of acute environmental stress. Yet, the infrequent fires which do occur are important in preventing or reducing woody plant encroachment (Cable 1965, 1972).

Bunchgrass communities appear to have a distinctly different evolutionary history than rhizomatous grasslands (Mack and Thompson 1982, Milchunas et al. 1988). They are characterized by fine fuel concentrations near the growing points of the dominant perennial species. Long fire-free intervals increase fuel accumulation and the incidence of mortality in the dominant bunchgrass species (Wright and Klemmedson 1965, Antos et al. 1983). Short fire-return intervals are detrimental to seedling establishment (Anderson and Bailey 1980), which is critically important in

Table 3. A comparison of minimum, maximum, and mean values recorded for rate of fire spread and fire intensity on initial seasonal burns and subsequent re-burns of the same plots in a northern mixed prairie, USA (Adapted from Steuter 1986). Summer fires had low rates of fire spread and intensity, and reduced post-fire fuel accumulation suggesting relatively small, highly stressed patches within the landscape.

Fire Season	Rate of Spread (m/min) min.	max.	$\bar{X}$	Fire Intensity (kW/m) min.	max.	$\bar{X}$
1st Summer (n =6)	1	6	3	201	868	540
1st Fall (n = 6)	24	86	53	1983	13716	6721
1st Spring (n = 3)	20	66	49	3185	10751	7440
2nd Summer (n = 6)	0	0	0	0	0	0
2nd Fall (n = 6)	8	33	18	565	3068	1459
2nd Spring (n = 6)	16	71	36	975	3980	2137
3rd Summer (n = 3)	2	3	2	233	404	338

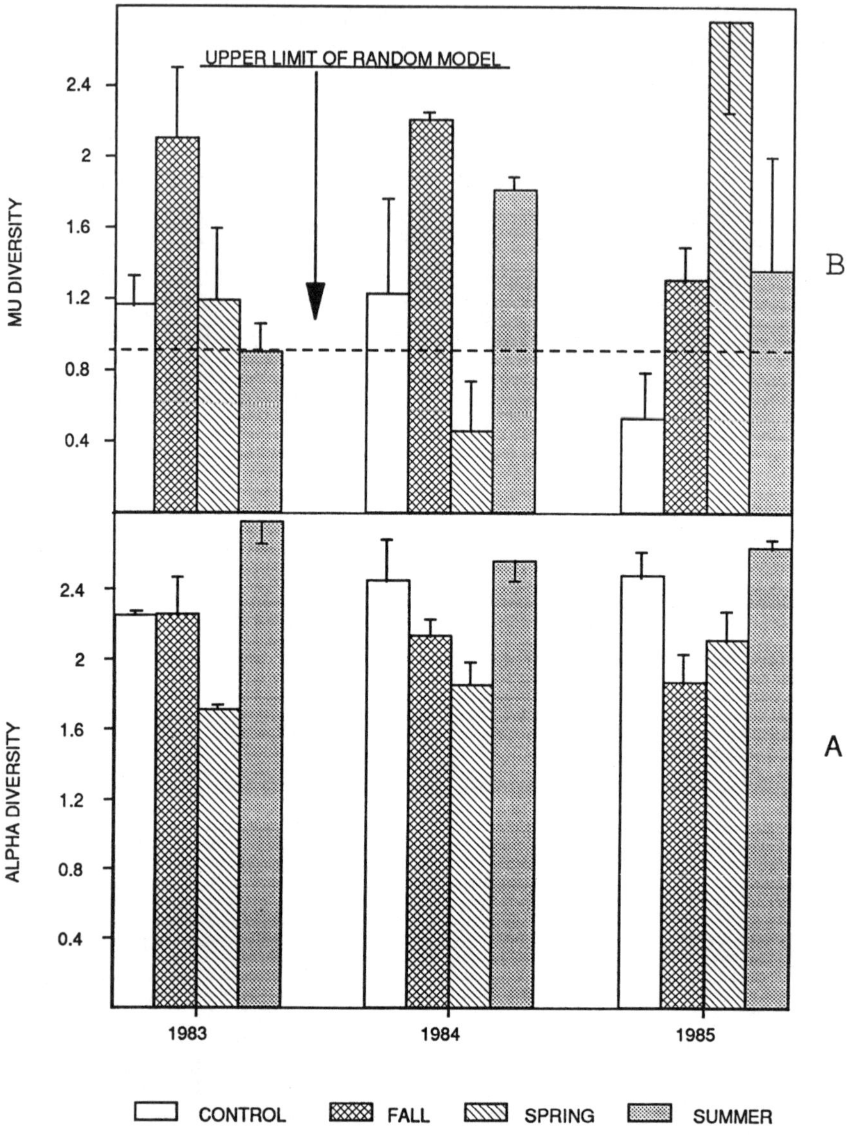

Figure 7. (A) Alpha diversity (Shannon-Wiener index), and (B) Mosaic diversity (MU) of the forb community on control, and spring, summer, and fall burned plots in a wet (1993), normal (1984), and dry (1985) year in the northern mixed prairie, USA (adapted from Biondini et al. 1989). Alpha diversity is highest on summer burn and unburned plots suggesting individual species adaptation to fire stress, while mosaic diversity was higher on fall or spring burns suggesting community adaptation to fire stress at the landscape scale.

determining community composition. Since small bunchgrass species and perennial forbs are less damaged by fire, they, along with species establishing from seed, are more important in the post-fire community (see review by Wright et al. 1979). These features of bunchgrass ranges have made them particularly susceptible to deterioration by exotic plant and animal species (Mack 1981) and a few resprouting shrubs (Wright et al. 1979). Significant shifts in post-fire structure may occur in arid bunchgrass communities. This suggests that fire is a relatively intense stress and that fires were infrequent or occurred under a specialized set of conditions.

Shrubland Ranges

Shrublands may occur in arid climates, in association with grasslands, and as sclerophyllous brush communities (e.g., chaparral). Fire regimes in arid shrublands and those associated with a grass matrix are controlled by fine fuels (Heirman and Wright 1973, Wright et al. 1979). Only in extremely hot, dry weather and heavy volatile fuel loads can fire spread be sustained by the woody canopies themselves (Bruner and Klebenow 1979). By contrast, the fire regime in chaparral is controlled by species composition and biomass accumulation within the shrub layer (Riggan et al. 1988). Chaparral is a volatile fuel type and fire intensities are usually high. In decadent stands, fire spread is limited only by local topography and short-term weather conditions.

Livestock management practices and the introduction of exotic plants have altered the fire regimes of many arid and semi-arid shrubland ranges. Consequently, these may be viewed as relatively new communities with a dynamic response to changes in the fire regime (Fig. 8). High shrub mortality, reduced species diversity, and long-term shifts in dominant species may result from fire (Wright et al. 1979, West and Hassan 1985), especially on desert ranges during periods of above-normal fine fuel loads (Brown and Minnich 1986). Fire size and frequency strongly influence community structure because replacement of the dominant pre-fire shrubs is controlled by the local presence or absence of a few prolific resprouters and the composition of the seed bank (Humphrey 1984, Hassan and West 1986).

Shrub species associated with the more productive grasslands generally resprout. These resprouting species are well adapted to maintaining their presence in communities exposed to fire frequencies less than 1 in 5 years (Neuenschwander et al. 1978, Steuter and Wright 1983, McFarland 1988, Bradstock and Myerscough 1988). In the absence of fire, these communities may develop dense overstories that significantly reduce herbaceous production (Ethridge et al. 1984). The lack of fine fuel reduces fire intensity and frequency, and the community itself changes (Bragg and Hulbert 1976, Archer 1990, Steuter et al. 1990). Once this occurs, fire alone cannot return the site to its earlier composition (Wright et al. 1976, Scifres 1980). Thus species diversity and herbage production are reduced in the absence of periodic fire stress.

Unlike shrub-dominated former grasslands, chaparral communities become increasingly flammable as the fire-free interval increases and herbaceous fuels decline (Riggan et al. 1988). A variety of shrub species have developed both sprouting and seeding strategies to maintain a presence in the post-fire community (Keeley and Zedler 1978, DeSouza et al. 1986, Keeley 1987). Fire increases species richness by periodically removing the light-dominating shrub canopies (Davis et al. 1988). Phytomass accumulates at an accelerating rate for up to 20 years after a

MOISTURE

EVOLUTIONARY FIRE FREQUENCY	ARID	TO	SUBHUMID
INCREASING	- only possible with increased fine fuels (introduced) - shrubs poorly adapted and so reduced - many native herbs also reduced - exotic species invasion may be enhanced - reduced species diversity	- shrub composition shift to resprouters - initially lower browse availability - some native herbs are reduced (bunchgrasses) - exotic species invasion may be enhanced - conversion to grassland possible on some sites - reduced species diversity	- loss of non-sprouting shrubs - increased browse availability - increased herb layer and fine fuels - increased stem density of resprouting shrubs - species diversity relatively stable
	Rare	30-70 years	20-100 years
DECREASING	- shrubs may increase in combination with change in grazing - species diversity relatively stable	- shrub biomass increase - herb layer decrease - fire size increase in volatile shrublands - non-volatile shrublands become "fire proof" - reduced species diversity	- shrub biomass increase - herb layer decrease - fire size increase in volatile shrublands - non-volatile shrublands become "fire proof" - reduced species diversity

Figure 8. Shrubland structure is influenced by increased or decreased fire frequencies relative to the evolutionary fire frequency and plant available moisture.

fire. Topoedaphic forces maintain distinct plant subassociations in the presence of fires which readily cross the topoedaphic boundaries (Malanson 1984). Although dominated by a different growth form and an order of magnitude difference in fire frequency (20 to 50 vs. 2 to 5 years between fires), fire seems as integral a part of chaparral communities as tallgrass communities.

Woodland Ranges

Woodland communities occur in association with semi-arid to humid grasslands, and as successional stages in cut or burned forest communities. Seral forest communities will not be addressed, although it should be recognized that in evolutionary time the stress from an intensified fire regime could result in a woodland community developing on some previously forested sites. In general, woodlands are characterized by a fire-tolerant tree overstory and a grassland understory (Bray 1958). Dominant tree species may be deciduous or coniferous, and the herb layer may be rhizomatous or caespitose. A shrub layer is variably developed depending on site conditions and fire frequency. The structure of most woodland ranges is closely tied to the prevailing fire regime (Fig. 9).

Often referred to as savannas, woodland communities also have a long evolutionary history of human and lightning ignitions (Gillion 1983). Recent evidence suggests that a distinct herb layer became adapted to the woodlands, including species generally not considered part of stable forests or prairies (Packard 1988). Local species distributions are primarily controlled by differences in light intensity resulting from tree canopies and soil disturbance (Bray 1958, Silvertown and Tremlett 1989), but interactions between available moisture, nutrients and fire determine regional plant distributions (Medina and Silva 1990). Platt et al. (1988b) describes the larger trees in long-leaf pine communities as the conduit which transforms an intense small-scale disturbance (lightning) into the less intense, but large-scale, ground fires necessary to maintain the community. This structural adaptation appears to maintain both humid southeastern and semi-arid western U.S. pine woodlands as open, uneven-aged stands with even-aged tree clusters (Wright 1978).

Humid woodlands are maintained as open communities by 1- to 5-year fire frequencies interacting with browsing and soil texture. Trees of semi-arid woodlands are generally poorly adapted to fire as juveniles (Wink and Wright 1973, Steuter and Britton 1983), and fire frequencies of 10 to 30 years allow tree recruitment without canopy closure. Fire exclusion results in humid woodlands becoming less flammable due to an increased deciduous cover and subsequent changes in fine fuel structure and moisture (Streng and Harcombe 1982). Woodlands characterized by conifers or other volatile fuels become increasingly susceptible to high-intensity crown fires if fire exclusion results in dense tree stocking and fine fuel accumulation (Wright 1978).

Woodlands are diverse communities due to the distribution of species between herb, shrub, and tree growth forms in a relatively high light environment (Fig. 9). Production in the herb layer is inversely related to tree stocking (e.g. Uresk and Severson 1989, Mitchell and Bartling 1991) and therefore directly related to fire frequency. Although the shrub response is site-specific, one would expect peak

EVOLUTIONARY FIRE FREQUENCY	MOISTURE: SEMI-ARID	TO	HUMID
INCREASING	- reduced tree cover - shift to resprouting shrubs - some native herbs reduced - exotic species invasion may be enhanced - conversion to grassland possible on some sites - loss of soil and nutrients - reduced species diversity	- reduced tree cover - increased stem density of resprouting shrubs - increased herb layer and fine fuels - conversion to grassland possible on some sites - loss of soil and nutrients possible on some sites - reduced species diversity	- tree cover relatively stable - shrub biomass reduced - herb layer relatively stable - species diversity relatively stable
	10–30 years	6–22 years	2–10 years
DECREASING	- increased tree cover - increase in non-sprouting and sprouting shrubs - reduced herb layer - loss of soil possible on some sites (loss of herb layer) - fire size and intensity increases in volatile and is less in non-volatile - species diversity reduced	- tree canopy closure possible - shrub layer increases but eventually is reduced - loss of soil possible on some sites (loss of herb layer) - fire size and intensity increases in volatile and becomes less in non-volatile - species diversity reduced	- change in community composition (forest) - understory woody plants become part of canopy - change in herb layer composition to shade tolerant types - very small fire size - species diversity may increase (different)

Figure 9. Woodland structure is influenced by increased or decreased fire frequencies relative to the evolutionary fire frequency and plant available moisture.

production in this layer under intermediate fire frequencies for a given community. Increasing the fire frequency of a woodland may result in a shift toward the shrubby growth form if resprouting species are present (Grimm 1984, Fox and Fox 1986, Rasmussen and Wright 1989).

Herbaceous nutrient concentrations can increase dramatically following fire in woodland communities, with the greatest increases in the oldest stands (Harris and Covington 1983). Increased light, reduced allelopathy, increased decomposition and mineralization, and altered species composition all contribute to nutrient increases in the herb layer following woodland fires. It appears that the herb layer in woodland communities acts as a sink for nutrients released from the woody overstory and thus reduces the loss of nutrients due to leaching (Kellman et al. 1985). Increased nutrient levels in herbage, and potential for losses due to run-off and leaching without a ground cover, suggests that heavy grazing following fire can lead to long-term declines in woodland productivity (Oswald and Covington 1983, Kellman et al. 1985).

Summary

The Known

The response of vegetation to fire has been intensively studied over the last 30 years. Research has focussed on the individual plant and plant community scales of a complex and interactive system. We know that individual plants have structural and functional characteristics at various life-history stages which make them more or less susceptible to fire stress. Seeds and other regenerative organs may be insulated from the heat of a fire by the soil, or specialized plant tissues. Although juveniles tend to be most susceptible to fire, their location or growth rates may afford some level of resistance. Dominant plant species in fire-prone habitats tend to have adaptations which allow adults to survive fire stress or have increased regeneration potential following fire. Some plant traits which confer drought or grazing resistance also reduce fire stress.

The array of individual plant responses to the fire environment in part defines the plant community. Changes in production, diversity, and structure have been interpreted as a measurement of fire stress at this scale. This assumes that fire has been a consistent enough force to have resulted in adaptation by the rest of the system, and that significant deviations from the fire regime produce significant changes in the community. Fire season and fire intensity are probably the two most significant components of the fire regime when managing at the scale of the plant community or lower.

The Unknown

Plant community and individual plant response to fire occurs within a broader spatial and temporal context than the individual fire event. Indeed, the fire regime, whether managed or unmanaged, is driven by forces which operate at the landscape level and above. The biological resources (plant and animal species) on rangelands consist of populations which also operate at the landscape level and above. These spatial effects of fire and other ecosystem processes are poorly understood. It may be even more difficult to deal with the temporal effects of fire management, since current fire management determines, in part, the future configuration of the resource. Both fire and grazers have been managed by humans over evolutionary time scales and we should expect to see evidence of adaptations in the ecosystem. However, the implications of community re-organization resulting from our technological abilities in landscape fragmentation, species introductions, and intensive management are emergent properties of the new, and still changing, rangeland resource. For example, recently-introduced species may significantly alter fire regimes and, subsequently, community structure and function. We should not assume that adaptations to fire in an extensively exploited landscape will mirror those of intensively managed rangelands.

Future Research Needs

It would be difficult, if not impossible, to enumerate all the information needed for appropriate management of specific sites and species. However, some important questions which address general principles include:

1) How do recently-introduced species influence fire regimes?
2) At what level in the ecosystem are adaptations fire-evolved (vs. other stresses), and is this level the same for different rangeland types?
3) How does intensive local management of the fire regime influence the viability and response of regional populations?
4) How do the increased rates of nutrient cycling which may result from a fire-grazing interaction impact resource values in other parts of the landscape?
5) What is an appropriate temporal scale for fire management planning?
6) What changes in fire regime can be expected as a result of changes in climate?

Acknowledgements

We would like to acknowledge the support of our respective institutions in this effort: The Nature Conservancy, Niobrara Valley Preserve, Johnstown, Nebraska 69214; and the University of Arizona, School of Renewable Natural Resources, Tucson, Arizona 85721. Critical reviews provided by M.E. Biondini, T.B. Bragg, C.M. Britton, K. Haworth, V.T. Parker, E.M. Steinauer, and R. Wallace were responsible for significant improvements in the chapter. Omissions and errors remain solely our responsibility.

Literature Cited

Abrahamson, W.G. 1984. Species responses to fire on the Florida Lake Wales Ridge. Amer. J. Bot. 71:35-43.

Allen, R.B. and T.R. Partridge. 1988. Effects of spring and autumn fires on the composition of *Chionochloa rigida* tussock grassland, New Zealand. Vegetatio 76:37-44.

Anderson, H.E. 1982. Aids to determing fuel models for estimating fire behavior. USDA For. Serv. Intermt. Res. Stn. Gen. Tech. Rep. INT-122, Ogden, UT.

Anderson, H.G., and A.W. Bailey. 1980. Effects of annual burning on grassland in the aspen parkland of east-central Alberta. Can. J. Bot. 58:985-996.

Andrews, P.L. 1986. BEHAVE: fire behavior prediction and fuel modeling system/burn subsystem, Part I. USDA For. Serv. Intermt. Res. Stn. Gen. Tech. Rep. INT-194, Ogden, UT.

Antos, J.A., B. McCune, and C. Bara. 1983. The effect of fire on an ungrazed western Montana grassland. Amer. Midl. Natur. 110:354-364.

Archibold, O.W. 1989. Seed banks and vegetation processes in coniferous forests. Pp. 107-122 *in* M.A. Leck, V.T. Parker, and R.L. Simpson (eds.) Ecology of Soil Seed Banks. Academic Press, San Diego, CA.

Archer, S. C., C.J. Scifres, C.R. Bassham, and R. Maggio. 1988. Autogenic succession in a subtropical savanna: conversion of grassland to thorn woodland. Ecol. Monogr. 58:111-127.

Armour, C.D., S.C. Bunting, and L.F. Neuenschwander. 1984. Fire intensity effects on the understory in Ponderosa pine forests. J. Range Manage. 37:44-49.

Ashton, D.H. 1986. Viability of seeds of *Eucalyptus obliqua* and *Leptospermum juniperinum* from capsules subjected to a crown fire. Aust. For. 49:28-35.

Auld, T.D. 1986. Population dynamics of the shrub *Acacia suaveolens* (Sm.) Willd.: Fire and the transition to seedlings. Aust. J. Ecol. 11:373-385.

Auld, T.D. 1987. Post-fire demography in the resprouting shrub *Angophora hispida* (Sm.) Blaxell: flowering, seed production, dispersal, seedling establishment and survival. Proc. Linn. Soc. N.S.W. 109:259-269.

Axelrod, D.I. 1985. Rise of the grassland biome, central North America. Bot. Rev. 51:163-201.

Bailey, A.W., and M.L. Anderson. 1978. Prescribed burning of a Festuca-Stipa grassland. J. Range Manage. 31:446-449.

Bailey, A.W. and M.L. Anderson. 1980. Fire temperatures in grass, shrub, and aspen forest communities of Central Alberta. J. Range Manage. 3: 37-40.

Bailey, A.W. and C.E. Poulton. 1968. Plant communities and environmental interrelationships in a portion of the Tillamook Burn, northwestern Oregon. Ecology 49:1-13.

Bates, C.G., H.C. Hilton, and T. Krueger. 1929. Experiments in the silvicultural control of natural reproduction of lodgepole pine in the central Rocky Mountains. J. Agric. Res. 38:229-243.

Beadle, N.C.W. 1940. Soil temperatures during forest fires and their effect on the survival of vegetation. J. Ecol. 28:180-192.

Belsky, A.J. 1989. Landscape patterns in a semi-arid ecosystem in East Africa. J. Arid Environ. 17:265-270.

Bentley, J.R. and R.L. Fenner. 1958. Soil temperatures during burning related to postfire seedbeds on woodland range. J. For. 56:737-774.

Biondini, M.E., A.A. Steuter, and C.E. Grygiel. 1989. Seasonal fire effects on the diversity patterns, spatial distribution and community structure of forbs in the northern mixed prairie, USA. Vegetatio 85:21-31.

Bradstock, R.A., and P.J. Myerscough. 1988. The survival and population response to frequent fires of two woody resprouters *Banksia serrata* and *Isopogon anemonifolius*. Aust. J. Bot. 36:415-431.

Bragg, T.B. 1982. Seasonal variation in fuel and fuel consumption by fires in a bluestem prairie. Ecology 63:7-11.

Bragg, T.B., and L.C. Hulbert. 1976. Woody plant invasion of unburned Kansas bluestem prairie. J. Range Manage. 29:19-24.

Bray, R.J. 1958. The distribution of savanna species in relation to light intensity. Can. J. Bot. 36:671-681.

Britton, C.M., and A.A. Steuter. 1983. Production and nutritional attributes of tobosagrass following burning. Southwest. Natur. 28:347-352.

Britton, C.M. and H.A. Wright. 1971. Correlation of weather variables to mesquite damage by fire. J. Range Manage. 24:136-141.

Britton, C.M., R.G. Clark, and F.A. Sneva. 1983. Effects of soil moisture on burned and clipped Idaho fescue. J. Range Manage. 36:708-710.

Britton, C.M., G.R. McPherson, and F.A. Sneva. 1990. Effects of burning and clipping on five bunchgrasses in eastern Oregon. Great Basin Nat. 50:115-120.

Brown, D.E., and R.A. Minnich. 1986. Fire and changes in creosote bush scrub of the western Sonoran desert, California. Amer. Midl. Natur. 116:411-422.

Bruner, A.D., and D.A. Klebenow. 1979. Predicting success of prescribed fires in pinyon-juniper woodland in Nevada. USDA For. Serv. Intermt. Res. Stn. Res. Pap. INT-219, Ogden, UT.

Bryant, F.C., G.K. Launchbaugh, and B.H. Koerth. 1983. Controlling mature ashe juniper in Texas with crown fires. J. Range Manage. 36:165-168.

Bunting, S.C., B.M. Kilgore, and C.L. Bushey. 1987. Guidelines for prescribed burning sagebrush-grass rangelands in the northern Great Basin. USDA For. Serv. Intermt. Res. Stn. Gen. Tech. Rep. INT-231, Ogden, UT.

Bunting, S.C., H.A. Wright, and W.H. Wallace. 1983. Seasonal variation in the ignition time of redberry juniper in west Texas. J. Range Manage. 36:169-171.

Burton, G.W. 1944. Seed production of several southern grasses as influenced by burning and fertilization. Amer. Soc. Agron. J. 36:523-529.

Cable, D.R. 1965. Damage to mesquite, Lehmann lovegrass, and black grama by a hot June fire. J. Range Manage. 18:326-329.

Cable, D.R. 1967. Fire effects on semi-desert grasses and shrubs. J. Range Manage. 20:170-176.

Cable, D.R. 1972. Fire effects on southwestern semidesert grass-shrub communities. Proc. Tall Timbers Fire Ecol. Conf. 12:109-127.

Cable, D.R. 1975. Influence of precipitation on perennial grass production in the semidesert southwest. Ecology 56:981-986.

Cairns, A.L.P. and O.T. de Villiers. 1986. Breaking dormancy of *Avena fatua* L. seed by treatment with ammonia. Weed Res. 26:191-197.

Chandler, C., P. Cheney, P. Thomas, L. Trabaud, and D. Williams. 1983. Fire in Forestry. John Wiley & Sons, New York.

Chapman, H.H. 1944. Fire and pine. Amer. For. 50:60-62, 91-93.

Cheplick, G.P. and T.A. Quinn. 1988. Subterranean seed production and population responses to fire in *Amphicarpum purshii* (Gramineae). J. Ecol. 76:263-273.

Christensen, N.L. 1985. Shrubland fire regimes and their evolutionary consequences. Pp. 86-100 *in* S.T.A. Pickett and P.S. White (eds.) The Ecology of Natural Disturbance and Patch Dynamics. Academic Press, Orlando, FL.

Christensen, N.L. and C.H. Muller. 1975. Effects of fire on factors controlling plant growth in *Adenostoma* chaparral. Ecol. Monogr. 45:29-55.

Clark, J.S. 1990. Landscape interactions among nitrogen, species composition, and long-term fire frequency. Biogeochemistry 11:1-22.

Clary, W.P. 1987. Herbage production and livestock grazing on pinyon-juniper woodlands. Pp. 440-447 *in* R.L. Everett (compiler) Proceedings—Pinyon-Juniper Conference. USDA For. Serv. Gen. Intermt. Res. Stn. Gen. Tech. Rep. INT-215, Ogden, UT.

Collins, S.L. 1987. Interaction of disturbances in tallgrass prairie: a field experiment. Ecology 68:1243-1250.

Collins, S.L., and S.C. Barber. 1985. Effects of disturbance on diversity in mixed-grass prairie. Vegetatio 64:87-94.

Connell, J.H. 1978. Diversity in tropical rain forests and coral reefs. Science 199:1302-1310.

Coppock, D.W., and J.K. Detling. 1986. Alteration of bison and blacktailed prairie dog grazing interaction by prescribed burning. J. Wildl. Manage. 50:452-455.

Cornelius, D.R. 1950. Seed production of native grasses under cultivation in eastern Kansas. Ecol. Monogr. 20:1-29.

Cornett, J.W. and J.M. Stewart. 1986. Palm burning and increased spadix production. Southwest. Natur. 31:552-553.

Curtis, J.T., and M.L. Partch. 1948. Effects of fire on the competition between bluegrass and certain prairie plants. Amer. Midl. Natur. 39:437-443.

Curtis, J.T. and M.L. Partch. 1950. Some factors affecting flower production in *Andropogon gerardi*. Ecology 31:488-489.

Cushwa, C.T., R.E. Martin, and R.L. Miller. 1968. The effects of fire on seed germination. J. Range Manage. 21:250-254.

Daubenmire, R. 1968. Ecology of fire in grasslands. Adv. Ecol. Res. 5:209-266.

Davis, F.W., D.E. Hickson, and D.C. Odion. 1988. Composition of maritime chaparral related to fire history and soil, Burton mesa, Santa Barbara County, California. Madrono 35:169-195.

Davis, K.M., B.D. Clayton, and W.C. Fischer. 1980. Fire ecology of Lolo National Forest habitat types. USDA For. Serv. Intermt. Exp. Stn. Gen. Tech. Rep. INT-79, Ogden, UT.

de Lange, J.H. and C. Boucher. 1990. Autecological studies on *Audouinia capitata* (Bruniaceae). I. Plant-derived smoke as a seed germination cue. S. Afr. J. Bot. 56:700-703.

DeBano, L.F. 1981. Water repellent soils: a state-of-the-art. USDA For. Serv. Pac. Southwest Exp. Stn. Gen. Tech. Rep. PSW-46, Berkeley, CA.

DeBano, L.F., P.H. Dunn, and C.E. Conrad. 1977. Fire's effect on physical and chemical properties of chaparral soils. Pp. 440-447 *in* Symposium on Environmental Consequences of Fire and Fuel Management in Mediterranean Ecosystems. USDA For. Serv. Gen. Tech. Rep. WO-3, Washington, D.C.

DeBano, L.F., R.M. Rice, and C.E. Conrad. 1979. Soil heating in chaparral fires: effects on soil properties, plant nutrients, erosion, and runoff. USDA For. Serv. Pac. Southwest Exp. Stn. Res. Pap. PSW-145, Berkeley, CA.

Denslow, J.S. 1980. Patterns of plant species diversity during succession under different disturbance regimes. Oecologia 46:18-21.

DeSouza, J., P.A. Silka, and S.D. Davis. 1986. Comparative physiology of burned and unburned *Rhus laurina* after chaparral wildfire. Oecologia 71:63-68.

Dix, R.L. 1960. The effects of burning on the mulch structure and species composition of grasslands in western North Dakota. Ecology 41:49-56.

Dwyer, D.D., and R.D. Pieper. 1967. Fire effects on blue grama-pinyon-juniper rangeland in New Mexico. J. Range Manage. 20:359-362.

Engle, D.M., and P.M. Bultsma. 1984. Burning of northern mixed prairie during drought. J. Range Manage. 37:398-401.

Esashi, Y. and A.C. Leopold. 1969. Dormancy regulation in subterranean clover seeds by ethylene. Plant Physiol. 44:1470-1472.

Ethridge, D.E., B.E. Dahl, and R.E. Sosebee. 1984. Economic evaluation of chemical mesquite control using 2,4,5-T. J. Range Manage. 37:152-156

Ewing, A.L., and D.M. Engle. 1988. Effects of late summer fire on tallgrass prairie microclimate and community composition. Amer. Midl. Natur. 120:212-223.

Fowells, H.A. 1965. Silvics of forest trees of the United States. USDA Hankb. No. 271, Washington, D.C.

Fox, M.D., and B.J. Fox. 1986. The effect of fire frequency on the structure and floristic composition of a woodland understorey. Aust. J. Ecol. 11:77-85.

Garren, K.H. 1943. Effects of fire on vegetation of the southeastern United States. Bot. Rev. 9:617-654.

Gibson, D.J. 1988. Regeneration and fluctuation of tallgrass prairie vegetation in response to burning frequency. Bull. Torr. Bot. Club 115:1-12.

Gibson, D.J., and L.C. Hulbert. 1987. Effects of fire, topography and year-to-year climatic variation on species composition in tallgrass prairie. Vegetatio 72:175-185.

Gill, A.M. 1975. Fire and the Australian flora: a review. Aust. For. 38:1-25.

Gill, A.M. 1977. Plants' traits adaptive to fires in the Mediterranean land ecosystems. Pp. 17-26 *in* H.A. Mooney and C.E. Conrad (tech. coord.) Symp. Environmental Consequences of Fire and Fuel management in Mediterranean Ecosystems. USDA For. Serv. Gen. Tech. Rep. WO-3, Washington, D.C.

Gill, A.M. 1981. Fire adaptive traits of vascular plants. Pp. 208-230 *in* H.A. Mooney, T.M. Bonnicksen, N.L. Christensen, J.E. Lotan and W.A. Reiners (eds.) Fire Regimes and Ecosytem Properties. USDA For. Serv. Gen. Tech. Rep. WO-26, Washington, D.C.

Gill, A.M. and F. Ingwerson. 1976. Growth of *Xanthorrhoea australis* R. Br. in relation to fire. J. Appl. Ecol. 13:195-203.

Gillion, D. 1983. The fire problem in tropical savannas. Pp. 617-641 *in* F. Bourliere (ed.) Tropical Savannas. Elsevier Sci. Publ. Co., Amsterdam.

Grimm, E.C. 1984. Fire and other factors controlling the big woods vegetation of Minnesota in the mid-nineteenth century. Ecol. Monogr. 53:291-311.

Hanes, T.L. 1971. Succession after fire in the chaparral of southern California. Ecol. Monogr. 41:27-52.

Hare, R.C. 1961. Heat effects on living plants. USDA For. Serv. Occas. Pap. S-183, Southern For. Exp. Stn., New Orleans, LA.

Harris, G.R., and W.W. Covington. 1983. The effect of a prescribed fire on nutrient concentration and standing crop of understory vegetation in ponderosa pine. Can. J. For. Res. 13:501-507.

Hartnett, D.C. and D.R. Richardson. 1989. Population biology of Bonama grandiflora (Convolvulaceae): effects of fire on plant and seed bank dynamics. Amer. J. Bot. 76:361-369.

Hassan, M.A., and N.E. West. 1986. Dynamics of soil seed pools in burned and unburned sagebrush semi-deserts. Ecology 67:269-272.

Heirman, A.L., and H.A. Wright. 1973. Fire in the medium fuels of West Texas. J. Range Manage. 26:331-335.

Henderson, R.A., D.L. Lovell, and E.A. Howell. 1983. The flowering responses of 7 grasses to seasonal timing of prescribed burns in remnant Wisconsin prairie. Pp. 7-10 *in* R. Brewer (ed.) Proc. Eighth North Amer. Prairie Conf., W. Michigan Univ.

Heyward, F. 1938. Soil temperatures during forest fires in the longleaf pine region. J. For. 36:478-491.

Hodgkinson, K.C. and R.E. Oxley. 1990. Influence of fire and edaphic factors on germination of the arid-zone shrubs *Acacia aneura*, *Cassia nemophila*, and *Dodonaea viscosa*. Aust. J. Bot. 38:269-279.

Hulbert, L.C. 1988. Causes of fire effects in tallgrass prairie. Ecology 69:46-58.

Humphrey, L.D. 1984. Patterns and mechanisms of plant succession after fire on *Artemisia*-grass sites in southeastern Idaho. Vegetatio 60:91-101.

Isaac, L.A. and H.G. Hopkins. 1937. The forest soil of the Douglas fir region and the changes wrought upon it by logging and slash burning. Ecology 18:264-279.

Keeley, J.E. 1987. Role of fire in seed germination of woody taxa in California chaparral. Ecology 68:434-443.

Keeley, J.E. 1991. Seed germination and life history syndromes in the California chaparral. Bot. Rev. 57:81-116.

Keeley, J.E., and P.H. Zedler. 1978. Reproduction of chaparral shrubs after fire: A comparison of sprouting and seeding strategies. Amer. Midl. Natur. 99:142-161.

Keeley, J.E., B.A. Morton, A. Pedrosa, and P. Trotter. 1985. Role of allelopathy, heat and charred wood in the germination of chaparral herbs and suffrutescents. J. Ecol. 73:445-458.

Keeley, S.C. and M. Pizzorno. 1986. Charred wood stimulated germination of two fire-following herbs of the California chaparral and the role of hemicellulose. Amer. J. Bot. 73:1289-1297.

Kellman, M., K. Miyanishi, and P. Hiebert. 1985. Nutrient retention by savanna ecosystems. II. Retention after fire. J. Ecol. 73:953-962.

Kimmins, J.P. 1987. Forest Ecology. Macmillan Publ. Co., New York.

Knapp, A.K. 1984. Post-burn differences in solar radiation, leaf temperature and water stress influencing production in a lowland tallgrass prairie. Amer. J. Bot. 71:220-227.

Knapp, A.K. and L.C. Hulbert. 1986. Production, density and height of flower stalks of three grasses in annually burned and unburned eastern Kansas tallgrass prairie: a four year record. Southwest. Natur. 31:235-241.

Koch, E. 1942. When the mountains roared: stories of the 1910 fire. USDA For. Serv. Misc. Rep., Idaho Panhandle National Forest.

Kucera, C.L. and J.H. Ehrenreich. 1962. Some effects of annual burning on central Missouri prairie. Ecology 43:334-336.

Kucera, C.L., R.C. Dahlman, and M. Koelling. 1967. Total net productivity and turnover on an energy basis for tallgrass prairie. Ecology 48:536-541.

Lamont, B.B. 1993. Plant senescence in fire-prone perennials. Trends Ecol. Evolut. 8:147.

Lamont, B.B. and M.J. Barker. 1988. Seed bank dynamics of a serotinous, fire-sensitive *Banksia* species. Aust. J. Bot. 36:193-203.

Lamont, B.B., S.W. Connell, and S.M. Berkl. 1991a. Seed bank and population dynamics of Banksia cuneata: the role of fire, time, and moisture. Bot. Gaz. 152:114-122.

Lamont, B.B., D.C. Le Maitre, R.M. Cowling, and N.J. Enright. 1991b. Canopy seed storage in woody plants. Bot. Rev. 57:277-317.

Launchbaugh, J.L. 1972. Effect of fire on shortgrass and mixed prairie species. Proc. Tall Timbers Fire Ecol. Conf. 12:129-151.

Lewcock, H.K. 1937. The use of acetylene to induce flowering in pineapple plants. Qd. Agric. J. 48:532-543.

Lotan, J.E. 1964. Regeneration of lodgepole pine: a study of slash disposal and cone opening. USDA For. Serv. Intermt. Res. Stn. Res. Note INT-16, Odgen, UT.

Lotan, J.E. 1974. Cone serotiny-fire relationships in lodgepole pine. Pp. 267-278 *in* Proc. Tall Timbers Fire Ecology Conf. No. 14, Tall Timbers Research Station, Tallahassee, FL.

Lotan, J.E., J.K. Brown, and L.F. Neuenschwander. 1985. Role of fire in lodgepole pine forests. Pp. 133-152 *in* D.E. Baumgartner, R.G. Krebill, J.T. Arnott, and G.F. Weetman (eds.) Lodgepole Pine: the Species and its Management, Symposium Proceedings. Cooperative Extension Service, Washington State Univ., Pullman.

Lovell, D.L., R.A. Henderson, and E.A. Howell. 1983. The response of forb species to seasonal timing of prescribed burns in remnant Wisconsin prairies. Pp. 11-15 *in* R. Brewer (ed.) Proc. Eighth North Amer. Prairie Conf., W. Michigan Univ.

Mack, R.N. 1981. Invasion of *Bromus tectorum* L. into western North America: an ecological chronicle. Agro-Ecosystems 7:145-165.

Mack, R.N., and J.N. Thompson. 1982. Evolution in steppe with few large, hooved mammals. Amer. Natur. 119:757-773.

Malanson, G.P. 1984. Fire history and patterns of Venturan subassociations of California coastal sage scrub. Vegetatio 57:121-128.

McFarland, D.C. 1988. Fire and the vegetation composition and structure of subtropical heathlands in south-eastern Queensland. Aust. J. Bot. 36:533-546.

McPherson, G.R., and H.A. Wright. 1989. Direct effects of competition on individual juniper plants: a field study. J. Appl. Ecol. 26:976-988.

Medina, E., and J.F. Silva. 1990. Savannas of northern South America: a steady state regulated by water-fire interactions on a background of low nutrient availability. J. Biogeography 17:403-413.

Milchunas, D.G., O.E. Sala, and W.K. Lauenroth. 1988. A generalized model of the effects of grazing by large herbivores on grassland community structure. Amer. Natur. 132:87-106.

Miller, T.E. 1982. Community diversity and interactions between the size and frequency of disturbance. Amer. Natur. 120:533-536.

Mitchell, J.E., and P.N.S. Bartling. 1991. Comparison of linear and nonlinear overstory-understory models for ponderosa pine. For. Ecol. Manage. 42:195-204.

Miyanishi, K. and M. Kellman. 1986. The role of fire in recruitment of two neotropical savanna shrubs, *Miconia albicans* and *Clidemia sericea*. Biotropica 18:224-230.

Moe, S.R., P. Wegge, and E.B. Kapela. 1990. The influence of man-made fires on large wild herbivores in Lake Burungi area in northern Tanzania. Afr. J. Ecol. 28:35-43.

Muller, C.H., R.B. Hanawalt, and J.K. McPherson. 1968. Allelopathic control of herb growth in the fire cycle of California chaparral. Bull. Torr. Bot. Club 95:225-231.

Mutch, R.W. 1970. Wildland fires and ecosystems—a hypothesis. Ecology 51:1046-1051.

Naveh, Z. 1975. Effects of fire in the Mediterranean region. Pp. 401-434 *in* T.T. Kozlowski and C.E. Ahlgren (eds.) Fire and Ecosystems. Academic Press, New York.

Nelson, R.M. and I.H. Sims. 1934. A method of measuring experimental forest fire temperatures. J. For. 32:488-490.

Neuenschwander, L.F., H.A. Wright, and S.C. Bunting. 1978. The effect of fire on a tobosa-grass-mesquite community in the Rolling Plains of Texas. Southwest. Natur. 23:315-338.

Nisbet, E.G. 1991. Leaving Eden—to protect and manage the earth. Pp. 90-106, Cambridge University Press, New York.

Norton, B.E. and J.W. McGarity. 1965. The effect of burning of native pasture on soil temperature in northern New South Wales. J. Brit. Grassl. Soc. 29:101-105.

Ojima, D.S., W.J. Parton, D.S. Schimel, and C.E. Owensby. 1990. Simulated impacts of annual burning in prairie ecosystems. Pp. 118-132 in S.L. Collins and L.L. Wallace (eds.), Fire in North American Tallgrass Prairies. University of Oklahoma Press, Norman.

Old, S.M. 1969. Microclimate, fire, and plant production in an Illinois prairie. Ecol. Monogr. 39:355-384.

Oswald, B.P., and W.W. Covington. 1983. Changes in understory production following a wildfire in southwestern ponderosa pine. J. Range Manage. 36:507-509.

Owensby, C.E., and K.L. Anderson. 1967. Yield response to time of burning in the Kansas Flinthills. J. Range Manage. 20:12-16.

Packard, S. 1988. Just a few oddball species: restoration and the rediscovery of the tallgrass savanna. Restor. and Manage. Notes 6:13-20.

Parker, V.T. and V.R. Kelly. 1989. Seed banks in California chaparral and other Mediterranean climate shrublands. Pp. 231-255 *in* M.A. Leck, V.T. Parker, and R.L. Simpson (eds.) Ecology of Soil Seed Banks. Academic Press, San Diego, CA.

Patton, B.D., M. Hironaka, and S.C. Bunting. 1988. Effect of burning on seed production of bluebunch wheatgrass, Idaho fescue, and Columbia needlegrass. J. Range Manage. 41:232-234.

Peart, M.H. 1984. The effects of morphology, orientation and position of grass diaspores on seedling survival. J. Ecol. 72:437-453.

Pieterse, P.J. and A.L.P. Cairns. 1986. The effect of fire on an *Acacia longifolia* seed bank in the south-western Cape. S. Afr. J. Bot. 52:233-236.

Platt, W.J., G.W. Evans, and M.M. Davis. 1988a. Effects of fire season on flowering of forbs and shrubs in longleaf pine forest. Oecologia 76:353-363.

Platt, W.J., G.W. Evans, and S.L. Rathburn. 1988b. The population dynamics of a long-lived conifer (*Pinus palustris*). Amer. Natur. 131:491-525.

Pratt, D.W., R.A. Black, and B.A. Zamora. 1984. Buried viable seed in a ponderosa pine community. Can. J. Bot. 62:44-52.

Quick, C.R. and A.S. Quick. 1961. Germination of *Ceanothus* seeds. Madrono 16:23-30.

Raison, R.J. 1979. Modification of the soil environment by vegetation fires, with particular reference to nitrogen transformations: a review. Plant and Soil 51:73-108.

Rasmussen, G.A., and H.A. Wright. 1988. Germination requirements of flameleaf sumac. J. Range Manage. 41:48-52.

Rasmussen, G.A., and H.A. Wright. 1989. Succession of secondary shrubs on ashe juniper communities after dozing and prescribed burning. J. Range Manage. 42:295-298.

Rasmussen, G.A., G.R. McPherson, and H.A. Wright. 1988. Economic comparison of aerial and ground ignition for rangeland prescribed fires. J. Range Manage. 41:413-415.

Redmann, R.E. 1978. Plant and soil water potentials following fire in a northern mixed grassland. J. Range Manage. 31:443-445

Reynolds, H.G. and J.W. Bohning. 1956. Effects of burning on a desert grass-shrub range in southern Arizona. Ecology 37:769-777.

Riggan, P.J., S. Goode, P.M. Jacks, and R.N. Lockwood. 1988. Interaction of fire and community development in chaparral of southern California. Ecol. Monogr. 58:155-176.

Risser, P.G., and W.J. Parton. 1982. Ecosystem analysis of the tallgrass prairie: nitrogen cycle. Ecology 63:1342-1351.

Risser, P.G., E.C. Birney, H.D. Blocker, S.W. May, W.J. Parton, J.A. Wiens. 1981. The True Prairie Ecosystem. Dowden, Hutchinson and Ross, Stroudsburg, PA.

Rogers, J.M., R.A. Susott, and R.G. Kelsey. 1986. Chemical composition of forest fuels affecting their thermal behavior. Can. J. For. Res. 16:721-726.

Rothermel, R.C. 1983. How to predict the spread and intensity of forest and range fires. USDA For. Serv. Intermt. Res. Stn. Gen. Tech. Rep. INT-143, Ogden, UT.

Rundel, P.W. 1981. Fire as an ecological factor. Pp. 501-538 *in* O.L. Lange et al. (eds.) Physiological Plant Ecology I: Responses to the Physical Environment. Springer-Verlag, Berlin.

Russell, J.D., A.R. Fraser, J.R. Watson, and J.W. Parsons. 1974. Thermal decomposition of protein in soil organic matter. Geoderma 11:63-66.

Sampson, A.W. 1944. Plant succession on burned chaparral lands in northern California. Calif. Agric. Exp. Stn. Bull. 685, Berkeley.

Savage, M.J., and K. Vermeulen. 1983. Microclimate modification of tall moist grasslands of Natal by spring burning. J. Range Manage. 36:172-174.

Schimel, D., M.A. Stillwell, and R.G. Woodmansee. 1985. Biogeochemistry of C, N, and P in a soil catena of the shortgrass steppe. Ecology 66:276-282.

Scifres C.J. 1980a. Fire and range vegetation of the Rio Grande Plains. Pp. 6-11 *in* L.D. White (ed.), Prescribed Range Burning in the Rio Grande Plains of Texas. Texas Agr. Ext. Serv. Publ. 88 p.

Scifres, C.J. 1980b. Integration of prescribed burning with other brush management methods: The systems concept of brush management. Pp. 44-50 *in* L.D. White (ed.), Prescribed Range Burning in the Rio Grande Plains of Texas. Texas Agr. Ext. Serv. Publ. 88 p.

Seastedt, T.R. 1988. Mass, nitrogen, and phosphorus dynamics in foliage and root detritus of tallgrass prairie. Ecology 69:59-65.

Sharrow, S.H., and H.A. Wright. 1977. Effects of fire, ash and litter on soil nitrate, temperature, moisture, and tobosagrass production in the Rolling Plains. J. Range Manage. 30:266-270.

Silvertown, J., and M. Tremlett. 1989. Interactive effects of disturbance and shade upon colonization of grassland: an experiment with *Anthriscus sylvestris* (L.) Hoffm., *Conium maculatum* L., *Daucus carota* L. and *Heracleum sphondylium* L. Funct. Ecol. 3:229:235.

Smith, D.W. and J.H. Sparling. 1966. The temperatures of surface fires in jack pine barren.

I. The variation in temperature with time. Can. J. Bot. 44:1285-1292.

Sneeuwjagt, R.J., and W.H. Frandsen. 1977. Behavior of experimental grass fires vs. predictions based on Rothermel's fire model. Can. J. For. Res. 7:357-367.

Snyder, J.R. 1984. The role of fire: Mutch ado about nothing. Oikos 43:404-405.

Snyder, J.R. 1986. The impact of wet season and dry season prescribed fires on Miami Rock Ridge Pineland, Everglades National Park. South Florida Res. Center Rep. SFRC-86/06.

Spalt, K.W. and W.E. Reifsnyder. 1962. Bark characteristics and fire resistance: a literature survey. USDA For. Serv. Southern Exp. Stn. Occas. Pap. S-193, New Orleans, LA.

Steuter, A.A. 1986. Fire behavior and standing crop characteristics on repeated seasonal burns - northern mixed prairie. Pp. 54-59 *in* A. Koonce (ed.), Prescribed Burning in the Midwest: State of the Art Symposium. Univ. Wisconsin, Stevens Point, WI.

Steuter, A.A. 1987. C3/C4 production shift on seasonal burns - northern mixed prairie. J. Range Manage. 40:27-31.

Steuter, A.A., and C.M. Britton. 1983. Fire-induced mortality of redberry juniper (*Juniperus pinchotii* Sudw.). J. Range Manage. 36:343-345.

Steuter, A.A., B. Jasch, J. Ihnen, and L.L. Tieszen. 1990. Woodland/grassland boundary changes in the middle Niobrara valley of Nebraska identified by delta 13C values of soil organic matter. Amer. Midl. Natur. 124:301-308.

Steuter, A.A., and H.A. Wright. 1983. Spring burning effects on redberry juniper-mixed grass habitats. J. Range Manage. 36:161-164.

Steward, F.R., S. Peter, and J.B. Richon. 1990. A method for predicting the depth of lethal heat penetration into mineral soils exposed to fires of various intensities. Can. J. For. Res. 20:919-926.

Stinson, K.J. and H.A. Wright. 1969. Temperature of headfires in the southern mixed prairie of Texas. J. Range Manage. 22:169-174.

Stock, W.D., and O.A.M. Lewis. 1986. Soil nitrogen and the role of fire as a mineralizing agent in a South African coastal fynbos ecosystem. J. Ecol. 74:317-328.

Stone, E.C. 1951. The stimulative effect of fire on the flowering of the golden Brodioea (*Brodioea ixodes* Wats. var. *lugens* Jeps.). Ecology 32:534-537.

Stone, E.C. and G. Juhren. 1953. Fire stimulated germination. Calif. Agric. 7(9):13-14.

Stott, P. 1986. The spatial pattern of dry season fires in the savanna forests of Thailand. J. Biogeog. 13:345-358.

Streng, D.R., and P.A. Harcombe. 1982. Why don't East Texas Savannas grow up to forest? Amer. Midl. Natur. 108:278-294.

Stur, W.W. and L.R. Humphreys. 1987. Seed production in *Brachiaria decumbens* and *Paspalum plicatulum* as influenced by system of residue removal. Aust. J. Agric. Res. 38:869-880.

Svejcar, T.J., and J.A. Browning. 1988. Growth and gas exchange of *Andropogon gerardii* as influenced by burning. J. Range Manage. 41:239-244.

Swamy, P.S. and P.S. Ramakrishnan. 1988. Effect of fire on growth and allocation strategies of *Mikania micrantha* under early successional environments. J. Appl. Ecol. 25:653-658.

Towne, G., and C.E. Owensby. 1984. Long-term effects of annual burning at different dates in ungrazed Kansas tallgrass prairie. J. Range Manage. 37:392-397.

Troumbis, A.Y., and L. Trabaud. 1989. Some questions about flammability in fire ecology. Acta Oecologica Plant. 10:167-175.

Turner, M.G., and S.P. Bratton. 1987. Fire, grazing, and landscape heterogeneity of a Georgia barrier island. Pp. 86-101 *in* M.G. Turner (ed.) Landscape Heterogeneity and Disturbance. Ecological Studies: Vol. 64, Springer-Verlag, New York.

Uresk, D.W., and K.E. Severson. 1989. Understory-overstory relationships in ponderosa pine forests, Black Hills, South Dakota. J. Range Manage. 42:203-208.

van de Venter, H.A. and A.D. Esterhuizen. 1988. The effect of factors associated with fire on seed germination of *Erica sessiflora* and *E. hebecalyx* (Ericaceae). S. Afr. J. Bot. 54:301-304.

van Wilgen, B.W. and M. Viviers. 1985. The effect of fire on serotinous Protaceae in the western Cape and the implications for fynbos management. S. Afr. For. J. 130:49-53.

Vinton, M.A., D.C. Hartnett, E.J. Finck, and J.M. Briggs. 1993. Interactive effects of fire, bison (*Bison bison*) grazing and plant community composition in tallgrass prairie. Amer. Midl. Nat. 129:10-18.

Vogl, R.J. 1973. Ecology of knobcone pine in the Santa Ana Mountains, California. Ecol. Monogr. 43:125-143.

Wells, P.V. 1970. Post-glacial vegetation history of the Great Plains. Science 167:1574-1582.

Went, F.W., G. Juhren, and M.C. Juhren. 1952. Fire and biotic factors effecting germination. Ecology 33:351-363.

West, N.E., and M.A. Hassan. 1985. Recovery of sagebrush-grass vegetation following wildfire. J. Range Manage. 38:131-134.

White, R.S., and P.O. Currie. 1983. Prescribed burning in the northern Great Plains: yield and cover responses of three forage species in the mixed grass prairie. J. Range Manage. 36:179-183.

Whitney G.G. 1986. Relation of Michigan's presettlement pine forests to substrate and disturbance history. Ecology 67:1548-1559.

Whittaker, E. 1961. Temperature in heath fires. J. Ecol. 49:709-715.

Wiens, J.A., C.S. Crawford, and J.R. Gosz. 1985. Boundary dynamics: a conceptual framework for studying landscape ecosystems. Oikos 45:421-427.

Wink, R.L., and Wright. 1973. Effects of fire on an ashe juniper community. J. Range Manage. 26:326-329.

Wright, E. 1931. The effect of high temperatures on seed germination. J. For. 29:679-687.

Wright, H.A. 1970. A method to determine heat-caused mortality in bunchgrass. Ecology 51:582-587.

Wright, H.A. 1971. Why squirreltail is more tolerant to burning than needle-and-thread. J. Range Manage. 24:277-284.

Wright, H.A. 1978. The effects of fire on vegetation in ponderosa pine forests: a state-of-the-art review. Texas Tech Univ. Range and Wildl. Info. Ser. No. 2, Lubbock.

Wright, H.A., and J.O. Klemmedson. 1965. Effects of fire on bunchgrasses of the sagebrush grass region in southern Idaho. Ecology 46:680-688.

Wright, H.A., L.F. Neuenschwander, and C.M. Britton. 1979. The role and use of fire in sagebrush-grass and pinyon-juniper plant communities: a state-of-the-art review. USDA For. Serv. Intermt. Res. Stn. Gen. Tech. Rep. INT-58, Ogden, UT.

Wright, H.A., and A.W. Bailey. 1980. Fire ecology and prescribed burning in the Great Plains – a research review. USDA For. Serv. Intermt. Res. Stn. Gen. Tech. Rep. INT-77, Ogden, UT.

Wright, H.A. and A.W. Bailey. 1982. Fire Ecology: United States and southern Canada. John Wiley & Sons, New York.

Zammit, C. and M. Westoby. 1987. Seedling recruitment strategies in obligate-seeding and resprouting Banksia shrubs. *Ecology* 68:1984-1992.

Zimmerman, G.T. and L.F. Neuenschwander. 1984. Livestock grazing influences on community structure, fire intensity, and fire frequency within the Douglas-fir/ninebark habitat type. J. Range Manage. 37:104-110.

The Relations of Phytophagous Invertebrates and Rangeland Plants

E. W. Evans and T. R. Seastedt

I. Introduction

"Range ecologists have given prominent position to the role of large herbivores and rodents in determining the composition of range vegetation. They have paid practically no attention to the role of thc much more selective grazers among the plant feeding insects which abound on those same ranges ..."

(Huffaker and Kennett 1959: 69)

Although there has been considerable research on rangeland insects and other phytophagous (plant-feeding) invertebrates in the thirty years since Huffaker and Kennett penned these lines, there remains a large element of truth to their assessment when applied to present range science. Those invertebrates, especially insects, subject to outbreak are of course well known for their major impact on rangelands (Table 1). Grasshoppers are perhaps the most widely appreciated of such species (e.g. Hewitt 1977, Hewitt and Onsager 1983). During outbreaks of the old world desert locust, *Schistocerca gregaria* (Forsk.), for example, swarms may consume daily the equivalent of 150,000 head of cattle (Bullen 1966). Other rangeland insects such as the harvester ants (*Pogonomyrmex* spp.) of western North America are also conspicuous in their general abundance and obvious impact. But a great many other, less conspicuous (yet often very abundant) species of phytophagous invertebrates inhabit rangelands (Watts et al. 1989). The biologies of these species and their impact on their host plants have generally received only limited study.

Here we highlight aspects of these biologies and impact within the conceptual framework of plant/herbivore interactions that has grown so enormously over the past thirty years (Feeny 1992, Herms and Mattson 1992, Stamp 1992). The emerging ideas, concepts, and generalizations are highly applicable to rangeland ecosystems, but remain to be more fully tested by well-directed field studies and experiments. Our aims in this brief overview are (1) to impress upon those unfamiliar with the subject, the importance of and great potential for studies focused upon the relations of rangeland plants with their associated phytophagous invertebrates, and (2) to provide lead-ins to a vast, diverse, and rapidly growing literature.

After briefly surveying the diversity of phytophagous invertebrates in rangelands, we examine characteristics of range plant individuals and species that influ-

Evans: Department of Biology, Utah State University, Logan, Utah 84322-5305; Seastedt: EPO Biology and Institute of Arctic and Alpine Research, University of Colorado, Boulder, Colorado 80309-0450.

Table 1. Examples of insects subject to outbreak in rangelands of western North America.

Order	Family: Species	Selected References
Grasshoppers (Orthoptera)	Acrididae:Migratory grasshopper *Melanoplus sanguinipes* (Fab.)	Parker et al. 1955 McAnelly and Rankin 1986a,b
	Tettigoniidae: Mormon cricket *Anabrus simplex* Haldeman	Wakeland 1959 MacVean 1987
Bugs (Hemiptera)	Miridae: Black grass bugs *Labops* and *Irbisia* spp.	Haws et al. 1973 Todd and Kamm 1974
Beetles (Coleoptera)	Scarabaeidae: White grubs *Phyllophaga* spp.	Ritcher 1966 Ueckert 1979
	Chrysomelidae: leaf beetles *Trirhabda* spp.	Pringle 1960 Hogue 1970
Moths (Lepidoptera)	Saturniidae: range caterpillar *Hemileuca oliviae* Cockerell	Watts and Everett 1976 Shaw et al. 1987
	Gelechiidae: sagebrush defoliator *Aroga websteri*	Gates 1964

ence their use by these invertebrates, and the responses of individual plants to such herbivory. We also consider impacts of invertebrate herbivory from community and ecosystem perspectives, including how phytophagous invertebrates may interact with each other, and with vertebrate herbivores, in affecting rangeland plants. We conclude by examining possible roles of the host plants themselves in contributing to the violent fluctuations in density of invertebrates prone to outbreak in rangelands.

Even in the absence of outbreaks, invertebrate herbivory is a chronic, universal phenomenon in rangelands. The costs and consequences of invertebrate herbivory on plant morphology and physiology are potentially more variable than those resulting from vertebrate herbivory (Fig. 1). As Huffaker and Kennett (1959) noted, invertebrates have the potential to be much more selective in exactly which plant parts are consumed. Moreover, invertebrates also play a larger role belowground than do vertebrates in most rangelands. The presence of numerous phytophagous invertebrate species both above and belowground requires that their combined impact, potentially rife with complex interactions, be considered (Fig. 2).

Rarely does invertebrate herbivory directly and immediately kill range plants, but host plants weakened by herbivory may fare poorly in competing for water and nutrients (e.g., insect herbivory may reduce ectomycorrhizae of rangeland plants; Del Vecchio et al. 1993) and in staving off disease. Furthermore invertebrates, particularly sucking species that inject saliva into their hosts while feeding, may transmit plant diseases. Aboveground herbivory may lower potential photosynthetic capacity in reducing leaf area. However, herbivory may also reduce self-shading, particularly in mesic grasslands, thereby ameliorating the negative impact on photosynthetic capacity. Belowground herbivory reduces the potential water and nutrient uptake capacity of roots. Above and belowground herbivory may result in vari-

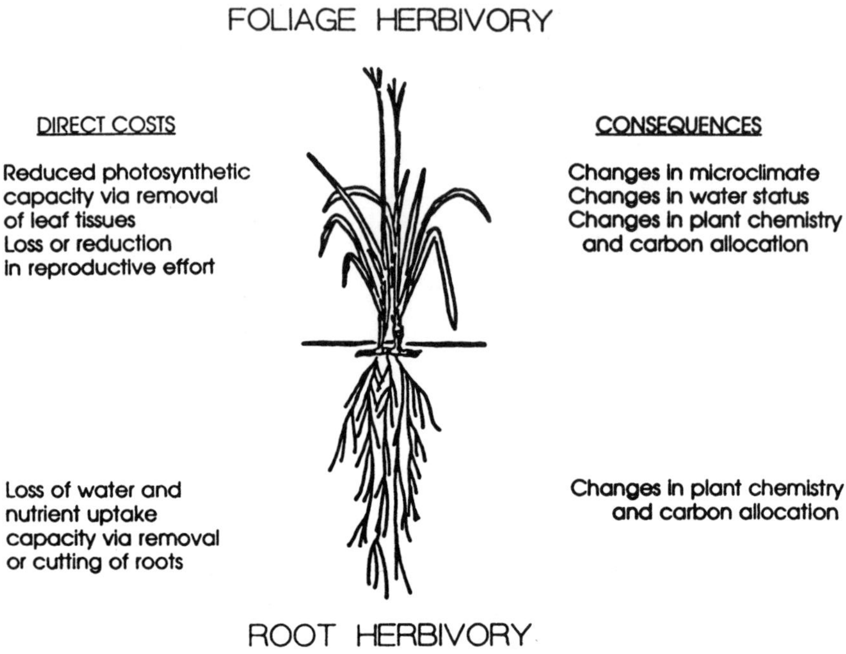

Figure 1. Foliage and root herbivory by invertebrates have diverse costs and consequences for rangeland plants.

ous shifts in energy and nutrient allocation by the host plant among sexual reproduction, tillering and shoot and root growth. The complexity of potential combined impacts of above and belowground invertebrate herbivory can be appreciated by considering the many kinds of phytophagous invertebrates inhabiting rangelands, and the diversity of ways in which they feed upon range grasses, forbs, and shrubs.

II. A Brief Survey of Rangeland Phytophagous Invertebrates

Aboveground, many chewing insects feed on leaves and stems, either externally, or internally as leaf miners, gall makers, and stem borers. Major groups include grasshoppers (Orthoptera: Acrididae and Tettigoniidae); walkingsticks (Orthoptera: Phasmatidae); leaf beetles (Coleoptera: Chrysomelidae); caterpillars of various moths and butterflies (Lepidoptera), flies (Diptera: e.g., leatherjackets, Tipulidae; grass stem maggots, Chloropidae; and gall midges, Cecidomyiidae); and sawflies (Hymenoptera: Cephidae). Another group, termites (Isoptera), are often thought of primarily as feeding on dead plant matter, but in some rangelands they may also consume large quantities of live plant material (Bodine and Ueckert 1975).

Many sap-feeding insects and mites (invertebrates with sucking mouthparts) also attack the phloem, xylem, and cell contents of leaves and stems. Particularly notable are members of the order Hemiptera, including Homoptera: leafhoppers (Cicadellidae), planthoppers (Fulgoroidea), aphids (Aphidae), and coccids (e.g., mealybugs and scales, Coccoidea); and Heteroptera: plant bugs (Miridae). Spider

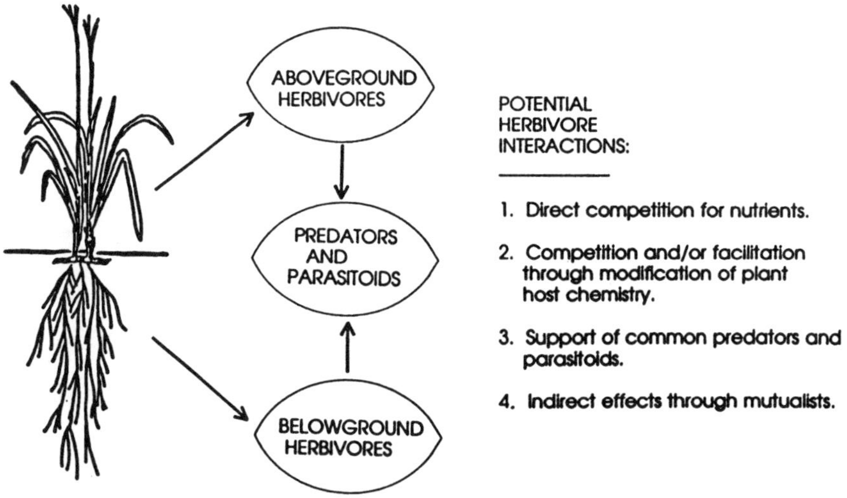

Figure 2. Invertebrates interact with each other both directly and indirectly in affecting rangeland plants.

mites (Acarina: Tetranychidae) may be the most destructive non-insect arthropods. Still other phytophagous insects feed especially on reproductive tissues of rangeland plants, including flowers and buds (e.g., blister beetles, Meloidae; thrips, Thysanoptera) as well as seeds (e.g., Hemiptera: seed bugs, Lygaeidae, and stinkbugs, Pentatomidae; Coleoptera: seed beetles, Bruchidae, and weevils, Curculionidae; Hymenoptera: seed chalcids, Eurytomidae, and ants, Formicidae). Watts et al. (1982, 1989) provide excellent introductions to the life histories of many of these groups.

Perhaps even more diverse, and certainly less studied, are the many species of belowground invertebrates that feed upon the roots and rhizomes of range plants. Phytophagous invertebrates are found throughout the soil profile of rangelands, roughly in proportion to the density of roots (Smolik and Lewis 1982, Freckman and Virginia 1990). Highest densities occur in the upper soil horizons. Almost all soil arthropods must complete their life cycles at or above the soil surface; hence, the final immature stages tend to be spent near the soil surface. Nematodes are free from this constraint, but remain most abundant in the upper soil horizons. Egg-laying in the arthropods is generally restricted to once per year at most; population dynamics are affected accordingly. In contrast, nematode reproduction is opportunistic, constrained by temperature and moisture conditions. Peak population densities are variable in their seasonal occurrence in arid regions (Smolik and Rogers 1976, Smolik and Lewis 1982), but occur in late spring and autumn in the more climatically predictable tallgrass prairie (Seastedt et al. 1987). These peaks may be correlated with times of relatively high resource quality (protein and amino acid concentrations) because of translocation and retranslocation from aboveground plant parts.

Identification of the trophic status of many soil invertebrates is difficult. Many arthropods and nematodes are opportunistic omnivores (Walter 1987a,b, 1988,

Moore et al. 1988). Given large constraints on both movement and resource quality in soil, many belowground consumers respond by ingesting most forms of organic matter encountered. Some invertebrates are attracted to roots to feed on plant pathogens (Curl et al. 1988). The benefit to plants of having a rhizosphere fauna capable of consuming pathogenic microbes may be outweighed by feeding on roots by these same organisms when preferred food items are not available.

Belowground arthropods in rangelands that are obligate or facultative root feeders include both sucking insects, especially cicadas and smaller Homoptera, and chewing insects, primarily larvae of Lepidoptera and the beetle families Scarabeidae, Elateridae, Chrysomelidae and Tenebrionidae (e.g., McDaniel 1971, Lloyd and Kumar 1977). European grasslands also experience substantial grazing by fly larvae (Curry 1987a).

Three major groups of root suckers can be found in rangelands. Xylem specialists consist of cicada nymphs (Cicadidae). In North America, cicadas are found in tallgrass and shortgrass prairies and in shrub-steppe, but are absent in the shortgrass prairie of eastern Colorado (Lloyd and Kumar 1977). While this group has been reported to cause damage in forests, no measurable reduction in productivity has yet been reported for cicadas in grasslands. Root phloem feeders include some relatively obscure species of Homoptera, for example the Megalodidae. These "ground pearls" have not yet been associated with major damage in rangelands, although they do at times significantly damage urban grass plantings.

Belowground arthropods probably consume only small to modest amounts of roots in most years. However, as with aboveground insects, occasional outbreaks by a few species result in great devastation of rangeland vegetation (Graber et al. 1931, Hewitt et al. 1974, Ueckert 1979, Weiner and Capinera 1980). The propensity for certain species for overgraze the vegetation appears related to their form of feeding. Root-chewing insects may overgraze the rangeland; root sucking insects and nematodes apparently do not. The most visibly destructive root-feeding insects of North American and Australian rangelands include a diverse assemblage of beetle larvae, especially larvae of the family Scarabeidae (June beetles, *Phyllophaga* spp.; Fig. 3). In North America, these insects were present in sufficient numbers to be recorded in the journals of westward pioneers as they crossed the great central grasslands:

> "A shower of bugs ... ("black beetles") ... fell like hail ... They could be heard all night, pattering against the tents like large drops of rain in a heavy shower." (Mattes 1969: 146)

The grasslands of Australia and New Zealand receive equal or more damage by beetle larvae than do the more mesic to xeric grasslands of North America (c.f. Hutchinson and King 1980). Surprisingly, recent reviews (Curry 1987a,b) indicate that managed European grasslands are less affected by these insects. Population regulatory mechanisms therefore appear to differ across sites. High soil moisture content appears to be negatively associated with survivorship (Davidson and Roberts 1968), and may provide a partial explanation for this pattern.

Phytophagous nematodes are another major group of belowground invertebrates; Smolik and Lewis (1982), for example, estimated that 2 to 6 million plant-

Figure 3. Root-feeding white grubs, or larvae of beetles of the genus *Phyllophaga* (Scarabaeidae) are among the relatively few species of rangeland invertebrates that are periodically subject to widespread and highly damaging outbreaks.

feeding nematodes per m^2 live in the top 60 cm of mixed grass prairie soil. The dominant phytophagous soil nematodes of the grasslands include species within the Tylenchida and Dorylaimida (e.g., Freckman 1982, Anderson et al. 1984). Phytophagous nematodes are similar to the Homopterans in their feeding habits, and both groups have been termed sap-feeders (Scott et al. 1979). An equally abundant component of the nematode fauna has been identified as consisting of facultative herbivores and/or fungivores. These organisms may simply consume whatever resource (live roots, dead roots, fungi) that they encounter. On average, nematodes consume more root biomass than any other invertebrate taxon and may compete with cattle as the most important rangeland consumer (Stanton et al. 1981, Stanton 1988). However, while nematodes impose high, chronic levels of herbivory on grass roots, the group has never been recorded to overgraze rangelands (c.f. Stanton et al. 1984, Stanton 1987).

In considering how such a vast set of above and belowground invertebrates affects rangeland plants, one must note that many invertebrates attack only a limited array of plants. Many grassland leafhoppers, for example, are highly host-specific (Whitcomb et al. 1994). Root-feeding invertebrates, such as nematodes and white grubs, also show definite host preferences (Davidson and Roberts 1968, Ingham and Detling 1984, Freckman and Virginia 1989). Even grasshoppers, infamous for their relatively broad feeding habits, are not indiscriminate feeders, but rather have marked, complexly determined feeding preferences (Isely 1938, Gangwere 1961, Mulkern et al. 1969, Joern 1979, Chapman 1990). Bernays and Graham (1988), Bernays (1991), and Appel and Martin (1992) provide current assessments of selective factors hypothesized to influence the evolution of diet breadth in phytophagous insects.

III. Plant Traits Influencing Invertebrate Herbivory

The great diversity of species of phytophagous invertebrates in rangelands can be misleading in suggesting that range plants represent a readily available food source that can be exploited with little difficulty. In fact, there are considerable evolutionary "hurdles" to feeding on seed plants, and relatively few major groups of invertebrates have evolved and radiated as herbivores (Southwood 1973, 1985, Mitter et al. 1988). In particular, physical and chemical features of plants generally pose major difficulties for herbivores, and phytophagous invertebrates are often specially adapted to overcome these difficulties.

It is frequently suggested that many plant traits affecting phytophagous invertebrates have, in fact, evolved as defense mechanisms against these organisms. It is extremely difficult to retrace the paths of evolution to test such hypotheses critically (but see Berenbaum 1983, Farrell et al. 1991, Farrell et al. 1992). It is nevertheless useful to consider how present-day attributes of plants influence the degree of herbivory that they sustain. We therefore review those physical and chemical traits of range plants, as well as spatial and temporal attributes, known to affect exploitation by phytophagous invertebrates.

We note at the outset that research is just beginning to reveal the wealth of genetically-based variation within most species of native rangeland plants that influences and is influenced by invertebrate herbivory (e.g., Moran 1981, Berenbaum et al. 1986, Fritz 1990, Fritz and Simms 1992, Vrieling 1991, Berenbaum and Zangerl 1992). A new and intriguing area of emphasis is the nature and intensity of insect herbivory in plant hybrid zones (Whitham 1989, Boecklen and Spellenberg 1990, Floate et al. 1993). Genetic variation is exploited to great benefit in agronomic plants, and may be similarly exploited in economically important rangeland plants (Campbell et al. 1984). In one application to rangeland, for example, Hewitt and Berdahl (1984) have tested for grasshopper food preferences among cultivars of alfalfa used for interseeding range grasslands.

Physical Features

Tissue toughness. A variety of physical and chemical attributes toughen the tissues of rangeland plants, making them difficult for chewing phytophagous invertebrates to clip or tear and process. Grasses in particular are characterized by high silica and lignin content. Graminivorous (grass-feeding) grasshoppers and caterpillars have mouthparts well-suited for these tough tissues (Isely 1944, Brown and Dewhurst 1975). In his pioneering studies, Isely (1944) noted that grasshopper mandibular morphology generally varies with diet in striking parallel fashion to variation in herbivorous mammals. Graminivorous species have relatively blunt incisors and a series of well-developed, parallel molar ridges, while forbivorous species have numerous sharply pointed incisor and molar cusps with poorly developed molar ridges; mixed feeders have mandibles intermediate in structure between these extremes (Chapman 1964).

The mouthparts of graminivorous grasshoppers are well-suited for grinding down tough, silicaceous foodstuffs to extract nutrients, while those of forbivorous grasshoppers are more generally adapted for incising and tearing off pieces of leaves. Patterson (1984) notes in addition that the specialized mouthparts of graminivorous grasshoppers enable them to split leaves between the parallel veins characteristic of grasses. High silica content can be highly abrasive even to mouthparts of adapted grasshoppers. The mandibles of grass-feeding grasshoppers generally show considerably more wear over time than do those of forb-feeding grasshoppers, and such wear may restrict food consumption of older individuals (Chapman 1964).

Morphological adaptations for feeding on tough plant tissues are also reflected in head allometry (relative head size). Mattson (1980) noted that the unusually large heads and associated mouthparts of many Hesperiid butterfly larvae may be adaptations giving mechanical advantage for feeding on fibrous, siliceous grasses. Bernays (1986) has subsequently demonstrated that among caterpillars and grasshoppers, grass-feeders have larger heads than other foliage feeders of comparable size. Similarly, allometric differences in relative head mass and differences in mandibular morphology also exist between Saturniid and Sphingid caterpillars (Bernays and Janzen 1988). The former ingest old, tough, tannin-rich leaves in relatively large, uniformly sized pieces, whereas the latter tear, crush, and thoroughly masticate soft, young leaf tissue before ingesting it. Head allometry can vary ontogenetically depending on environment; the grass-feeding noctuid caterpillar *Pseudaletia unipuncta* develops a relatively larger head and associated manbibles and mandibular musculature when reared on hard grass as opposed to soft artificial diet (Bernays 1986).

Amounts of lignin and silica undoubtedly influence belowground patterns of invertebrate herbivory as well. Large primary roots develop the high lignin and low nitrogen content characteristic of wood. We have seldom observed evidence of herbivory on these roots. Grass roots are higher in silica content than any other plant part (McNaughton et al. 1985, Seastedt et al. 1989). If high silica concentrations in grasses have indeed evolved as a defense against herbivory (McNaughton and Tarrants 1983), then the concentration of this material in roots implies that defense of roots is of major importance. Silica is deposited in discrete bundles (opal phytoliths) in roots. Sucking insects and nematodes appear capable of feeding between these bundles and thus circumventing the silica defense. Silica therefore may be deposited in roots at high concentrations to defend particularly against larger root chewers.

Pubescence. The degree of pubescence can also influence herbivory. Plant breeders have developed crop lines resistant to such insects as aphids, leaf beetles, and leafhoppers by selecting for increased density of hair-like appendages, or trichomes (Levin 1973, Norris and Kogan 1980). Both glandular and nonglandular trichomes (those that do not secrete chemical substances) can deter invertebrate feeding and oviposition, and may increase mortality, slow development, and reduce fecundity and adult body size. The dense, long nonglandular trichomes of the sagewort (*Artemisia ludoviciana*), for example, apparently deter most rangeland grasshoppers from feeding (Knutson 1982). When forced, the generalist

Melanoplus sanguinipes will consume the plant, but the trichomes apparently require so much energy to digest that the insect is unable to grow normally on the plant (Smith and Grodowitz 1983). The monophagous grasshopper *Hypochlora alba* (Dodge), in contrast, is adapted to this plant and can readily process the trichomes (Smith and Grodowitz 1983, Smith and Kreitner 1983).

Rangeland plants can vary in pubescence geographically, perhaps with important consequences to herbivores. For example, leaf pubescensce of major grasses varies throughout the tallgrass prairie biome (McMillan 1964). Rangeland plants may also vary in pubescence in response to previous injury from invertebrates; for example, grey alder (*Alnus incana* (L.)) produces new leaves with increased density of nonglandular trichomes following defoliation by beetles (Baur et al. 1991).

C3 and C4 plants. There are basic anatomical differences between plants possessing the Calvin cycle photosynthetic pathway (C3) (e.g., cool-season grasses) and plants possessing the dicarboxylic acid (C4) pathway (e.g., warm-season grasses). In particular, C4 plants differ from C3 plants in having bundle-sheath cells wherein proteins and carbohydrates are highly concentrated as constituents of cellular organelles and photosynthetic products (the latter are transported rapidly from the site of synthesis to these cells; Caswell and Reed 1976). Because these cells are very thick-walled and highly lignified, they may be indigestible for many herbivores, e.g., polyphagous grasshoppers (Caswell and Reed 1975, 1976). With this and related features of C4 plants in mind (e.g., C4 plants generally have lower foliar nitrogen concentrations), Caswell et al. (1973) hypothesized that C4 plants in general are inferior to C3 plants as food, and as a consequence, herbivores tend to avoid feeding upon C4 plants.

The basic hypothesis has been tested with field data from grasslands and old-fields by considering the ratio of C13/C12 isotopes in vegetation and consumers, particularly grasshoppers. The ratio of these isotopes differs in C3 vs C4 plants, while the ratio in an herbivore's body tissues generally is the same as that in its diet. Consequently, one can test for a preference for C3 plants by examining whether the ratio of C13/C12 in an herbivore's body tissues differs significantly from that present in the plant community. Such tests have only sometimes supported the hypothesis that phytophagous invertebrates feed disproportionately on C3 plants (Boutton et al. 1978, 1980, Pinder and Kroh 1987).

C4 grasses may not be inferior foodstuffs to insects particularly adapted to feeding on grasses (Bernays and Barbehehn 1987). Many rangeland grasshoppers feed heavily on warm-season C4 grasses (Fry et al. 1978, Joern 1979). It is possible, however, that they prefer C3 grasses if and when available. For example, in a carefully controlled field experiment, Heidorn and Joern (1984) demonstrated that the generalist graminivorous grasshopper *Ageneotettix deorum* prefers C3 to C4 grasses. In its natural habitat, however, the grasshopper feeds most heavily on the more abundant C4 grasses, particularly blue grama (*Bouteloua gracilis*). Thus, food preference may play a relatively minor role in food utilization, and other factors, including host plant availability and perhaps predation pressure, may be more important. In reviewing the literature, Pinder and Jackson (1988) conclude that as a group, free-ranging grasshoppers show no clear tendency to avoid C4 plants. The authors also point out, however, that too few tests of the Caswell et al.

(1973) hypothesis have been made to reach firm conclusions concerning the significance of C4 vs C3 anatomy and nutritional suitability to herbivores.

Plant Architecture. The intensity of herbivory experienced by a plant depends in part on the number of herbivore species attacking the plant. Among the most important determinants of herbivore diversity is the plant's growth form, or plant architecture (Lawton and Schroder 1977, Lawton 1983). Briefly, larger and structurally more complex plants support more insect species than do smaller, more simply constructed plants; for a given host geographic range, trees for example support more insect species than do shrubs, which in turn support more than do forbs and grasses (Lawton and Schroder 1977, Strong and Levin 1979, Southwood et al. 1979). The causes underlying these patterns may include a greater diversity of microhabitats and larger herbivore populations (less subject to extinction) on larger, more structurally complex host plants (Strong et al. 1984). Moran (1980), for example, assigned "architectural ratings" to prickly pear cacti of the genus *Opuntia*, based on plant height, number of cladodes, average cladode area, and development of woody stems (a fifth category of overall cladode complexity, as expressed for example in the degree to which spines and thorns are present, proved of little predictive value). Moran found a strong positive relationship between architectural complexity and the number of phytophagous insect species associated with individual species of *Opuntia*.

The relationship between root architecture and invertebrate herbivory remains unstudied, although root size and age are documented to influence the form and intensity of herbivory. The small size of many roots and their thinly spread distribution in the soil profile explain why smaller invertebrates are the dominant root feeders. Root architecture and spatial patterns may also influence population dynamics of root feeders, for example, phytophagous nematodes. Unlike arthropods, nematodes use soil water as their medium for existence and dispersal. Swimming or moving passively with gravity would appear to be a more successful dispersal mechanism than burrowing. At the same time, however, movement between roots during dry periods would be inhibited. This may explain why nematode outbreaks in rangeland are unknown.

Chemical Traits

Nutritional considerations. Perhaps the most challenging evolutionary hurdles to phytophagy (Southwood 1973, 1985) are the nutritional problems encountered by invertebrates in obtaining sufficient nutrients by feeding on plants. Although invertebrates have many complex dietary requirements (Dadd 1985, Bernays and Simpson 1990), studies in nutritional ecology of invertebrates have focused especially on nitrogen (protein) and to a lesser degree water (Scriber and Slansky 1981, Slansky and Rodriquez 1987, Bernays 1990). Except under extreme conditions (e.g., during outbreaks and/or drought), phytophagous invertebrates are generally considered not to be limited in growth or numbers by energy acquisition (but see Belovsky 1986). Seed-feeders, however, may frequently face food shortage (Slansky and Panizzi 1987) and their feeding habits may be modified accordingly. Kelrick et al. (1986), for example, report that seed preferences of harvester ants (*Pogonomyrmex occidentalis*) are highly correlated with calorie content and %

soluble carbohydrate (see Kelrick and MacMahon 1985 for implications for rangeland reseeding programs). Other dietary requirements, such as for minerals, vitamins, lipogenic factors, fatty acids and sterols, have generally not received much attention in ecological considerations of plant/invertebrate interactions (but see Arms et al. 1974, Bernays and Barbehehn 1987, Mattson and Scriber 1987, Tabashnik and Slansky 1987). The potential for complexity is heightened further as the basic suitability of plant tissue for invertebrates may be modified by microbes in numerous, diverse ways (Barbosa et al. 1991).

Nitrogen. Just as plant growth is often limited by supplies of inorganic nitrogen, so too is invertebrate growth limited by supplies of organic nitrogen. Whereas the percentage dry weight that is nitrogen in carbohydrate-rich plant tissue is generally 7% or less, the dry weight of protein-rich animal tissue in contrast is generally 7-14% nitrogen (Mattson 1980). The average protein content of plants is well below levels required for successful synthetic diets for insects (McNeill and Southwood 1978). Thus phytophagous invertebrates face a basic problem of extracting sufficient nitrogen from their diet, and many aspects of plant-invertebrate relations in rangelands reflect this central problem. Increasing concentrations of atmospheric carbon dioxide associated with global warming may further exacerbate the problem for rangleland insects, as range plants respond to the increase in carbon dioxide by producing tissues that are even more carbon-rich and nitrogen-poor (Fajer et al. 1989, Johnson and Lincoln 1990, 1991, Lincoln et al. 1993).

Nitrogen content of tissue varies widely within and among plants (McNeill and Southwood 1978, Mattson 1980). Actively growing plant tissues such as young leaves and reproductive tissues generally contain higher nitrogen concentrations than more quiescent tissues such as mature foliage or woody tissue. Short-lived, fast-growing species growing in favorable, N-rich soils generally have higher N content than long-lived, slow-growing species that can tolerate the stresses of xeric, N-poor soils (Mattson 1980). For example, C3 grasses usually have higher N content than the C4 grasses that flourish in poorer soils (Boutton et al. 1978). Foliage of woody plants generally has lower N concentrations than forb foliage (Scriber and Feeny 1979), and grass leaves have particularly low N content (Bernays and Berbehehn 1987).

The concentrations of different forms of N in plant tissue also vary widely. Succulent young tissue, for example, often has higher levels of non-protein nitrogen (e.g., in free amino acids, amines, or secondary compounds such as alkaloids) than does mature tissue in which much of the N is bound up as insoluble protein. Stresses imposed by temperature, moisture, fire, and simple physical damage, furthermore, may alter the balance of N in different forms within plant tissue (Mattson 1980, White 1984). Moisture stress may result in protein hydrolysis, for example, as the plant increases the concentration of free amino acids to maintain osmolarity. Through diverse effects on nitrogen dynamics in particular, and plant physiology and biochemistry in general, environmental stresses can profoundly affect the degree of invertebrate herbivory sustained by individual host plants (e.g., Louda and Collinge 1992; see also Price 1991a).

The importance of variations in N content to phytophagous insects can be appreciated by considering the efficiency with which these herbivores convert ingested plant material into additional biomass. In general, this efficiency increases from about 5% to 30–40% as the nitrogen content of the foodstuff increases from 1 to 7% (Mattson 1980). Thus, a phytophagous invertebrate must consume and process several to many units of plant biomass to produce each unit of its own biomass. White (1978, 1993) suggests that the difficulty of obtaining sufficient nitrogen in the diet, especially for young animals, is the single most important factor limiting the abundance of phytophagous invertebrates. The balance of nitrogenous compounds, as well as their summed quantity, is generally critical to the nutritional well-being of insects (e.g., Brodbeck and Strong 1987); subtle changes in amino acid content of host plants, for example, can influence insect growth, development, and fecundity (McNeill and Southwood 1978, Prestidge and McNeill 1983).

Phytophagous invertebrates exhibit a variety of adaptations for coping with the variable, low nitrogen content of their food plants (McNeill and Southwood 1978, Mattson 1980). Phytophagous insects feeding on materials of low N content often compensate by increasing their rate of food consumption, although in doing so they usually realize lowered efficiencies of digestion (due to the reduction in gut retention time) (Slansky and Feeny 1977, McNeill and Southwood 1978, Lincoln et al. 1986). For example, sap-sucking leafhoppers and spittlebugs feeding on xylem fluids with very low N content, consume 100-1000 times their body weight per day (Mittler 1967, Horsfield 1977, 1978). This is one to several orders of magnitude greater than the ingestion rates of leafhoppers feeding on phloem or parenchyma cell sap (Prestidge 1982a). Some consumers adapt phenologically such that their life cycle is synchronized with availability of N-rich plant tissues. In the tallgrass prairie, for example, the blister beetle *Epicauta fabricii* spends its larval life as a soil-dwelling predator. Adult emergence from the soil is timed such that the adults feed heavily on the young reproductive tissue of their chief host plants, wild indigo (*Baptisia australis* and *B. leucophaea*) (Evans et al. 1989, Evans 1990). Other consumers vary their feeding sites on and among plants as nitrogen quality and quantity change. The grass bug *Leptoterna dolabrata*, for example, moves among leaves and flower heads of *Holcus* species in acidic grasslands as they vary in nitrogen content (McNeill 1971, 1973). Other adaptations include prolonged periods of feeding and development, specialized digestive systems relying on activities of endo or ectosymbiotic microorganisms, occasional carnivory and opportunistic scavenging (e.g., Lockwood 1989), regulation of plant chemistry (e.g., induction of gall formation; Abrahamson and Weis 1987, Price et al. 1987, Larson and Whitham 1991), and evolution of larger body size (see discussion in McNeill and Southwood 1978 and Mattson 1980).

The nitrogen content of belowground plant tissue is particularly low. Roots lack the enzymatic apparatus associated with photosynthesis; much nitrogen is bound up in this apparatus aboveground. There are major nutritional problems posed for belowground invertebrates by low root nitrogen. As Stanton (1987) stated, selective feeders such as nematodes may have access to much higher than average nitrogen resources by selectively feeding on cell contents. However, generalist root chewers do not have this option, and low protein intake relative to fixed carbon ingestion results in very slow growth rates (Mattson 1980). Thus, many soil-dwelling herbivores have multi-year life cycles, unlike their aboveground counterparts. Cicadas are renowned for this characteristic, and rangeland cicadas have life cycles of about four years (Beamer 1928). Many of the white grubs (Scarabeidae) have facultative, 2–3-year life cycles. Should food resource quality be high, growth to the adult form can be completed in a shorter period of time (Ritcher 1966).

The importance of nitrogen content of host plants to invertebrates of rangelands has been investigated experimentally by applying nitrogenous fertilizer. In some cases, densities of leafhoppers and grasshoppers have been shown to increase in response to fertilization of grassland plots, at least in part in response to increasing levels of foliar nitrogen (Byers and Jung 1979, Prestidge 1982b, Heidorn and Joern 1987). The mechanisms responsible for these increases may be complicated, however, as the feeding responses of insects to increased nitrogen levels in host plants are not always straightforward (e.g., Joern and Alward 1988). The optimal level of foliar nitrogen may vary for different species of insects, for example, and therefore not all species may respond positively to increases in foliar nitrogen (Prestidge 1982a, 1982b). Lightfoot and Whitford (1987) report that on the whole, sap-sucking insects respond more dramatically than leaf-chewing insects to increased levels of nitrogen in creosotebush foliage. Kirchner (1977) documented only limited response of foliage arthropods to increased nitrogen content of shortgrass prairie vegetation, and Capinera and Thompson (1987) detected no changes in grasshopper densities and assemblage structure.

Dodd and Lauenroth (1979) found that nitrogen (and water) addition to shortgrass prairie had little effect on the belowground fauna, in contrast to the findings of Ingham and Detling (1984) and Seastedt et al. (1988) in mixed and tallgrass prairie. The addition of organic matter rich in nitrogen can increase growth and survivorship of beetle larvae, provided that soil moisture content does not increase (Davidson and Roberts 1968). In contrast, organic matter addition can reduce densities of plant-feeding nematodes (Norton 1979). Decomposing organic matter produces substances toxic to nematodes, and may support high densities of nematode predators as well (Stanton 1987).

Water. Phytophagous invertebrates rely on host plant water content to satisfy much of their moisture requirement. The water relations of host plants therefore can impact the plant/insect relationship. In the hot dry dustbowl years in Kansas, for example, grasshoppers caused unusual injury to prairie grasses, heavily attacking the inflorescences in an apparent attempt to obtain moisture (Wilbur 1936).

Scriber (1978) has suggested that leaf water content may often be even more fundamental than nitrogen availability in limiting growth and development of phytophagous insects. The water and nitrogen content of plant tissues are generally positively correlated, making it difficult to distinguish effects of these factors (Scriber and Feeny 1979, Mattson 1980, Scriber and Slanksy 1981). In studies of feeding performance in Lepidoptera, however, caterpillars' rates of gain in biomass and nitrogen were more strongly correlated with leaf water content than with leaf nitrogen (Scriber 1978, Scriber and Feeny 1979). Leaves of trees and shrubs typically have lower water content than leaves of forbs, and tree-feeders correspondingly have lower consumption and growth rates than forb-feeders. Larvae feeding on woody plants may be particularly sensitive to leaf water content while larvae feeding on the relatively succulent leaves of forbs may be more sensitive to variation in leaf nitrogen content (Scriber and Feeny 1979).

Comparisons among herbivore species feeding on a variety of plant species do not necessarily apply to responses of herbivores to varying water status of individual host plants. White (1969, 1974, 1976) in particular has emphasized the potential importance of the increase in free amino acids in plants subject to drought stress: because of the qualitative change in nitrogen content, host plants with low water content may serve as better sources of food for insects. The grasshopper *Melanoplus differentialis*, for example, prefers wilted over turgid sunflower leaves, and grows better on a diet including wilted tissue; chemical changes other than loss of water itself may explain the preference for and improved growth performance associated with wilted tissue (Lewis 1979, 1982, 1984). The desert locust, *Schistocerca gregaria*, also prefers wilted over turgid leaves in some host species, but not, however, in others (Bernays and Lewis 1986). Gall-forming Cecidomyiid flies vary in their response to water stress in creosote bush (*Larrea tridentata*); densities of these flies may be more sensitive to plant architectural rather than biochemical changes associated with water stress (Waring and Price 1990).

Seasonal variation. Both nitrogen and water content generally decline in leaf tissue as leaves age (Scriber and Slansky 1981). The life cycles of many phytophagous invertebrates are adapted to track these seasonal variations in food quality. Flush-feeders are timed to exploit the rapid but ephemeral burst of succulent new plant growth with high N content, while mature/senescence feeders are timed to exploit older, seemingly less nutritious (yet long-lasting) plant tissue (White 1978, 1993). The feeding activities of phytophagous invertebrates may also lead to seasonal changes in their host plants; shoot tissue of snakeweed (*Gutierrezia microcephala*) individuals fed upon by nymphs of the grasshopper *Hesperotettix viridis* had significantly lower nitrogen and water content by midsummer than did shoot tissue of individuals protected from grasshoppers (Parker 1984).

Allelochemicals. Vascular plants produce a great diversity of "secondary substances" with apparently no clear role in primary metabolism. Many of these compounds (e.g., alkaloids and terpenoids) act as allelochemicals inimical to phytophagous invertebrates (Fraenkel 1959, Whittaker and Feeny 1971, Rosenthal and Berenbaum 1991). Plant allelochemicals may inhibit invertebrate feeding (Bernays 1983, Chapman 1990) or interfere with digestion and incorporation of nutrients, thereby slowing growth and/or reducing fecundity (Reese 1979). They may also be toxic to herbivores, either killing them outright, as is the case in root exudates that kill certain plant-feeding nematodes (Sayre 1971), or weakening their resistance to pathogens and other stresses (Stubblevine and Langenheim 1977). A particularly unique mode of action is found in wild potatoes, *Solanum berthaultii*, that repel aphids by producing the insects' alarm pheromone (Gibson and Pickett 1983).

Individual allelochemicals may act in several ways and vary in activity among herbivores. Weis and Berenbaum (1989), for example, note that tannins, the polyphenols present in foliage of many woody plants, may act as feeding repellents, digestibility-reducing compounds, and/or toxins causing lesions of insect gut walls. Furthermore, allelochemicals within plants may often interact with each other in having complex and diverse effects on phytophagous invertebrates (Berenbaum 1985).

Host plant allelochemicals thus serve to "poison the platter" (Brattsten 1979) for rangeland invertebrates, which must balance their needs for nutrients against the need to avoid these noxious compounds (Bernays and Chapman 1978). While other factors may also be involved (Bernays 1988, Bernays and Graham 1989), the distinctive secondary chemical profiles of plants are generally thought to play a central role in the evolution of herbivore diets (e.g., Ehrlich and Murphy 1988).

Several major types of solutions to the problems posed by host plant chemistry can be identified in phytophagous invertebrates. They may feed selectively, thereby avoiding encounter. Many allelochemicals serve as feeding deterrents to the grass-feeding grasshopper *Locusta migratoria*, for example; as a result, this insect feeds readily only on grasses, which do not contain sufficient concentrations of allelochemicals to inhibit feeding (Bernays and Chapman 1977; Bernays and Barbehehn 1987). On given host plants, furthermore, insects may feed selectively on those tissues with particularly low concentrations of allelochemicals (e.g., Cornell 1989).

Phytophagous insects may also neutralize the chemical defenses of host plants before feeding (e.g., Tallamy 1986). Many plants incorporate allelochemicals in gummy resins or latex stored in secretory canal systems such as laticifers and ducts (Farrell et al. 1991). Held under pressure, these chemical defenses exude rapidly to contact and often engulf invertebrate herbivores as they bite into leaves. Many species of phytophagous insects sever or trench across leaf veins, thereby cutting off the flow of chemical defenses to more distal portions of the leaf on which the insect subsequently feeds (Dussourd and Eisner 1987, Dussourd and Denno 1994).

If allelochemicals are ingested, phytophagous invertebrates may possess various abilities to detoxify them or otherwise render them ineffective (Brattsten 1992). The allelochemical may be excreted rapidly with or without prior metabolic degradation, it may be stopped by special barriers or channeled into specialized storage tissue (Duffey 1980), or it may be unable to act because of special modification of the target organ (Martin and Martin 1984). Prominent amongst the metabolic machinery protecting the consumer are the mixed function oxidases (MFOs), which catalyze conversion of lipophilic toxins to more readily excretable hydrophilic substances, and which are capable of rapid induction as the need arises (Brattsten et al. 1977; but see Gould 1984).

The effectiveness with which phytophagous invertebrates have evolved physiological and biochemical protection against plant allelochemicals is seen in the many cases in which these consumers are adapted to feed on plants laced with highly toxic compounds (Brattsten 1979, Bernays 1983). Adapted herbivores may even evolve to use plant allelochemicals for their own benefit, often as long-distance cues in locating hosts or as feeding stimulants (Finch 1980). The tree locust *Anacridium melanorhodon* uses plant phenols in its own metabolism; the allelochemicals are bound into the insect's cuticle where they may stabilze protein. The insect, which feeds on tree leaves of low N content, is thus able to conserve amino nitrogen by not having to generate its own phenol via the tyrosine pathway (Bernays and Woodhead 1982). Although many insects are specially adapted to feed on particular plants, they may nevertheless be significantly affected by variation in allelochemical content and composition of their host plants (Berenbaum et al. 1989).

Plants may often derive chemical protection from invertebrates through their association with endophytes, widespread ascomycete fungi that grow within plant tissues and produce toxins. Livestock have long been known to be adversely affected by consuming grasses infected with endophytes (e.g., Clay 1988). There has been much recent interest in the seemingly mutualistic relationship of plants and endophytes, with a growing number of studies focusing on implications for invertebrate nutrition and growth (e.g., Carroll 1988, Clay 1988, Patterson et al. 1991, Knoch et al. 1993).

Spatial and Temporal Patterns ("Escape in Space and Time")

In addition to physical and chemical attributes of rangeland plants, their spatial and temporal characteristics can influence their colonization and use by phytophagous insects. On a broad spatial scale, the size of the geographic range of a host plant is an important determinant of the number of phytophagous insects associated with it (Southwood 1961, Strong 1979, Strong et al. 1984). Widely distributed plant species serve as hosts to greater numbers of phytophagous insects than do more geographically restricted plant species. This general pattern probably arises because more widely distributed plants generally occupy a greater diversity of habitats within their ranges, they are "exposed" to colonization and adoption by insect species over a greater area, and their larger range permits greater overall population sizes of associated phytophagous insects, which are therefore less subject to extinction. The number of phytophagous insect species occurring on the host plant at any location within its range (i.e., the local insect species richness), however, is not necessarily greater for a more widely distributed plant (Strong et al. 1984; see also Cornell 1993).

On a smaller spatial scale, the physical and biological attributes of a plant's immediate surroundings will also influence the degree to which it is discovered and fed upon by phytophagous invertebrates. Insects can be highly sensitive to microclimate, for example. They may prefer sunlit over shaded microenvironments or vice versa, and plants may be more or less subject to herbivory accordingly (e.g., Lincoln and Mooney 1984, Collinge and Louda 1989). Chrysomelid beetles of the genus *Chrysolina*, for example, avoid shaded areas when ovipositing on their host plant, Klamath weed (*Hypericum perforatum*). The beetles are highly effective in controlling the weed in the sunny, open rangelands of California but not in the more wooded, shadier rangelands of Australia (Clark 1953, Huffaker and Kennett 1959).

The species composition and density of adjacent vegetation can also markedly influence exploitation of individual host plants by their insect associates (Root 1973, 1975, Atsatt and O'Dowd 1976, Kareiva 1983, Stanton 1983). The spatial patterns of neighboring conspecifics (e.g., host plant patch size; Bach 1988a,b) can be of key importance. In tallgrass prairie, the intensity of attack on seed pods of individual milkweeds, *Asclepias viridis*, increases with increasing density of neighboring conspecific host plants, as the number and reproductive status of these nearby host plants influences the rates of immigration and emigration of the bug

Lygaeus kalmii onto and from individual milkweeds (Evans 1983). In contrast, the grasshopper *Hesperotettix viridis* is less likely to consume seedlings of snakeweed (*Gutierrezia microcephala*) when these grow in the immediate vicinity of conspecific adults (Parker 1982); the grasshopper generally moves directly among canopies of the mature shrubs and is thus less likely to discover nearby seedlings among mature shrubs than in the open. However, seedlings growing far from adult plants, for example seedlings sprouting after a rangeland fire, may at least temporarily escape invertebrate herbivory (Whelan and Main 1979).

Heterospecific neighbors may also influence an individual host plant's exposure to herbivores. Heterospecific neighbors may provide "associational resistance" (Tahvanainen and Root 1972) leading to reduced herbivory, for example by inhibiting discovery in masking chemical or visual cues for phytophagous invertebrates (e.g., Rausher 1981). The presence of heterospecific neighbors may in other cases promote herbivory. Parker and Root (1981) found that the grasshopper *Hesperotettix viridis* quickly consumed seedlings of tansyaster (*Machaeranthera canescens*) transplanted to stands of the grasshopper's major host plant, snakeweed (*Gutierrezia microcephala*); the grasshopper may be the major factor preventing *M. canescens* from establishing in these stands. The effects of neighboring plants on herbivory can be quite complex. In a two-year experimental field study, Holmes and Jepson-Innes (1989) found in one year (but not the next) that tillers of blue grama (*Bouteloua gracilis*) were less likely to be attacked by grasshoppers when their nearest neighbor was a tiller of three-awn grass (*Arisitida* spp.) rather than a conspecific tiller. In both years, however, those tillers that were attacked by grasshoppers were fed upon more heavily when they grew next three-awn grass rather than conspecific tillers.

Just as plants may "escape in space" from herbivory depending on the nature of their immediate surroundings, they may also "escape in time" to varying degrees depending on how their life cycle timing meshes with that of phytophagous invertebrates. In tallgrass prairie, for example, individual flowers of wild indigo (*Baptisia australis*) vary in their likelihood of escaping consumption by insects depending on when they open within a season (Evans et al. 1989). Populations of the moth *Heliodines nyctaginella* (Heliodinidae), build up over the season in tallgrass prairie, inflicting greatest damage to late-flowering individuals of their host plant, umbrellawort (*Mirabilis hirsuta*) (Kinsman and Platt 1984). Within plant populations, individuals may be consistently early or late among years in life cycle timing, with consequences for the severity and nature of insect herbivory that they experience (e.g., Crawley and Akhteruzzaman 1988).

Variation in weather often affects the timing of herbivore activity relative to that of plant availability, leading to greater loss to herbivores in some years than in others (Breedlove and Ehrlich 1968). Wild indigo suffered heavier losses to blister beetles in the warm spring of 1986 than in the cooler spring of 1984 when the beetles colonized the host plants relatively late in the flowering season (Evans et al. 1989). Poor weather may also limit the searching time available to phytophagous insects (e.g., Hayes 1981, Kingsolver 1989). Severe weather may even cause extinction of local populations of herbivores, as has been observed in butterfly populations (Ehrlich et al. 1972, 1980).

Plant traits influencing mutualists and natural enemies

Plant attributes may also influence the degree of herbivory indirectly by affecting relations with natural enemies of the herbivores and invertebrate mutualists (especially ants) of the plant (Price et al. 1980, Keeler 1989). Rangeland plants may possess structural modifications, for example, to attract and house mutualists. Particularly well-known are defense nectaries (generally synonymous with extrafloral nectaries; see Keeler 1989), or glands that secrete nectar but are not involved in pollination (Bentley 1977). The extrafloral nectar often contains amino acids in addition to sugar (Baker et al. 1978). Ants are attracted to the nectaries, and defend the plant and its nectaries by driving off herbivores. In the grasslands of the North American Great Plains, for example, ants visiting the sepal-surface defense nectaries of bush morning glory (*Ipomoea leptophylla*) significantly reduce flower damage by grasshoppers and seed losses to seed beetles (Keeler 1980). Secretion by defense nectaries is often greatest when the plant is growing rapidly and can benefit greatly from ant protection of succulent, young tissue (Beattie 1985). A variety of herbivore natural enemies (parasitoids, predatory wasps, ladybeetles) also visit defense nectaries and thus may contribute to reduced herbivory on plants bearing these glands (Beattie 1985).

The extreme in structural modification to accomodate ants as mutualists is seen in those rangeland plants housing ants (myrmecophytes). These plants, which house ants in hollow stems, swollen thorns, and galls, are apparently restricted to the tropics, perhaps because ants are unable to survive temperate zone winters in thin-walled plant domiciles (Keeler 1989). Best known are the acacias of tropical Africa and America whose mutualistic ants patrol the trees day and night (Janzen 1966, Hocking 1970). Although the mutualism may have evolved especially in response to grazing pressure from large mammals (both ant-acacias and large grazers are absent in Australia; Brown 1960), the mutualism serves to protect the plant from phytophagous invertebrates as well (Rehr et al. 1973).

Natural enemies may be affected by many of the same plant traits of importance to their phytophagous invertebrate prey (Table 2). Physical features such as the presence of wax or trichomes on leaf surfaces and the degree of leaf overlap may impair the foraging efficiency of these enemies (Shah 1982, Obrycki 1986, Kareiva and Perry 1989). The architectural complexity of the host plant (e.g., trees

Table 2. Traits of host plants and their neighbors that can affect herbivory indirectly through effects on herbivore natural enemies (see text for details and examples).

Physical features
morphology and structural complexity (e.g., shrub vs. herb)
nature of plant substrate (e.g., presence of trichomes)
Chemical attributes
low nutritive quality (prolonged herbivore development and exposure to enemies)
plant volatiles (attraction of natural enemies)
allelochemicals (deleterious effects on enemies from feeding on the herbivore)
Influences of neighboring plants
interference with host location by enemies
provision of alternate prey and/or nectar for enemies

and shrubs versus herbs and grasses) may also influence the number of parasitic insects attacking a herbivore (Hawkins 1988a, 1990); furthermore, the intensity of attack by parasitoids depends on the degree of exposure in the feeding niche of the herbivore (e.g., leaf-feeder vs stem-borer; Gross 1991). Indirect effects of natural enemies on their prey, for example, reduction in growth rate as the herbivore alters its behavior in reacting to natural enemies, may also depend on host plant morphology and architecture (Stamp and Bowers 1991).

Chemical traits of plants may influence herbivory in several ways through effects on natural enemy impact. Low nutritive quality may serve to reduce herbivory by lowering the growth rate of phytophagous invertebrates, thereby exposing them to increased risk of predation (Feeny 1975, Slansky and Feeny 1977, Moran and Hamilton 1980). Volatile plant allelochemicals may serve to attract natural enemies to either intact or damaged plants (Vinson 1985, Nordlund et al. 1988, Whitman 1988, Turlings et al. 1990). Plant allelochemicals may also reduce control by natural enemies, either by deterring them from attacking phytophagous invertebrates that sequester the compounds (Duffey 1980, Denno et al. 1990) or by harming them upon ingestion of prey (Campbell and Duffey 1979). Indeed, the ability to sequester plant allelochemicals as a defense against generalist predators may favor the evolution of host specialization in herbivores (crypsis may also be important; Bernays and Graham 1988).

Finally, neighboring plants may also influence the degree of protection a plant derives from natural enemies. Heterospecific neighbors, for example, may depress enemy activity by interfering with host location (Monteith 1960, Shahjahan and Streams 1973), or they may enhance enemy activity by providing alternate prey and/or nectar (Atsatt and O'Dowd 1976, Russell 1989). Price et al. (1980) and Rowell-Rahier and Pasteels (1992) stress the need to consider impacts of plant traits on natural enemies as well as phytophagous invertebrates, and highlight many potential "tritrophic" interactions.

IV. Responses of Individual Plants to Invertebrate Herbivory

Physiological responses: compensatory growth and reproduction

Range plants may exhibit various physiological responses to tissue damage by herbivores, including elevated rates of photosynthesis in undamaged and/or regrowth tissues; changes in assimilate allocation (e.g., increased production of new leaves, tillers or shoots, often at the expense of root growth); and perhaps increased efficiency of water use (McNaughton 1979, Caldwell et al. 1981, Hilbert et al. 1981, Belsky 1986, Detling 1988, Hendrix 1988, Trumble et al. 1993). These intrinsic responses may interact with extrinsic mechanisms (e.g., improved microclimate and increased soil nutrient availability for remaining tissues) to ameliorate the impact of herbivory (McNaughton 1983). One can distinguish several potential patterns in plant growth (and reproduction) associated with these responses. Such responses may lead to compensatory growth in damaged plants wherein primary production equals or exceeds that of undamaged plants (exact and overcompensation, respectively; Belsky 1986). Furthermore, even when undercompensation occurs (primary production of damaged plants is less than that of undamaged plants), partial compensation may occur if more biomass is produced than would be expected without such responses to herbivore damage (Belsky 1986).

The degree of compensation by a given plant may depend on many factors, including severity and timing of defoliation, distribution of damage over the host plant, microclimatic and weather conditions, availability of water, light, and nutrients, competition with other plants, and length of time available for recovery (Bardner and Taylor 1974, Boscher 1979, Hendrix 1979, Crawley 1983, Solomon 1983, Stamp 1984, Maschinski and Whitham 1989, Marquis 1992). The ability of tansy ragwort (*Senecio jacobaeae* L.) to compensate for damage from the cinnabar moth, for example, depends upon available soil moisture following defoliation (Cox and McEvoy 1983). The morphological and physiological factors so far implicated in interspecific differences among plants in compensatory ability are reviewed by Maschinski and Whitham (1989); these include growth form, patterns of life-history and carbon allocation, and rates of photosynthesis and growth.

Many studies of plant compensatory responses to herbivory have involved grazing by vertebrates or artificial clipping to mimic feeding by herbivores (most often vertebrate grazers). One must exercise caution in extrapolating these experimental results to interactions of rangeland plants and phytophagous invertebrates, as damage from vertebrates and artificial clipping rarely mimics the complex and intricate patterns of damage inflicted by invertebrates (Capinera and Roltsch 1980, Owen and Wiegert 1987, Hendrix 1988). In addition to these studies, a smaller number of studies indicate partial to exact compensation by the host plant in response to damage by phytophagous invertebrates. When stripped of its aboveground tissue by the cinnabar moth (*Tyria jacobeae* (L.), Arctiidae), tansy ragwort initiates regrowth shoots from the root crown; these produce numerous but small seeds (Islam and Crawley 1983). Thistles (*Jurinea mollis*) produces multiple rosettes when the basal rosette is damaged by caterpillars (Inouye 1982). Floral sex expression in late-maturing umbels of wild and cow parsnip (*Pastinaca sativa* L. and *Heracleum lanatum* Michx.) is altered by damage from the moth *Depressaria pastinacella* (Duponchel) (Oecophoridae) to primary umbels. As a result, more seeds are produced in late-maturing umbels of previously damaged than of undamaged plants (these seeds, however, are smaller than those from primary umbels) (Hendrix 1979, 1984, Hendrix and Trapp 1981).

Overcompensation and plant/herbivore mutualism

The possibility that host plant overcompensation may result in response to herbivory has led to the general hypothesis that herbivores at least sometimes may act as mutualists by benefitting plants (Owen and Wiegert 1987). In a speculative paper, Owen and Wiegert (1976; see also Owen 1978, 1980) offer several plausible scenarios in which insects may assist plants in obtaining soil nutrients. For example, plant growth is often limited more by availability of soil nitrogen than by photosynthetic capacity. In excreting sugar water (honeydew) upon ingestion of "surplus" plant carbohydrates, aphids may stimulate energy-limited, nitrogen-fixing microbes in the soil beneath the plant canopy, thereby increasing the availability of nitrogen to plant roots. In dry African savannas, xylem-feeding spittlebugs recycle through excretion large quantities of water to the soil surface, perhaps thereby enabling plant roots to take up otherwise unavailable nutients. When insects defoliate plants, the associated rain of leaf fragments and fecal material may provide organic material and nutrients to the soil during times when these are otherwise limiting. Insect herbivores may benefit host plants in other ways as well. Caterpillar feeding often serves to prune shrubs and trees, resulting in increased production of lateral branches (Whitham and Mopper 1985, Owen and Wiegert 1987).

The possibility that phytophagous invertebrates may sometimes benefit plants is intriguing, but convincing supporting evidence is lacking as yet. Grasshopper saliva, for example, has been suggested to stimulate grass growth (Dyer and Bokhari 1976), but carefully designed experiments have failed to support this hypothesis (Capinera and Roltsch 1980, Detling and Dyer 1981). The literature contains numerous cases where overcompensation in response to herbivory might be inferred. A critical review, however, indicates that unlike partial compensation in growth and reproduction, overcompensation by non-crop plants is not well-documented (Belsky 1986, 1987; see also McNaughton 1986 in response). Paige and Whitham (1987; but see also Mashinski and Whitham 1989, Aarssen and Irwin 1991) present an example of overcompensation in scarlet gilia (*Ipomopsis aggregata*), seemingly increasing plant fitness, in response to vertebrate herbivory; no similarly complete and convincing studies involving invertebrate herbivory yet exist.

Phytochemcial changes induced by herbivory

When a phytophagous invertebrate damages plant tissue, chemical changes may be triggered in the plant in response (Tallamy and Raup 1991). Many of these changes are seemingly defensive adjustments (Rhoades 1979, Edwards and Wratten 1983, 1985, Zangerl 1990, but see also Tuomi et al. 1984, 1990, Karban and Myers 1989, Bryant et al. 1991), analogous to the well-documented responses of plants to infection by microorganisms (Levin 1976; indeed, these adjustments may have evolved as defenses against infection following herbivore damage). Chemical changes in the host plant and adverse consequences for herbivores are outlined in Figure 4. It should be noted, however, that in some instances when host plants are lightly or moderately attacked by herbivores, these changes may be beneficial for later herbivores (Williams and Myers 1984).

It is difficult at present to assess the importance of induced phytochemical changes in interactions of plants with phytophagous invertebrates (e.g., implications for insect population dynamics, Rhoades 1985; patterns and significance of dispersion of insect damage among and within leaves, Silkstone 1987, Wratten et al. 1988). The experimental evidence so far amassed is in many cases flawed statistically and/or open to alternative explanations (Fowler and Lawton 1985, Bergelson et al. 1986, Haukioja and Neuvonen 1987). A good example is the possibility of interplant communication (as in so-called "talking trees"). Preliminary experimental evidence suggests that chemical changes may be induced in nearby host plants when herbivores damage a given plant (Baldwin and Schultz 1983, Rhoades 1983). The interplant communication presumably involves volatile chemical messengers emitted by the damaged host and detected by neighbors. This intriguing possibility still awaits rigorous experimental confirmation (Fowler

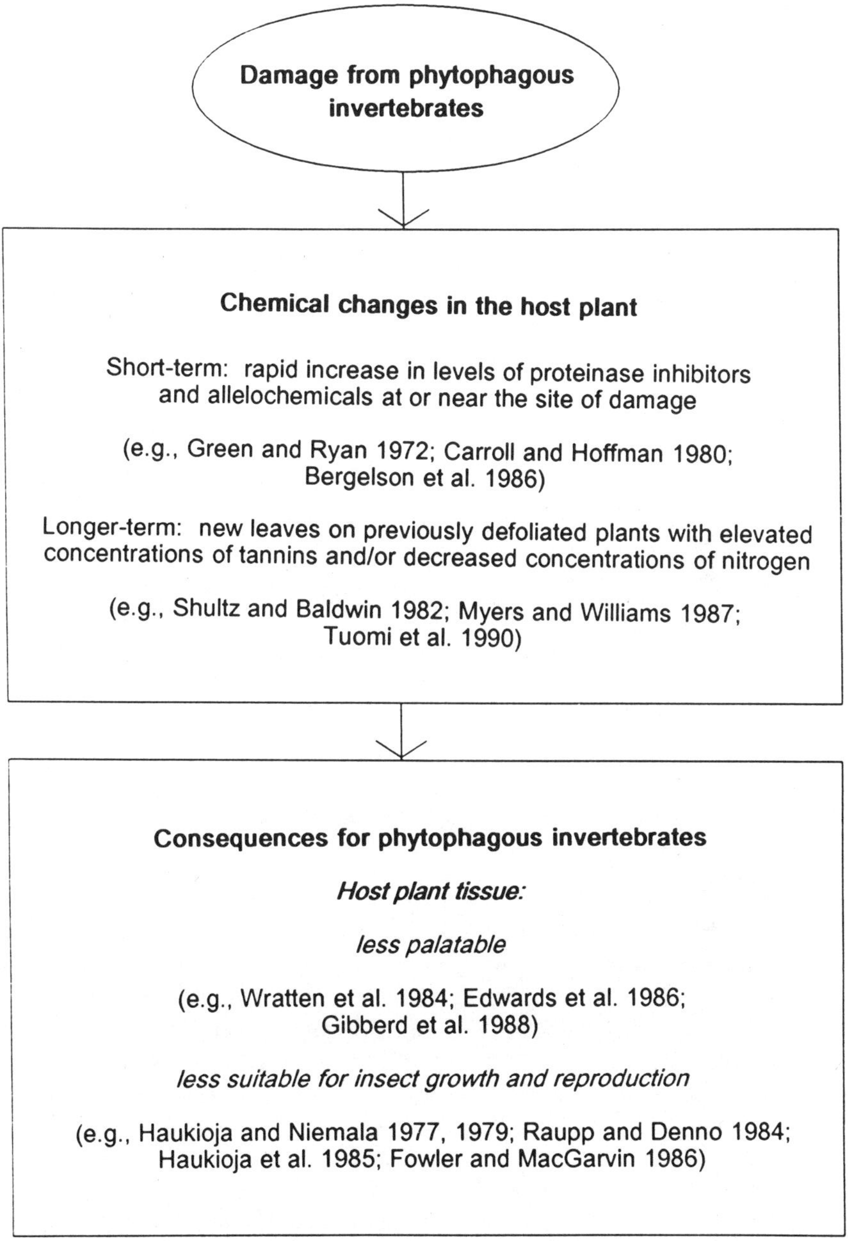

Figure 4. Chemical changes in host plants following damage from phytophous invertebrates, and consequences of these changes in the host plant for the herbivore.

and Lawton 1985). Recent experiments with big sagebrush (*Artemisia tridentata*) show that airborne methyl jasmonate from sagebrush leaves can induce synthesis of proteinase inhibitors in neighboring plants (Farmer and Ryan 1990).

Induced leaf abscission

Feeding by phytophagous insects often leads to early abscission of damaged leaves by the host plant (reproductive tissue may also be aborted in response to insect damage; Stephenson 1981). Faeth et al. (1981) drew attention to this phenomenon, pointing out that in some cases, this early abscission may result in considerable mortality to leaf-mining insects. Williams and Whitham (1986) documented that 98+% of leaf-galling aphids, *Pemphigus* spp., died when their host leaves (*Populus* spp.) were prematurely shed. These authors suggest that early abscission may be an induced plant defense in *Populus* against herbivory; Stiling and Simberloff (1989; see also Stiling et al. 1991), however, argue that premature leaf fall associated with insect herbivory is best interpreted as a special case of a more generalized response of leaves to physical damage. In other cases, however, leaf-miners and gallers may be adapted to early leaf abscission, either in maturing rapidly before leaves are shed and/or in manipulating plant physiology (e.g., cytokinin production) so that "green islands" of suitable host material surround the mine even after a leaf is shed (e.g., Kahn and Cornell 1983, Pritchard and James 1984a,b). Kahn and Cornell (1989) provide an example in which early leaf abscission has neglible and sometimes even positive effects on leaf-miner survivorship; the holly leaf-miner, *Phytomyza ilicicola* pupates before leaves are abscissed, and pupae in fallen leaves are less subject to parasitism than those in leaves not abscised. Thus, while the phenomenon of early leaf abscission in response to insect damage is widespread, its consequences for phytophagous insects seem quite variable from case to case.

V. General Patterns of Plant Defense Against Invertebrate Herbivores

Because they are capable of inflicting heavy damage to their host plants, phytophagous invertebrates may act as selective forces eliciting plant adaptations over evolutionary time. The widespread existence of chemical and morphological traits that act to protect plants against herbivory reinforces the notion that plants defend themselves against phytophagous invertebrates. Interest in the evolution of such plant/herbivore interactions was sparked by the seminal papers by Fraenkel (1959) and Ehrlich and Raven (1964). In the mid 1970s, a convergence in thinking by Feeny (1975, 1976) and Rhoades and Cates (1976; Cates and Rhoades 1977) resulted in broad generalizations to characterize the evolutionary ecology of plant defense. These generalizations came to be known as the plant apparency hypothesis.

Two ends of a continuum are highlighted in the hypothesis. At one extreme, there exist rare or ephemeral (e.g., early successional) plants (or plant parts) with limited exposure to herbivores (these plants "escape in space and time"). Such plants were hypothesized by Feeny, Rhoades, and Cates to defend themselves chemically by producing highly toxic compounds of low molecular weight (e.g., mustard oils). Because most insects have limited contact over evolutionary time with these plants, few insects become metabolically adapted to the toxins. Consequently, the plant must expend little energy to produce only small quantities of these "qualitative" defensive compounds that act in all or none fashion to deter non-adapted generalist herbivores. Such defenses are, however, fairly easily countered by adaptation in those relatively few insects evolving to specialize on that host plant.

In contrast, there exist other plants that are abundant and widespread over space and time, including such late successional species as oaks. Over evolutionary time, these species have much greater contact with many more herbivore species (they are highly "apparent" to insects) and must invest heavily in chemical defenses that, unlike qualitative defenses, cannot easily be overcome by counteradaptation. Consequently, these plants accumulate high concentrations of chemicals that act in dose-dependent manner by slowing insect growth and reproduction (e.g., by reducing or preventing digestion), rather than killing insects outright. These "quantitative defenses", which include tannins in woody plants and perhaps silicates in range grasses, are expensive to produce and/or maintain.

The plant apparency hypothesis stimulated tremendous interest in general patterns of plant defense. Close examination suggested needed modifications. The distinction between qualitative and quantitative defenses is not clearcut. Allelochemicals classed as qualitative defenses may act in dose-dependent manner, while quantitative defenses may sometimes also be highly toxic (e.g., see discussions in Barbosa 1988 and Karowe 1989). Furthermore, insects may adapt to quantitative defenses over evolutionary time as readily as they do to qualitative defenses; rangeland grasshoppers furnish excellent examples (Bernays et al. 1980, Bernays 1981). The roles of tannins as plant defenses have especially been scrutinized. Martin et al. (1987) conclude that tannins do not generally act as digestibility-reducing compounds in insects, as originally proposed, but rather act as toxins and feeding deterrents. The plant apparency hypothesis is also predicated on the assumption that generalist herbivores are less efficient in metabolizing qualitative defensive compounds than are more specialized herbivores, but this assumption is not well supported (Fox and Morrow 1981). Finally, it is difficult to define "apparency" operationally in applying the plant apparency hypothesis.

Mattson (1980) pointed out that many toxic, qualitative defensive compounds have high nitrogen content (e.g., alkaloids and glucosinolates), in contrast to many digestibility-reducing, quantitative defensive compounds (e.g., tannins). "A tempting conclusion is that N-rich plants and plant tissues employ mainly N-based allelochemics, whereas N-poor plants and plant tissues employ mainly non-N-based allelochemics" (Mattson 1980: 142). Coley et al. (1985) subsequently developed the "resource availability hypothesis", in which a continuum of plant defense patterns exists along a gradient of resource (light, water, nutrients) availability. At one extreme, plants growing in resource-poor sites are selected for slow growth to withstand stressful periods. Leaves are long-lived and as such, are important to protect. Consequently, large quantities of carbon-based allelochemicals such as tannins, lignins, and other quantitative defenses are produced. Although initially expensive to produce, these defensive compounds are relatively immobile with little turnover in leaf tissue, thus reducing the metabolic costs of maintenance. In contrast, plants growing in resource-rich sites are able to grow quickly. Leaves are short-lived and easily replaced; they are therefore only lightly protected by small quantities of mobile, inducible, nitrogen-based compounds. The metabolic cost of maintaining this chemical defense is relatively high, but the high turnover permits nutrient resorption before leaf fall and reversible commitment to defense.

The amounts and types of chemical defense thus are hypothesized by Coley et al. (1985) to be determined especially by the plant's nutrient status; nitrogen-fixing legumes of rangelands, for example, produce mobile, N-based alkaloids, whereas desert shrubs, limited more by availability of nitrogen than of light and ability to fix carbon, produce carbon-based terpenes. The plant's nutrient status is often but not always related to its apparency to herbivores in space and time. Fast-growing plant species in resource-rich environments generally experience greatest herbivory. Thus, as the basic plant-soil nutrient relationship varies, so does the nature of plant defense against herbivores, and consequently so do the basic patterns of herbivory in different habitats (see Herms and Mattson 1992 for a thorough, recent assessment of the evolutionary trade-offs of plant growth and defense).

VI. Plant Community and Ecosystem Impacts

Rangeland primary productivity

The impact of phytophagous invertebrates on primary productivity in rangelands is generally not well known, especially during non-outbreak periods. In grasslands and other ecosystems generally, the percentage of host plant productivity ingested by single species of insects has been reported to vary from <<1% to 19% (mean 3.5%; Wiegert and Petersen 1983). Detling (1988) summarizes studies indicating phytophagous invertebrates as a group may typically consume 15% or less of net primary productivity aboveground in grasslands and savannahs, and perhaps somewhat more belowground. Nematodes alone, for example, consumed or wasted 6-13% of belowground net primary productivity in the mixed grass prairie of South Dakota (Ingham and Detling 1984), and application of systemic nematicide to mixed grass prairie increased foliage production 25–60% (Smolik 1977, Ingham and Detling 1990). Scott et al. (1979) and Anderson (1987) concluded from reviewing the literature that root-feeding arthropods and nematodes consume 7–26% of net primary productivity under non-outbreak conditions. The complexities of estimating the combined impacts of individual species are illustrated by the intensive study by Menhinick (1967). Menhinick estimated that insect herbivores consumed about 1% of aboveground net primary productivity in a stand of lespedeza (*Lespedeza cuneata*), as compared to 2–10% in similar studies of salt marshes and forests.

Acridid grasshoppers are the one group of rangeland invertebrates whose ingestion of net primary productivity has been relatively well-studied (Hewitt 1977). The percentage of host production consumed by populations of individual species of rangeland grasshoppers varies from 1–3% (Bailey and Riegert 1973, Mispagel 1978) to 99% during outbreaks (Nerney and Hamilton 1969). Hewitt and Onsager (1983) estimate that on average at least 21–23% of available range vegetation is destroyed annually by grasshoppers (all species combined) in the western United States. Estimates from other grassland ecosystems where grasshoppers are generally less abundant range from 1 to 10% (Kohler et al. 1987).

A simple accounting of the calories of energy or grams of tissue that are extracted from rangeland plants by phytophagous invertebrates of course does not reflect fully the potential impact of these herbivores on primary productivity. Insect outbreaks may kill large numbers of plants, temporarily reducing primary productivity. However, the increase in available nutrients and light may lead to increased primary productivity in subsequent years (Mattson and Addy 1975, Romme et al. 1986). Primary productivity may also be more immediately stimulated at less destructive levels of invertebrate herbivory by the rapid recycling of key nutrients back to the soil in chewed leaf fragments and feces (Owen and Wiegert 1976). Lightfoot and Whitford (1990), for example, found that foliage arthropods of creosote bush enhanced nitrogen flux in desert rangelands, as these arthropods' excretions accounted for about 20% of the nitrogen flux from plant canopy to soil during the period of peak plant productivity. Not all of the readily-leached nutrients in insect frass (and remains) may be retained by the ecosystem, however (Swank et al. 1981, Hollinger 1986). Lightfoot and Whitford (1990) note that herbivores may also affect nutrient cycling rates by stimulating plant growth and nutrient uptake; by affecting plant litter production, nutrient content, and decomposition; and by increasing rates at which nutrients leach from intact foliage (see also Christensen et al. 1989a,b,c).

The process of herbivory may in some cases create a strong positive feedback loop between plants and herbivores. This phenomenon is best known from the "grazing lawns" exploited by large ungulates (McNaughton 1979, 1985). More recently, Holland and Detling (1990) have demonstrated that grazing acts to remove a bottleneck in the nitrogen cycle of rangelands. Foliage removal results in less carbon allocation to roots. This change, in turn, results in less root detritus production and a concurrent reduction in microbial immobilization of nitrogen. Inorganic nitrogen is therefore more available to plant roots, which is then used to create more photosynthetically active plants. The loop may be maintained as long as a plant root system is maintained that is capable of efficiently harvesting the available nitrogen.

Rangeland invertebrates may also influence primary productivity in other, relatively subtle ways. Grasshoppers may destroy seed heads, preventing range reseeding and thereby reducing future primary productivity (Hewitt 1977). Grasshoppers also act as "litter-making machines" in that they ingest only a fraction of the pieces they clip from grasses; the remainder falls to the ground and becomes litter (Mitchell and Pfadt 1974, Kohler et al. 1987). Such litter build-up can depress primary productivity in mesic grasslands, e.g., tallgrass prairie (Knapp and Seastedt 1986). In desert grasslands, however, litter build-up may increase primary productivity by increasing rainfall infiltration, and reducing rates of evaporation of soil moisture (Bodine and Ueckert 1975). In the shortgrass ecosystem of west Texas, the termite *Gnathamitermes tubiformans* significantly reduces the standing crop of grasses in wet years; it may also indirectly affect primary productivity by removing large quantities of litter (Bodine and Ueckert 1975). Finally, phytophagous invertebrates may subtly influence primary productivity by influencing the outcomes of interspecific competition among plants differing in patterns of growth, productivity, and reproduction.

Impacts on plant population and community dynamics

The impact of vertebrate herbivores on rangeland plant community patterns is well-documented (Ellison 1960, Detling 1988). Extensive field experiments with cattle, sheep, and goats in particular provide the basis for incorporating vertebrate herbivory with other mechanisms (for example, unusual climatic events, fire positive feedback) generating the diverse successional states and transitions now recognised to characterize many rangeland plant communities (Westoby et al. 1989). A similar potential of phytophagous invertebrates to influence the species composition of rangeland plant communities is strikingly illustrated by the classic case of biological control of the alien Klamath weed (*Hypericum perforatum*) in the far western United States. Following introduction to California in 1947, two leaf-feeding beetles (genus *Chrysolina*, Chrysomelidae) rapidly reduced the weed to a very small fraction of its former abundance, thus allowing a diversity of rangeland grasses and forbs to flourish (Huffaker and Kennett 1959). No comparable example involving native plants and insects has yet been documented in rangelands, however.

The impact of phytophagous invertebrates on plant population dynamics, in contrast to their impact on individual plant performance, is not well-studied and may in many cases be negligible (Crawley 1988, 1989, Andersen 1989). In the coastal sage scrub communities of California, insects inflict heavy damage on flowers and seeds of goldenweed (*Haplopappus squarrosus* and *H. venetus*). Exclusion of these insects led to greater seedling establishment in both species, but increased recruitment of juveniles only in the former species (Louda 1982a, 1982b, 1983). In the latter species, high and geographically variable seedling mortality (probably from vertebrate herbivory) was the key limiting process. Richman and Huddleston (1981) and Parker (1985) document the heavy impact of insect herbivores on the demography of the rangeland shrubs, *Gutierrezia* spp. Parker (1985) suggests that a well-plenished seed bank may permit host plant populations to persist by rapidly renewing population recruitment during periods of reduced herbivore numbers. Heavy insect defoliation is associated with and may be a major contributor to tree mortality in the eucalypt dieback syndrome of Australian rangelands (Lowman and Heatwole 1992). These studies suggest that phytophagous invertebrates may at least on occasion be influential in population dynamics of rangeland plants, but additional experimental work is needed.

The impact of phytophagous invertebrates on rangeland plant communities is likely to be generally more subtle than that observed in the dramatic case of Klamath weed, in which large numbers of individual plants were killed outright by the feeding activities of the beetle larvae. Less extensive but differential damage from herbivores may alter the competitive balance in interactions among rangeland plants, thus ultimately influencing the absolute and relative abundance of species in rangeland plant communities (Harper 1977, Whittaker 1979, Schowalter 1981, Gange and Brown 1989). Indeed, this is generally the case in successful cases of biological control (using introduced insects) of rangeland weeds (Harris 1988). For example, in a drought year when grasshopper populations were dense, the negative effect of intra and interspecific competition (with grasses) on surviviorship of snakeweed (*Gutierrezia microcephala*) was greatest when grasshoppers were allowed to feed on the plant (Parker and Salzman 1985).

The possible interaction of herbivory with interspecific competition among plants has also been investigated experimentally in grasslands and other herbaceous communities by applying pesticides to reduce numbers of invertebrates on all plant species in experimental plots. Studies of early successional British grasslands indicate that insect herbivores have significant and often complex impacts on plant community dynamics; the collective action of these invertebrates generally slows the rate with which perennial grasses replace annual forbs (Brown 1984, 1985, Gibson et al. 1987, Brown et al. 1988, Hendrix et al. 1988, Brown and Gange 1989a). Significant impacts are associated with belowground as well as aboveground insect herbivores (Brown and Gange 1989b). Similarly, an outbreak of leaf-feeding beetles (*Trirhabda* spp.) altered the course of old-field plant succession in eastern Canada (McBrien et al. 1983; see also Cain et al. 1991). Four consecutive years of pesticide application to non-outbreak populations of above and belowground invertebrates in mature tallgrass prairie, however, had little effect on plant community patterns (D. Gibson et al. 1990).

The impacts of harvester ants on western North American rangelands have received considerable attention in recent years. The ants remove plants, thus denuding the area immediately surrounding their gravel nest mounds. At one Great Basin study site near Reno, Nevada, Clark and Comanor (1975) estimated that as many as 157-226 million annual plants per hectare are cut each year by harvester ants. Surprisingly, such heavy herbivory may have little effect on primary production because of what Wight and Nichols (1966) refer to as the "border effect." These authors report that in the shrub steppe of western Wyoming, production of the dominant plant, Nuttall saltbush (*Atriplex nuttalii*), was not reduced by the plant's removal from ant mounds; plants growing at the perimeter of the denuded zone grew especially vigorously (perhaps because of increased water availabilty) such that production did not differ between local areas with and without ant mounds.

Some Sonoran Desert plants are particularly associated with harvester ant nests and realize an increase in reproductive output when developing from seeds dispersed to ant-nest refuse piles (Rissing 1986). The species composition of seedlings establishing on ant mounds in annual grasslands of California also differs significantly from that of surrounding vegetation (Hobbs 1985; see also Nowak et al. 1990). In addition to cutting grasses and forbs, harvester ants in the Mojave Desert defoliate mature shrubs (but make no apparent use of the plant tissue); such defoliation may elevate death rates of the shrubs, with which the ants may compete for water (Rissing 1988). Harvester ants also forage for seeds at considerable distances from their nests. A reduction in postdispersal seed predation by chemical exclusion of harvester ants dramatically increased soil densities of seeds at a Sonoran Desert site (Reichman 1979). Furthermore, densities (relative as well as absolute) of germinating annuals at this site as well as at a second site in the Chihuahuan Desert were also greatly affected by chemical exclusion of ants (Inouye et al. 1980, Davidson et al. 1984, 1985, Samson et al. 1992; but see also Brown and Heske 1990). As major sources of postdispersal seed mortality, other ant species may also critically affect plant population and community dynamics of rangeland plants (e.g., Andersen 1987).

Biological control of rangeland weeds

As noted above, our most dramatic examples of the capacity of insects to affect rangeland plant communities come from biological control programs against weeds. The spectacular success in using insects to suppress populations of weeds such as prickly pear (Dodd 1959) and Klamath weed (Huffaker and Kennett 1959) has generated great interest in biological control of rangeland weeds. By far the most widely practiced form of biocontrol on rangelands has been "classical" biological control, wherein exotic (non-native) phytophagous insects are introduced to attack exotic weeds (e.g., Batra 1982, Rosenthal et al. 1985, Wapshere et al. 1989). The basic rationale in such efforts is that noxious weeds often gain a foothold in natural communities, and indeed are frequently competitively dominant, because unlike the native flora, they lack natural enemies (Harris 1988). The standard procedure has been to redress the situation by searching for natural enemies in the weed's presumed country of origin. This approach continues to yield dramatic success stories (e.g., McEvoy et al. 1991). Recently Hokkanen and Pimentel (1984, 1989) have proposed that natural enemies with little or no history of coevolution with the weed will be more "virulent" and therefore more effective as biological control agents. Their thesis has been challenged vigorously (Goeden and Lok 1986). Relatively few efforts have been made to introduce insects against native weeds (but see Hokkanen and Pimentel 1989); the wisdom of making such attempts has also been a matter of some controversy (Harris 1990).

In classical biological control, insects are introduced against a weed only after extensive research over a period of years, including evaluation of the weed problem, field studies of the plant/insect relationship in the country of origin, and intensive host specificity testing (Batra 1982, Rosenthal et al. 1985, Harris 1991). Obviously, great care must be taken to identify host-specific insects that will largely or entirely confine their feeding to the weed in question. While there will always be some risk that even a highly host-specific insect, once introduced, will extend its host range beyond the weed to attack desirable plants, careful host specificity testing can minimize this risk (e.g., Harris and Zwolfer 1968, Wapshere 1974; but see also Howarth 1991). As Harris (1988, 1990) points out, the native vegetation will certainly be affected if the major alternatives to biological control, namely chemical control or no control, are selected in managing rangeland weed populations.

Table 3. Ecological factors that may affect the degree of success achieved in classical biological control of weeds by insects (for discussion, see Goeden and Louda 1976, Crawley 1986, 1987, Myers 1987, Myers et al. 1989, Harris 1991).

Climate in country of release vs. origin
Weather conditions at or following release
Size of initial release (number of insects released)
Genetic composition of insect and weed populations
Native predators and parasitiods that attack the insect
Competitive ability of the insect
Life history attributes of the insect (e.g., intrinsic rate of increase)
Life history strategy of the host plant
Life history stage of the host attacked by the insect

Despite the great care and effort that is invested in each biological control program, many introduced insects fail either to become established or to consume substantial quantities of target weeds (Harris 1988). A number of ecological factors may critically affect the success of biological control programs against weeds (Table 3). Increasing concern over undesirable effects of herbicides, coupled with continuing lack of cost-effectiveness of herbicidal treatment of large acreages of rangeland, will likely stimulate much more research and many new developments in biocontrol of rangeland weeds in the coming years. Of particular interest are integrated pest management programs in which biological control is integrated with other approaches (Wapshere et al. 1989, Harris 1991).

VII. Interactions and Inter-Relationships Among Herbivores

The impacts of phytophagous invertebrates on rangeland plant populations and communities may depend on interactions among these herbivores and their vertebrate counterparts. Thus the effect of a given level of herbivory from one source may depend upon the nature and degree of other sources of herbivory (e.g., Strauss 1991; Christensen and Whitham 1993). Gibson et al. (1987), for example, concluded that the effects of sheep and insect grazing on the plant community of a British limestone grassland were not always additive. In contrast, however, Brown and Gange (1989) reported a lack of interactions in above- and belowground insecticide treatments to an early successional plant community.

Herbivores may also directly or indirectly affect each others' numbers. As a result, the level of herbivory from a given source may vary with the nature and degree of other sources of herbivory. There has been considerable interest, for example, in how grazing by livestock may influence populations of phytophagous invertebrates (e.g., Morris 1967, 1968, 1969, 1971, Freckman et al. 1979, Hutchinson and King 1980, Smolik and Lewis 1982, Smolik and Dodd 1983). The most intensively studied relationship is that of cattle and grasshoppers. The influences of cattle on grasshopper numbers are complex in that grazing not only reduces available plant biomass, but also generally changes species composition of the plant community. This leads to corresponding change in the species composition of associated grasshopper assemblages as individual species rise or fall in density, and may lead to higher or lower overall grasshopper densities (e.g., Smith 1940, Holmes et al. 1979, Capinera and Sechrist 1982, Jepson-Innes and Bock 1989, Quinn and Walgenbach 1990).

Moderate levels of vertebrate grazing sometimes stimulate population growth of phytophagous invertebrates. Greatest densities of white grubs, for example, were associated with moderate levels of grazing by sheep (Hutchinson and King 1980, Roberts and Morton 1985). Seastedt (1985) suggests that a moderate level of grazing may improve food quality for belowground phytophagous invertebrates by lowering the carbon-to-nitrogen ratio of root tissues. Consequently, moderately grazed rangeland may support the greatest numbers of these invertebrates. This pattern, observed both in Australia and North America, has not been reported in European grasslands, perhaps because these grasslands are heavily fertilized, and hence grazing has minimal detrimental effect on the food quality of roots for belowground invertebrates. Patterns of invertebrate abundance in these grasslands appear to be largely related to plant productivity (Curry 1987b).

The effects of rangeland phytophagous invertebrates on each others' population dynamics are not well-studied. The importance of interspecific competition among phytophagous insects has been a subject of continuing controversy (e.g., Hairston et al. 1960, Lawton and Strong 1981). Few experiments have been conducted to resolve the debate (Connell 1983, Schoener 1983, Fritz et al. 1986). Strong et al. (1984) tentatively concluded upon reviewing the literature that interspecific competition is generally of little importance in the population dynamics of foliage-feeding insects. Studies of rangeland grasshoppers, for example, have generally failed to establish strong competitive interactions (e.g., Ueckert and Hansen 1971, Joern 1979, Joern and Lawlor 1980, Mulkern 1980, Joern and Klucas 1993), although such interactions may exist when grasshopper densities are high, for example in periods of outbreak (Belovsky 1986, Pfadt et al. 1988, Chase and Belovsky 1994). In experimental manipulation of densities of the most abundant grasshopper species at a tallgrass prairie site, Evans (1989) detected no impact on densities of other grasshoppers or of other phytophagous insects. However, recent studies of sedentary insects, in which individuals spend much of their lives on single plants (unlike free-ranging grasshoppers), indicate that interspecific competition may indeed often be significant (e.g., Faeth 1986, Harrison and Karban 1986, Karban 1986, 1989, Moran and Whitham 1990, Mopper et al. 1990).

Competition is only one of many interactions that may exist among invertebrate and vertebrate herbivores. In some cases, for example, facilitation among these herbivores may occur. In particular, the feeding activities of herbivores may enhance each others' foraging success (ultimately leading to increased density); such indirect mutualisms have been documented among vertebrate herbivores (e.g., Krueger 1986) and may also exist among invertebrates (e.g., Ritchie and Tilman 1993). Ingham and Detling (1984), for instance, report that heavy aboveground grazing by prairie dogs and other vertebrates (bison and pronghorn) facilitates grazing by root-feeding nematodes. In a second example, removal of granivorous rodents led to a short-term increase in numbers of harvester ants at a Sonoran desert site (Brown and Davidson 1977, Brown et al. 1979; see also Valone et al. 1994). But subsequently, ant densities declined on rodent-removal plots; apparently populations of large-seeded plants increased in the absence of rodents to dominate the small-seeded plants preferred by the ants (Davidson et al. 1984). Davidson (1985) provides another example of indirect mutualism among three species of harvester ants overlapping in diet; the largest species suppresses populations of the intermediate-sized species, thereby facilitating indirectly the smallest sized species. Phytophagous invertebrates may also influence each other indirectly as shared prey of predators and parasitoids (e.g., Hawkins 1988b, Belovsky and Slade 1993). Finally, indirect interactions among herbivores may arise from mutualisms, with important consequences for plant performance and attributes (e.g., Bach 1991, Vrieling et al. 1991). Messina (1981), for example, found that goldenrod (*Solidago altissima*) stems suffered less defoliation when harboring sap-feeding treehoppers tended by ants; in protecting the sap-feeders, the ants forced would-be defoliators from the host plant.

VIII. Invertebrate Outbreaks

Of the many phytophagous invertebrate species inhabiting rangelands, a relatively small number are subject to periodic outbreak. During outbreaks, these species can have enormous impact on rangeland plant populations and communities; clearly, the ability to predict such outbreaks is of great practical importance (e.g., Kemp 1987). Why only relatively few phytophagous invertebrates outbreak in ecosystems generally, and what causes such species to outbreak, are questions that have received considerable attention but remain largely unanswered (Barbosa and Shultz 1987). Generalizations are likely to be slow in coming given the complexity of potential interactions among such general factors as abiotic conditions (e.g., weather), food quantity and quality, impacts of natural enemies such as predators, parasitoids, and disease, and genetic changes in the phytophagous invertebrate population. Price (1991b), however, has taken an important initial step, drawing together many general patterns that may underly insect herbivore population dynamics; he emphasizes the importance of changing key characteristics of host plants along successional gradients.

The nature of the food supply also figures prominently in other recent hypotheses concerning population dynamics of phytophagous invertebrates. Myers and Post (1981), for example suggest that those populations of cinnabar moth that feed on tansy ragwort of particularly high nitrogen content are most prone to marked fluctuations in numbers; high food quality may permit the moths to periodically overexploit their food supply.

Changes in both host plant quantity and quality, especially as influenced by variation in weather, have also been hypothesized to trigger outbreaks of phytophagous invertebrates. White (1984, 1993, and previous papers cited therein) in particular has argued and provided data suggesting that many outbreaks of phytophagous invertebrates result from increased availability of nitrogen in drought-stressed host plants. In a thorough, recent assessment, Mattson and Haack (1987a,b) suggest a variety of other ways water stress from drought may result in greater susceptibility and suitability of host plants to their adapted consumers, and thus lead to outbreaks. They hypothesize, for example, that drought may enhance insect detoxification systems, thereby enabling these herbivores to cope with the often-increased levels of allelochemicals in drought-stressed host plants. They also hypothesize that insect chemoreceptors are especially well-designed to detect host plant chemical changes associated with drought. Particularly intriguing is the observation that only larvae of phytophagous insects possess amino acid receptors. Some of the amino acids to which these insects are sensitive (e.g., inositol and proline) are osmolytes that increase in concentration in host plant tissues during drought.

Mattson and Haack (1987a,b) also stress that not all host plant changes associated with drought may be beneficial to phytophagous invertebrates. In particular, significantly less plant tissue may be produced during a drought, and this tissue may be less suitable because it is available for less time (e.g., there may be premature leaf fall) and/or because its texture may change (leaves may become thicker, waxier, and/or more densely covered with trichomes). For example, the reduced quantities of green foliage associated with dry years may limit population growth of grasshoppers in arid rangelands (Dempster 1963, Nerney and Hamilton 1969, Capinera and Horton 1989). Finally, Mattson and Haack (1987a,b) point out that the impact of drought on invertebrate/plant relations is also likely to vary with the severity of the drought. For example, while free amino acid and sugar levels generally increase in plants during times of drought, severe and prolonged drought may lead to a decline in such levels. Invertebrate outbreaks eventually collapse, even in the face of continuing drought. The high levels of herbivory associated with outbreaks may lead to dramatic declines in host plant quantity and quality, which in turn may compromise the ability of the invertebrates to ward off disease and other natural enemies. English-Loeb (1990) illustrates the potential for non-linearity in the relationship between the degree of host drought stress and the response of herbivore populations. Larsson (1989) discusses how drought and other stresses may influence host plant physiology with net effects varying from positive to negative, depending on the feeding style of insects (e.g., suckers vs. chewers; see also Louda and Collinge 1992).

In addition to drought, another factor influencing host plant quality that has received considerable attention in recent years is the chemical response of host plants to herbivory. Recognition that chemical changes in host plants are often induced by herbivory has led to hypotheses that such chemical changes may in turn affect population dynamics of the herbivores. As herbivore outbreaks build, for example, induced chemical defenses in host plants may soon lead to herbivore population collapse even in the face of plentiful vegetation, and prevention of a new outbreak for some time thereafter (Haukioja 1980). Rhoades (1985) has suggested two alternative sets of traits ("offensive strategies") of herbivores in response to inducible plant defenses. Stealthy herbivores, with typically low population densities, may cause relatively little damage to their host plants, thereby inducing relatively little chemical defense. In contrast, opportunistic herbivores, with tendencies to outbreak, may take advantage of abiotic stress and other circumstances that weaken the host plant's defensive abilities; indeed, their mass-attack during outbreaks may be one important means of stressing their host plant. At present, one can only speculate on the potential importance of chemical changes in host plant quality in the population dynamics of rangeland phytophagous invertebrates. Haukioja and Neuvonen (1987) critically discuss experimental approaches to exploring this intriguing area of research further.

Aboveground grazing may influence host quality belowground such as to promote outbreaks of soil arthropods. Moderate to heavy grazing of foliage results in root die-back and the microbial colonization of senescing roots results in substantial increases in root nitrogen concentrations (Clark and Campion 1976). While the absolute amount of nitrogen present in roots of partially defoliated plants at any point in time is less than that of ungrazed plants, a portion of the nitrogen associated with senescing roots derives from microbes, which extract inorganic nitrogen from soil water (Aber and Mellillo 1980). Hence, there is a brief period when each senescing root, infused with microbes, contains more nitrogen (both in concentration and absolute amount) than did the living root. During this period, the senescing root may be a better food source for herbivores and omnivores (promoting higher growth and survivorship rates) than are living, healthy roots. Thus a pulse of root die-back, caused by foliage removal, may lead to the highest densities of belowground arthropod herbivores observed in grazed North American and Australian grasslands.

Once a pulse of root die-off is created by overgrazing of foliage, weather or other mechanisms, and *if* sufficient larvae are present in the soil due to low adult mortality, good survivorship of eggs, and so on, an eruption of belowground herbivores may result. Since many of these organisms have multi-year life cycles with damage to plants most commonly resulting from consumption by older larvae and nymphs, the system could exhibit cyclic behavior. The destruction of the resource base may ultimately terminate the cycle, but more likely a combination of weather and biotic controls supervene to control the herbivores (Daniels 1966, Ueckert 1979, Wiener and Capinera 1981).

IX. Summary

Our review has emphasized how many aspects of plant morphology and physiology may be intertwined with patterns of consumption by phytophagous invertebrates in rangelands. The devastating consequences of such consumption during outbreaks of locusts, grasshoppers, and other invertebrates are well-known, but the significance for range management of chronic, often "low-level" herbivory by numerous other, less conspicuous species of invertebrates remains unclear. The large body of literature on plant/invertebrate relations that has developed especially over the past three decades lays the groundwork for detailed studies to illuminate how stresses imposed by invertebrates interact with other environmental stresses such as drought, fire, nutrient deficiencies, and grazing by cattle and sheep, in affecting rangeland plants. Such studies are needed to improve our management of grazing lands by more successfully integrating invertebrate herbivory into our general understanding of patterns and processes of rangeland ecosystems.

Acknowledgements

We thank B.A. Haws and F.J. Messina for critically reviewing drafts of the manuscript.

Literature Cited

Aarssen, L.W., and D.L. Irwin. 1991. What selection: herbivory or competition? Oikos 60:261-262.

Aber, J.D., and J.M. Melillo. 1980. Litter decomposition: measuring state of decay and percent transfer into forest soils. Can. J. Bot. 58:416-421.

Abrahamson, W.G., and A.E. Weis. 1987. Nutritional ecology of arthropod gall makers, pp. 235-258. *In*: F. Slansky Jr. and J.G. Rodriquez (eds.), Nutritional ecology of insects, mites, spiders and related invertebrates. John Wiley and Sons, New York.

Andersen, A.N. 1987. Effects of seed predation by ants on seedling densities at a woodland site in SE Australia. Oikos 48:171-174.

Andersen, A.N. 1989. How important is seed predation to recruitment in stable populations of long-lived perennials? Oecologia 81: 310-315.

Andersen, D.C. 1987. Below-ground herbivory in natural communities: a review emphasizing fossorial animals. Quart. Rev. Biol. 62:261-286.

Anderson, R.V., R.E. Ingham, J.A. Trofymow, and D.C. Coleman. 1984. Soil mesofaunal distribution in relation to habitat types in a shortgrass prairie. Pedobiologia 26:257-261.

Appel, H.M., and M.M. Martin. 1992. Significance of metabolic load in the evolution of host specificity of *Manduca sexta*. Ecology 73:216-228.

Arms, K., P. Feeny, and R. Lederhouse. 1974. Sodium: stimulus for puddling behavior by tiger swallowtail butterflies, *Papilio glaucus*. Science 185: 372-374.

Atsatt, P.R., and D.J. O'Dowd. 1976. Plant defense guilds. Science 193:24-29.

Bach, C.E. 1988a. Effects of host plant patch size on herbivore density: patterns. Ecology 69:1090-1102.

Bach, C.E. 1988b. Effects of host plant patch size on herbivore density: underlying mechanisms. Ecology 69:1103-1117.

Bach, C.E. 1991. Direct and indirect interactions between ants (*Pheidole megacephala*), scales (*Coccus viridis*) and plants (*Pluchea indica*). Oecologia 87:233-239.

Bailey, C.G., and P.W. Riegert. 1973. Energy dynamics of *Encoptolophus sordidus costalis* (Scudder) (Orthoptera: Acrididae) in a grassland ecosystem. Can. J. Zool. 51:91-100.

Baker, H.G., P.A. Opler, and I. Baker. 1978. A comparison of the amino acid complements of floral and extrafloral nectars. Bot. Gaz. 139:322-332.

Baldwin, I.T., and J.C. Shultz. 1983. Rapid changes in tree leaf chemistry induced by damage: evidence for communication between plants. Science 221: 277-279.

Barbosa, P. 1988. Natural enemies and herbivore-plant interactions: influence of plant allelochemicals and host specificity, pp. 201-229. *In*: P. Barbosa and D.K. Letourneau (eds.), Novel aspects of insect-plant interactions. John Wiley and Sons, New York.

Barbosa, P. and J.C. Shultz. 1987. (eds.). Insect outbreaks. Academic Press, Inc., San Diego, California.

Barbosa, P., V.A. Krischnik, and C.G. Jones (eds.). 1991. Microbial mediation of plant-herbivore interactions. John Wiley and Sons, New York.

Bardner, R. and K.E. Fletcher. 1974. Insect infestations and their effects on the growth and yield of field crops. A review. Bull. Entomol. Res. 64:141-160.

Batra, S.W.T. 1982. Biological control in agroecosystems. Science 215:134-139.

Baur, R., S. Binder, and G. Benz. 1991. Nonglandular leaf trichomes as short-term inducible defense of the grey alder, *Alnus incana* (L.), against the chrysomelid beetle, *Agelastica alni* L. Oecologia 87:219-226.

Beamer, R.H. 1928. Studies on the biology of Kansas Cicadidae. University of Kansas Science Bull. 28:155-263.

Beattie, A.J. 1985. The evolutionary ecology of ant-plant mutualisms. Cambridge University Press, Cambridge, England.

Belovsky, G. 1986. Generalist foraging and its role in competitive interactions. Amer. Zool. 26:51-69.

Belovsky, G. and J.B. Slade. 1993. The role of vertebrate and invertebrate predators in a grasshopper community. Oikos 68: 193-201.

Belsky, A.J. 1986. Does herbivory benefit plants? A review of the evidence. Am. Nat. 127:870-892.

Belsky, A.J. 1987. The effects of grazing: confounding of ecosystem, community, and organism scales. Am. Nat. 129:777-783.

Bentley, B.L. 1977. Extrafloral nectaries and protection by pugnacious bodyguards. Ann. Rev. Ecol. Syst. 8:407-427.

Berenbaum, M.R. 1983. Coumarins and caterpillars: a case for coevolution. Evolution 37:386-410.

Berenbaum, M.R. 1985. Brementown revisited: interactions among allelochemicals in plants. Recent Advances in Phytochemistry 19:139-169.

Berenbaum, M.R., and A.R. Zangerl. 1992. Genetics of secondary metabolism and herbivore resistance in plants, pp. 415-438. *In* G.A. Rosenthal and M.R. Berenbaum (eds.), Herbivores: their interactions with secondary plant metabolites. 2nd ed. Vol. II. Academic Press, San Diego, CA

Berenbaum, M.R., A.R. Zangerl, and J.K. Nitao. 1986. Constraints on chemical coevolution: wild parsnips and the parsnip webworm. Evolution 40:1215-1228.

Berenbaum, M.R., A.R. Zangerl, and K. Lee. 1989. Chemical barriers to adaptation by a specialist herbivore. Oecologia 80:501-506.

Bergelson, J., S. Fowler, and S. Hartley. 1986. The effects of foliage damage on casebearing moth larvae, *Coleophora serratella*, feeding on birch. Ecol Entomol 11:241-250.

Bernays, E.A. 1981. Plant tannins and insect herbivores: an appraisal. Ecol Entomol 6:353-360.

Bernays, E.A. 1983. Nitrogen in defence against insects, pp. 321-344. *In*: J.A. Lee, S. McNeill, and I.H. Rorison (eds.), Nitrogen as an ecological factor. Blackwell Scientific Publications, Oxford.

Bernays, E.A. 1986. Diet-induced head allometry among foliage-chewing insects and its importance for graminivores. Science 231:495-497.

Bernays, E.A. 1989. Host range in phytophagous insects: the potential role of generalist predators. Evol. Ecol. 3:299-311.

Bernays, E.A. 1990. Water regulation, pp. 129-134. *In*: R.F. Chapman and A. Joern (eds.). Biology of grasshoppers. John Wiley and Sons, New York.

Bernays, E.A. 1991. On the narrow host ranges of phytophagous insects. Phytoparasitica 19:91-94.

Bernays, E.A., and R.F. Chapman. 1977. Deterrent chemicals as a basis of oligophy in *Locusta migratoria* (L.). Ecol. Entomol. 2:1-18.

Bernays, E.A., and R.F. Chapman. 1978. Plant chemistry and acridoid feeding behaviour, pp. 99-141. *In*: J.B. Harborne (ed.), Biochemical aspects of plant and animal coevolution. Academic Press, London.

Bernays, E.A., D.J. Chamberlain, and P. McCarthy. 1980. The differential effects of ingested tannic acid on different species of Acridoidea. Entomol. exp. appl. 28:158-166.

Bernays, E.A., and S. Woodhead. 1982. Plant phenols utilized as nutrients by a phytophagous insect. Science 216:201-203.

Bernays, E.A., and A.C. Lewis. 1986. The effect of wilting on palatability of plants to *Schistocerca gregaria*, the desert locust. Oecologia 70:132-135.

Bernays, E.A. and R. Barbehehn. 1987. Nutritional ecology of grass foliage-chewing insects, pp. 147-175. *In* F. Slansky Jr. and J.G. Rodriquez (eds.), Nutritional ecology of insects, mites, spiders, and related invertebrates. John Wiley and Sons, New York.

Bernays, E.A., and M. Graham. 1988. On the evolution of host specificity in phytophagous insects. Ecology 69:886-892.

Bernays, E.A., and D.H. Janzen. 1988. Saturniid and sphingid caterpillars: two ways to eat leaves. Ecology 69:1153-1160.

Bernays, E.A., and S.J. Simpson. 1990. Nutrition, pp. 105-127. *In* R.F. Chapman and A. Joern (eds.), Biology of grasshoppers. John Wiley and Sons, New York.

Bodine, M.C., and D.N. Ueckert. 1975. Effect of desert termites on herbage and litter in a shortgrass ecosystem in West Texas. J. Range Manage. 28:353-358.

Boecklen, W.J., and R. Spellenberg. 1990. Structure of herbivore communities in two oak (*Quercus* spp.) hybrid zones. Oecologia 85: 92-100.

Boscher, J. 1979. Modified reproduction strategy of leek *Allium porrum* in response to a phytophagous insect, *Acrolepiopsis assectella*. Oikos 33: 451-456.

Boutton, T.W., G. Cameron, and B. Smith. 1978. Insect herbivory on C3 and C4 grasses. Oecologia 36:21-32.

Boutton, T.W., B.N. Smith, and A.T. Harrison. 1980. Carbon isotope ratios and crop analyses of *Arphia* (Orthoptera: Acrididae) species in southeastern Wyoming grassland. Oecologia 45:299-306.

Brattsten, L.B. 1979. Biochemical defense mechanisms in herbivores against plant allelochemicals, pp. 199-270. *In* G.A. Rosenthal and D.H. Janzen (eds.), Herbivores: their interaction with secondary plant metabolites. Academic Press, New York.

Brattsten, L.B. 1992. Metabolic defenses against plant allelochemicals, pp. 175-242. *In* G.A. Rosenthal and M.R. Berenbaum (eds.). Herbivores: their interactions with secondary plant metabolites. 2nd ed. Vol II. Academic Press, San Diego, CA

Brattsten, L.B., C.F. Wilkenson, and T. Eisner. 1977. Herbivore-plant interactions: mixed-function oxidases and secondary plant substances. Science 196:1349-1352.

Breedlove, D.E. and P.R. Ehrlich. 1968. Plant-herbivore coevolution: lupines and lycaenids. Science 162:671-672.

Brodbeck, B. and D.R. Strong. 1987. Amino acid nutrition of herbivorous insects and stress to host plants, pp. 347-364. *In*: P. Barbosa and J.C. Schultz (eds), Insect outbreaks. Academic Press, San Diego, CA

Brown, E.S., and C.F. Dewhurst. 1975. The genus *Spodoptera* (Lepidoptera, Noctuidae) in Africa and the Near East. Bull. Entomol. Res. 65: 221-262.

Brown, J.H., and D.W. Davidson. 1977. Competition between seed-eating rodents and ants in desert ecosystems. Science 196:880-882.

Brown, J.H., D.W. Davidson, and O.J. Reichman. 1979. An experimental study of competition between seed-eating desert rodents and ants. Amer. Zool. 19:1129-1143.

Brown, J.H. and E.J. Heske. 1990. Control of a desert-grassland transition by a keystone rodent guild. Science 250:1705-1707.

Brown, V.K. 1984. Secondary succession: insect-plant relationships. BioScience 34:710-716.

Brown, V.K. 1985. Insect herbivores and plant succession. Oikos 44:17-22.

Brown, V.K., M. Jepsen, and C.W.D. Gibson. 1988. Insect herbivory: effects on early old field succession demonstrated by chemical exclusion methods. Oikos 52:293-302.

Brown, V.K., and A.C. Gange. 1989a. Differential effects of above- and below-ground insect herbivory during early plant succession. Oikos 54:67-76.

Brown, V.K., and A.C. Gange. 1989b. Herbivory by soil-dwelling insects depresses plant species richness. Functional Ecology 3:667-671.

Brown, W.L. 1960. Ants, acacias and browsing mammals. Ecology 41:587-592.

Bryant, J.P., I. Heitkonig, P. Kuropat, and N. Owen-Smith. 1991. Effects of severe defoliation on the long-term resistance to insect attack and on leaf chemistry in six woody species of the southern African savanna. Am. Nat. 137:50-63.

Bullen, F.T. 1966. Locusts and grasshoppers as pests of crops and pasture—a preliminary economic approach. J. Appl. Ecol. 3:147-168.

Byers, R.A., and G.A. Jung. 1979. Insect populations on forage grasses: effect of nitrogen fertilizer and insecticides. Environ. Entomol. 8:11-18.

Cain, M.L., W.P. Carson, and R.B. Root. 1991. Long-term suppression of insect herbivores increases the production and growth of *Solidago altissima* rhizomes. Oecologia 88:252-257.

Caldwell, M.M., J.H. Richards, D.A. Johnson, R.S. Nowak, and R.S. Dzurec. 1981. Coping with herbivory: photosynthetic capacity and resource allocation in two semiarid *Agropyron* bunchgrasses. Oecologia 50:14-24.

Campbell, B.C. and S.S. Duffey. 1979. Tomatine and parasitic wasps: potential incompatibility of plant antibiosis with biological control. Science 205:700-702.

Campbell, W.F., B.A. Haws, K.H. Asay, and J.D. Hansen. 1984. A review of black grass bug resistance in forage grasses. J. Range Manage. 37:365-369.

Capinera, J.L., and W.J. Roltsch. 1980. Response of wheat seedlings to actual and simulated migratory grasshopper defoliation. J. Econ. Entomol. 73:258-261.

Capinera, J.L., and T.S. Sechrist. 1982. Grasshopper (Acrididae) - host plant associations: response of grasshopper populations to cattle grazing intensity. Can. Entomol. 114:1055-1062.

Capinera, J.L., and D.C. Thompson. 1987. Dynamics and structure of grasshopper assemblages in shortgrass prairie. Can. Entomol. 119:567-575.

Capinera, J.L., and D.R. Horton. 1989. Geographic variation in effects of weather on grasshopper infestation. Environ. Entomol. 18:8-14.

Carroll, C.R., and C.A. Hoffman. 1980. Chemical feeding deterrent mobilized in response to insect herbivory and counteradaptation by *Epilachna tredecimnotata*. Science 209:414-416.

Carroll, G. 1988. Fungal endophytes in stems and leaves: from latent pathogen to mutualistic symbiont. Ecology 69:2-9.

Caswell, H. F. Reed, S.N. Stephenson, and P.A. Werner. 1973. Photosynthetic pathways and selective herbivory: a hypothesis. Am. Nat. 107:465-480.

Caswell, H., and F.C. Reed. 1975. Indigestibility of C4 bundle sheath cells by the grasshopper, *Melanoplus confusus*. Ann. Entomol. Soc. Amer. 68:686-689

Caswell, H., and F.C. Reed. 1976. Plant-herbivore interactions. The indigestibility of C4 bundle sheath cells by grasshoppers. Oecologia 26:151-156.

Cates, R.G., and D.F. Rhoades. 1977. Patterns in the production of antiherbivore defenses in plant communities. Biochem. Syst. Ecol. 5:185-193.

Chapman, R.F. 1964. The structure and wear of the mandibles in some African grasshoppers. Proc. Zool. Soc. Lond. 142:107-121.

Chapman, R.F. 1990. Food selection, pp. 39-72. *In*: R.F. Chapman and A. Joern (eds.), Biology of grasshoppers. John Wiley and Sons, New York.

Chase, J.M. and G.E. Belovsky. 1994. Experimental evidence for the included niche. American Naturalist 143:514-527.

Christiansen, K.M., and T.G. Whitham. 1993. Impact of insect herbivores on competition between birds and mammals for pinyon pine seeds. Ecology 74:2270-2278.

Christiansen, T.A., J.A. Lockwood, and J. Powell. 1989a. Mediation of nutrient cycling by arthropods in unmanaged and intensively managed mountain brush habitats. Great Basin Naturalist 49:134-139.

Christiansen, T.A., J.A. Lockwood, and J. Powell. 1989b. Effects of arthropods on root:shoot production in undisturbed and modified mountain shrub habitats. Great Basin Naturalist 49:456-460.

Christiansen, T.A., J.A. Lockwood, and J. Powell. 1989c. Litter decomposition by arthropods in undisturbed and intensively managed mountain brush habitats. Great Basin Naturalist 49:562-569.

Clark, F.B., and M. Campion. 1976. Distribution of nitrogen in root materials of blue grama. J. Range Manage. 29:256-258.

Clark, L.R. 1953. The ecology of *Chrysomela gemellata* Rossi and *C. hyperici* Forst, and their effect on St. John's wort in the Bright District, Victoria (Australia). Austr. J. Zool. 1:1-69.

Clark, W.H., and P.L. Comanor. 1975. Removal of annual plants from the desert ecosystem by western harvester ants, *Pogonomyrmex occidentalis*. Environ. Entomol. 4:52-56.

Clay, K. 1988. Fungal endophytes of grasses: a defensive mutualism between plants and fungi. Ecology 69:10-16.

Coley, P.D., J.P. Bryant, and F.S. Chapin III. 1985. Resource availability and plant antiherbivore defense. Science 230:895-899.

Collinge, S.K., and S.M. Louda. 1989. Influence of plant phenology on the insect herbivore/bittercress interaction. Oecologia 79:111-116.

Connell, J.H. 1983. On the prevalence and relative importance of interspecific competition: evidence from field experiments. Am. Nat. 122:661-696.

Cornell, H.V. 1989. Endophage-ectophage ratios and plant defense. Evol. Ecol. 3:64-76.

Cornell, H.V. 1993. Unsaturated patterns in species assemblages: the role of regional processes in setting local species richness, pp. 243-252. *In* R.E. Ricklefs and D. Schluter (eds.), Species diversity in ecological communities. Univ. Chicago Press, Chicago.

Cox, C.S., and P.B. McEvoy. 1983. Effect of summer moisture stress on the capacity of tansy ragwort (*Senecio jacobaea*) to compensate for defoliation by cinnabar moth (*Tyria jacobaeae*). J. Appl. Ecol. 20:225-234.

Crawley, M.J. 1983. Herbivory. The dynamics of animal-plant interactions. Blackwell Scientific Publications, Oxford.

Crawley, M.J. 1986. The population biology of invaders. Phil. Trans. R. Soc. Lond. B 314:711-731.

Crawley, M.J. 1987. What makes a community invasible?, pp. 429-453. *In*: A.J. Gray, M.J. Crawley, and P.J. Edwards (eds), Colonization, succession, and stability. Blackwell Scientific Publ., Oxford.

Crawley, M.J. 1988. The relative importance of vertebrate and invertebrate herbivores in plant population dynamics, pp. 45-71. *In*: E.A. Bernays (ed.), Insect-Plant Interactions, Volume I. CRC Press, Boca Raton, Florida.

Crawley, M.J. 1989. Insect herbivores and plant population dynamics. Ann. Rev. Entomol. 34:531-564.

Crawley, M.J., and M. Akhteruzzaman. 1988. Individual variation in the phenology of oak trees and its consequences for herbivorous insects. Functional Ecology 2:409-415.

Curl, E.A., R. Lartey, and C.M. Peterson. 1988. Interactions between root pathogens and soil microarthropods. Agriculture, Ecosystems, and Environment 24:249-261.

Curry, J.P. 1987a. The invertebrate fauna of grassland and its influence on productivity. I. The composition of the fauna. Grass and Forage Science 42:102-120.

Curry, J.P. 1987b. The invertebrate fauna of grassland and its influence on productivity. II. Factors affecting the abundance and composition of the fauna. Grass and Forage Science 42:197-211.

Dadd, R.H. 1985. Nutrition: organisms, pp. 313-390. *In*: G.A. Kerkut and L.I. Gilbert (eds.), Comprehensive insect physiology, biochemistry, and pharmacology. Volume 4. Pergamon Press, Oxford, England.

Daniels, N.E. 1966. The association of a rangeland grub, *Phyllophaga koehleriana* (Coleoptera: Scarabaeidae), with asilid larvae and with mites. Ann. Entomol. Soc. Amer. 59:1021.

Davidson, D.W., R.S. Inouye, and J.H. Brown. 1984. Granivory in a desert ecosystem: experimental evidence for indirect facilitation of ants by rodents. Ecology 65:1780-1786.

Davidson, D.W., D.A. Samson, and R.S. Inouye. 1985. Granivory in the Chihuahan desert: interactions within and between trophic levels. Ecology 66:486-502.

Davidson, D.W. 1985. An experimental study of diffuse competition in harvester ants. Am. Nat. 125:500-506.

Davidson, R.L., and R.J. Roberts. 1968. Species differences in scarab-pasture relationships. Entomological Research 58:315-324.

Del Vecchio, T.A., C.A. Gehring, N.S. Cobb, and T.G. Whitham. 1993. Negative effects of scale insect herbivory on the ectomycorrhizae of juvenile pinyon pine. Ecology 74:2297-2302.

Dempster, J.P. 1963. The population dynamics of grasshoppers and locusts. Biol. Rev. 38:490-529.

Denno, R.F., S. Larsson, and K.L. Olmstead. 1990. Role of enemy-free space and plant quality in host-plant selection by willow beetles. Ecology 71:124-137.

Detling, J.K. 1988. Grasslands and savannahs: regulation of energy flow and nutrient cycling by herbivores, pp. 131-154. *In* L.R. Pomeroy and J.J. Alberts (eds), Concepts of ecosystem ecology, a comparative view. Springer-Verlag, New York.

Detling, J.K., and M.I. Dyer. 1981. Evidence for potential plant growth regulators in grasshoppers. Ecology 62:485-488.

Dodd, A.P. 1959. The biological control of prickly pear in Australia, pp. 565-577. *In*: Biogeography and Ecology in Australia (Monographiae Biologicae, Vol. VIII). W. Junk, The Hague.

Dodd, J.L. and W.K. Lauenroth. 1979. Analysis of the response of a grassland ecosystem to stress, pp. 43-58. *In*: N.R. French (ed.), Perspectives in grassland ecology. Springer-Verlag, New York.

Duffey, S.S. 1980. Sequestration of plant natural products by insects. Ann. Rev. Entomol. 25:447-478.

Dussourd, D.E., and T. Eisner. 1987. Vein-cutting behavior: insect counterploy to the latex defense of plants. Science 237:898-901.

Dussourd, D.E., and R.F. Denno. 1994. Host range of generalist caterpillars: trenching permits feeding on plants with secretory canals. Ecology 75:69-78.

Dyer, M.I., and U.G. Bokhari. 1976. Plant-animal interactions: studies of the effects of grasshopper grazing on blue grama grass. Ecology 57:762-772.

Edwards, P.J. and S.D. Wratten. 1983. Wound induced defences in plants and their consequences for patterns of insect grazing. Oecologia 59:88-93.

Edwards, P.J. and S.D. Wratten. 1985. Induced plant defences against insect grazing: fact or artefact? Oikos 44:70-74.

Edwards, P.J., S.D. Wratten, and S. Greenwood. 1986. Palatability of British trees to insects: constitutive and induced defences. Oecologia 69:316-319.

Ehrlich, P.R., and R.H. Raven. 1964. Butterflies and plants: a study in coevolution. Evolution 18:586-608.

Ehrlich, P.R., D.E. Breedlove, P.F. Brussard, and M.A. Sharp. 1972. Weather and the "regulation" of subalpine populations. Ecology 53:243-247.

Ehrlich, P.R., D.D. Murphy, M.C. Singer, S.W. McKechnie, and L.E. Gilbert. 1980. Extinction, reduction and stability: the responses of checkerspot butterflies to the California drought. Oecologia 46:221-227.

Ehrlich, P.R., and D.D. Murphy. 1988. Plant chemistry and host range in insect herbivores. Ecology 69:908-909.

Ellison, L. 1960. Influence of grazing on plant succession of rangelands. Bot. Rev. 26:1-78.

English-Loeb, G.M. 1990. Plant drought stress and outbreaks of spider mites: a field test. Ecology 71:1401-1411.

Evans, E.W. 1983. The influence of neighboring hosts on colonization of prairie milkweeds by a seed-feeding bug. Ecology 64:648-653.

Evans, E.W. 1989. Interspecific interactions among phytophagous insects of tallgrass prairie: an experimental test. Ecology 70:435-444.

Evans, E.W. 1990. Dynamics of an aggregation of blister beetles (Coleoptera: Meloidae) attacking a prairie legume. J. Kans. Entomol. Soc. 63:616-625.

Evans, E.W., C.C. Smith, and R.P. Gendron. 1989. Timing of reproduction in a prairie legume: seasonal impacts of insects consuming flowers and seeds. Oecologia 78:220-230.

Faeth, S.H. 1986. Indirect interactions between temporally separated herbivores mediated by the host plant. Ecology 67:479-494.

Faeth, S.H., E.F. Connor, and D. Simberloff. 1981. Early leaf abscission: a neglected source of mortality for folivores. Am. Nat. 117:409-415.

Fajer, E.D., M.D. Bowers, and F.A. Bazzaz. 1989. The effects of enriched carbon dioxide atmospheres on plant-insect herbivore interactions. Science 243:1198-1200.

Farmer, E.E., and C.A. Ryan. 1990. Interplant communication: airborne methyl jasmonate induces synthesis of proteinase inhibitors in plant leaves. Proc. Natl. Acad. Sci. USA 87:7713-7716.

Farrell, B.D., D.E. Dussourd, and C. Mitter. 1991. Escalation of plant defense: do latex and resin canals spur plant diversification? Am. Nat. 138:881-900.

Farrell, B.D., C. Mitter, and D.J. Futuyma. 1992. Diversification at the insect-plant interface. BioScience 42:34-42.

Feeny, P. 1975. Biochemical coevolution between plants and their insect herbivores, pp. 3-19. *In*: L.E. Gilbert and P.H. Raven (eds.), Coevolution of animals and plants. University of Texas Press, Austin.

Feeny, P. 1976. Plant apparency and chemical defense. Recent Advances in Phytochemistry 10:1-40.

Feeny, P. 1992. The evolution of chemical ecology: contributions from the study of herbivorous insects, pp. 1-44. *In* G.A. Rosenthal and M.R. Berenbaum, Herbivores: their interactions with secondary plant metabolites. 2nd ed. Vol. II. Academic Press, San Diego, CA.

Finch, S. 1980. Chemical attraction of plant-feeding insects to plants. Appl. Biol. 5:67-143.

Floate, K.D., M.J.C. Kearsley, and T.G. Whitham. 1993. Elevated herbivory in plant hybrid zones: *Chrysomela confluens*, *Populus*, and phenological sinks. Ecology 74: 2056-2065.

Fowler, S.V., and J.H. Lawton. 1985. Rapidly induced defenses and talking trees: the devil's advocate position. Am. Nat. 126:181-195.

Fowler, S.V., and M. MacGarvin. 1986. The effects of leaf damage on the performance of insect herbivores on birch, *Betula pubescens*. J. Anim. Ecol. 55:565-573.

Fox, L.R., and P.A. Morrow. 1981. Specialization: species property or local phenomenon? Science 211:887-893.

Fraenkel, G.S. 1959. The raison d'etre of secondary plant substances. Science 129:1466-1470.

Freckman, D.W. 1982. Parameters of nematode contribution to ecosystems, pp. 81-97. *In*: D.W. Freckman (ed.), Nematodes in Soil Ecosystems. Univ of Texas Press, Austin.

Freckman, D.W., D.A. Duncan, and J.R. Larson. 1979. Nematode density and biomass in an annual grassland ecosystem. J. Range Manage. 32:418-422.

Freckman, D.W., and R.A. Virginia. 1989. Plant-feeding nematodes in deep-rooting desert ecosystems. Ecology 70:1665-1678.

Fritz, R.S. 1990. Effects of genetic and environmental variation on resistance of willow to sawflies. Oecologia 82:325-332.

Fritz, R.S., C.F. Sacchi, and P.W. Price. 1986. Competition versus host plant phenotype in species competition: willow sawflies. Ecology 67:1608-1618.

Fritz, R.S., and E.L. Simms (eds.). 1992. Plant resistance to herbivores and pathogens. Ecology, evolution, and genetics. The University of Chicago Press, Chicago.

Fry, B., A. Joern, and P.L. Parker. 1978. Grasshopper food web analysis: use of carbon isotope ratios to examine feeding relationships among terrestrial herbivores. Ecology 59:498-506.

Gange, A.C., and V.K. Brown. 1989. Insect herbivory affects size variability in plant populations. Oikos 56:351-356.

Gangwere, S.K. 1961. A monograph of food selection in Orthoptera. Trans. Amer. Entomol. Soc. (Philadelphia) 87:67-230.

Gates, D.H. 1964. Sagebrush infested by lead defoliating moth. J. Range Manage. 17:209-210.

Gibberd, R., P.J. Edwards, and S.D. Wratten. 1988. Wound-induced changes in the acceptability of tree-foliage to Lepidoptera: within-leaf effects. Oikos 51:43-47.

Gibson, C.W.D., V.K. Brown, and M. Jepson. 1987. Relationships between the effects of insect herbivory and sheep grazing on seasonal changes in an early successional plant community. Oecologia 71:245-253.

Gibson, D.J., C.C. Freeman, and L.C. Hulbert. 1990. Effects of small mammal and invertebrate herbivory on plant species richness and abundance in tallgrass prairie. Oecologia 84:169-175.

Gibson, R.W. and J.A. Pickett. 1983. Wild potato repels aphids by release of aphid alarm pheromone. Nature 302:608-609.

Goeden, R.D., and S.M. Louda. 1976. Biotic interference with insects imported for weed control. Ann. Rev. Entomol. 21:325-342.

Goeden, R.D., and L.T. Lok. 1986. Comments on a proposed "new" approach for selecting agents for the biological control of weeds. Can. Entomol. 118:51-58.

Gould, F. 1984. Mixed function oxidases and herbivore polyphagy: the devil's advocate position. Ecol. Entomol. 9:29-34.

Graber, L.F., C.L. Fluke, and S.T. Dexter. 1931. Insect injury of blue grass in relation to the environment. Ecology 12:547-566.

Green, T.R., and C.A. Ryan. 1972. Wound-induced proteinase inhibitor in plant leaves: a possible defense mechanism against insects. Science 175:776-777.

Gross, P. 1991. Influence of target pest feeding niche on success rates in classical biological control. Environ. Entomol. 20:1217-1227.

Hairston, N.G., F.E. Smith, and L.B. Slobodkin. 1960. Community structure, population control, and competition. Am. Nat. 94:421-425.

Harper, J.L. 1977. Population biology of plants. Academic Press, London.

Harris, P. 1988. Environmental impact of weed-control insects. BioScience 38: 542-548.

Harris, P. 1990. Environmental impact of introduced biological control agents, pp. 289-300. *In* M. Mackauer, L.E. Ehler, and J. Roland (eds), Critical issues in biological control. Intercept, Andover, Hants, UK.

Harris, P. 1991. Classical biocontrol of weeds: its definition, selection of effective agents, and administrative-political problems. Can. Entomol. 123:817-849.

Harris, P., and H. Zwolfer. 1968. Screening of phytophagous insects for biological control of weeds. Can. Entomol. 100:295-303.

Harrison, S., and R. Karban. 1986. Effects of an early-season folivorous moth on the success of a later-season species, mediated by a change in the quality of the shared host, *Lupinus arboreus* Sims. Oecologia 69:354-359.

Haukioja, E. 1980. On the role of plant defenses in the fluctuation of herbivore populations. Oikos 35:202-213.

Haukioja, E., and P. Niemala. 1977. Retarded growth of a geometrid larva after mechanical damage to leaves of its host tree. Ann. Zool. Fennici 14:48-52.

Haukioja, E., and P. Niemela. 1979. Birch leaves as a resource for herbivores: seasonal occurrence of increased resistance in foliage after mechanical damage of adjacent leaves. Oecologia 39:151-159.

Haukioja, E., and S. Neuvonen. 1987. Insect population dynamics and induction of plant resistance: the testing of hypotheses, pp. 411-432. *In* J.C. Schultz and P. Barbosa (eds.), Insects outbreaks. Academic Press, San Diego, CA.

Haukioja, E., J. Suomela, and S. Neuvonen. 1985. Long-term inducible resistance in birch foliage: triggering cues and efficacy on a defoliator. Oecologia 65:363-369.

Hawkins, B.A. 1988a. Species diversity in the third and fourth trophic levels: patterns and mechanisms. J. Anim. Ecol. 57:137-162.

Hawkins, B.A. 1988b. Foliar damage, parasitoids and indirect competition: a test using herbivores of birch. Ecol. Entomol. 13:301-308.

Hawkins, B.A. 1990. Global patterns of parasitoid assemblage size. J. Anim. Ecol. 59:57-72.

Haws, B.A., D.D. Dwyer, and M.G. Anderson. 1973. Problems with range grasses? Look for black grass bugs! Utah Science 34:3-9.

Hayes, J.L. 1981. The population ecology of a natural population of the pierid butterfly *Colias alexandra*. Oecologia 49:188-200.

Heidorn, T., and A. Joern. 1984. Differential herbivory on C3 versus C4 grasses by the grasshopper *Ageneotettix deorum* (Orthoptera: Acrididae). Oecologia 65:19-25.

Heidorn, T., and A. Joern. 1987. Feeding preference and spatial distribution of grasshoppers (Acrididae) in response to nitrogen fertilization of *Calamovilfa longifolia*. Functional Ecology 1:369-375.

Hendrix, S.D. 1979. Compensatory reproduction in a biennial herb following insect defloration. Oecologia 42:107-118.

Hendrix, S.D. 1984. Reactions of *Heracleum lanatum* to floral herbivory by *Depressaria pastinacella*. Ecology 65:191-197.

Hendrix, S.D. 1988. Herbivory and its impact on plant reproduction, pp. 246-263. *In*: J. Lovett Doust and L. Lovett Doust (eds), Plant reproductive ecology: patterns and strategies. Oxford University Press, New York.

Hendrix, S.D., and E.J. Trapp. 1981. Plant-herbivore interactions: insect induced changes in host plant sex expression and fecundity. Oecologia 49:119-122.

Hendrix, S.D., V.K. Brown, and A.C. Gange. 1988. Effects of insect herbivory on early plant succession: comparison of an English site and an American site. Biol. J. Linnean Soc. 35:205-216.

Herms, D.A., and W.J. Mattson. 1992. The dilemma of plants: to grow or defend. Quarterly Review of Biology 67:283-335.

Hewitt, G.B. 1977. Review of forage losses caused by rangeland grasshoppers. U.S. Department of Agriculture Miscellaneous Publication No. 1348.

Hewitt, G.B., and J.A. Onsager. 1983. Control of grasshoppers on rangeland in the United States - a perspective. J. Range Manage. 36: 202-207.

Hewitt, G.B., and J.D. Berdahl. 1984. Grasshopper food preferences among alfalfa cultivars and experimental strains adapted for rangeland interseeding. Environ. Entomol. 13:828-831.

Hilbert, D.W., D.M. Swift, J.K. Detling, and M.I. Dyer. 1981. Relative growth rates and the grazing optimization hypothesis. Oecologia 51:14-18.

Hobbs, R.J. 1985. Harvester ant foraging and plant species distribution in an annual grassland. Oecologia 67:519-523.

Hocking, B. 1970. Insect association with the swollen thorn acacias. Trans. R. Entomol. Soc. Lond. 122:211-255.

Hogue, S.M. 1970. Boisystematics of the genus *Trirhabda* LeConte of America North of Mexico. Dissertation. University of Idaho, Moscow.

Hokkanen, H., and D. Pimentel. 1984. New approach for selecting biological control agents. Can. Entomol. 116:1109-1121.

Hokkanen, H., and D. Pimentel. 1989. New associations in biological control: theory and practice. Can. Entomol. 121:829-840.

Holland, E.A., and J.K. Detling. 1990. Plant response to herbivory and belowground nitrogen cycling. Ecology 71:1040-1049.

Hollinger, D.Y. 1986. Herbivory and the cycling of nitrogen and phosphorus in isolated California oak trees. Oecologia 70:291-297.

Holmes, N.D., D.S. Smith, and A. Johnston. 1979. Effect of grazing by cattle on the abundance of grasshoppers on fescue grassland. J Range Manage. 32:310-311.

Holmes, R.D. and K. Jepson-Innes. 1989. A neighborhood analysis of herbivory in *Bouteloua gracilis*. Ecology 70:971-976.

Horsfield, D. 1977. Relationships between feeding of *Philaenus spumarius* (L.) and the amino acid concentration in the xylem sap. Ecol. Entomol. 2: 259-266.

Horsfield, D. 1978. Evidence for xylem feeding by *Philaenus spumarisu* (L.) (Homoptera: Cercopidae). Entomol. exp. appl. 24:95-99.

Howarth, F.G. 1991. Environmental impacts of classical biological control. Ann. Rev. Entomol. 36:485-509.

Huffaker, C.B., and C.E. Kennet. 1959. A ten-year study of vegetational changes associated with biological control of Klamath weed. J. Range Manage. 12:69-82.

Hutchinson, K.J., and K.L. King. 1980. The effects of sheep stocking level on invertebrate abundance, biomass and energy utilization in a temperate, sown grassland. J. Appl. Ecol. 17:369-387.

Ingham, R.E., and J.K. Detling. 1984. Plant-herbivore interactions in a North American mixed-grass prairie. III. Soil nematode populations and root biomass on *Cynomys ludovicianus* colonies and adjacent uncolonized areas. Oecologia 63:307-313.

Ingham, R.E., and J.K. Detling. 1990. Effects of root-feeding nematodes on aboveground net primary production in a North American grassland. Plant and Soil 121:279-281.

Inouye, D.W. 1982. The consequences of herbivory: a mixed blessing for *Jurinea mollis* (Asteraceae). Oikos 39:269-272.

Inouye, R.S., G.S. Byers, and J.H. Brown. 1980. Effects of predation and competition on survivorship, fecundity, and community structure of desert annuals. Ecology 61:1344-1351.

Isely, F.B. 1938. The relations of Texas Acrididae to plants and soils. Ecol. Monogr. 8:551-604.

Isely, F.B. 1944. Correlation between mandibular morphology and food specificity in grasshoppers. Ann. Entomol. Soc. Amer. 37:47-67.

Islam, A. and M.J. Crawley. 1983. Compensation and regrowth in ragwort (*Senecio jacobaea*) attacked by cinnabar moth (*Tyria jacobaeae*). J. Ecology 71:829-843.

Janzen, D.H. 1966. Coevolution of mutualism between ants and acacias in Central America. Evolution 20:249-275.

Jepson-Innes, K., and C.E. Bock. 1989. Response of grasshoppers (Orthoptera: Acrididae) to livestock grazing in southeastern Arizona: differences between seasons and subfamilies. Oecologia 78:430-431.

Joern, A. 1979. Feeding patterns in grasshoppers (Orthoptera: Acrididae): factors influencing diet specialization. Oecologia 38:325-347.

Joern, A. 1986. Experimental study of avian predation on coexisting grasshopper populations (Orthoptera: Acrididae) in a sandhills grassland. Oikos 46:243-249.

Joern, A., and L.R. Lawlor. 1980. Arid grassland grasshopper community structure: comparisons with neutral models. Ecology 61:591-597.

Joern, A., and R. Alward. 1988. Effect of nitrogen fertilization on choice among grasses by the grasshopper *Phoetaliotes nebrascensis* (Orthoptera: Acrididae). Ann. Entomol. Soc. Amer. 81:240-244.

Joern, A., and G. Klucas. 1993. Intra- and interspecific competition in adults of two abundant grasshoppers (Orthoptera: Acrididae) from a sandhills grassland. Environ Entomol 22;352-361.

Johnson, R.H., and D.E. Lincoln. 1990. Sagebrush and grasshopper responses to atmospheric carbon dioxide concentration. Oecologia 84:103-110.

Johnson, R.H., and D.E. Lincoln. 1991. Sagebrush carbon allocation patterns and grasshopper nutrition: the influence of CO_2 enrichment and soil mineral limitation. Oecologia 87:127-134.

Kahn, D.M., and H.V. Cornell. 1983. Early leaf abscission and folivores: comments and considerations. Am. Nat. 122:428-432.

Kahn, D.M., and H.V. Cornell. 1989. Leafminers, early leaf abscission, and parasitoids: a tritrophic interaction. Ecology 70:1219-1226.

Karban, R. 1986. Interspecific competition between folivorous insects on *Erigeron glaucus*. Ecology 67:1063-1072.

Karban, R. 1989. Community organization of *Erigeron glaucus* folivores: effects of competition, predation, and host plant. Ecology 70:1028-1039.

Karban, R., and J.H. Myers. 1989. Induced plant responses to herbivory. Ann Rev Ecol Syst 20:331-348.

Kareiva, P. 1983. Influences of vegetation texture on herbivore populations: resource concentration and herbivore movement, pp. 259-290. *In* R.F. Denno and M.S. McClure (eds), Variable plants and herbivores in natural and managed systems. Academic Press, New York.

Kareiva, P., and R. Perry. 1989. Leaf overlap and the ability of ladybird beetles to search among plants. Ecol. Entomol. 14:127-129.

Karowe, D.N. 1989. Differential effect of tannic acid on two tree-feeding Lepidoptera: implications for theories of plant anti-herbivore chemistry. Oecologia 80:507-512.

Keeler, K.H. 1980. The extrafloral nectaries of *Ipomoea leptophylla* (Convolvulaceae). Amer. J. Bot. 67:216-222.

Keeler, K.H. 1989. Ant-plant interactions, pp. 207-242. *In*: W.G. Abrahamson (ed.), Plant-animal interactions. McGraw-Hill, New York.

Kelrick, M.I., and J.A. MacMahon. 1985. Nutritional and physical attributes of seeds of some comon sagebrush-steppe plants; some implications for ecological theory and management. J. Range Manage. 38:65-69.

Kelrick, M.I., J.A. MacMahon, R.R. Parmenter, and D.V. Sisson. 1986. Native seed preferences of shrub-steppe rodents, birds, and ants; the relationships of seed attributes and seed size. Oecologia 68:327-337.

Kemp, W.P. 1987. Probability of outbreak for rangeland grasshoppers (Orthoptera: Acrididae) in Montana: applications of Markovian principles. J. Econ. Entomol. 80:1100-1105.

Kingsolver, J.G. 1989. Weather and the population dynamics of insects: integrating physiological and population ecology. Physiol. Zool. 62:314-334.

Kinsman, S., and W.J. Platt. 1984. The impact of a herbivore upon *Mirabilis hirsuta*, a fugitive prairie plant. Oecologia 65:2-6.

Kirchner, T.B. 1977. The effects of resource enrichment on the diversity of plants and arthropods in a shortgrass prairie. Ecology 58:1334-1344.

Knapp, A.K., and T.R. Seastedt. 1986. Detritus accumulation limits productivity of tallgrass prairie. BioScience 36:662-668.

Knoch, T.R., S.H. Faeth, and D.L. Arnott. 1993. Endophytic fungi alter foraging and dispersal by desert seed-harvesting ants. Oecologia 95;470-473.

Knutson, H. 1982. Development and survival of the monophagous grasshopper *Hypochlora alba* (Dodge) and the polyphagous *Melanoplus bivittatus* (Say) and *Melanoplus sanguinipes* (F.) on Louisiana sagewort, *Artemisia ludoviciana* Nutt. Environ. Entomol. 11:777-782.

Kohler, G., H.-P. Brodhun, and G. Schaller. 1987. Ecological energetics of Central European grasshoppers (Orthoptera: Acrididae). Oecologia 74:112-121.

Krueger, K. 1986. Feeding relationships among bison, pronghorn, and prairie dogs: an experimental analysis. Ecology 67:760-770.

Larson, K.C., and T.G. Whitham. 1991. Manipulation of food reserves by a gall-forming aphid: the physiology of sink-source interactions. Oecologia 88:15-21.

Larsson, S. 1989. Stressful times for the plant stress-insect performance hypothesis. Oikos 56:277-283.

Lawton, J.H. 1983. Plant architecture and the diversity of phytophagous insects. Ann. Rev. Entomol. 28:23-39.

Lawton, J.H., and D. Schroder. 1977. Effects of plant type, size of geographical range and taxonomic isolation on number of insect species associated with British plants. Nature 265:137-140.

Lawton, J.H., and D.R. Strong, Jr. 1981. Community patterns and competition in folivorous insects. Am. Nat. 118:317-338.

Levin, D.A. 1973. The role of trichomes in plant defense. Quart. Rev. Biol. 48:3-15.

Levin, D.A. 1976. The chemical defenses of plants to pathogens and herbivores. Ann. Rev. Ecol. Syst. 7:121-159.

Lewis, A.C. 1979. Feeding preference for diseased and wilted sunflower in the grasshopper, *Melanoplus differentialis*. Entomol. exp. appl. 26:202-207.

Lewis, A.C. 1982. Leaf wilting alters a plant species ranking by the grasshopper *Melanoplus differentialis*. Ecol. Entomol. 7:391-395.

Lewis, A.C. 1984. Plant quality and grasshopper feeding: effects of sunflower condition on preference and performance in *Melanoplus differentialis*. Ecology 65:836-843.

Lightfoot, D.C., and W.G. Whitford. 1987. Variation in insect densities on desert creosotebush: is nitrogen a factor? Ecology 68:547-557.

Lightfoot, D.C., and W.G. Whitford. 1990. Phytophagous insects enhance nitrogen flux in a desert creosotebush community. Oecologia 82:18-25.

Lincoln, D.E., and H.A. Mooney. 1984. Herbivory on *Diplacus aurantiacus* shrubs in sun and shade. Oecologia 64:173-176.

Lincoln, D.E., D. Covet, and N. Sionit. 1986. Response of an insect herbivore to host plants grown in carbon dioxide enriched atmospheres. Oecologia 69:556-560.

Lincoln, D.E., E.D. Fajer, and R.H. Johnson. 1993. Plant-herbivore interactions in elevated CO_2 environments. Trends in Ecology and Evolution 8:64-68.

Lloyd, J.E., and R. Kumar. 1977. Root feeding insects of shortgrass prairie and their response to grazing pressure and ecosystem stresses, pp. 267-272. *In*: J.K. Marshall (ed.), The belowground ecosystem. Range Science Dept Science Series 26, Colorado State University, Fort Collins, Colorado.

Lockwood, J.A. 1989. Ontogeny of cannibalism in rangeland grasshoppers (Orthoptera: Acrididae). J. Kans. Entomol. Soc. 62:534-541.

Louda, S.M. 1982a. Limitation of the recruitment of the shrub *Happlopappus squarrosus* (Asteraceae) by flower- and seed-feeding insects. J. Ecol. 70:43-53.

Louda, S.M. 1982b. Distribution ecology: variation in plant recruitment over a gradient in relation to insect seed predation. Ecol. Monogr. 52:25-41.

Louda, S.M. 1983. Seed predation and seedling mortality in the recruitment of a shrub, *Happlopappus venetus* (Asteraceae), along a climatic gradient. Ecology 64:511-521.

Louda, S.M., and S.K. Collinge. 1992. Plant resistance to insect herbivores: a field test of the environmental stress hypothesis. Ecology 73:153-169.

Lowman, M.D., and H. Heatwole. 1992. Spatial and temporal variability in defoliation of Australian eucalypts. Ecology 73:129-142.

MacVean, C.M. 1987. Ecology and management of Mormon Cricket, *Anabrus simplex* Haldeman, pp. 116-136. *In*: J.L. Capinera (ed.), Integrated pest management on rangeland: a shortgrass prairie perspective. Westview Press, Boulder, CO.

Marquis, R.J. 1992. A bite is a bite is a bite? Constraints on response to folivory in *Piper arieianum* (Piperaceae). Ecology 73:142-152.

Martin, J.S., M.M. Martin, and E.A. Bernays. 1987. Failure of tannic acid to inhibit digestion or reduce digestibility of plant protein in gut fluids of insect herbivores: implications for theories of plant defense. J. Chem. Ecol. 13:605-621.

Martin, M.M., and J.S. Martin. 1984. Surfactants: their role in preventing the precipitation of proteins by tannins in insect guts. Oecologia 61:342-345.

Maschinski, J., and T.G. Whitham. 1989. The continuum of plant responses to herbivory: the influence of plant association, nutrient availability, and timing. Am. Nat. 134:1-19.

Mattes, M.J. 1969. The great Platte River road. Nebraska State Historical Society 25, Lincoln, Nebraska. 583 pp.

Mattson, W.J., Jr. 1980. Herbivory in relation to plant nitrogen content. Ann. Rev. Ecol. Syst. 11:119-161.

Mattson, W.J., and N.D. Addy. 1975. Phytophagous insects as regulators of forest primary productivity. Science 190:515-522.

Mattson, W.J. and R.A. Haack. 1987a. The role of drought stress in provoking outbreaks of phytophagous insects, pp. 365-407. *In*: P. Barbosa and J.C. Schultz (eds.), Insect outbreaks. Academic Press, San Diego, CA.

Mattson, W.J., and R.A. Haack. 1987b. The role of drought in outbreaks of plant-eating insects. BioScience 37:110-118.

Mattson, W.J. and J.M. Scriber 1987. Nutritional ecology of insect folivores of woody plants: nitrogen, water, fiber, and mineral considerations, pp.105-146. *In*: F. Slansky Jr. and J.G. Rodriquez (eds.), Nutritional ecology of insects, mites, spiders, and related invertebrates. John Wiley and Sons, New York.

McAnelly, M.L., and M.A. Rankin. 1986a. Migration in the grasshopper *Melanoplus sanguinipes* (Fab.). I. The capacity for flight in non-swarming populations. Biol. Bull. 170:368-377.

McAnelly, M.L., and M.A. Rankin. 1986b. Migration in the grasshopper *Melanoplus sanguinipes* (Fab.). II. Interactions between flight and reproduction. Biol. Bull. 170:378-392.

McBrien, H., R. Harmsen, and A. Crowder. 1983. A case of insect grazing affecting plant succession. Ecology 64:1035-1039.

McDaniel, B. 1971. The role of invertebrates in the grassland biome, pp. 267-315. *In*: N.R. French (ed.), Preliminary analysis of structure and function in grasslands. Range Science Dept. Science Series 10. Colorado State University, Fort Collins, Colorado.

McEvoy, P., C. Cox, and E. Coombs. 1991. Successful biological control of ragwort, *Senecio jacobaea*, by introduced insects in Oregon. Ecol. Appl. 1:430-442.

McMillan, C. 1964. Ecotypic differentiation within four North American prairie grasses. I. Morphological variation within transplanted communities. Am. J. Bot. 51:1118-1138.

McNaughton, S.J. 1979. Grazing as an optimization process: grass-ungulate relationships in the Serengeti. Am. Nat. 113:691-703.

McNaughton, S.J. 1983. Compensatory plant growth as a response to herbivory. Oikos 40:329-336.

McNaughton, S.J. 1985. Ecology of a grazing system: the Serengeti. Ecol. Monogr. 55:259-294.

McNaughton, S.J. 1986. On plants and herbivores. Am. Nat. 128:765-770.

McNaughton, S.J., and J.L. Tarrants. 1983. Grass leaf silification: natural selection for inducible defense against herbivores. Proc. Natl. Acad. Sci. USA. 80:790-791.

McNaughton, S.J., J.L. Tarrants, M.M. McNaughton, and R.H. Davis. 1985. Silica as a defense against herbivory and a growth promoter in African grasses. Ecology 66:528-535.

McNeill, S. 1971. The energetics of a population of *Leptoterna dolabrata* (Heteroptera: Miridae). J. Anim. Ecol. 40:127-140.

McNeill, S. 1973. The dynamics of a population of *Leptoterna dolabrata* (Heteroptera: Miridae) in relation to its food resources. J. Anim. Ecol. 42:496-507.

McNeill, S. and T.R.E. Southwood. 1978. The role of nitrogen in the development of insect/plant relationships, pp. 77-98. *In*: J.B. Harborne (ed.), Biochemical aspects of plant and animal coevolution. Academic Press, London.

Menhinick, E.F. 1967. Structure, stability, and energy flow in plants and arthropods in a sericea lespedeza stand. Ecol. Monogr. 37:255-272.

Messina, F.J. 1981. Plant protection as a consequence of an ant-membracid mutualism: interactions on goldenrod (*Solidago* sp.). Ecology 62:1433-1440.

Mispagel, M.E. 1978. The ecology and bioenergetics of the Acridid grasshopper. *Bootettix punctatus* on creosotebush, *Larrea tridentata*, in the northern Mojave Desert. Ecology 59:779-788.

Mitter, C., B. Farrell, and B. Wiegmann. 1988. The phylogenetic study of adaptive zones: has phytophagy promoted insect diversification? Am. Nat. 132:107-128.

Mittler, T.E. 1967. Water tensions in plants - an entomological approach. Ann. Entomol. Soc. Amer. 60:1074-1076.

Mitchell, J.E., and R.E. Pfadt. 1974. A role of grasshoppers in a shortgrass prairie ecosystem. Environ. Entomol. 3:358-360.

Monteith, L.G. 1960. Influence of plants other than the food plants of their host on host-finding by tachinid parasites. Can. Entomol. 92:641-652.

Moore, J.C., D.W. Walter, and H.W. Hunt. 1988. Arthropod regulation of micro- and mesobiota in belowground detrital based food webs. Ann. Rev. Entomol. 33:419-439.

Mopper, S., T.G. Whitham, and P.W. Price. 1990. Plant phenotype and interspecific competition between insects determine sawfly performance and density. Ecology 71:2135-2144.

Moran, N. 1981. Intraspecific variability in herbivore performance and host quality: a field study of *Uroleucon caligatum* (Homoptera: Aphididae) and its *Solidago* hosts (Asteraceae). Ecol. Entomol. 6:301-306.

Moran, N., and W.D. Hamilton. 1980. Low nutritive quality as defense against herbivores. J. Theoret. Biol. 86:247-254.

Moran, N., and T.G. Whitham. 1990. Interspecific competition between root-feeding and leaf-galling aphids mediated by host-plant resistance. Ecology 71:1050-1058.

Moran, V.C. 1980. Interactions between phytphagous insects and their *Opuntia* hosts. Ecol. Entomol. 5:153-164.

Morris, M.G. 1967. Differences between the invertebrate faunas of grazed and ungrazed chalk grassland. I. Responses of some phytophagous insects to cessation of grazing. J. Appl. Ecol. 4:459-474.

Morris, M.G. 1968. Differences between the invertebrate faunas of grazed and ungrazed chalk grassland. II. The faunas of sample turves. J. Appl. Ecol. 5:601-611.

Morris, M.G. 1969. Differences between the invertebrate faunas of grazed and ungrazed chalk grassland. III. The heteropterous fauna. J. Appl. Ecol. 6: 475-487.

Morris, M.G. 1971. Differences between the invertebrate faunas of grazed and ungrazed chalk grassland. IV. Abundance and diversity of Homoptera-Auchenorrhyncha. J. Appl. Ecol. 8:37-52.

Mulkern, G.B. 1980. Population fluctuations and competitive relationships of grasshopper species (Orthoptera: Acrididae). Trans. Amer. Entomol. Soc. 106:1-41.

Mulkern, G.B., K.P. Pruess, H. Knutson, A.F. Hagen, J.B. Campbell, and J.D. Lambley. 1969. Food habits and preferences of grassland grasshoppers (Acrididae) of the North Central Great Plains. North Dakota State University Agricultural Experiment Station Bulletin Number 481.

Myers, J.H. 1987. Population outbreaks of introduced insects: lessons from the biological control of weeds, pp. 173-193. *In*: P. Barbosa and J.C. Schultz (eds), Insect outbreaks. Academic Press.

Myers, J.H., C. Higgins, and E. Kovacs. 1989. How many insect species are necessary for the biological control of insects? Environ. Entomol. 18:541-547.

Myers, J.H., and B.J. Post. 1981. Plant nitrogen and fluctuations of insect populations: a test with the cinnabar moth-tansy ragwort system. Oecologia 48:151-156.

Myers, J.H., and K.S. Williams. 1987. Lack of short or long term inducible defenses in the red alder-western tent caterpillar system. Oikos 48:73-78.

Nerney, N.J., and A.G. Hamilton. 1969. Effects of rainfall on range forage and populations of grasshoppers, San Carlo Apache Indian Reservation, Arizona. J. Econ. Entomol. 62:329-333.

Nordlund, D.A., W.J. Lewis, and M.A. Altieri. 1988. Influences of plant- produced allelochemicals on the host/prey selection behavior of entomophagous insects, pp. 65-95. *In*: P. Barbosa and D.K. Letourneau (eds.), Novel aspects of insect-plant interactions. John Wiley and Sons, New York.

Norris, D.M., and M. Kogan. 1980. Biochemical and morphological bases of resistance, pp. 23-61. *In*: F.G. Maxwell and P.R. Jennings (eds.), Breeding plants resistant to insects. John Wiley and Sons, New York.

Norton, D.C. 1979. Relationship of physical and chemical factors to populations of plant parasitic nematodes. Ann. Rev. Phytopathology 17:279-299.

Nowak, R.S., C.L. Nowak, T. DeRocher, N. Cole, and M.A. Jones. 1990. Prevalence of *Oryzopsis hymenoides* near harvester ant mounds: indirect facilitation by ants. Oikos 58:190-198.

Obrycki, J. 1986. The influence of foliar pubesscence on entomophagous species, pp. 61-83. *In*: D.J. Boethel and R.D. Eikenbary (eds.), Interactions of plant resistance and parasitoids and predators of insects. J. Wiley and Sons, New York.

Owen, D.F. 1978. Why do aphids synthesize melezitose? Oikos 31:264-267.

Owen, D.F. 1980. How plants may benefit from the animals that eat them. Oikos 35:230-235.

Owen, D.F. and R.G. Wiegert. 1976. Do consumers maximize plant fitness? Oikos 27:488-492.

Owen, D.F. and R.G. Wiegert. 1987. Leaf eating as mutualism, pp. 81-95. *In* P. Barbosa and J. Schultz (eds.), Insect outbreaks. Academic Press, San Diego, CA.

Paige, K.N., and T.G. Whitham. 1987. Overcompensation in response to mammalian herbivory: the advantage of being eaten. Am. Nat. 129:407-416.

Parker, M.A. 1982. Association with mature plants protects seedlings from predation in an arid grassland shrub, *Gutierrezia microcephala*. Oecologia 53:276-280.

Parker, M.A. 1984. Local food depletion and the foraging behavior of a specialist grasshopper, *Hesperotettix viridis*. Ecology 65:824-835.

Parker, M.A. 1985. Size-dependent herbivore attack and the demography of an arid grassland shrub. Ecology 66:850-860.

Parker, M.A., and R.B. Root. 1981. Insect herbivores limit habitat distribution of a native composite, *Machaeranthera canescens*. Ecology 62:1390-1392.

Parker, M.A., and A.G. Salzman. 1985. Herbivore exclosure and competitor removal: effects on juvenile survivorship and growth in the shrub *Gutierrezia microcephala*. J. Ecol. 73:903-913.

Parker, J.R., R.C. Newton, and R.L. Shotwell. 1955. Observations on mass flights and other activities of the migratory grasshopper. USDA Technical Bulletin 1109:1-46.

Patterson, B.D. 1984. Correlation between mandibular morphology and specific diet of some desert grassland Acrididae. Amer. Midl. Nat. 111:296-303.

Patterson, C.G., D.A. Potter, and F.F. Fannin. 1991. Feeding deterrency of alkaloids from endophyte-infected grasses to Japanese beetle grubs. Entomol. exp. appl. 61:285-289.

Pfadt, R.E., J.A. Lockwood, and T.M. Foppe. 1988. Diets of three common grasshoppers (Orthoptera: Acrididae) inhabiting desert grassland. Can. Entomol. 120:415-424.

Pinder III, J.E., and G.C. Kroh. 1987. Insect herbivory and photosynthetic pathways in old-field ecosystems. Ecology 68:254-259.

Pinder III, J.E., and P.R. Jackson. 1988. Plant photosynthetic pathways and grazing by phytophagous orthopterans. Amer. Midl. Nat. 120:201-211.

Prestidge, R.A. 1982a. Instar duration, adult consumption, oviposition and nitrogen utilization efficiencies of leafhoppers feeding on different quality food (Auchenorrhyncha: Homoptera). Ecol. Entomol. 7:91-101.

Prestidge, R.A. 1982b. The influence of nitrogenous fertilizer on the grassland Auchenor-ryncha (Homoptera). J. Appl. Ecol. 19:735-749.

Prestidge, R.A., and S. McNeill. 1983. The role of nitrogen in the ecology of grassland Auchenorryncha, pp. 257-281. *In*: J.A. Lee, S. McNeill, and I.H. Rorison (eds.), Nitrogen as an ecological factor. Blackwell Scientific, Oxford.

Price, P.W. 1991a. The plant vigor hypothesis and herbivore attack. Oikos 62:244-251.

Price. P.W. 1991b. Darwinian methodology and the theory of insect herbivore population dynamics. Ann. Entomol. Soc. Amer. 84:465-473.

Price, P.W., C.E. Bouton, P. Gross, B.A. McPheron, J.N. Thompson, and A.E. Weis. 1980. Interactions among three trophic levels: influence of plants on interactions between insect herbivores and natural enemies. Ann. Rev. Ecol. Syst. 11:41-65.

Price, P.W., G.W. Fernandes, and G.L. Waring. 1987. Adaptive nature of insect galls. Environ. Entomol. 16:15-24.

Pringle, W.L. 1960. The effect of a leaf feeding beetle on big sagebrush in British Columbia. J. Range Manage. 13:139-142.

Pritchard, I.M., and R. James. 1984a. Leaf mines: their effect on leaf longevity. Oecologia 64:132-139.

Pritchard, I.M. and R. James. 1984b. Leaf fall as a source of leaf miner mortality. Oecologia 64:140-141.

Quinn, M.A. and D.D. Walgenbach. 1990. Influence of grazing history on the community structure of grasshoppers of a mixed-grass prairie. Environ. Entomol. 19:1756-1766.

Raupp, M.J. and R.F. Denno. 1984. The suitability of damaged willow leaves as food for the leaf beetle, *Plagiodera versicolora*. Ecol. Entomol. 9:443-448.

Rauscher, M.D. 1981. The effects of native vegetation on the susceptibility of *Aristolochia reticulata* (Aristolochiaceae) to herbivore attack. Ecology 62:1187-1195.

Reese, J.C. 1979. Interactions of allelochemicals with nutrients in herbivore food, pp. 309-330. *In*: G.A. Rosenthal and D.H. Janzen (eds.), Herbivores: their interaction with secondary plant metabolites. Academic Press, New York.

Rehr, S.S., P.P. Feeny, and D.H. Janzen. 1973. Chemical defense in Central American non-ant-acacias. J. Anim. Ecol. 42:405-416.

Reichman, O.J. 1979. Desert granivore foraging and its impact on seed densities and distributions. Ecology 60:1085-1092.

Rhoades, D.F. 1979. Evolution of plant chemical defense against herbivores, pp. 3-54. *In* G.A. Rosenthal and D.H. Janzen (eds.), Herbivores: their interaction with secondary plant metabolites. Academic Press, New York.

Rhoades, D.F. 1983. Responses of alder and willow to attack by tent caterpillars and webworms: evidence for pheromonal sensitivity of willows, pp. 55-68. *In* P.A. Hedin (ed), Plant resistance to insects. American Chemical Society, Washington, D.C.

Rhoades, D.F. 1985. Offensive-defensive interactions between herbivores and plants: their relevance in herbivore population dynamics and ecological theory. Am. Nat. 125:205-238.

Rhoades, D.F., and R.G. Cates. 1976. Toward a general theory of plant antiherbivore chemistry. Recent Advances in Phytochemistry 10:168-213.

Richman, D.B., and E.W. Huddleston. 1981. Root feeding by the beetle *Crossidius pul-*

chellus LeCone and other insects on broom snakeweed (*Gutierrezia* spp.) in eastern and central New Mexico. Environ. Entomol. 10:53-57.

Ritcher, P.O. 1966. White grubs and their allies. Oregon State University Press, Corvallis, Oregon.

Rissing, S.W. 1986. Indirect effects of granivory by harvester ants: plant species composition and reproductive increase near ant nests. Oecologia 68:231-234.

Rissing, S.W. 1988. Seed-harvester ant association with shrubs: competition for water in the Mohave Desert? Ecology 69:809-813.

Ritchie, M.E., and D. Tilman. 1993. Predictions of species interactions from consumer-resource theory; experimental tests with grasshoppers and plants. Oecologia 94:516-527.

Roberts, R.J., and R. Morton. 1985. Biomass of larval Scarabaeidae (Coleoptera) in relation to grazing pressures in temperate, sown pastures. J Appl. Ecol. 22:863-874.

Romme, W.H., D.H. Knight, and J.B. Yavitt. 1986. Mountain pine beetle outbreaks in the Rocky Mountains: regulators of primary productivity? Am. Nat. 127:484-494.

Root, R.B. 1973. Organization of a plant-arthropod association in simple and diverse habitats: the fauna of collards (*Brassica oleraceae*). Ecol. Monogr. 43:1325-1340.

Root, R.B. 1975. Some consequences of ecosystem texture, pp. 83-97. *In*: S.A. Levin (ed.), Ecosystem analysis and prediction. Society for Industrial and Applied Mathematics, Philadelphia, Pennsylvania.

Rosenthal, G.A., and M.R. Berenbaum (eds.). 1991. Herbivores: their interactions with secondary plant metabolites. 2nd ed. Vol. I. Academic Press, San Diego, CA.

Rosenthal, S.S., D.M. Maddox, and K. Brunetti. 1985. Biological control methods, pp. 65-94. *In*: California Weed Conference. Principles of weed control in California. Thomson Publications, Fresno, CA.

Rowell-Rahier, M. and J.M. Pasteels. 1992. Third trophic level influences of plant allelochemicals, pp. 243-277. *In* G.A. Rosenthal and M.R. Berenbaum (eds.), Herbivores: their interactions with secondary plant metabolites. 2nd ed. Vol II. Academic Press, San Diego, CA.

Russell, E.P. 1989. Enemies hypothesis: a review of the effect of vegetational diversity on predatory insects and parasitoids. Environ. Entomol. 18:590-599.

Samson, D.A., T.E. Phillippi, and D.W. Davidson. 1992. Granivory and competition as determinants of annual plant diversity in the Chihuahuan desert. Oikos 65:61-80.

Sayre, R.M. 1971. Biotic influences in the soil environment, pp. 235-256. *In*: B.M. Zuckerman, W.F. Mai, and R.A. Rohde (eds). Plant parasitic nematodes. Academic Press, NY.

Schoener, T.W. 1983. Field experiments on interspecific competition. Am. Nat. 122:240-285.

Schultz, J.C., and I.T. Baldwin. 1982. Oak leaf quality declines in response to defoliation by gypsy moth larvae. Science 217:149-151.

Scott, J.A., N.R. French, and J.W. Leetham. 1979. Patterns in consumption in grasslands, pp. 89-105. *In*: N.R. French (ed.), Perspectives in grassland ecology. Springer-Verlag, New York.

Scriber, J.M. 1978. The effects of larval feeding specialization and plant growth form on the consumption and utilization of plant biomass and nitrogen: an ecological consideration. Entomol exp appl 24: 494-510.

Scriber, J.M., and P.P. Feeny. 1979. Growth of herbivorous caterpillars in relation to feeding specialization and to the growth form of their food plants. Ecology 60:829-850.

Scriber, J.M., and F. Slansky, Jr. 1981. The nutritional ecology of immature insects. Ann. Rev. Entomol. 26:183-211.

Seastedt, T.R. 1985. Maximization of primary and secondary productivity by grazers. Am. Nat. 126:559-564.

Seastedt, T.R., D.C. Hayes, and R.A. Ramundo. 1988. Maximization of belowground consumer densities by removal of foliage: empirical evidence, graphical and conceptual models. Oikos 51:243-248.

Seastedt, T.R., R.A. Ramundo, and D.C. Hayes. 1989. Silica, nitrogen and phosphorus dynamics of grazed tallgrass prairie. *In*: T.M. Bragg and J. Stubbendieck (eds). Prairie pioneers: ecology, history and culture. Proc. 11th North American Prairie Conference. Univ Nebraska.

Seastedt, T.R., T.C. Todd, and S.W. James. 1987. Experimental manipulations of arthropod, nematode and earthworm communities in a North American grassland. Pedobiologia 30:9-17.

Shah, M.A. 1982. The influence of plant surfaces on the searching behaviour of coccinellid larvae. Entomol. exp. appl. 31:377-380.

Shahjahan, M. and F.A. Streams. 1973. Plant effects on host-finding by *Leiophron pseudopallipes* (Hymenoptera: Braconidae), a parasitoid of the tarnished plant bug. Environ. Entomol. 2:921-925.

Shaw, P.B., D.B. Richman, J.C. Owens, and E.W. Huddleston. 1987. Ecology of the range caterpillar, *Hemileuca oliviae*, pp. 234-247. *In*: J.L. Capinera (ed.), Integrated pest management on rangeland: a shortgrass prairie perspective. Westview Press, Boulder, Co.

Silkstone, B.E. 1987. The consequences of leaf damage for subsequent insect grazing on birch (*Betula* spp.). A field experiment. Oecologia 74:149-152.

Slansky, F., Jr., and P.P. Feeny. 1977. Stabilization of the rate of nitrogen accumulation by larvae of the cabbage butterfly on wild and cultivated plants. Ecol. Monogr. 47:209-228.

Slansky, F., Jr., and A.R. Panizzi. 1987. Nutritional ecology of seed-sucking insects, pp. 283-320. *In*: F. Slansky Jr. and J.G. Rodriquez (eds.), Nutritional ecology of insects, mites, spiders, and related invertebrates. John Wiley and Sons, New York.

Slansky, F., Jr., and J.G. Rodriquez (eds.). 1987. Nutritional ecology of insects, mites, spiders, and related invertebrates. John Wiley and Sons, New York.

Smith, C.C. 1940. The effects of overgrazing and erosion upon the biota of the mixed grass prairie of Oklahoma. Ecology 21:381-397.

Smith, S.G.F., and G.U. Grodowitz. 1983a. Effects of nonglandular trichomes of *Artemisia ludoviciana* Nutt. (Asteraceae) on ingestion, assimilation, and growth of the grasshoppers *Hypochlora alba* (Dodge) and *Melanoplus sanguinipes* (F.) (Orthoptera: Acrididae). Environ. Entomol. 12:1766-1772.

Smith, S.G.F., and G.L. Kreitner. 1983b. Trichomes in *Artemisia ludoviciana* Nutt. (Asteraceae) and their ingestion by *Hypochlora alba* (Dodge) (Orthoptera: Acrididae). Amer. Midl. Nat. 119:118-123.

Smolik, J.D. 1977. Effect of nematicide treatment on growth of range grasses in field and glasshouse studies, pp. 257-260. *In*: J.K. Marshall (ed), The belowground ecosystem. Range Science Dept Science Series 26, Colorado State University, Fort Collins, CO.

Smolik, J.D., and L.E. Rogers. 1976. Effects of cattle grazing and wildfire on soil dwelling nematodes of the shrub-steppe ecosystem. J. Range Manage. 33: 657-663.

Smolik, J.D., and J.K. Lewis. 1982. Effect of range condition on density and biomass of nematodes in a mixed prairie ecosystem. J. Range Manage. 35:657-663.

Smolik, J.D., and J.L. Dodd. 1983. Effect of water and nitrogen, and grazing on nematodes in a shortgrass prairie. J. Range Manage. 36:744-748.

Solomon, B.P. 1983. Compensatory production in *Solanum carolinense* following attack by a host-specific herbivore. J. Ecol. 71:681-690.

Southwood, T.R.E. 1961. The number of species of insect associated with various trees. J. Anim. Ecol. 30:1-8.

Southwood, T.R.E. 1973. The insect/plant relationship - an evolutionary perspective, pp. 3-30. *In*: H.F. van Emden (ed.), Symp Royal Entomol Soc Lond: 6. Blackwell Scientific, Oxford.

Southwood, T.R.E. 1985. Interactions of plants and animals: patterns and processes. Oikos 44:5-11.

Southwood, T.R.E., V.K. Brown, and P.M. Reader. 1979. The relationships of plant and insect diversities in succession. Biol. J. Linn. Soc. 12:327-348.

Stamp, N.E. 1984. Effect of defoliation by checkerspot caterpillars (*Euphydryas phaeton*) and sawfly larvae (*Macrophya nigra* and *Tenthredo grandis*) on their host plants (*Chelone* spp.). Oecologia 63:275-280.

Stamp, N.E. 1992. Theory of plant-insect herbivore interactions on the inevitable brink of re-synthesis. Bull. Ecol. Soc. Amer. 73:28-34.

Stamp, N.E., and M.D. Bowers. 1991. Indirect effect on survivorship of caterpillars due to presence of invertebrate predators. Oecologia 88:325-330.

Stanton, M.L. 1983. Spatial patterns in the plant community and their effects upon insect search, pp. 125-157. *In*: S. Ahmad (ed.), Herbivorous insects: host-seeking behavior and mechanisms. Academic Press, New York.

Stanton, N.L. 1987. Nematodes in rangelands, pp. 291-311. *In*: J. Capinera (ed.), Integrated pest management on rangeland: a shortgrass prairie perspective. Westview Press, Boulder, CO

Stanton, N.L. 1988. The underground in grasslands. Ann. Rev. Ecol. Syst. 19:573-589.

Stanton, N.L., M. Allen, and M. Campion. 1981. The effect of the pesticide carbofuran on soil organisms and root and shoot production in shortgrass prairie. J. Appl. Ecol. 18:417-431.

Stanton, N.L., D. Morrison, and W.A. Laycock. 1984. The effect of phytophagous grazing on blue grama die-off. J. Range Manage. 37:447-450.

Stephenson. A.G. 1981. Flower and fruit abortion: proximate causes and ultimate functions. Ann. Rev. Ecol. Syst. 12:253-279.

Stiling, P. and D. Simberloff. 1989. Leaf abscission: induced defense against pests or response to damage? Oikos 55:43-49.

Stiling, P., D. Simberloff, and B.V. Brodbeck. 1991. Variation in rates of leaf abscission between plants may affect the distribution patterns of sessile insects. Oecologia 88:367-370.

Strauss, S.Y. 1991. Direct, indirect, and cumulative effects of three native herbivores on a shared host plant. Ecology 72:543-558.

Strong, D.R., Jr. 1979. Biogeographic dynamics of insect-host plant communities. Ann. Rev. Entomol. 24:89-119.

Strong, D.R., Jr., and D.A. Levin. 1979. Species richness of plant parasites and growth form of their hosts. Am. Nat. 114:1-22.

Strong, D.R., Jr., J.H. Lawton, and T.R.E. Southwood. 1984. Insects on plants. Community patterns and mechanisms. Blackwell Scientific Publications, Oxford.

Stubblevine, W.H., and J.H. Langenheim. 1977. Effects of *Hymenaea courbaril* leaf resin on the generalist herbivore *Spodoptera exigua* (beet armyworm). J. Chem. Ecol. 3:633-647.

Swank, W.T., J.B. Waide, D.A. Crossley, Jr., and R.L. Todd. 1981. Insect defoliation enhances nitrate export from forest ecosystems. Oecologia 51:297-299.

Tabashnik, B.E, and F. Slanksky Jr. 1987. Nutritional ecology of forb foliage-chewing insects, pp. 71-103. *In* F. Slansky Jr. and J.G. Rodriquez (eds.), Nutritional ecology of insects, mites, spiders, and related invertebrates. John Wiley and Sons, New York.

Tahvanainen, J.O., and R.B. Root. 1972. The influence of vegetational diversity on the population ecology of a specialized herbivore, *Phyllotreta cruciferae* (Coleoptera: Chrysomelidae). Oecologia 10:321-346.

Tallamy, D.W. 1986. Behavioral adaptations in insects to plant allelochemicals, pp. 273-290. *In*: L.B. Brattsten and S. Ahmad (eds.), Molecular aspects of insect/plant associations, Plenum Press, New York, USA.

Tallamy, D.W., and M.J. Raupp (eds.). 1991. Phytochemical induction by herbivores. John Wiley and Sons, New York.

Todd, J.G., and J.A. Kamm. 1974. Biology and impact of a grass bug *Labops hesperius* Uhler in Oregon rangeland. J. Range Manage. 27:453-458.

Trumble, J.T., D.M. Kolodny-Hirsch, and I.P. Ting. 1993. Plant compensation for arthropod herbivory. Ann. Rev. Entomol. 38:93-119.

Tuomi, J., P. Niemala, E. Haukioja, S. Siren, and S. Neuvonen. 1984. Nutrient stress: an explanation for plant anti-herbivore responses to defoliation. Oecologia 61:208-210.

Tuomi, J., P. Niemala, and S. Siren. 1990. The Panglossian paradigm and delayed inducible accumulation of foliar phenolics in mountain birch. Oikos 59:399-410.

Turlings, T.C., J.H. Tumlinson, and W.J. Lewis. 1990. Exploitation of herbivore-induced plant odors by host-seeking parasitic wasps. Science 250:1251-1253.

Ueckert, D.N. 1979. Impact of a white grub (*Phyllophaga crinita*) on a shortgrass community and evaluation of selected rehabilitation practices. J. Range Manage. 32:445-448.

Ueckert, D.N., and R.M. Hansen. 1971. Dietary overlap of grasshoppers on sandhill rangeland in northeastern Colorado. Oecologia 8:276-295.

Valone, T.J., J.H. Brown, and E.J. Heske. 1994. Interactions between rodents and ants in the Chihuahuan desert: an update. Ecology 75:252-255.

van der Meijden, E., and R.E. van der Waals-Kooi. 1979. The population ecology of *Senecio jacobea* in a sand dune system. I. Reproductive strategy and the biennial habitat. J. Ecol. 67:131-153.

Vinson, S.B. 1985. The behavior of parasitoids, pp. 417-469. *In*: G.A. Kerkut and L.I. Gilbert. Comprehensive Insect Physiology, Biochemistry and Pharmacology. Volume 9: Behaviour. Pergamon Press, Oxford, England.

Vrieling, K., W. Smit, and E. van der Meijden. 1991. Tritrophic interactions between aphids (*Aphis jacobaeae* Schrank), ant species, *Tyria jacobaeae* L., and *Senecio jacobaea* L. lead to maintainence of genetic variation in pyrrolizidine alkaloid concentration. Oecologia 86:177-182.

Wakeland, C. 1959. Mormon crickets in North America. USDA Tech. Bull. 1202:1-77.

Walter, D.E. 1987a. Trophic behavior of mycophagous microarthropods. Ecology 68:226-229.

Walter, D.E. 1987b. Belowground arthropods of semiarid grasslands, pp. 271- 290. *In*: J.L. Capinera (ed.), Rangeland pest management: a shortgrass prairie perspective. Westview Press, Boulder, Colorado.

Walter, D.E. 1988. Nematophagy by soil arthropods from the shortgrass steppe, Chihuahuan Desert and Rocky Mountains of the central United States. Agriculture, Ecosystems, and Environment 24:307-316.

Wapshere, A.J. 1974. A strategy for evaluating the safety of organisms for biological weed control. Ann. appl. Biol. 77:201-211.

Wapshere, A.J., E.S. Delfosse, and J.M. Cullen. 1989. Recent developments in biological control of weeds. Crop Protection 8:227-250.

Waring, G.L., and P.W. Price. 1990. Plant water stress and gall formation (Cecidomyiidae: *Asphondylia* spp.) on creosote bush. Ecol. Entomol. 15:87-95.

Watts, J.G., and T.D. Everett. 1976. Biology and behavior of the range caterpillar. Agric. Exp. Stn., New Mexico State University, Las Cruces, Bull. 646. 32 pp.

Watts, J.G., E.W. Huddleston, and J.C. Owens. 1982. Rangeland entomology. Ann. Rev. Entomol. 27:283-311.

Watts, J.G., G.B. Hewitt, E.W. Huddleston, H.G. Kinzer, R.J. Lavigne, and D.N. Ueckert. 1989. Rangeland entomology. Second edition. Range Science Series No. 2. Society for Range Management, Denver, CO.

Weiner, L.F., and J.L. Capinera. 1980. Preliminary study of the biology of the white grub *Phyllophaga fimbripes* (Le Conte) (Coleoptera: Scarabeidae). J. Kans. Entomol. Soc. 53:701-710.

Westoby, M., B. Walker, and I. Noy-Meir. 1989. Opportunistic management for rangelands not at equilibrium. J. Range Manage. 42:266-274.

Weigert, R.G., and C.E. Petersen. 1983. Energy transfer in insects. Ann. Rev. Entomol. 28:455-486.

Weis, A.E. and M.R. Berenbaum. 1989. Herbivorous insects and green plants, pp. 123-162. *In* W.G. Abrahamson (ed.), Insect-animal interactions. McGraw-Hill, New York.

Whelan, R.J., and A.R. Main. 1979. Insect grazing and post-fire plant succession in southwest Australian woodland. Austr. J. Ecol. 4:387-398.

Whitcomb, R.F., A.L. Hicks, H.D. Blocker, and D.E. Lynn. 1994. Biogeography of leafhopper specialists of the shortgrass prairie: evidence for the roles of phenology and phylogeny in determination of biological diversity. American Entomologist 40:19-34.

White, T.C.R. 1969. An index to measure weather-induced stress of trees associated with outbreaks of psyllids in Australia. Ecology 50:905-909.

White, T.C.R. 1974. A hypothesis to explain outbreaks of looper caterpillars, with special reference to populations of *Selidosema suavis* in a plantation of *Pinus radiata* in New Zealand. Oecologia 16:279-301.

White, T.C.R. 1976. Weather, food and plagues of locusts. Oecologia 22:119-134.

White, T.C.R. 1978. The importance of a relative shortage of food in animal ecology. Oecologia 33:71-86.

White, T.C.R. 1984. The abundance of invertebrate herbivores in relation to the availability of nitrogen in stressed food plants. Oecologia 63:90-105.

White, T.C.R. 1993. The inadequate environment. Nitrogen and the abundance of animals. Springer-Verlag, Berlin.

Whitham, T.G. 1989. Plant hybrid zones as sinks for pests. Science 244:1490-1493.

Whitham, T.G., and S. Mopper. 1985. Chronic herbivory: impacts on architecture and sex expression of pinyon pine. Science 228:1089-1091.

Whitman, D.W. 1988. Allelochemical interactions among plants, herbivores, and their predators, pp. 11-64. *In*: P. Barbosa and D.K. Letourneau (eds.), Novel aspects of insect-plant interactions. John Wiley and Sons, New York.

Whittaker, J.B. 1979. Invertebrate grazing, competition and plant dynamics, pp. 207-222. *In*: R.M. Anderson, B.D. Turner, and L.R. Taylor (eds), Population dynamics. Blackwell Scientific Publications, Oxford.

Whittaker, R.H., and P.P. Feeny. 1971. Allelochemics: chemical interactions between species. Science 171:757-770.

Wight, J.R., and F.T. Nichols. 1966. Effects of harvester ants on production of a saltbush community. J. Range Manage. 19:68-71.

Wilbur, D.A. 1936. Grasshopper injury to the inflorescence of pasture grasses. J. Kans. Entomol. Soc. 9:1-12.

Williams, A.G. and T.G. Whitham. 1986. Premature leaf abscission: an induced plant defense against gall aphids. Ecology 67:1619-1627.

Williams, K.S., and J.H. Myers. 1984. Previous herbivore attack of red alder may improve food quality for fall webworm larvae. Oecologia 63:166-170.

Wratten, S.D., P.J. Edwards, and I. Dunn. 1984. Wound-induced changes in the palatability of *Betula pubescens* and *B. pendula*. Oecologia 61:372-375.

Wratten, S.D., P.J. Edwards, and L. Winder. 1988. Insect herbivory in relation to dynamic changes in host plant quality. Biol. J. Linn. Soc. 35:339-350.

Zangerl, A.R. 1990. Furanocoumarin induction in wild parsnip: evidence for an induced defense against herbivores. Ecology 71:1926-1932.

Plant Responses to Defoliation: A Physiological, Morphological and Demographic Evaluation

David D. Briske and James H. Richards

Introduction

The importance of developing the fundamental principles of rangeland management from knowledge of the growth requirements and life history attributes of important plant species was recognized early in the development of the range science profession (Sampson 1914). Since then, a great deal has been learned concerning plant responses to a wide array of environmental variables and management practices. Plant responses to defoliation have been emphasized because of the extensive use of rangelands as a forage source for both wild and domestic herbivores. Knowledge of plant response to defoliation has been incorporated into various procedures and prescriptions to effectively manage both vegetation and herbivores to minimize the detrimental consequences of grazing and maintain plant and animal production on a sustainable basis.

Research activity in plant-herbivore interactions was stimulated to an unprecedented level with the introduction of the grazing optimization hypothesis in the mid-1970s (Dyer 1975, McNaughton 1976, 1979). This hypothesis proposes that maximum plant productivity occurs at an optimal intensity of grazing rather than in ungrazed vegetation. The grazing optimization hypothesis also implies that grazed plants must exhibit mechanisms enabling them to increase their growth rates above those of ungrazed plants if they are to replace biomass removed by herbivores. This alternative perspective of the production responses of plants to grazing has created a substantial amount of confusion and generated a great deal of controversy among ecologists and natural resource managers (e.g., McNaughton 1979, Belsky 1986, Painter and Belsky 1993, Belsky et al. 1993).

The introduction and wide acceptance of ecological hierarchy theory, at about the same time that the grazing optimization hypothesis was initially being debated, has further modified our interpretation of plant responses to grazing. Ecological hierarchy theory indicates that communities consist of a series of graded hierarchies or ecological scales that possess unique temporal and spatial characteristics (Belsky 1987, Brown and Allen 1989, Allen and Hoekstra 1990). Grasslands are very well suited to the application of the hierarchy theory. For example, individual grass plants consist of an assemblage of tillers, populations are defined by both plant density and tiller number per plant, and grassland communities are further

Department of Rangeland Ecology and Management, Texas A&M University, College Station, TX 77843-2126 and Department of Land, Air and Water Resources, University of California, Davis, CA 95616-8627.

composed of an aggregation of species populations (Archer and Smeins 1991, Briske 1991). The structure and function of lower hierarchical levels (i.e., tiller and plant) partially defines the higher levels of vegetation organization, while the higher hierarchical levels (i.e., communities) constrain the structure and function of lower levels in ways that are not readily predictable from an evaluation of the lower levels in isolation. This theory establishes that environmental variables and processes at the higher hierarchical levels (i.e., competition and nutrient cycling) can regulate plant growth following defoliation to an equal or greater extent than the direct effects of defoliation (McNaughton 1979, Archer and Smeins 1991).

The objective of this chapter is to present a comprehensive synthesis of the information addressing plant responses to grazing in a hierarchical context. Physiological and morphological responses of individual plants to defoliation are evaluated in chronological sequence beginning with plant function during "steady state" growth prior to defoliation, followed by the short-term effects of defoliation, and concluding with long-term processes contributing to the reestablishment of "steady-state" growth. Our knowledge of plant populations are assessed through an evaluation of tiller and plant demography in the absence of grazing followed by an assessment of the affects of grazing on these demographic processes. Grazing resistance is addressed by evaluating its component mechanisms of avoidance and tolerance and assessing the potential costs of these mechanisms. The chapter concludes with an assessment of how the effects of grazing are translated from individual plants through populations to affect productivity within plant communities.

Individual Plant Responses to Defoliation

Individual plant responses to defoliation encompass a large number of physiological and morphological processes spanning a broad range of temporal scales. The objective of this section is to evaluate individual plant responses in a chronological sequence beginning with plant function during "steady-state" growth prior to defoliation, followed by the short-term effects of defoliation, and conclude with long-term processes contributing to the reestablishment of "steady-state" growth.

Immediate Effects of Defoliation

Whole-plant photosynthesis

Whole-plant photosynthesis is instantaneously reduced in response to canopy removal by grazing or clipping. Large portions of the canopy of individual plants are frequently removed by grazing animals in a single grazing event (e.g., Jarman and Sinclair 1979, Norton and Johnson 1983, Anderson and Shumar 1986, Krueger 1986). Rapid, large-scale reductions in canopy area are less likely to occur as a result of chronic defoliation which is often associated with herbivorous insects (e.g., Coley 1983, Seastedt et al. 1983, Lowman 1984, see Evans and Seastedt this volume). These contrasting patterns of defoliation modify whole-plant photosynthesis in very different ways. Plants adjust to conditions of chronic defoliation and the associated reduction in whole-plant photosynthetic rates by altering resource allocation patterns and reducing relative growth rates. In contrast, a transient period of modified physiological function frequently accompanies plant defoliation by large herbivores followed by recovery of steady-state plant function. Relatively few studies have adequately monitored the immediate (<48 hours) effects of reduced whole-plant photosynthesis on plant function in response to defoliation. Knowledge of the immediate responses of plants to defoliation is critical for understanding the process of plant recovery following defoliation.

The reduction in whole-plant photosynthesis following defoliation is not necessarily proportional to leaf-area or biomass removal because of associated modification in canopy microclimate, the unequal photosynthetic contributions of leaves of various age and, in some cases, compensatory photosynthesis. For example, when mature, previously shaded leaves remain on the plant following defoliation, canopy photosynthesis is reduced to a greater extent than the proportion of leaf area removed because of the low photosynthetic capacity of the remaining leaves (Ludlow and Charles-Edwards 1980, Gold and Caldwell 1989b). A large decrease in the photosynthesis/transpiration ratio of the canopy (i.e., water-use efficiency) is also associated with this pattern of plant defoliation (Caldwell et al. 1983, Gold and Caldwell 1989b). Conversely, if a high proportion of relatively young leaves remain on the plant following defoliation, the reduction in canopy photosynthesis is more directly related to amount of leaf area removed. Consequently, canopy measurements of photosynthesis are more strongly correlated with the potential for regrowth than are measurements of single-leaf photosynthesis (Ludlow and Charles-Edwards 1980, Parsons et al. 1983a, 1983b, King et al. 1984, 1988, Nowak and Caldwell 1984, Gold and Caldwell 1989b). Estimates of canopy photosynthesis integrate the disparate photosynthetic activities of foliage elements, including leaf blades, sheaths and culms, their relative surface areas, and the influence of microclimatic variables in a way that is difficult to achieve with measurements of single-leaf photosynthesis.

Root growth, respiration, and nutrient uptake

Immediately following defoliation the effects of reduced canopy photosynthesis are rapidly propagated throughout the plant. Studies of numerous grazing-tolerant C_3 and C_4 forage grasses growing with high nutrient availability in controlled-environment or greenhouse conditions have demonstrated that root elongation essentially ceases within 24 hours after removal of approximately 50% or more of the shoot system (Fig. 1) (Crider 1955, Troughton 1957, Oswalt et al. 1959, Davidson and Milthorpe 1966a, Ryle and Powell 1975) and root mortality and decomposition may begin within 36–48 hours (Oswalt et al. 1959). Root respiration and nutrient acquisition are also reduced following defoliation, but to a lesser extent than root growth (Fig. 1) (Davidson and Milthorpe 1966a, Chapin and Slack 1979, Macduff et al. 1989). Root respiration begins to decline within hours of defoliation and it may decrease substantially within 24 hours (Davidson and Milthorpe 1966a, Clement et al. 1978, Thorgeirsson 1988).

Concomitant with the reduction in root respiration following defoliation is a rapid reduction in nutrient absorption (Fig. 1). Experiments conducted with perennial ryegrass (*Lolium perenne*) growing in nutrient solution demonstrated that the rate of nitrate (NO_3^-) absorption began to decline within 30 minutes following removal of 70% of shoot biomass. NO_3^- absorption decreased to less than 40% of

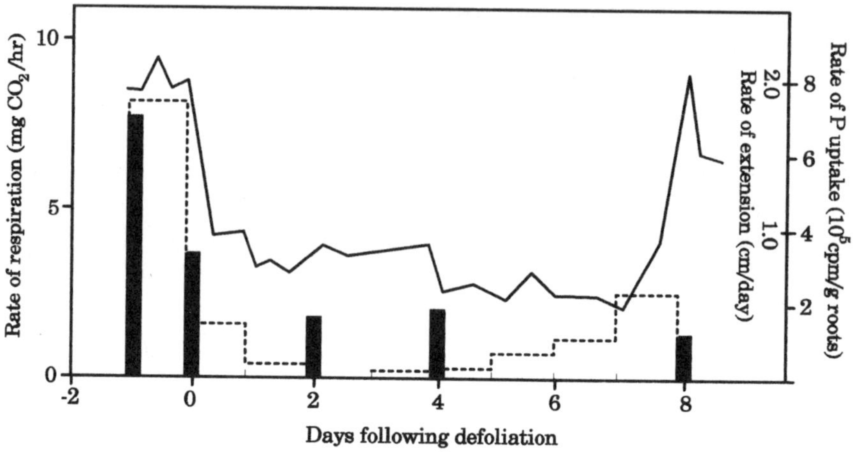

Figure 1. Root growth and function in orchardgrass plants defoliated to a height of 2.5 cm. Root respiration (solid line), root length extension (dashed line), and relative uptake of ^{32}P during 4-hour exposures (histograms) prior to and for 8 days following defoliation imposed on day 0. Plants were grown in solution culture to allow continuous physiological measurements (From Davidson and Milthorpe 1966b).

the pre-defoliation rate within 2 hours following defoliation (Clement et al. 1978). In these experiments, NO_3^- absorption continued to decline over the next 4–12 hours until it became negligible for 2 or 7 days before recovery began under high and low light intensities, respectively. NO_3^- absorption did not resume until a positive daily carbon balance was reestablished within the plant (Clement et al. 1978). Rapid reductions in root respiration and nutrient absorption following plant defoliation are proportional to the intensity of defoliation (Davidson and Milthorpe 1966a, Thorgeirsson 1988). Similarly, canopy shading or root severing causes large, rapid decreases in root respiration and nutrient absorption (Clarkson et al. 1974, Hansen and Jensen 1977, Massimino et al. 1981, Saglio and Pradet 1980, Aslam and Huffaker 1982, Thorgeirsson 1988, Bloom and Caldwell 1988, Macduff and Jackson 1992). These experiments clearly demonstrate the importance of current photosynthesis for the maintenance of root growth and function in rapidly growing plants.

The responses of plants adapted to infertile environments differ from those previously described for rapidly growing species adapted to fertile environments. For example, root growth was unaffected for 48 hours following a single, severe defoliation of two nutrient-limited tundra graminoids, cottongrass (*Eriophorum vaginatum*) and *Carex aquatilis* (Chapin and Slack 1979). Root growth was not reduced until the plants were subjected to two or more successive defoliations. Respiration and nutrient absorption were maintained, or even increased, following defoliation of these nutrient-limited plants (Chapin and Slack 1979). Defoliation has even been documented to increase NO_3^- absorption in perennial ryegrass within 8 hours when plants were grown with low fertility, rather than high fertility (Macduff et al. 1989). In addition to these short-term responses, long-term increases in specific root respiration rate and nutrient absorption capacity (V_{max}) have been found in both tropical and tundra graminoids growing under nutrient-limited conditions (Chapin and Slack 1979, Ruess et al. 1983, McNaughton and Chapin 1985).

Root growth and function can only be maintained by the continuous utilization of carbon within the plant (Chapin and Slack 1979). Therefore, the maintenance or increase in root activity following defoliation in these species was dependent upon carbohydrate translocation from remaining shoot tissues (e.g., stem bases, rhizomes) or the rapid reestablishment of canopy photosynthesis. Carbon allocation to roots may have continued following defoliation in these species because they were nutrient-limited and possessed especially strong root sinks. It is unlikely that an adequate amount of soluble carbohydrates existed in the roots to support root activity following defoliation. For example, the total nonstructural carbohydrate (TNC) concentration in roots of the tundra graminoid, cottongrass, at the time of defoliation (69 $mg.g^{-1}$ dry wt.; Chapin and Slack 1979) was well within the range of TNC concentrations observed for roots of temperate species in which root respiration and nutrient absorption rapidly declined following defoliation or root excision [e.g., barley (*Hordeum vulgare*), 166 $mg.g^{-1}$ sucrose, glucose and fructose, Bloom and Caldwell 1988; white clover (*Trifolium repens*), 81 $mg.g^{-1}$, Gordon et al. 1986; orchardgrass *(Dactylis glomerata)*, 32 $mg.g^{-1}$, Davidson and Milthorpe 1966a]. Tundra graminoids do not appear to possess unusually high concentrations of carbohydrate reserves in comparison with ecologically and genetically similar temperate populations (Chapin and Shaver 1989). Whether the immediate defoliation responses of plants adapted to infertile environments are a function of nutrient availability, unique adaptations to low nutrient supply (Chapin 1980), or both, has not been determined.

Nitrogen fixation

Biological nitrogen fixation in nodulated legumes is more sensitive to the reduction in leaf area or current photosynthesis following defoliation than are root growth or respiration (Hardy and Havelka 1976, Vance et al. 1979, Ryle et al. 1985). Removal of all expanded leaves from white clover plants resulted in a decline in nitrogenase-linked nodule respiration within 10-30 minutes, depending on plant size, and nodule respiration was reduced 80–90% within 2 hours (Ryle et al. 1985). Shading and chemical inhibition of photosynthesis also led to a rapid decline in nodule respiration in these experiments, but the subsequent decline was slower than that observed following defoliation. A short time lag existed between the time of defoliation and the reduction in nodule respiration based on the time required for the utilization of carbon that was in transit between the leaf and the nodules (Ryle et al. 1985).

The dependence of nitrogenase activity, in addition to nodule respiration, on current photosynthesis has also been documented in studies with white clover. In this case, the rapid decline in nodule respiration and nitrogen fixation was not related to the carbohydrate content of the nodules immediately following defoliation (Gordon et al. 1986). However, it has been demonstrated that oxygen diffusion into nodules is reduced after defoliaton causing an oxygen, rather than a carbohydrate, limitation to nitrogenase-linked respiration (Hartwig et al. 1987, 1990). These results indicate that a reduction in current photosynthesis was not directly responsible for the decrease in nitrogen fixation after defoliation. Other effects or signals associated with leaf removal apparently regulate oxygen diffusion into nodules. Although the regulatory mechanism is undefined, the rapid reduction in nodule respiration may be important for nodule maintenance during the period of reduced carbon availability after defoliation. In additon to these very rapid effects on nodule respiration, total root and nodule respiration, and carbohydrate content decline after 24–48 hours (Gordon et al. 1986, Culvenor et al. 1989a, 1989b). Apparently both current photosynthesis and an additional signal originating in the shoot system are necessary for continued nitrogen fixation within nodules.

Carbohydrate depletion

The previous investigations demonstrate that root growth, respiration, and nutrient absorption in rapidly growing plants are all dependent upon a continuous supply of labile carbohydrates produced in the shoot system. However, root growth and function may be partially buffered for 2–48 hours after defoliation by carbohydrate translocation from remaining shoot tissues (Fig. 1). The dependence of root growth and function on a continuous supply of carbohydrates produced in the shoots is further supported by investigations in which exogenously applied carbohydrates have prevented the usual decline in root respiration and nutrient absorption following defoliation (Saglio and Pradet 1980, Aslam and Huffaker 1982, Davidian et al. 1984). Without the addition of exogenous carbohydrates, the decline in root respiration and nutrient absorption is accompanied by the depletion of soluble carbohydrate pools in both attached and excised roots (Davidson and Milthorpe 1966a, Chapin and Slack 1979, Gordon et al. 1986, Bloom and Caldwell 1988, Thorgeirsson, 1988).

The rapid decline in soluble carbohydrates within roots commonly observed following defoliation (Jameson 1963, Deregibus et al. 1982) results from: (1) a reduction, although not necessarily a complete cessation, in the amount of photosynthetic carbon translocated from the shoot system and (2) a continuation of carbohydrate utilization by root respiration. Quantitative carbon balance studies have shown that the root system continues to function as a net sink for carbon immediately following defoliation (Davidson and Milthorpe 1966a, Ryle and Powell 1975, Muldoon and Pearson 1979, Richards and Caldwell 1985, Danckwerts and Gordon 1987, Thorgeirsson 1988). Therefore, it is unlikely that soluble carbohydrates in the root systems are mobilized to meet the carbon demands of the shoot system during regrowth as widely assumed. Continuous carbohydrate allocation from the shoot system or the mobilization of substrates other than TNC from source tissues within the plant is required to maintain minimal root activity following defoliation (Davidson and Milthorpe 1966a). In orchardgrass for example, the net loss of TNC from the root system following moderate or severe defoliation only accounted for a fraction of the total respiratory carbon loss from the roots. The inability of carbohydrate reserves within roots to balance root respiratory demands resulted in the utilization of compounds other than TNC. Several other studies have also produced data suggesting that alternative substrates which are not evaluated in TNC analyses, including hemicelluloses, proteins and organic acids, may be utilized by plants following severe defoliation (Muldoon and Pearson 1979, Chung and Trlica 1980, Dewald and Sims 1981, Richards and Caldwell 1985).

Carbohydrate pools within the crowns, including sheath bases, of crested wheatgrass (*Agropyron desertorum*) and bluebunch wheatgrass (*Pseudoroegneria spicata* ssp. *spicata*; syn: *Agropyron spicatum*) were equivalent to the amount of

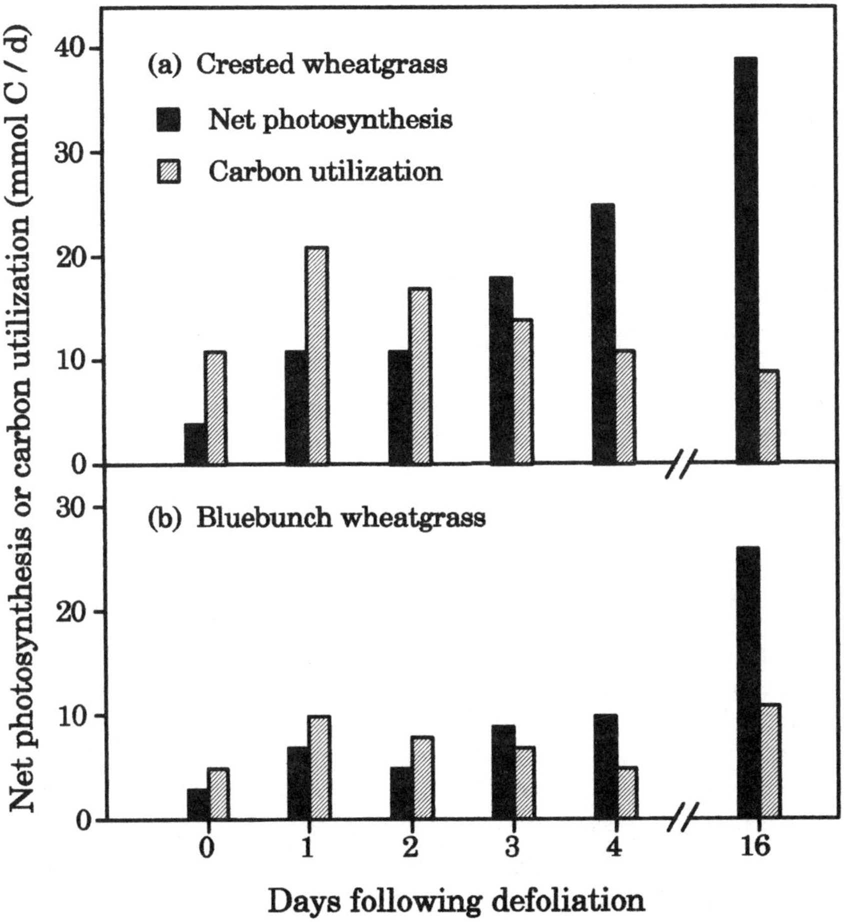

Figure 2. Daily net photosynthetic carbon uptake and carbon utilization for plants of (a) crested wheatgrass and (b) bluebunch wheatgrass on the day of defoliation (day zero) and 1, 2, 3, 4, and 16 days following defoliation. The daily availability of photosynthetic carbon exceeded the amount of carbon utilized for growth and respiration within 3 days of defoliation in both species (From Richards and Caldwell 1985).

carbohydrates produced in only 3 days of photosynthesis (Richards and Caldwell 1985). Consequently, plant growth was more dependent upon current photosynthesis than stored carbohydrates within 3 days following severe defoliation (Fig. 2). Similarly, leaf elongation or tillering were not proportionately related to carbohydrate concentrations in leaves, crowns or roots of tall fescue (*Festuca arundinacea*), orchardgrass (*Dactylis glomerata*) or reed canarygrass (*Phalaris arundinacea*) (Sambo 1983, Zarrough et al. 1984, Volenec and Nelson 1984) and carbohydrate concentrations were not correlated with tiller development in switchgrass (*Panicum virgatum*) (Anderson et al. 1989). These data collectively indicate that the contribution of carbohydrate reserves to plant growth following defoliation is considerably smaller than previously assumed.

A consistent, positive relationship between the size of carbohydrate pools and plant regrowth has not been definitively established (e.g., Richards and Caldwell 1985, Busso et al. 1990). In fact, an inverse relationship is frequently observed with carbohydrate pools increasing during periods of minimal plant growth and then decreasing during rapid plant growth (Davidson and Milthorpe 1966a, 1966b, Caldwell 1984). The potential for plant regrowth increased only when TNC concentrations in roots and crowns of crested wheatgrass and bluebunch wheatgrass increased to exceptionally high levels in response to an experimentally imposed drought (Fig. 3). Under all other conditions in a 2-year field study, crown and root TNC concentrations or pools were poor indicators of regrowth potential in early spring when these two grasses exhibit their greatest growth potential (Busso et al. 1990). In addition, neither carbohydrate concentrations nor pools account for the wide variation in grazing tolerance between crested wheatgrass and bluebunch wheatgrass (Richards and Caldwell 1985).

Carbohydrate pools play an important role in initiating plant growth when photosynthetic capacity is severely limited. However, the minimal amounts of carbohydrates stored in tiller bases (crowns and sheaths), the inaccessibility of root carbohydrates to support shoot growth, and the poor correlation between shoot growth and carbohydrate concentrations or pools limits their use as an effective index of shoot regrowth in perennial grasses and potentially other growth forms as well. The traditional interpretation of carbohydrate reserves appears to have been founded on an over-simplified, static conceptualization of the carbon economy of plants. The amount of residual photosynthetic area or the availability of active meristems following defoliation may be of equal or greater consequence in determining potential leaf regrowth of grasses and other forage species.

Resource allocation

Carbon. Carbohydrate availability within roots is reduced immediately after defoliation in response to: (1) a reduction in whole-plant photosynthesis and (2) the preferential allocation of photosynthetic carbon to active shoot sinks. For example, in sugarcane (*Saccharum officinarum*) the relative proportion of photosynthetically fixed ^{14}C translocated to the growing stem and leaves above a labeled, mature leaf was 2.7-fold greater in defoliated shoots within 2 hours after defoliation than it was in undefoliated shoots (Hartt et al. 1964). Two-three fold increases in the relative proportion of photosynthetic carbon allocated to rapidly growing shoots have been reported for barley within 24 hours after defoliation (Ryle and Powell 1975) (Fig. 4) and 2–10-fold increases have been documented in *Populus* x-*euramericana* (Bassman and Dickmann 1985). The relative proportion of photosynthetic carbon allocated to roots decreases as the proportion of photosynthetic carbon allocated to growing shoots increases following defoliation (Hartt et al. 1964, Marshall and Sagar 1965, Ryle and Powell 1975, Bassman and Dickmann 1985).

An increased proportion of photosynthetic carbon is exported from the remaining photosynthetic surfaces following defoliation which minimizes the absolute reduction in carbon allocation to the roots. Both the increase in carbon export from source tissues and greater relative carbon allocation to rapidly growing shoots are compensatory processes that promote canopy reestablishment by maintaining

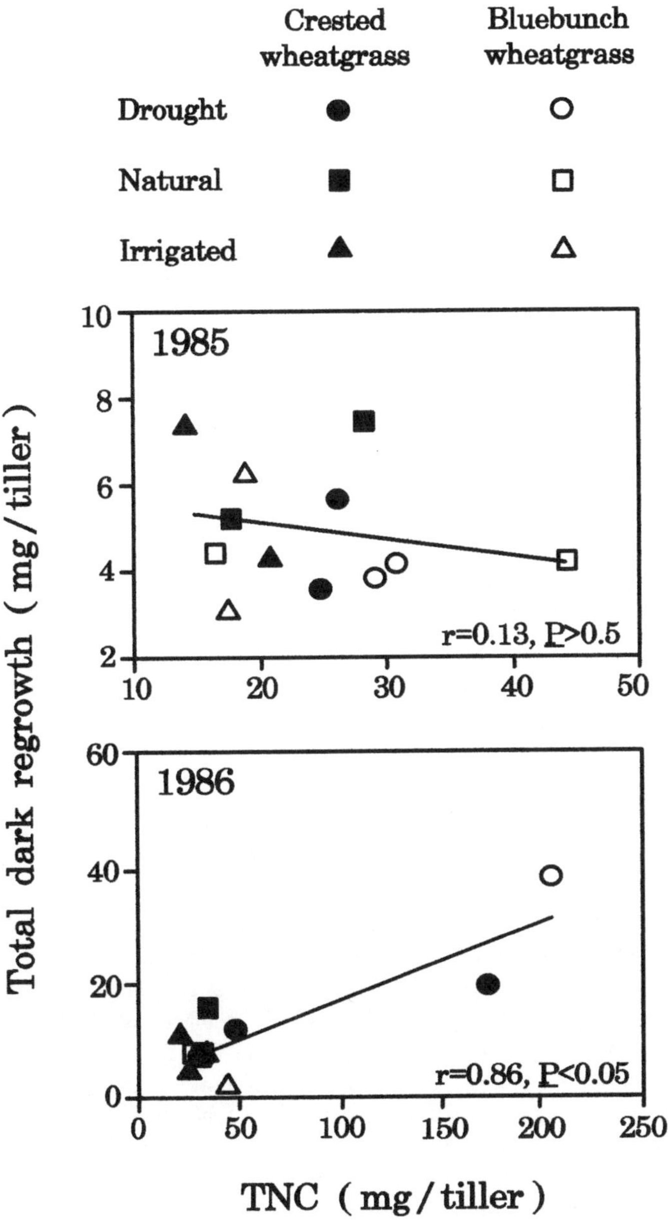

Figure 3. Relationship between root and crown TNC (total nonstructural carbohydrate) pools and regrowth of crested and bluebunch wheatgrass produced in the dark under irrigated, natural rainfed, and drought conditions. Regrowth produced in the dark quantifies the contribution of carbohydrate reserves to regrowth in the absence of current photosynthesis. A significant, positive relationship between TNC pools and regrowth was only observed in 1986 when severely water stressed plants had unusually large TNC pools and produced large amounts of regrowth. Under all other growth conditions in both years the relationship was not significant (From Busso et al. 1990).

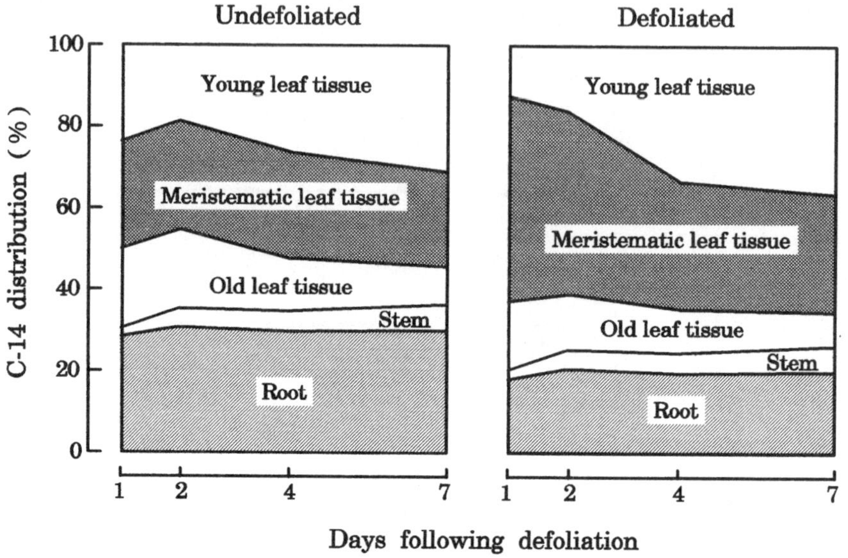

Figure 4. Relative allocation of ^{14}C to various components of defoliated and undefoliated barley plants. Plants were labelled for 25 minutes and harvested at 1, 2, 4, and 7 days. Defoliated plants were cut at the ligule of the 3rd leaf at the beginning of day 1. Allocation to meristematic leaf tissue (apical meristem, leaf primordia, and unexpanded leaves) increased and allocation to roots decreased at day 4 and 7 of regrowth (From Ryle and Powell 1975).

carbon availability for shoot meristems. Both compensatory processes may occur within hours following defoliation. For example, defoliation reduced whole plant photosynthesis in barley by 51% on the day of defoliation, but carbon allocation to the growing shoots was only reduced by 40% (Ryle and Powell 1975). Allocation of photosynthetic carbon to shoot sinks was proportionally greater in the defoliated plants than in the undefoliated plants for the next 7 days following defoliation. The increase in carbon allocation to shoot sinks occurred largely at the expense of carbon allocation to roots, which received both a lower total and relative amount of carbon than did the roots of undefoliated plants (Ryle and Powell 1975) (Fig. 4).

Carbon allocation between connected shoots or branches within a plant is also rapidly modified when a portion of the shoot system remains undefoliated or is defoliated to a lesser extent than the remainder of the plant (Marshall and Sagar 1965, 1968, Gifford and Marshall 1973, Welker et al. 1985). Carbon import from attached, undefoliated parental tillers increased 36–85% within 30 minutes following defoliation of a single daughter tiller of brownseed paspalum (*Paspalum plicatulum*) and little bluestem (*Schizachyrium scoparium*) (Welker et al. 1985). Carbon allocation from undefoliated parental tillers to defoliated daughter tillers increased to a maximum 10–84 hours following defoliation and then declined as daughter tillers reestablished their own photosynthetic capacity. Similar allocation patterns have been found in partially defoliated Italian ryegrass (*Lolium multiflorum*) plants (Marshall and Sagar 1968, Gifford and Marshall 1973) and in regularly mown Kentucky bluegrass (*Poa pratensis*) (Hull 1987). Carbon import from undefoliated, attached parental tillers can be maintained for a longer period or increased to a greater extent when the importing tiller is repeatedly defoliated (Gifford and Marshall 1973, Welker et al. 1985). The rapid resumption of carbon import by tillers following defoliation is dependent upon the maintenance of physiological integration among connected tillers during steady-state growth (Welker et al. 1985).

The allocation of carbon, nitrogen, and presumably other resources from undefoliated to defoliated tillers within a plant, may provide a potential mechanism of herbivory tolerance by facilitating tiller survival and growth following defoliation. For example, growth of tall fescue tillers increased progressively when either a greater percentage of tillers remained undefoliated or when the defoliation intensity was reduced (Matches 1966, Watson and Ward 1970). A portion of this growth response was attributable to resource allocation from nondefoliated to defoliated tillers. Similar conclusions have been drawn from investigations with several other graminoid species including *Dupontia fischeri* (Mattheis et al. 1976), *Carex aquatilis* (Archer and Tieszen 1986) and *Carex bigelowii* (Jónsdóttir and Callaghan 1989). Even though intertiller resource allocation functions as a mechanism to enhance herbivory tolerance, the primary function of parental resource support is very likely to increase the probability of juvenile tiller establishment in the competitive environment created by the high tiller density within established plants (Welker et al. 1991, Williams and Briske 1991).

In contrast to carbon allocation between roots and shoots, increased carbon export to defoliated tillers does not necessarily occur at the expense of carbon allocation to the root systems of undefoliated tillers. For example, increased carbon allocation to defoliated tillers of Italian ryegrass was accompanied by a large increase in carbon export from the undefoliated parental tillers so that the amount of carbon allocated to the root system of the parental tiller remained unchanged (Marshall and Sagar 1968). The pattern of intertiller resource allocation is dependent upon the relative leaf areas of the undefoliated parental tiller(s) and the attached, daughter tiller(s) remaining after defoliation (Gifford and Marshall 1973).

It is important to recognize that actively growing shoot meristems remained on the plants following defoliation to produce the patterns of carbon allocation previously described. Carbon allocation patterns in plants in which most sinks have been removed by defoliation will be modified substantially. The reorganization of carbon allocation priorities following defoliation is largely sink controlled, through either direct source-sink feedback or hormonal signals (Bucher et al. 1987a, Bassman and Dickmann 1985, Geiger and Fondy 1985). When actively growing shoot sinks are absent or limited, available carbon is allocated to alternative sinks, including roots (Richards 1984) and storage sites in the sheath and stem base in grasses (Bucher et al. 1987a, 1987b).

Increased carbon export from the remaining source leaves on defoliated plants can be interpreted as a response to decreased source/sink ratios. These short-term modifications in the pattern of carbon allocation probably result from competition among sinks of various strengths and locations in relation to the availability of carbon produced by source tissues. Long-term changes in allocation patterns, however, involve less direct interactions between sources and sinks and involve adaptive modifications most likely mediated by hormonal signals (Wareing et al. 1968, Geiger 1976, Geiger and Fondy 1985).

Nitrogen. Short-term changes in nitrogen (N) allocation within and between shoots have also been documented following defoliation. Previously absorbed N was allocated to regrowing leaves of perennial ryegrass within 2 days following defoliation (Fig. 5) (Ourry et al. 1988, 1990). The majority (80%) of the N was remobilized from remaining shoot tissues while the remainder was allocated from the root system. The contribution of remobilized N exceeded that of concurrent N absorption for 6 days following defoliation, after which concurrent absorption supplied the majority of N for regrowth.

N availability in the growth medium greatly affects the relative contribution of remobilized N to regrowth in perennial ryegrass following defoliation. The contribution of remobilized N to regrowth from the residual shoot system increased proportionally with decreasing N availability within the growth medium, while the amount of N contributed by the root system decreased with decreasing N availability (Millard et al. 1990, Ourry et al. 1990). For example, 69% of the N in shoot regrowth originated from remobilized N when plants were grown with low N availability, but only 40% of the N in regrowth originated from remobilized N when plants were grown with high N availability (Ourry et al. 1990). Similarly, in excess of 50% of the N in new leaves and tillers was derived from remobilized N in nondefoliated thickspike wheatgrass (*Agropyron dasystachyum*) plants grown with low N availability (Li et al. 1992). However, only 11% of the N in these tissues was derived from remobilized N when the plants were grown with high N availability. These data demonstrate, that in contrast to carbohydrate pools, N pools in the roots of grasses can be mobilized to support shoot growth following defoliation (Millard et al. 1990, Ourry et al. 1990). Similar conclusions have been drawn from investigations with herbaceous dicots including alfalfa (Hender-

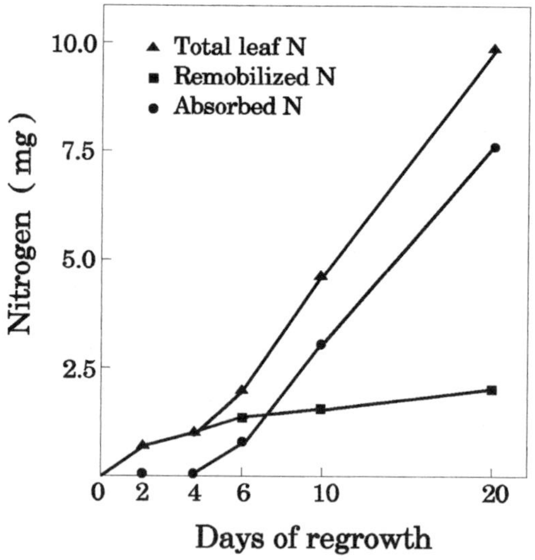

Figure 5. Total amount of N allocated to regrowing leaves of perennial ryegrass following defoliation to a height of 4 cm. The contribution of remobilized N from within the plant was greater than that of N absorption from the growth medium for 6 days following defoliation, after which N absorption supplied the majority of N for regrowth (From Ourry et al. 1988).

shot and Volenec 1993) and subterranean clover (*Trifolium subterraneum*) (Culvenor et al. 1989a).

Previously absorbed N is transported from roots to shoots in an organic form requiring simultaneous carbon transport (Ourry et al. 1989). The carbon associated with the nitrogenous compounds is included in estimates of the amount of TNC mobilized from roots to shoots in whole-plant labeling investigations. However, the reallocated organic N, and associated carbon, are primarily used for synthesis of proteins and other N-containing compounds within regrowing shoots rather than as substrate for respiration or synthesis of carbon-based compounds (e.g., celluose, hemicelluose). The biochemical identity of remobilized carbon and N in regrowing shoots must be determined to clarify the relative contributions of TNC and protein remobilization during regrowth.

N was preferentially concentrated in rapidly growing, defoliated daughter tillers which were attached to nondefoliated parental tillers that had been labelled with ^{15}N prior to defoliation (Welker et al. 1987, 1991). The absolute amounts of N imported by defoliated daughter tillers was much lower, however, than that of undefoliated tillers because of a reduction in the total N requirement created by the removal of shoot biomass. Nitrogen was preferentially allocated to defoliated daughter tillers until they reestablished a substantial amount of the leaf area which had been removed by defoliation. The patterns of N allocation within or between shoots of defoliated plants appear to be qualitatively similar to those for carbon allocation.

Photosynthetic capacity

The rapid modifications in carbon and nitrogen allocation patterns following defoliation are not paralleled by rapid increases in net photosynthetic rates of the remaining foliage. In fact, net photosynthetic rates are commonly reduced within hours for both partially defoliated and undefoliated leaves on plants subjected to partial canopy removal. The duration of the temporary decline in photosynthetic rate ranges from a few hours to 2 days. For example, net photosynthesis of undefoliated western wheatgrass (*Agropyron smithii*) tillers declined by 6–7% within 30 minutes following defoliation of all other attached tillers (Painter and Detling 1981) and simulated grasshopper grazing of leaf blades in this species reduced net photosynthesis by 31% after 1 hour (Detling et al. 1979a). Undefoliated leaves on defoliated alfalfa (*Medicago sativa*) plants had net photosynthetic rates 4–20% lower than comparable leaves on undefoliated plants, regardless of leaf or plant age, on the day of defoliation (Hodgkinson et al. 1972, Hodgkinson 1974). Nowak and Caldwell (1984) also documented a brief period of reduced (<30%) photosynthetic rates relative to comparable leaves on undefoliated crested wheatgrass plants after each of three sequential clippings. A similar response occurred in bluebunch wheatgrass plants, but photosynthetic rates of defoliated plants were only reduced below those of undefoliated plants after the first clipping. Net photosynthesis was reduced 30–80% in the basal portion of leaf blades in three species (*Deschampsia flexuosa*, *Nardus stricta*, *Juncus squarrosus*) from upland acidic, nutrient-poor soils within 10 hours following removal of the distal portion of the blades (Atkinson 1986). Dark respiration was not significantly altered by defoliation in these species. In a fourth species from the same site, net photosynthesis of sheep fescue (*Festuca ovina*) was not reduced until three sequential clippings had been imposed (Atkinson 1986). Both net photosynthesis and dark respiration increased within 20 hours of the first and second defoliation in sheep fescue. The mechanisms contributing to the rapid reduction in photosynthetic rates following defoliation have not been studied in detail, but may involve a direct wounding response (Macnicol 1976) or a response to physical disturbance (Ferree and Hall 1981).

A review of the immediate effects of defoliation on individual plants indicates that both the carbon and nitrogen economy are modified within 24–48 hours after defoliation and prior to the initiation of substantial regrowth. Soluble carbohydrate pools are depleted within several days following defoliation in response to continued respiratory demands, rather than as a result of allocation to regrowth that is produced over a period of weeks following defoliation. Carbon and nitrogen allocation priorities are rapidly modified to increase resource allocation to active meristematic sinks in the shoot. Rapid growth in the apical meristem region of existing shoots and new tillers produces sufficient sink strength to divert resources from other sinks, including root respiration and nutrient absorption, until leaf area expansion and whole-plant photosynthesis increase sufficiently to exceed the carbon demands of the actively growing shoot sinks. Consequently, root growth and function are substantially reduced within 24–48 hours after moderate or severe defoliation in response to a carbon limitation. An increase in photosynthesis and carbon export from remaining and regrowing leaves frequently minimizes the absolute reduction in carbon allocation to the roots. These sink-driven modifications are fundamental to plant recovery from defoliation and are well expressed in defoliation tolerant plants. When strong shoot sinks are absent or can not be rapidly activated after defoliation, recovery takes longer and defoliation tolerance is reduced. An understanding of the immediate effects of defoliation is critical because these initial processes establish the potential for subsequent plant recovery from defoliation.

Reestablishment of Whole-Plant Photosynthetic Capacity

Defoliation-induced changes in leaf photosynthetic capacity and carbon allocation that enable defoliated plants to compensate for foliage losses have been the focus of a great deal of attention. The initial manifestations of these compensatory processes, especially in allocation patterns, become apparent immediately following defoliation. However, adjustments in photosynthetic capacity occur over a period of several days following defoliation. Two commonly observed responses in defoliated plants that greatly affect the recovery rate of whole-plant photosynthetic capacity are: (1) increased photosynthetic capacity of remaining and regrowing foliage, and (2) increased growth rates of leaves and shoots. Increased rates of these two processes have a multiplicative effect on the rate of canopy reestablishment following defoliation. Although increased photosynthetic capacities of leaves during regrowth have been emphasized, the rates of both photosynthesis and leaf area expansion must be known to quantify the respective effects of either process for plant recovery.

Compensatory photosynthesis

Photosynthetic rates of foliage on defoliated plants are often higher than those of foliage of the same age on undefoliated plants. This response, which develops over a several-day period following defoliation, is termed compensatory photosynthesis and has been documented for both mature and expanding leaves remaining on the defoliated plant and for leaves that are produced during regrowth. Compensatory photosynthesis is a very consistent physiological response following defoliation that has been documented in a large number of species, including *Acer rubrum* (Heichel and Turner 1983), crested wheatgrass (Caldwell et al. 1981, Nowak and Caldwell 1984), western wheatgrass (Painter and Detling 1981, Detling and Painter 1983), sheep fescue (Atkinson 1986), fine thatching grass (*Hyparrhenia filipendula*) (Wallace et al. 1984), Italian ryegrass (Gifford and Marshall 1973), perennial ryegrass (Woledge 1977), alfalfa (Hodgkinson et al. 1972, Hodgkinson 1974), bean (*Phaseolus vulgaris*) (Wareing et al. 1968, Neales et al. 1971, Alderfer and Eagles 1976, Carmi and Koller 1979, von Caemmerer and Farquhar 1984), Monterey pine (*Pinus radiata*) (Sweet and Wareing 1966), bluebunch wheatgrass (Caldwell et al. 1981, Nowak and Caldwell 1984), *Quercus rubra* (Heichel and Turner 1983), red oatgrass (*Themeda triandra*) (Wallace et al. 1984), mesa dropseed (*Sporobolus flexuosus*) (Senock et al. 1991), and maize (*Zea mays*) (Wareing et al. 1968). Compensatory photosynthesis has also been documented for leaves that were produced following plant defoliation (Bassman and Dickmann 1982, Heichel and Turner 1983, Caldwell et al. 1981, Nowak and Caldwell 1984). The relative increase in photosynthetic rates in these cases was less than those observed for leaves existing on the plant prior to defoliation.

While increased photosynthesis of foliage on defoliated plants is frequently observed, rates less than or equal to those on undefoliated plants have also been reported (Ryle and Powell 1975, Poston et al. 1976, Detling et al. 1979a, Hall and Ferree 1976, Ferree and Hall 1981, Atkinson 1986, Boucher et al. 1987). The discrepancy between these studies and those reporting compensatory photosynthesis has not been adequately explained in most cases. There appears to be a tendency for minimal compensatory photosynthesis in leaves damaged by insects or clipped so that a long cut edge (wound) remained (Hall and Ferree 1976, Poston et al. 1976, Detling et al. 1979a, Ferree and Hall 1981). However, experimental aphid infestation of pea (*Pisum sativum*), bean (*Vicia faba*) and *Vigna unguiculata* induced compensatory photosynthesis to an extent that just balanced the photosynthate demand of the aphids (Hawkins et al. 1987). Repeated severe defoliation may also reduce photosynthetic rates, in even very grazing-tolerant species, such as *Kyllinga nervosa* (Wallace 1981, Wallace et al. 1985, Hodgkinson et al. 1989).

Compensatory photosynthesis originates from a rejuvenation of leaves and/or an inhibition of the decline in photosynthetic capacity that normally occurs as leaves age and senesce (Fig. 6). The degree and duration to which net photosynthetic rates are rejuvenated following defoliation varies greatly among species and among leaves on individual plants. Rejuvenation may only partially occur indicating that photosynthetic rates do not attain values comparable to those of newly developed leaves or photosynthetic rates of rejuvenated leaves may exceed those of newly developed leaves on undefoliated plants (Hodgkinson 1974, Heichel and Turner 1983). For example, in alfalfa the ability of mature leaves to increase photosynthetic rates to values equivalent to those of recently expanded leaves declined with age; 65-day-old leaves only increased to 66% of maximum capacity while younger leaves attained maximum photosynthetic rates (Fig. 6) (Hodgkinson

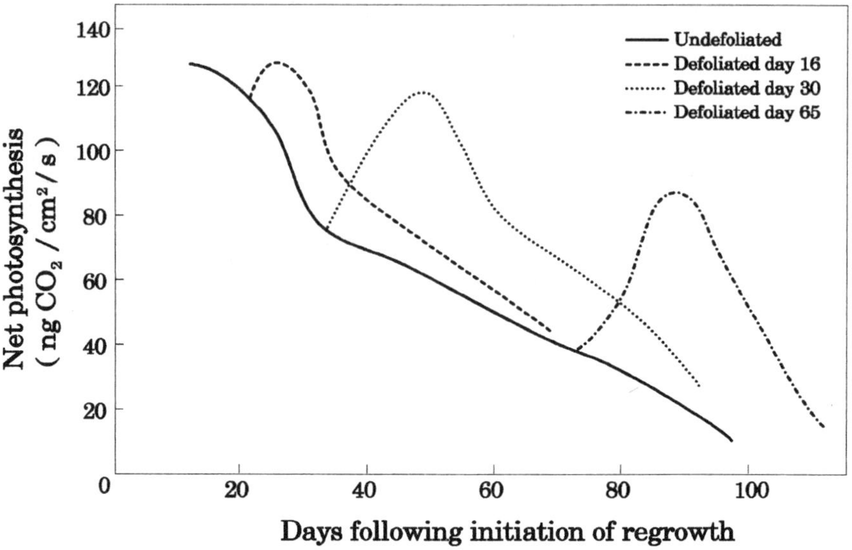

Figure 6. Net photosynthetic rate per unit leaf area for alfalfa leaves with increasing age on undefoliated plants and for undefoliated leaves on plants subjected to partial defoliation. Compensatory photosynthetic responses of leaves of three different ages are shown in comparison with leaves of the same chronological age on undefoliated plants. Maximum photosynthetic rates of leaves on undefoliated plants were achieved 10-20 days after leaf appearance and several days after maximum leaf expansion (From Hodgkinson 1974).

1974). However, because of the low initial rates of net photosynthesis in older leaves, proportional increases were greater in older than in younger leaves. Photosynthetic capacity exceeded the predefoliation rates by 181% in 65-day-old leaves compared to 12 and 74% in 16- and 30-day-old leaves, respectively (Hodgkinson 1974). Older leaves also required longer than younger leaves to achieve maximum recovery (5 versus 12 days; Hodgkinson 1974). Similar patterns were observed in perennial bunchgrasses by Nowak and Caldwell (1984).

Leaves on defoliated plants which do not rejuvenate may still exhibit higher photosynthetic rates than comparable leaves on undefoliated plants because the normal decline in photosynthetic capacity associated with aging is inhibited (Gifford and Marshall 1973). A reduced rate of leaf senescence is also expressed as an increase in mean leaf lifespan on defoliated plants (Jones et al. 1982, Nowak and Caldwell 1984).

Regulation. Modifications in microclimate and plant function are both potentially important processes regulating compensatory photosynthesis following defoliation. Leaves remaining on defoliated plants are often exposed to greater light intensity and altered light quality (Wallace 1990, Senock et al. 1991). Light intensity and quality influence the development of photosynthetic capacity in expanding leaves and leaf senescence in the absence of defoliation. Consequently, compensatory photosynthesis may occur in response to modifications in the light environment alone. The reallocation of carbon and nitrogen resources that accompanies compensatory photosynthesis may be a plant response to more closely optimize photosynthetic carbon gain in the altered light environment (Field 1983, Gutschick and Wiegel 1988). For example, compensatory photosynthesis of developing leaves of perennial ryegrass was inhibited when they were maintained in a shaded environment (Woledge 1977). This response indicates that the occurrence of compensatory photosynthesis was at least partially light dependent, rather than being completely determined by source-sink or hormonal changes in the plant. In many other cases, however, compensatory photosynthesis occurs independent of changes in the light environment (Wareing et al. 1968, Hodgkinson et al. 1972, Gifford and Marshall 1973, Hodgkinson 1974, Woledge 1977). These data indicate that compensatory photosynthesis can be induced by both changes in the light environment, and by physiological modifications in plant function following defoliation, or by a combination of these processes.

The relative importance of two potential physiological mechanisms of compensatory photosynthesis was assessed by Carmi and Koller (1979). They considered the role of increased carbon demand per unit photosynthetic area following defoliation (i.e., direct source-sink interaction), and the increased supply of hormones or other substances transported from the root system to the defoliated shoot system (i.e., increased root/shoot ratio). Compensatory photosynthesis in primary leaves of bean was not caused by direct source-sink interactions, but was dependent on substances transported in the xylem from the root system to the leaves on defoliated plants. Further support for this mechanism comes from studies demonstrating that modifications in photosynthetic capacity following defoliation are consistent with the effects of exogenously applied cytokinins (Wareing et al. 1968). Finally, the absence of immediate increases in photosynthesis following a decrease in the source/sink ratio supports the conclusion that compensatory photosynthesis is not the direct result of increased carbon demand (King et al. 1967, Geiger 1976, von Caemmerer and Farquhar 1984). Rather, indirect effects resulting from the increased root/shoot ratio, presumably mediated by cytokinins or other signals produced in the root, appear to affect leaf development and aging so that the photosynthetic apparatus is rejuvenated and/or its senescence is inhibited.

Mechanisms. Gas exchange and biochemical analyses of leaves exhibiting compensatory photosynthesis have documented a number of closely coordinated physiological and morphological modifications that result in increased photosynthetic capacity per unit leaf area. Many of the mechanistic components associated with compensatory photosynthesis have been consistently observed in numerous species, although every component has not been documented in each case.

Compensatory photosynthesis results from an increased photosynthetic capacity within the mesophyll, rather than improved stomatal conductance, with very few exceptions (e.g., Gifford and Marshall 1973). Increased leaf nitrogen content is the most frequently documented modification related to compensatory photosynthesis. Increased nitrogen content is consistent with increased mesophyll conductance and decreased internal CO_2 concentrations under ambient conditions and is associated with increased RNA, soluble protein, and chlorophyll content (e.g., Satoh et al. 1977, Nowak and Caldwell 1984, Yamashita and Fujino 1986). Dark respiration rates may also be higher in leaves exhibiting compensatory photosynthesis than in leaves on undefoliated plants which is characteristic of leaves with higher protein contents (e.g., Atkinson 1986). Direct measurements of increased RuBP carboxylase activity or amount, and electron transport capacity have also been correlated with compensatory photosynthesis as determined by gas exchange (Wareing et al. 1968, Neales et al. 1971, Jenkins and Woolhouse 1981, Aoki 1981, von Caemmerer and Farquhar 1984, Yamashita and Fujino 1986). Yamashita and Fujino (1986) also documented increased RNA content in leaves of defoliated mulberry (*Morus alba*) plants. These coordinated biochemical changes provide the basis for increased photosynthetic capacity in leaves on defoliated plants. Mature leaves on defoliated plants frequently develop increased specific leaf mass (mass per unit area) within 1–14 days after defoliation and expanding leaves tend to grow larger on defoliated than on undefoliated plants (Alderfer and Eagles 1976, Satoh et al. 1977, Carmi and Koller 1979, Bassman and Dickmann 1982). The coordination of these physiological and morphological changes and the time required for their development is consistent with the hypothesis that root produced substances and increased nitrogen allocation are important processes regulating the occurrence of compensatory photosynthesis.

Significance. Compensatory photosynthesis enables defoliated plants to fix more carbon than if photosynthetic rates were maintained at levels comparable to undefoliated plants. Unfortunately, few analyses have been done to determine the quantitative contribution of compensatory photosynthesis to the carbon budget of defoliated plants. Heichel and Turner (1986) calculated that the carbon assimilation associated with compensatory photosynthesis of two defoliated tree species was 19–67% greater than would have been expected if no compensatory photosynthesis had occurred. Gold and Caldwell (1990), however, present evidence indicating that recovery of whole-plant photosynthesis in crested wheatgrass subjected to various defoliation treatments was more closely correlated with the rate of leaf area expansion than with the photosynthetic rate per unit foliage area, which partially reflected compensatory photosynthesis. The compounding effect of increasing leaf area on canopy photosynthetic capacity and the transient nature of compensatory photosynthesis are at least partially responsible for these conflicting interpretations. These data suggest that the contribution of compensatory photosynthesis is variable among species and is influenced by a large number of plant and environmental variables.

Compensatory photosynthesis is a relatively consistent response of defoliated plants. The physiological basis for this response is a coordinated set of changes in the biochemical composition and morphological structure of leaves that is consistent with the reversal of developmental processes that normally occur during leaf aging. Compensatory photosynthesis appears to be an important component of compensatory growth in defoliated plants, yet it occurs in many defoliation-sensitive as well as defoliation-tolerant species and it often does not persist throughout the recovery period. The quantitative contribution of compensatory photosynthesis to plant recovery following defoliation cannot be evaluated without considering the simultaneous modifications in leaf area expansion and canopy microenvironment.

Canopy reestablishment

Compensatory leaf and tiller growth. Recovery of whole-plant photosynthetic capacity is dependent on the rate at which new photosynthetic surfaces are produced to an equal or greater extent than compensatory photosynthesis. The rate of canopy reestablishment, or the leaf replacement potential (Hyder 1972), is affected by environmental, physiological and morphological considerations. Reinvestment of current photosynthetic carbon in new leaf area by allocation to shoot meristems has a compounding effect on whole-plant photosynthetic capacity, making small increases in the rate of leaf expansion an important component of compensatory growth. The capacity for rapid canopy reestablishment is an important characteristic of defoliation tolerant plants (Davidson and Milthorpe 1966b, Caldwell et al. 1981, Richards and Caldwell 1985, Fankhauser and Volenec 1989, Hodgkinson et al. 1989).

Rates of leaf and tiller growth on defoliated plants may occasionally exceed those of undefoliated plants and can increase the rate of canopy reestablishment. More rapid leaf expansion on defoliated plants than on undefoliated plants has been observed in a number of species, including crested wheatgrass, tall fescue, *Kyllinga nervosa*, and kleingrass (*Panicum coloratum*) (Wallace 1981, Wolf and Parrish 1982, Wallace et al. 1985, Gold and Caldwell 1989a). In crested wheatgrass, grazed tillers had higher relative growth rates than tillers of ungrazed plants when grazing occurred prior to internode elongation in the spring (Olson and Richards 1988c). Enhanced leaf and tiller growth rates usually persist for only a few weeks and are not consistently expressed in all environmental conditions or phenological stages within the growing season (e.g., Volenec and Nelson 1983, Olson and Richards 1988c) (see Compensatory Growth section).

Developmental controls. The potential for canopy reestablishment in grasses is determined by the availability and activity of intercalary, apical and axillary meristems. The interaction among meristem type, environmental variables, and resource availability determines the rate of leaf area expansion in plants. For example, the availability of active intercalary and apical meristems following defoliation establishes high priority sinks which sequester currently produced photosynthetic carbon or reserve compounds located in stem and sheath bases (Richards and Caldwell 1985, Bucher et al. 1987a, 1987b, Schnyder and Nelson 1989, Danckwerts and Gordon 1987). Activity of these sinks induces the modified allocation patterns observed immediately following defoliation and contributes to the rapid replacement of the photosynthetic canopy. When active shoot meristems are removed, or when unfavorable environmental conditions limit growth, recovery is delayed or slowed (Muldoon and Pearson 1979, Richards and Caldwell 1985, Bucher et al. 1987a, 1987b). Recovery is most rapid in plants possessing abundant resources to support growth through a large number of active shoot meristems (Davidson and Milthorpe 1966b, Richards and Caldwell 1985, Hodgkinson et al. 1989, Busso et al. 1990).

Differences in herbivory tolerance between rhizomatous grasses and bunchgrasses, and among species within each group, is largely a function of meristem availability at the time of defoliation. Rhizomatous species frequently possess large numbers of active meristem throughout the growing season. The presence of these active shoot sinks rapidly sequesters available carbon and nitrogen within the plant increasing the relative rate of shoot growth following defoliation. For example, switchgrass plants exhibited reduced growth and reproduction when defoliated the same year they were initiated from seed, while established plants with well developed rhizomes increased growth and reproduction when subjected to a comparable defoliation intensity (Hartnett 1989). Rhizomatous species are most severely impacted by defoliation when tiller densities are at a seasonal low (Hull 1987). However, additional attributes such as intertiller resource allocation and defoliation avoidance characteristics (see Grazing Resistance section) also contribute to the grazing resistance of rhizomatous grasses in the field (Youngner 1972, Briske 1991).

Synchronous tiller development increases the susceptibility of bunchgrasses to a greater loss of active shoot meristems when grazed after internode elongation (Branson 1953, Westoby 1980, Olson and Richards 1988c). Synchronous tiller development also contributes to wide fluctuations in grazing tolerance with the progression of phenological plant development. For example, the grazing-sensitive, bluebunch wheatgrass, is quite tolerant of defoliation in the early spring when culmless, because active intercalary and apical meristems are located at or below ground level. However, defoliation tolerance decreases rapidly following internode elongation (Richards and Caldwell 1985, Busso et al. 1990). Seasonal variation in defoliation tolerance is much less pronounced in species with asynchronous tiller development. The most grazing tolerant bunchgrasses frequently possess asynchronous tiller development, producing a situation similar to that described for rhizomatous grasses. Examples of grazing tolerant bunchgrasses include buffelgrass (*Cenchrus ciliaris*), tanglehead (*Heteropogon contortus*), and guineagrass (*Panicum maximum*) (e.g., Hodgkinson et al. 1989, Mott et al. 1992).

Patterns of plant defoliation including the type, amount, and frequency of tissue removal, must be given careful consideration because the number and source of available meristems function as important determinants to plant recovery following grazing. The following sections review the developmental considerations that influence the rate of plant recovery following defoliation and the associated patterns of resource allocation among the various sinks on defoliated plants.

Intercalary meristems. Cellular division and expansion from a zone of meristematic tissue located at the base of the blade, sheath and internode collectively contribute to phytomer growth (Fig. 7) (Evans and Grover 1940, see Dahl, this volume). The initiation and maturation of intercalary meristems (i.e., meristems separated from the apical meristem by a region of nonmeristematic tissue) occurs in sequence beginning with the blade and terminating with the internode (Stubbendieck and Burzlaff 1971, Sims et al. 1971). Meristematic activity of the blade ceases at about the time the leaf tip emerges from the subtending sheaths, which is comparable to the time of ligule differentiation. Meristematic activity of the sheath ceases when the ligule emerges from the subtending sheaths (Sharman 1942). Leaf growth continues following the completion of cell division as recently divided cells in the blade and sheath expand. Although, intercalary meristem activity is short-lived and the total amount of biomass differentiated per individual meristem is relatively small, phytomer growth collectively determines the production potential of individual tillers.

Defoliation influences phytomer growth by primarily affecting cell expansion rather than cell division (Grant et al. 1981, Volenec and Nelson 1983). Defoliation at 2-week intervals reduced leaf elongation rate by 30% in tall fescue and was

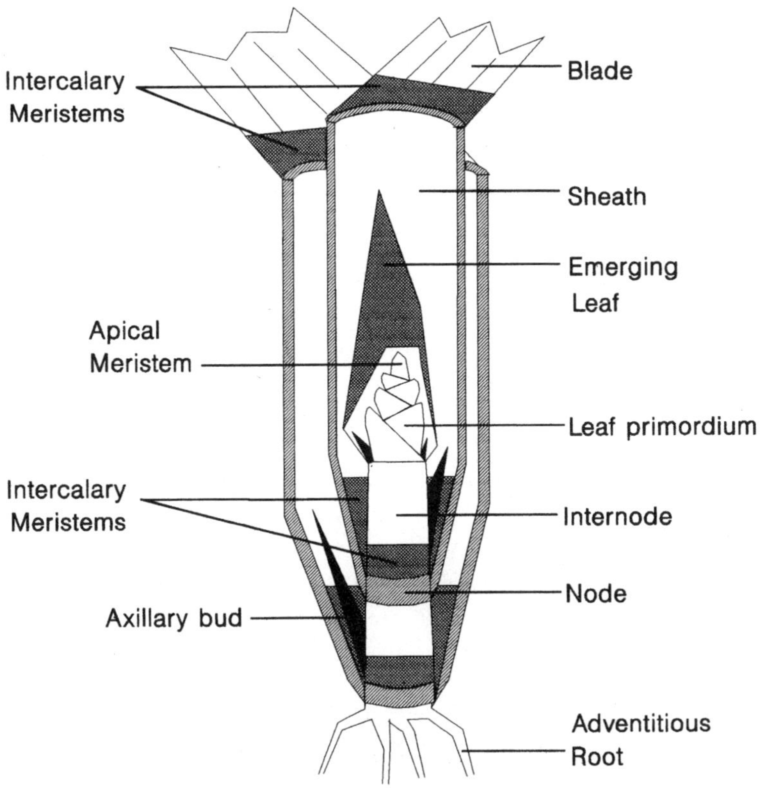

Figure 7. Longitudinal view of the crown of a grass tiller illustrating three meristematic sources contributing to growth and development in grasses. The magnitude of biomass production and rate of growth following defoliation differ for intercalary, apical and axillary meristems, but all three contribute to canopy reestablishment (Modified from Jewiss 1972).

associated with a reduction in the length of mature epidermal cells in comparison with plants defoliated at 6-week intervals (Volenec and Nelson 1983). The reduction in cellular elongation suggests that growth is restricted by substrate availability from either stored carbohydrates (Grant et al. 1981) or current photosynthesis (Richards and Caldwell 1985, Gold and Caldwell 1989a, 1989b). However, in some cases defoliation may also promote leaf elongation, as discussed previously. For example, improved water status, resulting from a reduction in transpirational area following defoliation, promoted cell expansion which contributed to an increased rate of leaf elongation in tall fescue plants (Wolf and Parrish 1982).

Apical meristems. Growth from apical meristems is dependent upon the successive differentiation of leaf primordia (i.e., blades and sheaths of differentiating phytomers). These primordia provide the meristematic source for leaf replacement following defoliation (Fig. 7). Initially, the entire primordium is meristematic, but cellular division is quickly restricted to individual intercalary meristems of the blade and sheath (Dahl, this volume, Sharman 1945). The interval between the appearance of successive leaf primordia is approximately 7 days in vegetative tillers under favorable environmental conditions (Anslow 1966). However, the rate of leaf appearance is known to be species specific and influenced by a wide range of environmental conditions.

Differentiation and growth of leaf primordia are relatively insensitive to defoliation based on the limited number of species that have been evaluated (Anslow 1966, Jones et al. 1982, Chapman et al. 1983). Severe defoliation reduces leaf appearance by directly removing apical meristems or indirectly by reducing photosynthetic carbon availability. Removal of more than two entire leaves per tiller is required to decrease the rate of leaf appearance in perennial ryegrass (Davies 1974, Grant et al. 1983). Defoliation may also increase leaf initiation by increasing irradiance on the remaining portions of the canopy thereby potentially increasing photosynthetic efficiency and increasing carbon availability for leaf replacement (e.g., Gold and Caldwell 1989b). An increase in leaf initiation rate with increasing irradiance (Anslow 1966) and leaf area per tiller (Grant et al. 1983) is well documented.

Grazing tolerance in grasses is partially attributable to the basal location of both the apical and intercalary meristems. The apical meristem remains near the soil surface continually differentiating leaf primordia until it undergoes floral induction in grasses that do not produce culmed vegetative shoots (Branson 1953, Dahl, this volume). Culm elongation occurs from the activation of intercalary meristems at the base of the several uppermost internodes. Although apical meristems are much more vulnerable to removal by large herbivores following floral induction, the capacity to differentiate subsequent leaf primordia has already been terminated (Sharman 1947). Culm elongation does occur in vegetative tillers in a number of species elevating the apical meristem while it is still capable of differentiating leaf primordia (Branson 1953, Dahl, this volume). Apical meristem removal in these species determines that leaf replacement must occur from immature intercalary meristems or tiller growth from axillary buds. Consequently, species producing a large proportion of reproductive or culmed vegetative tillers are best suited to intermittent defoliation rather than continuous grazing (Branson 1953, Hyder 1972). Intermittent defoliation provides a sufficient period of time for tillers to express maximum vegetative production prior to the termination of growth following floral induction or apical meristem removal.

Axillary buds. Axillary buds are rudimentary apical meristems which possess the capacity to differentiate into complete tillers. A single bud is differentiated from the apical meristem of the parent tiller as a component of each phytomer (Fig. 7). However, within an individual phytomer, axillary bud growth is preceded by growth of the blade, sheath and potentially the internode. If conditions are unfavorable for growth at the time of bud maturation, or if apical dominance is imposed, the buds may be inhibited for a period of time prior to growth. Axillary bud development in response to environmental conditions and defoliation are discussed in the tiller demography section.

Relative contribution of meristematic sources. Growth is dependent on all three meristematic sources in perennial grasses, but their relative contributions differ in magnitude and chronology (Briske 1986, 1991). Although intercalary meristems form the basis for phytomer growth, their contribution is limited to the relatively brief period of growth associated with the respective intercalary meristems of individual phytomers (Table 1). Apical meristems play a central role in the meristematic potential of grasses because they are the source of phytomer production including axillary buds. Axillary buds develop into mature apical meristems, which can differentiate successive axillary buds, thereby contributing to the persistence and sustainable productivity of perennial grasses.

Canopy reestablishment proceeds most rapidly from intercalary meristems because cell division has previously occurred within leaf primordia (Cook and Stoddart 1953, Hyder 1972, Briske 1991) (Table 1). Leaf replacement from apical meristems occurs at a slower rate because of the time required for differentiation and cell expansion within individual leaf primordia. Canopy reestablishment is slowest from axillary buds because of the time required for bud activation and leaf primordia differentiation. Although growth of these three meristematic sources can be considered independently, all three grow simultaneously during canopy reestablishment. The relative contribution of these meristematic sources to plant growth varies among species and is influenced by environmental variables and stage of phenological development (Coughenour et al. 1985a, 1985b, Olson and Richards 1988c). Knowledge of the relative availability and activity of meristematic sources is necessary to accurately interpret and potentially anticipate the regrowth potential of grasses.

Tiller and Plant Demography

Information addressing tiller and plant demography of important North American grasses is extremely limited. Demographic processes including recruitment, mortality, density, and size class distribution of tillers and plants, have important consequences for the persistence and sustainable productivity of species populations. The objectives of this section are to address the essential aspects of tiller and plant demography in ungrazed plants and populations and then evaluate the effects of grazing on these demographic processes.

Plant Organization and Persistence

The developmental morphology of grasses is based upon the successive differentiation of phytomers from apical meristems (Briske 1991, Dahl, this volume). However, an individual plant can also be viewed as a population of tillers because each tiller is composed of a series of phytomers differentiated from an individual apical meristem. Regardless of the growth unit recognized, the architecture and

Table 1. Growth rate following defoliation is most rapid from intercalary meristems,intermediate from apical meristems and slowest from axillary buds. However, the relative contribution of these meristematic sources to total biomass production occurs in the reverse order. (Adapted from Briske 1986).

	Meristematic Source		
Growth Response	Intercalary	Apical	Axillary
Growth Rate	Rapid	Intermediate	Slow
Total production	Low	Intermediate	High

productivity of plants is partially determined by the population dynamics of growth units (i.e., phytomers or tillers). This concept is clearly demonstrated by the relatively constant leaf number maintained by tillers throughout the growing season (Anslow 1966, Vine 1983). A relatively constant number of leaves is maintained because leaf production (recruitment) and leaf mortality occur at approximately comparable rates. Similarly, the relative rates of tiller recruitment and mortality establish the number of live tillers per plant or per unit area (White 1980).

Grasses can adjust to resource availability within their environment by altering plant density, tiller density or tiller weight because of their modular construction (Kays and Harper 1974). The demographic principle stating that a plant population can potentially yield a comparable level of production over a wide range of plant densities is known as the law of constant yield (Harper 1977, p. 154). This principle indicates that a plant population possesses sufficient plasticity to adjust to a particular level of resource availability through various combinations of plant size and density. In other words, resource availability within the environment, rather than plant growth potential, ultimately imposes an upper limit on population growth.

Population persistence and productivity may be maintained through either sexual or asexual (i.e., vegetative growth) reproduction, but available information suggests that vegetative growth (i.e., tillering) is the dominant form of reproduction in both semiarid and mesic grasslands (e.g. Belsky 1992). Investigations conducted in the tallgrass (Rabinowitz 1981, Johnson and Anderson 1986), midgrass (Kinucan 1987), and shortgrass (Coffin and Lauenroth 1989) prairies of North America consistently demonstrate a lack of correspondence between the existing vegetation of late-successional grassland communities and the species composition of the seed bank. Although caryopses are frequently produced, few appear to retain their viability within the soil for greater than one year (Thompson and Grime 1979, Pyke 1990). In addition, the number of seedlings recruited within established grasslands are frequently low and occur only sporadically during years of favorable moisture and temperature conditions (Wilson and Briske 1979, Salihi and Norton 1987, Pyke 1990, Jónsdóttir 1991).

Tiller recruitment from vegetative growth is more consistent than recruitment from seed because juvenile tillers import resources from parental tillers (Tripathi and Harper 1973, Welker et al. 1991). Consequently, vegetative growth is much less dependent upon immediate resource availability than is sexual reproduction. However, sexual reproduction is necessary for the maintenance of genetic diversity in populations and for population regeneration following plant mortality associated with large scale disturbances (O'Connor 1991). Additional research is needed to assess the relative contributions of vegetative and sexual reproduction to sustainable productivity and population maintenance in grasslands.

Tiller Recruitment and Mortality

Tiller recruitment may occur throughout the growing season, but maximum recruitment frequently occurs in the spring and/or autumn in both cool-season (Langer 1956, Langer et al. 1964, Robson 1968) and warm-season perennial grasses (Reidenbaugh 1983, Butler and Briske 1988). For example, crested wheatgrass, bluebunch wheatgrass (Mueller and Richards 1986), Arizona cottontop

(*Trichachne californica*) (Cable 1971), big bluestem (*Andropogon gerardii*) (McKendrick et al. 1975), and Nebraska sedge (*Carex nebraskensis*)(Ratliff and Westfall 1992) recruit a single tiller cohort in the autumn or spring, while Indiangrass (*Sorghastrum nutans*) (McKendrick et al. 1975) and little bluestem (Butler and Briske 1988) recruit two tiller cohorts in the late spring, early summer or autumn.

Tillers recruited early in the growing season frequently become florally induced and terminate their life cycle during the same growing season while tillers recruited later in the season frequently over-winter and resume growth the subsequent growing season. This pattern of tiller development results in greater tiller longevity, but longevity does not generally exceed two complete growing seasons in temperate perennial grasses (Langer 1956, Robson 1968, McKendrick et al. 1975, Butler and Briske 1988, Ratliff and Westfall 1992). Culms of Arizona cottontop survived three years in some cases, but the vegetative tissue they supported was comprised of younger tissues originating from elevated axillary buds along the culm (Cable 1971).

Tiller longevity is greater for grasses and sedges growing at higher latitudes than it is for grasses at lower latitudes. For example, tillers of *Agrostis stolonifera*, *Poa pratensis* ssp. *irrigata*, and *Festuca rubra* survived 3, 3, and 5 years, respectively, in a Baltic seashore meadow near Stockholm, Sweden (Jónsdóttir 1991). Tillers of *Dupontia fischeri* survived 4–5 years near Barrow, Alaska (Mattheis et al. 1976) and mean longevities of *Carex bigelowii* tillers have been estimated at 3–4.6 years (Carlsson and Callaghan 1990). The mechanism(s) promoting greater tiller longevities in high latitude graminoids in comparison with temperate grasses is unknown.

Mortality of vegetative tillers is frequently associated with reproductive development of parental tillers. Young vegetative tillers succumb to shading by the taller reproductive tillers and/or the cessation of resource allocation from parental tillers because of the additional resource demand created by culm and inflorescence development (Smith and Leinweber 1973, Ong 1978). Apparently the smallest tillers within the plant, which are partially dependent upon parental tillers for resources, are unable to effectively compete for resources when the plant encounters stress and are the first to undergo mortality (Ong 1978, Parsons et al. 1984). The longevity of reproductive tillers is established by the transition of the apical meristem from a vegetative structure to a reproductive structure. The capacity for vegetative growth is lost during this transition and subsequent vegetative growth can only be initiated from axillary buds (Sharman 1945, Langer 1972).

Mechanisms of Tiller Population Regulation

Tiller populations of numerous species, including *Cyperus esculentus* (Lapham and Drennan 1987), big galleta (*Hilaria rigida*) (Robberecht et al. 1983) and crested wheatgrass (Olson and Richards 1989), are known to be regulated by density-dependent mechanisms. Density-dependent mechanisms regulate population size by increasing recruitment and survival at low densities, while reducing recruitment and survival at high densities (Silvertown and Lovett Doust 1993, p. 136). Presumably competition influences resource availability and therefore tiller recruitment and/or mortality. Tiller populations of little bluestem in east central

Texas were regulated in similar proportion by competition from associated tillers within plants and by neighboring plants (Briske and Butler 1989). Tiller recruitment increased 57 and 71% following removal of neighboring plants and approximately one-half of the existing tillers within plants, respectively. Neither tiller survival nor reproductive development were affected as dramatically as recruitment by reduced competition in comparison with undisturbed plants. However, density-dependent regulation may only be a significant mechanism of population regulation in specific communities and during certain seasons of the year (Fowler 1986, Olson and Richards 1989). In addition, density-dependent mortality, rather than recruitment, has been documented to be an important population regulation mechanism in several herb species (Lovett-Doust 1981).

Numerous environmental variables are known to influence tillering in perennial grasses. Environmental variables which promote growth, including favorable temperatures, irradiance, water and nutrient availability, generally promote tillering as well (Langer 1963, Williams 1970a). Tillering is also affected by interactions among environmental variables, plant age and genotype. Even though environmental variables, either singly or in combination, influence various aspects of plant growth, they do not appear to directly regulate tillering (Phillips 1975). Presumably, environmental variables must interact with the mechanisms of apical dominance to regulate the dynamics of tiller populations.

Apical dominance defines the physiological processes by which the apical meristem exerts hormonal regulation over axillary bud growth (Murphy and Briske 1992). The theory was initially developed by Thimann and Skoog (1933, 1934) who demonstrated the regulatory influence of auxin on lateral branching in bean. The concept was extended to grasses when Leopold (1949) demonstrated that indoleacetic acid inhibited axillary bud growth in teosinte (*Euchlaena mexicana*) and barley. This theory of apical dominance, termed the direct inhibition theory, continues as the prevailing view of apical dominance in grassland ecology and management even though insight into the physiological mechanisms has advanced substantially (Murphy and Briske 1992). Although auxin produced in the apical meristem or young leaf primordia provides the principle correlative signal for bud inhibition, a second growth hormone, cytokinin, is also involved (Phillips 1975, Hillman 1984). Auxin is proposed to interfere with the metabolic function of cytokinin in the bud or prevent its transport into the bud thereby preventing bud growth. Two additional hormones, abscisic acid (Nojima et al. 1989) and ethylene (Harrison and Kaufman 1982), both of which are plant growth inhibitors, have also been associated with apical dominance in several grasses. These growth inhibitors may transmit the correlative signal of auxin into the axillary buds because auxin has not been observed to directly enter the buds (Tucker 1977, 1978). Although apical dominance is not thoroughly understood, recent advances indicate that the direct inhibition theory is no longer an appropriate interpretation of the phenomena (Murphy and Briske 1992).

Continued adherence to the direct inhibition theory of apical dominance in grassland ecology and management poses several disconcerting questions. How do plants initiate replacement tillers if the apical meristem is undisturbed? What mechanisms regulate tiller recruitment in specific seasonal patterns? How does tiller recruitment respond to changes in resource availability in the immediate environment? These considerations suggest that apical dominance may potentially respond or interact with environmental signals.

Radiation quality is an environmental signal documented to regulate a variety of developmental processes including seed germination, chlorophyll synthesis, leaf expansion, and internode elongation in a large number of vascular plants. The ratio of red:far-red (660:730 nm) radiation is correlated with the abundance of biologically active and inactive forms of phytochrome which can induce a broad range of photomorphogenetic effects (Smith and Morgan 1983). Radiation quality may also function as an environmental signal capable of regulating tiller recruitment in grasses (Fig. 8) (Kasperbauer and Karlen 1986, Casal et al. 1986, 1987, Simon and Lemaire 1987). Red radiation is attenuated more rapidly than far-red radiation as it

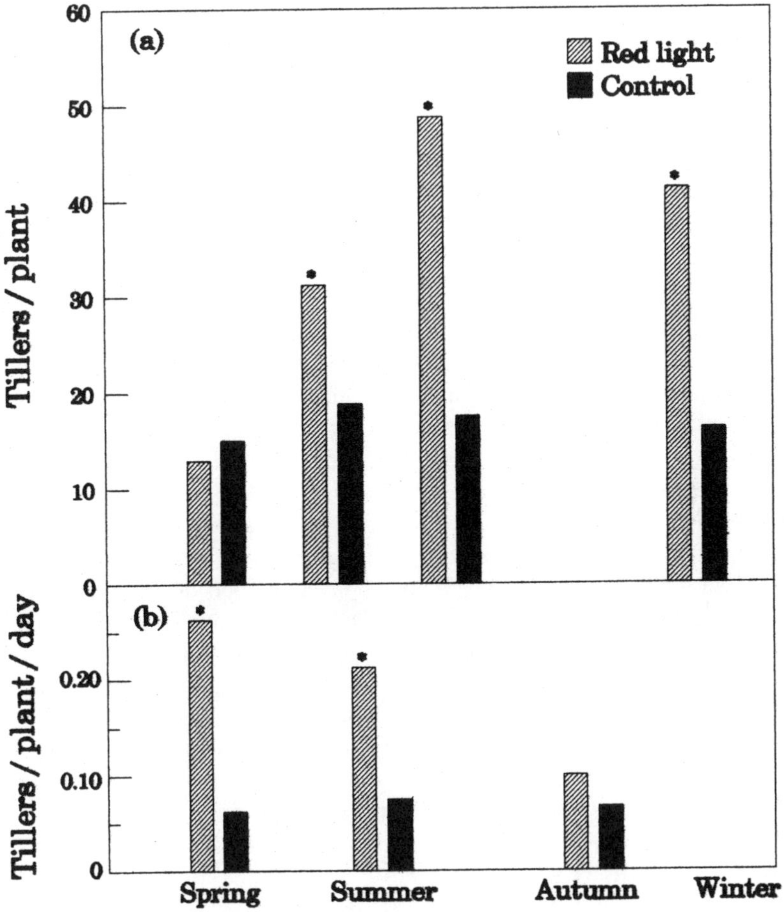

Figure 8. (a) Number of tillers per plant and (b) daily rates of tiller appearance per plant for Dallisgrass (*Paspalum dilatatum*) as affected by red light enrichment in the lower portion of the canopy. Asterisks indicate significant differences between red light enriched and control plants (From Deregibus et al. 1985).

passes through a plant canopy. The decreasing red:far-red ratio may function as a signal to reduce tiller recruitment before the carrying capacity of the environment it exceeded (Deregibus et al. 1985, Casal et al. 1986, 1987). Alternatively, prior to complete canopy development or following partial canopy removal by grazing, the red:far-red ratio may function as a signal to initiate tiller recruitment.

The hypothesis indicating that depressions in the red:far-red ratio directly suppresses tiller recruitment was tested with a single cohort of parental tillers on established little bluestem plants in the field (Murphy and Briske 1994). The red:far-red ratio was modified at the site of juvenile tiller initiation near plant bases throughout the daylight hours without increasing the amount of photosynthetically active radiation by either (1) supplementing red light beneath canopies to raise the naturally low red:far-red ratio or (2) supplementing far-red light beneath partially defoliated canopies to suppress the natural increase in the red:far-red ratio following defoliation. Neither supplemental red nor far-red radiation influenced the rate or magnitude of tiller recruitment despite the occurrence of tiller recruitment in all experimental plants (Fig. 9).

The hypothesis advocating direct regulation of tiller recruitment by the red:far-red ratio is further challenged by three categories of experimental evidence. First, reduced tiller recruitment is only one of several shade avoidance responses exhibited by tillers in response to the red:far-red signal (Murphy and Briske 1994). It is possible that a low red:far-red ratio reduces tiller recruitment indirectly by increasing carbon allocation to the blades at the expense of the axillary buds after tiller initiation has already occurred. Second, increases in the red:far-red ratio at the plant bases following partial canopy removal are relatively transient and do not override the associated constraints on tiller recruitment resulting from defoliation (Murphy and Briske 1992, 1994). Third, immature leaf blades appear to function as sites of red:far-red photoperception on individual tillers in addition to the previously recognized site at the tiller base (i.e., leaf sheaths or axillary buds) (Skinner and Simmons 1993, Murphy and Briske 1994). If blades are an important site of red:far-red photoperception, the ecological significance of the light quality signal is potentially minimized because the site of perception and a low red:far-red signal would spatially coincide only in young, juvenile tillers or seedlings located beneath or near established plant canopies. Therefore, depressions in the red:far-red ratio are probably of greater ecological significance as a signal of impending competition for light in vegetation canopies than they are as a density-dependent signal capable of directly regulating tiller recruitment. These findings demonstrate that additional research is required before definitive ecological or managerial conclusions can be established concerning the significance of radiation quality as an environmental signal capable of regulating tiller recruitment.

Axillary Bud Longevity

Axillary buds accumulate on the stem bases (crowns) of individual tillers as they are successively differentiated with each phytomer (Fig. 7). This population of buds comprises a bud bank and provides the potential for subsequent tiller recruitment within the population (Harper 1977, p. 108, Noble et al. 1979). The bud bank is depleted as buds grow out to form tillers, succumb to biotic or abiotic stresses or age and senesce. Unfortunately, the extent of bud longevity is largely

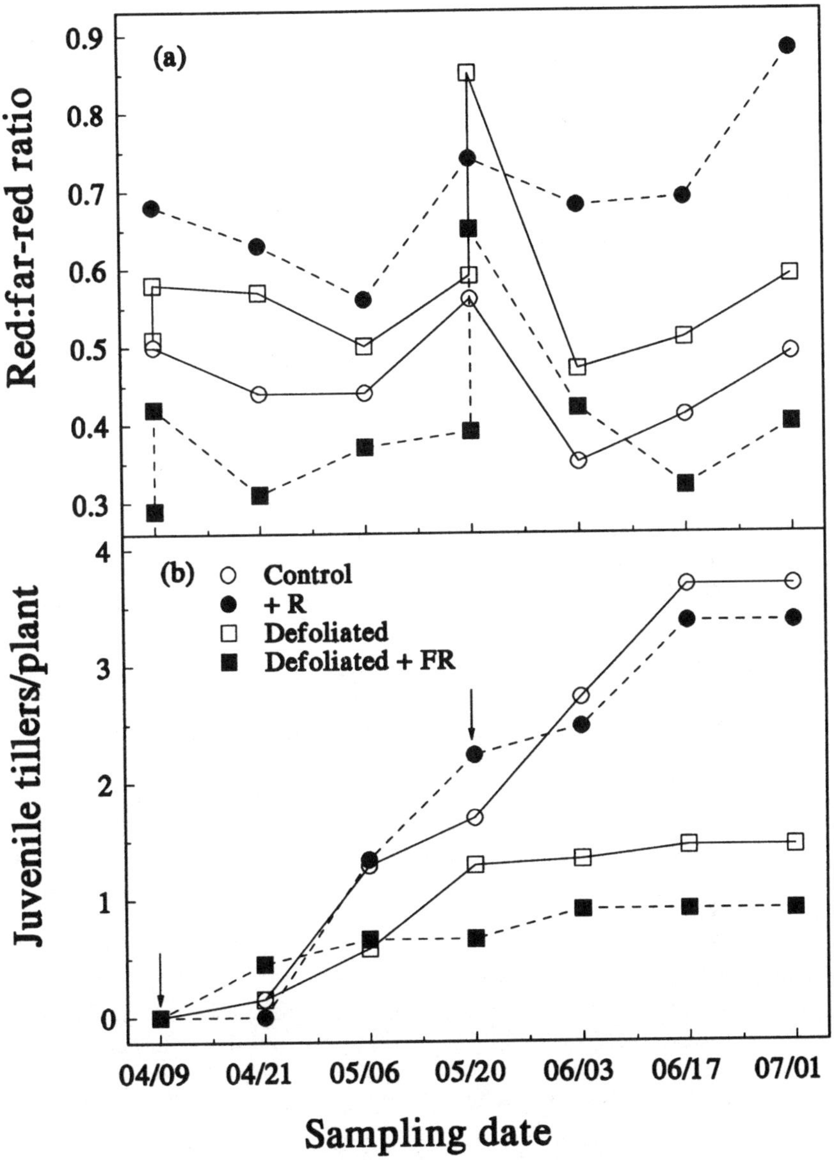

Figure 9. (a) Mean red:far-red ratios beneath canopies of little bluestem plants and (b) cumulative mean number of tillers recruited per plant in each of four treatments. Arrows indicate the dates of partial canopy removal to a height of 15 cm on plants assigned to defoliation treatments. The two values shown for the red:far-red ratios on 9 April and 20 May for plants in the defoliated and defoliated plus far-red treatments indicate the values measured immediately before and after partial canopy removal (From Murphy and Briske 1994).

unexplored and two contrasting viewpoints have developed concerning their contribution to population persistence. One perspective maintains that if buds do not develop shortly following maturation they quickly become senescent (Hyder 1972, 1974) while the alternative perspective suggests that buds express a greater longevity forming a reserve of meristematic potential (e.g., Cable 1971, McKendrick et al. 1975).

Axillary bud longevity has tremendous ecological consequences for sustainable productivity and persistence of grasses. Species populations which rely on short-lived axillary buds are entirely dependent upon on the successive recruitment of tillers for their persistence. If climatic variation or biotic stresses (e.g., grazing or pathogens) eliminates recruitment of one or two tiller generations, persistence would be jeopardized because meristematic potential could potentially be depleted within the population. Contrastingly, species populations which possess axillary buds with greater longevities are less susceptible to a loss of meristematic potential following the interruption of successive tiller recruitment.

The limited amount of available data suggests that a portion of the axillary buds survive as long as the parental tiller remains alive (Briske and Butler 1989). However, it is consistently buds which have most recently matured that grow out to form tillers even though several ontogenetically older buds may exist on the crown (Mitchell 1953, Busso et al. 1989). The longer buds remain inhibited and the further they become distanced from the apical meristem the less likely they are to form tillers (e.g., Mueller and Richards 1986). Therefore, the contribution of mature, inhibited buds to the persistence of grass populations is largely unknown.

Tiller Demography in Response to Grazing

An incomplete understanding of the mechanisms regulating tiller recruitment and the contribution of axillary bud banks severely limits our ability to anticipate the demographic responses of grasslands to grazing. Defoliation can potentially influence tillering by affecting substrate availability for bud growth, the degree of bud inhibition (i.e., apical dominance), the number of viable buds and microclimatic conditions for growth. The end result is that the conceptual model currently used to anticipate and manage tiller population dynamics in grazed systems requires additional development.

Tiller recruitment

The widely held assumption that apical meristem removal stimulates tillering in rangeland grasses is not consistently supported by the available information (Murphy and Briske 1992). Both Ellison (1960) and Jameson (1963), following a thorough evaluation of the early literature, concluded that defoliation does not consistently increase tillering in native, perennial grasses. Many of the recent research findings substantiate their conclusions. For example, two seasons of livestock grazing did not increase, and in the case of the intensive grazing decreased, the total number of tillers recruited by crested wheatgrass (Olson and Richards 1988b and 1988c) and little bluestem (Butler and Briske 1988) in comparison with ungrazed plants. In addition, four consecutive years of biweekly defoliation to stubble heights of 5, 10, and 15 cm decreased tiller densities proportionately in pinegrass (*Calamagrostis rubescens*) (Stout et al. 1981). The apparent complexity of the physiological mechanisms regulating axillary bud growth (i.e., apical dominance) and the large number of potentially intervening factors including, environmental variables, species specific responses, stage of phenological development, and the frequency and intensity of defoliation, further minimize the likelihood of a consistent tillering response to defoliation.

The time interval during which tillering is evaluated following defoliation may be the single greatest factor contributing to our misinterpretation of this response. Tillers are frequently initiated within 2–3 weeks of defoliation (Olson and Richards 1988c), and are much more obvious following partial removal of the canopy. However, this short-term "flush" of tiller initiation following defoliation may reduce subsequent tiller recruitment by a comparable number in comparison with maximum tiller recruitment in undefoliated plants (Fig. 10) (Butler and Briske 1988). In addition, tiller recruitment following spring grazing of crested wheatgrass did not contribute to tiller replacement the following growing season because of greater tiller mortality in grazed than in ungrazed plants during the winter (Olson and Richards 1988b). Therefore, in spite of an increased rate of tillering shortly following defoliation, defoliated grasses may not produce a greater number of tillers than undefoliated grasses when evaluated over one or more growing seasons. Defoliation may more appropriately be viewed as a means of altering the timing or seasonality of tiller recruitment, rather than increasing total tiller recruitment over the long-term, in many native range grasses. However, forage grasses adapted to mesic, fertile environments (e.g., perennial ryegrass) frequently do increase net tiller production in response to grazing (Jones et al. 1982, Tallowin et al. 1989). The mechanism(s) associated with these contrasting responses is unknown.

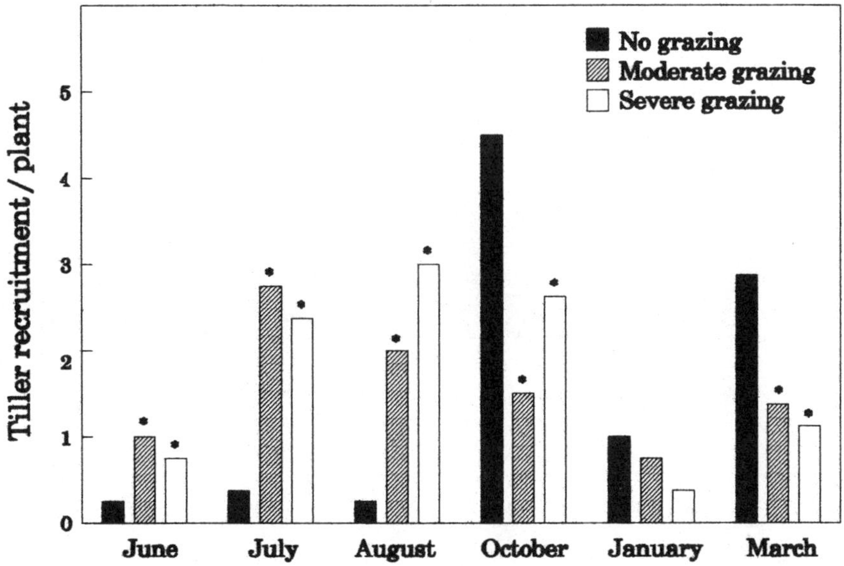

Figure 10. Tiller recruitment in little bluestem populations that had been moderately or severly grazed by cattle in a rotational grazing system for 4 years. Control populations had been protected from livestock grazing for 25 years. Values are a mean of 10 permanently marked tillers per each of 20 plants in each of the grazed populations and 10 plants in the ungrazed population. Asterisks indicate a significant difference between the grazed and ungrazed populations (From Butler and Briske 1988).

Increased tiller densities per plant or per unit area do not necessarily translate into greater productivity in accordance with the law of constant yield previously discussed (Kays and Harper 1974). An inverse relationship is commonly observed between individual tiller weight and tiller density (Grant et al. 1981). For example, continuously-grazed populations of perennial ryegrass maintained tiller densities 2 to 5 times greater than those of infrequently mowed populations (Jones et al. 1982). However, greater weights and leaf areas of individual tillers in the mown populations compensated for the low tiller density enabling both populations to produce comparable amounts of biomass annually.

Phenological development

The interaction between the phenological stage of plant development and plant defoliation is not well understood given its potential significance to grassland production and management; consequently, the available data are inconsistent and difficult to interpret. For example, defoliation during early vegetative growth exerts a negligible (Olson and Richards 1988b and 1988c) or slight stimulatory effect on tillering (Vogel and Bjugstad 1968). However, defoliation during vegetative growth promotes tiller recruitment in Arizona cottontop to a greater extent than any other phenological stage (Cable 1971). Defoliation at the time of culm elongation, but prior to inflorescence emergence, stimulates tillering in crested wheatgrass (Olson and Richards 1988c), little bluestem (Jameson and Huss 1959), timothy (*Phleum pratense*) and meadow fescue *Festuca pratensis* (Fig. 11). However,

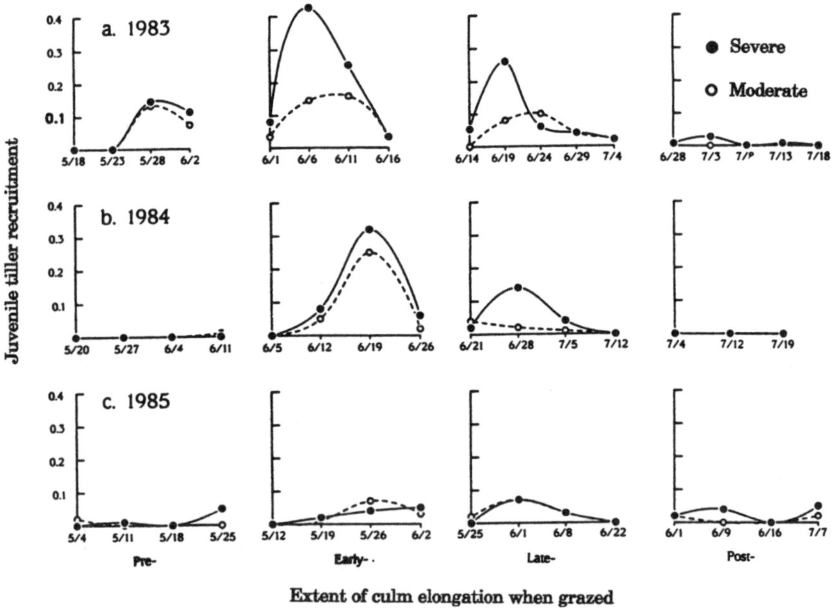

Figure 11. Rate of juvenile tiller recruitment (juvenile tillers/parent tiller/time interval) in response to the intensity and timing of a single grazing event. Crested wheatgrass plants were moderately or severely grazed by cattle at each of four stages of culm elongation during a 3-year period. Tiller counts were made at 5–8 day intervals following grazing in each experiment (From Olson and Richards 1988c).

defoliation at this phenological stage has been documented to suppresses tillering in big bluestem, little bluestem and Indiangrass (Vogel and Bjugstad 1968). Defoliation during inflorescence emergence stimulates tillering in the three tallgrasses mentioned previously (Jameson and Huss 1959, Vogel and Bjugstad 1968), but contradictory responses have been documented for big bluestem (Neiland and Curtis 1956), timothy and meadow fescue (Langer 1959).

It is tempting to speculate that a portion of the confusion associated with tiller initiation during the reproductive stage is based upon whether or not the apical meristem is removed by defoliation. However, defoliation did not consistently stimulate tillering in several wheatgrass species even when the apical meristem was removed (Branson 1956, Richards et al. 1988). Apical meristem removal was more consistently associated with increased tillering in several grasses of tropical origin. Apparently, physiological processes and/or environmental variables in addition to direct regulation by apical meristems are involved in the regulation of tillering in perennial grasses (Tainton and Booysen 1965, Richards et al. 1988, Murphy and Briske 1992).

Defoliation severity

The effects of defoliation frequency and intensity on tiller initiation are also difficult to generalize and easily confounded with phenological development and the seasonal progression of environmental variables. Tiller recruitment did not differ significantly between moderate and severe intensities of livestock grazing in crested wheatgrass (Olson and Richards 1988b) and little bluestem (Butler and Briske 1988). In some instances, the more severely grazed plants recruited a greater number of smaller tillers. Crested wheatgrass grazed twice during the spring (Olson and Richards 1988b) and intensively grazed perennial ryegrass display this pattern of tillering (Grant et al. 1983). However, more often, tiller recruitment decreases with increasing frequency and intensity of defoliation (Stout et al. 1981). It has been suggested that a stubble height of 2–3 cm is the optimal grazing intensity to maintain high tiller densities in populations of perennial ryegrass (Grant et al. 1983). More lenient grazing reduces tiller densities by decreasing tiller recruitment and increasing tiller mortality through self-shading while more severe grazing decreases tiller densities by limiting substrate availability following removal of excessive leaf area (Grant et al. 1983). This may serve as a useful generalization of the effect of defoliation intensity on tiller densities even though the optimal defoliation intensity will undoubtedly vary with species, stage of phenological development and associated environmental variables (Langer 1963).

System of grazing

The best documented examples of the effect of grazing systems on tiller dynamics are for perennial ryegrass. Continuous grazing maintains greater tiller densities than rotational grazing or infrequent mowing in contrast to native range grasses (Jones et al. 1982, Tallowin et al. 1989). Similarly, the more intensive, continuous grazing regimes produce the greatest number of tiller cohorts within the population (Parsons et al. 1984). Both responses are associated with a reduction in foliage density allowing more irradiance to penetrate into the canopy. This microenvironmental modification facilitates both tiller recruitment and daughter tiller survival

(Parsons et al. 1984, Tallowin et al. 1989). Even though continuously grazed populations maintain the highest tiller densities, they also possess tillers with the lowest weights (Grant et al. 1981, Jones et al. 1982). However, perennial ryegrass has undergone substantial genetic selection, is subject to intensive management practices (e.g., fertilization and livestock movement), and is restricted to mesic, temperate climates (e.g., United Kingdom and New Zealand). Consequently, it is questionable to what extent these results are directly applicable to native grasslands in extensively managed systems with more rigorous climatic regimes.

Plant Demography in Response to Grazing

Grazing-induced modifications in tiller recruitment, longevity and phenological development are eventually manifested in plant architecture and population structure. Population structure includes the attributes of density, size class distribution and age structure of plants and has important implications to grazing resistance, competitive ability and population persistence.

Population structure

The most pronounced modifications induced by grazing in native, bunchgrass populations are the reduction in individual plant basal area and increase in total plant density (Fig. 12). Both processes apparently result from the fragmentation of individual large plants within the population and may represent the initial and predominant process contributing to the decline of bunchgrass populations in

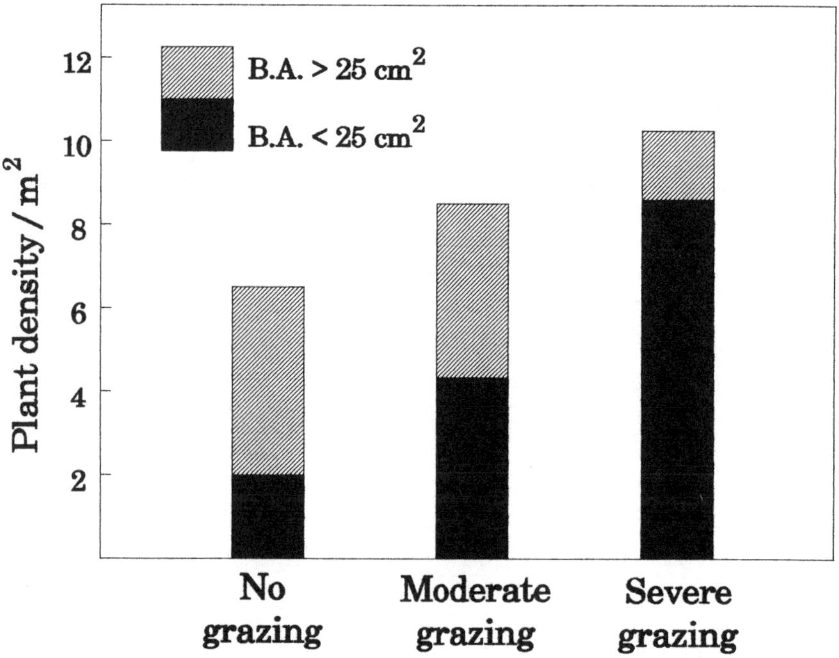

Figure 12. Densities of two size classes of little bluestem plants that were moderately or severely grazed by cattle in a rotational grazing system for 4 years. Control populations had been protected from livestock grazing for 25 years. Values are a mean of 20 plants in each of the grazed populations and 10 plants in the ungrazed population (From Butler and Briske 1988).

response to grazing (Butler and Briske 1988). Grazed populations of several perennial grasses have been documented to consist of individuals with smaller basal areas in comparison with ungrazed populations including Idaho fescue (*Festuca idahoensis*) (Pond 1960), crested wheatgrass (Hickey 1961), tussock grasses in Kazakhstan (Vorontzova and Zaugolnova 1985), grasses in the flooding pampas of Argentina (Sala et al. 1986), and dune grasses in Wales (Gibson 1988). Grazing induced population degradation eventually reduces tiller numbers and total basal area on a per unit area basis and potentially decreases the productivity and competitive ability of plants within the community (Fig. 13).

However, the ecological consequences of grazing-induced modifications in population structure are largely unexplored. Comparable tiller densities distributed in a low density of plants with large basal areas or a high density of plants with small basal areas did not significantly influence aboveground production of defoliated plants in comparison with undefoliated plants over a 2-year investigation (Briske and Anderson 1990). Apparently, both defoliated and undefoliated plants were equally capable of accessing and incorporating resources into production regardless of the spatial pattern of tiller distribution. However, tiller distribution did significantly affect the relative increase in tiller density and basal area for both defoliated and undefoliated plants. Relative increases in both variables were greatest for the high density of small plants and least for the low density of large plants in the second year of the investigation. A population structure consisting of a high density of small plants confers a greater potential for tiller recruitment and basal area expansion because a majority of tiller recruitment occurs on the plant periphery rather than in the interior (Butler and Briske 1988, Olson and Richards 1988a).

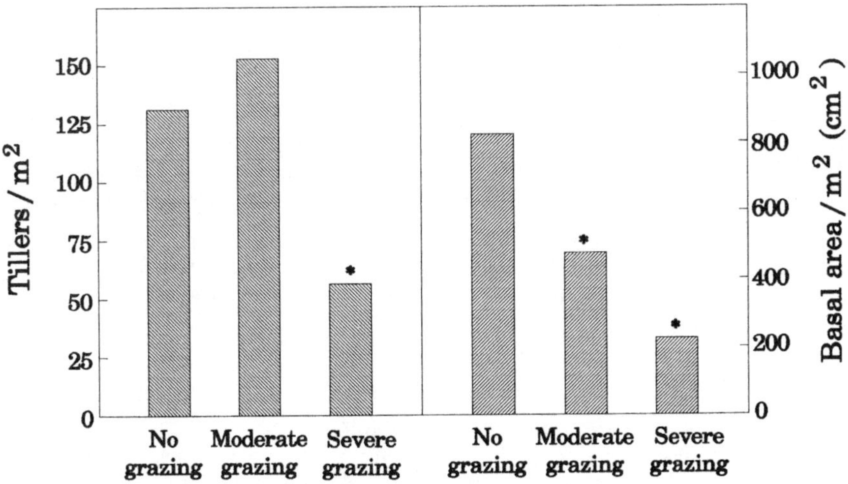

Figure 13. Tiller densities and total basal areas of little bluestem populations that had been moderately or severely grazed by cattle in a rotational grazing system for 4 years. Control populations had been protected from grazing for 25 years. Values are a mean of 20 plants in each of the grazed populations and 10 plants in the ungrazed population. Only plants with basal areas of 25 cm^2 or greater were monitored to calculate these variables. Asterisks indicate significant differences between grazed and ungrazed populations (From Butler and Briske 1988).

Although the spatial distribution of tillers did not influence production following defoliation, it represents only one aspect of grazing induced modifications in population structure. Grazing also modifies plant size class distribution and absolute tiller densities among the various species populations which comprise the community (Sala et al. 1986, Butler and Briske 1988). A large portion of the population growth of bunchgrasses can be ascribed to processes associated with individuals in the largest size classes (Moloney 1988). Eventually, small plants with few tillers will be placed at a competitive disadvantage in relation to their larger neighbors (Liddle et al. 1982, Fowler 1986, Olson and Richards 1989, Hartnett 1989). Similarly, populations possessing the greatest tiller densities may potentially acquire the greatest proportion of available resources within the community (Caldwell et al. 1987, Soriano et al. 1987). The population attributes of plant size class distribution and absolute tiller density appear to be of greater significance to herbivory tolerance in perennial bunchgrass populations than is the spatial distribution of tillers (Briske and Anderson 1990).

Plant longevity

Estimates of perennial grass longevity are limited and a portion of the available data document the occurrence of relatively short lifespans. Plant longevity estimates from the Santa Rita Experimental Range in Arizona (Canfield 1957), Jornada Experimental Range in New Mexico (Wright and Van Dyne 1976), and the U.S. Sheep Experimental Station in Idaho (West et al. 1979) are summarized in Table 2. Maximum longevities ranged from 4 years in curly mesquite (*Hilaria belangeri*) to 43 years in needle-and-thread (*Stipa comata*). Mean maximum plant longevities were shortest at the Santa Rita Experimental Range (8.4 years), intermediate at the Jornada Experimental Range (13.1 years), and longest at the Sheep Experimental Station (34.4 years).

Observations of relatively short lifespans may partially originate from the architectural pattern of bunchgrass growth and the charting procedures frequently used to monitor plant survival. Basal area expansion and subsequent plant fragmentation may yield estimates of premature plant mortality even though the genotype continues to survive in the form of one or more remnants of the original plant (West et al. 1979). Chronological estimates of the architectural development of tufted hairgrass (*Deschampsia caespitosa*) in northern Europe suggest that 25–60 years is required for plants to progress from seedlings to complete fragmentation of the plant (Gatsuk et al. 1980). Comparable evaluations of tussock grasses in Kazakhstan, including *Festuca*, *Koeleria*, and *Stipa* spp., suggest maximum plant longevities of 30–80 years (Vorontzova and Zaugolnova 1985, Zhukova and Ermakova 1985). However, the remnants of fragmented plants were assumed to have a limited probability of survival requiring that population densities be maintained by sexual reproduction. Chronological estimates of architectural development throughout the life history of North American grasses are not available.

The limited amount of information addressing the effects of grazing on plant longevity indicates that both increases and decreases in longevity may occur. Longevity of *Chloris acicularis*, the sole representative of the late-successional vegetation in New South Wales, Australia was reduced by grazing while longevity of the mid-successional dominant, *Danthonia caespitosa*, was unaffected

Table 2. Maximum longevity (years) of perennial grasses in ungrazed communities at the Santa Rita Experimental Range (Canfield 1957), Jornada Experimental Range (Wright and Van Dyne 1976), and the U.S. Sheep Experimental Station (West et al. 1979).

Santa Rita Experimental Range		Jornada Experimental Range		Sheep Experimental Station	
Species	*Longevity*	*Species*	*Longevity*	*Species*	*Longevity*
Aristida hamulosa Mesa threeawn	13	*Aristida divaricata* Poverty threeawn	9	*Koeleria cristata* Prairie junegrass	27
Bouteloua curtipendula Sideoats grama	7	*Aristida longiseta* Red threeawn	9	*Oryzopsis hymenoides* Indian ricegrass	19
Bouteloua eriopoda Black grama	14	*Bouteloua eriopoda* Black grama	27	*Poa spp.*[1] Bluegrasses	41
Bouteloua hirsuta Hairy grama	6	*Hilaria mutica* Tobosa grass	7	*Pseudoroegneria spicata* Bluebunch wheatgrass	42
Heteropogon contortus Tanglehead	6	*Muhlenbergia arenacea* Ear muhly	9	*Stipa comata* Needle-and-thread	43
Hilaria belangeri Curly mesquite	4	*Scleropogon brevifolius* Burro grass	12		
Trichachne californica Arizona cottontop	8	*Sporobolus flexuosus* Mesa dropseed	19		

[1] Includes *Poa secunda*, *Poa nevadensis*, and *Poa cusickii*.

(Williams 1970b). Longevities of three perennial grasses increased in relation to that of three-tipped sagebrush (*Artemisia tripartita*) which decreased in response to autumn grazing (West et al. 1979). Similarly, longevities of late-successional grasses decreased in response to grazing while longevities of the mid-successional species increased (Fig. 14) (Canfield 1957). These seemingly contradictory responses suggest that the late-successional grasses may have been grazed more severely than the associated mid-successional species. Selective grazing may have placed the late-successional plants at a competitive disadvantage with the less severely grazed mid-successional species thereby reducing their longevity. However, grazing did not significantly affect plant longevity in the investigations conducted on the Jornada Experimental Range (Wright and Van Dyne 1976) and Sheep Experimental Station (West et al. 1979).

A review of the population ecology literature for North American grasses confirms the conclusion of Crawley (1987) indicating that our understanding of herbivore-induced effects on population processes is poorly understood. Only limited data exist to substantiate that herbivory is a key regulatory process influencing the population dynamics of plants. Alternatively, the seemingly minimal impacts of herbivory on demographic processes may result from the complexity of the combined processes and associated interactions. The impacts of herbivory on population processes are difficult to discern because it may not directly induce plant mortality, but rather reduce growth, seed production (Verkaar 1987), and competitive ability (Briske 1991). These latter effects may only become evident in subsequent plant generations.

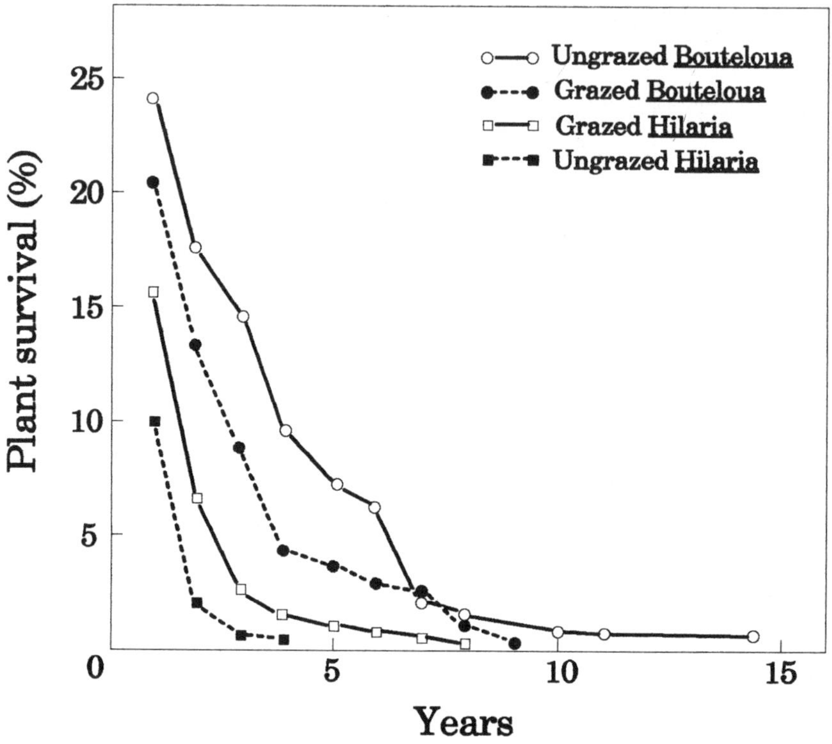

Figure 14. Maximum longevity of ungrazed and grazed black grama (*Bouteloua eriopoda*) and curly mesquite (*Hilaria belangeri*) plants on the Santa Rita Experimental Range near Tuson, Arizona. Grazing reduced the longevity of the late-successional dominant (black grama), but increased the longevity of the mid-successional species (curly mesquite) (Modified from Canfield 1957).

Grazing Resistance

A relative degree of grazing resistance can be assigned to numerous grasses based on the currently available information. Species characterized by decumbent canopy architectures (Arnold 1955, Hodgkinson et al. 1989, Painter et al. 1989), rapid rates of tiller recruitment (Richards et al. 1988) or high concentrations of secondary compounds (Coley et al. 1985, Georgiadis and McNaughton 1988, Briggs and Schultz 1990) are frequently more resistant than species which do not display these characteristics. Many characteristics associated with grazing resistance have been attributed to the selection pressure exerted by grazing and are frequently presented as evidence in support of the occurrence of coevolution between grasses and grazers (Georgiadis and McNaughton 1988). However, in both instances, the supporting evidence is limited and available data do not strongly confirm the role of grazers as selective agents in the evolutionary development of grass structure and function (Stenseth 1978, Herrera 1982, Simms and Rausher 1987, Belsky et al. 1993). Several environmental variables, other than herbivory, may have functioned as selective agents in the evolution of grasses

(Coughenour 1985). For example, the basal location of meristems may have evolved in response to drought or fire as well as herbivory. If either of these abiotic variables were the selective agent which initially contributed to the evolution of basal meristems, then the occurrence of this trait in relation to grazing avoidance would be considered an ex-adaptation (i.e., adaptation to an environmental variable other than the one which originally contributed to its selection; syn. preadaptation; Coughenour 1985). These considerations indicate that even though the concept of herbivory resistance is frequently used by ecologist and resource managers, it is based on a limited amount of quantitative information.

Avoidance and Tolerance Components

The concept of grazing resistance defines the relative ability of plants to persist in a grazed plant community. Grazing resistance can be divided into avoidance and tolerance components based on the general mechanism(s) conferring resistance (Briske 1986, 1991). Grazing avoidance includes mechanisms that reduce the probability and severity of grazing while grazing tolerance consists of mechanisms that facilitate growth following defoliation. Avoidance mechanisms include anatomical and architectural attributes in addition to secondary compounds which reduce tissue accessibility and palatability at several levels of grassland organization (Table 3). In Table 3, silicification refers to the occurrence of the mineral silica in epidermal cells (McNaughton et al. 1985) and interspecific association describes the protection afforded a species with a lesser expression of avoidance mechanisms when growing in close proximity to species with a greater expression of avoidance mechanisms (McNaughton 1978). Meristematic activity and compensatory physiological processes which increase growth following defoliation comprise the tolerance component (Table 4) (see Individual Plant Responses to Defoliation section).

The occurrence of varying degrees of grazing resistance among plants is well established by the predictable patterns of species selection by herbivores and species replacement in response to long-term herbivory (e.g., Dyksterhuis 1949, Heitschmidt et al. 1990). However, in most cases, it is uncertain whether herbivory induces species replacement through the selective utilization of plant species (i.e., avoidance mechanisms) or the occurrence of unequal growth capabilities among species following grazing (i.e., tolerance mechanisms). Although, both mechanisms are known to occur, the predominant mechanism, or relative combination of mechanisms, remains unknown for most species combinations and plant communities (e.g., van der Meijden et al. 1988). Selective grazing of the late-successional dominant, little bluestem, relative to mid-successional species, has been recognized as the predominant mechanism contributing to species replacement in the southern true prairie of Texas (Brown and Stuth 1993, Anderson and Briske, 1995). Alternatively, more rapid canopy reestablishment in crested wheatgrass than bluebunch wheatgrass following comparable levels of defoliation demonstrates that these two species possess different degrees of grazing tolerance (Caldwell et al. 1981, Richards and Caldwell 1985, Caldwell et al. 1987). In this comparison, the potential contribution of avoidance mechanisms is minimized by comparable canopy architectures, phenology, and leaf nitrogen content between the two species and the

Table 3. Mechanisms of grazing avoidance in grasses at the leaf, tiller, plant and community levels of grassland organization. The herbivore(s) and selected reference(s) are indicated for each mechanism.

	Mechanism	Herbivore	Reference
Leaf	Presence of vascular bundles	Grasshoppers	Caswell et al. 1973, Boutton et al. 1978, Heidorn and Joern 1984
		Cattle	Wilson et al. 1983, Akin et al. 1983
	Silicification	Slugs	Wadham and Parry 1981
		Stem-boring larvae	Moore 1984
		Serengeti ungulates	McNaughton et al. 1985, but see Shewmaker et al. 1989
		Prairie vole	Gali-Muhtasib et al. 1992
	Secondary compounds	Sheep	Marten et al. 1973, 1976
		Geese	Buchsbaum et al. 1984, but see Gauthier and Bédard 1990
		Cattle	Georgiadis and McNaughton 1988
Tiller	Height and angle	Clipping	Westoby 1980
		Prairie dogs	Detling and Painter 1983, Painter et al. 1989
		Cattle	Heitschmidt et al. 1990
	Height of apical meristem	Clipping	Branson 1953, Booysen et al. 1963, Westoby 1980
	Culm elongation/flowering	Cattle	Gammon and Roberts 1978
Plant	Accumulation of culms	Cattle, Deer	Willms et al. 1980
		Cattle	Norton and Johnson 1983, Truscott and Currie 1989, Ganskopp et al. 1992
	Large or small basal area	Cattle	Norton and Johnson 1983, Truscott and Currie 1989, Ganskoff and Rose 1992
Community			
	Interspecific association	Serengeti ungulates	McNaughton 1978
		Cattle	Davis and Bonham 1979
		Rabbits	Jaksić and Fuentes 1980

imposition of equivalent defoliation intensities. The observation that plant productivity is often not reduced in direct proportion to the severity of defoliation imposed also substantiates the occurrence of tolerance mechanisms which facilitate growth following defoliation (McNaughton 1983, Painter et al. 1989).

Recognition of herbivory avoidance or tolerance as the predominant mechanism inducing species composition shifts has important implications to grassland management. It is especially important in cases where the reduction and local extinction of late-successional dominants in severely grazed grasslands may have been misinterpreted as a consequence of inferior herbivory tolerance, rather than inferior herbivory avoidance, in relation to the associated mid-successional species. If the

Table 4. Mechanisms of grazing tolerance in grasses at the leaf, tiller, and plant levels of grassland organization. The physiological process(es) and selected reference(s) are indicated for each mechanism.

	Mechanism	Physiological Processes	Reference
Leaf	Compensatory photosynthesis	Increased nitrogen content	Hodgkinson 1974, Nowak and Caldwell 1984
		Increased carboxylase activity or amount	Yamashita and Fujino 1986, von Caemmerer and Farquhar 1984
		Increased electron transport capacity	Jenkins and Woolhouse 1981
		Increased stomatal conductance	Gifford and Marshall 1973, Wallace et al. 1984
	Rapid elongation rate	Intercalary meristem activity	Volenec and Nelson 1983, Coughenour et al. 1985a, 1985b
		Enhanced water status	Wolf and Parrish 1982, Toft et al. 1987
	Increased blade:sheath ratio	Increased photosynthetic rate	Detling et al. 1979b, Painter and Detling 1981
Tiller	Rapid leaf recruitment	Apical meristem activity	Youngner et al. 1976, Kotanen and Jeffries 1987
	Rapid tiller recruitment	Axillary bud activation	Stur and Humphreys 1988, Richards et al. 1988
	Increased C allocation to shoots	Source-sink activity	Ryle and Powell 1975, Richards 1984
	Increased N allocation to shoots	Source-sink activity	Ourry et al. 1988, Millard et al. 1990
Plant	Compensatory shoot growth	Potentially all processes above and a modified canopy age-structure	Dyer and Bokhari 1976, Georgiadis et al. 1989, Gold and Caldwell 1990, Oesterheld 1992
	Compensatory root activity	Nutrient reallocation to roots	Jaramillo and Detling 1988
		Increased sink activity of shoots	Polley and Detling 1988, Chapin and McNaughton 1989
	Inter-tiller resource allocation	Source-sink activity	Watson and Ward 1970, Welker et al. 1987, 1991, Jónsdóttir and Callaghan 1989
	Inherent competitive ability	Maintenance of resource acquisition	Mueggler 1972, Caldwell et al. 1985, 1987, Wallace and Macko 1993

late-successional dominants are perceived as being relatively intolerant of herbivory, the only viable management strategy to maintain dominance of the late-successional species is to reduce the severity of grazing within the community. Alternatively, if inferior herbivory avoidance by the late-successional dominants is the predominant mechanism inducing species replacement, then managerial decisions to regulate the uniformity of grazing among species may be implemented to maintain dominance of the late-successional species. Intensive grazing of mid-successional species when the relative expression of avoidance mechanisms is lowest, including the period of maximum vegetative plant height (Arnold 1955, Belsky 1992) and prior to reproductive culm development (Willms et al. 1980, Ganskopp et al. 1992), can minimize, although not eliminate, selective herbivory among grass species (Briske and Heitschmidt 1991, Brown and Stuth 1993).

Grazing Morphotypes

An indication of the relative contribution of avoidance and tolerance mechanisms to grazing resistance can be derived from the occurrence of morphotypic selection within species populations subjected to long-term grazing. Grazing or mowing selects against morphotypes possessing an erect canopy architecture in approximate proportion to the severity of defoliation (e.g., Scott and Whalley 1984). Grazing by domestic cattle has been demonstrated to function as a selection pressure capable of inducing architectural variation in perennial grass populations in an ecological time frame (ca <25 years; Peterson 1962, Briske and Anderson 1992). Selection has been documented to occur within much shorter time periods (2–12 years) in several perennial grasses subjected to intensive grazing by prairie dogs (*Cynomys ludovicianus*) in the mixed prairie (Detling and Painter 1983, Jaramillo and Detling 1988, Painter et al. 1993). Kemp (1937) was among the first to quantify that grazing selected against erect morphotypes of Kentucky bluegrass, orchardgrass and white clover, but similar responses have been documented for several perennial grasses (Table 5). Decumbent morphotypes, in contrast to erect morphotypes, are characterized by a large number of small tillers with reduced leaf numbers and blade areas. However, physiological variables that may potentially contribute to grazing tolerance, including photosynthetic rate, transpiration rate and water-use efficiency, did not differ significantly between erect and decumbent growth forms of western wheatgrass (Detling and Painter 1983). Limited evidence does exist to suggest that morphotypes subjected to long-term grazing may possess a greater degree of grazing tolerance in addition to greater grazing avoidance. Decumbent morphotypes of western wheatgrass with a history of grazing had 3-fold greater rates of nitrogen accumulation per unit root mass than did populations with no history of grazing (Polley and Detling 1988). Plants with a history of grazing had greater rates of shoot growth following defoliation which may have generated the greater demand for nitrogen. In addition, tillers of western wheatgrass and blue grama (*Bouteloua gracilis*) plants with a history of grazing produced more biomass and had greater survivorship following defoliation than did populations which had not been subjected to intensive selection by herbivores (Painter et al. 1989).

The inference drawn from morphotypic selection in grazed grass populations is that avoidance mechanisms, rather than tolerance mechanisms, are the predominant component of grazing resistance (Detling and Painter 1983, Jaramillo and Detling 1988). Decumbent morphotypes are better able to avoid grazing because less biomass is removed by herbivores and a greater number of meristems remain to facilitate growth following defoliation (Carman and Briske 1985) (see Canopy Reestablishment section). In contrast, morphotypes characterized by a small number of large tillers with large leaf areas are more competitive in environments

Table 5. Species populations in which defoliation has been documented to induce morphotypic selection. The herbivore(s), time interval grazing was imposed and reference(s) are indicated for each species.

Plant species	Herbivore	Time interval	Reference
Andropogon gerardii Big bluestem	prairie dogs/bison	2–100 yrs	Painter et al. 1993
Agropyron smithii Western wheatgrass	prairie dogs/bison	12 yrs	Detling and Painter 1983, Painter et al. 1989, 1993
Agrostis tenuis Colonial bentgrass	rabbits/sheep	unknown	Bradshaw 1959
Arrhenatherum elatius Tall oatgrass	sheep/cattle	unknown	Mahmoud et al. 1975
Bouteloua gracilis Blue grama	prairie dogs/bison	12 yrs 2–100 yrs	Jaramillo and Detling 1988, Painter et al. 1989
Dactylis glomerata Orchardgrass	sheep/rabbits sheep	unknown unknown	Gregor and Sansome 1926 Etherington 1984
Danthonia linkii *D. richardsonii* *D. racemosa* Oatgrasses	sheep	30 – >100 yrs	Scott and Whalley 1984
Lolium perenne Perennial ryegrass	sheep/rabbits	unknown	Gregor and Sansome 1926
Poa annua Annual bluegrass	clipping clipping/sheep	unknown unknown	Warwick and Briggs 1978 McNeilly 1981
Poa pratensis Kentucky bluegrass	livestock	37 yrs	Kemp 1937
Schizachyrium scoparium Little bluestem	cattle/bison cattle prairie dogs/bison	56 - >100 yrs 20 yrs 2 - 100 yrs	Carman and Briske 1985 Briske and Anderson 1992 Painter et al. 1993
Stipa comata Needle-and-thread	livestock	13 yrs	Peterson 1962

with dense canopies (but see Briske and Anderson 1992). Correspondingly, intraspecific competition is greatest within populations composed of erect morphotypes (Painter et al. 1989). However, grazing imposes a much greater detrimental impact on the competitive ability of erect morphotypes than of decumbent morphotypes because a greater portion of the canopy is removed.

The process by which grazing induces variation among growth forms within species populations has not been definitively established. However, a majority of the evidence indicates that grazing induced selection among growth forms is genetically based (McNeilly 1984, Scott and Whalley 1984, Carman and Briske 1985, Painter et al. 1989, 1993). The fact that grazing morphs of several midgrasses retained their decumbent stature for several growing seasons after being transplanted into a common garden or in a greenhouse in the absence of defoliation supports this interpretation (Polley and Detling 1988, Painter et al. 1993).

Paradoxically, there is little indication that grazing decreases genetic variability within species populations by selecting against the erect genotypes possessing limited avoidance mechanisms. A minimum of 66% of the individual plants subjected to electrophoretic analysis in both long-term grazed and ungrazed populations of little bluestem represented distinct genotypes even though plants in the grazed population displayed characteristics of a decumbent growth form in comparison with those of the ungrazed population (Carman and Briske 1985). Comparable genetic diversity between these two populations indicates that grazing does not necessarily eliminate the larger, slower tillering genotypes from grazed populations. Alternatively, grazing may influence the expression of relative dominance among genotypes variously adapted to herbivory, competition and associated environmental variables (McNeilly 1984, McNeilly and Roose 1984). Although a portion of the erect genotypes may be eliminated from a population following frequent, intense herbivory, the majority of genotypes less adapted to herbivory may exist in a subordinate position in the community or in areas less frequented by herbivores (Carman and Briske 1985). Eventually, when herbivory is eliminated or reduced, canopy expansion and mulch accumulation increase self-shading and potentially shift relative dominance toward the larger, fewer tillered genotypes. Genetic variation may also be maintained by the development of stable associations between and among interspecific neighbors (Aarssen and Turkington 1985) and the occurrence of disturbances which characterizes most grasslands (Collins and Barber 1985, Collins 1987, Coffin and Lauenroth 1988).

Role of Competition

Plants rarely respond to defoliation as isolated individuals in field settings. Defoliation affects the intensity of both intra-and interspecific competition in addition to directly influencing plant function (Caldwell 1984). For example, production of bluebunch wheatgrass plants subjected to 50% canopy removal just prior to culm elongation was equivalent to the production of undefoliated plants growing with full competition when associated vegetation within a 90 cm radius of the defoliated plants was clipped at ground level (partial competition) (Mueggler 1972) (Fig. 15). When competition from associated vegetation was removed by tilling within a 90 cm radius (no competition), defoliated plants produced three times the biomass of undefoliated plants growing with full competition. Similarly, defoliation of big bluestem to a height of 4 cm reduced biomass production 25% when plants were grown with a low density of nondefoliated neighbors, but production was reduced 90% by the same defoliation intensity when plants were grown with a high density of nondefoliated neighbors (Hartnett 1989). These data indicate that competition may influence growth following defoliation to an equal or greater extent than the direct effects of defoliation. The inherent competitive ability of a species within the community is an important determinant of its response to grazing.

Defoliation reduces the competitive ability of plants by reducing the effectiveness of resource acquisition within the environment. For example, the proportion of phosphorus isotope acquired by big sagebrush (*Artemisia tridentata*) from interspaces shared with bluebunch and crested wheatgrass increased by as much as 6-fold following a single defoliation removing 85% of the grass canopies (Caldwell et al. 1987). Significant increases in phosphorus absorption by big sagebrush

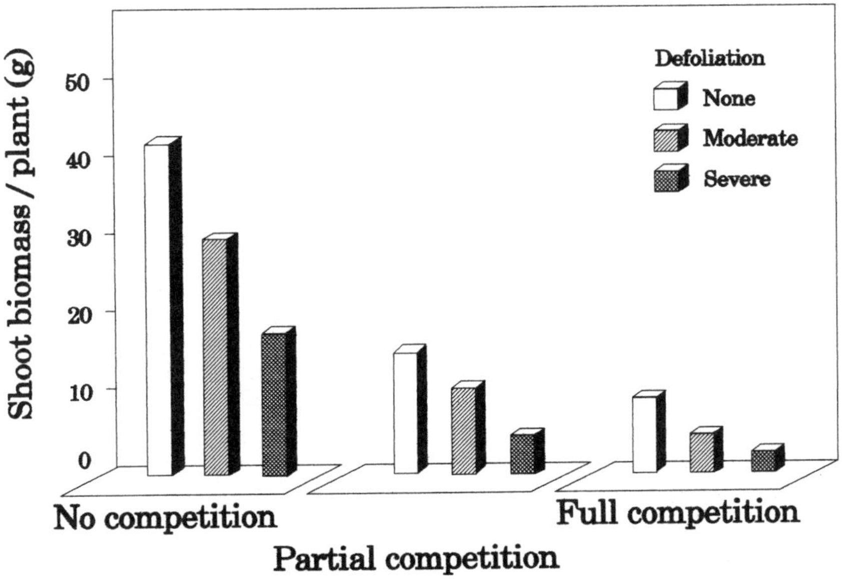

Figure 15. Response of bluebunch wheatgrass to three defoliation intensities in the presence of full, partial or no competition from association vegetation. Competition from associated species exerted a greater influence on growth following defoliation than did defoliation intensity (Modified from Mueggler 1972).

occurred within 2 weeks of grass defoliation, indicating that resource competition is rapidly modified by grazing. Similarily, the competitive ability of big bluestem for nitrogen was reduced by a single defoliation to a height of 10 cm regardless of whether the neighbors within a 50 cm radius were defoliated or not (Wallace and Macko 1993).

Caldwell (1984) has cautioned that many growth responses attributed to grazing resistance may actually be a consequence of the competition experienced by plants following defoliation. Although, inherent physiological and morphological mechanisms regulate plant growth following defoliation, differential resource acquisition among species may be of equal or greater significance (Fig. 15) (Mueggler 1972, Caldwell et al. 1985, Banyikwa 1988, Maschinski and Whitham 1989). The benefit a plant derives from defoliation of its neighbors coincident with its own defoliation has been termed competitive fitness (Belsky 1986). Larger statured grasses are frequently placed at a competitive disadvantage when grown with smaller statured grasses following defoliation to a comparable height because a greater proportion of biomass is removed from the larger plants (Arnold 1955, Belsky 1992).

The inherent mechanisms of grazing resistance are probably similar over the distributional range of a species, but the expression of relative resistance mechanisms will vary in relation to the grazing resistance and competitive ability of associated species. Species responses to grazing have been observed to vary depending upon the grazing intensity and topographic position within a mixed-grass prairie community (Archer and Smeins 1991). Relative abundance of western wheatgrass decreased regardless of grazing intensity or topographic position suggesting that it possessed limited grazing resistance relative to the associated species (Fig. 16). By contrast, the relative grazing resistance of buffalograss (*Buchloë dactyloides*) was relatively low in the swale, but relatively high on the ridge. Blue grama consistently expressed a high degree of grazing resistance except in the swale in response to severe grazing. The relative grazing resistance of buffalograss and sun sedge (*Carex eleocharis*) varied in response to (1) species specific patterns of livestock grazing, (2) site specific distribution of environmental variables, and (3) unique competitive interactions among the various plant species even though the inherent mechanisms of grazing resistance remained relatively constant (Archer and Smeins 1991).

Cost–Benefit Theory

The concept of grazing resistance is at least partially based on the assumption that the possession of grazing resistance mechanisms represent an associated cost to the plant (Simms and Rausher 1987, 1989). This assumption is most readily quantified in plants which produce secondary compounds to deter herbivory. For example, seedlings of the neotropical tree, *Cecropia peltata*, possessing high tan-

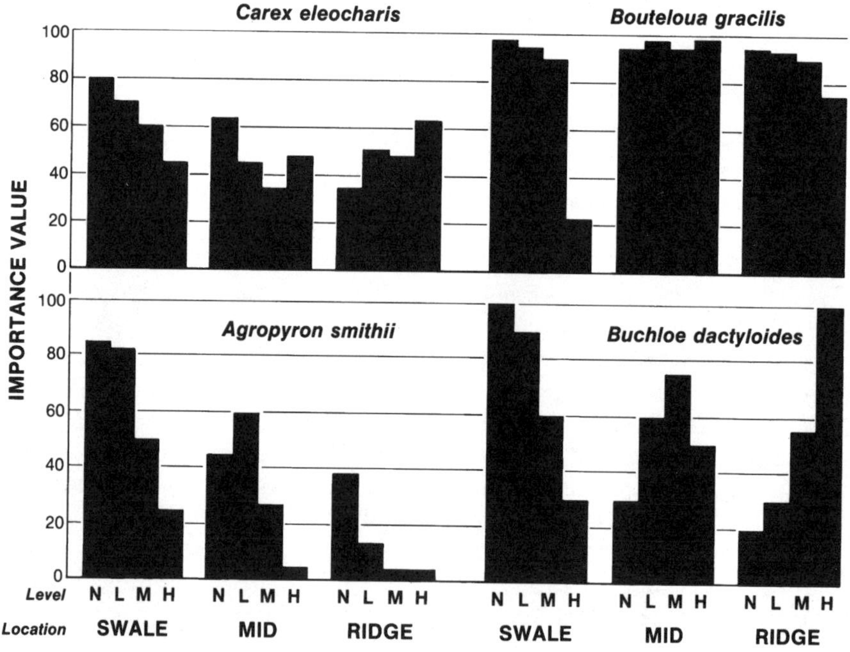

Figure 16. Mean importance values (relative cover x relative frequency) of four graminoids along mixed prairie toposequences (swale, mid-slope or ridge) subjected to different grazing intensities (N = none, L = light, M = moderate and H = heavy). The relative grazing resistance of sun sedge and buffalograss varied with topographic position and grazing intensity while the grazing resistance of western wheatgrass and blue grama remained relatively constant (From Archer and Tieszen 1986).

nin concentrations, experienced less insect herbivory, but displayed lower growth rates in the absence of herbivory than did seedlings which possessed lower tannin concentrations (Coley 1986). The difference in growth rate between these two groups of seedlings was attributed to the cost of producing secondary compounds and can be interpreted as the cost of herbivory resistance. Similar conclusions have been drawn from other investigations in which secondary compounds were variously produced by plants to deter herbivores (Windle and Franz 1979, Dirzo and Harper 1982). Mechanical defenses are also associated with large costs to the plant, but the evidence is much less extensive than it is for chemical defenses (van der Meijden 1988, Björkman and Anderson 1990, but see Ågren and Schemske 1993).

Grazing resistance presumably evolved as individuals in a population allocated various amounts of resources to resistance mechanisms and thereby experienced differing degrees of fitness. Since natural selection is driven by the relative contribution of individuals to the gene pool of subsequent generations within a population, fitness estimates must be based on sexual reproduction and associated fitness related characteristics rather than solely on estimates of biomass production. Grazing resistance is potentially maximized in an evolutionary context at the point where a plant realizes the greatest increase in fitness for the amount of resources allocated to resistance mechanisms. In other words, grazing resistance is maximized when the cost of resistance approximates the benefits of resistance (Coley et al. 1985, Simms and Rausher 1987) (Fig. 17). Therefore, plants do not become

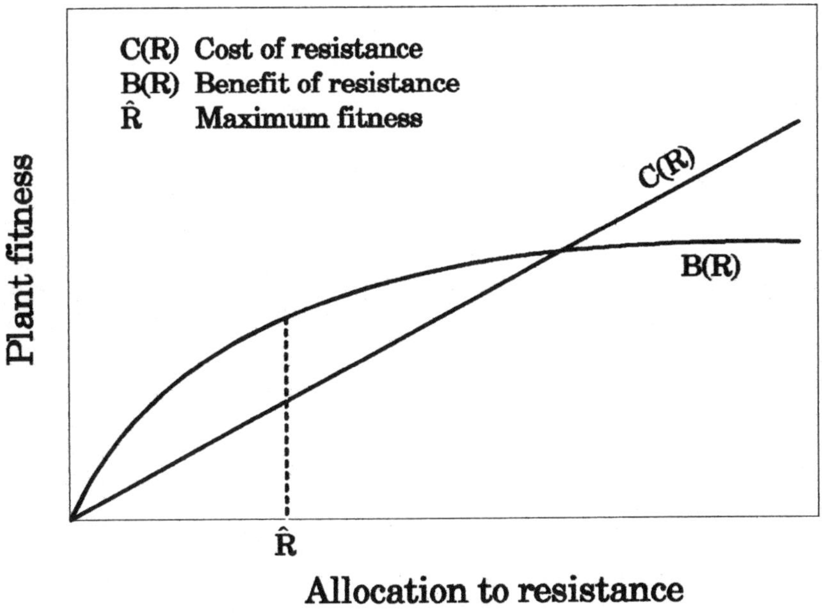

Figure 17. Schematic presentation of the costs and benefits of grazing resistance when treated as a quantitative trait. Plants experience maximum fitness at a point where the greatest grazing resistance is attained with minimum resource allocation to resistance. Plants do not become completely resistant to herbivores because at some point the cost of resistance exceeds the benefits to plant fitness (From Simms and Rausher 1987).

completely resistant to herbivores because the cost of resistance must, at some point, exceed the benefit conveyed by the resistance mechanisms (Pimentel 1988). The potential trade-off between the costs and benefits of herbivory resistance occur because competition is the predominant selective agent constraining the evolution of herbivory resistance (Herms and Mattson 1992).

The cost of herbivory resistance is influenced by both chemical (i.e., turnover rates and redistribution) and ecological considerations (i.e., resource availability and plant growth rate) (Skogsmyr and Fagerstrom 1992). Therefore, the cost of herbivory resistance will vary considerably given the wide range of defense mechanisms that are known to occur and the relative degree to which they are expressed. Trade-offs between herbivory resistance and competitive ability are assumed to be greatest in resource-rich environments (Coley et al. 1985, Herms and Mattson 1992). Alternatively, herbivory resistance is assumed to be more compatible with competitive ability, and therefore, less costly, in resource-limited environments. The evolutionary models of herbivory resistance are reviewed by Herms and Mattson (1992).

An accurate evaluation of herbivory resistance requires an estimate of not only the fitness conferred by the associated resistance mechanism(s), but also some indication of the extent to which herbivory reduces fitness in individuals which do not possess resistance mechanisms. Unfortunately, the influence of resistance mechanisms on fitness related characteristics are difficult to quantify between grazed and ungrazed populations. The concept of herbivory resistance requires additional conceptual development and quantification in order to develop a greater understanding of how specific resistance mechanisms influence plant responses to herbivory.

Compensatory Growth

Grazing is generally perceived to have a detrimental impact on grassland productivity by natural resource managers. Ellison (1960) concluded from a thorough review of the early literature that the beneficial effects of grazing on herbaceous plant growth, if any, were difficult to discern. However, this perspective was markedly altered by Chew (1974) and Mattson and Addy (1975) who suggested that herbivores may influence the structure and function of ecosystems to a greater extent than indicated by their biomass or the amount of plant material they consume. These authors suggested that herbivores may act as regulators of energy flow and nutrient cycling through a series of feedback loops on vegetation function. Owen and Wiegert (1976) further hypothesized that herbivores may increase plant fitness and McNaughton (1976, 1979) proposed that grazing could potentially optimize grassland production. These viewpoints stimulated a lively debate which continued through much of the following decade and increased the popularity of plant-animal interactions.

Grazing Optimization Hypothesis

The grazing optimization hypothesis originated from this contemporary view of plant-animal interactions (Dyer 1975, McNaughton 1976, 1979). The hypothesis states that primary production increases above that of ungrazed vegetation as grazing intensity increases to an optimal level followed by a decrease at greater grazing

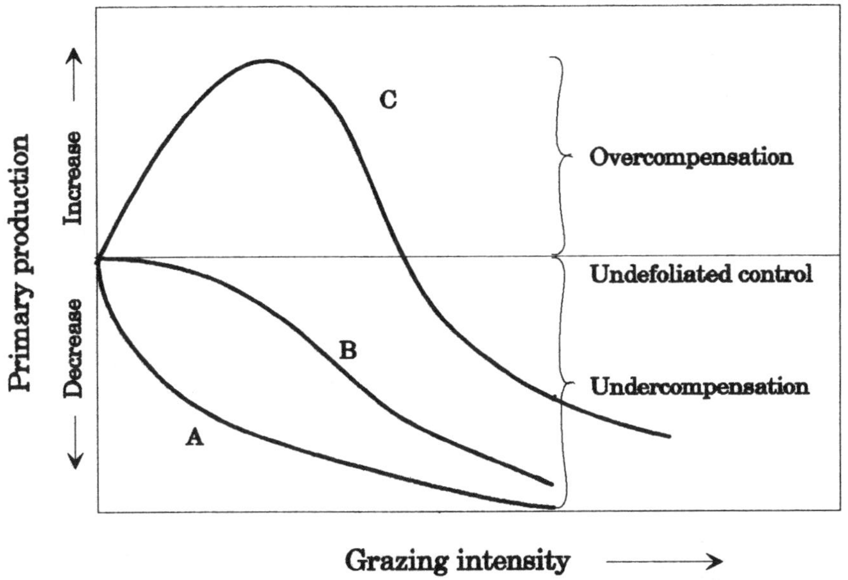

Figure 18. Three potential responses of primary production to increasing grazing intensity as proposed by the grazing optimization hypothesis. Primary production may: (a) decrease with increasing grazing intensity, (b) remain unaffected until intermediate levels of grazing intensity are attained and then decrease, or (c) increase with increasing grazing intensity to an optimal level and then decrease at greater grazing intensities. The concepts of over- and undercompensation are also illustrated (From Belsky 1986).

intensities (Fig. 18). This perspective represents a dramatic departure from the alternative view that grazing decreases primary production or that production remains constant up to a moderate intensity of grazing beyond which production decreases. Unfortunately, the general applicability of the grazing optimization hypothesis has not yet been established.

Ambiguity of the term "compensatory growth" has added considerable confusion to the debate over whether or not grazing increases plant productivity (Belsky 1986). Compensatory growth has been used to describe plant responses ranging from a partial replacement of removed biomass to production exceeding that of ungrazed plants. It is often difficult to discern which definition of "compensatory growth" is being used when reviewing the literature. Belsky (1986) proposed specific terms to clarify these growth responses (Fig. 18). Overcompensation refers to instances where cumulative total weight, including biomass removed by defoliation, exceeds the cumulative total weight of undefoliated plants. Partial compensation defines instances where more growth is produced by defoliated plants than would be expected if no increase in growth rate, had occurred.

Supporting Evidence

Much of the evidence in support of the grazing optimization hypothesis has been derived from containerized plant investigations conducted under controlled conditions. Overcompensation has been documented in certain instances with

Kyllinga nervosa, an African C_4 sedge, defoliated to a height of 4 cm at daily intervals (McNaughton 1979), *K. nervosa* defoliated to a height of 2 cm at 2 day intervals when grown with limited water and high nitrogen availability (McNaughton et al. 1983), *K. nervosa* defoliated to a height of 5 cm at 7 day intervals when grown with moderate nitrogen availability (Wallace et al. 1985), and *Sporobolus kentrophyllus* when infrequently defoliated to a height of 6 cm and grown with limited water and high nitrogen availability (Georgiadis et al. 1989). Complete compensation has also been reported in *Briza subaristata* and *Stipa bavioensis* in response to an exponential increase in relative growth rate of shoots with increasing defoliation intensity (Oesterheld 1992). Although these data substantiate the occurrence of overcompensation, they also demonstrate the specific growth conditions necessary to induce its response (Hilbert et al. 1981, Olson et al. 1989, Maschinski and Whitham 1989, Alward and Joern 1993, Belsky et al. 1993).

Examples of overcompensation by individual plants in field settings are less numerous and less definitive. Grazing by native ungulates or clipping increased total plant production of the biennial herb, scarlet gilia (*Ipomopsis aggregata*), 2 and 1.5 times, respectively, within the same growing season in comparison with undefoliated plants (Paige and Whitham 1987). Although, the intensities of both grazing and clipping were severe (95% of the shoot biomass was removed), defoliation was restricted to the phenological stage of plant development associated with stem elongation following vegetative growth, but prior to inflorescence production. Defoliation also increased flower and fruit production and did not detrimentally affect seed weight or germinability in comparison with undefoliated plants. Root biomass of defoliated plants was comparable or greater than that of undefoliated plants demonstrating that overcompensation did not result from the reallocation of belowground resources. Ungrazed plants of this species may have been meristem limited (i.e., Watson 1984) and defoliation may have activated growth of additional meristems which contributed to the increase in biomass yield and sexual reproduction.

However, Bergelson and Crawley (1992) evaluated 14 populations of scarlet gilia and found no evidence for overcompensation following defoliation throughout much of its range. They hypothesized that shade played a substantial role in the occurrence of compensatory growth reported by Paige and Whitham (1987). Defoliation may have released buds from apical dominance and promoted branching to a greater extent in shaded than in full sun habitats. Conflicting reports of compensatory growth within the same species clearly illustrates the involvement of specific environmental variables and our limited understanding of the mechanisms contributing to the process.

Overcompensation has also been documented in the Intermountain shrub bitterbrush (*Purshia tridentata*) under field conditions (Bilbrough and Richards 1993). Spring shoot growth of plants subjected to simulated winter browsing exceeded that of unbrowsed control plants following removal of terminal buds only, 60%, or 100% of the previous year's growth from all branches within a plant, but only the 60% defoliation intensity was significant. Overcompensation partially resulted from defoliation-induced meristem activation based on an increase in the frequency of short shoot and node production and the greater length of long shoots in defoliated plants.

Big bluestem and switchgrass on the Konza Prairie both compensated for bison grazing by increasing their relative growth rates in comparison with ungrazed plants (Vinton and Hartnett 1992). However, big bluestem tillers grazed repeatedly during the previous year had lower relative growth rates, tiller weights and tiller survival the following year than did ungrazed tillers. Overcompensation was only observed in switchgrass when all of the tillers within a plant were defoliated (Hartnett 1989). However, overcompensation in switchgrass plants was associated with a reduction in biomass partitioning to rhizomes. Therefore, the short-term increase in shoot growth in response to defoliation may occur at the expense of perennating organs and potentially compromise plant production and survival over the long-term (Hartnett 1989, Vinton and Hartnett 1992).

Although not a direct documentation of overcompensation, relative growth rates of crested wheatgrass tillers grazed by cattle were greater than for tillers in ungrazed plants when grazing occurred prior to internode elongation in the spring (Olson and Richards 1988c, Olson et al. 1989). Compensatory growth was only maintained for a 2–3 week period after which the relative growth rates of tillers on defoliated plants decreased to rates less than those of ungrazed plants. Compensatory growth was not observed in this species when spring precipitation was below normal or when grazing occurred during culm elongation or inflorescence emergence.

Definitive examples of overcompensation in plant communities are also limited, but information is becoming available for a greater number of diverse systems. McNaughton (1979) was among the first to indicate that moderate grazing intensities imposed by native ungulates doubled aboveground productivity of grasslands dominated by *Andropogon greenwayi* in the Serengeti Plains of east-central Africa. Although grazing intensity was the predominant variable affecting overcompensation, it occurred to the greatest extent when soil water was limiting. In another study, production of mixed prairie vegetation was 41% greater in a rotational grazing system in comparison to that of ungrazed vegetation during the second year of a 2-year investigation (Heitschmidt et al. 1982). Overcompensation primarily resulted from compensatory growth of warm-season grasses during the autumn when soil water was readily available. Aboveground herbaceous production on a diverse set of sites grazed by elk (*Cervus elaphus*) and bison (*Bison bison*) in Yellowstone National Park was found to be 47% greater than on comparable ungrazed sites over a 2-year period (Frank and McNaughton 1993). Overcompensation was documented in both vegetative and reproductive tissues of *Bouteloua gracilis* and *B. hirsuta* in response to grazing by grasshoppers (*Ageneotettix deorum*) in the mixed-prairie of southwestern Nebraska (Alward and Joern 1993). Overcompensation occurred most consistently under conditions of reduced competition on sites in which each species was naturally most abundant.

Perhaps the most definitive examples of overcompensation within plant communities are those where the production of intertidal vegetation near Hudson Bay was monitored following grazing by lesser snow geese (*Chen caerulescens caerulescens*). Production increases of 60–80% and 35–80% occurred in grazed monocultures of *Carex subspathacea* (Cargill and Jefferies 1984a) and grazed stands of *C. subspathacea* and *Puccinellia phryganodes* (Cargill and Jefferies 1984b), respectively, in comparison with ungrazed vegetation during the same growing season (Fig. 19). This system is extremely nitrogen limited and the effect of grazing without the addition of nitrogen in goose feces did not result in overcompensation (Hik and Jefferies 1990, Hik et al. 1991). Compensatory growth of *P. phryganodes* was dependent upon the rapid recycling of N, a low to moderate grazing intensity, and grazing early in the growing season. The potential for compensatory growth decreased as the growing season and phenological development of the vegetation progressed. Additional field investigations of primary production responses to grazing in a variety of ecological systems are required to more thoroughly assess the general applicability of herbivore optimization.

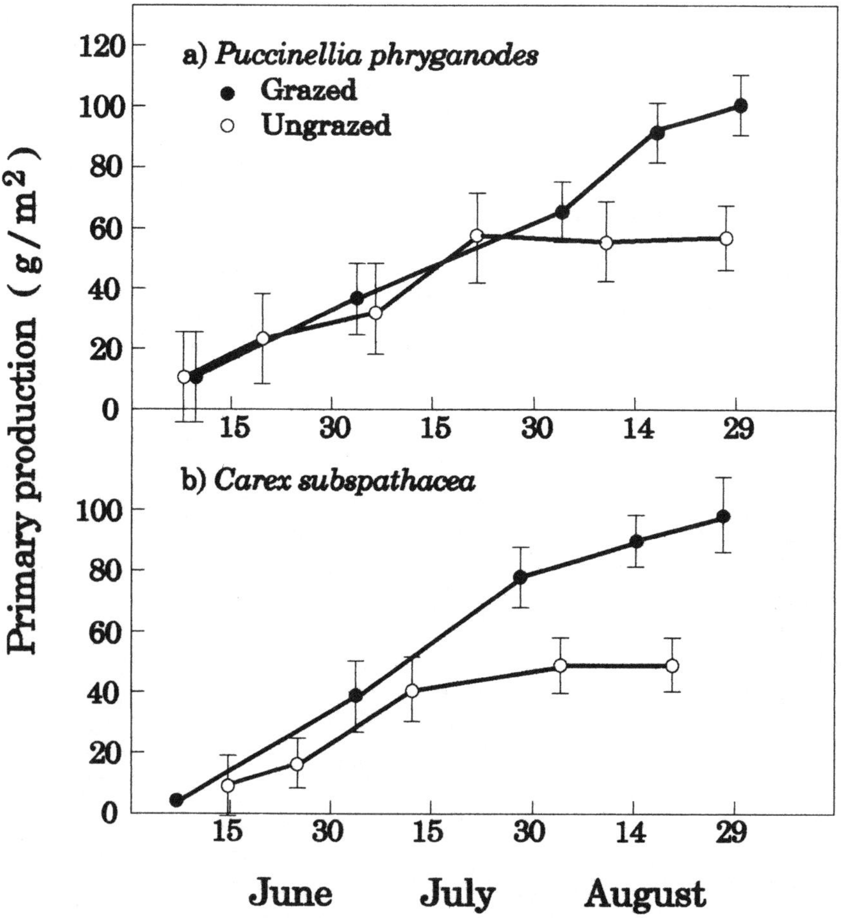

Figure 19. Cumulative net aboveground primary production of grazed and ungrazed swards of (a) *Puccinellia phryganodes* and (b) *Carex subspathacea* in response to early season grazing by geese near Hudson Bay, Canada. Productivity of both the grazed and ungrazed swards were similar early in the season, but grazed swards maintained a high growth rate throughout the season in comparison with ungrazed swards (From Cargill and Jefferies 1984b).

Causal Mechanisms

A wide variety of mechanisms have been proposed to explain compensatory growth responses. Intrinsic mechanisms, which are associated with herbivore-induced physiological processes, have received the greatest attention (McNaughton 1979, 1983) while extrinsic mechanisms, those involving herbivore-mediated environmental modifications, have been less thoroughly investigated, but may be of equal or greater importance (Table 6). For example, accelerated rates of nutrient cycling play an essential role in the occurrence of compensatory growth within plant communities (Ruess and McNaughton 1987, Hik and Jefferies 1990) thus verifying the assumption of Ellison (1960) that if vegetation benefits from herbivory it occurs as a result of ecosystem processes rather than individual plant function. Herbivory increases the rate of mineralization thereby increasing the rate at which nutrients become available for reabsorption within the system (Woodmansee et al. 1981, Floate 1981, Holland et al. 1992). Overcompensation in grasslands grazed during only the winter when plants are dormant also document the occurrence of extrinsic mechanisms (Frank and McNaughton 1993). However, intrinsic mechanisms must be the predominant explanation for compensatory growth in individual plant investigations, especially those in which defoliation is imposed by clipping (e.g., Paige and Whitham 1987, Bilbrough and Richards 1993).

Table 6. Two categories of mechanisms potentially contributing to compensatory plant growth following defoliation. Intrinsic mechanisms are associated with herbivore-induced physiological processes while extrinsic mechanisms result from herbivore-mediated environmental modifications (Modified from McNaughton 1983).

Mechanism	Reference
Intrinsic category	
Accelerated photosynthetic rate per leaf area	Detling et al. 1979b, Nowak and Caldwell 1984, Wallace 1990
Accelerated nutrient absorption per unit root mass	Ruess et al. 1983, Polley and Detling 1988, Chapin and McNaughton 1989
Greater resource allocation to shoots	Hartt et al. 1964, Ryle and Powell 1975, Richards 1984
Increased tiller initiation	Jameson and Huss 1959, Oesterheld and McNaughton 1988, Richards et al. 1988, but see text
Improved water status	Wolf and Parrish 1982, Toft et al. 1987, Briske and Anderson 1990
Extrinsic category	
Increased irradiance on remaining leaf tissue	Woledge 1977, Caldwell et al. 1983, Wallace 1990 Senock et al. 1991
Conservation of soil water following leaf area removal	Caldwell et al. 1983, Wraith et al. 1987 Svejcar and Christiansen 1987, but see Naeth et al. 1991, Miller and Rose 1992
Accelerated rate of nutrient cycling	Floate 1981, Hik and Jefferies 1990, Holland et al. 1992
Increased activity of decomposer organisms	Ruess and McNaughton 1987, Holland and Detling 1990

Both intrinsic and extrinsic mechanisms presumably contribute to compensatory growth, but their relative contribution varies with species, defoliation regime, associated environmental variables and the ecological scale investigated (e.g., Brown and Allen 1989). For example, compensatory growth in *Ipomopsis arizonica* decreased in response to increasing interspecific competition, decreasing nutrient availability, and grazing during the later portion of the growing season (Maschinski and Whitham 1989). Brief periods of intensive herbivory confined to the early portion of the growing season appear to be an important requirement for the occurrence of compensatory growth in several diverse systems (McNaughton 1976, Paige and Whitham 1987, Hik et al. 1991, Frank and McNaughton 1993). This pattern of grazing is most representative of the spatiotemporal patterns produced by migratory herbivores (Frank and McNaughton 1993).

Collectively, the currently available data does not support the conclusion that overcompensation is a widely occurring phenomena in grazed systems. Belsky (1986) reviewed 48 reports in the literature referencing the response of aboveground production to grazing and found that 34 documented a decrease in production, 5 reported no change and 9 reported an increase in production. Similarily, Milchunas and Lauenroth (1993) concluded that overcompensation occurred in only 17% of a total of 276 data sets representing 159 communities. In addition, the increase in aboveground production observed in the grazed communities was quite small relative to production in ungrazed communities. Differences in aboveground production between grazed and ungrazed communities, for all communities combined, were significantly and negatively related to increasing (1) years of grazing, (2) intensity of biomass utilization, and (3) community productivity. Nevertheless, these three variables only explained 25% of the variation among the communities evaluated. When only grassland or grassland and shrubland communities were evaluated, the differences in aboveground production between grazed and ungrazed communities were more sensitive to changes in ecosystem variables (i.e., production and history of grazing) than to variables associated with grazing (i.e., intensity of utilization, years of grazing). However, various compensatory mechanisms may increase plant growth rates following defoliation, but only partially compensate for the total amount of biomass removed (Belsky 1986). Partial compensation determines that plant productivity is not suppressed in direct proportion to the frequency and intensity of defoliation imposed, even though total production does not exceed that of ungrazed vegetation (McNaughton 1983, 1985).

It is important to recognize that much of the data collected in support of the grazing optimization hypothesis were from grazed systems where herbivore density and movement were not directly regulated by humans (e.g., McNaughton 1979, 1985, Paige and Whitham 1987, Hik and Jefferies 1990, Frank and McNaughton 1993). In these systems, primary production and herbivore density often fluctuate widely in response to annual climatic variation. Conversely, herbivore density and movement are controlled in intensively managed systems and precautions are taken to minimize deleterious consequences on animal production. Consequently, the intensity of grazing in these systems may frequently exceed the optimal intensity required to consistently stimulate primary production as indicated by the grazing optimization hypothesis (Briske and Heitschmidt 1991,

Oesterheld et al. 1992, McNaughton 1993). These differences between plant-animal systems may also partially explain why the grazing optimization hypothesis originated with researchers working in extensively managed, rather than intensively managed systems and why the hypothesis receives limited support from most natural resource managers.

The potential benefits of compensatory growth should not obscure the effects of grazing on the ecological processes governing the sustainability of grazed systems (Briske 1993). Overgrazing and site degradation frequently result from the attempt of humans to maintain a desired number of animals or to produce a sufficient amount of animal products for subsistence agriculture or economic profitability, rather than from the inherent instability of grazed systems. While compensatory plant growth may potentially optimize primary production in some systems given the appropriate environmental conditions, it very likely has a negligible effect on the threshold limits defining system stability. In this context, the sustainability of grazed systems is a more fundamental ecological issue than is grazing optimization.

Summary

The following dichotomous model and summary statements are intended to emphasize the most important conclusions concerning the morphological, physiological and demographic responses of plants to grazing. Responses of plants to grazing span temporal scales beginning with the disruption of steady-state growth within minutes, to the expression of compensatory physiological processes within days, followed by the progressive development of regrowth over a period of weeks, and conclude with demographic processes spanning months and years. Morphological, physiological and demographic responses to defoliation encompass the ecological scales of tillers, plants and populations within grassland communities.

Dichotomous Model for Analyzing Regrowth Responses

The ability of individual plants to regrow following defoliation is a multifacetted process involving the meristematic potential and compensatory physiological processes of individual plants, intra- and interspecific competition among plants, and abiotic variables within the environment. The inherent morphological and physiological attributes of individual plants establish the potential for regrowth while the intensity of plant competition, resource availability and environmental conditions constrain the extent to which this regrowth potential is realized. The following dichotomous model is presented as a concise summary of the plant attributes and environmental variables that must be considered when predicting and/or analyzing regrowth responses of individual plants following defoliation.

I. Active meristems intact (dependent upon plant architecture, phenology, and defoliation intensity and pattern). [If not, go to II.]
   A. Environmental conditions appropriate for growth (consider both abiotic, e.g., light intensity, temperature, water and nutrient availability, and biotic variables, e.g., competitive interactions and pathogens). [If not, go to I. B.]

1. High plant capacity for compensatory photosynthesis and/or reallocation of photosynthetic carbon to new shoot tissues (dependent upon sink strength and number of active apical and intercalary meristems).
   These conditions lead to rapid canopy reestablishment; rapid recovery of photosynthetic capacity and recharge of carbohydrate and nutrient pools; root growth and function begins to recover within 2–4 days after defoliation; dependence on stored carbohydrates and nutrients is limited to several days.
2. Low plant capacity for compensatory photosynthesis and/or reallocation of photosynthetic carbon.
   Canopy reestablishment occurs more slowly than I.A.1. because less current photosynthetic carbon is reinvested in shoot regrowth. Predefoliation root/shoot ratios are reestablished more slowly.

B. Environmental conditions are suboptimal for growth (consider both abiotic, e.g., light intensity, temperature, water and nutrient availability, and biotic variables, e.g., competitive interactions and pathogens).
   The potential for canopy reestablishment is great, but it is not realized. In some cases (e.g., defoliation during winter) environmental conditions may both prevent regrowth and induce quiescence so that stored carbohydrates and nutrients remain in the plant until optimal environmental conditions for growth occur.

II. Active meristems removed (regrowth dependent on activation of shoot sinks, i.e., axillary bud growth).

A. Quiescent axillary meristems absent or necrotic. [If present, go to II. B.]
   Canopy reestablishment does not occur; plants respire available substrate and die.

B. Quiescent axillary meristems present.

1. Rapid activation of axillary meristems. [If slow, go to II. B.2.]
   Canopy reestablishment is relatively rapid, but less than in I.A.1. because of time lag required for bud activation. Regrowth is dependent on available carbohydrate pools at the time growth is initiated and on the reinvestment of photosynthetic carbon into shoot growth. Presence of residual leaf area or resource reallocation from within the plant are critical for rapid regrowth.
2. Slow activation of axillary meristems.
   Potential loss of stored carbohydrates and nutrients within the plant and potential loss of water, nutrient and light resources to competitors.
   (1) "Dormant" season:
   Low metabolic activity of plants and inactivity of competitors preserves stored carbohydrates and nutrients within plants and access to environmental resources, respectively, until quiescence or dormancy is alleviated. Regrowth rate may be rapid depending on the number and sink strength of activated shoot meristems and on the availability of carbohydrates to promote the initial growth of leaves.
   (2) Growing season:
   High metabolic activity of plants reduces resources to a level that may severely limit regrowth rates when shoot meristems are acti-

vated. Competitors may preempt environmental resources and minimize the potential for canopy reestablishment and plant survival.

Summary statements

1. Steady-state plant growth is immediately disrupted by defoliation in response to a substrate limitation imposed by a reduction in photosynthetic area. A reduction in whole-plant photosynthesis and preferential carbon allocation to active shoot sinks reduces root growth and nutrient absorption as root carbohydrates are depleted. The extent to which these physiological processes are suppressed and their potential rates of recovery directly affect the grazing resistance, competitive ability and productivity of defoliated plants.
2. The reduction of carbohydrate pools within root systems following defoliation primarily results from the combined effects of a reduction in the amount of photosynthetic carbon allocated to the root system and the continued utilization of carbohydrates in root respiration, rather than the remobilization of carbohydrates to support shoot regrowth. Although, carbohydrate pools within the remaining shoot system are important for initiating plant growth when photosynthetic capacity is severely limited, the relative inaccessibility of root carbohydrates to support shoot growth, the limited amount of carbohydrates stored in tiller bases, and poor correlations between shoot regrowth and carbohydrate concentrations or pools limits their use as an effective index of shoot regrowth in perennial grasses and potentially other growth forms as well.
3. Compensatory photosynthesis is a very consistent physiological response following defoliation that has been documented in numerous species. The process involves the rejuvenation of leaves and/or the postponement of the normal decline in photosynthetic capacity that occurs as leaves age and senesce. Although the occurrence of compensatory photosynthesis is partially influenced by increased light availability following partial canopy removal, indirect mechanisms presumably mediated by cytokinins or other root produced signals, appears to be the predominant mechanism inducing the response. Compensatory photosynthesis occurs in defoliation-sensitive as well as defoliation-tolerant species and it often does not persist throughout the recovery period. Therefore, the rate of leaf area expansion following defoliation plays a greater role than does the increase in photosynthetic rate per unit leaf area in the reestablishment of whole-plant photosynthetic capacity.
4. Canopy reestablishment is a function of the source, number, and activity of remaining meristems; the availability of photosynthetic carbon, and the amount of stored carbohydrates and nutrients. Growth following defoliation is most rapid from intercalary meristems, intermediate from apical meristems and slowest from axillary buds. Canopy reestablishment is delayed when a large portion of the active meristems are removed by defoliation, or when suboptimal environmental conditions or intense competition limit growth, photosynthesis, or nutrient acquisition. A series of compensatory processes, including increased resource allocation to shoots, increased photosynthetic rates, and enhanced rates of nutrient absorption by roots, may occur within several days of defoliation to increase relative growth rates over those of undefoliated plants.

5. The widely held assumption that apical meristem removal stimulates tiller initiation in grasses is not consistently supported by either traditional or contemporary experimental data demonstrating that the process is more complex than the direct inhibition of axillary buds by auxin produced in the apical meristem. Although tiller recruitment may increase for a short period following defoliation, tiller recruitment or longevity is often reduced over the course of one or more growing seasons in comparison with undefoliated plants. Defoliation-induced reductions in tiller recruitment and longevity suppress tiller numbers per plant and per unit area when evaluated over one or more growing seasons in even the most grazing resistant rangeland grasses.
6. Radiation quality has been hypothesized to function as an environmental signal capable of regulating tiller recruitment in grasses. Reductions in the red:far-red ratio, resulting from greater attenuation of red than far-red wavelengths as radiation passes through a plant canopy, may function as a signal to reduce tiller recruitment before tiller densities exceed the carrying capacity of the environment. However, supplemental red or far-red radiation applied to the base of plants does not consistently modify the rate or magnitude of tiller recruitment under field conditions. Reductions in the red:far-red ratio are probably of greater ecological significance as a signal of impending competition for light than they are as a density-dependent signal capable of directly regulating tiller recruitment. These findings demonstrate that additional research is required before definitive ecological or managerial conclusions can be established concerning the significance of radiation quality as an environmental signal capable of regulating tiller recruitment.
7. Grazing induced modifications of tiller demography are eventually manifested in plant demography and population structure. Individual plant basal areas decrease and plant densities increase following the fragmentation of individual bunchgrass plants subjected to grazing. Total basal area and tiller number per unit area may be substantially reduced by continued severe grazing. Grazing-induced modifications in population structure reduce the potential for population maintenance and decrease competitive ability thereby increasing resource availability for more grazing resistant species populations within the community. Grazing appears to decrease plant longevity by placing the most severely grazed plants at a competitive disadvantage with associated species which are grazed less severely.
8. Grazing or mowing is known to select against morphotypes possessing an erect canopy architecture and increase the proportion of decumbent growth forms, commonly referred to as grazing morphs, within species populations. Decumbent growth forms are characterized by a large number of small tillers with reduced leaf numbers and blade areas which are better able to avoid grazing because less biomass is removed by herbivores and a greater number of meristems remain to facilitate growth following defoliation. A majority of the evidence indicates that grazing-induced selection among growth forms is genetically based. Selection has been documented to occur within <25 years in response to grazing by cattle and in 2–12 years in response to intensive grazing by prairie dogs.

9. Grazing resistance consists of avoidance and tolerance mechanisms which reduce the probability and/or severity of grazing and increase the rate of growth following grazing, respectively. Although plants possess components of both mechanisms, avoidance mechanisms appear to dominate the relative expression of grazing resistance among grasses. Grazing resistance is maximized in an evolutionary context when a plant attains the greatest increase in fitness for the amount of resources allocated to resistance mechanisms. However, the relative costs and benefits associated with avoidance and tolerance mechanisms require additional investigation if the concept of grazing resistance is to be of greater value in anticipating and predicting plant responses to defoliation.
10. Plants rarely respond to defoliation as isolated individuals in the field. Competition from undefoliated plants has been demonstrated to constrain growth of defoliated plants to as great an extent as the direct effects of defoliation. Defoliation reduces the competitive ability of plants by decreasing resource acquisition from the environment. The significant influence of competitive interactions on canopy reestablishment indicate that grazing tolerance (i.e., regrowth ability following defoliation) and competitive ability are distinct processes regulating the regrowth capacity of plants.
11. Compensatory growth following defoliation has been documented in both individual plants and plant communities. Intrinsic mechanisms, which are associated with herbivore-induced physiological processes of individual plants, and extrinsic mechanisms, which involve herbivore-mediated environmental modifications, both contribute to compensatory growth, but their relative contributions appear to vary with species or community, associated environmental variables, defoliation regime, and the ecological scale investigated. Compensatory growth appears to be promoted by limited competition, increased nutrient availability, and grazing only during the early portion of the growing season. Current information indicates that grazing only infrequently increases total growth of grazed plants above that of ungrazed plants in managed systems. However, greater growth rates of grazed plants very likely prevent growth from being suppressed in direct proportion to the severity of defoliation imposed even though complete compensation does not occur.
12. Plant communities are composed of several hierarchical scales possessing unique temporal and spatial characteristics. Lower scales (i.e., tillers and plants) contribute to the structure of higher scales (i.e., populations and communities) while higher scales can modify responses of the scales below in ways which are not predictable from the investigation of lower scales in isolation. Consequently, it is essential that several hierarchical scales be evaluated simultaneously to develop an accurate interpretation of plant and vegetation responses to grazing.

Literature Cited

Aarssen, L.W., and R. Turkington. 1985. Biotic specialization between neighbouring genotypes in *Lolium perenne* and *Trifolium repens* from a permanent pasture. J. Ecol. 73:605-614.

Ågren, J., and D.W. Schemske. 1993. The cost of defense against herbivores: an experimental study of trichome production in *Brassica rapa*. Am. Nat. 141:338-350.

Akin, D.E., J.R. Wilson, and W.R. Windham. 1983. Site and rate of tissue digestion in leaves of C_3, C_4 and C_3/C_4 intermediate *Panicum* species. Crop Sci. 23:147-155.

Alderfer, R.G., and C.F. Eagles. 1976. The effect of partial defoliation on the growth and photosynthetic efficiency of bean leaves. Bot. Gaz. 137:351-355.

Allen, T.F.H., and T.W. Hoekstra. 1990. The confusion between scale-defined levels and conventional levels of organization in ecology. J. Veg. Sci. 1:5-12.

Alward, R.D., and A. Joern. 1993. Plasticity and overcompensation in grass responses to herbivory. Oecologia 95:358-364.

Anderson, B., A.G. Matches, and C.J. Nelson. 1989. Carbohydrate reserves and tillering of switchgrass following clipping. Agron J. 81:13-16.

Anderson, J.E., and M.L. Shumar. 1986. Impacts of black-tailed jackrabbits at peak population densities on sagebrush-steppe vegetation. J. Range Manage. 39:152-155.

Anderson, V.J., and D.D. Briske. 1995. Herbivore-induced species replacement in grasslands: Is it driven by herbivory tolerance or avoidance? Ecol. Appl. 5:(in press).

Anslow, R.C. 1966. The rate of appearance of leaves on tillers of the Gramineae. Herb. Abst. 36:149-155.

Aoki, S. 1981. Effects of plucking of young tea plants on their photosynthetic capacities in the mature and overwintered leaves. Japan J. Crop Sci. 50:445-451.

Archer, S., and F.E. Smeins. 1991. Ecosystem-level processes. p. 109-139. *In*: R.K. Heitschmidt and J.W. Stuth (eds.) Grazing Management: an ecological perspective. Timber Press, Portland, Oregon.

Archer, S.R., and L.L. Tieszen. 1986. Plant response to defoliation: hierarchical considerations. p. 45-59. *In*: O. Gudmundsson (ed.) Grazing research at northern latitudes. Plenum Publ. Corp.

Arnold, J.F. 1955. Plant life-form classification and its use in evaluating range conditions and trend. J. Range Manage. 8:176-181.

Aslam, M., and R.C. Huffaker. 1982. In vivo nitrate reduction in roots and shoots of barley (*Hordeum vulgare* L.) seedlings in light and darkness. Plant Physiol. 70:1009-1013.

Atkinson, C.J. 1986. The effect of clipping on net photosynthesis and dark respiration rates of plants from an Upland Grassland, with reference to carbon partitioning in *Festuca ovina*. Ann. Bot. 58:61-72.

Banyikwa, F.F. 1988. The growth response of two east African perennial grasses to defoliation, nitrogen fertilizer and competition. Oikos 51:25-30.

Bassman, J.H., and D.I. Dickmann. 1982. Effects of defoliation in the developing leaf zone on young populus x-euramericana plants. I. Photosynthetic physiology, growth, and dry weight partitioning. For. Sci. 28:599-612.

Bassman, J.H., and D.I. Dickmann. 1985. Effects of defoliation in the developing leaf zone on young populus x-euramericana plants. II. Distribution of 14-C-photosynthate after defoliation. For. Sci. 31:358-366.

Belsky, A.J. 1986. Does herbivory benefit plants? A review of the evidence. Am. Nat. 127:870-892.

Belsky, A.J. 1987. The effects of grazing: confounding of ecosystem community, and organism scales. Am. Nat. 129:778-783.

Belsky, A.J. 1992. Effects of grazing, competition, disturbance and fire on species composition and diversity in grassland communities. J. Veg. Sci. 3:187-200.

Belsky, A.J., W.P. Carson, C.L. Jensen, and G.A. Fox. 1993. Overcompensation by plants: herbivore optimization or red herring? Evol. Ecol. 7:109-121.

Bergelson, J., and M.J. Crawley. 1992. Herbivory and *Ipomopsis aggregata*: The disadvantages of being eaten. Am. Nat. 139:870-882.

Bilbrough, C.J., and J.H Richards. 1993. Growth of sagebrush and bitterbrush following simulated winter browsing: mechanisms of tolerance. Ecology 74:481-492.

Björkman, C., and D.B. Anderson. 1990. Trade-off among antiherbivore defenses in a South American blackberry (*Rubus bogotensis*). Oecologia 85:247-249.

Bloom, A.J., and R.M. Caldwell. 1988. Root excision decreases nutrient absorption and gas fluxes. Plant Physiol. 87:794-796.

Booysen, P. DE V., N.M. Tainton, and J.D. Scott. 1963. Shoot-apex development in grasses and its importance in grassland management. Herb. Abst. 33:209-213.

Boucher, T.J., D.G. Pfeiffer, J.A. Barden, and J.M. Williams. 1987. Effects of simulated insect injury on net photosynthesis of potted grapevines. Hort. Sci. 22:927-928.

Boutton, T.W., G.N. Cameron, and B.N. Smith. 1978. Insect herbivory on C_3 and C_4 grasses. Oecologia 36:21-32.

Bradshaw, A.D. 1959. Population differentiation in *Agrostis tenuis* sibth. I. Morphological differentiation. New Phytol. 58:208-227.

Branson, F.A. 1953. Two new factors affecting resistance of grasses to grazing. J. Range Manage. 6:165-171.

Branson, F.A. 1956. Quantitative effects of clipping treatments on five range grasses. J. Range Manage. 9:86-88.

Briggs, M.A., and J.C. Schultz. 1990. Chemical defense production in *Lotus corniculatus* L. II. Trade-offs among growth, reproduction and defense. Oecologia 83:32-37.

Briske, D.D. 1986. Plant response to defoliation: morphological considerations and allocation priorities. p. 425-427. *In*: P.J. Joss, P.W. Lynch and O.B. Williams (eds.) Rangelands: A Resource Under Siege. Cambridge Univ. Press.

Briske, D.D. 1991. Developmental morphology and physiology of grasses. p. 85-108. *In*: R.K. Heitschmidt and J.W. Stuth (eds.) Grazing Management: an ecological perspective. Timber Press, Portland, Oregon.

Briske, D.D. 1993. Grazing optimization: a plea for a balanced perspective. Ecol. Appl. 3:24-26.

Briske, D.D., and J.L. Butler. 1989. Density-dependent regulation of ramet populations within the bunchgrass *Schizachyrium scoparium*: interclonal versus intraclonal interference. J. Ecol. 77:963-974.

Briske, D.D., and R.K. Heitschmidt. 1991. An ecological perspective. p. 11-26. *In*: R.K. Heitschmidt and J.W. Stuth, (eds.) Grazing management: an ecological perspective. Timber Press, Portland, Oregon, USA.

Briske D.D., and V.J. Anderson. 1990. Tiller dispersion in populations of the bunchgrass *Schizachyrium scoparium*: implications for herbivory tolerance. Oikos 59:50-56.

Briske, D.D., and V.J. Anderson. 1992. Competitive ability of the bunchgrass *Schizachyrium scoparium* as affected by grazing history and defoliation. Vegetatio 103:41-49.

Brown, B.J., and T.F.H. Allen. 1989. The importance of scale in evaluating herbivory impacts. Oikos 54:189-194.

Brown, J.R., and J.W. Stuth. 1993. How herbivory affects grazing tolerant and sensitive grasses in a central Texas grassland: integrating plant response across hierarchical levels. Oikos 67:291-298.

Bucher, H.P., F. Machler, and J. Nöesberger. 1987a. Sink control of assimilate partitioning in meadow fescue (*Festuca pratensis* Huds.). J. Plant Physiol. 129:469-477.

Bucher, H.P., F. Machler, and J. Noesberger. 1987b. Storage and remobilization of carbohydrates in meadow fescue (*Festuca pratensis* huds.). J. Plant Physiol. 130:101-109.

Buchsbaum, R., I. Valiela, and T. Swain. 1984. The role of phenolic compounds and other plant constituents in feeding by Canada geese in a coastal marsh. Oecologia 63:343-349.

Busso, C.A., R.J. Mueller, and J.H. Richards. 1989. Effects of drought and defoliation on bud viability in two caespitose grasses. Ann. Bot. 63:477-485.

Busso, C.A., J.H. Richards, and N.J. Chatterton. 1990. Nonstructural carbohydrates and spring regrowth of two cool-season grasses: Interaction of drought and clipping. J. Range Manage. 43:336-343.

Butler, J.L., and D.D. Briske. 1988. Population structure and tiller demography of the bunchgrass *Schizachyrium scoparium* in response to herbivory. Oikos 51:306-312.

Cable, D.R. 1971. Growth and development of Arizona cottontop (*Trichachne californica* [Benth.] Chase). Bot. Gaz. 132:119-145.

Caldwell, M.M. 1984. Plant requirements for prudent grazing. p. 117-152. *In*: Developing strategies for Rangeland Management. Westview Press, Boulder, Colorado.

Caldwell, M.M., J.H. Richards, D.A. Johnson, R.S. Nowak, and R.S. Dzurec. 1981. Coping with herbivory: photosynthetic capacity and resource allocation in two semiarid *Agropyron* bunchgrasses. Oecologia 50:14-24.

Caldwell, M.M., T.J. Dean, R.S. Nowak, R.S. Dzurec, and J.H. Richards. 1983. Bunchgrass architecture, light interception, and water-use efficiency: assessment by fiber optic point quadrats and gas exchange. Oecologia 59:178-184.

Caldwell, M.M., D.M. Eissenstat, J.H. Richards, and M.F. Allen. 1985. Competition for phosphorus: differential uptake from dual-isotope-labeled soil interspaces between shrub and grass. Science 229:384-386.

Caldwell, M.M., J.H. Richards, J.H. Manwaring, and D.M. Eissenstat. 1987. Rapid shifts in phosphate acquisition show direct competition between neighbouring plants. Nature 327:615-616.

Canfield, R.H. 1957. Reproduction and life span of some perennial grasses of southern Arizona. J. Range Manage. 10:199-203.

Cargill, S.M., and R.L. Jefferies. 1984a. Nutrient limitation of primary production in a sub-arctic salt marsh. J. Appl. Ecol. 21:657-668.

Cargill, S.M., and R.L. Jefferies. 1984b. The effects of grazing by lesser snow geese on the vegetation of a sub-arctic salt marsh. J. Appl. Ecol. 21:669-686.

Carlsson B.Å., and T.V. Callaghan. 1990. Programmed tiller differentiation, intraclonal density regulation and nutrient dynamics in *Carex bigelowii*. Oikos 58:219-230.

Carman, J.G., and D.D. Briske. 1985. Morphologic and allozymic variation between long-termgrazed and non-grazed populations of the bunchgrass *Schizachyrium scoparium* var. *frequens*. Oecologia 66:332-337.

Carmi, A., and D. Koller. 1979. Regulation of photosynthetic activity in the primary leaves of bean (*Phaseolus vulgaris* L.) by materials moving in the water-conducting system. Plant Physiol. 64:285-288

Casal, J.J., R.A. Sanchez, and V.A. Deregibus. 1986. The effect of plant density on tillering: the involvement of R/FR ratio and the proportion of radiation intercepted per plant. Environ. Exp. Bot. 26:365-371.

Casal, J.J., R.A. Sanchez, and V.A. Deregibus. 1987. Tillering responses of *Lolium multiflorum* plants to changes of red/far-red ratio typical of sparse canopies. J. Exp. Bot. 38:1432-1439.

Caswell, H., F. Reed, S.N. Stephenson, and P.A. Werner. 1973. Photosynthetic pathways and selective herbivory: a hypothesis. Am. Nat. 107:465-480.

Chapin III, S.F. 1980. The mineral nutrition of wild plants. Ann. Rev. Ecol. Syst. 11:233-260.

Chapin III, F.S., and G.R. Shaver. 1989. Lack of latitudinal variations in graminoid storage reserves. Ecology 70:269-272.

Chapin III, S.F., and S.J. McNaughton. 1989. Lack of compensatory growth under phosphorus deficiency in grazing-adapted grasses from the Serengeti Plains. Oecologia 79: 551-557.

Chapin III, S.F., and M. Slack. 1979. Effect of defoliation upon root growth, phosphate absorption and respiration in nutrient-limited tundra graminoids. Oecologia 42:67-79.

Chapman, D.F., D.A. Clark, C.A. Land, and N. Dymock. 1983. Leaf and tiller growth of *Lolium perenne* and *Agrostis* spp. and leaf appearance rates of *Trifolium repens* in set-stocked and rotationally grazed hill pastures. N.Z. J. Agri. Res. 26:159-168.

Chew, R.M. 1974. Consumers as regulators of ecosystems: an alternative to energetics. Ohio J. Sci. 74:359-370.

Chung, H.H., and M.J. Trlica. 1980. 14C distribution and utilization in blue grama as affected by temperature, water potential and defoliation regimes. Oecologia 47:190-195.

Clarkson, D.T., M. Shone, and A. Wood. 1974. The effect of pretreatment temperature on the exudation of xylem sap by detached barley root systems. Planta 121:81-92.

Clement, C.R., M.J. Hopper, L.H.P. Jones, and E.L. Leafe. 1978. The uptake of nitrate by *Lolium perenne* from flowing nutrient solution: II. Effect of light, defoliation, and relationship to CO_2 flux. J. Exp. Bot. 29:1173-1183.

Coffin, D.P., and W.K. Lauenroth. 1988. The effects of disturbance size and frequency on a shortgrass plant community. Ecology 69:1609-1617.

Coffin, D.P., and W.K. Lauenroth. 1989. Spatial and temporal variation in the seed bank of a semiarid grassland. Amer. J. Bot. 76:53-58.

Coley, P.D. 1983. Herbivory and defensive characteristics of tree species in a lowland tropical forest. Ecol. Mono. 53:209-233.

Coley, P.D. 1986. Cost and benefits of defense by tannins in a neotropical tree. Oecologia 70:238-241.

Coley, P.D., J.P. Bryant, and F.S. Chapin III. 1985. Resource availability and plant antiherbivore defense. Science 230:895-899.

Collins, S.L. 1987. Interaction of disturbances in tallgrass prairie: a field experiment. Ecology 68:1243-1250.

Collins, S.L., and S.C. Barber. 1985. Effects of disturbance on diversity in mixed-grass prairie. Vegetatio 64:87-94.

Cook, C. W., and L.A. Stoddart. 1953. Some growth responses of crested wheatgrass following herbage removal. J. Range Manage. 6:267-270.

Coughenour, M.B. 1985. Graminoid responses to grazing by large herbivores: adaptations, expectations, and interacting processes. Ann. Missouri Bot. Gard. 72:852-863.

Coughenour, M.B., S.J. McNaughton, and L.L. Wallace. 1985a. Responses of an African tall-grass (*Hyparrhenia filipendula* stapf.) to defoliation and limitations of water and nitrogen. Oecologia 68:80-86.

Coughenour, M.B., S.J. McNaughton, and L.L. Wallace. 1985b. Responses of an African graminoid (*Themeda triandra* Forsk.) to frequent defoliation, nitrogen and water: a limit of adaptation to herbivory. Oecologia 68:105-110.

Crawley, M. J. 1987. Herbivores and plant population dynamics. p.367-392. *In*: A.J. Davy, M.J. Hutchings, and A.R. Watkinson (eds.). Plant population ecology. Blackwell Sci. Pub., Oxford, England

Crider, F.J. 1955. Root-growth stoppage resulting from defoliation of grass. U.S.D.A. Tech. Bull. 1102.

Culvenor, R.A., I.A. Davidson, and R.J. Simpson. 1989a. Regrowth by swards of subterranean clover after defoliation. 1. Growth, non-structural carbohydrate and nitrogen content. Ann. Bot. 64:545-556.

Culvenor, R.A., I.A. Davidson, and R.J. Simpson. 1989b. Regrowth by swards of subterranean clover after defoliation. 2. Carbon exchange in shoot, root and nodule. Ann. Bot. 64:557-567.

Danckwerts, J.E., and A.J. Gordon. 1987. Long-term partitioning, storage and remobilization of ^{14}C assimilated by *Lolium perenne* (cv. Melle). Ann. Bot. 59:55-66.

Davidian, J., A. Soler, and C. Grignon. 1984. Development of H^+ extrusion by barley roots after their excision. Physiol. Veg. 22:163-170.

Davidson, J.L., and F.L. Milthorpe. 1966a. Leaf growth in *Dactylis glomerata* following defoliation. Ann. Bot. 30 (118):173-184.

Davidson, J.L., and F.L. Milthorpe. 1966b. The effects of defoliation on the carbon balance in *Dactylis glomerata*. Ann. Bot. 30:185-198.

Davies, A. 1974. Leaf tissue remaining after cutting and regrowth in perennial ryegrass. J. Agric. Sci., Camb. 82:165-172.

Davis, J.H., and C.D. Bonham. 1979. Interference of sand sagebrush canopy with needle and thread. J. Range Manage. 32:384-386.

Deregibus, V.A., M.J. Trlica, and D.A. Jameson. 1982. Organic reserves in herbage plants: Their relationship to grassland management. p. 315-344. *In*: M.J. Recheigl (ed.), Handbook of agricultural productivity. Vol. I. Plant productivity. CRC Press, Inc.

Deregibus, V.A., R.A. Sanchez, J.J. Casal, and M.J. Trlica. 1985. Tillering responses to enrichment of red light beneath the canopy in a humid natural grassland. J. Appl. Ecol. 22:199-206.

Detling, J.K., and E.L. Painter. 1983. Defoliation responses of western wheatgrass populations with diverse histories of prairie dog grazing. Oecologia 57:65-71.

Detling, J.K., M.I. Dyer, and D.T. Winn. 1979a. Effect of simulated grasshopper grazing on CO_2 exchange rates of western wheatgrass leaves. J. Econ. Ent. 72:403-406.

Detling, J.K., M.I. Dyer, and D.T. Winn. 1979b. Net photosynthesis, root respiration, and regrowth of *Bouteloua gracilis* following simulated grazing. Oecologia 41:127-134.

Dewald, C.L., and P.L. Sims. 1981. Seasonal vegetative establishment and shoot reserves of eastern gamagrass. J. Range Manage. 34:300-304.

Dirzo, R., and J.L. Harper. 1982. Experimental studies on slug-plant interaction. IV. The performance of cyanogenic and acyanogenic morphs of *Trifolium repens* in the field. J. Ecol. 70:119-138.

Dyer, M.I. 1975. The effects of red winged blackbirds (*Agelaius phoeniceus* L.) on biomass production of corn grains (*Zea mays* L.). J. Appl. Ecol. 12:719-726.

Dyer, M.I., and U.G. Bokhari. 1976. Plant-animal interactions: studies of the effects of grasshopper grazing on blue grama grass. Ecology 57:762-772.

Dyksterhuis, E.J. 1949. Condition and management of range land based on quantitative ecology. J. Range Manage. 2:104-115.

Ellison, L. 1960. Influence of grazing on plant succession of rangelands. Bot. Rev. 26:1-78.

Etherington, J.R. 1984. Relationship between morphological adaptation to grazing, carbon balance and waterlogging tolerance in clones of *Dactylis glomerata* L. New Phytol. 98:647-658.

Evans, M.W., and F.O. Grover. 1940. Developmental morphology of the growing point of the shoot and the inflorescence in grasses. J. Agric. Res. 61:481-520.

Fankhauser Jr., J.J., and J.J. Volenec. 1989. Root vs. shoot effects on herbage regrowth and carbohydrate metabolism of alfalfa. Crop Sci. 29:735-740.

Ferree, D.C., and F.R. Hall. 1981. Influence of physical stress on photosynthesis and transpiration of apple leaves. J. Amer. Soc. Hort. Sci. 106:348-351.

Field, C. 1983. Allocating leaf nitrogen for the maximization of carbon ga in: leaf age as a control on the allocation program. Oecologia 56:341-347.

Floate, M.J.S. 1981. Effects of grazing by large herbivores on nitrogen cycling in agricultural ecosystems. In: F.E. Clark and T. Rosswall (eds.) Terrestrial Nitrogen Cycles. Ecol. Bull. (Stockholm) 33:585-601.

Fowler, N.L. 1986. Density-dependent population regulation in a Texas grassland. Ecology 67:545-554.

Frank, D.A., and S.J. McNaughton. 1993. Evidence for the promotion of aboveground grassland production by native large herbivores in Yellowstone National Park. Oecologia. 96:157-161.

Gali-Muhtasib, H.U., C.C. Smith, and J.J. Higgins. 1992. The effect of silica in grasses on the feeding behavior of the prairie vole, *Microtus ochrogaster*. Ecology 73:1724-1729.

Gammon, D.M., and B.R. Roberts. 1978. Patterns of defoliation during continuous and rotational grazing of the Matopos Sandveld of Rhodesia. 1. Selectivity of grazing. Rhod. J. Agric. Res. 16:117-131.

Ganskopp D., and J. Rose. 1992. Bunchgrass basal area affects selection of plants by cattle. J. Range. Manage. 45:538-541.

Ganskopp, D., R. Angell, and J. Rose. 1992. Response of cattle to cured reproductive stems in a caespitose grass. J. Range Manage. 45:401-404.

Gatsuk, L.E., O. V. Smirnova, L.I. Vorontzova, L.B. Zaugolnova, and L.A. Zhukova. 1980. Age states of plants of various growth forms: a review. J. Ecol. 68:675-696.

Gauthier, G., and J. Bédard. 1990. The role of phenolic compounds and nutrients in determining food preference in greater snow geese. Oecologia 84:553-558.

Geiger, D.R. 1976. Effects of translocation and assimilate demand on photosynthesis. Can. J. Bot. 54:2337-2345.

Geiger, D. R., and B.R. Fondy. 1985. Responses of export and partitioning to internal and environmental factors. p. 177-194 *In*: Jeffcoat, B., A.F. Hawkins and A.D. Stead *eds*. Regulation of sources and sinks in crop plants. British Plant Growth Regulator Group Monograph No., 12, Long Ashton.

Georgiadis, N.J., R.W. Ruess, S.J. McNaughton, and D. Western. 1989. Ecological conditions that determine when grazing stimulates grass production. Oecologia 81:316-322.

Georgiadis, N.J., and S.J. McNaughton. 1988. Interactions between grazers and a cyanogenic grass, *Cynodon plectostachyus*. Oikos 51:343-350.

Gibson, D.J. 1988. The relationship of sheep grazing and soil heterogeneity to plant spatial patterns in dune grassland. J. Ecol. 76:233-252.

Gifford, R.M., and C. Marshall. 1973. Photosynthesis and assimilate distribution in *Lolium multiflorum* Lam. following differential tiller defoliation. Aust. J. Biol. Sci. 26:517-526.

Gold, W. G., and M. M. Caldwell. 1989a. The effects of the spatial pattern of defoliation on regrowth of a tussock grass. I. Growth responses. Oecologia 80:289-296.

Gold, W.G., and M.M. Caldwell. 1989b. The effects of the spatial pattern of defoliation on regrowth of a tussock grass. II. Canopy gas exchange. Oecologia 81:437-442.

Gold, W.G., and M.M. Caldwell. 1990. The effects of the spatial pattern of defoliation on regrowth of a tussock grass. III. Photosynthesis, canopy structure and light interception. Oecologia 82:12-17.

Gordon, A.J., G.J.A. Ryle, D.F. Mitchell, K.H. Lowry, and C.E. Powell. 1986. The effect of defoliation on carbohydrate, protein and leghaemoglobin content of white clover nodules. Ann. Bot. 58:141-154.

Grant, S.A., G.T. Barthram, and L. Torvell. 1981. Components of regrowth in grazed and cut *Lolium perenne* swards. Grass Forage Sci. 36:155-168.

Grant, S.A., G.T. Barthram, L. Torvell J. King, and H.K. Smith. 1983. Sward management, lamina turnover and tiller population density in continuously stocked *Lolium perenne*-dominated swards. Grass Forage Sci. 38:333-344.

Gregor, J.W., and F.W. Sansome. 1926. Experiments on the genetics of wild populations. Part I. grasses. J. Gent. 17:349-364.

Gutschick, V.P., and F.W. Wiegel. 1988. Optimizing the canopy photosynthetic rate by patterns of investment in specific leaf mass. Am. Nat. 132:67-86.

Hall, F.R., and D.C. Ferree. 1976. Effects of insect injury simulation on photosynthesis of apple leaves. J. Econ. Entom. 69:245-248.

Hansen, G.K., and C.R. Jensen. 1977. Growth and maintenance respiration in whole plants, tops, and roots of *Lolium multiflorum*. Physiol. Plant. 39:155-164.

Hardy, R.W.F., and U.D. Havelka. 1976. Photosynthate as a major factor limiting nitrogen fixation by field-grown legumes with emphasis on soybeans. p. 421-439. *In*: P.S. Nutman (ed.) Symbiotic nitrogen fixation in plants, International Biological Programme. Vol. 7. Cambridge Univ. Press.

Harper, J.L. 1977. Population biology of plants. Academic Press, London.

Harrison, M.A., and P.B. Kaufman. 1982. Does ethylene play a role in the release of lateral buds (tillers) from apical dominance in oats? Plant Physiol. 70:811-814.

Hartnett, D.C. 1989. Density- and growth stage-dependent responses to defoliation in two rhizomatous grasses. Oecologia 80:414-420.

Hartt, C.E., H.P. Kortschak, and G.O. Burr. 1964. Effects of defoliation, deradication, and darkening the blade upon translocation of C^{14} in sugarcane. Plant Physiol. 39:15-22.

Hartwig, U., B.C. Boller, and J. Nösberger. 1987. Oxygen supply limits nitrogenase activity of clover nodules after defoliation. Ann. Bot. 59:285-291.

Hartwig, U., B.C. Boller, B. Baur-Höch, and J. Nösberger. 1990. The influence of carbohydrate reserves on the response of nodulated clover to defoliation. Ann. Bot. 65:97-105.

Hawkins, C.D.B., M.J. Aston, and M.I. Whitecross. 1987. Short-term effects of aphid feeding on photosynthetic CO_2 exchange and dark respiration in legume leaves. Physiol. Plant. 71:379-383.

Heichel, G.H., and N.C. Turner. 1983. CO_2 assimilation of primary and regrowth foliage of red maple (*Acer rubrum* L.) and red oak (*Quercus rubra* L.): response to defoliation. Oecologia 57:14-19.

Heidorn, T., and A. Joern. 1984. Differential herbivory on C_3 versus C_4 grasses by the grasshopper *Ageneotettix deorum* (*Orthoptera*: *acrididae*). Oecologia 65:19-25.

Heitschmidt, R.K., D.D. Briske, and D.L. Price. 1990. Pattern of interspecific tiller defoliation in a mixed-grass prairie grazed by cattle. Grass Forage Sci. 45:215-222.

Heitschmidt, R.K., D.L. Price, R.A. Gordon, and J.R. Frasure. 1982. Short duration grazing at the Texas Experimental Ranch: effects on above ground primary production and seasonal growth dynamics. J. Range Manage. 35:367-372.

Herms, D.A., and W.J. Mattson. 1992. The dilemma of plants: to grow or defend. Q. Rev. Biol. 67:283-335.

Hendershot, K.L., and J.J. Volenec. 1993. Nitrogen pools in taproots of *Medicago sativa L.* after defoliation. J. Plant Physiol. 141:129-135.

Herrera, C.M. 1982. Grasses, grazers, mutualism, and coevolution: a comment. Oikos 38:254-259.

Hickey, W.C. Jr. 1961. Growth form of crested wheatgrass as affected by site and grazing. Ecology 42:173-176.

Hik D.S., and R.L. Jefferies. 1990. Increases in the net above-ground primary production of a salt-marsh forage grass: a test of the predictions of the herbivore-optimization model. J. Ecol. 78:180-195.

Hik, D.S., H..A. Sadul, and R.L. Jefferies. 1991. Effects of the timing of multiple grazings by geese on net above-ground primary production of swards of *Puccinellia phryganodes*. J. Ecol. 79:715-730.

Hilbert, D.W., D.M. Swift, J.K Detling, and M.I. Dyer. 1981. Relative growth rates and the grazing optimization hypothesis. Oecologia 51:14-18.

Hillman, J.R. 1984. Apical dominance. p. 127-148. *In*: M.B. Wilkins (ed.) Advanced plant physiology. Pitman Publ. London.

Hodgkinson, K.C., N.G. Smith, and G.E. Miles. 1972. The photosynthetic capacity of stubble leaves and their contribution to growth of the lucerne plant after high level cutting. Aust. J. Agri. Res. 23:225-238.

Hodgkinson, K.C. 1974. Influence of partial defoliation on photosynthesis, photorespiration and transpiration by lucerne leaves of different ages. Aust. J. Plant Physiol. 1:561-578.

Hodgkinson, K.C. M.M. Ludlow, J.J. Mott, Z. Baruch. 1989. Comparative responses of the savanna grasses *Cenchrus ciliaris* and *Themeda triandra* to defoliation. Oecologia 79:45-52.

Holland, E.A., and J.K. Detling. 1990. Plant response to herbivory and belowground nitrogen cycling. Ecology 71:1040-1049.

Holland, E.A., W.J. Parton, J.K. Detling, and D.L. Coppock. 1992. Physiological responses of plant populations to herbivory and their consequences for ecosystem nutrient flow. Am. Nat. 140:685-706.

Hull, R.J. 1987. Kentucky bluegrass photosynthate partitioning following scheduled mowing. J. Amer. Soc. Hort. Sci. 112:829-834.

Hyder, D.N. 1972. Defoliation in relation to vegetative growth. p. 302-317. *In*: V.B. Youngner, and C.M. McKell (eds.) The biology and utilization of grasses. Academic Press, New York.

Hyder, D.N. 1974. Morphogenesis and management of perennial grasses in the U.S. p. 89-98. *In*: Plant morphogenesis as the basis for scientific management of range resources. USDA Misc. Publ. 1271.

Jaksic, F.M., and E.R. Fuentes. 1980. Why are native herbs in the Chilean Matorral more abundant beneath bushes: microclimate or grazing? J. Ecol. 68:665-669.

Jameson, D.A. 1963. Responses of individual plants to harvesting. Bot. Rev. 29:532-594.

Jameson, D.A., and D.L. Huss. 1959. The effect of clipping leaves and stems on number of tillers, herbage weights, root weights, and food reserves of little bluestem. J. Range Manage. 12:122-126.

Jaramillo, V.J., and J.K. Detling. 1988. Grazing history, defoliation, and competition: effects on shortgrass production and nitrogen accumulation. Ecology 69:1599-1608.

Jarman, P.J., and A.R.E. Sinclair. 1979. Feeding strategy and the pattern of resource-partitioning in ungulates. p. 130-163. *In*: A.R.E. Sinclair and M. Norton-Griffiths (eds.). Sergengeti dynamics of an ecosystem. Univ. Chicago Press.

Jenkins, G.I., and H.W. Woolhouse. 1981. Photosynthetic electron transport during senescence of the primary leaves of *Phaseolus vulgaris* L. J. Exp. Bot. 32:467-478.

Jewiss, O. R. 1972. Tillering in grasses—its significance and control. J. Br. Grassld. Soc. 27:65-82.

Johnson, R.G., and R.C. Anderson. 1986. The seed bank of a tallgrass prairie in Illinois. Am. Midl. Nat. 115:123-30.

Jones, M.B., B. Collett, and S. Brown. 1982. Sward growth under cutting and continuous stocking managements: sward canopy structure, tiller density and leaf turnover. Grass Forage Sci. 37:67-73.

Jónsdóttir, G. Á. 1991. Tiller demography in seashore populations of *Agrostis stolonifera, Festuca rubra* and *Poa irrigata*. J. Veg. Sci. 2:89-94.

Jónsdóttir, I.S., and T.V. Callaghan. 1989. Localized defoliation stress and the movement of ^{14}C - photoassimilates between tillers of *Carex bigelowii*. Oikos 54:211-219.

Kasperbauer, M.J., and D.L. Karlen. 1986. Light-mediated bioregulation of tillering and photosynthate partitioning in wheat. Physiol. Plant. 66:159-163.

Kays S., and J.L Harper. 1974. The regulation of plant and tiller density in a grass sward. J. Ecol. 62:97-105

Kemp, W.B. 1937. Natural selection within plant species as exemplified in a permanent pasture. J. Hered. 28:329-333.

King, R.W., I.F. Wardlaw, and L.T. Evans. 1967. Effect of assimilate utilization on photosynthetic rate in wheat. Planta 77:261-276.

King, J., E.M. Sim, and S.A. Grant. 1984. Photosynthetic rate and carbon balance of grazed ryegrass pastures. Grass Forage Sci. 39:81-92.

King, J., E.M. Sim, G.T. Barthram, S.A. Grant and L. Torvell. 1988. Photosynthetic potential of ryegrass pastures when released from continuous stocking management. Grass Forage Sci. 43:41-48.

Kinucan, R.J. 1987. Influence of soil seed bank, seed rain, inhibition competition and site disturbance on successional processes within three long-term grazing regimes on the Edwards Plateau, Texas. Ph.D. Diss. Texas A&M Univ.

Kotanen, P., and R.L. Jefferies. 1987. The leaf and shoot demography of grazed and ungrazed plants of *Carex subspathacea*. J. Ecol. 75:961-975.

Krueger, K. 1986. Feeding relationships among bison, pronghorn, and prairie dogs: an experimental analysis. Ecology 67:760-770.

Langer, R.H.M. 1956. Growth and nutrition of timothy (*Phleum pratense*). I. The life history of individual tillers. Ann. Appl. Biol. 44:166-187.

Langer, R.H.M. 1959. A study of growth in swards of timothy and meadow fescue. II. The effects of cutting treatments. J. Agric. Sci. 52:273-281.

Langer, R.H.M. 1963. Tillering in herbage grasses. Herb. Abst. 33:141-148.

Langer, R.H.M. 1972. How Grasses Grow. Edward Arnold Ltd., London.

Langer R.H.M., S.M. Ryle, and O.R. Jewiss. 1964. The changing plant and tiller populations of timothy and meadow fescue swards. I. Plant survival and the pattern of tillering. J. Appl. Ecol. 1:197-208

Lapham, J., and D.S.H. Drennan. 1987. Intraspecific regulation of populations of the clonal herb, *Cyperus esculentus*. J. Appl. Ecol. 24:1011-1024.

Leopold, A.C. 1949. The control of tillering in grasses by auxin. Amer. J. Bot. 36:437-440.

Li, Y.S., R.E. Redmann, and C. van Kessel. 1992. Nitrogen budget and ^{15}N translocation in a perennial wheatgrass. Funct. Ecol. 6:221-225.

Liddle, M.J., C.S.J. Budd, and M.J. Hutchings. 1982. Population dynamics and neighborhood effects in establishing swards of *Festuca rubra*. Oikos 38:52-59.

Lovett-Doust L. 1981. Intraclonal variation and competition in *Ranunculus repens*. New Phytol. 89:495-502.

Lowman, M.D. 1984. An assessment of techniques for measuring herbivory: Is rainforest defoliation more intense than we thought? Biotropica 16:264-268.

Ludlow, M.M, and D.A. Charles-Edwards. 1980. Analysis of the regrowth of a tropical grass/legume sward subjected to different frequencies and intensities of defoliation. Aust. J. Agric. Res. 31:673-692.

Macduff, J.H., S.C. Jarvis, and A. Mosquera. 1989. Nitrate nutrition of grasses from steady-state supplies in flowing solution culture following nitrate deprivation and/or defoliation: II. Assimilation of NO^{-}_3 and short-term effects on NO^{-}_3 uptake. J. Exp. Bot. 40:977-984.

Macduff, J.H., and S.B. Jackson. 1992. Influx and efflux of nitrate and ammonium in Italian ryegrass and white clover roots: comparisons between effects of darkness and defoiation. J. Exp. Bot. 43:525-535.

Macnicol, P.K. 1976. Rapid metabolic changes in the wounding response of leaf discs following excision. Plant Physiol. 57:80-84.

Mahmoud, A., J.P. Grime, and S.B. Furness, S.B. 1975. Polymorphism in *Arrhenatherum elatius* (L.) Beauv. Ex J. & C. Presl. New Phytol. 75:269-276.

Marshall, C., and G.R. Sagar. 1965. The influence of defoliation on the distribution of assimilates in *Lolium multiflorum* Lam. Ann. Bot. 29:365-370.

Marshall, C., and G.R. Sagar. 1968. The distribution of assimilates in *Lolium multiflorum* Lam. following differential defoliation. Ann., Bot. 32:715-719.

Marten, G.C., R.F. Barnes, A.B. Simons, and F.J. Wooding. 1973. Alkaloids and palatability of *Phalaris arundinacea* L. grown in diverse environments. Agron. J. 65:199-201.

Marten, G.C., R.M. Jordan, and A.W. Hovin. 1976. Biological significance of reed canary-grass alkaloids and associated palatability variation to grazing sheep and cattle. Agron. J. 68:909-914.

Maschinski, J., and T.G. Whitham. 1989. The continuum of plant responses to herbivory-the influence of plant association, nutrient availability, and timing. Am. Nat. 134:1-19.

Massimino, D., M. André, C. Richaud, A. Daguenet, J. Massimino and J. Vivoli. 1981. The effect of a day at low irradiance of a maize crop. I. Root respiration and uptake of N, P and K. Physiol. Plant. 51:150-155.

Matches, A.G. 1966. Influence of intact tillers and height of stubble on growth responses of tall fescue (*Festuca arundinacea* Schreb.). Crop Sci. 6:484-487.

Mattheis, P.J., L.L. Tieszen, and M.C. Lewis. 1976. Responses of *Dupontia fischeri* to simulated lemming grazing in an Alaskan arctic tundra. Ann Bot. 40:179-197

Mattson, W.J., and N.D. Addy. 1975. Phytophagous insects as regulators of forest primary production. Science 190:515-522.

McKendrick, J.D., C.E. Owensby, and R.M. Hyde. 1975. Big bluestem and indiangrass vegetative reproduction and annual reserve carbohydrate and nitrogen cycles. Agro-Ecosystems 2:75-93.

McNaughton, S.J. 1976. Serengeti migratory wildebeest: facilitation of energy flow by grazing. Science 191:92-94.

McNaughton, S.J. 1978. Serengeti ungulates: feeding selectivity influences the effectiveness of plant defense guilds. Science 199:806-807.

McNaughton, S.J. 1979. Grazing as an optimization process: grass-ungulate relationships in the Serengeti. Am. Nat. 113:691-703.

McNaughton, S.J. 1983. Compensatory plant growth as a response to herbivory. Oikos 40:329-336.

McNaughton, S.J. 1985. Ecology of a grazing ecosystem: the Serengeti. Ecol. Monogr. 55:259-294.

McNaughton, S.J. 1993. Grasses and grazers, science and management. Ecol. Appl. 3:17-20.

McNaughton, S.J., and F.S. Chapin III. 1985. Effects of phosphorus nutrition and defoliation on C_4 graminoids from the Serengeti Plains. Ecology 66:1617-1629.

McNaughton, S.J., J.L. Tarrants, M.M. McNaughton, and R.H. Davis. 1985. Silica as a defense against herbivory and a growth promotor in African grasses. Ecology 66:528-535.

McNaughton, S.J., L.L. Wallace, and M.B. Coughenour. 1983. Plant adaptation in an ecosystem context: effects of defoliation, nitrogen, and water on growth of an African C_4 sedge. Ecology 64:307-318.

McNeilly, T. 1981. Ecotypic differentiation in *Poa annua*: interpopulation differences in response to competition and cutting. New Phytol. 88:539-547.

McNeilly, T. 1984. Ecotypic differentiation in *Poa annua*: within population variation in response to competition and cutting. New Phytol. 96:307-316.

McNeilly, T., and M.L. Roose. 1984. The distribution of perennial ryegrass genotypes in swards. New Phytol. 98:503-513.

Milchunas, D.G., and W.K. Lauenroth. 1993. Quantitative effects of grazing on vegetation and soils over a global range of environments. Ecol. Monogr. 63:327-366.

Millard, P., R.J. Thomas, and S.T. Buckland. 1990. Nitrogen supply affects the remobilization of nitrogen for the regrowth of defoliation *Lolium perenne* L. J. Exp. Bot. 41:941-947.

Miller, R.F., and J.A. Rose. 1992. Growth and carbon allocation of *Agropyron desertorum* following autumn defoliation. Oecologia 89:482-486.

Mitchell, K.J. 1953. Influence of light and temperature on the growth of ryegrass (*Lolium* spp.) II. The control of lateral bud development. Physiol. Plant. 6:425-43.

Moloney, K.A. 1988. Fine-scale spatial and temporal variation in the demography of a perennial bunchgrass. Ecology 69:1588-1598.

Moore, D. 1984. The role of silica in protecting Italian ryegrass (*Lolium multiflorum*) from attack by dipterous stem-boring larvae (*Oscinella frit* and other related species). Ann. Appl. Biol. 104:161-166.

Mott, J.J., M.M. Ludlow, J.H. Richards, and A.D. Parsons. 1992. Effects of moisture supply in the dry season and subsequent defoliation on persistence of the savanna grasses *Themeda triandra*, *Heteropogon contortus* and *Panicum maximum*. Aust. J. Agric. Res. 43:241-260.

Mueggler, W.F. 1972. Influence of competition on the response of bluebunch wheatgrass to clipping. J. Range Manage. 25:88-92.

Mueller, R.J., and J.H. Richards. 1986. Morphological analysis of tillering in *Agropyron spicatum* and *Agropyron desertorum*. Ann. Bot. 58:911-921.

Muldoon, D.K., and C.J. Pearson. 1979. Morphology and physiology of re-growth of the tropical tallgrass hybrid *Pennisetum*. Ann. Bot. 43:719-728.

Murphy, J.S., and D.D. Briske. 1992. Regulation of tillering by apical dominance: Chronology, interpretive value, and current perspectives. J. Range Manage. 45:419-429.

Murphy, J.S., and D.D. Briske. 1994. Density-dependent regulation of ramet recruitment by the red:far-red ratio of solar radiation: A field evaluation with the bunchgrass *Schizachyrium scoparium*. Oecologia 97:462-469.

Naeth, M.A., D.S. Chanasyk, R.L. Rothwell, and A.W. Bailey. 1991. Grazing impacts on soil water in mixed prairie and fescue grassland ecosystems of Alberta. Can. J. Soil Sci. 71:313-325.

Neales, T.F., K.J. Treharne, and P.F. Wareing. 1971. A relationship between net photosynthesis, diffusive resistance, and carboxylating enzyme activity in bean leaves. p. 89-97. *In*: M.D. Hatch, C.B. Osmond, and R.O. Slatyer (eds.) Photosynthesis and Photorespiration. New York:Wiley-Interscience.

Neiland, B.M., and J.T. Curtis. 1956. Differential responses to clipping of six prairie grasses in Wisconsin. Ecology 37:355-365.

Noble, J.C., A.D. Bell, and J.L. Harper. 1979. The population biology of plants with clonal growth I. The morphology and structural demography of *Carex arenaria*. J. Ecol. 67:983-1008.

Nojima, H., S. Fuketa, A. Isoda, and Y. Takasaki. 1989. Levels of abscisic acid in lateral buds during regrowth of *Sorghum bicolor* M. p. 505-506. *In*: Proc. XVI Int. Grassld. Cong., Nice, France.

Norton, B.E., and P.S. Johnson. 1983. Pattern of defoliation by cattle grazing crested wheatgrass pastures. p. 462-464. *In*: J.A. Smith, and V.W. Hayes (eds.) Proc. XIV Int. Grassld. Cong. Westview Press, Boulder, Colorado.

Nowak, R.S., and M.M. Caldwell. 1984. A test of compensatory photosynthesis in the field: implications for herbivory tolerance. Oecologia 61:311-318.

O'Connor, T.G. 1991. Local extinction in perennial grasslands: a life-history approach. Am. Nat. 137:753-773.

Oesterheld, M. 1992. Effect of defoliation intensity on aboveground and belowground relative growth rates. Oecologia. 92:313-316.

Oesterheld, M., O.E. Sala, and S.J. McNaughton. 1992. Effects of animal husbandry on herbivore carrying capacity at a regional scale. Nature 356:234-236.

Oesterheld M., and S.J. McNaughton. 1988. Intraspecific variation in the response of *Themeda triandra* to defoliation: the effect of time of recovery and growth rates on compensatory growth. Oecologia 77:181-186.

Olson, B.E., and J.H. Richards. 1988a. Spatial arrangement of tiller replacement in *Agropyron desertorum* following grazing. Oecologia 76:7-10.

Olson, B.E., and J.H. Richards. 1988b. Annual replacement of the tillers of *Agropyron desertorum* following grazing. Oecologia 76:1-6.

Olson, B.E., and J.H. Richards. 1988c. Tussock regrowth after grazing: intercalary meristem and axillary bud activity of *Agropyron desertorum*. Oikos 51:374-382.

Olson, B.E., and J.H. Richards. 1989. Crested wheatgrass growth and replacement following fertilization, thinning and neighbor removal. J. Range Manage. 42:93-97.

Olson, B.E., R.L. Senft, and J.H. Richards. 1989. A test of grazing compensation and optimization of crested wheatgrass using a simulation model. J. Range Manage. 42 :458-467.

Ong, C.K. 1978. The physiology of tiller death in grasses. 1. The influence of tiller age, size and position. J. Brit. Grassld. Soc. 33:197-203.

Oswalt, D.L., A.R. Bertrand, and M.R. Teel. 1959. Influence of nitrogen fertilization and clipping on grass roots. Soil Sci. Soc. Proc. 23:228-230.

Ourry, A., J. Boucaud, and J. Salette. 1988. Nitrogen mobilization from stubble and roots during re-growth of defoliated perennial ryegrass. J. Exp. Bot. 39:803-809.

Ourry, A., J. Bigot, and J. Boucaud. 1989. Protein mobilization from stubble and roots, and proteolytic activities during post-clipping re-growth of perennial ryegrass. J. Plant Physiol. 134:298-303.

Ourry, A., J. Boucaud, and J. Salette. 1990. Partitioning and remobilization of nitrogen during regrowth in nitrogen-deficient ryegrass. Crop Sci. 30:1251-1254.

Owen, D.F., and R.G. Wiegert. 1976. Do consumers maximize plant fitness? Oikos 27:488-492.

Paige, K.N., and T.G. Whitham. 1987. Overcompensation in response to mammalian herbivory: the advantage of being eaten. Am. Nat. 129:407-416.

Painter, E.L., and A.J. Belsky. 1993. Application of herbivore optimization theory to rangelands of the western United States. Ecol. Appl. 3:2-9.

Painter, E.L., and J.K Detling. 1981. Effects of defoliation on net photosynthesis and regrowth of western wheatgrass. J. Range Manage. 34:68-71.

Painter, E.L., J.K. Detling, and D.A. Steingraeber. 1989. Grazing history, defoliation, and frequency-dependent competition: effects on two North American grasses. Amer. J. Bot. 76:1368-1379.

Painter, E.L., J.K. Detling, and D.A. Steingraeber. 1993. Plant morphology and grazing history: Relationships between native grasses and herbivores. Vegetatio 106:37-62.

Parsons, A.J., E.L. Leafe, B. Collett, and W. Stiles. 1983a. The physiology of grass production under grazing. I. Characteristics of leaf and canopy photosynthesis of continuously-grazed swards. J. Appl. Ecol. 20:117-126.

Parsons, A.J., E.L. Leafe, B. Collett, P.D. Penning, and J. Lewis. 1983b. The physiology of grass production under grazing. II. Photosynthesis, crop growth and animal intake of continuously-grazed swards. J. Appl. Ecol. 20:127-139.

Parsons, A.J., B. Collett, and J. Lewis. 1984. Changes in the structure and physiology of perennial ryegrass sward when released from a continuous stocking management: implications for the use of exclusion cages in continuously stocked swards. Grass Forage Sci. 39:1-9.

Peterson, R.A. 1962. Factors affecting resistance to heavy grazing in needle-and-thread grass. J. Range Manage. 15:183-189.

Phillips, I.D.J. 1975. Apical dominance. Ann. Rev. Plant Physiol. 26:341-367.

Pimentel, D. 1988. Herbivore population feeding pressure on plant hosts: feedback evolution and host conservation. Oikos 53:289-302.

Polley, H.W., and J.K. Detling. 1988. Herbivory tolerance of *Agropyron smithii* populations with different grazing histories. Oecologia 77:261-267.

Pond, F.W. 1960. Vigor of Idaho Fescue in relation to different grazing intensities. J. Range Manage. 13:28-30.

Poston, F.L., L.P. Pedigo, R.B. Pearce, and R.B. Hammond. 1976. Effects of artificial and insect defoliation on soybean net photosynthesis. J. Econ. Ent. 69:109-112.

Pyke, D.A. 1990. Comparative demography of co-occurring introduced and native tussock grasses: persistence and potential expansion. Oecologia 82:537-543.

Rabinowitz, D. 1981. Buried viable seeds in a North American tallgrass prairie: the resemblance of their abundance and composition to dispersing seeds. Oikos 36:191-195.

Ratliff, R.D., and S.E. Westfall. 1992. Nebraska sedge (*Carex nebraskensis* Dewey): Phenology and life history at Tule Meadow, Sierra National Forest, California. J. Range. Manage. 45:210-215.

Reidenbaugh, T.G. 1983. Tillering and mortality of the salt marsh cordgrass, *Spartina alterniflora*. Amer. J. Bot. 70:47-52.

Richards, J.H. 1984. Root growth response to defoliation in two *Agropyron* bunchgrasses: field observations with an improved root periscope. Oecologia 64:21-25.

Richards, J.H., and M.M. Caldwell. 1985. Soluble carbohydrates, concurrent photosynthesis and efficiency in regrowth following defoliation: A field study with *Agropyron* species. J. Appl. Ecol. 22:907-920.

Richards, J.H., R.J. Mueller, and J.J. Mott. 1988. Tillering in tussock grasses in relation to defoliation and apical bud removal. Ann. Bot. 62:173-179.

Robberecht R., B.E. Mahall, and P.S. Nobel. 1983. Experimental removal of intraspecific competitors—effects on water relations and productivity of a desert bunchgrass, *Hilaria rigida*. Oecologia 60:21-24.

Robson, M.J. 1968. The changing tiller population of spaced plants of S.170 tall fescue (*Festuca arundinacea*). J. Appl. Ecol. 5:575-590.

Ruess, R.W., and S.J. McNaughton. 1987. Grazing and the dynamics of nutrient and energy regulated microbial processes in the Serengeti grasslands. Oikos 49:101-110.

Ruess, R.W., S.J. McNaughton, and M.B. Coughenour. 1983. The effects of clipping, nitrogen source and nitrogen concentration on the growth responses and nitrogen uptake of an east african sedge. Oecologia 59:253-261.

Ryle, G.J., and C.E. Powell. 1975. Defoliation and regrowth in the graminaceous plant: the role of current assimilate. Ann. Bot. 39:297-310.

Ryle, G.J.A., C.E. Powell, and A.J. Gordon. 1985. Short-term changes in CO_2 evolution associated with nitrogenase activity in white clover in response to defoliation and photosynthesis. J. Exp. Bot. 36:634-643.

Saglio, P.H., and A. Pradet. 1980. Soluble sugars, respiration, and energy charge during aging of excised maize root tips. Plant Physiol. 66:516-519.

Sala, O.E., M. Oesterheld, R.J.C. Leon, and A. Soriano. 1986. Grazing effects upon plant community structure in subhumid grasslands of Argentina. Vegetatio 67:27-32.

Salihi, D.O., and B.E. Norton. 1987. Survival of perennial grass seedlings under intensive management of semiarid rangelands. J. Appl. Ecol. 24:145-153.

Sambo, E.Y. 1983. Leaf extension rate in temperate pasture grasses in relation to assimilate pool in the extension zone. J. Exp. Bot. 34:1281-1290.

Sampson, A.W. 1914. Natural revegetation of rangelands based upon growth requirements and life history of the vegetation. J. Agric. Res. 3:93-147.

Satoh, M., P.E. Kriedemann, and B.R. Loveys. 1977. Changes in photosynthetic activity and related processes following decapitation in mulberry trees. Physiol. Plant. 41:203-210.

Schnyder, H., and C.J. Nelson. 1989. Growth rates and assimilate partitioning in the elongation zone of tall fescue leaf blades at high and low irradiance. Plant Physiol. 90:1201-1206.

Scott, A.W., and R.D.B. Whalley. 1984. The influence of intensive sheep grazing on genotypic differentiation in *Danthonia linkii. D. richardsonii* and *D.racemosa* on the New England Tablelands. Aust. J. Ecol. 9:419-429.

Seastedt, T.R., D.A. Crossley, Jr., and W.W. Hargrove. 1983. The effects of low-level consumption by canopy arthropods on the growth and nutrient dynamics of black locust and red maple trees in the southern Appalachians. Ecology 64:1040-1048.

Senock, R.S., W.B. Sisson, and G.B. Donart. 1991. Compensatory photosynthesis of *Sporobolus flexuosus* (Thurb.) rydb. following simulated herbivory in the northern Chihuahuan desert. Bot. Gaz. 152:275-281.

Sharman, B.C. 1942. Developmental anatomy of the shoot of *Zea mays* L. Ann. Bot. 6:245-282.

Sharman, B.C. 1945. Leaf and bud initiation in the Gramineae. Bot. Gaz. 106:269-289.

Sharman, B.C. 1947. The biology and developmental morphology of the shoot apex in the Gramineae. New Phytol. 46:20-34.

Shewmaker, G.E., H.F. Mayland, R.C. Rosenau, and K.H. Asay. 1989. Silicon in C-3 grasses: effects on forage quality and sheep preference. J. Range Manage. 42:122-127.

Silvertown, J.W., and J. Lovett Doust. 1993. Introduction to plant population biology. Blackwell Sci. Pub., Oxford.

Simms, E.L., and M.D. Rausher. 1987. Costs and benefits of plant resistance to herbivory. Am. Nat. 130:570-581.

Simms, E.L., and M.D. Rausher. 1989. The evolution of resistance to herbivory in *Ipomoea purpurea*. II. Natural selection by insects and costs of resistance. Evolution 43:573-585.

Simon, J.C., and G. Lemaire. 1987. Tillering and leaf area index in grasses in the vegetative phase. Grass Forage Sci. 72:373-380.

Sims, P.L., L.J. Ayuko, and D.N. Hyder. 1971. Developmental morphology of switchgrass and sideoats grama. J. Range Manage. 24:357-360.

Skinner, R.H., and S.R. Simmons. 1993. Modulation of leaf elongation, tiller appearance and tiller senescence in spring barley by far-red light. Plant, Cell, Environ. 16:555-562.

Skogsmyr I., and T. Fagerström. 1992. The cost of anti-herbivory defence: an evaluation of some ecological and physiological factors. Oikos 64:451-457.

Smith, A.E., and C.L. Leinweber. 1973. Incorporation of ^{14}C by little bluestem tillers at two stages of phenological development. Agron. J. 65:908-910.

Smith, H., and D.C. Morgan. 1983. The function of phytochrome in nature. p. 491-517. *In*: W. Shropshire and H. Mohr (eds.) Photomorphogenesis, Encycl. Plant Physiol. NS16B Springer-Verlag, New York.

Soriano, A. R.A. Golluscio, and E. Satorre. 1987. Spatial heterogeneity of the root system of grasses in the patagonian arid steppe. Bull. Torrey Bot. Club 114:103-108.

Stenseth, N.C. 1978. Do grazers maximize individual plant fitness? Oikos 31:299-306.

Stout, D.G., J. Hall, B. Brooke, and A. McLean. 1981. Influence of successive years of simulated grazing (clipping) of pinegrass growth. Can. J. Plant Sci. 61:939-943.

Stubbendieck J., and D.F. Burzlaff. 1971. Nature of phytomer growth in blue grama. J. Range Manage. 24:154-156.

Stur, W.W., and L.R. Humphreys. 1988. Defoliation and burning effects on the tillering of *Brachiaria decumbens*. J. Appl. Ecol. 25:273-277.

Svejcar, T., and S. Christiansen. 1987. Grazing effects on water relations of Caucasian bluestem. J. Range Manage. 40:15-18.

Sweet, G.B., and P.F. Wareing. 1966. Role of plant growth in regulating photosynthesis. Nature 210:77-79.

Tainton, N.M., and P. DE V. Booysen. 1965. Growth and development in perennial veld grasses. I. *Themeda triandra* tillers under various systems of defoliation. S. Afr. J. Agric. Sci. 8:93-110.

Tallowin, J.R.B., J.H.H. Williams, and F.W. Kirkham. 1989. Some consequences of imposing different continuous-grazing pressures in the spring on tiller demography and leaf growth. J. Agric. Sci., Camb. 112:115-122.

Thimann, K.V., and F. Skoog. 1933. Studies on the growth hormone of plants. 3. The inhibiting action of the growth substance on bud development. Proc. Nat. Acad. Sci. USA 19:714-716.

Thimann, K.V., and F. Skoog. 1934. On the inhibition of bud development and other functions of growth substances in *Vicia faba*. Proc. Roy. Soc. Lond. B114:317-339.

Thompson, K., and J.P. Grime. 1979. Seasonal variation in the seed banks of herbaceous species in ten contrasting habitats. J. Ecol. 67:893-921.

Thorgeirsson, H. 1988. The modeling and measurement of respiratory carbon use and net carbon gain of two Agropyron bunchgrasses. Ph.D. Dissertation, Utah State University, Logan.

Toft, N.L., S.J. McNaughton, and N.J. Georgiadis. 1987. Effects of water stress and simulated grazing on leaf elongation and water relations of an east African grass, *Eustachys paspaloides*. Aust. J. Plant Physiol. 14:211-226.

Troughton, A. 1957. The underground organs of herbage grasses. Commonwealth Bureau of Pastures and Field Crops. Hurley, Bul. 44. pp.163.

Tripathi, R.S., and J.L. Harper. 1973. The comparative biology of *Agropyron repens* (L.) Beauv. and *A. caninum* (L.) Beauv. I. The growth of mixed populations established from tillers and from seeds. J. Ecol. 61:353-368.

Truscott, D.R., and P.O. Currie. 1989. Cattle preferences for a hybrid grass: chemical and morphological relationships. J. Range Manage. 42:22-27.

Tucker, D.J. 1977. Hormonal regulation of lateral bud outgrowth in the tomato. Plant Sci. Lett. 8:105-111.

Tucker, D.J. 1978. Apical dominance in the tomato: the possible roles of auxin and abscisic acid. Plant Sci. Lett. 12:273-278.

Vance, C.P., G.H. Heichel, D.K. Barnes, J.W. Bryan, and L.E. Johnson. 1979. Nitrogen fixation, nodule development, and vegetative regrowth of alfalfa (*Medicago sativa* L.) following harvest. Plant Physiol. 64:1-8.

van der Meijden, E., M. Wijn, and H.J.Verkaar. 1988. Defence and regrowth, alternative plant strategies in the struggle against herbivores. Oikos 51:355-363.

Verkaar, H. J. 1987. Population dynamics - the influence of herbivory. New Phytol. 106:49-60.

Vine, D.A. 1983. Sward structure changes within a perennial ryegrass sward: leaf appearance and death. Grass Forage Sci. 38:231-242.

Vinton, M.A., and D.C. Hartnett. 1992. Effects of bison grazing on *Andropogon gerardii* and *Panicum virgatum* in burned and unburned tallgrass prairie. Oecologia 90:374-382.

Vogel, W.G., and A.J. Bjugstad. 1968. Effects of clipping on yield and tillering of little bluestem, big bluestem, and Indiangrass. J. Range Manage. 21:136-140.

Volenec, J.J., and C.J. Nelson. 1983. Responses of tall fescue leaf meristems to N fertilization and harvest frequency. Crop Sci. 23:720-724.

Volenec J. J., and C.J. Nelson. 1984. Carbohydrate metabolism in leaf meristems of tall fescue. II. Relationship to leaf elongation rates modified by nitrogen fertilization. Plant Physiol. 74:595-600.

von Caemmerer, S., and G.D. Farquhar. 1984. Effects of partial defoliation, changes in irradiance during growth, short-term water stress and growth at enhanced p(CO_2) on the photosynthetic capacity of leaves of *Phaseolus vulgaris* L. Planta 160:320-329.

Vorontzova, L.I., and L.B. Zaugolnova. 1985. Population biology of steppe plants. p. 143-178. *In*: J. White (ed). The population structure of vegetation. Dr W. Junk Publishers, Dordrecht.

Wadham, M.D., and D.W. Parry. 1981. The silicon content of *Oryza sativa* L. and its effect on the grazing behaviour of *Agriolimax reticulatus* Muller. Ann. Bot. 48:399-402.

Wallace, L.L. 1981. Growth, morphology and gas exchange on mycorrhizal and non-mycorrhizal *Panicum coloratum* L., a C_4 grass species, under different clipping and fertilization regimes. Oecologia 49:272-278.

Wallace, L.L. 1990. Comparative photosynthetic responses of big bluestem to clipping versus grazing. J. Range Manage. 43:58-61.

Wallace, L.L., and S.A. Macko. 1993. Nutrient acquisition by clipped plants as a measure of competitive success: the effects of compensation. Funct. Ecol. 7:326-331.

Wallace, L.L., S.J. McNaughton, and M.B. Coughenour. 1984. Compensatory photosynthetic responses of three African graminoids to different fertilization, watering, and clipping regimes. Bot. Gaz. 145:151-156.

Wallace, L.L., S.J. McNaughton, and M.B. Coughenour. 1985. Effects of clipping and four levels of nitrogen on the gas exchange, growth, and productivity of two east African graminoids. Amer. J. Bot. 72:222-230.

Wareing, P.F., M.M. Khalifa, and K.J. Treharne. 1968. Rate-limiting processes in photosynthesis at saturating light intensities. Nature 220:453-457.

Warick, S.I., and D. Briggs. 1978. The genecology of lawn needs. II. Evidence for disruptive selection in *Poa annua* L. in a mosaic environment of bowling green lawns and flower beds. New Phytol. 81:725-737.

Watson, M.A. 1984. Developmental constraints: effect on population growth and patterns of resource allocation in a clonal plant. Am. Nat. 123:411-426.

Watson, V.H., and C.Y. Ward. 1970. Influence of intact tillers and height of cut on regrowth and carbohydrate reserves of Dallisgrass (*Paspalum dilatatum* Poir.). Crop Sci. 10:474-476.

Welker, J.M., D.D. Briske, and R.W. Weaver. 1987. Nitrogen-15 partitioning within a three generation tiller sequence of the bunchgrass *Schizachyrium scoparium*: response to selective defoliation. Oecologia 24:330-334.

Welker, J.M., D.D. Briske, and R.W. Weaver. 1991. Intraclonal nitrogen allocation in the bunchgrass *Schizachyrium scoparium*: an assessment of the physiological individual. Funct. Ecol. 5:433-440.

Welker, J.M., E.J. Rykiel, Jr., D.D. Briske, and J.D. Goeschl. 1985. Carbon import among vegetative tillers within two bunchgrasses: assessment with carbon-11 labelling. Oecologia 67:209-212.

West, N.E., K.H. Rea, and R.O. Harniss. 1979. Plant demographic studies in sagebrush-grass communities of southeastern Idaho. Ecology 60:376-388.

Westoby, M. 1980. Relations between genet and tiller population dynamics: survival of *Phalaris tuberosa* tillers after clipping. J. Ecol. 68:863-869.

White, J. 1980. Demographic factors in populations of plants. p. 21-48. *In*: O.T. Solbrig (ed.) Demography and evolution in plant populations. Univ. Calif. Press.

Williams, D.G., and D.D. Briske. 1991. Size and ecological significance of the physiological individual in the bunchgrass *Schizachyrium scoparium*. Oikos 62:41-47.

Williams, R.D. 1970a. Tillering in grasses cut for conservation, with special reference to perennial ryegrass. Herb. Abst. 40:383-388.

Williams, O.B. 1970b. Population dynamics of two perennial grasses in Australian semi-arid grassland. J. Ecol. 58:869-875.

Willms, W., A.W. Bailey, and A. Mclean. 1980. Effect of burning or clipping *Agropyron spicatum* in the autumn on the spring foraging behaviour of mule deer and cattle. J. Appl. Ecol. 17:69-84.

Wilson, A.M., and D.D. Briske. 1979. Seminal and adventitious root growth of blue grama seedlings on the central plains. J. Range Manage. 32:209-213.

Wilson, J.R., R.H. Brown, and W.R. Windham. 1983. Influence of leaf anatomy on the dry matter digestibility of C_3, C_4, and C_3/C_4 intermediate types of *Panicum* species. Crop Sci. 23:141-146.

Windle, P.N., and E.H. Franz. 1979. The effects of insect parasitism on plant competition: greenbugs and barley. Ecology 60:521-529.

Woledge, J. 1977. The effects of shading and cutting treatments on the photosynthetic rates of ryegrass leaves. Ann. Bot. 41:1279-1286.

Wolf, D.D., and D.J. Parrish. 1982. Short-term growth responses of tall fescue to changes in soil water potential and to defoliation. Crop Sci. 22:996-999.

Woodmansee, R.G., I. Vallis, and J.J. Mott. 1981. Grassland nitrogen. *In*: F.E. Clark and T. Rosswall (eds.) Terrestrial nitrogen cycles. Ecol. Bull. (Stockholm) 33:443-462.

Wraith, J.M., D.A. Johnson, R.J. Hanks, and D.V. Sisson. 1987. Soil and plant water relations in a crested wheatgrass pasture: response to spring grazing by cattle. Oecologia 73:573-578.

Wright, R.G., and G.M. Van Dyne. 1976. Environmental factors influencing semidesert grassland perennial grass demography. Southwest. Nat. 21:259-274.

Yamashita, T., and A. Fujino. 1986. Effects of pruning of young and old shoots on ribulose bisphosphate carboxylase and other constituents in leaves of the mulberry tree (*Morus alba* L.). J. Exp. Bot. 37:1836-1841.

Youngner, V.B. 1972. Physiology of defoliation and regrowth.p. 292-303. *In*: V.B. Youngner, and C.M. McKell (eds.) The biology and utilization of grasses. Academic Press.

Youngner, V.B., F. Nudge, and R. Ackerson. 1976. Growth of Kentucky bluegrass leaves and tillers with and without defoliation. Crop Sci. 16:110-113.

Zarrough, K.M., C.J. Nelson, and D.A. Sleper. 1984. Interrelationships between rates of leaf appearance and tillering in selected tall fescue populations. Crop Sci. 24:565-569.

Zhukova, L.A., and I.M. Ermakova. 1985. Structure and dynamics of coenopopulations of some temperate grasses. p. 179-205. *In*: J. White (ed.) The Population structure of vegetation. Dr W. Junk Publishers, Dordrecht.